Gravel-Bed Rivers

Gravel-Bed Rivers

Processes and Disasters

Edited by

Daizo Tsutsumi
Disaster Prevention Research Institute, Kyoto University, Kyoto, Japan

and

Jonathan B. Laronne
Department of Geography and Environmental Development,
Ben Gurion University of the Negev, Beer Sheva, Israel

WILEY Blackwell

This edition first published 2017

Registered Office
John Wiley & Sons Ltd, The Atrium, Southern Gate, Chichester, West Sussex, PO19 8SQ, UK

Editorial Office
9600 Garsington Road, Oxford, OX4 2DQ, UK

For details of our global editorial offices, customer services, and more information about Wiley products visit us at www.wiley.com.

Wiley also publishes its books in a variety of electronic formats and by print-on-demand. Some content that appears in standard print versions of this book may not be available in other formats.

Library of Congress Cataloging-in-Publication Data applied for:

Hardback : 9781118971406

Cover Design: Wiley
Cover Image: Photograph by Daizo Tsutsumi

Set in 10/12pt Warnock by SPi Global, Pondicherry, India

Printed and bound by CPI Group (UK) Ltd, Croydon, CR0 4YY

C9781118971406_220524

Contents

List of Contributors

Mikaël Attal
School of GeoSciences
University of Edinburgh, Edinburgh, UK

Alberto Bellin
Department of Civil
Environmental and Mechanical Engineering
University of Trento, Trento, Italy

Walter Bertoldi
Department of Civil
Environmental and Mechanical Engineering
University of Trento, Trento, Italy

Astrid Blom
Faculty of Civil Engineering and Geosciences
Delft University of Technology
Delft, Netherlands

D. Nathan Bradley
US Bureau of Reclamation
Sedimentation and River Hydraulics Group
Denver, Colorado, USA

Ryan W. Bradley
Department of Geography
Simon Fraser University, Burnaby
British Columbia, Canada

James Brasington
The School of Geography
Queen Mary, University of London
London, UK

Paul A. Carling
Geography and Environment
University of Southampton
Southampton, UK

Yu-Shiu Chen
Disaster Prevention Research Center
National Cheng Kung University
Tainan City, Taiwan, ROC

Philippe Claudin
Physique et Mécanique des Milieux Hétérogènes (PMMH), ESPCI Paris - PSL Research University, Univ. P. M. Curie - Sorbonne Universités, Univ. D. Diderot - Sorbonne Paris Cité, Paris, France.

Mathias J. Collins
NOAA, National Marine Fisheries Service
Restoration Center
Gloucester, MA, USA

Thibault Datry
IRSTEA, UR MALY
Centre de Lyon-Villeurbanne
Villeurbanne, France

Olivier Devauchelle
Institut de Physique du Globe de Paris, Sorbonne Paris Cité, Univ. Paris-Diderot, Paris, France

William E. Dietrich
Department of Earth and Planetary Science
University of California at Berkeley
Berkeley, CA, USA

Amy E. East
US Geological Survey, Pacific Coastal and Marine Science Center
Santa Cruz, CA, USA

Cristian Escauriaza
Departamento de Ingeniería Hidráulica y Ambiental
Pontificia Universidad Católica de Chile
Santiago, Chile

Taymaz Esmaeili
Civil Engineering Department
Islamic Azad University of Gorgan
Aq Qala, Iran

Siobhan L. Fathel
Department of Earth and Environmental Sciences, and Department of Civil and Environmental Engineering, Vanderbilt University
Nashville, TN, USA

Koichi Fujita
The National Institute for Land and Infrastructure Management (NILIM)
The Japanese Ministry of Land, Infrastructure, Transport and Tourism (MLIT)
Tsukuba, Ibaraki, Japan

Masaharu Fujita
Disaster Prevention Research Institute
Kyoto University, Kyoto, Kyoto, Japan

David Jon Furbish
Department of Earth and Environmental Sciences and Department of Civil and Environmental Engineering
Vanderbilt University, Nashville, TN, USA

Alessandro B. Gitto
Department of Geography
Simon Fraser University
Burnaby, British Columbia, Canada

Gordon E. Grant
USDA Forest Service
Pacific Northwest Research Station
Corvallis, OR, USA

Marwan A. Hassan
Department of Geography
The University of British Columbia
Vancouver, British Columbia, Canada

Dan Haught
Department of Geography
Simon Fraser University, Burnaby
British Columbia, Canada

Ricardo R. Hernandez Moreira
Department of Civil and Environmental Engineering
University of South Carolina at Columbia
South Carolina, USA

Murray Hicks
National Institute of Water and Atmospheric Research
Christchurch, New Zealand

Kimberly M. Hill
St. Anthony Falls Laboratory and the Department of Civil, Environmental and Geo-Engineering, University of Minnesota
Minneapolis, Minnesota, USA

Yuki Hiramatsu
Department of Civil and Environmental Engineering
Waseda University, Tokyo, Japan

Rebecca A. Hodge
Department of Geography
Durham University, Durham, UK

Sameh Kantoush
Disaster Prevention Research Institute
Kyoto University
Uji, Kyoto, Japan

Florent Lachaussée
Laboratoire FAST
Univ. Paris-Sud, CNRS
Université Paris-Saclay
Orsay, France

Eric Lajeunesse
Institut de Physique du Globe de Paris
Sorbonne Paris Cité
Univ. Paris-Diderot
Paris, France

Djoko Legono
Department of Civil and
Environmental Engineering
Universitas Gadjah Mada
Yogyakarta, Indonesia

Kevin W. Lewis
Department of Earth
and Planetary Sciences
Johns Hopkins University
Baltimore, MD, USA

Christopher S. Magirl
US Geological Survey
Arizona Water Science Center
Tucson, AZ, USA

Jon J. Major
US Geological Survey
Volcano Science Center
Cascades Volcano Observatory
Vancouver, WA, USA

Lorenzo Marchi
Research Institute for
Geo-hydrological Protection
National Research Council of Italy (CNR IRPI)
Padova, Italy

Alessandra Marzadri
Center for Ecohydraulics Research
University of Idaho
Boise, Idaho, USA

Peter A. Nelson
Department of Civil and
Environmental Engineering
Colorado State University
Fort Collins, Colorado, USA

Giyoung Ock
Division of Basic Research
National Institute of Ecology
Seocheon, South Korea

Jim E. O'Connor
US Geological Survey
Geology, Minerals, Energy
and Geophysics Science Center
Portland, OR, USA

Marisa C. Palucis
Division of Geological and
Planetary Sciences
California Institute of Technology
Pasadena, CA, USA

Chris Paola
Department of Earth Sciences
St. Anthony Falls Laboratory
University of Minnesota
Minneapolis, Minnesota, USA

A.N. (Thanos) Papanicolaou
Department of Civil and
Environmental Engineering
Hydraulics and Sedimentation Laboratory
The University of Tennessee, Knoxville
Tennessee, USA

Adam Pamudji Rahardjo
Department of Civil and
Environmental Engineering
Universitas Gadjah Mada
Yogyakarta, Indonesia

Colin D. Rennie
Department of Civil Engineering
University of Ottawa, Ottawa, Canada

Dieter Rickenmann
Mountain Hydrology and Mass Movements
Swiss Federal Institute for Forest
Snow and Landscape Research WSL
Birmensdorf, Switzerland

Frances Rivera-Hernandez
Geology Department
University of California at Davis
Davis, CA, USA

Virginia Ruiz-Villanueva
Dendrolab.ch, Institute of Geological Sciences
University of Bern, Bern, Switzerland
and Institute for Environmental Sciences
University of Geneva, Geneva
Switzerland

Mark W. Schmeeckle
School of Geographical Sciences and
Urban Planning, Arizona State University
Tempe, Arizona, USA

Masato Sekine
Department of Civil and
Environmental Engineering
Waseda University
Tokyo, Japan

Chjeng-Lun Shieh
Disaster Prevention Research Center
National Cheng Kung University
Tainan City, Taiwan, ROC

Annunziato Siviglia
Laboratory of Hydraulics
Hydrology and Glaciology VAW
ETH Zurich, Zurich, Switzerland

Guglielmo Stecca
National Institute of Water and
Atmospheric Research
Christchurch, New Zealand
and Department of Civil and
Environmental Engineering
University of Trento
Trento, Italy

Muhammad Sulaiman
Department of Civil and Environment
Engineering, Gadjah Mada University
Yogyakarta, Indonesia

Tetsuya Sumi
Disaster Prevention Research Institute
Kyoto University
Uji, Kyoto, Japan

Dawn Y. Sumner
Geology Department
University of California at Davis
Davis, CA, USA

Hiroshi Takebayashi
Disaster Prevention Research Institute
Kyoto University, Kyoto
Kyoto, Japan

Danielle Tan
Department of Mechanical Engineering
National University of Singapore
Singapore

Daniele Tonina
Center for Ecohydraulics Research
University of Idaho
Boise, Idaho, USA

Achilleas G. Tsakiris
Department of Civil and
Environmental Engineering
Hydraulics and Sedimentation Laboratory
The University of Tennessee
Knoxville, Tennessee, USA

Daizo Tsutsumi
Disaster Prevention Research Institute
Kyoto University
Kyoto, Japan

Desiree D. Tullos
Department of Biological and
Ecological Engineering
Oregon State University
Corvallis, OR, USA

Jeremy G. Venditti
Department of Geography
Simon Fraser University
Burnaby, British Columbia, Canada

Damià Vericat
Fluvial Dynamics Research Group
Department of Environment and
Soil Sciences
University of Lleida, Lleida
Catalonia, Spain
and Forest Sciences Centre of Catalonia, Solsona
Catalonia, Spain

Enrica Viparelli
Department of Civil and Environmental
Engineering, University of South Carolina at
Columbia, South Carolina, USA

Vaughan R. Voller
Department of Civil, Environmental and
Geo-Engineering, St. Anthony Falls Laboratory
University of Minnesota, Minneapolis
Minnesota, USA

Joseph M. Wheaton
Department of Watershed Sciences
Utah State University
Logan, Utah, USA

Andrew C. Wilcox
Department of Geosciences
University of Montana
Missoula, MT, USA

Richard D. Williams
School of Geographical and Earth Sciences
University of Glasgow
UK

Rebecca M. E. Williams
Planetary Science Institute
Tucson, AZ, USA

Preface

The *8th International Gravel-Bed Rivers (GBR) Workshop* was held in Kyoto (first day) and Takayama, Japan between 14 and 18 September 2015. Kyoto was the capital of Japan during the period AD 794–1869 and is the center of Japanese culture, with its many traditional Buddhist temples, Shinto shrines together with their unique gardens, and Zen gardens that are adorned with gravel that is as lovely as the plants. Kyoto University is among the oldest in Japan. Takayama is a small city surrounded by mountains. Due to its location and cultural uniqueness, tradition and cuisine are preserved in this city, which is traversed by gravel-bed rivers. The workshop was attended by 116 (of which, 17 were PhD students) gravel-bed river enthusiasts from Austria, Canada, Chile, China, France, Germany, Indonesia, Israel, Italy, Japan, Netherlands, New Zealand, Norway, Spain, Switzerland, Taiwan, United Kingdom, and United States.

In further keeping with the GBR tradition, the workshop was designed to present an authoritative review of recent progress in understanding the morphology of and processes operating in gravel-bed rivers. Accordingly, the workshop was constructed around a series of invited keynote presentations that tackled the principal themes selected for the meeting. Each theme included two presentations – the chapters comprising this book. Sessions for each theme incorporated considerable time for discussions. These discussions appear at the end of each chapter. The workshop included 61 posters (of which 15 were first-authored by PhD students). Ample time was given for questioning the poster presenters and exchanging information. Several among the poster presenters are preparing manuscripts for a special issue of the journal Earth Surface Processes and Landforms.

The workshop's main theme was Gravel-Bed Rivers and Disasters. In keeping with previous workshops, session topics covered a wide area, only some directly relevant to the workshop main theme of disasters. Because most measurements and data for gravel-bed rivers are available almost exclusively for small to intermediate (bankfull) flows, it is timely to pay more attention to large, often disastrous floods, more so in this age of climate change and urbanization. For instance, in Japan such floods accounted for 54% of the dead and missing in natural disasters during the period 2010–2014: altogether 5512 sediment disasters (landslides, slope failures, or debris flows) took place and 254 people were killed or missing in Japan (MLIT; http://www.mlit.go.jp/river/sabo/jirei/h26dosha/150331_H26saigai.pdf, 2015). We had chosen this theme because Japanese gravel-bed rivers cause many and frequent disasters (as in other tectonically and/or mountainous countries), but also because the Disaster Prevention Research Institute, Kyoto University played host to the workshop. The topics of this theme included gravel transport and deposition by large floods, sediment supply and availability, and integrated channel management.

Workshop themes included those that have become classics for the quinquennial meetings. So, gravel transport processes dealt with flow and transport near the river bed, theoretical considerations of bedload transport and bedload transport quantification by both direct and surrogate methods.

Represented were the two different approaches of physical and morphodynamic modeling. In demonstrating the continued awareness of the relevance of ecology to gravel-bed rivers, one session was devoted to in-stream habitat issues and the significance of vegetation. Dam removal and sediment flushing were included because these topics have gained interest and momentum by researchers and practitioners. As geomorphic, hydraulic and fluvial sedimentologic studies are often insufficiently interdisciplinary, the workshop included two other relevant issues: sediment porosity and modeling of deposit stratigraphy and bed cover. Also, for first time in the history of these meetings, the workshop hosted a topic on gravels away from Earth and explored evidence of past activity on Mars.

Field trips have continued to be central to successive workshops and that reported here was no exception. The pre-workshop trips took us to the urban Kamo River in Kyoto, guided by Yasuhiro Takemon, and to the Ujigawa Hydraulics Laboratory, Disaster Prevention Research Institute (DPRI), Kyoto University, where researchers from outside the university and the country are invited to take advantage of the flumes. The Gamada River (demonstrated by Michinobu Nonaka) and the Kamikoch area (guided by Hiroshi Suwa) were visited mid-workshop, demonstrating various devices used to monitor bedload and debris flows in DPRI's long-running Hodaka Observatory, only 4 km from the crater of the active Mount Yakedake volcano. The observatory invites researchers to collaborate or merely stay for short or extended periods at no cost. The post-workshop field trip was devoted to the Kurobe River, with its source in Mount Washiba (2924 m a.s.l.), part of the Hida Mountain Range of the Japanese Alps, carving the deep Kurobe gorge. It emerges from the mountains at Unazuki and forms an alluvial fan that debouches directly to the Sea of Japan. The guide (Tetsuya Sumi) demonstrated evidence of the ultra-high sediment yields from the Japanese Alps and the means by which dams are used to retain and flush the sediments.

Novelties of this workshop were the two-day special short-courses offered free to PhD students and postdoctoral fellows. Moreover, live web-streaming was accessible and presentations as well as discussions were recorded and uploaded to a freely available website (https://www.youtube.com/channel/UC8oW0AbmlhcJYwvtTBrAAsA).

There are many people to thank for the success of the workshop. First, our sponsors: the Japan World Exposition 1970 Commemorative Fund; Hida-Takayama Tourism and Convention Bureau; Sabo and Landslide Technical Center; Sabo Frontier Foundation; Association for Disaster Prevention Research; Nippon Koei; Civil Engineering and Eco-Technology Consultants Co. Ltd; the Kyoto University Foundation; and the DPRI, Kyoto University. We thank our supporters: Takayama City, the Jinzu and Matsumoto Sabo (meaning 'sand') offices, and Kurobe River office of the Japan Ministry of Land, Infrastructure, Transportation and Tourism.

Peer reviews of the invited papers were completed prior to the workshop. The papers were accepted between 17.03.2015 and 16.09.2015; the last reply to discussions was submitted on 1.03.2016. Special thanks go to more than 60 reviewers who helped to ensure the quality of the manuscripts. We thank the discussants of the papers and the presenters of the posters. Thanks go to Shusuke Miyata for organizing the workshop web site, the live videoconference and its recording, to Tatsuko Fujita for the workshop logo, to Harumi Miura for management of registration issues, preparation of the workshop, and tours for companions, to Cecilia Laronne Corrado for her help in the registration and support for attendees, to the staff of the Japan Travel Bureau for their assistance and management of the workshop, to the staff of the Hotel Associa Takayama Resort for their help in preparing the meeting place and accommodations, and to K. Lab Ltd for the booklet of the program and abstracts. Hiroshi Takebayashi and Joe Wheaton voluntarily offered short courses respectively on 'Modeling river flow, sediment transport and morphodynamics' and 'Introduction to geomorphic change,' to Ian Reid for English edits, to the PhD students and postdoctoral fellows, for which students Jun Nishiura and Satoru Masuda ensured a welcome. Appreciation goes to the excellent suggestions made by members of the Scientific Advisory Board: Jochen Aberle, Peter Ashmore, James Brasington, Shinji Egashira, Chris Paola, Klement Tockner and Guido Zolezzi. Many thanks to Rachael Ballard, Shummy Metilda, Vinodhini Mathiyalagan, Delia Sandford, Audrie Tan, Teresa Netzler, Bella Talbot and Andrew Harrison, Lyn Nesbitt-Smith, staff at John Wiley & Sons, for their invaluable help during the entire editing process and to Mike Church for his suggestions and introduction to Wiley.

Professor Richard Hey (who could not participate), Professor James Bathurst and Professor Colin Thorne were guests invited and honored for their foresight in initiating the Gravel-Bed River series of workshops, beginning with the first in Newtown, Wales, in 1980. Thanks go to James and Colin, who, as banquet speakers, entertained everyone with the history of GBR workshops and their importance. Professor Takahisa Mizuyama was praised during the dinner for his central role in developing Japanese research on gravel bed rivers during the past two decades.

We trust that this book, like its predecessors, will become part of an authoritative record of advances in knowledge and understanding of gravel-bed rivers. And we wish the hosts of the next meeting, GBR9, to be held in Chile, as much success as we have enjoyed.

Daizo Tsutsumi
Kyoto, Japan
and
Jonathan B. Laronne
Beer Sheeva, Israel

1

Computational Models of Flow, Sediment Transport and Morphodynamics in Rivers

Cristian Escauriaza, Chris Paola, and Vaughan R. Voller

1.1 Introduction

Models of various kinds have long been essential tools for investigating the dynamics of fluvial systems, providing over a wide range of time and space scales fundamental insights on the coupled interactions between turbulent flows, sediment transport, and the morphodynamic evolution of rivers. From a scientific standpoint, models can effectively help us understand phenomena in nature and investigate the response of rivers subject to varying environmental and anthropogenic forcings.

Models are simplified, often mathematical, representations of fundamental conservation laws of nature. They typically comprise a set of equations or rules with variables that describe the state of the system and contain, either implicitly or explicitly, a definition of the time and length scales that characterize the dynamics in the application of interest. Beyond the predictions they provide, model development in itself provides a creative portal in advancing scientific understanding and explaining key dynamics. Models in general must be tested against measurements. Ideally, since in general neither models nor observations are without errors, this is done iteratively by adapting the equations and solution methods, as well as measurement protocols, such that both parts of the process add to our understanding.

Despite advances in computational power and numerical strategies to solve equations, predictions of river flows and sediment dynamics still pose a significant challenge for models of river flows. The large variety and complexity of available simulations do not necessarily mean that our understanding of the essential processes that explain the river dynamics has advanced at the same pace. Even for a "basic" river flow problem, i.e., only water and noncohesive sediment, the challenges are formidable. The main challenges are the intrinsic complexity of turbulent flow, especially under time-varying conditions and over complex boundaries, and the intractability of calculating the sediment flux precisely, even by performing a particle-by-particle force balance. We consider these well known problems in this chapter but we would also like to call attention to one that has received less attention: that of deciding the most appropriate model for a given scientific question. In complex systems such as rivers in which there are many levels of phenomena of interest, some questions may be answered with quite simple – but perhaps unconventional – modeling methods, while others require all the fine detail and resolution that we can muster.

In this chapter we review some significant advances in numerical models of fluvial environments, connect the development of models with new insights that have been unveiled by their use, and look at some of the less conventional approaches that have been applied to river behavior. We hope this

Gravel-Bed Rivers: Processes and Disasters, First Edition. Edited by Daizo Tsutsumi and Jonathan B. Laronne.

chapter will also serve as a guide to discuss and help select the appropriate model for different processes of interest in rivers, emphasizing challenges that future research should address. The chapter is organized as follows: in section 1.2 we provide a brief review of the most common modeling frameworks and describe examples of successful applications in river flows; in section 1.3 we discuss the reasoning behind the selection of a modeling approach, and suggest guidelines for choosing the best approach for the problem of interest; in section 1.4 we focus on future challenges, especially the prospects for using more detailed, computationally intensive models to inform less detailed ones that can potentially cover greater time/space scales and/or be run multiple times; in section 1.5 we summarize ideas and formulate questions to advance toward a synthesis of models that can contribute to understanding natural processes in rivers.

1.2 Numerical Simulations in Rivers

Models in fluvial systems range from highly simplified and/or abstracted approximations that retain a minimal set of physical laws supplemented with observation-based rules, to detailed direct numerical simulations of the equations that describe multiphase turbulent flows with complex boundaries. As discussed throughout this section, the relevance of turbulent motions as mechanisms of sediment transport and erosion, and the scales at which the river is studied have a great influence on the development and hence selection of models that represent physical processes in the flow and the evolution of the bed.

Our aim in this section is to provide a brief "field guide" to modeling approaches with application examples, highlighting the main assumptions and insights they have provided for specific problems in rivers. In a very broad sense, modeling approaches can be envisaged as being laid out on a spectrum. At one end of this spectrum are highly abstracted and simplified models, including for example, descriptive models and highly generalized balances, e.g., the Lane (1955) model for river slope or the classical cellular automaton (CA) models (Wolfram 2002). At the other end of the spectrum are models that we refer to as high-fidelity models. The most detailed of these models perform direct simulations that capture all of the relevant scales of the underlying physics, resolving down to the smallest length scale in the system – the Kolmogorov microscale, at which viscosity dissipates kinetic energy – or the size of individual sediment particles. Most models in use today lie somewhere in the middle of this spectrum. Figure 1.1 shows examples produced by approaches that are close to the ends of the modeling spectrum: from rule-based models to understand the features of drainage networks, to small-scale turbulence simulations coupled to the dynamics of individual particles to capture the details of hydrodynamic mechanisms of sediment transport.

1.2.1 Level 0: Reduced Complexity Models

At the simplified end of the model-complexity spectrum sit the so-called reduced complexity models (RCM). RCM simulations in geomorphology focus on the interaction of sediment and water or air, and are generally aimed at pattern formation and self-organization. These models are based on a discrete grid of cells in a two-dimensional plane. Their simplest form, which so far has seen little application in morphodynamics modeling, is the CA approach championed by Wolfram (2002). In their purest state, CA models generate emergent behavior of a process on a grid of cells through rules that switch a state variable on or off based only on local conditions. In slightly more complex RCM models, the simple binary CA settings are replaced by more complex rules that typically represent what the modeler considers to be the important dynamics. Models typically include mass conservation for fluid and sediment, a simplified description of sediment motion, and a highly simplified formulation of the flow hydrodynamics.

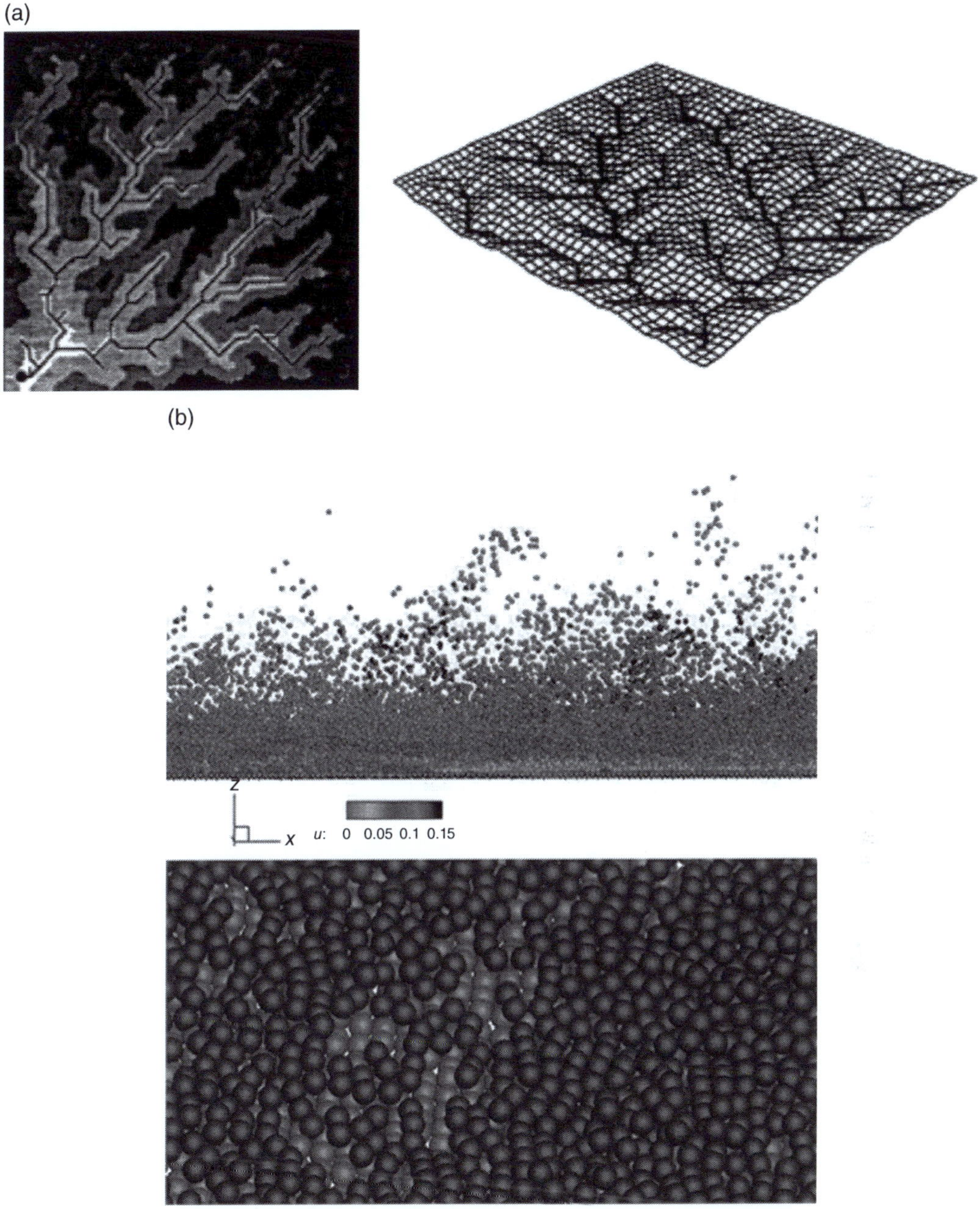

Figure 1.1 Models range from simplified approximations of conservation laws to very detailed simulations of the entire spectrum of eddy scales in turbulent flows. (a) Simple model of channel network and hillslope evolution by Willgoose, Bras, and Rodríguez-Iturbe (1991). (b) Direct numerical simulations of the turbulent boundary layer coupled with a Lagrangian model of sediment transport. Particles are modeled as pointwise elements whose dynamics are computed from the instantaneous flow field that resolves all the turbulent scales (González *et al.* 2016). (*Sources*: Willgoose 1991. Reproduced with permission of John Wiley & Sons.)

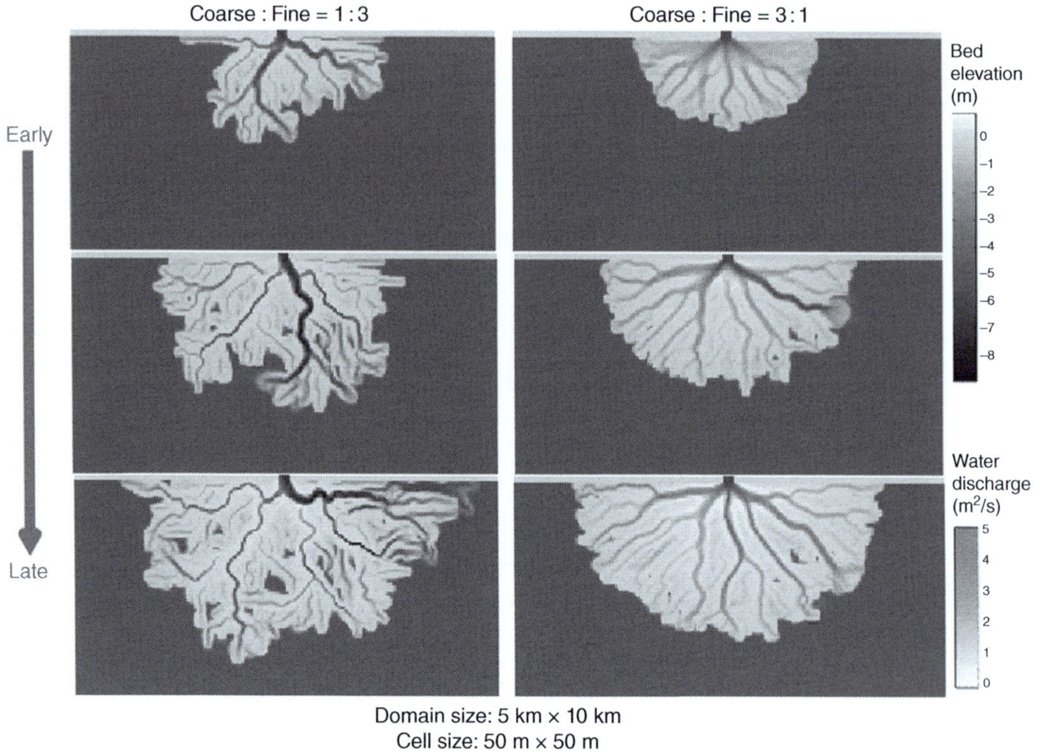

Figure 1.2 Reduced complexity model of river delta formation based on random walks and a set of mathematical rules in a discretized plane (Liang, Voller, and Paola 2015a). (*Source*: Liang 2015. http://www.earth-surf-dynam.net/3/67/2015/esurf-3-67-2015.pdf. Used under CC BY SA 3.0. https://creativecommons.org/licenses/by/3.0/.)

Usually the aim of RCMs is not to predict specific morphodynamic cases, but rather to explore general behaviors, commonly emphasizing pattern formation. A key underlying hypothesis, not always stated explicitly in such models, is that the dynamics at the scales of interest are sufficiently independent of the underlying fine-scale details that the latter can be represented in highly simplified, summary forms (e.g., Werner 1999). There is no formalism for developing RCMs akin to the more conventional models, which are based on appropriate discrete forms of the governing conservation equations. In RCMs, usually the only conservation laws that are strictly enforced are those for mass (volume) of water and sediment. Since most of the complexity in river morphodynamics arises from conserving fluid momentum, "shortcuts" are usually applied here. The rules that form these models can come from a variety of sources: analogy (e.g., the electrical analog used by Seybold, Andrade, and Herrmann (2007) and Seybold *et al.* (2009) to model deltas); synthetic summaries of observed behavior (e.g., the bedform models of Werner and Kocurek (1997), and the models of eolian ripples by Anderson and Bunas (1993) and Haff and Anderson (1993)); semi-empirical models (e.g., stream-power erosion laws used to model erosional landscapes); models abstracted from more detailed analyses at smaller scales (e.g., the meander model of Howard (1992)); adaptations of generalized behavioral models, such as random walks (Overeem, Syvitski, and Hutton 2005); and/or rough approximations of the governing equations, such as the Murray and Paola (1997) braided river model or the delta model of Liang, Voller, and Paola (2015), shown in Figure 1.2. RCMs have been successful in capturing important emergent dynamics at large spatial scales, for example, braided rivers (Murray and Paola 1994, 1997) and deltas (Seybold, Andrade, and Herrmann 2007), and erosional landscapes, starting with the work of Willgoose *et al.* (1991), and leading to models such as CHILD (Tucker *et al.* 2001) and CAESAR (Coulthard *et al.* 2013). Werner (1999) and Murray (2007, 2013) have provided summaries of the rationale for these approaches.

The RCM approach has also received its share of criticism (see Doeschel-Wilson and Ashmore 2005; Nicholas 2010). Broadly speaking the criticisms fall into two categories, which are both related to the unavoidable fact that a RCM leaves out and/or drastically simplifies some of the relevant mechanics: first, that the behavior they predict does not match reality; and second, that RCMs are so approximate that they can be adjusted to produce any result the modeler desires.

On one hand, the criteria for evaluating RCMs – comparison with observations – is no different than for any other model. On the other, it is unreasonable to expect a model that leaves out most of the governing dynamics to reproduce nature in detail. Generally, RCMs should be measured by how much their return in reproducing observed emergent behavior (e.g., pattern dynamics) exceeds their input in process complexity. The second criticism, when it applies, can be seen as a bug or as a feature. Seen as a feature, a shortfall in the physics or predictive quality can be mitigated by successive iteration that modifies the rules and transport treatments until the desired level of model agreement with reality is achieved; this can be viewed either as "tuning" of the model to produce the desired result or as part of the normal scientific process, in which we improve our guesses by comparison of RCM simulations with observations. We will return to this point and advocate a new approach to synthesize across-model approaches in section 1.4, but for now we continue our survey of different modeling approaches, moving across the spectrum towards increasing complexity and higher levels of fidelity and resolution.

1.2.2 Level 1: Diffusion Models

The idea that sedimentary processes at landscape scales are broadly diffusive is intuitively natural and has been in the literature for many decades (e.g., Scheidegger 1970). Derivations of diffusion equations from the governing equations of mass and momentum conservation are by now numerous (Begin, Meyer, and Schumm 1981; Gill 1983; Ribberink and van der Sande 1985; Paola 2000), and they are applied fairly routinely to investigate large-scale landscape evolution (e.g., Martin and Church 1997; Whipple and Tucker 2002; Tucker and Hancock 2010), especially when the focus is on coupling surface processes to tectonics. Diffusion models have also been able to describe the processes producing grain-size changes in the stratigraphy of alluvial basins (Paola, Heller, and Angevine 1992), as shown in Figure 1.3. The reader can find details of the diffusion approach in the works cited above. The formal derivation of the diffusion approximation for landscapes (e.g., Paola 2000) starts with the shallow-water equations (SWEs), discussed in the next section. The additional basic assumptions and conditions for the diffusive approximation to be valid include the following:

1) Length scale of the problem $L >> L_{bw}$ where L_{bw} is the backwater length, defined as the ratio between the representative magnitudes of the water depth h and the slope S ($L_{bw} = h/S$). This means that spatial acceleration and pressure-gradient terms can be neglected, calculating the bed shear stress as the product of the depth, slope and water specific weight.
2) Discharge variability can be represented via a simple intermittency that considers the system either to be "on," i.e., in flood, or "off." This amounts to replacing the hydrograph with a single characteristic flood.
3) Sediment transport can be represented as a simple function of total bed shear stress via one of the classical formulas (e.g., Engelund and Hansen 1972) with no complications from bed armoring, topography, etc. The effects of bedforms and bars, if present, can be represented via a constant drag coefficient.

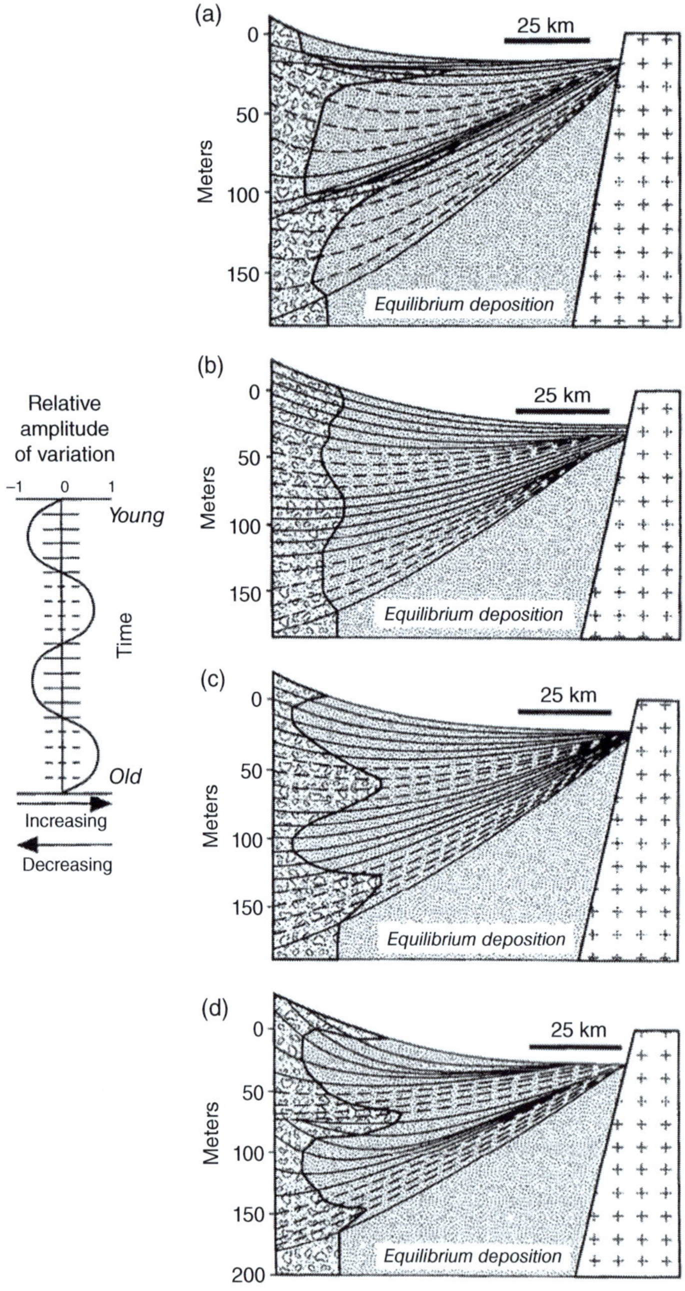

Figure 1.3 Gravel–sand front in the diffusion model of basin stratigraphy. Slow variation of sediment flux, subsidence, gravel fraction and diffusivity produce different cross-sections (see details in Paola, Heller, and Angevine 1992). (*Source*: Paola, Heller, and Angevine 1992. Reproduced with permission of Basin Research.)

1.2.3 Level 2: Models Based on Solving the Shallow-Water Equations

The shallow-water (St. Venant) equations represent the starting point for the family of traditional, discrete-element-based river models. The simplest version of the SWE is the well-known depth–slope product force balance for nonaccelerated flow that underpins the diffusion approximation discussed above. More complete versions of the SWE generally require numerical solution for flows with spatial and/or temporal acceleration. Numerical models in rivers at larger scales commonly use the SWEs to represent the flow hydrodynamics in one or two dimensions. The general unsteady nonlinear SWEs incorporate the conservation of mass and momentum, which are obtained by vertically integrating the Reynolds equations over water depth, with the following assumptions: (i) pressure distribution is hydrostatic; (ii) horizontal spatial scales are significantly larger than the vertical wave height scales; (iii) viscous stresses and Coriolis accelerations are negligible; and (iv) boundary stresses can be represented by algebraic equations originally derived for uniform flows. Assuming horizontal Cartesian axes x and y, the nondimensional two-dimension equations in conservative form can be written as,

$$\frac{\partial Q}{\partial t}+\frac{\partial F}{\partial x}+\frac{\partial G}{\partial y}=S \tag{1.1}$$

where the vector of hydrodynamic variables that contains the water depth h, and the depth-averaged velocity components u and v is written as $Q=\begin{bmatrix} h & uh & vh \end{bmatrix}^T$. The flux vectors that contain the mass and momentum balance are defined as follows,

$$F=\begin{bmatrix} uh \\ u^2h+h^2/(2Fr^2) \\ uvh \end{bmatrix};\quad G=\begin{bmatrix} vh \\ vuh \\ v^2h+h^2/(2Fr^2) \end{bmatrix} \tag{1.2}$$

which are written in terms of the characteristic Froude number of the flow, Fr. The source term S contains the local bed slope and the friction slope calculated with the Manning, Chezy, or Darcy–Weisbach friction factor equation. These friction momentum sources that account for interactions with the boundaries usually incorporate effects of other momentum exchange processes when the model is applied to large spatial scales, such as local curvature effects and additional drag by subgrid elements (Hodges 2013). This is accomplished by increasing the friction terms in the SWE, to account for processes acting at smaller scales than the size of the discretization elements of the domain.

Numerical solutions of the SWE, coupled with sediment transport formulas and the Exner equation for sediment mass balance (Paola and Voller 2005), have been used for many years to predict the sediment transport and morphodynamics in long-term processes ranging from months to years. For processes occurring at larger spatial and temporal scales, morphodynamic models have sometimes neglected the unsteady and inertial terms to represent the long-term evolution of the bed by considering steady, gradually varied flow or even local uniform flow in one-dimension formulations, using a constant or variable water discharge. The limiting case is the basic depth–slope product force balance mentioned above that readily yields the bed shear stress, and is coupled with empirical transport equations to estimate the sediment flux. These SWE models without the unsteady and inertial terms have successfully predicted the progradation of deltas and alluvial fans (e.g., Parker *et al.* 1998; Kostic and Parker 2003a,b; Papanicolaou, Bdour, and Wicklein 2004; Kim *et al.* 2009; Contreras *et al.* 2015), the evolution of a river after dam removal (Cui *et al.* 2006a,b), the morphodynamics of rivers and bed composition in cases with mixed sediments (e.g., Cui, Parker, and Paola 1996; Cui *et al.* 2003; Verhaar *et al.* 2008; Viparelli *et al.* 2010, 2011; Stecca, Siviglia, and Blom 2014), and recently to simulate the evolution of river meanders with bank erosion and deposition (Asahi *et al.* 2013). As explained in the previous section, it is common to use one-dimension formulations of the SWE for landscape evolution and RCMs (Nicholas, Thomas, and Quine 2006; van de Wiel *et al.* 2007), and they also have been employed in models that use quasi-Lagrangian approaches for long-term sedimentary processes (e.g., Tetzlaff and Harbaugh 1989). These one-dimension models represent processes over the river cross-section in averaged form, explaining the overall long-term evolution of rivers at the reach scale or larger (i.e., at length scales > > channel width), and for long time scales (typically averaging over many flood events). Fasolato *et al.* (2011) also analyzed the validity of the one-dimension steady uniform flow approximation for morphodynamic simulations comparing numerical solutions with analytical formulas. They suggested that this approximation is appropriate for increasing spatial scales and longer time scales as the Froude number decreases.

As pointed out by Ferguson (2003) and Ferguson and Church (2009), the main shortcoming of the simplified steady or unsteady cross-section averaged one-dimension models is that they can misrepresent the sediment transport flux in many situations. One major difficulty arises in the distribution of flow properties across the width, especially in complex cross-sections or in cases where the stresses are nonuniformly distributed and/or close to the critical shear stress that determines the initiation of sediment motion (see Paola 1996; Paola *et al.* 1999; and El kadi Abderrezzak and Paquier 2009, for more details). Two-dimensional models have also been used to simulate sediment transport processes by computing an entrainment function or a bedload flux vector tangential to the bed surface. The suspended sediment is considered by coupling the flow to advection-diffusion equations for sediment concentration, which require an assumed vertical velocity distribution to estimate the bed stresses. Papanicolaou *et al.* (2008) summarized the key capabilities and applications of models that couple the hydrodynamics and sediment transport for river flows.

The SWEs have also been employed in the simulation of rapid and extreme flooding events, such as tsunami propagation, riverine flooding, and dam-break flows. These conditions require inclusion of time variation, as well as shock-capturing numerical schemes that can reproduce complex discontinuities such as shockwaves or wet–dry interfaces by solving the so-called Riemann problem at the interfaces of the elements of the discretization (Toro 2001). Fully two-dimensional formulations have been used to simulate the effects of rapid floods with sharp free-surface gradients over rough and complex topographies (Figure 1.4 ; de Almeida *et al.* 2012; Guerra *et al.* 2014), and the dynamics of debris flows in mountain rivers, including bed erosion and deposition (Armanini, Fraccarollo, and Rosatti 2009).

1.2.4 Level 3: Unsteady Reynolds-averaged Navier–Stokes

Unsteady Reynolds-averaged Navier–Stokes (URANS) equations form the entry point for three-dimensional modeling of turbulent flows. The main purpose of three-dimensional models in rivers is to capture the influence of turbulence at the scales of the processes that are being analyzed and the impact of the flow on morphodynamics and sediment transport. Turbulence is the collective set of phenomena that occurs in flows with a Reynolds number above some threshold, characterized by unsteadiness, irregularity, greatly enhanced mixing, and eddies of a wide range of scales. Velocity components, pressure, and other flow variables fluctuate irregularly in space and time, rapidly and effectively transferring momentum, mass, and heat. As discussed in this chapter, the simulation of all scales in complex real-life turbulent flows might be not only computationally very expensive but also impractical. In most cases, then, turbulent flows are analyzed by equations that explicitly resolve the average velocity and pressure fields, whereas the effects of the fluctuations produced by the local dynamics of small unsteady vortices in the flow are represented through statistical or deterministic models.

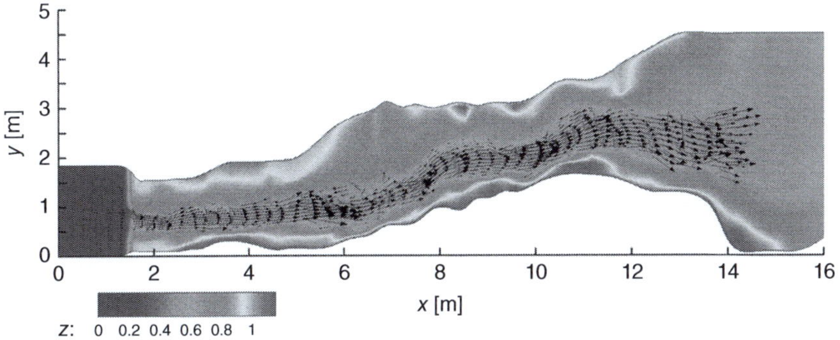

Figure 1.4 Snapshot of the velocity field during the propagation of a dam-break flow over a complex bathymetry using the SWE model of Guerra *et al.* (2014). The two-dimensional model yields the flooded area and instantaneous depth-averaged velocities at every point, incorporating the inundation of previously dry zones of the terrain. (*Source*: Guerra 2014. Reproduced with permission of American Society of Civil Engineers.) (See insert for color representation of this figure.)

Most mathematical descriptions of turbulent flows are based on the similarity hypotheses of Kolmogorov (Pope 2000), which allow us to represent the nonlinear transfer of kinetic energy or energy cascade from large to small eddy scales as a statistically universal regime. Small scales are assumed to be homogeneous and isotropic, and energy is assumed to be continuously dissipated by viscosity at the smallest scales. A common statistical description of turbulent flows is based on a procedure known as Reynolds decomposition, in which we separate the mean and fluctuating parts of velocity components and pressure as follows,

$$u_i = U_i + u'_i$$
$$p = P + p'$$

where the capital letters and the lower-case letters designated with a prime correspond to the mean and zero-mean fluctuating parts, respectively.

When this decomposition is applied to the equations of conservation of mass and momentum, which are then ensemble-averaged, we obtain the well-known unsteady Reynolds-averaged Navier–Stokes equations (URANS). The time derivatives of the mean hydrodynamic field are preserved when the flow is assumed to be nonstationary. For a homogeneous incompressible flow, the URANS equations can be written in nondimensional form as follows,

$$\frac{\partial U_i}{\partial x_i} = 0 \tag{1.3}$$
$$\frac{\partial U_i}{\partial t} + U_j \frac{\partial U_i}{\partial x_j} = -\frac{\partial P}{\partial x_i} + \frac{1}{Re}\frac{\partial^2 U_i}{\partial x_j \partial x_j} - \frac{\partial}{\partial x_j}\langle u'_i u'_j \rangle$$

where Re is the Reynolds number defined in terms of the characteristic velocity and length scales of the problem and the fluid viscosity, and the angle brackets denote ensemble averaging.

In these equations, the effects of the fluctuations on the mean flow appear as derivatives of unknown second-order statistical moments of the velocity fluctuations. These are the Reynolds stresses represented by the tensor $\langle u'_i u'_j \rangle$, which contains correlation terms that cannot be found by deriving evolution equations to close the system. Equations for second-order moments are always expressed in terms of additional unknown higher-order statistics, so additional simplifications are needed to close Eq. (1.3) and solve for the mean flow field.

Over the past century or so, many solutions have been proposed for this closure problem. The most common relations to represent the turbulent transport of momentum by the Reynolds stresses, such as the $k-\varepsilon$ and $k-\omega$ turbulence models (Wilcox 1994), assume isotropy of the smaller scales, i.e., translational and rotational invariance of the fluctuating field. These models are based on the Boussinesq approximation, which makes an analogy between turbulent stresses and molecular processes by defining a variable known as the eddy or turbulent viscosity, denoted as ν_t. The eddy viscosity is not a material quantity that can be directly measured in experiments, but rather a rough theoretical description of the statistical effects of the fluctuating field acting on the mean flow. In these isotropic models, the Reynolds stresses are represented as a function of the eddy viscosity and mean flow properties as follows,

$$\langle u'_i u'_j \rangle = \frac{2}{3} k \delta_{ij} - 2\nu_t S_{ij} \tag{1.4}$$

where the first term is the Kronecker delta tensor δ_{ij} and the turbulent kinetic energy $k = \frac{1}{2}\langle u'^2_i \rangle$. The second term contains the eddy viscosity multiplied by the mean rate-of-strain tensor S_{ij}, in analogy to the calculation of viscous stresses in Newtonian fluids.

Statistical turbulence models are designed to represent the effects of turbulent fluctuations in the flow by deriving equations to compute the instantaneous eddy viscosity at every point of the discretization in the spatial domain as a function of space, mean velocity gradients, etc. These equations have a physical basis but the eddy viscosity is so dependent on specific local flow properties that in many complex flows it must be nearly hand-fitted to the constraints at hand (for details on URANS models the reader is referred to Wilcox (1994), Pope (2000) and Durbin and Pettersson-Reif (2011)). Depending on the turbulence models, we might also have to determine the boundary conditions employed in the calculation, as some models require the use of wall functions at solid boundaries, while in others we impose directly the magnitudes of the turbulent variables at the walls.

Numerous investigations have demonstrated that URANS models can reasonably predict the mean velocity and turbulent kinetic energy distributions in space with moderate computational costs, providing new insights of three-dimensional mean flows in rivers at real-life Reynolds numbers (see for instance Lane *et al.* 1999; Ge and Sotiropoulos 2005; Ge *et al.* 2005; Sandbach *et al.* 2012). Typical spatial scales of the simulations range from laboratory experiments (10^0 m) to long river reaches using coarse meshes (10^3 m), and time scales vary from seconds to hours to represent the relevant features of the mean three-dimensional flow field, depending also on whether the averaged flow is considered steady (stationary) or unsteady (nonstationary).

URANS models are also capable of predicting mean velocities and sediment concentrations in laboratory flows (e.g., Wu, Rodi, and Wenka 2000; Olsen 2003) and river sections (e.g., Sinha, Sotiropoulos, and Odgaard 1998; Ferguson *et al.* 2003; Rodriguez *et al.* 2004), including flows around hydraulic structures with complex geometries such as bridge piers or abutments (Ge and Sotiropoulos 2005; Ge *et al.* 2005), and of estimating bed scour (Nagata *et al.* 2005; Roulund *et al.* 2005; Duc and Rodi 2008).

However, URANS models have important limitations as well. For example the turbulent stresses calculated with the eddy viscosity directly affect the numerical solution for the mean flow, as each specific turbulence model selected to perform the simulations can change the predictive ability of the URANS model. Isotropic turbulence models are usually chosen by their simplicity in implementation and robustness of their numerical solutions, but they add a considerable viscous-like dissipation to the computed flow fields. As a result, URANS turbulence models produce in many cases overly smooth solutions that cannot resolve the unsteady large-scale vortical structures that characterize high Reynolds-number flows. In addition, the use of wall functions based on logarithmic or power-law profiles at solid boundaries produces discrepancies in some situations, for example, in highly unsteady flows near the wall or when strong adverse pressure gradients are present. The most complicated problems arise in massively separated flows, recirculating flows, helical flows, or curvature-induced secondary flows. In all these situations, which are very common in rivers, there is an important turbulence anisotropy generated by large-scale coherent structures. Consequently, URANS models that assume isotropic small-scale turbulence are difficult to apply and error-prone.

Reynolds-stress transport equations, algebraic, or nonlinear models (Speziale 1998; Durbin and Pettersson-Reif 2011) can overcome some of these limitations by calculating directly the components of the Reynolds stress tensor or modeling turbulence anisotropy. However, they generally require solving more complex equations in finer grids, which increases the cost of the simulations. This approach also may not resolve the unsteadiness of large-scale structures and still introduces a large turbulent dissipation through ν_t. As discussed in the following section, advanced large-eddy simulations (LES) or hybrid URANS–LES models can resolve the rich dynamics of complex turbulent flows, but at the cost of increasing significantly the computational effort of the simulations.

1.2.5 Level 4: Large-Eddy Simulations

Even though URANS models can provide a reliable representation of three-dimensional averaged velocities and pressure, when large-scale turbulence unsteadiness and anisotropy are present in the fundamental mechanisms driving the phenomena in question, more advanced, albeit computationally expensive models are required to resolve the flow.

Large-eddy simulation (LES) is an approach to modeling complex turbulent flows dominated by large-scale coherent structures. The LES models can provide accurate descriptions of the flow fields, yielding detailed solutions by resolving directly all the turbulent scales at or above the size of the computational grid. Using URANS as a starting point, we could describe the Reynolds averaging process as a form of spatial filtering (averaging). In URANS, the filter is large enough to encompass the whole spectrum of turbulent length scales, i.e., the scale is in a sense that of the whole flow. In LES, the basic idea is the to reduce the size of this spatial filter as much as possible, so that in effect the filtering (averaging) is carried out only on the smallest turbulent scales of the flow. A spatial filter is applied to the instantaneous flow, as described by the Navier–Stokes equations, and the governing equations are expressed in terms of the filtered variables. The effect of the residual flow field that remains unresolved after filtering the Navier–Stokes equations is captured via a subgrid-scale (SGS) model. The selected SGS model represents the energy transfer and dissipation between the resolved and unresolved flow fields, and gives an approximation of the stress tensor corresponding to the subgrid scales of turbulent motion. In URANS, for comparison, the eddy viscosity along with auxiliary equations that are coupled to it could be thought of as the SGS model.

In most LES turbulence models that do not consider backscatter (energy transfer from small to large scales) the so-called Smagorinsky approach is used, in which the same URANS concept of eddy viscosity is used to represent the unresolved scales. Specifically, the SGS stresses τ_{ij} are obtained from the instantaneous magnitude of ν_t, multiplied by the filtered (resolved) rate of strain tensor $\widetilde{S}_{ij}$, as follows,

$$\frac{\tau_{ij}}{\rho} = \frac{1}{3}\frac{\tau_{kk}}{\rho}\delta_{ij} - 2\nu_t\widetilde{S}_{ij} \tag{1.5}$$

where the eddy viscosity is a nonlinear function of the flow and the local grid size (for more details and other LES modeling approaches the reader is referred to Pope (2000), Meneveau and Katz (2000) and Sagaut (2006)). The fundamental assumption underlying LES is that the smallest scales of turbulence – the SGS – are more isotropic and more universal than the large (resolved) scales. It can be thus anticipated that LES performed on coarse computational meshes at high Reynolds numbers will not capture the dynamics of turbulence well because, in effect, it is not that different from a URANS model.

The simulation of near-wall flows is a major challenge in LES, since resolving the detailed dynamics at the fine scales that become important near the wall may not be possible at full-scale Reynolds numbers due to the computational cost. The details of the coherent structure dynamics of the boundary layer are typically important only for near-bed particle transport and small bedforms in sand. Wall-modeling approximations for the boundary conditions at solid walls can be used to resolve in detail the most energetic scales of motion away from solid boundaries. This is done either by the use of a simplified wall model to obtain the eddy viscosity near the solid boundaries (Piomelli and Balaras 2002), or by the use of hybrid URANS–LES turbulence models (Spalart 2000).

Hybrid models have been proposed to decrease the computational costs of LES, overcoming the limitations of URANS simulations for high Reynolds number flows in complex geometries. The basic idea is to use a URANS turbulence closure to model regions near solid walls, which allows a coarser grid resolution near the boundaries, while LES is used in the rest of the domain. A URANS model equation is retained throughout the boundary layer, and far from the solid boundaries the same model equation becomes the SGS eddy-viscosity equation, balancing turbulence production and dissipation, and allowing the energy cascade down to a length scale proportional to the grid size. These hybrid approaches, such as detached-eddy simulations (DES; Spalart 2009), can capture the large-scale coherent structures of the flow, resolving the most energetic scales of motion away from solid boundaries.

Recent LES and hybrid URANS–LES models have shown to be very effective in capturing complex turbulence dynamics and advancing the fundamental understanding of turbulent flows in natural streams (Kang *et al.* 2011), and around complex hydraulic structures (McCoy Constantinescu, and Weber 2008; Escauriaza and Sotiropoulos 2011c), as discussed in the next section (see also the recent discussion by Stoesser 2014).

1.2.6 Level 5: Direct Numerical Simulations

Direct numerical simulations (DNS) resolve all the scales in a turbulent flow through the numerical solution of the Navier–Stokes equations, using a grid resolution of the same order of magnitude as the Kolmogorov scale. This approach provides the maximum of physical fidelity as well as space–time resolution of the flow field, but is limited as it has an extremely large computational cost at even small Reynolds numbers. Computational domains usually represent a very small section of the flow and boundary conditions need to be approximated at the resolution of the simulation, which may involve assuming nonphysical periodic conditions in the cross-stream and longitudinal directions.

This high level of description of the entire spectrum of turbulent scales is necessary only when the finest-grained structure of the flow is important for the problem at hand, or in cases where we believe the approximations required for one of the above methods lead to incorrect results at the scales of interest. For problems with sediment transport, a true DNS would also require resolving each grain and its surrounding flow field – a very formidable computational requirement if the smallest grain diameter is less than the Kolmogorov scale. The DNS approach has mainly been employed in highly detailed models of sediment transport, such as Lagrangian particle models that integrate the dynamics of individual sediment grains in a boundary layer flow with high resolution.

DNS models involving sediment to date, however, use a simplified description of the particles, representing them by pointwise elements with hydrodynamic forces computed from analytical or empirical equations to account for the flow effects on individual grains. These terms are linearly added to the particle linear momentum or angular momentum equations, which are integrated in time coupled with the trajectory for each particle. Models also account for grain interactions with the walls and collisions with other particles in flows with high particle concentrations, in what is called the discrete element method (DEM) (see Crowe, Sommerfeld, and Tsuji 1998).

Soldati and Marchioli (2012) analyzed the advantages and limitations of Lagrangian particle transport models in DNS for turbulent boundary layer flows compared to LES models. The influence of particle inertia and vertical sediment concentration leads to different requirements for the models to describe adequately the interactions between particles and near-wall turbulent coherent structures. A similar approach was employed in the recent work of Kidanemariam and Uhlmann (2014) and Ji *et al.* (2014), which represent the interparticle collisions and hydrodynamic forces with different degrees of detail on a DNS, identifying mechanisms of sediment transport at low Reynolds numbers in simplified domain configurations at small length scales. A different model for the sediment description was adopted in the DNS carried out by Yue *et al.* (2013). Their Eulerian approach for the sediment concentration helped to characterize the turbulent boundary layer flow and near-wall sediment transport by incorporating rheology changes and local stratification. DNS showed the effects on the turbulence damping near the wall induced by fine sediments. More complex applications of DNS models are currently intractable, due to the amount of data that would have to be analyzed even in simple flows.

1.3 Choosing the Right Modeling Approach

As presented in the previous sections, there is a wide range of applications and alternatives of numerical models that can be used to simulate river flows with different degrees of detail in terms of the representation of the conservation equations and description of morphology, mean flow field, and turbulence. The answer to the question "which model is best?" depends on the problem at hand, as the decision to move toward very realistic, albeit complex, modeling techniques is not necessarily an improvement from a broader scientific point of view. Here we discuss the ability of models to advance our knowledge and we also provide a few strategies to help in selecting the best approach, depending on the processes that control the specific problem.

1.3.1 Models as Research Tools

In research, the main purpose of numerical models is to gain insight about physical processes by solving a set of equations devised to predict the evolution of variables and describe the dynamics of the system. The definition of the system variables such as velocities, pressure, mass fluxes, concentrations, bed elevations, etc., and dimensionless parameters that appear in the equations are closely linked to time and length scales at which processes are being observed and analyzed. Therefore, variables in numerical models always imply an a priori definition of the scales of interest, except in LES-type models, where the smallest scales of the resolved flow field are established by the grid size and the instantaneous computed local velocity. This range of scales defines the scope and applicability of the model, which is a major issue as river flows are characterized by interconnected subsystems and interacting processes that cover a wide range of scales (Church 2008).

The nondimensional parameters in the equations (Reynolds, Froude, Shields, Richardson, etc.) characterize the physical system by incorporating the essential factors that govern the problem, and by assisting in identifying the importance of the dynamic variables in the model. Knowledge of the relations among these parameters can help to simplify the equations or select the appropriate modeling framework for specific problems – for example, including or excluding backwater effects based on *Fr* for large-scale models.

Once established, a model can help to predict the evolution of the variables for practical purposes, explore the causes and consequences of different mechanisms that control the flow, and help to pose new research questions and design new experiments. Therefore, models not only answer applied questions involving predicting and evaluating possible scenarios, but they also yield quantitative insights to understand aspects of the system dynamics, for example:

1) They can reveal patterns and self-organization. Models can uncover complex interactions of processes that occur in a hierarchical-scale system, from river networks to turbulence, and characterize statistically the dynamics of the variables in nonlinear systems, for example by showing that certain components of the system exhibit consistent scaling. Although high fidelity models can also reveal characteristic forms of self-organization (e.g., coherent structures in turbulence and bedform development), large-scale self-organization and pattern formation are often the targets of RCM approaches.
2) Through numerical models we can identify threshold values of dimensionless parameters that characterize transitions in the dynamics of the system.
3) We can recognize transport mechanisms and predominant coherent structures. Using simulations we can understand the controlling mechanisms of transport and mixing in the flow that give shape to large-scale patterns.
4) Models provide information on the interactions of the flow with the boundaries and sediment transport in mobile beds.
5) Numerical simulations produce different information compared to experiments. Once thoroughly tested and verified, and depending on the variables and resolution of the model, they can be used to investigate parts of the system or processes that are experimentally inaccessible or that cannot be directly measured.

In addition to the objectives and analysis of the results of a given model, there are two important technical aspects to model development that we do not discuss in this chapter. First, we do not attempt to summarize the large literature on the numerical strategies to solve the equations with a prescribed accuracy. We do not address the discretization techniques to create adequate computational grids with a specified resolution for complex and arbitrary geometries in aquatic environments, although these can significantly affect the results and calculation time of the simulations (e.g., Ferziger and Perić 2002). The second aspect not reviewed here is techniques for testing a given model, which help define its applicability and limitations. Testing is performed either (for deterministic models) by comparing directly the numerical results with individual experimental or field data measured or expressed at the same resolution (which also have a defined degree of precision), or by assessing if the results match the statistics of observed quantities with good accuracy. Since stochastic behavior is common in geomorphic systems and models of them, we would like to stress that the basic process of testing models that produce stochastic output is no different from testing deterministic models. The difference is that far more data are needed for model testing when the target is a probability distribution rather than a set of discrete predictions (e.g., mean velocity profile, average meander length). Sometimes a comprehensive validation of a model under certain conditions can allow us to extend its applicability, testing new scenarios to explore physical problems for which experimental data are not yet available. In these cases, once a model has been validated, we can trust that it is representing the system with good accuracy, and we can apply it to new cases to understand the dynamics under different or new conditions.

We want to state clearly our summary overall philosophy here: more is not better, i.e., very detailed and complex numerical models are not necessarily better for all situations. The most pragmatic basis to decide about how to represent a given real system is to decide if the cost and significant amount of additional data that is necessary for implementing a more detailed model will really provide new insights and produce meaningfully different results (note that we are excluding simulations built for practical prediction where life and property are at stake; arguably, in those cases we always want as much detail and fidelity as we can obtain). At a more fundamental level, we can ask if there is value in using a simpler model even if the computational costs of higher fidelity are not an issue. On one level it would seem the answer is obviously no – is it not always better to capture more of the underlying physics, remain closer to the fundamental governing equations, and obtain more detailed answers? We believe that, tantalizing though it is, there are good reasons to reject this view. The motivation for doing so becomes clearer if we ask instead "is there value in identifying and understanding what is the *minimal set* of processes needed to produce a given phenomenon?" We would not, for example, learn much by finding that a complete model (assuming we could build one!) of every process that could possibly contribute to shaping rivers, resolved down to the level of the finest grains, or perhaps to the cellular workings of organisms, reproduced the dynamics we observe. Nature has already performed that experiment for us. The questions are: "What do I learn from this model?" and "At which level(s) of complexity do I get more back than I put in?" We tackle these and related questions in the next section.

1.3.2 Heuristics: How to Select a Modeling Approach

At present there is no systematic, reliable method to define the model category appropriate for a specific phenomenon in nature. In practice, intuition and experience and some amount of trial and error are relied upon. Moreover, it seems safer to include more detail – there is a "precautionary principle" at work in that it is hard to see how including unnecessary detail does any harm, whereas leaving out something essential clearly does. For these reasons, people often assume that it is always better to include more detail. As stated above, one of our main goals in this chapter is to dispute this assumption. In our view, making a model more complex than necessary for the problem at hand not only slows the calculations but also slows the growth of understanding.

Complex models should not only produce more information and details of the flow and morphodynamics, but they should also produce new insights. One simple but useful test is to ask if the results of a high-fidelity model, degraded to a coarser resolution, improve the description and results obtained with a simpler model at that same coarse resolution.

The first decision to be made is whether to try a rule-based (RCM) approach versus some form of more traditional solver for the coupled flow–sediment equations. Although there is a continuum between these worlds, there is a fundamental difference between them: as stated above, in many RCMs the only conservation laws that are strictly enforced are those for mass balance (for water and sediment). These offer only limited constraint; model behavior is often governed by heuristic rules that come from a variety of sources, as discussed in section 1.2.2.

Werner (1999) proposed some general guidelines for when RCMs (broadly defined) are viable. He argued that in many morphodynamic systems the behavior at the scales at which patterns like bars and bedforms develop is largely decoupled from the details of the underlying dynamics. The recent models of Escauriaza and Sotiropoulos (2011a) and Khosronejad and Sotiropoulos (2014) show, on the other hand, that if one is interested in the details of formation of bedforms such as current ripples by the action of near-bed turbulence, a high level of turbulence resolution is required. While there is room for debate about the extent to which higher-level dynamics is independent of the details of lower-level dynamics across morphodynamic systems, we think that asking the question is a good way to begin thinking about whether a given system is amenable to the RCM approach or not. It is certainly a necessary condition for the RCM approach; whether it is sufficient remains to be seen.

One indicator that nature provides for independence of the processes governing a particular morphologic process is evidence of scale independence. Fractal geometry is in turn a common indicator of scale independence. This is discussed extensively in Paola *et al.* (2009) as a basis for understanding why laboratory experiments that are not true scale models of large-scale systems nonetheless seem to capture more of their behavior than would be expected based on classical dynamical scaling. Inasmuch as an experiment is a form of simplified "model" of the natural system, the same argument can be applied to deciding whether a simplified modeling approach might work. Scale independence is one indicator of Werner's (1999) condition of independence of large-scale dynamics from the details of small-scale processes.

At present, the only general alternative to RCM is to solve the coupled flow and sediment equations in one or more of the levels of approximation discussed in the previous section. Here the criteria for deciding on the approach are better developed. Specifically, SWE coupled with sediment transport and morphodynamic models effectively average fluctuations of the river channel morphology and flow variables in space over vertical and horizontal scales, preserving the relevant information of mechanisms that most likely contribute to the transport and river dynamics at large scales, such as the water discharge, channel slope, grain size, and sediment supply. The effects of turbulent coherent structures can be represented only in an averaged sense. For long times and large space scales, the use of a steady one-dimensional local uniform formulation with time variation represented only by a simple intermittency can take us surprisingly far, as the numerous papers applying the diffusion approximation to landscapes attest (section 1.2.1). For problems with length scales below the backwater length, the advective terms can become important, which require a backwater formulation for the SWE.

SWE- type models provide a fast, inexpensive technique to obtain a good approximation of the free surface dynamics when it is not affected by three-dimensional flow features. As pointed out by Keylock *et al.* (2005), problems depending mostly on bulk quantities, such as the effect of discharge on flood risk, do not benefit from having a well-resolved turbulent field with LES, or a similarly detailed approach. In some large-scale problems, however, it can be useful to understand the small-scale processes, as we discuss in the next section.

As we move toward higher-fidelity morphodynamic models, the most important aspect becomes the influence of turbulence on the problem at hand. If the details of the turbulent flow dynamics, especially features like coherent structures, leave a significant imprint on the processes of interest, we need models that can capture the major features of the three-dimensional flow at the scales that are relevant on the real system. URANS models with isotropic turbulence are simple approaches to represent three-dimensional flow fields in rivers, and more complex methods can be incorporated when turbulence anisotropy is significant, by using Reynolds-stress transport models or turbulence-resolving methods such as LES. Cases where anisotropy is large include important river phenomena such as secondary flows and strong flow separation. The dissipation introduced by URANS turbulence models hampers the numerical solution as we increase the grid resolution. Flow fields computed with URANS turbulence models will converge to a grid-independent solution, and they might not be able to resolve the unsteadiness of dynamically rich coherent structures.

When the flow unsteadiness and instantaneous turbulence fluctuations at smaller scales play a significant role on the dynamics of the flow field, the best alternative is to employ LES or hybrid formulations. These simulations, however, can become computationally expensive, as very fine grids are required at high Reynolds numbers to resolve the physical mechanisms that are important for transport and mixing processes in fluvial environments. It is important to note that boundary conditions in

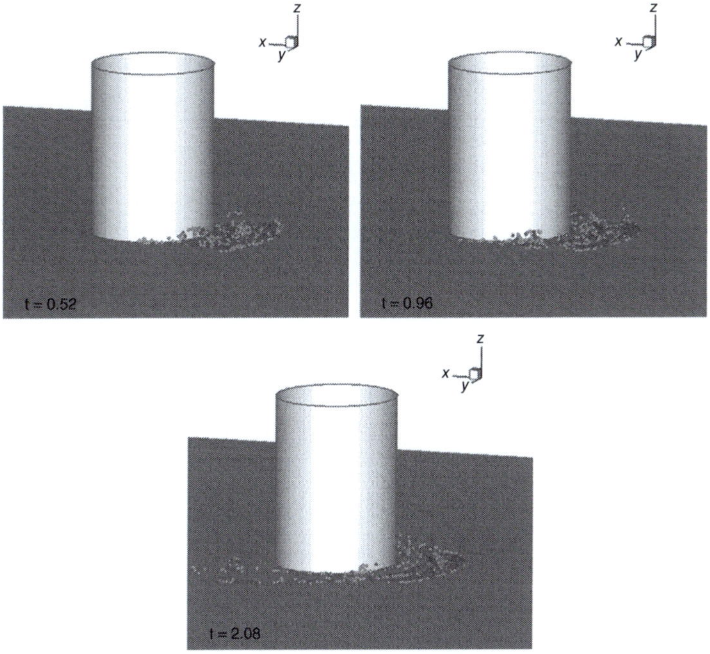

Figure 1.5 Instantaneous images of particle transport driven by the turbulent horseshoe vortex system resolved with DES, using the model of Escauriaza and Sotiropoulos (2011b). The model reproduces the intermittency of bedload transport near the threshold of motion controlled by the large-scale coherent structures of the flow. (*Source*: Escauriaza 2011. Reproduced with permission of Cambridge University Press.)

these three-dimensional models can also introduce additional errors to the numerical solutions, including the geometry of the solid boundaries, and the inlet velocity fields and turbulence variables, which are not necessarily known at the resolution of the simulations.

Two recent examples can illustrate the advantages of using highly resolved models in fluvial research. Hybrid URANS–LES and LES formulations have provided relevant new insights on fundamental mechanisms of large-scale coherent structures in turbulent flows past bed-mounted obstacles, and also in flow dynamics in meandering channels. The hybrid formulation known as DES (Spalart 2009) resolves essentially all turbulence scales produced by the flow around obstacles installed in a channel or a stream. In response to the adverse pressure gradient, a highly dynamic system of necklace-like vortical structures known as the turbulent horseshoe vortex is formed around the obstacle. Simulations have shown that intense velocity fluctuations generated by the coherent structures near the bed are characterized by low-frequency bimodal oscillations, quasi-periodic vortex shedding, vortex merging and wall interaction, and hairpin secondary structures due to a centrifugal instability that controls the vortex dynamics (see Paik, Escauriaza, and Sotiropoulos (2007) and Escauriaza and Sotiropoulos (2011c) for more details). These processes increase the instantaneous shear stress near the obstacle at levels sufficiently high to initiate bedload transport and scour. The insights obtained from these simulations allowed the development of a Lagrangian sediment transport model shown in Figure 1.5, which replicated numerically, and explained the multifractal properties of transport of sediment near the threshold of motion (Escauriaza and Sotiropoulos 2011b; Link *et al.* 2012). This Lagrangian model inspired a new Eulerian bedload transport model that was the first to capture the dynamics of bedforms produced by large-scale coherent structures in fine sand (Escauriaza and Sotiropoulos 2011a).

Other situations when LES is likely to be the best option are cases where the dynamics of large-scale vortices induces a strong anisotropy of the turbulent field. The flow in meander bends is characterized by secondary motions produced by imbalances in the transverse pressure gradient and the centrifugal force, generating at least one large streamwise helical rotating vortex, with flow velocities in the outer bank direction near the surface and inward near the bed. This main secondary flow is commonly accompanied by an outer-bank counter-rotating structure, and flow separation in the inner bank that produces a recirculation zone and a shear layer. The complex geometry of these flows and interactions among various turbulent scales in natural streams has been the subject of extensive research in recent decades. Recently, Kang and Sotiropoulos (2012) performed numerical simulations using steady-state RANS and LES in a natural meandering channel, employing a very detailed natural bathymetry with a fine grid discretization (approximately 4.9×10^7 nodes). They compared the mean flow and turbulence statistics (turbulent kinetic energy and bed shear-stress) for both approaches, and velocity profiles measured with an acoustic Doppler velocimeter (ADV). Even though the mean velocity profiles of the RANS simulation appeared closer to the measured velocity components, the comparison of the three-dimensional geometry and flow statistics between the models showed a totally different picture. Flow features such as the peak velocity near the outer bank, the inner recirculation zone, and most importantly the outer-bank streamwise structure were not captured by the RANS simulation, but were in the LES, showing the inability of RANS to resolve highly anisotropic flow fields.

These examples underscore the role of high-resolution LES models in cases when the limitations of URANS require a more complex modeling approach. They also illustrate an important general issue of modeling complex multiscale systems: having "almost high enough" fidelity in a model may actually be worse than having low fidelity. Even though they looked more realistic, the RANS secondary flows in Kang and Sotiropoulos's work were not much more accurate than rough approximations based on, for example, a simple radial force balance around the bend (Rozovskii 1961), but the large amount of rendered detail gave a false sense of fidelity.

These examples illustrate cases where, having done the modeling, we can see that the added detail of a high fidelity turbulence simulation was essential to understanding the phenomenon in question. Are there guidelines that would allow us to determine a priori how much detail is required?

One general suggestion is, for any approach, to aim to resolve a scale at least 10 times smaller than the smallest scale we think affects the dynamics. This intuitive idea has the purpose of decreasing the responsibility of the "modeled scales" to represent the effects of unresolved phenomena. Here we can learn from the experience of turbulence modeling, where unresolved scales acquire greater importance at increasing Reynolds numbers. The mathematical representations are based on hypotheses such as Boussinesq's with the introduction of the eddy viscosity to account for the influence of fluctuations on the variables of the model at the scales of interest. Analogous, though less well constrained, approximations have been applied to modeling landscapes (Passalacqua *et al.* 2006). In both of these cases (turbulence and erosional landscapes) the fractal nature of the phenomenon helps to guide the modeling of the unresolved scales, as discussed in the final section. Unfortunately, this feature is absent from many other landscape processes of interest.

Returning to turbulence, since most of the statistical or SGS parameterizations were designed for specific flows and scales and cannot be generalized, they wrongly approximate the dynamics of the unresolved small scales when the problem of interest is too close to that resolution in general situations. Experimental and field observations coupled with dimensional analysis can also guide this process by helping to identify the nondimensional parameters that affect the outcome of the simulations, and to quantify the relation between external forcings and internal processes of the system.

We propose five other major guidelines for selecting a modeling approach:

1) The RCM approach, probably the most controversial and untested of the approaches we discuss, is best applied to studies of self-organization and pattern formation under conditions in which there is reason to think that the essential dynamics of the self-organization process is not sensitive to the details of the underlying mechanics. Werner (1999) and Murray (2007, 2013) provide more analysis of this line of thinking. The RCM approach is especially well suited to exploratory studies where the goal is to find out if a given process set can even qualitatively capture the phenomenon in question. It is worth noting that even a heuristic, "hand-crafted" model of the right formative process is more useful than a fully worked out formal theory of the wrong one, as exemplified by the various models proposed for beach-cusp formation (e.g., Coco and Murray 2007).
2) If the focus is on large space and time scales where bulk quantities such as discharge, slope, grain size distribution, etc., dominate the morphology of the river system, then fairly simple formulations may provide a good deal of insight. This level of simplicity includes RCMs, the most highly averaged deterministic models, e.g., diffusive models that use a local uniform approximation, through to the SWE with basic backwater profiles. For bulk flood routing, the system of nonlinear SWEs in highly unsteady flows is a good approximation.
3) When three-dimensional effects of boundaries or initial conditions influence the dynamics of the system, and especially when the hydrostatic pressure assumption fails, the problem requires a three-dimensional formulation for the flow.
4) URANS is the most expedient alternative for three-dimensional flows, except when dynamically rich vortical structures that produce highly anisotropic flows (e.g., shear layers, flow separation, strong curvature, etc.) are the fundamental mechanisms for the problem of interest, and where they occur at a spatial scale close to the smallest scale that can be resolved with the discretization. In these cases, LES or URANS–LES would be the best options to resolve the time-dependent flow structure.
5) Wall-resolving LES or DNS are required when the fine-grained structure of turbulence is the driving mechanism of the phenomenon of interest, i.e., individual particle transport or, as we will see below, formation of small-scale bedforms in sandy sediment.

One way this general strategy can fail is in situations when a simple view of the scales of the problem breaks down. The worst case – and the antithesis of the basic assumption behind RC modeling discussed above – is where the dynamical details at smaller scales significantly affect the large-scale dynamics in ways that cannot be embodied in synthetic rules. In turbulence modeling, an example would be small-scale features that give rise, through a cascade of interactions, to large-scale coherent structures that locally dominate the bed stresses, flow velocities, and sediment fluxes. An example is the effect on turbulence structure of the formation and evolution of bedforms over an initial flat bed. These processes have been recently captured by the simulations of Khosronejad and Sotiropoulos (2014), showing that the bed friction changes considerably over time as bedforms develop, and in particular that turbulence processes initiating near the bed – itself quite complex and evolving with time – can interact with one another to generate large, coherent structures that then dominate stresses and sediment fluxes. Of course, the general idea that self-organized morphologic features such as bedforms can strongly affect turbulence structure is not new; what is emerging from these new models is a new appreciation of the complexity of the turbulence produced, the strength of its influence on bed evolution, and the difficulty of representing these effects without fully solving for the coupled turbulent flow and sediment-transport fields. At this point, we cannot say whether morphodynamics problems such as these are irreducibly complex, or whether we simply have not been smart enough to find ways to capture the complexity using simpler approaches (e.g., variants of the rule-based bedform models cited earlier: Werner and Kocurek 1997; Anderson and Bunas 1993; Haff and Anderson 1993). In the next section we focus on specific incremental steps that could be taken to improve high-fidelity models, but we suggest that, in a broader view, the way forward is to attempt new syntheses of these approaches.

1.4 Next Steps in Modeling

1.4.1 High-Fidelity Modeling

In several respects we are far from reaching a thorough understanding of important processes in rivers. New challenges seem to emerge in flows near interfaces, which include the interaction of sediment grains with the coherent structures of the turbulent boundary layer, and problems that appear at increasingly larger orders of magnitude and overlapped dynamical scales, such as grain roughness effects, bedform development and their influence on hydraulic resistance, large-scale secondary flow structures, river-channel size and shape, and the morphology and planform geometry of rivers.

Near the bed, the global sediment transport measured over a long period of time might depend on small-scale processes that occur at the scale of the particles and the coherent structures of the turbulent boundary layer at the interface. In unsteady flows, most of the models that account for bedload in three-dimensional turbulent flows have used empirical formulas to compute the bulk sediment flux, which have been developed in a totally different context, using field or experimental data and assuming uniform flow. As pointed out by Frey and Church (2009), these large errors of bedload transport formulas can be corrected in the future by representing directly the dynamics of grains in the flow and grain–grain interactions. Significant progress has been made recently by developing Lagrangian sediment transport models that couple a detailed description of the near-bed flow with the integration of the dynamics of a large number of grains (see, e.g., the recent work of Escauriaza and Sotiropoulos (2011b), Kidanemariam and Uhlmann (2014), Ji *et al.* (2014), Schmeeckle (2014), and González *et al.* (2016)). As discussed in the DNS-particle model section, the grain forces have so far been highly simplified using a DEM approach, with analytical expressions and coefficients to represent the hydrodynamic effects and collisions in a sort of "reduced complexity model" for the particle representation, without even considering the effects of the particle volume on fluid continuity (most models are coupled only through momentum, and not on the mass conservation equation).

Lagrangian models of sediment transport have yet to be developed to be able to explain sediment–turbulence interactions and consequences on the total bedload flux, but they have already helped to elucidate important dynamics of particle motion on complex turbulent flows, mainly coupled to LES or DNS formulations. These Lagrangian approaches require detailed descriptions of the flow since the smallest scales in a river flow near the bed correspond to the grain size and the turbulent structures that can effectively interact with the particles. Turbulent boundary layers are dominated by small coherent structures that produce intermittent velocity fluctuations and give rise to sequences of near-wall fluid ejections. Groups of so-called hairpin and streamwise vortices are known to develop at the wall and travel downstream, merging and generating low-speed streaks, and producing both turbulent kinetic energy and Reynolds stresses of high magnitude. Does this mean that in Lagrangian models of sediment transport we always need to perform DNS when studying the interaction of the boundary layer and sediment grains? Numerical investigations (e.g., Crowe *et al.* 1998; Escauriaza and Sotiropoulos 2009, 2011b; Soldati and Marchioli 2012) have demonstrated that nondimensional parameters, such as the Stokes number that relates the particle characteristic time scale with turbulence motions, indicate that very small eddies compared to particle time and length scales may have a limited effect on the grain dynamics. These results show that nondimensional parameters can be used to define the model resolution and the importance of small-scale fluctuations on transport. Coarsely resolved LES compared to the particle size can also substitute higher-resolution models by stochastic approaches that incorporate the SGS effects on particle motions (e.g., Kuerten and Vreman 2005; Soldati and Marchioli 2012). With analyses based on concepts from statistical mechanics and nonlinear dynamics (e.g., Furbish and Schmeeckle 2013), future LES models coupled with sediment transport will advance our knowledge to discover the small-scale processes that have larger implications on sediment transport and bed roughness.

Significant progress has also been made recently on the investigation of bedform dynamics using Eulerian models of sediment transport to reproduce the formation of ripples in turbulent flows as a consequence of turbulent sweep events near the bed. Models have been able to understand the mechanisms that generate the initial perturbations that later organize as ripples, reproducing the same statistical properties of real bedforms in experiments (Escauriaza and Sotiropoulos 2011a; Khosronejad and Sotiropoulos 2014).

Models of the free surface dynamics coupled to the flow, on the other hand, have been less explored. Two-dimensional formulations with the SWEs compute directly the dynamics of the free surface in subcritical and supercritical flows, but require specific functions to represent wave breaking and dissipation, which are especially important in the dynamics of steep, coarse-bedded rivers with Froude numbers near unity. In three-dimensional models it is very difficult to quantify the effects of the free surface in general situations. For low Froude numbers ($Fr < 0.35$), it is generally accepted that the dynamics of the free surface has little effect on the three-dimensional structure of the turbulent flow, and the "rigid-lid" symmetric boundary condition provides a good approximation. For flows at larger *Fr*, there are complex models that capture the water–air interface with advanced techniques such as volume-of-fluid (e.g., Ramamurthy, Qu, and Vo 2007) or level-set approaches (e.g., Yue, Lin and Patel 2005). No model has yet been developed to consider the interaction of the resolved and modeled turbulence scales with the free surface, especially in near-critical flows where the effects of waves, air entrainment, and breaking can strongly influence the dynamics of turbulence.

The parameterization of bed roughness is an additional topic in which high fidelity models can contribute to improve representations at larger scales. Rameshwaran, Naden, and Lawless (2011) and Zeng and Li (2012), for example, have recently demonstrated that three-dimensional RANS simulations perform better with wall models based on a spatially averaged roughness that considers the form-induced momentum due to drag, which can also be included in spatial correlations terms that emerge from the double-average approach presented by Nikora *et al.* (2007). A similar methodology can be applied on vegetated streams, in which the roughness depends on the area obstructed by vegetation and properties of the canopy as explained by Nepf (2012).

New theories of nutrient uptake processes, mobilization and stabilization of contaminants in the water and sediments, and hyporheic flow connections might also benefit from high fidelity approaches. Most of the models discussed above consider the channel bed is impermeable, i.e., that there is no flow normal to the bed surface. In reality, however, both standing waves on the free surface (Qian, Voller, and Stefan 2008) or flow over bedforms (e.g., Packman, Salehin, and Zaramella 2004) can create a pressure distribution that is sufficient to divert a fraction of the flow into and through the subsurface of the bed. Such hyporheic flows have a dramatic impact on both the movement of solutes and temperature of the system, in addition to modifying the turbulent and flow shear conditions at the bed surface. While specific models of hyporheic flows under an imposed pressure field have been developed (e.g., Packman, Salehin, and Zaramella 2004; Trauth *et al.* 2014; Marzadri *et al.* 2014), models that explicitly couple morphodynamic channel flow models with subsurface flows are still in early development. Future advances in the interactions of streamflow with the subsurface will allow us to gain knowledge of the mass exchange processes through models that couple the river flow with the bed.

1.4.2 Model Synthesis

Throughout the preceding sections, the unstated assumption has been that the appropriate level of model 'fidelity' for the problem at hand must be chosen, with reduced-complexity, typically rule-based models at one limit and fully resolved (DNS) models of turbulence and particle motion at the other. The guidelines given above provide, we hope, some insight as to how to choose the appropriate level for a given problem. But given that each approach has its advantages, it seems obvious that it would be better to avoid choosing one specific style, instead opting for a more holistic approach that blends modeling styles. Some would argue that it is always better to use the most detailed model available for a given problem, but we strongly disagree. Even with very fast computers, there will be problems, like running models over planetary time and space scales, or running thousands of model realizations in Monte-Carlo type studies of possible future scenarios, in which high-fidelity models are impractical. More importantly, we believe there will always be an intellectual payoff, in insight and in satisfaction, for finding the simplest way of representing a given set of dynamics in a model. At the same time, even if the aim is for radical simplification, it seems silly not to use the best computational tools available to develop the representations used in the simplified model. Yet, surprisingly, to our knowledge this has not been systematically pursued. Thus we conclude with some starting thoughts for working across the spectrum, combining high-fidelity and reduced-complexity models.

At present high fidelity approaches are typically applied to model problems chosen to illustrate the phenomenon of interest. How may the choice of model runs be different if instead the aim was to use them to elucidate a set of rules that could be used to build a much simpler (RC type) model? Given finite computing resources and time, how may an optimal set of high-fidelity model runs be designed to best embody the important physics being represented? May a thoughtfully chosen set of model experiments suggest better ways of implementing rules than, for example, brute-force curve fitting to model output (or observational data)? In a partial example from the opposite end of the sedimentary spectrum of gravel-bed rivers, Edmonds and Slingerland (2007) used a fairly high fidelity model (Delft3D with a URANS level turbulence treatment) to study the development of mouth bars in river deltas. Part of the justification for this, they argued, is that mouth bars are a critical element in creating the characteristic large-scale distributary pattern because they nucleate channel bifurcations. Results such as these seem to call out to be systematically mined for rules that could be applied in RCMs.

A complementary approach would be to use RCMs to explore "sensitive points" in the dynamics, for example, by varying rules and initial conditions to determine what aspects are most important for determining overall behavior. Findings from a reduced-complexity delta model (Liang, Voller, and Paola 2015; Liang *et al.* 2015), again from the low-gradient deltaic world, provide a partial example of this. The authors found that a key aspect of delta evolution (avulsion) was so sensitive to the behavior of the free surface that fairly detailed backwater dynamics were required for the model to function properly, even at a qualitative level.

The above ideas indicate how modeling styles of different complexity can be used in concert to add physical insight or accuracy to model predictions and calculations. An alternative, drawing from a classical modeling paradigm of integrating across modeling complexity, is to use the method of "progressive refinement." In essence, this involves starting with a reduced complexity model and, based on its predictions, then systematically adding model refinements (complexity) to the point where the problems of interest are resolved. In the past this progressive refinement approach has been restricted to models of a given class, for example, models built around numerical discretization of partial differential equations (PDEs). In this context model refinement (increase in complexity) involves such things as moving from constant to variable parameters, replacing diffusion transport with advective diffusion transport or, as detailed above, improving the resolution of turbulence modeling. Here, however, we suggest an important extension to the classical practice of model refinement, advocating that it not be confined to a given class of model on the spectrum but rather to stretch across a number of classes. That is, we are envisaging a process that, for a particular modeling purpose, may start out as a pure CA treatment and yet end up having a significant component based on classical high-fidelity methods like the numerical solution of PDEs. The research question that arises in this refinement is how do we merge from one level of model complexity to the next?

Our proposed holistic viewpoint of constructing and using models should also include the possibility of reversing the refinement procedure to allow for progressive coarsening. Here our vision is that the starting point is a high fidelity model that provides a detailed reproduction of the physics. The understanding of the physical mechanisms that underlie these phenomena are then "snagged out" by systematically deconstructing the complexity; this could be a way, for example, of constructing rules for RCMs as discussed above. At each step, the residual from the previous level of complexity informs the key elements in the resulting lower order model.

1.5 Concluding Questions

We would like to close this chapter with two synthesis questions that we hope will help to improve future model development:

1) Can lessons learned from the successive approximations developed over the years in turbulence research be applied to the RCM approach and is turbulence a good metaphor for understanding methods and limitations for abstracting small-scale dynamics in large-scale models?

   We believe that some fundamental concepts used in turbulence modeling can help improve models developed for complex nonlinear morphodynamical systems. Initial developments have successfully advanced our understanding on the effects that small-scale interactions produce on the dynamics of larger scales. The landscape evolution model of Passalacqua *et al.* (2006), for example, incorporated a subgrid-scale parameterization to account for the effects of small-scale fluctuations on the resolved erosion, in analogy to LES formulations. Paola (1996) considered the inherent variability of unresolved spatial scales to estimate the stochastic changes of the shear stress, and improve the description of sediment discharge and morphological features of braided streams.

   To describe the large-scale dynamics of Earth-surface systems using similar ideas to turbulence modeling, important issues will need to be addressed:

   - Self-similarity: Earth-surface systems exhibit scale-invariance for a wide range of spatial scales of many observable variables or statistical moments that describe these systems. Similar to the kinetic energy mostly transferred to smaller and more isotropic eddies and eventually dissipated into heat in turbulent flows, we will need to understand the connections among the scales in these other systems, and formulate hypotheses of the transfer mechanisms (analogous to vortex stretching in turbulence).
   - Parameter space: It is also important to identify the nondimensional parameters that can be used to characterize the transitions and dynamics of these systems. Models can be utilized to carry out systematic studies and explore the importance of different effects on the dynamics, varying the scales and also the magnitudes of the parameters. In analogy to the Reynolds number in turbulent flows, parameters such as the ratio of width to depth in braided rivers, Shields, Rouse number, or the granularity (Paola *et al.* 2009) have demonstrated that they can be used to characterize scale-invariance in geomorphic systems. It would be interesting to explore if the scale-invariance

observed in these physical phenomena is universal, and how it can be employed to improve models for rivers and landscape evolution (Haff, 2007).

2) Can we develop systematic methods for using high-fidelity models to inform RCMs?
   This would be the best way to resolve the greatest weakness of RCM but would require not only new levels of cooperation between modeling "camps" that have been quite disparate but also new ways of formulating problems to determine where the more synthetic models would most benefit from a concentrated dose of focused, high-fidelity modeling. Likewise, a well-formulated low-fidelity model may provide a means for bringing into clear focus the critical conditions in which the full power of a high-fidelity model is needed. Similar ideas have been proposed recently by Keylock (2015) to advance turbulence modeling strategies and approximations near the bed. When the anisotropy and intermittency of unresolved velocity fluctuations play a significant role on the dynamics of the flow, we can use detailed simulations or experiments to characterize these features and improve our description of boundary layer flows with LES, especially in complex flows over bedforms or vegetation.

Acknowledgments

This work has been supported by Fondecyt grant 1130940 and CONICYT/FONDAP Grant 15110017. We thank Professor Emmanuel Mignot for useful suggestions and Dr. Man Liang for providing the images of RCM fluvial delta formation. Comments and suggestions by Professor Thanos Papanicolaou and by an anonymous reviewer greatly improved the manuscript.

References

Anderson, RS and Bunas, KL 1993. Grain size segregation and stratigraphy in aeolian ripples modelled with a cellular automaton. *Nature* 365, 740–743.

Armanini, A, Fraccarollo, L, and Rosatti, G 2009. Two-dimensional simulation of debris flows in erodible channels. *Comput. Geosci.* 35, 993–1006.

Asahi, K, Shimizu, Y, Nelson, J, and Parker, G 2013. Numerical simulation of river meandering with self-evolving banks. *J. Geophys. Res.* 118, 2208–2229.

Begin, ZB, Meyer, DF, and Schumm SS 1981. Development of longitudinal profiles of alluvial channels in response to base-level lowering. *Earth Surf. Proc. Land.* 6, 49–68.

Church M 2008. Multiple scales in rivers. In *Gravel-bed Rivers VI: from Process Understanding to River Restoration* (eds H Habersack, H. Piegay, M Rinaldi). Elsevier: Amsterdam; 3–32.

Coco G and Murray AB 2007. Patterns in the sand: from forcing templates to self-organization. *Geomorphology*, 91, 271–290.

Contreras, MT, Müllendorff, D, Pastén, P *et al.* 2015. Potential accumulation of contaminated sediments in a reservoir of a high-Andean watershed: morphodynamic connections with geochemical processes. *Water Resour. Res.* 51, 3181–3192.

Coulthard, T J, Neal, JC, Bates, PD *et al.* 2013. Integrating the LISFLOOD-FP 2D hydrodynamic model with the CAESAR model: implications for modelling landscape evolution. *Earth Surf. Proc. Land.* 38, 1897–1906.

Crowe, CT, Sommerfeld, M, and Tsuji, Y 1998. *Multiphase Flows with Droplets and Particles.* CRC Press LLC: Montpelier, VT.

Cui, Y, Parker, G, and Paola, C 1996. Numerical simulation of aggradation and downstream fining. *J. Hydraul. Res.* 34, 185–204.

Cui, Y, Parker, G, Pizzuto, JE, and Lisle, TE 2003. Sediment pulses in mountain rivers. Part 2. Comparison between experiments and numerical predictions.*Water Resour. Res.* 39, 1240.

Cui, Y, Parker, G, Braudrick, C *et al.* 2006a. Dam Removal Express Assessment Models (DREAM). Part 1: model development and validation. *J. Hydraul. Res.* 44, 291–307.

Cui, Y, Braudrick, C, Dietrich, WE *et al.* 2006b. Dam Removal Express Assessment Models (DREAM). Part 2: sample runs/sensitivity tests. *J. Hydraul. Res.* 44, 308–323.

De Almeida, GAM, Bates, P, Freer, JE, and Souvignet, M 2012. Improving the stability of a simple formulation of the shallow water equations for 2-D flood modeling. *Water Resour. Res.* 48, W05528.

Doeschel-Wilson, AB and Ashmore, PE 2005. Assessing a numerical cellular braided stream model with a physical model. *Earth Surf. Process. Landf.* 30, 519–540.

Duc, BM and Rodi, W 2008. Numerical simulation of contraction scour in an open laboratory channel. *J. Hydraul. Eng.* 134, 367–377.

Durbin, PA and Pettersson-Reif, BA 2011. *Statistical Theory and Modeling for Turbulent Flows*, 2nd edn. John Wiley & Sons, Ltd.

Edmonds, DA and Slingerland, RL 2007. Mechanics of river mouth bar formation: Implications for the morphodynamics of delta distributary networks. *J. Geophys. Res.* 112, F02034.

El kadi Abderrezzak, K, and Paquier, A 2009. One-dimensional numerical modeling of sediment transport and bed deformation in open channels. *Water Resour. Res.* 45, W05404.

Engelund, F and Hansen, E 1972. *A Monograph on Sediment Transport.* Teknik Vorlag: Copenhagen.

Escauriaza, C and Sotiropoulos, F 2009. Trapping and sedimentation of inertial particles in three-dimensional flows in a cylindrical container with exactly counter-rotating lids. *J. Fluid Mech.* 641, 169–193.

Escauriaza, C and Sotiropoulos, F 2011a. Initial stages of erosion and bed form development in a turbulent flow around a cylindrical pier. *J. Geophys. Res.* 116, F03007.

Escauriaza, C and Sotiropoulos, F 2011b. Lagrangian model of bed-load transport in turbulent junction flows. *J. Fluid Mech.* 666, 36–76.

Escauriaza, C and Sotiropoulos, F 2011c. Reynolds number effects on the coherent dynamics of the turbulent horseshoe vortex system. *Flow Turbul. Combust.* 86, 231–262.

Fasolato G, Ronco P, Langendoen EJ, and Di Silvio G 2011. Validity of uniform flow hypothesis in one-dimensional morphodynamic models. *J. Hydraul. Eng.* 37, 183–195.

Ferguson, RI 2003. The missing dimension: effects of lateral variation on 1-D calculations of fluvial bedload transport. *Geomorphology* 56, 1–14.

Ferguson, RI and Church, M 2009. A critical perspective on 1-D modeling of river processes: gravel load and aggradation in lower Fraser River. *Water Resour. Res.* 45, W11424.

Ferguson, RI, Parsons, DR, Lane, SN, and Hardy, RJ 2003. Flow in meander bends with recirculation at the inner bank. *Water Resour. Res.* 39, 1322.

Ferziger, JH and Peric, M 2002. *Computational Methods in Fluid Dynamics.* Springer-Verlag: New York.

Frey, P and Church, M 2009. How river beds move. *Science* 325, 1509–1510.

Furbish, DJ and Schmeeckle, MW 2013. A probabilistic derivation of the exponential-like distribution of bed load particle velocities. *Water Resour. Res.* 49, 1–15.

Ge, L and Sotiropoulos, F 2005. 3D unsteady RANS modeling of complex hydraulic engineering flows. Part I: numerical model. J. Hydraul. Eng. 131, 800–808.

Ge, L, Lee, SO, Sotiropoulos, F, and Sturm, TW 2005. 3D unsteady RANS modeling of complex hydraulic engineering flows. Part II: model validation and flow physics. *J. Hydraul. Eng.* 131, 809–820.

Gill, MA 1983. Diffusion model for degrading channels. *J. Hydraul. Res.* 21, 369–378.

González, C, Richter, DH, Bolster, D *et al.* 2016. Characterization of bedload intermittency near the threshold of motion using a Lagrangian sediment transport model. *Environ. Fluid Mech.* DOI:10.1007/s10652-016-9476-x.

Guerra, M, Cienfuegos, R, Escauriaza, C *et al.* 2014. Modeling rapid flood propagation over natural terrains using a well-balanced scheme. *J. Hydraul. Eng.* 140, 04014026.

Haff, PK 2007. The landscape Reynolds number and other dimensionless measures of Earth surface processes. *Geomorphology* 91,178–185.

Haff, PK and Anderson RS 1993. Grain scale simulations of loose sedimentary beds – the example of grain-bed impacts in aeolian saltation. *Sedimentology* 40, 175–198.

Hodges, B 2013. Challenges in continental river dynamics. *Environ. Modell. Softw.* 50, 16–20.

Howard, AD 1992. Modeling channel migration and floodplain sedimentation in meandering streams. In *Lowland Flooplain Rivers: Geomorphological Perspectives* (eds P Carling and G Petts). John Wiley & Sons, Ltd: Chichester.

Ji, C, Munjiza, A, Avital, E *et al.* 2013. Direct numerical simulation of sediment entrainment in turbulent channel flow. *Phys. Fluids* 25, 056601.

Kang, S and Sotiropoulos, F 2012. Numerical modeling of 3D turbulent free surface flow in natural waterways. *Adv. Water Resour.* 40, 23–36.

Kang, S, Lightbody, A, Hill, C, and Sotiropoulos, F 2011. High-resolution numerical simulation of turbulence in natural waterways. *Adv. Water Res.* 34, 98–113.

Keylock, CJ 2015. Flow resistance in natural, turbulent channel flows: The need for a fluvial fluid mechanics. *Water Resour. Res.* 51, 4374–4390.

Keylock, CJ, Hardy, RJ, Parsons, DR *et al.* 2005. The theoretical foundation and potential for large-eddy simulation (LES) in fluvial geomorphic and sedimentological research. *Earth-Sci. Rev.* 71, 271–304.

Kidanemariam, AG and Uhlmann, M 2014. Direct numerical simulation of pattern formation in subaqueous sediment. *J. Fluid Mech.* 750, 1–13.

Kim, W, Dai, A, Muto, T, and Parker, G 2009. Delta progradation driven by an advancing sediment source: Coupled theory and experiment describing the evolution of elongated deltas. *Water Resour. Res.* 45, W06428.

Khosronejad, A and Sotiropoulos, F 2014. Numerical simulation of sand waves in a turbulent open channel flow. *J. Fluid Mech.* 753, 150–216.

Kostic, S and Parker G 2003a. Progradational sand–mud deltas in lakes and reservoirs. Part 1. Theory and numerical modeling. *J. Hydraul. Res.* 41, 127–140.

Kostic, S and G. Parker G 2003b. Progradational sand–mud deltas in lakes and reservoirs. Part 2. Experiment and numerical simulation. *J. Hydraul. Res.* 41, 141–152.

Kuerten, JGM and Vreman, AW 2005. Can turbophoresis be predicted by large-eddy simulation?' *Phys. Fluids* 17, 011701.

Lane, EW 1955. Design of stable channels. *Trans. ASCE* 120, 1234–1279.

Lane, SN, Bradbrook, KF, Richards, KS *et al.* 1999. The application of computational fluid dynamics to natural river channels: three-dimensional versus two-dimensional approaches. *Geomorphology* 29, 1–20.

Liang, M, Voller, VR, and Paola, C 2015. A reduced-complexity model for river delta formation – Part 1: Modeling deltas with channel dynamics. *Earth Surf. Dynam.* 3, 67–86.

Liang, M, Geleynse, M, Edmonds, DA, and Passalacqua, P 2015. A reduced-complexity model for river delta formation – Part 2: Assessment of the flow routing scheme. *Earth Surf. Dynam.* 3, 87–104.

Link, O, González. C, Maldonado, M, and Escauriaza, C 2012. Coherent structure dynamics and sediment particle motion around a cylindrical pier in developing scour holes. *Acta Geophys.* 60, 1689–1719.

Martin, Y and Church, M 1997. Diffusion in landscape development models: on the nature of basic transport relations. *Earth Surf. Proc. Land.* 22, 273–279.

Marzadri, A, Tonina, D, McKean, JA *et al.* 2014. Multi-scale streambed topographic and discharge effects on hyporheic exchange at the stream network scale in confined streams. *J. Hydrol.* 519, 1997–2011.

McCoy, A, Constantinescu, G, and Weber, LJ 2008. Numerical investigation of flow hydrodynamics in a channel with a series of groynes. *J. Hydraul. Eng.* 134, 157–172.

Meneveau, C and Katz, J 2000. Scale-invariance and turbulence models for large-eddy simulation. *Annu. Rev. Fluid Mech.* 32:1–32.

Murray, AB 2007. Reducing model complexity for explanation and prediction. *Geomorphology* 90, 178–191.

Murray, AB 2013. Which models are good (enough), and when? In *Treatise on Geomorphology* (ed. ACW Baas), Vol. 2, *Quantitative Modeling of Geomorphology*. Academic Press: San Diego, CA; 50–58.

Murray, AB and Paola, C 1994. A cellular model of braided rivers. *Nature* 371, 54–57.

Murray, AB and Paola, C 1997. Properties of a cellular braided-stream model. *Earth Surf. Proc. Land.* 22, 1001–1025.

Nagata, N, Hosoda, T, Nakato, T, and Muramoto, Y 2005. Three-dimensional numerical model for flow and bed deformation around river hydraulic structures. *J. Hydraul. Eng.* 131, 1074–1087.

Nepf, HM 2012. Hydrodynamics of vegetated channels. *J. Hydraul. Res.* 50, 262–279.

Nicholas, AP 2010. Reduced-complexity modeling of free bar morphodynamics in alluvial channels. *J. Geophys. Res.* 115, F04021.

Nicholas, AP, Thomas, R, and Quine, TA 2006. Cellular modelling of braided river form and process. In *Braided Rivers: Process, Deposits, Ecology and Management* (eds GH Sambrook, JL Best, CS Bristow and GE Petts). Blackwell Publishing Ltd: Oxford.

Nikora, V, McLean, S, Coleman, S *et al.* 2007. Double averaging concept for rough-bed open-channel and overland flows: application. *J. Hydraul. Eng.* 33, 884–895.

Olsen, NRB 2003. Three-dimensional CFD model of self-forming meandering channel. *J. Hydraul. Eng.* 129, 366–372.

Overeem, I, Syvitski, JPM, and Hutton, EWH 2005. Three-dimensional numerical modeling of deltas. In *River Deltas: Concepts, Models and Examples* (eds JP Bhattacharya and L Giosan). *SEPM Spec. Publ.* 83, 13–30.

Packman, AI, Salehin, M, and Zaramella, M 2004. Hyporheic exchange with gravel beds: basic hydrodynamic interactions and bedform-induced advective flows. *J. Hydraul. Eng.* 130, 647–656.

Paik, J, Escauriaza, C, and Sotiropoulos, F 2007. On the bimodal dynamics of the turbulent horseshoe vortex system in a wing-body junction. *Phys. Fluids* 19, 045107.

Paola, C 1996. Incoherent structure: turbulence as a metaphor for stream braiding. In *Coherent Flow Structures in Open Channels* (eds P Ashworth, SJ Bennett, JL Best and SJ McLelland), John Wiley & Sons, Ltd: Chichester; 705–723,

Paola, C 2000. Quantitative models of sedimentary basin filling. *Sedimentology* 47, 121–178.

Paola, C and Voller, VR 2005. A generalized Exner equation for sediment mass balance. *J. Geophys. Res.* 110, F04014.

Paola, C, Heller, PL, and Angevine, CL 1992. The large-scale dynamics of grain-size variation in alluvial basins 1: theory. *Basin Res.* 4, 73–90.

Paola, C, Parker, G, Mohrig, DC, and Whipple, KX 1999. The influence of transport fluctuations on spatially averaged topography on a sandy, braided fluvial fan. In *Numerical Experiments in Stratigraphy: Recent Advances in Stratigraphic and Sedimentologic Computer Simulations* (eds JW Harbaugh, WL Watney, EC Rankey *et al.*). *SEPM Spec. Publ.* 62, 211–218,

Paola, C, Straub, K, Mohrig, D, and Reinhardt, L 2009. The "unreasonable effectiveness" of stratigraphic and geomorphic experiments. *Earth-Sci. Rev.* 97, 1–43.

Papanicolaou, AN, Bdour, A, and Wicklein, E 2004. One-dimensional hydrodynamic/sediment transport model applicable to steep mountain streams. *J. Hydraul. Res.* 42, 357–375.

Papanicolaou AN, Elhakeem M, Krallis G *et al.* 2008. Sediment transport modeling review-current and future developments. *J. Hydraul. Eng.* 134, 1–14.

Parker, G, Paola, C, Whipple, KX, and Mohrig D 1998. Alluvial fans formed by channelized fluvial and sheet flow. *I: theory. J. Hydraul. Eng.* 124, 985–995.

Passalacqua, P, Porté-Agel, F, Foufoula-Georgiou, E, and Paola, C 2006. Application of dynamic subgrid-scale concepts from large-eddy simulation to modeling landscape evolution. *Water Resour. Res.* 42, W06D11.

Piomelli, U and Balaras, E 2002. Wall-layer models for large- eddy simulations. *Ann. Rev. Fluid Mech.* 34, 349–374.

Pope, SB 2000. *Turbulent Flows*. Cambridge University Press.

Qian, Q, Voller, VR, and Stefan HG 2008. A vertical dispersion model for solute exchange induced by underflow and periodic hyporheic flow in a stream gravel bed. *Water Resour. Res.* 44, W07422.

Ramamurthy, AS, Qu, J, and Vo, D 2007. Numerical and experimental study of dividing open-channel flows. *J. Hydraul. Eng.* 133, 1135–1144.

Rameshwaran, P, Naden, PS, and Lawless, M 2011. Flow modelling in gravel-bed rivers: rethinking the bottom boundary condition. *Earth Surf. Process. Land.* 36, 1350–1366.

Ribberink, JS and van der Sande, JTM 1985. Aggradation in rivers due to overloading: analytical approaches. *J. Hydraul. Res.* 23, 273–284.

Rodriguez, JF, Bombardelli, FA, Garcia, MH *et al.* 2004. High-resolution numerical simulation of flow through a highly sinuous river reach. *Water Resour. Manag.* 18, 1573–1650.

Roulund, A, Sumer, BM, Fredsøe F, and Michelsen, J 2005. Numerical and experimental investigation of flow and scour around a circular pile. *J. Fluid Mech.* 534, 351–401.

Rozovskii, IL 1961. *Flow of Water in Bends of Open Channels*. No. OTS 60-51133, translation by Y. Prushansky, Office of Technical Service, US Department of Commerce.

Sagaut, P 2006. *Large Eddy Simulation for Incompressible Flows*, 3rd edn. Springer-Verlag, Berlin/New York.

Sandbach, SD, Lane, SN, Hardy, RJ *et al.* 2012. Application of a roughness-length representation to parameterize energy loss in 3D numerical simulations of large rivers. *Water Resour. Res.* 48, W12501.

Scheidegger, AE 1970. *Theoretical Geomorphology*. Springer-Verlag: Berlin.

Schmeeckle, MW 2014. Numerical simulation of turbulence and sediment transport of medium sand. *J. Geophys. Res.* 119, 1240–1262.

Seybold, H, Andrade, JS, and Herrmann HJ 2007. Modeling river delta formation. *Proc. Natl. Acad. Sci.* 104, 16804–16809.

Seybold, H, Molnar, P, Singer, HM *et al.* 2009. Simulation of birdfoot delta formation with application to the Mississippi Delta. *J. Geophys. Res.* 114, F03012.

Sinha, S, Sotiropoulos, F, and Odgaard, A 1998. Three-dimensional numerical model for flow through natural rivers. *J. Hydraul. Eng.* 124, 13–24.

Soldati, A and Marchioli, C 2012. Sediment transport in steady turbulent boundary layers: Potentials, limitations, and perspectives for Lagrangian tracking in DNS and LES. *Adv. Water Res.* 48, 18–30.

Spalart, PR 2000. Strategies for turbulence modeling and simulations. *Int. J. Heat Fluid Flow.* 21, 252–263.

Spalart, PR 2009. Detached-eddy simulation. *Ann. Rev. Fluid Mech.* 41, 181–202.

Speziale, CG 1998. Turbulence modeling for time-dependent RANS and VLES: a review. *AIAA J.* 36, 173–184.

Stecca, G, Siviglia, A, and Blom, A 2014. Mathematical analysis of the Saint-Venant-Hirano model for mixed-sediment morphodynamics. *Water Resour. Res.* 50, 2014WR015251.

Stoesser, T 2014. Large-eddy simulation in hydraulics: *Quo Vadis*? *J. Hydraul. Res.* 52, 441–452.

Tetzlaff, DM and Harbaugh, JW 1989. *Simulating Clastic Sedimentation*. Van Nostrand Reinhold: New York.

Toro, EF 2001, *Shock-Capturing Methods for Free-Surface Shallow Flow*. John Wiley & Sons, Ltd: Chichester.

Trauth, N, Schmidt, C, Vieweg, M *et al.* 2014. Hyporheic transport and biogeochemical reactions in pool–riffle systems under varying ambient groundwater flow conditions. *J. Geophys. Res. Biogeosci.* 119, 910–928.

Tucker, GE and Hancock, GR 2010. Modeling landscape evolution. *Earth Surf. Proc. Land.* 35, 28–50.

Tucker, GE, Lancaster, S, Gasparini, N, and Bras, R 2001. The Channel–Hillslope Integrated Landscape Development model (CHILD). In *Landscape Erosion and Evolution Modeling* (eds RS Harmon and WW Doe III). Kluwer: New York.

Van de Wiel, MJ, Coulthard, TJ, Macklin, MG, and Lewin, J 2007. Embedding reach-scale fluvial dynamics within the CAESAR cellular automaton landscape evolution model. *Geomorphology*, 90, 283–301.

Verhaar, PM, Biron, PM, Ferguson, RI, and Hoey, TB 2008. A modified morphodynamic model for investigating the response of rivers to short-term climate change. *Geomorphology* 101, 674–682.

Viparelli, E, Sequeiros, O, Cantelli, A *et al.* 2010. Modeling of river morphodynamics with creation/consumption of grain size stratigraphy. Part 2: numerical model. *J. Hydraul. Res.* 48, 727–741.

Viparelli, E, Gaeuman, D, Wilcock, P, and Parker, G 2011. A model to predict the evolution of a gravel bed river under an imposed cyclic hydrograph and its application to the Trinity River. *Water Resour. Res.* 47, W02533.

Werner,BT 1999. Complexity in natural landform patterns. *Science* 284, 102–104.

Werner, BT and Kocurek, G 1997. Bed-form dynamics: Does the tail wag the dog? *Geology* 25, 771–774.

Whipple, K and Tucker, G 2002. Implications of sediment-flux dependent river incision models for landscape evolution. *J. Geophys. Res.* 107 (B2), 2039.

Wilcox, DC 1994. *Turbulence Modeling for CFD*. DCW Industries, Inc.: Lake Arrowhead, CA.

Willgoose, G, Bras, RL, and Rodríguez-Iturbe, I 1991. A coupled channel network growth and hillslope evolution model: 1. *Theory. Water Resour. Res.* 27, 1671–1684.

Wolfram, S 2002, *A New Kind of Science*. Wolfram Media Inc.: Champaign, IL.

Wu, W, Rodi, W, and Wenka, T 2000. 3D numerical model of flow and sediment transport in open channels. *J. Hydraul. Eng.* 126, 4–15.

Yue, W, Lin CL, and Patel VC 2005. Large eddy simulation of turbulent open-channel flow with free surface simulated by level set method. *Phys. Fluids* 17, 025108.

Zeng, C and Li, CW 2012. Modeling flows over gravel beds by a drag force method and a modified S–A turbulence closure. *Adv. Water Res.* 46, 84–95.

Discussion

Discussion by Philippe Frey

This discussion is about free surface effects, which are likely to become more important on steep slope mountain streams. Effects include very low relative submergence consequences on turbulent flow patterns, flow resistance and bedload transport, emerging boulder issues but also the free surface acting as a barrier for bedload grain trajectories. Could the authors comment on issues of this free surface effect and on the associated experimental, modelling, numerical, and theoretical challenges?

Discussion by Colin Rennie

You provided an excellent summary of the range of numerical models available for morphodynamic modelling of gravel bed rivers, presenting Kang and Sotiropoulos (2012), who compared fully converged isotropic RANS and LES models to flow measurements in a meander bend collected with an acoustic Doppler velocimeter. LES was able to resolve secondary flow and turbulent vortices whereas RANS was not. However, RANS yielded better predictions of mean velocities. We have also made this observation when comparing fully converged anisotropic RANS and LES models of bend flow (Kashyap 2012), and have speculated that larger LES errors in mean velocity may be due to inexact locations and sizes of predicted flow structures. Possibly, such errors may arise from LES sensitivity to boundary conditions. The consequence may be incorrect spatial distribution of boundary shear stress, which could have implications for accuracy of morphodynamic predictions. Discussion of this point would be appreciated.

Reference

Kashyap, S 2012. *A 3-D numerical study of flow, coherent structures, and mechanisms leading to scour in a high curvature 135° channel bend with and without submerged groynes*. PhD thesis, University of Ottawa.

Reply by the Authors

P. Frey and C. Rennie make interesting points discussing the details of modeling techniques and future challenges to advance our scientific understanding of fluvial systems. The effects of the free surface, which we have discussed in section 1.4.1 of this chapter, is one of the topics that will require new ideas to study flows that have not been fully explored with numerical simulations. As we have emphasized for all models in general, the definition of the level of complexity of the model depends on the scientific questions of the research problem. To simulate flows at larger time and length scales, we will need to simplify the representation of complex surface waves, and processes such as (i) wave breaking, (ii) energy dissipation, (iii) air entrainment, and (iv) free surface and turbulence interactions. Similar approaches will have to be adopted to account for flow resistance, in cases where low relative submergence and obstacles are important. High fidelity models that we highlight in this chapter have recently provided new insights on some of these topics, which can help advancing these tasks. Future research should address questions on the effects of free surface phenomena and flow resistance to improve current descriptions, investigating self-invariance in the case of resistance, and imprints of all these processes on flow statistics. The development of models will also require new and more detailed experimental data to improve the descriptions, which in many cases is difficult to obtain in complex free-surface shallow flows, especially for the four processes described above.

We also agree with the observation by C. Rennie regarding the ability of RANS or LES to resolve highly anisotropic flows, such as the meander bends studied by Kang and Sotiropoulos (2012) and Kashyap, (2012). It is important to note that in addition to the potential effects of the boundary conditions, there are other aspects of the model that can affect the results of the simulations. When turbulence is important, previous knowledge about the physics of the flow and the hypotheses behind the turbulence closures can become critical to the selection and performance of the model. For the implementation of LES in river flows, for example, the relation between the spatial resolution of the discretization and the turbulence model can have a significant impact on the results of the simulation. As the discretization becomes coarser, the SGS model has the responsibility of representing a larger portion of the turbulent kinetic energy of the flow, especially if energetic large-scale coherent structures cannot be resolved in a coarser grid. Additional factors need to be considered such as the treatment of solid walls and the resolved near-wall flow, the free surface, and the inlet boundary conditions, which in many cases require unsteady conditions to capture the turbulence dynamics. A systematic analysis of these factors in the models can be used to explore the importance of some particular aspect of the problem, such as the effects of the unresolved turbulence on the global dynamics of the system.

2

Boulder Effects on Turbulence and Bedload Transport

A.N. (Thanos) Papanicolaou and Achilleas G. Tsakiris

2.1 Boulders in the Riverine Continuum

Bedforms are believed to be derived from the complex feedback interactions between flow structures such as vortices and sediment and constitute an integral component of the riverine continuum spanning from the headwaters to the river outlet (e.g., Church 2006; Frey and Church 2011; Jerolmack and Mohrig 2005; Kuhnle 1996; Lanzoni and Tubino 1999; Müller and Gyr 1996; Singh *et al.* 2012). Bedforms have been studied extensively in the literature and are well documented to alter the near-bed turbulent flow patterns as well as affect the magnitude and spatiotemporal characteristics of sediment transport (e.g., Bennett and Bridge 1995; Bridge 2009; Curran and Tan 2014; MacVicar and Roy 2007; Nelson *et al.* 1995; Strom and Papanicolaou 2007; Strom *et al.* 2004). Because of their importance in the fluvial system, classification or taxonomy tools, such as phase or regime diagrams have been developed to delineate the flow–sediment conditions governing the formation and disintegration of each bedform type.

For the lower river segments, which are typically comprised of sand sized bed surface sediment, classification tools are relatively well-established in the literature (e.g., Best 1996; Bridge 2009). In contrast, a hierarchical bedform classification system has yet to be developed for upper river segments, where bed surface sediment is coarser, flow is transcritical, and movement of sediment is intermittent (Ghilardi, Franca, and Schleiss 2014a, 2014b; Singh *et al.* 2009; Strom *et al.* 2004). In these systems, bedform formation and evolution is highly controlled by the interaction of flow and ubiquitous roughness elements (e.g., Grams and Wilcock 2007; Parker 2008; Wilcock and Kenworthy 2002). An outcome of this interaction is believed to be the generation of cluster bedforms, which are sought to form and evolve during the rising or falling limb of a hydrograph (e.g., Billi 1988; Brayshaw 1984; Reid, Frostick, and Brayshaw 1992).

A particle cluster, or simply cluster, is considered to be a group of distinguishable sediment grains found atop the surrounding gravel bed surface. A universal definition of clusters is at present not available. Nevertheless, a typical cluster consists of an obstacle clast, also known as the anchor particle, against which a stoss-side accumulation of imbricated finer particles develops and behind which a wake tail grows (e.g., Brayshaw 1984; Lacey and Roy 2007; Laronne and Carson 1976; Laronne, Garcia, and Reid 2001; Reid, Frostick and Brayshaw 1992; Strom and Papanicolaou 2008). Papanicolaou and Kramer (2005) recently challenged this paradigm by observing that the deposition location of the

Gravel-Bed Rivers: Processes and Disasters, First Edition. Edited by Daizo Tsutsumi and Jonathan B. Laronne.

incoming material in relation to the anchor particle (stoss or wake) was found to vary depending on the relative submergence of the anchor particle. More recently, Strom and Papanicolaou (2008) and Tsakiris and Papanicolaou (2008) provided qualitative information about different cluster shapes and spacing. Further, Papanicolaou, Tsakiris, and Strom (2012) employed fractal theory as a means of classifying cluster shapes and relating them to flow characteristics (Griffiths 1967; Hendrick, Ely, and Papanicolaou 2010; Papanicolaou and Schuyler 2003).

Compilation of available bed-state data found in the lower and upper regimes and obtained by the ingenious work of Best (1996), Montgomery and Buffington (1997), Chin (1999), Wohl and Merritt (2008) as well as of Thompson and Croke (2008) for established bedforms, in combination with preliminary data on cluster evolution from the experimental and field studies of the Papanicolaou group in Figure 2.1, provide a first attempt to "place" clusters within the river continuum (from Hendrick, Ely, and Papanicolaou 2010; Papanicolaou and Kramer 2005; Papanicolaou and Schuyler 2003; Papanicolaou *et al.* 2003; Strom and Papanicolaou 2008, 2009; Strom *et al.* 2004).

The data compiled in Figure 2.1 reveal that the flow–sediment regimes corresponding to particle cluster formation are quite comparable to those of pool-riffle and plain-bed bedforms and within the limit of step-pools in the upper portions of the riverine continuum. In other words, particle clusters are small-scale microtopographic features, which are superimposed on the larger scale bedforms, such as the pool–riffles, plane beds and step-pools.

Because the formation and presence of clusters is interwoven with the availability of anchor particles, we describe in more detail below the influence of these anchor particles on the approach flow and cluster morphology. Anchor particles modify their surrounding flow field by dampening the level of turbulent intensity in the streamwise direction and creating pockets of reduced flow velocity in their

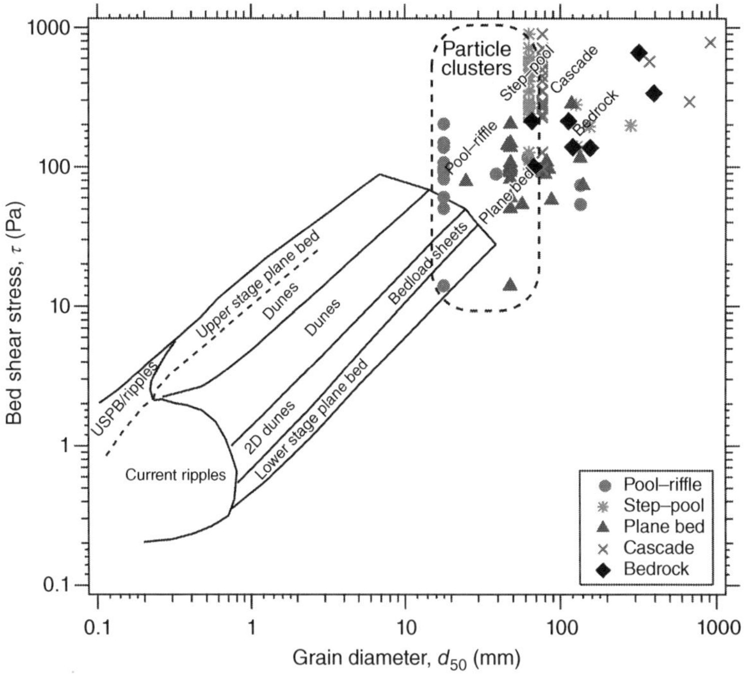

Figure 2.1 A phase diagram for sand and gravel bed streams from compilation of field and laboratory measurements.

upstream and downstream locations (e.g., Curran and Tan 2014; Hajimirzaie *et al.* 2014; Lacey and Roy 2007; Lawless and Robert 2001; Papanicolaou *et al.* 2012; Strom and Papanicolaou 2007; Yager *et al.* 2012). The degree of modification of the flow field and type of vortices found around the anchor particles is strongly dependent on the anchor particle relative submergence (e.g., Flammer *et al.* 1970; Shamloo, Rajaratnam, and Katopodis 2001). The anchor particle relative submergence, H/d_c, is defined as the ratio of the approach flow depth to the characteristic diameter of the anchor particle. The anchor particle relative submergence, H/d_c, encapsulates the effects of flow depth and anchor particle length scale on flow modification around anchor particles. These modifications on flow vortex structures and depositional patches remain unclear (e.g., Lacey and Rennie 2012; Papanicolaou and Kramer 2005; Shamloo, Rajaratnam, and Katopodis 2001; Suszka 1991). In streams with poorly sorted coarse bed material, i.e., having geometric standard deviation, $\sigma_g = (d_{84}/d_{16})^{0.5} > 1.50$ (e.g., Bathurst 1978; 1987; Wittenberg, Laronne and Newson 2007), H/d_c affects the depositional patterns around the anchor particles. Low relative submergence conditions lead to the formation of discrete clusters with deposited sediment in the stoss (upstream) side of the anchor particle. Instead, high relative submergence conditions result in sediment deposited in the wake (downstream) region of the anchor particle (Papanicolaou and Kramer 2005).

The anchor particles can exhibit a range of sizes (Figure 2.2), varying from coarse gravel to coarser sediment fractions, such as coarse cobbles and boulders (e.g., L'Amoreaux and Gibson 2013; Strom and Papanicolaou 2008; Wittenberg, Laronne, and Newson 2007). The focus herein is placed on the coarser cobble- and boulder-sized anchor particles, which act as stand-alone anchor particles, and promote the formation of discrete clusters around them. Such coarse anchor particles have been documented to occupy between 2% and 25% of the plan view area of the bed surface of reaches of natural rivers (e.g., Bathurst 1978; 1987; Byrd, Furbish, and Warburton 2000; Ghilardi, Franca, and Schleiss 2014a, 2014b; Nitsche *et al.* 2011, 2012; Yager, Kirchner, and Dietrich 2007). Depending on their availability in the reach plan view area, or equivalently on their characteristic spacing, λ, the anchor

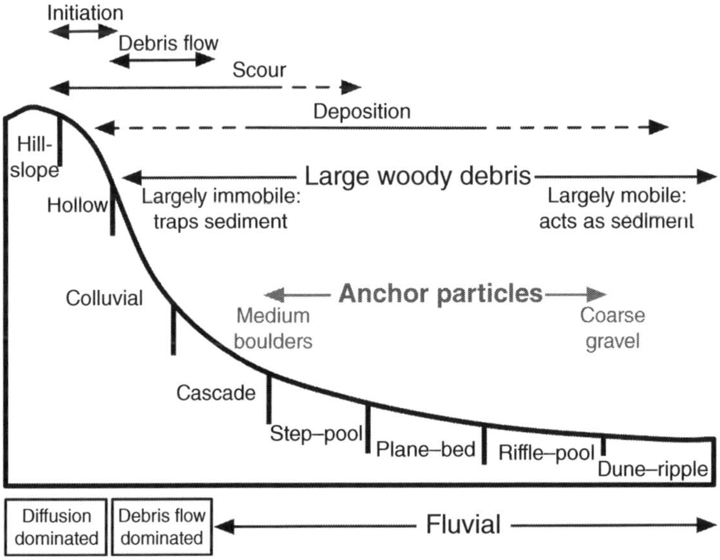

Figure 2.2 Conditions where anchor particles are found to occur in the stream continuum.

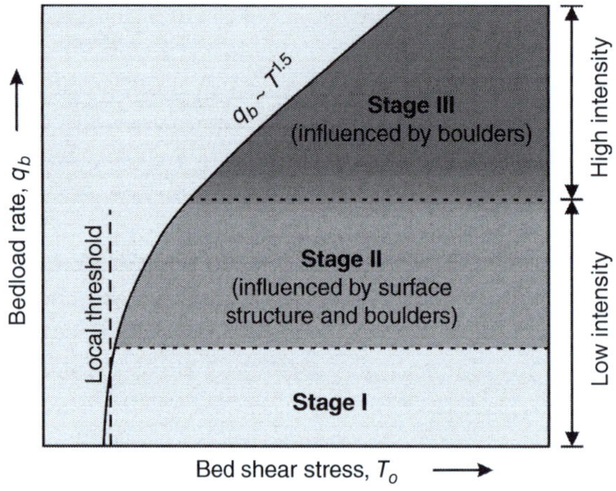

Figure 2.3 Conceptual schematic of the conditions under which surface structure organization and boulders influence sediment transport in coarse-grained streams. (*Source*: M. Hassan)

particles, and hence the clusters that form around them, may correspond to isolated ($\lambda/d_c \geq 6$), wake interaction ($2 \leq \lambda/d_c < 6$) or skimming ($\lambda/d_c < 2$) roughness conditions (e.g., Ghilardi, Franca, and Schleiss 2014c; Morris 1955; Papanicolaou *et al.* 2002; Tan and Curran 2012).

Overall, clusters provide enhanced bed stability to bed particles, due to particle interlocking and by minimizing the exposure of these particles to the flow (Piedra, Haynes, and Hoey 2012; Reid, Frostick, and Brayshaw 1992; Strom *et al.* 2004; Wittenberg and Newson 2005; Yager *et al.* 2012). Clusters have been shown to delay bedload movement of sediment for bed shear stresses up to 3.5 times the critical Shields parameter (Figure 2.3), which correspond to general movement, i.e., Stage II, transport conditions (Frey and Church 2011; Kuhnle 1996; Strom *et al.* 2004). At higher, Stage III, transport conditions, clusters do not enhance the stability of bed surface material, but disintegrate thus acting as sources of bedload. Even at Stage III transport conditions, however, the coarser anchor particles still offer hiding effects to the incoming sediment (Papanicolaou and Kramer 2005) and affect flow and bedload transport patterns (Figure 2.3).

2.2 Scope and Objectives of the Study

The focus of this study is placed on the coarser anchor particles, which for simplicity will be referred to as boulders, and their implications on their surrounding mean and turbulent flow, as well as on the bedload transport patterns. In particular, we will first examine the modification of the mean flow field characteristics around these boulders, including the time-averaged (mean) streamwise velocity profiles, and the spatial variability of the bed shear stress. In addition, we will critically investigate the role of relative submergence on the topology of the vortex structures that form around the boulders. The insights of the mean flow field and the turbulent vortex structures around a boulder will be complemented with observations of mobile sediment depositional patterns. The latter will allow inferences to be drawn about the controls that boulders exert on bedload transport patterns in their vicinity through the modification of the mean and turbulent flow fields.

Based on the earlier discussion in section 2.1 a key impediment for improving the understanding of the hydrodynamics and bedload transport within a reach with ubiquitous boulders is the variable relative submergence that boulders may exhibit during a year. Under bankfull flow conditions, the boulders become fully submerged, with $H/d_c > 1$ (e.g., Saldi-Caromile *et al.* 2004; Wohl 2000; Yager,

Kirchner, and Dietrich 2007). In contrast, throughout the remainder of the year, when lower flow conditions prevail, boulders, due to their large size, remain partially submerged, with $H/d_c < 1$. The limited water depth, uneven water surface, and elevated water aeration levels (i.e., bubbles) make the acquisition of reliable mean and turbulent flow measurements in the vicinity of partially submerged boulders at best very challenging (e.g., Strom and Papanicolaou 2007).

Under high relative submergence conditions, some insight into the mean flow patterns around a wall-mounted, spherical boulder has been provided by Franca (2005), Franca, Ferreira, and Lemmin (2008), Dey *et al.* (2011), Papanicolaou *et al.* (2012), and Tsakiris, Papanicolaou and Lauth (2014), Tsakiris *et al.* (2014). These researchers reported that the mean flow approaching the boulder decelerates and separates at the crest of the boulder, thus forming a near-wake and a far-wake region. The near-wake region immediately downstream of the boulder is characterized by recirculation that typically leads to a reversal of the time-averaged (mean) streamwise flow velocity, $\bar{u}$. The mean flow reattaches and begins to gradually accelerate in the far-wake region, albeit with the boulder effects still evident on the $\bar{u}$ profile. Under low relative submergence conditions, the limited evidence available suggests that the $\bar{u}$ velocity profiles follow an "S" or linear shape (Byrd, Furbish, and Warburton 2000; Ferro 2003; Franca and Lemmin 2009; Lawless and Robert 2001; Nikora *et al.* 2001).

The complexity of the flow field around a boulder at both high and low relative submergence conditions implies, in turn, that the bed shear stress distribution around the boulder is also highly variable (e.g., Dey *et al.* 2011; Papanicolaou *et al.* 2012; Sadeque, Rajaratnam, and Loewen 2009; Tsakiris *et al.* 2014). To accurately capture the bed shear stress variability around the boulder, the Boundary Characteristics Method (BCM) is considered in section 2.3.1 of this chapter (Afzalimehr and Anctil 2000; Hinze 1975; Papanicolaou *et al.* 2012; Tsakiris *et al.* 2014).

The current knowledge on vortex structure topology around boulders for high and, particularly, low relative submergence conditions is with few exceptions based on qualitative observations. Such observations are mostly derived from turbulent flow visualizations. For instance, through dye injections around a fully submerged hemisphere mounted atop a flat smooth bed. Shamloo, Rajaratnam, and Katopodis (2001) observed the formation of a detached shear layer in the boulder recirculation region, and reported the formation of a horseshoe vortex at the stoss side of the hemisphere. For the low relative submergence condition, Shamloo, Rajaratnam, and Katopodis (2001) reported the formation of a von Kárman vortex street downstream of the boulder. Boyer and Roy (1991), who investigated the turbulent flow structures around a partially submerged hemispherical obstacle mounted atop a rough bed under subcritical and supercritical conditions, reported a similar topology of vortices. In fact, Boyer and Roy (1991) observed an increase of the width of the von Kárman vortex street as flow conditions transitioned from subcritical to supercritical.

The effects of boulders on the effective area that is available within a reach for the incoming sediment to mobilize have in general not been accounted for by traditional bedload transport formulae. As a result, such formulae break down when applied at mountain stream reaches with ubiquitous boulders and yield bedload flux predictions that are in error by several orders of magnitude. Recently, Yager, Kirchner, and Dietrich (2007) notably proposed a modification (see equation 2.1) applicable to traditional bedload flux formulae based on the excess bed shear stress, such as those of Meyer-Peter and Müller (1948), Fernandez Luque and Van Beek (1976), Parker, Klingeman, and McLean (1982), and Smart (1984). The Yager, Kirchner, and Dietrich (2007) approach in parts addressed the drawbacks of the aforementioned formulae by accounting for: (i) the reduction in the bed shear stress that is applied by the flow on the mobile sediment, due to the form drag that is borne by the immobile boulders; and (ii) the reduction of the effective area that mobile incoming sediment could utilize within a stream reach, due to the area occupied by the immobile boulders (Figure 2.4a vs. 4b). The Yager, Kirchner, and Dietrich (2007) study first included a decomposition of the reach-averaged total

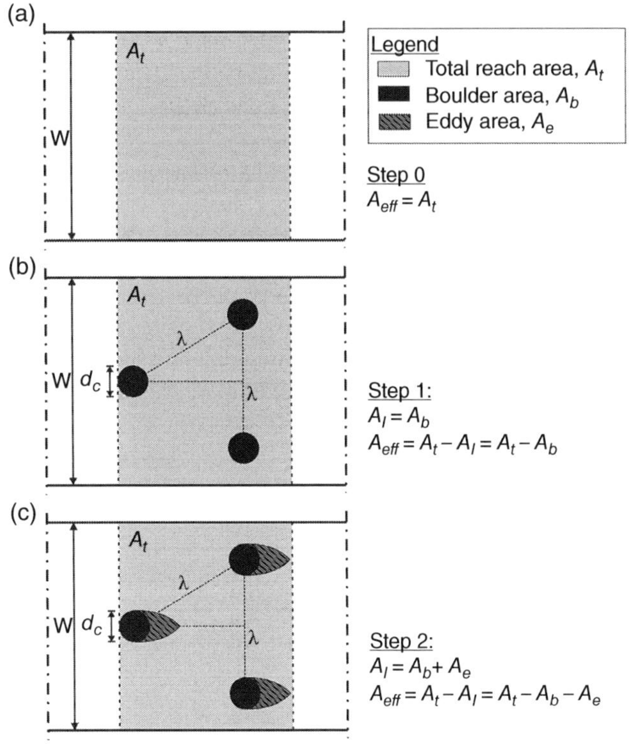

Figure 2.4 Various ways to account for the boulder influence zone: (a) when no boulders are present; (b) based on geometry; and (c) combination of geometry and hydrodynamics. (From Papanicolaou *et al.* 2012 and modified from Yager, Kirchner and Dietrich 2007.)

bed shear stress, $<\tau_o>$ into its components applied to the mobile sediment, $<\tau_{os}>$, and to the immobile boulders $<\tau_{of}>$ for providing the dimensionless bedload rate of the mobile sediment $q^*_{s,m}$

$$q^*_{s,m} = C\left(\left\langle\tau^*_{os}\right\rangle - \tau^*_{cr}\right)^{1.5} \cdot \frac{A_{eff}}{A_t} \tag{2.1}$$

where $\left\langle\tau^*_{os}\right\rangle = \frac{\left\langle\tau^*_o\right\rangle A_t - \left\langle\tau^*_{of}\right\rangle A_I}{A_{eff}}$ and $q^*_{s,m}$ is the dimensionless bedload rate of the mobile sediment; C is a constant; $\left\langle\tau^*_{os}\right\rangle$ is the dimensionless bed shear stress applied to the mobile sediment; τ^*_{cr} is the dimensionless critical bed shear stress for incipient motion of the mobile sediment; A_{eff} is the effective plan view bed area that contributes to the mobile sediment supply (Figure 2.4b); A_t is the total bed plan view area (Figure 2.4b); $\left\langle\tau^*_{of}\right\rangle$ is the dimensionless bed shear stress applied to the immobile boulders; A_I is the plan view bed area that remains immobile and does not contribute to the mobile sediment supply (Figure 2.4b).

Second, the proposed modifications incorporated a reduction to the mobile sediment supply due to the presence of the immobile boulders, by excluding the area of the immobile boulders, $A_I = A_b$ in (Figure 2.4b), from the plan view area that contributes to the mobile sediment supply. This effective area reduction proposed by Yager, Kirchner, and Dietrich (2007) was based solely on geometric considerations, such as the diagonal spacing between the boulders, λ/d_c, and the boulder diameter, d_c (Figure 2.4b).

More recently, Papanicolaou *et al.* (2012) enhanced the two modifications of Yager, Kirchner, and Dietrich (2007) through a combined geometrical and physically based approach. First, Papanicolaou *et al.* (2012) suggested that the improved formulae should account for the variability of the bed shear stress applied to the mobile sediment around the boulder. Second, Papanicolaou *et al.* (2012) recognized that the reduction in the effective area, A_{eff}, must also consider the area of the vortex structures developing around the boulders (Figure 2.4c). Equivalently, the effective area, A_{eff}, should not only depend on the diagonal spacing between the boulders, λ/d_c, and the boulder diameter, d_c, but also on the hydrodynamic effects that boulders have on the flow for different λ/d_c and H/d_c. Employing observations of mobile sediment transport patterns from the study of Papanicolaou and Kramer (2005), Papanicolaou *et al.* (2012) noted a correspondence of the area occupied by the vortex structures around the boulder with the area of mobile sediment accumulations around the boulders. However, the observations of Papanicolaou *et al.* (2012) were of rather preliminary nature and more systematic insight into the correspondence of the area of the vortex structures with the mobile sediment deposition area is needed.

The overarching goal of this study is to evaluate the effects of the coarse anchor particles or boulders on their surrounding turbulent flow and on the bedload transport flux. In doing so, we will address the following four specific objectives:

1) Provide an improved method for estimating the spatial distribution of the bed shear stress or equivalently of the friction velocity, u_*, and an improved insight into the mean, i.e., time-averaged, characteristics of the flow around a boulder under high and low relative submergence conditions.
2) Identify the dominant vortex structures that develop around boulders under high and low relative submergence conditions and quantify their plan view area.
3) Assess the mobile sediment deposition patterns around the boulders at high and low relative submergence conditions and establish their correspondence with the vortex structure topology developing around the boulder.
4) Examine the morphology of the patches that develop around the boulders.

2.3 Dataset Selection and Methodology

2.3.1 Dataset Selection

To address the study objectives, the flow measurements conducted by the Papanicolaou group were synthesized. A cornerstone of the experimental synthesis was that it was designed to address the two improvements on the Yager, Kirchner, and Dietrich (2007) approach brought forth by Papanicolaou *et al.* (2012) through two sets of experiments. The key experimental (or numerical) parameters considered in each of these experimental runs along with the source are summarized in Table 2.1.

The first set of experiments aimed to include the spatial variability of the bed shear stress applied by the flow on the mobile sediment around the immobile boulders, due to the form drag that is generated by the immobile boulders. In this first set of experiments, we provided estimates of the spatial distribution of the friction velocity, u_*, and thus of the bed shear stress, $<\tau_o>$, around a boulder, thereby improving the Yager, Kirchner, and Dietrich (2007) study that considered a constant value for u_*. In addition, the experiments of Set 1 also yielded the mean flow characteristics around the boulder. This first set of experiments was performed for a single boulder as a baseline condition for the u_* estimation.

Table 2.1 Sources and key parameters for the datasets synthesized in this study.

Set	Source	Dataset designation	Bed configuration	Bed roughness d_b (mm)	Relative submergence H/d_c (–)	Dimensionless bed shear stress $<\tau_o^*>$ (–)	Flow stage $<\tau_o^*>/\tau^*_{cr}$ ‡ (–)	Mobile sediment diameter d_{50} (mm)	Intended use
1	Tsakiris *et al.* (2014)	Run 1	Single boulder	19.1	3.50	0.021	2.50	19.1 (assumed)	**1** Mean flow field **2** $u*$ distribution at HRS
1	Present study	Run 2	Single boulder	19.1	0.80	0.021	2.50	19.1 (assumed)	**1** Mean flow field **2** $u*$ distribution at LRS
2	Papanicolaou *et al.* (2012)	Array – HRS	Boulder array	19.1	3.50	0.021	2.50	19.1 (assumed)	Collective array form drag
2	Papanicolaou *et al.* (2012)	Array – LES	Boulder array	0 (smooth)	3.50	0.021	2.50	19.1 (assumed)	Mean vortex structure (HRS)
2	Papanicolaou, Dermisis, and Elhakeem (2011)	Array – LRS	Boulder array	19.1	0.80	0.056	2.50	1.90	Mean vortex structure (LRS)
2	Papanicolaou and Kramer (2005)	Series A†	Boulder array	19.1	3.50	0.017–0.030	2.00–3.50	19.1	Sediment transport (HRS)
2	Papanicolaou and Kramer (2005)	Series B†	Boulder array	19.1	0.80	0.017–0.030	2.00–3.50	19.1	Sediment transport (LRS)

$<\tau_o^*> = \rho u_*^2/[g(\rho_s - \rho)d_{50}]$, where $\rho = 999$ kg/m^3 is the water density and $\rho_s = 2550$ kg/m^3 the mobile sediment density.

‡ Critical dimensionless bed shear stress for incipient motion of material with size d_{50}.

† Series of four runs reported in Table 2.3.

The second set of experimental runs was performed to introduce a hydrodynamic aspect in the estimation of the effective area, A_{eff}, within a boulder array for mobile sediment conveyance. The second set of experiments allowed estimating the A_{eff} such that it accounted for the effects of the vortex topology generated around the boulders on reducing the conveyance of incoming grains, in addition to the geometry of the boulder array. This second set of experiments involved an array of boulders, as it was more meaningful to examine the vortex topology and its implications on the conveyance of the incoming sediment grains when an array of boulders was present.

To allow integration of the experimental findings and explore their utility towards the development of a bedload formula that accounts for both the effects of the vortex structure topology on the A_{eff}, as well as for the spatial variability of u_* in the boulder vicinity, all experimental runs were performed for the same reach-averaged dimensionless bed shear stress level value, $< \tau_o^* > / \tau^*_{cr}$. This is a condition that replicated general sediment mobilization (e.g., Frey and Church 2011; Pitlick and Van Steeter 1998; Strom *et al.* 2004). Furthermore, to examine the role of the variable relative submergence, H/d_c, on the bed shear stress variability around the boulder and the vortex topology in conjunction with the sediment transport patterns and ultimately on A_{ff}, the experimental Sets 1 and 2 included runs conducted under both high and low relative submergence conditions. The selected high and low relative submergence values, namely $H/d_c = 3.50$ and 0.80, respectively, were consistent for both experimental sets to allow further integration of the experimental findings (Table 2.1).

2.3.2 Experimental Setup and Measurement Techniques

The experimental setup and procedures for each experimental run in Table 2.1 are briefly presented in the following. The reader is referred to the respective studies reported in Table 2.1 for additional details.

All experiments of the datasets reported in Table 2.1 were conducted in a 21 m long, 0.91 m wide, and 0.53 m deep, water-recirculating flume with glass sidewalls that enabled experimental observations from the side. The test section was 5.2 m in length and began 10.9 m downstream of the flume entrance, where the flow was fully developed, as verified through preliminary measurements. The flume bed was well packed, with five layers of spherical glass particles of diameter $d_b = 19.1$ mm placed in a hexagonal arrangement, which created a flat, rough, porous bed. The boulder (anchor particle) for the first set of experiments was replicated by a sphere with diameter $d_c = 0.055$ m, which was rigidly mounted atop the flat rough bed at the flume centerline at the middle of the test section. For the second set of experiments, an array of identical spherical boulders with $d_c = 0.055$ m was placed with diagonal spacing $\lambda/d_c = 6.0$, which corresponded to isolated roughness conditions (e.g., Morris 1955; Papanicolaou *et al.* 2002).

The flow depth, H, for the experiments conducted under high relative submergence (HRS) conditions was $H = 0.193$ m, yielding a relative submergence $H/d_c = 3.50$ (Table 2.2). For all experiments conducted under low relative submergence (LRS) conditions the flow depth was $H = 0.044$ m, and the relative submergence $H/d_c = 0.80$ (Table 2.2). The flow depth, H, for all experiments reported in Table 2.2 was uniform throughout the test section by adjusting a tailgate at the downstream end of the flume. The flow depth, H, uniformity was verified via regularly spaced rulers placed on the flume glass sidewalls and point gages. For the single boulder experiments (designated as Run 1 and Run 2 in Table 2.2) the flow for the HRS run was subcritical ($Fr = 0.69$), whereas the LRS run was conducted under supercritical flow conditions ($Fr = 1.25$). In the experimental runs of Set 2, the flow was subcritical with $Fr = 0.57$ and $Fr = 0.54$ (Table 2.2). The flow aspect ratio for all experiments was larger than 4.5 (Table 2.2), to minimize the effects of secondary currents (e.g., Ferro 2003; Nezu and Nakagawa 1993).

Table 2.2 Summary of the key experimental parameters for the flow runs of experimental Sets 1 and 2.

Source	Dataset designation	Flow depth H (m)	Relative submergence H/d_c (–)	Bed slope S_o (%)	Flow discharge Q (m^3/s)	Bulk velocity U_{bulk} (m/s)	Froude number Fr (–)	Uniform friction velocity u_* (m/s)	Uniform bed shear stress $<\tau_o>$ (Pa)	Dimensionless bed shear stress $<\tau_o^*>$ (–)
Tsakiris *et al.* (2014)	Run 1	0.193	3.50	0.33	0.167	0.95	0.69	0.079	6.24	0.021
Present study	Run 2	0.044	0.80	1.43	0.033	0.82	1.25	0.079	6.24	0.021
Papanicolaou *et al.* (2012)	Array – HRS	0.193	3.50	0.3	0.137	0.78	0.57	0.076	5.67	0.021
Papanicolaou, Dermisis, and Elhakeem (2011)	Array – LRS	0.044	0.80	0.36	0.014	0.35	0.54	0.039	1.55	0.056

$U_{bulk} = Q/BH$, where $B = 0.91$ m is the flume width.
$Fr = U_{bulk}/(gH)^{0.5}$, where $g = 9.81$ m/s^2 is the acceleration of gravity.
$u_* = (gHS_o)^{0.5}$.
$<\tau_o> = \rho u_*^2$, where $\rho = 999$ kg/m^3 is the water density.
$<\tau_o^*> = <\tau_o>/[g(\rho_s - \rho)d_{50}]$, where $\rho_s = 2550$ kg/m^3 is the assumed mobile sediment density for Runs 1 and 2 and $d_{50} = 0.0191$ m is the median diameter.

The uniform, reach-averaged dimensionless bed shear stress is herein defined as $<\tau_o^*> = <\tau_o>/[g(\rho_s - \rho)d_{50}]$, where $<\tau_o> = \rho g H S_o$ was the uniform, reach-averaged, bed shear stress, $\rho = 999\ \mathrm{kg/m^3}$ was the water density, and $g = 9.81\ \mathrm{m/s^2}$ was the acceleration of gravity. Although no mobile sediments were present during run 1, run 2, the array HRS, and the array LRS flow experiments (Table 2.1), the entrainable material was assumed to have median diameter $d_{50} = 19.1$ mm and density $\rho_s = 2550\ \mathrm{kg/m^3}$, to match the d_{50} and ρ_s of the bedload transport experiments conducted for the purposes of experimental Set 2.

The parameter τ^*_{cr} denoted the dimensionless critical bed shear stress for incipient motion of the assumed entrainable material with $d_{50} = 19.1$ mm. The τ^*_{cr} was based upon the concept of the probability of entrainment (Papanicolaou *et al.* 2002), which was considered equal to 2%. The τ^*_{cr} was thus determined in a separate series of experiments, where 100 unisize spherical particles with $d = 19.1$ mm were placed at isolated conditions atop a flat, rough bed comprising identical particles. The τ^*_{cr} corresponded to the flow condition at which 2 of the 100 (2%) preplaced particles were entrained (Papanicolaou *et al.* 2002). These experiments yielded $\tau^*_{cr} = 0.0085$, which is lower than the typical values of τ^*_{cr} between 0.02 and 0.07 for coarse natural sediments (e.g., Buffington and Montgomery 1997; Recking 2009). The smaller τ^*_{cr} value used herein is attributed to the minimal interlocking of the spherical mobile particles with the identical bed particles, as well as to the full exposure of the mobile particles to the flow (e.g., Buffington and Montgomery 1997; Vollmer and Kleinhans 2007; Wiberg and Smith 1987).

Flow measurements around the single boulder for both Runs 1 and 2 of experimental Set 1 in Table 2.1 were acquired via a two-component (2D) laser Doppler velocimeter (LDV) manufactured by TSI™. The 2D LDV system was operated in coincidence mode, in order to measure streamwise, u, and vertical, w, instantaneous velocities along the x and z directions, respectively. A positive shift varying from 5 to 200 kHz was applied to both color channels to distinguish bidirectional motion. At every measurement point, 10 000 simultaneous u and w measurements were collected at an average data rate between 30 Hz and 250 Hz. The locations in plan view and along the vertical, z, direction, where the LDV measurements were performed for Run 1 (HRS) and Run 2 (LRS), are indicated in Figure 2.5. More details of the LDV system and measurements are given in Tsakiris *et al.* (2014).

Flow measurements in the Papanicolaou *et al.* (2012) experiments were acquired via a two-component (2D) acoustic Doppler velocimeter (ADV). The ADV was used to collect continuous measurements over a period of 2 min at 25 Hz resulting in 3000 simultaneous u and v velocity component measurements. The reader is referred to Papanicolaou *et al.* (2012) for a detailed presentation of the ADV measurements and to Tsakiris *et al.* (2014) for a comparison between the ADV and the LDV measurement techniques.

The experimental measurements of Papanicolaou *et al.* (2012) were supplemented with a large eddy simulation (LES), for elucidating the vortex structure topology that forms around a fully submerged boulder from its interaction with the approach turbulent flow. The LES simulated the geometry setup and flow conditions of the Papanicolaou *et al.* (2012) experiment, but employed a smooth, impermeable boundary in place of the flat, rough, porous bed used in the experiment. Despite the use of a smooth boundary in the LES simulations, the LES may be used to elucidate the key vortex structures around the boulder at least in qualitative terms. The specific details of the LES numerical setup are provided in Papanicolaou *et al.* (2012).

Papanicolaou, Dermisis, and Elhakeem (2011) experimentally determined the vortex structure topology forming around a boulder under LRS conditions by using the large-scale particle image velocimetry (LSPIV) technique (e.g., Hauet *et al.* 2008; Muste *et al.* 2004). In summary, the LSPIV technique considers recordings of sequential images depicting tracer particles seeded at the flow free surface. The analysis of the successive images yields the "instantaneous" streamwise and lateral

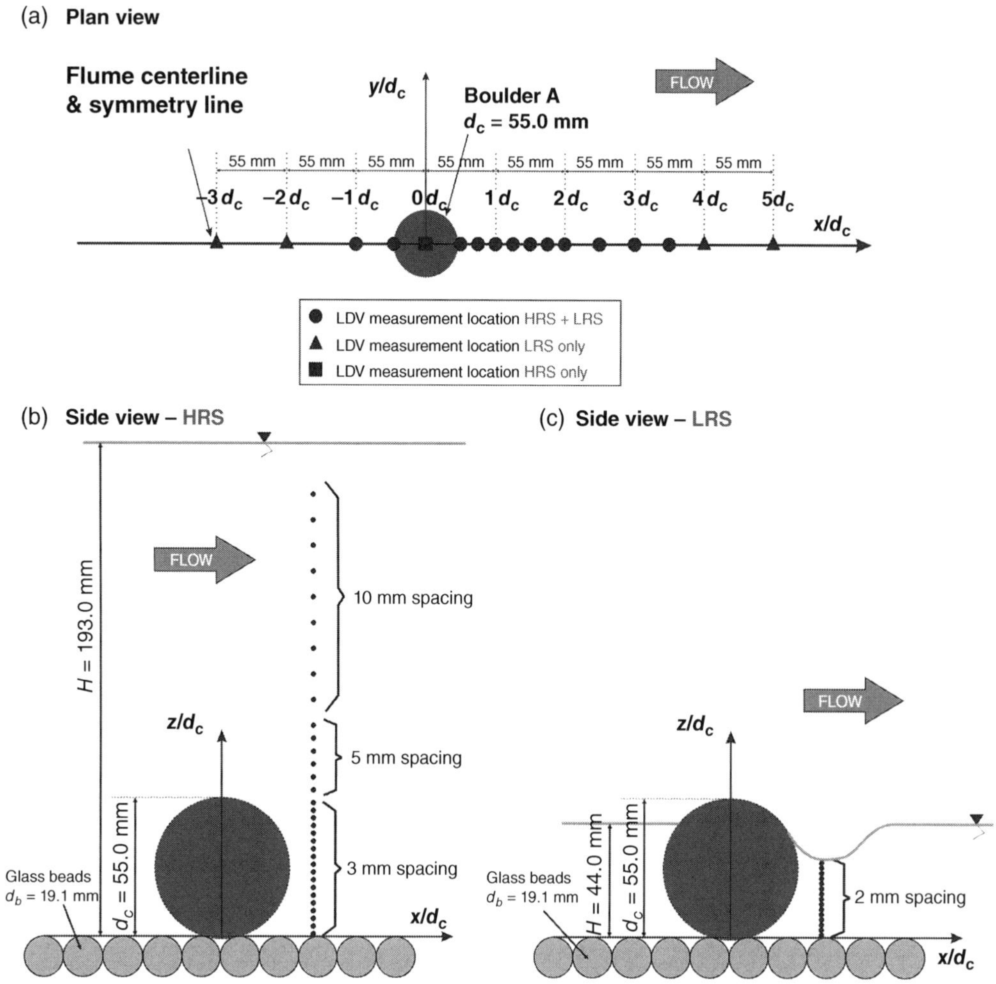

Figure 2.5 Locations of the ADV and LDV measurements: (a) in plan view; (b) side view for the high relative submergence (HRS) conditions; and (c) side view for the low relative submergence (LRS) conditions.

velocity components, u and v respectively, within the water free surface. Because the flow depth in Papanicolaou, Dermisis, and Elhakeem (2011) is shallow, i.e., $H = 0.044$ m (Table 2.2), and smaller than the boulder diameter ($H/d_c = 0.80$), it is believed that the surface velocity patterns can capture the developing vortex structures around the boulder with reasonable accuracy.

The flow experiments were complemented with observations of the depositional patches formed around boulders placed within an array under HRS and LRS conditions during experiments performed with mobile sediment in Papanicolaou and Kramer (2005). The reader is referred to Papanicolaou and Kramer (2005) as well as to Kramer (2005) for a presentation of these experiments in great detail. Similarly to the flow experimental sets, two series of experiments were performed, namely series A with high relative submergence, $H/d_c = 3.50$, and a series B with low relative submergence, $H/d_c = 0.80$ (Table 2.3). For each of the series A and B, four experimental runs were performed with ratio $< \tau_o^* > / \tau^*_{cr} = 2.00$, 2.50, 3.00, and 3.50 (Table 2.3). The results for $< \tau_o^* > / \tau^*_{cr} = 2.50$ were combined

Table 2.3 Summary of the key experimental parameters for the sediment transport runs of experimental set 2 (from Papanicolaou and Kramer 2005).

Series	Number	Flow depth H (m)	Relative submergence H/d_c (–)	Bed slope S_o (%)	Flow discharge Q (m^3/s)	Bulk velocity U_{bulk} (m/s)	Uniform friction velocity u_* (m/s)	Dimensionless bed shear stress $<\tau_o^*>$ (–)	Froude number Fr	Reynolds number $Re \times 10^{-5}$
A	1	0.193	3.5	0.30	0.13	0.76	0.08	0.017	0.56	5.871
A	2	0.193	3.5	0.33	0.13	0.78	0.08	0.021	0.57	6.020
A	3	0.193	3.5	0.36	0.14	0.82	0.08	0.026	0.60	6.194
A	4	0.193	3.5	0.42	0.15	0.84	0.08	0.030	0.61	6.417
B	1	0.044	0.8	1.59	0.02	0.49	0.08	0.017	0.75	0.867
B	2	0.044	0.8	1.86	0.02	0.54	0.09	0.021	0.81	0.941
B	3	0.044	0.8	2.12	0.03	0.70	0.10	0.026	1.06	1.226
B	4	0.044	0.8	2.39	0.03	0.74	0.10	0.030	1.13	1.313

$U_{bulk} = Q/BH$, where $B = 0.91$ m is the flume width.
$u_* = (gHS_o)^{0.5}$, where $g = 9.81$ m/s^2 is the acceleration of gravity.
$<\tau_o^*> = <\tau_o>/[g(\rho_s - \rho)d_{50}]$, where $\rho_s = 2440$ kg/m^3 the mobile sediment density for Runs 1 and 2 and $d_{50} = 0.0191$ m is the median diameter.
$Fr = U_{bulk}/(gH)^{0.5}$.
$Re = 4U_{\rm bulk} \times H/\nu$, where $\nu = 1 \times 10^{-6}$ m^2/s the water kinematic viscosity.

with the vortex topology for the experiments of Set 2 with the boulder array present, in order to examine the effects of the vortex structure topology on the sediment depositional patterns around a boulder.

The experimental setup of Papanicolaou and Kramer (2005) was identical to the setup of Papanicolaou *et al.* (2012), with the array of boulders placed atop a flat, rough bed comprising well-packed glass spheres with $d_b = 19.1$ mm. The boulders of the array were placed with diagonal spacing $\lambda/d_c = 6.0$, which corresponds to isolated roughness conditions (e.g., Ghilardi, Franca, and Schleiss 2014c; Morris 1955; Papanicolaou *et al.* 2002). The mobile sediment used in the experiments of Papanicolaou and Kramer (2005) comprised of a mixture of glass spheres with five different sizes, namely $d = 8.0$, 15.9, 19.1, 22.2, and 25.4 mm. The mobile sediment mixture had a median diameter $d_{50} = 19.1$ mm, which was also the same as the assumed diameter for the mobile sediment assumed by Papanicolaou *et al.* (2012). For each of the four runs of the experimental Series A and B, an amount of sediment mixture, predetermined through preliminary experimentation at identical flow conditions, was supplied upstream of the boulder array and allowed to be entrained by the flow action.

2.3.3 Data Post-Processing

The streamwise instantaneous velocity measurements collected via the LDV or the ADV were averaged over the measurement duration to yield the time-averaged (mean) streamwise velocity, $\bar{u}$ as:

$$\bar{u} = \frac{1}{N} \cdot \sum_{i=1}^{N} u_i \tag{2.2}$$

where the overbar denotes time-averaging; N is the number of instantaneous velocity measurements, with $N = 10\ 000$ and $N = 3000$ for the LDV and the ADV measurements respectively; and u_i is the instantaneous velocity recorded. A time-averaged streamwise velocity, $\bar{u}$, profile was also collected after removing the boulder at identical flow conditions with Runs 1 and 2, in order to provide benchmark $\bar{u}$ profiles atop a flat rough bed.

To address specific objective **1**, the friction velocity, u_*, was estimated herein using the boundary characteristics method, which is referred to as the BCM (Afzalimehr and Anctil 2000; Hinze 1975; Papanicolaou *et al.* 2012). The BCM allows estimation of the u_* from the mean flow characteristics assuming that the defect law is valid in the outer flow layer, i.e., above $z/H > 0.20$. The BCM offers two advantages towards the calculation of u_*. First, no limiting assumption is made for the mean streamwise velocity profile shape in the near-bed region ($z/H < 0.20$). The outer flow is generally not affected greatly by the presence of deeply submerged large roughness elements and the defect law in the outer flow layer is in general valid (e.g., Dey *et al.* 2011; Franca, Ferreira, and Lemmin 2008; Papanicolaou *et al.* 2012). In contrast, the frequently employed Clauser method (Clauser 1956), which also estimates u_* from mean velocity profiles (e.g., Ferro 2003; Kironoto and Graf 1994) requires that the velocity profile in the inner flow layer ($z/H < 0.20$) follows the logarithmic law (Clauser 1956; Wei, Schmidt, and McMurtry 2005). However, in the presence of boulders, the velocity profile significantly deviates from the logarithmic law in the inner flow layer and the aforementioned assumption is violated (e.g., Dey *et al.* 2011; Papanicolaou *et al.* 2012; Shamloo, Rajaratnam, and Katopodis 2001). Second, the BCM does not require turbulence measurements, as opposed to the Reynolds shear stress method (e.g., Biron *et al.* 2004; Rowiński, Aberle, and Mazurczyk 2005), which necessitates the estimation of the streamwise and vertical velocity fluctuations, u' and w', respectively, for calculating the Reynolds bed shear stress, $\tau_{\text{Re}} = -\rho\left(\overline{u'w'}\right)$.

At each measurement location (Figure 2.5), the friction velocity $u_{*,BCM}$ from the BCM was calculated as:

$$u_{*,BCM} = \frac{(\delta_d - \delta_\theta) \cdot u_e}{K \cdot \delta_d} \tag{2.3}$$

where $K = 2/\kappa = 4.87$ (Afzalimehr and Anctil 2000) a constant; $\kappa = 0.41$ the von Karman constant in clear water flow; u_e is the maximum time-averaged velocity at a measurement location; and δ_d and δ_θ are the displacement and momentum thicknesses respectively of the boundary layer given by equations 2.4a and 2.4b, respectively, as:

$$\delta_d = \int_0^H \left(1 - \frac{\bar{u}}{u_e}\right) dz \tag{2.4a}$$

$$\delta_\theta = \int_0^H \frac{\bar{u}}{u_e} \cdot \left(1 - \frac{\bar{u}}{u_e}\right) dz \tag{2.4b}$$

The $u_{*,BCM}$ was also compared to the uniform friction velocity $u_{*,unif} = (gHS_o)^{0.5}$ (Table 2.1).

The transverse instantaneous vorticity, ω_y, within a streamwise-vertical, x–z, plane along the boulder centerline and the vertical instantaneous vorticity, ω_z within a streamwise-lateral, x–y, plane through the boulder center were calculated for the LES of Papanicolaou *et al.* (2012), respectively, as:

$$\omega_y = \frac{\partial u}{\partial z} - \frac{\partial w}{\partial x} \tag{2.5a}$$

$$\omega_z = \frac{\partial v}{\partial x} - \frac{\partial u}{\partial y} \tag{2.5b}$$

where v and w are the instantaneous lateral and vertical velocity components. The time-averaged streamwise vorticity, $\bar{\omega}_x$ within lateral–vertical, y–z, planes was calculated as:

$$\bar{\omega}_x = \frac{\partial \bar{w}}{\partial y} - \frac{\partial \bar{v}}{\partial z} \tag{2.6}$$

The time-averaged vertical vorticity component, $\bar{\omega}_z$ was estimated from the LSPIV measurements of Papanicolaou, Dermisis, and Elhakeem (2011) by taking the time average of equation (2.5b).

2.4 Mean Flow Field Around a Single, Wall-Mounted Boulder

2.4.1 Mean Streamwise Flow Velocity Field

The mean streamwise velocity, $\bar{u}$, profiles along the centerline of the single boulder for the HRS conditions (Run 1 of Set 1) are shown in Figure 2.6. Our measurements support the near/far wake model for the mean flow field downstream of an anchor particle (e.g., Dey *et al.* 2011; Franca 2005; Franca and Lemmin 2009; Papanicolaou *et al.* 2012). In particular, the $\bar{u}$ profiles indicate the presence of a flow recirculation region in the boulder near-wake, between locations $x/d_c = 0.50$ and $x/d_c = 1.25$, where $\bar{u}$ attains negative values. Further downstream, at $x/d_c = 1.50$, $\bar{u}$ becomes positive throughout the depth, indicating that the flow has reattached. The flow reattachment length is therefore $x/d_c = 1.40$, which compares favorably with previous similar studies (e.g., Dey *et al.* 2011; Sadeque, Rajaratnam, and Loewen 2009).

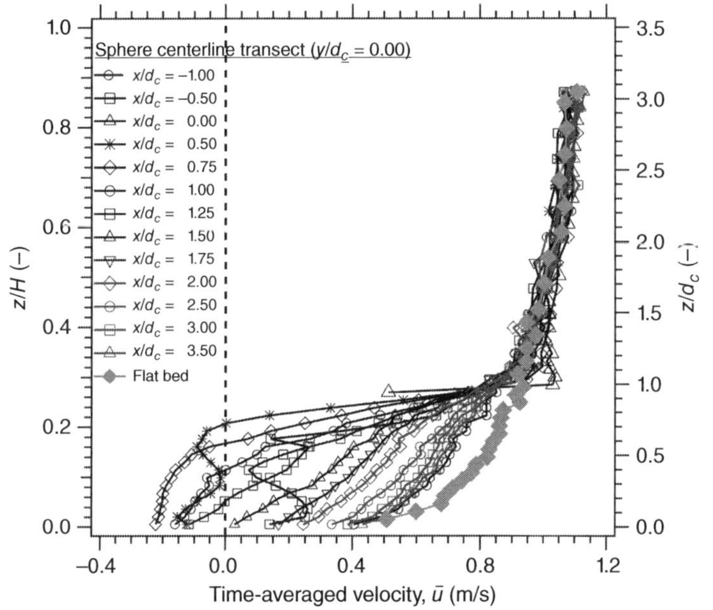

Figure 2.6 Time-averaged, streamwise velocity profiles around the single boulder and atop the flat rough bed under high relative submergence (HRS) conditions.

The $\bar{u}$ profiles in the vicinity of the boulder illustrated in Figure 2.6 attain overall an "S" shape, which markedly deviates from the logarithmic $\bar{u}$ profile for the flat rough bed. This "S" shape of the $\bar{u}$ profiles is indicative of the effects that the boulder has on its surrounding mean flow and more specifically the significant deceleration of the time-averaged velocity below $z/d_c = 1.00$ (Franca and Lemmin 2009). This deceleration is more pronounced in the boulder near-wake region, with negative $\bar{u}$ values and flow reversal. Further downstream of the boulder, i.e., for $x/d_c > 1.50$ the effects of the boulder on the $\bar{u}$ profiles attenuate and the $\bar{u}$ profile shapes have smaller deviations from the $\bar{u}$ profile corresponding to the flat rough bed.

Alternately, these deviations of the $\bar{u}$ profile shape in the vicinity of the boulder may be viewed as the outcome of the local form drag, which is generated by the presence of the boulder. Immediately downstream of the boulder, within the near-wake region, $x/d_c = 0.50$ and $x/d_c = 1.25$, the form drag generated by the boulder is maximized. This additional form drag decelerates the flow, thus causing a deficit in the mean $\bar{u}$ velocity and the flow reversal. Further downstream of the boulder, i.e., for $x/d_c > 1.50$, the form drag generated by the boulder attenuates, leading to increasingly smaller deficit in $\bar{u}$. As a result, the deviations of the $\bar{u}$ profile shape from the $\bar{u}$ profile corresponding to the flat rough bed gradually diminish.

Furthermore, the $\bar{u}$ profiles illustrated in Figure 2.6 highlight the vertical extent of the boulder effects on the surrounding mean flow field. In particular, the $\bar{u}$ velocity profiles for all locations along the centerline transect at $y/d_c = 0.00$ show an overall collapse above $z/d_c = 1.00$. This suggests that the effects of the boulder on the surrounding time-averaged flow field remain confined below the boulder crest, i.e., below $z/d_c = 1.00$. A similar finding has also been reported for flows around protruding roughness elements, such as transverse ribs and cubes (e.g., Coleman *et al.* 2007; Florens, Eiff, and Moulin 2013; Nikora *et al.* 2007; Stoesser and Nikora 2008) and natural coarse sediments (e.g., Aberle, Koll, and Dittrich 2008; Mignot, Hurther, and Barthelemy 2009; Pokrajac, McEwan, and Nikora 2008; Dey and Das 2012). These researchers noted an increase in the spatial heterogeneity of the flow below the roughness elements crests. The increased flow spatial heterogeneity below the roughness element crests was reflected in the nonzero values of the form-induced (or dispersive) stresses, which result from the application of the double-averaging framework to the momentum equations (Coleman *et al.* 2007; Nikora *et al.* 2001; Nikora *et al.* 2007).

For the LRS conditions (Run 2 of Set 1), the mean streamwise velocity profiles along the centerline of the boulder are shown in Figure 2.7. Remarkably, under these shallow, supercritical ($Fr <$ 1) flow conditions, the shape of the $\bar{u}$ profiles does not resemble the logarithmic velocity profile even at the furthest locations from the boulder, i.e., at $x/d_c = -3.00$ and $x/d_c = 5.00$. Upstream of the boulder, at locations $x/d_c = -3.00$, -2.00, and -1.00 the $\bar{u}$ profile approaches an "S" shape, indicating rapid deceleration of the flow in the near-bed region (Ferro 2003; Franca and Lemmin 2009). The magnitude of $\bar{u}$ reduces gradually from $x/d_c = -3.00$ to $x/d_c = -2.00$ and to $x/d_c = -1.00$ suggesting that the flow decelerates as the boulder is approached. Downstream of the boulder, between $x/d_c = 0.75$ and $x/d_c = 1.00$ the $\bar{u}$ profile shape is overall linear. More importantly, there is a sharp increase in the magnitude of $\bar{u}$ from $x/d_c = 0.75$ to $x/d_c = 1.00$ throughout the flow depth. This sharp increase suggests the presence of a high-speed stream at $x/d_c = 0.75$ to $x/d_c = 1.00$. This high speed stream is likely an outcome of the flow depth constriction immediately downstream of the boulder, where a standing wave forms in these supercritical flow conditions (e.g., Papanicolaou and Kramer 2005; Peakall, Ashworth, and Best 1996; Shamloo, Rajaratnam, and Katopodis 2001).

Between $x/d_c = 1.25$ and $x/d_c = 2.00$ the $\bar{u}$ profile shape also appears to follow an "S" shape (Figure 2.6), indicating that the boulder triggers some retardation to the velocity in the near-bed

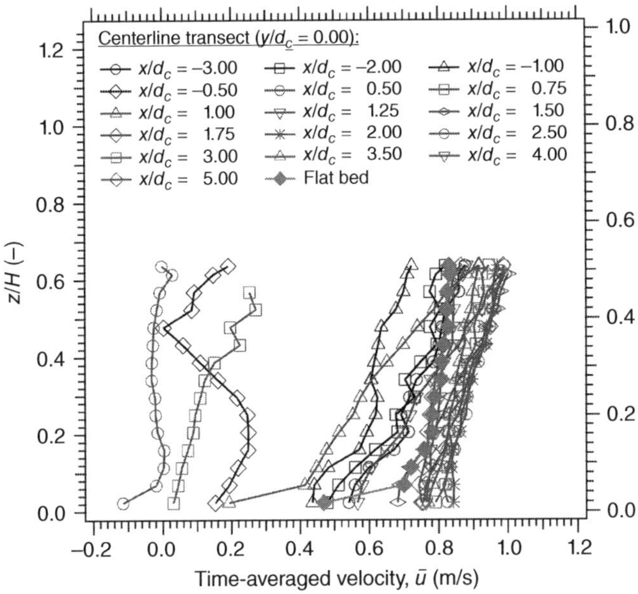

Figure 2.7 Time-averaged, streamwise velocity profiles around the single boulder and atop the flat rough bed under low relative submergence (LRS) conditions.

region, below $z/d_c = 0.15$. Further downstream of $x/d_c = 2.50$, the $\bar{u}$ profiles become linear and overall collapse, thus suggesting that the effects of the boulder are not significant.

2.4.2 Distribution of u_* Around the Boulder

The distribution of the friction velocity, $u_{*,BCM}$, calculated by the BCM for the HRS conditions of Run 1 of Set 1 is shown in Figure 2.8a. The results in Figure 2.8a show that the BCM performs well at depicting the variability of u_* while highlighting the effects of the form drag, due to the presence of the boulder. Upstream of the boulder at $x/d_c = -1.0$ the BCM yields comparable u_* values with the uniform method. The $u_{*,BCM}$ increases sharply around the boulder between $x/d_c = -1.00$ and $x/d_c = 0.75$, where the $u_{*,BCM}$ exhibits its largest value, which is nearly three times larger than the $u_{*,unif}$. Thereafter, i.e., downstream of $x/d_c = 0.75$, the $u_{*,BCM}$ decreases gradually up to $x/d_c = 3.50$, where the $u_{*,BCM}$ attains a comparable value to the $u_{*,BCM}$ at $x/d_c = -1.00$.

Also, the distribution of $u_{*,BCM}$ along the centerline transect at $y/d_c = 0.00$ (Figure 2.8a) may be interpreted by considering the decomposition of the local total bed shear stress, τ_o, namely τ_{os} and τ_{of} due to the skin friction and form drag generated by the boulder, respectively. Further away from the anchor particle, i.e., at $x/d_c = -1.00$ and $x/d_c = 3.50$, the influence of the boulder on the mean flow is not significant, as also illustrated in Figure 2.6, and $\tau_o = \rho u_*^2 \approx \tau_{os}$. Within close proximity of the anchor particle, τ_{of} becomes important at the local scale and it is added to the existing τ_{os} due to the smallscale bed roughness. Hence, τ_o and thus the $u_{*,BCM}$ increases from $x/d_c = -1.00$ to $x/d_c = 0.75$. At $x/d_c = 0.75$, the τ_{of} generated by the form drag on the boulder is maximized, hence leading to a maximum value of $u_{*,BCM}$ at this location. Further downstream of the boulder, i.e., for $x/d_c > 0.75$, the form roughness effects attenuate, hence τ_{of} decreases as also does $\tau_o = \rho u_*^2 = \tau_{os} + \tau_{of}$. This analysis of Figure 2.8a verifies in parallel the finding in Figure 2.6 that the form drag generated by the

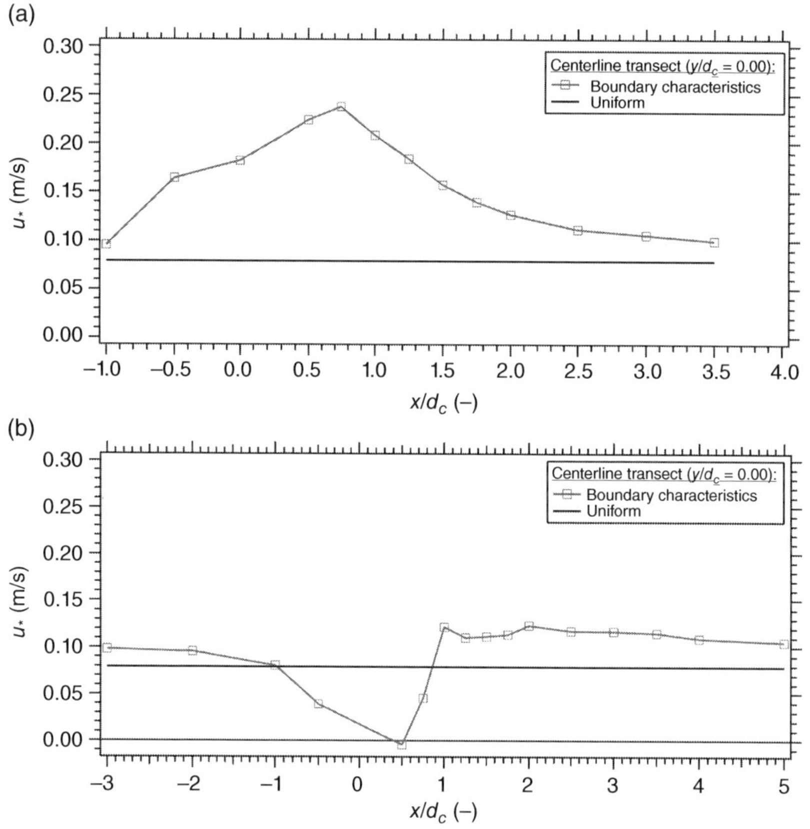

Figure 2.8 Spatial distribution of the friction velocity u_* calculated from the boundary characteristics method (BCM) along the boulder centerline transect for (a) high relative submergence (HRS) and (b) low relative submergence (LRS) conditions.

boulder is reflected in the mean $\overline{u}$ profiles. In particular, the $u_{*,BCM}$ is maximized at $x/d_c = 0.75$, which coincides with the location where the largest deviations in the $\overline{u}$ profile shape are observed (Figure 2.6). Further downstream, i.e., for $x/d_c > 1.75$, where the form drag generated by the anchor particle attenuates, the shape of the $\overline{u}$ profiles shows smaller deviations from the $\overline{u}$ profile corresponding to the flat rough bed (Figure 2.6).

The distribution of the friction velocity, $u_{*,BCM}$, calculated via the BCM under the LRS conditions of Run 2 within the centerline of the boulder is shown in Figure 2.8b. Further upstream of the boulder, between $x/d_c = -3.00$ to $x/d_c = -1.00$, the $u_{*,BCM}$ has comparable magnitude to the $u_{*,unif}$ and does not vary significantly. Downstream of $x/d_c = -1.00$ the $u_{*,BCM}$ shows a strong decrease and attains a nearly zero value at the downstream edge of the boulder at $x/d_c = 0.50$. The $u_{*,BCM}$ then rapidly increases from the zero value at $x/d_c = 0.50$ to $u_{*,BCM} = 0.13$ at $x/d_c = 1.00$, and remains approximately constant further downstream.

The $u_{*,BCM}$ distribution in Figure 2.8b is indicative of the total bed shear stress, $\tau_o = \rho u_*^2$ applied by the flow to the bed. Upstream of the boulder, i.e., between $x/d_c = -3.00$ and $x/d_c = -1.00$ the boulder effects are not significant and the $u_{*,BCM}$ remains essentially constant. In turn, this implies, that between $x/d_c = -3.00$ and $x/d_c = -1.00$ the majority of the flow bed shear stress is applied on the flat rough bed, or $\tau_o = \rho u_*^2 \approx \tau_{os}$. This finding was also observed in Figure 2.8a for the HRS condition of Run 1. At close proximity of the boulder, part of the total applied flow shear stress is balanced by the form drag generated by the anchor particle. As a result, the $u_{*,BCM}$ decreases sharply between $x/d_c = -1.00$ and $x/d_c = 0.50$. This $u_{*,BCM}$ decrease suggests that the boulder provides shelter to incoming mobile sediment from the flow action on its upstream side. This finding is in accordance with the observations of Papanicolaou and Kramer (2005), who documented a preferential deposition of incoming mobile sediment under LRS conditions on the upstream side of the boulder (see section 2.7). Downstream of the boulder, the sharp increase of $u_{*,BCM}$ is indicative of the rapid flow acceleration. The boulder ceases to offer sheltering effects to mobile particles on its downstream side, i.e., between $x/d_c = 0.50$ and $x/d_c = 1.00$, thus allowing their entrainment from these locations. This finding is also in agreement with the observations of Papanicolaou and Kramer (2005), who documented the absence of mobile sediment depositional patches in the region downstream of the boulders under LRS conditions (see section 2.7). Further downstream, i.e., for $x/d_c > 1.00$, the $u_{*,BCM}$ displays small variability.

The distribution of $u_{*,BCM}$ in Figure 2.8b exhibits a correspondence with the $\bar{u}$ profiles in Figure 2.7. Upstream of the boulder, the slight reduction of the $\bar{u}$ profile magnitude between $x/d_c = -3.00$ and $x/d_c = -1.00$ and its sharp reduction between $x/d_c = -1.00$ and $x/d_c = 0.50$ is captured by the slight and sharp reduction in the $u_{*,BCM}$ values in these locations respectively. Similarly, the sharp flow acceleration downstream of the boulder is captured by the sharp increase in the magnitude of $\bar{u}$ profile magnitude and in the $u_{*,BCM}$ values between $x/d_c = 0.50$ and $x/d_c = 1.00$. Further downstream, i.e., for $x/d_c > 1.00$, the magnitudes of both the $\bar{u}$ profiles and the $u_{*,BCM}$ do not show significant variability.

Overall, Figure 2.8b shows that the BCM can reliably capture the variability of the u_* and the local bed shear stress, $\tau_o = \rho u_*^2$, in the vicinity of the boulder and its effects on the surrounding flow field under LRS conditions. The determination of the variable $u_{*,BCM}$ in Figure 2.8 was based on the assumption that the defect law is valid in the outer flow layer (e.g., Afzalimehr and Anctil 2000; Hinze 1975; Papanicolaou *et al.* 2012). Under the LRS conditions of Run 2 of Set 2, the $\bar{u}$ profiles attain "S" or linear shapes, and it is rather unlikely that they follow a defect law distribution. This finding implies that the BCM is a robust method for estimating the u_* and its assumption that the defect law is valid in the outer flow layer is not critical (e.g., Afzalimehr and Anctil 2000; Hinze 1975). The robustness of the BCM for estimating the u_* was also reported by Papanicolaou *et al.* (2012) for HRS conditions.

2.5 Mean Vortex Structure Around a Wall-Mounted Boulder

The transverse instantaneous vorticity, ω_y, within a streamwise-vertical, x–z, plane along the boulder centerline and the vertical instantaneous vorticity, ω_z within a streamwise-lateral, x–y, plane through the boulder center were calculated by the LES are illustrated in Figure 2.9a and b, respectively. Figure 2.9c and d shows the time-averaged streamwise vorticity, $\bar{\omega}_x$ within lateral-vertical, y–z, planes at $x/d_c = 1.00$ and 2.00 downstream of the boulder center. Combined, the plots (a–d) of Figure 2.9 allow elucidating the three-dimensional vortex structure field that develops around the boulder under HRS conditions. The flow around the boulder is characterized by large-scale vortex structures, which are predominantly generated by the flow separation from the boulder (Figure 2.9a and b). The flow separates at the boulder crest, forms a recirculation zone and reattaches at approximately $x/d_c = 1.40$, as also observed by the mean streamwise velocity, $\bar{u}$, profiles (Figure 2.6).

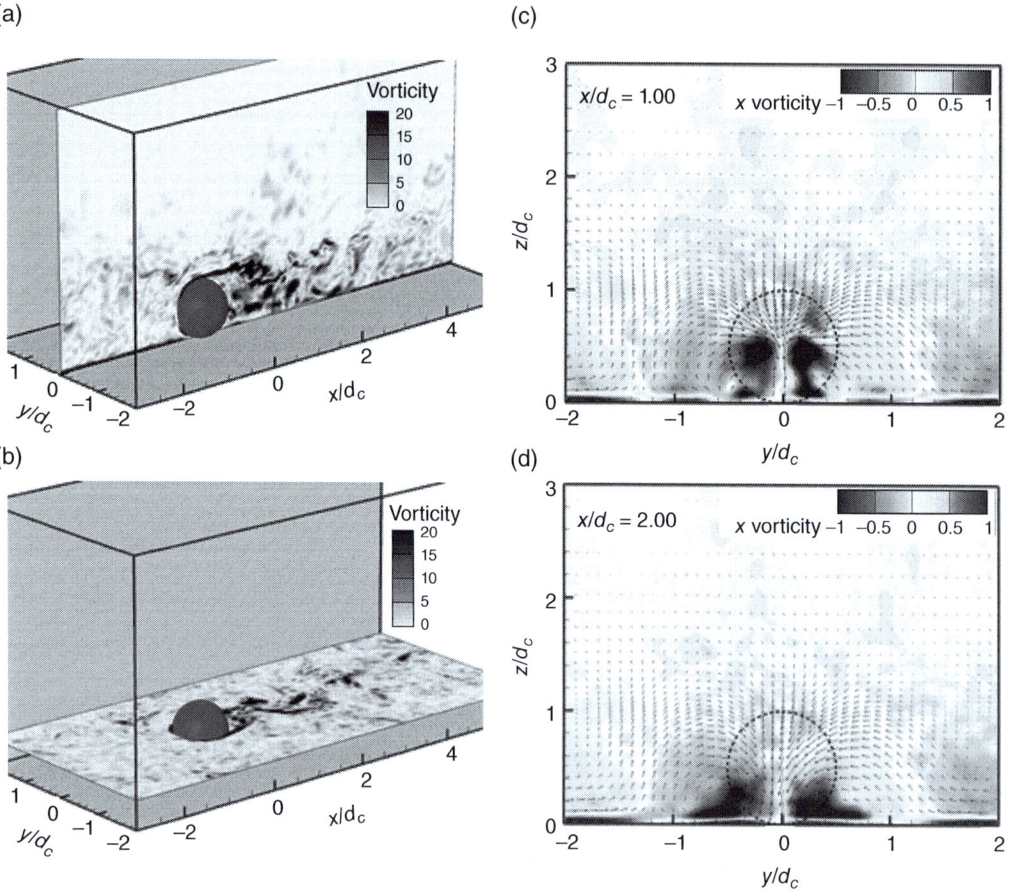

Figure 2.9 (a) Instantaneous lateral vorticity, ω_y, distribution within the streamwise-vertical, x–z, plane through the boulder centerline; (b) instantaneous vertical vorticity, ω_z, distribution within the streamwise-transverse, x–y, plane through the boulder center; (c) time-averaged streamwise vorticity, $\bar{\omega}_x$, distribution within the transverse-vertical, y–z, plane at $x/d_c = 1.00$; (d) time-averaged streamwise vorticity, $\bar{\omega}_x$, distribution within the transverse-vertical, y–z, plane at $x/d_c = 2.00$. (From Papanicolaou *et al.* 2012.) (See insert for color representation of this figure.)

A detailed examination of the time-averaged streamwise vorticity, $\bar{\omega}_x$ patterns at $x/d_c = 1.00$ (Figure 2.9c) provides deeper insight into the vortex structures that develop around the boulder. There appear to be four dominant vortex structures, which are indicated by the ω_y vorticity concentrations downstream of the boulder. First, a pair of counter-rotating vortices, i.e., a pair of $\bar{\omega}_x$ concentrations with opposite sign, are observed near the boulder periphery, at approximately $z/d_c = 0.90$ and $y/d_c = \pm 0.20$ (Figure 2.9c). This vortex pair near the periphery of the boulder has the same sense of rotation and location as the upper vortices documented by Hajimirzaie *et al.* (2014) in their study of a wall-mounted sphere within a boundary layer flow.

A second pair of strong counter-rotating vortices is illustrated by the concentration of ω_y vorticity near the middle of the boulder at $z/d_c = 0.50$ extending between $x/d_c = 0.50$ to approximately $x/d_c = 2.00$ (Figure 2.9a). This ω_y concentration corresponds to a second pair of strong counter-rotating vortices near the boulder center at approximately $z/d_c = 0.50$ and $y/d_c = \pm 0.20$ (Figure 2.9c) with opposite sense of rotation comparatively to the upper vortices. This pair of strong counter-rotating vortices resembles in terms of direction of rotation and location relative to the obstacle the tip vortices observed by Hajimirzaie *et al.* (2014).

A third pair of near-bed $\bar{\omega}_x$ concentrations are observed at approximately $z/d_c < 0.20$ and $y/d_c = \pm 0.40$ in Figure 2.9c. These $\bar{\omega}_x$ concentrations indicate the presence of near-bed vortices with similar sense of rotation as the tip vortices forming near the boulder center. The near-bed vortices in Figure 2.9c bear strong resemblance to the lobe vortices observed by Hajimirzaie *et al.* (2014), which were found to be related to the like-signed tip vortices.

Finally, at $y/d_c = \pm 1.00$ and $z/d_c = 0.10$ two concentrations of $\bar{\omega}_x$ can be observed with similar sign to the tip and near-bed vortices (Figure 2.9c). These $\bar{\omega}_x$ concentrations are the limbs of the well-known horseshoe vortex, which is formed by the roll-up of the boundary layer at the upstream face of the spherical boulder, due to the adverse pressure gradient (e.g., Ballio, Bettoni, and Franzetti 1998; Hajimirzaie *et al.* 2014). The horseshoe vortex can also be discerned wrapping around the spherical boulder in the instantaneous ω_z vorticity contours of Figure 2.9b.

Figure 2.9d shows that the three vortex systems identified in Figure 2.9a and c coalesce at $x/d_c = 2.00$ and form a pair of so-called trailing vortices observed near the bed (e.g., Hajimirzaie *et al.* 2014; Pattenden, Turnock, and Zhang 2005). These trailing vortices were found to be robust features that form further downstream of obstacles of different shape and in a variety of flows (e.g., Hajimirzaie *et al.* 2014; Hajimirzaie, Wojcik, and Buchholz 2012; Okamoto 1982; Pattenden, Turnock, and Zhang 2005). Furthermore, the vertical vorticity, ω_y, and streamwise $\bar{\omega}_x$ vorticity concentrations in Figure 2.9b and c suggest that the lateral extent of the vortex systems generated by the boulder is approximately between $y/d_c = -0.75$ to $y/d_c = 0.75$. Papanicolaou *et al.* (2012) reported a similar extent of the boulder effects on the mean flow characteristics.

The flow field around a partially submerged boulder within an array is dominated by a pair of counter-rotating vortices, as shown in Figure 2.9c of Papanicolaou, Dermisis, and Elhakeem (2011). These counter-rotating vortices are illustrated in that figure by regions of time-averaged, vertical vorticity, $\bar{\omega}_z$ with similar magnitude but opposite sign. The counter-rotating vortex pair downstream of the anchor particle bears strong resemblance to the von Kárman vortex street visualized downstream of hemispherical obstacles by Shamloo, Rajaratnam, and Katopodis (2001) and Boyer and Roy (1991). From this striking similarity it is thus conjectured that the vortex structures downstream of a partially submerged boulder is a von Kárman vortex street. In plan view, Papanicolaou, Dermisis, and Elhakeem (2011) found the von Kárman vortex street to extend approximately $3d_c$ downstream of the anchor particle in the streamwise direction, and $2d_c$ in the transverse direction.

2.6 Collective Effects of the Boulder Array

As noted in section 2.2, boulders and coarser anchor particles in natural rivers are typically organized in arrays and have been reported to occupy a significant (2–25%) portion of the plan view area in mountainous stream reaches (e.g., Bathurst 1987; Byrd, Furbish, and Warburton 2000; Ghilardi, Franca, and Schleiss 2014c; Nitsche *et al.* 2011, 2012; Papanicolaou *et al.* 2012; Papanicolaou, Dermisis, and Elhakeem 2011; Yager, Kirchner, and Dietrich 2007). The role that an array of boulders collectively plays on the flow field around an individual boulder within the array has been overlooked. Perhaps only Papanicolaou, Dermisis, and Elhakeem (2011) suggested that the presence of the boulder array even at isolated roughness conditions (diagonal spacing $\lambda/d_c = 6.0$ between the anchor particles) collectively generates additional form drag, which essentially decelerates the flow within the reach. In their study of sediment movement within an array of boulders at LRS conditions, Papanicolaou, Dermisis, and Elhakeem (2011) observed, for the very first time, a sediment depositional formation upstream of the array. This formation, referred to as a "sand ridge" in their paper, was found to bear up to 20% of the total bed shear stress, $<\tau_o>$, applied by the flow. For comparison, the 40 individual boulders comprising the array in the Papanicolaou, Dermisis, and Elhakeem (2011) experiment were found to bear only 6% of the $<\tau_o>$.

Following the findings of Papanicolaou, Dermisis, and Elhakeem (2011), Tsakiris *et al.* (2014) provided some additional insight into the role of the form drag collectively generated by the boulder array under HRS conditions. In doing so, Tsakiris *et al.* (2014) compared the mean streamwise velocity, $\bar{u}$, profiles (Figure 2.6) and the $u_{*,BCM}$ distribution (Figure 2.8a) around the single boulder (Run 1 of Table 2.1) with the corresponding $\bar{u}$ profiles and the $u_{*,BCM}$ distribution around a boulder within an array from the experiment of Papanicolaou *et al.* (2012), i.e., the Array-HRS run in Table 2.1.

The mean streamwise velocity, $\bar{u}$, profiles measured at corresponding locations around a boulder with and without the presence of a boulder array are shown in Figure 2.10. In Figure 2.10 the $\bar{u}$ velocity profiles are normalized with the respective bulk flow velocity, U_{bulk}, to account for the different U_{bulk} of the two experimental scenarios. Overall, the $\bar{u}/U_{bulk}$ profiles for the single boulder appear to be parallel shifted in comparison to the corresponding $\bar{u}/U_{bulk}$ profiles for the boulder array. This parallel shift suggests that the boulder array globally decelerates the mean streamwise velocity by nearly 20–25% compared to the single boulder condition. The deceleration observed in Figure 2.10 is attributed to the form drag, which is generated collectively by the boulder array present in the Papanicolaou *et al.* (2012) experiment.

In addition, below $z/d_c = 1.00$, at locations $x/d_c = 2.00$ and 3.00, the $\bar{u}/U_{bulk}$ profiles for the single boulder show quick recovery to the shape of the $\bar{u}/U_{bulk}$ profile at $x/d_c = -1.00$ upstream of the

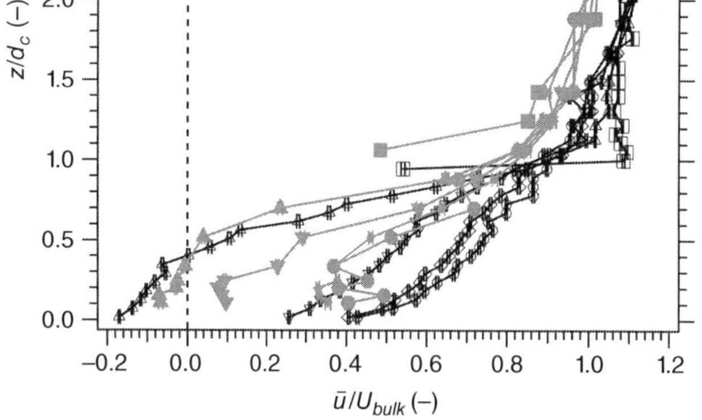

Figure 2.10 Time-averaged (mean) streamwise velocity, $\bar{u}$, distribution within the centerline transect of a boulder with and without an array present. (Adapted from Tsakiris *et al.* 2014.)

boulder. In contrast, especially at $x/d_c = 2.00$ and at $x/d_c = 3.00$, the $\overline{u}/U_{bulk}$ profiles around the boulder within the array attain smaller $\overline{u}/U_{bulk}$ values than the corresponding $\overline{u}/U_{bulk}$ around the single boulder. This slower recovery of the $\overline{u}/U_{bulk}$ implies that the global mean flow deceleration, due to the presence of the boulder array, amplifies the local effects of the boulder on its surrounding mean flow field.

The friction velocity, $u_{*,BCM}$, calculated via the BCM at each measurement location in the vicinity of the single boulder and the representative boulder within the boulder array is presented in Figure 2.11. The uniform flow friction velocity, $u_{*,unif} = (gHS_o)^{0.5} = 0.079\,\text{m/s}$, which is the same for Run 1 (Table 2.1) and for the experiment of Papanicolaou *et al.* (2012), is also included for comparison purposes. Most notably, Figure 2.11 shows that the $u_{*,BCM}$ in the vicinity of the boulder with the array present, i.e., at locations $x/d_c = 0.00$, 1.00, and 2.00, is consistently lower than the $u_{*,BCM}$ around the single boulder. The lower $u_{*,BCM}$ values around the boulder with the array present compared to the $u_{*,BCM}$ for the single boulder are attributable to the form drag, which results from the collective effects of the boulder array. In other words, the boulder array collectively bears a portion of the total bed shear stress applied by the flow. As a result, the portion of the total bed shear stress applied to each individual boulder within the array and hence the $u_{*,BCM}$ (Figure 2.11) is reduced. In a mathematical context, the reach-averaged total bed shear stress applied by the flow, $<\tau_o>$, may be expressed as (e.g., Yager, Kirchner and Dietrich 2007):

$$\langle \tau_o \rangle = \langle \tau_{os} \rangle + \langle \tau_{of} \rangle \tag{2.7}$$

where, $<\tau_{os}>$ is the bed shear stress applied as skin friction onto the rough bed consisting of the well-packed glass beads, and $<\tau_{of}>$ is the component borne by the single boulder as form drag. When the array of anchor particles is present, the $<\tau_o>$ should be decomposed as follows (Papanicolaou, Dermisis, and Elhakeem 2011):

$$\langle \tau_o \rangle = \langle \tau_{os} \rangle + \langle \tau_{of} \rangle + \langle \tau_{oa} \rangle \tag{2.8}$$

where $<\tau_{oa}>$ denotes the component of the total bed shear stress borne as form drag collectively generated by the boulder array. For a given $<\tau_o>$ applied by the flow, the collective form drag generated by the array and expressed in equation (2.8) at term $<\tau_{oa}>$ will reduce the sum $<\tau_{os}> + <\tau_{of}>$, when a single boulder is present (e.g., Papanicolaou *et al.* 2012; Papanicolaou,

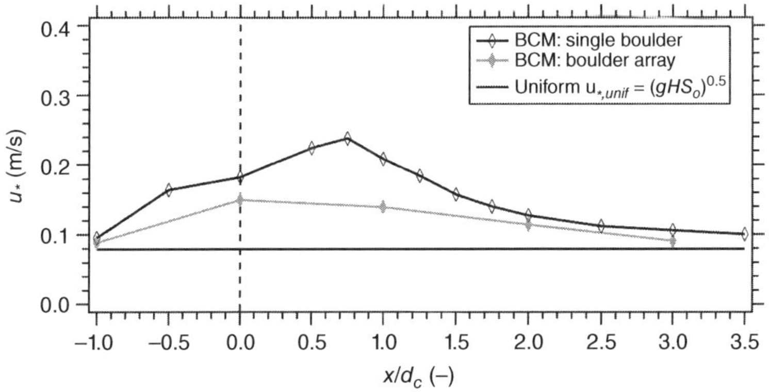

Figure 2.11 Comparison of the friction velocity, u_*, distribution from the boundary characteristics method (BCM) along the centerline of a boulder with and without an array present. (Adapted from Tsakiris *et al.* 2014.)

Dermisis, and Elhakeem 2011). This reduction is reflected in the smaller $u_{*,BCM}$ predicted in the vicinity of the boulder within the array (Figure 2.11).

The additional form drag generated collectively by the array of boulders should be accounted for when predicting bedload rates in stream reaches with ubiquitous boulders (e.g., Papanicolaou, Dermisis, and Elhakeem 2011; Rickenmann 2001; Yager, Kirchner, and Dietrich 2007). The $u_{*,BCM}$ in the vicinity of an individual boulder is reduced by 8–45% when the array is present. Consequently, not accounting for the boulder array effects would lead to an overestimation of the bed shear stress by 18–300%. This overestimation would, in turn, lead to a more substantial overestimation of the bedload rate, which in typical bedload relations is given as an exponential function of the bed shear stress with an exponent greater than unity.

2.7 Sediment Transport Within a Boulder Array

2.7.1 Qualitative Assessment of Flow and Sediment Deposition Patterns

Before examining the salient sediment depositional features developed in the experiments of Papanicolaou and Kramer (2005), it is of interest to relate the type of depositional features with the bulk experimental parameters based on prior observations. To this end, the bulk flow velocity, U_{bulk}, and the corresponding flow depth, H, for each of experimental run in Table 2.2 are compared with the H versus U_{bulk} data compiled by Peakall, Ashworth, and Best (1996) for river physical models (Figure 2.12). Peakall, Ashworth, and Best (1996) noted that under supercritical flow conditions (i.e., with $Fr > 1$), low-relief bedforms (e.g., ridges) and rhomboidal standing waves form. As the Fr

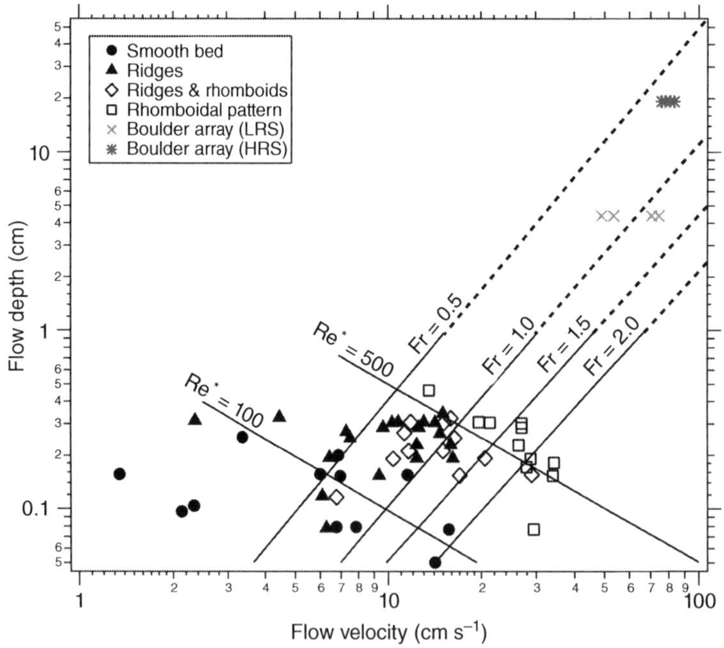

Figure 2.12 Conditions for the formation of low relief bedforms and water surface effects. (Adapted from Peakall, Ashworth, and Best 1996.)

decreases and flow becomes subcritical, the standing waves attenuate, while the ridges give way to a flat bed. Although these observations of Peakall, Ashworth, and Best (1996) were based on experiments conducted with sand-sized mobile sediment, they appear to qualitatively hold for the coarser material considered in the experiments of Papanicolaou and Kramer (2005). In particular, for the experimental runs with $<\tau_o^*>/\tau^*_{cr} = 3.0$ and 3.5 of Series B (Table 2.2), where flow was supercritical with $Fr > 1.0$, low-relief sediment patches formed upstream of the boulders and standing waves were observed at the boulder locations (Figure 2.13c and d). For the experiments of Series B conducted at $<\tau_o^*>/\tau^*_{cr} = 2.0$ and 2.5 with $Fr < 1.0$, only a depression of the water free surface was observed at the same location relative to the boulder with the standing waves for the runs with $<\tau_o^*>/\tau^*_{cr} = 3.0$ and 3.5 (Figure 2.13a and 13b). Similarly, Boyer and Roy (1991) observed distinct water surface patterns around a partially submerged boulder under subcritical ($Fr < 1.0$) and supercritical ($Fr > 1.0$) flow conditions. Peakall, Ashworth, and Best (1996), in the context of their scaling study, attributed the formation of the standing waves and low-relief bedforms in supercritical flows to smaller skin friction roughness relative to that predicted by scaling laws. In the experiments of Kramer (2005) reported

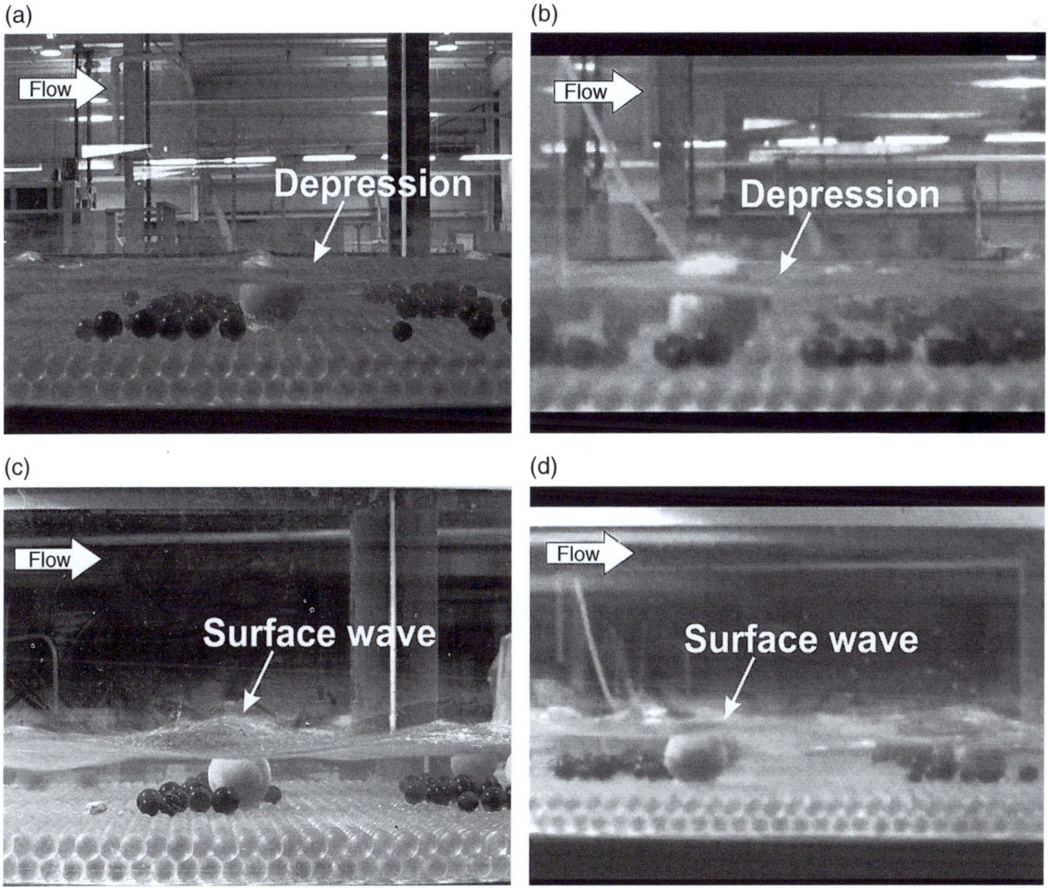

Figure 2.13 Side view observations of mobile sediment deposition patterns around partially submerged boulders at various flow stages: (a) $<\tau_o^*>/\tau^*_{cr} = 2.00$; (b) $<\tau_o^*>/\tau^*_{cr} = 2.50$; (c) $<\tau_o^*>/\tau^*_{cr} = 3.00$; (d) $<\tau_o^*>/\tau^*_{cr} = 3.50$. (From Papanicolaou and Kramer 2005.)

herein (Table 2.2), the smaller skin friction roughness that is associated with the presence of the standing waves and the sediment patch formation is attributable to the presence of the boulder array. The boulder array bears a portion of the total applied bed shear stress by the flow as form drag (Figure 2.11), which effectively reduces the bed skin friction.

2.7.2 Sediment Deposition Patterns around Boulders

Under HRS conditions (Series A of Table 2.3), the incoming mobile sediment tends to preferentially deposit in the wake region that forms downstream of a boulder within the array, as shown in Figures 5.1–5.4 of Kramer (2005) for $<\tau_o^*>/\tau^*_{cr} = 2.0$, 2.5, 3.0 and 3.5, respectively. Within the boulder wake region, the mean flow significantly decelerates or even reverses direction (Figure 2.6). Moreover, the component of the total bed shear stress applied onto the mobile sediment, τ_{os}, is minimized (Figure 2.8a and 11). At the same time, a complex vortex field, comprising of upper, tip and near-bed vortices, forms within the boulder wake region (Figure 2.9). These mean flow patterns and the vortex structures promote the deposition of incoming mobile sediment within the recirculation region forming downstream of a fully submerged boulder.

For the runs of Series B conducted under LRS conditions (Table 2.3), plan view images of the boulder array after the end of the test runs are shown in figures 5.17–5.20 of Kramer (2005) for $<\tau_o^*>/\tau^*_{cr} = 2.0$, 2.5, 3.0, and 3.5, respectively. Similarly, the sand sized sediment in Papanicolaou, Dermisis, and Elhakeem (2011) preferentially deposited as sand patches upstream of the partially submerged boulders of the array. Therefore, under LRS conditions, mobile sediment tends to deposit at the stoss (upstream) side of the boulders, and not at the downstream side of the boulders, as is the case under HRS conditions. The mean flow downstream of a partially submerged boulder accelerates (Figure 2.7) and a von Kárman vortex is formed (Figure 2.9c in Papanicolaou, Dermisis and Elhakeem (2011)). At the same time, upstream of a partially submerged boulder the flow decelerates (Figure 2.7), thus creating a low bed shear stress region, indicated by the reduced $u_{*,BCM}$ in Figure 2.8b. Overall, the interaction of a partially submerged boulder with the approach flow creates favorable conditions for the deposition of incoming sediment at the stoss and not in the wake region of a partially submerged boulder.

A quantitative look of the mobile sediment amount trapped by the array boulders under HRS and LRS conditions is offered in Figure 2.14. Figure 2.14 illustrates the plan view area occupied by the deposited sediment, A_{sd}, as a function of $<\tau_o^*>/\tau^*_{cr}$. The A_{sd} was determined through analysis of the images in figures 5.1–5.4 and figure 5.17–5.20 of Kramer (2005) for the experimental Series A and B of Table 2.3, respectively. The total immobile area of the bed, A_I, (Figure 2.4c) was also calculated by adding the plan view area of the boulders to the A_{sd}. Under HRS conditions, the trapping efficiency of the boulder array, which is quantified by A_{sd}, increases monotonically with $<\tau_o^*>/\tau^*_{cr}$. For the LRS condition, the A_{sd} increases gradually from $<\tau_o^*>/\tau^*_{cr} = 2.0$ to 2.5, but shows a sharp increase to $<\tau_o^*>/\tau^*_{cr} = 3.0$, where A_{sd} attains its maximum value. As $<\tau_o^*>/\tau^*_{cr}$ further increases to 3.5, the A_{sd} exhibits a sharp drop and becomes comparable to the A_{sd} for $<\tau_o^*>/\tau^*_{cr} = 2.0$. This sharp decrease of A_{sd} from $<\tau_o^*>/\tau^*_{cr} = 3.0$ to 3.5 suggests that at high flow conditions, where flow becomes supercritical, the partially submerged boulders offer reduced protection to the incoming mobile sediment from the flow shearing action. This pattern illustrated in Figure 2.14 for the LRS condition qualitatively resembles the findings of Papanicolaou *et al.* (2003) and Strom *et al.* (2004), who investigated the organization of unisize bed surface material into surface patches or clusters. The formation of clusters for $<\tau_o^*>/\tau^*_{cr} < 2.00$ added to the bed stability by trapping incoming sediment, as is the case with the boulders herein $<\tau_o^*>/\tau^*_{cr} < 3.00$. However, as the $<\tau_o^*>/\tau^*_{cr}$ increased above 2.00, the clusters disintegrated and ceased to offer additional bed stability, which is a similar observation for the partially submerged boulder array for $<\tau_o^*>/\tau^*_{cr} > 3.00$ (Figure 2.14).

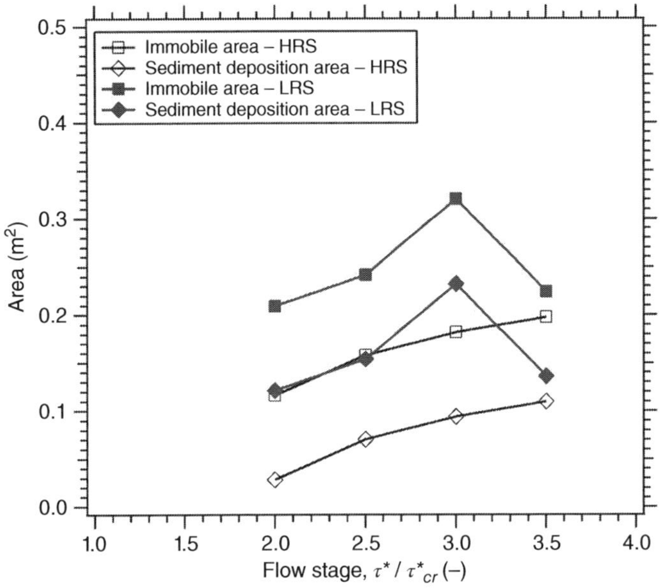

Figure 2.14 Area occupied by the deposited sediment and total immobile bed area as a function of the flow stage, $<\tau_o^*>/\tau^*_{cr}$, for high and low relative submergence conditions (HRS and LRS).

2.7.3 Correspondence of Sediment Deposition and Vortex Structure Areas

In the following, we evaluate the hypothesis of Papanicolaou *et al.* (2012) that there is a correspondence of the vortex structures developing around the boulders with the deposition of mobile sediment. To this end, the area of the mobile sediment patches deposited around the boulders is compared to the area, A_e, of the vortex structures, or eddies, which develop around the boulders under HRS and LRS conditions. Preliminary estimates of A_e were provided for boulders under HRS and LRS conditions from the LES of Papanicolaou *et al.* (2012) and from the LSPIV measurements of Papanicolaou, Dermisis, and Elhakeem (2011) in section 2.5 of this chapter. Atop flat smooth beds without any boulders present, it has been found that the average length, l_x, and width, l_y, of the turbulent eddy structures scale with the flow inner variables, namely the water kinematic viscosity, ν, and the uniform friction velocity, $u_{*,unif}$ (e.g., Cao 1997; Grass 1971; Kim, Kline, and Reynolds 1971; Kline *et al.* 1967; Wallace, Eckelmann, and Brodkey 1972). The proposed correlations developed from experimental data are the following:

$$l_x = 40\frac{\nu}{u_{*,unif}} \tag{2.9a}$$

and

$$l_y = 10\frac{\nu}{u_{*,unif}} \tag{2.9b}$$

From a physical point of view, equation (2.9) indicates that the eddy dimensions l_x and l_y are an outcome of the balance between the energy offered by the flow shear and the viscous dissipation. The flow shear action in equation (2.9) is quantified by $u_{*,unif}$ and the viscous dissipation through the water kinematic viscosity, ν. In the presence of a boulder, the flow shear variability around the boulder is reliably captured by the $u_{*,BCM}$, as opposed to $u_{*,unif}$ (Figure 2.8). In fact, in the highly turbulent flows considered herein, energy is predominantly dissipated by the large eddy structures generated by the boulder (Figure 2.9), as opposed to the viscous effects (e.g., Bridge 2009; Nezu and Nakagawa 1993). Therefore, to account for the effects of the turbulent structures in equation (2.9), we consider the turbulent eddy viscosity, ν_T, in place of ν. The dimensions l_x and l_y of the turbulent eddy structures around the boulder would thus be given as:

$$l_x = 40\frac{\nu_T}{u_{*,BCM}} \tag{2.10a}$$

and

$$l_y = 10\frac{\nu_T}{u_{*,BCM}} \tag{2.10b}$$

For the purposes of the semi-theoretical calculations herein, the estimation of ν_T is based on the Boussinesq hypothesis, which mathematically may be represented as:

$$\tau_{Re} = -\overline{\rho u' w'} \approx \rho \nu_T \frac{\partial \bar{u}}{\partial z} \Rightarrow \nu_T = \frac{-\overline{u' w'}}{\frac{\partial \bar{u}}{\partial z}} \tag{2.11}$$

Equation (2.11) indirectly accounts for the presence of the boulder effects on its surrounding flow field, as it accounts for the effects of the boulder on the Reynolds shear stress, τ_{Re}, and the mean streamwise velocity profiles around it. Herein, the ν_T and l_x and l_y are calculated from the LDV measurements around the single boulder under HRS ($H/d_c = 3.50$) and LRS ($H/d_c = 0.80$) conditions for $<\tau_o^*>/\tau^*_{cr} = 2.5$ conducted for the purposes of test Runs 1 and 2 of experimental Set 1 (Table 2.1). For the HRS conditions of test Run 1, an average value of ν_T within the boulder wake region yields an average eddy size of $l_x = 0.1378$ m and $l_y = 0.034$ m. The predicted streamwise length of the turbulent eddy structures, $l_x = 0.1378\text{ m} = 2.50d_c$, is in remarkable agreement with the qualitative assessment of the vortex structure length in Figure 2.9. The lateral extent of the vortices $l_y = 0.034\text{ m} = 0.63d_c$ is smaller than the qualitative assessment of $1.50d_c$ in Figure 2.9. This discrepancy is likely attributed to the simplified calculation of ν_T, which does not account for the lateral Reynolds stress component, $-\rho(\overline{v'w'})$. The area of the turbulent eddy structures around the boulder is therefore: $A_e = l_x l_y = 0.0047\text{ m}^2$. The total eddy area downstream of all 12 boulders in the array is thus $A_e = 0.056\text{ m}^2$, which is good agreement with the area, $A_{sd} = 0.07\text{ m}^2$, occupied by the mobile sediment (Figure 2.14).

For the LRS conditions of test Run 2 of Set 1 (Table 2.1), the average eddy length and width are calculated from equation 2.10a and b to be $l_x = 0.002$ m and $l_y = 0.001$ m. The l_x and l_y for the LRS conditions are significantly different from the extent of the vortex structures estimated by Papanicolaou, Dermisis, and Elhakeem (2011) and presented in section 2.5. This discrepancy for the LRS conditions is attributed to the strong three-dimensional flow around the boulder. To account for the three-dimensional flow, the ν_T estimation for the LRS conditions should include the lateral component of the Reynolds shear stress.

2.8 Morphology of Depositional Patches Around Boulders

The analysis in section 2.7.2 suggests that the location of mobile sediment deposition relative to an anchor particle (stoss or wake) is intimately related to the vortex topology that develops from the interaction of the boulder with the approach flow. In turn, the vortex topology is a function of the boulder relative submergence. Under LRS conditions, the mobile sediment deposits upstream of the anchor particle, contrary to the currently accepted cluster paradigm, that incoming particles are entrapped downstream of the anchor particle within its wake.

Similar to the sediment deposition observations in the laboratory experiments of Papanicolaou and Kramer (2005), the sediment accumulations or patches around boulders and other anchor particles in natural coarse-grained streams also exhibit distinct morphologies. To date several efforts for developing classification schemes for morphology of these have been made (e.g., De Jong 1995; Hendrick, Ely, and Papanicolaou 2010; Strom and Papanicolaou 2008; Wittenberg 2002; Wittenberg and Newson 2005). Yet these efforts lack the necessary objectivity, as most of the morphology classification schemes were predominantly based on site-specific visual observations during field campaigns and definition sketches. Such empirical tools hinder the establishment of a classification scheme for clusters, thus ultimately delaying their incorporation in a universal regime diagram for the gravel-bed-river continuum (e.g., Figure 2.1). Simultaneously, the lack of a classification scheme for cluster morphology also impedes comparisons between the morphology of patches observed in different mountain streams with their counterparts developed in the laboratory.

To address this shortcoming, Papanicolaou, Tsakiris, and Strom (2012) recently employed principles of fractal theory to develop a quantitative measure for describing the morphology of the patches that develop around boulders. Through a systematic fractal analysis of artificial images, the areal fractal dimension, D_{AT}, of the plan view image of these patches was identified as the most suitable measure for the patch morphology. The reader is referred to Papanicolaou, Tsakiris, and Strom (2012) for an in-depth derivation of D_{AT} and the methodology for determining D_{AT} via the box-counting method. The D_{AT} provided an objective measure of cluster morphologies documented in natural coarse-grained streams by Strom and Papanicolaou (2008). In fact each cluster morphology examined corresponded to a unique range of D_{AT} values. Two main categories of cluster morphologies were identified by the D_{AT}, namely the in-line, which corresponded to D_{AT} closer to 1; and the "non-linear" cluster morphologies where D_{AT} attained values closer to 2. Furthermore, because the D_{AT} is a scale-independent index, it allowed comparisons between the morphologies at different scales. In particular, through the use of D_{AT}, Papanicolaou, Tsakiris, and Strom (2012) illustrated a correspondence between the morphology of the patches recorded in the field and of patches developed in the laboratory from uni-size sediment. Analysis of the patches with different morphologies that resulted for the experimental series in the experiments of Kramer (2005) would assist in better understanding their underlying forming processes in natural coarse-grained streams.

2.9 Concluding Remarks

The focus of this study was to isolate the hydrodynamic and sediment transport processes that led to the formation of the small-scale bedforms found in the upper segments of streams known as clusters. Existing hydraulic data from field investigations were combined with bed state observations in order to demarcate the flow–sediment conditions under which clusters are present in the riverine continuum. This compilation showed that the flow–sediment regimes corresponding to cluster formation are quite comparable to those of pool–riffles and plane beds and in the limit of those of step-pools in the upper portions of the riverine continuum.

We systematically examined the effects that boulder-sized anchor particles have on their surrounding flow field and sediment movement patterns that promote cluster formation. More specifically, this study: (i) promoted a fundamental understanding of the form drag generated by individual boulders in their vicinity, as well as collectively by an array of boulders; and (ii) documented the vortex structure topology that develops around a boulder and assessed its influence on the amount of mobile sediment trapped by the boulders. The boulder relative submergence was a key parameter in this investigation. The relative submergence was considered to regulate not only the mean flow patterns around a boulder but also the vortex structure topology. We further considered that the sediment movement patterns around the boulders were intimately related to the topology of macro-eddies.

The key feature of the mean flow field around a boulder at HRS conditions was a mean flow recirculation region in the boulder near-wake, which extended $1.00d_c$ downstream of the boulder and was characterized by a deceleration and even reversal of the flow. Within the near-wake region, the BCM predicted an increase in the bed shear stress resulting from the additional form drag generated by the interaction of the boulder with the flow. This additional form drag effectively reduced the bed shear stress available for transporting the mobile sediment within the boulder wake region. Under LRS conditions, the velocity profile attained an "S"-shaped profile in the proximity of, and linear shape further downstream of, a partially submerged boulder. Immediately downstream of the boulder, the flow was characterized by intense acceleration and high flow shear, which was attributed to the presence of a standing wave that constricted the mean flow. In contrast, upstream of the boulder, a zone of flow deceleration and low flow shear was documented.

The topology of the vortex structures, which develop within the wake region of a fully submerged boulder, was assessed in qualitative terms from a LES performed by Papanicolaou *et al.* (2012). The LES highlighted four key vortex structures developing within the boulder wake region, namely the upper, tip, near-bed, and the horseshoe vortices. The vortex structures were found to extend for $2d_c$ in the streamwise direction and $1.5d_c$ in the transverse. For the LRS conditions, a von Kárman vortex street was the dominant vortex structure in the boulder wake extending $3d_c$ downstream of the boulder and $1.5d_c$ in the lateral.

The form drag generated collectively by an array of isolated boulders was estimated through systematic comparisons of the mean streamwise velocity profiles and $u_{*,BCM}$ distributions around a boulder with and without the array present (Tsakiris *et al.* 2014). A boulder array was found to decelerate the mean flow within the array and reduce the bed shear stress applied to each individual boulder by 8–300%.

The analysis of the mean streamwise velocity patterns and the vortex structures developing around the boulders was complemented with observations of the mobile sediment depositional patterns within the array for HRS and LRS conditions. A reversal of the sediment deposition location relative to the boulder was found to occur between HRS and LRS conditions. In particular, under HRS conditions deposition occurred within the low shear boulder near-wake region. In contrast, under LRS conditions, mobile sediment particles preferentially deposited at the upstream, i.e., lee, face of the boulders. This location of sediment deposition for the LRS coincided with the low bed shear stress region, highlighted by small $u_{*,BCM}$ values from the flow analysis. This correspondence of the mobile sediment deposition areas around the boulders with the low shear regions around a boulder for HRS and LRS conditions provides compelling evidence that the interaction of the boulders with the approach flow regulates the sediment depositional patterns in their vicinity, and in turn cluster morphology.

A quantitative assessment of the amount of the mobile sediment accumulating around the boulders for different flow stages, $<\tau_o{}^*>/\tau^*{}_{cr}$, was performed through image analysis. This quantitative assessment was combined with vortex topology scaling laws, which provided preliminary, semi-theoretical estimates of the area occupied by the vortex structures around the boulders. The calculated vortex structure area around a boulder showed remarkably good agreement with the sediment deposition area derived from image analysis for the HRS condition. This remarkable agreement further solidified the intimate relation between vortex structure topology and the sediment depositional areas around the boulders. At the same time, the agreement between the sediment depositional and the vortex structure area around a boulder provided a hydrodynamically based estimate of the effective bed area for mobile sediment conveyance, A_{eff}, through a boulder array.

This study provided a fundamental examination of the control that boulder-sized anchor particles exert on their surrounding turbulent flow and bedload transport processes, which contribute to the formation of clusters in gravel-bed rivers. The observed relation between the sediment depositional patterns around the boulders with the vortex topology under different degrees of boulder submergence has important implications for classifying cluster microforms within the river continuum from a geomorphological perspective, but also from an engineering perspective, when designing habitat restoration structures and sediment control best-management practices.

Notation and Abbreviations

A_b	Plan view area of the boulders
ADV	Acoustic Doppler velocimeter
A_e	Plan view area of the eddies (vortices) around the boulders
A_{eff}	Effective plan view bed area contributing to the mobile sediment supply
A_I	Plan view area of the bed that remains immobile
A_{sd}	Plan view area of sediment deposited around the boulders
B	Flume width
BCM	Boundary characteristics method
C	Constant in equation (2.1)
D_{AT}	Total areal fractal dimension
d	Mobile sediment diameter
d_{16}	Diameter of the grains coarser than the 16% of the bed material
d_{50}	Diameter of the grains coarser than the 50% of the bed material (median diameter)
d_{84}	Diameter of the grains coarser than the 84% of the bed material
d_b	Diameter of the glass beads comprising the flat rough bed
d_c	Boulder (clast) diameter
Fr	Froude number
g	Acceleration of gravity
H	Flow depth
HRS	High Relative Submergence
i	Index in equation (2.2)
K	Constant in equation (2.3)
LDV	Laser Doppler velocimeter

LES	Large eddy simulation
LRS	Low relative submergence
LSPIV	Large-scale particle image velocimetry
l_x	Characteristic length of the eddies (vortices) along the streamwise direction
l_y	Characteristic length of the eddies (vortices) along the transverse direction
N	Number of velocity measurements in equation (2.2)
Q	Flow discharge
$q_{s,m}{}^*$	Dimensionless bedload rate of the mobile sediment
S_o	Bed slope
$u_{*,BCM}$	Friction velocity calculated from the BCM
$u_{*,unif}$	Uniform friction velocity
u_*	Friction velocity
u	Instantaneous streamwise (along x direction) velocity
$\bar{u}$	Time-averaged (mean) streamwise (along x direction) velocity
u'	Streamwise velocity fluctuation
U_{bulk}	Bulk flow velocity
u_e	Maximum time-averaged, streamwise velocity along the depth
v	Instantaneous transverse (along y direction) velocity
v'	Transverse velocity fluctuation
w	Instantaneous vertical (along z direction) velocity
w'	Vertical velocity fluctuation
x	Streamwise direction
y	Transverse direction
z	Vertical direction
δ_d	Boundary layer displacement thickness given by equation (2.4a)
δ_θ	Boundary layer momentum thickness given by equation (2.4b)
κ	Von-Kárman constant
λ	Diagonal spacing between boulders of an array
ν	Water kinematic viscosity
ν_T	Turbulent eddy viscosity
ρ	Water density
ρ_s	Sediment density
σ_g	Geometric standard deviation
τ^*_{cr}	Critical dimensionless bed shear stress for incipient motion
τ_{cr}	Critical bed shear stress for incipient motion

$<\tau_o^*>$ Dimensionless, reach-averaged, total bed shear stress

$<\tau_o>$ Reach-averaged, total bed shear stress

$<\tau_{oa}>$ Reach-averaged, bed shear stress borne collectively by the boulder array

$<\tau_{of}^*>$ Dimensionless, reach-averaged, total bed shear stress applied as form drag on the boulders

$<\tau_{of}>$ Reach-averaged, total bed shear stress applied as form drag on the boulders

$<\tau_{os}^*>$ Dimensionless, reach-averaged, bed shear stress applied as skin friction

$<\tau_{os}>$ Reach-averaged, bed shear stress applied as skin friction

τ_{Re} Reynolds shear stress

ω_y Instantaneous transverse vorticity given by equation (2.5a)

ω_z Instantaneous vertical vorticity given by equation (2.5b)

$\bar{\omega}_x$ Time-averaged, streamwise vorticity given by equation (2.6)

Acknowledgments

The authors would like to thank all who contributed to the studies of the Papanicolaou group, including Dr. Kyle Strom, Dr. D. Dermisis, Dr. Mohamed Elhakeem, Casey Kramer, Adam Schuyler, Dr. Adam Maxwell, Doug Knapp, Brandon Billing as well as Dr. Lisa Ely and Dr. Ross Hendrick. Furthermore, thanks are due to Dr. James Buchholz and Dr. Seyed Hajirmirzaie, for the discussions on vortex topology, as well as Professors Vladimir Nikora and V. Kindratenko for the discussions on fractals. The helpful discussions with Professors J. Laronne and Ian Reid as well as other members of the gravel-bed community are gratefully acknowledged. The authors also wish to acknowledge the financial support provided for these studies, including US Geological Survey Program 104, Contract No. 02HQGR0139, and Grant No. 3995–7941, the US National Science Foundation NSF Hydrologic division under NSF grant EAR-0208358, NSF Integrative Graduate Education and Research Training (IGERT) grant under NSF Grant No. DGE9972887 and Grant No. CBET-1033732. The comments provided by Dr. Mario Franca and Professor Dieter Rickenmann have substantially improved the quality of the manuscript.

References

Aberle, J, Koll, K, and Dittrich, A 2008. Form induced stresses over rough gravel-beds. *Acta Geophysica* 56(3), 584–600. DOI: 10.2478/s11600-008-0018-x.

Afzalimehr, H and Anctil, F 2000. Accelerating shear velocity in gravel-bed channels. *Hydrological Sciences Journal* 45(1), 113–124. DOI: 10.1080/02626660009492309.

Ballio, F, Bettoni, C, and Franzetti, S 1998. A survey of time-averaged characteristics of laminar and turbulent horseshoe vortices. *Journal of Fluids Engineering* 120(2), 233–242.

Bathurst, JC 1978. Flow resistance of large-scale roughness. *Journal of the Hydraulics Division* 104(12), 1587–1603.

Bathurst, JC 1987. Critical conditions for bed material movement in steep, boulder-bed streams. In *Erosion and Sedimentation in the Pacific Rim* (eds RL Beshta, T Blinn, and G Grant *et al.*). Publication 165, International Association of Hydrological Sciences Publication: Wallingford, UK; 309–318.

Bennett, SJ and Bridge, JS 1995. An experimental study of flow, bedload transport and bed topography under conditions of erosion and deposition and comparison with theoretical models. *Sedimentology* 42(1), 117–146.

Best, J 1996. The fluid dynamics of small-scale alluvial bedforms. in *Advances in Fluvial Dynamics and Stratigraphy* (eds PA Carling and MR Dawson). John Wiley & Sons: New York; 67–125.

Billi, P 1988. A note on cluster bedform behaviour in a gravel-bed river. *Catena* 15(5), 473–481. DOI: 10.1016/0341-8162(88)90065-3.

Biron, PM, Robson, C, Lapointe, MF, and Gaskin, SJ 2004. Comparing different methods of bed shear stress estimates in simple and complex flow fields. *Earth Surface Processes and Landforms* 29(11), 1403–1415. DOI: 10.1002/esp.1111.

Boyer, C and Roy, AG 1991. Morphologie du lit autour d'un obstacle soumis à un écoulement en couche mince. *Géographie physique et Quaternaire* 45(1), 91–99.

Brayshaw, AC 1984. Characteristics and origin of cluster bedforms in coarse-grained alluvial channels. in *Sedimentology of Gravels and Conglomerates* (eds EH Koster and FJ Steel). Memoir 10, Canadian Society of Petroleum Geologists: Calgary; 77–85.

Bridge, JS 2009. *Rivers and Floodplains: Forms, Processes, and Sedimentary Record.* John Wiley and Sons: Chichester.

Buffington, JM and Montgomery, DR 1997. A systematic analysis of eight decades of incipient motion studies, with special reference to gravel-bedded rivers. *Water Resources Research* 33(8), 1993–2029.

Byrd, TC, Furbish, DJ, and Warburton, J 2000. Estimating depth-averaged velocities in rough channels. *Earth Surface Processes and Landforms* 25(2), 167–173.

Cao, Z 1997. Turbulent bursting-based sediment entrainment function. *Journal of Hydraulic Engineering* 123(3), 233–236.

Chin, A 1999. On the origin of step-pool sequences in mountain streams. *Geophysical Research Letters* 26 (2), 231–234.

Church, M 2006. Bed material transport and the morphology of alluvial river channels. *Annual Review of Earth Planetary Science* 34, 325–354.

Clauser, FH 1956. The turbulent boundary layer. *Advances in Applied Mechanics* 4, 1–51.

Coleman, S, Nikora, V, McLean, S, and Schlicke, E 2007. Spatially averaged turbulent flow over square ribs. *Journal of Engineering Mechanics* 133(2), 194–204.

Curran, JC and Tan, L 2014. The effect of cluster morphology on the turbulent flows over an armored gravel bed surface. *Journal of Hydro-Environment Research* 8(2), 129–142.

De Jong, C 1995. Temporal and spatial interactions between river bed roughness, geometry, bedload transport and flow hydraulics in mountain streams: examples from Squaw Creek (Montana, USA) and Lainbach, Schmiedlaine (Upper Bavaria, Germany). *Berliner Geographische Abhandlungen* 59.

Dey, S and Das, R 2012. Gravel-bed hydrodynamics: double-averaging approach. *Journal of Hydraulic Engineering* 138(8), 707–725. DOI: 10.1061/(asce)hy.1943-7900.0000554.

Dey, S, Sarkar, S, Bose, SK, Tait, S, and Castro-Orgaz, O 2011. Wall-wake flows downstream of a sphere placed on a plane rough wall. *Journal of Hydraulic Engineering* 137(10), 1173–1189. DOI: 10.1061/(asce) hy.1943-7900.0000441.

Fernandez Luque, R and Van Beek, R 1976. Erosion and transport of bed-load sediment. *Journal of Hydraulic Research* 14(2), 127–144.

Ferro, V 2003. ADV measurements of velocity distributions in a gravel-bed flume. *Earth Surface Processes and Landforms* 28(7), 707–722.

Flammer, GH, Tullis, JP, and Mason, ES 1970. Free surface, velocity gradient flow past hemisphere. *Proceedings of the American Society of Civil Engineers, Journal of the Hydraulics Division* 96(HY7), 1485–1502.

Florens, E, Eiff, O, and Moulin, F 2013. Defining the roughness sublayer and its turbulence statistics. *Experiments in Fluids* 54(4). DOI: 10.1007/s00348-013-1500-z.

Franca, M 2005. Flow dynamics over a gravel riverbed. In *Proceedings of the XXXI IAHR Congress,* Seoul, Korea; 6542–6551.

Franca, MJ and Lemmin, U 2009. The simultaneous occurrence of logarithmic and S-shaped velocity profiles in gravel-bed river flows. *Archives of Hydro-Engineering and Environmental Mechanics* 56(1–2), 29–41.

Franca, MJ, Ferreira, RM, and Lemmin, U 2008. Parameterization of the logarithmic layer of double-averaged streamwise velocity profiles in gravel-bed river flows. *Advances in Water Resources* 31(6), 915–925.

Frey, P and Church, M 2011. Bedload: a granular phenomenon. *Earth Surface Processes and Landforms* 36(1), 58–69.

Ghilardi, T, Franca, MJ, and Schleiss, AJ 2014a. Bed load fluctuations in a steep channel. *Water Resources Research* 50(8), 6557–6576. DOI: 10.1002/2013wr014449.

Ghilardi, T, Franca, MJ, and Schleiss, AJ 2014b. Period and amplitude of bedload pulses in a macro-rough channel. *Geomorphology* 221, 95–103.

Ghilardi, T, Franca, MJ, and Schleiss, AJ 2014c. Sediment transport in steep channels with large roughness elements. In *Proceedings of River Flow 2014* (eds A Schleiss, G de Cesare, MJ Franca, and M Pfister). CRC Press/Balkema; 899–907.

Grams, PE and Wilcock, PR 2007. Equilibrium entrainment of fine sediment over a coarse immobile bed. *Water Resources Research* 43(10). DOI: 10.1029/2006wr005129.

Grass, A 1971. Structural features of turbulent flow over smooth and rough boundaries. *Journal of Fluid Mechanics* 50(02), 233–255.

Griffiths, JC 1967. *Scientific Method in Analysis of Sediments.* McGraw-Hill Book Co: New York.

Hajimirzaie, SM, Wojcik, CJ, and Buchholz, JH 2012. The role of shape and relative submergence on the structure of wakes of low-aspect-ratio wall-mounted bodies. *Experiments in Fluids* 53(6), 1943–1962.

Hajimirzaie, SM, Tsakiris, AG, Buchholz, JHJ, and Papanicolaou, AN 2014. Flow characteristics around a wall-mounted spherical obstacle in a thin boundary layer. *Experiments in Fluids* 55(6). DOI: 10.1007/s00348-014-1762-0.

Hauet, A, Kruger, A, Krajewski, WF *et al.* 2008. Experimental system for real-time discharge estimation using an image-based method. *Journal of Hydrologic Engineering* 13(2), 105–110.

Hendrick, RR, Ely, LL, and Papanicolaou, A 2010. The role of hydrologic processes and geomorphology on the morphology and evolution of sediment clusters in gravel-bed rivers. *Geomorphology* 114(3), 483–496.

Hinze, J 1975. *Turbulence.* McGraw-Hill, New York, NY USA.

Jerolmack, D and Mohrig, D 2005. Interactions between bed forms: Topography, turbulence, and transport. *Journal of Geophysical Research: Earth Surface (2003–2012)* 110(F2. DOI: 10.1029/2004JF000126.

Kim, H, Kline, S, and Reynolds, W 1971. The production of turbulence near a smooth wall in a turbulent boundary layer. *Journal of Fluid Mechanics* 50(01), 133–160.

Kironoto, B and Graf, WH 1994. Turbulence characteristics in rough uniform open-channel flow. *Proceedings of the ICE – Water, Maritime and Energy* 106(4), 333–344.

Kline, S, Reynolds, W, Schraub, F, and Runstadler, P 1967. The structure of turbulent boundary layers. *Journal of Fluid Mechanics* 30(04), 741–773.

Kramer, C 2005. *The role of relative submergence on cluster microtopography and bedload predictions in mountain streams.* Thesis, University of Iowa.

Kuhnle, RA 1996. Unsteady transport of sand and gravel mixtures. In *Advances in Fluvial Dynamics and Stratigraphy* (eds PA Carling and MR Dawson). John Wiley & Sons; New York;183–201.

L'Amoreaux, P and Gibson, S 2013. Quantifying the scale of gravel-bed clusters with spatial statistics. *Geomorphology* 197, 56–63.

Lacey, RJ and Rennie, CD 2012. Laboratory investigation of turbulent flow structure around a bed-mounted cube at multiple flow stages. *Journal of Hydraulic Engineering* 138(1), 71–84.

Lacey, RWJ and Roy, AG 2007. A comparative study of the turbulent flow field with and without a pebble cluster in a gravel bed river. *Water Resources Research* 43(5). DOI: 10.1029/2006wr005027.

Lanzoni, S and Tubino, M 1999. Grain sorting and bar instability. *Journal of Fluid Mechanics* 393, 149–174.

Laronne, J and Carson, M 1976. Interrelationships between bed morphology and bed-material transport for a small, gravel-bed channel. *Sedimentology* 23(1), 67–85.

Laronne, JB, Garcia, C, and Reid, I 2001. Mobility of patch sediment in gravel bed streams: patch character and its implications for bedload. In *Gravel-Bed Rivers V* (ed. P Mosley). New Zealand Hydrological Society: Wellington; 249–289.

Lawless, M and Robert, A 2001. Three-dimensional flow structure around small-scale bedforms in a simulated gravel-bed environment. *Earth Surface Processes and Landforms* 26(5), 507–522.

MacVicar, BJ and Roy, AG 2007. Hydrodynamics of a forced riffle pool in a gravel bed river: 1. Mean velocity and turbulence intensity. *Water Resources Research* 43(12). DOI: 10.1029/2006wr005272.

Meyer-Peter, E and Müller, R 1948. Formulas for bed-load transport. In *2nd Meeting of the International Association for Hydraulic Structure Research*, IAHR, Stockholm; 39–64.

Mignot, E, Hurther, D, and Barthelemy, E 2009. On the structure of shear stress and turbulent kinetic energy flux across the roughness layer of a gravel-bed channel flow. *Journal of Fluid Mechanics* 638, 423. DOI: 10.1017/s0022112009990772.

Montgomery, DR and Buffington, JM 1997. Channel-reach morphology in mountain drainage basins. *Geological Society of America Bulletin* 109(5), 596–611.

Morris, HM 1955. Flow in rough conduits. *Transactions of the American Society of Civil Engineers* 120(1), 373–398.

Müller, A and Gyr, A 1996. Geometrical analysis of the feedback between flow, bedforms and sediment transport. In *Coherent Flow Structures in Open Channels* (eds P Ashworth, SJ Bennett, JL Best, and S McLelland). John Wiley & Sons: New York; 237–247.

Muste, M, Xiong, Z, Schöne, J, and Li, Z 2004. Validation and extension of image velocimetry capabilities for flow diagnostics in hydraulic modeling. *Journal of Hydraulic Engineering* 130(3), 175–185.

Nelson, JM, Shreve, RL, McLean, SR, and Drake, TG 1995. Role of near-bed turbulence structure in bed load transport and bed form mechanics. *Water Resources Research* 31(8), 2071–2086.

Nezu, I and Nakagawa, H 1993. *Turbulence in Open Channels*. IAHR/AIRH Monograph, Balkema: Rotterdam.

Nikora, V, Goring, D, McEwan, I, and Griffiths, G 2001. Spatially averaged open-channel flow over rough bed. *Journal of Hydraulic Engineering* 127(2), 123–133.

Nikora, V, McEwan, I, McLean, S *et al.* 2007. Double-averaging concept for rough-bed open-channel and overland flows: theoretical background. *Journal of Hydraulic Engineering* 133(8), 873–883.

Nitsche, M, Rickenmann, D, Turowski, JM *et al.* 2011. Evaluation of bedload transport predictions using flow resistance equations to account for macro-roughness in steep mountain streams. *Water Resources Research* 47(8). DOI: 10.1029/2011wr010645.

Nitsche, M, Rickenmann, D, Kirchner, JW *et al.* 2012. Macroroughness and variations in reach-averaged flow resistance in steep mountain streams. *Water Resources Research* 48(12). DOI: 10.1029/2012wr012091.

Okamoto, S 1982. Turbulent shear flow behind hemisphere-cylinder placed on ground plane. In *Turbulent Shear Flows 3* (eds LJS Bradbury, F Durst, BE Launder *et al.*). Springer Verlag: Berlin; 171–185.

Papanicolaou, A and Kramer, C 2005. The role of relative submergence on cluster microtopography and bedload predictions in mountain streams. In *Proceedings of the International Symposium "River, Coastal and Estuarine Morphodynamics"* (eds G Parker and MH Garcia). Taylor and Francis: Urbana, IL.

Papanicolaou, A and Schuyler, A 2003. Cluster evolution and flow-frictional characteristics under different sediment availabilities and specific gravity. *Journal of Engineering Mechanics* 129(10), 1206–1219.

Papanicolaou, AN, Dermisis, DC, and Elhakeem, M 2011. Investigating the role of clasts on the movement of sand in gravel bed rivers. *Journal of Hydraulic Engineering – ASCE* 137(9), 871–883. DOI: 10.1061/(Asce)Hy.1943-7900.0000381.

Papanicolaou, AT, Tsakiris, AG, and Strom, KB 2012. The use of fractals to quantify the morphology of cluster microforms. *Geomorphology* 139, 91–108.

Papanicolaou, A, Diplas, P, Evaggelopoulos, N, and Fotopoulos, S 2002. Stochastic incipient motion criterion for spheres under various bed packing conditions. *Journal of Hydraulic Engineering* 128(4), 369–380.

Papanicolaou, AN, Strom, K, Schuyler, A, and Talebbeydokhti, N 2003. The role of sediment specific gravity and availability on cluster evolution. *Earth Surface Processes and Landforms* 28(1), 69–86. DOI: 10.1002/esp.427.

Papanicolaou, A, Kramer, C, Tsakiris, A, Stoesser, T, Bomminayuni, S, and Chen, Z 2012. Effects of a fully submerged boulder within a boulder array on the mean and turbulent flow fields: implications to bedload transport. *Acta Geophysica* 60(6), 1502–1546.

Parker, G 2008. Transport of gravel and sediment mixtures. In *Sedimentation Engineering: Theory, Measurements, Modeling and* (ed. MH Garcia). Manuals and Reports on Engineering Practice No. 110, ASCE: Reston, VA; 165–251.

Parker, G, Klingeman, PC, and McLean, DG 1982. Bedload and size distribution in paved gravel-bed streams. *Journal of the Hydraulics Division* 108(4), 544–571.

Pattenden, R, Turnock, S, and Zhang, X 2005. Measurements of the flow over a low-aspect-ratio cylinder mounted on a ground plane. *Experiments in Fluids* 39(1), 10–21.

Peakall, J, Ashworth, P, and Best, J 1996. Physical modelling in fluvial geomorphology: principles, applications and unresolved issues. In *The Scientific Nature of Geomorphology: Proceedings of the 27th Binghampton Symposium in Geomorphology* (eds BL Rhoads and C Thorn). John Wiley & Sons; New York; 221–253.

Piedra, MM, Haynes, H, and Hoey, TB 2012. The spatial distribution of coarse surface grains and the stability of gravel river beds. *Sedimentology* 59(3), 1014–1029.

Pitlick, J and Van Steeter, MM 1998. Geomorphology and endangered fish habitats of the upper Colorado River 2. Linking sediment transport to habitat maintenance. *Water Resources Research* 34(2), 303–316.

Pokrajac, D, McEwan, I, and Nikora, V 2008. Spatially averaged turbulent stress and its partitioning. *Experiments in Fluids* 45(1), 73–83. DOI: 10.1007/s00348-008-0463-y.

Recking, A 2009. Theoretical development on the effects of changing flow hydraulics on incipient bed load motion. *Water Resources Research* 45(4). DOI: 10.1029/2008wr006826.

Reid, I, Frostick, L, and Brayshaw, A 1992. Microform roughness elements and the selective entrainment and entrapment of particles in gravel-bed rivers. In *Dynamics of Gravel-Bed Rivers* (eds P Billi, RD Hey, CR Thorne and P Tacconi). John Wiley & Sons: Chichester; 253–275.

Rickenmann, D 2001. Comparison of bed load transport in torrents and gravel bed streams. *Water Resources Research* 37(12), 3295–3305.

Rowiński, P, Aberle, J, and Mazurczyk, A 2005. Shear velocity estimation in hydraulic research. *Acta Geophysica Polonica* 53(4), 567–583.

Sadeque, M, Rajaratnam, N, and Loewen, M 2009. Effects of bed roughness on flow around bed-mounted cylinders in open channels. *Journal of Engineering Mechanics* 135(2), 100–110.

Saldi-Caromile, K, Bates, K, Skidmore, P *et al.* 2004. *Stream Habitat Restoration Guidelines: Final Draft.* Washington State Department of Ecology and the US Fish and Wildlife Service: Olympia, Washington.

Shamloo, H, Rajaratnam, N, and Katopodis, C 2001. Hydraulics of simple habitat structures. *Journal of Hydraulic Research* 39(4), 351–366.

Singh, A, Fienberg, K, Jerolmack, DJ *et al.* 2009. Experimental evidence for statistical scaling and intermittency in sediment transport rates. *Journal of Geophysical Research* 114(F1). DOI: 10.1029/2007jf000963.

Singh, A, Foufoula-Georgiou, E, Porté-Agel, F, and Wilcock, PR 2012. Coupled dynamics of the co-evolution of gravel bed topography, flow turbulence and sediment transport in an experimental channel. *Journal of Geophysical Research* 117(F4). DOI: 10.1029/2011jf002323.

Smart, GM 1984. Sediment transport formula for steep channels. *Journal of Hydraulic Engineering* 110, 267–276.

Stoesser, T and Nikora, V 2008. Flow structure over square bars at intermediate submergence: Large Eddy Simulation study of bar spacing effect. *Acta Geophysica* 56(3). DOI: 10.2478/s11600-008-0030-1.

Strom, KB and Papanicolaou, AN 2007. ADV measurements around a cluster microform in a shallow mountain stream. *Journal of Hydraulic Engineering* 133(12), 1379–1389.

Strom, K and Papanicolaou, A 2008. Morphological characterization of cluster microforms. *Sedimentology* 55(1), 137–153.

Strom, K and Papanicolaou, A 2009. Occurrence of cluster microforms in mountain rivers. *Earth Surface Processes and Landforms* 34(1), 88–98.

Strom, K, Papanicolaou, A, Evangelopoulos, N, and Odeh, M 2004. Microforms in gravel bed rivers: formation, disintegration, and effects on bedload transport. *Journal of Hydraulic Engineering* 130(6), 554–567.

Suszka, L 1991. Modification of transport rate formula for steep channels. In *Lecture Notes in Earth Sciences* (eds A Armanini and G di Silvio). Springer-Verlag: Berlin; 59–70.

Tan, L and Curran, JC 2012. Comparison of turbulent flows over clusters of varying density. *Journal of Hydraulic Engineering* 138(12), 1031–1044. DOI: 10.1061/(asce)hy.1943-7900.0000635.

Thompson, C and Croke, J 2008. Channel flow competence and sediment transport in upland streams in southeast Australia. *Earth Surface Processes and Landforms* 33(3), 329–352.

Tsakiris, A and Papanicolaou, AN 2008. A fractal approach for characterizing microroughness in gravel streams. *Archives of Hydro-Engineering and Environmental Mechanics* 55(1–2), 29–43.

Tsakiris, AG, Papanicolaou, AN, and Lauth, TJ 2014. Signature of bedload particle transport mode in the acoustic signal of a geophone. *Journal of Hydraulic Research* 52(2), 185–204. DOI: 10.1080/00221686.2013.876454.

Tsakiris, AG, Papanicolaou, ANT, Hajimirzaie, SM, and Buchholz, JHJ 2014. Influence of collective boulder array on the surrounding time-averaged and turbulent flow fields. *Journal of Mountain Science* 11(6), 1420–1428. DOI: 10.1007/s11629-014-3055-8.

Vollmer, S and Kleinhans, MG 2007. Predicting incipient motion, including the effect of turbulent pressure fluctuations in the bed. *Water Resources Research* 43(5). DOI: 10.1029/2006wr004919.

Wallace, JM, Eckelmann, H, and Brodkey, RS 1972. The wall region in turbulent shear flow. *Journal of Fluid Mechanics* 54(01), 39–48.

Wei, T, Schmidt, R, and McMurtry, P 2005. Comment on the Clauser chart method for determining the friction velocity. *Experiments in Fluids* 38(5), 695–699.

Wiberg, PL and Smith, JD 1987. Calculations of the critical shear stress for motion of uniform and heterogeneous sediments. *Water Resources Research* 23(8), 1471–1480.

Wilcock, PR and Kenworthy, ST 2002. A two-fraction model for the transport of sand/gravel mixtures. *Water Resources Research* 38(10), 12-1–12-12. DOI: 10.1029/2001wr000684.
Wittenberg, L 2002. Structural patterns in coarse gravel river beds: typology, survey and assessment of the roles of grain size and river regime. *Geografiska Annaler: Series A, Physical Geography* 84A(1), 25–37.
Wittenberg, L and Newson, MD 2005. Particle clusters in gravel-bed rivers: an experimental morphological approach to bed material transport and stability concepts. *Earth Surface Processes and Landforms* 30(11), 1351–1368.
Wittenberg, L, Laronne, J, and Newson, M 2007. Bed clusters in humid perennial and Mediterranean ephemeral gravel-bed streams: the effect of clast size and bed material sorting. *Journal of Hydrology* 334(3), 312–318.
Wohl, E and Merritt, DM 2008. Reach-scale channel geometry of mountain streams. *Geomorphology* 93(3), 168–185.
Wohl, EE 2000. *Mountain Rivers*. American Geophysical Union: Washington, DC.
Yager, EM, Kirchner, JW, and Dietrich, WE 2007. Calculating bed load transport in steep boulder bed channels. *Water Resources Research* 43(7). DOI: 10.1029/2006wr005432.
Yager, EM, Dietrich, WE, Kirchner, JW, and McArdell, BW 2012. Prediction of sediment transport in step-pool channels. *Water Resources Research* 48(1). DOI: 10.1029/2011wr010829.

Discussion

Discussion by J.C. Bathurst

The flume study of Flammer *et al.* (1970) is often overlooked but shows how boulder drag may vary as a function of relative roughness and Froude number. This insight should be incorporated in the development and testing of new models of the interaction between boulders and flow hydraulics.

Discussion by P. Frey

This discussion is about free-surface effects, which are likely to become more important on steep-slope mountain streams. Effects include very low relative submergence consequences on turbulent flow patterns, flow resistance and bedload transport, emerging boulders issues, and also the free surface acting as a barrier for bedload grain trajectories. Could the authors comment on these free-surface effect issues and on the associated experimental, modeling, numerical, and theoretical challenges.

Reply by the Authors

The authors agree with the discussion by James Bathurst that the overlooked work of Flammer *et al.* (1970) provides comprehensive insight into the parameters influencing the drag force on an obstacle, placing emphasis on the free-surface effects, quantified by the relative submergence and Froude number. The authors have included the reference in their chapter. Through careful experimentation, Flammer *et al.* (1970) delineate three flow regimes, in terms of relative submergence and Froude number values, where the free surface has pronounced, moderate or negligible effects on boulder drag. The establishment of these regimes allows identifying the flow conditions where free-surface effects on boulder drag become important, as well as the resulting boulder-drag magnitude variation, which can improve flow resistance and sediment-transport flux estimates. Furthermore, the Flammer *et al.* (1970) flow regime classification promotes the educated design of new studies by selecting the ranges of the experimental and/or numerical parameters suitable for reproducing (or eliminating) the free-surface effects on boulder drag.

Our findings support the observation of Philippe Frey that at low relative submergence conditions, the free surface acts as a "barrier" that confines the flow in a shallow layer. This barrier constricts the transported grain trajectories and influences their transport mode. The free-surface effects promote secondary currents downstream of the boulder leading to highly three-dimensional, highly turbulent nonequilibrium flow. The pressure distribution deviates from the hydrostatic, limiting the applicability of many one-, two-, and quasi-three-dimensional models. The numerical solution of the few suitable for such flows models is a further challenge, requiring, e.g., shock-capturing schemes, a moving boundary free-surface treatment and grid refinement, for preventing numerical instabilities. Depending on the Froude number value the shocks generated by the boulder–flow interaction may be pseudo-stationary or "rolling," which necessitates the need for new closure schemes for the stress terms in the Navier–Stokes equations. The experimental investigation of free-surface effects still remains a formidable task, due to the inherent complexity of accurately reproducing them at the laboratory scale, along with the limited accessibility of nonintrusive measurement instruments, and the oftentimes poor-quality measurements from the accompanying aeration effects. The laboratory challenge of mapping the flow with particle image velocimetry remains and the authors have proposed a new set-up arrangement for mapping the eddies shedding off the partially submerged structures. This study provides a first step to assess these impacts.

3

Granular Flows Applied to Gravel-Bed Rivers: Particle-Scale Studies of the Mobilization of a Gravel Bed by the Addition of Fines

Kimberly M. Hill and Danielle Tan

3.1 Introduction

In recent decades, researchers have demonstrated that the addition of fine particles to the gravel bed of a river can increase the mobility of the bed significantly, beyond what one might expect from simple observations or what established transport formulas can predict (Jackson and Beschta 1984; Ikeda 1984; Iseya and Ikeda 1984; Wilcock, Kenworthy, and Crowe 2001; Wilcock and Crowe 2003; Venditti *et al.* 2010a; Hill *et al.* 2011). In some cases, the most obvious manifestation of this effect is in the increase of overall bedload transport rates of a gravel riverbed upon introduction of fine particles (e.g. Beschta and Jackson, 1979; Venditti *et al.* 2010b). In other cases, the same dynamics can manifest in a reduction in stress needed to transport the same quantity of material, once fine particles are introduced. In this context, the mobilization effect can significantly alter the morphology of the channel bed, as demonstrated, for example, by Jackson and Beschta (1984) among a sequence of pools and riffles.

Consequently, this mobilization effect can degrade and undermine restoration projects where bank stability is affected by fine particles running off the land in flood events. This is of even greater concern where the presence of fine particles in streams can be enhanced by anthropogenic actions, from increased logging to agricultural development (e.g., Wilcock, Kenworthy, and Crowe 2001). On the other hand, the mobilization effect has the potential to aid in other restoration efforts. For example, in imbricated, static river beds downstream from dams, mobilization of static gravel beds could serve to rejuvenate salmonid spawning habitat (e.g., Venditti *et al.* 2010a). There have been a number of recent efforts to understand this phenomenology, as we discuss in the next two sections.

3.1.1 Governing Parameters Revealed Through Experimental Flume Studies

While the addition of sand has been demonstrated to greatly increase the mobility and flux of gravel in a bed for several decades (e.g., Beschta and Jackson 1979), recent systematic studies in experimental flumes have helped bring to light the role of certain parameters in determining the details of that mobilization. These parameters include the quantity – i.e., the concentration – of the finer particles in the bed (Wilcock, Kenworthy, and Crowe 2001; Wilcock and Crowe 2003); the relative size of the particles in the coarse bed compared with that of the mobilizing fine particles (Venditti *et al.* 2010a,b; Hill *et al.* 2011), and the absolute particle size (Hill *et al.* 2011) on the quantitative details of the bed mobilization.

Gravel-Bed Rivers: Processes and Disasters, First Edition. Edited by Daizo Tsutsumi and Jonathan B. Laronne.

Wilcock, Kenworthy, and Crowe (2001) performed a series of flume experiments with bimodal sediments where the size distribution of the coarse fraction (ranging in size from 2.0 to 64 mm) was kept constant from one run to the next, but the concentration of the sand particles was varied from 6.2 to 34%. In doing so, they demonstrated that the mobility of the mixtures monotonically increased with concentration of sand. Further, they showed that the relative mobility of the different size fractions changed with the addition of sand. Specifically, they found that the difference in mobility between coarse and fine fractions increased with increasing concentrations of sand. Wilcock and Crowe (2003) developed a surface-based transport model based on these data and, in doing so, illustrated the importance of the fraction of sand in the bed surface in determining the relative and absolute transport rates of the bedload material.

Venditti *et al.* (2010a, b) demonstrated that the mobilization of gravel particles by sand could be extended to larger (fine) particles. Venditti *et al.* (2010a) performed experiments in which small gravel particles, median grain size D_{50} = 3 mm, were added in pulses to an armored, nearly imbricated, gravel bed, $D_{50,surf}$ = 16.3 mm. They found these pulses increased surface transport rates, increased the coarse particles in bedload transport, and decreased the typical size of the particles on the bed. In Venditti *et al.* (2010b), they demonstrated that increasing the grain size of the fine particle pulses and decreasing the size tended to reduce the mobilization effects. Together with experiments of Wilcock and co-workers (e.g., Wilcock and McArdell 1993, 1997; Wilcock, Kenworthy, and Crowe 2001; Wilcock and Crowe 2003), these results suggest that rather than the absolute size of the fine particles, the relative grain size of coarse and fine sediment fractions governs the mobilization of coarse particles by the introduction of fine particles.

Hill *et al.* (2011) performed experiments in a laboratory flume to investigate the importance of relative and absolute grain size using three different gravel beds each with several experimental sets of different fine particles (Table 3.1). In their experiments, they allowed the slope to change with the addition of fine particles and in this way investigated the mobilization effect through the stress required to transport bed material of a particular grain size distribution. They performed experiments with three different gravel beds ($D_{50,coarse}$ = 3.8 mm, $D_{50,coarse}$ = 7.0 mm, and $D_{50,coarse}$ = 12.0 mm) where they fed the coarse gravel at a steady rate until the slope no longer changed and other steady-state indicators were established. Then, they maintained the same feed rate of the coarse particles and doubled the total feed rate using a second sediment feeder adjacent to the first to introduce fine particles. In addition to the coarse particle sizes, the particles sizes used for $D_{50,fine}$ included 0.08, 0.11, 0.2, 0.55, 1.21, and 2.5 mm (so that the grain size ratio $D_* = D_{50,coarse}/D_{50,fine}$ varied from $D_* \approx 1$ to 150).

As they increased D_* for each particular gravel bed, they found each system transitioned between three regimes.

Table 3.1 Ranges of steady state experimental conditions listed for each set of experiments performed by Hill *et al.* (2011), as described in more detail in Hill *et al.* (2016).

Coarse D_{50} (mm)	Fines D_{50} (mm)	Final bed slope, S_b (m/m)	Flow depth H (m)	Bed shear stress τ (N/m^2)	Friction velocity u_τ (m/s)	Flow velocity U (m/s)	Froude number $F = U/\sqrt{gH}$
3.8	0.08–3.8	0.0069–0.0077	0.12–0.15	7.53–9.86	0.0868–0.0993	0.762–0.923	0.633–0.845
7.0	0.08–7.0	0.0065–0.011	0.11–0.13	7.778–12.17	0.091–0.111	0.86–1.02	0.763–0.968
12	0.08–12	0.012–0.017	0.10–0.17	13.62–17.65	0.1182–0.1329	0.965–1.076	0.903–1.063

1) For D_* close to one, (for the systems they investigated, they found this for $D_* = 1$ only) the bed slope increased to supply the additional shear stress needed to transport the additional load of similar-sized particles.
2) For intermediate values of D_* (up to $D_* = 10$), the increased fines content increased the apparent mobility of the bed, and the slope decreased. The bed surface consisted of patchy regions of fine and coarse grains.
3) For the largest particle size ratios (the smallest fines), the additional fines caused essentially no change in bed slope; virtually no fine particles were visible at the bed surface, though the subsurface became clogged with fine sediment.

They demonstrated that the results most relevant to this chapter, namely, the mobilization effects of regime **2**, could be well-characterized by considering D_* alone. The transition to regime **3**, which they characterized as a transition to washload, was better characterized by the Rouse number of the fine particles and other considerations, as detailed in Hill *et al.* (2011).

3.1.2 Potential Physical Mechanisms Governing the Mobilizing Behaviors

There have been a number of suggestions proposed for the physical mechanisms responsible for the mobilization effect. These can be broken down roughly into three categories: (i) a change in the near-bed flow conditions that increases the effective drag on all bed particles by the fluid; (ii) a change in the bed itself that increases the likelihood that bed particles are entrained into bedload; and (iii) a change in the bed that decreases the likelihood that bedload particles are distrained into the bed.

Research into the mechanisms related to changes in near-bed flow conditions includes considerations of the effect of near-bed suspended particles, near-bed velocities, and turbulent fluctuations. In considering the effect of increased sand delivery on riffle degradation in pool–riffle sequences, Jackson and Beschta (1984) cited previous work that suggested that increased concentrations of sediment near the streambed increased the transport capacity of the system to transport riffle sediments. Ikeda (1984) proposed that the introduction of sand into pockets surrounding coarse bed particles results in hydraulic smoothing of the bed, subsequently increasing near-bed velocities so that protruding particles experienced more lift and drag (e.g., as modeled by Kirchner *et al.* 1990). In their experiments, Venditti *et al.* (2010a) measured increased near-bed velocities and decreased turbulent kinetic energies as a pulse of fine particles passed over and mobilized their gravel bed, supporting these conjectures.

Other mechanisms include geometric effects associated with the filling of pockets in the bed giving rise to increased entrainment rates of coarse particles into bedload and decreased distrainment rates. Ikeda (1984) suggested that the introduction of sand into the bed surface changes the pocket geometry, specifically via the reduction in the pivot angle a particle must rotate to be entrained into the bed. Others (Whiting *et al.* 1988; Kirchner *et al.* 1990; Buffington, Dietrich, and Kirchner 1992) suggested that as the fraction of fine particles increases in the bed surface, the larger particles protrude more, possibly due to the filling of pockets in the bed (e.g., Whiting *et al.* 1988). This would result in the greater exposure of these particles to near-bed lift and drag forces (as modeled by Kirchner *et al.* 1990). Ikeda (1984) also suggested that the infilling of pockets of the bed is likely responsible for a decreased propensity for larger particles to be disentrained from the flow. The bed becomes smoother, resulting in less collisional energy loss of the particles in bedload transport, and the particles are less likely to find a wake behind a protruding bed particle where there is a significantly reduced force from the driving fluid. This is supported by more recent measurements of travel distances of tracer particles in an experimental flume by Hill, DellAngelo, and Meerschaert (2010).

All of these suggestions are physically viable, and likely all contribute to the mobilization of bedload particles by increased presence of fine particles in the bed, though the contributions may vary depending on the parameters of the system. Obvious considerations include the stage of bedload transport and the particle sizes in the mixture, though there may be others including particle properties and nonuniformities in the system, such as bedforms and sinuosity. It is difficult to isolate the different mobilizing effects to determine which dominate(s) under particular circumstances, though it would be useful for modeling the response of a river and interpretation of historical deposits where the mobilization is relevant.

Herein, we investigate the possibility that the mobilization of bed particles by the addition of fine particles can be driven partially or entirely through particle interactions alone, independent of fluid effects. We do so through the investigation of tools obtained from what might be called the "granular physics" community, that is, the research that focuses on understanding how relatively simple particle-scale interactions drive bulk behavior of large conglomerations of macroscopic particles. In the next section we investigate how constitutive relationships derived for what is often called "liquid-like" or "dense" particle systems (as we explain in the next section) may provide insight to these mobilizing effects. We demonstrate that while the systems are somewhat different from those more typical of bedload transport, some useful insight can be gained. In section 3 we present computational simulations with some fluid effects to drive a simplified form of bedload transport, but for which particle–particle effects dominate the behavior of particle systems. We use these to investigate how the effect of size alone, independent of turbulent fluctuations, can drive a range of behaviors associated with different sized particles. We conclude by discussing the strengths and weaknesses for applying this simulation framework toward an understanding of problems like this associated with particle transport in gravel-bed rivers.

3.2 Insights from Rheological Models of Dry Dense Granular Flows

In this section we consider related investigations for transport behaviors of dry dense granular mixtures. As we will show, these results demonstrate that increasing the concentration of small particles in a mixture results in a well-defined increase in mobility that can be quantitatively expressed through recently established constitutive relationships (e.g., Rognon *et al.* 2007; Yohannes and Hill 2010; Tripathi and Khakhar 2011).

Admittedly, the transport of the particles in dense granular flows can be somewhat different than that in bedload transport. For example, in bedload transport, typically the particle movement is restricted primarily to the surface grains (at most 1–2 grain diameters from the bed surface). (Exceptions to this include (i) sheet flows where, under conditions of relatively high shear stresses, the bedload layer devolves into a collisional sliding layer of grains that can be several grains thick (Wilson 1966), and (ii) bedload sheets, highly mobile and very low-relief bedforms, a few grain diameters thick and up to several meters long (Whiting *et al.* 1988; Recking *et al.* 2009).) Typically, in dense flows, the particles moving with significant velocities can be 10 grain diameters or more from the bed surface, with a well-defined and, typically, spatially variable shear rate. Additionally, for particles in bedload transport (except in sheet flows and bedload sheets), most particle motion appears associated directly with a fluid velocity that is either partially or primarily directly responsible for their motion. In contrast, most particles in dense flow are being driven by neighboring particles and their own weight rather than by external fluid forcing. Indeed, while Frey and Church (2011) demonstrated many examples where insight from granular flows could be applied to sediment transport in rivers, they pointed out limitations to the application of results from granular physics to particles in intermittent bedload transport. Nevertheless, we consider the analogous problem – the effect of the presence of fine particles on particle transport – in dense granular flows for insight into the problem of mobilization of gravel beds through the introduction of fine particles.

To clarify, by "dense" granular flows, we consider those flows of sufficient density and intermediate shear rates, where particles interact by both collisional and frictional contacts. These are often called "liquid" granular flows to distinguish them from solid-like deformations in granular materials and gas-like granular flows (e.g., Jaeger, Nagel, and Behringer 1996). In solid-like deformations, the constitutive behavior is essentially rate independent and well-characterized using models based on plasticity, while in gas-like flows, the behaviors are dominated by relatively isolated collisions between particles and well-characterized using models based on kinetic theory of dense gases. Until recently there has been no agreement on a modeling framework for dense sheared granular materials, even those comprised of monosized spherical particles. Within the last decade or so, compilations of a wide range of experimental and computational research (e.g., GDR MiDi 2004) have provided the data for relatively reliable and surprisingly simple relationships for quantifying some flow behaviors of these systems, which we describe in the next section.

3.2.1 Rheology of Dense Granular Flows

While there is no universal physics-based model for dense granular flows, it is generally agreed that the rheology of relatively hard spherical particles should depend in some way on the following nondimensional form of the shear rate $\dot{\gamma}$:

$$I = \frac{\dot{\gamma} D}{\sqrt{p/\rho_p}} \tag{3.1}$$

referred to as the inertial number (e.g., Forterre and Pouliquen 2008). Here, D is the particle size, ρ_p is the density of the particles rather than the density of the agglomeration of particles and pore spaces, and p is the average pressure or normal stress. The value of a system's inertial number has been shown to indicate the nature of the flow. This is reviewed elsewhere (briefly in Yohannes and Hill 2010 and in more detail elsewhere, e.g., Forterre and Pouliquen 2008), but we summarize briefly here. Systems of very small values of I (roughly, $I \leq 10^{-3}$) behave as solid-like granular materials, and their rheological properties are rate independent. Systems with relatively high values of I (roughly, $I \geq 0.2$) (e.g., daCruz *et al.* 2005) generally behave as a dissipative gas, and kinetic theory of dense gas predicts the behavior well (e.g., Jenkins and Mancini 1987). Systems with intermediate values of I are in the dense flow regime, where the rheology appears explicitly dependent on I.

For example, for quasi two-dimensional steady dense flows, the rheology is efficiently described using an effective friction coefficient $\mu^* = \tau/p$, where τ is the local averaged streamwise shear stress. Depending on the boundary conditions, the relationship between μ^* and I has been found to describe well monosized particulate flows using simple relationships. These include linear (e.g., da Cruz *et al.* 2005) and power-law (e.g., Hatano 2007) relations and over the broadest range of flow conditions by more complicated relationships (e.g., GDR MiDi 2004; Jop, Forterre, and Pouliquen 2006; Forterre and Pouliquen 2008) such as

$$\mu^* = \mu_1 + (\mu_2 - \mu_1)/(I_0/I + 1) \tag{3.2}$$

Here, μ_1, μ_2, and I_0 are all, strictly speaking, fitting parameters, but μ_1 and μ_2 are considered effective friction coefficients associated with quasistatic deformation and rapid collisional flows, respectively.

This framework has been shown to be relatively successful for a wide range of monosized particle systems, within some limits, as described by Forterre and Pouliquen (2008).

3.2.2 Rheology of Dense Granular Mixtures

For the relationships described by Equations (3.1) and (3.2) to be applicable to the bimodal systems of the focus of the chapter, there are two apparent ambiguities. First, it is not immediately obvious what particle size is appropriate for the representative particle size, D, Equation (3.1). There is no single central particle size that is represented in the mixture for a bimodal sediment bed. Second, it is not obvious how to determine the coefficients μ_1, μ_2, and I_0, and, indeed, whether they should vary with components in the mixture. Recent simulations point to potential answers to these questions (Rognon *et al.* 2007; Yohannes and Hill 2010; Tripathi and Khakhar 2011).

Rognon *et al.* (2007) performed two-dimensional simulations of binary mixtures of different-sized circular particles down a rough plane inclined at an angle θ relative to the horizontal. As is typical of these gravity-driven flows, the large particles segregated toward the free surface, and the small particles, away, not unlike armoring in appearance. Nevertheless, when they assumed a linear relationship between $\bar{I}$ and μ^*, where $\bar{I}$ is defined by replacing the particle size D in Equation (3.1) with the area-averaged particle size $\bar{D}$,

$$\bar{I} = \frac{\dot{\gamma}\bar{D}}{\sqrt{p/\rho_p}}, \tag{3.3}$$

they found an expression that is independent of depth:

$$\mu^*{}_R = a_1(D_*,f_L) + a_2(D_*,f_L)\bar{I}. \tag{3.4}$$

In this equation, a_1 and a_2 are fitting parameters, both potentially functions of D_* and f_L; D_* is the ratio between the diameters of the large and small particles, and f_L is the fraction of large particles in the mixture.

More recently, Yohannes and Hill (2010) investigated this same question using three-dimensional simulations of binary mixtures of different-sized (10 mm and 20 mm) spherical particles. The geometry was somewhat different: they sheared the particles in a plane Couette cell in the absence of gravity. These boundary conditions resulted in minimal segregation over the duration of the simulations, provided they maintained a relatively high solid fraction $f \approx 0.6$. They varied the volume-averaged concentration of small particles from 0% to 100%. As done by Rognon *et al.* (2007), they calculated $\bar{D}$ based on a volume-averaged calculation of the particles in the mixture.

Even though $\bar{D}$ was not represented in their mixtures, Yohannes and Hill (2010) found all of their mixtures behaved according to Equation (3.2), substituting $\bar{I}$ in Equation (3.3) for the inertial number I and using only one set of parameters, $(\mu_1, \mu_2, I_0) = (0.35, 0.64, 0.25)$. Within error, these constants are equivalent to those found for a wide range of other computational and experimental results (e.g., Forterre and Pouliquen 2008, and references within). Through other analysis of their simulated data, Yohannes and Hill (2010) showed that this rather remarkable data collapse could be related to collisional time scales of the problem, a topic beyond that of this chapter. Regardless, the data are supported by three-dimensional simulations of segregating particle mixtures on an inclined plane by

Tripathi and Khakhar (2011), where they showed a further generalization was possible for mixtures of particles differing in both size and density:

$$\bar{I}_{TK} = \frac{\dot{\gamma}\bar{D}}{\sqrt{p/\bar{\rho}_p}}, \tag{3.5}$$

where $\bar{D}$ is defined as for Equation (3.3) and $\bar{\rho}_p$ is the volume-averaged particle density in the mixture.

3.2.3 Rheology of Dense Granular Mixtures Applied to the Mobility Problem of Bedload Mixtures

To understand the implications of these results for our bimodal bed mobility problem, we first investigate the effect of varying $\bar{D}$ in equations (3.3) and (3.5) on $\dot{\gamma}$, then we consider the results in the context of simplified bedload conditions. First, we note that since $\bar{D}$ is based on a volume-averaged calculation of two different sized particles, for two different mixtures in which the size of the coarse particles is the same, if we either increase the concentration of fine particles in a mixture or decrease the size of the fine particles relative to the coarse particles, $\bar{D}$ decreases.

Results obtained by Yohannes and Hill (2010) and Tripathi and Khakhar (2011) indicate that, for two different mixtures subjected to the same applied stress and pressure, $\bar{I}$ is the same:

$$\bar{I} = I_0 \times (\mu^* - \mu_1)/(\mu_2 - \mu^*). \tag{3.6}$$

Based on this, and Equation (3.3) (or more generally, Equation (3.4)), we can express $\dot{\gamma}$ as a simple function of $\bar{D}$:

$$\dot{\gamma} = \bar{I} \times \sqrt{p/\bar{\rho}_p}/\bar{D} = I_0/\bar{D} \times \sqrt{p/\bar{\rho}_p} \times (\mu^* - \mu_1)/(\mu_2 - \mu^*). \tag{3.7}$$

In other words, for our special case in which we hold the size of the coarse particles, and the pressure and stress the same, Equations (3.2) and (3.3) indicapte that $\dot{\gamma}$ increases as the fraction of fines in the mixture decreases and/or the size of the fine particles decreases.

To apply these predictions directly to systems of bedload transport, we consider a case where the slope is relatively small so that the shear stress on the bed is due primarily to the fluid, and slope effects are minimal. Over the relatively small thickness of the particles in bedload transport, the shear stress on the particles is approximately equal to the shear stress at the top of the bed τ_b. To further simplify the problem, we consider sediment of a uniform material density ρ_p and solid fraction f, so that the pressure associated with the weight of the bed particles can be expressed by:

$$p = \left(\rho_p - \rho_f\right) fg\zeta. \tag{3.8}$$

Therefore,

$$\mu^* = \tau_b / \left[\left(\rho_p - \rho_f\right) fg\zeta\right], \tag{3.9}$$

decreases with increasing distance from the bed surface, ζ (see Figure 3.1). Then, to find the transport rate, we first express $\dot{\gamma}$ in terms of the variation of the local average downstream particle velocity $\bar{u}_p(\zeta)$ with distance from the bed surface:

$$\dot{\gamma} = d\bar{u}_p/d\zeta = \begin{cases} I_0/\bar{D} \times \sqrt{\left(1 - \rho_f/\rho_p\right) fg\zeta} \times (\mu^* - \mu_1)/(\mu_2 - \mu^*), & \mu^* > \mu_1 \\ 0 & \mu^* \le \mu_1 \end{cases}. \tag{3.10}$$

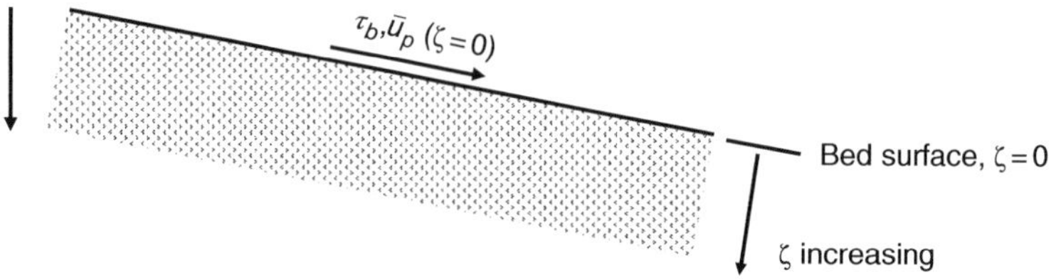

Figure 3.1 Sketch of bed to consider application of Equation (3.7) to bedload mixtures. The bedslope is exaggerated in this sketch.

To find the local average downstream velocity, we integrate Equation (3.10):

$$\bar{u}_p(\zeta) = \bar{u}_p(0) + I_0/\bar{D} \times \sqrt{\left(1 - \rho_f/\rho_p\right) fg} \times \int_{\zeta=0}^{\zeta} \left[\sqrt{\zeta} \times (\mu^* - \mu_1)/(\mu_2 - \mu^*)\right] d\zeta. \tag{3.11}$$

In cases where the stress τ_b is constant, μ^* is a simple function of ζ. Thus, the integral in Equation (3.11) is also a function only of ζ To somewhat simplify the expression, we introduce a new variable $L_1(\zeta)$ with units of length, where:

$$(L_1(\zeta))^{3/2} = \int_{\zeta=0}^{\zeta} \left[\sqrt{\zeta} \times (\mu^* - \mu_1)/(\mu_2 - \mu^*)\right] d\zeta \tag{3.12}$$

We can then express the transport rate as $q_b = \int_{\zeta=0}^{\zeta_{mdx}} f \bar{u}_p(\zeta) d\zeta$, i.e., if f is independent of depth in the transported layer

$$q_b = f \bar{u}_p(0) \times \zeta_{max} + f I_0/\bar{D} \times \sqrt{\left(1 - \rho_f/\rho_p\right) fg} \times \int_{\zeta=0}^{\zeta_{max}} (L_1(\zeta))^{3/2} d\zeta. \tag{3.13}$$

ζ_{max} is the distance from the bed at which $\dot{\gamma}$ and $\bar{u}_p$ vanish, where $\mu^* \to \mu_1$. We use this in Equation (3.9),

$$\zeta_{max} = \tau_b / \left[\left(\rho_p - \rho_f\right) fg\mu_1\right]. \tag{3.14}$$

While there are limits to applicability of the granular rheology framework in the slow or creeping regime (Forterre and Pouliquen 2008), the contribution of the transport in this region is minimal, so we extend the application of the rheology to the bottom of the bedload layer to keep the discussion relatively simple.

To determine the effect of varying $\bar{D}$ on q_b, we might first consider two cases where we keep τ_b the same, but vary $\bar{D}$ from one to the next. It is a reasonable guess that the near-surface particle velocity $\bar{u}_p(0)$ scales with the shear velocity and, therefore, with $\sqrt{\tau_b}$, in which case q_b depends only on $\bar{D}^{-1}$. If, on the other hand, we consider a case where we keep water discharge the same and vary $\bar{D}$ only from one next, we need to understand the effect of varying $\bar{D}$ on $\bar{u}_p(0)$ and τ_b. Here, we consider a couple of simplifying assumptions regarding the predictions of Equation (3.13). We might expect under certain circumstances (e.g., for small quantities of fine particles on the bed surface and a constant water supply) that τ_b is dependent only on the size of the coarse particles in the bed. As we illustrate in the next section, this is reasonable in cases where the log law of the wall holds and the effective friction length scale of the bed is that of the larger particles in the bed. This is likely an oversimplification, particularly given the observations of a patchy surface for these mixtures; here we use it to develop a qualitative picture of the phenomenology rather than to develop a detailed quantitative prediction. In this case, for two systems where the coarse particles are the same size and the fine particles vary, the second term in Equation (3.13) is still inversely dependent on $\bar{D}$. For an alternative consideration of $\bar{u}_p(0)$, we borrow from measurements of surface flows and granular materials for which near-bed velocities increase with percentage of smaller particles in the mixture (e.g., Hill and Zhang 2008). We can extend this to our considerations here and suggest that $\bar{u}_p(0)$ increases with decreasing $\bar{D}$, that is decreasing fine particle size or increasing quantity of fine particles in the mixture.

Based on these calculations and our simplifying assumptions, we would expect q_b to increase with decreasing the size of fine particles in the mixture for coarse–fine particle size ratios not too much greater than 1. We would further expect the enhanced mobility of these mixtures – what one might call a lubrication effect – to level off for substantial particle size ratios as $\bar{D} \rightarrow D_{50,\,coarse} \times f_{coarse}$, where f_{coarse} is the fraction of the particles in bedload transport that are coarse. Both effects were measured for experiments using bimodal mixtures of particles in bedload transport by Hill *et al.* (2011).

To summarize, theoretical predictions based on a rheological model for the liquid-like behavior of dense granular mixtures are consistent with observations of enhanced mobility in the bedload transport of bimodal mixtures. However, we are cautious in applying the predictions directly to bedload systems with so many simplifying assumptions required. Rather than this, we use this example to point out that under the right conditions, a conglomeration of particles can be treated as a continuum where response to applied shear stress resembles that of bimodal mixtures in bedload transport. It is likely most applicable to relatively high transport rates for bedload when the particles are moving in sheet flow. On the other hand, the manner in which the stress is partitioned between the fluid (water) and solid (particle) phase might vary, so that the assumption of a constant shear stress on the particles throughout the depth is likely an oversimplification. It is also interesting to consider how this rheology may apply to other modes of bedload transport, a topic beyond the scope of this chapter, but one for future research.

Since experimental systems do not allow adequate exploration, alone, for isolating the mechanisms behind the effect of enhanced mobility with the addition of fine particles, and theoretical development is somewhat limited in this regime, we turn to computational simulations to provide a virtual laboratory for numerical experiments that can help improve our intuition of the importance of particle–particle contacts on this problem. In the next section we describe the use of relatively simple discrete element simulations to address this problem.

3.3 Discrete Element Model Simulations of Bimodal Mixtures in Bedload Transport

To computationally study the effect of particle contacts on the effect of fine particles on increasing the mobility of a coarse bed, we use the discrete element method (DEM), first proposed by Cundall and Strack (1979). In the DEM method, every particle can be tracked at each time step of the computation. As such, details of the segregation and transport arise spontaneously, so they can be ideal for studying systems for which no constitutive relations exist.

Since the initial proposal of the use of DEM to study conglomerations of macroscopic particles, models based on the DEM have been applied to a variety of bedload problems. Forrest and Haff (1992) and Jiang and Haff (1993) were among the first to apply DEM to model bedload transport. Forrest and Haff (1992) used it to investigate the stratigraphy of wind ripples, and Jiang and Haff

(1993) used it to investigate certain transport dynamics of the bedload transport problem, including particle entrainment and deposition and prediction of transport rates. Jiang (1995) and McEwan, Jefcoate, and Willetts (1999) used DEM simulations to investigate details of the bed layer such as thickness and velocity profiles for particles in bedload transport. Schmeeckle and Nelson (2003) developed a parameterized approach to incorporating fluid effects on the bed and found their DEM simulations roughly reproduced the shear stress versus bedload transport relationship of Fernandez Luque and van Beck (1976).

While DEM simulations are capable of investigating a wide range of phenomenology, they are limited in a purely physics-based representation of fluid effects. Thus considering particle–fluid systems requires some level of simplification. There have been more recent developments of this DEM framework that have increased significantly in sophistication. For example, Ji *et al.* (2013) fully simulated the fluid using a large eddy simulation (LES)/direct numerical simulation (DNS) for the fluid, a DEM approach for the particles, and an immersed boundary method for coupling the two. They used this method to numerically investigate the entrainment and movement of coarse particles on the bed of an open channel, for example, to determine what details of the turbulent statistics correlate with particle transport details. Fukuoka *et al.* (2014) investigated the effect of particle shape on particle transport numerically using a mixture model within conservation equations for the combined fluid-particle motion based on local particle concentration in the mixture. They used essentially a Smagorinsky model as the subgrid turbulence model for this fluid. Then, the motion of individual particles modelled using a DEM framework was determined according to calculated local forces from the fluid on the particles. While the representation of the turbulence can be quite sophisticated as in the simulations by Ji *et al.* (2013), the details of the interparticle interaction mediated by fluids are limited by the resolution of the fluid grid, particularly for irregularly shaped particles as simulated by Fukuoka, Fukuda, and Uchida (2014). Additionally, even simulations of dry granular flows can be computationally expensive, so the size of simulations is limited. For our first numerical investigation into mobilization of a coarse bed by the addition of fines, we focused on whether interparticle interactions alone may have the capability of this dynamic. To accomplish this, rather than a simulation that captures dynamic fluid effects in detail, we developed a simple simulation to try to capture and isolate the effects of particle–particle interaction on these mobilization dynamics. We describe some of the simplifications we make in our simulations below.

3.3.1 Our DEM model

Our basic particle–particle contact model is based on an established mechanistic model for representing particle–particle interactions of spheres. It uses Hertz–Mindlin contact theory with a damping component specified by Tsuji, Tanaka, and Ishida (1992) and Coulomb sliding friction:

$$F_n = -k_n\delta_n^{3/2} - \eta_n\delta_n^{1/4}\dot{\delta}_n, \text{ and} \tag{3.15}$$

$$F_t = \min\left\{-k_t\delta_n^{1/2}\delta_t - \eta_t\delta_n^{1/4}\dot{\delta}_t, \mu F_n\right\} \tag{3.16}$$

where F_n and F_t are the contact forces in the normal and tangential directions to the contact plane between two contacting particles, δ_n and δ_t are the corresponding deformations, $\dot{\delta}_n$ and $\dot{\delta}_t$ are the corresponding temporal derivatives of those deformations, and k_n, η_n, k_t, and η_t are stiffness and damping coefficients. In this chapter we derive them in terms of the particle properties, as defined in Tables 3.2 and 3.3 according to the respective sizes, Young's moduli, shear moduli, masses, and restitution coefficients of the particles in contact. We used four different particle sizes for our mixtures, ranging from 7 mm to 1.75 mm to create bimodal mixtures that typically resulted in enhanced mobilization effects

Table 3.2 Expressions for stiffness coefficients (k_n and k_t) and damping coefficients (η_n and η_t) in the DEM model (Equations 3.15 and 3.16).‡

k_n	$\frac{4}{3}\sqrt{R_{\text{eff}}}E_{\text{eff}}$
k_t	$8\sqrt{R_{\text{eff}}}G_{\text{eff}}$
η_n	$\alpha\sqrt{m_{\text{eff}}k_n}$
η_t	$\alpha\sqrt{m_{\text{eff}}k_t}$

Source: Tsuji, Tanaka, and Ishida (1992). Reproduced with permission of Elsevier. $\alpha = 0.07$ is related to a coefficient of restitution ~0.9 according to details in Tsuji, Tanaka, and Ishida (1992).
‡Subscripts n and t indicate the directions normal and tangential to the contacting surfaces, respectively.

Table 3.3 Expressions for effective radius R_{eff}, elastic modulus E_{eff}, shear modulus G_{eff}, and mass m_{eff} of two contacting particles in our DEM model.‡

R_{eff}	$\left(\frac{1}{R_1}+\frac{1}{R_2}\right)^{-1}$
E_{eff}	$\left(\frac{1-\nu_1^2}{E_1}+\frac{1-\nu_2^2}{E_2}\right)^{-1}$
G_{eff}	$\left(\frac{2(1+\nu_1)(2-\nu_1)}{E_1}+\frac{2(1+\nu_2)(2-\nu_2)}{E_2}\right)^{-1}$
m_{eff}	$\left(\frac{1}{m_1}+\frac{1}{m_2}\right)^{-1}$

‡Defined using the two particles' radii (R_1 and R_2), Poisson's ratios (ν_1 and ν_2), elastic modulii (E_1 and E_2), and masses (m_1 and m_2).

in experimental runs (Hill *et al.* 2011). In each of the grain sizes we used a 10% polydispersity to minimize crystallization in the bed as it is sheared. The particle properties we used are similar to a relatively hard rock, like granite.

Prior to introducing the particles into the simulations, we prepare a channel that is 105 mm long by 21 mm wide and sufficiently high so that the particles never encounter the top boundary. We roughen the bottom of the channel by gluing 7-mm spheres in a random array that covers roughly 45% of the boundary. The horizontal boundaries are periodic in both downstream and transverse directions.

We initialize the bed by first releasing 300 particles 7 mm in diameter with random (small) velocities and allow them to settle. As they drop, the forces on each particle are due to any other particles in which they are in contact according to Equations (3.15) and (3.16) and gravity. Once they have stopped moving, they form a bed approximately 46 mm above the roughened channel bottom. We then initiate our model fluid flow, described presently, and let the bed reach steady-state conditions.

We incorporate fluid effects in a simple manner to investigate the effect of particle size alone on the transport properties. Specifically, we impose a steady velocity profile uniform in the downstream direction that varies, normal to the bed, as the log of the height above the bed, similar to the "log law of the wall" as in the inertial region of a hydraulically rough turbulent boundary layer:

$$\frac{u_{fx}(y)}{u_\tau} = \frac{1}{\kappa}\ln\left(\frac{y}{y_0}\right) \tag{3.17}$$

In this equation, we use the standard notation that u_τ is the shear velocity, which scales with a shear stress at the bed τ_b according to $u_\tau = \sqrt{\tau_b/\rho}$; y_0 is the position at which our velocity is zero (typically, a short distance below the surface of the bed that scales with the coarse particles in the bed); $\kappa = 0.41$ is the von Karman constant. For a hydraulically rough flow, typical of gravel-bed rivers, y_0 scales with a roughness length scale of the system, k_s, as $y_0 = k_s/30$ (e.g., Wiberg and Smith 1987; Pope 2000).

We chose parameters for Equation (3.17) based on experiments performed in a laboratory flume by Hill *et al.* (2011) and maintain the same fluid parameters for all simulations we report here. Namely, $u_\tau = 0.09\,\text{m/s}$, so that if $\rho \approx 1000\,\text{kg/m}^3$, $\tau_b \approx 8.1\,\text{N/m}^2$. For D_{50} of the coarse fraction (7 mm), the Shields number $\tau_* = \tau_b/\left[\left(\rho - \rho_f\right)gD_{50}\right] \approx 0.071$. We scaled k_s with the larger of the grains in all of our mixtures somewhat arbitrarily as $k_s = 3.5 \times 7\,\text{mm} = 24.5\,\text{mm}$. The fluid velocity is nonzero at the top of the bed, that is the top of the bed is not a no-slip boundary for the fluid. We set $y = y_0$ (where the fluid velocity $u_{fx}(y = y_0) = 0$) at a distance $2 \times D_{50}$ of the coarse fraction (7 mm) from the top of the bed, and allow this "boundary" to move up and down as particles are deposited and entrained, as illustrated in Figure 3.2. We include the details of this calculation in Section 3.2 with the rest of the numerical procedures.

We set up the simulations to allow the fluid to interact with the particles via a one-way coupling. That is, the fluid provides a force on the particles, but the presence of the particles on the bed or in the water above the bed does not affect the baseline fluid velocity in our simulations, except in the manner in which upstream particles can shield downstream particles from the effect of the flow which we explain shortly. The forces of the fluid on each particle include three forces: (i) a buoyancy force, (ii) a drag force, and (ii) an "added mass" force to account for the fact that an accelerating (or decelerating) particle must accelerate (or deflect) fluid around it. The choice to include both average

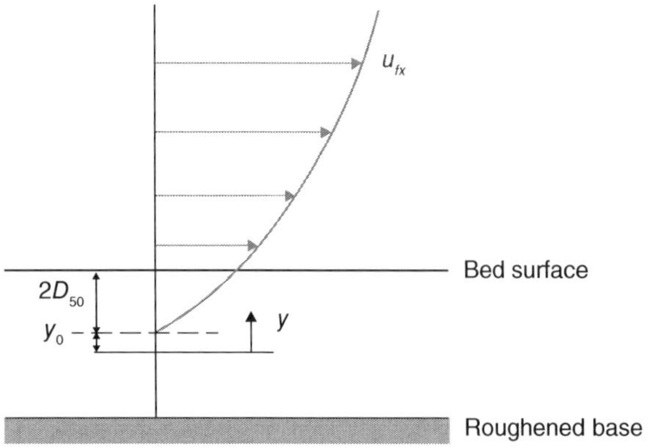

Figure 3.2 Sketch of flow profile of the simulated bedload transport.

downstream forces (drag and added mass) and neglect vertical forces associated with lift was primarily to focus efforts on understanding the potential involvement of particle–particle interactions in governing the mobilization effects of fine particles. We note that others such as Schmeeckle and Nelson (2003) and Calantoni, Todd Holland and Drake (2004) have neglected the lift force in part to simplify the simulation and in part because the form an average lift force should take is not clear.

We calculate the buoyancy force as $\vec{F}_B = -\rho_f \forall_p \vec{g}$, where $\forall_p$ is the volume of the particular particle, and $\vec{g}$ is gravitational acceleration. We calculate the added mass force as $\vec{F}_A = -\rho_f \forall_p c_m (d\vec{u}_p/dt)$. Here, $\vec{u}_p$ is the velocity of the particle, and c_M is the coefficient of added mass; we use $c_M = 0.5$ as in Batchelor (1967). We calculate the drag force of the fluid on each particle as $\vec{F}_D = -\rho_f c_D A |\vec{u}_{rel}| \vec{u}_{rel}$. In this equation, c_D is the coefficient of drag, and we follow Drake and Calantoni (2001) in using $c_D = 24/Re_s + 4/\sqrt{Re_s} + 0.4$, where $Re_s = |\vec{u}_{rel}| D/\nu$ is a relevant particle Reynolds number. D is the particle diameter, and ν is the fluid viscosity. $\vec{u}_{rel}$ is the local effective velocity of the fluid relative to the particle, calculated in such a way to account for "blocking" associated with upstream particles. That is, we do not modify our basic expression for the fluid velocity profile. Instead, we use an empirical relationship to account for the effect upstream grains have on shielding downstream grains from fluid drag effects, according to that derived from experiments by Schmeeckle and Nelson (2003) and described therein. The expression for this relative velocity can be broken into downstream, vertical, and transverse components, $\vec{u}_{rel} = u_{rel,x}\hat{x} + u_{rel,y}\hat{y} + u_{rel,z}\hat{z}$. Since the fluid velocity is steady and only in the downstream direction,

$$u_{rel,y} = -u_{p,y};\, u_{rel,z} = -u_{p,z}, \tag{3.18}$$

and the downstream component can be written:

$$u_{rel,y}(y) = \left[u_{fx}(y) - u_{px}\right] \times erf\left(\frac{l_d - h_p}{5h_p}\right) \tag{3.19}$$

That is, the drag on a bed particle downstream of a protruding particle is reduced according to the downstream distance between the two particles (l_d) and the height of the top of the protruding particle above the bed (h_p). We use the results by Schmeeckle and Nelson (2003) demonstrating that the distance at which the blocking effects diminish scales with the height of the top of the protruding particle above the bed as $5h_p$. It is interesting to consider whether the results are sensitive to this parameter, though it is an issue beyond the scope of this chapter.

The full equation we use to calculate the force on each particle includes the weight of that particle as well as the forces of all other particles in contact with that particle:

$$\vec{F}_p = \left(\rho_s - \rho_f\right)\forall_s \vec{g} - \rho_f \forall_s c_m \frac{d\vec{u}_p}{dt} + \frac{1}{2}\rho_f c_D A |\vec{u}_{rel}| \vec{u}_{rel} + \sum_j \left(\vec{F}_{jn} + \vec{F}_{jt}\right) \tag{3.20}$$

where the last term on the right side sums over the forces due to all j particles in contact with the particle.

3.3.2 Simulation Procedures and Results

As mentioned, to initialize the bed for each simulation, we release 300 7-mm particles with random velocities into our channel and wait until the particles are moving with negligible velocities. They form an initial bed that is typically 46 mm tall.

We then "turn on" the fluid flow. As mentioned above, the fluid flow is effectively a steady velocity profile in the horizontal direction that follows a log-law. We set the height at which the velocity becomes zero, y_0, at a predefined distance below the top of the bed, y_s. Specifically, $y_s = y_0 + 2 \times D_{50,coarse}$, i.e., $y_s = y_0 + 14\,\text{mm}$. We determine the location of the top of the bed dynamically, that is, at the beginning of each time step, by considering the average height of the top of the particles in the bed. We define bed particles as all particles with less than the near-bed fluid velocity, that is, any particles whose speed, v_p,is less than the fluid speed at the top of the bed, in Equation 3.17, $u_f(y_s) = (u_\tau/\kappa) \times \ln(y_s/y_0)$, i.e., $u_f(y_s) \approx 0.64\,\text{m/s}$. This seems a bit high, as it may be expected that this should be closer to the shear velocity, 0.09 m/s. There are several factors that might contribute to this. It may result from scaling of the roughness coefficient ($k_s \sim 3.5D_{50}$) or the depth in the bed ($y_s - y_0$) at which the velocity is zero. Increasing the former or decreasing the latter would decrease $u_f(y_s) \approx 0.64\,\text{m/s}$. It would be interesting to understand the sensitivity of the results to $y_s - y_0$, but we suspect this governs the absolute transport of the particles, and not the relative effects of varying the content or size of fine particles. This is under investigation. At each time step, we calculate the average vertical position of these highest bed particles and set that as the top of the bed, y_s. Then, for calculating the drag on the particles during that time step we set $y_s - y_0 = 2 \times 7\,\text{mm}$.

When the "fluid" flow is initiated in the simulations, all particles are at rest, so y_s is defined based on the top of all of the particles. As particles are entrained, most of the 7-mm particles remain bed particles, but the bed surface (and corresponding details of the velocity profile) moves up as the bed expands. We continue the simulations with just the 7-mm particles until the system reaches a steady state. The transport per unit width is approximately $q_b \sim 2.4 \times 10^{-4}\,\text{m}^2/\text{s}$, a bit high compared to natural gravel under the same conditions or that calculated based on the Meyer–Peter Müller formula for bedload transport (e.g., Wong and Parker 2006): $q_b \sim 3 \times 10^{-5}\,\text{m}^2/\text{s}$. There are a number of potential reasons for this discrepancy. These largely include considerations of the simplification of the model framework of the fluid, the parameter choices (such as roughness lengthscale) within this framework, and the asphericity of the particles, the last of which has been shown to have a significant effect in the mobility of particles (e.g., Thomas and Bray 1999).

We then introduce the finer particles by releasing them from the top of the domain. We release a volume of small particles equal to 1/8 the volume of large particles in the established bed. The particles are accelerated by the fluid as they descend until they impact the bed and become incorporated into it. As is the case for the coarser particles, we consider the finer particles part of the bedload once their downstream velocity decreases to less than that of the near-bed fluid velocity. As the finer particles are incorporated into the bed, we monitor the relative transport rate of the two components and the absolute transport rate of the mixture as they vary with time. Since the downstream boundary conditions are periodic, the evolution of the system in time resembles the evolution of the state of a river bed as the particles travel downstream, rather than the evolution of the bed at a single location in a flume that receives a steady grain size distribution input from upstream. Effectively, both the "supply" and subsequent state of the bed evolve in time.

In our simulations, this evolution manifests in a few ways. When the fine particles first impact the bed, they move with a speed similar to that of the near-bed velocity, so transport rates initially increase substantially, even beyond what may be expected from previous reports of mobilization of a coarse bed by fine particles. The bed transport rate quickly decreases, but remains significantly higher than for the case with coarse bed particles alone. After additional time, the particles segregate. Specifically, the large particles move up toward the surface and the smaller particles segregate downward. As the small particles segregate downward, they no longer participate in bedload transport. Correspondingly, as the large

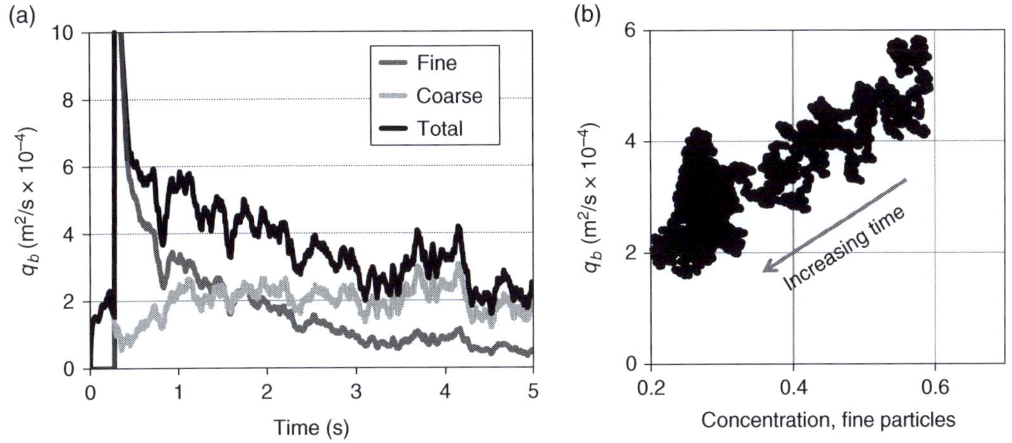

Figure 3.3 Data from simulations shortly after 1.75-mm particles are introduced into the flow above the 7-mm steady state gravel bed. (a) Time dependence of q_b for the fine and coarse particle mixture components. (b) A parametric plot of q_b for the mixture at each time step plotted versus fine particle concentration at each corresponding time step from $t = 1$ s to 5 s following the introduction of fine particles. As mentioned in the text, after the initial introduction of fines to the surface of the bed, the finer particles gradually become segregated beneath the top layer where particle velocities are highest. In (b), time flows roughly from right to left as indicated in this plot.

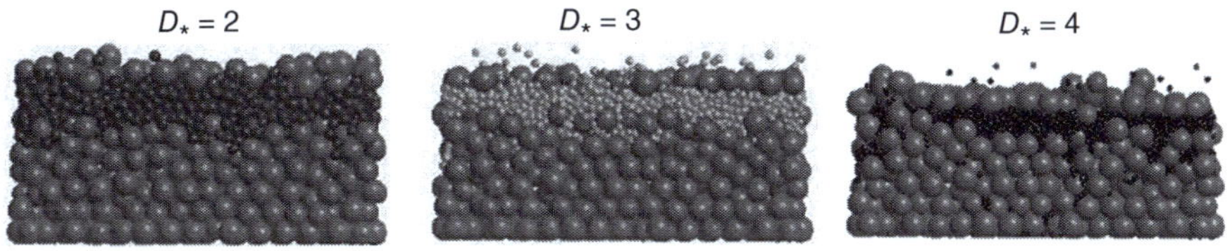

Figure 3.4 Snapshots from the near steady-state conditions from the end of each simulation. The grain size ratio is provided above. (See insert for color representation of this figure.)

particles segregate upward they participate increasingly in the bedload transport (Figure 3.3a). As the concentration of large particles in bedload transport increases, the total transport rate decreases (Figure 3.3b). Snapshots from the simulations from each of the grain size ratios are shown in Figure 3.4.

Because of the issues associated with boundary conditions, we cannot make a direct comparison with flume experiments of Hill *et al.* (2011). Indeed, since we have periodic boundary conditions, the slope cannot decrease, and the bed cannot degrade. Instead, for each system of fine particles i, we calculate the average value of $q_{b,i}$ for the times during which the transport of fine particles in bed load is approximately equal to that of the coarse particles in bed load, the conditions under which Hill *et al.* (2011) performed their experiments. We normalize each of these by the average value of the particles in bedload transport when the bed consists of 7-mm particles alone, $q_{b,7}$. We consider the results in their nondimensional form as $q_* = q_{b,7}/q_{b,i}$ and the particle size ratio $D_* = D_7/D_{i,f}$. We plot $1/q_*$ versus D_* in Figure 3.5 to illustrate the increasing mobilization of the bed with smaller fine

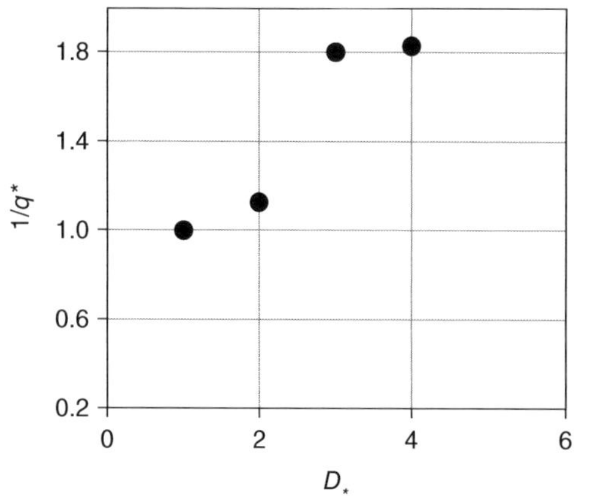

Figure 3.5 Transport results from the four different mixtures described in the text. The results are plotted for the average transport rate of each system when the fraction of bedload transport was approximately the same for each component in the mixture. $q_* = q_{b,7}/q_{b,i}$ and $D_* = D_7/D_{i,f}$.

particles. We note this effect levels off for only moderately high values of D_*, somewhat smaller than that suggested by theoretical considerations discussed immediately following Equation (3.14) and also smaller than that observed experimentally by Hill *et al.* (2011).

3.4 Conclusions

Our simple considerations of the influence of particle size alone on the mobilization of a gravel bed by the addition of fine particles has provided us with some insights to the problem as well as our naïve approaches to the questions at hand. Application of the rheological framework that has recently emerged using computational and experimental data from dense granular flows (e.g., Forterre and Pouliquen 2008) to the mobilization problem reproduces certain trends provided fine particle concentrations are not too high. This suggests an additional mechanism in the mobilization dynamics than has been previously considered that involves the behavior of the particles in bulk resulting from particle–particle interactions. While the relationships are purely empirically based, it motivates the question of how broadly models based on dense granular flows may be applied beyond sheared flows and sheet flows to lower modes of bedload transport.

The DEM simulations involving a simple representation of fluid forcing provide further evidence of the importance of particle–particle interactions in controlling the mobilization effect and other dynamics in bedload transport of a bimodal mixture. We also found some unexpected additional similarities, namely, for the finest particles we used, there was a moderate level of particles present above the bed even without turbulent fluctuations. This appeared to be the result of momentum exchange of particles during collisions, a topic currently under investigation. The similarities between the simulation results and analogous experiments (Hill *et al.* 2011) implies a potential mechanism for exploring the questions raised through the theoretical investigation of the mobilization phenomenology based on dense granular flows. On the other hand, certain differences in transport rates between physical and numerical experiments indicate there are limitations of the modeling efforts described here that need to be explored for a more complete understanding of the phenomenology. Some may be related to the importance of asphericity in governing the transport behavior (e.g., Fukuoka, Fukuda, and Uchida 2014), though as pointed out in the description of the model development, there a number of other simplifications, the effect of which need to be investigated in more detail, a topic of ongoing research.

Notation and Abbreviations

a_1	A fitting parameter in an expression proposed by Rognan *et al.* (2007) to relate the effective friction coefficient in a mixture to $\bar{I}$ (see Equation 3.4)
a_2	A fitting parameter in an expression proposed by Rognan *et al.*, (2007) to relate the effective friction coefficient in a mixture to $\bar{I}$ (see Equation 3.4)
c_D	Coefficient of drag in the DEM model; we follow Drake and Calantoni (2001) in using $c_D = 24/Re_s + 4/\sqrt{Re_s} + 0.4$, where Re_s is a relative particle Reynolds number
c_m	Coefficient of added mass in the DEM model; we use a value of 0.5 as in Batchelor (1967)
D	Grain size
DEM	Discrete element model
D_*	Grain size ratio of coarse particles to fine particles in a bimodal grain size distribution: $D_{50,coarse}/D_{50,fine}$ in laboratory experiments summarized in Table 3.1 and $D_7/D_{i,f}$ in DEM simulations
D_7	Average grain size of the coarse fraction of particles used in the DEM simulations
D_{50}	Median grain size (7 mm)
$D_{50,coarse}$	Median grain size of the coarse fraction of a bimodal grain size distribution
$D_{50,fine}$	Median grain size of the fine fraction of a bimodal grain size distribution
$D_{50,surf}$	Median grain size of the surface
$\bar{D}$	Mass-averaged grain size of a mixture of different sized particles
$D_{i,f}$	Average grain size of the fine particles of size *i* mm in a bimodal mixture in the DEM simulations
E_{eff}	Effective elastic modulus associated with two contacting particles in the DEM model (see Table 3.3)
E_i	Elastic modulus of particle species *i*, where $i = 1$ or 2
erf	Error function
$\vec{F}_A$	$\vec{F}_A = \rho_f \forall_p c_m (d\vec{u}_p/dt)$ added mass force accounting for inertial effects of the fluid in the DEM model.
$\vec{F}_B$	Buoyancy force $\vec{F}_B = -\rho_f \forall_p \vec{g}$ in the DEM model
$\vec{F}_D$	Drag force of the fluid on each particle $\vec{F}_D = \rho_f c_D A \lvert\vec{u}_{rel}\rvert \vec{u}_{rel}$ in the DEM model
F_n	Component of interparticle force normal to an interparticle contact plane in the DEM model
F_t	Component of interparticle force tangential to an interparticle contact plane in the DEM model
f	Solid fraction in a conglomeration of particles
f_{coarse}	The fraction of the particles in bedload transport that are coarse

f_L	Fraction of large particles in a mixture of different sized particles
F	Froude number of the open channel flow: $F = U/\sqrt{gH}$
G_{eff}	Effective shear modulus associated with two contacting particles in the DEM model (see Table 3.3)
G_{i}	Shear modulus of particle species *I*, where $i = 1$ or 2
g	Gravitational acceleration
H	Flow depth
h_p	Height of the top of a particle protruding from the bed, above the top of the bed (as in Equation 3.19)
I	Inertial number defined for a conglomeration of particles of the same size D: $I = \dot{\gamma}D/\sqrt{p/\rho_p}$
$\bar{I}$	Inertial number defined for a mixture of particles of different sizes based on the mass-averaged average particle size $\bar{I} = \dot{\gamma}\bar{D}/\sqrt{p/\rho_p}$
$\bar{I}_{TK}$	Inertial number defined for a mixture of particles of different sizes and density based on the mass-averaged particle size and density: $\bar{I} = \dot{\gamma}\bar{D}/\sqrt{p/\rho_p}$
I_0	Fitting parameter for a relationship proposed between μ^* and I (see Equation 3.2)
k_n	Coefficient of stiffness associated with compressive deformation of a particle in the DEM model (see Table 3.2)
k_s	Roughness length scale
k_t	Coefficient of stiffness associated with shear deformation of a particle in the DEM model (see Table 3.2)
$L_1(\zeta)$	Friction length scale: $L_1(\zeta) = \left\{\int_{\zeta=0}^{\zeta}\left[\sqrt{\zeta} \times (\mu^* - \mu_1)/(\mu_2 - \mu^*)\right]d\zeta\right\}^{-3/2}$
l_d	Downstream distance between an upstream particle protruding from the bed and a downstream particle (as in Equation 3.19)
m_{eff}	Effective mass associated with two contacting particles in the DEM model (see Table 3.3)
m_i	Mass of a particle species *i*, where $i = 1$ or 2
p	Average pressure or normal stress
q_b	Volumetric sediment transport per unit width
$q_{b,i}$	Volumetric sediment transport of particles of a mixture of particles of average size 7 mm and average size *i* mm in the DEM model (e.g., contributing to data plotted in Figure 3.5)
q_*	The ratio of volumetric transport rate of 7-mm particles alone to the transport rate of a mixture of 7-mm particles and particles of size *i* mm: $q_* = q_{b,7}/q_{b,i}$ (contributing to data plotted in Figure 3.5)
R_{eff}	Effective radius associated with two contacting particles in the DEM model (see Table 3.3)
R_{i}	Radius of a particle species *i*, where $i = 1$ or 2

Re_s	A particle Reynolds number based on the velocity of the particle relative to the local fluid velocity: $Re_s = \lvert \vec{u}_{rel} \rvert D/\nu$
S_B	Bed slope; $S_B = sin\theta$, where θ is the angle at which the bed surface is inclined relative to the horizontal direction
$\vec{u}_p$	Instantaneous velocity of a particle in the DEM model
$u_{p,i}$	The i-component of the instantaneous velocity of a particle; $i = x$, y, or z
$\vec{u}_{rel}$	The local effective velocity of the fluid relative to the particle, calculated in such a way to account for "blocking" associated with upstream particles (see Equations 3.18 and 3.19)
$u_{rel,i}$	The i-component of the local effective velocity of the fluid relative to a particle; $i = x$, y, or z (see Equations 3.18 and 3.19)
y_0	Position in the bed at which the fluid velocity is zero in the log-law in the DEM model (see Equation 3.17)
y_s	Height of the bed surface in the DEM model
u_{fx}	Downstream local fluid velocity
u_τ	Friction velocity; $u_\tau = \sqrt{\tau_B/\rho}$, where τ_B is the average shear stress on the bed by a shearing fluid, and ρ is the fluid density
U	Depth-averaged flow velocity
α	An empirically determined coefficient related to the coefficient of restitution associated with an interparticle collision in the DEM model
δ_n	Representative deformation normal to the contact plane of an interparticle contact in the DEM model
δ_t	Representative deformation tangential to the contact plane of an interparticle contact in the DEM model
$\dot{\gamma}$	Shear rate
ζ	Distance upward from the stationary sediment bed
ζ_{max}	Distance from the bed at which $\dot{\gamma}$ and $\bar{u}_p$ vanish
η_n	Damping coefficient associated with compressive deformation of a particle (see Table 3.2)
η_t	Damping coefficient associated with shear deformation of a particle (see Table 3.2)
κ	Von Karman's constant ($\kappa = 0.41$)
ρ	Fluid density
ρ_p	Material density of the sediment particles
$\bar{\rho}_p$	Mass-averaged material density of a mixture of sediment particles
μ	Interparticle friction coefficient
μ^*	Effective friction coefficient in a particle bed: $\mu^* = \tau/p$
${\mu^*}_R$	An expression proposed by Rognan *et al.* (2007) to relate the effective friction coefficient in a mixture to $\bar{I}$ (see Equation 3.4)

μ_1	Fitting parameter for a relationship proposed between μ^* and I, often associated with the quasistatic limit of the flow (see Equation 3.2)
μ_2	Fitting parameter for a relationship proposed between μ^* and I, often associated with the rapid limit of the flow (see Equation 3.2)
θ	Angle at which the bed surface is inclined relative to the horizontal
τ	Local shear stress
τ_B	Average shear stress on the bed by a shearing fluid
τ_*	Shields number: $\tau_* = \tau_B / \left[\left(\rho - \rho_f\right) g D_{50}\right]$
$\forall_p$	Volume of a particular particle

Acknowledgments

We gratefully acknowledge support provided for this research in part by the University of Minnesota in part by the National Center for Earth Surface Dynamics (NCED), a NSF Science and Technology Center funded under agreement EAR-0120914, and in part by NSF grant CBET-0756480. We thank Professor Chris Paola (University of Minnesota) for helpful discussions. We are also grateful to Professor Philippe Frey (IRSTEA, University Grenoble Alpes), an anonymous reviewer, and to the editors Professors Daizo Tsutsumi (Kyoto University) and Jonathan Laronne (Ben-Gurion University of the Negev) for helpful comments and suggestions.

References

Batchelor, GK 1967. *An Introduction to Fluid Dynamics.* Cambridge University Press: New York.

Beschta, RL and Jackson, WL 1979. The intrusion of fine sediments into a stable gravel bed. *Journal of the Fisheries Research Board of Canada* 36(2), 204–210.

Buffington, JM, Dietrich, WE, and Kirchner, JW 1992. Friction angle measurements on a naturally formed gravel streambed: Implications for critical boundary shear stress. *Water Resources Research* 28), 411–425.

Calantoni, J, Todd Holland, K, and Drake, TG 2004. Modelling sheet-flow sediment transport in wave-bottom boundary layers using discrete-element modelling. *Philosophical Transactions of the Royal Society of London, Series A* 362, 1987–2001.

Cundall, PA and Strack, ODL 1979. A discrete numerical model for granular assemblies. *Géotechnique* 29 (1), 47–65.

DaCruz, F, Emam, S, Prochnow, M *et al.* 2005. Rheophysics of dense granular materials: discrete simulation of plane shear flows. *Physical Review E* 72, 021309.

Drake, TG and Calantoni, J 2001. Discrete particle model for sheet flow sediment transport in the nearshore. *Journal of Geophysical Research: Oceans* 106(19), 19,859–19,868.

Fernandez Luque, R and vanBeck, R 1976. Erosion and transport of bed-load sediment. *Journal of Hydraulic Research* 14, 127–144.

Forrest, SB and Haff, PK 1992. Mechanics of wind ripple stratigraphy. *Science* 255(5049), 1240–1243.

Forterre, Y and Pouliquen, O 2008. Flows of dense granular media. *Annual Reviews of Fluid Mechanics* 40, 1–24.

Frey, P and Church, M 2011. Bedload: a granular phenomenon. *Earth Surface Processes and Landforms* 36 (1), 58–69.

Fukuoka, T, Fukuda, T, and Uchida, T 2014. Effects of sizes and shapes of gravel particles on sediment transports and bed variations in a numerical movable-bed channel. *Advances in Water Resources* 72, 84–86.

GDR MiDi 2004. On dense granular flows. *The European Physical Journal E* 14, 341–365.

Hatano, T 2007. Power-law friction in closely packed granular materials. *Physical Review E* 75, 060301(R).

Hill, KM and Zhang, J 2008. Kinematics of densely flowing granular mixtures. *Physical Review E* 77, 061303.

Hill, KM, DellAngelo, L, and Meerschaert MM 2010. Heavy tailed travel distance in gravel bed transport: an exploratory enquiry. *Journal of Geophysical Research: Earth Surface* 115, F00A14.

Hill, KM, Baumgardner, S, Gaffney J *et al.* 2011. The effect of fines on a gravel bed: from lubrication to washload. *Fall Meeting of the American Geophysical Union*, San Francisco, CA.

Hill, KM, Gaffney, J, Baumgardner, S, Wilcock, P, and Paola, C 2016. Experimental study of the effect of grain sizes in a bimodal mixture on bed slope, bed texture, and the transition to washload. *Water Resources Research* 53, 923–941.

Ikeda, H 1984. Flume experiments on the superior mobility of sediment mixtures. *Annual report of the Institute of Geoscience* 10, 53–56.

Iseya, F and Ikeda, H 1987. Pulsations in bedload transport rates induced by a longitudinal sediment sorting: A flume study using sand and gravel mixtures. *Geografiska Annaler: Series A, Physical Geography* 69(1), 15–27.

Jackson, WL and Beschta, RL 1984. Influences of increased sand delivery on the morphology of sand and gravel channels. *Journal of the American Water Resources Association* 20(4), 527–533.

Jaeger, HM, Nagel, SR, and Behringer, RP 1996. The physics of granular materials. *Physics Today* 49, 32–39.

Jenkins, JT and Mancini, F 1987. Balance laws and constitutive relations for plane flows of a dense, binary mixture of smooth, nearly elastic, circular disks. *Journal of Applied Mechanics* 54, 27–34.

Ji, C, Munjiza, A, Avital, E *et al.* 2013. Direct numerical simulation of sediment entrainment in turbulent channel flow. *Physics of Fluids* 25, 056601.

Jiang, Z 1995. The motion of sediment–water mixtures during intense bedload transport: computer simulations. *Sedimentology* 42, 935–945.

Jiang, Z and Haff, PK 1993. Multiparticle simulation methods applied to the micromechanics of bed load transport. *Water Resources Research* 29, 399–412.

Jop, P, Forterre, Y, and Pouliquen, O 2006. A constitutive law for dense granular flows. *Nature* 441, 727–730.

Kirchner, JW, Dietrich, WE, Iseya, F, and Ikeda, I 1990. The variability of critical shear stress, friction angle, and grain protrusion in water-worked sediments. *Sedimentology* 37, 647–672.

McEwan, IK, Jefcoatc, BJ, and Willetts, BB 1999. The grain–fluid interaction as a self-stabilizing mechanism in fluvial bed load transport. *Sedimentology* 46, 407–416.

Pope, S 2000. *Turbulent Flows*, 1st revised edn. Cambridge University Press.

Recking, A, Frey, P, Paquier, A, and Belleudy P 2009. An experimental investigation of mechanisms involved in bed load sheet production and migration. *Journal of Geophysical Research – Earth Surface* 114, F03010.

Rognon, PG, Roux, J-N, Naaim, M, and Chevoir, F 2007. Dense flows of bidisperse assemblies of disks down an inclined plane. *Physics of Fluids* 19, 058101.

Schmeeckle, MW and Nelson, JM 2003. Direct numerical simulation of bedload transport using a local, dynamic boundary condition. *Sedimentology* 50, 279–301.

Thomas, P and Bray 1999. Capturing nonspherical shape of granular media with disk clusters. *Journal of Geotechnical and Geoenvironmental Engineering* 125(3), 169–178.

Tripathi, A and Khakhar, DV 2011. Rheology of binary granular mixtures in the dense flow regime. *Physics of Fluids* 23, p. 113302.

Tsuji, Y, Tanaka, T, and Ishida, I 1992. Lagrangian numerical simulation of plug flow of cohesionless particles in a horizontal pipe. *Powder Technology* 71, 239–250.
Venditti, JG, Dietrich, WE, Nelson, PA *et al.* 2010a. Mobilization of coarse surface layers in gravel-bedded rivers by finer gravel bed load. *Water Resources Research* 46, W07506.
Venditti, JG, Dietrich, WE, Nelson, PA *et al.* 2010b. Effect of sediment pulse grain size on sediment transport rates and bed mobility in gravel bed rivers. *Journal of Geophysical Research: Earth Surface* 115, F03039.
Whiting, PJ, Dietrich, WE, Leopold, LB *et al.* 1988. Bedload sheets in heterogeneous sediment. *Geology* 16), 105–108.
Wiberg, PL and Smith, JD 1987. Calculations of the critical shear stress for motion of uniform and heterogeneous sediments. *Water Resources Research* 23(8), 1471–1480.
Wilcock, PR and Crowe, JC 2003. Surface-based transport model for mixed-size sediment. *Journal of Hydraulic Engineering* 129(2), 120–128.
Wilcock, PR, and McArdell BW 1993. Surface-based fractional transport rates: Mobilization threshold and partial transport of a sand–gravel sediment. *Water Resources Research* 29(4), 1297–1312.
Wilcock, PR and McArdell BW 1997. Partial transport of a sand/gravel sediment. *Water Resources Research* 33(1), 235–245.
Wilcock, PR, Kenworthy, ST, and Crowe, JC 2001. Experimental study of the transport of mixed sand and gravel. *Water Resources Research* 37(12), 3349–3358.
Wilson, KC 1966. Bed load transport at high shear stresses. *Journal of Hydraulic Engineering* 92(6), 49–59.
Wong, M and Parker, G 2006. Reanalysis and correction of bed-load relation of Meyer-Peter and Müller using their own database. *Journal of Hydraulic Engineering* 132(11), 1159–1168.
Yohannes, B and Hill, KM 2010. Rheology of dense granular mixtures: particle-size distributions, boundary conditions, and collisional time scales. *Physical Review E* 82, 061301.

Discussion

Discussion by Philippe Frey on K. Hill and D. Tan and D. Furbish *et al.*

While entrainment of bedload particles has been extensively studied in relation to incipient motion in turbulent flow, this is not the case for disentrainment. Yet, disentrainment conditions should *a priori* be as important as entrainment to understand and predict bedload transport and associated morphological change. More specifically, focusing on the granular bed and on granular interactions, with uniform and nonuniform material leading to complex morphological structures, how could (i) the stochastic approach and (ii) the discrete element method help better parameterize disentrainment for a better understanding of bedload transport.

Discussion by Francesco Ballio

My question arises from personal experience in the difficulty of defining "bed" (still particles?) and "transport layer" (moving particles) whenever describing the problem at a particle scale. Concepts appear to be clear, but whenever univocal criteria/definitions are sought, it turns out that it is not so easy to find these definitions.

You define bed surface (which is key for determining the velocity flow field) "by considering the average height of the top of the particles in the bed. We define bed particles as all particles with less than the near-bed fluid velocity, that is, any particles whose speed, v_p, is less than the fluid speed at the top of the bed".

Please explain two matters in this regard.

1) It cannot be that the height of the top of all particles in the bed is taken for averaging, because you want to define the surface. So, I guess you somehow consider only the first layer. But how does one decide whether or not a particle belongs to the first layer?
2) How exactly is the "near-bed fluid velocity" defined?

Replies

We thank Professor Frey for raising this interesting point. We agree this is important for understanding transport and may prove to be even more important and useful in improving models for sediment transport and, ultimately, associated evolving bed morphology under unsteady, nonuniform conditions. Since the discrete element method captures the motion of each particle in a simulated flow, it has the potential of developing robust statistics on disentrainment under a variety of fluid transport conditions and evolving grain size distributions in the bed. Potential weaknesses lie in any lagging representation of the relevant flow dynamics and particle–fluid interactions and associated time and cost of more physically robust simulations. That being said, as simulations become more sophisticated and processors increasingly faster, these limitations are becoming less onerous and the model representations are becoming more robust and satisfying (e.g., Kidanemariam and Uhlmann, 2014).

The clarifying questions by Professor Ballio are helpful and we are grateful in raising these points. It is helpful to think of these questions in two stages.

1) Before we "turned on" the fluid in the simulations, the 7-mm particles were at rest in the bed. We divided the horizontal plane into a grid of squares of approximately two particle diameters on each side. Then, we considered the top particle in each square. We averaged the height of the particle in each square to define an average y-position for the top of the bed. Then, we set the vertical position of the fluid velocity profile (Equation 3.17) relative to the top of the bed $u_{fx}(y = y_0) = 0$) at a distance $2 \times D_{50}$ of the coarse fraction (7 mm).
2) Once we initiate the fluid effects, we define the near-bed fluid velocity $u_f(y_s)$ according to the log law and this distance: $u_f(y_s) = (u_\tau/\kappa) \times ln(y_s/y_0)$, i.e., $u_f(y_s) \approx 0.64\, m/s$. Again, we consider particles according to their position in a stationary horizontal grid of squares, though from this time onward, we consider only particles with velocities less than the near-bed fluid velocity ($\approx 0.64\, m/s$). In each square, the particle with the highest position whose velocity is less than that near-bed fluid velocity is used to calculate a representative bed height. We averaged the height of these top "bed particles" to define the top of the bed at each time step.

Reference

Kidanemariam, AG and Uhlmann, M 2014. Direct numerical simulation of pattern formation in subaqueous sediment. *Journal of Fluid Mechanics* 750, R2.

4

Particle Motions and Bedload Theory: The Entrainment Forms of the Flux and the Exner Equation

David Jon Furbish, Siobhan L. Fathel, and Mark W. Schmeeckle

4.1 Introduction

As Colombini (2014) points out in his brief review of the study of the stability of erodible streambeds, the Exner equation (Exner 1925) – a basic description of conservation of sediment – has for decades been a central element in studies of sediment transport and river morphodynamics. With respect to bedload transport, the Exner equation has two versions. Perhaps most commonly this equation is written such that the rate of change in the local streambed elevation is equal to the divergence of the sediment flux, assuming that moving particles remain more-or-less in contact with the streambed and are included within the measure of streambed elevation. In contrast, the entrainment form of the Exner equation relates the rate of change in the streambed elevation to the difference in the local rates of sediment entrainment and deposition (e.g., Tsujimoto 1978; Parker, Paola, and Leclair 2000; Furbish *et al.* 2012), where streambed elevation is defined with respect only to particles at rest.

Both versions of the Exner equation can be expressed in the form of a Master equation, a general probabilistic expression of conservation which, in the present context, describes the difference between the rate at which particles arrive at a local coordinate position from all other possible positions and the rate at which particles leave this local coordinate position to other positions (e.g., Risken 1984; Ebeling and Sokolov 2005; Furbish *et al.* 2012). In the case of the divergence form of the Exner equation, the central elements of the formulation involve the particle activity (the number of particles in motion per unit streambed area) and the probability distribution of particle displacements occurring during a small interval of time $\mathrm{d}t$, leading to a Fokker–Planck-like description of particle advection and diffusion in the limit of $\mathrm{d}t \to 0$ (Furbish *et al.* 2012) – analogous to formulations of these quantities in the field of statistical mechanics. In the case of the entrainment form of the Exner equation (the focus of this chapter), the central elements of the formulation involve the particle entrainment rate and the joint probability distribution of particle hop distances and associated travel times (Furbish *et al.* 2012), or, with simplifications, the entrainment rate and the distribution of just the particle hop distances (e.g., Parker, Paola and Leclair 2000; Pelosi *et al.* 2014).

These formulations of the Exner equation are important because they provide a framework for examining details of sediment motions and transport from a probabilistic perspective. Indeed, as

Gravel-Bed Rivers: Processes and Disasters, First Edition. Edited by Daizo Tsutsumi and Jonathan B. Laronne.

noted by Fathel, Furbish, and Schmeeckle (2015), there has been a recent, renewed interest in probabilistic descriptions of bedload particle motions pertaining to the ingredients of the particle flux (e.g., Parker, Paola, and Leclair 2000; Ancey *et al.* 2008; Ancey 2010; Furbish *et al.* 2012; Furbish, Ball, and Schmeeckle 2012; Ancey and Heyman 2014; Armanini, Cavedon, and Righetti 2014; Heyman 2014), as well as the behavior of tracer particles (e.g., Nikora *et al.* 2002; Schumer, Meerschaert, and Baeumer 2009; Foufoula-Georgiou and Stark 2010; Bradley, Tucker, and Benson 2010; Ganti *et al.* 2010; Voller and Paola 2010; Hill, DellAngelo, and Meerschaert 2010; Ball 2012; Martin, Jerolmack, and Schumer 2012; Lajeunesse *et al.* 2013; Hassan *et al.* 2013; Pelosi *et al.* 2014; Martin, Purohit, and Jerolmack 2014; Zhang *et al.* 2014; Hassan and Bradley 2015). This interest stems from the recognition that the concepts and language of probability are particularly well suited to the problem of describing bedload particle motions and transport, building from early ideas provided by Einstein (1937, 1942, 1950). Moreover, this effort is aimed at a deeper understanding of the transport of sediment and sediment-borne substances, the influence of stochastic sediment motions on bedform evolution, and, in turn, the effect of changes in streambed elevation on tracer particle motions. In addition, new measurement techniques, including high-speed imaging, are providing an unprecedented opportunity to observe and measure particle motions, and to test and constrain theories of particle behavior during transport, including advanced numerical renditions of this behavior (e.g., Schmeeckle 2014a, b; Kidanemariam and Uhlmann 2014) – techniques and numerical methods that were unavailable to H. Einstein and his contemporaries.

The probabilistic version of the entrainment form of the Exner equation has two important attributes. First, it embodies the idea of "nonlocal" transport. As elaborated below, this means that the local rate of change in the streambed elevation at a position x at time t may be viewed as the outcome of conditions influencing the entrainment of particles from upstream positions at earlier instants in time as well as their subsequent travel to and arrival at x at time t. This point of view focuses on the possibility of explicitly addressing the effects of nonuniform conditions (e.g., entrainment) on transport rates and the divergence of these rates. Second, and equally important, this probabilistic version of the Exner equation does not assume at the outset that sediment transport satisfies the assumptions of a continuum behavior. This provides an opportunity to consider the effects of rarefied conditions that typically occur at low to moderate transport stages, including the extent to which continuum formulations of streambed evolution involving continuously differentiable functions are adequate.

The purpose of this chapter is to describe the ingredients of the probabilistic version of the entrainment form of the Exner equation, and to examine implications regarding its use. Because of the probabilistic basis of this equation, we start in section 4.2 with a summary of the idea of ensemble behavior, wherein the probability distributions describing characteristic features of particle motions (i.e., particle velocities, hop distances, and travel times) are to be considered a 'signature' of the transport process for given sediment and macroscopic flow conditions. Here we outline the relationship between ensemble behavior and rarefied transport conditions. In section 4.3 we describe the idea of nonlocal transport, as this provides important context for the probabilistic formulation. We then present the entrainment form of the flux, and, in turn, the entrainment form of the Exner equation. These forms highlight how the entrainment rate and the distributions of particle hop distances (measured start-to-stop) and associated travel times are at the heart of these probabilistic descriptions of transport and conservation. In section 4.4 we focus on a description of the joint distribution of hop distances and travel times, and the statistically expected relationship between these quantities for comparison with experimental measurements. In section 4.5 we offer an interpretation of the actual meaning of continuous functions applied to conditions of rarefied transport. For simplicity we focus on transport of a single particle size throughout.

4.2 Sediment Ensembles and Rarefied Conditions

4.2.1 Ensembles and Ensemble Distributions

Because of the importance of the idea of ensemble behavior in the developments below, here we elaborate previous descriptions of sediment ensembles, paying particular attention to key points regarding particle hop distances and associated travel times added by Fathel, Furbish, and Schmeeckle (2015) to the descriptions provided by Furbish *et al.* (2012) and Furbish and Schmeeckle (2013).

The idea of ensemble behavior starts by envisioning bedload particles moving over an area B on a streambed subjected to steady macroscopic flow conditions, where we momentarily assume that the streambed is planar (e.g., Lajeunesse, Malverti, and Charru 2010; Roseberry, Schmeeckle and Furbish 2012), albeit possibly involving small, stationary fluctuations in elevation (e.g., Wong *et al.* 2007). Here we are focusing on motions of particles that are entrained from the streambed and then return to it after traveling some distance, so of particular interest are motions measured parallel to the surface. Of less concern are the details of motions normal to the surface (although these details matter in certain problems). From this point of view, bedload transport is intrinsically a two-dimensional problem.

We initially choose B to be sufficiently large that the total number N_a of active particles within B may be considered fixed from one instant to the next. We then envision all possible instantaneous configurations of the N_a active particles as defined by their xy positions within B at a fixed time. This imagined set of possible configurations constitutes an ensemble of active particle positions. Moreover, each active particle in each possible configuration possesses an instantaneous velocity $\mathbf{u}_p = \mathbf{i}u_p + \mathbf{j}v_p$. We now envision the ensemble as consisting of all possible instantaneous states defined by the joint occurrence of particle positions and velocities u_p and v_p.

Let us now imagine randomizing the time series of particle states in order to remove any temporal correlation in particle configurations and velocities from one instant to the next. At this point we may imagine that each of the states represents an instant in an independent system subjected to the same macroscopic flow conditions. Indeed, we may imagine that the great number of states represents a great number of independent but nominally identical systems at a fixed time, in which case the ensemble of configurations and velocities is like that envisioned by Gibbs (1902) for gas particle systems.

Of particular interest are the distributions of particle configurations and velocities. These may vary from one member of the ensemble to another. Indeed, our conceptualization of an ensemble of velocity states is the basis for describing the *most probable* velocity states of the particles compatible with the extant fluid–particle physics, leading to the ensemble distribution of velocities whose moments coincide with the expected ensemble values. In this point of view, each member of the ensemble represents a sample drawn from the ensemble distribution. Moreover, this ensemble distribution – an exponential-like distribution (Furbish and Schmeeckle 2013) – is to be considered a *signature* of the transport process for specified sediment and macroscopic flow conditions.

Turning to the distributions of particle hop distances L_x and associated travel times T_p, let t_p denote the travel time ("age") of a particle measured from the instant that its motion begins, and let $l_x(t_p)$ denote the associated travel distance. Note that $t_p \le T_p$ and $l_x \le L_x$, where $t_p \to T_p$ and $l_x(T_p) \to L_x$ only when motion ceases. If $f_{t_p}(t_p)$ [T^{-1}] denotes the probability distribution of particle ages t_p, then at any instant conservation of probability requires that

$$\frac{\partial f_{t_p}(t_p)}{\partial t} = -\frac{\partial f_{t_p}}{\partial t_p} - D_f, \tag{4.1}$$

which is a simple rendering of the Master equation. The first term on the right-hand side describes advection of probability over the domain of t_p due to particle aging at a rate $\mathrm{d}t_p/\mathrm{d}t = 1$, and D_f [T^{-2}] denotes a "failure rate" associated with deposition of particles with age $t_p = T_p$; note that D_f may be a function of t_p or f_{t_p}, or both. With steady conditions where entrainment and deposition are balanced, then at all times, $\mathrm{d}f_{t_p}/\mathrm{d}t_p = -D_f$. The solution of this requires a fixed boundary condition $f_{t_p}(0) = E_f$ [T^{-1}] associated with steady entrainment of particles with starting age $t_p = 0$. Moreover, in this situation $f_{t_p}(t_p) \rightarrow f_{T_p}(T_p)$ represents the distribution of expected age states of the particles compatible with the extant fluid–particle physics, that is, the ensemble distribution of particle travel times. By a similar development we may assert that the steady distribution $f_{l_x}(l_x) \rightarrow f_{L_x}(L_x)$ represents the ensemble distribution of hop distances. These quantities are jointly distributed as $f_{L_x,T_P}(L_x, T_p)$.

As described by Fathel, Furbish, and Schmeeckle (2015), we may envision an ensemble of independent systems wherein particle motions begin within an area $b\mathrm{d}x$ during the interval from t to $t + \mathrm{d}t$ at the fixed time t, allowing each independent system to evolve long enough for the N particles starting during $\mathrm{d}t$ to complete their motions. As in the example above involving particle velocities, we may consider this set of travel distances L_x for each system as being a sample of size N drawn from the ensemble distribution $f_{L_x}(L_x)$. For a given width b, N varies from one system of the ensemble to the next. Moreover, in general, any one sample has the appearance of an irregular histogram of hop distances rather than a smooth distribution. Then, upon collecting all such samples from all members of the ensemble and pooling them, the resulting distribution converges to its ensemble form. Or, with respect to time, we may imagine letting an individual system evolve. During successive intervals $\mathrm{d}t$, successive sets of N particles begin their hops from within $b\mathrm{d}x$, each set leading after a finite period of time to a histogram of travel distances L_x. Assuming the process acting on the particles does not change over time, then with sufficient time T we may assume the accumulated samples converge to the ensemble distribution $f_{L_x}(L_x)$, which is invariant with respect to time. Similarly we may assume these accumulated samples yield the ensemble distribution $f_{T_p}(T_p)$. As with ensemble distributions of instantaneous quantities (e.g., particle velocities or ages), these ensemble distributions are to be considered a signature of the transport process for specified sediment and macroscopic flow conditions.

With respect to an individual realization, here we re-emphasize that, at any instant the velocities of the active particles within an elementary area $b\mathrm{d}x$ represent a sample drawn from the ensemble distribution of velocities. Similarly, during any small interval $\mathrm{d}t$, the particles entrained from an elementary area $b\mathrm{d}x$ represent, upon completion of their hops, a sample drawn from the ensemble distribution of hop distances and travel times. In the presence of patchy, intermittent sediment motions, the sizes of these samples vary with position and from one instant to the next. This means that, depending on the size of the area $b\mathrm{d}x$, it is not necessarily defensible to assume at the outset that quantities such as the particle entrainment rate, the particle activity, and the average particle velocity, as well as derivative quantities such as the particle flux, can be adequately represented as continuously differentiable two-dimensional fields. Rather, as elaborated in the penultimate section of this chapter, to describe these quantities as continuous functions requires suitable ensemble averaging, wherein they are envisioned as representing *statistically expected* values, analogous to the use of Reynolds-averaged quantities in studies of turbulence. Herein resides the significance of paying attention to the ensemble distributions of particle velocities, hop distances, and travel times. Namely, the parametric quantities of these distributions represent the *appropriate* expected (average) values compatible with the sediment and macroscopic flow conditions. For further context we turn to experimental measurements to illustrate how low to moderate transport rates involve rarefied conditions.

4.2.2 Rarefied Versus Continuum Conditions

Consider the measurements of particle motions reported by Roseberry, Schmeeckle, and Furbish (2012), which involve high-speed imaging of coarse sand particles moving as bedload within an experimental flume under uniform, steady transport conditions. Details of the experimental conditions are provided in this cited paper, and involve four sets of measurements at different transport rates (see section 4.4.2). In each of the experiments the measurement window size dB is such that the number of particles within dB varies over time (from frame to frame). The time-averaged particle number activity n_a varies over nearly three orders of magnitude, from $n_a = 0.0447$ [number cm^{-2}] at the lowest transport rate to $n_a = 21.3$ [number cm^{-2}] at the highest transport rate. In turn the average particle velocity $\overline{u_p}$ respectively varies only by about 40%, from $\overline{u_p} = 3.45$ cm s^{-1} to $\overline{u_p} = 5.46$ cm s^{-1}. (This contrast is consistent with the idea that variations in the transport rate, including short-term fluctuations, are dominated by variations in the particle activity, not by variations in particle velocity (Schmeeckle and Furbish 2007; Ancey *et al.* 2008; Ancey 2010; Roseberry, Schmeeckle and Furbish 2012; Heyman and Ancey 2014; Heyman *et al.* 2014).

Momentarily setting aside the occurrence of patchiness of active particles, let us consider the issue of rarefied versus continuum conditions. Namely, let us define a mean spacing of active particles as $L_a = \sqrt{1/n_a}$. In turn, we define a one-dimensional measure of the uniformity in n_a as $\lambda_a \sim n_a/(\mathrm{d}n_a/\mathrm{d}x)$ (Furbish 1997, p. 29). If we assume that L_a is a suitable analogue of the mean-free path λ of active particles, then the Knudsen number, $Kn = \lambda/\lambda_a \sim L_a/\lambda_a$, indicates that a continuum description of active particles is permissible when $Kn \ll 1$, that is, $L_a \ll \lambda_a$. Asserting that this condition is satisfied with $Kn \sim 0.01$, then a continuum approximation is permissible with $\lambda_a \sim 100L_a$. In the experiments above, L_a varies from about 5 cm ($n_a = 0.0447$ cm^{-2}) to 0.2 cm ($n_a = 21.3$ cm^{-2}), suggesting that a continuum approximation is reasonable with smoothing (averaging) at a scale of at least as large as ~ 5 m to 20 cm. Viewed at smaller scales, these experiments represent rarefied conditions.

Similarly, Fathel, Furbish, and Schmeeckle (2015) report that (see section 4.4.2) over an experimental sampling area of d$B = 45.8$ cm^2, patchy sediment motions yield a number activity n_a that fluctuates from a low of 0.37 [number cm^{-2}] to a high of 7.31 [number cm^{-2}] with a time average of 2.58 [number cm^{-2}] over a 5 s interval sampled at a rate of 250 frames per second. As in the example above, these values of particle activity represent rarefied conditions. In all cases the averaging area needed to envision the particle activity as being a continuous field is much larger than important scales of interest, for example, the scale of ripples and small bedforms. At these scales, both intensive properties (e.g., the particle number activity) and derivative quantities (e.g., the particle flux) are spatially and temporally discontinuous. We return to these points in section 4.5.

4.3 Entrainment Forms of the Flux and the Exner Equation

4.3.1 Nonlocal Behavior

As described below, the probabilistic versions of the entrainment forms of the flux and the Exner equation embody the idea of "nonlocal" transport conditions. We thus start with an outline of this idea.

Let x denote a coordinate axis along which a great number of sediment particles intermittently move, each as a series of hops of varying distance L_x and duration T_p, with rest periods of unspecified duration between successive hops. Now choose a particular position x. The rate at which particles are crossing x at a particular instant in time t represents the instantaneous flux $q_x(x, t)$ at this position. Observe, moreover, that these particles which are crossing x at time t started their motions (hops) from many upstream positions $x' = x - l_x$ and previous times $t' = t - t_p$. This means that the instantaneous flux $q_x(x, t)$ can be a consequence of factors or events that occurred "far" from x at "much" earlier times, inasmuch as these factors or events control the onset of the motions of those particles that eventually cross x at time t. That is, the instantaneous flux at x is not necessarily determined by factors occurring at x.

This nonlocal view of transport is quite different from a local view. For example, consider a precise definition of the instantaneous flux. Let A denote a vertical surface of width b positioned at x through which particles are moving with varying velocities, where A is normal to the x coordinate. The instantaneous solid volume flux q_x associated with A is then defined as

$$q_x(x,t) = \frac{1}{b}\int_A M_A \mathbf{u}_p \cdot \mathbf{n} \mathrm{d}A, \tag{4.2}$$

where $\mathbf{u}_p$ is the discontinuous particle velocity field viewed at the surface A, and $\mathbf{n}$ is the unit vector normal to A. With $u_p = \mathbf{u}_p \cdot \mathbf{n}$, then $M_A = 1 - H(u_p)H(-u_p)$ denotes a "mask" projected onto A such that $M_A = 1$ where $u_p \neq 0$ and $M_A = 0$ elsewhere (Furbish *et al.* 2012). Here, $H(u_p)$ is the Heaviside step function defined by $H(u_p) = 0$ for $u_p < 0$ and $H(u_p) = 1$ for $u_p \geq 0$.

The instantaneous velocity component u_p of a particle intersecting A at time t derives from the time integral of all forces acting on it from the moment of entrainment at time $t' < t$. Namely,

$$u_p(t) = \frac{1}{m_p}\int_{t'}^{t} F_x(w)\mathrm{d}w, \tag{4.3}$$

where m_p denotes the particle mass, $F_x(t)$ denotes the net instantaneous force acting on the particle, and w is a variable of integration. The history represented by the integral in (4.3) is unique to each particle intersecting A. It may be possible to know the distribution of instantaneous velocities (e.g., Furbish and Schmeeckle 2013) for statistically homogeneous, steady behavior over b (and in the upstream vicinity of A). This distribution of velocities then reflects the amalgamation of the histories represented by (4.3), specifically, the details of all particle motions associated with the near-bed turbulence, particle–particle interactions, and particle–bed interactions. In this sense, the number of particles intersecting A and their velocities may be viewed as being determined locally, inasmuch as the turbulence conditions, etc., in the upstream vicinity of A are statistically indistinguishable from those at A. That is, with a translation of the surface A along x, the number of particles intersecting A and the associated distribution of velocities do not change.

With nonuniform conditions in the streamwise direction, the velocities of particles intersecting A at time t again involve a set of unique histories prior to t. But these histories may involve upstream near-bed turbulence conditions and rates of entrainment (or disentranment) that are statistically different from those at A. For example, particles arriving at A at time t may be more numerous, or on average originate from farther upstream, or on average move faster, than would otherwise occur with statistically uniform (upstream) conditions matching those at A. It then becomes less clear that the number of particles intersecting A and their velocities can be related to local conditions at A. We now turn to a probabilistic formulation of this idea.

4.3.2 Instantaneous Flux

Let $f_{L_x,T_p}(L_x,T_p;x',t')$ denote the joint ensemble distribution of particle hop distances L_x and associated travel times T_p for particles whose motions start at position x' at time t'. These quantities generally are correlated (Roseberry, Schmeeck, and Furbish 2012; Fathel, Furbish, and Schmeeckle 2015). We now define a conditional probability density as

$$f_{L_x|T_p}\left(L_x|T_p;x',t'\right) = \frac{f_{L_x,T_p}\left(L_x,T_p;x',t'\right)}{f_{T_p}\left(T_p;x',t'\right)}, \tag{4.4}$$

where $f_{T_p}\left(T_p;x',t'\right)$ is the marginal distribution of $f_{L_x,T_p}\left(L_x,T_p;x',t'\right)$ representing the travel times T_p. In turn we define a conditional survival function as $R_{L_x|T_p}\left(L_x|T_p;x',t'\right) = 1 - F_{L_x|T_p}\left(L_x|T_p;x',t'\right)$, where

$$F_{L_x|T_p}\left(L_X|T_p;x',t'\right) = \int_0^{L_x} f_{L_x|T_p}\left(L'_x|T_p;x',t'\right)\mathrm{d}L'_x \tag{4.5}$$

is a conditional cumulative distribution function, and the prime on L_x denotes this as the variable of integration. We now observe that, so long as $f_{L_x|T_p}\left(L_x|T_p;x',t'\right)$ possesses finite first and second moments (i.e., it is not heavy tailed), the conditionally expected hop distance, $L_{x1}\left(T_p;x',t'\right) = \mathrm{E}\left(L_x|T_p;x',t'\right)$, is given by (Furbish and Haff 2010 Furbish and Roering 2013)

$$L_{x1}\left(T_p;x',t'\right) = \int_0^{\infty} R_{L_x|T_p}\left(L_x|T_p;x',t'\right)\mathrm{d}L_x, \tag{4.6}$$

and the conditionally expected second moment, $L_{x2}\left(T_p;x',t'\right) = \mathrm{E}\left(L_x^2|T_p;x',t'\right)$, is given by

$$L_{x2}\left(T_p;x',t'\right) = 2\int_0^{\infty} R_{L_x|T_p}\left(L_x|T_p;x',t'\right)\mathrm{d}L_x. \tag{4.7}$$

These definitions become useful below.

Let $E(x',t')$ denote the volumetric rate of particle entrainment per unit sediment-surface area at the position x' at time t'. Then, during a small interval $\mathrm{d}t'$ the volume of particles entrained from the streambed area $b\mathrm{d}x'$ is $b\mathrm{d}x'\mathrm{d}t'E(x',t')$. Of this total volume of entrained particles, the smaller volume of particles that possess a travel time within the interval T_p to $T_p + \mathrm{d}T_p$ is $b\mathrm{d}x'$ $\mathrm{d}t'E(x',t')f_{T_p}\left(T_p;x',t'\right)\mathrm{d}T_p$, and the particle volume possessing this travel time which also moves at least as far as L_x is

$$b\mathrm{d}x'\mathrm{d}t'E(x',t')f_{T_p}\left(T_p;x',t'\right)\mathrm{d}T_pR_{L_x|T_p}\left(L_x|T_p;x',t'\right). \tag{4.8}$$

The particle volume $\mathrm{d}V$ crossing the position x during a small interval $\mathrm{d}t$ is

$$\mathrm{d}V = b\mathrm{d}t\int_{-\infty}^{t}\int_{-\infty}^{x} E(x',t')f_{T_p}\left(T_p;x',t'\right)R_{L_x|T_p}\left(L_x|T_p;x',t'\right)\mathrm{d}x'\mathrm{d}t', \tag{4.9}$$

in which $T_p = t - t'$ and $L_x = x - x'$ with $\mathrm{d}T_p = \mathrm{d}t$. That is, (4.9) accumulates the small volume of particles $\mathrm{d}V$ entrained from all possible positions x' at previous times t' which move at least as far as x by time t during the interval $\mathrm{d}t$. Then, because the flux $q_x(x,t) = (1/b)\mathrm{d}V/\mathrm{d}t$,

$$q_x(x,t) = \int_{-\infty}^{t}\int_{-\infty}^{x} E(x',t')f_{T_p}\left(T_p;x',t'\right)R_{L_x|T_p}\left(L_x|T_p;x',t'\right)\mathrm{d}x'\mathrm{d}t'. \tag{4.10}$$

This is an expression of the "entrainment form" of the instantaneous flux. It requires knowledge of the entrainment rate as this varies with position and time, the distribution of particle travel times, which also may vary with position and time, and the distribution of hop distances conditional on the travel time.

Note that equation (4.10) is a nonlocal expression of the flux. It explicitly acknowledges that particles contributing to the flux at x at time t started their motions at upstream positions $x' = x - L_x$ at previous times $t' = t - T_p$. Further note that this expression is scale independent.

Although (4.10) is a precise probabilistic description of the flux, it is not necessarily useful in this form. Let us therefore consider an approximation of (4.10) having the form of a Fokker–Planck-like equation, which provides a valuable perspective on the idea of nonlocal transport.

With $x' = x - L_x$ and $t' = t - T_p$, and noting that $\mathrm{d}x' = \mathrm{d}x = \mathrm{d}L_x$ and $\mathrm{d}t' = \mathrm{d}T_p$, then a change of variables leads to

$$q_x(x,t) = \int_0^\infty \int_0^\infty E\left(x - L_x, t - T_p\right) f_{T_p}\left(T_p; x - L_x, t - T_p\right) R_{L_x}\left(L_x \middle| T_p; x - L_x, t - T_p\right) \mathrm{d}L_x \mathrm{d}T_p. \tag{4.11}$$

Assuming that E and $f_{T_p}\left(T_p\right)$ change negligibly (or only very slowly) over time, we expand the integrand in (4.11) as a Taylor series to first order only with respect to x, and, with appropriate factoring obtain

$$\begin{aligned} q_x(x,t) &= E(x,t) \int_0^\infty f_{T_p}\left(T_p; x, t\right) \mathrm{d}T_p \int_0^\infty R_{L_x|T_p}\left(L_x \middle| T_p; x, t\right) \mathrm{d}L_x \\ &- \frac{\partial}{\partial x}\left[E(x,t) \int_0^\infty f_{T_p}\left(T_p; x, t\right) \mathrm{d}T_p \int_0^\infty L_x R_{L_x|T_p}\left(L_x \middle| T_p; x, t\right) \mathrm{d}L_x\right]. \end{aligned} \tag{4.12}$$

Using (4.6) and (4.7) and the law of the unconscious statistician to obtain expected values of these quantities as a function of the travel time T_p, and letting an overbar denote these averaged quantities,

$$q_x(x,t) = E(x,t)\overline{L_x}(x,t) - \frac{1}{2}\frac{\partial}{\partial x}\left(E\overline{L_x^2}\right), \tag{4.13}$$

which is the Fokker–Planck-like approximation of (4.10). Here it is important to recognize that $\overline{L_x^2}$ is the raw variance of the hop distances, that is, the second moment measured about the origin, rather than the ordinary variance measured about the average.

Note that the diffusive term in (4.13) is an approximation of the nonlocal behavior embodied in (4.10). In effect this term "looks" at conditions upstream of x, and takes into account the effect of variations in these conditions relative to those at x. This term is not "diffusive" in the classic sense of molecular diffusion. Although it describes effects of spatial variations in active particle numbers (via E) and deviations in particle velocities about the average (via $\overline{L_x^2}$), as in molecular diffusion, it is more accurate to think of this term as modulating the advective term in (4.13). Moreover, it is possible to show that this approximation is very good so long as $\overline{L_x} \ll \lambda_1$, where the characteristic length scale λ_1 is defined as

$$\lambda_1 = \frac{\left|\mathrm{d}\left(E\,\overline{L_x^2}\right)/\mathrm{d}x\right|}{\left|\mathrm{d}^2\left(E\,\overline{L_x^2}\right)/\mathrm{d}x^2\right|}, \tag{4.14}$$

which is a measure of the distance over which the gradient $\mathrm{d}\left(E\overline{L_x^2}\right)/\mathrm{d}x$ changes. Inasmuch as the numerator in (4.14) is constant over a distance $\lambda_1 \gg \overline{L_x}$, then in the limit of $\lambda_1 \to \infty$, (4.13) becomes an exact approximation of (4.10).

Note that for uniform, quasi-steady conditions (4.13) yields the definition of the flux introduced by Einstein (1950), namely,

$$q_x = E\overline{L_x}, \tag{4.15}$$

which has been elaborated more recently by Wilcock (1997), Parker, Paola, and Leclair (2000), Seminara, Solari, and Parker (2002), Wong *et al.* (2007), Ganti *et al.* (2010), Furbish *et al.* (2012) and others.

It is important to emphasize that, in asserting at the outset that the distributions of hop distances and travel times are ensemble distributions, we are envisioning the instantaneous flux given by (4.13) as an abstraction such that q_x and the associated moments $\overline{L_x}$ and $\overline{L_x^2}$ are, as ensemble averaged quantities, continuously differentiable with respect to space and time. In fact, as described in section 4.2.1 above, for any particular realization, these quantities represent, from one instant to the next, samples drawn from the ensemble distributions. We elaborate this idea in section 4.5.

4.3.3 The Exner Equation

In its simplest form the Exner equation is (Tsujimoto, 1978)

$$c_b \frac{\partial \eta(x,t)}{\partial t} = -E(x,t) + D(x,t). \tag{4.16}$$

Here, $c_b = 1 - \phi$ is the volumetric particle concentration of the sediment bed where ϕ is the porosity, and $D(x, t)$ is the volumetric rate of particle deposition per streambed area. Assuming for simplicity that all particle motions are downstream, we now write the deposition rate in terms of the entrainment rate giving (Furbish *et al.* 2012)

$$c_b \frac{\partial \eta(x,t)}{\partial t} = -E(x,t) + \int_{-\infty}^{t} \int_{-\infty}^{x} E(x',t') f_{L_x,T_p}(x-x',t-t';x',t')\,\mathrm{d}x'\,\mathrm{d}t'. \tag{4.17}$$

Like the entrainment form of the flux (4.10), this nonlocal expression explicitly acknowledges that particles contributing to the local rate of change in the streambed elevation at position x at time t started their motions at upstream positions x' at previous times t'. Simplifications of (4.17) (e.g., Parker, Paola, and Leclair 2000; Pelosi *et al.* 2014) assume negligible travel times T_p or involve temporal averaging (Furbish and Roering 2013) such that the integral with respect to time is neglected and the joint probability distribution $f_{L_x,T_p}(L_x, T_p; x', t')$ is replaced with the distribution of hop distances, $f_{L_x}(L_x; x', t')$. In either case, it is important to note that this formulation involves the distribution of hop distances measured start-to-stop, not a distribution of travel distances involving multiple hops and intervening rest periods. To invoke the latter is not mass conserving.

Like the entrainment form of the flux, (4.17) can be approximated as a Fokker–Planck-like equation (Furbish *et al.* 2012), which amounts to taking the divergence of (4.13). Namely, a change of variables as used in the preceding section followed by a Taylor expansion to second order leads to

$$c_b \frac{\partial \eta(x,t)}{\partial t} = -\frac{\partial}{\partial x}\left(E\overline{L_x}\right) + \frac{1}{2}\frac{\partial^2}{\partial x^2}\left(E\overline{L_x^2}\right). \tag{4.18}$$

Applications of the entrainment form of the Exner equation principally have focused on tracer particle motions (e.g., Parker, Paola. and Leclair 2000; Ganti *et al.* 2010; Furbish *et al.* 2012; Pelosi *et al.* 2014), in which (4.17) is modified to include an expression of the fraction of particles that are tracers.

For completeness we note that, as Ancey and Heyman (2014) point out, the volumetric particle activity $\gamma = V_p n_a$ satisfies a statement of conservation of the form

$$\frac{\partial \gamma}{\partial t} = -\frac{\partial q_x}{\partial x} + E(x,t) - D(x,t). \tag{4.19}$$

This represents a bridge between the divergence and entrainment forms of the Exner equation, an idea that is examined separately.

4.4 Distributions of Hop Distances and Travel Times

4.4.1 Disentrainment Rates

The preceding descriptions of the entrainment forms of the flux and the Exner equation highlight the central roles of the entrainment rate $E(x, t)$ and the joint probability distribution $f_{L_x,T_p}(L_x, T_p)$ of particle hop distances L_x (measured start-to-stop) and associated travel times T_p. Clearly, there is a need to more deeply understand how these features of particle motions arise from the probabilistic physics of coupled fluid–particle behavior, including particle disentrainment. (It is noteworthy that the "Sed-Mech Working Group" composed of nearly 20 researchers around the world, during its inaugural September 2014 meeting in Lausanne, Switzerland, selected the topic of particle entrainment as the focus of its next meeting in 2015.) Here we focus on the distributions of particle hop distances and travel times, starting with the probabilistic (kinematic) basis of particle disentrainment. Also, in anticipation of comparisons with experimental measurements involving uniform, quasi-steady ("equilibrium") transport conditions, we hereafter omit the dependence of probability functions on x' and t' in the functional notation.

The temporal disentrainment rate is defined as (Furbish and Haff 2010; Furbish *et al.* 2012)

$$P_{T_p}(T_p) = \frac{f_{T_p}(T_p)}{1 - F_{T_p}(T_p)} = \frac{f_{T_p}(T_p)}{R_{T_p}(T_p)}, \tag{4.20}$$

and the spatial disentrainment rate is

$$P_{L_x}(L_x) = \frac{f_{L_x}(L_x)}{1 - F_{L_x}(L_x)} = \frac{f_{L_x}(L_x)}{R_{L_x}(L_x)}, \tag{4.21}$$

where $F_{T_p}(T_p)$ and $F_{L_x}(L_x)$ are cumulative distribution functions, and $R_{T_p}(T_p)$ and $R_{L_x}(L_x)$ are the associated survival functions. These disentrainment rates are not to be confused with particle deposition rates, although they reflect particle deposition. Rather, these disentrainment rates represent the particles that cease motion within an interval T_p to $T_p + \mathrm{d}T_p$ (or within L_x to $L_x + \mathrm{d}L_x$) expressed as a proportion of those which have remained in motion ("survived") to T_p (or to L_x).

In turn, the distributions of the travel times T_p and the hop distances L_x are expressed in terms of the disentrainment rates as

$$f_{T_p}(T_p) = P_{T_p}(T_p) e^{-\int_0^{T_p} P_{T_p}(w)\,\mathrm{d}w}, \tag{4.22}$$

and

$$f_{L_x}(L_x) = P_{L_x}(L_x)e^{-\int_0^{L_x} P_{L_x}(w)\,\mathrm{d}w}, \tag{4.23}$$

where w is a variable of integration.

The disentrainment rates P_{T_p} and P_{L_x} are physically related inasmuch as $f_{T_p}(T_p)$ and $f_{L_x}(L_x)$ are the marginal distributions of the joint probability density $f_{L_x,T_p}(L_x, T_p)$, where, on average, L_x varies with T_p as $L_x \sim T_p^{\alpha}$. Here, α is an exponent which thus far has been estimated from experiments (Roseberry, Schmeeckkle, and Furbish 2012; Fan *et al.* 2014; Fathel, Furbish, and Schmeeckle 2015), but which, as described below, must have a theoretical basis. As described in the next section, experiments involving high-resolution particle tracking suggest that the temporal disentrainment rate P_{T_p} is a fixed value and that $\alpha \sim 2$. Here we explore the reasons and implications.

A fixed temporal disentrainment rate is consistent with the idea that the rate of particle deposition is proportional to the number of particles in motion and "available" to be disentrained (e.g., Heyman *et al.* 2014). At any instant the active particles within an area B are in all possible states of entrainment, acceleration, deceleration, and disentrainment. Some proportion of this number of particles is in a state of imminent disentrainment (and then does so during any small interval dt). The simplest defensible assumption is that this proportion of particles in a state of imminent disentrainment is fixed for all T_p.

Specifically, whereas particles having relatively large travel times on average experience higher velocities at some point during their travel than particles having small travel times (Fathel, Furbish, and Schmeeckle 2015), particles in a state of imminent disentrainment, regardless of their travel history, are on average likely to possess a similar distribution of relatively small velocities. We thus assume that, of the particles that have at any instant survived to an "age" of $t_p = T_p$, a fixed proportion are deposited during T_p to $T_p + \mathrm{d}T_p$.

Using (4.22), this hypothesis of a fixed temporal disentrainment rate P_{T_p} yields an exponential distribution of travel times T_p, namely,

$$f_{T_p}(T_p) = \frac{1}{\overline{T_p}} e^{-T_p/\overline{T_p}}, \tag{4.24}$$

where $\overline{T_p}$ is the average travel time such that the temporal disentrainment rate $P_{T_p} = 1/\overline{T_p}$ (Fathel *et al.* 2015). We also note that, upon letting $\gamma = V_p n_a$ [L] denote the volumetric particle activity, where V_p is the particle volume, then with reference to (4.1), $E_f = 1/\overline{T_p} = E/\gamma$ and $D_f = P_{T_p} f_{T_p}$, where, under steady conditions, $E_f = P_{T_p}$.

We now assume that particle hop distances L_x vary with travel times T_p as

$$L_x = aT_p^2 + \varepsilon, \tag{4.25}$$

where, as described below, a is interpreted as a characteristic acceleration and ε denotes a stochastic deviation about the expected hop distance L_x. Taking averages, $\overline{L_x} = a\overline{T_p^2}$. Further assuming that T_p indeed is exponentially distributed according to (4.24), then $\overline{T_p^2} = 2\overline{T_p}^2$, or $\overline{L_x} = 2a\overline{T_p}^2$. The distribution $f_{L_x}(L_x)$ of hop distances L_x is then given by

$$f_{L_x}(L_x) = \left|\frac{\mathrm{d}}{\mathrm{d}L_x}\left[g^{-1}(L_x)\right]\right| f_{T_p}\left[g^{-1}(L_x)\right], \tag{4.26}$$

where, neglecting the deviation ε, the function $L_x = g(T_p) = aT_p^2$ so that the inverse function $g^{-1}(L_x)$ is $T_p = \sqrt{L_x/a}$. These relations then yield

$$f_{L_x}(L_x) = \frac{A}{2\,L_x^{1/2}} e^{-AL_x^{1/2}}, \tag{4.27}$$

where $A = \left(2/\overline{L_x}\right)^{1/2}$. This has the form of a Weibull distribution with shape parameter $k = 1/2$ and scale parameter $\lambda = \overline{L_x}/2$. The spatial disentrainment rate is given by

$$P_{L_x}(L_x) = \frac{A}{2\,L_x^{1/2}}, \tag{4.28}$$

indicating that this rate decreases with increasing travel distance.

More generally, if T_p is exponentially distributed and the expected relationship $L_x \sim T_p^\alpha$ with $\alpha > 1$, then the expected distribution of L_x will have the form of a Weibull distribution. We examine this hypothesis in the next section using experimental data. Meanwhile we note that, because this formulation gives the distribution (4.27) of hop distances L_x in terms of the expected relationship $L_x = aT_p^2$, it does not account for effects of the residual variance associated with ε in (4.25).

To clarify these points in mechanical terms we turn to the relationship (4.25) between hop distances and travel times. Let $x_p(t)$ denote the streamwise position of a particle at time t relative to its starting position at time $t = 0$. According to Newton's second law,

$$L_x = x_p(T_p) = \frac{1}{m_p}\int_0^{T_p}\int_0^t F_x(t')\,\mathrm{d}t'\,\mathrm{d}t, \tag{4.29}$$

where m_p denotes the particle mass, $F_x(t)$ denotes the net instantaneous streamwise force acting on the particle, and t' is a variable of integration. The inner integral in (4.29) involving $\mathrm{d}t'$ represents an impulse over t, and can be replaced with a time average. Namely, letting an overbar denote an average over t,

$$L_x = \frac{1}{m_p}\int_0^{T_p} \overline{F}_x t\,\mathrm{d}t. \tag{4.30}$$

In turn, letting $\overline{a} = \overline{F}_x/m_p$ denote an average acceleration during t such that $\overline{a}\,t = u_p(t)$ is the velocity at t, we average this again over the travel time T_p to give

$$L_x = \overline{\overline{a}t}\,T_p. \tag{4.31}$$

We now assume that $\overline{\overline{a}t} \approx \bar{\bar{a}}\,T_p/2$ and then take an average over the ensemble of particles possessing the travel time T_p to give

$$\langle L_x\rangle = \frac{1}{2}\langle\bar{\bar{a}}\rangle T_p^2, \tag{4.32}$$

where the angle brackets denote the ensemble averaging. We now let $a = \langle\bar{\bar{a}}\rangle/2$ denote a characteristic acceleration as in (4.25), where the brackets are omitted for simplicity and it becomes clear that ε represents a deviation about the expected travel distance $L_x = \langle L_x\rangle$ for particles with travel time T_p. Note that a actually may depend on T_p rather than being a fixed value. We assess this below with experimental data. Also note that this scaling is equivalent to assuming that $\langle\overline{u_p}\rangle = \langle L_x\rangle/T_p \sim aT_p$.

4.4.2 Experimental Measurements

Here we focus on experimental measurements reported by Fathel, Furbish, and Schmeeckle (2015) concerning the hop distances and travel times of coarse sand particles transported as bedload. The data were collected in the River Dynamics Laboratory at Arizona State University using an 8.5 m × 0.3 m recirculating flume. As reported by Roseberry, Schmeeckle, and Furbish (2013) and Furbish, Ball, and Schmeeckle (2012), these experiments involved four flow conditions. (The Froude number varied from 0.30 to 0.35, and the Shields number varied from 0.034 to 0.063.) Fathel, Furbish, and Schmeeckle (2015) reanalyzed Run 3, in which the time-averaged velocity at 1 cm above the bed was 31.0 cm s^{-1}. The bed material used in the experiment is a uniform, coarse sand with an average diameter D_{50} of 0.05 cm.

The sediment bed is photographed through a plexiglass "sled" window placed on the surface of the water, which prevents distortion of the images caused by light refraction from fluctuations in the water surface. The flow is sufficiently deep (12.5 cm) that the plexiglass window does not interfere with the flow at the bed surface. The high-speed imagery captures a 7.57 cm (streamwise) by 6.05 cm (cross-stream) area of the bed, with a corresponding resolution of 1280 × 1024 pixels. The images are taken at a rate of 250 frames per second. Particle tracking over 5 s of imagery was performed using the open-source software ImageJ. This involved marking the centroid of each particle as it moves through successive frames, with subpixel resolution. Approximately 95% of the particles in motion were tracked, amounting to more than 140 000 unique spatial positions from more than 10 000 active particles. The 5% of particles that were not tracked either left the field of view early or proved too difficult to follow through the course of the video. Particle hops were counted only if they started and stopped within the field of view. We identified approximately 4000 complete hops (measured start-to-stop).

Particles were tracked only after they began to move, excluding particles that wiggled back and forth within their pockets, but otherwise did not move. This manual method provides accurate measurements of all motions, involving both short and long displacements, with minimal bias. Individual hops were measured from incipient motion to a state of rest where both the streamwise and cross-stream velocities are zero.

Fathel, Furbish, and Schmeecckle (2015) conclude from their analyses that the distribution of the streamwise particle velocities and the distribution of the magnitude of the cross-stream velocities both are exponential-like, consistent with previous experiments (Lajeunesse, Malverti, and Charru 2010; Roseberry, Schmeeckle, and Furbish 2012) and with theory (Furbish and Schmeeckle 2013). The data confirm the presence of a "light" tail in the streamwise velocities, where the largest particle velocities are limited by near-bed fluid velocities (Roseberry, Schmeeckle, and Furbish 2012; Furbish and Schmeeckle 2013). Distributions of streamwise and cross-stream particle accelerations are Laplace-like and centered on zero means, consistent with uniform, quasi-steady ("equilibrium") transport conditions (Furbish, Roseberry, and Schmeeckle 2012; Furbish and Schmeeckle 2013).

Turning specifically to streamwise particle hop distances L_x and associated travel times T_p, a bivariate plot of L_x versus T_p (Figure 4.1) represents a sample drawn from the joint distribution $F_{L_x,T_p}(L_x,T_p)$. If $x=0$ denotes the upstream position of the sampling window and $x=X$ denotes its downstream position, then the hop distances L_x (and the associated travel times T_p) of particles starting at positions $0<x'<X$ are spatially censored if $L_x>X-x'$. That is, these particles leave the sampling window before completing their full hops. As Fathel, Furbish and Schmeeckle (2015) note, whereas some proportion of all hop distances and associated travel times is censored, longer hop distances (and travel times) are preferentially censored. The visual effect of this is a preferential reduction in the largest hop distances and travel times, giving the appearance of a less steep overall relationship

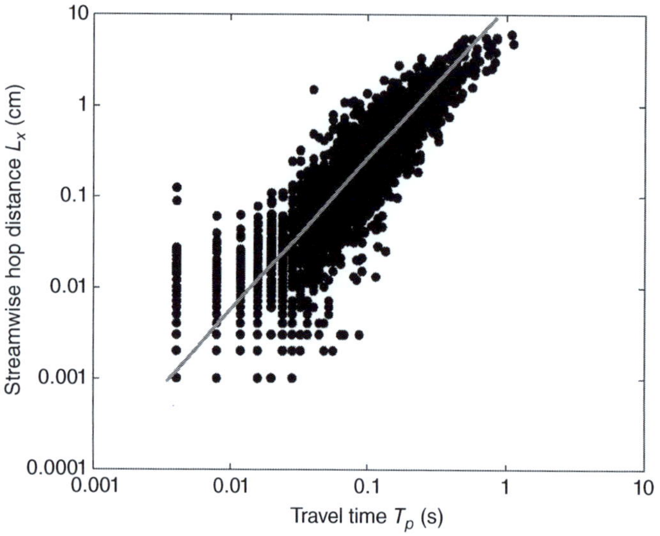

Figure 4.1 Plot of measured particle hop distances L_x versus travel times T_p with eye-fit line having a log–log slope = 2.

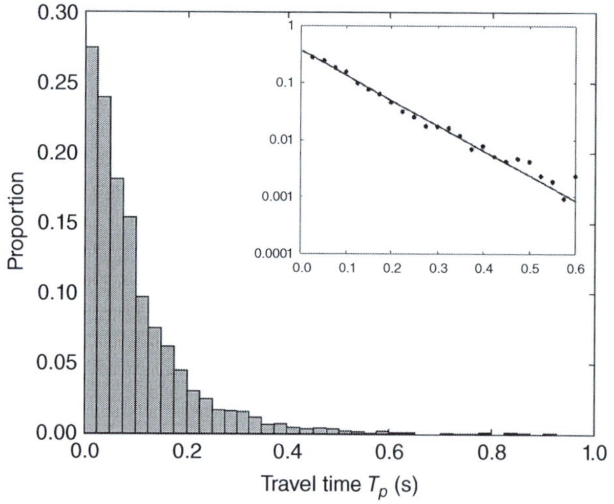

Figure 4.2 Marginal distribution of particle travel times T_p showing exponential fit (inset).

between L_x and T_p (Figure 4.1). With this in mind, we suggest that the data are consistent with a slope of $\alpha \sim 2$, as described in the previous section (rather than $\alpha \sim 5/3$, as initially suggested by Roseberry, Schmeeckle, and Furbish (2012) and reproduced in the simulations of Fan *et al.* (2014)).

In turn, the marginal distribution $f_{T_p}(T_p)$ of travel times T_p is exponential-like (Figure 4.2) and the marginal distribution $f_{L_x}(L_x)$ of hop distances L_x is Weibull-like (Figure 4.3). As described in the previous section, these are consistent with a fixed temporal disentrainment rate P_{T_p} and a decreasing spatial disentrainment rate $P_{L_x}(L_x)$. Estimated values of the shape and scale parameters of the Weibull

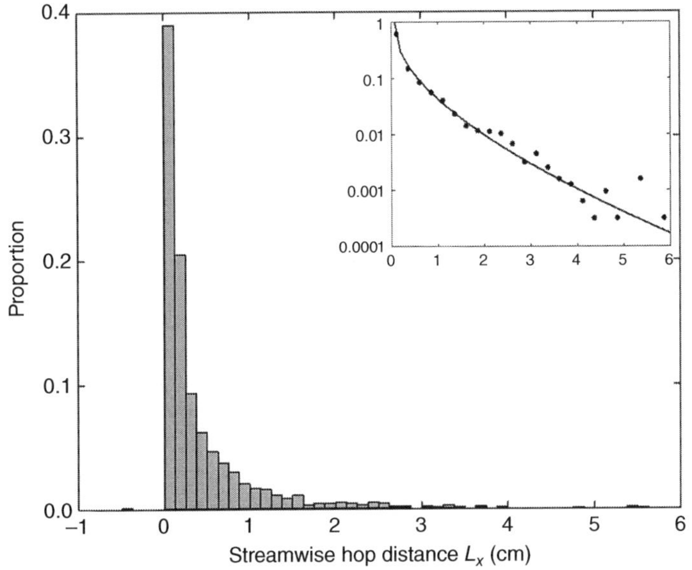

Figure 4.3 Marginal distribution of particle hop distances L_x showing Weibull fit (inset).

distribution are $k = 0.67$ (rather than $k = 0.5$) and $\lambda = 0.33$ (rather than $\lambda = \overline{L_x}/2 = 0.23$). Nonetheless, that the data possess a Weibull-like form with $k < 1$ is consistent with the expected relationship between hop distances and travel times. These estimates of k and λ likely reflect that the transformation (4.26) neglects the residual variance of deviations ε about the expected hop distance L_x.

The majority of sediment "hops" involve short positive displacements. If particles gain a sufficiently high velocity, they tend to travel farther in a shorter amount of time than similar slower moving particles. Thus, with increasing travel times, hop distances on average disproportionately increase as particle velocities increase (Fathel, Furbish, and Schmeeckle 2015).

4.5 The Meaning of Continuous Functions Applied to Conditions of Rarefied Transport

Here we return to the Exner equation (4.17) and its approximation (4.18), and place these within the context of an individual realization. In (4.17) the entrainment rate $E(x', t')$ is an ensemble expected value, and the ensemble joint distribution $f_{L_x,T_p}(L_x, T_p; x', t')$ is considered to be a smooth function over the $L_x - T_p$ domain. However, for a given position x in an individual realization, what matters is the *actual* joint distribution that occurs in relation to variations in the entrainment rate over all positions $x' < x$ and times $t' < t$. Let $f^a_{L_x,T_p}(x-x', t-t'; x', t')$ denote this actual joint distribution of L_x and T_p. As a sample drawn from the ensemble distribution $f_{L_x,T_p}(x-x', t-t'; x', t')$, this actual distribution may be irregular. We now define a deviation about the ensemble distribution as $f^d_{L_x,T_p} = f^a_{L_x,T_p} - f_{L_x,T_p}$. Note that this is not a single value, but rather represents deviations for all L_x and T_p. Also note that $f^d_{L_x,T_p}$ in general has both positive and negative values; it is not a probability density, and by definition integrates to zero. Moreover, this function may be discontinuous. Letting

$E^a(x', t')$ represent the actual entrainment rate, we similarly define a deviation in the entrainment rate about its ensemble value as $E^d = E^a - E$.

Replacing the ensemble quantities in (4.17) with the actual entrainment rate $E^a = E + E^d$ and the actual distribution $f^a_{L_x,T_p} = f_{L_x,T_p} + f^d_{L_x,T_p}$ then leads to

$$c_b \frac{\partial \eta(x,t)}{\partial t} = -\frac{\partial}{\partial x}\left(E\overline{L_x}\right) + \frac{1}{2}\frac{\partial^2}{\partial x^2}\left(E\overline{L_x^2}\right) + \varepsilon_\eta \tag{4.33}$$

where ε_η represents noise that is given by

$$\begin{aligned}\varepsilon_\eta = &\int_{-\infty}^{t}\int_{-\infty}^{x} E(x',t') f^d_{L_x,T_p}(x-x', t-t'; x', t')\,\mathrm{d}x'\,\mathrm{d}t' \\ &+ \int_{-\infty}^{t}\int_{-\infty}^{x} E^d(x',t') f^a_{L_x,T_p}(x-x', t-t'; x', t')\,\mathrm{d}x'\,\mathrm{d}t'.\end{aligned} \tag{4.34}$$

The first two terms on the right-hand side of (4.33) represent the Fokker–Planck-like approximation of the ensemble expected behavior, that is, the idealized behavior of a system for which f_{L_x,T_p} is continuous over L_x and T_p, and E and f_{L_x,T_p} are continuous over x' and t'. The term ε_η represents a random deviation about this expected behavior. Because the integrands in (4.34) contributing to ε_η involve irregular if not discontinuous quantities, we cannot expand them as Taylor series (as in section 4.3) to obtain continuous expressions analogous to the first two terms on the right-hand side of (4.33). In turn, the integral expressions in (4.34) symbolically involve piecewise integration over these integrands. Furthermore, whereas $f^d_{L_x,T_p}\left(L_x, T_p; x', t'\right)$ integrates to zero for all x' and t', the first integration in (4.34) of the product $E(x',t')f^d_{L_x,T_p}\left(L_x, T_p; x', t'\right)$ may be finite. Similarly, the second integration in (4.34) may be finite. One may think of these integral quantities as representing covariance-like terms that vary with position x and time t.

For a given entrainment rate E^a, by definition an increasing width b normal to x involves an increasing number of particles contributing to the realization of the joint distribution $f^a_{L_x,T_p}$. If conditions are statistically homogeneous along b, the integral terms in (4.34) presumably vanish with sufficiently large b (and number of moving particles) such that (4.33) converges to the ensemble form (4.18). At smaller scales (of b), the ensemble part of (4.33) is an abstraction representing the statistically expected behavior. Conversely, for a given width b, the integral terms in (4.34) presumable decrease in magnitude with increasing entrainment rate E^a, although an approximate continuum-like behavior is perhaps achieved only at very high particle activities.

Here it is useful to remind ourselves of the interpretation of Reynolds averaging. In practice this involves appealing to time-averaged quantities, although formally Reynolds-averaged quantities are ensemble averages (e.g., Monin and Yaglom 1971). A description of the Reynolds-averaged pressure and velocity fields within a flow domain based on the Reynolds-averaged continuity and Navier–Stokes equations, of which the well-known logarithmic velocity profile is an example, does not actually exist at any instant within an individual realization. Rather, such a description represents the *statistically expected* behavior within the flow domain. Similarly, the behavior described by the continuously differentiable equation (4.18) is an expected behavior, not an actuality in any realization.

Indeed, momentarily focusing on the particle flux (rather than on its divergence, $\partial\eta/\partial t$) under nominally uniform, quasi-steady transport, Heyman (2014) uses the model of Ancey *et al.* (2008) to describe the volume-averaged flux in terms of a probability distribution whose parametric values depend on the averaging (window) size, where there is a finite probability of a zero flux in the absence of moving particles. Heyman (2014) further shows experimentally how the volume-averaged flux systematically varies with sampling window size where, as this size decreases, the flux becomes increasingly intermittent (discontinuous in time). In relation to this, Ballio, Nikora, and Coleman (2014) provide experimental measurements of the sediment particle flux using the precise surface-integral definition (4.2). These measurements clearly illustrate the time-discontinuous nature of the flux under rarefied conditions.

More generally, these points remind us that, under conditions of rarefied, intermittent transport, formulations of the particle flux and its divergence using continuously differentiable equations are abstractions that represent the statistically expected behavior of ensemble-value fields – or that we are envisioning (for any particular realization) spatial or temporal averaging at scales that may be larger than important scales of interest. An important challenge consists of clarifying how the uncertainties in these formulations due to rarefied conditions translate to predictions of bed-surface evolution, and the conditions and scales for which continuum averaging is appropriate.

4.6 Conclusions

The entrainment forms of the flux and the Exner equation provide a probabilistic framework for examining details of sediment motions and transport. Our principal conclusions are as follows.

The central elements of the entrainment forms of the flux and the Exner equation involve the particle entrainment rate and the joint probability distribution of particle hop distances and associated travel times measured start-to-stop, not a distribution of travel distances involving hops and intervening rest periods (e.g., during a flood). To appeal to the latter in the integral form of the Exner equation is not mass conserving.

When expressed in integral (convolution) form, the Exner equation may be viewed as a Master equation that is both "nonlocal" and scale independent. If the joint distribution of hop distances and travel times has finite moments, this integral form can be approximated as a Fokker–Planck-like equation possessing advective and diffusive parts. The diffusive term is an approximation of the nonlocal behavior embodied in the integral form of the Exner equation, and, in effect, modulates the advective term.

Theory and experimental measurements suggest that the relationship between particle hop distances L_x and travel times T_p has the form $L_x = aT_p^2 + \varepsilon$, where a is a characteristic acceleration and ε is a stochastic deviation about the expected hop distance L_x. Moreover, an exponential distribution of travel times T_p and a Weibull distribution of hop distances L_x with shape parameter $k < 1$ reflect a clear internal consistency concerning expected particle behavior.

The probabilistic formulation presented here leads to the idea that the Exner equation, written in the continuous form of, say, a Fokker–Planck-like equation, represents *statistically expected* behavior, analogous to the interpretation of Reynolds-averaged quantities in the study of turbulent flows. The behavior of any particular realization involves uncertainty about this expected behavior in relation to patchy, intermittent sediment motions. More generally, and particularly under conditions of rarefied transport, continuum descriptions of the particle flux and its divergence may be specious. (For example, see the insightful admonitions of Haff (1983) regarding the use of continuum equations for granular flows.)

Notation

a	Characteristic acceleration [L T^{-2}]
A	Vertical surface normal to particle flux [L^2]
b	Width of surface A [L]
B	Area on streambed [L^2]
c_b	Volumetric particle concentration of sediment bed
D	Volumetric rate of particle deposition per unit streambed area [L T^{-1}]
E	Volumetric rate of particle entrainment per unit streambed area [L T^{-1}]
E^a	Actual volumetric rate of particle entrainment per unit streambed area [L T^{-1}]
E^d	Deviation in volumetric rate of particle entrainment per unit streambed area about the ensemble value [L T^{-1}]
f_{l_x}	Probability distribution of particle travel distances l_x [L^{-1}]
f_{L_x}	Probability distribution of particle travel distances L_x [L^{-1}]
F_{L_x}	Cumulative probability distribution of particle travel distances L_x
f_{t_p}	Probability distribution of particle travel times t_p [T^{-1}]
f_{T_p}	Probability distribution of particle travel times T_p [T^{-1}]
F_{T_p}	Cumulative probability distribution of particle travel times T_p
f_{L_x,T_p}	Joint ensemble distribution of particle hop distances L_x and travel times T_p [L^{-1} T^{-1}]
$f^a_{L_x,T_p}$	Actual joint distribution of particle hop distances L_x and travel times T_p [L^{-1} T^{-1}]
$f^d_{L_x,T_p}$	Deviation about the joint ensemble distribution of particle hop distances L_x and travel times T_p [L^{-1} T^{-1}]
$f_{L_x\|T_p}$	Conditional distribution of particle hop distances L_x for given travel time T_p [L^{-1}]
$F_{L_x\|T_p}$	Conditional cumulative distribution of particle hop distances L_x for given travel time T_p
F_x	Net instantaneous force acting on a particle [M L T^{-2}]
H	Heaviside step function
k	Shape parameter in Weibull distribution
Kn	Knudsen number
L_a	Spacing of active particles [L]
l_x	Particle travel distance [L]
L_x	Particle hop distance [L]
L_{x1}	Conditionally expected hop distance defined by (4.6) [L]
L_{x2}	Conditionally expected second moment of hop distances defined by (4.7) [L^2]
m_p	Particle mass [M]
M_A	Mask projected onto surface A

n_a	Particle number activity [L^{-2}]
P_{L_x}	Spatial disentrainment rate defined by (4.21) [L^{-1}]
P_{T_p}	Temporal disentrainment rate defined by (4.20) [T^{-1}]
q_x	Volumetric particle flux [L^2 T^{-1}]
R_{L_x}	Survival function of particle hop distances L_x
$R_{L_x\|T_p}$	Conditional survival function of particle hop distances L_x for given travel time T_p
R_{T_p}	Survival function of particle travel times T_p
u_p, v_p	Streamwise and cross-stream particle velocities [L T^{-1}]
V	Total particle volume [L^3]
V_p	Particle volume [L^3]
x, y	Streamwise and cross-stream coordinates [L]
t	Time [T]
t_p	Particle travel time (“age”) [T^{-1}]
T_p	Particle travel time [T^{-1}]
X	Sampling window size [L]
α	Exponent in relation between hop distances L_x and travel times T_p
γ	Volumetric particle activity [L]
ε	Stochastic deviation about the expected hop distance [L]
ε_η	Noise in rate of change in streambed elevation defined by (4.34) [L T^{-1}]
η	Sediment surface elevation [L]
λ	Mean-free path [L]; scale parameter in Weibull distribution
λ_a	Length scale of uniformity in particle activity [L]
λ_1	Length scale defined by (4.14) [L]
ϕ	Porosity of sediment bed

Acknowledgments

We thank Christophe Ancey for reviewing our work and acknowledge support by the National Science Foundation (EAR-1226076, EAR-1226288).

References

Ancey, C 2010. Stochastic modeling in sediment dynamics: Exner equation for planar bed incipient bed load transport conditions. *Journal of Geophysical Research – Earth Surface* 115, F00A11. DOI:10.1029/2009JF001260.

Ancey, C and Heyman, J 2014. A microstructural approach to bed load transport: mean behavior and fluctuations of particle transport rates. *Journal of Fluid Mechanics* 744, 129–168. DOI: 10.1017/jfm.2014.74.

Ancey, C, Davidson, AC, Böhm, T *et al.* 2008. Entrainment and motion of coarse particles in a shallow water stream down a steep slope. *Journal of Fluid Mechanics* 595, 83–114.

Armanini, A, Cavedon, V, and Righetti, M 2014. A probabilistic/deterministic approach for the prediction of the sediment transport rate. *Advances in Water Resources.* DOI: 10.1016/j.advwatres.2014.09.008.

Ball, AE 2012. *Measurements of bed load particle diffusion at low transport rates.* Junior thesis, Vanderbilt University, Nashville, TN.

Ballio, F, Nikora, V, and Coleman, SE 2014. On the definition of solid discharge in hydro-environment research and applications. *Journal of Hydraulic Research.* DOI: 10.1080/00221686.2013.869267.

Bradley, DN, Tucker, GE, and Benson, DA 2010. Fractional dispersion in a sand bed river. *Journal of Geophysical Research – Earth Surface* 115, F00A09. DOI: 10.1029/2009JF001268.

Colombini, M 2014. A decade's investigation of the stability of erodible stream beds. *Journal of Fluid Mechanics* 756, 1–4. DOI: 10.1017/jfm.2014.391.

Ebeling, W and Sokolov, IM 2005. *Statistical Thermodynamics and Stochastic Theory of Nonequilibrium Systems.* Series on Advances in Statistical Mechanics 8, World Science: Hackensack, NJ.

Einstein, HA 1937. *Bedload transport as a probability problem.* PhD thesis, Mitteilung der Versuchsanstalt fur Wasserbau Eidgenössische Technische Hochsch: Zürich.

Einstein, HA 1942. Formula for the transportation of bed load. *Transactions of the American Society of Civil Engineers* 107, 561–597.

Einstein, HA 1950. *The Bed-Load Function for Sediment Transportation in Open Channel Flows.* Technical Bulletin 1026, Soil Conservation Service, US Department of Agriculture: Washington, DC.

Exner, FM 1925. Uber die Wechselwirkung zwischen Wasser und Geschiebe in Flüssen. *Sitzungsberichte der Akademie der Wissenschaften in Wien Abteilung IIA* 134, 166–204.

Fan, N, Zhong, D, Wu, B *et al.* 2014. A mechanistic–stochastic formulation of bed load particle motions: From individual particle forces to the Fokker–Planck equation under low transport rates. *Journal of Geophysical Research – Earth Surface* 119. DOI: 10.1002/2013JF002823.

Fathel, SL, Furbish, DJ, and Schmeeckle, MW 2015. Experimental evidence of statistical ensemble behavior in bed load sediment transport. *Journal of Geophysical Research – Earth Surface* 120. DOI: 10.1002/2015JF003552.

Foufoula-Georgiou, E and Stark, C 2010. Introduction to special section on stochastic transport and emergent scaling on Earth's surface: rethinking geomorphic transport – stochastic theories, broad scales of motion and nonlocality. *Journal of Geophysical Research – Earth Surface* 115, F00A01. DOI: 10.1029/2010JF001661.

Furbish, DJ 1997. *Fluid Physics in Geology.* Oxford University Press, New York.

Furbish, DJ and Haff, PK 2010. From divots to swales: hillslope sediment transport across divers length scales. *Journal of Geophysical Research – Earth Surface* 115, F03001. DOI: 10.1029/2009JF001576.

Furbish, DJ and Roering, JJ 2013. Sediment disentrainment and the concept of local versus nonlocal transport on hillslopes. *Journal of Geophysical Research* 118, DOI: 10.1002/jgrf.20071.

Furbish, DJ and Schmeeckle, MW 2013. A probabilistic derivation of the exponential-like distribution of bed load particle velocities. *Water Resources Research* 49, 1537–1551. DOI: 10.1002/wrcr.20074.

Furbish, DJ, Ball, AE, and Schmeeckle, MW 2012. A probabilistic description of the bed load sediment flux: 4. Fickian diffusion at low transport rates. *Journal of Geophysical Research – Earth Surface* 117. DOI: 10.1029/2012JF002356.

Furbish, DJ, Roseberry, JC, and Schmeeckle, MW 2012. A probabilistic description of the bed load sediment flux: 3. The particle velocity distribution and the diffusive flux. *Journal of Geophysical Research – Earth Surface* 117, F03033. DOI: 10.1029/2012JF002355.

Furbish, DJ, Haff, PK, Roseberry, JC, and Schmeeckle, MW 2012. A probabilistic description of the bed load sediment flux: 1. *Theory. Journal of Geophysical Research* 117, F03031. DOI: 10.1029/2012JF002352.

Ganti, V, Meerschaert, MM, Foufoula-Georgiou, E *et al.* 2010. Normal and anomalous diffusion of gravel tracer particles in rivers. *Journal of Geophysical Research – Earth Surface* 115, F00A12. DOI: 10.1029/2008JF001222.

Gibbs, JW 1902. *Elementary Principles in Statistical Mechanics.* Yale University Press: New Haven.

Haff, PK 1983.Grain flow as a fluid–mechanical phenomenon. *Journal of Fluid Mechanics* 134, 401–430.

Hassan, M, Voepel, H, Schumer, R *et al.* 2013. Displacement characteristics of coarse fluvial bed sediment. *Journal of Geophysical Research – Earth Surface* 118, 155–165. DOI: 10.1029/2012JF002374.

Hassan, MA and Bradley, DN 2016. Geomorphic controls on tracer particle dispersion in gravel bed rivers. In *Gravel-Bed Rivers: Processes and Disasters* (eds D. Tsutsumi and JB Laronne). John Wiley, & Sons: Chichester, UK.

Heyman, J 2014. *A study of the spatio-temporal behaviour of bed load transport rate fluctuations.* PhD thesis, École Polytechnique Fédérale de Lausanne.

Heyman, J and Ancey, C 2014. Tracking bed load particles in a steep flume: methods and results. In *River Flow 2014* (eds A. J. Schleiss, G. de Cesare, M. J. Franca, and M. Pfister). CRC Press; 909–916. DOI: 10.1201/b17133-123.

Heyman, J, Ma, HB, Mettra, F, and Ancey, C 2014. Spatial correlations in bed load transport: Evidence, importance, and modelling. *Journal of Geophysical Research – Earth Surface.* DOI: 10.1002/2013JF003003.

Hill, K, DellAngelo, L, and Meerschaert, MM 2010. Heavy-tailed travel distance in gravel bed transport: an exploratory enquiry. *Journal of Geophysical Research – Earth Surface* 115. DOI: 10.1029/2009JF001276.

Kidanemariam, AG and Uhlmann, M 2014. Direct numerical simulation of pattern formation in subaqueous sediment. *Journal of Fluid Mechanics.* DOI: 10.1017/jfm.2014.284.

Lajeunesse, E, Malverti, L, and Charru, F 2010. Bed load transport in turbulent flow at the grain scale: experiments and modeling. *Journal of Geophysical Research – Earth Surface* 115, F04001. DOI: 10.1029/2009JF001628.

Lajeunesse, E, Devauchelle, O, Houssais, M, and Seizilles, G 2013. Tracer dispersion in bedload transport. *Advances in Geosciences* 37. DOI: 10.5194/adgeo-37-1-2013.

Martin, RL, Jerolmack, DJ, and Schumer, R 2012. The physical basis for anomalous diffusion in bed load transport. *Journal of Geophysical Research – Earth Surface* 117, F01018. DOI: 10.1029/2011JF002075.

Martin, R, Purohit, P, and Jerolmack, D 2014. Sedimentary bed evolution as a mean-reverting random walk: Implications for tracer statistics. *Geophysical Research Letters.* DOI: 10.1002/2014GL060525.

Monin, AS and Yaglom, AM 1971. *Statistical Fluid Mechanics: Mechanics of Turbulence.* Massachusetts Institute of Technology Press: Cambridge, MA.

Nikora, V, Habersack, H, Huber, T, and McEwan, I 2002. On bed particle diffusion in gravel bed flows under weak bed load transport. *Water Resources Research* 38, 1081. DOI: 10.1029/2001WR000513.

Parker, G, Paola, C, and Leclair, S 2000. Probabilistic Exner sediment continuity equation for mixtures with no active layer. *Journal of Hydraulic Engineering* 126, 818–826.

Pelosi, A, Parker, G, Schumer, R, and Ma, H-B 2014. Exner-based Master equation for transport and dispersion of river pebble tracers: derivation, asymptotic forms, and quantification of nonlocal vertical dispersion. *Journal of Geophysical Research – Earth Surface* 119, 1818–1832. DOI: 10.1002/2014JF003130.

Risken, H 1984. *The Fokker–Planck Equation.* Springer, Berlin.

Roseberry, JC, Schmeeckle, MW, and Furbish, DJ 2012. A probabilistic description of the bed load sediment flux: 2. Particle activity and motions. *Journal of Geophysical Research – Earth Surface* 117, F03032. DOI: 10.1029/2012JF002353.

Schmeeckle, MW 2014a. Numerical simulations of turbulence and sediment transport of medium sand. *Journal of Geophysical Research – Earth Surface*. DOI: 10.1002/2013JF002911.
Schmeeckle, MW 2014b. The role of velocity, pressure, and bed stress fluctuations in bed load transport over bed forms: Numerical simulation downstream of a backward-facing step. *Earth Surface Dynamics Discussions* 3, 105–112. DOI: 10.5194/esurf-3-105-2015.
Schmeeckle, MW and Furbish, DJ 2007. A Fokker–Planck model of bedload transport and morphodynamics. Paper presented at the *Stochastic Transport and Emerging Scaling on Earth's Surface (STRESS) Work Group Meeting*, National Center for Earth-Surface Dynamics, University of Illinois and Desert Research Institute, Lake Tahoe, NV.
Schumer, R, Meerschaert, MM, and Baeumer, B 2009. Fractional advection-dispersion equations for modeling transport at the Earth surface. *Journal of Geophysical Research – Earth Surface* 114, F00A07. DOI: 10.1029/2008JF001246.
Seminara, G, Solari, L, and Parker, G 2002. Bed load at low Shields stress on arbitrarily sloping beds: Failure of the Bagnold hypothesis. *Water Resources Research* 38, 1249. DOI: 10.1029/2001WR000681.
Tsujimoto, T 1978. *Probabilistic model of the process of bed load transport and its application to mobile-bed problems*. PhD thesis, Kyoto University, Kyoto, Japan.
Voller, VR, and Paola, C 2010. Can anomalous diffusion describe depositional fluvial profiles?. *Journal of Geophysical Research – Earth Surface* 115, F00A13. DOI: 10.1029/2009JF001278.
Wilcock, PR 1997. Entrainment, displacement and transport of tracer gravels. *Earth Surface Processes and Landforms* 22, 1125–1138.
Wong, M, Parker, G, DeVries, P *et al.* 2007. Experiments on dispersion of tracer stones under lower-regime planebed equilibrium bed load transport. *Water Resources Research* 43, W03440. DOI: 10.1029/2006WR005172.
Zhang, Y, Martin, RL, Chen, D *et al.* 2014. A subordinated advection model for uniform bedload transport from local to regional scales. *Journal of Geophysical Research – Earth Surface*. DOI: 10.1002/2014JF003145.

Discussion

Discussion by Astrid Blom

The authors have worked on the modelling of the streamwise dispersion of grains and have presented models that are very interesting yet seem to lack one physical mechanism inducing streamwise dispersion: the temporary burial of particles within the bed. Crickmore and Lean (1962a, b) did pioneering work on this subject. Can the authors elaborate on the relevance of this mechanism in the field (and in the laboratory), and, if relevant, if and how they intend to incorporate the mechanism in their future modeling efforts of streamwise dispersion? Regarding temporary particle burial: do the authors expect to gain from insights due to the recent work by Pelosi and co-authors?

References

Crickmore, MJ and Lean GH 1962a. The measurement of sand transport by means of radioactive tracers. *Proceedings of the Royal Society London A* 266, 402421.
Crickmore, MJ and Lean GH 1962b. The measurement of sand transport by the time-integration method with radioactive tracers. *Proceedings of the Royal Society London A* 270, 2747.

Pelosi, A, Parker G, Schumer, R, and Ma, H-B 2014. Exner-based Master equation for transport and dispersion of river pebble tracers: derivation, asymptotic forms, and quantification of nonlocal vertical dispersion. *Journal of Geophysical Research – Earth Surface* 119, 18181832. DOI: 10.1002/2014JF0 03130.

Discussion by Francesco Ballio

Any statistic on moving particles (as well on still ones, but this is not the case here) depends on how we distinguish between movement and stillness, that is, the definition of movement. This can be done on minimum velocity, on minimum displacement, etc. I find it difficult to find unequivocal definitions which are supported by evidence of their significance with respect to the process (why this particle yes and that not? Do they really behave differently?).

Would the authors clarify and justify their choice on this point and, if relevant, show that results are not affected by conventional thresholds in the definition of "moving" (e.g., by sensitivity analysis). Please also notice the relevant chapter by Lajeuness *et al.*

Discussion by Cristian Escaurizia

This work provides an excellent account of the current approaches that have furthered our understanding of bedload transport, and how the statistical analysis of sediment motion is used to improve our predictions on the transport flux and bed evolution. The particle trajectories and travel times obtained from detailed experiments and numerical simulations can help us describe the continuous equations at a macroscopic level, by computing the statistics of velocities, and time and spatial scales of transport. At times scales at which we lose correlation, the Exner equation is represented as a Fokker–Planck relation, in which we can separate the ensemble distributions and the actual distribution functions obtained from a realization.

I would like the authors to comment about the conditions when long correlations and nonlocal transport are present, and how we could incorporate the effect of fluctuations on the bed evolution when particles experience long periods of rest, and transport becomes intermittent, influencing also the characteristic timescales, and producing transport that can no longer be represented by a diffusive term.

Discussion by Philippe Frey

While entrainment of bedload particles has been extensively studied in relation to incipient motion in turbulent flow, this is not the case for disentrainment. Yet, disentrainment conditions should a priori be as important as entrainment to understand and predict bedload transport and associated morphological change. More specifically, focusing on the granular bed and on granular interactions, with uniform and nonuniform material leading to complex morphological structures, how could (i) the stochastic approach and (ii) the discrete element method help better parameterize disentrainment for a better understanding of bedload transport.

Response by the Authors

In response to Astrid Blom, to be clear, the diffusion (dispersion) of particles as described in our chapter concerns all active particles, in which case rest times do not enter the formulation. With respect to tracer particles, indeed rest times, whether involving burial or not, almost certainly dominate spreading behavior in natural systems at timescales much longer than the characteristic timescale of motions measured start-to-stop. The work of Pelosi *et al.* (2014) and others clearly points to the idea that burial and exhumation of particles associated with fluctuations in bed elevation strongly influence the nature of rest times, which means that describing tracer particle spreading cannot be decoupled from descriptions of bed-surface evolution and associated particle-size sorting. A significant challenge consists of characterizing variations in spreading behavior associated with the multiplicity of individual realizations about the expected ensemble behavior, particularly for conditions of rarefied transport, including relatively small tracer concentrations.

In response to Francesco Ballio, we are not sure an unequivocal definition of particle movement is possible using current imaging techniques. Noting that our measurements are not influenced by apparent motions due to light refraction associated with water-surface fluctuations, we neglect the `jiggling' of particles within their pockets. We include all *net* displacements measurable at our resolution and apply no filtering. (One particle diameter represents about 10 pixels, so the resolution of motion is about 0.05 mm.) We prefer this simple kinematic definition of net displacements, recognizing the existence of irreducible noise associated with resolution, rather than parsing the effects that imposed (normally low-pass) filtering has on subsequent calculations (e.g., differencing to obtain velocities) or on the forms of the distribution of motions (e.g., hop distances, velocities).

In response to Cristian Escaurizia, as in the above, it is important to distinguish between transport of all active particles and transport (and spreading) of tracer particles. The diffusive term in the Fokker–Planck-like equation is in fact an approximation of the "nonlocal" behavior embodied in the convolution form of the Exner equation. Moreover, nonlocal behavior does not require the presence of a heavy-tailed distribution of hop distances, although with a heavy-tailed distribution a nonfractional diffusive term may not be appropriate. For transport of active particles, rest times do not enter the formulation. For tracer particles, rest times (perhaps involving effects of bed-surface fluctuations with burial and exhumation) do matter. The occurrence of patchy, intermittent motions means that the continuous form of the Exner equation, in its convolution form or as approximated by a Fokker–Planck-like equation, must be viewed as describing the ensemble expected behavior, not the precise behavior during any individual realization.

In response to Phillippe Frey, we concur with the assessment that disentrainment should be considered on equal footing with entrainment in order to aim at a deeper understanding of transport. The stochastic approach provides a clear probabilistic framework for describing disentrainment in its formulation of both the spatial and temporal disentrainment rates (which are not independent) as a basis for specifying the joint distribution of particle hop distances and travel times – a centerpiece of the entrainment forms of the flux and the Exner equation. Whereas this probabilistic formulation currently can be constrained and parameterized using high-resolution measurements of particle motions, it contains no physics. Our hope is that the formulation will provide impetus for efforts aimed at physical descriptions of disentrainment (on par with efforts given to the initial motion problem), as well as serve to constrain and test such descriptions. We anticipate that the role of coupled fluid–granular behavior will be key.

5

Revisiting the Morphological Approach: Opportunities and Challenges with Repeat High-Resolution Topography

Damià Vericat, Joseph M. Wheaton, and James Brasington

5.1 Introduction

The morphology of gravel bed rivers reflects the relative roles of four fundamental geomorphic processes: the erosion, transport, deposition, and storage of sediment. Fluvial geomorphologists have long focused on understanding the controls on bed-material transport, as it is this coarse fraction of the total sediment load that provides the fundamental building blocks of channel morphology. Attempts to link quantitatively the observed morphology of channels to formative transport processes have been approached from two directions. The forward method (after Church 2006) involves using established physical principles to deduce constitutive equations to estimate sediment transport and predict patterns of sedimentation that build channel morphology. By contrast, the inverse approach, involves using observations of channel morphology and, specifically, the change in form through time, to infer rates of sediment transport. While the forward approach offers an attractive grounding in physically based relationships, the inverse approach rests on measurements of channel morphology that, at least superficially, are readily quantifiable. Moreover, the spatially explicit nature of the inverse approach offers the opportunity to gain insights into longitudinal variations in form and transport that cannot be established readily from the at-a-point or section predictions associated with the forward approach.

As outlined in the next section, the inverse method is based on the sediment continuity equation and dates back to Exner's (1925) formulation for the motion of a travelling wave. This original approach (and similar later formulations, e.g. Wittmann 1927; Hubbell 1964) related bed-material discharge explicitly to the celerity of bedforms, notably dunes. Most applications of the inverse or, as also known, 'morphological' method (after Ashmore and Church, 1998), however, involve accounting for more general changes in net sediment storage derived from repeat measurements of channel topography. Customarily, these measurements have been based on cross-section surveys (Brewer and Passmore 2002) and, not surprisingly given the complexity of channel topography, have been shown to exhibit significant dependency to survey technique (precision and reproducibility) and the spatial and temporal frequency of sections (e.g. Lane, Chandler, and Richards 1994; Fuller *et al.* 2002). The advent of survey methods capable of generating dense topographic data that, in turn, enable the construction of continuous digital elevation models (DEMs) have therefore been greeted with great excitement by the fluvial research community, and has stimulated a large and growing body of research emphasizing the potential opportunities afforded by what is increasingly recognized as a fundamental data revolution (Lane, Westaway, and Hicks 2003).

Gravel-Bed Rivers: Processes and Disasters, First Edition. Edited by Daizo Tsutsumi and Jonathan B. Laronne.

Early research in this theme was largely restricted to deriving storage budgets from DEMs of small reaches, intensively surveyed using electronic tacheometry and differential global positioning systems (GPS; e.g., Lane, Richards, and Chandler 1995; Brasington, Rumsby, and Mcvey 2000; Eaton and Lapointe 2001). Rapid progress in geomatics has, however, facilitated dramatic enhancements in the density and extent of data acquisition, illustrated most obviously by the advent of channel models derived from light detection and ranging (LiDAR; Westaway, Lane, and Hicks 2003). This technological development has seen typical datasets describing channel morphology to grow from a few tens of cross-sections, perhaps comprising a few hundred survey observations, to airborne LiDAR surveys incorporating millions of points, and now mobile terrestrial laser scan data sets with even a greater density of observations; an increase of seven orders of magnitude (Brasington, Vericat, and Rychkov 2012). Rescher (1996) argued that the advancement of modern science is intimately intertwined with such technological escalation. He contends that advances in technology are essential to provide new observations and perspectives to deepen and generalize our view of reality. He illustrates this with reference to Eddington's parable of the experimental fisherman, who trawls the ocean with a single net of fixed mesh to study and measure the diversity of life in the oceans. However, in common with all empiricists, technology limits his insights and all fish smaller than his mesh slip through, leading to a distorted and partial view of aquatic life.

Given the subtle amplitude of channel morphology and deformation, the refinement of our observations has had tangible implications for the quality of models of channel change (Brasington, Langham, and Rumsby 2003; Brasington, Vericat, and Rychkov 2012). Moreover, the technological overheads associated with high-frequency monitoring of topography are also diminishing with time, most notably through advances in low-cost methods such as structure-from-motion (SfM) stereopsis (e.g. Fonstad *et al.* 2013; Javernick, Brasington, and Caruso 2014). This raises the prospect of widely available high-quality topographic data over the coming decade, and the tangible possibility to acquire datasets at temporal frequencies tuned to the driving processes of the natural environment.

This chapter seeks to provide a timely review of the application and scope of the morphological method following the advent of this high-resolution topography (HRT) revolution. Specifically, we seek to summarize the tools that are now available to apply the morphological method for utilizing information from repeat HRT surveys. However, we also argue that while this data revolution enables ever more precise measurement of topography, problems with the inverse method persist, so that perhaps the more significant opportunities of this empirical bounty lie in the richness of the spatial patterning of channel change we can now quantify. These spatial models offer fundamental insights into the process mechanisms (erosion, deposition, and change in storage) that underpin channel morphodynamics in different fluvial contexts and forcing regimes, and provide vital data needed to parameterize and test numerical models suitable for generalizing empirical results. We contend that there is a pressing need to refocus the energy of our community away from tantalizing methodological 'tinkering' and instead employ the broader space–time perspective these data now offer to address basic questions about how gravel-bed rivers function.

5.2 The Morphological Approach: a Primer

The morphological approach has been reviewed thoroughly elsewhere and we refer readers to Ashmore and Church (1998) and Brewer and Passmore (2002), as well as Church (2006) for more thorough treatments. Ashmore and Church (1998) presented their review of the morphological approach to estimating bed-material transport at the 4th Gravel Bed Rivers Workshop in Colorado.

It is therefore fitting that four workshops later, we revisit their 'new paradigm' in the light of recent advances, both technological and theoretical. Somewhat presciently, Ashmore and Church (1998) recognized, albeit cautiously, the potential promise of 'distributed, terrain-sensitive surfaces' (i.e. DEMs) in reference to the initial efforts by Lane *et al.* (1994). Here, we briefly review the basics of the morphological approach before emphasizing the dividends that HRT might offer.

The morphological approach is typically described as an inverse-solution to estimating sediment transport, in contrast to direct measurements and hydraulically-based predictions of sediment transport. The starting point for this approach is sediment continuity (i.e. conservation of mass), in which changes in topography through time ($\partial z/\partial t$) can be formally linked to bed-material transport per unit channel width (q_b):

$$\partial q_b/\partial x + \partial q_b/\partial y + (1-p)\partial z/\partial t + \partial C_b/\partial t = 0 \tag{5.1}$$

where p is sediment deposit porosity and C_b is the concentration per unit bed area of sediment in motion. As Church (2006) highlights, the solution to Equation 5.1 typically involves approximating this two-dimensional formulation to a simplified one-dimensional case, by neglecting the lateral overbank sediment transport, and assuming that temporal variations in sediment concentration are averaged over the typically long sampling intervals used in repeat topography. Under these assumptions, Equation 5.1 is reduced to:

$$\partial Q_b/\partial x + (1-p)\partial z/\partial t = 0 \tag{5.2}$$

in which Q_b refers to the bed-material transport as a cross-sectionally averaged quantity. Equation 5.2 represents a one-dimensional form of the Exner equation fundamental to morphodynamic models (see Paola and Voller (2005) for a helpful review). Recasting this in finite difference form yields the more familiar statements of continuity in a sediment budgeting context:

$$\Delta Q_b/\Delta x + (1-p)\Delta z/\Delta t = 0 \tag{5.3}$$

$$\text{i.e.}(1-p)\Delta V = (Q_{b_{in}} - Q_{b_{out}})\Delta t \tag{5.4}$$

in which $Q_{b_{in}}$ and $Q_{b_{out}}$ represent bed-material flux (in mineral volumes) into and out of a river reach with no significant tributary inputs, and ΔV is the net change of the volume of sediment stored in the reach. If we only consider conservation of mass in volumetric terms,

$$\Delta V = V_{in} - V_{out} \tag{5.5}$$

where V_{in} and V_{out} are the volumetric equivalents of $Q_{b_{in}}$ and $Q_{b_{out}}$ over the same period Δt. It is worth noting that Equation 5.5 is also an expression for a simple sediment budget for a reach (see review by Brewer and Passmore 2002). It follows that any term can be solved for, if the other two are known. In practice, all three quantities are typically highly uncertain (Grams *et al.* 2013) and closure (i.e. measuring all three terms) is very rarely attempted, much less successfully achieved (e.g. Erwin *et al.* 2012). However, the morphological method is primarily concerned with estimating sediment flux. Therefore, the integration form of Equation 5.3 or 5.4 over space (i.e. the control volume of our reach) and time (i.e. time between surveys) is required to solve Q_b.

Early attempts to apply these relationships sought to quantify the rates of transport associated with mobile bedforms in sand-bedded rivers (e.g. Exner 1925; Simmons, Richardson, and Nordin 1965). While the generic form used is arguably applicable, mobile bedforms (i.e. dunes) in gravel-bed rivers are relatively rare, such that application of the morphological method is of limited interest in this context. By contrast, we focus here on the two primary ways the inverse method is used to estimate sediment flux in gravel-bed rivers.

The first approach involves sediment budgeting principles, to infer the value of either V_{in} or V_{out} of a reach (i.e. a reference sediment flux), and measuring ΔV from repeat topography, to then back-calculate the other unknown term (i.e. V_{in} or V_{out}). The second does not require sediment budgeting, and instead relies directly on the quantification of erosion from topographic change, combined with information about the virtual velocity of bed material (i.e. its path length), to estimate the bed-material flux. Figure 5.1 illustrates application of morphological data with repeat HRT and the definition of terms.

5.2.1 Transport Rates from Sediment Budgeting

The specification of transport rates from explicit channel sediment budgeting requires knowledge of one boundary flux, which is used to provide the upper or lower boundary condition to a longitudinal numerical solution to Equation 5.3. Typically, this has been approached using one of the following two strategies.

1) Define the study region to incorporate an unambiguous zero-flux boundary, e.g. a distal gravel–sand transition (e.g. Griffiths 1979; McLean and Church, 1999) or an upstream lake or reservoir (Merz, Pasternack, and Wheaton 2006) where sediment transport is known to approach to zero.
2) Discretize the study reach into a streamwise series of subreaches and specify the boundary flux into or out of one section under one of the following assumptions, from which a numerical solution can be obtained for all other reaches:
   a) assume a zero-flux boundary into a net-erosional subreach (from a net-depositional reach), such that overall transport rates represent lower-bound estimates (i.e. the minimum possible flux, e.g. Eaton and Lapointe 2001);
   b) use actual continuous measurements of flux across one boundary (e.g. Grams *et al.* 2013);
   c) use an estimate of the time-integrated net flux computed either from an empirical (e.g. Erwin *et al.* 2012) or theoretical (e.g. Anderson and Pitlick 2014) sediment transport rating curve/model across any given boundary.

It is worth mentioning that for strategy **2** the length of the subreaches needs to be larger than the expected step-length in order to prevent the sediment being exported from one reach passing through the following reach. Figure 5.2 and Table 5.1 illustrate application of strategy **2a** in a braided gravel river. Beyond the specification of boundary conditions, clear definition of the control volume and fluxes is critical. As Ashmore and Church (1998) note, a significant potential danger arises in assuming that the volumetric budget comprises solely the bedload flux. They note that observed volumes of erosion (V_e) or deposition (V_d), calculated topographically, are likely to incorporate significant volumes of fine sand and/or fine gravel that pass through the conservation volume not as bedload (i.e. in saltation or even in suspension in strong currents). Failure to account for this labile fraction is likely to bias bed-material flux estimates.

5.2.2 Transport Rates from Path Lengths

An alternative to explicit sediment budgeting is to use estimates of the velocity of bed-material load in conjunction with measurements of the volume of eroded bed material. In the context of repeat topographic surveys, we can define the particle path length along a streamline (i.e. L) as the mean distance travelled by sediment particles over a time T. This definition gives rise to the concept of a virtual

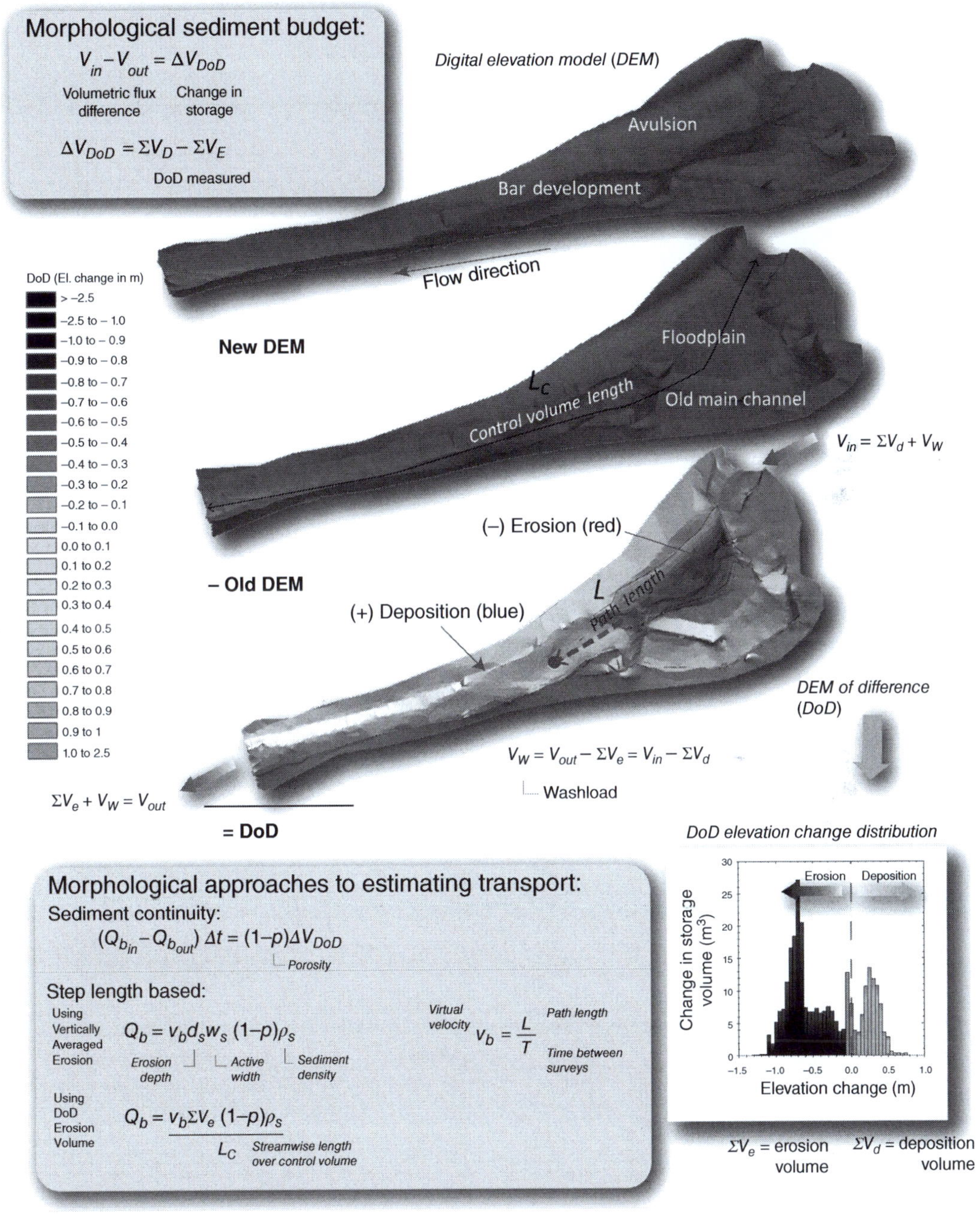

Figure 5.1 Definition diagram for terms related to the morphological approaches to estimate bed-material transport in fluvial systems.

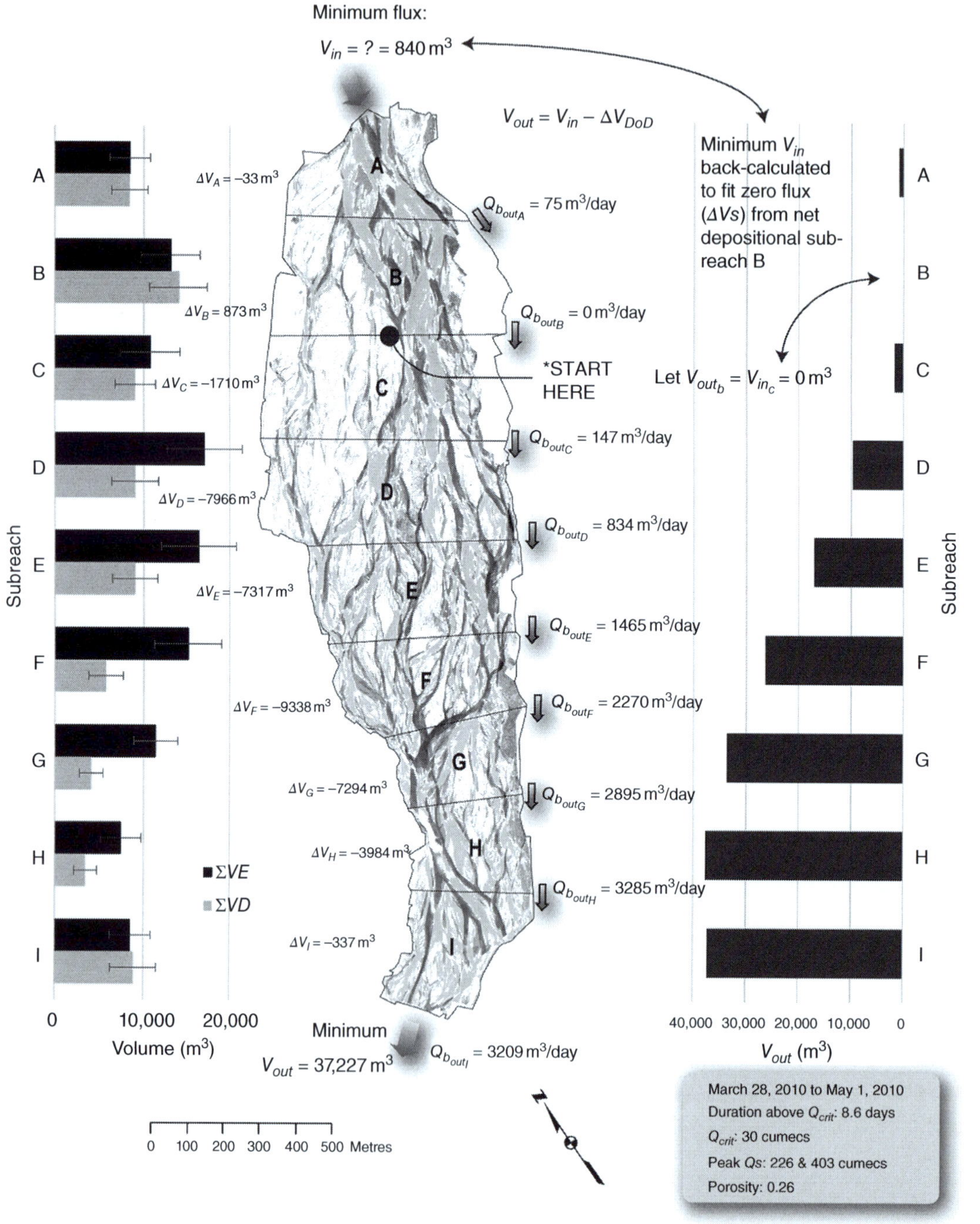

Figure 5.2 Example illustration of application of morphological approach using sediment budgeting principles and assuming a no-flux boundary to back-calculate out a minimum V_{in} at the top of reach to support observed net changes (ΔV). Example from Rees River, New Zealand for two events exceeding critical discharge. See Table 5.1 for data.

Table 5.1 Example calculations using sediment budgeting principles and following the morphological approach to estimate a minimum $Q_{b_{out}}$ by assuming a no-flux boundary at the transition from reach B to C. That value is used to back out a minimum V_{in} of 840 m^3 (i.e. $\Delta V_B + \Delta V_A$) at the top of the reach. See Figure 5.2 for illustration.

Subreach	ΣV_e m^3	ΣV_d m^3	ΔV m^3	V_{in} m^3	V_{out} m^3	$Q_{b\text{-}out}$ m^3/day	Cumulative ΔV m^3
A	8496 ± 2278	8463 ± 2040	−33 ± 3058	840	873	75	−33
B	13 146 ± 3363	14 019 ± 3323	873 ± 4728	873	0	0	840
C	10 800 ± 3358	9090 ± 2272	−1710 ± 4054	0	1709	147	−869
D	17 021 ± 4408	9055 ± 2639	−7966 ± 5137	1709	9675	834	−8835
E	16 371 ± 4382	9055 ± 2556	−7317 ± 5074	9675	16 992	1465	−16 152
F	15 129 ± 3914	5791 ± 1934	−9338 ± 4366	16 992	26 330	2270	−25 490
G	11 400 ± 2502	4151 ± 1371	−7249 ± 2828	26 330	33 579	2895	−32 739
H	7477 ± 2254	3493 ± 1303	−3984 ± 2603	33 579	37 563	3238	−36 723
I	8546 ± 2327	8883 ± 2609	337 ± 3496	37 563	37 227	3209	−36 387
Total	108 387 ± 28.786	72 001 ± 19 992	−36 387 ± 35 343	840	37 227	3209	−36 387

ΣV_e, total erosion volume in a given reach; ΣV_d, total sedimentation volume in a given reach; ΔV, net change of the volume of sediment stored in the reach; V_{in}, volume of bed material into a reach; V_{out}, volume of bed material out of a reach, $Q_{b\text{-}out}$, bed-material flux out of a reach.

velocity: $v_b = L/T$. Thus, revisiting the integration of Equation 5.4, Church (2006) showed that bed-material flux Q_b can be approximated as:

$$Q_b = v_b d_s w_s (1-p)\rho_s \tag{5.6}$$

where d_s is the scour depth, w_s is the active width of the channel bed, and ρ_s is the sediment density. Equation 5.6 is ideally suited to one-dimensional applications in which depths of erosion can be estimated from scour chains (e.g. Laronne *et al.* 1994) or burial depths from tracer studies (e.g. Hassan and Ergenzinger 2003). Scour chains have been used to obtain discrete observations of erosion or deposition at a point (see the pioneering study by Emmett and Leopold (1965) and subsequent studies by Hollingshead (1971) Ergenzinger, Schmidt, and Busskamp (1989), Hassan (1990), Laronne *et al.* (1994), Haschenburger (1996) Wilcock *et al.* (1996), Vericat, Batalla, and Garcia (2008), Vázquez-Tarrío and Menéndez-Duarte (2014), and Chapuis *et al.* (2015)). This approach is normally limited to small spatial scales (e.g. bar-scale studies) and does not typically provide information on the spatial distribution of topographic changes.

For applications using repeat HRT, we can specify d_s in Equation 5.6 from a spatially averaged erosion volume (e.g. $\Sigma V_e / A_e$; where A_e is the area experiencing erosion or area of interest). Alternatively the erosion volume ΣV_e could be used directly if Equation 5.6 is modified so that:

$$Q_b = (v_b \Sigma V_e (1-p)\rho_s)/L_c \tag{5.7}$$

in which we introduce a characteristic length scale L_c to normalize the virtual velocity v_b and keep Q_b dimensionally consistent (i.e. a volumetric rate). L_c needs to be longer than the step lengths L, and thus implies that the reach or control volume from which ΣV_e is determined must be large enough to allow typical step lengths to occur.

Regardless of whether Equation 5.6 or 5.7 is used, the estimate of the path length L itself is critical. At least two methods exist for estimating L: (i) tracer studies (e.g. Figure 5.3); and (ii) linking transport pathways and modes of morphological adjustment to the characteristic length scale of fluvial units (e.g. Figure 5.1). Exhaustive reviews of tracing techniques are presented in Sear *et al.* (2000) and Hassan and Ergenzinger (2003). Pyrce and Ashmore (2003) synthesized and reanalysed results from published tracing experiments in gravel-bed rivers to identify the variety of possible path-length distributions for differing channel morphology, channel dimensions, bed particle size, and flow magnitude. They also looked for occurrences of path length coinciding with the length scale of the morphology. Their results showed that path-length distributions may be variable (from positively skewed to multimodal; e.g. Church and Hassan 1992). Variability in path-length distributions may be related to multiple factors (e.g. event magnitude, system size) so it is very difficult to establish general models. However, gravel-bed rivers often have regularly spaced erosion (potentially corresponding with concavities such as pools) and deposition (potentially corresponding to bars) units and, consequently, it is reasonable to expect that particle path lengths should be related to this morphological scale (Pyrce and Ashmore, 2003).

One way to overcome the limitations of tracers in providing a representative estimate of the travelling distance in a given period is by investigating the distribution of sediment sources (erosional features) and sinks (depositional features). This approach was first applied by Neill (1971) for meandering rivers; he considered the travel distance as equal to one half of the meander wavelength under the assumption that material is typically transported from outer-bank meander bends to point bars. This assumption is clearly morphology specific. Carson and Griffiths (1989), Ferguson and Ashworth (1992), and Goff and Ashmore (1994) presented an alternative approach for dynamic braided rivers that involved matching successive loci of erosion and deposition (see figure 7.3 in Ashmore and Church 1998). If topographic surveys are performed frequently (e.g. event-based), successive erosional and depositional areas of similar volumes can be distinguished and the distance between the centroids used to provide an estimate of the transport lengths.

While apparently simple, Ashmore and Church (1998) highlighted the difficulties of matching erosion–deposition volumes in rivers with spatially complex topographic changes. They suggested that estimating mean travel distance could, by contrast, be based on erosional/depositional features such as bar-pool or step-pool spacing. Pyrce and Ashmore (2003) also indicated that travel distances are often related to channel morphology, especially during highly competent flows. They concluded that there is a need for deliberate systematic investigation of the relation between path-length distributions and channel morphology to establish the types of channels and particle mobility conditions in which particle transfer is, or is not, primarily stochastic. Following this, a recent flume study by Kasprak *et al.* (2015) demonstrated the importance of channel bars in acting as sedimentation zones for particles in braided rivers. These results have the potential to improve simplified relationships between channel morphology and sediment transport distances. Finally, the estimate of the virtual velocity (v_b) also requires knowledge of the duration over which the particles moved. As shown by Milan (2013), the competence duration (T) of a given event is controlled by the grain size of the bed-material flux. An under- or overestimation of T can have a direct impact on estimation of v_b and, consequently, it will skew the estimation of the bed-material flux (Q_b).

5.3 Applying a Morphological Approach with HRT

Fundamental to either of the approaches to using a morphological method as described above is the measurement of topographic change (i. e. $\Delta z/\Delta t$). Channel changes between two periods can be determined by a variety of repeat surveys, ranging from simple one-dimensional vertical depths based

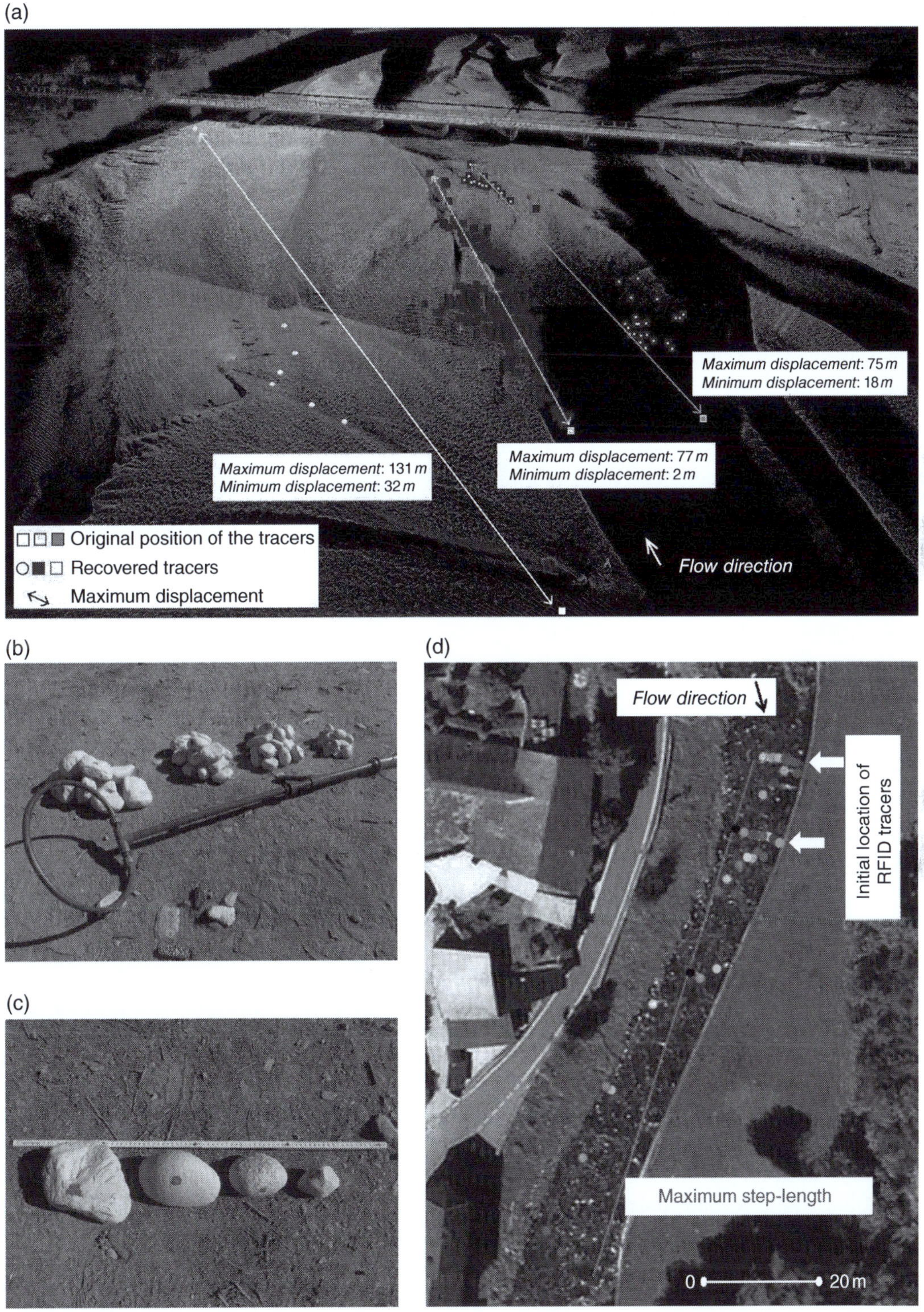

Figure 5.3 Examples of tracer deployment for estimating path lengths to drive morphological approach. (a) Painted tracers in dry exposed bars: example of displacement of painted tracers on a gravel bar in the River Rees (New Zealand) after a small magnitude flood event. (b) A mobile radio frequency identification system (RFID) to track the displacement of particles in a gravel-bed river: Mobile antenna and RFID tracers grouped by size classes. (c) Four individual tracers in which the location of the RFID transponders can be seen. (d) Example of particle step lengths after a spring flood in the River Noguera Pallaresa (Southern Pyrenees, Spain).

on scour chains to complex high-density three-dimensional point clouds (e.g. Lague, Brodu, and Leroux 2013). Fundamental to the choice of technique is the density of observations (spatial resolution). Coarser resolution sampling (e.g. lower point density, sparse cross sections, coarse grid sizes) can negatively bias topographic changes. Beyond these conditions, all studies of topographic change require measurements within a robust frame of reference (coordinate system) that facilitates accurate reoccupation of position. This can be achieved using monumented survey marks or, as is increasingly popular, by establishing control on-the-fly from established continuous GPS networks. Such control networks may need to be established outside the active survey area, and may then be used to position temporary control within the active channel on a survey-by-survey basis (Columbia Habitat Monitoring Program 2012). As one-dimensional approaches to this problem have been reviewed thoroughly elsewhere, here we focus on approaches that utilize raster DEM differencing methods based on HRT; defined as topographic surveys that exceed metre resolution data (i.e. > 1 point/m^2, see Passalacqua *et al.* (2015) for a broader discussion).

5.3.1 Digital Elevation Models and DEMs of Difference

If DEMs from repeat surveys are prepared appropriately as orthogonal rasters, with consistent projections and datums (see Passalacqua *et al.* 2015), models representing the topography at two epochs can be subtracted to produce a DEM of Difference (DoD; see Williams 2012). As the DoD calculation is a simple subtraction of elevations on a cell-by-cell basis, it is critical that rasters align and that unnecessary resampling artefacts are not introduced. Under this assumption, simple raster algebra can be used to derive the ΔV term in Equation 5.4:

$$\Delta V_{DoD} = \Sigma V_d - \Sigma V_e \tag{5.8}$$

where ΣV_d is the total volume of deposition in the same reach, and ΣV_e is the total volume of erosion in a reach. Both ΣV_d and ΣV_e are calculated directly from the DoD by taking the $\Delta z/\Delta t$ and multiplying by the respective cell area to derive sediment volumes.

While the calculation of Equation 5.8 is essentially the same when applied to cross-section or point data, the use of high-resolution DEMs provides a near continuous sampling of topography that results in robust continuum statistics. This approach therefore dramatically reduces the potential for bias that might arise when sampling channel change volumes with more typical low-density cross-section surveys (i.e. major source of historical survey data). This effect is visualized in Figure 5.4, which compares these two approaches using field data obtained from the River Feshie (Scotland) in 2006 and 2007 respectively. Here, the three randomly placed cross sections in Figure 5.4a illustrate the possible bias that may be incorporated into reach-averaged volumetric estimates when the frequency of sections is significantly lower that the spatial distribution of scour-fill units, shown here by the DoD in Figure 5.4b and c. The spatial distribution of erosion and deposition reflects different processes (e.g. bank erosion, avulsion) that are only partially sampled by these sparse sections. The dependency of the volumetric budget estimated from such low-frequency cross-section-based surveys will be strongly dependent on the style (length-scale and complexity) of morphological change, which will vary according to the fluvial context. Lane, Westaway, and Hicks (2003) analysed this effect by simulating the reliability of average bed-level estimates derived for a 3-km braided reach of the Waimakariri River, New Zealand. In their experiment, they determined the average bed-level for the entire reach based on a 1-m resolution DEM and compared this to estimates based on averaging simulated cross-sections extracted from the DEM at progressively larger and larger spacing. They found that the difference between the reference DEM-based bed-level and the section-based estimates diverged rapidly as section spacing increased beyond the scale of typical morphological units (100–200 m; see

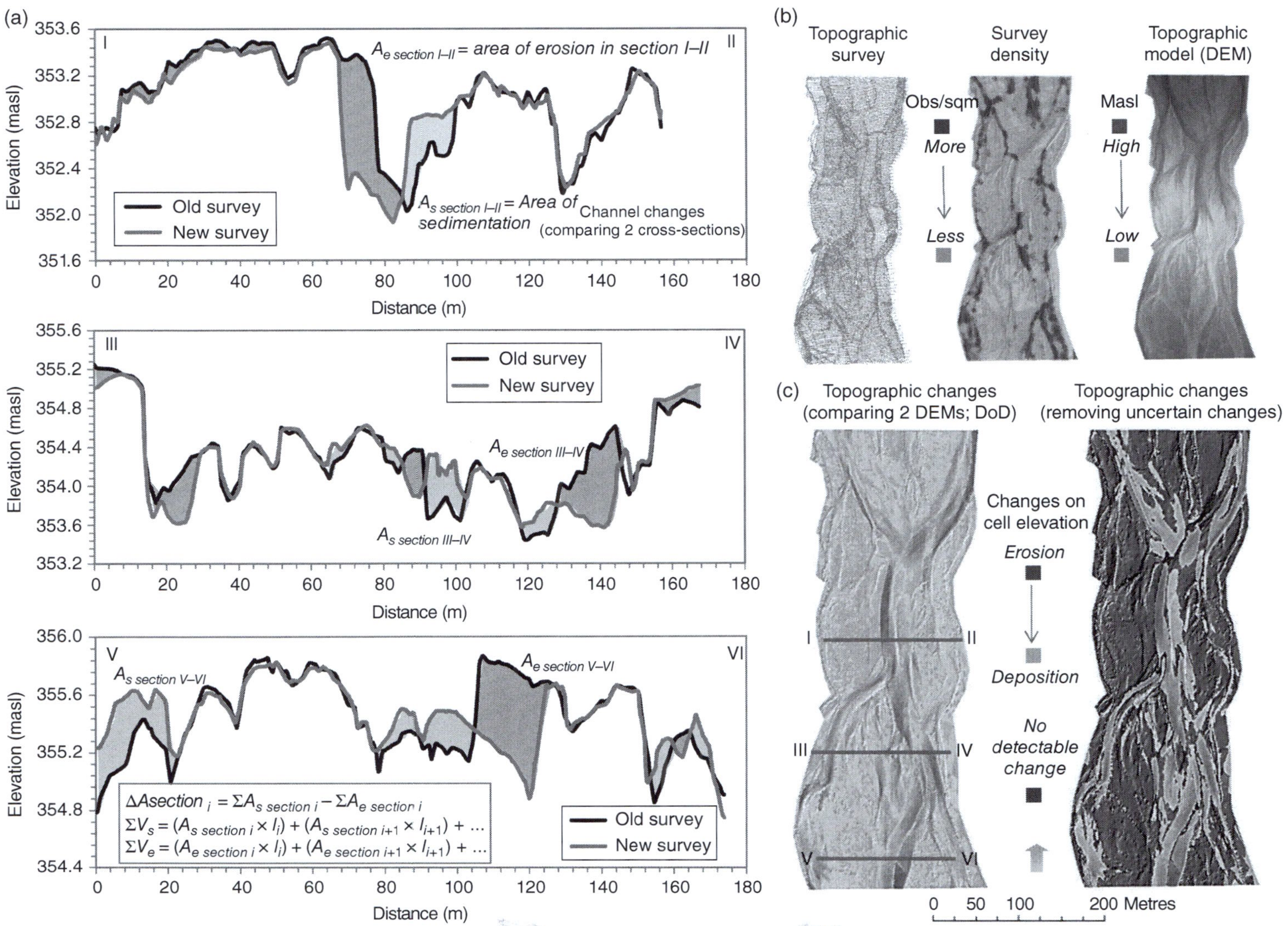

Figure 5.4 Illustrative comparison of (a) traditional one-dimensional cross-sectional information versus (b) a HRT data set and DEM. (c) DoD-based information on topographic changes between two different periods (2006–2007) in the River Feshie (Scotland). Note that the location of the cross-sections in (a) are represented in (c).

discussion in Hicks 2012). The implication is that many standard survey protocols (e.g. the 500–800-m spacing used by the Canterbury Regional Council on the Waimakariri) are likely to incorporate significant uncertainty into sediment budget calculations and that ultimately, robust sampling is likely to require section spacing that is more or less comparable to a continuous DEM.

5.3.1.1 Estimation of DEM Uncertainty

The example described above indicates how uncertainty in the underlying topographic data can potentially propagate into derived sediment budgets. For cross-section surveys, measurements are typically well resolved in the vertical dimension, and therefore, in such a case uncertainties are largely a result of the low spatial sampling frequency. However, while HRT datasets dramatically resolve this spatial bias they are nonetheless still subject to positional and vertical uncertainties (errors) that result from a complex combination of instrument errors (both random and systematic), sampling issues, interpolation, and processing artefacts and blunders (e.g. Bangen *et al.* 2014a).

Given the low vertical relief that characterises fluvial topography and many channel deformation processes (e.g. bedload sheets), it is important to develop a framework to separate the topographical signal of actual channel change from spurious DEM differences that arise from uncertainties in the before and after surfaces (Brasington, Rumsby, and Mcvey 2000). As DEM uncertainty is likely to be spatially variable, it follows that this analysis ideally requires a distributed approach to quantifying DEM uncertainty and accounting for artefacts when estimating volumetric changes (Brasington, Langham, and Rumsby 2003; Wheaton *et al.* 2010; Milan *et al.* 2011). These uncertainties need to be carefully considered since, as recently demonstrated (Erwin *et al.* 2012), topographic models may not be sufficient to determine the sign of the net change in a given reach, even using the most recent survey techniques and data analysis (e.g. Heritage and Hetherington 2007; Milan *et al.* 2011; Grams *et al.* 2013).

There are three strategies typically employed to estimating DEM uncertainties (elevation errors, i.e. δz): (i) treat them as spatially uniform; (ii) allow uncertainties to vary in given regions (e.g. according to survey methods or surface cover classes); or (iii) model uncertainties at the grid-cell scale. Representing DEM uncertainties as elevation errors is statistically convenient as the change detection problem can be treated using established error propagation theory (see below). Passalacqua *et al.* (2015) provide a more complete review of these methods and in particular the specific situations that warrant using the simpler versus more complex methods. Here we briefly differentiate these methods.

5.3.1.1.1 Spatially Uniform Methods

Treating DEM uncertainty as spatially uniform is by far the most simple and also most commonly adopted strategy. Herein, elevation errors are assumed not to be a function of location but approximated as a constant +/− elevation value across a DEM. Typical methods for estimating representative DEM error values are to compute simple statistics (e.g. range, standard deviation, root-mean-square error) using standard cross-validation methods such as hold-out or leave-one-out cross-validation or reoccupation of check points (Federal Geospatial Data Committee 1998; Brasington, Langham, and Rumsby 2003). A common, but arguably flawed, practice is to use manufacturer reported estimates of instrument precision (e.g. point quality metrics in differential GPS surveys) as an approximation of DEM error. Such estimates are generally far too optimistic and neglect the effects of spatial sampling, interpolation, and potential blunders. Other approaches include spot elevation checks against independent data or measurements of notionally higher accuracy, and/or use of fiduciary check points on unchanging surfaces (e.g. Kaplinski *et al.* 2014).

5.3.1.1.2 Spatially Variable Methods

A more sophisticated, though still simple, form of error modelling is to use spatially uniform approximations within specific regions. For example, Lane, Westaway, and Hicks (2003) used check-point data to calculate errors in wet (submerged) versus dry (exposed) areas of the channel. Similarly, Kaplinski *et al.* (2014) used a hybrid survey approach to produce a DEM from a mix of multibeam and singlebeam sonar, total station and LiDAR data, and differentiated spatially uniform error estimates for each survey method, as well as for smooth and rough areas within each method. Differences in surface texture, vegetation, and survey methods are all good reasons to use different estimates of elevation error and the critical thing is that clear, mutually exclusive boundaries can be derived for these different regions (e.g. Williams *et al.* 2013).

Spatially variable error models can also be derived using independent models of elevation error that describe how errors vary on a cell-by-cell basis. There are many different approaches to building spatially variable error models, including full error budgeting, modelling variance of independent repeat surveys of unchanging surfaces, statistical models of error, bootstrap tests, interpolation error estimates, and fuzzy inference systems. Not all of these methods are appropriate for all surveys. Full error budgeting is typically not practical and often results in overly conservative estimates (e.g. CUBE – combined uncertainty and bathymetry estimator; Calder and Mayer 2001, 2003). Traditional total station, real-time kinematic (RTK) GPS and even many airborne LiDAR surveys often lack the point density to hold back a portion of the data to use for building reliable statistical models of error. Where data density is adequate (e.g. terrestrial laser scanning (TLS) and SfM), statistical methods that look at elevation variance within a moving window or bins of a defined cell size (Milan *et al.* 2011) can be applied.

Wheaton *et al.* (2010) proposed the use of fuzzy inference systems (FIS) as a pragmatic approach to model spatially variable errors. Fuzzy inference systems are convenient frameworks for taking the information that is known (e.g. point density, slope, etc.) and producing an appropriate output (i.e. a spatial distributed δz in this case) based on expert-based rule tables (Jang and Gulley 2007; Wheaton *et al.* 2010). In this approach, both input and output are represented with continuous variables and use fuzzy membership functions to translate them into categorical data. The key to fuzzy inference systems is that membership of a given category can overlap and, therefore, multiple rules can apply to any combination of input, which gives rise to a much richer model output. A simple example is shown in Figure 5.5, whereby spatially variable (at the grid-size scale), continuous inputs of point density (a proxy for sampling), slope (a proxy for topographic complexity), and surface roughness (a proxy for subgrid topographic variability) are combined in a FIS to predict total vertical DEM uncertainty.

5.3.1.2 Error Propagation and DoD Thresholding

Given an error model (δz_{DEM_i}), the significance of observed vertical changes calculated between two DEMs can then be readily established. Both Lane and Chandler (2003) and Brasington, Langham, and Rumsby (2003) suggest using traditional error propagation theory to support the identification of a minimum level of detection (LOD_{min}) to threshold significant vertical changes from noise.

This approach involves first defining the propagated error (i. e. δz_{DoD}) associated with vertical differencing, which can be taken as the sum of the errors of both DEMs (i and $i+1$) in quadrature (see Taylor 1997);

$$\delta z_{DoD} = \sqrt{\delta z_{DEM_i}{}^2 + \delta z_{DEM_{i+1}}{}^2} \tag{5.9}$$

Each elevation difference (i.e. $z_i - z_{i+1}$) can then, on a cell-by-cell basis, be converted to a t statistic by considering the magnitude of change relative to the propagated error;

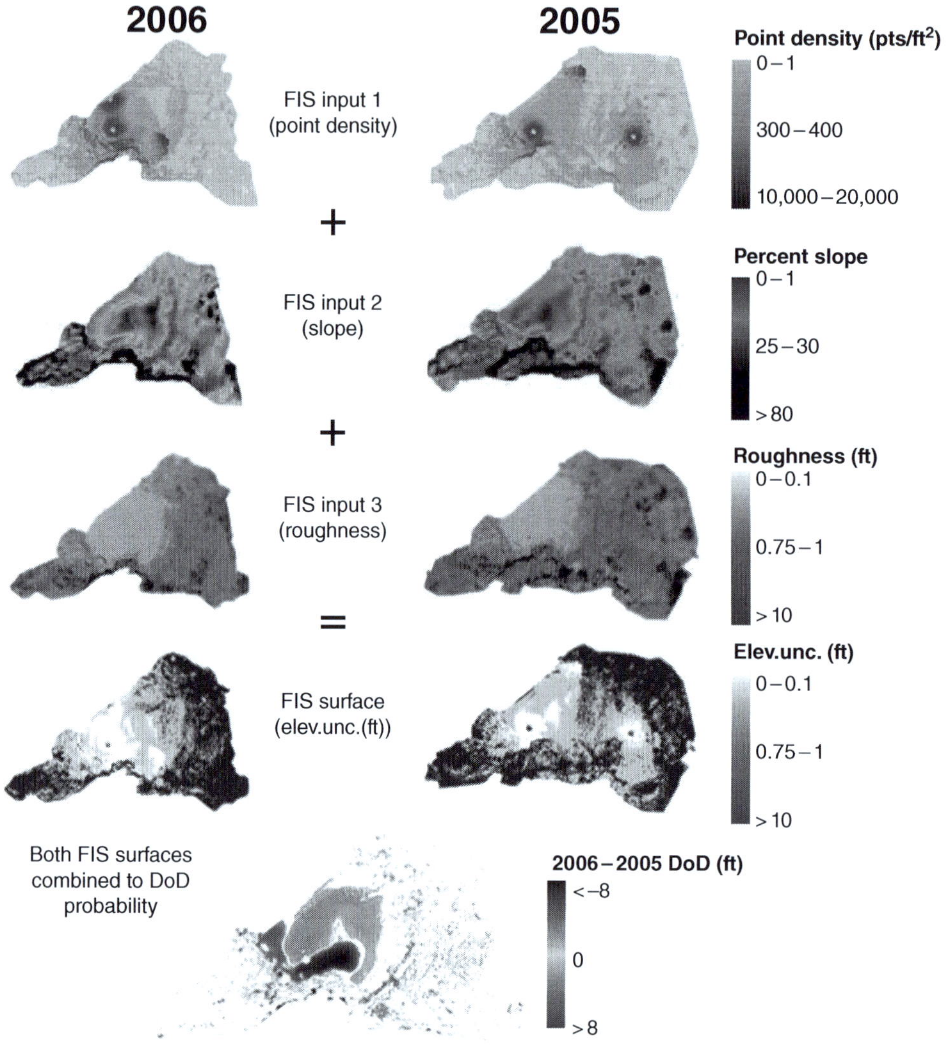

Figure 5.5 An example of spatially variable inputs (point density, slope and roughness) derived from point clouds and used as inputs into a fuzzy inference system (FIS) model of elevation uncertainty (elev. unc.) to produce a spatially variable error models (at the grid-size scale or by cell) for each DEM. (*Source*: Leary, Wheaton, and DeMeurichy (2012). Reproduced with the permission of Joe Wheaton.)

$$t = \frac{Z_i - Z_{i+1}}{\sqrt{\delta Z_{DEMi}{}^2 + \delta Z_{DEMi+1}{}^2}} \tag{5.10}$$

This approach generates a spatial map of *t-scores* that can be used to threshold the vertical changes as predicted by the DoD. For example, under a two-tailed test, a *t-score* of 1.96 or greater (where the degrees of freedom are taken to be large, i.e. $n_x \times n_y > 30$) would indicate changes that are significant at the 95% confidence limit. In more general terms, this approach enables the definition of a flexible $\mathrm{LoD_{min}}$. Given that ΣV_e and ΣV_d estimates are likely to be underestimates of total volumetric changes due to possible compensating changes between surveys (i.e. unobserved cut and fill; Lindsay and Ashmore, 2002), there is a strong argument for using slightly more relaxed confidence intervals ~80% (Mapstone 1995; Bradford, Korman, and Higgins 2005).

Once a threshold level of detection has been established, it can be applied to calculate the volumetric sediment budget in a variety of ways. The most restrictive and least common approach is to subtract the portion of elevation change below the threshold from the total change (Milan *et al.* 2011). More commonly, the LoD_{min} threshold is used simply to identify cells where the elevation difference is taken to be true and the volumetric changes for these cells are preserved and summed to generate the net budget. Finally, although rarely implemented, it is possible to weight the values of cells below the threshold and include only part of their values.

5.3.1.3 Geomorphic Change Detection: Workflow and Software

The basic geomorphic change detection workflow described above and illustrated in Figure 5.6 has become a well-accepted and standardized method for DEM differencing and volumetric budgeting (Williams 2012; Passalacqua *et al.* 2015). The main steps in this basic workflow comprise:

1) building orthogonal DEMs;
2) estimating error surfaces independently associated with each DEM;
3) calculating the DEM difference error propagation on a cell-by-cell basis;
4) considering the change detection problem and thresholding to differentiate changes at a given confidence level;
5) determining the zonal statistics and geomorphic interpretation to summarize the 'significant' changes into erosion (ΣV_e) and deposition (ΣV_d) estimates and calculate the net budget between periods (ΔV_{DoD}).

This approach and closely related variants have been variously described in Lane, Westaway, and Hicks (2003), Brasington, Langham, and Rumsby (2003), Wheaton *et al.* (2010), Milan *et al.* (2011), and James *et al.* (2012). The workflow can be implemented in a variety of GIS and scripting software environments and the Geomorphic Change Detection (GCD) software (http:///gcd.joewheaton.org) provides a project-based software platform to undertake such analyses (Wheaton *et al.* 2010; Williams 2012). The most common version of the software is available as an ArcGIS 10.X add-in, but command line, stand-alone, web applications (http://zcloudtools.org and http://opentopograhy.org) also exist.

5.3.2 Extracting DEMs from Hyperscale Point Clouds

Advances in survey technologies have made it progressively easier to acquire three-dimensional point clouds that capture terrain at extreme resolutions (mm–cm spacing) and over increasingly large spatial extents (Bangen *et al.* 2014b). These developments have stemmed from a range of evolving technologies that include terrestrial laser scanning (TLS; e.g. Milan, Heritage, and Hetherington 2007; Williams *et al.* 2011, 2013; Brasington, Vericat, and Rychkov 2012); structure-from-motion photogrammetry (James and Robson 2012; Westoby *et al.* 2012; Fonstad *et al.* 2013; Javernick, Brasington, and Caruso 2014; Woodget *et al.* 2014; Smith *et al.* 2014; Smith and Vericat 2015; Dietrich 2015; Smith, Carrivick, and Quincey 2015) and multi-beam sonar (Parsons *et al.* 2005; Grams *et al.* 2013; Kaplinski *et al.* 2014). Brasington, Vericat, and Rychkov (2012) used the term hyperscale, in an analogy to the remote sensing community's reference to hyperspectral cf. multispectral imaging (> 4 bands), to highlight that such high-density point clouds facilitate multiple different spatial scales of analysis. Although their acquisition over larger spatial extents was initially limited, mobile platforms (e.g. mobile TLS applied to fluvial studies; Wang *et al.* 2013) and the proliferation of consumer-grade

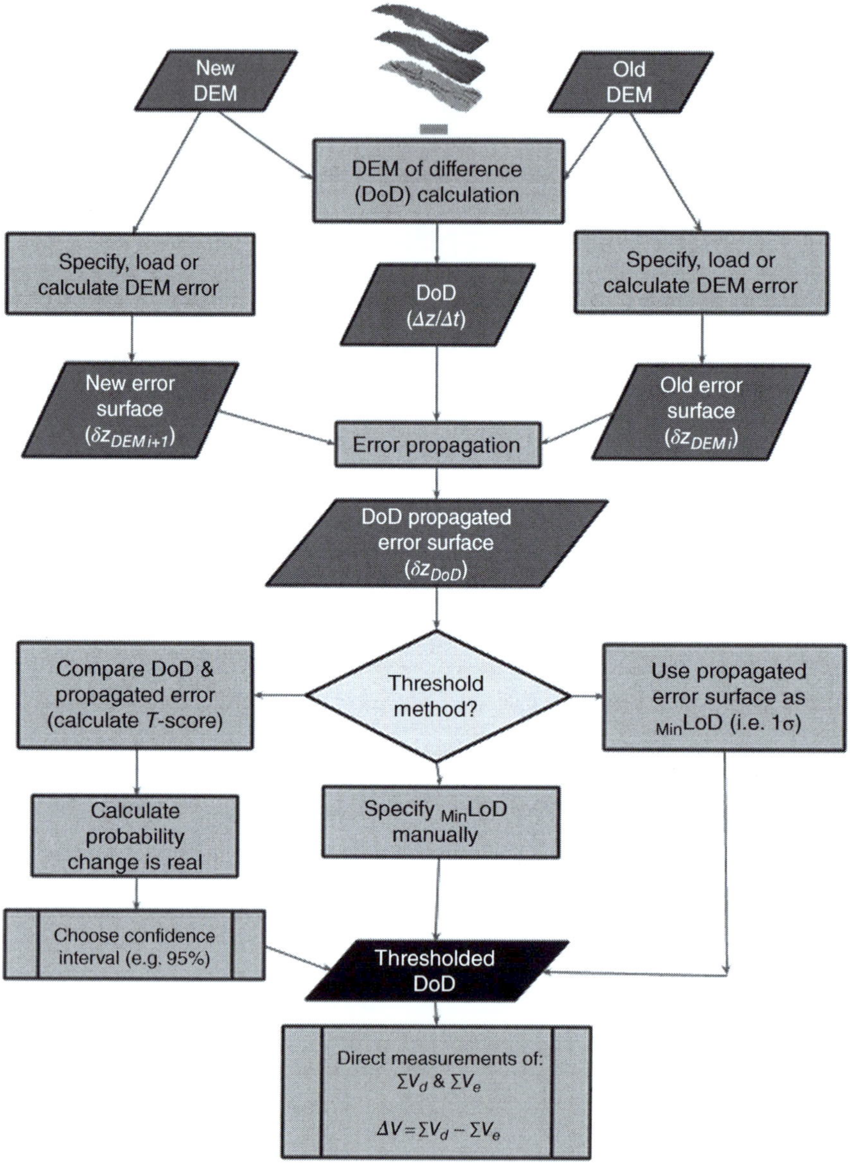

Figure 5.6 A standardized geomorphic change detection workflow facilitated within the geomorphic change detection (GCD) software. The outputs of ΣV_d and ΣV_e are used to derive ΔV; ΔV can be used for 'budget-based' application of the morphological approach, whereas ΣV_e can be used for path-length-based application. (Modified from Wheaton *et al.*, 2013.)

unmanned aerial vehicles (Woodget *et al.* 2015) have allowed data acquisitions over increasingly larger reaches. Many of these new survey technologies also have the advantage of being 'non-contact' and enable the extraction of high-density point clouds (i.e. hundreds to thousands of points per square metre).

Passalacqua *et al.* (2015) review the key steps used to convert point clouds to reliable bare-earth DEMs, which typically involves unifying and georegistering point clouds acquired from different perspectives, vegetation and surface object removal, and manual cloud editing. The automation of this process is complicated further by the large data volumes involved, so that the adoption of dense point clouds for change detection has been comparatively slow. However, some practical tools to support workflows are starting to emerge, such as PySESA (Buscombe, Grams, and Smith 2015), CloudCompare (Girardeau-Montaut 2011) and CANUPO (Brodu and Lague 2012; Lague, Brodu, and Leroux 2013). The Topographic Point Cloud Analysis Toolkit (ToPCAT) offers a computationally efficient geospatial toolkit to perform intelligent decimation of Tb-sized point cloud datasets that creates GIS-compatible data models that retain statistical information about the subgrid elevation distribution (Brasington, Vericat, and Rychkov 2012). Figure 5.7 illustrates the ToPCAT workflow and the outputs generated. This decimation algorithm is available in a distributable form as a simple standalone toolbox and is also implemented in the GCD software described previously. Performance tests on the distribution version of the code by Brasington, Vericat, and Rychkov (2012) revealed its suitability for desktop deployment without the need for high-performance computing. ToPCAT has been successfully used in several studies to provide microscale information about surface topography (Williams *et al.* 2011, 2013, 2014, 2015; Bangen *et al.* 2014b; Javernick, Brasington, and Caruso 2014; Vericat, Smith, and Brasington 2014; Storz-Peretz and Laronne 2015; Smith and Vericat 2015).

5.3.3 What More Might Hyperscale Data Reveal?

While simply being able to produce DEMs from hyperscale point clouds is one challenge, the real question is whether DEMs derived from point clouds up to two orders of magnitude denser than typical HRT data offer any advantages from a change detection perspective and whether this helps matters for estimating transport rates morphologically. As an illustrative example, Figure 5.8 compares the patterns and estimates of channel change based on TLS and RTK-GPS-derived DEMs of the River Feshie acquired in 2006 and 2007 (see Wheaton *et al.* (2010) and Brasington, Vericat, and Rychkov (2012) for further information about data acquisition). For the TLS data, thresholds for DoDs were established using the locally detrended standard deviation of elevation to define a representative DEM uncertainty on a cell-by-cell basis (derived using ToPCAT; see Figure 5.7). For the RTK-GPS surveys, a uniform standard deviation of error was estimated by using a bootstrap test, using 5% of the GPS points as check data. These errors were used to define a 1σ LoD_{min} threshold (i.e. a 68% confidence interval), below which changes were interpreted as indistinguishable from noise.

The resulting contrast in DoDs are shown in Figure 5.8. The most notable differences are observed in the degree of low-magnitude elevation changes that the higher precision, hyperscale TLS data are able to reveal. For example, there is 35% more estimated erosion in the TLS ($\Sigma V_{e_{GPS}} = -7141\text{m}^3$ vs. $\Sigma V_{e_{TLS}} = -10\,992\,\text{m}^3$) and 35% more estimated deposition in the TLS ($\Sigma V_{d_{GPS}} = 4618\,\text{m}^3$ = vs. $\Sigma V_{d_{TLS}} = 7144\,\text{m}^3$). Although these overall changes in erosion and deposition volumes do not have an impact on the direction of change for the net signal, differences exist between the two techniques. These differences reveal fine-scale sheets of deposition on the braid plain and bar top, as well as lower-magnitude sculpting of the bar surfaces (e.g. minor chute dissection and head cutting that the GPS survey did not pick up; Figure 5.8b).

The impact of using these values in the morphological approach depends on whether transport rates are estimated using sediment budgeting principles (i.e. using the net changes ΔV_{DoD}) or are path-length based (i.e. using just the ΣV_e). In the above example, overall flux estimates will be lower with the lower-precision GPS data for both because the magnitudes of ΔV_{DoD} and ΣV_e quantities are both

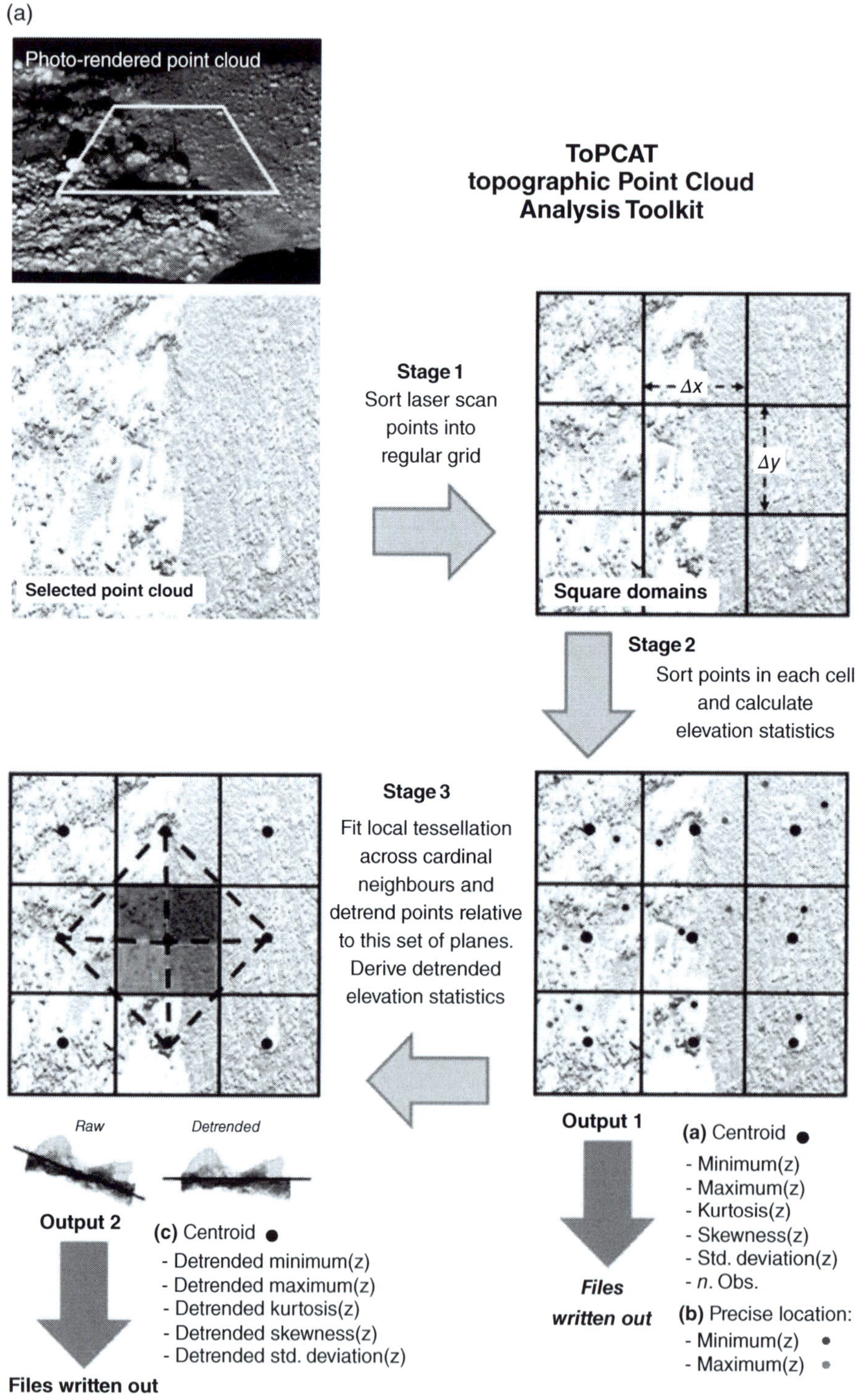

Figure 5.7 Example of processing hyperscale point cloud data to drive morphological approach. (a) The ToPCAT workflow for filtering three-dimensional point clouds and deriving useful derivatives. (Figure reproduced from Brasington, Vericat, and Rychkov 2012) (b) Applying ToPCAT in the Upper River Cinca (www.morphsed.es): from the point cloud to a 1 m grid DEM and roughness map. The roughness map is based on the detrended standard deviation (dSD) of elevation, while the DEM is based on the minimum elevation (*z*) in each square domain using ToPCAT (see a).

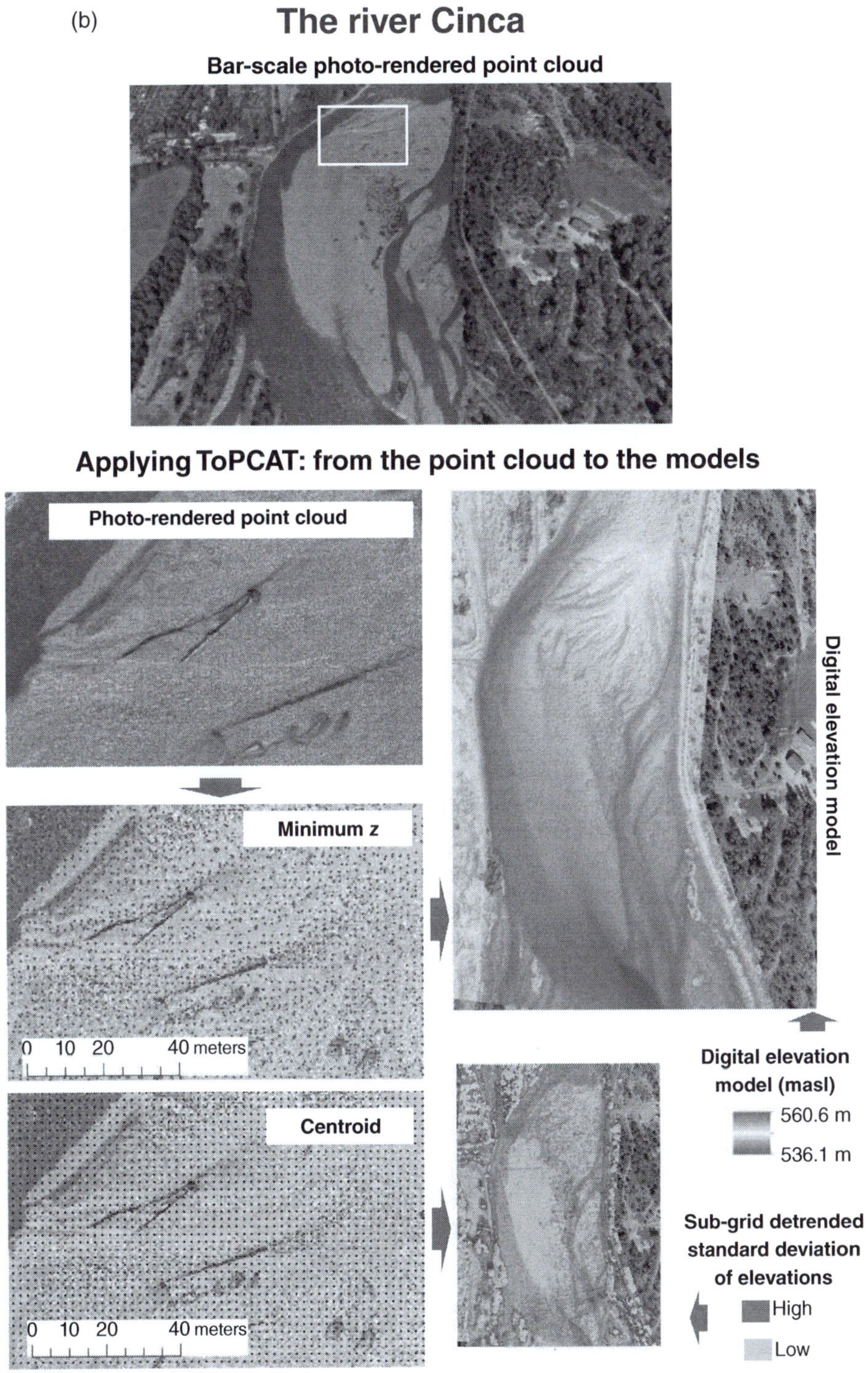

Figure 5.7 (*Continued*)

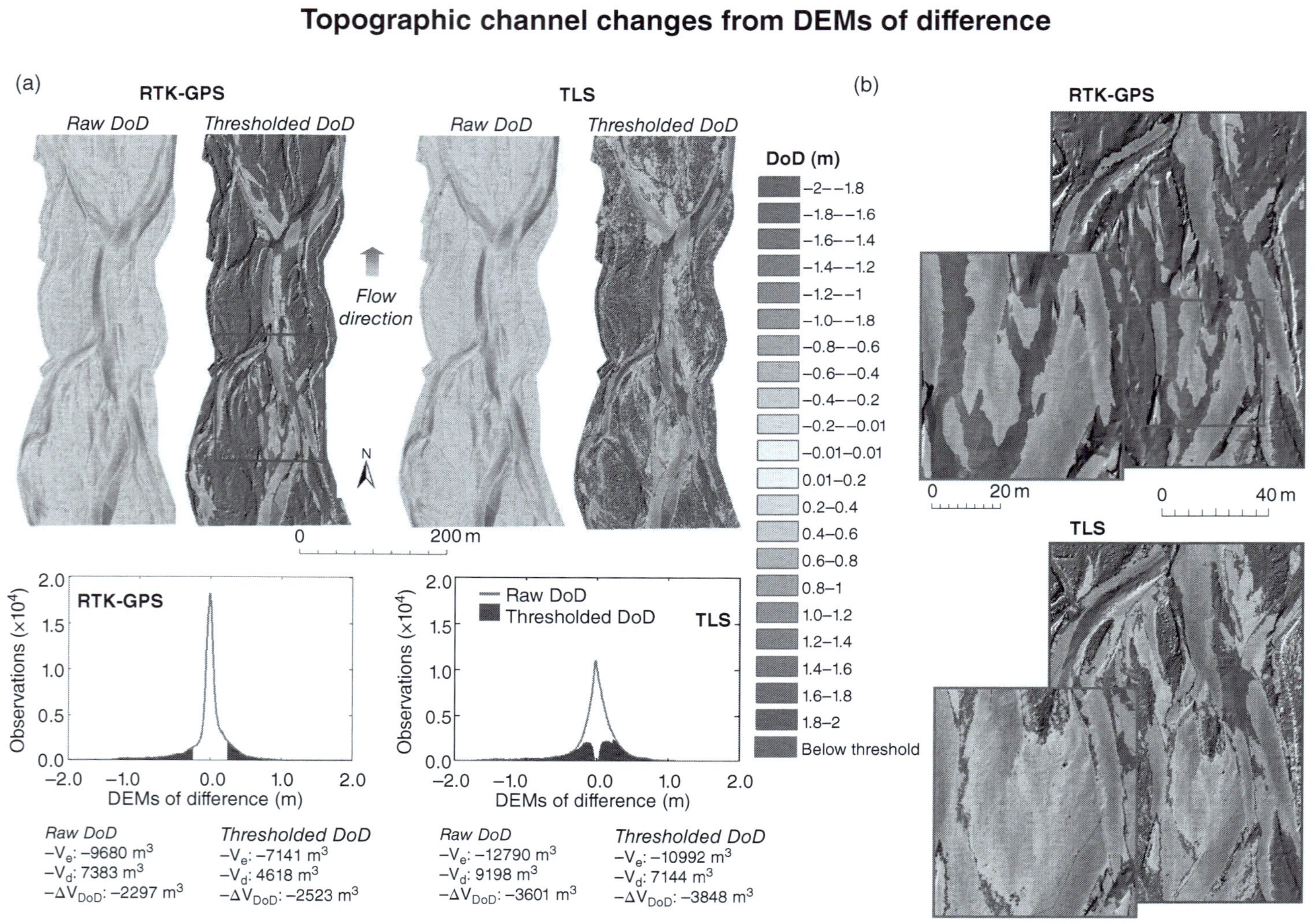

Figure 5.8 Illustration of additional erosion and deposition mechanisms identified with higher precision, hyperscale TLS HRT data versus traditional RTK-GPS HRT. (a) Comparison of reach-scale DoDs derived from RTK-GPS versus TLS DEMs for same time periods (2006–2007) from the Feshie River (Scotland). (b) Detailed close-up of changes on a gravel bar. (See insert for color representation of this figure.)

lower. However, this is partly an artefact of this budget having only a weak degradational trend in the context of the total volume of change (i.e. $\Sigma V_e + \Sigma V_d$). In general, the impacts of the lower precision data will be an even smaller minimum transport rate.

5.3.4 Case Study: High Frequency Morphodynamics of the Braided Rees River

While technical advances in geomatics have enabled progressive increases in the spatial frequency of survey data, the calculation of net sediment budgets is also known to be negatively biased by local compensation of scour and fill that occurs between surveys and is unaccounted for within the topographic budget (Lindsay and Ashmore 2002). In this section we use a case study of morphodynamics on the braided Rees River (New Zealand) to illustrate an application of the morphological method with event-based, hyperscale topographic monitoring. A principal goal of this research was to use the morphological approach to support the development of a bed-material rating curve for a large braided river in which direct measurements of flux during floods are practically impossible. As such, the case study provides a dual illustration of the utility of the morphological method and how hyperscale terrain data can support that goal.

The study period comprised an annual flood cycle, during which 10 flood events were captured. DEMs were derived using a combination of mobile TLS and optical-depth mapping, and DoDs produced and thresholds set using a spatially variable LoD_{min} (for further details see Williams *et al.* (2014) and the illustrative example in Figure 5.2 and Table 5.1). Figure 5.9 reveals that a strong net erosional signal is evident for the periods with the largest events, which together account for over 50% of the total erosion during the annual period. In addition to volumetric trends, erosional and depositional patterns can be quantified in terms of area and average depths. Figure 5.9b reveals that the average local depth of erosion and sedimentation is largely consistent between events, despite the widely varying totals of erosion and sedimentation (Figure 5.9a). This points towards a similarity of processes driving the overall pattern of bed mobility, but as indicated by the changing area between events, the volumes are a function of this pattern being dispersed over wider areas as the wetted width increases with discharge (a reflection of the local relief of the channels). This mirrors a similar observation to Ashmore and Sauks' (2006) observations of hydraulic geometry on the Sunwapta River (Canada) in that increasing discharge is accommodated largely by increases in wetted width, and so that no significant increases in mean bed shear stress would be expected and that total transport may scale simply with width (similarly patterns were also observed in an earlier study by Mosley in 1982). Additionally, results from the Rees also show that the average depth of erosion and deposition is approximately half the bankfull channel depth, implying that local stalling of bed-material sheets is likely to reduce local conveyance and encourage channel widening or avulsion within 1–2 events.

The distribution of erosion and deposition across the floodplain of the River Rees for the entire study period is presented in Figure 5.10a. The net annual DoD reveals that 65% of the bed appears to have undergone either significant erosion or deposition through the flood season. However, when these changes are integrated sequentially, event-by-event, the actual areas found to have experienced erosion or deposition were 71% and 72% respectively. This mismatch is explained by the fact that a significant proportion of the braidplain (~50%) experienced repeated cut and fill cycles (Figure 5.10b), where in some cases, the total net bed deformation summed to zero. These results reveal the importance of the frequency of topographic surveys for the estimates of channel changes. Compensating erosion and deposition might mask processes if the frequency of surveys is not appropriate, as has been revealed in previous studies and clearly presented here in Figure 5.10b.

Effective scaling of morphological change into estimates of bed-material transport requires information on the typical step length of particle transport or a reference sediment flux in a given

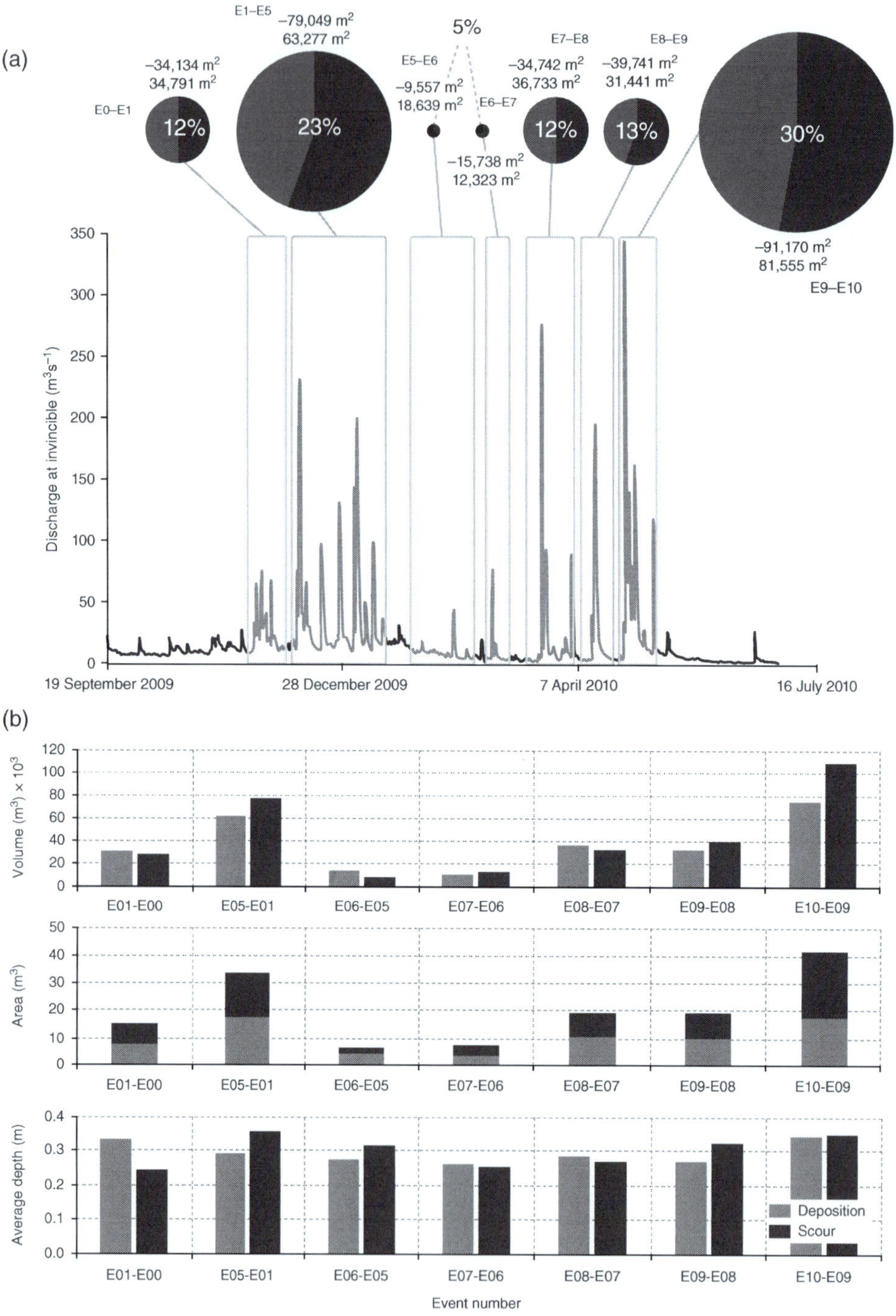

Figure 5.9 Event-scale DoD from the River Rees (New Zealand) during a flood season (2009–2010). (a) A hydrograph over the study period, with boxes bracketing the epochs of analysis and pie charts showing the relative proportions of erosion and deposition for each period. The percentages indicate the proportion of event-based geomorphic change to the total during the study period. (b) Contrast of the event-based volumes, areas, and average thickness of erosion and deposition.

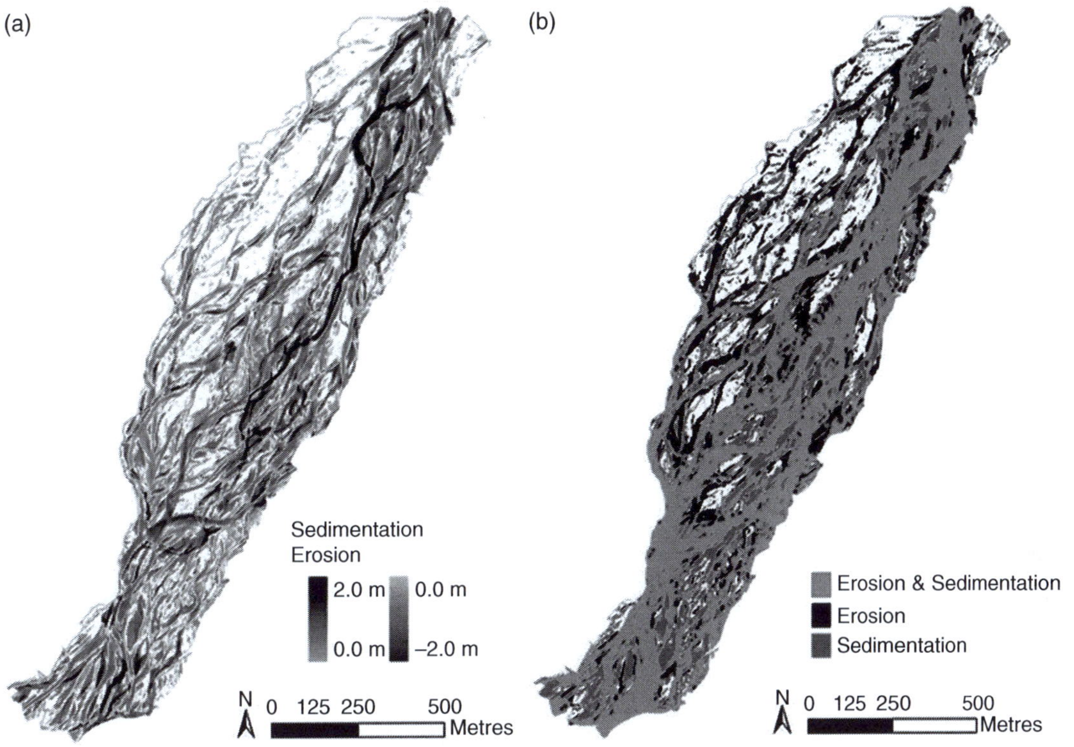

Figure 5.10 (a) Net annual thresholded DoD from the River Rees and (b) classification of regions that experienced just erosion, just sedimentation, versus a more complicated, event-transgressive mix of erosion and sedimentation during a flood season (2009–2010). (See insert for color representation of this figure.)

section (e.g. as illustrated for two of the Rees events in Figure 5.2). However, as reviewed in section 5.2, both requirements are complicated by the difficulties of field data acquisition and have numerous uncertainties that directly affect estimates derived from the morphological approach. In the following sections we present a proxy for the step length of particle transport based on the multi-event HRT database acquired for the River Rees and developed by Redolfi (2014). We then apply this method to the entire data set to estimate flood-scale bed-material transport and investigate the relationship between transport flux and measures of flow intensity.

5.3.4.1 Flood-Based Estimates of Particle Step Lengths

As discussed above, Pyrce and Ashmore (2003) highlighted the variability in the relationship between path-length distributions and channel morphology in gravel-bed rivers; although they note that at higher flows travel distances may be related to specific morphological units (e.g. bar-pool spacing). From a flume study, Kasprak *et al.* (2015) showed results supporting the idea that step-lengths are highly related to bar spacing and bar heads were consistent sinks for tracer particles. Establishing path-length data from long-term field monitoring is constrained by logistics and in wider gravel-bed rivers the recovery rate of tracers is low and decreases with time due to the spatial extent of dispersal and depth of burial (Laronne and Duncan 1989, 1992). As an attempt to circumvent these limitations, Redolfi (2014) presented a proxy for measuring flood-based step lengths of particle transport by characterizing the size-distribution of discrete areas of erosion using image processing (Figure 5.11a). Briefly, discrete areas of erosion were first isolated using a segmentation of the DoD (i.e. classifying cells that represent erosion, deposition, and no significant change). Erosional features were quantified by fitting an ellipse to each erosion patch from which the major axes could be computed. The length of this ellipse was then weighted according to the volume of erosion to yield the proxy of the step length (i.e. L; see Redolfi (2014) for more details). Analysis of this metric through the annual series of events registered in the River Rees is presented in Figure 5.11b. The results reveal that this method scales strongly with flow magnitude, varying from around 50 m for the low-intensity events registered in the period E05–E06 and E06–E07, through more than 250 m for the high-intensity events E01–E05 and E09–E10 (Figure 5.11b). This approach differs from directly matching centres of erosion and deposition as the complex patterns of channel observed precluded simple identification of sediment dispersal patterns. While a crude proxy, the scale dependence of this relationship with discharge is encouraging and may provide a means to estimate a lower bound estimate of particle step length, and opens the possibility to relate step lengths to flow magnitude using data intrinsic to a DoD.

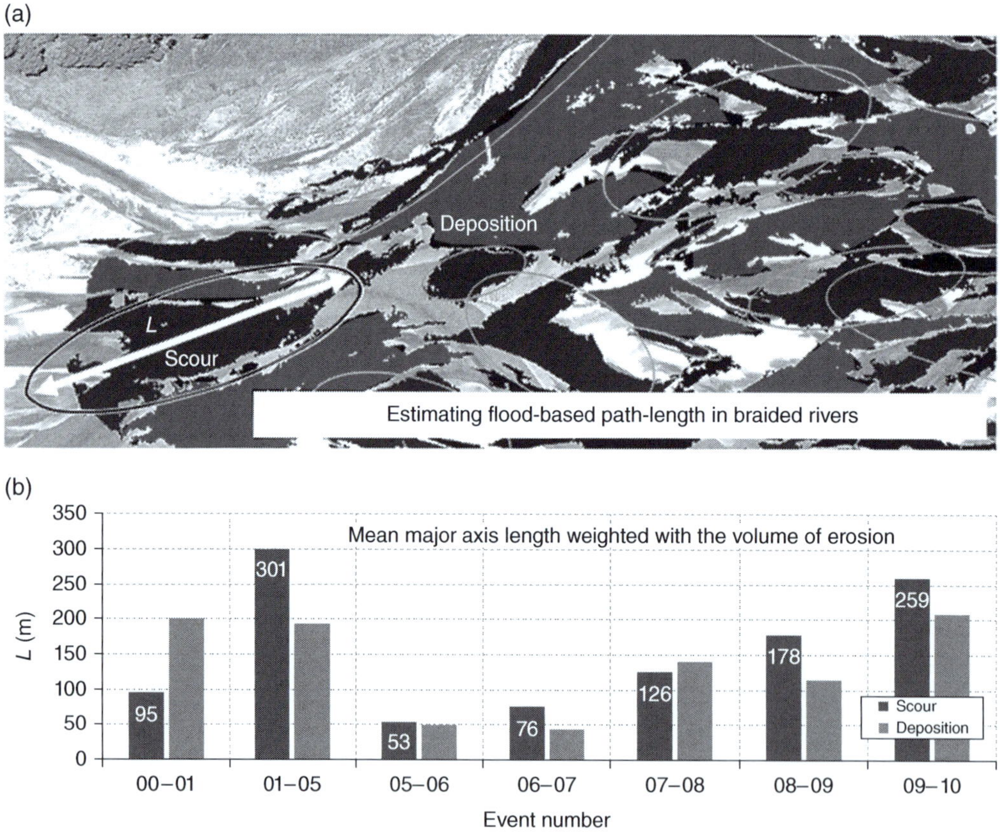

Figure 5.11 (a) Example of estimating flood-based path-lengths (L) of particle transport by characterizing the size-distribution of discrete areas of erosion using image processing (see text for more details). This approach was first developed by Redolfi (2014). (b) Assessing flood-based particle step lengths by applying the approach presented in A to the River Rees data set (Figure 5.9).

This approach is still being critically reviewed and it would require a comparison with results obtained from tracers. Painted and radio frequency identification system (RFID) tracer data obtained in the River Rees are being analysed to compare the distributions of both path-length models.

5.3.4.2 Reach-Scale Bed-Material Flux and Flow Intensity

Event-based total bed-material flux in the River Rees was estimated using the step-length weighting model introduced in the previous section and the volume of erosion derived from Figure 5.11b (recall the step length formulation shown in Figure 5.1). To obtain an estimate of the bed-material transfer, mean flood-based particle step length was multiplied by total erosion-patch volume. This estimate is turned into a section-flux by dividing by the reach length and the event duration. However, the competence duration of each event may have a direct influence on the flux estimate. This approach yields a storm-by-storm bed-material flux that can then be related to measures of flow intensity. Simple rating relationships for reach-scale transport are shown in Figure 5.12, where measures of flow intensity have been normalized so that the range 0–1 corresponds to the maximum and minimum observed intensity during the study period. These results indicate that the fit to peak discharge and flow duration have strongly positive relationships, but with significant scatter, particularly for the high-intensity events. However, by integrating the flow through time into a combined measure that characterizes the dimensionless total energy expenditure (~ stream power) during each event, demonstrates a much stronger relationship to the estimated transport flux (Figure 5.12a). Weighting this index by an exponent reveals that the strength of the relationship increases as the exponent is raised from unity (Figure 5.11a) through to 2 (Figure 5.12b), a pattern that is mirrored in flume experiments (e.g. Bertoldi, Ashmore, and Tubino 2009). These results show that by combining a proxy for measuring flood-based step lengths of particle and high-resolution topographic changes, bed-material fluxes can be estimated, compared, and correlated with different measures of flow intensity.

5.4 Discussion

5.4.1 New Technologies, Old Problems

Since the 4th Gravel Bed Rivers Workshop when Ashmore and Church (1998) reviewed the morphological method we have gone from an era of one-dimensional abstractions of rivers with cross sections, longitudinal profiles, planform maps, and occasionally all three (e.g. Brewer and Passmore 2002), to spatially continuous DEMs (2.5-dimensional surfaces; e.g. Lane and Chandler 2003), to three-dimensional terrains and high-resolution point clouds (e.g. Lague, Brodu, and Leroux 2013). We term this period starting in the mid- to late 1990s, the high-resolution topography revolution. The geomorphology community did not start this revolution, but has benefited from technological advances in both data acquisition (e.g. SfM, TLS) and post-processing that have been driven principally by other applications and industries (Bangen *et al.* 2014b). While continuous topographic surveys have been tractable for decades, if not centuries (e.g. plane-table survey since at least the 1550s; photogrammetry with stereoscopes since at least the 1950s), it was not until the 1990s that ground-based surveys with total-stations became cost-effective and practical so that researchers were able to implement topographic surveys of gravel-bed rivers (e.g. Lane, Chandler, and Richards 1994; Milne and Sear, 1997). At approximately the same time, GIS and remote sensing software enabled the ready conversion of these data into DEMs, and these datasets proliferated following the declassification of precise GPS. The first commercial LiDAR followed shortly thereafter. This confluence of technological,

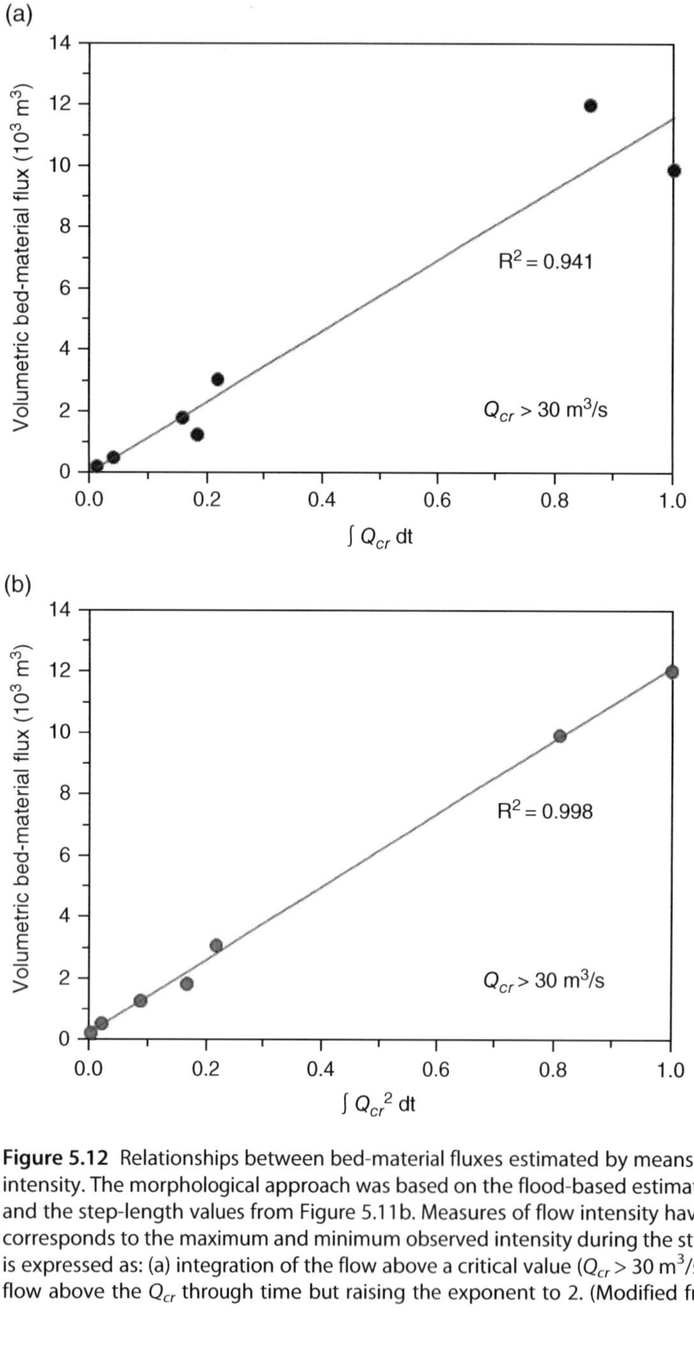

Figure 5.12 Relationships between bed-material fluxes estimated by means of the morphological approach and flow intensity. The morphological approach was based on the flood-based estimates of bed erosion presented in Figure 5.9 and the step-length values from Figure 5.11b. Measures of flow intensity have been normalized so that the range 0–1 corresponds to the maximum and minimum observed intensity during the study period (i.e. E00 to E10). Flow intensity is expressed as: (a) integration of the flow above a critical value ($Q_{cr} > 30$ m^3/s) through time and (b) integration of the flow above the Q_{cr} through time but raising the exponent to 2. (Modified from Refolfi 2014.)

remote sensing, and surveying advances spawned the HRT revolution that so many geomorphologists duly followed. What could be a more perfect battle to fight? This was a revolution that required field work, but also allowed us to quantify landscapes, landforms, and their changes like never before (Roering *et al.* 2013; Tarolli 2014).

What is perhaps surprising is that while the morphological approach has been established for over 50 years, the application of the approach with HRT data (e.g. Anderson and Pitlick 2014) remains relatively rare in the literature. Nonetheless, methods to produce high quality DEMs from HRT and supporting change detection analyses have matured (Passalacqua *et al.* 2015), therefore applications of these data to the morphological approach need to be evaluated. One major impediment to utilizing HRT data for application of the morphological approach is a lack of basic geomatics training and practical surveying knowledge (Passalacqua *et al.* 2015). However, this should be possible to address. Most of the undergraduate and graduate programmes that produce researchers who study fluvial geomorphology and gravel-bed rivers do not provide adequate (or in many cases any) training in geomatics and surveying. Many papers already published contain a disproportionate amount of text explaining what was done to overcome simple surveying blunders and mistakes, such as datum busts, miss-matched flight lines, inconsistent control networks between surveys, etc. Many of these errors are avoidable and standard practices and workflows exist to streamline the post-processing and change detection analyses.

Application of the morphological approach in many fluvial environments is complicated by long-standing challenges in simply surveying those environments. In some non-wadeable rivers, techniques for acquiring bathymetry, such as spectral-depth correlation from imagery (e.g. Legleiter, Roberts, and Lawrence 2009), green LiDAR (e.g. McKean *et al.* 2009) and multibeam echosounding (e.g. Hazel *et al.* 2006), are not always feasible. As such, errors associated with extracting the topography of wet channels can be considerably greater than for dry beds, where more accurate methodologies and workflows can be applied (Williams *et al.* 2013).

5.4.2 Channel Morphodynamics: Beyond Quantifying Flux

As we begin to emerge from this HRT revolution, it is important to refocus our efforts around our basic geomorphic curiosities. Arguably, we have become overly focused on measuring and estimating the same old values at the continuum scale: net change in storage (ΔV) and sediment transport rate (Q_b). While these metrics provide useful insights into the processes shaping gravel-bed rivers, they are average expressions of morphodynamic processes that we know vary significantly over both space and time. We contend that the full value of repeat topography lies in using the rich spatial patterning these data reveal, to better understand river morphodynamics and longitudinal patterns of sediment transport in the context of other specific erosion and deposition mechanisms (i.e. the dominance of lateral vs. vertical adjustments, or catastrophic changes, e.g. avulsions). Morphodynamics are defined as the changes to morphology (i.e. topography) as a result of these processes through time and are thus reflected in snap-shots of topography and sediment in storage, which ultimately give rise to distinctive assemblages of geomorphic units (Wheaton *et al.* 2015). There remains much to learn about how longitudinal variations in sediment trasport rates, revealed by the morphological method, give rise to specific assemblages of these geomorphic units (river styles) and, in turn, how these units control transport rates. There are also other mechanisms by which topography can be modified either naturally (e.g. tectonic uplift, isostasy, compaction, deflation) and by humans (e.g. gravel mining or extraction, fill, compaction, tilling) or ecosystem engineers (e.g. beaver, willows), but these are not necessarily the direct result of morphodynamics (Merz, Pasternack, and Wheaton 2006). Wheaton *et al.* (2013) defined a morphodynamic signature as 'a distinct mechanism of erosion or deposition that leads to a consistent morphological response'. Moreover, Wheaton *et al.* (2013) and Williams *et al.* (2015) have demonstrated how repeat topography can be used to classify net changes in topography (i.e. net erosion and sedimentation) into specific morphodynamic signatures such as braiding mechanisms, bank erosion, bar-edge trimming, channel incision, confluence pool scour, overbank gravel/cobble sheets, and lateral bar development (e.g. Figure 5.13).

5.4.3 Future Opportunities and Challenges

While HRT is beginning to open doors to facilitate (and extend) the morphological method, enabling more accurate, better resolved, more spatially extensive estimates of fluvial processes and mechanisms of change, these efforts have barely started to help us better understand how gravel-bed rivers really work. Although there is scope for continued improvements in topographic data acquisition, we need to be very careful what we ask, because we just might get it. What if we could have HRT point clouds for every river we wanted at subevent frequency? The result could be data overload that might entail more methodological reflection and redefinition of workflows, instead of a focus on what these data actually reveal about the links between morphology and transport. Arguably, technological progress means that this prospect is neither fanciful nor does it lie in the too distant future. So there is surely some methodological urgency in building algorithms for post-processing HRT data, estimating errors, performing uncertainty analysis, and providing reliable change detection outputs in a manner that is

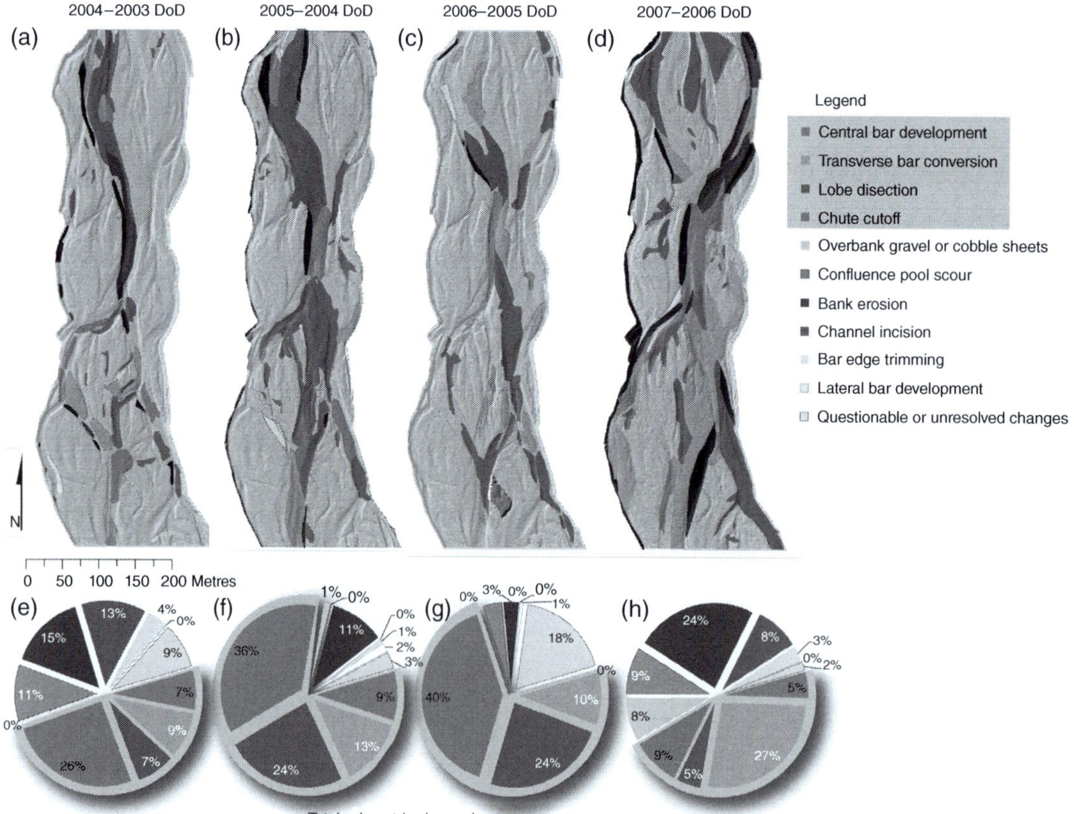

Figure 5.13 Example of using budget segregation of DoD to quantify morphodynamic signatures. (Figure reproduced from Wheaton *et al.* 2013.) (See insert for color representation of this figure.)

not just possible with analytical brute force, but in a scalable, computationally efficient manner that transcends scale and the volume of data we are quickly amassing. Efforts to make algorithms more scalable and accessible from a variety of web-based, super-computer-backed and cloud-computing solutions (e.g. http://zcloudtools.org) are scratching the surface in terms of building this capacity and we believe are the future to building the capacity our community needs.

5.5 Conclusions

The inverse approach to estimating bed-material transport, known also as the morphological approach, was first stated formally nearly a century ago. Nonetheless, application of this method has been relatively rare and limited by the high labour demand and costs associated with undertaking slow topographic surveys. The morphological approach was originally implemented with planform mapping, scour chains, and sparse monumented cross sections. As we transition from the battle to acquire better topographic data, to a reality in which HRT surveys are commonplace, the morphological approach now has more relevance than ever before as a viable and pragmatic tool. Repeat HRT surveys can be effectively utilized now to support change detection, and the application of the data using the morphological approach to estimate bed-material fluxes is now possible. When coupled, in the traditional sense, with a reference sediment flux or particle step lengths, reliable estimates of net bed-material transport can be made without direct measurement and local sediment budgets may also be elucidated. Moreover, new workflows, as demonstrated here for the Rees River, can generate bed-material flux estimates based entirely on DoDs, through simultaneous measurements of erosion volumes that also support the definition of proxies for particle step lengths. Such methods clearly require further evaluation, but are ultimately no more conjectural or subject to more severe uncertainties than direct methods of sediment transport estimation, which moreover cannot be readily applied in large rivers.

The adoption of HRT data analysis must, however, address the all-to-familiar amateur surveying blunders, post-processing mistakes, and ineffective study designs. Information and methods exist to avoid these costly mistakes and as a community we need to pay more attention to basic data acquisition protocols in order to generate more effective data to address our key research questions. Such progress would enable wider application of the modern morphological approach not just to estimate sediment flux and net changes, but also to better understand the underlying morphodynamic mechanisms responsible for shaping gravel-bed rivers. With the modern morphological approach, we are well positioned to exploit the explosion of repeat HRT survey data to generate and test the next generation of hypotheses and theories about how gravel-bed rivers work.

Perhaps more important than just being able to measure topography at ever increasing temporal and spatial resolution just because we can, is the question of what can all that data tell us that we did not already know about gravel-bed rivers? What are the major theoretical advances we will be able to make because of such rich empirical topographic time series? Will this lead to new insights about the morphodynamic signatures and mechanisms by which gravel-bed rivers organize themselves in response to fluvial disturbance? Will we be able to take this newly found theoretical understanding and be able to build the next generation of predictive morphodynamic models of gravel-bed-river evolution? Will these be models so computationally complex we can barely model a handful of floods? Or will the empirical insights into the mechanisms by which gravel-bed rivers adjust reveal some clever computational short cuts we can take to better model these dynamics and focus on the most essential processes? At the moment these are questions that we can pose but not yet fully answer.

Acknowledgements

The authors wish to thank the organizers of the Gravel-Bed Rivers Workshop for inviting this contribution and being so patient with our manuscript submission. The work presented in this chapter was supported by different research projects and contracts funded by a UK NERC Grant NE/G005427/1 (ReesScan), the Spanish Ministry of Economy and Competiveness and the European Regional Development Fund Scheme (MorphSed, www.morphsed.es; CGL2012-36394), research contracts funded by Endesa Generación S.A. (Spain), the National Science Foundation (Awards 1147942, 1226145 and 1226127), the US Geological Survey Grand Canyon Monitoring and Research Center (Awards: 08WRAG0053, G12AC20346), Idaho Power Company (IPC 4568), and Eco Logical Research, Inc (USU Award 100652). The ReesScan project was also supported by the NERC Geophysical Equipment Facility (award GEF 892), the New Zealand's Foundation for Science and Technology Grant C01X0308 and Spain's Ministry of Science and Innovation's Jose Castillejo travel fund. The first author is funded by a Ramon y Cajal Fellowship (RYC-2010-06264). James Brasington wrote this chapter with the support of a Leverhulme Trust International Academic Fellowship (IAF-2014-038). Many thanks to all members of the ReesScan team, different members from the Fluvial Dynamics Research Group (RIUS) at the University of Lleida and the members of Utah State University Department of Watershed Sciences, and Ecogeomorphology Topographic Analysis Laboratory for their valuable assistance during the different stages of the research presented here. The authors wish to thank the discussion authors and Mike Church for helpful conversations at the GBR conference, which helped us strengthen the manuscript during revisions. We are especially grateful to Jonathan Laronne and Daizo Tsutsumi for their patience, encouragement, comments and excellent editorial advice throughout the process that helped us complete the chapter. Authors are indebted to Chris Gibbins for reviewing the first version of this manuscript and Murray Hicks for a thoughtful and constructive review. We would also like to thank the authors of the discussions provided at the end of this chapter for their constructive comments.

References

Anderson, S and Pitlick, J 2014. Using repeat lidar to estimate sediment transport in a steep stream. *Journal of Geophysical Research – Earth Surface* 119, 621–643.

Ashmore, P and Sauks, E 2006. Prediction of discharge from water surface width in a braided river with implications for at-a-station hydraulic geometry. *Water Resources Research* 42(3), W03406.

Ashmore, PE and Church, M 1998. Sediment transport and river morphology: a paradigm for study. In *Gravel Bed Rivers in the Environment* (eds PC Klingeman, RL Beschta, PD Komar, and JB Bradley). Water Resources Publications LLC; Highlands Ranch, CO; 115–148.

Bangen, SG, Wheaton, JM, Bouwes, N *et al.* 2014a. Crew variability in topographic surveys for monitoring wadeable streams: a case study from the Columbia River Basin. *Earth Surface Processes and Land Forms* 39(15), 2070–2086.

Bangen, SG, Wheaton, JM, Bouwes, N *et al.* 2014b. A methodological intercomparison of topographic survey techniques for characterizing wadeable streams and rivers. *Geomorphology* 206, 343–361.

Bertoldi, W, Ashmore, P, and Tubino, M 2009. A method for estimating the mean bed load flux in braided rivers. *Geomorphology* 103(3), 330–340.

Bradford, MJ, Korman, J, and Higgins, PS 2005. Using confidence intervals to estimate the response of salmon populations (*Oncorhynchus* spp.) to experimental habitat alterations. *Canadian Journal of Fish and Aquatic Sciences* 62(12), 2716–2726.

Brasington, J, Langham, J, and Rumsby, B 2003. Methodological sensitivity of morphometric estimates of coarse fluvial sediment transport. *Geomorphology* 53(3–4), 299–316.

Brasington, J, Rumsby, BT, and Mcvey, RA 2000. Monitoring and modelling morphological change in a braided gravel bed river using high resolution GPS-based survey. *Earth Surface Processes and Landforms* 25(9), 973–990.

Brasington, J, Vericat, D, and Rychkov, I 2012. Modeling river bed morphology, roughness, and surface sedimentology using high resolution terrestrial laser scanning. *Water Resources Research* 48(11), 1–18.

Brewer, PA and Passmore, DG 2002. Sediment budgeting techniques in gravel bed rivers. In *Sediment Flux to Basins: Causes, Controls and Consequences* (eds S. Jones and L. Frostick). Special Publication 191, Geological Society: London; 97–113.

Brodu, N and Lague, D 2012. 3D terrestrial LiDAR data classification of complex natural scenes using a multi-scale dimensionality criterion: applications in geomorphology. *ISPRS Journal of Photogrammetry and Remote Sensing* 68, 121–134.

Buscombe, D, Grams, PE, and Smith, SM 2015. Automated riverbed sediment classification using low-cost sides-can sonar. *Journal of Hydraulic Engineering.* DOI:10.1061/(ASCE)HY.1943-7900.0001079 , 06015019.

Calder, BR and Mayer, LA 2001. *Robust Automatic Multibeam Mathymetric Processing.* Center for Coastal and Ocean Mapping and Joint Hydrography Center: Durham, NH; 20 pp. Available from http://www.thsoa.org/hy01/3_4.pdf

Calder, BR and Mayer, LA 2003. Automatic processing of high-rate, high-density multibeam echosounder data. *Geochemistry, Geophysics, Geosystems* 4(6), 1–22.

Carson, MA and Griffiths, GA 1989. Gravel transport in the braided Waimakariri River: mechanisms, measurements and predictions. *Journal of Hydrology* 109, 201–220.

Chapuis, M, Dufour, S, Provansal, M *et al.* 2015. Coupling channel evolution monitoring and RFID tracking in a large, wandering, gravel-bed river: Insights into sediment routing on geomorphic continuity through a riffle–pool sequence. *Geomorphology* 231, 258–269.

Church, M 2006. Bed material transport and the morphology of alluvial river channels. *The Annual Review of Earth and Planetary Science* 34, 325–354.

Church, M and Hassan, MA 1992. Size and distance of travel of unconstrained clasts on a streambed. *Water Resources Research* 28, 299–303.

Columbia Habitat Monitoring Program 2012. *Scientific Protocol for Salmonid Habitat Surveys Within the Columbia Habitat Monitoring Program.* Integrated Status and Effectiveness Monitoring Program, Terraqua, Inc.: Wauconda, WA, 188 pp. Available from http://champmonitoring.org/Program/RetreiveProgramDocumentFile/1/Protocol%20Documents/1113376526

Dietrich, JT 2015. Riverscape mapping with helicopter-based structure-from-motion photogrammetry. *Geomorphology* 252, 144–157.

Eaton, B and Lapointe, M 2001. Effects of large floods on sediment transport and reach morphology in the cobble-bed Sainte Marguerite River. *Geomorphology* 40(3), 291–309.

Emmett, WW and Leopold, LB 1965. Downstream pattern of river bed scour and fill. *Proceeding Federal Interagency Conference on Sedimentation*, Jackson, MS; 399–409.

Ergenzinger, P, Schmidt, K-H, and Busskamp, R 1989. The pebble transmitter system (PETS): first result of a technique for studying coarse material erosion, transport and deposition. *Zeitschrift für Geomorphologie* 33, 509–508.

Erwin, SO, Schmidt, JC, Wheaton, JM, and Wilcock, PR 2012. Closing a sediment budget for a reconfigured reach of the Provo River, Utah, United States. *Water Resources Research* 48, W10512.

Exner, FM 1925. *Über die Wechselwirkung zwischen Wasser und Geschiebe in Flüssen* (in German). *Sitzungsberichte der Akademie der Wissenschaften in Wien Abteilung IIA* 134, 165–203.

Federal Geospatial Data Committee 1998. *Geospatial Positioning Accuracy Standards. Part 3: National Standard for Spatial Data Accuracy*, FGDC-STD-007.3-1998: Federal Geospatial Data Committee; Reston, VA.

Ferguson, RI and Ashworth, P 1992. Spatial patterns of bedload transport and channel change in braided and near braided rivers. In *Dynamics of Gravel-Bed Rivers* (eds P Billi, RD Hey, CR Thorne and P Taconni). John Wiley & Sons: Chichester, UK; 477–492.

Fonstad, M, Dietrich, JP, Courville, B *et al.* 2013. Topographic structure from motion: a new development in photogrammetric measurement. *Earth Surface Processes and Land Forms* 38(4), 421–430.

Fuller, IC, Passmore, DG, Heritage, G *et al.* 2002. Annual sediment budgets in an unstable gravel bed river: the River Coquet, northern England. In *Sediment Flux to Basins: Causes, Controls and Consequences* (eds S Jones and L Frostick), Special Publication 191, Geological Society: London; 115–131.

Girardeau-Montaut, D 2011. *CloudCompare Open Source Project.* OpenSource Project, available from www.cloudcompare.net

Goff, JR and Ashmore, P 1994. Gravel transport and morphological change in braided Sunwapta River, Alberta, Canada. *Earth Surface Processes and Landforms* 19(3), 195–212.

Grams, PE, Topping, DJ, Schmidt, JC *et al.* 2013. Linking morphodynamic response with sediment mass balance on the Colorado River in Marble Canyon: issues of scale, geomorphic setting, and sampling design. *Journal of Geophysical Research: Earth Surface* 118(2), 361–381.

Griffiths, GA 1979. Recent sedimentation history of the Waimakariri River, New Zealand. *Journal of Hydrology (NZ)* 18, 6–28.

Haschenburger, JK 1996, *Scour and fill in a gravel bed channel: observations and stochastic models.* Doctoral Dissertation, University of British Columbia, Vancouver.

Hassan, MA 1990. Scour, fill, and burial depth of coarse material in gravel bed streams. *Earth Surface Processes and Landforms* 15(4), 341–356.

Hassan, MA and Ergenzinger, P 2003. Use of tracers in fluvial geomorphology. In *Tools in Fluvial Geomorphology* (eds GM Kondolf and H Piégay). John Wiley & Sons: Chichester, UK.

Hazel, JE, Topping, DJ, Schmidt, JC, and Kaplinski, M. 2006. Influence of a dam on fine-sediment storage in a canyon river. *Journal of Geophysical Research – Earth Surface* 111, F01025.

Heritage, G and Hetherington, D 2007. Towards a protocol for laser scanning in fluvial geomorphology. *Earth Surface Processes and Landforms* 32(1), 66–74.

Hicks, DM 2012. Remotely sensed topographic change in gravel riverbeds with flowing channels. In *Gravel-Bed Rivers: Processes, Tools, Environments* (eds M Church, PM Biron, and AG Roy). John Wiley & Sons: Chichester, UK; 303–314.

Hollingshead, AB 1971. Sediment transport measurements in gravel river. *Journal of the Hydraulics Division* 97(11), 1817–1834.

Hubbell, DW 1964. *Apparatus and Techniques for Measuring Bedload.* Water Supply Paper 1748, US Geological Survey, 74 pp. Available from http://pubs.er.usgs.gov/publication/wsp1748

James, LA, Hodgson, ME, Ghoshal, S, and Latiolais, MM 2012. Geomorphic change detection using historic maps and DEM differencing: the temporal dimension of geospatial analysis. *Geomorphology* 137(1), 181–198.

James, MR and Robson, S 2012. Straightforward reconstruction of 3D surfaces and topography with a camera: accuracy and geoscience application. *Journal of Geophysical Research – Earth Surface* 117, F03017.

Jang, JSR and Gulley, N 2007. *Fuzzy Logic Toolbox 2: User Guide.* Matlab: Natick, MA; 299 pp. Available at: http://www.mathworks.com/access/helpdesk/help/pdf_doc/fuzzy/fuzzy.pdf

Javernick, L, Brasington, J, and Caruso, B 2014. Modelling the topography of shallow braided rivers using Structure-from-motion photogrammetry. *Geomorphology* 213, 166–182.

Kaplinski, M, Hazel, JE, Grams, PE, and Davis, PA. 2014. *Monitoring Fine-Sediment Volume in the Colorado River Ecosystem, Arizona – Construction and Analysis of Digital Elevation Models.* Open File Report

2014-1052, US Geological Survey, Grand Canyon Monitoring Research Center: Flagstaff, AZ; 29 pp. Available from http://dx.doi.org/10.3133/ofr20141052

Kasprak, A, Wheaton, JM, Ashmore, PE *et al.* 2015. The relationship between particle travel distance and channel morphology: results from physical models of braided rivers. *Journal of Geophysical Research: Earth Surface* 120, 55–74.

Lague, D, Brodu, N, and Leroux, J 2013. Accurate 3D comparison of complex topography with terrestrial laser scanner: application to the Rangitikei canyon (NZ). *ISPRS journal of Photogrammmetry and Remote Sensing* 82, 10–26.

Lane, SN and Chandler, JH 2003. The generation of high quality topographic data for hydrology and geomorphology: new data sources, new applications and new problems. *Earth Surface Processes and Landforms* 28(3), 229–230.

Lane, SN, Chandler, JH, and Richards, KS 1994. Developments in monitoring and modeling small-scale river bed topography. *Earth Surface Processes and Landforms* 19(4), 349–368.

Lane, SN, Richards, KS, and Chandler, JH 1995. Morphological estimation of the time-integrated bed-load transport rate. *Water Resources Research* 31(3), 761–772.

Lane, SN, Westaway, RM, and Hicks, DM 2003. Estimation of erosion and deposition volumes in a large, gravel bed, braided river using synoptic remote sensing. *Earth Surface Processes and Landforms* 28(3), 249–271.

Laronne, JB and Duncan, MJ 1989. *Constraints on Duration of Sediment Storage in a Wide, Gravel Bed River, New Zealand.* Publication 184, International Association of Hydrological Sciences; Wallingford, UK; 165–172.

Laronne, JB and Duncan, MJ 1992. Bedload transport paths and gravel bar formation. In *Dynamics of Gravel Bed Rivers* (eds P Billi, RD Hey, CR Thorne, and P Tacconi). John Wiley & Sons: Chichester, UK.

Laronne, JB, Outhet, DN, Carling, PA, and McCabe, TJ 1994. Scour chain employment in gravel bed rivers. *Catena* 22(4), 299–306.

Leary, RJ, Wheaton, JM, and DeMeurichy, KD 2012. *Recommended Geomorphic Change Detection Procedures for Repeat TLS Surveys from Hells Canyon, Idaho.* Ecogeomorphology and Topographic Analysis Laboratory, Utah State University; Prepared for Idaho Power: Logan, UT. Available from http://etal.usu.edu/IPC/HellsCanyon_TLS_ChangeDetection_Report_ETAL.pdf

Legleiter, CJ, Roberts, DA, and Lawrence, RL 2009. Spectrally based remote sensing of river bathymetry. *Earth Surface Processes and Landforms* 34(8), 1039–1059.

Lindsay, JB and Ashmore, PE 2002. The effects of survey frequency on estimates of scour and fill in a braided river model. *Earth Surface Processes and Landforms* 27(1), 27–43.

Mapstone, BD 1995. Scalable decision rules for environmental impact studies: effect size, type I, and type II errors. *Ecological Applications* 5(2), 401–410.

McKean, J, Nagel, D, Tonina, D *et al.* 2009. Remote sensing of channels and riparian zones with a narrow-beam aquatic-terrestrial LIDAR. *Remote Sensing* 1(4), 1065–1096.

McLean, DG and Church, M 1999. Sediment transport along lower Fraser River – 2. Estimates based on the long-term gravel budget. *Water Resources Research* 35(8), 2549–2559.

Merz, J, Pasternack, G, and Wheaton, JM 2006. Sediment budget for salmonid spawning habitat rehabilitation in a regulated river. *Geomorphology* 76(1–2), 207–228.

Milan, DJ 2013. Virtual velocity of tracers in a gravel-bed river using size-based competence duration. *Geomorphology* 198, 107–114.

Milan, DJ, Heritage, GL, and Hetherington, D 2007. Application of a 3D laser scanner in the assessment of erosion and deposition volumes and channel change in a proglacial river. *Earth Surface Processes and Landforms* 32(11), 1657–1674.

Milan, DJ, Heritage, GL, Large, ARG, and Fuller, IC 2011. Filtering spatial error from DEMs: Implications for morphological change estimation. *Geomorphology* 125(1), 160–171.

Milne, JA and Sear, D 1997. Modelling river channel topography using GIS. *International Journal of Geographical Information Science* 11(5), 499–519.

Mosley, MP 1982. Analysis of the effect of changing discharge on channel morphology and instream uses in a braided river, Ohau River, New Zealand. *Water Resources Research* 18(4), 800–812.

Neill, CR 1971. River bed transport related to meander migration rates. *Journal of the Waterways, Harbours and Coastal Engineering Division ASCE* 97, 783–86.

Paola, C and Voller, VR 2005. A generalized Exner equation for sediment mass balance. *Journal of Geophysical Research* 110, F04014.

Parsons, DR, Best, JL, Orfeo, O *et al.* 2005. Morphology and flow fields of three-dimensional dunes, Rio Paraná, Argentina: results from simultaneous multibeam echo sounding and acoustic Doppler current profiling. *Journal of Geophysical Research* 110(F4).

Passalacqua, P, Belmont, P, Staley, D *et al.* 2015. Analyzing high resolution topography for advancing the understanding of mass and energy transfer through landscapes: a review. *Earth Science Reviews* 148, 174–193.

Pyrce, RS and Ashmore, PE 2003. The relation between particle path length distributions and channel morphology in gravel bed streams: a synthesis. *Geomorphology* 56(1–2), 167–187.

Redolfi, M 2014, *Sediment transport and morphology of braided rivers: steady and unsteady regime.* PhD thesis, Trento University and Queen Mary University of London; 316 pp.

Rescher, N 1996. *Process Metaphysics: an Introduction to Process Philosophy.* SUNY; 215 pp.

Roering, JJ, Mackey, BH, Marshall, JA *et al.* 2013. 'You are HERE': connecting the dots with airborne lidar for geomorphic fieldwork. *Geomorphology* 200, 172–183.

Sear, DA, Lee, MWE, Oakey, RJ *et al.* 2000. Coarse sediment tracing technology in littoral and fluvial environments: review. In *Tracers in Geomorphology* (ed. IDL Foster). John Wiley & Sons: Chichester, UK; 21–55.

Simmons, DB, Richardson, EV, and Nordin Jr, CF 1965. *Bedload Equation for Ripples and Dunes.* US Geological Survey Professional Paper 462-H, 1–9.

Smith, MW and Vericat, D 2015. From experimental plots to experimental landscapes: topography, erosion and deposition in sub-humid badlands from Structure-from-motion photogrammetry. *Earth Surface Processes and Landforms.* DOI:10.1002/esp.3747.

Smith, MW, Carrivick, JL, and Quincey, DJ 2015. Structure from Motion photogrammetry in physical geography. *Progress in Physical Geography.* DOI:10.1177/0309133315615805.

Smith, MW, Carrivick, JL, Hooke, J, and Kirkby, MJ 2014. Reconstructing flash flood magnitudes using 'Structure-from-motion': a rapid assessment tool. *Journal of Hydrology* 519, 1914–1927.

Storz-Peretz, Y and Laronne JB 2015. Morphotextural characterization of dryland braided channels. *Geological Society of America Bulletin* 125, 1599–1617.

Tarolli, P 2014. High-resolution topography for understanding Earth surface processes: opportunities and challenges. *Geomorphology* 216, 295–312.

Taylor, J 1997. *An Introduction to Error Analysis: the Study of Uncertainties in Physical Measurements,* 2nd edn. University Science Books: Sausalito, CA; 327 pp.

Vázquez-Tarrío, D and Menéndez-Duarte, R 2014. Bedload transport rates for coarse-bed streams in an Atlantic region (Narcea River, NW Iberian Peninsula). *Geomorphology* 217, 1–14.

Vericat, D, Batalla, RJ, and Garcia C 2008. Bed-material mobility in a large river below dams. *Geodinamica Acta* 21(1–2), 3–10.

Vericat, D, Smith, MW, and Brasington, J 2014. Patterns of topographic change in sub-humid badlands determined by high resolution multi-temporal topographic surveys. *Catena* 120, 164–176.

Wang, Y, Liang, X, Flener, C *et al.* 2013. 3D Modeling of coarse fluvial sediments based on mobile laser scanning data. *Remote Sensing* 5, 4571–4592.

Westaway, RM, Lane, SN, and Hicks, DM 2003. Remote survey of large-scale braided rivers using digital photogrammetry and image analysis. *International Journal of Remote Sensing* 24, 795–816.

Westoby, MJ, Brasington, J, Glasser, NF *et al.* 2012. 'Structure-from-motion' photogrammetry: a low-cost, effective tool for geoscience applications. *Geomorphology* 179, 300–314.
Wheaton, JM, Brasington, J, Darby, S, and Sear, D 2010. Accounting for uncertainty in DEMs from repeat topographic surveys: improved sediment budgets. *Earth Surface Processes and Landforms* 35(2), 136–156.
Wheaton, JM, Brasington, J, Darby, SE *et al.* 2013. Morphodynamic signatures of braiding mechanisms as expressed through change in sediment storage in a gravel bed river. *Journal of Geophysical Research – Earth Surface* 118(2), 759–779.
Wheaton, JM, Fryirs, K, Brierley, G *et al.* 2015. Geomorphic mapping and taxonomy of fluvial landforms. *Geomorphology* 248, 273–295.
Wilcock, PR, Barta, AF, Shea, CC *et al.* 1996. Observations of flow and sediment entrainment on a large gravel bed river. *Water Resources Research* 32(9), 2897–2909.
Williams, RD 2012. DEMs of difference. In *Geomorphological Techniques*. British Society for Geomorphology; 17 pp.
Williams, RD, Brasington, J, Vericat, D *et al.* 2011. Monitoring braided river change using terrestrial laser scanning and optical bathymetric mapping. In *Developments in Earth Surface Processes* (eds J Mike, PP Smith, and SG James). Elsevier: Amsterdam; 507–532.
Williams, RD, Brasington, J, Hicks, M *et al.* 2013. Hydraulic validation of two-dimensional simulations of braided river flow with spatially continuous aDcp data. *Water Resources Research* 49(9), 5183–5205.
Williams, RD, Brasington, J, Vericat, D, and Hicks, DM 2014. Hyperscale terrain modelling of braided rivers: fusing mobile terrestrial laser scanning and optical bathymetric mapping. *Earth Surface Processes and Landforms* 39(2), 167–183.
Williams, RD, Rennie, CD, Brasington, J *et al.* 2015. Linking the spatial distribution of bedload transport to morphological change during high-flow events in a shallow braided river. *Journal of Geophysical Research – Earth Surface* 120(3), 604–622.
Wittmann, Rgbt 1925. Der Einfluss der Korrecktion des Rheins zwischen Basel und Mannheim auf die Geschiebebewegung des Rheins. *Deutsche Wasserwirtschaft* 22, 245–261.
Woodget, AS, Carbonneau, PE, Visser, F, and Maddock IP 2015. Quantifying submerged fluvial topography using hyperspatial resolution UAS imagery and structure from motion photogrammetry. *Earth Surface Processes and Landforms* 40(1), 47–64.

Discussion

Discussion by Helmut Habersack

The improvement of the morphological approach by repeat high-resolution topography measurements seems to be very promising. This is especially true for rivers with significant morphodynamics, like braided rivers, where cross-sectional sediment monitoring, e.g. with basket samplers, is impossible due to fast bed changes. However, experiences of working with highly dynamic gravel bed rivers like the Waimakariri River in New Zealand or the Sunwapta River in Canada showed that the comparison of repeat topography measurements allows the calculation of volumetric changes (erosion vs. sedimentation) but due to the unknown travel distance and active layer, not necessarily the determination of bedload transport rate. The missing link can be found by combining repeated high-resolution topography with tracer studies. In this case especially the travel distance from erosion to deposition of single particles (e.g. about 180 m in single small floods/freshes at the Waimakariri River) between the dates of topography measurements and if possible additionally the burial depths/active layer are essential to derive results comparable to direct bedload monitoring.

Discussion by Jonathan Laronne

The DoD is based on very advanced error analyses. Changes in topographic elevation smaller than these errors bring about a delineation of areas where no change is detected – although it may have likely occurred. For these DoD = 0 areas:

1) When analysing temporal changes for periods of time > 2 HRT DEMS with resultant more than one DoD, areas for which DoD are non-varying need not be identical. The procedure to determine the DoD of a second DEM pair in comparison to a first DEM pair will produce results for areas of supposedly non-varying topography. However, this is an error caused by the accuracy of the method.
2) As the number of DoDs increases with additional acquisition of DEMs, this type of error may increase.

Have you considered how to deal with this issue, particularly so because river bed areas of little change can be very interesting?

Discussion by David Milan

The combination of high quality repeat topographic data, providing information on spatial patterns of erosion and deposition, in conjunction with tracer data, providing clast step lengths, offers great potential to investigate spatial patterns in bedload transport rates. The calculation of bedload transport rate using this technique relies on the accurate determination of virtual velocity, calculated from the step length divided by the duration of competence. No allowance is usually made for differences in competence duration according to grain-size. However, Milan (2013) provides an approach to account for this, demonstrating that relatively smaller clasts can have lower virtual velocity in comparison to larger clasts, due to their longer competence duration. Developments in active tracer technology should improve estimates of grain-size-specific competence duration and thus sediment transport rates calculated using this methodology.

Discussion by Murray Hicks

My comment concerns use of the length of erosion zones as a proxy for step length when using morphological change from DoDs to estimate bedload flux. Even though this offers a tantalizingly simple approach, my concern is that it is appears to have no physical basis nor has been validated off any data. One source of validation data could use tracer cobbles to directly measure step length over events captured with DoDs. Another (potentially very powerful) approach could use two-dimensional/three-dimensional morphodynamical model runs over events to produce: (i) DoDs and from those statistics of the dimensions of erosion zones; (ii) data on step lengths by using numerical tracers; and (iii) direct measures of bedload flux for comparison with the estimates based on the morphological method. There would appear to be the opportunity to pursue this using the existing Delft3D model of the Rees study reach (e.g. Williams *et al.* 2015).

Discussion by Philippe Belleudy

When using frequent topographical levellings it is not always possible to have them in the same (low) flow conditions. Yet the flooding conditions of bars may not be comparable and lead to false conclusions concerning lateral erosion or aggradation between two topographic measurements. Does your system deal with the rendering of morphological changes that could artificially result from this effect?

Reply by the Authors

We thank the authors of the discussions for their constructive comments. Two main issues arise from their discussions: opportunities given by tracers and uncertainty of the morphological approach.

H. Habersack and D. Milan both commented about the use of tracer studies to estimate virtual velocity. Habersack points out that gravel tracer studies in combination with repeat HRT surveys can provide the 'missing link'. Although compelling this may seem, the establishment of robust path-length models remains a challenging task for geomorphologists (e.g. Pyrce and Ashmore 2003). The potential of active tracers (e.g. radio-tracers) is evident and may provide very valuable data on the dynamics of single pebbles during an entire competent event (e.g. Habersack 2001; Liedermann *et al.* 2013). However, the cost of this method limits their application at broader spatial and temporal scales. Milan (2013) also highlights the importance grain size plays in competence duration and therefore in the estimation of virtual velocity. Although this may improve the estimates of the morphological approach, it remains challenging to segment the results obtained by differencing DEMs by the different grain size fractions of the virtual velocity models (see Equation 5.7).

While we agree with both Habersack and Milan's enthusiasm for the potential of tracer studies, our broader point is that tracer studies may not be the only way to estimate a characteristic length scale L_c (i.e. path length). It may be possible to estimate that L_c term from topographic signatures such as bar spacing, confluence–diffluence spacing, meander wavelength (e.g. Neill 1971), or even, potentially, from a morphodynamic signature such as the length scale of the long axis of erosional units as recently developed by Redolfi (2014). Within this context, M. Hicks argues that this last approach needs specific validation. We absolutely agree that more empirical study and validation comparing distributions of such DoD derived length scales with actual path length distributions from tracer studies are needed. Figure 5.3a shows an example of some of the available tracer data from the Rees that could be used to validate the approach. The authors, Redolfi, and others are in the process of analysing all data from painted tracers obtained along one flood season at multiple bars in the Rees. We chose not to include this validation data in the chapter as we did not have the space to properly flush this idea out and feel it deserves its own study. Hicks also argues that while the approach is 'tantalizingly simple...it appears to have no physical basis'. We concur that this theory has not yet been properly theorized or vetted in the literature. We postulate that the physical basis for the length of erosional units working as a potential proxy for step length has less to do with the distance over which the topography is eroded and more to do with a geometric correlation with the topography and features being eroded. In many gravel-bed rivers, the erosional units' length scale is a direct reflection of the length scale of the geomorphic units they are eroding. The extent of the long axis of that length scale of erosional units appears to scale with event magnitude and/or duration. The length of erosional units and connectivity of units is sensitive to the DoD thresholding methods used, and this sensitivity needs to be explored further, as does their relevance outside braided and wandering gravel-bed rivers (e.g. single thread). Finally, Hicks points out another exciting approach for testing the validity of the erosional unit approximation would be to test them in two-dimensional/three-dimensional morphodynamics models, by measuring the erosion units produced and comparing them with transport distances that in principle could be known, because they are modelled. We agree this should be explored, but caution against believing that the physical basis for morphodynamics models in gravel-bed rivers should be trusted any more than a field approximation. Most two-dimensional/three-dimensional morphodyanmic models have still not been adequately vetted for gravel-bed rivers. Moreover, the standard off-the-shelf morphodynamics packages focus only on 'bed erosion', and where they include a lateral migration module, to date they have been woefully inadequate at producing realistic bank-erosion and bar-edge-trimming mechanisms.

The comments by J. Laronne and P. Belleudy are related to various aspects of the uncertainty of the morphological approach. Belleudy simply asks if the DoD method is sensitive to the flow stage at which the surveys are performed. In principle, the answer is not necessarily. For many gravel-bed rivers, a variety or even a hybrid of survey techniques (e.g. Williams *et al.* 2014) will facilitate complete survey coverage over a wide range of flows. The effects of wet channels on geomorphic change detection depend on the accuracy and precision of the survey technologies employed to obtain and post-process the topographic data (Bangen *et al.* 2014a; Passalacqua *et al.* 2015). As discussed in the chapter (see section 5.3), all studies of topographic change require measurements registered to a robust topographic control network that facilitates accurate reoccupation of position. Therefore, neither the position of the wet channel in each of the surveys to compare, nor the level of the water during surveys should be a limitation if the method employed to obtain the topography under the water (i.e. bathymetry) has the same accuracy as the method for the dry areas. Unfortunately, methods to obtain HRT that provide high accuracies in dry areas (e.g. TLS) often do not provide any useable data for submerged areas, and methods employed for wet channels have less accuracy under most circumstances. The different accuracy between methods needs to be taken into account when the uncertainty and the errors of DEMs is assessed and propagated.

Finally, Laronne raised the interesting question of the influence of the areas thresholded out of a DoD analysis (i.e. where $\text{DoD} < \text{LoD}_{\text{min}}$) on the estimate of sediment flux. Briefly, when the errors are propagated on a cell-by-cell basis, where the vertical changes in those cells are smaller than its LoD_{min} it is not considered zero change, but instead is considered indistinguishable from noise. Thus, it is important that these 'thesholded' areas are not equated with no change. They simply reflect the precision limits of the survey methods, error modelling, and change detection methods. Calculated sediments fluxes will generally be underestimated, as a result of thresholding, because the estimates do not incorporate the small changes in these areas. This manifests itself differently depending on whether the budget-based approach and the net ΔV term or a virtual velocity approach and only reliant on the ΣV_e (volume of erosion) term from the DoD is being used. However, the greater uncertainty is probably related to the temporal frequency of surveys. These areas (as well as the rest of the DoD) could be subjected to local compensation of scour and fill processes that occur between surveys (e.g. Lindsay and Ashmore 2002). That is, a surface could experience significant erosion and deposition between two points in time, but the measured net change is negligible because the starting and ending elevations are similar. The impact of this would also be an underestimate of total sediment flux. This impact can be minimized by having shorter time steps between surveys and generally reserving the morphological approach for event-based or more frequent DEM datasets, or by coupling DoD results with data from scour chains, although it remains an understudied problem and should be further explored.

Overall, the morphological approach is positioned to be more fully exploited in the future and the comments here provide some interesting caveats to consider moving forward cautiously.

References

Habersack, HM 2001. Radio-tracking gravel particles in a large braided river in New Zealand: a field test of the stochastic theory of bed load transport proposed by Einstein. *Hydrological Processes* 15, 377–391.

Liedermann, M, Tritthart, M, and Habersack, H 2013. Particle path characteristics at the large gravel-bed River Danube: results from a tracer study and numerical modelling. *Earth Surface Processes and Landforms* 38, 512–522.

6

Geomorphic Controls on Tracer Particle Dispersion in Gravel-Bed Rivers

Marwan A. Hassan and D. Nathan Bradley

6.1 Introduction

A longstanding problem in the study of the sediment transport in gravel-bed streams is the detailed description of the processes of individual particle displacements in relation to bed texture, flow regime, and channel morphology. Bed-surface composition and channel morphology are direct results of individual particle movement and, in turn, are controls on particle displacement. An accurate description of fluvial sediment dispersion in streams is essential for predicting and understanding channel stability, spatial and temporal variations in sediment mobility, changes in bed texture and structure, depth of the active layer, aquatic habitat, and landscape evolution. Detailed observation of the dispersion of individually identifiable tracer particles is likely to improve the physical description of the sediment transport in streams and provide the information needed for the development of better predictive models.

While the technical difficulties of tracking large numbers of tracer particles impose a major limitation on particle tracking efforts, extensive research has been conducted using tracer particles. In spite of these efforts, there are still major gaps in our knowledge. We remain at the level where we are examining the basic features of grain movement such as the relative frequency of grain displacements, the mode of movement, and the travel speed and distance in relation to particle properties and flow characteristics. Major research gaps include the role of channel morphology and bed texture, the depth of the active layer and vertical mixing of particles, particle storage timescales, relations between burial depth and particle mobility, spatial and temporal patterns of bed mobility, and the roles of mobile and immobile particles within the active layer. Long-term tracer monitoring programs in a wide range of fluvial environments aimed at exploring the effects of channel morphology, bed texture, and flow characteristics on particle mobility are needed to fill the gaps in our knowledge. Such monitoring programs should be aimed at obtaining detailed descriptions of particle movement at a broad range of spatial and temporal scales.

This chapter is intended to present a comprehensive, critical review of the field of fluvial grain kinematics in order to evaluate progress made over the years and identify gaps in our knowledge. The chapter covers the following topics: the relationship between grain kinematics and bedload discharge, basic concepts and theory, stochastic transport models, tracer experiments examining the fundamental properties of sediment dispersion (step length, rest period, and travel distance),

Gravel-Bed Rivers: Processes and Disasters, First Edition. Edited by Daizo Tsutsumi and Jonathan B. Laronne.

and the interaction of moving particles with the channel morphology (virtual velocity, depth of active layer, vertical mixing, and burial depth).

6.2 Bedload Estimates Using Tracers

From a practical point view, one of the main reasons to study the movement of individual particles in streams is to estimate bedload transport. While the motion of individual particles is highly stochastic, the transport rate can be estimated from the average motion of the tracer population. The lengths of individual particle displacements provide a formal definition of the transport rate and a link between the Lagrangian and Eulerian frameworks (Einstein 1937; Crickmore and Lean 1962; Hubbell and Sayre 1964). Wong *et al.* (2007; see also Crickmore *et al.* 1990) detailed three methods for calculating sediment transport using tracer data: the first is based on the continuity of moving bedload particles, the second on a conservation relation, and the third on a conservation relation with virtual velocity.

In the first method, the volumetric bedload transport rate per unit width (q_b) has the form

$$q_b = \xi_b U_{bm} \tag{6.1}$$

where ξ_b is the volume of the grains in motion per unit area and U_{bm} is the mean mobile grain velocity. Because ξ_b has units of length, it can also be interpreted as the thickness of the transport layer. Similar relations have been used by others to estimate sediment transport in streams (Wiberg and Dungan Smith1989; Sekine and Kikkawa 1992; Niño and García 1994). This method is limited by the ability to estimate the thickness of the moving bedload layer if a large proportion of the fine fraction of the bedload moves into suspension at relatively high shear stresses.

In the second method, the volumetric sediment transport rate is the product of the volume of sediment entrained per unit time per unit area, E_b, and the mean particle step length, L_s:

$$q_b = E_b L_s \tag{6.2}$$

Nakagawa and Tsujimoto (1976) developed one of the early relations of this kind, utilizing the theoretical results of Einstein (1937) to obtain values for the entrainment rate and step length. According to Wong *et al.* (2007), the difficulty of defining the average step length in a way that can be consistently measured makes this method difficult to apply. Wilcock (1997) further developed this method, addressing aspects of data collection, analysis, and interpretation of tracer gravel observations. He suggested the following relation to calculate the fractional mass transport rate q_{bi} from tracer data:

$$q_{b_i} = M_{ai}\left(\frac{N_i}{T}\right)L_{1i} \tag{6.3}$$

where M_{ai} is the entrained mass of size fraction i per unit bed area over time period T, N_i is the number of times an individual grain of fraction i is entrained during T, and L_{1i} is the travel distance of the grains during time period T. Wilcock (1997) linked the particle entrainment (M_{ai}) to partially and fully mobile stages of sediment transport, developed expressions to calculate sediment entrainment for both stages, and tested some of the concepts using both laboratory and field data.

The third method calculates the transport rate (in newtons per second) from the virtual velocity of the tracers and the depth of the active layer. The virtual velocity of the tracers is determined from change in the position of the center of mass of the spatial distribution of tracers, or from the travel distance of individual particles. Hubbell and Sayre (1964) and Sayre and Hubbell (1965) used this method in the form

$$q_b = \gamma_s(1-p)d_x\frac{L_{1i}}{t} \tag{6.4}$$

where t is the time interval, d_x the average depth over which the tracer particles are distributed beneath the bed surface, γ_s is the bulk specific weight of the bed material, and p is porosity. The term (L_{1i}/t) is the virtual velocity of grains as first introduced by Einstein (1937). Although Sayre and Hubbell (1965) used the relation for the estimation of sand, the relation is applicable for gravel-bed streams if the mean travel distance of individual particles is used instead of the centroid of the tracer plume.

The accuracy of the estimate of the sediment transport rate by the methods described above is determined by the quality of the information on particle entrainment, velocity, travel distance, virtual velocity, and depth of the active layer. The difficulty in quantifying the characteristics of particle dispersion has to do with the intermittent nature of movement of individual particles in streams. Particles spend most of their time at rest, moving only during infrequent floods. During a flood, entrainment, transport, and deposition are highly stochastic processes that are difficult to treat in a deterministic manner. This recognition led Einstein (1937) and many others to develop probabilistic models based on the idea that the motion of a single particle is a series of steps with intervening rests. The probability distributions that describe the length of steps and the duration of rests can be combined to estimate sediment transport if the forms and parameters of the distributions can be determined. Stochastic transport models provide a second link between the particle-focused Lagrangian framework and the flux-based Eulerian framework.

A simple example of a stochastic transport model is the gamma distribution. The total distance traveled by a particle that takes a series of n steps, with random lengths drawn from an exponential distribution, is a random variable that follows a gamma distribution. The parameters of the distribution are the number of steps and the mean step length. The gamma distribution describes both the probability of finding a single particle in a particular place, and the spatial concentration profile (in one dimension) of a population of particles. While the gamma distribution does not explicitly include a time component, flood duration can be used to compute the velocity term in equation 6.1 to estimate flux.

A more sophisticated transport model, called a compound Poisson process, allows the number of steps to be a random variable, which is equivalent to adding exponentially distributed rests between the steps. The compound Poisson Einstein–Hubbell–Sayre (EHS) model (Einstein 1937; Hubbell and Sayre 1964; Sayre and Hubbell 1965) treated steps and rests as exponentially distributed random variables. The mean tracer velocity is the ratio of the two model parameters, the mean step length λ and the mean resting time τ. A critical assumption of this velocity is that steps are so brief relative to the duration of the rests that they can be treated as if they are instantaneous.

The various methods reviewed above to estimate flux from tracers treat the effects of bed texture and morphology superficially. In the first method, the term ξ_b can be interpreted as an active-layer thickness. In the second, bed-texture effects are implicit in the entrainment rate term. The third method includes mixing depth and porosity. The EHS stochastic model incorporates assumptions about bed texture and morphology in the choice of the exponential distributions. The exponential distributions imply steady-in-time deposition (in the case of steps) and erosion (in the case of rests) probabilities and spatially uniform bed conditions. For example, steady deposition probability (per unit time) coupled with steady particle velocity results in an exponential distribution of step length. Similarly, steady erosion probability during a rest results in an exponential distribution of rest duration.

The EHS model has inspired many similar models that allow the steps and rests to follow a variety of probability distributions, (e.g. Paintal 1971; Yang and Sayre 1971; Niño *et al.* 2002; Papanicolaou *et al.* 2002; Marion, Zaramella, and Bottacin-Busolin 2008). A critical challenge in applying stochastic models is defining what constitutes a step or a rest. Is a step simply a displacement followed by a pause, no matter how brief the pause? Or is a step a mostly continuous episode of motion with a few brief pauses? Einstein (1937) defined a step as particle displacement followed by an immobile period that is much longer than the time the particle was in motion. By this definition, can the travel distance during an entire flood be considered a step? What are the appropriate spatial and temporal scales to describe particle motion? These questions deserve further consideration.

6.3 Scales of Particle Motion

Nikora *et al.* (2001; see also Nikora *et al.* 2002) addressed these questions by introducing three "ranges" to describe the spatial and temporal scales of sediment particle motion and the nature of particle dispersion at each scale. The first addresses motion at the shortest scales and is called the *local* range. A single saltation (Nikora *et al.* 2001; Seminara, Solari, and Parker 2002), or in the case of rolling and sliding the motion between two "significant" collisions with the bed (Nikora *et al.* 2002), fall within the local range. A collision with a particle of similar or larger size that causes a change of direction is considered significant. Displacements within the local range are not steps so long as the pauses between displacements are similar in duration to the intervals of motion.

Nikora *et al.* (2001) assigned Einstein's step to the *intermediate* range. A step is a series of one or more local displacements followed by rest that is long relative to the time in motion. Similarly, Seminara, Solari, and Parker (2002) defined a step as a series of several (~18) saltations. The length of steps and duration of rests are presumably functions of the local hydraulic conditions (e.g. the local shear stress), the bed texture, and grain size.

The EHS stochastic transport model and related models are based on probability distributions of step length and resting time. The particle motion in these models is composed of many intermediate trajectories with many intervening rests. The total transport is built from the distributions of the intermediate-scale components, the steps and rests. Nikora *et al.* (2001) assigned these models to the *global* range. The global range may be as short as a single flood, if particles move multiple times, or much longer if particles move infrequently. The global range conflates two potentially disparate immobile timescales, the duration of an intermediate-scale rest and a much longer timescale related to flood frequency and particle storage in features such as bars and riffles.

6.4 Types of Tracer Experiments and a Review of Results

Tracking uniquely labeled tracer stones provides the best means to observe the kinematics of individual gravel clasts. As a small sample of the bed material, tracers are assumed to represent the whole bed and to provide information on particle displacement. Tracers provide an opportunity to study the general characteristics of sediment movement and reveal the long-term action of the fluvial system under varying flow and sediment supply regimes, as well as the influence of bed composition (texture and structure), bedforms, and channel morphology. Reviews of the gravel tracking technology may be found in Sear *et al.* (2000), Hassan and Ergenzinger (2003), Bradley and Tucker (2012) and Hassan and Roy (2016).

Table 6.1 Active tracer experiments using radio transmitter equipped rocks that measured step length or resting time.

Study	Site	Tracers size range (mm)	Step length	Resting time
Chacho, Emmett, and Burrows (1994)	Toklat	60–100	Not reported	Exponential
Ergenzinger and Schmidt (1990) Schmidt and Ergenzinger (1992)	Lainbach	85–130	Exponential	Exponential
Habersack (2001)	Waimakariri	42–62	Gamma	Exponential
McNamara and Borden (2004)	Reynolds	76–92	Gamma	Gamma or exponential if longer rests were excluded

Tracer technology generally falls into two categories: "active" and "passive" tracers. An active tracer is typically a battery-powered radio transmitter embedded in a natural or artificial clast that broadcasts to one or more receivers stationed on the riverbank. Active tracer experiments are limited to particles large enough to hold a transmitter and battery. The receivers detect the location of the tracer and whether it is in motion or at rest. Active tracers observe transport over the local and intermediate ranges and therefore active tracer studies (summarized in Table 6.1) have generally focused on measuring step lengths and resting times. These studies have provided substantial evidence that step lengths are exponentially or gamma distributed and that resting times are exponentially distributed.

A significant advantage of active tracers is that transport is observed in real time and can be correlated with fluctuations in the flood hydrograph, linking the hydraulic conditions to the particle transport (McNamara and Borden 2004). However, active tracers are expensive and are limited by the battery life. The experiments performed to date have used a small number of tracers (fewer than 40 in the studies summarized in Table 6.1) and have been relatively short, capturing motion over short periods and providing limited information about transport at the global scale and how tracers interact with the bed and the channel morphology over the long term. However, a recent study by Liedermann, Tritthart, and Habersack (2013) observed active tracers in a large river for a full year, bridging the gap between the intermediate range observed by most active tracer experiments, and the global range more typically investigated using passive tracers.

Passive tracers are either fine sediment that is tagged in some way so that the concentration of the tracer sediment in the bed can be measured, or individually labeled (i.e., numbered) gravel clasts. Sand has been labeled with radioactive material, allowing *in situ*, continuous detection of particles within the active layer (surface and subsurface) over large areas and a range of time intervals. The experimental results of Hubbell and Sayre (1964) and Sayre and Hubbell (1965) supported Einstein's model for particle travel distance and provided support for EHS-style stochastic transport models. Subsequent work in sand-bed rivers motivated similar models that used a gamma distribution for the step length (Grigg 1970; Yang and Sayre 1971; Nakagawa and Tsujimoto 1976). However, concerns about introducing radioactive material into fluvial environments have prevented recent use of this method to track fine sediment. Consequently, most tracer experiments have focused on larger material, mainly gravel.

Gravel tracers are usually modified in some way to make them easier to find, such as painting the rocks a distinctive color, inserting an iron or magnetic core, or using passive integrated transponder (PIT) tags. Passive tracers have a number of advantages over active tracers; they are generally inexpensive, making large experiments possible, and are not limited by battery life but by how long the identifying marks last. They can be smaller than active tracers because no battery is necessary, but most experiments have been limited to gravel. Passive tracers observe transport at the global scale, the cumulative motion in all the floods between recovery campaigns.

The recent use of PIT tagged fluvial tracers (Nichols 2004; Lamarre, Macvicar, and Roy 2005; Lamarre and Roy 2008b; Bradley and Tucker 2012; Liébault *et al.* 2012; Olinde and Johnson 2013; Schneider *et al.*, 2010; Phillips, Martin, and Jerolmack 2013) has overcome the chief disadvantages of earlier passive tracers: decaying labels, low recovery rates, and the need to dig the tracers out of the bed in order to identify them. The last destroys features such as bed texture that may be relevant to entrainment rates. The high recovery rates of PIT tagged gravel – > 90% after several years (Bradley and Tucker 2012) – have made tracking large amounts of gravel more practical than ever. In addition, PIT tagged gravel can be detected up to burial depths of about 0.5 m (Lamarre, Macvicar, and Roy 2005; Chapuis *et al.* 2014; Bradley and Tucker 2012), but the PIT tag technique does not provide sorely needed information on burial depths below this. Recently, Tsakiris *et al.* (2015) developed a semi-anlaytical expression to relate low-frequency passive RFID signals with burial depth. They reported that the received input voltage decays inversely with the cube of the burial depth in all mediums (e.g., air, water, sediment). Furthermore, the orientation angle of the excitation antenna and the sediment composition around transponder miniature antennas can have a significant effect on the magnitude of the input voltage and its decay rate in relation to burial depth. Additional testing of the Tsakiris *et al.* (2015) method could provide a solution for the determination of burial depth and is therefore worth pursuing. A second disadvantage is that current PIT tag recovery techniques are limited to wadeable streams and are not suitable for large rivers (Liedermann, Tritthart, and Habersack 2013). A final disadvantage of PIT tagged gravel is that only cumulative transport over the course of a flood is observed. Recent work with accelerometer-equipped tracers and in-stream RFID antennas provided observations of the local and intermediate scales (Olinde and Johnson 2013) that could overcome this limitation.

As with studies using active tracers, a key motivator for passive tracer experiments has been to test stochastic transport models. Early work in gravel-bed rivers showed that the gamma probability function and the EHS model fit the data reasonably well for relatively short single peaked flow events during which the number of steps that particles take is relatively low (on the order of 2–3) (Hassan, Church, and Schick 1991; Hassan and Church 1992). Bradley and Tucker (2012) argued that the Yang and Sayre (1971) model, which uses a gamma distribution of step lengths, is more flexible than the EHS model because the gamma distribution can model more diverse transport than the exponential. The gamma distribution can be peaked, as opposed to the monotonically decreasing exponential distribution, and can decrease monotonically faster than the exponential. It also closely approximates the EHS compound Poisson distribution, so transport over multipeak events such as a snowmelt flood or a season of floods can be modeled by treating the total transport during a single peak as a single step. The weakness in this interpretation is that it conflates two different immobile timescales – the rests between actual intermediate steps and the global-scale intervals between flood peaks – in a single exponential distribution.

Subsequent research in both sand- and gravel-bed rivers has emphasized the difficulty of applying EHS-style stochastic models to describe the movement of gravel in streams (Stelczer, 1981). This is in part because the EHS-style models are formulated on the assumption that steps and rests are each statistically independent, identically distributed random variables, which implies steady uniform flow and bed conditions and a well-defined distribution of particle size. These assumptions may be approximately correct on the timescale of a single flood, but less so over longer timescales as transport is affected by channel morphology, bed texture, and tracer burial.

6.5 Practical Relations for Travel Distance

Although linking the parameters of the step-length and resting-time distributions to flow conditions, channel morphology, and bed composition remains an open research question, empirical relations to estimate the mean travel distances have been developed. Observations of particle displacement have led to the development of two types of practical relations. The first type relates the mean travel distance to several measures of the flow strength: peak stream power (Hassan and Church 1992; Hassan, Church, and Ashworth 1992; Schneider *et al.* 2014), time-integrated excess stream power (Haschenburger 2013; Schneider *et al.* 2014), excess energy (this chapter), and an impulse framework that describes the total momentum imparted to a particle by the flow (Phillips, Martin, and Jerolmack 2013; Phillips and Jerolmack 2014). The second relates the mean travel distance during individual flow events to the mean particle size of unconstrained clasts on the riverbed (Church and Hassan 1992; Hassan and Church 1992; Hassan, Church, and Ashworth 1992; Gintz, Hassan, and Schmidt 1996).

Hassan and Church (1992) and Hassan, Church, and Ashworth (1992) developed a functional relation between the mean travel distance and excess stream power at the flood peak for tracer particles. They presented two relations: one for gravel clasts on a gravel bed and the second for gravel on a sand bed. The mean travel distance increases with excess peak stream power, but the wide amount of scatter in the data is typical of the variability found in measurements of sediment transport in streams and reflects the important role that channel morphology plays in sediment transport (Figure 6.1a). For example, tracer particles seeded in Rio Cordon step-pool stream plot below the riffle–pool streams, implying a shorter distance of travel for a given excess stream power (Figure 6.1a). However, the step-pool morphology of the Lainbach (Germany) and the Erlenbach (Switzerland) yielded different results. The mean travel distance for the Lainbach plots close to the upper envelop of the riffle–pool streams (Figure 6.1). This could be due to the poorly developed step-pool morphology in the Lainbach. In the case of the Erlenbach, the mean travel distance plots within the riffle–pool range (Figure 6.1a). Travel distances of gravel on a sand bed are about two orders of magnitude greater than those of gravel on a gravel bed. Recently, Phillips and Jerolmack (2014) found that transport distances from different streams (with different linear relations to impulse) can be treated similarly if impulse is scaled by a dimensionless friction factor related to the grain size and relative submergence of major roughness elements.

Haschenburger (2013) and Schneider *et al.* (2014) related tracer transport to excess stream power integrated over time. Haschenburger (2013) found a power-law relationship between the time-integrated excess stream power and the mean tracer displacement, and linear relationships between the time-integrated excess stream power and the position of the central tracer, the leading edge of the tracer plume, and the standard deviation of position. In all cases, the coefficient of determination of the relationship was high ($R^2 \geq 0.98$). Schneider *et al.* (2014) also found power-law scaling between mean displacement and integrated excess stream power, but with different scaling for small and large floods and with somewhat lower R^2 values. Using Halfmoon Creek data (collected by Bradley and Tucker 2012), we related the mean travel distance to the excess energy. Similar to the relation developed for the excess stream power, the coefficient of determination for the relationship was high (Figure 6.1c).

Phillips, Martin, and Jerolmack (2013), inspired by the laboratory experiments of Diplas *et al.* (2008) and Lajeunesse, Malverti, and Charru (2010), related tracer transport to nondimensional impulse in the form of time-integrated excess shear velocity (Figure 6.1b). Impulse is a measure of the momentum imparted to a particle by a flow. They found a linear relationship ($R^2 = 0.98$) between mean dimensionless tracer displacement (distance divided by particle size) and dimensionless impulse

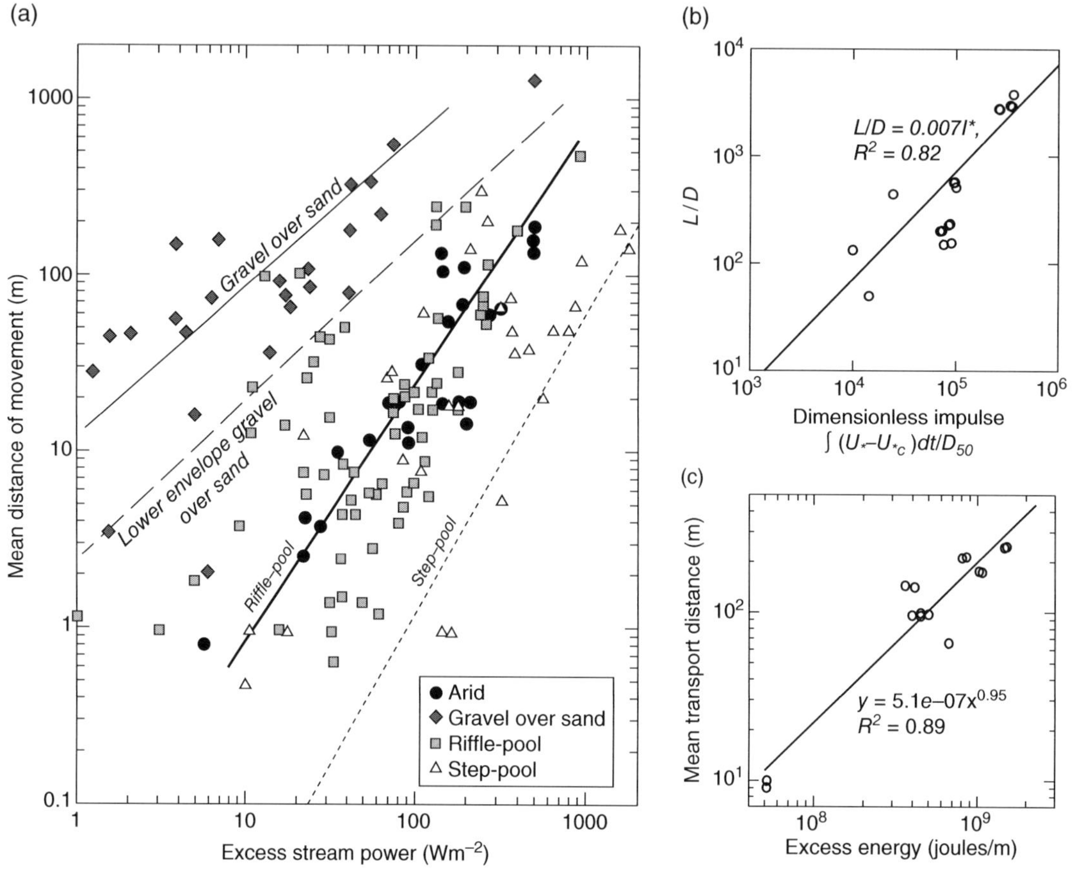

Figure 6.1 (a) Mean travel distance as a function of excess stream power. The figure is based on published data reported in Hassan, Church, and Ashworth (1992), Gintz, Hassan, and Schmidt (1996), Sear (1996), Lenzi (2004), and Schneider *et al.* (2014). (b) Scaled travel distance (*L* is the mean travel distance and *D* is the particle size) versus the nondimensional impulse in the form of time-integrated excess shear velocity (U_* is the shear velocity, U_{*c} is the threshold shear velocity, and D_{50} is the median size of the bed sediment). (After Phillips *et al.* (2013), with permission of American Geophysical Union). (c) Mean travel distance as a function of excess energy for PIT tagged gravel-travel distances during snowmelt events in Halfmoon Creek, near Leadville, CO (see Bradley and Tucker (2012) for general information about the study).

for tracers in the Mameyes River in Puerto Rico. A nearly identical relationship between mean tracer displacement and the product of average excess shear velocity and flood duration shows that cumulative impulse is proportional to the time above the transport threshold if the distribution of excess shear velocity is thin-tailed and consequently has a well-defined average. Phillips and Jerolmack (2014) point out that stream discharge (and by implication, stream power) may range over many orders of magnitude and may not be thin-tailed, while shear velocity does not vary by much more than a factor of two for most rivers.

Field observations have also shown that the mean travel distance of particles during competent floods tends to decrease only weakly with increasing particle size (e.g., Church and Hassan 1992; Phillips and Jerolmack 2014). Mean travel distance is better correlated with particle size relative to the size

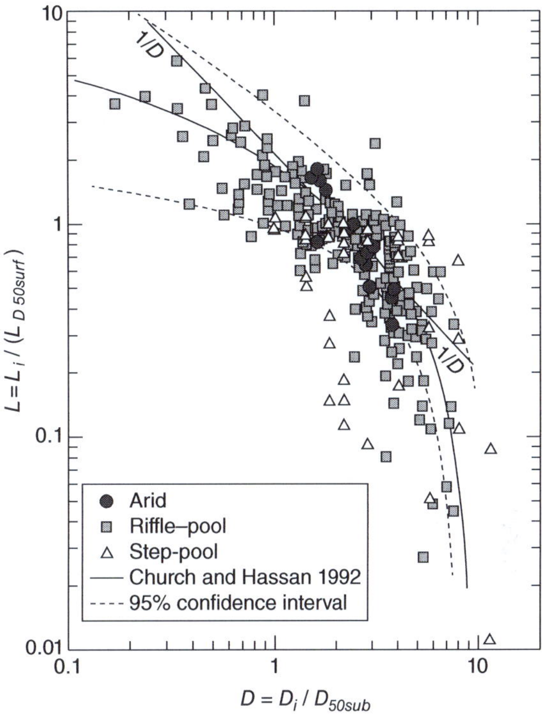

Figure 6.2 Scaled travel distance (*L*) of individual particles as a function of scaled particle size (*D*). The travel distance of each size fraction (L_i) is scaled by the mean travel distance of the fraction that includes the median size of the bed-surface sediment (LD_{50surf}). The particle size of each fraction (D_i) is scaled by the median size of the subsurface material (D_{50sub}). The figure is based on published data reported in Hassan and Church (1992), Ferguson and Wathen (1998), Lenzi (2004), MacVicar and Roy (2011), Milan (2013), Haschenburger (2013), and Vázquez-Tarrío and Menéndez-Duarte (2015).

of the bed material. Church and Hassan (1992) and Hassan and Church (1992) related the scaled mean travel distance to the scaled particle size and showed a decrease in travel distance with increasing particle size (Figure 6.2). For particles larger than the median size of the bed material, the mean travel distance declines rapidly as the particle size increases. Wilcock (1997) confirmed the relation while exploring the relationship between entrainment, flux and travel distance in a flume. He showed that the partial sediment transport regime implies a nonlinear relation of travel distance to particle size. Wilcock (1997) asserted that the hyperbolic functional relation applies when material finer than the median size of the bed material is fully mobile and the coarser fractions are partially mobile. Ferguson and Wathen (1998) and Ferguson *et al.* (2002) showed some scatter in the relation between the scaled travel distance and scaled particle size in an extensive dataset, but overall, the relation holds. Recently, MacVicar and Roy (2011) showed that a power-law function describes the decline in the travel distance with particle size for free-surface particles in the forced riffle–pool morphology of Mores Creek, Canada. However, the Mores data plot near the upper envelope of the riffle–pool streams. Of further interest is the travel distance obtained for the step-pool morphology of Rio Cordon (Lenzi 2004). The data plot in two groups: below and above the riffle–pool streams (Figure 6.2).

6.6 Virtual Velocity

The effects of morphology, bed texture, and burial on tracer transport are expressed largely on the duration of the immobile periods. Immobile periods include brief rests during a flood, and longer term stationary periods that span multiple floods. Recognizing the importance of stationary periods on total transport, Einstein introduced the concept of virtual velocity. The virtual velocity $V_b = L/T$ is the total displacement, L, over the duration of the competent flow, T. This is distinguished from the actual velocity $V_s = ds/dt$ that the grains move during a single step of length ds and duration dt. The virtual velocity differs from the actual velocity in that the time over which it is measured includes both mobile and immobile periods; $T = t_{immobile} + t_{mobile}$.

Over relatively short time periods, field observations show that virtual velocity among a population of tracer particle varies widely. Sayre and Hubbell (1965) reported that the virtual velocity of radioactive sand exhibited a wide amount of scatter. Stelczer (1981) related the particle virtual velocity to the mean flow velocity close to the bed, and to the bed composition. His results were consistent with those of Sayre and Hubbell in terms of a wide range of values, which he attributed to reduced travel distance due to the exchange between the surface and subsurface. Hassan, Church, and Ashworth (1992) and Haschenburger and Church (1998) related the virtual velocity and the total travel distance to both peak stream power and the time-integrated excess stream power and found a weak correlation due to the large amount of scatter in the data that they attributed to bed-surface structures, sampling problems, and varying magnitude of sediment supply and flow. Phillips, Martin, and Jerolmack (2013) found that tracers in the Mameyes River traveled at a constant mean virtual velocity over 2 years. Wong *et al.* (2007) explored the relation between particle virtual velocity and flow characteristics in a series of flume experiments with uniform sediment and a plane bed to avoid the complexity of field data. Their experiments showed that although step length decreases weakly with increasing excess Shields stress, the virtual velocity increases due to more frequent particle entrainment. They also found that, contrary to some field results (see below), the virtual velocity was independent of the duration of competent flow. However, reanalysis of their data revealed that the virtual velocity depends on flow duration as well as shear stress (Voepel, Schumer, and Hassan 2013). This is probably due to particle trapping and burial that slows the particle virtual velocity as the flow duration increases.

Over longer periods, most field observations show that the virtual velocity of tracer particles slows over time. Ferguson and Wathen (1998) tracked particles along six reaches in the Allt Dubhaig in Scotland for 2 years and found that the virtual velocities of small particles were higher than those of large particles, and that 90% of the variation in the virtual velocity of tracer particles could be explained by relative particle size and reach Shields stress. Ferguson and Hoey (2002) resurveyed the positions of the same tracer population 8 years after installation and found that the virtual velocity over the full 8 years of the experiment averaged just 50% of the virtual velocity during the first 2 years. They also found a much weaker dependence on shear stress than in the first 2 years of the experiment, which they attributed to vertical mixing of tracers into the streambed and storage in locations of lower mobility such as gravel bars. Haschenburger (2011b) tracked tracer gravels in Carnation Creek, Vancouver Island, over 18 years and found that virtual velocity decreased rapidly with increasing cumulative stream power, eventually reaching an equilibrium value of about ~28% of the initial virtual velocity. She also attributed the long-term slowing of virtual velocity to tracer burial. Burial depth is likely to determine the timing and the duration of movement of particles within the active layer, as deeply buried particles are likely to move later during the event and for shorter duration (Hassan and Church 1994; DeVries 2002; Wong *et al.* 2007).

Recent research has focused on how the amount of variability in the virtual velocity of a population of sediment particles affects the mode of dispersion of a sediment plume. In particular, the possibility that excessive variability in virtual velocity leads to anomalous dispersion (also known as non-Fickian or fractional dispersion) as opposed to normal dispersion: a sediment plume undergoing anomalous dispersion spreads out along a stream either faster or slower than a normally dispersive plume. The spread of the tracer plume is measured by the scaling of the variance of particle position with time,

$\sigma^2 \sim t^\gamma$. The value of γ indicates the mode of dispersion. A normally dispersive plume spreads linearly, $\gamma = 1$, and an anomalous plume spreads nonlinearly, $\gamma \neq 1$. Ganti *et al.* (2010) discussed anomalous dispersion in the context of sediment transport, and Metzler and Klafter (2004) provide a general review of the physics of anomalous dispersion.

Anomalous dispersion arises when the probability distribution of virtual velocity has a heavy tail, meaning that extreme values occur with higher probability than predicted by thin-tailed probability distributions. Heavy-tailed distributions have a cumulative exceedance function with a power-law tail that decays as $x^{-\alpha}$ with $\alpha < 2$, and a divergent second moment, meaning that the sample variance continues to increase with sample size due to the high probability of extreme values. This is in contrast to thin-tailed distributions such as the exponential or gamma distribution, where the sample variance converges on a stable value once the sample size is sufficiently large. Newman (2005) provides a thorough and accessible discussion of heavy-tailed probability distributions.

A heavy-tailed virtual velocity distribution can arise as a result of either heavy tailed step-length or heavy tailed resting-time distributions. Hill, Dellangelo, and Meerschaert (2010) showed how heavy tailed step-length distributions can arise from a gamma distribution of grain sizes and grain-size-dependent exponential distributions of step length. Bradley, Tucker, and Benson (2010) showed an example of anomalous sediment dispersion in a sand tracer experiment that they attributed to heavy-tailed step lengths. More recently, Martin, Jerolmack, and Schumer (2012) and Phillips, Martin, and Jerolmack (2013) showed examples of anomalous gravel dispersion in both laboratory and field experiments that appears to arise as a consequence of heavy-tailed resting times. Nikora *et al.* (2001) argued that heavy-tailed resting times should result in anomalous subdiffusion ($\gamma < 1$) but Martin, Jerolmack, and Schumer (2012) and Phillips, Martin, and Jerolmack (2013) both observed anomalous superdiffusion ($\gamma > 1$). Superdiffusion arises from heavy-tailed rests when the steps are strongly biased in one direction, as is the case with downstream gravel transport (Weeks and Swinney 1998). The superdiffusive sand plume discussed by Bradley, Tucker, and Benson (2010) could have arisen as a consequence of heavy-tailed rests, though they did not consider that possibility. Hassan *et al.* (2013) compiled data from 64 gravel-bed field tracer experiments to determine probabilistic characteristics of gravel travel distance and waiting time. They reported that travel distances of mobile particles were in most cases thin-tailed, and concluded that continuous time tracking of tracers is necessary to capture exact step and waiting-time distributions.

6.7 Burial Depth and Vertical Mixing

Rivers mix sediment vertically, moving surface particles into the subsurface and displacing subsurface material upwards through scour and fill. The burial depth of tracer particles is controlled by the thickness of the active layer, the magnitude and duration of the flow, the texture and structure of the bed sediment, and the channel morphology (Hassan and Church 1994; Haschenburger 1999; Wong *et al.* 2007). The vertical dispersion of tracer particles affects both the vertical and longitudinal dispersion of sediment, and involves both active mobile gravels and passive stationary material (Hassan 1992; Haschenburger 2011a,b). The location of the tracer particles relative to the bed surface determines the entrainment probability, the length of the waiting time (Voepel, Schumer, and Hassan 2013; Martin, Jerolmack, and Schumer 2012; Martin, Purohit, and Jerolmack 2014), the travel distance, and the virtual velocity. The vertical exchange between the bed material and the mobilized bedload also modifies the characteristics of the bed texture (Hassan and Church 1994; Hoey and Ferguson 1994; Haschenburger 2013).

In contrast to the extensive literature on particle travel distance and step length, limited research has been focused on vertical mixing and burial depth, despite the fact that it is a fundamental control on virtual velocity. Field observations are limited to a few streams. Hassan (1990) reported that deeply buried particles are less mobile than ones on the surface, that burial is deeper in bars than in pools and riffles, and that substantial proportions of the particles become buried after one or two events (see Schick, Lekach, and Hassan 1987; Hassan 1992). Liébault *et al.* (2012) found that 80% of their tracer population became buried after 17 competent floods. Hassan and Church (1994) measured tracer particle burial in four different streams and reported that single-peaked and short floods tended to yield exponential distributions of burial depth, whereas multipeak and long-duration events resulted in uniform distributions of burial depth. They found no simple relation between particle size and burial depth. Gottesfeld *et al.* (2004) found that the concentration of buried tracers in small salmonid spawning streams in British Columbia decreased quasi-exponentially with depth and that the burial depth increased with flood discharge.

Haschenburger (2011a; see also Haschenburger 2013) evaluated the depth of burial and vertical exchange of magnetically tagged tracers recovered 11 times over a 17 year flood series in Carnation Creek. The data set is unique in terms of its long duration and relatively large number of tracers. She found that the vertical mixing of particles in Carnation Creek was very rapid; within one to eight events the mean burial depth reached approximately two times the thickness of the bed surface layer (defined as the local D_{90}), and small particles were mixed faster than large ones (Haschenburger 2011a). The exact time to reach vertical mixing equilibrium was largely determined by the timing of large floods after tracer deployment. After the tracer particles were well-mixed into the bed, they showed a pattern of exchange with the surface layer broadly consistent with multi-layer vertical exchange models (see below) (Schick, Lekach, and Hassan 1987; Hassan and Church 1994; Ferguson and Hoey 2002). Haschenburger (2011b) observed that the vertical component of tracer dispersion was constrained by the depth of the active layer and therefore streamwise dispersion of coarse sediment in gravel-bed streams dominates over the vertical dispersion. Nevertheless, vertical mixing is an essential part of the dispersion of fluvial sediment through its effects on particle entrainment (Haschenburger 2011b). Wong *et al.* (2007) reported an increase in the depth of exchange and bed mobility as the flow magnitude or duration increased in a flume using uniform sediment under constant flow conditions; an outcome supported by field observations (Schick, Lekach, and Hassan 1987; Hassan 1992; Haschenburger 1999). Phillips, Martin, and Jerolmack (2013) observed a pattern of tracer dispersion in the Mameyes River that implied heavy-tailed particle resting times that they attributed to tracer burial based on laboratory experiments (Martin, Jerolmack, and Schumer 2012).

Although burial and vertical mixing had been used to explain low recover rates of painted particles (e.g. Leopold, Emmett, and Myrick 1966), Schick, Lekach, and Hassan (1987) were the first to describe burial depth and vertical mixing in gravel-bed streams with a Markov model of particle exchange between the surface and subsurface layers. The model consisted of a surface layer and a subsurface layer to simulate vertical exchange of sediment due to scour and fill. It showed that after just a few (~1–5) sediment-mobilizing events, the exchange rate and the number of particles in each layer approached an equilibrium value. The time to reach equilibrium exchange decreased with the fraction of tracers mobilized, which in turn was largely determined by the amount of bed-surface armoring and the magnitude of the mobilizing events. Haschenburger (2013) applied the model to extensive tracer data sets from Carnation Creek and the Allt Dubhaig.

Hassan and Church (1994) developed a ten-layer sediment exchange model based on field observations of the distribution of burial depth of coarse material. The model is based on the assumptions that the burial depth of individual particles in single-peaked short events is exponentially distributed and that the probabilities for particles to move to another layer (the transition matrix) remain constant from event to event. As designed, the model yielded an exponential distribution of tracer burial depth for single-peaked short events and armored surfaces. After many events however, the distribution of burial depth tended towards a uniform distribution.

Ferguson and Hoey (2002) developed a three-layer Markov model of vertical particle exchange to simulate the slowdown of tracers due to vertical mixing as a compromise between the two-layer model of Schick, Lekach, and Hassan (1987) and the ten-layer model of Hassan and Church (1994). The model includes a surface layer, an active layer, and an inactive layer, with each layer corresponding to successively deeper tracer burial. The model was calibrated to tracer displacements after 2 and 8 years in the Allt Dubhaig, Scotland, and showed that mixing equilibrium was reached between 3 and 5 years. In an "environment like Scotland with no shortage of floods" (an average of about one flood a month), this corresponds to around 30 to 50 competent events (Ferguson and Hoey 2002).

6.8 Depth of the Active Layer

The active-layer thickness limits the depth of vertical mixing and controls the estimate of bedload flux based on the mean displacement distance of a population of tracers (DeVries 2002; Haschenburger and Church 1998; Hassan and Ergenzinger 2003; Liébault and Laronne 2008; Houbrechts *et al.* 2012; Laronne *et al.* 1992; Wong *et al.* 2007). Field and experimental research has linked the depth of the active layer to the magnitude of the flow (Hassan *et al.* 1992; Wilcock and McArdell 1997; Haschenburger and Church 1998; Wong *et al.* 2007; Houbrechts *et al.* 2012; Schneider *et al.* 2014). Field observations have also shown that the active layer is relatively thin, and that in most cases is smaller than the thickness of the bed surface layer as represented by the 90th percentile of grain size, D_{90} (Hassan 1990; Haschenburger 1999; Houbrechts *et al.* 2012), and generally does not exceed $2D_{90}$ (Wilcock *et al.* 1996; Wilcock and McArdell 1997; Haschenburger and Church 1998; DeVries 2002).

Wilcock *et al.* (1996) related the thickness of the active layer of the Trinity River in northern California to the local bed shear stress. They reported that particle exchange associated with bedload transport occurred to a depth of $1D_{90}$ of the bed material at a dimensionless shear stress of 0.035. A curve fit to their data suggested that the exchange depth should asymptotically approach a limiting value of approximately $1.7D_{90}$ at higher shear stresses. Wilcock and McArdell (1997) showed that the depth of the vertical mixing in a flume has a limiting value of about $2D_{90}$. Haschenburger and Church (1998) measured active-layer thickness using both the depth of tracer burial depths and the disturbance depth of scour chains and found that both measures indicated that the depth of the active layer increased with stream power and did not exceed $2D_{90}$.

DeVries (2002) used scour chains to study the temporal and spatial variability in the maximum disturbance depth of the beds of seven rivers in Washington State. He reported that the maximum disturbance depth was limited to about $2D_{90}$, consistent with the results of Wilcock *et al.* (1996), but found no clear trend of increasing disturbance depth with increasing peak Shields stress above a threshold stress of about 0.035. Furthermore, he reported that only a fraction of the streambed experiences a disturbance depth of $2D_{90}$ and the remainder of the channel experiences much less than that, with mean values ranging between $1D_{90}$ and $1.5D_{90}$ when the entire bed was mobilized. The average disturbance depth can be less than D_{90} when the bed is only partially mobilized. In the most active parts of the channel, disturbance depths were approximately uniformly distributed, which is consistent with the results of the one-dimensional model for vertical mixing developed by Hassan and

Church (1994), who determined that mixing-depth distributions tended to uniformity with increased transport activity and areal extent of distribution.

Houbrechts *et al.* (2012), using one of the most extensive tracer data sets, showed that the depth of the active layer was positively correlated with the peak-specific stream power of the mobilizing event. Although the field data included several relatively large floods (return period > 10 years), the thickness of the active layer was less than the D_{50} of the bed material in almost every case. The relatively low thickness of the active layer in comparison to other published values (e.g. DeVries 2002; Lenzi, Mao, and Comiti 2006; Rovira and Kondolf 2008) was attributed to the flatness of the schist pebbles that comprise the bed material, and to the imbricated and armored bed surface. In most cases, the thickness of the active layer was comparable to the *c*-axis of the bed material.

6.9 Morphology

Although the role of channel morphology (e.g., bars, riffles, pools) on particle dispersion has long been recognized (Takayama 1965; Leopold, Emmett, and Myrick 1966; Laronne and Carson 1976; Hassan, Church, and Schick 1991; Sear 1996; Beechie 2001; Pyrce and Ashmore; 2003a, 2005), the traditional view that particle displacement is a purely stochastic process subject to hydraulic control is still the dominant one. However, tracer studies have yielded evidence that for larger flows capable of mobilizing most of the stream bed, particle displacement is not purely stochastic, and the hydraulic control on particle mobility becomes secondary to the effects of channel morphology and bed-surface structures (Church and Hassan 1992; Pyrce and Ashmore 2003a).

Characteristics of the channel morphology (e.g., pool–bar spacing, riffle–pool spacing, spacing between deposition zones) have been used as an estimate of the mean travel distance of bedload particles in efforts to estimate bedload transport. For example, Neill (1987) used the distance between scour pool and point bar in a meander bend to estimate the mean travel distance of particles and to calculate the sediment transport rate. Other morphologic surrogates for travel distance include the spacing between riffle and bar (Ashmore and Church 1998), the distance between successive sedimentation zones (Church *et al.* 1986; McLean 1990), the distance between riffles, and channel width (Beechie 2001).

While there is evidence to suggest that stochastic models such as the gamma distribution or the EHS compound Poisson model are suitable for small, single-peak flow events, low rates of sediment transport, or channels lacking well-developed morphology (e.g. Hassan *et al.* 1991; see Tables 6.1 and 6.2 for other examples), the models are not suitable for large, multipeak flow events or in channels with well-developed morphology. The failure of simple stochastic models to describe particle displacements has been highlighted for different cases by Hassan and Church (1992). They reported poor performance of the models for gravel transport in Carnation Creek, a stream with well-developed pool–bar channel morphology and moderate to high sediment transport rates, in Harris Creek, a stream with long-duration snowmelt events, complex surface structures, and channel morphology with relatively low sediment transport rates, and in Nahal Og, a semi-arid stream with multipeak events and relatively high transport rates. For these streams, channel features such as bars and pools are likely to significantly influence the travel distance of particles, especially during large events in which the bed is fully mobile.

The influence of bars on travel distance and sediment mobility has been examined using data from Allt Dubhaig (Wathen *et al.* 1995) and Halfmoon Creek (Bradley and Tucker 2012). After several competent flow events, a large proportion of the tagged particles remained trapped within the main bar in

Table 6.2 Tracer experiments, methods, and transport models.

Study	Technique	Transport model
Hassan, Church, and Schick (1991)	Magnetically tagged gravel	EHS or gamma distribution
Hassan and Church (1992)	Magnetically tagged gravel and distinct lithologies	EHS or gamma distribution
Schmidt and Ergenzinger (1992)	Iron and magnetically gravel particles	Exponential distribution
Schmidt and Gintz (1995)	Magnetically tagged gravel	Gamma distribution, but a poor fit
Gintz, Hassan, and Schmidt (1996)	Magnetically tagged gravel	EHS or gamma distribution
Pyrce and Ashmore (2003b)	Fluorescent grains	Cauchy or combined gamma–Cauchy
Bradley and Tucker (2012)	PIT tagged gravel	Gamma–exponential (GEM)
Liébault *et al.* (2012)	PIT tagged gravel	Power-law steps
Phillips, Martin, and Jerolmack (2013), Phillips and Jerolmack (2014)	PIT tagged gravel	Exponential steps with heavy-tailed rests

Allt Dubhaig (Figure 6.3a). Results from a large PIT-tagged gravel study in Halfmoon Creek, a snowmelt-driven stream near Leadville, CO (see Bradley and Tucker (2012) for general information about the study), clearly show that tracers have a tendency to accumulate on gravel bars (Figure 6.3b), and that the accumulation happens during higher flow years (2010 and 2011 in Figure 6.3c). Liébault *et al.* (2012) also recently documented the critical influence of bars on particle dispersion in a wandering gravel-bed stream. Furthermore, these results confirmed early work on the influence of bar morphology on particle dispersion in streams (e.g., Mosley 1978; Kondolf and Matthews 1986).

The role of riffle–pool morphology on sediment entrainment, mobility, sorting, and flow has been the focus of many studies. Field tracer studies (summarized in Sear 1996) show that the travel distances of tracers seeded in pools are larger than those seeded in riffles (Figure 6.4a). Furthermore, at the bankfull flow, the travel distances in pool-head and pool-tail are higher than on riffles (Ashworth 1987; Sear 1996). The influence of riffles and pools on bed mobility was examined using data from East Creek, Canada (Figure 6.4b). In the low-flow year the majority of the channel is under partial mobility or no mobility, with only a few areas being fully mobile. Areas of partial and full mobility are found across the entire width of the riffle–bar unit, while they are concentrated in the center of the channel and exit slope in the pool. In the high flow year, full mobility is reached everywhere on the bed for the chosen subunits. In 2012–2013, when flows were nearest bankfull, patterns of mobility match the patterns of bed shear stress computed by the model. Partial and full mobility is concentrated in the right-hand side of the riffle and on the bar. In the pool, mobility in concentrated down the center of the channel and in the exit slope of the pool.

Pyrce and Ashmore (2003a) provided the most comprehensive attempt to explore the role of channel morphology on travel distance of tracer particles. They synthesized and reanalyzed published data from a wide range of river environments (e.g. differing flow regimes, channel morphology, and bed-surface composition). Their analyses yielded uni-, bi-, and multimodal distributions of travel distance, including modes that correspond to channel features such as bars. They concluded that gamma and exponential distributions with a short modal path length are typical of small events, while bimodal and

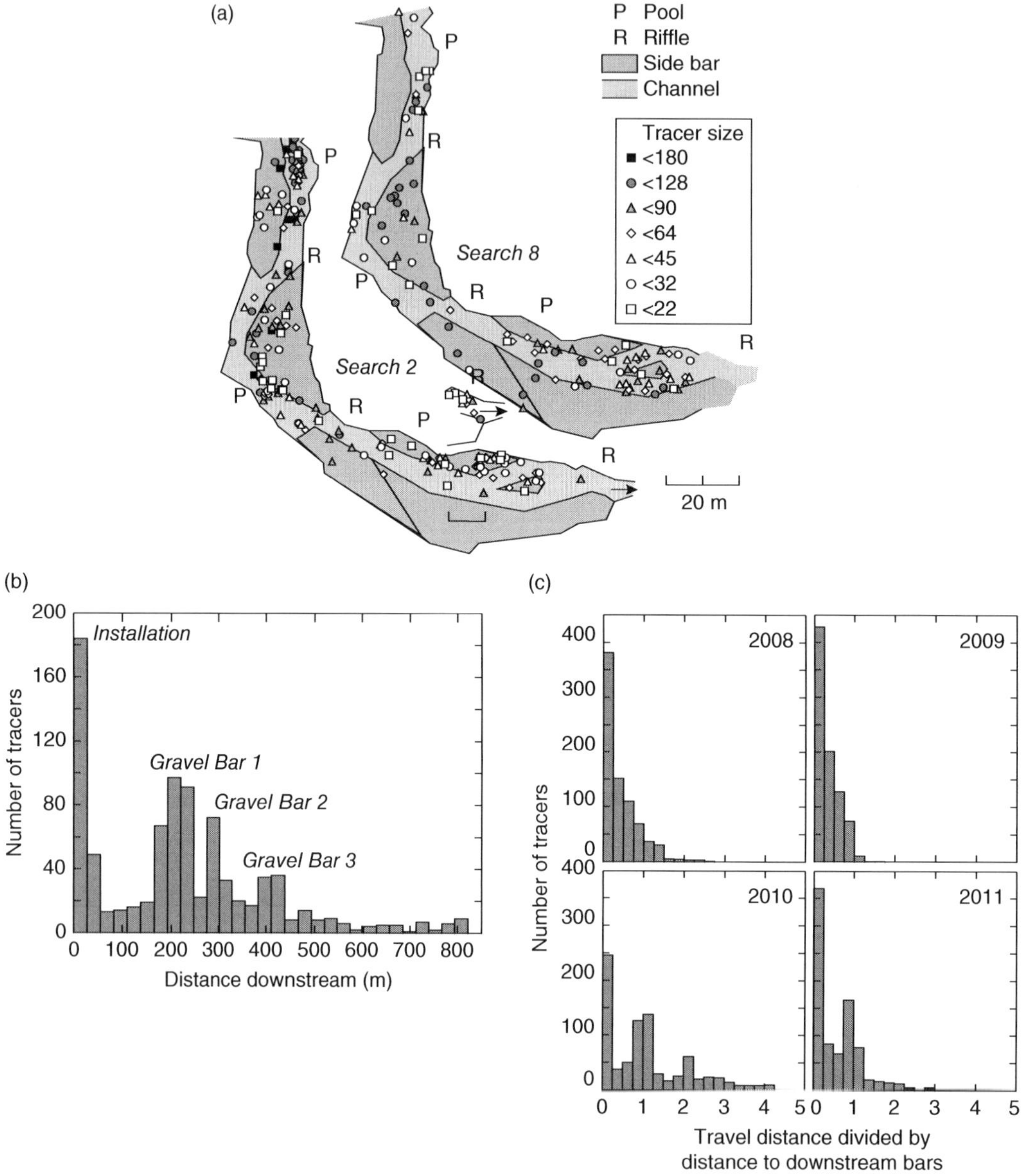

Figure 6.3 (a) Tagged particle movement plots grouped by size classes during a few events in Allt Dubhaig (data from Wathen *et al.* 1995). (b) PIT tagged gravel-travel distances during snowmelt events in Halfmoon Creek. (c) Travel distances during four flood seasons in Halfmoon Creek showing that accumulation happens during higher flow years (2010 and 2011).

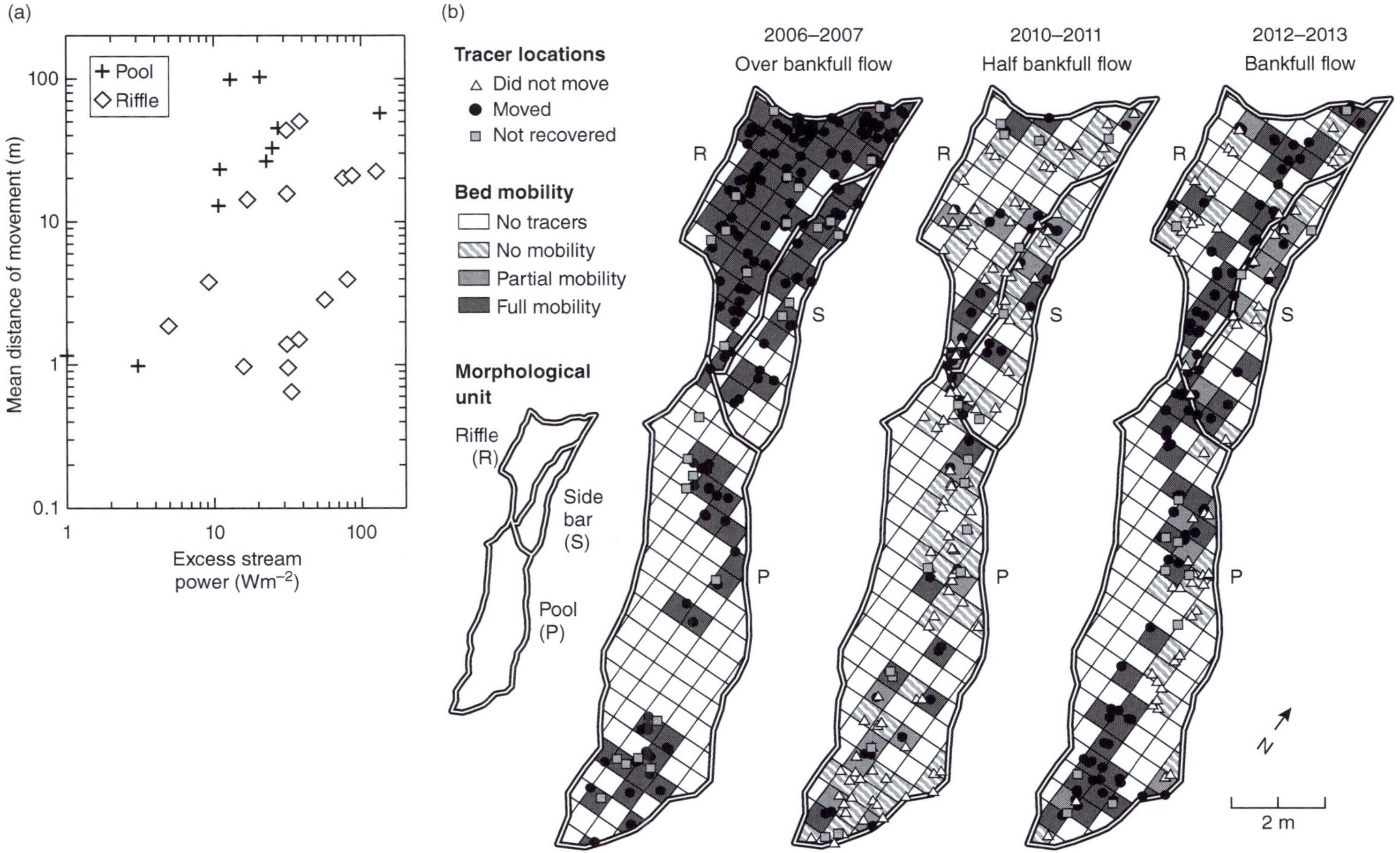

Figure 6.4 (a) Mean travel distance of particles seeded in pool and riffles. (Modified from Sear 1996). (b) Sediment mobility in a riffle–pool sequence in East Creek, Canada. An 0.5 m by 0.5 m grid was overlaid on top of the bed. This grid size was chosen because 0.5 m was selected as the threshold for tracer movement. The relative mobility in each grid was calculated as the fraction of grains that moved divided by the total number of tracer stones in that grid. Stones that were not recovered are excluded from the analysis.

multimodal travel distances are associated with large floods and well-developed channel morphology. However, they concluded that available field data are inadequate to thoroughly explore the role of channel morphology on particle displacement.

Pyrce and Ashmore (2003b, 2005) studied the influence of bars on the travel distance of particles in a flume. They reported a departure from the two-parameter gamma distribution attributed to the modulating effect of channel bedforms on the travel distance of tracer particles. They demonstrated that the travel distances of particles at high flows (channel forming discharges) follow a Cauchy distribution centered on the distance to the next bar downstream and the travel distances followed the gamma distributions at low flows. The large proportion of tracer particles that were deposited on the first point bar downstream of the seeding area emphasized the role that channel morphology exerts on travel distance. Furthermore, Pyrce and Ashmore (2005) linked bar development in a low-sinuosity channel to the pattern of particle displacement. In the early stages of the bar formation, sediment moved from scour zones across the point-bar head to the bar apex, while at later stages, less sediment moved across the bar and the locus of the movement shifted to the thalweg. Recently, Kasprak *et al.* (2014) used high-resolution elevation models derived from structure-from-motion photogrammetry in a flume setting to study downstream dispersion of tracer particles in braided rivers. They reported that about 80% of the particles were found in association with compound, lateral, diagonal, and point bars. About 70% of recovered particles were found in bar heads and bar margins, and only 3% were found on the bar tail.

The role of the step-pool morphology on particle dispersion has been the focus of several studies (Gintz, Hassan, and Schmidt 1996; Lenzi 2004; Lamarre and Roy 2008b). Travel-distance distributions of particles yielded irregular skewed results with secondary modes in most cases. The secondary mode is attributed to trapping by channel features such as bars and pools. Lamarre and Roy (2008b) reported travel-distance distributions that follow the gamma distribution and that the dominant morphological length-scale of the bed controls the travel distance of the tagged clasts. In most cases, however, the path length they reported was larger than the step-pool spacing, indicating that particles traverse several step-pool pairs during a flow event.

In summary, the influence of channel morphology on particle dispersion is still an open research question. To fill this gap in our knowledge there is a need for new studies that are designed to isolate the role of channel morphology and particle dispersion. In spite of the early recognition of the relative importance of channel morphology on particle dispersion, there is still a need for field and experimental studies on the effects of channel morphology, bed composition, and flow regime on the travel distance of coarse bed material.

6.10 Bed Texture

Bed-surface armoring and structures such as pebble clusters, transverse ribs, and stone cells are known to impact the rate of particle entrainment (Cin 1968; Laronne and Carson 1976; Brayshaw, Frostick, and Reid 1983; Church, Hassan, and Wolcott 1998; Oldmeadow and Church 2006), and therefore the duration of the rest period. Particles trapped in between large stones as a part of bed-surface structures are likely to have longer rest periods (Laronne and Carson 1976; Church, Hassan, and Wolcott 1998; Hassan *et al.* 2008), while the opposite is true in the case of an unstructured bed (Stelczer 1981; Hassan *et al.* 2008). Surface structures may also influence particle travel distance by enhancing deposition probability (Lamarre and Roy 2008a). Although tracer particles could provide detailed information on the influence of bed-surface structures on sediment mobility and bed stability, few attempts have been made to relate the displacement of individual particles to bed-surface arrangement.

Laronne and Carson (1976) attempted to explain the lack of a clear relation between travel distance, particle size and shape, and water discharge in terms of packing of alluvial sediment on bed surfaces. Based on detailed mapping of the deposition environment of the tracer particles, they identified an array of grain-packing arrangements including pebble clusters, vertically in-filled cavities, and imbricated structures. These structures increased the force required to initiate motion to approximately two to six times the submerged particle weight. They concluded that particle travel distance is smallest for grains in tight structural arrangements and greatest in open and unfilled structures. Their results are consistent with studies focusing on pebble clusters where it has been shown that particles seeded into pebble clusters were entrained less frequently and moved shorter distances than those that were seeded on the bed surface (Cin 1968; Brayshaw, Frostick, and Reid 1983; Reid, Frostick, and Brayshaw 1992)

Oldmeadow and Church (2006) investigated the influence of reticulate bed-surface stone cells (Church *et al.* 1998) on channel stability by manually destroying all the structures in an experimental reach. Particle entrainment and mobility in the disturbed reach was much greater than in an adjacent undisturbed control reach. Lamarre and Roy (2008a) used PIT tagged tracers to confirm these results. They examined the development and influence of bed-surface structures on sediment mobility in a relatively steep cobble-bedded channel. Tagged particles were found deposited around "keystones" – particles that were part of the sedimentary structures that developed rapidly and remained largely intact even during relatively large flood events.

In summary, there is a need for both field and experimental work on the impacts of bed-surface structures on particle dispersion at both the grain and reach scales. Linking particle path length to bed-surface arrangement would provide essential information on particle dispersion and aid in the development of predictive models for sediment transport.

6.11 Closing Remarks

In this chapter we reviewed dispersion patterns and characteristics of step length, rest period, travel distance, burial depth, and vertical mixing of coarse bed material in gravel-bed streams. Gaps in our knowledge were identified, and a need for future research was highlighted. The spatial and temporal variability of individual particle dispersion in floods highlights the complexity of sediment transport in gravel-bed rivers and the need to better understand the interactions among flow, channel morphology, and particle characteristics.

The lack of detailed information on the movement of individual particles is for the most part caused by the difficulties in finding suitable measuring techniques to deploy in the field. Ideally, the technique should allow for the tracing of large samples (thousands of stones), have the ability to provide real-time data on the tracer's location and to estimate the burial without disturbing the particles, and a long life expectancy that would allow the monitoring of the movement for long periods of time. Such a technique should also be suitable for use in both large and small rivers. In spite of the technological advances in the decades since Einstein (1937), such a technique does not exist.

The three ranges of spatial and temporal scales (local, intermediate, and global) of sediment particle motion and particle dispersion could be adopted as a framework for the study of particle movement in streams. We propose that the global range might be meaningfully split into a "flood range," in which motion consists of steps and rests, and a "geomorphic range," in which the storage of particles in sedimentary deposits exerts a strong control on the total travel distance of a particle over a series of floods. Transport at geomorphic timescales is largely a function of the hydrologic conditions (e.g., flood frequency and magnitude) and channel morphology.

To address transport at the geomorphic scale, stochastic models need to consider that the interaction with bed texture and channel morphology leads to two widely different immobile timescales: that of a rest between steps, and that of a potentially much longer immobile timescale associated with particle burial and channel morphology. The impulse framework (Phillips, Martin, and Jerolmack 2013) is promising, but needs to be tested over a longer periods and in different types of streams as its prediction of a constant mean virtual velocity is inconsistent with studies that show that tracers slow down over the long term (e.g. Ferguson *et al.* 2002; Haschenburger 2011b). Recently, it has been suggested that heavy-tailed probability distributions without characteristic scales may better describe step length than more traditional exponential or gamma models (Ancey *et al.* 2006; Bradley, Tucker, and Benson 2010; Ganti *et al.* 2010; Hill, Dellangelo, and Meerschaert 2010). Subsequent work has indicated that heavy-tailed resting times coupled with thin-tailed step lengths are more likely to be responsible for patterns of tracer dispersion that could also be attributed to heavy-tailed step lengths (Martin, Jerolmack, and Schumer 2012; Phillips, Martin, and Jerolmack 2013; Martin, Purohit, and Jerolmack 2014). It is possible that treating the distribution of resting times as a heavy-tailed random variable may bridge the gap between the flood range and the geomorphic range. Models of this type have been applied to solute transport in porous media (e.g., Schumer *et al.* 2003), but the possibility of heavy-tailed resting times in sediment transport has only recently received attention (Martin, Jerolmack, and Schumer 2012; Bradley and Tucker 2013; Phillips, Martin, and Jerolmack 2013; Voepel, Schumer, and Hassan 2013).

Although burial depth and vertical mixing have been recognized as a fundamental part of the process of sediment movement and dispersion in streams, little research has been focused on these topics. The burial depth controls particle entrainment, travel distance, virtual velocity, and movement duration and is therefore fundamental for the estimation of sediment flux in streams. Also, the burial depth can provide information on the depth of the active layer, vertical sorting of sediment and residence time. For progress in our understanding of particle kinematics, there is a need to conduct field and theoretical work on the topic. Similarly, there is a need to conduct more research aiming at the quantification of the role of channel morphology on sediment dispersion in streams.

Given the recent theoretical progress in the study of sediment transport in streams, there is a need to design new studies focused on testing theory and providing information needed for developing new sediment-transport models. Long-term experiments designed to explore the role of flood magnitude, duration, and history on sediment motion are essential for real progress on the topic. Similarly, channel morphology, bed composition, burial depth, and the active layer need to be explored more extensively. This can be achieved by combined field measurements and laboratory experiments.

Acknowledgments

The research was funded by the Natural Sciences and Engineering Research Council of Canada through a Research Grant to M. Hassan. D. N. Bradley was supported by a Bureau of Reclamation Science and Technology Program grant. Trevor Hoey and Rob Ferguson provided original data from the Allt Dubhaig. Elli Papangelakis produced Figure 6.3. Eric Leinberger prepared the final version of all figures. Douglas Jerolmack, an anonymous reviewer, Jonathan Laronne, and Daizo Tsutsumi furnished helpful reviews, which we appreciate.

References

Ancey C, Böhm, T, Jodeau, M, and Frey, P 2006. Statistical description of sediment transport experiments. *Physical Review E* 74. DOI:org/10.1103/PhysRevE.74.011302.

Ashmore, P and Church, M 1998. Sediment transport and river morphology: a paradigm for study. in *Gravel-Bed Rivers in the Environment* (eds PC Klingeman, RL Beschta, PD Komar, and JB Bradley). Water Resources: Highlands Ranch, CO; 115–148.

Ashworth, PJ 1987. *Bedload transport and channel change in gravel-bed rivers.* PhD thesis, University of Stirling, UK.

Beechie, TJ 2001. Empirical predictors of annual bed load travel distance, and implications for salmonid habitat restoration and protection. *Earth Surface Processes and Landforms* 26, 1025–1034.

Bradley, DN and Tucker, GE 2012. Measuring gravel transport and dispersion in a mountain river using passive radio tracers. *Earth Surface Processes and Landforms* 37, 1034–1045.

Bradley, DN and Tucker, GE 2013. The storage time, age, and erosion hazard of laterally accreted sediment on the floodplain of a simulated meandering river. *Journal of Geophysical Research: Earth Surface* 118, 1308–1319.

Bradley, DN, Tucker, GE, and Benson, DA 2010. Fractional dispersion in a sand bed river. *Journal of Geophysical Research: Earth Surface* 115, 33–52.

Brayshaw, AC, Frostick, LE, and Reid, I 1983. The hydrodynamics of particle clusters and sediment entrapment in coarse alluvial channels. *Sedimentology* 30, 137–143.

Chacho, EF Jr, Emmett, WW, and Burrows, RL 1994. Monitoring gravel movement using radiotransmitters. *Hydraulic Engineering* 94, 785–789.

Chapuis, M, Bright, CJ, Hufnagel, J, and Macvicar, B 2014. Detection ranges and uncertainty of passive Radio Frequency Identification (RFID) transponders for sediment tracking in gravel rivers and coastal environments. *Earth Surface Processes and Landforms* 39, 2109–2120.

Church, M and Hassan, MA 1992. Size and distance of travel of unconstrained clasts on a streambed. *Water Resources Research* 28, 299–303.

Church, M, Hassan, MA, and Wolcott, JF 1998. Stabilizing self-organized structures in gravel-bed stream channels: field and experimental observations. *Water Resources Research* 34, 3169–3179.

Church, M, Miles, M, Rood, KM, and Miles, M 1986. *Sediment Transfer Along Mackenzie River: a Feasibility Study.* Report IWD-WNR(NWT)-WRB-SS-87-1, Environment Canada, Inland Waters Directorate, Western and Northern Region, Sediment Survey Section: Ottawa, Canada.

Cin, RD 1968. Pebble clusters: their origin and utilization in the study of palaeocurrents. *Sedimentary Geology* 2, 233–241.

Crickmore, M and Lean, G 1962. The measurement of sand transport by means of radioactive tracers. *Proceedings of the Royal Society of London. Series A. Mathematical and Physical Sciences* 266, 402–421.

Crickmore, MJ, Tazioli, OS, Appleby, PG, and Oldfield, F 1990. *The Use of Nuclear Techniques in Sediment Transport and Sedimentation Problems.* Technical Documents in Hydrology, IHP-III Project 5.2, UNESCO: Paris; 170 pp.

DeVries, P 2002. Bedload layer thickness and disturbance depth in gravel bed streams. *Journal of Hydraulic Engineering* 128, 983–991.

Diplas, P, Dancey, CL, Celik, AO *et al.* 2008. The role of impulse on the initiation of particle movement under turbulent flow conditions. *Science* 322, 717–720.

Einstein, H 1937. *Bedload Transport as a Probability Problem.* Mitteilung der Versuchsanstalt fuer Wasserbau an der Eidgenössischen Technischen Hochschule: Zürich; 110 pp.

Ergenzinger, P and Schmidt, K-H 1990. Stochastic elements of bedload transport in a step-pool mountain river. In *Hydrology in Mountainous Regions 11.* Publication No. 194, IAHS: Wallingford, UK; 39–46.

Ferguson, R and Hoey, T 2002. Long-term slowdown of river tracer pebbles: Generic models and implications for interpreting short-term tracer studies. *Water Resources Research* 38, 1142. DOI: 10.1029/2001WR000637.

Ferguson, RI, and Wathen, SJ. 1998. Tracer-pebble movement along a concave river profile: virtual velocity in relation to grain size and shear stress. *Water Resources Research* 34(8), 2031–2038.

Ferguson, R, Bloomer, D, Hoey, T, and Werritty, A 2002. Mobility of river tracer pebbles over different timescales. *Water Resources Research* 38, 1045. DOI:10.1029/2001WR000254.

Ganti, V, Meerschaert, MM, Foufoula-Georgiou, E *et al.* 2010. Normal and anomalous diffusion of gravel tracer particles in rivers. *Journal of Geophysical Research: Earth Surface* 115, F00A12. DOI:10.1029/2008JF001222.

Gintz, D, Hassan, MA, and Schmidt, K-H 1996. Frequency and magnitude of bedload transport in a mountain river. *Earth Surface Processes and Landforms* 21, 433–445.

Gottesfeld, AS, Hassan, MA, Tunnicliffe, JF, and Poirier, RW 2004. Sediment dispersion in fish bearing streams: the influence of floods and spawning salmon, *Journal of the American Water Resources Association* 40, 1071–1086.

Grigg, N 1970. Motion of single particles in alluvial channels. *Journal of the Hydraulics Division* 96, 2501–2518.

Habersack, HM 2001. Radio-tracking gravel particles in a large braided river in New Zealand: a field test of the stochastic theory of bed load transport proposed by Einstein. *Hydrological Processes* 15, 377–391.

Haschenburger, J 2013. Tracing river gravels: insights into dispersion from a long-term field experiment. *Geomorphology* 200, 121–131.

Haschenburger, JK 1999. A Probability model of scour and fill depths in gravel-bed channels. *Water Resources Research* 35, 2857–2869.

Haschenburger, JK 2011a. Vertical mixing of gravel over a long flood series. *Earth Surface Processes and Landforms* 36, 1044–1058.

Haschenburger, JK 2011b. The rate of fluvial gravel dispersion. *Geophysical Research Letters* 38, L24403. DOI:10.1029/2011GL049928.

Haschenburger, JK and Church, M 1998. Bed material transport estimated from the virtual velocity of sediment. *Earth Surface Processes and Landforms* 23, 791–808.

Hassan, MA 1990. Scour, fill, and burial depth of coarse material in gravel bed streams. *Earth Surface Processes and Landforms* 15, 341–356.

Hassan, MA 1992. Structural controls of the mobility of coarse material in gravel-bed channels. *Israel Journal of Earth Sciences* 41, 105–122.

Hassan, MA and Church, M 1992. The movement of individual grains on the streambed. in *Dynamics of gravel-bed rivers* eds P Billi, RD Hey, CR Thorne, and P Tacconi, P. Wiley, Chichester;159–175.

Hassan, MA and Church, M 1994. Vertical mixing of coarse particles in gravel bed rivers: a kinematic model. *Water Resources Research* 30, 1173–1186.

Hassan, MA and Ergenzinger, P 2003. Use of tracers in fluvial geomorphology. In *Tools in Fluvial Geomorphology* (eds GM Kondolf and H Piegay). John Wiley & Sons: Chichester; 397–423.

Hassan, MA and Roy, AG 2016. Coarse particle tracing in fluvial geomorphology. In *Tools in Fluvial Geomorphology*, 2nd edn (eds GM Kondolf and H Piégay). John Wiley & Sons: Chichester; 306–323.

Hassan, MA, Church, M, and Ashworth, PJ 1992. Virtual rate and mean distance of travel of individual clasts in gravel-bed channels. *Earth Surface Processes and Landforms* 17, 617–627.

Hassan, MA, Church, M, and Schick, AP 1991. Distance of movement of coarse particles in gravel bed streams. *Water Resources Research* 27, 503–511.

Hassan, MA, Smith, BJ, Hogan, DL *et al.* 2008. Sediment storage and transport in coarse bed streams: scale considerations. In *Gravel-Bed Rivers VI: From Process Understanding to River Restoration* (eds H Habersack, H Piegay, and M Rinaldi). Elsevier BV: Amestrdam; 473–496.

Hassan, MA, Voepel, H, Schumer, R *et al.* 2013. Displacement characteristics of coarse fluvial bed sediment. *Journal of Geophysical Research* 118. DOI:10.1029/2012JF002374.

Hill, KM, Dellangelo, L, and Meerschaert, MM 2010. Heavy-tailed travel distance in gravel bed transport: An exploratory enquiry. *Journal of Geophysical Research: Earth Surface* 115, F00A14. DOI:10.1029/2009JF001276.

Hoey, TB and Ferguson, R 1994. Numerical simulation of downstream fining by selective transport in gravel bed rivers: model development and illustration. *Water Resources Research* 30, 2251–2260.

Houbrechts, G, Van Campenhout, J, Levecq, Y *et al.* 2012. Comparison of methods for quantifying active layer dynamics and bedload discharge in armoured gravel-bed rivers. *Earth Surface Processes and Landforms* 37, 1501–1517.

Hubbell, D and Sayre, W 1964. Sand transport studies with radioactive tracers. *Proceedings of the American Society of Civil Engineers* 90, 39–68.

Kasprak, A, Wheaton, JM, Ashmore, PE *et al.* 2014. The relationship between particle travel distance and channel morphology: Results from physical models of braided rivers. *Journal Geophysical Research: Earth Surface* 120, 10.1002/2014JF003310.

Kondolf, MG and Matthews, WVG 1986. Transport of tracer gravels on a coastal California river. *Journal of Hydrology* 85, 265–280.

Lajeunesse, E, Malverti, L, and Charru F 2010. Bed load transport in turbulent flow at the grain scale: experiments and modeling. *Journal of Geophysical Research* 115, F04001. DOI:10.1029/2009JF001628.

Lamarre, H and Roy, AG 2008a. A field experiment on the development of sedimentary structures in a gravel-bed river. *Earth Surface Processes and Landforms* 33, 1064–1081.

Lamarre, H and Roy, AG 2008b. The role of morphology on the displacement of particles in a step–pool river system. *Geomorphology* 99, 270–279.

Lamarre, H, Macvicar, B, and Roy, A 2005. Using passive integrated transponder (PIT) tags to investigate sediment transport in gravel-bed rivers. *Journal of Sedimentary Research* 75, 736–741.

Laronne, J and Carson, M 1976. Interrelationships between bed morphology and bed-material transport for a small, gravel-bed channel. *Sedimentology* 23, 67–85.

Laronne, JB, Outhet, D, Duckham, J, and Mccabe, T 1992, Determining event bedload volumes for evaluation of potential degradation sites due to gravel extraction, NSW, Australia. In *Erosion and Sediment Transport Monitoring Programmes in River Basins: Proceedinings of the Oslo Symposium* (eds J Bogen, DE Walling, and TJ Day). IAHS Publications: Wallingford, UK; 87–94.

Lenzi, MA 2004. Displacement and transport of marked pebbles, cobbles and boulders during floods in a steep mountain stream. *Hydrological Processes* 18, 1899–1914.

Lenzi, MA, Mao, L, and Comiti, F 2006. When does bedload transport begin in steep boulder-bed streams? *Hydrological Processes* 20, 3517–3533.

Leopold, LB, Emmett, WW, and Myrick, RM 1966. Channel and hillslope processes in a semiarid area, New Mexico. *US Geololgical Survey Professional Paper* 352-G, 193–253.

Liébault, F and Laronne, JB 2008. Evaluation of bedload yield in gravel-bed rivers using scour chains and painted tracers: the case of the Esconavette Torrent (southern French Prealps). *Geodinamica Acta* 21, 23–34.

Liébault, F, Bellot, H, Chapuis, M *et al.* 2012. Bedload tracing in a high-sediment-load mountain stream. *Earth Surface Processes and Landforms* 37, 385–399.

Liedermann, M, Tritthart, M, and Habersack, H 2013. Particle path characteristics at the large gravel-bed River Danube: results from a tracer study and numerical modelling. *Earth Surface Processes and Landforms* 38, 512–522.

MacVicar, BJ and Roy, AG 2011. Sediment mobility in a forced riffle–pool. *Geomorphology* 125, 445–456.

Marion, A, Zaramella, M, and Bottacin-Busolin, A 2008. Solute transport in rivers with multiple storage zones: The STIR model. *Water Resources Research* 44, W10406. DOI:10.1029/2008WR007037.

Martin, RL, Jerolmack, DJ, and Schumer, R 2012. The physical basis for anomalous diffusion in bed load transport. *Journal of Geophysical Research: Earth Surface* 117, 1018. DOI:10.1029/2011JF002075.

Martin, RL, Purohit, P K, and Jerolmack, DJ 2014. Sedimentary bed evolution as a mean-reverting random walk: implications for tracer statistics. *Geophysical Research Letters* 41, 6152–6159.

McLean, DG 1990. *The relation between channel instability and sediment transport on lower Fraser River.* PhD thesis, University of British Columbia, Canada.

McNamara, JP and Borden, C 2004. Observations on the movement of coarse gravel using implanted motion-sensing radio transmitters. *Hydrological Processes* 18, 1871–1884.

Metzler, R and Klafter, J 2004. The restaurant at the end of the random walk: recent developments in the description of anomalous transport by fractional dynamics. *Journal of Physics A: Mathematical and General* 37, R161–R208.

Milan, DJ 2013. Sediment routing hypothesis for pool–riffle maintenance. *Earth Surface Processes and Landforms* 38, 1623–1641.

Mosley, MP 1978. Bed material transport in the Tamaki River near Dannevirke, North Island, New Zealand. *New Zealand Journal of Science* 21, 619–626.

Nakagawa, H and Tsujimoto, T 1976. Experimental study on evolution of sand waves from an initially flattened bed. *Annuals, Disaster Prevention Research Institute, Kyoto University*;189–309.

Neill, CR 1987. Sediment balance considerations linking long-term transport and channel processes. In *Sediment Transport in Gravel Bed Rivers* (eds CR Thorne, JC Bathurst, and RD Hey). John Wiley & Sons: Chichester, UK; 225–240.

Newman, MEJ 2005. Power laws, Pareto distributions and Zipfs law. *Contemporary Physics* 46, 323–351.

Nichols, M 2004. A radio frequency identification system for monitoring coarse sediment particle displacement. *Applied Engineering in Agriculture* 20, 783–787.

Nikora, V, Heald, J, Goring, D, and Mcewan, I 2001. Diffusion of saltating particles in unidirectional water flow over a rough granular bed. *Journal of Physics A: Mathematical and General* 34, L743. DOI.org/10.1088/0305-4470/34/50/103.

Nikora, V, Habersack, H, Huber, T, and Mcewan, I 2002. On bed particle diffusion in gravel bed flows under weak bed load transport. *Water Resources Research* 38. DOI:10.1029/2001WR000513.

Niño, Y and García, M 1994. Gravel saltation: 2. Modeling. *Water Resources Research* 30, 1915–1924.

Niño, Y, Atala, A, Barahona, M, and Aracena, D 2002. Discrete particle model for analyzing bedform development. *Journal of Hydraulic Engineering* 128, 381–389.

Oldmeadow, DF and Church, M 2006. A field experiment on streambed stabilization by gravel structures. *Geomorphology* 78, 335–350.

Olinde, L and Johnson, J 2013. Characterizing coarse bedload transport during floods with RFID and accelerometer tracers, in-stream RFID antennas and HEC-RAS modeling. *AGU Fall Meeting Abstracts*, San Francisco.

Paintal, A 1971. A stochastic model of bed load transport. *Journal of Hydraulic Research* 9, 527–554.

Papanicolaou, A, Diplas, P, Evaggelopoulos, N, and Fotopoulos, S 2002. Stochastic Incipient Motion Criterion for Spheres under Various Bed Packing Conditions. *Journal of Hydraulic Engineering* 128, 369–380.

Phillips, CB and Jerolmack, DJ 2014. Dynamics and mechanics of bed-load tracer particles. *Earth Surface Dynamics* 2, 513–530.

Phillips, CB, Martin, RL, and Jerolmack, DJ 2013. Impulse framework for unsteady flows reveals superdiffusive bed load transport. *Geophysical Research Letters* 40, 1328–1333.

Pyrce, RS and Ashmore, PE 2003a. The relation between particle path length distributions and channel morphology in gravel-bed streams: a synthesis. *Geomorphology* 56, 167–187.

Pyrce, RS and Ashmore, PE 2003b. Particle path length distributions in meandering gravel-bed streams: results from physical model. *Earth Surface Processes and Landforms* 28, 951–966.

Pyrce, RS and Ashmore, PE 2005. Bedload path length and point bar development in gravel-bed river models. *Sedimentology* 52, 839–857.

Reid, I, Frostick, L, and Brayshaw, A 1992. Microform roughness elements and the selective entrainment and entrapment of particles in gravel-bed rivers. In *Dynamics of Gravel-Bed Rivers* (eds P Billi, RD Hey, CR Thorne, and P Tacconi). John Wiley & Sons: Chichester, UK; 253–275.

Rovira, A and Kondolf, GM 2008. Bed mobility on the Deschutes River, Oregon: tracer gravel results. *Geodinamica Acta* 21, 11–22.

Sayre, W and Hubbell, D 1965. Transport and dispersion of labeled bed material, North Loup River, Nebraska. *US Geololgical Survey Professional Paper* 433-C, 48 pp.

Schick, A, Lekach, J, and Hassan, MA 1987. Vertical exchange of coarse bedload in desert streams. In *Desert Sediments: Ancient and Modern* (eds LE Frostick and I Reid). Special Publication 35, Geological Society: London; 7–16.

Schmidt, K-H and Ergenzinger, P 1992. Bedload entrainment, travel lengths, step lengths, rest periods – studies with passive (iron, magnetic) and active (radio) tracer techniques. *Earth Surface Processes and Landforms* 17, 147–165.

Schmidt, HK and Gintz, D 1995. Results of bedload tracer experiments in a mountain river. In *River Geomorphology* (ed. EJ Hickin). John Wiley & Sons: Chichester, UK; 145–158.

Schneider, J, Hegglin, R, Meier, S *et al.* 2010. *Studying Sediment Transport in Mountain Rivers by Mobile and Stationary RFID Antennas*. Bundesanstalt für Wasserbau: Braunschweig, Germany.

Schneider, JM, Turowski, JM, Rickenmann, D *et al.* 2014. Scaling relationships between bed load volumes, transport distances, and stream power in steep mountain channels. *Journal of Geophysical Research: Earth Surface* 119, 533–549.

Schumer, R., Benson, DA, Meerschaert, MM, and Baeumer B 2003. Fractal mobile/immobile solute transport. *Water Resources Research* 39, 1296. DOI:10.1029/2003WR002141.

Sear, D 1996. Sediment transport processes in pool–riffle sequences. *Earth Surface Processes and Landforms* 21, 241–262.

Sear, DA, Lee, MWE, Oakley, RJ *et al.* 2000. Coarse sediment tracing technology in littoral and fluvial environments: a review. In *Tracers in the Environment* (ed. I Foster). John Wiley & Sons: Chichester, UK; 21–55.

Sekine, M and Kikkawa, H 1992. Mechanics of saltating grains. II. *Journal of Hydraulic Engineering* 118, 536–558.

Seminara, G, Solari, L, and Parker, G 2002. Bed load at low Shields stress on arbitrarily sloping beds: Failure of the Bagnold hypothesis. *Water Resources Research* 38, 31-1–31-16. DOI:10.1029/2001WR000681.

Stelczer, K 1981. *Bed Load Transport: Theory and Practice*. Water Resource Publications: Ft. Collins, CO.

Takayama, S 1965. Bedload movement in torrential mountain streams. *Tokyo Geographical Paper* 9, 169–188.

Tsakiris, AG, Papanicolaou, AN, Moustakidis, IV, and Abban, BK 2015. Identification of the burial depth of radio frequency identification transponders in riverine applications. *Journal of Hydraulic Engineering*. DOI:10.1061/(ASCE)HY.1943-7900.0001001.

Vazquez-Tarrio, D and Menendez-Duarte, R 2015. Bedload transport rates of coarse-bed streams in an Atlantic region (Narcea River, NW Iberian Peninsula). *Geomorphology* 217, 1–14.

Voepel, H, Schumer, R, and Hassan, MA 2013. Sediment residence time distributions: theory and application from bed elevation measurements. *Journal of Geophysical Research: Earth Surface* 118, 2013JF002816.

Wathen, SJ, Ferguson, RI, Hoey, TB, and Werritty, A 1995. Unequal mobility of gravel and sand in weakly bimodal river sediments. *Water Resources Research* 31, 2087–2096.

Weeks, E and Swinney, H 1998. Anomalous diffusion resulting from strongly asymmetric random walks. *Physical Review E* 57, 4915–4920.

Wiberg, PL and Dungan Smith, J 1989. Model for calculating bed load transport of sediment. *Journal of Hydraulic Engineering* 115, 101–123.

Wilcock, PR 1997. Entrainment, displacement and transport of tracer gravels. *Earth Surf. Processes Landforms* 22(12), 1125–1138.

Wilcock, PR and Mcardell, BW 1997. Partial transport of a sand/gravel sediment. *Water Resources Research* 33, 235–245.

Wilcock, PR, Barta, AF, Shea, CC *et al.* 1996. Observations of flow and sediment entrainment on a large gravel-bed river. *Water Resources Research* 32, 2897–2909.

Wong, M, Parker, G, Devries, P *et al.* 2007. Experiments on dispersion of tracer stones under lower-regime plane-bed equilibrium bed load transport. *Water Resources Research* 43, W03440. DOI:10.1029/2006WR005172.

Yang, C and Sayre, W 1971. Stochastic model for sand dispersion. *Journal of the Hydraulics Division* 97, 265–288.

Discussion

Discussion by Francesco Comiti

The insights into how bedload transport processes are affected by channel morphology are indeed very valuable. One aspect that I find insufficiently addressed is the role of bedload transport duration, which is clearly relevant when analyzing displacement lengths. My concern is on the plot showing how mean transport distances are shorter in step-pool than in riffle–pool or larger gravel-bed rivers. How sure can we be that the evident stratification – for similar excess stream power – stems from morphological reasons, and not hydrological ones? In fact, step-pool streams feature on average smaller drainage areas than the other reach types, implying that when other conditions are equal, flashier storm flows with shorter duration for bedload transport should therefore give rise to shorter travel distances. My feeling is that we should better compare different morphologies in terms of virtual velocities (despite the uncertainties related to the determination of the actual transport duration) rather than travel distances.

Reply by the Authors

We thank Francesco Comiti for his interest in our work and his discussion on the subject. We assume that the comment referred to Figure 6.1a and 2, because Figure 6.1b and 1c include time in the analysis. We agree that virtual velocity would be a better representation of the travel distances, however, we do not have the available data to calculate the virtual velocity for most of the cases in the figures.

7

Bedload Transport Measurements with Geophones, Hydrophones, and Underwater Microphones (Passive Acoustic Methods)

Dieter Rickenmann

7.1 Introduction

7.1.1 The Need for Surrogate Measuring Techniques

Prediction of bedload transport is important for hazard assessment and in engineering applications such as the planning and implementation of flood protection measures. Furthermore, a wide range of scientific disciplines such as geomorphology, sedimentology, and river engineering rely on bedload transport predictions for modelling work. Despite more than 100 years of active research, knowledge of bedload transport processes is still limited. This is especially true for coarse-bedded gravel rivers and torrents, i.e., high-gradient mountain streams. Bedload transport is typically very variable for given stream-flow conditions. Continuous and reliable measurements of bedload transport are difficult and challenging to obtain, but important to advance our understanding of sediment transport in mountain streams.

Many bedload transport formulae for gravel-bed rivers have been developed or calibrated using data from experiments in laboratory flumes. Measured transport rates in natural streams can differ substantially from values predicted with such formulae (e.g., Bathurst, Graf, and Cao 1987; Gomez and Church 1989; Rickenmann 2001; Barry, Buffington, and King 2004). Many of the characteristics of steep streams can only be incompletely reproduced in the laboratory, such as large-scale roughness and a broad sediment size distribution. In steep headwater streams, hydraulically based functional relationships may overpredict sediment transport rates by one order of magnitude or more (Lenzi, D'Agostino, and Billi 1999; Rickenmann 2001, 2012; Hassan *et al.* 2005; Chiari and Rickenmann 2011, Nitsche *et al.* 2011).

Direct measurements of bedload transport in gravel-bed streams are typically carried out using physical traps and samplers (Ryan, Porth, and Troendle 2005; Gray, Laronne, and Marr 2010). Vortex and conveyor-belt samplers allow collecting the entire grain size distribution of the bedload particles transported over the entire stream width (Emmett 1980; Hayward 1980; Tacconi and Billi 1987); however, such installations are rare and rather expensive. Trough or pit samplers can typically sample smaller volumes of accumulated bedload particles, and the samples represent bedload transport characteristics at a given stream cross-section, either covering the entire width or parts thereof. In principle, vortex and conveyor-belt samplers can be operated over longer time periods, if the samples

Gravel-Bed Rivers: Processes and Disasters, First Edition. Edited by Daizo Tsutsumi and Jonathan B. Laronne.

can be either stored away or continuously analysed during the measurement. Bedload samplers retain particles by the sampler screen and include net-frame samplers, such as the Helley–Smith sampler (Helley and Smith, 1971; Childers, 1999), and bedload traps, such as the portable bedload traps developed by Bunte *et al.* (2004). These samplers collect particles small enough to enter the nozzle but larger than the mesh size of the collection bag. The samplers are typically operated manually by a single person, and thus operation at high flows may be difficult. Bedload samplers are flexible with regard to the selection of sampling locations both along and across a stream channel. However, sampling duration depends on the volume of the sampler and on the transport rate. Typically, samples are therefore quite limited regarding the temporal and spatial coverage of bedload transport.

Bedload transport measurements often show a considerable spatial and temporal variability (Mühlhofer 1933; Leopold and Emmett 1997; Childers 1999). Even for a constant discharge or shear stress, bedload transport rates may vary over several orders of magnitude (Bathurst, Graf, and Cao 1987; Gomez 1991; Leopold and Emmett 1997; Ryan and Dixon 2008; Recking 2010). Direct measurements of bedload transport in the field are rare, particularly in steep streams, and often episodical, because such measurements are laborious, expensive and technically challenging, especially at higher discharges and in large and steep streams (Rickenmann 2001; Bunte *et al.* 2004; Barry, Buffington, and King 2004; Ryan, Porth, and Troendle 2005; Gray, Laronne, and Marr 2010).

As a result of the difficulties and challenges involved with direct bedload measurements and the large natural variability of transport rates, the necessary data to systematically test, calibrate, and validate laboratory derived equations for natural streams are not currently available. Indirect methods to measure bedload transport can provide useful high-resolution data of value for fluvial sediment transport studies. Non-invasive techniques have the additional advantage to minimize local and temporal changes in the flow field near the sensor. The need for surrogate measuring techniques has been discussed recently, for example, at the International Bedload Surrogate Monitoring Workshop, held in April 2007 in Minneapolis, USA (Gray, Laronne, and Marr 2010) and at the International Workshop of Acoustic and Seismic Monitoring of Bedload and Mass Movements (Rickenmann *et al.* 2013).

7.1.2 Passive Acoustic Methods

The earliest acoustic monitoring of bedload transport in rivers was probably made by Mülhofer (1933), who also performed bedload basket measurements in the Inn River of Austria. A microphone was placed in a watertight metallic box set on the riverbed and connected to headphones. A similar measuring technique was later used in France by Bradeau (1951) to detect the critical discharge for the beginning of bed transport.

Passive acoustic measurements of bedload transport essentially record naturally generated signals, i.e., the sound or vibration induced by moving bedload particles. Interparticle collisions among moving particles or between moving particles and the streambed are the main source of sediment-related noise detected by hydrophones, i.e., underwater microphones. These sensors are placed in a river environment with a supporting structure (e.g. close to one bank) and record acoustic waves or the underwater sound generated by interparticle collisions during bedload movement (Barton *et al.* 2010; Geay 2013). Thus underwater microphones represent an integrated measurement of the bedload transport activity related to a certain bed area in the neighbourhood of the sensor.

Other measuring devices record the collisions of sediment particles with an impact structure, such as a pipe, plate, or column, and use different types of sensors, such as geophones, accelerometers, and piezoelectric sensors to measure the resulting vibrations of the structure. In contrast to hydrophones, impact sensors record bedload transport activity at the location of the streambed where they are deployed. Using arrays of impact plates or impact pipes allows monitoring the spatial variability, for example, across the entire width of the stream. The same is true in principle for impact columns; however, this technique has the disadvantage of introducing an obstacle near the streambed thereby influencing significantly the local flow hydraulics.

Passive acoustic methods are potentially very attractive, as they can provide continuous recordings, there may be no or only minimal interference with the flow or bedload movement, and they are typically associated with low maintenance costs. Some of these methods have been systematically developed and improved over recent decades. Being an indirect or surrogate measuring technique, passive acoustic measurements require direct measurements of bedload transport for calibration, preferably at the same field site and synchronized with the surrogate measuring system. Several research groups have recently started analysing the acoustic signal of more sensitive seismic sensors typically placed away from river courses, to infer information on bedload transport (e.g. Burtin *et al.* 2008; Schmandt *et al.* 2013). This technology is not discussed here.

The objective of this chapter is to present an overview of passive acoustic methods for bedload transport measurement or monitoring in streams, with a focus on those that have been tested and calibrated in the field. I will consider the Swiss plate geophone (Rickenmann *et al.* 2014), the Japanese pipe microphone (Mizuyama *et al.* 2010a,b), and other impact plate systems (Reid, Lane, and Berney 2007; Krein *et al.* 2008; Downing 2010; Tsakiris, Papanicolaou, and Lauth 2014) in section 7.2, and underwater microphones (Barton *et al.* 2010; Geay 2013) in section 7.3. This is followed by a comparative discussion of important issues related to system calibration (section 7.4) and of some operational aspects of different measuring systems (section 7.5). A section with conclusions and an outlook completes the review.

7.2 Particle Impact Systems

7.2.1 Swiss Plate Geophone

The Swiss plate geophone system essentially consists of a steel plate equipped with a geophone sensor underneath and mounted flush with the streambed. The steel plate has standard dimensions of $L \times B \times T = 36\ \text{cm} \times 49.6\ \text{cm} \times 1.5\ \text{cm}$, where L is the downstream length, B the transversal width, and T the thickness of the steel plate. Typically, a 20DX geophone from Geospace Technologies (Houston, TX) in a PC801 LPC Land-case is fixed in an aluminum case to the steel plate. Moving bedload particles cause impact shocks that are transmitted by the plate to the geophone sensor. The sensor has a magnet in a coil as an inductive element. The magnet moves with the steel plate and induces a current in the coil that is proportional to the velocity of the magnet or the velocity of the displacement of the plate (Rickenmann *et al.* 2012).

Often several steel plates are fixed next to each other into a steel canal laterally across the streambed. Each plate then covers a unit stream width of 0.5 m. The steel plates are acoustically isolated by rubber elements from each other and from the steel canal, to minimize recording of vibrations of neighboring plates or of particles impacting on the steel canal. The steel canal may be fixed on the downstream side of a check dam (Figure 7.1a) or may be placed upstream of a bed sill made of concrete and large blocks with a slightly increased local channel slope to prevent downstream erosion (Figure 7.1b). The setup in both cases also helps to avoid bedload deposition on the plates.

Rickenmann and McArdell (2007, 2008) reported the first successful calibration of the Swiss plate geophone system for a slightly different measuring system using a piezoelectric bedload impact sensor (PBIS) (instead of a geophone) at the Swiss Erlenbach stream and the Austrian Pitzbach stream (Bänziger and Burch 1990; Rickenmann *et al.* 1997). From 2000 onwards, the home-made PBIS

(a)

(b)

Figure 7.1 (a) Array with six steel plates and geophone sensors mounted in a steel canal on the downstream side of a check dam (Erlenbach, Switzerland, location at upper bridge). (b) Array with 12 steel plates and geophone sensors mounted in a steel canal on the upstream side of a cemented block ramp (Ruetz at Mutterbergalm, Austria). The steel pillar is used to position a metal basket in line with one steel plate for calibration measurements. The steel canal is inclined towards the orographic left-hand side to improve flow gauge measurements at small discharges.

was replaced by geophone sensors at the Erlenbach because these sensors are commercially available, are probably more reliable, and may also have a longer lifetime than the previously installed piezoelectric sensors (Rickenmann *et al.* 2012). In the meantime, the Swiss plate geophone system has been installed at more than 20 field sites, primarily in smaller and steeper streams in Switzerland and Austria but also in a few larger rivers in Austria and at the Elwha River in the United States (Rickenmann *et al.* 2014)[1].

During normal-flow monitoring conditions, a preprocessing of the geophone signal provides summary values; due to data storage limitations, the raw signal is not recorded. Typically, the following geophone summary is recorded during a predefined time interval such as 1 min or 15 min (Rickenmann *et al.* 2014): (i) whenever the voltage exceeds a preselected threshold value of typically 0.1 V, the shock is recorded as an impulse and the summed impulse counts are stored (*SumIMP*); (ii) the maximum value of the signal amplitude (*MaxA*); (iii) since the geophone signal is proportional to the velocity of the deformation of the steel plate, the momentum transmitted by the particles onto the plate should be related to the acoustic energy of the signal, which is proportional to the value obtained by summing the squared amplitudes values (*SumIQA*). The sampling frequency at most field sites is set to 10 kHz.

A comparative analysis of calibration measurements was made for five field sites, including the Erlenbach (with bedload transport induced by summer rainstorms), the Fischbach, the Ruetz and the Rofenache (with regular bedload transport during the summer months due to snow and glacier meltwater), and the Nahal Esthemoa (an ephemeral stream in Israel) (Rickenmann *et al.* 2014). The best performing calibration function for all streams was a linear relation between *SumIMP* and the bedload mass M transported over a plate:

$$SumIMP = k_b M \tag{7.1}$$

where the calibration coefficient k_b [kg^{-1}], indicating the number of registered impulses per transported mass, varied from site to site. For the Erlenbach, the Fischbach, and the Ruetz stream, the squared correlation coefficient r^2 between *SumIMP* and M for $D > 20$ mm was 0.97, 0.97, and 0.85, respectively. Calibration relations could also be established between *MaxA* and M, and between *SumIQA* and M. This was not surprising since strong correlations exist between *SumIMP*, *MaxA*, and *SumIQA*, which indicate a similar system behaviour at the different field sites (Rickenmann *et al.* 2014). The variation of the calibration coefficient k_b among field sites was found to partly depend on the mean flow velocity. Based on complementary flume experiments (Etter 1996; Krein *et al.* 2008; Turowski and Rickenmann 2009; Böckli 2011) it was concluded that the geophone signal response is mainly influenced by particle size, mean flow or particle velocity, and bed roughness, which all influence the type of particle motion over the steel plate (Rickenmann *et al.* 2014).

The comparative analysis also indicated that the amplitude of the geophone signal contains information on the transported particle sizes (Rickenmann *et al.* 2014). For the Erlenbach, the Fischbach, and the Ruetz stream, the squared correlation coefficient r^2 between the maximum transported particle size and the maximum recorded amplitude was 0.68, 0.85, and 0.76, respectively. Using the raw signal recorded during the calibration measurements with the moving baskets at the

1 The PBIS system was originally developed in the early 1980s by researchers at the Swiss Federal Research Institute WSL and the Laboratory of Hydraulics, Hydrology and Glaciology of ETH Zurich. While all active systems now have geophone sensors, the mounting system of the steel plates at the field sites is essentially similar to the original design. For brevity, this surrogate measuring system is called Swiss plate geophone system, in analogy to the Japanese pipe microphone (hydrophone) system.

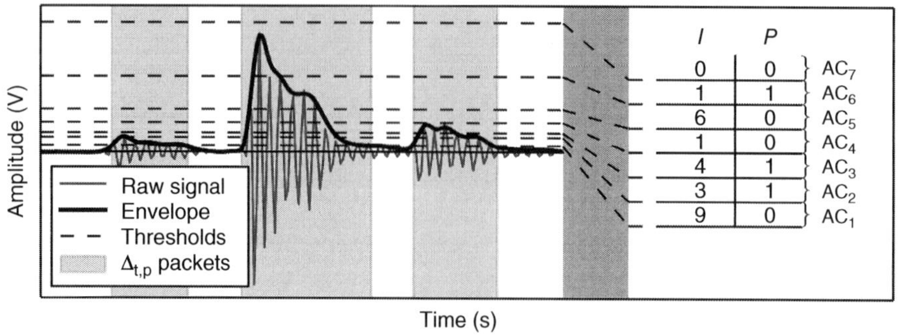

Figure 7.2 Illustration of the counting of impulses (*I*) and packets (*P*) for the Swiss plate geophone measurements. The threshold lines separate different amplitude classes. The envelope of the signal is calculated with a Hilbert transform. (*Source* : Wyss *et al.* 2016. Reproduced with permission of American Society of Civil Engineers.)

Erlenbach, Wyss *et al.* (2016) presented a refined method to estimate bedload transport by grain-size fraction from the geophone signal. The key element of the approach is to identify individual signal packets (periods of increased signal intensities) characterized by a maximum amplitude, which are thought to be due to a single particle impact on the plate. The signal packets were determined by an envelope function (Figure 7.2), using a Hilbert transform. Based on these signal packets, the relative frequencies were determined for different signal amplitude classes. In this method the calibration coefficient α represents a relative number of registered packets for all transported particles for a given grain-size class. The earlier method relating summed impulses *SumIMP* to total mass *M* (eq. 7.1) and using a detection threshold of 0.1 V at the Erlenbach was found to be more reliable for values of *M* larger than about a few 10 kg (Rickenmann *et al.* 2012). With the newer method the lowest detection threshold used was 0.018 V, and also smaller masses down to about 1 kg were estimated with a similar accuracy as for larger masses (Wyss *et al.* 2016). For particle sizes *D* increasing from 10 to 20 to 30 mm, the calibration coefficient α for the Erlenbach increases from about 0.005 to 0.09 to 0.15, indicating a more than 10-fold larger detection probability for particles with $D = 20$ mm as compared to those with $D = 10$ mm. For $D = 20$ mm about every tenth particle is detected on average at the Erlenbach. The bedload transport rates at the Erlenbach varied in the range from 0.0001 kg/s/m to 10 kg/s/m.

The calibration measurements with the Swiss plate geophones using eq. (7.1) indicate an uncertainty (standard error) between estimated and measured geophone summary values of 18% in the Swiss Erlenbach torrent and between 45% and 65% in three Austrian glacial streams (Rickenmann *et al.* 2014). For a similar system with PBISs used earlier in the Erlenbach, the error in estimating total sediment loads in the retention basin was assessed to be approximately a factor of two (Rickenmann and McArdell 2007). The uncertainty of the calibrated fractional bedload mass estimation using the packet-identification-based amplitude classes is around 10% for most of the analysed bedload samples (Wyss *et al.* 2016).

7.2.2 Japanese Pipe Microphone (Hydrophone)

Taniguchi *et al.* (1992) reported the first measurements in Japan with a microphone attached to one end of a metal pipe to detect bedload transport. A similar system had been developed earlier and independently in Poland, and Froehlich (2003, 2010) later summarized the related experiences. Recently, more systematic studies with such a system were performed in Japan (Mizuyama *et al.* 2010a,b). The

Figure 7.3 Japanese pipe microphone installed at a check dam in the Ashi-arai-tani River.(Photograph courtesy of Daizo Tsutsumi.)

pipe has a diameter of 42 mm at the Polish field sites (Froehlich 2003) and of 48 mm at the Japanese field sites, and it is typically placed transverse to the flow direction across the streambed, ideally buried by half its diameter at a stable bed section such as a check dam or a sill (Figure 7.3) (Mizuyama *et al.* 2010b). The pipe is filled with air, and the device is called here a pipe microphone, in contrast to the terminology used in Froehlich (2003) and in Mizuyama, Fujita, and Nonaka (2003).

Earlier feasibility tests under laboratory conditions suggested a linear relationship between the number of pulses within a given amplitude range and bedload volume (Taniguchi *et al.* 1992). Note that the counting of 'pulses' is different from the counting of 'impulses' as used with the Swiss plate geophone system (section 7.2.1). The raw signal is first treated with a band-pass filter, and a wave detector determines the envelope of the signal. In the earlier study of Taniguchi *et al.* (1992) the envelope was compared with a given amplitude threshold level to determine the number of waves or wave peaks exceeding the threshold level to arrive at the pulse counts. For the analysis of the more recent measurements with the same system, the pulse count depends on the amplification of the envelope wave signal (using six amplification factors: 2, 4, 16, 64, 256, and 1024), which is then compared to a constant threshold amplitude (of 2 V), thus determining the number of pulses to be recorded for six different channels corresponding to the different amplification levels (Mizuyama *et al.* 2010a,b; D. Tsutsumi and M. Nonaka, pers. comm. 2014). In comparison with the Swiss plate geophone system, the number of pulses roughly correspond to the number of packets, and the pulse counts resulting from the use of different amplification factors is analogous to the packet counts for different amplitude classes with the Swiss plate geophone system.

Based on a series of flume experiments with particle sizes in the range 2–16 mm, Mizuyama *et al.* (2010a) examined the suitability of different pipe lengths ranging from 0.3 to 1 m and different pipe shapes for bedload transport detection. They found that longer pipe segments result in a dampening of acoustic signals, particularly when bedload flux is high. For elevated bedload discharge the number of pulses temporarily decreased or dropped to zero. This dampening was thought to be due to numerous particles making it impossible to distinguish individual particle impacts based on number of pulses, and in addition the high bedload discharge caused a reverberant sound. The location of particle impacts affects the number of recorded collisions, increasing towards the root (where the microphone is fixed), thus limiting the allowable pipe length. There is a limiting grain size of about 8 mm that can be detected by this system (Mizuyama *et al.* 2010b). The use of a filter to dampen microphone sensitivity was found necessary for the successful detection of higher bedload fluxes.

A field calibration study with pipe microphones was undertaken in six steep, gravel-bed mountainous rivers in Japan, where automatic and continuously monitoring bedload samplers allowed for an independent determination of bedload transport (Mizuyama *et al.* 2010b). The measurement locations included sandy-gravel reaches such as the Nishi-takiga-tani to bouldery reaches in other rivers. The pipes were fixed at the bottom of a stable cross-section of a check dam or a bed sill. Both the total length of the pipe and the length of the acoustically sensitive pipe segment varied between the sites. The stainless steel pipe as used in all the rivers had a diameter of 48 mm and a 2 mm or 3 mm thick wall.

The measurements from the gravel-bedded Nishi-takiga-tani river showed a reasonable and linear correlation between pulses and bedload mass accumulated over 30–60 min intervals for intermediate bedload discharges (>0.01 kg/s/m), with a squared correlation coefficient $r^2 = 0.84$. Similarly, the observations for the Ashi-arai-tani river provided a fair and linear correlation between pulses and bedload flux for 5-min averaged data when bedload discharge was intermediate (0.01–0.1 kg/s/m), with a squared correlation coefficient $r^2 = 0.78$, but for low bedload discharges (<0.01 kg/s/m) the averaging time needed to be increased. The best correlation was obtained in both cases by using the pulses of channel 5, i.e. a fourfold amplification of the wave envelope signal. At high bedload discharges (>0.1 kg/s/m) the acoustic system tends to become saturated. The relation between pulse counts of the lower sensitivity channels and water level during high floods on the Joganzi River with elevated bedload discharge indicates that calibration may also be possible under these conditions.

7.2.3 Other Impact Plate Systems

Krein *et al.* (2008) report on measurements with an impact plate system consisting of a piezoelectric hydrophone mounted on the bottom side of a stainless steel plate with $L \times B \times T = 50\ \text{cm} \times 50\ \text{cm} \times 3\ \text{mm}$, which is fixed flush with the flume bottom or the stream bed. They used a shallow-water omnidirectional transducer (International Transducer Corp. ITC-4001 A) containing a flexural disc transducer with Channelite-5400 ceramics mounted in a rugged corrosion-resistant housing. Its recording range lies between 0.1 and 20 kHz with a stable frequency response, and the sensor is able to record acoustic signals in the sonic frequency range (kHz) with high signal-to-noise ratio (Barrière *et al.* 2015). In a series of flume tests, four different grain-size classes were considered, namely fine sand (0.063–0.2 mm), medium sand (0.2–0.63 mm), coarse sand (0.63–2 mm), and gravel (2–4 mm). The results showed that the entire signal spectrum shifted into the higher frequency range with decreasing grain sizes. Krein *et al.* (2008) also performed experiments with single stones to investigate the effect of different particle shapes on the signal response. Many of the flat stones could not be detected because they did not have contact with the surface of the steel plate. Saltation of the stones was not observed during the tests. Flat stones showed a higher variability of the movement patterns than angular stones.

In the past, identification of characteristic frequencies appeared to be possible only for uniformly sized particle mixtures. Barrière *et al.* (2015) performed further flume experiments with the hydrophone plate system of Krein *et al.* (2008), using again both uniform grain sizes and mixed grain classes. Grain sizes varied in the range 2 mm < D < 63 mm. The raw signal was sampled with a rate of 96 kHz.

Barrière *et al.* (2015) presented a novel method to extract grain size information from acoustic measurements of bedload transport in the gravel size particle range. The method is based on an advanced signal processing technique called 'first arrival atomic decomposition'. The basic idea is to identify signal packets pertaining to a single particle impact based on a continuous wavelet analysis. The original signal is then reconstructed or approximated by a superposition of several theoretical signal functions, with known frequency and amplitude characteristics. Based on the analysis of the flume experiments, the particle size D_{50} was found to increase with increasing amplitude A and decreasing frequency f, and the following calibration relation was developed:

$$D_{50} = 5.06 \times 10^4 A^{0.39} / f^{0.86} \qquad (7.2)$$

Having used the median values of A and f for the development of this relation, the squared correlation coefficient is $r^2 = 0.95$. The results appear to be interesting because they allow a promising analysis of the raw signal of grain size mixtures being transported over the measuring system.

Tsakiris, Papanicolaou, and Lauth (2014) report on flume experiments with an impact plate system using a geophone sensor mounted on the bottom of a 0.91 m long, 0.60 m wide and 0.30 m deep acrylic box in a cavity underneath the flume bed. A stainless steel impact plate with dimensions $L \times B \times T = 15$ cm × 15 cm × 3 mm was attached flush with the top surface of the acrylic box. For the geophone sensor they used the 'DSP-06 Particle Monitor' manufactured by ClampOn, Inc. The sensor sampled the acoustic energy generated by the acoustic force due to the impacts of mobile bedload particles on the impact plate at four different frequency ranges. Frequency 1 covering the range between 100 and 200 kHz and frequency 4 covering the range from 380 to 480 kHz were used in the study. Spherical glass particles with a diameter of 15.9 mm were used, and bedload transport rates were independently measured with an overhead video camera, ranging from about 0.5×10^4 kg/s/m to about 15×10^4 kg/s/m. The dominant transport mode was found to be rolling and also saltation at higher transport stages.

Several steps were involved in the analysis of the measurements (Tsakiris, Papanicolaou, and Lauth 2014). After removal of the background noise in the geophone output, spikes were isolated in the geophone output time series. Then the power spectral density of the geophone acoustic signal spikes was calculated, which was the basis to determine the geophone spike impulse Ip, representing a relative energy produced by the acoustic force due to the elastic impacts of the spheres onto the geophone plate. A reasonable calibration relation was found between dimensionless bedload-transport rate and geophone spike impulse Ip, with a logarithmic function and a squared correlation coefficient of $r^2 = 0.78$, valid for higher transport stages (τ^*/τ^*_{cr} larger than about 1.4), which included saltating particle motion. For lower transport stages, however, with only rolling particle motion, no good calibration relation for bedload flux was obtained.

Richardson, Benson, and Carling (2003) and Reid, Lane, and Berney (2007) describe a further impact plate system. The so-called sediment impact sensor consists of a steel plate with $L \times B \times T = 13$ cm × 15 cm × 6 mm, which is mounted flush with the channel bed. A small Tinytag Plus TGPR1201 Count Input Data logger is directly attached to the metal plate and it includes an accelerometer (Figure 7.4). The installed memory allows 55 days of logged information at 5 min intervals, with a 12 month battery life. For the study of Reid, Lane, and Berney (2007) the logger was set to detect only grains greater than 20 mm in diameter. By inserting the logger into bedrock or into large boulders, the device can survive more intense transport events. The loggers can record between 0 and 255 impacts within a given time interval and a maximum of three impacts per second. Since the probability of more than one particle hitting the plate in less than a third of a second increases with shear stress (discharge) and transport

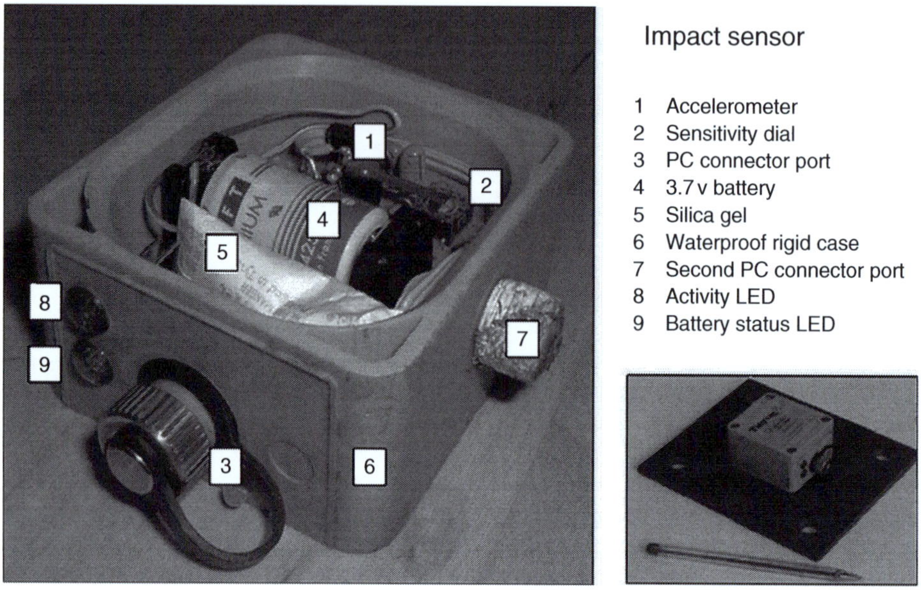

Figure 7.4 Sediment impact sensor (accelerometer), including data logger and battery, contained in a waterproof box (left-hand photograph), which is attached to a steel plate (right-hand photograph). (*Source*: Reid 2007. Reproduced with permission of Elsevier.)

rate, this results in a conservative estimate of transport intensity at higher flows and a more reliable estimate at low to moderate flows (Reid, Lane, and Berney 2007).

The sediment impact sensor was used by Reid, Lane, and Berney (2007), Raven *et al.* (2009), and Raven, Lane, and Ferguson (2010) to provide information on sediment transfer processes in a 5.6 km reach of an upland gravel-bed river in northern England. The main channel sensors were often saturated during larger storm events. The problem of saturation could be resolved by lowering the time-step to 1 min (from 5 min), but this would require a download every 10 days, which was deemed not to be practical. Therefore, data were converted from actual records to a relative saturation by dividing each record by 255 to scale data between 0 and 1, representing a transport intensity. Full saturation of 1 is equivalent to 255 impacts in 300 s, which gives a mean transport rate of 0.85 impacts per second (Raven, Lane, and Ferguson 2010).

Downing *et al.* (2003) and Downing (2010) developed the so-called gravel-transport sensor (GTS). It consists of a 1.6 mm thick steel pressure plate, fixed to polyvinylidene fluoride film, which is mounted on an aluminium block. The impact plate has a height of 5.4 cm and a width of 10.5 cm, is set vertically above the streambed, and inclined 45° to the main flow direction. Gravel particles colliding with the plate generate an electric charge proportional to the impact force, and the integral of the charge is proportional to the momentum of the particles. Polyvinylidene fluoride film is used because of its high sensitivity, excellent formability, low acoustic sensitivity, and low elastic modulus. Transducers constructed in this way are sensitive to particle momentum in the range from 0.00002 kg m/s (a 5 mm stone moving at 0.12 m/s) to 0.08 kg m/s (a 100 mm stone moving at 0.06 m/s). The data recording unit includes an analogue signal processor, a microcontroller and a solid-state data memory. Following a charge amplifier, a high-pass filter removes low-frequency noise, e.g. from flow turbulence, hydrostatic pressure fluctuations, and bedload noise. The further signal processing includes the detection of threshold exceedance, the pulse integration, and counting.

For the GTS, transported bedload mass is determined from the product of digital counts, which are proportional to momentum, and a conversion factor, divided by the sum of the size-fraction-weighted grain velocities (Downing 2010). The conversion factor is instrument specific and the bedload size distribution must be known and must remain the same during transport events. Field tests with the GTS were performed in Caspar Creek in the United States in conjunction with a Birkbeck weighing bedload trap. For two flow events of 58 h and 40 h duration with total transported bedload masses of 27 and 74 kg collected in the trap, the empirical conversion factors differed by about 20%. Conversion factors for computing mass flux from GTS measurements were also determined with video images obtained in a flume for 13.4 mm, 21.4 mm, and 40 mm gravel respectively, and were more than ten times larger than those determined from the field study. Conversion factors were found to vary from one instrument to another because of electronic gain and bedload size. The main limitation of this measuring system is that the size distribution of the bedload must be known and must remain the same during a transport event (Downing 2010). In contrast to the other techniques discussed here, the gravel-transport sensor interacts with the flow, and so a calibration involving the hydraulic efficiency is required.

7.3 Underwater Microphones

Underwater microphones have been used for various applications in the marine environment, identifying both anthropogenic and natural sources of ambient noise in the ocean (Hildebrand 2009). Etter (2012) compiled a comprehensive review of advanced applications for underwater acoustic modeling. Measurements of self-generated noise by moving sediments in a tidal environment have been reported, e.g. by Thorne (1990) and Bassett, Thomson, and Polagye (2013). However, there were only very few applications of underwater microphones to monitor bedload transport in rivers over recent decades. The measuring principle of underwater microphones (hydrophones) in rivers is similar to the passive acoustic techniques discussed in section 7.2. However, instead of recording the noise produced by particles impacting on a rigid body, underwater microphones pick up the noise generated by bedload particles colliding with each other or with those resting on the bed surface.

Barton *et al.* (2010) used a Geospace MP-18 hydrophone encased in a perforated PVC pipe filled with pieces of permeable foam to measure bedload transport rates. The hydrophone was connected to an adjustable-gain amplifier, which allowed for on-site adaptation to the sampling environment. One-minute long samples of the amplified and filtered voltage were recorded with a sampling rate of 62.5 kHz. To test the acoustic system, independent bedload measurements were carried out from a tethered cataraft using pressure-difference samplers in a study reach of the Trinity River, California. Hydrophone measurements were undertaken using a case housing the instrument anchored in the channel on the riverbank. Total acoustic power was calculated from the hydrophone data ranging from 0.01 to 14.8 kHz over 1-min intervals. The acoustic energy of grain collisions correlated linearly with the bedload data, with a correlation coefficient of $r = 0.76$. Bedload transport rates varied from 1.4 to 5.5 kg/s.

A similar hydrophone system was tested in the Arc en Maurienne mountain river in France (Camenen *et al.* 2012). During a reservoir flushing event, direct bedload transport measurements were made using a large (152 mm) Helley–Smith type bedload sampler. Passive acoustic measurements were carried out with an omnidirectional Brüel and Kjaer hydrophone type 8105, which was positioned 50 cm above the riverbed. The acoustic signal of the hydrophone was sampled at a rate of 100 kHz. Sound pressure levels (SPL) were determined using 30 s recording duration over a 0–22.5 kHz frequency band. Although SPL showed some correlation with bedload transport rates, the considerable temporal variability of bedload flux based on the Helley–Smith sampling contradicted the generally steady hydrophone signals. It was likely that the presence of large particles hampered the direct bedload transport measurements with the Helley–Smith sampler.

Extending the study by Camenen *et al.* (2012), Geay (2013) used an omnidirectional Brüel and Kjaer hydrophone type 8103 at the Drau mountain river in Austria, where independent bedload transport measurements were made with an array of 40 Swiss plate geophone sensors (Habersack, Seitz, and Liedermann 2010). The hydrophone was fixed onto a stainless steel structure 30 cm above the riverbed. The energy of the acoustic measurements, expressed by the root mean square values of the hydrophone raw signal, showed a good correlation with the geophone signal energy, both considering only the three nearest geophones or all geophones covering the entire channel width; the squared correlation coefficient r^2 was 0.89 and 0.86, respectively (Geay 2013). In the study at the Drau River, it was difficult to distinguish between sources of flowing water and bedload-generated noise; the energy level measured by the hydrophone also showed a strong correlation with water depth. Hourly variations of the monitored variables were highly intercorrelated and discrimination of the processes was not obvious. Several weeks of continuous data monitoring allowed observation of hysteresis effects, which indicated that the ambient noise of the Drau River was dominated by bedload self-generated noise. The analysis of the hydrophone signal along with the 40 geophone sensors spanning the entire channel width showed that the deployed system is sensitive to bedload motion not only locally, but also over distances of several metres (Geay 2013). A further comparison of the Brüel and Kjaer hydrophone type 8103 was made with video observations of bedload transport at the Rhine River, and the sound pressure level of the hydrophone signal for the frequency range 10–70 kHz correlated well with the video analysis of moving particles (Geay 2013).

7.4 Important Findings Related to System Calibration

7.4.1 Calibration for Total Mass Flux

A number of different acoustic techniques have been employed in field settings where independent bedload transport measurements were available in mountain streams. In most of these studies it was shown that the acoustic signal provides a very plausible measure of bedload transport intensity. In section 7.2 quantitative calibration relations are reported with r^2 values larger than about 0.7 between a summary value of the acoustic signal and total bedload mass or flux for the Swiss plate geophone (for three sites) and for the Japanese pipe microphone (for two sites). For the pipe microphone system an appropriate channel sensitivity had to be selected, and the validity of the calibration relations was restricted to intermediate bedload discharges (0.01–0.1 kg/s/m) (Mizuyama *et al.* 2010b). The linear calibration relation for the Swiss plate geophone at the Erlenbach was found to be applicable over a large range of bedload discharges of roughly 0.01–10 kg/s/m (Rickenmann *et al.* 2012; Wyss *et al.* 2016). A comparison of the impulse counts of the two systems is illustrated for a bedload-transporting flow event at the Erlenbach in Figure 7.5.

Measuring bedload transport of uniform spherical glass particles (D = 16 mm) in a flume with a geophone impact plate system, Tsakiris, Papanicolaou, and Lauth (2014) also determined a reasonable calibration relation between dimensionless bedload transport rate and geophone spike impulse Ip, valid for a flow stage with particles also transported in saltation mode.

Using underwater microphones and impact plate hydrophones, several laboratory experiments have demonstrated that the energy of the resulting acoustic signal is correlated with bedload flux through a power law relation (Thorne 1985, 1986; Voulgaris, Wilkin, and Collins 1995; Bogen and Møen 2003;

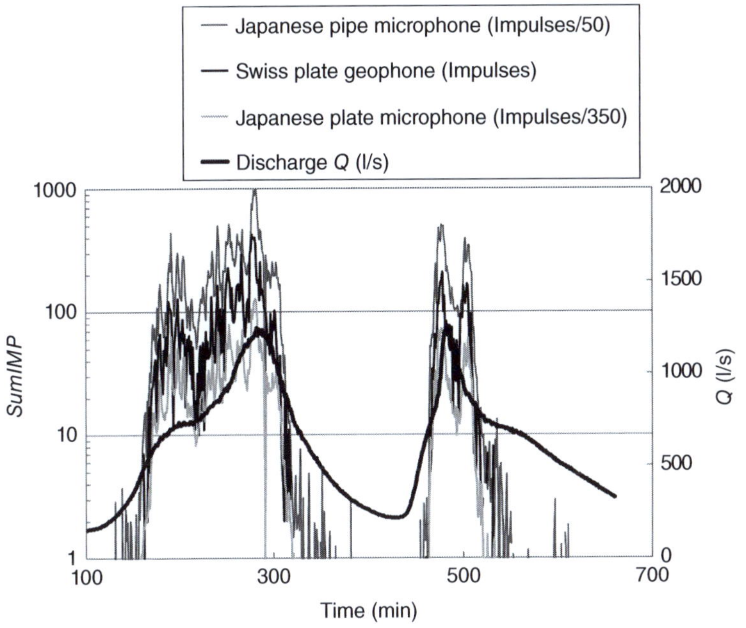

Figure 7.5 Bedload-transporting flow event of 29 July 2013 at the Erlenbach, showing acoustic measurements along with discharge over time. The plate microphone (a variant of the pipe microphone) was fixed under the same steel plate as one geophone, thus covering only a width of 0.5 m. The plate geophone data are the values of two neighboring plates, covering 1 m. Also the pipe microphone covers 1 m of channel width. The impulse count is made according to the procedure implemented for the Swiss plate geophone system (*SumIMP*). The scaling factors for the microphone data were chosen arbitrarily, to better illustrate the similarity of the signal responses.

Møen *et al.* 2010). As reported in section 7.3, this was confirmed for bedload transport measurements in a river by Barton *et al.* (2010), suggesting a linear relation between acoustic power and the bedload transport (with $r = 0.76$). Similarly, Geay (2013) provided indirect confirmation of such a relation, by showing a strong correlation between acoustic energy of an underwater microphone and the geophone square integrals (*SumIQA* of the three nearest sensors) for three bedload-transport periods of several days in a mountain river (with $r^2 = 0.89$).

It is evident from field investigations with different acoustic techniques that calibration with independent bedload transport data is necessary (Gray, Laronne, and Marr 2010; Barton and Pittman 2010). If sufficient calibration measurements are available from a number of field sites, and if a standard design of the measuring installation is used, it may be possible to identify site-specific factors that affect the calibration relation (Rickenmann *et al.* 2014). Even if no direct calibration is possible or available for passive acoustic bedload transport measurements, using such a technique may provide interesting qualitative information, for example, on the beginning and cessation of bedload transport and on the temporal dynamics of changing bedload transport intensities (e.g. Raven *et al.* 2009; Raven, Lane, and Ferguson 2010; Beylich and Laute 2014; Mao *et al.* 2014).

7.4.2 Identification of Grain Size (Classes)

Ideally, a surrogate bedload transport measuring technique should not only be calibrated for total mass flux but should also be able to identify different grain sizes or grain-size classes. For the Japanese pipe microphone, Taniguchi *et al.* (1992) found a correlation between the maximum amplitude and the momentum of the particle impact in laboratory tests. Based on flume experiments with their impact plate hydrophone, Krein *et al.* (2008) determined a weak correlation between the mass of the single particle and the maximum signal amplitude (with r ranging from 0.36 to 0.32). Beylich and Laute (2014) used an impact plate with an accelerometer (similar to the system of Reid, Lane, and Berney (2007) mentioned in section 7.2.3, but with a larger steel plate of dimensions $L \times B \times T = 46\ \text{cm} \times 34\ \text{cm} \times 3\ \text{mm}$), and they found from flume experiments that the mean acceleration increased with increasing grain-size classes but also depended on transport rate. Calibration measurements with the Swiss plate geophone at three field sites showed a fair correlation between maximum transported particle size and maximum amplitude of the geophone signal (with r^2 ranging from 0.68 to 0.85; see section 7.2.1). Using the raw signal of calibration measurements at the Swiss Erlenbach stream and an amplitude-based classification of the signal packets, a method was developed to successfully estimate bedload transport rates by grain-size fractions, distinguishing seven grain-size classes with median diameters ranging from 12.3 mm to 83 mm (Wyss *et al.* 2016). Similarly, a study using the Japanese pipe microphone system analysed the pulses recorded with different channels, corresponding to different amplification levels, and succeeded in predicting bedload transport by grain-size fraction for median diameters ranging from 9 mm to 36 mm, which was better for flume than for field data (Mao *et al.* 2016). This approach is in fact analogous to the packet counts for different amplitude classes with the Swiss plate geophone system used by Wyss *et al.* (2016).

Regarding underwater microphone and impact plate hydrophone measurements, several laboratory studies demonstrated that a characteristic frequency of the sediment related noise is inversely related to the grain diameter for relatively uniform-sized particles (Thorne 1986; Etter 1996; Krein *et al.* 2008; Møen *et al.* 2010; Belleudy, Valette, and Graff 2010). However, it is more difficult to infer grain size distribution for a mixture of particles (Thorne 1986; Carson *et al.* 2008; Uher and Beneš 2012). The frequency spectrum is mainly altered by sediment characteristics including size, shape, and mineralogy (Thorne and Foden, 1988). Tsakiris, Papanicolaou, and Lauth (2014) reported that rotary vibrations of the mobile particles rolling atop their impact plate produced a higher characteristic frequency response than spheroid vibrations, which are produced by the minute distortions of the spherical particle shape at the moment of the impingement. Thorne (1985) found that the characteristic frequency of a spheroid vibration is roughly half that of a rotary vibration.

According to a theoretical model applied by Thorne (1985, 2014) to collisions of gravel particles of different sizes (self-generated noise), it follows that the main factor determining the characteristic frequency f emitted after an impact is the particle size. According to this model, the frequency f is inversely proportional to particle size (diameter) D. The impact velocity only has a minor influence. For typical material properties of gravel sediments the following simplified relation can be given (Thorne 2014) for an assumed impact velocity of 0.5 m/s, where f is in Hz and D is in mm:

$$f = 141/D \tag{7.3}$$

It is interesting to compare characteristic frequencies determined in different experimental studies (Figure 7.6). Thorne (1986) determined the centroid frequency with a hydrophone for gravel particles within a size range 0.31–25 mm, which were agitated in a wooden rotating drum submerged in water, where the gravel particles produced self-generated noise due to interparticle collisions. For the gravel data of Thorne (1986, 2014), the equivalent sphere diameter is given in Figure 7.6. Belleudy, Valette, and Graff (2010) poured gravel particles into a cavity in a water basin and used a hydrophone to measure the shock frequencies of particles with diameters from 15 to 110 mm. Using particles from the Erlenbach streambed, Etter (1996) performed flume experiments with a Swiss impact plate system, using a PBIS (instead of a geophone) at that time. He determined the median frequency based on

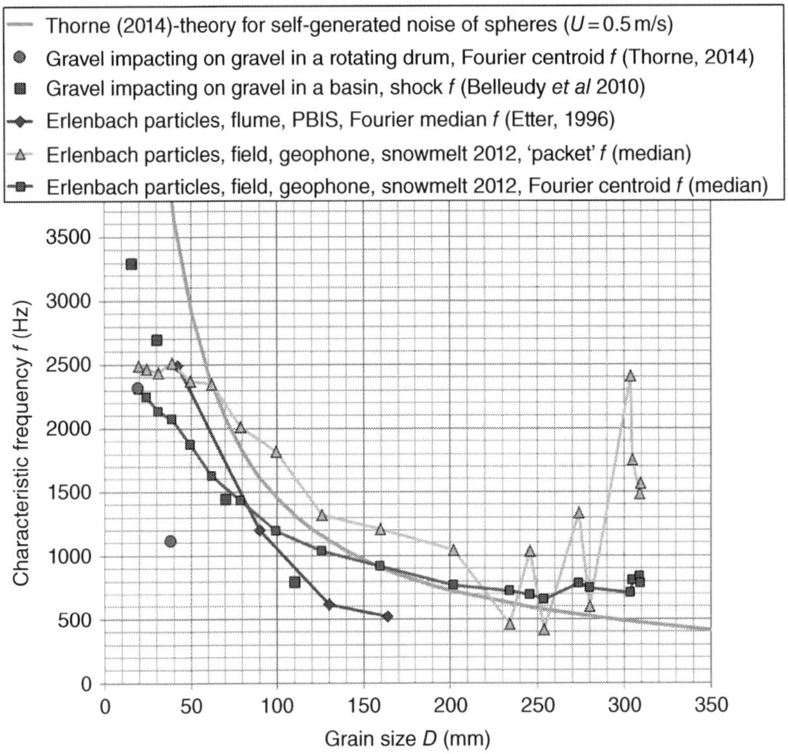

Figure 7.6 Characteristic frequency f versus grain size D. Experimental data from different studies (see text for details). In this plot, a single value of each characteristic frequency is shown: most measurements indicated a large variability around an average (typically the median) value. The theoretical equation (7.3) of Thorne (2014) is plotted here for an assumed impact velocity of 0.5 m/s. For the Erlenbach snowmelt 2012 experiments, for $D > 200$ mm only one particle per grain size was used.

a Fourier transform of the raw signal, for grain sizes from 42 to 164 mm. For this study we analysed field measurements at the Erlenbach with the Swiss plate geophone, where particles with known grain sizes were inserted into the flow during snowmelt runoff in 2012. We tested both a Fourier transform and a simplified approach to estimate a characteristic frequency of the signal due to a single particle impact, by dividing the number of full oscillations by the packet duration $\Delta_{t,p}$, as illustrated in Figure 7.2. Despite different measuring and analysis techniques used, and additional factors other than particle diameter (or mass) influencing the characteristic frequency, the experimental data indicate a similar trend for the change in f with D for a certain range of particle sizes (Figure 7.6). For the Swiss plate geophone measurements, this preliminary analysis suggests that it may be possible to improve grain-size identification in the approximate range 20 mm < D < 200 mm by using frequency in addition to amplitude information. We applied the theoretical model of Thorne (2014) to the Swiss plate geophone system by replacing one gravel particle by the mass and material properties of the steel plate. The result is a vanishing frequency dependency for impacting gravel particles with D larger than about 200 mm, which is in agreement with the data from the snowmelt experiments at the Erlenbach shown in Figure 7.6.

As is indicated by eq. (7.2) in section 7.2.3 based on the flume experiments of Barrière *et al.* (2015), grain-size identification may be improved by taking into account both amplitude and frequency information of the acoustic signal. These experiments included grain sizes ranging from 2 mm

$< D <$ 63 mm, and the characteristic frequency was determined with a novel method based on the 'first arrival atomic decomposition'.

7.5 Some Operational Aspects to be Considered For Different Systems

Both hydrophones and impact sensors permit in principle a continuous monitoring of bedload transport activity in time. However, if the raw signal is to be recorded continuously there are practical limits with regard to storing the signal from multiple sensors over longer time periods. Appropriate procedures for online processing of the raw signal and retrieving summary values may partly solve this technical problem (e.g. Geay 2013; Rickenmann *et al.* 2014; Barrière *et al.* 2015), but on-site signal processing may be limited by power supply.

Underwater microphone measurements represent integrated recordings of sediment-related noise covering an (unknown) area of the riverbed. An advantage of the integrative measurement is that it is not local and may provide information on average bedload fluxes, while a disadvantage is that the signal response is dependent on the location of the acoustic sources (Geay 2013). Using multiple hydrophones at different places may help to constrain major source areas of sediment-related noise (Bassett, Thomson, and Polagye 2013). These sensors are also sensitive to any other noise present in the river, e.g., generated due to turbulence (hydrodynamic noise) or due to vessel traffic (Bassett, Thomson, and Polagye 2013). Especially in natural gravel-bed rivers with large roughness and elevated turbulence levels (mountain streams), the distinction between sediment-related noise and noise due to the turbulence of water flow may become difficult (P. Belleudy, pers. comm. 2012).

The Japanese pipe mantle has a thickness of 2–3 mm (Mizuyama *et al.* 2010b), and thus potentially may be damaged by large boulder impacts. Installation costs are relatively low. The pipe is ideally half buried in the bed, but upstream abrasion may eventually change the upstream surface area exposed to particle impacts. The Swiss geophone system with 15 mm thick steel plates appears to be more robust in highly energetic environments including the transport of large boulders in steep streams. A disadvantage of this system are the relatively high installation costs, requiring not only a stable cross-section (e.g. check dam or sill) but also an embedded steel frame to hold the plates.

Installation costs are relatively low for underwater microphones, pipe microphones, the sediment impact plate, and the gravel transport sensor. Maintenance costs are likely to be low for sensors that are incorporated in a solid and watertight structure mounted flush with the stream bed, such as the Swiss plate geophone system. Higher maintenance costs may be expected for sensors that are fully or partly exposed to permanent water flow and which might be damaged by larger impacting particles.

7.6 Conclusions

Some important conclusions from the International Bedload Surrogate Monitoring Workshop held in April 2007 (Gray, Laronne, and Marr 2010) were confirmed at the International Workshop of Acoustic and Seismic Monitoring of Bedload and Mass Movements held in September 2013 (Rickenmann *et al.* 2013), and they are still valid: (i) indirect bedload measuring methods have the advantage of providing continuous records of bedload transport activity both in time and over a cross-section; (ii) controlled laboratory experiments are important for a better understanding of the factors influencing the calibration of these measuring methods; and (iii) additional field calibration of the sensors is still necessary to obtain a reasonable measuring accuracy.

More detailed field calibration efforts have been made, particularly for impact sensors, i.e., the Swiss plate geophone and the Japanese pipe microphone. For both techniques it was found that an approximately linear relationship exists between number of impulses or pulses, respectively, and bedload mass or flux (Mizuyama *et al.* 2010b; Rickenmann *et al.* 2012, 2014). While the lower limit of grain-size detection is 4–8 mm for the Japanese pipe microphone (Mizuyama *et al.* 2010a), this lower limit is about 10–20 mm for the Swiss plate geophone (Rickenmann *et al.* 2012, 2014; Wyss *et al.* 2016).

The performance of the Japanese pipe microphone system was found to be: (i) acceptable for low bedload discharges, given a sufficiently long averaging; (ii) good for intermediate bedload discharges; (iii) but for large bedload discharges multichannel outputs need to be compared to ensure that no signal-damping occurs (Mizuyama *et al.* 2010b). The Swiss plate geophone appears to be robust for a rather wide range of bedload fluxes (Wyss *et al.* 2016). In addition, using the raw signal of the calibration measurements at the Erlenbach, a refined method was developed to successfully estimate the mass of grain-size fractions, for median diameters ranging from 12.3 to 83 mm (Wyss *et al.* 2016).

Apart from the calibration for total mass flux, some studies indicate that identification of grain-size classes may be possible based on frequency and or amplitude information of the acoustic signal (Thorne 1986; Barrière *et al.* 2015; Wyss *et al.* 2016). If such methods of analysing the acoustic signal can be successfully applied and verified at different field sites, this would represent a major step forward regarding a more widespread use of surrogate bedload transport measuring techniques.

The underwater microphones also show encouraging results when acoustic energy is compared with independent bedload measurements in rivers. A major advantage is the relatively easy installation. However, the quantitative calibration appears to be more challenging than for impact-type acoustic techniques, since the integration area of underwater microphones placed at some height above the riverbed is difficult to determine.

Acknowledgement

I thank Carlos Wyss for the analysis of the Erlenbach snowmelt 'experiments' presented in Figure 7.6 and for comments on a draft manuscript.

References

Bänziger, R and Burch, H 1990. Acoustic sensors as indicators for bed load transport in a mountain torrent. In *Hydrology in Mountainous Regions I* (eds Lang, H and Musy, A). Publication 193, IAHS: Wallingford, UK; 207–214.

Barrière, J, Krein, A, Oth, A, and Schenkluhn, R 2015. An advanced signal processing technique for deriving grain size information of bedload transport from impact plate vibration measurements. *Earth Surface Processes and Landforms* 40, 913–924.

Barry, JJ, Buffington, JM, and King, JG 2004. A general power equation for predicting bed load transport rates in gravel bed rivers. *Water Resources Research* 40, W10401. DOI: 10.1029/2004WR003190.

Barton, JS, and Pittman, SA 2010. Passive-transducer hydroacoustics. In *Sedimentology of Aqueous Systems* (eds Poleto, C and Charlesworth, S). Wiley-Blackwell: Oxford, UK; 65–69.

Barton, JS, Slingerland, RL, Pittman, S, and Gabrielson, TB 2010. Monitoring coarse bedload transport with passive acoustic instrumentation: a field study. In *Bedload-Surrogate Monitoring Technologies* (eds Gray, JR, Laronne, JB, and Marr, JDG). Scientific Investigations Report SIR 2010-5091, US Geological Survey; 38–51. http://pubs.usgs.gov/sir/2010/5091/papers

Bassett, C, Thomson, J, and Polagye, B 2013. Sediment-generated noise and bed stress in a tidal channel. *Journal of Geophysical Research Oceans* 118, 2249–2265. DOI:10.1002/jgrc.20169.

Bathurst, JC, Graf, WH, and Cao, HH 1987. Bed load discharge equations for steep mountain rivers. In *Sediment Transport in Gravel-Bed Rivers* (eds Thorne, CR, Bathurst, JC, and Hey, RD). John Wiley & Sons: New York; 453–477.

Belleudy, P, Valette, A, and Graff, B 2010. Passive hydrophone monitoring of bedload in river beds: first trials of signal spectral analyses. In *Bedload-Surrogate Monitoring Technologies* (eds Gray, JR, Laronne, JB, and Marr, JDG). Scientific Investigations Report SIR 2010-5091, US Geological Survey; 67–84. http://pubs.usgs.gov/sir/2010/5091/papers

Beylich, AA and Laute, K 2014. Combining impact sensor field and laboratory flume measurements with other techniques for studying fluvial bedload transport in steep mountain streams. *Geomorphology* 218, 72–87.

Böckli, M 2011. *Laborversuche zum Verhalten von Geophon-Sensoren*. Master Thesis, Swiss Federal Institute of Technology ETH, Zürich, and Swiss Federal Research Institute WSL, Birmensdorf; 56 pp.

Bogen, J and Møen, K 2003. Bed load measurements with a new passive acoustic sensor. In *Erosion and Sediment Transport Measurement in Rivers: Technological and Methodological Advances* (eds Bogen, J, Fergus, T, and Walling, DE). Publication 283, IAHS: Wallingford, UK; 181–192.

Bradeau, G 1951. Quelques techniques pour l'étude et la mesure de débit solide. *La Houille Blanche* 6, 243–252.

Bunte, K, Abt, SR, Potyondy, JP, and Ryan, SE 2004. Measurement of coarse gravel and cobble transport using portable bedload traps. *Journal of Hydraulic Engineering* 130(9), 879–893.

Burtin, A, Bollinger, L, Vergne, J *et al.* 2008. Spectral analysis of seismic noise induced by rivers: a new tool to monitor spatiotemporal changes in stream hydrodynamics. *Journal of Geophysical Research* 113, B05301. DOI:10.1029/2007JB005034.

Camenen, B, Jaballah, M, Geay, T *et al.* 2012. Tentative measurements of bedload transport in an energetic alpine gravel bed river. In *River Flow 2012* (ed Munoz, M). Taylor & Francis Group: London; 379–386.

Carson, G, Mulholland, AJ, Nordon, A *et al.* 2008. Particle sizing using passive ultrasonic measurement of vessel wall vibrations. *Journal of Sound and Vibration* 317, 142–157.

Chiari M and Rickenmann D 2011. Back-calculation of bedload transport in steep channels with a numerical model. *Earth Surface Processes and Landforms* 36, 805–815.

Childers, D 1999. *Field Comparisons of Six Pressure Difference Bedload Samplers in High-Energy Flow*. Water-Resources Investigations Report 92–4068, US Geological Survey; 59 pp.

Downing, J 2010. Acoustic gravel-momentum sensor. In *Bedload-Surrogate Monitoring Technologies* (eds Gray, JR, Laronne, JB, and Marr, JDG). Scientific Investigations Report SIR 2010-5091, US Geological Survey; 352–366. http://pubs.usgs.gov/sir/2010/5091/papers.

Downing, J, Farley, PJ, Bunte, K *et al.* 2003. Acoustic gravel-transport sensor: description and field tests in Little Granite Creek, Wyoming, USA. In *Erosion and Sediment Transport Measurement in Rivers: Technological and Methodological Advances* (eds Bogen, J, Fergus, T, and Walling, DE). Publication 283, IAHS: Wallingford, UK; 193–200.

Emmett, WW 1980. *A Field Calibration of the Sediment-Trapping Characteristics of the Helley–Smith Bed-Load Sampler*. Professional Paper 1139, US Geological Survey; 44 pp.

Etter, M 1996. *Zur Erfassung des Geschiebetransportes mit Hydrophonen*. Diploma Thesis, Institute of Geography, University of Berne, Berne, and Swiss Federal Research Institute WSL, Birmensdorf; 110 pp.

Etter, PC 2012. Advanced applications for underwater acoustic modeling. *Advances in Acoustics and Vibration*, Article ID 214839; 28 pp. DOI:10.1155/2012/214839.

Froehlich, W 2003. Monitoring bed load transport using acoustic and magnetic devices. In *Erosion and Sediment Transport Measurement in Rivers: Technological and Methodological Advances* (eds Bogen, J, Fergus, T, and Walling, DE). Publication 283, IAHS: Wallingford, UK; 201–210.

Froehlich W 2010. Monitoring of bed load transport within a small drainage basin in the Polish flysch Carpathians. In *Bedload-Surrogate Monitoring Technologies* (eds Gray, JR, Laronne, JB, and Marr, JDG). Scientific Investigations Report SIR 2010-5091, US Geological Survey; 185–194. http://pubs.usgs.gov/sir/2010/5091/papers

Geay, T 2013. *Mesure acoustique passive du transport par charriage dans les rivières.* Doctoral Thesis, University of Grenoble, Grenoble, France.

Gomez B 1991. Bedload transport. *Earth-Science Reviews* 31, 89–132.

Gomez, B and Church, M 1989. An assessment of bedload sediment transport formulae for gravel bed rivers. *Water Resources Research* 25(6, 1161–1186.

Gray, JR, Laronne, JB, Marr, JDG (eds) 2010. *Bedload-Surrogate Monitoring Technologies.* Scientific Investigations Report SIR 2010-5091, US Geological Survey. http://pubs.usgs.gov/sir/2010/5091

Habersack, H, Seitz, H, and Liedermann, M 2010. Integrated automatic bedload transport monitoring. In *Bedload-Surrogate Monitoring Technologies* (eds Gray, JR, Laronne, JB, and Marr, JDG). Scientific Investigations Report SIR 2010-5091, US Geological Survey; 281–235. http://pubs.usgs.gov/sir/2010/5091/papers

Hassan, MA, Church, M, Lisle, TE *et al.* 2005. Sediment transport and channel morphology of small, forested streams. *Journal of the American Water Resources Association* 41, 853–876.

Hayward, JA 1980. *Hydrology and Stream Sediments in a Mountain Catchment.* Special Publication 17, Tussock Grasslands and Mountain Lands Institute: Canterbury, New Zealand; 235 pp.

Helley, EJ and Smith, W 1971. *Development and Calibration of a Pressure-Difference Bedload Sampler.* Open-File Report 73-108, US Geological Survey; 38 pp.

Hildebrand, JA 2009. Anthropogenic and natural sources of ambient noise in the ocean. *Marine Ecology Progress Series* 395, 5–20.

Krein, A, Klinck, H, Eiden, M *et al.* 2008. Investigating the transport dynamics and the properties of bedload material with a hydro-acoustic measuring system. *Earth Surface Processes and Landforms* 33, 152–163.

Lenzi, MA, D'Agostino, V, and Billi, P 1999. Bedload transport in the instrumented catchment of the Rio Cordon, part I: analysis of bedload records, conditions and threshold of bedload entrainment. *Catena* 36, 171–190.

Leopold, LB and Emmett, WW 1997. *Bedload and River Hydraulics – Inferences from the East Fork River, Wyoming.* Professional Paper 1583, US Geological Survey; 52 pp.

Mao, L, Carrillo, R, Escauriaza, C, and Iroume, A 2016. Flume and field-based calibration of surrogate sensors for monitoring bedload transport. *Geomorphology* 253, 10–21.

Mizuyama, T, Fujita, M, and Nonaka, M 2003. Measurement of bed load with the use of hydrophones in mountain rivers. In *Erosion and Sediment Transport Measurement in Rivers: Technological and Methodological Advances* (eds Bogen, J, Fergus, T, and Walling, DE). Publication 283, IAHS: Wallingford, UK; 222–227.

Mizuyama, T, Oda, A, Laronne, JB *et al.* 2010a. Laboratory tests of a Japanese pipe geophone for continuous acoustic monitoring of coarse bedload. In *Bedload-Surrogate Monitoring Technologies* (eds Gray, JR, Laronne, JB, and Marr, JDG). Scientific Investigations Report SIR 2010-5091, US Geological Survey; 319–335. http://pubs.usgs.gov/sir/2010/5091/papers

Mizuyama, T, Laronne, JB, Nonaka, M. et al 2010b. Calibration of a passive acoustic bedload monitoring system in Japanese mountain rivers. In *Bedload-Surrogate Monitoring Technologies* (eds Gray, JR,

Laronne, JB, and Marr, JDG). Scientific Investigations Report SIR 2010-5091, US Geological Survey; 296–318. http://pubs.usgs.gov/sir/2010/5091/papers

Mao, L, Dell'Agnese, A, Huincache, C *et al.* 2014. Bedload hysteresis in a glacier-fed mountain river. *Earth Surface Processes and Landforms* 39, 964–976.

Møen, KM, Bogen, J, Zuta, JF *et al.* 2010. Bedload measurement in rivers using passive acoustic sensors. In *Bedload-Surrogate Monitoring Technologies* (eds Gray, JR, Laronne, JB, and Marr, JDG). Scientific Investigations Report SIR 2010-5091, US Geological Survey; 336–351. http://pubs.usgs.gov/sir/2010/5091/papers

Mühlhofer, L 1933. Untersuchungen über die Schwebstoff und Geschiebeführung des Inn nächst Kirchbichl (Tirol). *Die Wasserwirtschaft* 1(6), 23 pp.

Nitsche, M, Rickenmann, D, Turowski, JM *et al.* 2011. Evaluation of bedload transport predictions using flow resistance equations to account for macro-roughness in steep mountain streams. *Water Resources Research* 47, W08513. DOI:10.1029/2011WR010645.

Raven, EK, Lane, SN, and Ferguson, R 2010. Using sediment impact sensors to improve the morphological sediment budget approach for estimating bedload transport rates. *Geomorphology* 119, 125–134.

Raven, EK, Lane, SN, Ferguson, RI, and Bracken, LJ 2009. The spatial and temporal patterns of aggradation in a temperate, upland, gravel-bed river. *Earth Surface Processes and Landforms* 34, 1181–1197.

Recking, A 2010. A comparison between flume and field bed load transport data and consequences for surface based bed load transport prediction. *Water Resources Research* 46, W03518. DOI:10.1029/2009WR008007.

Reid, SC, Lane, SN, and Berney, JM 2007. The timing and magnitude of coarse sediment transport events within an upland gravel-bed river. *Geomorphology* 83, 152–182.

Richardson, K, Benson, I, and Carling, PA 2003. An instrument to record sediment movement in bedrock channels. In *Erosion and Sediment Transport Measurement in Rivers: Technological and Methodological Advances* (eds Bogen, J, Fergus, T, and Walling, DE). Publication 283, IAHS: Wallingford, UK; 228–235.

Rickenmann, D 2001. Comparison of bed load transport in torrents and gravel bed streams. *Water Resources Research* 37(12), 3295–3305.

Rickenmann, D 2012. Alluvial steep channels: flow resistance, bedload transport and transition to debris flows. In *Gravel Bed Rivers: Processes, Tools, Environment* (eds Church, M, Biron, P, and Roy, A). John Wiley & Sons: Chichester, UK; 386–397.

Rickenmann, D and McArdell, BW 2007. Continuous measurement of sediment transport in the Erlenbach stream using piezoelectric bedload impact sensors. *Earth Surface Processes and Landforms* 32, 1362–1378.

Rickenmann, D and McArdell, BW 2008. Calibration measurements with piezoelectric bedload impact sensors in the Pitzbach mountain stream. *Geodinamica Acta* 21, 35–52.

Rickenmann, D, Hofer, B, and Fritschi, B 1997. Geschiebemessung mittels Hydrophon. *Österreichische Wasser- und Abfallwirtschaft* 49(11/12), 219–228.

Rickenmann, D, Turowski, JM, Fritschi, B *et al.* 2012. Bedload transport measurements at the Erlenbach stream with geophones and automated basket samplers. *Earth Surface Processes and Landforms* 37, 1000–1011.

Rickenmann, D, Laronne, JB, Turowski, JM, and Vericat, D (eds) 2013. *International Workshop of Acoustic and Seismic Monitoring of Bedload and Mass Movements*, Abstracts. Birmensdorf, Switzerland, Swiss Federal Research Institute WSL; 109 pp.

Rickenmann, D, Turowski, JM, Fritschi, B *et al.* 2014. Bedload transport measurements with impact plate geophones: comparison of sensor calibration in different gravel-bed streams. *Earth Surface Processes and Landforms* 39, 928–942.

Ryan, SE and Dixon, MK 2008. Spatial and temporal variability in stream sediment loads using examples from the Gros Ventre Range, Wyoming, USA. In *Gravel-Bed Rivers VI: From Process Understanding to River Restoration* (eds Habersack, H, Piégay, H, and Rinaldi, M). Developments in Earth Surface Processes Vol. 11, Elsevier: Amsterdam; 387–407.

Ryan, SE, Porth, LS, and Troendle, CA 2005. Coarse sediment transport in mountain streams in Colorado and Wyoming, USA. *Earth Surface Processes and Landforms* 30, 269–288.

Schmandt, B, Aster, RC, Scherler, D *et al.* 2013. Multiple fluvial processes detected by riverside seismic and infrasound monitoring of a controlled flood in the Grand Canyon. *Geophysical Research Letters* 40, 4858–4863. DOI:10.1002/grl.50953.

Tacconi, P and Billi, P 1987. Bed load transport measurements by the vortex-tube trap on Virginio Creek, Italy. In *Sediment Transport in Gravel-Bed Rivers* (eds Thorne, CR, Bathurst, JC, and Hey, RD). John Wiley & Sons: New York; 583–616.

Taniguchi, S, Itakura, Y, Miyamoto, K, and Kurihara, J 1992. A new acoustic sensor for sediment discharge measurement. In *Erosion and Sediment Transport Monitoring Programmes in River Basins* (eds Bogen, J, Walling, DE, and Day, TJ). Publication 210, IAHS: Wallingford, UK; 135–142.

Thorne, PD 1985. The measurement of acoustic noise generated by moving artificial sediments. *Journal of the Acoustic Society of America* 78, 1013–1023.

Thorne, PD 1986. Laboratory and marine measurements on the acoustic detection of sediment transport. *Journal of the Acoustic Society of America* 80(3), 899–910.

Thorne, PD 1990. Seabed generation of ambient noise. *Journal of the Acoustic Society of America* 87(1), 149–153.

Thorne, PD 2014. An overview of underwater sound generated by inter-particle collisions and its application to the measurements of coarse sediment bedload transport. *Earth Surface Dynamics* 2, 531–543.

Thorne, PD and Foden, DJ 1988. Generation of underwater sound by colliding spheres. *Journal of the Acoustic Society of America* 84(6), 2144–2152.

Tsakiris, AG, Papanicolaou, AN, and Lauth, TJ 2014. Signature of bedload particle transport mode in the acoustic signal of a geophone. *Journal of Hydraulic Research* 52(2), 185–204.

Turowski, JM and Rickenmann, D 2009. Tools- and cover-effect in bedload transport observations in the Pitzbach, Austria. *Earth Surface Processes and Landforms* 34, 26–37.

Uher, M and Beneš, P 2012. Measurement of particle size distribution by the use of acoustic emission method. In *IEEE International Instrumentation and Measurement Technology Conference* (I2MTC), 13–16 May, Graz, Austria; 1194–1198.

Voulgaris, G, Wilkin, MP, and Collins, MB 1995. The *in situ* passive acoustic measurement of shingle movement under waves and currents: instrument (TOSCA) development and preliminary results. *Continental Shelf Research* 15, 1195–1211.

Wyss, CR, Rickenmann, D, Fritschi, B *et al.* 2016. Measuring bedload transport rates by grain-size fraction using the Swiss plate geophone signal at the Erlenbach. *Journal of Hydraulic Engineering* 142(5), DOI:10.1061/(ASCE)HY.1943-7900.0001090.

Discussions

Discussion by Hiroshi Suwa

Reviewing many publications, this paper describes the achievements in developing the technique for monitoring bedload transport with acoustic sensors. It summarizes that bedload flux and grain sizes have been calibrated from sensor signals. These calibrations are successfully applied by counting number of impulses and intensity of signals. Data on wave frequency may also be useful for evaluating bedload transport. Also, the power spectra obtained from geophone signals in natural streams is expected to supply information not only on bedload texture but also on the temporal change in bedload grain-size.

Discussion by Kimberly Hill

At the beginning of this presentation, there was a demonstration of how the sound of a collision between two spheres varied with the sizes of the spheres colliding. Later, there was a discussion of how amplitude, rather than frequency, might be used to identify the size of the colliding particles. It surprised me that frequency was not included in this discussion. There has been considerable theoretical work as to how resonant frequencies of an object depend on its size (e.g., Lamb, 1882). Also, it seems that resonant frequency of rocks might not vary with transport conditions while collisional amplitude might. I am curious if there has been any work investigating how grain size may be reflected in the frequencies of the acoustic signals and, if not, why this direction may not have been pursued.

Reference

Lamb, H 1882. On the vibrations of an elastic sphere. *Proceedings of the London Mathematical Society* s1–13(1), 189–212.

Discussion by Philippe Belleudy

I take the opportunity of adding details on the hydrophone method as it is mentioned in this chapter. (i) The hydrophone method is integrative of a large surface of the bed. (ii) The transmission properties of the sound from the riverbed is considerably affected by the bouncing of the signal on flow and bed surfaces, resulting in a filtering of high frequencies in the case of shallow rivers. The determination of the grain size of the bedload by a dominant frequency in the signal must take in consideration such an effect. Yet, the hydrophone method, which still requires calibration with more classic methods, is particularly adapted to large rivers where other methods are more difficult to implement.

Discussion by Helmut Habersack and Andrea Kreisler

The bedload monitoring with plate geophones at the Urslau stream (Austria) during a flood with a return period of 100 years that occurred in 2013 shows that within the rising limb of the hydrograph the geophone impulses suddenly vanished to zero. After the flood the geophones responded once again to bedload particle collisions. Since geophones register the vibration of the steel plate when bedload particles collide on them, the disappearance of the signal is interpreted to suggest that gravel sheets may have been formed and migrated over the plates during this extreme event, hindering the geophones to record bedload. Therefore, we conclude that only an integrated bedload monitoring system with additional monitoring techniques, e.g., slot samplers, allows monitoring the full spectrum of discharges. For very large floods, where bedload dunes or bedload sheets are formed, the functioning of the plate geophones seems to be limited, depending on the boundary conditions of the measuring station.

Discussion by Jonathan Laronne

As this workshop is focused amongst others on disasters, hence on large floods, please clarify to what extent can the passive acoustic techniques developed hitherto be applied to large floods. This is particularly relevant because much bedload is mobilized in them and they have considerable influence on the morphology of gravel bedded rivers. The monitoring of bedload by microphones in the water column (commonly termed hydrophones) appears to be very advantageous for large floods because the sensor need not be deployed on the riverbed like geophones.

Reply by the Author

I thank the discussers for their comments and questions. Hiroshi Suwa indicated that frequency and power-spectra information of the acoustic signal may be useful for grain-size identification. Kimberly Hill addressed a similar point suggesting that identification of particle size may be more robust using frequency information rather than signal amplitude, which may depend more on transport conditions. In the chapter reference is made to several studies using acoustic sensors, wherein grain size was successfully determined by using frequency information of the acoustic signal. This was similarly confirmed in a field study deploying seismometers close to a river channel, showing that the power spectra and dominant frequencies tended to change with grain size (Burtin *et al.* 2011). Furthermore, changes in the power spectra with grain size were also found in flume and field experiments measuring rock impact on a gravel bed with a hydrophone and geophone sensor (Huang, Shieh, and Yin 2004; Huang *et al.* 2007). Reference is also made in the chapter to the study of Barrière *et al.* (2014) who used both frequency and amplitude information together for grain-size identification with an acoustic method. Figure 7.6 indicates that also for the Swiss plate geophone system (SPGS) the characteristic frequency is sensitive to grain size over a certain range of particle sizes. An ongoing flume study with the SPGS indicates that a combination of frequency and amplitude information results in relatively robust predictions of grain size but is not necessarily much more precise than grain-size identification by amplitude information alone (Wyss *et al.* 2016a), and that a flume-based interpretation of the geophone signal can be used in principle to predict bedload transport at field sites without direct calibration measurements (Wyss *et al.* 2016b).

The comments of Philippe Belleudy present some additional information on the use of underwater microphones (hydrophones), and I have nothing to add in this regard. Helmut Habersack and Andrea Kreisler hinted to a monitoring site with Swiss plate geophones in an Austrian stream where it is hypothesized that gravel sheets may have resulted in a cover effect over the steel plates, as a possible reason for non-recorded bedload transport activity during higher flows. For the field calibration measurements with the Swiss plate geophones in the Nahal Esthemoa stream in Israel, we found a relatively weak signal response as compared to other field sites with calibration measurements (Rickenmann *et al.* 2014), and we similarly hypothesized that (a temporary) deposition of gravel sheets may have been the reason for this.

Jonathan Laronne asked to what extent passive acoustic measuring techniques may also be useful during large floods and sediment disasters. In my view, impact plates and pipe hydrophones are expected to be useful also under such conditions, as long as they are not covered by sediments and they are not damaged or destroyed. In this regard, underwater microphones (hydrophones) placed at some distance above the streambed are less likely to be affected by such problems. Swiss plate geophones are now installed at more than 20 field sites, and they have not been damaged so far. Furthermore, if they are placed at a fixed point in the longitudinal profile such as at a check dam, sediment deposits are very unlikely. It can also be mentioned that Swiss plate geophones were installed at two water intakes in steep headwater streams in Switzerland where they also serve as an alarm system, and in case of very high bedload transport or debris-flow-like conditions a gate at the water intake could be closed by remote control.

References

Burtin, A, Cattin, R, Bollinger, L *et al.* 2011. Towards the hydrologic and bed load monitoring from high-frequency seismic noise in a braided river: the 'torrent de St Pierre', French Alps. *Journal of Hydrology* 408, 43–53.

Huang, CJ, Shieh, CL, and Yin, HY 2004. Laboratory study of the underground sound generated by debris flows. *Journal of Geophysical Research* 109, F01008. DOI:10.1029/2003JF000048.

Huang, CJ, Yin, HY, Chen, CY *et al.* 2007. Ground vibrations produced by rock motions and debris flows. *Journal of Geophysical Research* 112, F02014. DOI:10.1029/2005JF000437.

Wyss, CR, Rickenmann, D, Fritschi, B *et al.* 2016a. Laboratory flume experiments with the Swiss plate geophone bedload monitoring system. Part I: Impulse counts and particle size identification. *Water Resources Research*, 52, 7744–7759, DOI:10.1002/2015WR018555.

Wyss, CR, Rickenmann, D, Fritschi, B *et al.* 2016b. Laboratory flume experiments with the Swiss plate geophone bedload monitoring system. Part II: Application to field sites with direct bedload samples. *Water Resources Research*, 52, 7760–7778, DOI:10.1002/2015WR019283.

8

Calibration of Acoustic Doppler Current Profiler Apparent Bedload Velocity to Bedload Transport Rate

Colin D. Rennie, Damià Vericat, Richard D. Williams, James Brasington, and Murray Hicks

8.1 Introduction

Bed material transport has long been recognized as a key gravel-bed river process that determines channel form (Mulhoffer 1933). At all spatial and temporal scales, river morphology is determined by the spatial and temporal pattern of sediment transfer along a river reach through local erosion and deposition. For example, local differences in transport explain the development of meandering planforms due to erosion on the outer bank of bends and deposition of sediments on point bars. Similarly, bars develop in braiding and wandering rivers due to local deposition of sediments in areas of flow divergence where flow decelerates and loses transport capacity, following entrainment and transport from some upstream point of flow convergence and acceleration.

Most river management problems require understanding of bed material transport, including channel change, flooding, fish habitat, and infrastructure protection. Unfortunately, accurate measurement of bed material transport has perpetually confounded researchers. The most reliable direct measurement methods have involved channel-spanning bedload traps (e.g. Wolman 1954; Klingeman and Milhous, 1971; Rickenmann *et al.* 2012). This requires a permanent installation at a single cross-section, and allows for quantification of total bedload. However, these installations are necessarily limited to relatively small rivers, and cannot be used to estimate the bedload spatial distributions that determine channel morphodynamics. Moreover, in ecologically sensitive reaches the installation of hard structures to monitor bedload can be considered disruptive to natural ecosystem functioning.

The conventional approach to this problem is to deploy small bedload samplers at discrete points on the river bed. However, the presence of the sampler disturbs the local flow field and therefore the local bed material transport. Pressure-difference bedload samplers, such as the BL-84, attempt to overcome this problem by employing a flared exit to induce suction through the sampler to achieve a sampling efficiency near one. Still, sampling efficiencies need to be calibrated and are often grain-size dependent (Hubbell 1987; Edwards and Glysson 1999). Furthermore, bedload samplers can have large sampling bias and errors (e.g. Vericat, Church, and Batalla 2006), induced by gouging of the river bed during deployment and/or incorrect alignment of the sampler with the flow (Gaudet, Roy, and Best 1994). This is particularly problematic in large rivers, which require deployment of the samplers from a

Gravel-Bed Rivers: Processes and Disasters, First Edition. Edited by Daizo Tsutsumi and Jonathan B. Laronne.

boat – which impedes careful placement of the sampler on the bed. Besides these issues with measurement inaccuracy, bedload is notoriously difficult to measure with conventional samplers because: (i) real spatiotemporal variability of bedload requires a large number of samples distributed in both time and space for adequate representation of bedload transport rate; (ii) deploying a sampler near the bed is dangerous and/or difficult during large, channel-forming discharges with high water velocities and bedload transport rates; and (iii) a great deal of effort is required to collect and process samples. Consequently, very few attempts have been made to map the spatial distribution of bed material transport using conventional samplers (e.g. Davoren and Mosley 1986; Dietrich and Whiting 1989; Bridge and Gabel 1992; Ferguson *et al.* 1992).

The inverse or morphological approach has been seen as an alternative to study bed material transport (see the reviews by Ashmore and Church (1989) and Church (2006)). The inverse approach provides an estimation of the distribution of bed material transport at multiple spatial and temporal scales. This approach infers the links between channel hydraulics and morphology; however, some technical issues in relation to key variables for its application remain, such as reliable particle step lengths or a reference sediment flux. Moreover, this approach provides estimates of bed material transport that differ from what is often sampled in the field as either bedload or sediment transported in suspension. Technological progress in topography has opened new opportunities to the morphological approach, as is reviewed in this book (Chapter 5). Recent publications (e.g. Wheaton *et al.* 2013) have demonstrated how this approach may provide new insights into morphodynamic mechanisms in shaping fluvial channels. However, this approach fails in quantifying sediment fluxes during flood events.

As a result of these difficulties, there has been increasing interest in adopting surrogate approaches to bedload transport measurement (Gray, Laronne, and Marr 2010; Gray *et al.* 2010; Burtin *et al.* 2011; Rickenmann *et al.* 2013). Acoustic techniques are particularly attractive for river studies due to their noninvasive character and the relatively low attenuation and high resolution obtainable with high frequency sound in turbid water. Acoustic methods can be passive or active. Passive methods use hydrophones to "listen" to the intensity of acoustic noise generated by collisions amongst mobile bed material, whereas active methods analyse the effects of bedload on returns from an incident sound source. This chapter will focus on use of acoustic Doppler current profilers (aDcps) for active measurement of bedload. In particular, the bias in aDcp bottom tracking (Doppler sonar) induced by near-bed particle motion is used as a measure of apparent bedload velocity (Rennie, Millar, and Church 2002). In this chapter, the technique is explained and previous attempts to calibrate apparent bedload velocity for measurement of bedload transport rate are reviewed. An extensive new data set obtained on the braided gravel-bed Rees River, New Zealand, is used to expand and evaluate aDcp apparent bedload velocity calibration. In particular, the influence of bedload particle grain size on measured apparent bedload velocity is examined in detail.

8.2 aDcp Apparent Bedload Velocity

8.2.1 Estimation

An aDcp utilizes range-gated monostatic acoustic backscatter along multiple beams to measure spatial averages of the three principal water velocity components in individual "bins" throughout a vertical column of water (see Brumley, Cabrera, and Terray 1991; Gordon 1996; and Simpson 2001 for aDcp theory and methodology). These instruments are increasingly used to measure fluvial water velocity distributions and discharge (e.g. Dinehart and Burau 2005; Rennie and Church 2010;

Guerrero and Lamberti 2011; Jamieson *et al.* 2011a; Tsubaki *et al.* 2012; Petrie *et al.* 2013; Venditti *et al.* 2014; Vermeulen, Sassi, and Hoitink 2014).

Velocities are measured relative to the reference frame of the aDcp. If the instrument is mounted on a moving boat, corrections must be made for boat velocity in order to obtain absolute water velocities. The boat velocity is determined either by bottom tracking or by use of differential global positioning system (DGPS) data. Boat velocity by bottom tracking is typically more accurate than by DGPS, although accuracy of real-time kinematic DGPS (RTK-DGPS) can approach that of bottom tracking (Rennie and Rainville 2006, 2008; Rennie, Rainville, and Kashyap 2007). However, bottom tracking for water velocity measurement is limited to immobile bed conditions. Bottom tracking involves sending an acoustic pulse along each of multiple beams to the river bed. The velocity component parallel to each beam can be determined from the Doppler shift in frequency of the return pulse of each beam backscattered from the river bed. The Doppler shift is assumed to be due to boat velocity, but if the river bed material is moving then bottom tracking (BT) is biased by the bed velocity, and the frequency shift is due to both the boat velocity and the sediment motion. Under mobile bed conditions, the apparent velocity of the bed sediment (v_a) can be determined if the actual boat velocity is known by DGPS (Rennie, Millar, and Church 2002):

$$v_a = v_{DGPS} - v_{BT} \tag{8.1}$$

Typical aDcps provide one bottom track ping for each water velocity profile measurement, thus achievable v_a sampling rates are on the order of 1 Hz.

It is important to emphasize that v_{DGPS} is collected in Earth coordinates, whereas v_{BT} is collected in the reference frame of the instrument. The aDcp compass must be calibrated and an estimate of local magnetic declination is required to rotate v_{BT} to Earth coordinates for calculation of v_a.

8.2.2 Sampling Volume

The sampling volume for a four-beam aDcp consists of four independent volumes located at the impingement points of the four beams. The size of the insonified sampling area of the acoustic beams on the bed was described in Rennie, Millar, and Church (2002). The sampling volume was briefly described in Rennie, Millar, and Church (2002), and further detail was provided in Rennie and Millar (2004). The sampling location depends on beam geometry, distance to the bed, and length of the bottom track acoustic pulse (Rennie, Millar, and Church 2002; Rennie and Millar 2004). Three beams are required to resolve a three-dimensional Cartesian velocity from individual beam velocities, but most modern aDcps employ four acoustic beams. The redundant fourth beam provides an additional velocity component, allowing for two estimates of vertical velocity, from which error estimates can be derived. The estimated Cartesian bottom track velocity thus depends on either three or four individual beam velocities, each of which is positioned at a different location on the bed. Due to beam tilt from vertical (ϕ), the insonified bed areas have quasi-elliptical shapes; Rennie, Millar, and Church (2002) provided equations to estimate these dimensions. As seen in Figure 8.1, each beam samples a different location on the bed, and the sampling locations are located around the circumference of a circle with diameter approximately equal to the distance between the aDcp and the bed (depending on ϕ).

The sampling location is a volume as opposed to an area, because suspended scatterers above the bed can contribute to the biased bottom track signal if the reflected sound pulse is coincident with the incident pulse (Figure 8.2). The maximum height of the sampling volume above the bed is [$0.5\ l_p \sin(90 - \phi - \theta)$], where l_p is the length of the acoustic pulse, and θ is the half-intensity beam width due to beam spreading (Rennie, Millar, and Church 2002). The default pulse length for aDcps made by RD Instruments, such as the Rio Grande ADCP™, is 20% of the vertical range to the bed from the aDcp

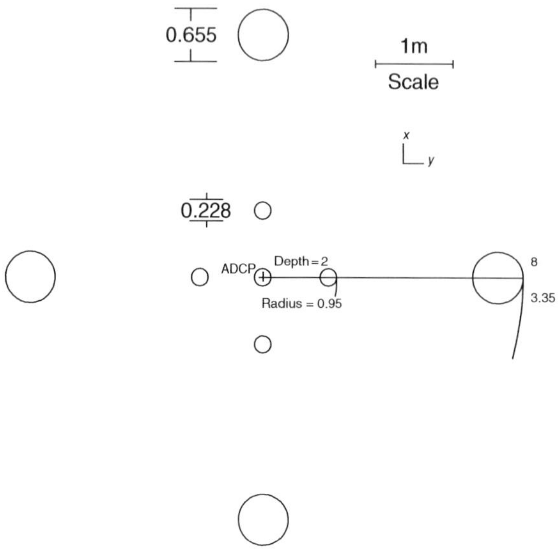

Figure 8.1 Plan view of bottom track sampling areas for a four-beam 1200 kHz Rio Grande ADCP at two different depths (2 m and 8 m). All dimensions in meters.

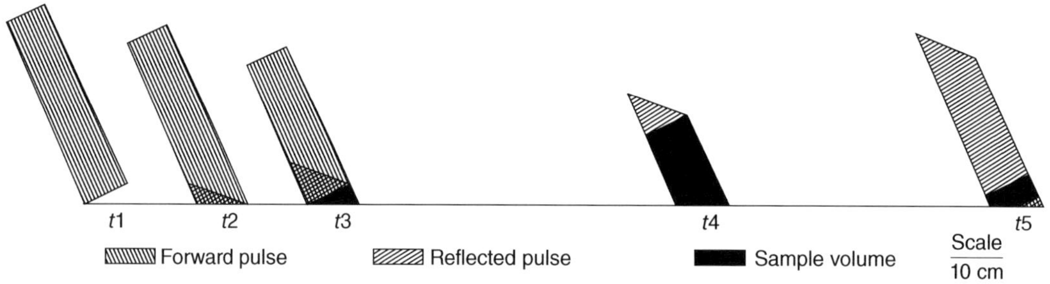

Figure 8.2 Profile of dynamic beam sample volume for a pulse of length 60 cm reflecting off a flat bed 3.05 m below a non-tilted 1.5 MHz Sontek ADP. Beam spread is considered, but the far-field approximation of plane incident and reflected waves is employed. Time is progressing from left to right (horizontally scaled 5x for clarity), such that the locations of incident and reflected pulse are shown at various times (t1 through t5) during pulse reflection. The sample volume is specified where suspended scatterers can contribute to the acoustic return, i.e. where the incident and reflected pulse are coincident and both the leading and trailing edge of the incident pulse contribute to the reflected pulse. Note the sample volume changes dynamically during pulse reflection. (*Source*: Rennie and Millar 2004. Reproduced with permission of John Wiley & Sons.)

(i.e., &R20 ADCP configuration command). Consequently, both bedload and near-bed suspended particles within roughly the lower 9% of the water column can contribute to the v_a signal. However, very fine suspended sediment should not contribute greatly to the v_a signal. This is because, at the acoustic frequencies of aDcps used for river studies, clays and silts scatter in the Rayleigh regime, whereas coarse sands and gravels scatter with orders of magnitude more intensity in the geometric specular regime (Moate and Thorne 2012). Geometric scattering occurs for particles with a diameter greater than $\frac{2c}{\pi f}$, where c is the speed of sound (~1500 m/s in water) and f is the acoustic frequency. For example, for a 1.2 MHz aDcp, particles with diameter greater than about 0.8 mm are most emphasized.

It should also be acknowledged that the precise location of reflection of a bottom track pulse and the corresponding location of measurement has some uncertainty. Bottom track processing finds the velocity of the stable bed level, which is determined based on location of the peak acoustic intensity. If the bedload layer is highly concentrated and thereby attenuates the acoustic pulse, the peak intensity could possibly occur near the top of the bedload layer, resulting in positive bias of the measured bedload velocity. However, this should be less of an issue in gravel-bed rivers than sand-bed rivers, due to the presence of coarse bedload in a thin sporadic bedload layer.

In order to resolve a Cartesian velocity from the individual beam velocities, the homogeneity assumption of equal bedload velocity in each insonified beam area must be applied. Bedload transport involves the intermittent entrainment and transport of material from the riverbed, and is a function (in part) of the composition of the surface particles. Gravel beds tend to entrain and transport material more stochastically than do sand beds, due to their larger and typically less sorted particle size, and transport is spatially and temporally heterogeneous. Thus, the bedload homogeneity assumption is likely violated in gravel-bed rivers. In sand-bed rivers the transport may be more homogeneous locally, but due to transport of bed material as bedforms (ripples and dunes), heterogeneous transport can be expected at these scales in sand-bed environments.

Beam velocities are converted to a Cartesian velocity using a coordinate transform. For a four beam aDcp, with beam 3 facing forward (x direction), beam 4 facing aft ($-x$ direction), beam 2 positive y (counterclockwise from x when facing the transducer), and beam 1 negative y (clockwise from x when facing the transducer):

$$\hat{v}_x = \frac{b_3 - b_4}{2\sin(\phi)} \tag{8.2a}$$

$$-\hat{v}_y = \frac{b_1 - b_2}{2\sin(\phi)} \tag{8.2b}$$

$$\hat{v}_z = \frac{-b_3 - b_4}{2\cos(\phi)} = \frac{-b_1 - b_2}{2\cos(\phi)} \tag{8.2c}$$

where b_i are the individual along-beam velocity components and $\hat{v}_x$, $\hat{v}_y$, and $\hat{v}_z$ are the Cartesian components with z positive upwards (i.e., into the page when facing transducers for a down-looking aDcp). Due to beam tilt from vertical (ϕ), both horizontal and vertical bedload velocities in the insonified area will contribute to the individual beam velocities:

$$b_3 = v_{3x}\sin(\phi) - v_{3z}\cos(\phi) \tag{8.3a}$$

$$b_4 = -v_{4x}\sin(\phi) - v_{4z}\cos(\phi) \tag{8.3b}$$

$$b_1 = -v_{1y}\sin(\phi) - v_{1z}\cos(\phi) \tag{8.3c}$$

$$b_2 = v_{2y}\sin(\phi) - v_{2z}\cos(\phi) \tag{8.3d}$$

Consequently, it can be shown that unequal velocities in beam pairs will result in erroneous estimates of $\hat{v}_x$, $\hat{v}_y$, and $\hat{v}_z$:

$$\hat{v}_x = \frac{v_{3x} + v_{4x}}{2} + \frac{v_{4z} - v_{3z}}{2}\cot(\phi) \tag{8.4a}$$

$$\hat{v}_y = \frac{v_{1y} + v_{2y}}{2} + \frac{-v_{2z} + v_{1z}}{2}\cot(\phi) \tag{8.4b}$$

$$\hat{v}_z = \frac{v_{3z} + v_{4z}}{2} + \frac{v_{4x} - v_{3x}}{2}\tan(\phi) = \frac{v_{1z} + v_{2z}}{2} + \frac{-v_{2y} + v_{1y}}{2}\tan(\phi) \tag{8.4c}$$

For the four-beam system, violation of the assumption of homogeneity results in contamination of horizontal velocity components by only vertical velocity components, which are usually small.

The error velocity arises from the redundant estimates of $\hat{v}_z$, and accounts for both errors in individual beam velocities and heterogeneity between beams. The vertical velocity difference can be converted to a horizontal velocity error (v_e) using the following transform (Teledyne RD Instruments 2010):

$$v_e = \frac{b_1 - b_2 + b_3 - b_4}{2\sqrt{2}sin\theta} \tag{8.5}$$

8.2.3 Errors

Stationary time series of individual single ping v_a samples have large variance, thus averaging is required (Rennie, Millar, and Church 2002). The variance is due to both error and real temporal variability of bedload. Rennie and Millar (2007) developed a statistical model to deconvolve the signal and noise. The model assumes a compound Poisson-gamma distribution for the spatially averaged bedload velocity signal and zero-mean normally distributed errors. For v_a time series collected from two different gravel-bed rivers, the estimated signal distributions were highly positively skewed, indicating that spatially averaged bedload velocity was mostly near zero interspersed with sporadic high values. The errors were large, of similar magnitude to the estimated signal. The errors arise due to decorrelation of the bottom track signal (i.e., increased bandwidth of the bottom track Doppler spectrum) in the presence of variable particle velocities and multiple scattering in the bedload layer, as well as inhomogeneity of bedload velocity between beams.

Rennie and Church (2010) employed the results of the deconvolution model to develop a method to estimate uncertainty associated with using individual v_a samples to represent the local mean within a spatial survey. The output of the deconvolution model for a stationary time series was used to estimate real temporal variability of spatially averaged bedload, and errors arising from the compass, DGPS, and v_{BT} were considered. The bottom track error velocity (v_e) was used for v_{BT} error. Compass induced errors in v_a ($\sigma_{c,\varepsilon}$) were calculated using (8.6), which was based on previous work of Joyce (1989) and Gaeuman and Jacobson (2005):

$$\begin{aligned} \sigma_{c,\varepsilon,E} &= v_{a,E}(\cos(\zeta) - 1) + v_{DGPS,E}(1 - \cos(\zeta)) + \sin(\zeta)(v_{a,N} - v_{DGPS,N}) \\ \sigma_{c,\varepsilon,N} &= v_{a,N}(\cos(\zeta) - 1) + v_{DGPS,N}(1 - \cos(\zeta)) + \sin(\zeta)(v_{DGPS,E} - v_{a,E}) \end{aligned} \tag{8.6}$$

where ζ is the counter-clockwise compass error angle, and E and N represent true east and north components. The error in v_a due to compass inaccuracy increases with true boat velocity, depends on the orientation of v_a with respect to v_{DGPS}, and results in an error in both magnitude and direction. For the Fraser River spatial survey, an assumed 1° compass error yielded $\sigma_{c,\varepsilon}$ as high as 3 cm/s. It is worth noting that compass error itself varies during a spatial survey, as it depends upon the direction of travel of the boat (Marsden 2012), and higher errors are often observed at turning points where boat accelerations induce dynamic compass errors (Gaeuman and Jacobson 2005). Lastly, v_{DGPS} errors were estimated based on evaluation of the RTK-GPS to be 3.1 cm/s on average (Rennie and Rainville 2006). Despite these uncertainties due to measurement error, the uncertainty model suggested that real temporal variability of bedload contributed more to uncertainty than measurement error.

It remains unknown if v_a is an unbiased estimate of the bedload transport velocity spatially averaged between the insonified beam areas of the aDcp (v_b). Individual particle velocities should be weighted by backscatter strength, which depends upon particle size, but may also differ between mobile particles and stationary bed (Rennie and Villard 2004; Gaeuman and Jacobson 2006; Gaeuman and Rennie 2006; Jamieson, Rennie, and Ramooz 2008; Blanckaert *et al.* 2013). An intensive effort to validate aDcp v_a time series with synchronous videography of instantaneous bedload particle velocities is required to resolve this issue.

If the actual mean bedload particle velocity (v_b) is known, then bedload transport rate can be calculated using the kinematic relation:

$$g_b = \rho_s v_b d_a (1 - \lambda) \tag{8.7}$$

where ρ_s is the sediment granular density, d_a is the thickness of the active bedload layer, and λ is the porosity of the active layer (related to bedload concentration). The relations between v_a, v_b, and g_b thus depend on a number of factors: the concentration and size of the particles moving in the bedload layer; the bedload layer thickness; and the backscatter strength from a unit area of moving particles relative to the echo strength from a unit area of unobstructed bed. An analytical solution relating v_a to g_b has yet to be realized. Two possible approaches are available to deal with this problem: (i) a site-specific calibration to relate v_a to g_b can be developed from concurrent aDcp and physical bedload samples collected in a given river reach (see below); or (ii) assuming a consistent linear relation between v_a and g_b (see Table 8.1), v_a may be utilized uncalibrated as a measure of the relative intensity of bedload transport throughout the reach (e.g. Rennie and Church 2010; Jamieson *et al.* 2011a; Williams *et al.* 2015).

8.3 Previous Calibration Efforts

8.3.1 Fraser River

The first attempt to correlate apparent bedload velocity with actual bedload transport rate was conducted in 2000 in the gravel reach (D_{50} = 25 mm; McLean, Church, and Tassone 1999) of Fraser River, Canada at the Agassiz Bridge cross-section (Rennie, Millar, and Church 2002). The Fraser River at the study site is a large wandering gravel-bed river (mean annual flood discharge 8760 m^3/s), with sufficient sediment load to induce mid-channel islands, although the channel is single thread at the bridge site. Pressure-difference bedload samplers were deployed from the bridge: a VuV sampler for the gravel fraction and a Helley–Smith (HS) sampler for the sand fraction. At most sampling stations, multiple samples were collected with each sampler. Bedload samples were weighed and sieved in the laboratory, and sampling efficiencies were respectively assumed to be 0.33 and 1.5 for the VuV and HS, from which a time-averaged fractional bedload transport rate (g_b) was determined for each sampling station.

A 1.5 MHz Sontek Acoustic Doppler Profiler (ADP) aDcp was deployed from a boat positioned immediately downstream of the bedload samplers. Deployment of the bedload sampler from the bridge in the fast current was challenging, resulting in some uncertainty in position of the bedload sampler on the bed, thus positions of the aDcp and bedload sampler may have been offset by a few meters. Stationary time series of v_a were collected throughout the bedload sampling at a given sampling station, with sampling durations ranging from 2 to 91 min. It was observed that v_a time series had high variance, and included many single ensemble negative values. Bedload transport in this gravel-bed river was likely in the partial transport regime, and thus spatially sporadic with

Table 8.1 Summary of previous aDcp apparent bedload velocity linear regression calibrations for estimation of bedload transport rate.

Type	Location	*N*	r^2	Calibration	aDcp	Bottom track pulse length	D_{50} bed material (mm)	D_{50} bedload (mm)
Gravel bed	Fraser River[1]	9	0.89	$g_b = 1.2v_a - 0.037$	Sontek 1.5 MHz ADP	0.2 m	25	0.34–50
	Trinity River, all data[2]	115	0.30	N/A	RDI 1.2 MHz RioGrande ADCP	N/A	65	N/A
	Trinity River, <60% sand[2]	~69	0.32	N/A	RDI 1.2 MHz RioGrande ADCP	N/A	65	N/A
	Trinity River, >60% sand[2]	~46	0.74	N/A	RDI 1.2 MHz RioGrande ADCP	N/A	65	N/A
	Trinity River, sand fraction only[2]	115	0.63	N/A	RDI 1.2 MHz RioGrande ADCP	N/A	65	N/A
	Rees River (2009/10)[3]	36	0.91	$g_b = 1.4v_a - 0.018$	Sontek S5 3.0 MHz	Proprietary, probably variable	6.5	1–13
	SAFL[4]	21	0.71	$g_b = 10v_a - 0.18$	RDI 1.2 MHz RioGrande ADCP	20% or 30% of depth	11.3	N/A
Sand bed	Fraser River, Sea Reach[5]	68	0.76	$g_b = 0.057v_a - 0.0007$	Sontek 1.5 MHz ADP	0.6 m	0.25	0.25
	Fraser River, Canoe Pass 2000[5]	49	0.38	$g_b = 0.23v_a + 0.001$	Sontek 1.5 MHz ADP	0.2 m	0.35	0.35
	Fraser River, Canoe Pass 2001[5]	15	0.42	$g_b = 0.090v_a - 0.0003$	Sontek 1.5 MHz ADP	0.6 m	0.35	0.35
	Missouri River 2004[6]	80	0.33	$g_b = 0.135{v_a}^{1.027}$	RDI 0.6 MHz RioGrande ADCP	20% of depth	0.55	0.25–1.25
	Missouri River 2005[6]	107	0.72	$g_b = 0.132{v_a}^{0.955}$ approximate equivalent $g_b = 0.13v_a + 0.004$	RDI 0.6 MHz RioGrande ADCP	20% of depth	0.55	0.55
	Assiniboine River[7]	22	0.63	$g_b = 3.3v_a$	RDI 1.2 MHz RioGrande ADCP	20% of depth	N/A	0.07–0.49
	SAFL[4]	11	0.90	$g_b = 2.1v_a - 0.024$	RDI 1.2 MHz RioGrande ADCP	20% of depth	1.05	1.05
	SAFL[4]	22	0.76	$g_b = 1.6v_a - 0.0026$	RDI 1.2 MHz RioGrande ADCP	20% or 30% of depth	1.05	1.05

SAFL, Saint Anthony Falls Laboratory.
[1]Rennie, Millar, and Church 2002; [2]Gaeuman and Pittman (2010); [3]Brasington *et al.* 2011; [4]Ramooz and Rennie 2010; [5]Rennie and Villard 2004; [6]Gaeuman and Jacobson 2006; [7]Rennie 2013

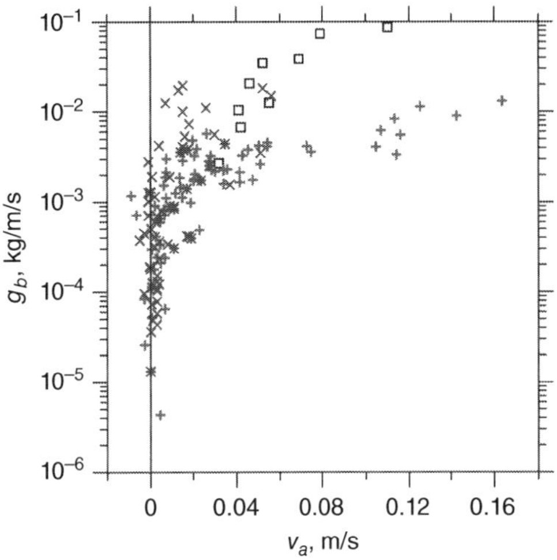

Figure 8.3 Fraser River bedload transport rate (g_b) versus mean apparent bedload velocity (v_a). Symbols: □ Agassiz (gravel-bed); x Canoe Pass 2000 (sand-bed); *Canoe Pass 2001 (sand-bed); + Sea Reach (sand-bed) (Rennie and Villard 2004). See Table 8.1 for linear regression calibrations.

heterogeneous transport between beams. Regardless, collection of long time series allowed for sufficient averaging, and mean values of v_a displayed good correlation with mean g_b (Figure 8.3 and Table 8.1; $g_b = 1.2v_a - 0.037$; $r^2 = 0.89$; $n = 9$).

Calibration was also attempted using the same aDcp in Fraser River sand-bed estuarine distributaries; Canoe Pass in 2000, and Canoe Pass and Sea Reach in 2001 (Rennie and Villard 2004). In this case only the HS sampler was utilized, with sampling efficiency assumed equal to one, and both the aDcp and HS were deployed from the same boat. Samples were collected on the stoss side of dunes to maximize homogeneity of transport between beams. Sample durations were 5 min. Some v_a time series had substantial proportions of bad v_{BT} data; sampling stations with less than 70% good v_{BT} data were excluded from analysis. The Fraser River calibration data are shown in Figure 8.3. Good correlation was achieved for samples collected in Sea Reach (Table 8.1; $g_b = 0.057v_a - 0.0007$; $r^2 = 0.76$; $n = 68$). Correlations were less good for Canoe Pass data, possibly because greater flow depths resulted in signal attenuation that approached the limit of feasible bottom tracking for this particular aDcp. The presence of large dunes (dune heights up to 3 m) in Canoe Pass may also have resulted in a more spatially complex bedload transport field and thus greater heterogeneity of transport between beams (see also Gaueman and Jacobson 2007; Jamieson *et al.* 2011b). Based on the observation that $v_a = \pm 1.0$ cm/s was recorded over a large range of low bedload transport rates (Figure 8.3), it was suggested precision (lower detection limit) of v_a measurement is ±1.0 cm/s.

8.3.2 Missouri River

A similar calibration effort was conducted on the Missouri River, United States, using an RD Instruments four-beam 600 kHz Rio Grande ADCP (Gaeuman and Jacobson 2006, 2007). The Missouri River in the study reach is a large single-thread sand-bed river (flood discharge of 4670 m^3/s; Gaeuman and Jacobson 2006), with a navigation channel maintained by means of wing dikes along the channel banks. HS and BL-84 bedload samplers were respectively used in 2004 and 2005. Stationary v_a time-series periods ranged from 2 to 10 min. Relatively poor correlation was observed between g_b and v_a samples, particularly for the 2004 HS data, which had an r^2 of 0.33 (Table 8.1). The 2005 BL-84 data had a higher r^2 of 0.72, although this was increased by a data cluster of a dozen low g_b (and low v_a)

samples. Alternatively, v_a was correlated to bedload transport rate estimates obtained from dune tracking ($r^2 = 0.65$). It was concluded that physical bedload sampling was unreliable in this large sand-bed river, and that dune tracking and v_a provided better representations of bedload (Gaeuman and Jacobson 2007).

8.3.3 Trinity River

A controlled dam release on the Trinity River in 2006 was utilized as an opportunity for v_a versus g_b calibration on the Trinity River, California (Gaeuman and Pittman 2010). Data were collected at a total of 11 verticals across one section on six different days between May 25 and June 12 when discharge ranged from 149 m^3/s to 280 m^3/s. The gauging section was rectangular in shape within a straight reach. Surface bed material at the section was coarse gravel and cobbles ($D_{50} = 0.065$ m, $D_{90} = 0.170$ m) with 5–10% sand. A 1200 kHz Rio Grande ADCP was deployed from a cataraft attached to a cableway, and a cable suspended Toutle River (TR-2) sampler collected bedload approximately 6 m downstream of the ADCP. The ADCP sample durations ranged from 2.5 to 30 min, with an average of 6 min. Individual bedload sample durations ranged from 30 to 60 s, and between one and three samples were aggregated at each vertical. Underwater videography confirmed that the majority of transport occurred towards the right-hand side of the channel, although the coarsest fraction (16–128+ mm) was more distributed across the section. The video images were also used to determine that suspended sand did not influence v_a. The Trinity River data showed strong correlation between v_a and the sand fraction of g_b, but the correlation was weaker between v_a and the coarse gravel and cobble fractions of g_b (Table 8.1). Also, for a given observed v_a the population of samples with less than 60% sand had greater g_b. Gaeuman and Pittman (2010) suggested that v_a was strongly influenced by sand bedload but not by transported gravel and cobble.

8.3.4 Assiniboine River

A calibration exercise was conducted on the Assiniboine River in Manitoba, Canada by Manitoba Hydro field staff (Rennie 2013). The Assiniboine River flows east out of Saskatchewan through southern Manitoba to its confluence with the Red River in the City of Winnipeg. The Assiniboine River drains 162,000 km^2, has an average flow of 50 m^3/s, and a peak recorded instantaneous flow of 610 m^3/s (Hydat database, Environment Canada Water Survey, gauge 05 MJ001 Assiniboine River at Headingley). The river morphology is single-thread meandering. A total of 22 concurrent 5 min v_a and g_b samples were collected between November 8–10, 2011 at six stations within a sand-bed reach near the Headingley gauge. A BL-84 bedload sampler was used, and bedload samples were processed in the laboratory for total mass and particle grain size distribution.

A 1200 kHz Rio Grande ADCP was used, integrated with a high quality Real Time Kinematic GPS (RTK-GPS). The default bottom track pulse length was 20% of the flow depth. A local magnetic declination was utilized, and the ADCP compass was calibrated *in situ* to within 1° for every measurement using Method 3 described in the WinRiver manual (RD Instruments). Nonetheless, at each site cross-sectional ADCP data from repeat transects suggested drift of the compass, and the compass calibration was optimized in post-processing by minimizing difference between GPS and bottom track boat tracks (Rennie and Rainville 2006). Despite generally low observed v_a, a reasonable calibration was obtained (Table 8.1), although the regression coefficient was higher than for other sand-bed rivers. This may have been due to less influence of near-bed suspended sediment ("water bias") on the observed v_a signal.

8.3.5 Saint Anthony Falls Laboratory

A calibration experiment was performed at the Saint Anthony Falls Laboratory in 2006 (Jamieson, Rennie, and Ramooz 2008; Ramooz and Rennie 2010) as part of the SedT StreamLab06 project (Singh *et al.* 2013). A gravel-bed (D_{50} = 11.3 mm) and a sand-bed (D_{50} = 1.0 mm) were recirculated in the Main Test Channel (2.75 m wide, 1.8 m deep, 85 m long). Both a 600 kHz Rio Grande ADCP and a 1200 kHz Rio Grande ADCP, under a range of bottom track pulse lengths, were tested. During testing the ADCP was placed 1.7 m upstream of a channel-spanning bedload trap that sampled bedload transport rate automatically using weigh pans. Bedload transport rate measured in the middle weigh pan was compared to v_a. Sample periods for each test case ranged from 30 to 180 min, which was long enough for passage of at least one dune wavelength. The ADCP was situated 1.0–1.2 m above the sand bed, depending upon passage of bedforms, and 1.05 m above the gravel bed. These ranges were at the lower limit for reliable bottom tracking using the ADCPs tested.

The experimental results confirmed that v_a is increased by near-bed suspension if longer bottom track pulse lengths are used. The best results were obtained with the 1200 kHz ADCP using a bottom track pulse length at the default value (&R20), i.e., 20% of range to the bottom. This configuration yielded the highest correlation with measured transport rates in the sand-bed runs, and was least sensitive to positive bias at low transport rates in the gravel-bed runs. Good correlative relations were obtained for the sand-bed tests. Using the data obtained with the 1200 kHz ADCP using bottom track pulse lengths set by &R20 and &R30, the relation was $g_b = 1.6v_a - 0.0026$; $r^2 = 0.76$; $n = 22$ (Figure 8.4, Table 8.1). The gravel-bed tests using the same ADCP configuration also showed reasonable correlation ($g_b = 10.1v_a + 0.18$; $r^2 = 0.71$; $n = 21$; Figure 8.4, Table 8.1); however, v_a values of up to 3 cm/s were obtained at low bedload transport rates broadly ranging from 0.001 to 0.2 kg/m/s, which exceeded the estimated v_a precision for the previous Fraser River data. This suggested that v_a was influenced by near-bed suspended scatterers, possibly related to detritus in the Mississippi River flume water that was unfiltered during the gravel-bed tests. Although synchronous high speed videography was obtained, the video was of insufficient quality for complete validation of v_a. Nonetheless video image analysis demonstrated that v_a increased with increasing percentage of the bed in transport (Jamieson, Rennie, and Ramooz 2008).

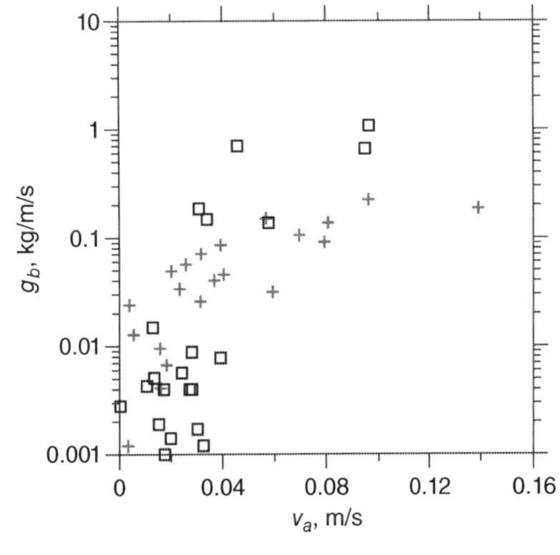

Figure 8.4 Laboratory calibration for both gravel-bed (□) and sand-bed (+) using 1200 kHz Rio Grande ADCP with either &R20 or &R30 bottom track pulse length setting. See Table 8.1 for linear regression calibrations.

8.4 Rees River Survey: New Fractional Calibration Data

The strong influence of bedload sediment grain size on the relation between v_a and g_b shown by these previous calibration efforts in both the field and laboratory motivated collection of an entirely new data set from the Rees River. A deeper examination of this influence was performed in this shallow braided river.

8.4.1 Study Site

The Rees River drains a 420 km^2 catchment in Central Otago to the east of New Zealand's Southern Alps. Cook, Quincey, and Brasington (2013) described the geomorphology of the Rees River. No long-term discharge record exists, but the highest discharge recorded in a 2-year period was 475 m^3/s (Williams *et al.* 2013). As it descends out of the mountains before draining into Lake Wakatipu, the Rees River deposits a substantial sediment load, resulting in a wide, slowly aggrading gravel-bed braided river. Rees River braiding morphodynamics have been studied in detail as part of the the ReesScan Project initiated by Brasington. The study involved sequential terrestrial laser scanning, fused with optical-empirical bathymetric mapping, of a 2.5 km long by 0.7 m wide braided reach before and after each flood event in the 2009/10 flood season (Williams *et al.* 2011, 2014). Terrestrial laser scanning and intensive aDcp spatial surveying were also conducted in a ~300 m reach of the primary and a secondary anabranch during multiple flood events in 2011 (Williams *et al.* 2013, 2015). Based on differencing digital elevation models obtained during each survey, time-averaged bed material transport was evaluated using the morphological sediment budgeting method. However, this method alone does not account for throughput of bed material. Consequently, an attempt was made to measure bedload transport across the downstream end of the study reach in order to close the sediment budget and estimate bed material transport through the inverse or morphological approach. The intent was to use calibrated aDcp bedload transport velocity for this purpose.

8.4.2 Calibration Measurements

A hand-held HS sampler was used to collect bedload samples, and one of two Sontek Riversurveyor aDcps (an S5 and an M9) integrated with a RTK-GPS was used to collect v_a data. Flow was always shallow (<0.6 m), thus both instruments functioned as a pulse incoherent four-beam 3 MHz aDcp that collected single ping data at 1 Hz. A total of 144 synchronous v_a and g_b samples were collected during the 2009/10 (36 samples) and 2011 campaigns (108 samples). Sample durations were short, ranging from 0.5 to 5 min, due to relatively high bedload transport rates. This was despite all sampling being conducted at low to moderate flow stage when the river was wadable. Even at low flow stage it was possible to find bedload transport conduits in this braided gravel-bed river. Bedload samples were weighed and sieved in the laboratory. Bedload materials in transport varied from sand to small cobbles. Bed material grain size was measured on a braid bar in the vicinity of the bedload sampling (Williams *et al.* 2015). The subsurface distribution (bulk sampling) had $D_{16} = 0.84 \pm 0.15$ mm, $D_{50} = 7.5 \pm 1.6$ mm, $D_{84} = 27.9 \pm 9.7$ mm, and $D_{90} = 40.5 \pm 21.9$ mm. The surface distribution (pebble count) had $D_{16} = 10.4 \pm 5$ mm, $D_{50} = 19.9 \pm 10.4$ mm, and $D_{84} = 35.2 \pm 19.2$ mm. Sampled bedload D_{50} ranged from 0.2 to 13 mm (Table 8.2).

8.4.3 Analysis

Initial assessment of the 2009/10 aDcp calibration dataset was reported in Brasington *et al.* (2011). For the present chapter, all available Rees River data were compiled and analyzed. Each aDcp time series

Table 8.2 Rees River bedload velocity linear regression calibrations for estimation of bedload transport rate.

Location	Subpopulation	N	r^2	Calibration	aDcp	Bottom track pulse length	D_{50} bed material (mm)	D_{50} bedload (mm)
Rees River	All data	144	0.29	$g_b = 1.5v_a + 0.086$	Sontek Riversurveyor 3.0 MHz	Proprietary, probably variable	6.5	0.2–13
	Sand with gravel	9	0.96	$g_b = 1.3v_a - 0.063$	Sontek Riversurveyor 3.0 MHz	Proprietary, probably variable	6.5	0.2–0.6
	Gravel with sand	87	0.45	$g_b = 1.8v_a + 0.041$	Sontek Riversurveyor 3.0 MHz	Proprietary, probably variable	6.5	0.7–13
	Gravel	17	0.98	$g_b = 7.0v_a + 0.022$	Sontek Riversurveyor 3.0 MHz	Proprietary, probably variable	6.5	2.3–9.7
Combined Fraser and Rees data	Sand	71	0.74	$g_b = 0.054v_a - 0.00073$				0.3–0.4
	Sand with gravel	11	0.96	$g_b = 1.3v_a - 0.057$				0.2–0.6
	Gravel with sand	89	0.45	$g_b = 1.9v_a + 0.038$				0.7–16
	Gravel	18	0.93	$g_b = 6.8v_a + 0.0074$				2.3–28

was processed using in-house Matlab code. Single ping apparent bedload velocity in the east and north directions were averaged to determine mean (v_a) and standard deviation of (σ_{va}) of apparent bedload velocity. The percentage of good v_{BT} pings (%$_{\mathrm{BT}}$), and flow depth (H) were also recorded. Depth-averaged water velocity (U) was also determined, unless flow depths were insufficient to obtain a water velocity bin. The bedload samples were processed to determine transport rate (g_b), representative percentiles of their grain size distributions (i.e. D_{50}, D_{16}, D_{84}) the percentage of sand (i.e., < 1 mm, %$_{\mathrm{S}}$), and the standard deviation of the bedload materials ($\sigma = \sqrt{D_{84}/D_{16}}$, assuming a log-normal distribution). Finally, bed shear velocity (u^*) was estimated using the Keulegan form of the depth-averaged log-law (van Rijn 1984): $u^* = \frac{U}{5.75 \log\left(\frac{12H}{3D_{84}}\right)}$.

Comparison of v_a and g_b using all 144 samples resulted in relatively low correlation ($r^2 = 0.29$; Table 8.2), which was similar to the correlation observed for the entire Trinity River data set ($r^2 = 0.30$; Table 8.1). The available Rees River data set was examined for correlation by stepwise multiple linear regression analysis using all 11 variables described above. It was found that v_a was significantly correlated to σ_{va}, g_b, and %$_{\mathrm{BT}}$, with the least-squares model fit of:

$$v_a = 0.23 + 0.092g_b + 0.46\sigma_{va} - 0.0028\%_{\mathrm{BT}} \qquad \left(R^2 = 0.64\right) \tag{8.8}$$

Significant positive correlation was observed between v_a and g_b. Interestingly, v_a was also positively correlated with σ_{va}, which suggests v_a signal variance increased with transport rate (cf., Rennie and

Millar 2007; Rennie, Rainville, and Kashyap 2007). This may have occurred due to increased real variance of transport as transport rates increased, which can be observed in the flume data of Gibbs (1973). However, the negative correlation with $\%_{BT}$ may also suggest that σ_{va} increased with decreasing v_a signal quality, which occurs with degradation of bottom tracking as bedload transport rate increases.

Meanwhile, g_b was significantly correlated to σ, u^*, D_{50}, and v_a individually, but the stepwise multiple regression model for g_b utilized only v_a and $\%_S$:

$$g_b = 0.18 + 2.1v_a - 0.0024\%_s - 0.026v_a\%_s \qquad \left(R^2 = 0.36\right) \tag{8.9}$$

Thus, utilization of additional parameters in multiple regression yielded only a marginal improvement in prediction of g_b as a function of v_a.

Alternatively, as a result of an optimization exercise, a data filtering and classification procedure was adopted. First, data were excluded if $\%_{BT} < 69$, which closely corresponded to the 70% threshold used by Rennie and Villard (2004). Second, data were grouped into four grain-size classes: entirely sand ($\%_s = 100$), sand with gravel ($D_{50} \leq 0.6$ mm, $\%_S < 100$), gravel with sand ($D_{50} > 0.6$ mm, $\%_S > 10$), and gravel ($D_{50} > 0.6$ mm, $\%_S \leq 10$). Note that the Rees River data set did not include any samples that were entirely sand. The sand threshold of 0.6 mm (medium sand) was selected as it better separated the data set and more closely corresponded to the transition to geometric scattering ($D = 0.3$ mm) for a 3 MHz aDcp. Although σ_{va} was correlated to v_a (as also found by Rennie, Rainville, and Kashyap 2007), σ_{va} did not prove useful as a data filtering or clustering parameter to improve the relation between g_b and v_a.

This data filtering and clustering exercise improved the observed correlations substantially (Table 8.2, Figure 8.5). In particular, both the gravel and the sand with gravel data sets displayed very good correlations between g_b and v_a, with r^2 of 0.96 and 0.98, respectively, although it should be acknowledged that the gravel data set was limited in that it had only three samples with v_a greater

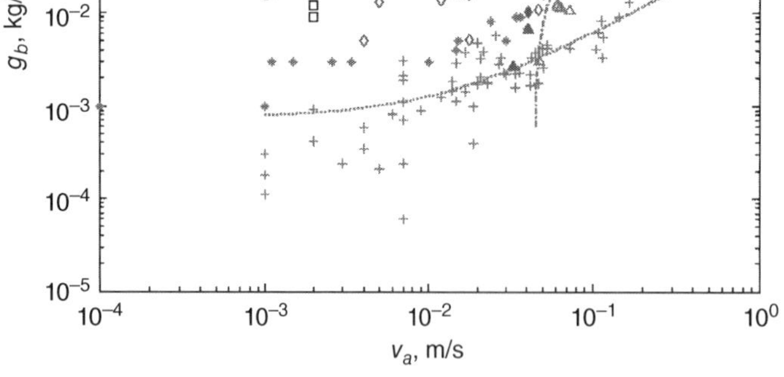

Figure 8.5 Rees River (R), Fraser River (F), and Assiniboine River (A) bedload transport rate (g_b) versus apparent bedload velocity (v_a) as a function of bedload grain size. Refer to the text for data clustering procedures, and Table 8.2 for linear regression calibrations.

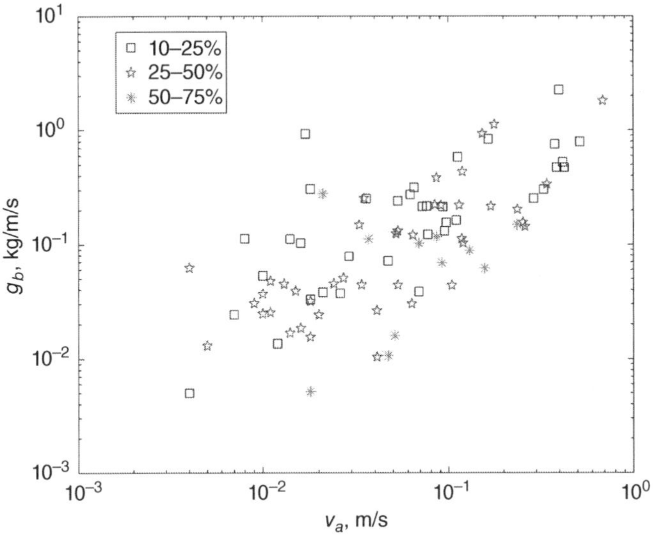

Figure 8.6 Gravel with sand data (combined Rees River and Fraser River data sets), subdivided based on the percentage of sand in the mixture.

than 1 cm/s and the regression was influenced heavily by the two largest samples. The large set of data defined as gravel with sand showed more scatter (r^2 of 0.45). As observed by Gaeuman and Pittman (2010), it is evident that very different relations are obtained depending on the size of sediment in transport. Similar values of apparent bedload velocity were obtained for much lower transport rates of sand than gravel. This confirms a previous assessment of Fraser River data (Rennie and Villard 2004). There is a continuous trend of increasing transport for a given v_a as grain size increases from sand, to sand with gravel, to gravel with sand, to gravel.

For comparative purposes the same procedure was applied to the Fraser River data collected at Agassiz and in Sea Reach and Canoe Pass in 2001, and these data are also plotted on Figure 8.5. Calibration relations obtained with the combined data set are provided in Table 8.2. Encouragingly, the classified Fraser River data reasonably plot within the same classification clusters as the Rees River data. Again, the weakest relation was obtained for the gravel with sand grouping (mixture of gravel with greater than 10% sand). Further subdivision of this grouping shows some trend of decreasing g_b with increasing percentage sand in the mixture (Figure 8.6), but subgroup distinction was insufficient to warrant individual subgroup calibration regressions.

It should be expected that calibration relations between g_b and v_a would have a zero intercept, whereas the relations presented in Tables 8.1 and 8.2 have nonzero intercepts. Rennie, Millar, and Church (2002) argued that a negative intercept term may arise if suspended sediments introduce water bias to v_a without bedload transport. It can also be seen in Figure 8.5 that the sign of the intercept term determines the concavity of a linear fit on a log–log plot.

8.5 Discussion

The g_b versus v_a calibrations made prior to the latest Rees study have been site specific, depending upon bedload grain-size distribution and instrument operating parameters (Rennie and Villard

2004), and have shown that in general gravel-bed rivers have higher transport rate (g_b) for a given v_a than sand-bed rivers (Table 8.1). There are several possible explanations. This could partially be because suspended fine sediments, or sand suspended above the bedload sampler, can be transported within the bottom track sample volume and bias bottom tracking without increasing measured bedload (please refer to calculations and discussion in Rennie and Millar (2004)). However, suspended sand observed above the bedload sampler did not result in a v_a signal at the Trinity River (Gaeuman and Pittman 2010). Instead, Gaeuman and Pittman (2010) suggested that v_a may be insensitive to transported gravel and cobble, thus only sand bedload increases v_a, whereas g_b increases dramatically with gravel. However, it is difficult to explain theoretically why a gravel particle in transport should be acoustically invisible. Backscatter is proportional to particle cross-sectional area in the geometric scattering regime, and less than sectional area for smaller particles in the Rayleigh regime. For aDcps typically used in rivers (i.e., operating frequencies > 1 MHz), backscatter from gravel should be emphasized in comparison to fine sands. It may be argued that the discrepancy arises from the difference between scattering and pure reflection. Given the angle of incidence of an aDcp bottom track beam, a perfectly smooth bottom would reflect little backscatter. It thus could be thought that large, relatively flat bed particles would backscatter less than small bed particles. However, sediment beds should be sufficiently rough to yield isotropic scattering, and particles in transport present even more irregular geometry, thus scatter from both large and small bedload particles should be isotropic. Furthermore, the Rees River data set includes a few samples with no sand yet significant v_a, which suggests that gravel alone can indeed influence v_a.

Alternatively, it is reasonable to suggest that gravel simply carries more mass than sand for a given v_a. Mass depends on particle volume (a^3, where a is particle radius), whereas v_a depends on particle surface area (a^2), thus mass increases more rapidly with particle size than v_a. Higher g_b for a given v_a in gravel-bed rivers implies that a set of gravel bedload particles carries greater mass than a set of sand bedload particles with the same planar surface area moving at the same speed. This suggests that the product of bedload layer depth and concentration (i.e., $\rho_s d_a(1-\lambda)$ in Equation 8.6) is less for sand transport than gravel transport. It is worth noting that a large calibration coefficient (7.0; Table 8.2) was obtained for the Rees River gravel subpopulation, similar to the calibration coefficient (10; Table 8.1) obtained for the gravel-bed experiments at Saint Anthony Falls Laboratory. The coefficient serves as a proxy for $\rho_s d_a(1-\lambda)$, and a large coefficient for gravel transport indicates high g_b for a given v_a. Previous field calibrations in gravel-bed rivers have included sand in the bedload mixture (Table 8.1), thus smaller calibration coefficients were obtained.

The expanded Rees River dataset presented here includes a broad range of bedload grain-size mixtures, with median bedload particle sizes (D_{50}) ranging from 0.2 mm to 13 mm. In effect, the Rees River dataset includes both sand and gravel transport within the same reach. It is not surprising, then, that correlation between g_b and v_a was relatively low for the entire Rees River dataset (Table 8.2). Separation of the dataset into more homogeneous subpopulations based on bedload grain-size classes results in higher correlation between g_b and v_a. Similar to the Trinity River observations (Gaeuman and Pittman 2010), it appears that the amount of fine to medium sand in transport has a large influence on the observed v_a signal. In the Rees River data, distinct subpopulations were identified for sand mixed with more than 10% gravel, gravel with more than 10% sand, and pure gravel. The Rees River dataset did not include any bedload samples that were entirely sand, but combination of the Fraser River and Rees River data confirmed that pure sand constitutes another distinct subpopulation. It is noteworthy that the sand-bed Assiniboine River data, while displaying higher g_b for a given v_a than the sand-bed Fraser River data (Figure 8.5), presumably due to less water bias of suspended sediments, still plot below Fraser River and Rees River data that include gravel in the bedload mixture.

This strong dependence of the v_a signal on the particle size distribution of bedload complicates application of the v_a method in rivers with a mixed sand-gravel sediment load. While it is possible to develop individual calibrations for different grain-size classes (Table 8.2), it is difficult to apply these calibrations for prediction of g_b because the bedload grain size for a given v_a measurement must be known *a priori*. As a case in point, Williams *et al.* (2015) present maps of v_a distributed throughout a Rees River braid anabranch during five different stages in a flood season. One of these maps has been reproduced in Figure 8.7. Assuming a constant relation between g_b and v_a, the spatial distribution of v_a provides an indication of the relative intensity of bedload throughout the reach. Given the dependence of v_a on grain size, and that the size distribution of bedload may vary spatially, it should be acknowledged that these maps provide only an approximate estimate of the relative intensity of transport. Still, these spatially distributed apparent bedload velocity data offer rich opportunities to evaluate morphodynamic processes through the identification of regions of bed disturbance as well as bedload transport pathways between identified zones of erosion and deposition (Rennie and Church 2010; Rennie 2013; Williams *et al.* 2015). These maps are thus useful for many aspects of river management, including assessments of channel change, habitat stability and availability, and infrastructure security.

For some management issues it is useful to quantify the bedload mass in transport. For the Rees River study, bedload transport rates are required to close sediment budgets obtained from the morphological method. This requires conversion of the spatially distributed v_a to spatially distributed g_b

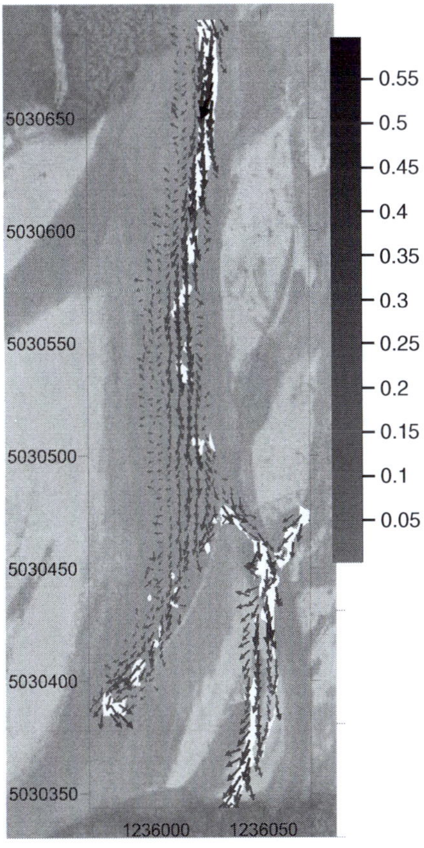

Figure 8.7 Interpolated spatially distributed apparent bedload velocity (m/s) in Rees River braided river anabranch. Data collected January 27–29, 2011. Locations with statistically significant bedload velocity are marked brighter. Grid coordinates (m) in New Zealand Transverse Mercator. The underlying aerial photograph was taken on February 27, 2011, after a high-flow event that caused morphological change within the study area. Maximum v_a was 0.55 m/s. Conversion of v_a to g_b depends on which Rees River calibration relation is utilized (see Table 8.2). Maximum g_b calculated as follows: the sand with gravel relation ($g_b = 1.3v_a - 0.063$), the gravel with sand relation ($g_b = 1.8v_a + 0.041$), and the gravel relation ($g_b = 7.0v_a + 0.022$) are 0.65, 1.0, and 3.9 kg/m/s, respectively. (*Source*: Williams *et al.* 2015. Reproduced with permission of John Wiley & Sons.) (See insert for color representation of this figure.)

using a g_b–v_a calibration relation. Which of the calibrations in Table 8.2 is most appropriate to use? The correct choice for which calibration relation to use depends upon the bedload grain-size distribution during the spatial survey. It is even possible that multiple calibrations must be employed due to the spatiotemporal variation in bedload size distribution. If a constant bedload grain-size distribution is allowed, assuming that transport was mostly sand requires use of the sand with gravel calibration, whereas assuming transport was almost entirely gravel requires the gravel calibration. As explained in the Figure 8.7 caption, the maximum transport rate varies by a factor of six depending upon which assumption is employed. Although this may seem to be a large degree of uncertainty, it is less than the errors often obtained with bedload transport formulae (Gomez and Church 1989; Habersack and Laronne 2002).

It is reasonable to suppose that bedload grain size increases with flow intensity, thus it may be tempting to use the sand with gravel relation during a low flow condition. Yet, in the Rees River we have observed gravel moving in narrow bedload transport pathways at even the lowest flows (Williams *et al.* 2015). Likewise, it may be tempting to use the pure gravel relation for high flow events. However, at very high flows it is reasonable to expect that all bed material sizes are transported in an equal mobility condition (Parker and Klingeman 1982; Wilcock 1992), which would suggest that the gravel with sand relation is most appropriate. Intriguingly, while displaying the lowest correlation, the gravel with sand relation was most similar to the calibration relation developed using all available data (Table 8.2). Thus, it may be that for a given reach a relation developed with all available data is most suitable for use, despite the higher likelihood of low correlation for that relation due to inclusion of a broad range of bedload grain-size distributions. In order to further evaluate this hypothesis, we are currently processing an additional g_b–v_a calibration dataset collected on the Rio Cinca, Spain as part of the MorphSed project (www.morphsed.es).

8.6 Conclusions

This chapter summarizes previous calibration experiments to relate a measured apparent bedload velocity (v_a) to a bedload transport rate (g_b), and presented an extensive new data set collected at the braided gravel-bed Rees River, New Zealand. These calibration experiments involved simultaneous measurement of v_a with an aDcp and g_b with a pressure-difference bedload sampler. Calibration relations have proven to be site specific, varying as a function of bedload grain-size distribution and aDcp operating parameters such as operating frequency and bottom track pulse length. The Rees River data set included a broad range of bedload grain-size distributions, which allows for detailed analysis of the influence of bedload grain size on measured v_a. It is demonstrated that gravel bedload is associated with greater g_b for a given measured v_a than sand bedload, and that based on bedload grain size (sand, sand with > 10% gravel, gravel with > 10% sand, and gravel). This suggests that an *a priori* estimation of the bedload grain-size distribution is required for application of g_b–v_a calibration relations in rivers with mixed sand–gravel bed material.

Notation

aDcp	Acoustic Doppler current profiler
ADCP	Acoustic Doppler Current Profiler (RD Instruments aDcp)
ADP	Acoustic Doppler Profiler (Sontek aDcp)
BT	Bottom tracking

DGPS	Differential global positioning system
&R20	Default bottom track pulse length of 20% of range to bed
$\%_{BT}$	Percentage of good v_{BT} pings
$\%_S$	Percentage of sand
b_i	Along beam velocity for beam i (m/s)
c	Speed of sound (~1500 m/s in water)
d_a	Thickness of the active bedload layer (m)
D_i	Ith percentile grain size (m)
f	Acoustic frequency (Hz)
g_b	Bedload transport rate (kg/m/s)
H	Flow depth (m)
l_p	Length of the acoustic pulse (m)
r^2	Coefficient of determination
U	Depth averaged water velocity
$u*$	Bed shear velocity (m/s)
v_a	Apparent bedload velocity (m/s)
v_b	Spatially averaged bedload particle velocity (m/s)
v_{BT}	Boat velocity by bottom tracking (m/s)
v_{DGPS}	Boat velocity by DGPS (m/s)
$\hat{v}_x$, $\hat{v}_y$, $\hat{v}_z$	Calculated Cartesian velocity component (x, y, z directions; z positive upwards) (m/s)
v_{ij}	Particle velocity in beam i in Cartesian direction j (m/s)
v_e	Error velocity (m/s)
ϕ	Beam tilt from vertical (°)
λ	Porosity of the active bedload layer
ρ_s	Sediment granular density (kg/m^3)
σ_{va}	Standard deviation of apparent bedload velocity (m/s)
σ	Standard deviation of bedload grain size distribution (m)
θ	Half-intensity beam width (°)
ζ	Counter-clockwise compass error angle (°)

Acknowledgements

The authors wish to thank Helen Reid for field assistance on the Rees River in 2011, and Jo Hoyle for grain size analyses of the 2011 Rees River bedload samples. The authors also wish to thank David Gaeuman and Ryota Tsubaki for very helpful reviews that lead to an improved manuscript, and Shinji Egashira and Colin Thorne for providing very interesting discussions. The Rees River field campaign was funded by United Kingdom NERC grant NE/G005427/1 and NERC Geophysical Equipment Facility Loan 892, as well as NSERC and CFI (Canada) grants to Rennie.

References

Ashmore, PE and Church, MA 1998. Sediment transport and river morphology: a paradigm for study. In *Gravel Bed Rivers in the Environment* (eds PC Klingeman, RL Beschta, BD Komar, and JB Bradley). Water Resources Publications: Oregon; 115–139.

Blanckaert, K, Rennie, CD, Heyman, J, and Chen, Q 2013. 3D-ADVP measurements of bedload transport. In *35th IAHR World Congress*, Chengdu, China.

Brasington, J, Rennie, CD, Vericat, D *et al.* 2011. Monitoring braided river morphodynamics with an acoustic Dopler current profiler. In *34th IAHR World Congress*, Brisbane.

Bridge, JS and Gabel, SL 1992. Flow and sediment dynamics in a low sinuosity, braided river: Calamus River, Nebraska sandhills. *Sedimentology* 39, 125–142.

Brumley, BH, Cabrera, KLD, and Terray, EA 1991. Performance of a broad-band Acoustic Doppler Current Profiler. *IEEE Journal of Oceanic Engineering* 16(4), 402–407.

Burtin, A, Cattin, R, Bollinger, L *et al.* 2011. Towards the hydrologic and bed load monitoring from high-frequency seismic noise in a braided river: the "Torrent de St Pierre", French Alps. *Journal of Hydrology* 408(1–2), 43–53.

Church, M 2006. Bed material transport and the morphology of alluvial river channels. *Annual Review of Earth and Planetary Sciences* 34, 325–354.

Cook, SJ, Quincey, DJ, and Brasington, J 2013. Geomorphology of the Rees Valley, Otago, New Zealand. *Journal of Maps* 10(1), 136–150.

Davoren, A, and Mosley, MP 1986. Observations of bedload movement, bar development and sediment supply in the braided Ohau River. *Earth Surface Processes Landforms* 11, 643–652.

Dietrich, WE and Whiting, P 1989. Boundary shear stress and sediment transport in river meanders of sand and gravel. In *River Meandering* (eds S Ikeda and G Parker). American Geophysical Union: Washington, DC; 1–50.

Dinehart, RL and Burau, JR 2005. Repeated surveys by acoustic Doppler current profiler for flow and sediment dynamics in a tidal river. *Journal of Hydrology* 314(1–4), 1–21.

Drake TG, Shreve, RL, Dietrich, WE *et al.* 1988. Bedload transport of fine gravel observed by motion-picture photography. *Journal of Fluid Mechanics* 192, 193–217.

Edwards, TK and Glysson, GD 1999. *Field Methods for Measurement of Fluvial Sediment.* US Geological Survey: Denver, CO.

Ferguson, RI, Ashmore, PE, Ashworth, PJ *et al.* 1992. Measurements in a braided river chute and lobe. 1. Flow pattern, sediment transport, and channel change. *Water Resources Research* 28(7), 1877–1886.

Gaeuman, D and Jacobson, RB 2005. *Aquatic Habitat Mapping with an Acoustic Doppler Current Profiler: Considerations for Data Quality.* US Geological Survey Open-File Report 2005–1163.

Gaeuman, D and Jacobson, RB 2006. Acoustic bed velocity and bed load dynamics in a large sand bed river. *Journal of Geophysical Research-Earth Surface* 111(F2), F02005.

Gaeuman, D and Jacobson, RB 2007. Field assessment of alternative bed-load transport estimators. *Journal of Hydraulic Engineering* 133(12), 1319–1328.

Gaeuman, D and Pittman, S 2010. Relative contributions of sand and gravel bedload transport to acoustic Doppler bed velocity magnitudes in the Trinity River, California. In *Bedload-Surrogate Monitoring*

Technologies (eds JR Gray, JB Laronne, and JDG Marr). Scientific Investigations Report 2010-5091, US Geological Survey (available at http://pubs.usgs.gov/sir/2010/5091).

Gaeuman, D and Rennie, CD 2006. A comparison of two field studies of acoustic bed velocity: Grain size and instrument frequency effects. In *8th Federal Interagency Sedimentation Conference.*

Gaudet, JM, Roy, AG, and Best, JL 1994. Effect of orientation and size of Helley–Smith sampler on its efficiency. *Journal of Hydraulic Engineering* 120(6), 758–766.

Gibbs CJ 1973. *Model study of the basket-type bed-load sampler.* MSc thesis, Department of Civil Engineering, University of Alberta.

Gomez B and Church, M 1989. An assessment of bed load sediment transport formulae for gravel bed rivers. *Water Resources Research* 25(6), 1161–1186.

Gordon, RL 1996. *Acoustic Doppler Current Profiler Principles of Operation: A Practical Primer (2nd Edition for Broadband ADCPs).* RD Instruments: San Diego, CA.

Gray, JR, Laronne, JB, and Marr, JDG 2010. *Bedload-Surrogate Monitoring Technologies.* Scientific Investigations Report 2010-5091, US Geological Survey (available at http://pubs.usgs.gov/sir/2010/5091).

Gray, JR, Gartner, JW, Barton, JS *et al.* 2010. Surrogate technologies for monitoring bed-load transport in rivers. In *Sedimentology of Aqueous Systems* (eds C Poleto and S Charlesworth). Wiley-Blackwell: Hoboken, NJ.

Guerrero, M and Lamberti, A 2011. Flow field and morphology mapping using ADCP and multibeam techniques: survey in the Po River. *Journal of Hydraulic Engineering* 137(12), 1576–1587.

Habersack, HM and Laronne, JB 2002. Evaluation and improvement of bed load discharge formulas based on Helley-Smith sampling in an alpine gravel bed river. *Journal of Hydraulic Engineering* 128(5), 484–499.

Hubbell, DW 1987. Bed load sampling and analysis. In *Sediment Transport in Gravel-Bed Rivers* (eds C. R Thorne, JC Bathurst, and RD Hey). John Wiley & Sons: Chichester, UK; 89–118.

Jamieson, E, Rennie, CD, and Ramooz, R 2008. Verification of *in-situ* ADCP bedload transport measurements using digital video analysis. In *River Flow 2008,* International Association for Hydraulic Research, Wallingford, UK.

Jamieson, EC, Rennie, CD, Jacobson, RB, and Townsend, RD 2011a. 3-D flow and scour near a submerged wing dike: ADCP measurements on the Missouri River. *Water Resources Research* 47, W07544.

Jamieson, EC, Rennie, CD, Jacobson, RB, and Townsend, RD 2011b. Evaluation of ADCP bed velocity in a large sand bed river: moving versus stationary boat conditions. *Journal of Hydraulic Engineering* 137, 1064–1071.

Joyce, TM 1989. On *in situ* "calibration" of shipboard ADCPs. *Journal of Atmospheric and Oceanic Technology* 6, 169–172.

Klingeman, PC and Milhous, RT 1971. Oak Creek vortex bed-load sampler. *EOS, Transactions of the American Geophysical Union* 52(5), 434.

Marsden, R 2012. Performance of several ADCP internal magnetic compasses. In *Hydraulic Measurements and Experimental Methods 2012,* ASCE and IAHR, Snowbird, UT.

McLean, DG Church, M, and Tassone, B 1999. Sediment transport along lower Fraser River: 1. Measurements and hydraulic computations. *Water Resources Research* 35(8), 2533–2548.

Moate, BD and Thorne, PD. Scattering from suspended sediments having different and mixed mineralogical compositions: comparison of laboratory measurements and theoretical predictions. *Journal of the Acoustical Society of America* 133(3), 1320–1334.

Mulhoffer, L 1933. Untersuchungen uber der schwebstoff und geschiebefurung des Inn nachst Kirchbichl, Tirol [Investigations into suspended load and bedload of the River Inn, near Kirchbichl, Tirol]. *Wasserwirtschaft* (1–6).

Parker, G and Klingeman, PC 1982. On why gravel bed streams are paved. *Water Resources Research* 18(5), 1409–1423.
Petrie, J, Diplas, P, Gutierrez, M, and Nam, S 2013. Combining fixed- and moving-vessel acoustic Doppler current profiler measurements for improved characterization of the mean flow in a natural river. *Water Resources Research* 49, 1409–1423. DOI:10.1002/wrcr.201396.
Ramooz, R and Rennie, CD 2010. Laboratory measurement of bedload with an ADCP. In *Bedload-Surrogate Monitoring Technologies* (eds JR Gray, JB Laronne, and JDG Marr). Scientific Investigations Report 2010-5091, US Geological Survey (available at http://pubs.usgs.gov/sir/2010/5091).
Rennie, CD 2013. Technical Memo: ADCP Bedload Measurement in Manitoba Rivers. Report for Manitoba Hydro.
Rennie, CD and Church, M 2010. Mapping spatial distributions and uncertainty of water and sediment flux in a large gravel bed river reach using an acoustic Doppler current profiler. *Journal of Geophysical Research-Earth Surface* 115, F03035.
Rennie, CD and Millar, RG 2004. Measurement of the spatial distribution of fluvial bedload transport velocity in both sand and gravel. *Earth Surface Processes and Landforms* 29(10), 1173–1193.
Rennie, CD and Millar, RG 2007. Deconvolution technique to separate signal from noise in gravel bedload velocity data. *Journal of Hydraulic Engineering* 133(8), 845–856.
Rennie, CD and Rainville, F 2006. Case study of precision of GPS differential correction strategies: influence on aDcp velocity and discharge estimates. *Journal of Hydraulic Engineering* 132(3), 225–234.
Rennie, CD and Rainville, F 2008. Improving precision in the reference velocity of ADCP measurements using a Kalman filter with GPS and bottom track. *Journal of Hydraulic Engineering* 134(9), 1257–1266.
Rennie, CD and Villard, PV 2004. Site specificity of bed load measurement using an acoustic Doppler current profiler. *Journal of Geophysical Research-Earth Surface* 109(F3), F03003.
Rennie, CD, Millar, RG, and Church, MA 2002. Measurement of bed load velocity using an acoustic Doppler current profiler. *Journal of Hydraulic Engineering* 128(5), 473–483.
Rennie, CD, Rainville, F, and Kashyap, S 2007. Improved estimation of ADCP bedload velocity using a real time Kalman filter. *Journal of Hydraulic Engineering* 133(12), 1337–1344.
Rickenmann, D, Turowski, JM, Fritschi, B *et al.* 2012. Bedload transport measurements at the Erlenbach Stream with geophones and automated basket samplers. *Earth Surface Processes and Landforms* 37(9), 1000–1011.
Rickenmann, D, Laronne, JB, Turowski, JM, and Vericat, D 2013. *International Workshop of Acoustic and Seismic Monitoring of Bedload and Mass Movements – Abstracts.* Swiss Federal Institute for Forest, Snow and Landscape WSL, Birmensdorf, Switzerland.
Simpson, MR 2001. *Discharge Measurements Using a Broad-Band Acoustic Doppler Current Profiler.* Open File Report 01-1, US Geological Survey: Sacramento, CA.
Singh, A, Czuba, JA, Foufoula-Georgiou, E *et al.* 2013. StreamLab collaboratory: experiments, data sets, and research synthesis. *Water Resources Research* 49(3), 1746–1752.
Teledyne RD Instruments 2010. *ADCP Coordinate Transformation: Formulas and Transformations.* Poway, CA.
Tsubaki, R, Kawahara, Y, Muto, Y, and Fujita, I 2012. New 3-D flow interpolation method on moving ADCP data. *Water Resources Research* 48, W05539.
Van Rijn, LC 1984. Sediment transport. Part I: bed load transport. *Journal of Hydraulic Engineering* 110(10), 1431–1456.
Venditti, JG, Rennie, CD, Bomhof, J *et al.* 2014. Flow in bedrock canyons. *Nature* 513(7519), 534–537.
Vericat, D, Church, M, and Batalla, RJ 2006. Bed load bias: comparison of measurements obtained using two (76 and 152 mm) Helley–Smith samplers in a gravel bed river. *Water Resources Research* 42(1), 1–13.
Vermeulen, B, Sassi, MG, and Hoitink, AJF 2014. Improved flow velocity estimates from moving-boat ADCP measurements. *Water Resources Research* 50(5), 4186–4196.

Wheaton, JM, Brasington, J, Darby, SE *et al.* 2013. Morphodynamic signatures of braiding mechanisms as expressed through change in sediment storage in a gravel-bed river. *Journal of Geophysical Research – Earth Surface* 118(2), 759–779.
Wilcock, PR 1992. Experimental investigation of the effect of mixture properties on transport dynamics. In *Dynamics of Gravel-Bed Rivers* (eds P Billi, RD Hey, CR Thorne, and P Tacconi). John Wiley & Sons: Chichester, UK; 109–130.
Williams, RD, Brasington, J, Vericat, D *et al.* 2011. Monitoring braided river change using terrestrial laser scanning and optical bathymetric mapping. In *Geomorphological Mapping: Methods and Applications* (eds M Smith, P Paron, and J Griffiths). Elsevier: Amsterdam; 507–532.
Williams, RD, Brasington, J, Hicks, M *et al.* 2013. Hydraulic validation of two-dimensional simulations of braided river flow with spatially continuous aDcp data. *Water Resources Research* 49(9), 5183–5205.
Williams, RD, Brasington, J, Vericat, D, and Hicks, DM 2014. Hyperscale terrain modelling of braided rivers: fusing mobile terrestrial laser scanning and optical bathymetric mapping. *Earth Surface Processes and Landforms* 39(2), 167–183.
Williams, RD, Rennie, CD, Brasington, J *et al.* 2015. Linking the spatial distribution of bedload transport to morphological change during high-flow events in a shallow braided river. *Journal of Geophysical Research*, 120, 604–622. DOI:10.1002/2014JF003346, 2015.
Wolman, MG 1954. A method of sampling coarse river-bed material. *Transactions of the American Geophysical Union* 35(6), 951–956.

Discussion

Discussion by Shinji Egashira

Many bedload formulas have been proposed, and they have a functional form such as

$$g_b \sim \tau_b{}^n (\tau_b : \text{bed shear stress}, n = 3/2 \sim 5/2)$$

According to the formula derived from mechanics of a continuum body, n is specified as 5/2. However, 3/2 is given to n for most of the existing bedload formulas. The bedload transport rate g_b can be defined as

$$g_b = \sigma V_s v_s$$

in which V_s is the bedload absolute volume, v_s is the average velocity of all bedload particles and σ is the mass density of a sediment particle.

You provide interesting results on the relation of g_b and v_a as illustrated in Figures 8.5 and 8.6. Studies on bedload formulas show that V_s is proportional to the bed shear stress, and it is considered that the apparent velocity v_a measured by the aDcp must be proportional to the average velocity v_s. Can you show results on the relation between the apparent velocity of bedload and independent hydraulic variables? Such results help us not only to develop a measuring tool for bedload, but also to consider a reasonable functional form for a bedload formula.

Discussion by Colin Thorne

I have for many years held that the unit transport rate for coarse sediment (i.e. gravel and cobbles) moving as bedload in a gravel-bed river rarely if ever exceeds 1 kg/s per meter of the active bed width. I must stress that this applies only to *true bedload as defined by Brigadier Bagnold* – that is mobile, gravel and cobbles whose weight is supported *solely* by solid–solid contact with the bed. It does not include sand moving in saltation, debris flows, or sediment carried by a non-Newtonian fluid.

In the authors' opinion, do their measured data support my proposition that gravel moving as true bedload transport rates rarely exceed 1 kg/s per meter bed width? Also, given reported rates much higher than this in the literature, is there a maximum gravel/cobble transport rate per unit bed width and, if so, how can this be explained?

Response by the Authors

We thank Professor Egashira for enquiring about estimation of bedload transport rate (g_b) using the observed apparent bedload velocity (v_a) and the kinematic relation for bedload (Equation 8.7). This possibility was examined at length using the Fraser River data set in Rennie, Millar, and Church (2002) and Rennie and Villard (2004). Rennie, Millar, and Church (2002) also plotted v_a as a function of bed shear stress (τ), and Rennie and Villard (2004) compared bedload transport rate estimated using v_a with the kinematic model to predictions from conventional models based on τ. Also, with data from the Fraser and Rees Rivers as reported earlier in this chapter, we observe relationships between v_a and nondimensional bed shear stress, $\tau^* = \frac{\tau}{\gamma(S_s-1)D_{50}}$, which indexes transport stage. The Fraser River sand data set demonstrates the tightest trend, with $v_a \propto \tau*^{2.3}$ and $r^2 = 0.76$. The exponent is greater than the expected value of 1.5 for bedload transport rate, but within the range specified by Dr. Egashira.

Professor Thorne raises the very interesting proposition that bedload of coarse gravel/cobble rarely exceeds 1 kg/s/m. Indeed, our data set lends some support to this proposition in that our observed bedload transport rates do not exceed 2 kg/s/m. This can be attributed to a physical limitation of contact bedload. Consider bedload consisting of a single layer of 10 mm diameter particles of density $\rho_s = 2650\ \text{kg/m}^3$. The velocity ($v_b$) of rolling gravel bedload particles is on the order of two or three times the shear velocity (Drake *et al.* 1988; Nino, Garcia, and Ayala 1994), and v_b of gravel bedload particles is generally observed to be < 1 m/s (Drake *et al.* 1988; Schmidt and Ergenzinger 1992, Chacho, Burrows, and Emmett *et al.* 1994; Nino, Garcia, and Ayala 1994, Habersack 2001; Lajeunesse, Malverti, and Charru 2010; Blanckaert *et al.* 2013). This constrains $g_b < 27(1-\lambda)$. The porosity of the bedload layer should be greater than that of a static bed ($\lambda > 0.4$). Transport in most gravel-bed rivers occurs near the threshold of motion, even at bankfull conditions (cf. Rennie and Millar 2000). Near the threshold of motion only a small percentage (~1%) of the bed surface is mobile at any given time (Drake *et al.* 1988; Lajeunesse, Malverti, and Charru 2010), thus $\lambda > 0.99$, in which case $g_b < 0.27$ kg/s/m. The percentage of mobile bed surface and thus transport rate increases with transport stage. Lajeunesse, Malverti, and Charru (2010) did not observe percentages greater than 50% ($\lambda > 0.7$) even in fully mobile conditions with $\tau^* = 0.24$, which in our thought experiment would constrain $g_b < 8.0$ kg/s/m. Bagnold (1966) predicted analytically that a traction carpet would occur at $\tau^* = 0.3$. Thus, it appears that transport rates of gravel greater than 1.0 kg/s/m could possibly occur in thicker bedload layers, either dilated traction carpets or fully mobile transport of very coarse beds. However, lowland perennial gravel-bed rivers are unlikely to generate the shear stresses required for these conditions except possibly during extreme events (Rennie and Millar 2000), thus $g_b > 1$ kg/s/m is not often observed. Exceptions include transport events with broken or absent armour layers, such as infrequent large floods in mountain torrents where transport rates of up to 36.5 kg/s/m have been observed (Rickenmann *et al.* 2014), and flash-flood events in ephemeral arid rivers where bedload transport rates up to 10 kg/s/m have been reported (Reid and Laronne 1995; Cohen and Laronne 2005; Cohen, Laronne, and Reid 2010; Rickenmann *et al.* 2014).

References

Bagnold, RA 1966. *An Approach to the Sediment Transport Problem from General Physics. Physiographic and Hydraulic Studies of Rivers*. Geological Survey Professional Paper 422.1, US Government Printing Office: Washinton, DC.

Chacho, EFJ, Burrows, RL, and Emmett, WW 1989. Detection of coarse sediment movement using radio transmitters. In *International Association for Hydraulic Research, XXIII Congress, Proceedings of Technical Session B: Fluvial Hydraulics*, Ottawa.

Cohen, H and Laronne, JB 2005. High rates of sediment transport by flashfloods in the Southern Judean Desert, Israel. *Hydrological Processes* 19, 1687–1702.

Cohen, H, Laronne, JB, and Reid I 2010. Simplicity and complexity of bed load response during flash floods in a gravel bed ephemeral river: a 10 year field study. *Water Resources Research* 46, W11542. DOI:10.1029/2010WR009160.

Habersack, HM 2001. Radio-tracking gravel particles in a large braided river in New Zealand: a field test of the stochastic theory of bed load transport proposed by Einstein. *Hydrological Processes* 15, 377–391.

Lajeunesse, E, Malverti, L, and Charru, F 2010. Bed load transport in turbulent flow at the grain scale: Experiments and modeling. *Journal of Geophysical Research* 115, F04001. DOI:10.1029/2009JF001628.

Nino, Y, Garcia, M, and Ayala, L 1994. Gravel saltation 1. Experiments. *Water Resources Research* 30(6), 1907–1914.

Reid, I and Laronne, JB 1995. Bed load sediment transport in an ephemeral stream and a comparison with seasonal and perennial counterparts. *Water Resources Research* 31(3), 773–781.

Rennie, CD and Millar, RG 2000. Spatial variability of stream bed scour and fill: a comparison of scour depth in chum salmon (*Oncorhynchus keta*) redds and adjacent bed. *Canadian Journal of Fisheries and Aquatic Science* 57(5), 928–938.

Rickenmann, D, Turowski, JM, Fritschi, B *et al.* 2014. Bedload transport measurements with impact plate geophones: comparison of sensor calibration in different gravel-bed streams. *Earth Surface Processes and Landforms* 39, 928–942.

Schmidt, KH and Ergenzinger, P 1992. Bedload entrainment, travel lengths, step lengths, rest periods – studied with passive (iron, magnetic) and active (radio) tracer techniques. *Earth Surface Processes and Landforms* 17, 147–165.

9

Modeling Surface–Subsurface Exchange of Heat and Nutrients

Daniele Tonina, Alessandra Marzadri, and Alberto Bellin

9.1 Introduction

When we look at streams, we only see the portion of their water that flows as surface water (Figure 9.1) (Vaux 1962, 1968). However, a part of stream water flows as pore water within the streambed material interstices (Bencala and Walters 1983; Bencala 1993). Streambed sediments are mostly granular material, which is permeable (Kennedy *et al.* 2009). As a result, stream water enters the streambed as downwelling fluxes and exits the streambed sediment as upwelling fluxes (Thibodeaux and Boyle 1987). These flow cells delineate a band of streambed sediment beneath and besides streams known as hyporheic zone (Edwards 1998). This zone may extend below the stream, defining the fluvial hyporheic zone, beneath emerged bars, known as the parafluvial hyporheic zone, between meanders or along paleochannels referred to as floodplain hyporheic zone (Figure 9.1) (Tonina and Buffington 2009a). This zone is divided operatively into biological, hydraulic, and geochemical methods (Tonina 2012). The first identifies the hyporheic zone based on the presence of hyporheic-specific organisms (Orghidan 1959; Williams and Hynes 1974), the second identifies the surface enveloping all flow paths of stream water entering and exiting the streambed material (Cardenas, Wilson, and Zlotnik 2004; Tonina and Buffington 2007, 2011), and the third the extracted pore water with a concentration higher than 10% of a selected stream water constituent (Figure 9.1 inset; Triska *et al.* 1989). Although, these methods may provide different hyporheic zone extensions they all agree on a common point that hyporheic flows are primarily surface water (Gooseff 2010).

Hyporheic fluxes have important functions for contaminant attenuation, nutrient cycling, and ecological processes (Stuart 1953; Stanford and Ward 1988; Hendricks 1993; Triska, Duff, and Avanzino 1993a, b; Payn *et al.* 2009; Boano *et al.* 2014). Streamwater temperature cools and warms at different time scales, which affect the thermal regime of the hyporheic zone (Fanelli and Lautz 2008; Marzadri, Tonina, and Bellin 2012a, 2013a;). Hyporheic water temperature is a key variable, because it regulates organisms' metabolism, growth, behavior, and biogeochemical reactions (Hynes 1970; Hendricks and White 1991; Bruno *et al.* 2009). Most biochemical transformations are microbially mediated and their reaction rates are temperature dependent. They chiefly take place in the streambed sediment rather than in the water columns, because of the higher density of microbial communities adhered to the streambed particles than suspended in the surface water (Master, Shavit, and Shaviv 2005). As surface

Gravel-Bed Rivers: Processes and Disasters, First Edition. Edited by Daizo Tsutsumi and Jonathan B. Laronne.

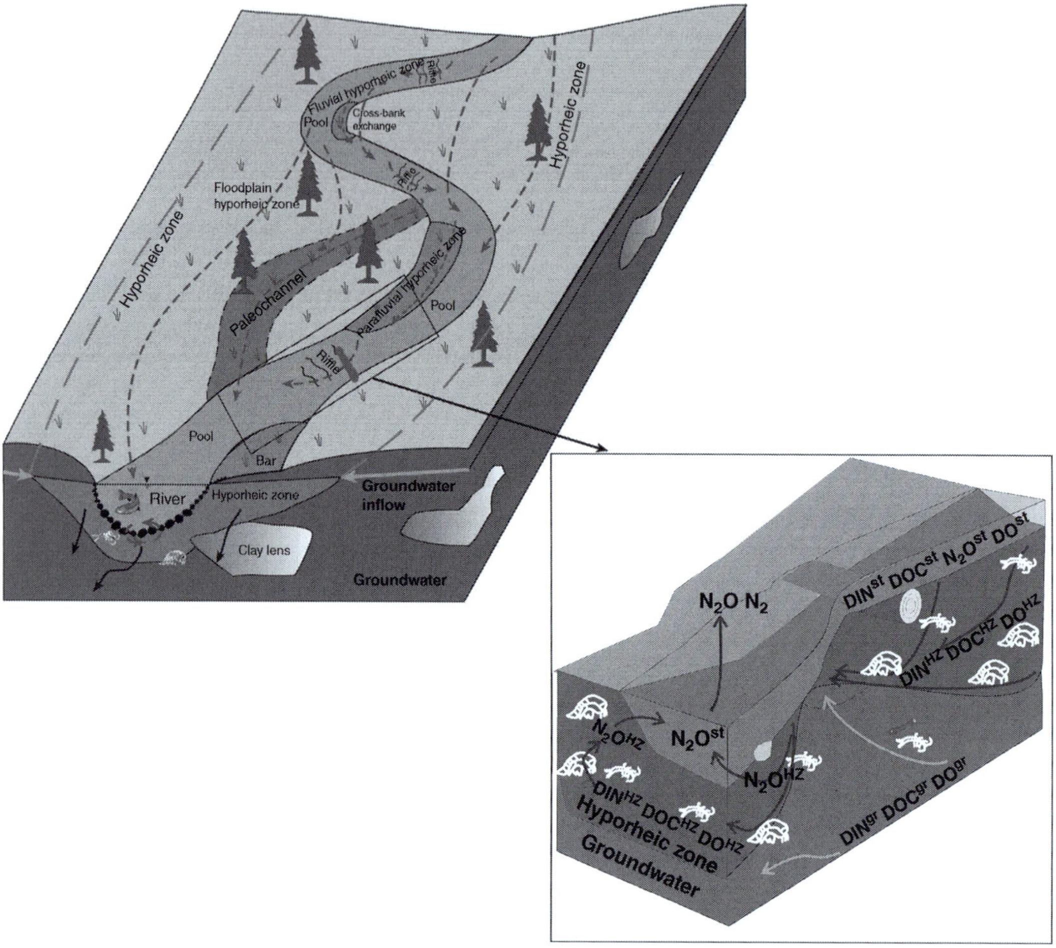

Figure 9.1 Fluvial, parafluvial, and floodplain hyporheic zones are depicted at the channel-unit scale around riffles and at channel-reach scale between meanders. The close-up shows the microscale fluvial hyporheic zone delineated by the biological method, shown by the presence of hyporheos, macroinvertebrates leaving in the hyporheic zone shown as white and green icons, geochemical method, shown by the dark purple color, which indicates more than 10% stream water, and hydraulic method, shown by the deepest red lines. Pathways of reactive solutes, dissolved organic carbon (DOC), dissolved inorganic nitrogen (DIN), and dissolved oxygen (DO), are also shown, with superscript *gr* for groundwater constituents, superscript *st* for stream and HZ for hyporheic zone (Modified from Tonina and Buffington 2009a). (See insert for color representation of this figure.)

water downwells in the streambed sediment, reactive solutes undergo biogeochemical reactions, which affect the quality of both surface and subsurface waters (Zarnetske *et al.* 2011; Boano *et al.* 2014). Among the most important biochemical processes occurring in streams are those of the nitrogen cycle, which includes nitrification and denitrification (Wondzell and Swanson 1996; Alexander, Smith, and Schwarz 2000; Beaulieu *et al.* 2011; Mulholland *et al.* 2008; Storey, Williams, and Fulthorpe 2004), because reactive dissolved inorganic nitrogen (DIN) is a potential contributor to the greenhouse effect and a primary cause of eutrophication of water bodies and periods of water column hypoxia. Nitrification transforms ammonium into nitrate or nitrite, which then through denitrification transforms to nitrogen gases: dinitrogen gas and nitrous oxide (Galloway *et al.* 2004). The latter is a greenhouse gas 300 times more potent than carbon dioxide; it has a long life lasting more than 100 years once in the stratosphere, where it reacts with ozone with negative effects on the ozone layer (Galloway *et al.* 2004, 2008). Streams, which were typically nitrogen limited, have seen an increase of DIN concentrations since the green revolution (Carpenter *et al.* 1998). The Haber–Bosch process allowed the mass production of synthetic reactive nitrogen, which is a fundamental component of fertilizers used in agriculture (Erisman *et al.* 2008). Excess DIN from agricultural land may reach streams via groundwater, run-off, and atmospheric deposition (Galloway *et al.* 2003). Here, it is a source of water body eutrophication and it undergoes biogeochemical reactions, which convert it to nitrogen gas (Mulholland and DeAngelis 2000; Mulholland *et al.* 2008; Wagenschein and Rode 2008).

These processes, which involve a combination of physical, chemical, and biogeochemical factors along the hyporheic flow paths, generate chemical and physical gradients that sustain a rich ecotone composed of macroinvertebrate and fish besides microbial colonies within the hyporheic zone (Bjornn and Reiser 1991; Gibert *et al.* 1994; Edwards 1998; Gibert and Deharveng 2002). For instance, salmonids spend between a few weeks to several months within streambed gravels, as incubating embryos and frys after hatching. Factors such as hyporheic flow rates, dissolved oxygen concentrations, and thermal regime of the hyporheic zone influence the habitat quality of their egg nests called redds (Bjornn and Reiser 1991; Tonina and Buffington 2009b). Consequently, hyporheic fluxes are a fundamental part of the whole river ecosystem (Malard *et al.* 2002; Stanford and Ward 1993).

Hyporheic flows stem from different mechanisms that include: variations of near-bed energy heads (Elliott and Brooks 1997b; Tonina and Buffington 2007; Tonina 2012); changes of alluvium thickness and valley confinement; sediment hydraulic conductivity heterogeneity (Vaux 1968); sediment transport (Packman and Brooks 2001); turbulence; and concentration gradients (Boano *et al.* 2009). Near-bed energy head variations, which are the main factor controlling hyporheic exchange, depend on streambed morphology (Elliott and Brooks 1997a; Stonedahl, Harvey, and Packman 2013). Piezometric head is the main component of the energy head controlling hyporheic exchange in streams characterized by pool–riffle morphology; while dynamic head plays a major role in streams with dune-like bedforms. A combination of the two (piezometric and dynamic) heads has been suggested for step-pool streams or for pool–riffle morphology with large amplitude and low-flow conditions (Tonina and Buffington 2007; Hassan *et al.* 2014). These head distributions form near-bed areas of high energy, which drives stream water into the sediment, and areas of low energy, which pumps pore water out of the sediment (Elliott and Brooks 1997b). Variations of streambed thickness and valley confinement also drive hyporheic flow. Stream water is driven into the sediment as the volume of sediment increases, moving from thin to thick sediment thickness or from confined to unconfined valley (Tonina and Buffington 2011). For instance, reductions in alluvium volume, as it occurs entering a confined valley or encountering thin alluvium layers, pump pore water out from the sediment. Hydraulic heterogeneity variations also regulate fluxes of surface water moving in and out of the sediment due to the capacity of the sediment to modulate the water movement (Cardenas, Wilson, and Zlotnik 2004; Salehin, Packman, and Paradis 2004; Marion *et al.* 2008). As pore water encounters less permeable sediment the excess portion is directed outside the streambed, whereas as stream flows encounter more permeable sediment stream water downwells into the streambed (Vaux 1968). Sediment transport may induce hyporheic flows as scour processes release pore waters into the stream and depositional processes entrain stream water within the sediment (Packman and Brooks 2001). Flow turbulence causes temporal and spatial fluctuations of near-bed pressure and velocity, which also drive hyporheic flows, though limited to a thin interface layer (Higashino and Stefan 2008).

The relative importance of these mechanisms in inducing hyporheic flows depends on stream morphology, surface hydrology, and ambient groundwater flows, both underflow (longitudinal groundwater flow component) and basal (vertical groundwater flow component) flows (Cardenas and Wilson 2007; Buffington and Tonina 2009; Stonedahl, Harvey, and Packman 2013; Fox, Boano, and Amon 2014). The most studied mechanism, and probably the most important, is the stream flow–bedform interaction, which causes near-bed head variations (Boano *et al.* 2014). In this chapter, we focus on the interaction between surface flow and two common bedforms: two-dimensional dune-like and three-dimensional pool–riffle morphologies. Our goal is to present a set of empirical relations to characterize the global biogeochemical and thermal behavior of the hyporheic zone at the bedform scale from a few standard field measurements, which include streamflow mean velocity, hydraulic depth, mean bed slope and streambed grain-size distribution. For the two-dimensional dune-like bedforms, we also analyze the effect of the ambient groundwater flow on the redox and thermal regimes of the hyporheic zone. To address our goal, we select the set of analytical solutions presented in previous studies (Elliott and Brooks 1997b; Packman and MacKay 2003; Marzadri *et al.* 2010, 2015) to characterize the hyporheic flow field induced by these two bedforms under different boundary conditions. These solutions adopt a Lagrangian approach to quantify the residence-time distribution of stream water within the streambed sediment and hyporheic fluxes. We then use two dimensionless numbers to characterize the biogeochemical response and thermal regime of the hyporheic zone at the bedform scale. They are the oxygen Damköhler number, Da_O (Marzadri, Tonina, and Bellin 2012b) and the temperature Damköhler number, Da_T (Marzadri, Tonina, and Bellin 2013a, 2013b), respectively. The interpretation and analysis of the data collected by the second Lotic Intersite Nitrogen eXperiment (LINX II) with Da_O highlights the importance of the streambed morphology in modulating the transformations of reactive solutes occurring in rivers and nitrous oxide production (Marzadri *et al.* 2014a).

9.2 Hyporheic Hydraulics

Hyporheic flows are typically modeled as a Darcian flow (Freeze and Cherry 1979; Elliott and Brooks 1997b):

$$\mathbf{u} = -\frac{\mathbf{K}}{n} \cdot \nabla h \tag{9.1}$$

where $\mathbf{u}$ is the seepage velocity, $\mathbf{K}$ is the tensor of the hydraulic conductivity, K, n is the sediment porosity, and h is the energy head. Only a few examples are available for heterogeneous K and unsteady surface flow (Cardenas, Wilson, and Zlotnik 2004; Salehin, Packman, and Paradis 2004; Boano, Revelli, and Ridolfi 2007; Hester, Young, and Widdowson 2013). Most analyses assume a homogeneous and isotropic K and stationary surface flows because of the simplification of the governing equation to (Elliott and Brooks 1997b; Marion *et al.* 2002):

$$\nabla^2 h = 0 \tag{9.2}$$

Analytical solutions of equation (9.2) have been obtained for three-dimensional pool–riffle (Marzadri *et al.* 2010) and two-dimensional dune-like (Elliott and Brooks 1997b; Packman and Brooks 2001; Marion *et al.* 2008; Marzadri *et al.* 2015) bedforms under several boundary conditions. Here, we present the solutions of the hyporheic flow for the following boundary conditions. The groundwater valley slope is specified as a hydraulic head drop equal to the mean streambed slope, s_0, between the

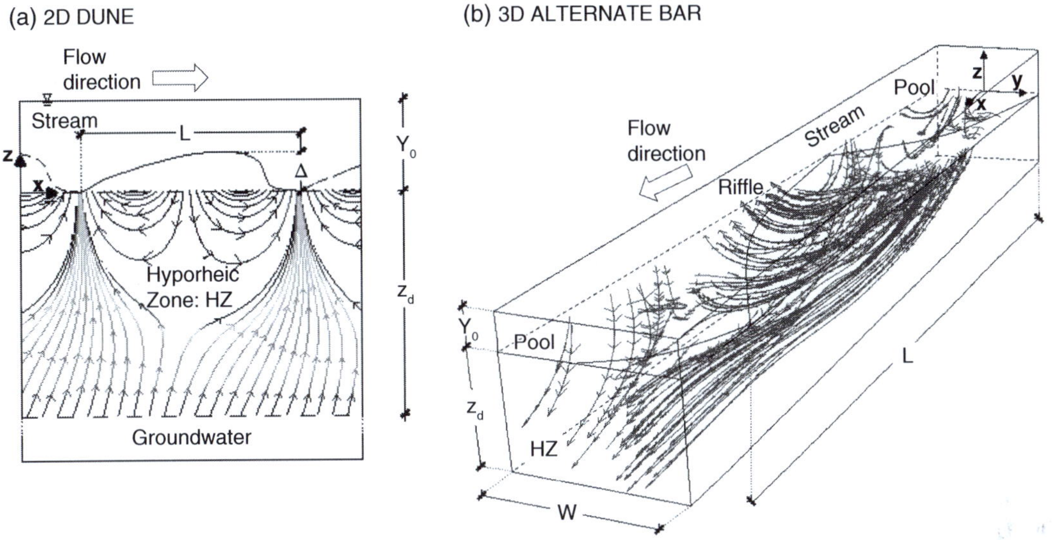

Figure 9.2 Hyporheic flow field shown by flow lines in a stream characterized by (a) two-dimensional dune morphology with positive ambient groundwater basal flow, stream gaining conditions, and (b) three-dimensional pool-and-riffle morphology with neutral case. This figure was created using ParaView (Henderson, Ahrens, and Law 2004).

upstream and downstream ends of the reach. This head difference causes the ambient groundwater underflow $u_{gw} = K \cdot s_0$. The banks of the river are typically considered as an impervious boundary (Figure 9.2) (Trauth *et al.* 2013), although groundwater heads may be specified (Cardenas 2007; Marzadri, Tonina, and Bellin 2013a). Impervious boundary conditions are set at the alluvium bottom for pool–riffle morphology, whereas a groundwater ambient basal flow, v_{gw}, is specified at a given depth, d_b for dune morphology (Boano *et al.* 2007; Hester, Young, and Widdowson 2013; Fox, Boano, and Amon 2014; Marzadri *et al.* 2015). The head distribution at the water–sediment interface is typically specified by solving the surface hydraulics numerically (Tonina and Buffington 2009b; Trauth *et al.* 2013), from measured values in flume or field experiments (Gooseff *et al.* 2006; Tonina and Buffington 2007; Tonina, Marzadri, and Bellin 2015), or from analytical solutions (Elliott and Brooks 1997a; Marion *et al.* 2002; Marion, Zaramella, and Bottacin-Busolin 2008; Marzadri *et al.* 2010). This distribution is applied at the mean streambed elevation, considering the streambed as a planar surface rather than at the actual bathymetric elevation to allow analytical solutions of the flow equation (9.2), with negligible effects on hyporheic hydraulics for these morphologies (Elliott and Brooks 1997a; Marzadri *et al.* 2010). Here, we adopt the analytical pressure-head distribution predicted by the Colombini, Seminara, and Tubino (1987) model for pool–riffle morphology. Their theory assumes that the pool–riffle bedform is in equilibrium with the surface hydraulics.

Under these assumptions, the mean hyporheic downwelling flow induced by pool–riffle morphology can be estimated with this equation (Tonina 2012):

$$\bar{q}^* = 41.108\, Y_D^{*\,-0.732};\ R^2 = 0.96$$
$$\bar{q} = \frac{\bar{q}^* K}{C_z^4} \qquad (9.3)$$

where Y_D^* is the ratio between the hydraulic depth, Y_0, and the bedform amplitude, Δ, and C_z is the dimensionless Chezy number calculated as:

$$C_z = 6 + 2.5\ln\left(\frac{1}{2.5d_s}\right) \tag{9.4}$$

with d_s the grain relative submergence, which is the ratio between d_{50}, the median grain size of the streambed material, and Y_0. These solutions are for fully submerged bars regardless of discharge.

For the dune-like bedform, we use the model proposed by Elliott and Brooks (1997a, b) to predict the near-bed pressure distribution. This model provides a sinusoidal near-bed pressure distribution over dune-like bedforms. The amplitude of the near-bed dynamic energy induced by the flow–dune bedform interaction, h_m, depends on mean flow velocity, V, mean water depth Y_0, and dune amplitude Δ. Its value can be estimated with the following equation (Shen, Fehlman, and Mendoza 1990):

$$h_m = \frac{0.28\,V^2}{2g}\begin{cases}\left(0.34\,Y_D^*\right)^{-3/8} & \mathit{if}\ Y_D^{*-1} < 0.34\\ \left(0.34\,Y_D^*\right)^{-3/2} & \mathit{if}\ Y_D^{*-1} > 0.34\end{cases} \tag{9.5}$$

where g is the acceleration of gravity (Elliott and Brooks 1997b). With the above boundary conditions and an ambient groundwater basal velocity, v_{gw}, which is positive for groundwater flow upwelling to the stream (stream gaining conditions) and negative for hyporheic flow entrained into the groundwater system (stream losing conditions), the mean hyporheic downwelling flow for dune morphology in both gaining, $\bar{q}_G$, and losing conditions, $\bar{q}_L$, is derived from flow-field analysis by Marzadri *et al.* (2015) for a finite alluvium and from mass balance analysis by Fox, Boano, and Amon (2014) for an infinite alluvium:

$$\begin{aligned}\bar{q}_G &= \frac{u_m}{\pi}\sqrt{1-\frac{v_{gw}^2}{u_m^2}} - \frac{v_{gw}}{\pi}\arccos\left(\frac{v_{gw}}{u_m}\right)\\ \bar{q}_L &= \frac{u_m}{\pi}\sqrt{1-\frac{v_{gw}^2}{u_m^2}} - \frac{v_{gw}}{\pi}\arccos\left(\frac{v_{gw}}{u_m}\right) + v_{gw}\\ u_m &= K\lambda h_m \tanh(\lambda d_b)\end{aligned} \tag{9.6}$$

where d_b is the alluvium depth, where the ambient groundwater flow v_{gw} is applied, λ is the bedform wave number, $2\pi/L$ with L the bedform length (or wavelength), and $u_m = h_m{\cdot}K{\cdot}\lambda{\cdot}\tanh(\lambda{\cdot}d_b)$ represents the maximum downwelling Darcy velocity for the neutral case $v_{gw} = 0$. Equation (9.6) coincides with the solution proposed by Packman and Brooks (2001) for a hyporheic zone of finite thickness d_b when $v_{gw} = 0$. When $|v_{gw}| > u_m$, with $|\ |$ indicating absolute value, hyporheic flow exchange does not form because groundwater upwells over the entire bedform, $v_{gw} > 0$, or stream water enters the groundwater system, $v_{gw} < 0$. Thus the range of values of the dimensionless ambient basal groundwater flow, v_{gw}^* ($v_{gw}^* = |v_{gw}|/u_m$), for which we do have hyporheic flow, is bounded between 0 and 1.

These flows may be large enough to facilitate the deposition of fine sediments (Einstein 1968; Garcia and Parker 1991), mostly in the range of colloids (Ren and Packman 2004a, b, 2007), in streambeds. Mechanical straining or electrical charges may trap the delivered particles in the streambed thereby forming siltation (Schälchli 1995), which in its extreme form may result as colmation (Brunke 1999). Presence of fine sediment may increase the mobility of large particles by the hiding/sheltering effect, which reduces mobility of small particles and promotes the mobility of large sediment (Andrews 1983; Andrews and Parker 1987).

9.3 Hyporheic Residence Time

Once the flow field is known, the residence time distribution (RTD) of surface water within the hyporheic zone can be quantified by using the particle tracking technique (e.g., Tonina and Bellin 2007) described in detail in Marzadri *et al.* (2010). We release a number of particles (NP) uniformly at the downwelling area, large enough to stabilize the cumulative density function CDF (NP ranges between 2500 and 80200 according to the type of morphology and the stream dimension). Only those particles that upwell at the streambed are considered as part of the hyporheic flow. The probability density function (PDF) of the RTD is well approximated by a log-normal distribution for both pool–riffle (Marzadri *et al.* 2010; Tonina and Buffington 2011; Marzadri, Tonina, and Bellin 2012b; Trauth *et al.* 2013) and dune-like morphologies (Marzadri *et al.* 2015; Wörman *et al.* 2002).

9.3.1 Pool–Riffle Morphology

The mean, τ_m, and variance, σ^2, of the RTD of pool–riffle morphologies for fully submerged bars are related to the bedform morphology through the dimensionless amplitude, Y_D^*. Both moments in their dimensionless forms are well represented in terms of the following power functions (Marzadri *et al.* 2010):

$$\begin{cases} \tau_m^* = 1.32\left(Y_D^*\right)^{0.68}; R^2 = 0.98; \\ \sigma^{2*} = 2.67\left(Y_D^*\right)^{1.37}; R^2 = 0.99 \end{cases} \tag{9.7}$$

where $\tau_m^* = \tau_m/t_f$, and $\sigma^{2*} = \sigma^2/t_f^2$ are represented by scaling the first two moments according to the following characteristic advective time scale $t_f = L/(K{\cdot}s_0{\cdot}C_z)$, which depends on sediment hydraulic conductivity (K), stream slope (s_0), bedform wavelength (L), and dimensionless Chezy coefficient (C_z). Similarly, the dimensionless median residence time, τ_{50}^*, which indicates the time for which 50% of the stream water entered the hyporheic zone is still within the sediment, is well represented by the following exponential function (Marzadri, Tonina, and Bellin 2012b):

$$\tau_{50}^* = 0.21\exp\left(1.22Y_D^*\right); \quad R^2 = 0.99 \tag{9.8}$$

9.3.2 Dune-Like Morphology

At the single bedform scale the hyporheic RTD is well modeled as log-normal distribution (Wörman *et al.* 2002) even with the presence of ambient groundwater flow (Marzadri *et al.* 2015). The dimensionless mean, τ_m^*, the dimensionless standard deviation, σ^*, and the dimensionless median residence time, τ_{50}^*, for hyporheic flow induced by dune-like bedforms change according to the magnitude of the dimensionless basal groundwater flow, v_{gw}^*.

The mean residence time decreases for both losing and gaining conditions as v_{gw}^* increases (Figure 9.3), because of the reduction of the hyporheic flow cells (see also Figure 9.2a). The two trends are well represented in terms of the following logarithmic functions valid for $v_{gw}^* > 0.01$:

$$\begin{cases} \tau_m^* = -3.261\ln\left(v_{gw}^*\right) + 4.6812; R^2 = 0.99 \text{ (losing)}; \\ \tau_m^* = -3.166\ln\left(v_{gw}^*\right) + 4.1208; R^2 = 0.99 \text{ (gaining)} \end{cases} \tag{9.9}$$

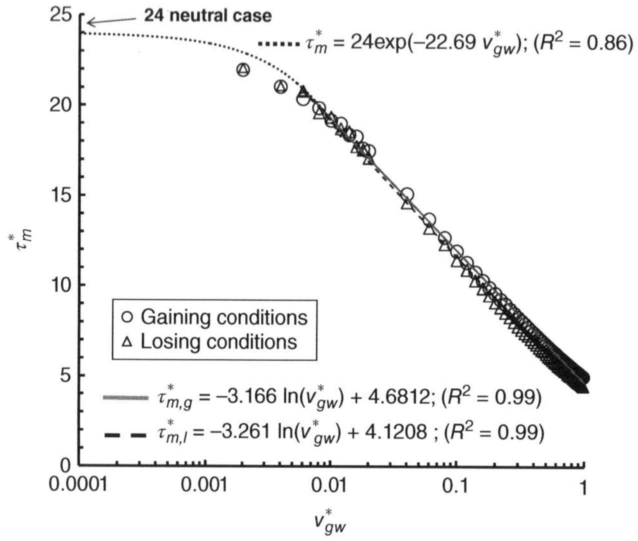

Figure 9.3 The dimensionless hyporheic mean residence time, τ^*_m, for dune-like bedforms under losing and gaining conditions as a function of the dimensionless groundwater ambient velocity $v^*_{gw} = v_{gw}/u_m$, where $u_m = h_m K \lambda \tanh(\lambda d_b)$ represents the maximum porewater Darcy velocity for the neutral case. Simulations are based on parameters reported in the Appendix.

The mean residence time is slightly higher with gaining than losing conditions because in the former case all the downwelled water returns to the stream while in the latter case part of the exchanged water is lost. This is in agreement with recent experimental finding by Fox, Boano, and Amon (2014). Both conditions tend to the same neutral case value of $\tau^*_m = 24$ as $v^*_{gw} \to 0$ following an exponential curve (Figure 9.3). The data show that a small basal flow has a large impact on the mean hyporheic residence time. For instance $v_{gw} = 1\%$ of u_m ($v^*_{gw} = 0.01$) reduces τ^*_m by 20%, as also observed numerically by Boano, Revelli, and Ridolfi (2008) and Cardenas and Wilson (2007). For v^*_{gw} tending to 1, τ^*_m tends to 5 and then drops to a value of 0 for $v^*_{gw} = 1$. The sudden drop is caused because few particles still sustain a small hyporheic zone with v^*_{gw} near 1. However, the hyporheic exchange ceases as soon as the basal groundwater flow is larger than u_m.

Similarly, the dimensionless standard deviation, σ^*, of RTD decreases with larger dimensionless basal velocity v^*_{gw}, under both gaining and losing conditions (Figure 9.4). This trend is due to the reduction of the hyporheic flow cells (see also Figure 9.2a), which shallows and thins, reducing the dispersion of streamline residence time near the mean value as v^*_{gw} increases. The value of σ^*, for losing and gaining conditions, is well represented in terms of the following power functions valid for $v^*_{gw} > 0.01$:

$$\begin{cases} \sigma^* = 5.4882\left(v^*_{gw}\right)^{-0.554}; R^2 = 0.99 \text{ (losing)}; \\ \sigma^* = 5.958\left(v^*_{gw}\right)^{-0.568}; \ R^2 = 0.99 \text{ (gaining)} \end{cases} \tag{9.10}$$

As for the mean, the standard deviation of the RTD is also slightly higher for gaining than losing conditions (Figure 9.4). This difference between gaining and losing conditions is negligible for $v^*_{gw} < 0.01$, after this value σ^* tends to the neutral case of 123.49 for $v^*_{gw} \to 0$ with the same exponential function for both cases. As for the mean, σ^* tends to 5 for $v^*_{gw} \to 1$ and it abruptly changes to 0 for v^*_{gw} equal to 1. For

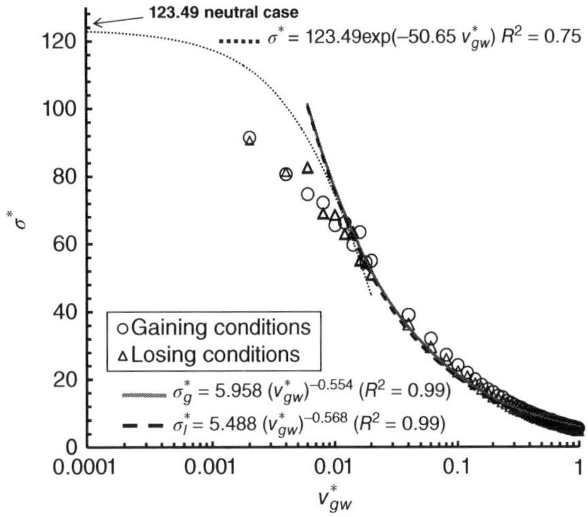

Figure 9.4 Dimensionless standard deviation, σ^*, of the RTD under losing and gaining conditions as a function of the dimensionless groundwater ambient velocity $v^*_{gw} = v_{gw}/u_m$, where $u_m = h_m K\lambda \tanh(\lambda d_b)$ represents the maximum porewater Darcy velocity for the neutral case. Simulations are based on parameters reported in the Appendix. The best fit of the data with two power equations is also shown.

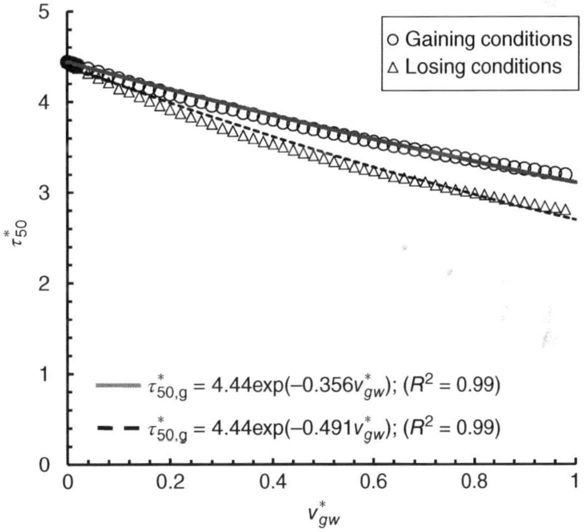

Figure 9.5 Dimensionless hyporheic median residence time, τ^*_{50}, for losing and gaining conditions as a function of the dimensionless groundwater ambient velocity $v^*_{gw} = v_{gw}/u_m$, where $u_m = h_m K\lambda \tanh(\lambda d_b)$ represents the maximum downwelling Darcy velocity for the neutral case. Simulations are based on parameters reported in the Appendix.

$v^*_{gw} = 1$ the hyporheic zone is sustained by very few particles and it completely disappears when $v_{gw} > u_m$ (i.e. $v^*_{gw} > 1$).

Similar to τ^*_m, τ^*_{50} is higher for gaining than losing conditions and it decreases with larger dimensionless basal velocity $v^*_{gw} = v_{gw}/u_m$, but slowly (Figure 9.5). This slower decrease could be due to the smaller influence of the tails on τ^*_{50} than on τ^*_m (Kirk 2008). Their values are well predicted by the following 2 exponential functions:

$$\begin{cases} \tau^*_{50} = 4.44\ \exp\left(-0.491 v^*_{gw}\right); R^2 = 0.99\ \text{(losing)}; \\ \tau^*_{50} = 4.44\ \exp\left(-0.356 v^*_{gw}\right);\ R^2 = 0.99\ \text{(gaining)} \end{cases} \tag{9.11}$$

For for $v^*_{gw} = 0$, the solution matches that derived by Elliott and Brooks (1997b) and for $v^*_{gw} = 1$ it tends to 2.79 and 3.17 for losing and gaining conditions respectively. Then its value drops to 0, as for the other moments.

These results show that groundwater ambient basal flow has a very important effect on hyporheic flows and hyporheic moments. The impact on the first two moments is strong even at small basal flows, i.e. $v^*_{gw} = 0.01$, which is 1% of the maximum hyporheic downwelling flow. These are strongly affected by the tails of the distributions. This result is in agreement with those presented by Stonedahl *et al.* (2012), who performed a series of simulations to analyze the effect of groundwater discharge to a small stream. They observed that "hyporheic exchange is very sensitive to small amounts of groundwater discharge." Similarly, Trauth *et al.* (2013) showed the same behavior numerically for τ^*_m for a pool–riffle morphology. On the other hand, the median residence time shows a less strong dependency on v^*_{gw} than τ^*_m and s^*. The weak dependence of τ^*_{50} on v^*_{gw} ensures that τ^*_{50} is a robust metric in many cases, because the intensity of the ambient groundwater flow is difficult to establish in the field. Additionally, τ^*_{50} for dune-like bedforms has been shown to have negligible dependency on low heterogeneity of sediment hydraulic conductivity (typical of sand streams) (Tonina *et al.* 2016).

9.4 Damköhler Numbers

Biogeochemical processes and heat transport within the hyporheic zone have been studied by solving the advection diffusion equation with or without the reaction transformations for biogeochemical processes and heat transport, respectively, once the flow field has been resolved (Bardini *et al.* 2012; Zarnetske *et al.* 2012; Marzadri, Tonina, and Bellin 2013b; Briggs, Lautz, and Hare 2014; Marzadri *et al.* 2014b; Briggs *et al.* 2015). These solutions provide concentration and temperature distributions within the hyporheic zone. However, most of the times,we are interested in the overall response of the hyporheic zone at the bedform scale rather than what is the chemistry at a particular location. Here, we use two dimensionless numbers, the oxygen Damköhler number, Da_O (Marzadri, Tonina, and Bellin 2012b) to quantify the chemical conditions and the thermal Damköhler number, Da_T (Marzadri, Tonina, and Bellin 2013a) to assess the thermal regime of the hyporheic zone. Both numbers are defined as the ratio between two time scales.

9.4.1 The Biogeochemical Damköhler Number

The oxygen Damköhler number, Da_O, is the ratio between the median hyporheic residence time, τ_{50}, and the time required by biogeochemical reactions and microbial respiration to consume dissolved oxygen from its surface water concentration, DO_0, to a threshold concentration, $DO_{\lim}$, below which the environment is anaerobic, $\tau_{\lim}$.

$$Da_O = \frac{\tau_{50}}{\tau_{\lim}} = \frac{\tau_{50}}{\frac{1}{K_{RN}} \ln\left(\frac{DO_0}{DO_{\lim}}\right)} \tag{9.12}$$

K_{RN} is the reaction rate that combines the effects of nitrification and biomass respiration (Rutherford *et al.* 1995; Marzadri, Tonina, and Bellin 2013a). Zarnetske *et al.* (2012) also propose a special case of this Damköhler number. Their Damköhler number has a value of 1 for the logarithm, which assumes that the threshold concentration, $DO_{\lim}$, is $1/e$ smaller than DO_0.

Da_O has been suggested to be an index of prevailing redox conditions within the hyporheic zone and of the production of nitrogen gases (Marzadri, Tonina, and Bellin 2011, 2012b; Marzadri *et al.* 2014a). Da_O values less than one are associated with prevailing aerobic conditions, whereas values larger than

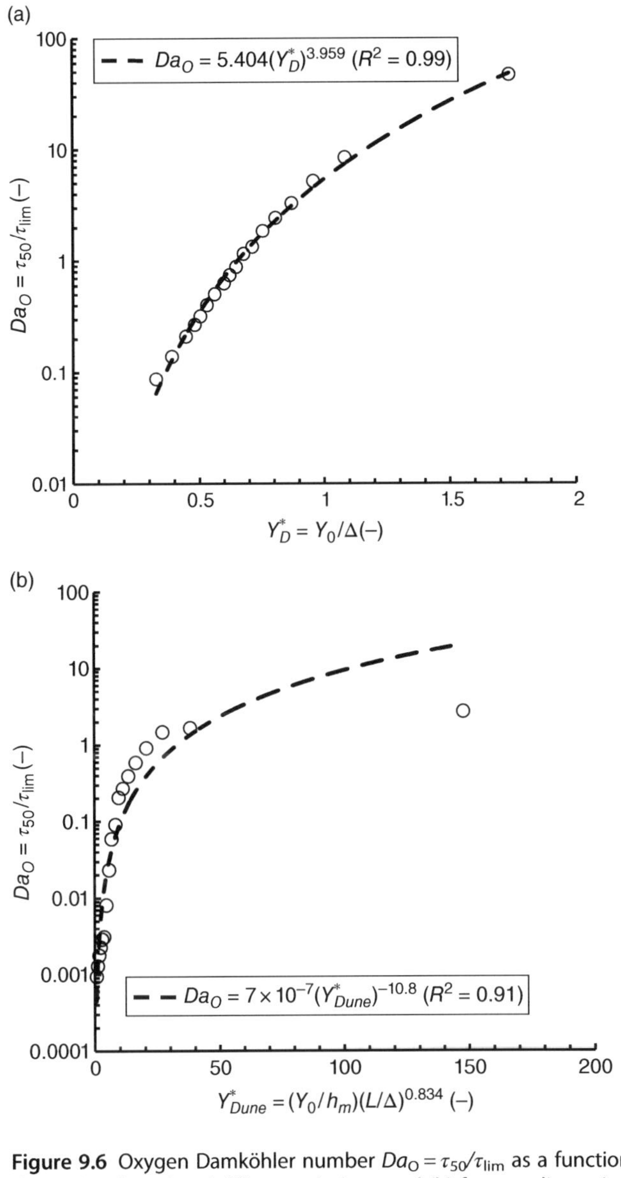

Figure 9.6 Oxygen Damköhler number $Da_O = \tau_{50}/\tau_{\lim}$ as a function of the dimensionless depth (a) for three-dimensional pool-and-riffle morphology and (b) for two-dimensional dune morphology. Simulations are based on parameters reported in the Appendix. The best fit of the data with the power equations is also shown.

one are associated with prevailing anaerobic conditions within the streambed material. When $\tau_{\lim}$ is constant, then an increase of stream size implies an increase in anaerobic conditions, as shown in Figure 9.6. The following power law functions for pool–riffle and dune-like morphologies well represent this trend:

$$\begin{cases} Da_O = 5.40\left(Y_D^*\right)^{3.96}; & R^2 = 0.99 \ \text{pool and riffle} \\ Da_O = 7\times10^{-7}\left(Y_{Dune}^*\right)^{-10.8}; & R^2 = 0.91 \ \text{dune} \end{cases} \tag{9.13}$$

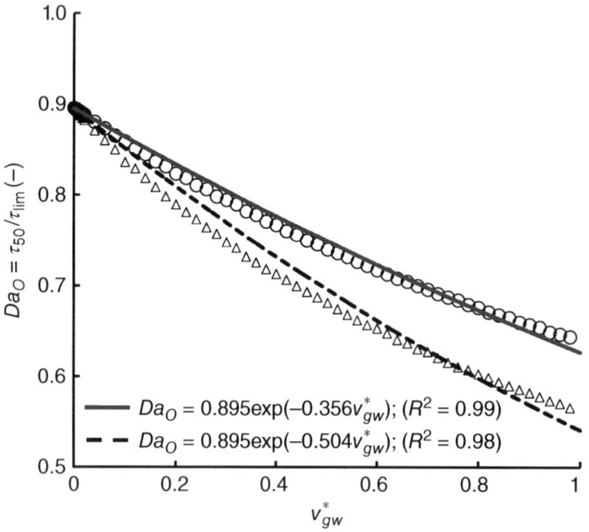

Figure 9.7 Oxygen Damköhler number, Da_O, for losing and gaining conditions as a function of the dimensionless ambient groundwater basal velocity $v^*_{gw} = v_{gw}/u_m$, where $u_m = h_m K\lambda \tanh(\lambda d_b)$ represents the maximum pore water Darcy velocity for the neutral case. Simulations are based on parameters reported in the Appendix.

Solutions for pool–riffle are for fully submerged bars, when discharge has been shown to play a negligible effect compared to the topography on the piezomentric head variations (Marzadri *et al.* 2010). Thus Da_O depends mostly on Y^*_D. Conversely, dynamic head variations, which drive hyporheic flows in dune-like bedforms, depend on both flow hydraulics, through h_m, and dune shape, via Δ and L. Thus, Da_O for dune morphology is shown to depend on a new dimensionless depth Y^*_{Dune} defined as the product between Y^*_D and the aspect ratio of the dune $(L/\Delta)^{0.834}$. In both cases, Da_O increases with both dimensionless depths, which typically increase with stream size. Moreover, the ambient groundwater basal flow reduces hyporheic residence time by reducing the dimension of the hyporheic flow cells. This effect, in turn, increases the probability of prevailing aerobic conditions ($Da_O < 1$) with a reduction of the Da_O as shown by the following equations for losing and gaining (Figure 9.7) conditions in the case of dune-like morphology:

$$\begin{cases} Da_O = -0.315\, v^*_{gw} + 0.853; \quad R^2 = 0.98 \text{ (losing)}; \\ Da_O = -0.248\, v^*_{gw} + 0.873; \quad R^2 = 0.99 \text{ (gaining)} \end{cases} \tag{9.14}$$

The decrease is due only to change in hyporheic hydraulics. Thus, the effect of v^*_{gw} on Da_O can be accounted in this equation by replacing the intercept with the Da_O value of the neutral case.

9.4.2 Thermal Damköhler Number

The thermal Damköhler number, Da_T, is an index that quantifies the control of stream morphology on the influence of daily stream water temperature oscillations on the hyporheic thermal regime. It is defined as the ratio between τ_{50} and a thermal residence time limit, which is an index of the temporal scale of temperature attenuation, $\tau_{T\lim}$ (Marzadri, Tonina, and Bellin 2013b):

$$Da_T = \frac{\tau_{50}}{\tau_{T\lim}} = \frac{\tau_{50}}{\left(\sqrt{r}\cos(\theta/2) - \frac{C_T}{2D}\right)^{-1}} \tag{9.15}$$

with:

$$\begin{cases} r = \dfrac{1}{D}\sqrt{\dfrac{C_T^4}{16D^2} + \omega^2}; \quad \theta = \text{arctg}\left(\dfrac{4\omega D}{C_T^2}\right); \\ D = \dfrac{n\alpha_L C_w}{C|u|} + \dfrac{K_T}{C|u|^2} \end{cases} \tag{9.16}$$

Here, $\tau_{T\text{lim}}$ is the time necessary to reduce the daily temperature oscillations of the stream water by the factor *1/e* (Marzadri, Tonina, and Bellin 2013b) within the hyporheic zone, α_L is the longitudinal component of the dispersivity tensor, C_w is the volumetric heat capacity of water, C is the effective volumetric heat capacity of the sediment–water matrix, K_T is the effective thermal conductivity and ω is the period of oscillation of temperature signal ($\omega = 2{\cdot}\pi{\cdot}d^{-1}$). Similar to Da_O, Da_T increases according to the bedform dimension because of the increase of the median hyporheic residence time:

$$\begin{cases} Da_T = 3.86\left(Y_D^*\right)^{5.17}; & R^2 = 0.97 \quad \text{pool and riffle} \\ Da_T = 0.0233\,Y_{Dune}^* + 0.002; & R^2 = 0.91 \quad \text{dune} \end{cases} \tag{9.17}$$

Values of $Da_T < 1$ indicate that daily fluctuations of stream water temperature significantly affect the hyporheic thermal regime at the bedform scale (Figure 9.8). This has important consequences on hyporheic organisms, as they need to adapt to these changes (Bruno *et al.* 2009). Hydraulic conductivity may also change due to temperature oscillations (Constantz *et al.* 2001; Constantz, Thomas, and Zellweger 1994; Constantz and Thomas 1997). Consequently, we expect faster flows when the warm water wave enters the sediments than during night when cold water would reduce the hydraulic conductivity of the sediments. Biogeochemical reaction rates also increase with temperature, suggesting higher production with warmer than colder conditions (Marzadri, Tonina, and Bellin 2013b). Conversely, values of $Da_T > 1$ indicate that most of the hyporheic zone is not influenced by the daily temperature oscillations of the surface water (Figure 9.8). Thus, mean daily temperatures provide sufficient information to characterize the hyporheic thermal regime at the bedform scale for this case. Note that at the local scale (i.e., evaluating the thermal Damköhler number by replacing τ_{50} with the streamline residence time τ) we can observe a different pattern of hot spots with both thermally dynamic or stable conditions (Marzadri, Tonina, and Bellin 2013b).

9.5 Role of Stream Morphology on Nitrous Oxide Emissions

Recently, Marzadri *et al.* (2014a) used the oxygen Damkhöler, Da_O, to demonstrate the role of stream morphology on the emissions of nitrous oxide, N_2O, from 72 headwater streams located across nine different US biomes as part of the 2nd Lotic Intersite Nitrogen eXperiment: LINXII (Mulholland *et al.* 2008; Beaulieu *et al.* 2011). They evaluated: (i) the median residence time, τ_{50}, as a function of stream morphology (Mulholland *et al.* 2008); (ii) the residence time limit, τ_{lim}, as a function of the stream biogeochemical parameters for DIN dynamics (Beaulieu *et al.* 2011); (iii) the nitrous oxide emissions from the hyporheic zone (Beaulieu *et al.* 2011). The analysis is represented in terms of $FN_2O_{HZ}^*$ to remove the influence of dissolved inorganic nitrogen loads, which may depend on nearby land use, on N_2O production. $FN_2O_{HZ}^*$ is defined as the ratio between the N_2O mass flux (FN_2O_{HZ}) and the total stream mass flux of ammonium ($V\,[NH_4^+]$) and nitrate ($V\,[NO_3^-]$), which are the potential sources of N_2O.

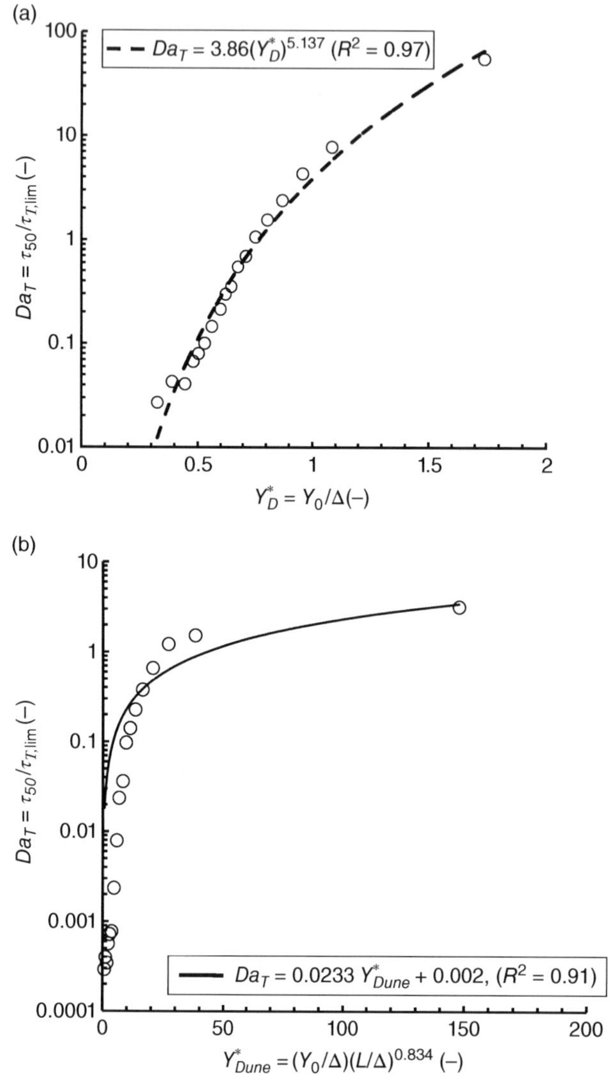

Figure 9.8 Thermal Damköhler number $Da_T = \tau_{50}/\tau_{T,\mathrm{lim}}$ as a function of the dimensionless depths for (a) three-dimensional pool-and-riffle morphology and (b) two-dimensional dune morphology. Simulations are based on parameters reported in the Appendix.

With this information, $FN_2O^*_{HZ}$ increases with $Da_O = \tau_{50}/\tau_{\mathrm{lim}}$ (Figure 9.9). LINX II stream reaches have negligible groundwater or surface water inputs (Beaulieu *et al.* 2011), thus τ_{50} is referred to the neutral case (i.e., $v_{gw} = 0$). By accounting for the correct morphology, streams with pool–riffle or dune-like morphology fall on the same relationship (Figure 9.9), which is well approximated by the following power function (Marzadri *et al.* 2014a):

$$FN_2O^*_{HZ} = 1.4 \times 10^{-7}(Da_O)^{0.22}; R^2 = 0.53 \tag{9.18}$$

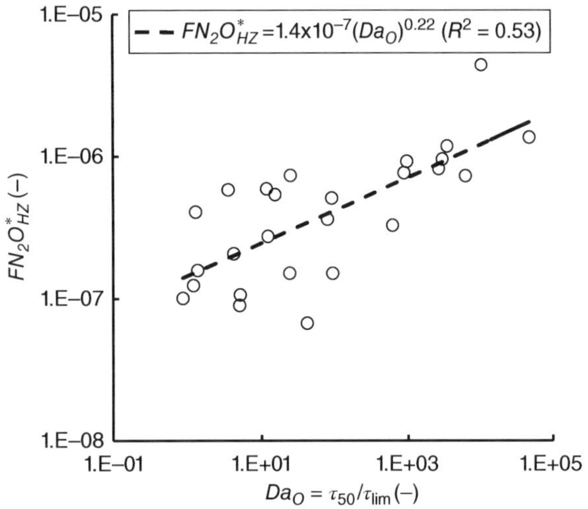

Figure 9.9 Dimensionless flux of nitrous oxide, $FN_2O^*_{HZ} = FN_2O_{HZ}/(V([NH_4^+] + [NO_3^-]))$, as a function of the oxygen Damköhler number, $Da_O = \tau_{50}/\tau_{lim}$, for the LINXII stream characterized by dune or pool-and-riffle morphology. Notice that all concentrations are expressed as N. The best fit of the data with a power law equation is also shown. (Modified from Marzadri *et al.* (2014a).)

The functional dependence of N_2O flux on Da_O, underscores the importance of stream morphology when interpreting experimental results and upscaling N_2O emissions from the reach to the network scale.

9.6 Conclusions and Research Needs

Stream morphology is well recognized to play a strong influence on both biogeochemical processes and thermal regime within the hyporheic zone. Two Damköhler numbers have been suggested as indicators of the prevailing redox conditions (Da_O) and of the prevailing thermal regime (Da_T) within the streambed sediments. Application of Da_O and Da_T for increasing stream dimensions shows different roles between the hyporheic zones of streams characterized by dune and by pool–riffle morphologies. Those trends of variations are well represented in terms of four power functions that relate these Damköhler numbers with the dimensionless water depths for pool–riffle, Y^*_D, and for dune-like, Y^*_{Dune}, bedforms. Both Y^*_D and Y^*_{Dune} values can be quantified easily from a few standard reach-scale hydromorphological data measured in the field. These measurements include the hydraulic depth, mean flow velocity, mean streambed slope and grain size distribution. Thus, these indices are useful tools to quantify the impact that redox conditions and thermal regime may have on chemical processes involving oxygen, nutrients and contaminants, besides the lifetime of many aquatic organisms and the rate of photosynthesis by aquatic plants.

The Damköhler approach used here can be applied to any hyporheic zone for which the median hyporheic residence time is known. Here, we applied the definitions of Damköhler numbers to derive a set of empirical relations that give us the possibility to characterize the biogeochemical and thermal behavior of the hyporheic zone at the global, bedform, scale. This provided the possibility to compare how bedforms with a prevailing dune-like two-dimensional pattern act with respect to bedforms with a prevailing three-dimensional pattern, pool–riffle. Future research in this direction is needed, because models that estimate the residence time distribution for other types of morphologies, such as step-pool, cascade and plane-bed, from accessible reach-scale hydromorphological information, like reach-scale water depth, channel width, distance between steps and high of the steps, are still missing.

Notation

C	Effective volumetric heat capacity of the sediment–water matrix, $Jm^{-3\circ}C^{-1}$
C_w	Volumetric heat capacity of water, $Jm^{-3\circ}C^{-1}$
C_z	Dimensionless Chezy number (–)
d_{50}	Median grain size of the streambed material, m
Da_O	Oxygen Damköhler number, the ratio between two times τ_{50} and $\tau_{\lim}$ (–)
Da_T	Thermal Damköhler number (–)
DIN	Dissolved inorganic nitrogen
L	Bedform wavelength, m
d_b	Depth of the impervious boundary layer or the depth at which the ambient groundwater basal flow is applied, m
$DO_{\lim}$	Dissolved oxygen threshold concentration below which the environment is anaerobic, mg/L
d_s	Grain relative submergence; ratio between d_{50} and Y_0 (–)
$FN_2O^*_{HZ}$	The ratio between FN_2O_{HZ} and V $[NH_4^+] + [NO_3^-]$ with concentrations expressed as N (–)
FN_2O_{HZ}	N_2O mass flux, mg N_2O–N $m^{-2}h^{-1}$
g	Acceleration of gravity, m/s^2
h	Energy head, m
h_m	Amplitude of the near-bed dynamic head induced by the flow–dune bedform interaction, m
$\mathbf{K}$	The tensor of the hydraulic conductivity, m/s
K	Hydraulic conductivity, m/s
K_{RN}	Reaction rate that combines the effects of nitrification and biomass respiration, 1/day
K_T	Effective thermal conductivity, $Wm^{-1\circ}C^{-1}$
n	Sediment porosity (–)
PDF	Probability density function (–)
$\bar{q}$	Mean hyporheic downwelling flux, m/s
$\bar{q}^*$	Dimensionless mean hyporheic downwelling flux (–)
$\bar{q}_G$	Mean hyporheic downwelling flux with groundwater gaining condition, m/s
$\bar{q}_L$	Mean hyporheic downwelling flux with groundwater losing condition, m/s
RTD	Residence time distribution (–)
s_0	Mean streambed slope, which is assumed equal to the valley scale groundwater energy slope (–)
$\mathbf{u}$	The seepage velocity vector, m/s

u_{gw}	Groundwater ambient underflow, m/s
u_m	Maximum downwelling Darcy velocity for the neutral case $v_{gw} = 0$, m/s
V	Cross-sectional averaged flow velocity of stream water
V $[NH_4^+]$	Stream mass flux of ammonium, mg NH_4^+–N $m^{-2}h^{-1}$
V $[NO_3^-]$	Stream mass flux of nitrate, mg NO_3^+–N $m^{-2}h^{-1}$
v_{gw}^*	Dimensionless groundwater ambient basal flow (–)
v_{gw}	Groundwater ambient basal flow, m/s
Y_D^*	Dimensional depth, ratio between Y_0 and Δ
Y_{Dune}^*	Dimensionless depth defined as the product between Y_D^* and the aspect ratio of the dune $(L/D)^{0.834}$ (–)
Y_0	Stream hydraulic depth, m
Δ	Bedform amplitude, m
α_L	Longitudinal component of the dispersivity tensor, m
λ	The bedform wave number, 1/m
σ^2	Variance of the residence time of the hyporheic zone at the bedform scale, s^2
σ^{2*}	Dimensionless variance of hyporheic residence time (–)
τ_{50}^*	Dimensionless median hyporheic residence time (–)
τ_m^*	Dimensionless mean hyporheic residence time (–)
τ_{50}	Median hyporheic residence time, s
$\tau_{\lim}$	Time required by biogeochemical reactions and microbial respiration to consume dissolved oxygen to $DO_{\lim}$, s
τ_m	Mean residence time of the hyporheic zone, s
$\tau_{T\lim}$	Thermal residence time limit, which is an index of the temporal scale of temperature attenuation, s
ω	Period of oscillation of temperature signal ($\omega = 2{\cdot}\pi/day$), 1/day

Acknowledgments

This research is partially supported by the Deadwood River Project, US Forest Service Award 009421-01. We thank Professor J. L. Tank for her assistance in analyzing the LINXII data and the LINXII group for providing empirical data that supported the result proposed in this manuscript. We also thank Professor A. Packman for his insights and valuable comments on an earlier version.

Appendix

To analyze the role of ambient groundwater basal flow, v_{gw}, on the RTD moments of hyporheic zone induced by dune-like bedform, a set of simulations was run for a dune configuration reported in Table 9.1 with v_{gw} varying between 0 and 5.82×10^{-6} m/s with 1.25×10^{-8} m/s increments.

To analyze the role of stream morphology on the thermal and oxygen Damköhler numbers, we compare the results for rivers with two-dimensional dunes under neutral conditions (no ambient groundwater basal flow) with those with three-dimensional pool and riffle, whose bedforms are in equilibrium with the water discharge (Q_{stream}). We use 18 different streams, whose morphodynamic parameters are summarized for dunes in Table 9.1 and for pool–riffle in Table 9.2 to span the natural range of variability of streambed slopes, discharges, and bedform dimension. In all the simulations, the number of released particles (NP) has been chosen to ensure the stability of the calculated moments.

To show the behaviors of Da_O we performed a set of simulations with $DO_{lim} = 2$ mg/L, which is the typical limit value for anoxic condition (Rosamond, Thuss, and Schiff 2012), $K_R = 0.053$ day^{-1} and $K_N = 1.998$ day^{-1} obtained applying the Arrhenius equation (Fogler 1999) on the data reported in Rutherford (1994) and Sjiodin, Lewis, and Saunders (1997), respectively, during a typical winter condition in Western North America ($T = 6\,°C$).

Table 9.1 Morphodynamic parameters of two-dimensional dunes.

Test	V (m/s)	Y_0 (m)	Δ (m)	Y_D^* (–)	s_0 (%)	L (m)	Q_{stream} (m^{3}s^{-1})
1	0.42	0.028	0.085	0.33	0.48	0.106	0.005
2	0.44	0.031	0.150	0.39	0.46	0.137	0.007
3	0.46	0.036	0.120	0.45	0.44	0.183	0.010
4	0.48	0.041	0.105	0.48	0.42	0.251	0.015
5	0.50	0.048	0.095	0.50	0.40	0.358	0.024
6	0.53	0.051	0.097	0.53	0.38	0.534	0.041
7	0.57	0.093	0.166	0.56	0.32	0.699	0.058
8	0.80	0.196	0.327	0.60	0.25	0.849	0.076
9	1.00	0.382	0.615	0.62	0.20	1.045	0.099
10	1.11	0.524	0.810	0.65	0.18	1.468	0.156
11	1.33	0.906	1.337	0.68	0.15	1.659	0.183
12	1.43	1.11	1.564	0.71	0.14	1.885	0.218
13	1.54	1.39	1.844	0.75	0.13	2.153	0.261
14	1.67	1.77	2.197	0.81	0.12	2.477	0.231
15	1.82	2.30	2.640	0.87	0.11	2.867	0.382
16	2.00	3.06	3.195	0.96	0.10	3.344	0.469
17	2.00	3.06	2.824	1.08	0.10	3.933	0.582
18	2.00	3.06	1.760	1.74	0.10	8.359	1.592

V, mean stream velocity; Y_0, mean flow depth; Δ, dune height; Y_D^*, dimensionless mean flow depth; s_0, streambed slope; L dune length; Q_{stream}, stream water discharge per unit width.

Table 9.2 Morphodynamic parameters of the pool–riffle bedform. In all simulations the Shields' number is constant $\theta = 0.08$ as the median grain size $d_{50} = 0.01$ m.

Test	W (m)	Y_0 (m)	Δ (m)	Y_D^* (-)	s_0 (%)	L (m)	Q_{stream} (m^3s^{-1})
1	1.30	0.05	0.15	0.33	2.64	8.76	0.057
2	1.73	0.07	0.17	0.39	1.98	11.52	0.111
3	2.17	0.08	0.19	0.45	1.58	14.23	0.185
4	2.48	0.10	0.20	0.48	1.39	16.16	0.251
5	2.74	0.11	0.21	0.50	1.25	17.77	0.314
6	3.06	0.12	0.22	0.53	1.12	19.75	0.404
7	3.47	0.13	0.23	0.56	0.99	22.26	0.537
8	4.00	0.15	0.26	0.60	0.86	25.53	0.738
9	4.33	0.17	0.27	0.62	0.79	27.57	0.883
10	4.73	0.18	0.28	0.65	0.73	29.98	1.072
11	5.20	0.20	0.29	0.68	0.66	32.87	1.325
12	5.78	0.22	0.31	0.71	0.59	36.40	1.675
13	6.50	0.25	0.33	0.75	0.53	40.82	2.174
14	7.43	0.29	0.33	0.81	0.46	46.52	2.902
15	8.67	0.33	0.38	0.87	0.40	54.13	4.101
16	10.40	0.40	0.42	0.96	0.33	64.82	6.121
17	13.00	0.50	0.46	1.08	0.26	80.97	9.977
18	26.00	1.00	0.56	1.74	0.13	163.14	45.037

W is stream width; Y_0, mean flow depth; Δ, dune height; Y_D^*, dimensionless mean flow depth; s_0, streambed slope; L dune length; Q_{stream}, stream water discharge per unit width.

In the simulations for Da_T, we assume $\alpha_L = 0.01$ m and $C_w = 4.2 \times 10^6$ ($Jm^{-3}°C^{-1}$), $C = 2.76 \times 10^6$ ($Jm^{-3}°C^{-1}$),and $K_T = 2.58$ ($Wm^{-1}°C^{-1}$) are taken according to the data reported by Gariglio, Tonina, and Luce (2013).

References

Alexander, RB, Smith, RA, and Schwarz, GE 2000. Effect of stream channel size on the delivery of nitrogen to the Gulf of Mexico. *Nature* 403.

Andrews, ED 1983. Entrainment of gravel from naturally sorted riverbed material. *Geological Society of American Bulletin* 94, 1225–1231.

Andrews, ED and Parker, G 1987. Formation of a coarse surface layer as the response to gravel mobility. In *Sediment Transport in Gravel-Bed Rivers* (eds CR Thorne, JC Bathurst, and RD Hey). John Wiley & Sons: Chichester, UK.

Bardini, L, Boano, F, Cardenas, BM *et al.* 2012. Nutrient cycling in bedform induced hyporheic zones. *Geochimica Et Cosmochimica Acta* 84, 47–61. DOI:10.1016/j.gca.2012.01.025.

Beaulieu, JJ, Tank, JL, Hamilton, SK *et al.* 2011. Nitrous oxide emission from denitrification in stream and river networks. *Proceedings of the National Academy of Sciences of the United States of America* 108(1), 214–219. DOI:10.1073/pnas.1011464108.

Bencala, KE 1993. A perspective on stream-catchment connections. *Journal of the North American Benthological Society* 12(1), 44–47.

Bencala, KE and Walters, RA 1983. Simulation of solute transport in a mountain pool-and-riffle stream: a transient storage model. *Water Resources Research* 19(3), 718–724.

Bjornn, TC and Reiser, DW 1991. Habitat requirements of salmonids in streams. In *Influence of Forest and Rangeland Management on Salmonid Fishes and Their Habitats* (ed. WR Meehan) Special Publication 19, American Fisheries Society: Bethesda, MD; 83–138.

Boano, F, Revelli, R, and Ridolfi, L 2007. Bedform-induced hyporheic exchange with unsteady flows. *Advances in Water Resources* 30, 148–156.

Boano, F, Revelli, R, and Ridolfi, L 2008. Reduction of the hyporheic zone volume due to the stream-aquifer interaction. *Geophysical Research Letters* 35, L09401. DOI:10.1029/2008GL033554.

Boano, F, Harvey, JW, Marion, A *et al.* 2014. Hyporheic flow and transport processes: Mechanisms, models, and biogeochemical implications. *Reviews of Geophysics* 52, 603–679. DOI:10.1002/2012RG000417.

Boano, F, Packman, AI, Cortis, A *et al.* 2007. A continuous time random walk approach to the stream transport of solutes. *Water Resources Research* 43, W10425. DOI:10.1029/2007WR006062.

Boano, F, Poggi, D, Revelli, R, and Ridolfi, L 2009. Gravity-driven water exchange between streams and hyporheic zones. *Geophysical Research Letters* 36, L20402. DOI:10.1029/2009GL040147.

Briggs, MA, Lautz, L, and Hare, DK 2014. Residence time control on hot moments of net nitrate production and uptake in the hyporheic zone. *Hydrological Processes* 28(11), 3741–3751. DOI:10.1002/hyp.9921.

Briggs, MA, Day-Lewis, FD, Zarnetske, JP, and Harvey, JW 2015. A physical explanation for the development of redox microzones in hyporheic flow. *Geophysical Research Letters* 42. DOI:10.1002/2015GL064200.

Brunke, M 1999. Colmation and depth filtration within streambeds: retention of particles in hyporheic interstices. *International Review of Hydrobiology* 84(2), 99–117.

Bruno, MC, Maiolini, B, Carolli, M, and Silveri, L 2009. Impact of hydropeaking on hyporheic invertebrates in an Alpine stream (Trentino, Italy). *International Journal of Limnology* 45, 157–170.

Buffington, JM and Tonina, D 2009. Hyporheic exchange in mountain rivers II: Effects of channel morphology on mechanics, scales, and rates of exchange. *Geography Compass* 3(3), 1038–1062.

Cardenas, MB 2007. Potential contribution of topography-driven regional groundwater flow to fractal stream chemistry: residence time distribution analysis of Toth flow. *Geophysical Research Letters* 34, L05403. DOI:10.1029/2006GL029126.

Cardenas, MB and Wilson, JL 2007. Exchange across a sediment–water interface with ambient groundwater discharge. *Journal of Hydrology* 346, 69–80. DOI:10.1016/j.jhydrol.2007.08.019.

Cardenas, MB, Wilson, JF, and Zlotnik, VA 2004. Impact of heterogeneity, bed forms, and stream curvature on subchannel hyporheic exchange. *Water Resources Research* 40, W08307. DOI:10.1029/2004WR003008.

Carpenter, SR, Caraco, NF, Correll, DL *et al.* 1998. Nonpoint pollution of surface waters with phosphorus and nitrogen. *Ecological Applications* 8(3), 559–568.

Colombini, M, Seminara, G, and Tubino, M 1987. Finite-amplitude alternate bars. *Journal Fluid Mechanics* 181, 213–232.

Constantz, J and Thomas, CL 1997. Stream bed temperature profiles as indicators of percolation characteristics beneath arroyos in the middle Rio Grande basin, USA. *Hydrological Processes* 11, 1621–1634.

Constantz, J, Thomas, CL, and Zellweger, G 1994. Influence of diurnal variations in stream temperature on streamflow loss and groundwater recharge. *Water Resources Research* 30(12), 3253–3264.
Constantz, J, Stonestrom, D, Stewart, AE *et al.* 2001. Analysis of streambed temperatures in ephemeral channels to determine streamflow frequency and duration. *Water Resources Research* 37(2), 317–328.
Edwards, RT 1998. The hyporheic zone. In *River Ecology and Management: Lessons from the Pacific Coastal Ecoregion* (eds RJ Naiman and RE Bilby). Springer-Verlag: New York; 399–429.
Einstein, HA 1968. Deposition of suspended particles in a gravel bed. *Journal of Proceedings of the American Society of Civil Engineers Journal of Hydraulics Division* 94(5), 1197–1205.
Elliott, A and Brooks, NH 1997a. Transfer of nonsorbing solutes to a streambed with bed forms: laboratory experiments. *Water Resources Research* 33(1), 137–151.
Elliott, A and Brooks, NH 1997b. Transfer of nonsorbing solutes to a streambed with bed forms: theory. *Water Resources Research* 33(1), 123–136.
Erisman, JW, Sutton, MA, Galloway, JN *et al.* 2008. How a century of ammonia synthesis changed the world. *Nature Geosciences* 1(10). DOI:10.1038/ngeo325.
Fanelli, RM and Lautz, L 2008. Water, heat and solute fluxes through the hyporheic zone of small dams. *Ground Water* 46(5), 671–687.
Fogler, HS 1999. *Elements of Chemical Reaction Engineering*. Prentice –Hall: Englewood Cliffs, NJ.
Fox, A, Boano, F, and Amon, S 2014. Impact of losing and gaining streamflow conditions on hyporheic exchange fluxes induced by dune-shaped bed forms. *Water Resources Research* 50, 1895–1907. DOI:10.1002/2013WR014668.
Freeze, RA and Cherry, JA 1979. *Groundwater*, Prentice Hall: Englewood Cliffs, NJ.
Galloway, JN, Aber, JD, Erisman, JW *et al.* 2003. The nitrogen cascade. *Bioscience* 53(4), 341–356.
Galloway, JN, Dentener, FJ, Capone, DG *et al.* 2004. Nitrogen cycles: past, present, and future. *Biogeochemestry* 70, 153–226.
Galloway, JN, Townsend, AR, Erisman, JW *et al.* 2008. Transformation of the nitrogen cycle: recent trends, questions, and potential solutions. *Science* 320(5878), 889–892.
Garcia, M and Parker, G 1991. Entrainment of bed sediment into suspension. *Journal of Hydraulic Engineering* 117(4), 414–435.
Gariglio, F, Tonina, D, and Luce, CH 2013. Spatio-temporal variability of hyporheic exchange through a pool-riffle-pool sequence. *Water Resources Research* 49(11), 7185–7204. DOI:10.1002/wrcr.20419.
Gibert, J and Deharveng, L 2002. Subterranean ecosystems: a truncated functional biodiversity. *Bioscience* 52(6), 473–481.
Gibert, J, Stanford, JA, Dole-Olivier, M-J, and Ward, JV 1994. Basic attributes of groundwater ecosystems and prospects for research. In *Groundwater Ecology* (eds J Gilbert, DL Danielpol, and JA Standford) Academic Press: San Diego, CA; 8–40.
Gooseff, MN 2010. Defining hyporheic zones – advancing our conceptual and operational definitions of where stream water and groundwater meet. *Geography Compass* 4(8), 945–955. DOI:10.1111/j.1749-8198.2010.00364.x.
Gooseff, MN, Anderson, JK, Wondzell, SM *et al.* 2006. A modelling study of hyporheic exchange pattern and the sequence, size, and spacing of stream bedforms in mountain stream networks, Oregon, USA. *Hydrological Processes* 20(11), 2443–2457.
Hassan, MA, Tonina, D, Beckie, RD, and Kinnear, M 2014. The effects of discharge and slope on hyporheic flow in step-pool morphologies. *Hydrological Processes* 29(3), 419–433. DOI:10.1002/hyp.10155.
Henderson, A, Ahrens, J, and Law, C 2004. *The ParaView Guide*. Kitware Inc.: Clifton Park, NY.
Hendricks, SP 1993. Microbial ecology of hyporheic zone: a perspective integrating hydrology and biology. *Journal of the North American Benthological Society* 12(1), 70–78.

Hendricks, SP and White, DS 1991. Physico-chemical patterns within a hyporheic zone of a northern Michigan river, with comments on surface water patterns. *Canadian Journal of Fisheries and Aquatic Sciences* 48, 1645–1654.

Hester, ET, Young, KI, and Widdowson, MA 2013. Mixing of surface and groundwater induced by riverbed dunes: implications for hyporheic zone definitions and pollutant reactions. *Water Resources Research* 49, 5221–5237.

Higashino, M and Stefan, HG 2008. Mixing of surface and groundwater induced by riverbed dunes: implications for hyporheic zone definitions and pollutant reactions. *Journal of Hydraulic Research* 46(3), 291–300. DOI:10.3826/jhr.2008.2769.

Hynes, HBN 1970. *The Ecology of Running Waters*. University Press, Liverpool.

Kennedy, CD, Genereux, DP, Corbett, DR, and Mitasova, H 2009. Spatial and temporal dynamics of coupled groundwater and nitrogen fluxes through a streambed in an agricultural watershed. *Water Resources Research* 45, W09401. DOI:10.1029/2008WR007397.

Kirk, RE 2008, *Statistics: An Introduction*. Thomson Higher Education: Belmont, CA.

Malard, F, Tockner, K, Dole-Olivier, M-J, and Ward, JV 2002. A landscape perspective of surface–subsurface hydrological exchanges in river corridors. *Freshwater Biology* 47, 621–640.

Marion, A, Zaramella, M, and Bottacin-Busolin, A 2008. Solute transport in rivers with multiple storage zones: The STIR model. *Water Resources Research* 44, W10406. DOI:10.1029/2008WR007037.

Marion, A, Bellinello, M, Guymer, I, and Packman, AI 2002. Effect of bed form geometry on the penetration of nonreactive solutes into a streambed. *Water Resources Research* 38(10), 1209. DOI:10.1029/2001WR000264.

Marion, A, Packman, AI, Zaramella, M, and Bottacin-Busolin, A 2008. Hyporheic flows in stratified beds. *Water Resources Research* 44, W09433. DOI:10.1029/2007WR006079.

Marzadri, A, Tonina, D, and Bellin, A 2011. A semianalytical three-dimensional process-based model for hyporheic nitrogen dynamics in gravel bed rivers. *Water Resources Research* 47, W11518. DOI:10.1029/2011WR010583.

Marzadri, A, Tonina, D, and Bellin, A 2012a. Effects of hyporheic fluxes on streambed pore water temperature. In *10th International Conference on Hydroinformatics HIC 2012*, IAHR, Hamburg, Germany.

Marzadri, A, Tonina, D, and Bellin, A 2012b. Morphodynamic controls on redox conditions and on nitrogen dynamics within the hyporheic zone: application to gravel bed rivers with alternate-bar morphology. *Journal of Geophysical Research* 117. DOI:10.1029/2012JG001966.

Marzadri, A, Tonina, D, and Bellin, A 2013a. Effects of stream morphodynamics on hyporheic zone thermal regime. *Water Resources Research* 49(4), 2287–2302. DOI:10.1002/wrcr.20199.

Marzadri, A, Tonina, D, and Bellin, A 2013b. Quantifying the importance of daily stream water temperature fluctuations on the hyporheic thermal regime: Implication for dissolved oxygen dynamics. *Journal of Hydrology* 507, 241–248. DOI:10.1016/j.jhydrol.2013.10.030.

Marzadri, A, Tonina, D, Bellin, A, Vignoli, G, and Tubino, M 2010. Effects of bar topography on hyporheic flow in gravel-bed rivers. *Water Resources Research* 46, p. W07531. DOI:10.1029/2009WR008285.

Marzadri, A, Tonina, D, Bellin, A, and Tank, JL 2014a. A hydrologic model demonstrates nitrous oxide emissions depend on streambed morphology. *Geophysical Research Letters* 41(15), 5484–5491. DOI:10.1002/2014GL060732.

Marzadri, A, Tonina, D, McKean, JA, Tiedemann, MG, and Benjankar, R 2014b. Multi-scale streambed topographic and discharge effects on hyporheic exchange at the stream network scale in confined stream. *Journal of Hydrology* 519, 1997–2011. DOI:10.1016/j.jhydrol.2014.09.076.

Marzadri, A, Tonina, D, Bellin, A, and Valli, A 2015. Mixing interfaces, fluxes, residence times and redox conditions of the hyporheic zones induced by dune-like bedforms and groundwater flows. *Advances in Water Resources*. DOI:10.1016/j.advwatres.2015.12.014.

Master, Y, Shavit, U, and Shaviv, A 2005. Modified isotope pairing technique to study N transformations in polluted aquatic systems: theory. *Environmental Science and Technology* 39, 1749–1756.

Mulholland, PJ and DeAngelis, DL 2000. Surface–subsurface exchange and nutrient spiraling. In *Streams and Ground Waters* (eds JB Jones and PJ Mulholland). Academic Press: San Diego, CA; 149–168.

Mulholland, PJ, Helton, AM, Poole, GC *et al.* 2008. Stream denitrification across biomes and its response to anthropogenic nitrate loading. *Nature* 452(13).

Orghidan, T 1959. Ein neuer lebensraum des unterirdischen wassers: Der hyporheische biotop. *Archiv für Hydrobiologie* 55, 392–414.

Packman, AI and Brooks, NH 2001. Hyporheic exchange of solutes and colloids with moving bed forms. *Water Resources Research* 37(10), 2591–2605.

Packman, AI and MacKay, J 2003. Interplay of stream-subsurface exchange, clay particle deposition, and streambed evolution. *Water Resources Research* 39(4). DOI:10.1029/2002WR001432.

Payn, RA, Gooseff, MN, McGlynn, BL *et al.* 2009. Channel water balance and exchange with subsurface flow along a mountain headwater stream in Montana, United States. *Water Resources Research* 45, W11427. DOI:10.1029/2008WR007644.

Ren, J and Packman, AI 2004a. Modeling of simultaneous exchange of colloids and sorbing contaminants between streams and streambeds. *Environmental Science and Technology* 38(10), 2901–2911. DOI: 10.1021/es034852l.

Ren, J and Packman, AI 2004b. Stream–subsurface exchange of zinc in the presence of silica and kaolinite colloids. *Environmental Science and Technology* 38(24), 6571–6581. DOI:10.1021/es035090x.

Ren, J and Packman, AI 2007. Changes in fine sediment size distributions due to interactions with streambed sediments. *Sedimentary Geology* 202, 529–537. DOI:10.1016/j.sedgeo.2007.03.021.

Rosamond, MS, Thuss, SJ, and Schiff, SL 2012. Dependence on riverine nitrous oxide emissions on dissolved oxygen levels. *Nature Geosciences* 5, 715–718.

Rutherford, JC 1994. *River Mixing*. John Wiley & Sons, Chichester, UK.

Rutherford, JC, Boyle, JD, Elliott, AH *et al.* 1995. Modeling benthic oxygen uptake by pumping. *Journal of Environmental Engineering* 121(1), Paper No. 7216.

Salehin, M, Packman, AI, and Paradis, M 2004. Hyporheic exchange with heterogeneous streambeds: laboratory experiments and modeling. *Water Resources Research* 40(11). DOI:10.1029/2003WR002567.

Schälchli, U 1995. Basic equations for siltation of riverbeds. *Journal of Hydraulic Engineering* 121(3), 274–287. DOI:10.1061/(ASCE)1234–5678(1995)121:3(274).

Shen, HV, Fehlman, HM, and Mendoza, C 1990. Bed form resistances in open channel flows. *Journal of Hydraulic Engineering* 116(6), 799–815.

Sjodin, AL, Lewis, WM Jr., and Saunders, JF III 1997. Denitrification as a component of the nitrogen budget for a large plains river. *Biogeochemistry* 39, 327–342.

Stanford, JA and Ward, JV 1988. The hyporheic habitat of river ecosystems. *Nature* 335, 64–65.

Stanford, JA and Ward, JV 1993. An ecosystem perspective of alluvial rivers: connectivity and the hyporheic corridor. *Journal of the North American Benthological Society* 12(1), 48–60.

Stonedahl, SH, Harvey, JW, and Packman, AI 2013. Interaction between hyporheic flow produced by stream meanders, bars and dunes. *Water Resources Research* 49, 5450–5461. DOI:10.1002/wrcr.20400.

Stonedahl, SH, Harvey, JW, Detty, J *et al.* 2012. Physical controls and predictability of stream hyporheic flow evaluated with a multiscale model. *Water Resources Research* 48, W10513. DOI:10.1029/2011WR011582.

Storey, RG, Williams, DD, and Fulthorpe, RR 2004. Nitrogen processing in the hyporheic zone of a pastoral stream. *Biogeochemestry* 69, 285–313.

Stuart, TA 1953. Water currents through permeable gravel and their significance to spawning salmonids, etc. *Nature* 172, 407–408.

Thibodeaux, LJ and Boyle, JD 1987. Bedform-generated convective transport in bottom sediment. *Nature* 325(22), 341–343.

Tonina, D 2012. Surface water and streambed sediment interaction: the hyporheic exchange. In *Fluid Mechanics of Environmental Interfaces* (eds C Gualtieri and DT Mihailović). CRC Press, Taylor & Francis Group: London; 255–294.

Tonina, D and Bellin, A 2007. Effects of pore-scale dispersion, degree of heterogeneity, sampling size, and source volume on the concentration moments of conservative solutes in heterogeneous formations. *Advances in Water Resources*. DOI:10.1016/j.advwatres.2007.08.009.

Tonina, D and Buffington, JM 2007. Hyporheic exchange in gravel-bed rivers with pool-riffle morphology: laboratory experiments and three-dimensional modeling. *Water Resources Research* 43, W01421.

Tonina, D and Buffington, JM 2009a. Hyporheic exchange in mountain rivers I: Mechanics and environmental effects.*Geography Compass* 3(3), 1063–1086.

Tonina, D and Buffington, JM 2009b. A three-dimensional model for analyzing the effects of salmon redds on hyporheic exchange and egg pocket habitat. *Canadian Journal of Fisheries and Aquatic Sciences* 66, 2157–2173. DOI:10.1139/F09-146.

Tonina, D and Buffington, JM 2011. Effects of stream discharge, alluvial depth and bar amplitude on hyporheic flow in pool-riffle channels. *Water Resources Research* 47, W08508. DOI:10.1029/2010WR009140.

Tonina, D, Marzadri, A, and Bellin, A 2015. Benthic uptake rate due to hyporheic exchange: the effects of streambed morphology for constant and sinusoidally varying nutrient loads. *Water* 7(2), 398–419. DOI:10.3390/w7020398.

Tonina, D, De Barros, FPJ, Marzadri, A and Bellin A 2016. Does streambed heterogeneity matter for hyporheic residence time distribution in sand-bedded streams? *Advances in Water Resources*, 96, 120–126, doi: j.advwatres.2016.07.009.

Trauth, N, Schimdt, C, Maier, U *et al.* 2013. Coupled three-dimensional stream flow and hyporheic flow model under varying stream and ambient groundwater flow conditions in pool-riffle system. *Water Resources Research* 49(9. DOI:10.1002/wrcr.20442.

Triska, FJ, Duff, JH, and Avanzino, RJ 1993. Patterns of hydrological exchange and nutrient transformation in the hyporheic zone of a gravel bottom stream: examining terrestrial-aquatic linkages. *Freshwater Biology* 29, 259–274.

Triska, FJ, Duff, JH, and Avanzino, RJ 1993. The role of water exchange between a stream channel and its hyporheic zone in nitrogen cycling at the terrestrial-aquatic interface. *Hydrobiologia* 251, 167–184.

Triska, FJ, Kennedy, VC, Avanzino, RJ *et al.* 1989. Retention and transport of nutrients in a third-order stream in Northwestern California: hyporheic processes. *Ecology* 70(6), 1893–1905.

Vaux, WG 1962. *Interchange of Stream and Intragravel Water in a Salmon Spawning Riffle.* Fisheries No 405, US Fish and Wildlife Service, Bureau of Commercial Fisheries: Washington, DC.

Vaux, WG 1968. Intragravel flow and interchange of water in a streambed. *Fishery Bulletin* 66(3), 479–489.

Wagenschein, D and Rode, M 2008. Modelling the impact of river morphology on nitrogen retention – a case study of the Weisse Elster River (Germany). *Ecological Modelling* 211(1–2), 224–232. DOI:10.1016/j.ecolmodel.2007.09.009.

Williams, DD and Hynes, HBN 1974. The occurrence of benthos deep in the substratum of stream. *Freshwater Biology* 4, 233–256.

Wondzell, SM and Swanson, FJ 1996. Seasonal and storm dynamics of the hyporheic zone of 4-th order mountain stream. 2: nitrogen cycling. *North America Benthological Society* 15, 20–34.
Wörman, A, Packman, AI, Johansson, H, and Jonsson, K 2002. Effect of flow-induced exchange in hyporheic zones on longitudinal transport of solutes in streams and rivers. *Water Resources Research* 38(1), 15. DOI:10.1029/2001WR000769.
Zarnetske, JP, Haggerty, R, Wondzell, SM, and Baker, MA 2011. Dynamics of nitrate production and removal as a function of residence time in the hyporheic zone. *Journal of Geophysical Research* 116, G01025. DOI:10.1029/2010JG001356.
Zarnetske, JP, Haggerty, R, Wondzell, SM *et al.* 2012. Coupled transport and reaction kinetics control the nitrate source-sink function of hyporheic zones. *Water Resources Research* 48, W11508. DOI:10.1029/2012WR011894.

Discussion

Discussion by Guido Zolezzi

Many results on the role of channel morphology on hyporheic flows and related geochemical processes presented by Tonina *et al.* refer to relatively high flow conditions, i.e., conditions under which the channel bathymetry is fully wet or might even be partially transporting gravel in the deepest and fastest flow regions. From an ecological point of view, most field studies that highlight the importance of hydrological and solute exchanges between surface water and the hyporheic region stress the relevance of such processes and provide information on its characteristics during low-flow conditions, with partially dry bar surfaces. In many river systems, these are the most durable flow conditions and have fundamental effects for the biota. Moving from these considerations, I would like the authors to discuss the following points:

1) To what extent can results on hyporheic processes obtained under relatively high flow conditions inform on the characteristics of the same processes at much lower flow stages?
2) How relevant from an ecological point of view are hyporheic exchanges during flow conditions under which gravel transport and channel morphological changes occur?

Discussion by David Furbish

To what extent does the formulation take into account the relatively long travel times of hyporheic flow associated with deeper flow paths in thick alluvium systems or within the third dimension, for example, involving channel curvature (e.g., meander bends)? To what extent is it actually necessary to take these long times into account, inasmuch as they represent a small proportion of the distribution of travel times?

Response by the Authors

Whereas the solutions for dune-like bedforms were derived at flows with low if any sediment transport, those for pool–riffle bedforms were derived at the formation discharge, which is typically near bankfull flows in most unregulated streams (Tubino 1991). These flows are rare, once a year or less in most systems (Goodwin 2004), but they may be quite common and persist for several months in meadow systems (McKean and Tonina 2013), which may even show static pool–riffles (Maturana *et al.* 2013). Low flows are more frequent than high flows and potentially more ecologically important. Thus, we tested the robustness of our solutions for discharges smaller than the forming discharge (Marzadri *et al.* 2010). Our results show that our solutions are also applicable at low-flow conditions with small errors as long as the bedforms are fully submerged. In case of partially submerged bedforms, the empirical equations proposed in the work of Tonina and Buffington (2011) could be used. In the case of bedform migration the turnover mechanisms could be used to adjust the hyporheic residence time and fluxes (Elliott and Brooks 1997a; Packman and Brooks 2001).

Ahmerkamp *et al.* (2015) suggest that bedform migration has a significant impact on porewater transport and benthic fluxes that in turn influence both oxygen uptake and denitrification, especially in small bedforms such as dunes or ripples. Hyporheic processes often occur simultaneously (i.e., pumping and turnover), and their relative importance on the overall exchange depend on both flow regime and stream conditions. Here, we analyze the effect of the pumping process, which is considered one of the most important mechanisms in controlling the interaction between surface and subsurface environments during the most persistent flow regime.

Hyporheic exchange occurs at different length scales, e.g., fluvial, parafluvial, and floodplain hyporheic zone and consequently time scales (Tonina 2012). Even within the same topographical features, e.g., a dune or a pool–riffle bedform, we do have a distribution of residence times. These probability density functions are skewed toward short residence times (Haggerty, Wondzell and Johnson 2002; Wörman *et al.* 2002). Our model uses the median residence time, τ_{50}, which is a robust moment even under uncertain long-tail values due to boundary conditions such as alluvium thickness and groundwater table, as indexes for hyporheic processes. We also selected the dominant morphology such that the exchange occurring between meanders is neglected in sinuous pool–riffle streams because the dominant topography is expected to mask the effect of single features at the large reach scale. Although these nondominant features may influence the tails of the residence-time distribution, their effect on τ_{50} is expected to be small. Tonina and Buffington (2009b) observed the small relevance of local features in the case of a single redd, which have significant local effects but negligible at the large reach scale.

References

Ahmerkamp, S, Winter, C, Janssen, F *et al.* 2015. The impact of bedform migration on benthic oxygen fluxes. *Journal of Geophysical Research: Biogeoscience* 120.

Goodwin, P 2004. Analytical solutions for estimating effective discharge. *Journal of Hydraulic Engineering* 130(8).

Haggerty, R, Wondzell, SM, and Johnson, MA 2002. Power-law residence time distribution in hyporheic zome of a 2nd-order mountain stream. *Geophysical Research Letters* 29(13), 4.

Maturana, O, Tonina, D, Mckean, JA *et al.* 2013. Modeling the effects of pulsed versus chronic sand inputs on salmonid spawning habitat in a low-gradient gravel-bed river. *Earth Surface Processes and Landforms* 39(7), 877–889.

McKean, JA and Tonina, D 2013. Bed stability in unconfined gravel-bed mountain streams: with implications for salmon spawning viability in future climates. *Journal of Geophysical Research: Earth Surface* 118, 1–14.

Tubino, M 1991. Growth of alternate bars in unsteady flow. *Water Resources Research* 27(1), 37–52.

10

Ecological Effects of Flow Intermittence in Gravel-Bed Rivers

Thibault Datry

10.1 Introduction

Rivers with intermittent flow regimes, those that do not flow permanently, are more abundant and widespread than previously thought, and may even make up more than half the global river network. Due to the nature of their streambed and the intensive hyporheic exchanges they experience, gravel-bed rivers (GBRs) are often temporally or spatially intermittent. Yet, this aspect of their flow regimes has been relatively overlooked. In this chapter I will use the recent developments of intermittent river ecology, a maturing and expanding discipline, to review various aspects of flow intermittence in GBRs. When possible, the specificities of GBRs and how these can make them different from other intermittent rivers will be highlighted. A first section will focus on defining terms and examine the quantitative importance and the distribution of these rivers. In a second section, I will use an eco-hydrological perspective at different spatial scales to illustrate the ecological implications of flow intermittence in GBRs. In a third section, I will expand on two aspects which, in my opinion, are really exciting and promising to explore in intermittent GBRs: the concept of disturbance ecology and the interaction between aquatic and terrestrial components in these systems. I will finish this review by identifying some of the current research needs and perspectives. Two important points deserve to be mentioned here. First, because I have been working in various gravel-bed rivers that are flowing intermittently, many examples and results shown in this review are from my past and current research, whether it has been published or not. Obviously, these are examples among various others, since this research field is really blooming. Second, due to my expertise, many examples will concern aquatic and terrestrial invertebrate and fish communities, but clearly, other groups of organisms (e.g., plant, microbial communities) have been studied, as well as the effects of flow intermittence on ecosystem processes, which I will only briefly mention here.

10.2 Flow Intermittence in GBRs from a Hydrological Perspective

10.2.1 What is Flow Intermittence?

Flow intermittence occurs when the flow of water in a river or a stream is not continuous (i.e., perennial, permanent) but ceases for a certain period. Flow intermittence comprises *flow cessation*

Gravel-Bed Rivers: Processes and Disasters, First Edition. Edited by Daizo Tsutsumi and Jonathan B. Laronne.

periods, where flow ceases but surface water is still present in disconnected pools (Figure 10.1C,D), and *dry* periods, where surface water disappears completely from the river channel(s) (Figure 10.1E,F). *Flow cessation* periods are preceded by low-flow periods, characterized by a lateral contraction of the flowing channel(s) leading to the emersion of the once-submerged lateral bars, littoral and riparian zones (Figure 10.1B; Boulton 2003). During *dry* periods, the sediments beneath and adjacent to the streambed (also called the hyporheic zone, White 1993) can be saturated with water, contain some water but be unsaturated, or be completely dry. Water-saturated streambed sediments may or may not be in contact (i.e., perched systems) with the underlying groundwater aquifer (Malard, Ward, and Robinson 2000).

Flow intermittence refers to any period or situation described above, but generally these periods occur sequentially over time: discharge decreases, flow ceases, disconnected pools are lost, and the benthic then hyporheic sediments dry (Figure 10.1A–F). The sequence is reversed with flow resumption in the river channel(s), but not necessarily passing through all the previous steps, as rewetting is generally more sudden (Figure 10.1G; Corti and Datry 2012).

Figure 10.1 Illustration of flow intermittence in a GBR showing a typical sequence of flow changes, from (A) bankfull, (B) low flow, (C, D) flow cessation, (E, F) dry, and (G) rewetting periods. All photographs are from the Albarine River, in France. (Adapted from Datry, Larned, and Tockner (2014).)

Flow intermittence is a defining feature of *intermittent rivers*, which have surface water that ceases to flow at some point(s) in space and time along their course, thus including all waterways that do not flow perennially/permanently (Acuña *et al.* 2014). This generic term includes many scientific appellations, such as temporary, seasonal, episodic, interrupted, or non-perennial streams and rivers, as well as more than 40 vernacular words used in various countries to refer to intermittent rivers (i.e., see table 2 in Steward *et al.* 2012). Although not all of these rivers are GBRs, due to the nature of their streambed, many GBRs are intermittent (Arscott, Tockner, and Ward 2005; Larned *et al.* 2011; Datry *et al.* 2014a, see below).

10.2.2 Intermittent Rivers: Importance and Distribution

10.2.2.1 Intermittent Rivers are Widespread and Abundant

Intermittent rivers are not limited to arid regions, and occur on all continents, including Antarctica (Dodds 1997; Larned *et al.* 2010; Steward *et al.* 2012). In fact, several examples at regional or continental scales suggest that intermittent rivers are likely to comprise at least half the global river network (Acuña *et al.* 2014; Datry, Larned, and Tockner 2014), although a large-scale and precise estimate of total length and discharge of intermittent rivers is still lacking.

Intermittent rivers are dominating features of hyper-arid, arid, semi-arid, and dry-subhumid regions, which represent a third to half of Earth's land surfaces (Tooth 2000; Whitford 2002; Millennium Ecosystem Assessment 2005). For example, more than 70% of the 3.5 million kilometres of river channels measured at the 1:250 000 scale are considered intermittent in Australia (Sheldon *et al.* 2010), along with 66–94% of the river length in the Southwestern United States (i.e., Arizona, New Mexico, Utah, Nevada, Colorado, and California; Levick *et al.* 2008). Intermittent rivers also make up the majority of rivers and streams in the Mediterranean regions (Sabater and Tockner 2010; Bonada and Resh 2013).

Intermittent rivers are also abundant in temperate, humid, boreal, alpine, and polar areas, although they have been much less studied in these regions (Leigh *et al.* 2016). Across all of the USA, using the National Hydrography Dataset (i.e., NHD, 1:100 000), it has been estimated that 59% of rivers and streams are intermittent (Nadeau and Rains 2007). Intermittent rivers have recently been reported to be common in Canada (Buttle *et al.* 2012), and in France, where 25–40% of the total river length is intermittent (Snelder *et al.* 2013; Figure 10.2). Many alpine, boreal, and polar rivers are intermittent (McKnight *et al.* 1999; Malard *et al.* 2003; Rüegg and Robinson 2004; Stanish *et al.* 2012; Datry, Larned, and Tockner 2014). As a detailed example, the western and very humid part of France includes many intermittent rivers (Figure 10.2); during summer, as much as 80% of some river catchments in this region dry or cease to flow (Datry *et al.* 2016).

10.2.2.2 Most Low-Order Streams are Intermittent

First-order streams absent from most topographic maps make up > 70% of river network lengths, and are typically intermittent (Meyer *et al.* 2001; Lowe and Likens 2005; Fritz *et al.* 2013; Lehnerand Grill 2013; Datry, Larned, and Tockner 2014). The previous estimates are therefore probably low because they are based on national hydrological networks (e.g., the NHD of the United States), generally derived from topographic maps, which do not include first-order streams (i.e., headwater streams), which are most prone to flow intermittence. For example, the US NHD does not reflect the total extent of intermittent rivers because it does not include stream segments less than 1 mile (~1.6 km) in length and is based on 1:100 000 scale topographic maps (Hansen 2001; Colson *et al.* 2008; Benstead and Leigh 2012; Fritz *et al.* 2013). Headwater streams are very difficult to detect by traditional mapping techniques, including airborne photography and satellite images (Benstead and Leigh 2012). For

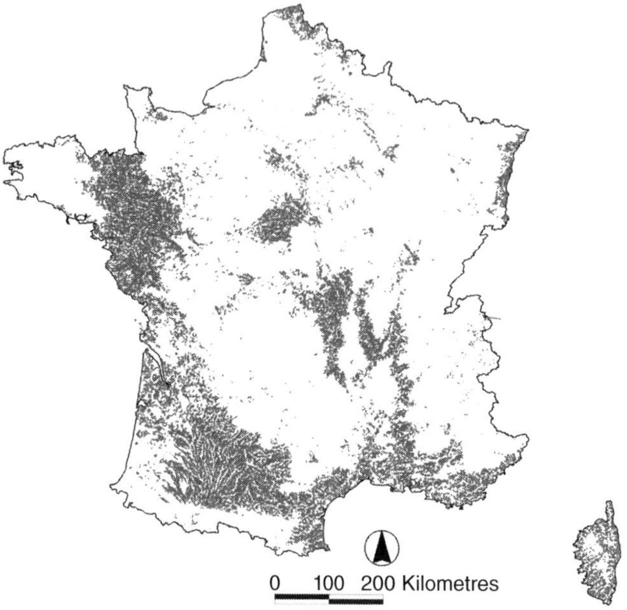

Figure 10.2 Distribution of river segments in France that are predicted to be intermittent based on an environmental classification of flow regimes at 628 sites. (After Snelder *et al.* 2013.)

example, in the Upper Little Tennessee River in the southeastern United States, the total stream channel length is threefold greater than indicated by the NHD (Benstead and Leigh 2012). When accounting for these headwater streams, the overall proportion of intermittent rivers on Earth may even exceed that of perennial ones.

That most low-order streams are prone to flow intermittence is illustrated in a recent, global estimate of the total stream surface area. Using the US NHD and extrapolation across the COSCAT (costal segmentation and related catchment) network segmentation framework (Meybeck, Dürr, and Vörösmarty 2006), global stream and river surface areas were estimated to be 624 000 km^2 and 84 000 km^2, of which ~14% was estimated to be intermittent (Raymond *et al.* 2013). This indicates that although intermittent rivers contribute greatly to river length and discharge, their importance is less in terms of surface area, which reflects the fact that flow intermittence is more likely to occur in stream orders lower than 6. For the COSCATs below 60°N, Raymond *et al.* (2013) predicted that 69, 56, 49, 42, and 34% of stream orders 1–5 were intermittent, all climatic regions considered.

10.2.2.3 Many GBRs are Intermittent

Many GBRs are spatially and/or temporally intermittent due to the nature of their streambed, which is very porous and the strong surface–groundwater interactions they experience (Doering *et al.* 2007; Vorste *et al.* 2016). Since large-scale quantification and classification of intermittent rivers are scarce, defining how many GBRs are intermittent and locating where they are is a difficult task. Nevertheless, many of the studies on intermittent rivers were carried out on GBRs in recent decades. These rivers include the Selwyn River in the Canterbury Plains of the South Island of New Zealand (Larned *et al.* 2008), the Tagliamento River in northern Italy (Tockner *et al.* 2003; Doering *et al.* 2007), the Albarine River in eastern France (Datry 2012) and many others (e.g., Sager 1983; Fowler 2004; Konrad 2006;

Table 10.1 Number of daily flow-gauging stations with >30 years of records, starting after 1970, for four countries. Stations with intermittent flow were identified as showing more than five consecutive days with no flow per annum.

	Australia	France	Spain	United States
Total	424	1030	118	112
Perennial	150	966	90	23
Intermittent	274	64	28	89
% intermittent	65	7	24	79

Source: after Hoeve and De Vries (2014).

Izbicki 2007). In a recent large-scale programme examining geomorphological and ecological dynamics of GBRs in southeastern France (Piégay *et al.* 2009; Belletti *et al.* 2013, 2014; Capderrey *et al.* 2013; Vorste *et al.* 2016), many reaches were found to be intermittent due to large-scale downwelling of surface water into the underlying aquifer (see page 273).

10.2.2.4 Flow Gauging Networks and Intermittent Rivers

Another approach used to examine the large-scale distribution of intermittent rivers, including intermittent GBRs, involves exploring flow-gauge time series at national or global scales, many of which are available online (e.g., Bundesanstalt für Gewässerkunde, www.bafg.de; US Geological Service, http://waterdata.usgs.gov/nwis). Unfortunately, because intermittent rivers are of limited economic importance, they are rarely gauged, and the position of flow-gauging stations across river networks is strongly biased towards perennial rivers or reaches (e.g., Datry, Arscott, and Sabater 2011; Snelder *et al.* 2013; Prat *et al.* 2014). For example, in Australia there is one streamflow gauging station per 350 000 km^2 in the arid zone compared with one per 3200 km^2 in humid regions (McMahon 1979). In spite of this strong bias, exploring these data also provides an indication of the abundance of intermittent rivers. In four countries for which flow-gauge data were available (i.e., Australia, France, United States, and Spain), daily flow time series with more than 30 years of continuous records and starting after 1970 (to ensure data reliability) were selected (Hoeve and De Vries 2014). Time series with intermittent flow were then identified as those showing more than five consecutive days of no-flow annually, this criterion being used to disregard false '0' due to radar or logger failure, for example. Using this approach, between 7 and 79% of the flow gauging stations had intermittent flow, which is high given the distribution bias these stations have towards larger streams (Hoeve and De Vries 2014, Table 10.1). Intermittent rivers can occur everywhere and are not restricted to particular regions or areas (Figure 10.3).

10.2.3 Origins, Causes of Flow Intermittence, and Trends

Flow intermittence can have natural and non-natural causes. Below are listed the various processes which can lead to flow intermittence. However, often, multiple causes interact and disentangling the natural and non-natural causes of flow intermittence is rarely easy.

10.2.3.1 Natural Causes

Different processes generate natural flow intermittence in streams and rivers, acting individually or in combination: transmission loss (infiltration of surface water into porous streambeds),

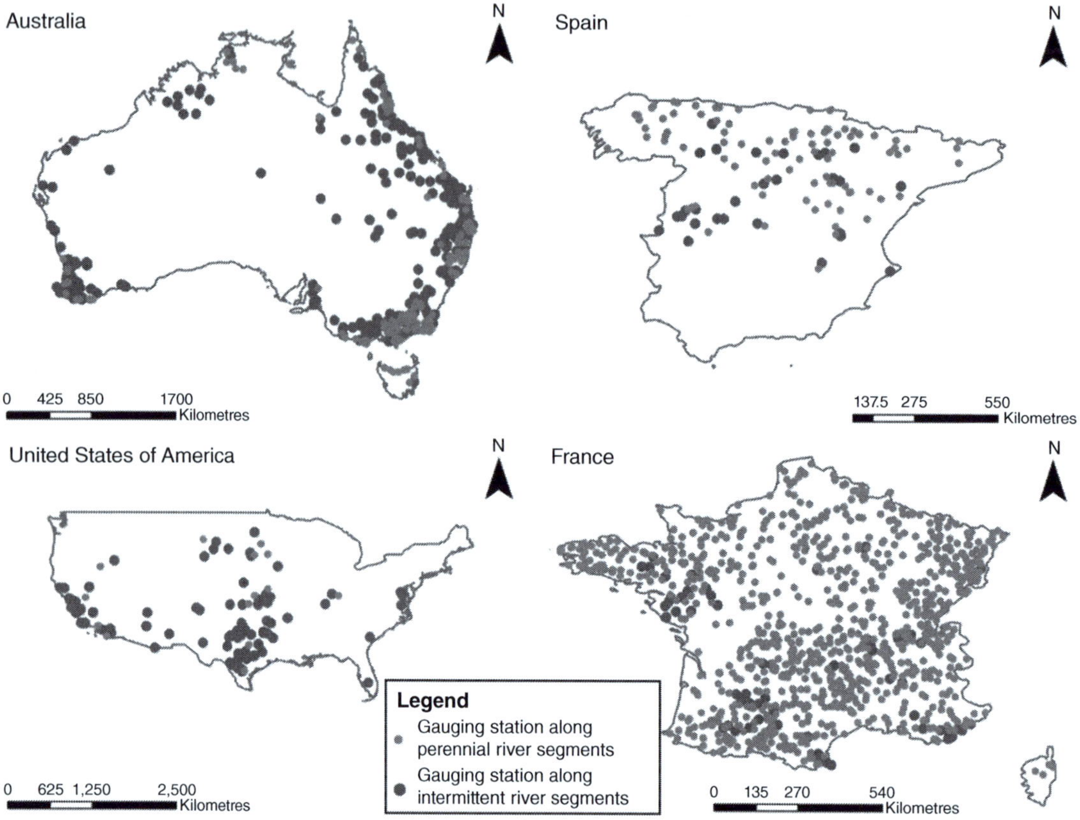

Figure 10.3 Classification of selected gauging stations across Australia, France, United States, and Spain. (Modified after Hoeve and De Vries (2014).) (See insert for color representation of this figure.)

evapotranspiration, downward shifts in groundwater tables, hillslope runoff recession, and freeze-up (Larned *et al.* 2010; Table 10.2). To date GBRs being intermittent due to freeze-up, soil freezing, water recession, and evapotranspiration have received least attention, while those in which intermittence is driven by transmission losses are among the most studied (Table 10.2 and Figure 10.4).

10.2.3.2 Non-natural Causes

Non-natural flow intermittence can be due to one or more of the following human activities: alteration of land-use patterns, flow regulation, surface- or groundwater extraction, and reduced precipitation and increased evaporation resulting from climate change (Palmer *et al.* 2008; Steward *et al.* 2012; Figure 10.5). The Colorado (United States) and Yellow (China) Rivers are notable examples of how human activities can induce intermittence in large perennial rivers. The effects of climate change are likely more subtle, but to date and to our knowledge, there is no published evidence that climate change alone has driven a perennial river to become intermittent; there is little doubt that this will change in the coming years.

Table 10.2 Types of naturally intermittent rivers and processes controlling flow intermittence.

Type of intermittent river*	Processes causing flow intermittence	Example	Reference
Snowmelt and glacial meltwater streams	Cessation of melting/ ablation, freezing of stream surface, and shallow subsurface water	Dry Valley streams, Antarctica; Val Roseg, and Macun catchments, Switzerland	Bronge (1996), McKnight *et al.* (1999), Tockner and Malard (2003), Conovitz, MacDonald, and McKnight (2006), Robinson and Matthaei (2007), Stanish *et al.* (2012)
Perched and semi-perched alluvial rivers	Transmission losses, depletion of bank storage, and floodplain aquifer	Albarine River, France; Selwyn River, New Zealand; Tagliamento River, Italy; Cooper Creek, Australia, Kuiseb River, Namibia; Mohave River, United States; Shashane and Wenlock Rivers, Zimbabwe	Jacobson *et al.* (2000), Hamilton and Gehrke (2005), Lange (2005), Mansell and Hussey (2005), Konrad (2006), Doering *et al.* (2007), Izbicki (2007), Larned *et al.* (2008), Datry, Corti, and Philippe (2012)
Non-perched rivers in arid and semi-arid regions	Depletion of surface water and shallow groundwater by direct evaporation and evapotranspiration	Sycamore Creek, United States; Oued Chelif, Algeria	Stanley, Fisher, and Grimm (1997), Arab *et al.* (2004)
Zero-order and headwater streams	Cessation of overland flow, depletion of saturated soil water or hillslope aquifer, macropore recession	Maybeso Creek, United States; Riera de Fuirosos, Spain; Stillman Creek watershed, United States	Butturini *et al.* (2002), Gomi, Sidle, and Richardson (2002), Hunter, Quinn, and Hayes (2005)
Permafrost streams	Soil freezing, soil water and wetland recession	Granger Creek and Wolf Creek, Canada	Metcalfe and Buttle (2001), Carey and Quinton (2005)
Karstic streams	Transmission loss, cessation of spring discharge	River Lathkill and River Wye tributaries, United Kingdom; Coulazou River, France; Alme River, Germany	Meyer qnd Meyer (2000), Wood *et al.* (2005), Jourde *et al.* (2007)
Lake outlets	Lake level drops below outlet elevation	Shadow Lake, Canada	Mielko and Woo (2006)

Source: modified after Larned *et al.* (2010).

*These types are not mutually exclusive; for example, alluvial rivers in desert regions are often perched or semi-perched, and water loss is jointly controlled by transmission loss and evaporation (Knighton and Nanson 1994; Jacobson *et al.* 2000).

10.2.3.3 Temporal Trends

Evidence indicates that the total number and length of intermittent rivers and intermittent GBRs are increasing due to anthropogenic causes. Water abstraction and impoundment have caused many formerly perennial rivers to become intermittent in the past 50 years, including large rivers such as the Nile, Indus, Yellow, Amu and Syr Darya, Rio Grande and Colorado (Postel 2000; Meybeck 2003). Most of the once-perennial rivers in arid and semi-arid regions are now intermittent (Gleick 2003). Indeed, the once-perennial Colorado River, which flows 1450 miles (2334 km) from its headwaters in the Rocky Mountains of Colorado to the Sea of Cortez (Gulf of California) in northwestern Mexico, has flowed intermittently in most years since the completion of the Glen Canyon Dam in 1963

Figure 10.4 The Asse River, a naturally intermittent GBR from Provence (France). (Photograph: B. Dumont.)

Figure 10.5 The Jankho Kala River, an artificial intermittent GBR from central Bolivia. (Photograph: T. Datry.)

(Gleick 2003). In March 2014, water in the Colorado River reached its final, natural destination, for the first time in 16 years thanks to agreement between the United States and Mexico (Witze 2014). Another well-documented example of a large river becoming progressively more intermittent due to dams and water abstractions is the Yellow River, China (Changming and Shifeng 2002; Fu *et al.* 2004). Today, this 5464 km long river (which was perennial until 1972) is intermittent, connects to the sea only infrequently, and has an extended drying segment which exceeds 500 km (Changming and Shifeng 2002). There is a direct relationship between this flow regime shift and the construction of 12 major dams, along with a dramatic increase in water abstraction from 7.4 10^6 m^3 in 1949 to >30 million m^3 in 1988, for human and agricultural uses (Changming and Shifeng 2002). Analyses of discharge trends in the Yellow River indicate that the situation is worsening each year (Changming and Shifeng 2002).

In the near future, the number of intermittent GBRs will increase in regions where severe climatic drying and/or water appropriation occurs (Larned *et al.* 2010; Döll and Schmied 2012; Cipriani *et al.* 2014). This will include many of the presently perennial rivers, whose flow regimes will become intermittent in all regions affected by drying trends. Globally, perennial to intermittent flow regime shifts are projected until the 2050s for northeastern Australia, Brazil, California, the Caribbean, southern Africa, western Africa, and around the Mediterranean Sea (Döll and Schmied 2012). Conversely, due to a warmer climate, some naturally intermittent GBRs will likely become perennial in Siberia, part of Canada and Alaska due to warmer winters (Döll and Schmied 2012).

Naturally intermittent GBRs are also at risk of experiencing longer and more frequent flow cessation and dry periods. For example, the intermittent Albarine River, a GBR in eastern France (Datry, Corti, and Philippe 2012), a hydrological modelling (using various rainfall–runoff models) coupled to climate change scenarios indicated that annual flow intermittence will increase on average (all climate change scenarios considered) by 5% by 2050 (Cipriani *et al.* 2014). Due to decreased annual precipitation (on average, 6%) and increased summer temperatures (on average, by 2 °C), dry periods will last 2–4 days longer than in the most recent decade (Cipriani *et al.* 2014). Similarly, at the intermittent flow-gauging stations selected across Australia, France, Spain and USA (see above), trends towards greater flow intermittence were detected for 64 records during the period 1970–2013, indicating that the number and frequency of both flow cessation and dry periods are increasing (Hoeve and De Vries 2014; Table 10.3). Conversely, 21 time series showed a trend towards less flow intermittence (negative slope), and 370 had no detectable trend (Hoeve and De Vries 2014).

It is worth noting that, contrary to the extension of non-natural flow intermittence described above, some naturally intermittent rivers and intermittent GBRs have become perennial due to controlled

Table 10.3 Number of time series (with intermittent flow), per cent significant trends (using Mann–Kendall correlation analyses adjusted for autocorrelation when necessary), and number of positive and negative slopes for annual duration of no-flow periods in four countries.

	Number of time series	Significant trend (%)	Number of positive slopes	Number of negative slopes
Australia	274	27	49	14
France	64	5	2	1
Spain	28	21	4	2
United States	89	15	9	4

releases from dams and weirs, discharge of agricultural, industrial, and municipal effluents, and inter-basin transfers (Hassan and Egozi 2001; Steward *et al.* 2012; Datry, Larned, and Tockner 2014).

10.3 Flow Intermittence in GBRs: an Ecohydrological Perspective

Flow intermittence is an extreme form of flow variability, which is the most important determinant of patterns and processes in river ecosystems (Power *et al.* 1995; Larned *et al.* 2010). River networks experience contraction and expansion cycles in response to variation in surface-and groundwater levels (Stanley, Fisher, and Grimm 1997; Malard, Ward, and Robinson 2000; Lake, 2003). Contraction can lead to flow cessation, and/or to dry periods, while expansion initiates rewetting and connections to floodplains. These cycles (Figure 10.1) are accompanied by dramatic habitat changes occurring at various spatial scales; such changes have direct and indirect consequences for both aquatic and terrestrial communities.

10.3.1 Flow Intermittence and Habitat Dynamics in GBRs at the Local Scale

The sequence of flow changes that occur during a contraction–expansion cycle (Figure 10.1) provokes dramatic habitat changes at the local scale (Boulton 2003). When flow recession commences from a bankfull period, fringing vegetation and gravel bars from the littoral and riparian zones become isolated (Figure 10.1A,B). This removes an important habitat for aquatic insects and fish, which feed, grow, and find refuge in the vegetation and low-current zones (Wright 1984; Ormerod *et al.* 1987; Poizat and Pont 1996; Boulton 2003). Conversely, this lateral contraction creates new habitats for terrestrial and riparian communities (Greenwood and McIntosh 2010; Corti and Datry 2012) and terrestrial successions are initiated.

The lateral contraction of river channels (Figure 10.1C,D) is an important stage for aquatic organisms, because shifts from lotic to lentic conditions rapidly eliminate rheophilic invertebrates and fish species (Boulton and Lake 1992; Stanley *et al.* 1994; Pires, Cowx, and Coelho 1999; Anna *et al.* 2009). The loss of the downstream flow of water also halts drift and subsequent recolonization pathways for upstream organisms (Williams and Hynes 1976). Disconnected pools can experience severe change in water physico-chemistry, such as increased temperatures and decreasing dissolved oxygen, during their contraction. For example, the contraction of three disconnected pools in the Albarine River (France) in July 2011 was accompanied by an increase of temperature from 15 to 25 °C, and a concomitant decrease in dissolved oxygen concentrations from 11 to 4.5 mg/L (Figure 10.6). Such changes induce the rapid death of many fish species (e.g., Larimore, Childers, and Heckrotte 1959). Additionally, organisms stranded in pools are subjected to increased biotic interactions including predation, and changes in resource availability place additional pressures on biotic communities (Boulton 2003; Lake 2003). Although not well understood, the abiotic and biotic changes associated with pool disconnection and drying may trigger anhydrobiosis and entrance into dormant forms (cysts, eggs, larvae) of organisms able to produce desiccation-resistant forms (Williams 2006; Datry, Corti, and Philippe 2012; Stubbington and Datry 2013). Some insect larvae and crustaceans can survive weeks or months without surface water in these forms. However, unsurprisingly, the number of aquatic taxa decreases rapidly when the channel becomes fragmented into pools, and species dominance in communities shifts from lotic to lentic (Boulton and Lake 1992; Bonada, Rieradevall, and Prat 2007; Anna *et al.* 2009). Moreover, the number and abundance of invertebrates able to survive in desiccation-resistant forms in dry bed sediments decreases gradually with increasing dry period duration

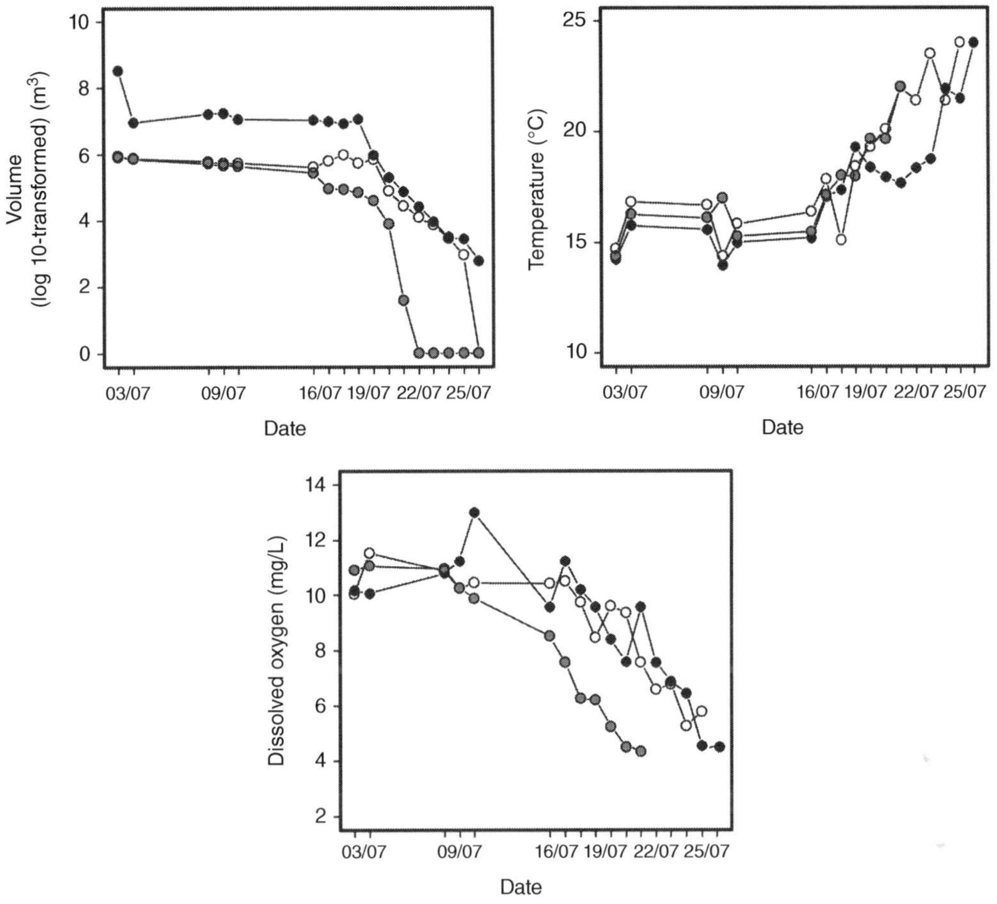

Figure 10.6 Volume, temperature, and dissolved oxygen changes occurring during summer 2011 and progressive drying in three pools of the Albarine River (France).

(Larned, Datry, and Robinson *et al.* 2007; Datry, Corti, and Philippe 2012; Stubbington and Datry 2013).

Disconnected pools in GBRs are generally connected by hyporheic flow (Boulton 2003), except in cases of perched conditions, where an unsaturated zone separates the streambed from the underlying aquifer (e.g., Malard, Ward, and Robinson 2000; Datry and Larned 2008). Hyporheic communities beneath these pools may also experience important habitat changes during these shifts, due to the infiltration of surface water with higher temperature, lower dissolved oxygen, and altered chemistry (Boulton and Lake 1992; Datry and Larned 2008). Hyporheic communities may also experience intense biotic interactions during these periods, due to the penetration of surface organisms into the hyporheic interstices (Stubbington, Wood, and Reid 2011). Indeed, the hyporheic zone of GBRs can be used as a refuge by predominantly benthic invertebrates, and even by small fish, escaping from the harsh conditions developing in disconnected surface pools (Del Rosario and Resh 2000; Boulton 2003; Stubbington 2012; Kawanishi *et al.* 2013). Recent data also indicate that this hyporheic zone subsequently contributes to the resilience of surface communities, demonstrating that at least some organisms return to the surface upon rewetting (Kawanishi *et al.* 2013).

The lotic–lentic shift is also a critical stage for terrestrial communities from the riparian zone, and upland areas. Aquatic organisms stranded in pools represent an important source of food for birds, mammals, and insects such as beetles (Baxter, Fausch, and Saunders 2005; Paetzold and Tockner 2005; Steward *et al.* 2012). This also concerns humans, since fishing in pools can be very common during these periods (Kappelman *et al.* 2014). Pools are also used as a source of water by various large mammals in desert regions (Steward *et al.* 2012). The responses of riparian communities to lotic–lentic shifts occurring in river channels are still poorly explored in temperate systems (but see Steward *et al.* 2011; Corti and Datry 2016).

Although some disconnected pools can persist for months to years in some intermittent rivers (Hamilton and Gehrke 2005; Sheldon *et al.* 2010), they typically dry completely, particularly in GBRs, due to transmission losses, drainage, and evaporation (Figure 10.1E,F). The loss of surface water represents a critical stage for aquatic organisms (Boulton 2003) and many die, unless they can migrate into the hyporheic zone, produce desiccation-resistant forms (see above), emerge as adults, or escape to refuges by crawling (Lytle and Poff 2004). The resultant large quantities of dead and dying organisms on dry streambeds are consumed by terrestrial scavengers such as invertebrates (spiders, beetles), birds, and mammals (Williams and Hynes 1976; Steward *et al.* 2012), and this may create a pulse of carbon, nutrients, and energy into the adjacent terrestrial ecosystems during the early stages of drying (Boulton 2003; Baxter, Fausch, and Saunders 2005; Greenwood and McIntosch 2010).

As the duration of the dry period increases, streambeds become increasingly terrestrial, showing a decrease in the number of aquatic species able to survive as resistance forms (Larned, Datry, and Robinson 2007; Datry, Corti, and Philippe 2012; Hershkovitz and Gasith 2013). Gradually, the water table within the hyporheic zone decreases (Boulton 2003; Datry 2012) and terrestrial plant and animal communities develop (Steward *et al.* 2012; Corti and Datry 2012; Figure 10.1F). Terrestrial phases in dry GBRs are still poorly studied and represent a *terra incognita* (Datry, Arscott, and Sabater 2011; Steward *et al.* 2012); thus, many aspects about their constituent communities and ecological processes remain largely unexplored. Last, the unsaturated zone below GBRs could also be a critical habitat for terrestrial invertebrates both during wet and dry phases (Langhans and Tockner 2014).

Flow resumption can take the form of a sudden rewetting event with an advancing wetted front, when driven by high discharge following rainfall (e.g., Jacobson *et al.* 2000; Cohen and Laronne 2005; Corti and Datry 2012), instigating a rapid reversal of the sequence from terrestrial to flowing periods (Figure 10.1G). Rewetting can also occur more steadily, when rainfall is localized to upstream catchments or when rewetting is driven by rising groundwater levels (Stanley, Fisher, and Grimm 1997; Tockner, Malard, and Ward 2000). From an aquatic perspective, rewetting triggers the breaking of dormancy and the hatching of juveniles from cysts and eggs (Datry, Corti, and Philippe 2012), the upward migration of benthic invertebrates that sought refuge in the hyporheic zone (Boulton 2003; Hose, Jones, and Lim 2005; Fenoglio *et al.* 2007; Vorste *et al.* 2016), the colonization of newly rewetted habitats by drifting and flying organisms (Williams and Hynes 1976), and the germination of aquatic seeds (Larned *et al.* 2010). From a terrestrial perspective, large numbers of terrestrial organisms are entrained downstream in the advancing wetting fronts (Corti and Datry 2012; Rosado, Morais and Tockner 2014), others escape into the riparian zone (Boumezzough and Musso 1983; Lambeets *et al.* 2008). Entrained organisms provide a food source for scavengers and predators from aquatic food webs, and some (e.g., some species of spiders, beetles, and ants) use these advancing fronts as recolonization pathways to colonize downstream riparian zones (Corti and Datry 2012; Jacobson and Jacobson 2013; Rosado, Morais and Tockner 2014).

10.3.2 Flow Intermittence and Habitat Dynamic in GBRs at the Reach Scale

Flow cessation, dry and rewetting periods create unique longitudinal and lateral dynamics of habitat change at the reach scale (Larned *et al.* 2011; Jaeger and Olden 2012; Datry *et al.* 2014a). This is particularly true in GBRs, which alternate bounded and unbounded reaches driving reach-scale extensive surface–groundwater exchanges, and in turn, flow intermittence patterns (Malard *et al.* 2002; Doering *et al.* 2007; Tonina and Buffington 2009; Capderrey *et al.* 2013; Vorste *et al.* 2016). Notably, retreating and advancing wetted fronts generate longitudinal patterns in dry duration and frequency, and hence flow intermittence, along reaches. In turn, flow intermittence gradients shape the longitudinal distribution of communities and ecosystem processes in intermittent GBRs (e.g., Larned *et al.* 2011; Bogan, Boersma, and Lytle 2013; Datry *et al.* 2014a).

Retreating wetted fronts (RWFs) occur during the transitions from low flow to flow cessation or dry periods (Figure 10.1B–D), while advancing wetting fronts (AWFs) occur during the shifts from flow cessation or dry period to low-flow or bankfull discharge periods (Figure 10.1G). These fronts occur laterally (lateral contraction/expansion of the wet section of the streambed), vertically (drying/rewetting of the hyporheic sediments), and longitudinally. Longitudinal movements of these fronts can be oriented downstream, upstream, or both. Downstream RWFs occur, for example, in the case of flow intermittence generated at a lake outlet where the level is dropping (Mielko and Woo 2006), or in headwater streams (Meyer *et al.* 2007; Datry *et al.* 2014a). Upstream RWFs happen, for example, when drying is caused by transmission losses into the gravels (Larned *et al.* 2008; Datry 2012). In some cases, when drying occurs in the central reaches of river networks, RWFs occur in both directions (Larned *et al.* 2008). Similarly, AWFs can occur downstream in cases of rainfall-induced rewetting periods (Corti and Datry 2012; Figure 10.1G), upstream when rising groundwater levels intercept the streambed, or in both directions (Tockner, Malard, and Ward 2000).

The velocity of RWFs and AWFs can vary greatly. For example, in the case of the Albarine River (France) in which drying is caused by transmission losses and rewetting by rainfall, RWFs have been measured moving downstream the channel at a rate of 0.01–0.1 km/h (personal observations), while AWFs advanced at a rate of 0.1–0.5 km/h (Corti and Datry 2012; Figure 10.7). In other systems, including GBRs, RWFs have been recorded retreating as fast as 5 km/h and AWFs advancing as rapidly as 7 km/h (Reid, Laronne, and Powell 1998; Larned *et al.* 2010).

Depending on the river system and the annual number of contraction–expansion cycles, RWFs and AWFs can vary substantially in terms of origin, velocity, frequency, and timing. Some systems are very seasonal with relatively well-predictable periods of flow contraction and expansion, such as glacial systems (Malard *et al.* 2002, 2003). Others experience multi-annual dry periods and rewetting occurs only once every 10 years (Jacobson *et al.* 2000). In contrast, some systems can be very flashy and experience multiple RWFs and AWFs each year (Arscott, Tockner, and Ward 2005; Corti and Datry 2012).

At the reach scale, RWFs and AWFs generate longitudinal patterns of flow intermittence. Such patterns have been particularly well described by Larned *et al.* (2011) across four river systems located in different countries, using a statistical hydrological model based on synoptic point gauging data coupled to long-term flow recorders (Schmidt *et al.* 2009; Larned *et al.* 2011). This empirical longitudinal flow model (ELFMOD) has been used to generate longitudinal flow statistics along river

Figure 10.7 Sequence of an advancing wet front in the Albarine River (France). (Photographs: R. Corti.)

lengths, such as annual number of no-flow events, annual duration of no-flow events, annual average dry-reach length, annual average number of dry reaches, or annual flow permanence (Figure 10.8). The longitudinal patterns of these different statistics are not smooth and continuous, but show rather abrupt changes in gradients (Figure 10.8), which are typically caused by hydrogeomorphic discontinuities, the most obvious being the input from a tributary. For example, the sharp increase in flow permanence in the central reach of the Selwyn River is caused by the confluence with a permanent tributary (Figure 10.8; Larned *et al.* 2008). However, hydrogeomorphic discontinuities can be less evident when associated with changes in geology, alluvial deposits, or faults (Larned *et al.* 2011). For example, the gradual decrease in flow permanence that occurs along the length of the downstream reach of the Albarine River is caused by progressive widening of an alluvial valley (Datry 2012). When the river course moves from the bounded to the unbounded alluvial reach, rapid transmission losses occur into the underlying aquifer, the water table of which is located 5–15 m below the streambed. These perched conditions lead to a rapid longitudinal decrease in flow permanence (Figure 10.8; Datry 2012).

Flow intermittence patterns at the reach-scale may differ in GBRs to those in other intermittent rivers. The alternation of bounded and unbounded reaches is a typical feature of GBRs (Baxter and Hauer 2000; Malard *et al.* 2002; Tonina and Buffington 2009; Capderrey *et al.* 2013). The typical sequence of canyons, gorges, and other bounded reaches following wide alluvial floodplains have been compared to 'beads on a string' (Stanford and Ward 1993). These hydrogeomorphic discontinuities control large-scale hyporheic exchange, and ultimately longitudinal patterns of flow intermittence (Figures 10.8 and 10.9). Unbounded reaches are generally characterized by thick, wide deposits of alluvium, in which surface water downwells to recharge the alluvial aquifer (Malard *et al.* 2002, 2003; Tonina and Buffington 2009; Capderrey *et al.* 2013). Perched conditions often prevail in the upstream areas of unbounded reaches and transmission losses can be very high (Malard *et al.* 2002, 2003; Datry and Larned 2008). Such conditions can lead to the complete drying of the river channel or to the drying of many braids (Capderrey *et al.* 2013; Vorste *et al.* 2016). In contrast, at transitions between unbounded and bounded reaches, bedrock outcrops may force groundwater to upwell into river channels (Malard *et al.* 2000; Tonina and Buffington 2009; Capderrey *et al.* 2013). This favours reach-scale flow permanence and creates disruptions in longitudinal flow permanence gradients (Capderrey *et al.* 2013; Vorste *et al.* 2016; Figures 10.8 and 10.9).

RWFs, AWFs, and the longitudinal patterns of flow intermittence they generate at reach scales have various consequences for ecological processes and patterns in GBRs. The species diversity of aquatic communities generally decreases along longitudinal gradients of increasing flow intermittence (Datry, Larned, and Tockner 2014). Very consistent species-diversity–flow-intermittence relationships have been reported for aquatic invertebrates in multiple GBRs from different biogeographic regions (Datry *et al.* 2014a; Figure 10.10). Moreover, species-poor communities from the more intermittent sites were nested subsets of richer communities from the less intermittent and perennial sites (Datry *et al.* 2014a). These patterns and many others reported from various intermittent rivers (e.g., Meyerhoff and Lind 1987; Delucchi and Peckarsky 1989; Feminella 1996; Bonada, Rieradevall, and Prat 2007) suggest that communities in GBRs are organized as metacommunities along longitudinal flow intermittence gradients (Larned *et al.* 2010). Metacommunities are sets of local communities connected in the landscape by dispersal (Leibold *et al.* 2004). In GBRs, local communities seem to be structured longitudinally along gradients of flow intermittence, with AWFs allowing occasional exchanges of genes, organisms, and energy between them (Larned *et al.* 2010). Following this, community composition may be primarily a result of species dispersal abilities, frequency and duration of flowing/dry periods, and distances between aquatic habitats. Yet, testing and development of the metacommunity theory in GBRs is in its infancy, and our understanding may increase substantially in the coming years.

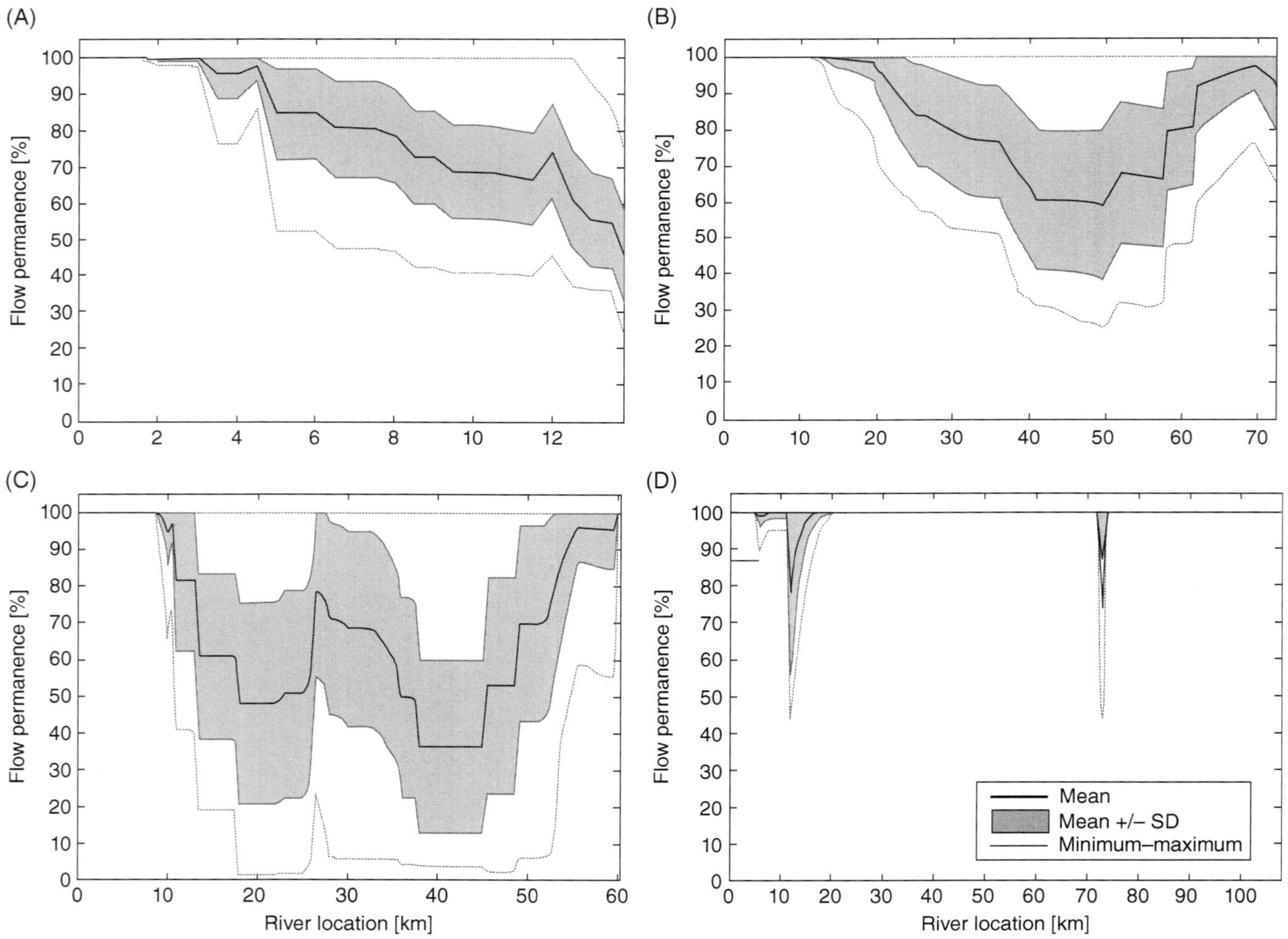

Figure 10.8 Variation in flow permanence (defined as the proportion of the year with flowing water) along intermittent reaches of four GBRs. (A) The Albarine River (France), (B) the Selwyn River (New Zealand), (C) the Orari River (New Zealand), and (D) the Methow River (United States). (Modified after Larned *et al.* (2011).)

Figure 10.9 Satellite image showing large-scale downwelling (left-upper side, dry channels) and upwelling (right-bottom side, flowing channels) patterns along a GBR (Rio Tapacari) in central Bolivia. (*Source*: Google Earth.)

Much less research has been conducted on reach-scale patterns of terrestrial communities in GBRs, but some results indicate they could be organized following a different spatial scheme to aquatic communities (Corti and Datry 2016). During AWFs and subsequent flowing periods, terrestrial organisms from dry riverbeds may seek refuge laterally in the riparian and upland areas (Steward *et al.* 2011; Corti and Datry 2014). Other organisms may use AWFs as dispersal pathways to colonize downstream reaches (Corti and Datry 2012; Rosado, Morais, and Tockner 2014). During RWFs, organisms may colonize dry riverbeds from these source habitats. Therefore, it is unlikely that terrestrial community patterns show longitudinal structures comparable to those described for aquatic communities. Instead, dry riverbeds may show patchy and discontinuous longitudinal community patterns, controlled by the species diversity and composition of terrestrial communities from riparian and upland habitats. A source–sink metacommunity model with an organization constrained laterally rather than longitudinally may characterize terrestrial communities in intermittent rivers (Corti and Datry 2016).

The cycling of organic matter (OM) and nutrients is controlled by RWFs and AWFs in intermittent rivers, and is therefore clearly discontinuous and longitudinally structured. Compared with perennial rivers, intermittent rivers accumulate large amounts of OM and nutrients during flow cessation and dry periods. Pools are sites of riparian leaf litter and sediment accumulation (Acuña and Tockner 2010) and, as discussed, contain high densities of refuge-seeking invertebrates and fish. Heterotrophic respiration and low solubility in disconnected pools cause rapid reductions in dissolved oxygen concentrations, and hypoxic conditions can promote ammonium and phosphate release from sediments (Williams 2006; Von Schiller *et al.* 2011, 2015). Following RWFs, dissolved nutrients are converted to evaporates, and terrestrial plants gradually replace aquatic macrophytes and algae (Larned *et al.* 2010). As shown earlier, various and abundant terrestrial communities, particularly of arthropods (Paetzold

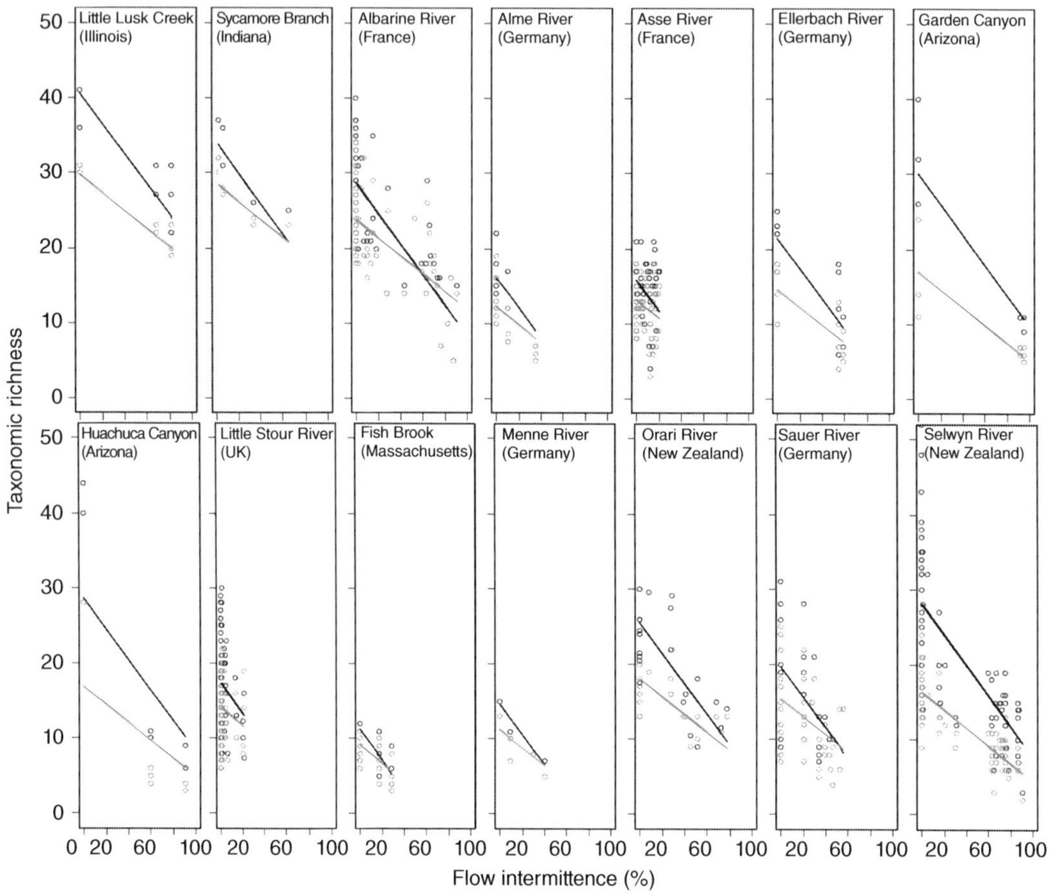

Figure 10.10 Average mixed-effects models for 14 different GBRs of family-level (black) and genera-level (grey) taxonomic richness and annual flow intermittence (percentage of the year without surface water) as a fixed slope, showing a congruent decrease in the number of taxa with increasing flow intermittence, regardless of the taxonomic resolution. Linear regressions performed on each dataset individually indicated significant relationships between taxonomic richness and flow intermittence in 11 and 10 rivers out of 14, when considering family- and genera-level taxonomic richness, respectively. (From Datry *et al.* (2014a).)

and Tockner 2005; Steward *et al.* 2011; Corti, Larned, and Datry *et al.* 2013; Corti and Datry 2012) colonize dry river channels. When dry periods extend into leaf-fall periods, very large quantities of riparian leaf litter accumulate along dry channels (Acuña and Tockner 2010; Corti and Datry 2012). This material is preconditioned by solar radiations, anoxia, and/or desiccation (Dieter *et al.* 2011). The subsequent decomposition of leaf litter occurs at much higher rates under flowing than intermittent or dry conditions (e.g., Gurtz and Tate 1988; Boulton 1991; Corti *et al.* 2011; Datry *et al.* 2011).

During AWFs, accumulated OM, nutrients, and organisms are entrained, and eventually deposited downstream. For example, terrestrial arthropod density and taxonomic richness, and OM concentration were several orders of magnitude higher in one AWF than in upstream perennial reaches in a French GBR, the Albarine (Corti and Datry 2012; Figure 10.11). The OM concentrations reached

Figure 10.11 Density and taxonomic richness of (A) terrestrial and (B) aquatic invertebrates entrained at Rewetted (a) and Perennial (b) sites indicating statistical significant differences based on Wilcoxon tests (density is individuals/m^3). Box plots show the median (solid line), 25th and 75th percentiles (box ends), and 10th and 90th percentiles (whiskers). (C) Mean (±1 SD) concentrations of particulate organic matter (POM) entrained during the advance of the wetted front. The first five points are located in the perennial section (km 13–15), whereas the seven others are in the AWFs. (Modified after Corti and Datry (2012).)

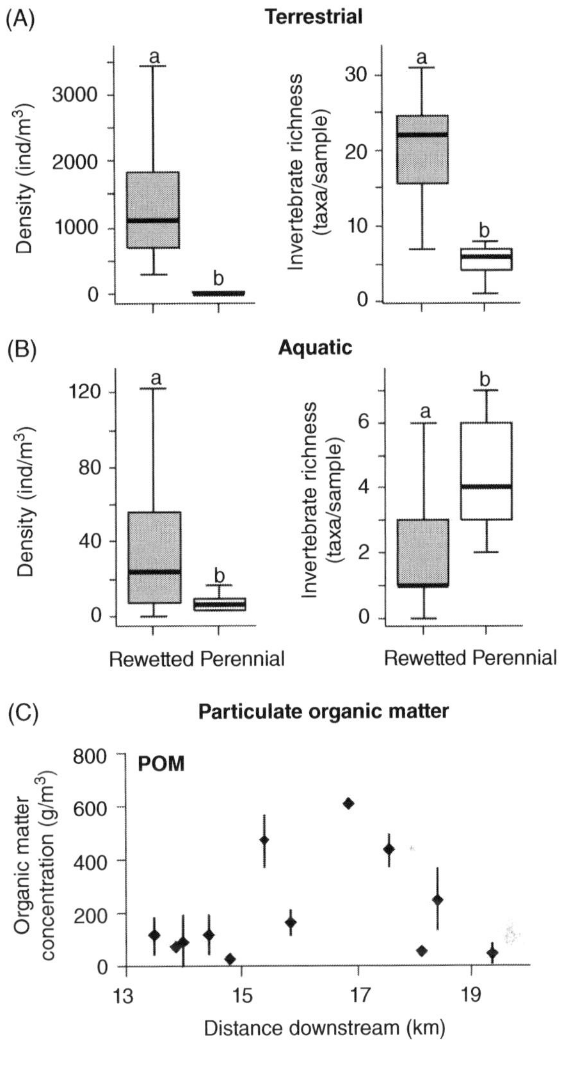

600 g/m^3 in the AWF but only <50 g/m^3 in the perennial upstream reach; similarly, invertebrate abundance was on average 200-fold higher in the AWF than in the perennial reach (Figure 10.11A,B; Corti and Datry 2012). Striking differences in OM concentrations between AWFs and perennial reaches have been reported in other systems, including GBRs (e.g., Jacobson *et al.* 2000; Larned *et al.* 2010). Dissolution of sediment evaporites and leaching of OM also cause associated spikes in solute and nutrient concentrations in AWFs (Jacobson *et al.* 2000; Bernal, Butturini, and Sabater 2005; Obermann *et al.* 2009), and can lead to hypoxic blackwater events (Hladyz *et al.* 2011; Datry, Larned, and Tockner 2014). AWFs move this material downstream where flow attenuation results in deposition within and at the margins of river channels (Jacobson *et al.* 2000). These depositional areas are not necessarily located far downstream, as no longitudinal increase in OM concentrations or invertebrate abundance were detected along the 7-km course of an AWF (Figure 10.11C; Corti and Datry 2012). The ecological consequences of these large OM and nutrient pulses are still poorly understood, but

could have positive as well as negative effects on downstream receiving waters. Deposited OM is a fundamental carbon source for fungi, micro-organisms, invertebrates, and fish, and can stimulate food webs, but it can also generate large-scale anoxic plumes and can therefore cause fish and invertebrate kills (Hladyz *et al.* 2011; Datry, Larned, and Tockner 2014).

10.3.3 Flow Intermittence and Habitat Dynamics in GBRs at the Catchment Scale

The sequence of flow changes occurring during a contraction–expansion cycle generates habitat changes at the catchment scale (Stanley, Fisher, and Grimm 1997; Jaeger and Olden 2012; Datry *et al.* 2016). These changes generate mosaics of habitat conditions at the river catchment scale. Shifting habitat mosaics are a defining feature of all intermittent rivers, including GBRs: mosaics of lotic (flowing), lentic (not flowing, disconnected pools), and terrestrial (dry) habitats are created across these river networks by cycles of expansion and contraction, as driven by surface discharge and groundwater-level variations (e.g., Stanley, Fisher, and Grimm 1997; Larned *et al.* 2010; Datry *et al.* 2016). The flow cessation, dry and rewetting periods lead to the simultaneous presence of all or any combination of lotic, lentic, and terrestrial habitats in the landscape (Larned *et al.* 2010; Datry, Larned, and Tockner 2014; Jaeger, Olden, and Pelland 2014; Datry *et al.* 2016). Because surface discharge and groundwater levels vary continuously, the spatial arrangement, temporal turnover, and connectivity of these three habitats constantly shift. Therefore, much like alluvial floodplains (e.g., Arscott, Tockner, and Ward 2005; Stanford, Lorang, and Hauer 2005), intermittent GBRs are 'shifting habitat mosaics' and are amenable to landscape ecological approaches.

One well-illustrated example of the habitat changes that occur at the scale of river networks comes from western France, where flow states are monitored along > 4000 km of a river network by a group of citizens (Datry *et al.* 2016). Although this example does not come from GBRs, it is still relevant to explore flow intermittence patterns and habitat changes at the network scale. Observations were initiated in 2004 by Carg'eau (Collectif Associatif Régional pour la Gestion de l'EAU, www.eau-poitou-charentes.org), an association of fishermen, environmentalists, and water users, to monitor and highlight the extent of summer dry events in the region, which are prone to conflicts around water uses: the observations cover > 4000 km of river length across 20 river catchments (Datry *et al.* 2016). For each observation, flow state is assigned to one of the following categories: *flowing,* when there is flow across the entire channel width; *low flow,* when flow is limited to less than 50% of the channel width; *no flow,* when the channel is composed of disconnected pools without surface flow in between; and *dry,* when the channel is entirely devoid of surface water (Datry *et al.* 2016). Observations are made along the entire channel network (from headwaters to catchment outlets) when possible, depending on accessibility, which is influenced by the availability of, for example, bridges, roads, and/or walking paths (Datry *et al.* 2016).

The analysis of these observations revealed the multiple and dynamic habitat changes that intermittent rivers experience at the network scale (Datry *et al.* 2016; Figure 10.12). Dry patches dominated most observation dates, although lotic patches had the longest average length and occupied the greatest proportion of channel length. At the start of each summer, catchments were almost entirely composed of lotic patches but more lentic and dry patches appeared along the network as summer progressed. Despite this seemingly consistent temporal pattern, habitat types shifted between and within catchments, resulting in a high level of spatiotemporal variability (Figure 10.13; Datry *et al.* 2016).

The ecological implications of such network-scale habitat changes are theoretically numerous but they have been poorly explored in empirical studies. For instance, these changes may shape the organization of metapopulations and communities in GBRs (Jaeger, Olden, and Pelland 2014;

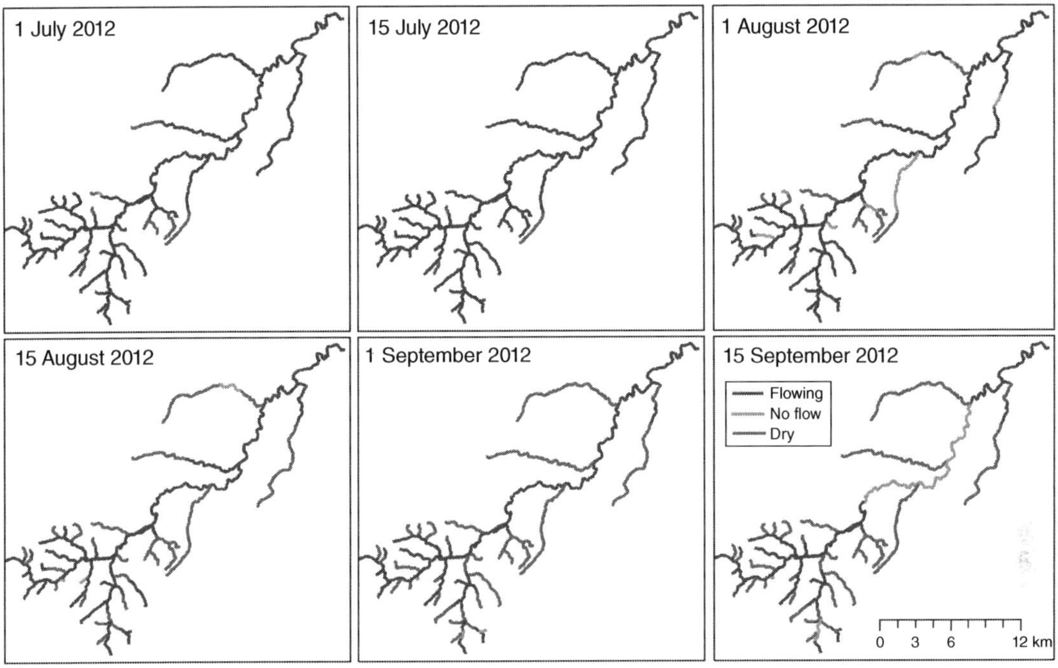

Figure 10.12 Temporal dynamics in patch composition and configuration of the Thouaret catchment during summer 2012: lotic patches; lentic patches; terrestrial patches. (Modified after Datry *et al.* (2016).) (See insert for color representation of this figure.)

Cañedo-Argüelles *et al.* 2015). In an attempt to explore the possible responses of organisms at the population level, the metapopulation capacity of four hypothetical species with contrasting features has been modelled in five of these catchments (Datry *et al.* 2016). Populations of small fish were the most affected by habitat fragmentation caused by drying. In contrast, large fish and invertebrates with good aerial dispersal capacities were mostly influenced by habitat size rather than fragmentation (Datry *et al.* 2016). In a previous study, Colvin *et al.* (2009) found that the number of fish species in intermittent rivers decreased as distance to perennial waters increased, so regardless of size, fish may be vulnerable to lotic habitat fragmentation. Insects were less sensitive to habitat fragmentation than fishes, which may be because local extinctions resulting from drying and habitat disconnection are compensated by colonization from perennial reaches via aerial dispersal (Múrria *et al.* 2010; Chester and Robson 2011; Swan and Brown 2014; Cañedo-Argüelles *et al.* 2015) or emergence from desiccation-resistant forms (Datry, Corti, and Philippe 2012; Stubbington and Datry 2013; Hershkovitz and Gasith 2013). Although these studies differed in landscape configurations (e.g., average patch size, distances among patches), which could alter these conclusions, the use of empirical estimates of population densities and dispersal distances applied to real landscapes (catchments of the Poitou-Charentes, western France) increases the confidence in these findings.

These habitat changes have also persuasive effects on OM and nutrient cycling across river networks (Acuña and Tockner 2010; Larned *et al.* 2010; Raymond *et al.* 2013; Datry, Larned, and Tockner 2014; Von Schiller *et al.* 2015). The respective proportion and spatial arrangement of the three habitat types across the network may notably control the net export, processing and storage of OM and nutrients.

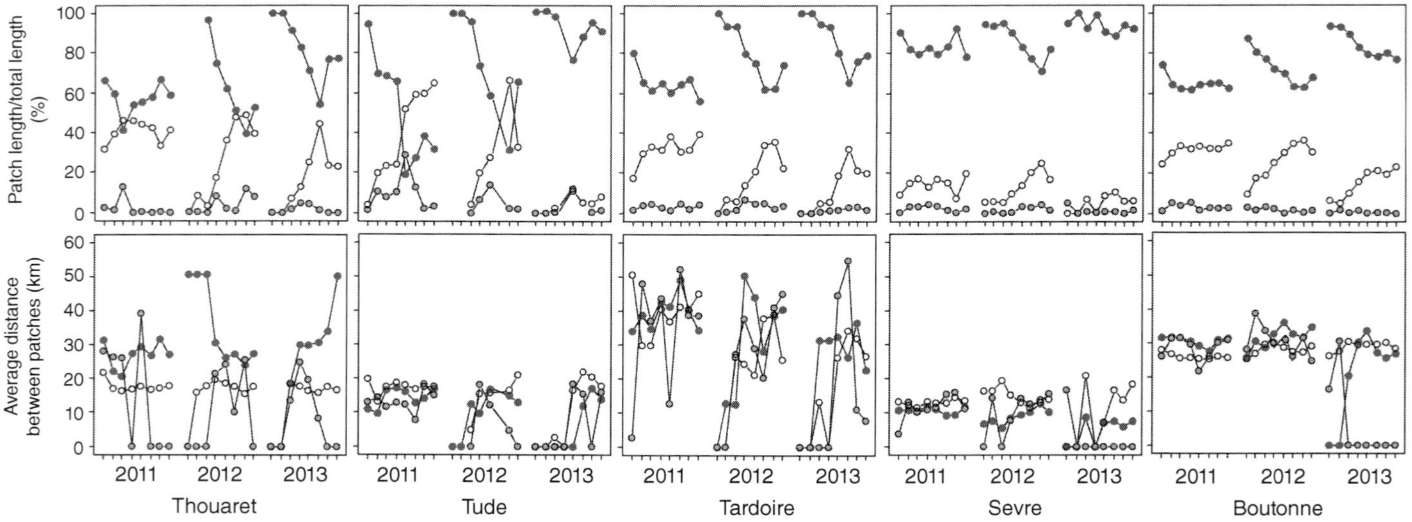

Figure 10.13 Temporal dynamics of two landscape metrics (percentage of the catchment composed of a given patch type, and average distance between patches) for five studied catchments in western France and year of study: dark grey - lotic patches; light grey - lentic patches; open circles - terrestrial patches. (Modified after Datry *et al*. (2016).)

A metaecosystem approach (Loreau, Mouquet, and Holt 2003; Gravel *et al.* 2010; Larned *et al.* 2010; Datry *et al.* 2016) would improve our understanding of OM and nutrient cycling in intermittent rivers, including GBRs.

River systems are among the most obvious dendritic dispersal networks occurring in the landscape (Fagan 2002; Swan and Brown 2014). Their hierarchical structure shows branching patterns, which result in isolated, exterior branches (headwaters), and converging, well-connected interior branches (mainstems) (Fagan 2002; Finn *et al.* 2011; Swan and Brown 2014). The spatial location of flow cessation and drying events interact with these hierarchical structures to influence river communities and processes (Lake 2003; Datry *et al.* 2014a). In a simple view, five different spatial patterns of flow intermittence occur at the catchment scale (Figure 10.14). In arid (Stanley, Fisher, and Grimm 1997) or karstic (Meyer and Meyer 2000) areas, entire river networks can remain dry for weeks or months (Figure 10.14A). In these systems, aquatic processes and community succession are halted across the entire catchment until flow resumption, and the resistance of micro- and macro-organisms to desiccation may be crucial to their persistence in these systems. Many river networks have intermittent headwaters and perennial downstream reaches (Meyer *et al.* 2007; Benstead and Leigh 2012; Figure 10.14B). The contribution of low-order streams may represent up to 70% of river network lengths (Lowe, Likens, and Power 2006). Upon flow resumption, the contribution of these 'small but mighty' channels to catchment-scale biodiversity and organic matter cycling is substantial (Meyer *et al.* 2007; Benstead and Leigh 2012; Swan and Brown 2014). Alluvial GBRs with high transmission losses are generally intermittent in their most downstream reaches (Figure 10.14C; Tockner *et al.* 2003; Larned *et al.* 2008; Datry 2012). For example, the most downstream section (11 km) of the Albarine River (France) is intermittent, showing a longitudinal gradient of increasing flow intermittence, while the upstream section (40 km) is perennial (Datry 2012). Communities in such intermittent sections generally return quickly to their pre-drying levels upon rewetting, as drifting organisms can quickly recolonize newly rewetted habitats. Within two weeks of flow resumption, benthic and hyporheic communities of the Albarine River had returned to their pre-drying stages (Datry 2012). Central, mid-reach drying reaches have also been reported in several GBRs, such as the Tagliamento (Arscott *et al.* 2010) and the Selwyn rivers (Figure 10.14D; Larned *et al.* 2008). Flow intermittence in these rivers is generated by transmission losses into the underlying aquifer, with downstream reaches being perennial due to large-scale upwelling flows of groundwater that occur near coastal areas underlain by confined aquifers (Tockner, Malard, and Ward 2000; Larned *et al.* 2008, 2010). Although these central intermittent reaches can sometime alter the longitudinal distribution of communities (Bogan

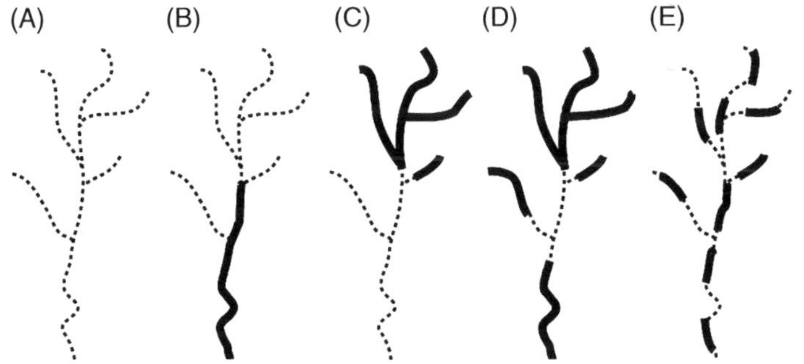

Figure 10.14 Various types of theoretical spatial patterns of drying: (A) complete drying, (B) headwater drying, (C) downstream drying, (D) mid-stream drying, and (E) point drying.

et al. 2013), these generally return to their pre-drying levels within a few weeks (e.g., Arscott *et al.* 2010). Last, flow cessation and dry periods can occur punctually throughout the network, as described previously for the Poitou-Charente region (Figures 10.12 and 10.14E). In such cases, the response of communities and processes to flow intermittence may be catchment-specific, depending mostly on the mosaic configuration (Datry *et al.* 2016).

10.4 Intermittent GBRs as Coupled Aquatic–Terrestrial Disturbed Ecosystems

10.4.1 Disturbance Ecology and Intermittent GBRs

For both aquatic and terrestrial habitats, species inhabiting intermittent GBRs experience disturbances related to flow cessation, drying, and the subsequent inundation that occurs when flow resumes (Figure 10.15). Unsurprisingly, the concept of disturbance ecology is very relevant to the study of intermittent GBRs. Disturbance is defined here *sensu* Lake (2000), as the cause of a perturbation of an ecosystem (and not defined as the response of the perturbation as it is sometime presented; Lake 2000). Although it could be argued that organisms in intermittent rivers are adapted to their flow regimes, a drying event for the aquatic biota or a rewetting event for the terrestrial biota, are disturbances. Beyond the endless debate disturbance ecology has been generating (see Lake 2000), disturbance ecology represents a rich and relevant framework to consider flow intermittence in rivers. To allow comparisons across systems or over time, a disturbance can be characterized by both spatial and temporal scale (see sections 10.2.2 and 10.2.3; Resh *et al.* 1988; Poff and Ward 1989; Larned *et al.* 2011). Concerning the spatial scale, determining where drying or rewetting occur and quantifying what proportion of the catchment is being affected by drying or rewetting is crucial, along with characterizing the frequency, duration, timing, and predictability of the event. Although early work on intermittent rivers has largely overlooked these aspects by using categorical comparisons (e.g. permanent vs. intermittent), more recent research has quantified both spatial and temporal scales at which flow cessation or drying occurred, allowing for cross-system comparisons (e.g., Datry *et al.*

Figure 10.15 Photographs showing aquatic invertebrates stranded in a dry pool (left) and terrestrial invertebrates entrained in an AWF (right). (Photographs: R. Corti.)

2014a; Stubbington and Datry 2013). Disturbance ecology will also help guide research aiming at comparing the effects of natural versus non-natural flow intermittence on riverine communities.

For aquatic organisms to tolerate desiccation, and for terrestrial organisms to cope with inundation, two types of mechanisms are involved: resistance and resilience. Resistance reflects the capacity of communities to persist unchanged during a disturbance, and involves a range of physiological adaptations allowing organisms to survive within dry riverbed sediments or disconnected pools (Lytle and Poff 2004; Williams 2006; Chester and Robson 2011; Hershkovitz and Gasith 2013), or for terrestrial organisms to resist submersion during flowing periods (Lytle ad Poff 2004; Corti and Datry 2012; Hershkovitz and Gasith 2013). Species from several aquatic invertebrate groups (e.g., oligochaetes, copepods, and dipterans) can persist for years in dry river sediments as cysts, cocoons, or in a state of diapause as larvae or adults (Williams 2006). Conversely, terrestrial organisms can have adaptations that allow them to escape AWFs, or to resist submersion (Boumezzough and Musso 1983; Adis and Junk 2002; Lytle and Poff 2004; Lambeets *et al.* 2008). In field and laboratory experiments, some riparian beetles such as Carabidae or Staphylinidae and some arachnid species of the Araneae Order (e.g., *Pardosa agricola, Chlaenius velutinus,* and *Pterostichus lasserei*) exhibit different survival strategies to resist submersion. Sheltering and escaping are the two dominant strategies during slow increases in water level (Boumezzough and Musso 1983; Lambeets *et al.* 2008), while crawling, floatation, and respiration through air-bubbles allow some taxa to survive up to 20 h of submersion when water levels increase rapidly. The unsaturated zone could also be used as a refuge for terrestrial invertebrates during short flowing periods (Langhans and Tockner 2014). In general, the contribution of resistance strategies to community persistence has been considered to be low, but it may have been underestimated (Stubbington and Datry 2013; Hershkovitz and Gasith 2013).

Resilience reflects the ability of communities to return to their pre-disturbance state, in terms of structure, composition, and function. This involves life-history traits and behavioural/morphological adaptations related to dispersal capacity, short life cycles, and high fecundity (Williams 2006). High resilience has been found to characterize aquatic communities in intermittent GBRs (e.g., Sagar 1983; Fowler *et al.* 2004; Arscott *et al.* 2010; Datry 2012). This was particularly apparent in eight intermittent GBRs of southern France, where communities from reaches experiencing drying were similar to perennial ones less than 3 weeks after rewetting (Vorste *et al.* 2016). The authors concluded that such high resilience was the result of: (i) a selection from the regional species pool of species with traits promoting rapid recovery, as braided GBRs are known to be highly disturbed systems; and (ii) the presence of an extensive hyporheic zone in which organisms could find refuges during drying. In general, resilience of aquatic communities may be intimately linked to the distance to perennial refuges and other sources of colonists (Larned *et al.* 2010; Chester and Robson 2011; Datry *et al.* 2014b). Resilience of terrestrial communities to inundation has not been quantified in intermittent GBRs or other rivers, but evidence suggests that it may also be very high. First, refuges and sources of colonists are located in the adjacent riparian zones and, therefore, colonization of dry riverbeds and gravel bars can be fast and is not probably a limiting factor (Datry *et al.* 2014b; Corti and Datry 2016). Second, AWFs are used as dispersal pathways by terrestrial invertebrates to colonize downstream terrestrial habitats (Corti and Datry 2012). Lastly, the resilience of riparian, terrestrial communities to river flooding is known to be high in other types of rivers (e.g., Adis and Junk 2002; Lambeets *et al.* 2008). Further research will help quantify and understand how terrestrial communities persist in intermittent GBRs.

10.4.2 Aquatic and Terrestrial Communities, and their Interactions, in Intermittent GBRs

Because aquatic and terrestrial conditions alternate or coexist in intermittent GBRs, their combined biotic communities are particularly rich and comprise aquatic, semi-aquatic, and terrestrial organisms

(Williams 2006; Larned *et al.* 2010; Datry, Larned, and Tockner 2014; Corti and Datry 2016). These aquatic and terrestrial components interact at any time, although the quantification of these interactions and their ecological significance is an open field of research (Leigh *et al.* 2016; Datry *et al.* 2016). Riparian predators (e.g., spiders, bats, birds) prey heavily on adult aquatic insect species that emerge from river channels (Nakano and Murakami 1999; Baxter, Fausch, and Saunder 2005), whereas some fish and invertebrates rely largely on terrestrial invertebrates as a food source (Baxter, Fausch, and Saunders 2005). The interaction between aquatic and terrestrial communities may peak during flow cessation and dry periods, when terrestrial species advantageously feed on the dying aquatic invertebrates and fish stranded in pools or on dry sediments. These interactions may then decrease as the dry period continues, unless disconnected pools persist (Hamilton and Gehrke 2005; Sheldon *et al.* 2010). Strong interactions may also occur during flow resumption, as AWFs entrain many terrestrial organisms downstream (Corti and Datry 2012; Rosado, Morais, and Tockner 2014). Yet, most of these aspects remain theoretical and empirical quantifications of these interactions, which likely peak both temporally and spatially along GBR corridors. Our ability to manage GBRs effectively and to conserve their constituent communities will depend on our capacity to understand and predict these interactions.

The intermittent flow regimes of many GBRs, as well as the extensive terrestrial and aquatic habitats they harbour, make them model systems to advance empirical and conceptual developments of aquatic–terrestrial interactions in river systems (Tockner *et al.* 2003). Notably, the relationships between aquatic and terrestrial communities are so interlinked in these systems, that a distinction between them may limit the understanding of each (e.g., Corti and Datry 2016). In fact, invertebrates in intermittent GBRs may be considered as a metacommunity of two communities, one terrestrial and one aquatic, each with contrasted spatial and temporal dynamics and interacting intensively, albeit punctually, in space and time. Empirical studies that simultaneously examine terrestrial and aquatic components in intermittent GBRs are appearing (e.g., Corti and Datry 2016), and community models to understand such dynamics are likely to follow. Such conceptual developments will in turn benefit various coupled aquatic–terrestrial systems, such as wetlands, river floodplains, or even stormwater and artificial recharge basins.

10.5 Flow Intermittence in GBRs: Research Needs and Open Questions

10.5.1 How Many, Where, How, and Why?

As emphasized throughout this chapter, various research gaps and needs are currently limiting both our perception and our ecological understanding of intermittent GBRs. Perhaps one of the most striking aspects is that there is currently no large-scale quantification of intermittent GBRs. In 2015, it is still unknown how much of the world's GBRs are spatially or temporally intermittent, where these are located, why they run dry, and what characterizes their flow regime. This gap originates from the lack of hydrological data collected in intermittent GBRs, in particular in their intermittent reaches. A direct consequence of this knowledge gap is that researchers and water managers systematically overlook these rivers (Datry, Larned, and Tockner 2014; Acuña *et al.* 2014). Most conceptual and empirical models developed in geomorphology, hydrology, and ecology are therefore not suitable for intermittent GBRs (Datry, Larned, and Tockner 2014). Very few management practices or water policies recognize rivers with intermittent flow (Acuña *et al.* 2014). Gathering hydrological (or other types of) data to quantify, locate, explain, and describe the flow regime of intermittent GBRs is the most urgent task to improve our perception and understanding of these systems. This will notably help to answer several important questions. What is the proportion of GBRs that are intermittent? Are human practices and climate change increasing this proportion? Are natural and anthropogenic intermittent GBRs similar (hydrologically, ecologically)?

10.5.2 The Vertical Dimension: Surface- and Groundwater Interactions in Intermittent GBRs

Another aspect requiring urgent attention is the vertical dimension in intermittent GBRs. Although such research has increased in the past decade (e.g., Tonina and Buffington 2009; Gray and Harding 2009; Capderrey *et al.* 2013; Vorste *et al.* 2016), many questions remain. Is the hyporheic zone of intermittent GBRs a refuge during disturbances (e.g., drying, flooding, scouring)? Do terrestrial invertebrates also use the hyporheic zone of intermittent GBRs during dry phases? What are the implied hydrogeomorphic drivers? From the opposite end of the vertical gradient, and clearly the least studied, how do large-scale downwelling and upwelling patterns influence communities and ecosystem processes at the water table of alluvial aquifers beneath intermittent GBRs? To what extent does groundwater abstraction alter flow intermittence patterns in intermittent GBRs? What are the effects of the shifts between saturated and unsaturated conditions in the vadose/hyporheic zone?

10.5.3 Diversity Patterns and Community Dynamics at the Catchment Scale in Intermittent Rivers

While knowledge about local responses of aquatic communities to flow cessation, drying, and rewetting has increased much in recent decades (Leigh *et al.* 2016), responses of the communities at larger scales are very rarely considered (but see Jaeger, Olden, and Pelland 2014; Widder *et al.* 2014; Datry *et al.* 2016). How beta diversity and community structure (e.g., nestedness and turnover) change during drying–rewetting cycles? Local decreases in taxonomic richness (alpha diversity) during drying may be accompanied by an increase in beta diversity at higher spatial scales. On the contrary, if organisms are highly adapted to drying, both alpha and beta diversity may decline with increasing drying. How community nestedness and species turnover respond to drying at the catchment scale is unknown, and also the underlying metacommunity processes are not explored. Improving our knowledge on these aspects is particularly required to increase the efficiency of management plans and biomonitoring programmes (Heino 2013).

10.5.4 Evolutionary Perspectives in Intermittent Rivers

On a geological timescale, rivers and streams are thought to be stable ecosystems in the landscape (Dijkstra, Monaghan, and Pauls 2014). In contrast, standing waters such as ponds, pools, wetlands, and even lakes are considered as more unstable (they appear or disappear in the landscape). As a consequence of natural selection, it has been proposed that the ability of organisms to disperse would be stronger, and the range size of individuals higher, in standing than in running waters, and some evidence from recent genetic studies supports this idea (e.g., Finn *et al.* 2006; Hughes, Schmidt, and Finn 2009; Dijkstra, Monaghan, and Pauls 2014). Low dispersal may favour speciation through allopatry in rivers: some river invertebrates harbour more genetic differentiation than lentic species (Marten, Braendle, and Brandl *et al.* 2006; Papadopoulou *et al.* 2008; Dijkstra, Monaghan, and Pauls 2014). However, it is now becoming clear that flow intermittence can strongly limit the dispersal of organisms in river systems, including those with subaerial stages (e.g., Jaeger, Olden, and Pelland 2014; Datry *et al.* 2016; Cañedo-Argüelles *et al.* 2015). Therefore, the current evolutionary perspective of river organisms might be revisited accounting for the fact that intermittent rivers are much more abundant than initially expected, and that limited dispersal could have been an evolutionary factor in lotic systems too. What are the evolutionary consequences of flow intermittence on the diversification of riverine species?

10.5.5 Aquatic–Terrestrial Interactions in Intermittent GBRs

Terrestrial communities and aquatic–terrestrial interactions are poorly explored in intermittent GBRs, although these may be model systems to advance our general understanding of these aspects, including both empirical and conceptual developments. Where and when do these interactions peak? Is it possible to consider aquatic and terrestrial communities as a single metacommunity? What is the significance of these interactions at the catchment scale, and for the terrestrial biodiversity of adjacent ecosystems?

10.5.6 What are the Costs of Not Managing and Protecting Intermittent GBRs?

Intermittent GBRs perform several ecosystem services (Larned *et al.* 2010; Acuña *et al.* 2014), even during dry phases (Steward *et al.* 2012). They have high biodiversity values, convey water, sediments, solutes, organisms across the landscape, and sustain flow and other services to downstream reaches and rivers. They contribute to the longitudinal connectivity of river networks (Bogan, Boersma, and Lytle 2013), exchange laterally with floodplain and riparian ecosystems (Tockner *et al.* 2003), stimulate biogeochemical transformations in the hyporheic zone (Datry and Larned 2008), and recharge aquifers (Larned *et al.* 2010; Boulton 2014). Drying in GBRs contributes to the dynamic and diversity of habitat mosaic (Sheldon, Boulton, and Puckridge 2002; Datry, Larned, and Tockner 2014). Finally, dry gravel-bed riverbeds can have considerable cultural, social, and spiritual value (Steward *et al.* 2012). Intermittent GBRs therefore clearly deserve comprehensive protection and efficient management (Acuña *et al.* 2014). While the Millennium Ecosystem Assessment (2005) has estimated the value of ecosystem services in perennial rivers, no study has yet quantified the costs for societies of not achieving the protection and management of intermittent GBRs and rivers in general. There is little doubt that such work will increase their recognition by scientists and societies.

Acknowledgements

I was supported by the French Foundation for Research on Biodiversity and the French National Agency for Water and Aquatic Environments in the context of the CESAB project 'Intermittent River Biodiversity Analysis and Synthesis' (IRBAS; http://irbas.cesab.org). I thank the many colleagues and friends who have influenced my research on intermittent rivers, particularly Scott Larned, Andrew Boulton, Klement Tockner, Cliff Dahm, Rachel Stubbington, Ken Fritz, Mike Bogan, Cath Leigh, and Roland Corti. I also thank the members of the IRBAS group for stimulating discussions and exchanges during recent workshops. Roland Corti, Ross Vander Vorste, and Rachel Stubbington provided valuable comments on early drafts. The helpful comments and suggestions of two anonymous reviewers have benefitted this chapter.

References

Acuña, V and Tockner, K 2010. The effects of alterations in temperature and flow regime on organic carbon dynamics in Mediterranean river networks. *Global Change Biology* 16, 2638–2650.

Acuña, V, Datry, T, Marshall, J *et al.* 2014. Why should we care about temporary waterways? *Science* 343, 1080–1082.

Adis, J. and Junk, WJ 2002.Terrestrial invertebrates inhabiting lowland river floodplains of Central Amazonia and Central Europe: a review. *Freshwater Biology* 47, 711–731.

Anna, A, Yorgos, C, Konstantinos, P, and Maria, L 2009. Do intermittent and ephemeral Mediterranean rivers belong to the same river type? *Aquatic Ecology* 43(2), 465–476.

Arab, A, Lek, S, Lounaci, A, and Park, YS 2004. Spatial and temporal patterns of benthic invertebrate communities in an intermittent river (North Africa). *Annales de Limnologie-International Journal of Limnology* 40(4), 317–327.

Arscott, DB, Tockner, K, and Ward, JV 2005. Lateral organization of aquatic invertebrates along the corridor of a braided floodplain river. *Journal of the North American Benthological Society* 24(4), 934–954.

Arscott, DB, Larned, ST, Scarsbrook, MR, and Lambert, P 2010. Aquatic invertebrate community structure along an intermittence gradient: Selwyn River, New Zealand. *Journal of the North American Benthological Society* 29, 530–545.

Baxter, CV and Hauer, FR 2000. Geomorphology, hyporheic exchange, and selection of spawning habitat by bull trout (*Salvelinus confluentus*). *Canadian Journal of Fisheries and Aquatic Sciences* 57(7), 1470–1481.

Baxter, CV, Fausch, KD, and Saunders, WC 2005. Tangled webs: reciprocal flows of invertebrate prey link streams and riparian zones. *Freshwater Biology* 50(2), 201–220.

Belletti, B, Dufour, S, and Piégay, H 2013. Regional variability of aquatic pattern in braided reaches (example of the French Rhône basin). *Hydrobiologia,* 712(1), 25–41.

Belletti, B, Dufour, S, and Piégay, H 2014. Regional assessment of the multi-decadal changes in braided riverscapes following large floods (example of 12 reaches in south east of France). *Advances in Geosciences* 37(37), 57–71.

Benstead, JP and Leigh, DS 2012. An expanded role for river networks. *Nature Geoscience* 5(10), 678–679.

Bernal, S, Butturini, A, and Sabater, F 2005. Seasonal variations of dissolved nitrogen and DOC: DON ratios in an intermittent Mediterranean stream. *Biogeochemistry* 75(2), 351–372.

Bogan, MT, Boersma, KS, and Lytle, DA 2013. Flow intermittency alters longitudinal patterns of invertebrate diversity and assemblage composition in an arid-land stream network. *Freshwater Biology* 58 (5), 1016–1028.

Bonada, N. and Resh, VH 2013. Mediterranean-climate streams and rivers: geographically separated but ecologically comparable freshwater systems. *Hydrobiologia* 719(1), 1–29.

Bonada, N, Rieradevall, M., and Prat, N 2007. Macroinvertebrate community structure and biological traits related to flow permanence in a Mediterranean river network. *Hydrobiologia* 589(1), 91–106.

Boulton, AJ 1991. Eucalypt leaf decomposition in an intermittent stream in south-eastern Australia. *Hydrobiologia* 211(2), 123–136.

Boulton, AJ 2003. Parallels and contrasts in the effects of drought on stream macroinvertebrate assemblages. *Freshwater Biology* 48(7), 1173–1185.

Boulton, AJ 2014. Conservation of ephemeral streams and their ecosystem services: what are we missing? *Aquatic Conservation: Marine and Freshwater Ecosystems* 24(6), 733–738.

Boulton, AJ and Lake, PS 1992. The ecology of two intermittent streams in Victoria, Australia. *Freshwater Biology* 27(1), 99–121.

Boumezzough, A, and Musso, JJ 1983. Etude des communautés animales ripicoles du bassin de la rivière Aille (Var-France). I. Aspects biologiques et éco-éthologiques. *Ecologia Mediterranea* 9, 31–56.

Bronge, C 1996. Hydrographic and climatic changes influencing the proglacial Druzhby drainage system, Vestfold Hills, Antarctica. *Antarctic Science* 8(04), 379–388.

Buttle, JM, Boon, S, Peters, DL *et al.* 2012. An overview of temporary stream hydrology in Canada. *Canadian Water Resources Journal* 37(4), 279–310.

Butturini, A, Bernal, S, Sabater, S, and Sabater, F 2002. The influence of riparian-hyporheic zone on the hydrological responses in an intermittent stream. *Hydrology and Earth System Sciences Discussions* 6(3), 515–526.

Cañedo-Argüelles, M, Boersma, KS, Bogan, MT *et al.* 2015. Dispersal strength determines meta-community structure in a dendritic riverine network. *Journal of Biogeography* 42(4), 778–790.

Capderrey, C, Datry, T, Foulquier, A *et al.* 2013. Invertebrate distribution across nested geomorphic features in braided-river landscapes. *Freshwater Science* 32(4), 1188–1204.

Carey, SK and Quinton, WL 2005. Evaluating runoff generation during summer using hydrometric, stable isotope and hydrochemical methods in a discontinuous permafrost alpine catchment. *Hydrological Processes* 19(1), 95–114.

Changming, L and Shifeng, Z 2002. Drying up of the Yellow River: its impacts and counter-measures. *Mitigation and Adaptation Strategies for Global Change* 7(3), 203–214.

Chester, ET and Robson, BJ 2011. Drought refuges, spatial scale and recolonisation by invertebrates in non-perennial streams. *Freshwater Biology*, 56(10), 2094–2104.

Cipriani, T, Tilmant, F, Branger, F *et al*, 2014. Impact of climate change on aquatic ecosystems along the Asse River Network. In *FRIEND-Water 2014, Hydrology in a Changing World: Environmental and Human Dimensions*, February 2014, Hanoi, Vietnam (IAHS Publication 36X, 2014).

Cohen, H and Laronne, JB 2005. High rates of sediment transport by flashfloods in the southern Judean Desert, Israel. *Hydrological Processes* 19(8), 1687–1702.

Colson, T, Gregory, J, Dorney, J, and Russell, P 2008. Topographic and soil maps do not accurately depict headwater stream networks. *National Wetland Newsletter* 30, 25–28.

Colvin, R, Giannico, GR, Li, J, Boyer, KL, and Gerth, WJ 2009. Fish use of intermittent watercourses draining agricultural lands in the Upper Willamette River Valley, Oregon. *Transactions of the American Fisheries Society*, 138(6), 1302–1313.

Conovitz, PA, MacDonald, LH, and McKnight, DM 2006. Spatial and temporal active layer dynamics along three glacial meltwater streams in the McMurdo Dry Valleys, Antarctica. *Arctic, Antarctic, and Alpine Research*, 38(1), 42–53.

Corti, R. and Datry, T 2012. Invertebrates and sestonic matter in an advancing wetted front travelling down a dry river bed (Albarine, France). *Freshwater Science*, 31(4), 1187–1201.

Corti, R and Datry, T 2014. Drying of a temperate, intermittent river has little effect on adjacent riparian arthropod communities. *Freshwater Biology* 59(4), 666–678.

Corti, R. and Datry, T 2016. Terrestrial and aquatic invertebrates along a gradient of flow permanence (Albarine, France). *Freshwater Biology* 61, 1308–1320.

Corti, R, Larned, ST, and Datry, T 2013. A comparison of pitfall-trap and quadrat methods for sampling ground-dwelling invertebrates in dry riverbeds. *Hydrobiologia* 717(1), 13–26.

Corti, R, Datry, T, Drummond, L, and Larned, ST 2011. Natural variation in immersion and emersion affects breakdown and invertebrate colonization of leaf litter in a temporary river. *Aquatic Sciences* 73(4), 537–550.

Datry, T 2012. Benthic and hyporheic invertebrate assemblages along a flow intermittence gradient: effects of duration of dry events. *Freshwater Biology* 57(3), 563–574.

Datry, T and Larned, S. T 2008. River flow controls ecological processes and invertebrate assemblages in subsurface flowpaths of an ephemeral river reach. *Canadian Journal of Fisheries and Aquatic Sciences*, 65(8), 1532–1544.

Datry, T, Arscott, DB, and Sabater, S 2011. Recent perspectives on temporary river ecology. *Aquatic Sciences – Research Across Boundaries* 73(4), 453–457.

Datry, T, Corti, R, and Philippe, M 2012. Spatial and temporal aquatic–terrestrial transitions in the temporary Albarine River, France: responses of invertebrates to experimental rewetting. *Freshwater Biology* 57(4), 716–727.

Datry, T, Larned, ST, and Tockner, K 2014. Intermittent rivers: a challenge for freshwater ecology. *BioScience*. DOI: 10.1093/biosci/bit027.

Datry, T, Corti, R, Claret, C, and Philippe, M 2011. Flow intermittence controls leaf litter breakdown in a French temporary alluvial river: the 'drying memory'. *Aquatic Sciences* 73(4), 471–483.

Datry, T, Larned, ST, Fritz, KM *et al.* 2014a. Broad-scale patterns of invertebrate richness and community composition in temporary rivers: effects of flow intermittence. *Ecography* 37(1), 94–104.

Datry, T, Corti, R, Belletti, B, and Piégay, H 2014b. Ground-dwelling arthropod communities across braided river landscape mosaics: a Mediterranean perspective. *Freshwater Biology* 59(6), 1308–1322.

Datry T, Pella H, Leigh C, *et al.* 2016. A landscape approach to advance intermittent river ecology. *Freshwater Biology* 61, 1200–1213.

Del Rosario, RB and Resh, VH 2000. Invertebrates in intermittent and perennial streams: is the hyporheic zone a refuge from drying?. *Journal of the North American Benthological Society* 19(4), 680–696.

Delucchi, CM and Peckarsky, BL 1989. Life history patterns of insects in an intermittent and a permanent stream. *Journal of the North American Benthological Society* 8(4), 308–321.

Dieter, D, von Schiller, D, Garcia-Roger, EM *et al.* 2011. Preconditioning effects of intermittent stream flow on leaf litter decomposition. *Aquatic Sciences* 73(4), 599–609.

Dijkstra, KDB, Monaghan, MT, and Pauls, SU 2014. Freshwater biodiversity and aquatic insect diversification. *Annual Review of Entomology* 59, 143–163.

Dodds, WK 1997. Distribution of runoff and rivers related to vegetative characteristics, latitude, and slope: a global perspective. *Journal of the North American Benthological Society* 16(1), 162–168.

Doering, M, Uehlinger, U, Rotach, A *et al.* 2007. Ecosystem expansion and contraction dynamics along a large Alpine alluvial corridor (Tagliamento River, Northeast Italy). *Earth Surface Processes and Landforms* 32(11), 1693–1704.

Döll, P and Schmied, HM 2012. How is the impact of climate change on river flow regimes related to the impact on mean annual runoff? A global-scale analysis. *Environmental Research Letters* 7(1), 014037.

Fagan, WF 2002. Connectivity, fragmentation, and extinction risk in dendritic metapopulations. *Ecology* 83(12), 3243–3249.

Feminella, JW 1996. Comparison of benthic macroinvertebrate assemblages in small streams along a gradient of flow permanence. Journal of the North American Benthological Society 651–669.

Fenoglio, S, Bo, T, Cucco, M, and Malacarne, G 2007. Response of benthic invertebrate assemblages to varying drought conditions in the Po river (NW Italy). *Italian Journal of Zoology* 74(2), 191–201.

Finn, DS, Theobald, DM, Black, WC, and Poff, NL 2006. Spatial population genetic structure and limited dispersal in a Rocky Mountain alpine stream insect. *Molecular Ecology* 15(12), 3553–3566.

Finn, DS, Bonada, N, Múrria, C, and Hughes, JM 2011. Small but mighty: headwaters are vital to stream network biodiversity at two levels of organization. *Journal of the North American Benthological Society* 30 (4), 963–980.

Fowler, RT 2004. The recovery of benthic invertebrate communities following dewatering in two braided rivers. *Hydrobiologia* 523(1–3), 17–28.

Fritz, KM, Hagenbuch, E, D'Amico, E *et al.* 2013. Comparing the extent and permanence of headwater streams from two field surveys to values from hydrographic databases and maps. *Journal of the American Water Resources Association* 49(4), 867–882.

Fu, G, Chen, S, Liu, C, and Shepard, D 2004. Hydro-climatic trends of the Yellow River basin for the last 50 years. *Climatic Change* 65(1–2), 149–178.

Gleick, PH 2003. Global freshwater resources: soft-path solutions for the 21st century. *Science* 302(5650), 1524–1528.

Gomi, T, Sidle, RC, and Richardson, JS 2002. Understanding processes and downstream linkages of headwater systems headwaters differ from downstream reaches by their close coupling to hillslope processes, more temporal and spatial variation, and their need for different means of protection from land use. *BioScience* 52(10), 905–916.

Gravel, D, Mouquet, N, Loreau, M, and Guichard, F 2010. Patch dynamics, persistence, and species coexistence in metaecosystems. *The American Naturalist* 176(3), 289–302.

Gray, D and Harding, J.S 2009. Braided river benthic diversity at multiple spatial scales: a hierarchical analysis of β diversity in complex floodplain systems. *Journal of the North American Benthological Society* 28(3), 537–551.

Greenwood, MJ and McIntosh, AR 2010. Low river flow alters the biomass and population structure of a riparian predatory invertebrate. *Freshwater Biology* 55(10), 2062–2076.

Gurtz, ME and Tate, CM 1988. Hydrologic influences on leaf decomposition in a channel and adjacent bank of a gallery forest stream. *American Midland Naturalist* 120(1), 11–21.

Hamilton, SK and Gehrke, PC 2005. Australia's tropical river systems: current scientific understanding and critical knowledge gaps for sustainable management. *Marine and Freshwater Research* 56(3), 243–252.

Hansen, WF 2001. Identifying stream types and management implications. *Forest Ecology and Management* 143(1), 39–46.

Hassan, MA and Egozi, R 2001. Impact of wastewater discharge on the channel morphology of ephemeral streams. *Earth Surface Processes and Landforms* 26(12), 1285–1302.

Heino, J 2013. The importance of metacommunity ecology for environmental assessment research in the freshwater realm. *Biological Reviews* 88(1), 166–178.

Hershkovitz, Y and Gasith, A 2013. Resistance, resilience, and community dynamics in mediterranean-climate streams. *Hydrobiologia* 719(1), 59–75.

Hladyz, S, Watkins, SC, Whitworth, KL, and Baldwin, DS 2011. Flows and hypoxic blackwater events in managed ephemeral river channels. *Journal of Hydrology* 401(1), 117–125.

Hoeve, J and de Vries, S 2014. *Characterizing Spatial and Temporal Patterns in Hydrological Data to Simulate Flow Intermittence.* Final Report, University of Twente, Nederlands, and Irstea, France; 55 pp.

Hose, GC, Jones, P, and Lim, RP 2005. Hyporheic macroinvertebrates in riffle and pool areas of temporary streams in south eastern Australia. *Hydrobiologia* 532(1–3), 81–90.

Hughes, JM, Schmidt, DJ, and Finn, DS 2009. Genes in streams: using DNA to understand the movement of freshwater fauna and their riverine habitat. *BioScience* 59(7), 573–583.

Hunter, MA, Quinn, T, and Hayes, MP 2005. Low flow spatial characteristics in forested headwater channels of southwest Washington. *Journal of the American Water Resources Association* 41, 503–516.

Izbicki, JA 2007. Physical and temporal isolation of mountain headwater streams in the Western Mojave Desert, Southern California. *Journal of the American Water Resources Association* 43(1), 26–40.

Jacobson, PJ and Jacobson, KM 2013. Hydrologic controls of physical and ecological processes in Namib Desert ephemeral rivers: implications for conservation and management. *Journal of Arid Environments* 93, 80–93.

Jacobson, PJ, Jacobson, KM, Angermeier, PL, and Cherry, DS 2000. Variation in material transport and water chemistry along a large ephemeral river in the Namib Desert. *Freshwater Biology* 44(3), 481–491.

Jaeger, KL and Olden, JD 2012. Electrical resistance sensor arrays as a means to quantify longitudinal connectivity of rivers. *River Research and Applications* 28(10), 1843–1852.

Jaeger, KL, Olden, JD, and Pelland, NA 2014. Climate change poised to threaten hydrologic connectivity and endemic fishes in dryland streams. *Proceedings of the National Academy of Sciences* 111(38), 13894–13899.

Jourde, H, Roesch, A, Guinot, V, and Bailly-Comte, V 2007. Dynamics and contribution of karst groundwater to surface flow during Mediterranean flood. *Environmental Geology* 51(5), 725–730.
Kappelman, J, Tewabe, D, Todd, L *et al.* 2014. Another unique river: a consideration of some of the characteristics of the trunk tributaries of the Nile River in northwestern Ethiopia in relationship to their aquatic food resources. *Journal of Human Evolution* 77, 117–131.
Kawanishi, R, Inoue, M, Dohi, R *et al.* 2013. The role of the hyporheic zone for a benthic fish in an intermittent river: a refuge, not a graveyard. *Aquatic Sciences* 75(3), 425–431.
Knighton, AD and Nanson, GC 1994. Waterholes and their significance in the anastomosing channel system of Cooper Creek, Australia. *Geomorphology* 9(4), 311–324.
Konrad, CP 2006. Location and timing of river–aquifer exchanges in six tributaries to the Columbia River in the Pacific Northwest of the United States. *Journal of Hydrology* 329(3), 444–470.
Lake, PS 2000. Disturbance, patchiness, and diversity in streams. *Journal of the North American Benthological Society* 19(4), 573–592.
Lake, PS 2003. Ecological effects of perturbation by drought in flowing waters. *Freshwater Biology* 48(7), 1161–1172.
Langhans, SD and Tockner, K 2014. Is the unsaturated sediment a neglected habitat for riparian arthropods? Evidence from a large gravel-bed river. *Global Ecology and Conservation* 2, 129–137.
Lambeets, K, Vandegehuchte, ML, Maelfait, JP, and Bonte, D 2008. Understanding the impact of flooding on trait-displacements and shifts in assemblage structure of predatory arthropods on river banks. *Journal of Animal Ecology* 77(6), 1162–1174.
Lange, J. 2005. Dynamics of transmission losses in a large arid stream channel. *Journal of Hydrology* 306, 112–126.
Larimore, RW, Childers, WF, and Heckrotte, C 1959. Destruction and re-establishment of stream fish and invertebrates affected by drought. *Transactions of the American Fisheries Society* 88(4), 261–285.
Larned, ST, Datry, T, and Robinson, CT 2007. Invertebrate and microbial responses to inundation in an ephemeral river reach in New Zealand: effects of preceding dry periods. *Aquatic Sciences* 69(4), 554–567.
Larned, ST, Hicks, DM, Schmidt, J *et al.* 2008. The Selwyn River of New Zealand: a benchmark system for alluvial plain rivers. *River Research and Applications* 24(1), 1–21.
Larned, ST, Datry, T, Arscott, D.B, and Tockner, K 2010. Emerging concepts in temporary-river ecology. *Freshwater Biology* 55(4), 717–738.
Larned, ST, Schmidt, J, Datry, T *et al.* 2011. Longitudinal river ecohydrology: flow variation down the lengths of alluvial rivers. *Ecohydrology* 4(4), 532–548.
Lehner, B and Grill G 2013. Global river hydrography and network routing: baseline data and new approaches to study the world's large river systems. *Hydrological Processes* 27(15), 2171–2186.
Leibold, MA, Holyoak, M, Mouquet, N *et al.* 2004. The metacommunity concept: a framework for multi-scale community ecology. *Ecology Letters* 7(7), 601–613.
Leigh, C, Boulton, AJ, Courtwright, JL *et al.* 2016. Ecological research in intermittent rivers: an historical review. *Freshwater Biology* 61,1181–1199.
Levick, LR, Goodrich, DC, Hernandez, M *et al.* 2008. *The Ecological and Hydrological Significance of Ephemeral and Intermittent Streams in the Arid and Semi-Arid American Southwest.* US Environmental Protection Agency, Office of Research and Development.
Loreau, M, Mouquet, N, and Holt, RD 2003. Meta-ecosystems: a theoretical framework for a spatial ecosystem ecology. *Ecology Letters* 6(8), 673–679.
Lowe, WH and Likens, GE 2005. Moving headwater streams to the head of the class. *BioScience* 55(3), 196–197.
Lowe, WH, Likens, GE, and Power, ME 2006. Linking scales in stream ecology. *BioScience* 56(7), 591–597.

Lytle, DA and Poff, NL 2004. Adaptation to natural flow regimes. *Trends in Ecology and Evolution* 19(2), 94–100.

Malard, F, Ward, JV, and Robinson, CT 2000. An expanded perspective of the hyporheic zone. *Verhandlungen der Internationalen Vereinigung für Theoretische und Angewandte Limnologie* 27, 431–437.

Malard, F, Tockner, K, Dole-Olivier, MJ, and Ward, JV 2002. A landscape perspective of surface–subsurface hydrological exchanges in river corridors. *Freshwater Biology* 47(4), 621–640.

Malard, F, Galassi, D, Lafont, M *et al.* 2003. Longitudinal patterns of invertebrates in the hyporheic zone of a glacial river. *Freshwater Biology* 48(10), 1709–1725.

Mansell, MG and Hussey, SW 2005. An investigation of flows and losses within the alluvial sands of ephemeral rivers in Zimbabwe. *Journal of Hydrology* 314(1), 192–203.

Marten, A, Braendle, M, and Brandl, R 2006. Habitat type predicts genetic population differentiation in freshwater invertebrates. *Molecular Ecology* 15(9), 2643–2651.

McKnight, DM, Niyogi, DK, Alger, AS *et al.* 1999. Dry valley streams in Antarctica: ecosystems waiting for water. *BioScience* 49(12), 985–995.

McMahon, TA 1979. Hydrological characteristics of arid zones. In *Hydrology of Areas of Low Precipitation, Proceedings of the Canberra Symposium*. Publication No. 128, IAHS-AISH: Wallingford, UK; 105–123.

Metcalfe, RA and Buttle, JM 2001. Soil partitioning and surface store controls on spring runoff from a boreal forest peatland basin in north-central Manitoba, Canada. *Hydrological Processes* 15(12), 2305–2324.

Meybeck, M 2003. Global analysis of river systems: from Earth system controls to Anthropocene syndromes. *Philosophical Transactions of the Royal Society of London. Series B: Biological Sciences* 358 (1440), 1935–1955.

Meybeck, M, Dürr, HH, and Vörösmarty, CJ 2006. Global coastal segmentation and its river catchment contributors: a new look at land–ocean linkage. *Global Biogeochemical Cycles* 20(1). DOI: 10.1029/2005GB002540.

Meyer, A and Meyer, EI 2000. Discharge regime and the effect of drying on macroinvertebrate communities in a temporary karst stream in East Westphalia (Germany). *Aquatic Sciences* 62(3), 216–231.

Meyer, JL, Wallace, JB, Press, MC *et al.* 2001. Lost linkages and lotic ecology: rediscovering small streams. In *Ecology: Achievement and Challenge: the 41st Symposium of the British Ecological Society*, Orlando, FL, 10–13 April 2000. Sponsored by the Ecological Society of America, Blackwell Science; 295–317.

Meyer, JL, Strayer, DL, Wallace, JB *et al.* 2007. The contribution of headwater streams to biodiversity in river networks. *Journal of the American Water Resources Association* 43(1), 86–103.

Meyerhoff, RD and Lind, OT 1987. Factors affecting the benthic community structure of a discontinuous stream in Guadelupe Mountains National Park, Texas. *Internationale Revue der gesamten Hydrobiologie und Hydrographie* 72(3), 283–296.

Mielko, C and Woo, MK 2006. Snowmelt runoff processes in a headwater lake and its catchment, subarctic Canadian Shield. *Hydrological Processes* 20(4), 987–1000.

Millenium Ecosystem Assessment 2005. *Ecosystems and Human Well-being: Wetland and Water. Synthesis*. Available at http://www.millenniumassessment.org/proxy/Document

Múrria, C, Bonada, N, Ribera, C, and Prat, N 2010. Homage to the Virgin of Ecology, or why an aquatic insect unadapted to desiccation may maintain populations in very small, temporary Mediterranean streams. *Hydrobiologia* 653(1), 179–190.

Nadeau, T-L and Rains, MC 2007. Hydrological connectivity of headwaters to downstream waters: introduction to the featured collection. *Journal of the American Water Resources Association* 43(1), 1.

Nakano, S and Murakami, M 2001. Reciprocal subsidies: dynamic interdependence between terrestrial and aquatic food webs. *Proceedings of the National Academy of Sciences* 98(1), 166–170.

Obermann, M, Rosenwinkel, KH, and Tournoud, MG 2009. Investigation of first flushes in a medium-sized mediterranean catchment. *Journal of Hydrology* 373(3), 405–415.

Ormerod, SJ, Wade, KR, and Gee, AS 1987. Macro-floral assemblages in upland Welsh streams in relation to acidity, and their importance to invertebrates. *Freshwater Biology* 18(3), 545–557.

Paetzold, A and Tockner, K 2005. Effects of riparian arthropod predation on the biomass and abundance of aquatic insect emergence. *Journal of the North American Benthological Society* 24(2), 395–402.

Palmer, MA, Reidy Liermann, CA, Nilsson, C *et al.* 2008. Climate change and the world's river basins: anticipating management options. *Frontiers in Ecology and the Environment* 6(2), 81–89.

Papadopoulou, A, Bergsten, J, Fujisawa, T *et al.* 2008. Speciation and DNA barcodes: testing the effects of dispersal on the formation of discrete sequence clusters. *Philosophical Transactions of the Royal Society of London B: Biological Sciences* 363(1506), 2987–2996.

Piégay, H, Alber, A, Slater, L, and Bourdin, L 2009. Census and typology of braided rivers in the French Alps. *Aquatic Sciences* 71(3), 371–388.

Pires, AM, Cowx, IG, and Coelho, MM 1999. Seasonal changes in fish community structure of intermittent streams in the middle reaches of the Guadiana basin, Portugal. *Journal of Fish Biology* 54(2), 235–249.

Poff, NL and Ward, JV 1989. Implications of streamflow variability and predictability for lotic community structure: a regional analysis of streamflow patterns. *Canadian Journal of Fisheries and Aquatic Sciences* 46(10), 1805–1818.

Poizat, G and Pont, D 1996. Multi-scale approach to species–habitat relationships: juvenile fish in a large river section. *Freshwater Biology* 36(3), 611–622.

Postel, SL 2000. Entering an era of water scarcity: the challenges ahead. *Ecological Applications*, 10(4), 941–948.

Power, ME, Parker, G, Dietrich, WE, and Sun, A 1995. How does floodplain width affect floodplain river ecology? A preliminary exploration using simulations. *Geomorphology* 13(1), 301–317.

Prat N, Gallart F, Von Schiller D *et al* 2014. The MIRAGE Toolbox: an integrated assessment tool for temporary streams. *River Research and Applications*. DOI:10.1002/rra.2757.

Raymond, PA, Hartmann, J, Lauerwald, R *et al.* 2013. Global carbon dioxide emissions from inland waters. *Nature* 503(7476), 355–359.

Reid, I, Laronne, JB, and Powell, DM 1998. Flash-flood and bedload dynamics of desert gravel-bed streams. *Hydrological Processes* 12(4), 543–557.

Resh, VH, Brown, AV, Covich, AP *et al.* 1988. The role of disturbance in stream ecology. *Journal of the North American Benthological Society* 7(4), 433–455.

Robinson, CT and Matthaei, S 2007. Hydrological heterogeneity of an alpine stream–lake network in Switzerland. *Hydrological Processes* 21(23), 3146–3154.

Rosado, J, Morais, M, and Tockner, K 2014. Mass dispersal of terrestrial organisms during first flush events in a temporary stream. *River Research and Applications* 31(7), 912–917.

Rüegg, J and Robinson, CT 2004. Comparison of macroinvertebrate assemblages of permanent and temporary streams in an Alpine flood plain, Switzerland. *Archiv für Hydrobiologie* 161(4), 489–510.

Sabater, S and Tockner, K 2010. Effects of hydrologic alterations on the ecological quality of river ecosystems. In *Water Scarcity in the Mediterranean*. Springer: Berlin, Heidelberg; 15–39.

Sagar, PM 1983. Invertebrate recolonisation of previously dry channels in the Rakaia River. *New Zealand Journal of Marine and Freshwater Research* 17(4), 377–386.

Schmidt, J, Larned, ST, Arscott, D, and Diettrich, JC 2009. Hydrological indices for quantifying ecologically relevant flow conditions in intermittent alluvial plain rivers. *IAHS Publications* 20, 94.

Sheldon, F, Boulton, AJ, and Puckridge, JT 2002. Conservation value of variable connectivity: aquatic invertebrate assemblages of channel and floodplain habitats of a central Australian arid-zone river, Cooper Creek. *Biological Conservation* 103(1), 13–31.

Sheldon, F, Bunn, SE, Hughes, JM *et al.* 2010. Ecological roles and threats to aquatic refugia in arid landscapes: dryland river waterholes. *Marine and Freshwater Research* 61(8), 885–895.
Snelder, TH, Datry, T, Lamouroux, N *et al.* 2013. Regionalization of patterns of flow intermittence from gauging station records. *Hydrological and Earth System Sciences* 17, 2685–2699.
Stanford, JA and Ward, JV 1993. An ecosystem perspective of alluvial rivers: connectivity and the hyporheic corridor. *Journal of the North American Benthological Society* 12(1), 48–60.
Stanford, JA, Lorang, MS, and Hauer, FR 2005. The shifting habitat mosaic of river ecosystems. *Internationale Vereinigung fur Theoretische und Angewandte Limnologie Verhandlungen*, 29(1), 123–136.
Stanish, LF, Kohler, TJ, Esposito, RM *et al.* 2012. Extreme streams: flow intermittency as a control on diatom communities in meltwater streams in the McMurdo Dry Valleys, Antarctica. *Canadian Journal of Fisheries and Aquatic Sciences* 69(8), 1405–1419.
Stanley, EH, Fisher, SG, and Grimm, NB 1997. Ecosystem expansion and contraction in streams. *BioScience* 47(7), 427–435.
Stanley, EH, Buschman, DL, Boulton, AJ *et al.* 1994. Invertebrate resistance and resilience to intermittency in a desert stream. *American Midland Naturalist* 131(2), 288–300.
Steward, AL, Marshall, JC, Sheldon, F *et al.* 2011. Terrestrial invertebrates of dry river beds are not simply subsets of riparian assemblages. *Aquatic Sciences* 73(4), 551–566.
Steward, AL, von Schiller, D, Tockner, K *et al.* 2012. When the river runs dry: human and ecological values of dry riverbeds. *Frontiers in Ecology and the Environment* 10(4), 202–209.
Stubbington, R 2012. The hyporheic zone as an invertebrate refuge: a review of variability in space, time, taxa and behaviour. *Marine and Freshwater Research* 63(4), 293–311.
Stubbington, R and Datry, T 2013. The macroinvertebrate seedbank promotes community persistence in temporary rivers across climate zones. *Freshwater Biology* 58(6), 1202–1220.
Stubbington, R, Wood, PJ, and Reid, I 2011. Spatial variability in the hyporheic zone refugium of temporary streams. *Aquatic Sciences* 73(4), 499–511.
Swan, CM and Brown, BL 2014. Using rarity to infer how dendritic network structure shapes biodiversity in riverine communities. *Ecography* 37(10), 993–1001.
Tockner, K and Malard, F 2003. Channel typology. In *Ecology of a Glacial Floodplain* (eds JV Ward and U Uehlinger). Kluwer Academic Publishers: Dordrecht; 57–73.
Tockner, K, Malard, F, and Ward, JV 2000. An extension of the flood pulse concept. *Hydrological Processes* 14(16–17), 2861–2883.
Tockner, K, Ward, JV, Arscott, DB *et al.* 2003. The Tagliamento River: a model ecosystem of European importance. *Aquatic Sciences* 65(3), 239–253.
Tonina, D and Buffington, JM 2009. Hyporheic exchange in mountain rivers I: mechanics and environmental effects. *Geography Compass* 3(3), 1063–1086.
Tooth, S 2000. Process, form and change in dryland rivers: a review of recent research. *Earth-Science Reviews* 51(1), 67–107.
Von Schiller, D, Acuña, V, Graeber, D *et al.* 2011. Contraction, fragmentation and expansion dynamics determine nutrient availability in a Mediterranean forest stream. *Aquatic Sciences* 73(4), 485–497.
Von Schiller, D, Graeber, D, Ribot, M *et al.* 2015. Hydrological transitions drive dissolved organic matter quantity and composition in a temporary Mediterranean stream. *Biogeochemistry* 123(3), 429–446.
Vorste, RV, Corti, R, Sagouis, A, and Datry, T 2016. Invertebrate communities in gravel-bed, braided rivers are highly resilient to flow intermittence. *Freshwater Science* 35(1), 164–177.
White, DS 1993. Perspectives on defining and delineating hyporheic zones. *Journal of the North American Benthological Society* 12(1), 61–69.
Whitford, WG 2002. *Ecology of Desert Systems.* Academic Press.

Widder, S, Besemer, K, Singer, GA *et al.* 2014. Fluvial network organization imprints on microbial co-occurrence networks. *Proceedings of the National Academy of Sciences* 111(35), 12799–12804.

Williams, DD 2006. *The Biology of Temporary Waters*. Oxford University Press.

Williams, DD and Hynes, HBN 1976. The recolonization mechanisms of stream benthos. *Oikos* 27(2), 265–272.

Witze, A 2014. Water returns to arid Colorado River delta. *Nature News* 507(7492).

Wood, PJ, Gunn, J, Smith, H, and Abas-Kutty, A 2005. Flow permanence and macroinvertebrate community diversity within groundwater dominated headwater streams and springs. *Hydrobiologia* 545(1), 55–64.

Wright, JF 1984. The chironomid larvae of a small chalk stream in Berkshire, England. *Ecological Entomology* 9(2), 231–238.

11

Catastrophic Deposition of Gravel from Outbreak Floods

Paul A. Carling

11.1 Introduction

This chapter is a condensation of Carling (2013) and focuses on general stratigraphic associations that typify outbreak flood characteristics rather than the site characteristics, flood generating mechanisms, and the detail of the sedimentology. Those readers interested in the additional detail should refer to Carling (2013).

Catastrophic freshwater flooding can occur at a variety of scales and is due to different generating mechanisms, including excessive precipitation and/or ice and snow melt. However, large-scale events are often associated with the failure of man-made or natural dams and so are related to the dispersal of impounded water. The latter events include: landslides into lakes causing overtopping, failure of debris dams, tectonic adjustments, and geothermal activity such as volcanic eruptions beneath ice-sheets. Perhaps most well studied are (i) those due to ice-dam mechanical failures and (ii) those due to geothermal activity beneath ice-sheets – the jökulhlaups of Iceland in particular. The biggest floods are associated with failures of Quaternary lakes impounded by glaciers or ice-sheet lobes. Given the past extensive research focus on Quaternary proglacial lakes, this review draws heavily but not exclusively on the deposits of such examples, although important commonalities can be drawn with deposits generated by other failure mechanisms.

Although possible examples of catastrophic Quaternary freshwater flooding have been known for decades, it is only in recent years that it has become evident that such spectacular events were circumpolar; associated with the termination of the Pleistocene Epoch and the decay of polar and continental ice sheets. Flooding was a global phenomenon (Figure 11.1; Baker 2013), with extraordinary spatial extent and repeated frequency (Baker 1997, 2007, 2013; Mangerud *et al.* 2004). Ice-marginal and subglacial palaeolakes were often associated with catastrophic flooding due to natural ice-dam failure (Margold *et al.* 2011). Those floods with a *peak* discharge of greater than $10^6\ m^3\ s^{-1}$ (1 Sverdrup) are termed 'megafloods' (Baker 2002); and the best studied are examples from the Northern Hemisphere (O'Connor and Baker 1992; O'Connor 1993; Clarke *et al.* 2004; Miyamoto *et al.* 2006a,b; Alho, Baker, and Smith 2010; Carling *et al.* 2010). Some Holocene Icelandic floods may have exceeded 1 Sv (Tómasson 1973), so Icelandic jökulhlaups are also considered.

Maizels (1997a) and Marren and Schuh (2009) reviewed small-scale catastrophic flooding in the context of Holocene Icelandic jökulhlaups. They concluded that there are sedimentary signatures that are unique to jökulhlaups, but most signatures can be found in other fluvial deposits as well. Two key

Gravel-Bed Rivers: Processes and Disasters, First Edition. Edited by Daizo Tsutsumi and Jonathan B. Laronne.

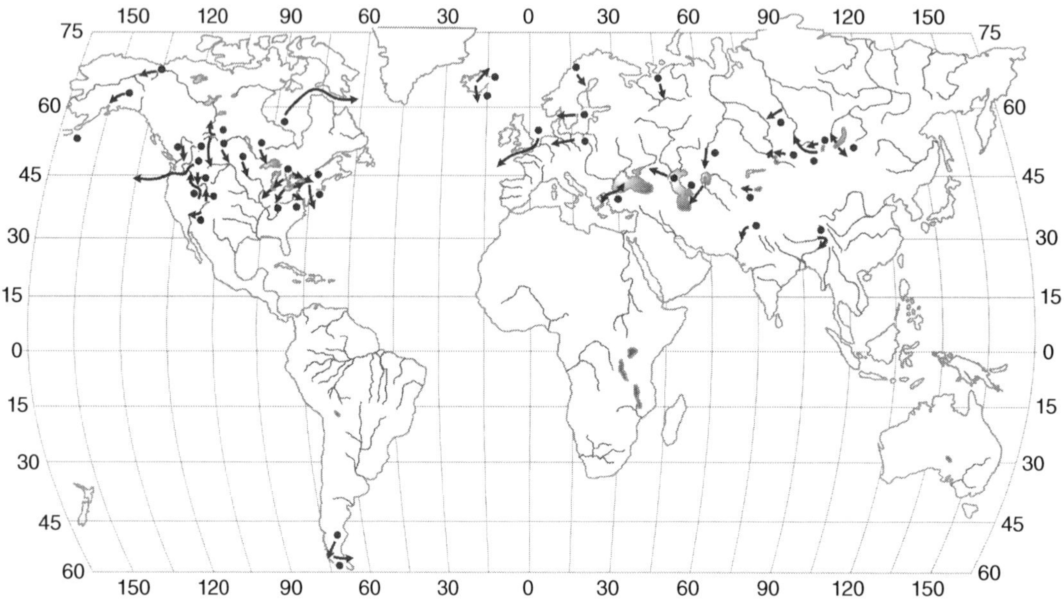

Figure 11.1 Locations for 41 regions where terrestrial megaflooding or related large-scale flooding has been recorded. (Redrawn with modification and permission from Baker (2013).)

issues were identified (Marren and Schuh 2009); one was that the range of criteria for identifying catastrophic flood deposition was *context-specific*. Thus if a glacial context could be determined using criteria other than the sedimentary signatures then a jökulhlaup interpretation was likely to be correct, in contrast to other fluvial contexts. The second issue is one of scale. Marren and Schuh (2009) indicate that the thickness and extent of catastrophic flood sediment successions is likely to be greater than the thickness of successions of non-catastrophic deposits in the same location. These authors also indicated that the catastrophic deposits might be coarser and also deposited to higher elevations than more usual waterlain deposits.

By their very nature, the floods were extremely unsteady, macroturbulent, and of very high power. Modern jökulhlaups typically develop sustained power of around 3000–5000 W m^{-2}, which can suspend 100–200 mm sized clasts (Russell and Knudsen 1999a), whereas short-duration peak flows can be associated with power of 40 000 W m^{-2}, which can suspend 2-m boulders (Russell 2005). The largest Quaternary floods had powers of 10^5–10^6 W m^{-2} sustained over large distances and for durations of several days. These latter flows suspended large blocks and comminuted rock debris to produce fine suspended granules. Thus, with megafloods it is not possible to preclude the possibility of deposition of gravel from high-concentration suspensions (Lord and Kehew 1987; Mutti *et al.* 1996).

11.2 Depositional Context

Coarse and fine sediments associated with outburst floods are deposited in a number of distinct contexts. Coarser sediments tend to be deposited in the upper portions of the flood tract whereas very fine sediment may form distal slackwater sequences. Considering the coarser components, sediment deposits are chiefly found in flow expansions along the course of the main flood. The resultant barforms can be classified (Figure 11.2) for convenience (Carling *et al.* 2009a).

1) 'Expansion bars' (Figure 11.2a) occur downstream of locations where the flow suddenly widens, and they consist of a 'deltaic' deposit of coarse traction sediment, the surface of the bar may be characterised by longitudinal ridges and swales. Usually, the surface is dissected during the falling flow to leave a field of longitudinal barforms. The downstream termination of the expansion bar may be steep (e.g. O'Connor 1993; his figure 57).
2) 'Point bars' (Figure 11.2b) may form in similar location and manner to lesser-scale channel flows, although the scale is much larger. Deposition occurs on the inner bend of the flow and the outer margin can be very steep owing to lateral cutting of the coarse deposits by later lesser floods.
3) 'Eddy bars' (Figure 11.2c) form due to flow separation along the channel margins, such that flow recirculates within embayments in the channel margins. At the simplest a bar may fill a small re-entrant completely. However, eddying also occurs within the entrances to tributary valleys such that deposition may occur far up the tributary. In the latter case, the bar will fine and lose height up the tributary (O'Connor 1993; Benito and O'Connor 2003).
4) 'Pendant' bars (Figure 11.2d) form downstream of isolated obstacles located within the main flow, typically small hills and rocky outcrops. A bar is deposited in the separation zone downflow of the obstacle, although a shallow depression due to turbulent scour may occur close to the obstacle (O'Connor 1993, p. 46).

At the simplest, expansion bars form at the base of the main flow and tend to be composed of very coarse gravel. The other bars form at the margins of the main flow and can display distinct gradation from large boulders proximal to the main flow through to finer sediments distally (O'Connor 1993; Benito and O'Connor 2003). The simple sequences detailed below tend to represent single-channel floodways, but more complex depositional sequences can occur (Miyamoto *et al.* 2006b) in those cases where flood tracts become anastomosed (e.g. Bretz 1925a; Bretz, Smith, and Neff 1956), such that flood waves progressing in different channels within the network arrive at a given location at different times.

11.3 A Framework for Description of Megaflood Sedimentary Successions

Many megaflood depositional successions exist in locations which have been subject to 'normal' proglacial stream sedimentary processes, including annual meltwater events. These processes were active both before and after the megaflood(s) and sedimentary evidence for these normal processes has to be discriminated from the megaflood successions (Marren 2002a). Studies of normal braidplain sedimentary facies are legion (e.g. Miall 1977, 1978) and the key characteristics have been summarized by Maizels (1997a) as her 'Outwash Type I' to discriminate them from large-flood deposits.

The complete vertical lithostratigraphic stack of sediments associated with a given megaflood might be termed a 'succession' and may be of the order of tens of metres thick. This succession might contain information on the waxing and waning of a single flood or several floods. The unit of investigation at the next lesser scale can be termed a 'sequence' (up to several metres thick) such that several sequences make up the succession. A sequence has specific stratigraphic motifs that might be ascribed to defined hydraulics of deposition. Thus, the number and nature of the sequences provide the large-scale information on flood

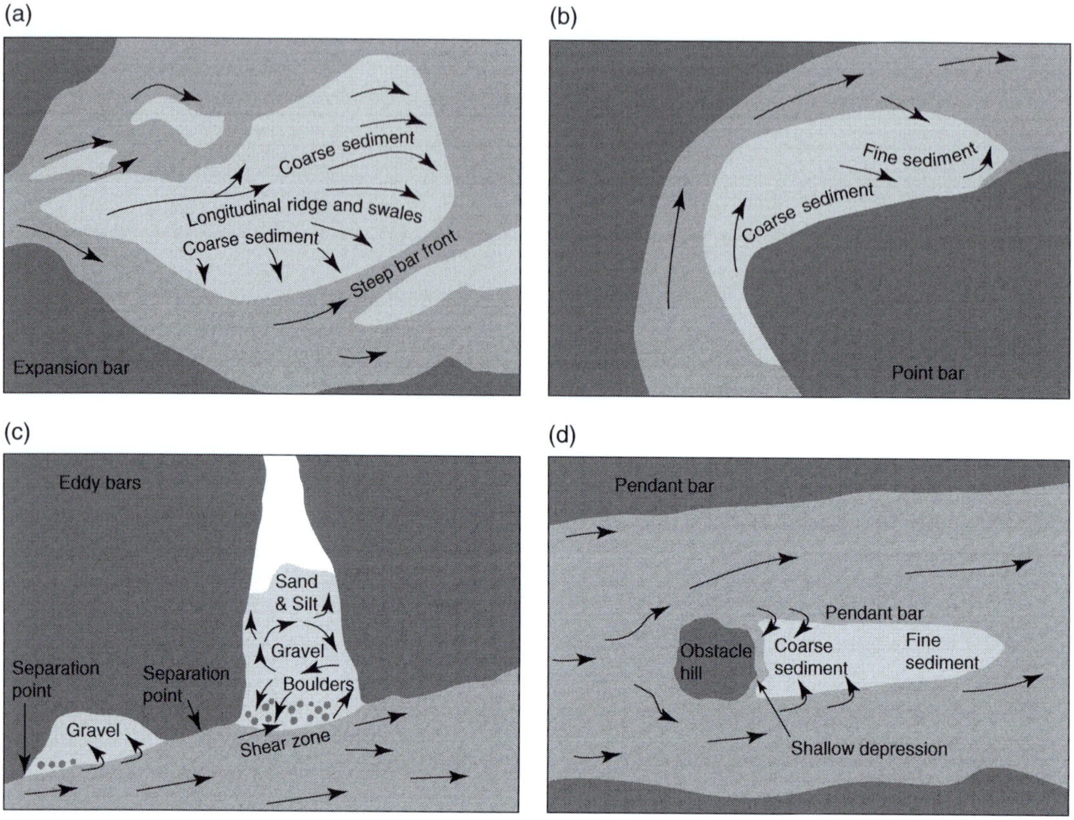

Figure 11.2 Planview cartoons of typical barform locations along the course of megafloods. (a) Expansion bars occur where the channel suddenly expands and flow power is reduced. (b) Point bars occur on the inside of channel bends in the same fashion as occurs in lower-energy channels. (c) Eddy bars occur where flow recirculates in channel-side re-entrants, which include back-flooded tributary valleys. (d) Pendant bars occur on the downflow side of obstructions to flow in the main channel, such as isolated hills.

hydrograph(s) progression. Sequences are composed of individual beds (centimetres to metres thick), which, if sufficiently thin (e.g. <0.01 m), conveniently can be termed laminae. If 'facies' is defined as — a body of intimate sedimentary characteristics that form under certain conditions of sedimentation – the term readily can be related to small-scale fluid processes and so facies characteristics are the smallest scale of investigation and at this scale the fluid dynamics become evident.

11.4 Typical Sequences Within a Succession

To orientate the reader the key succession distilled from the literature is summarized at this point and justified latterly. Megaflood successions typically exhibit sequence variants on the following theme from the base to the top of each succession: (i) basal coarse parallel-bedded units; (ii) large-scale clinoforms; (iii) horizontally bedded laminations; (iv) ripple and dune cross-lamination; (v) silt beds; and (vi) debris-flow units. The successions are almost entirely aggradational, the lack of major discontinuities being striking. The beds are often thickest proximal to the entrance to the depositional embayment and thin and fine into the embayment (Bretz 1959; Benito and O'Connor 2003; Carling *et al.* 2009b). Individual similar sequences might become progressively thinner and finer higher in the succession (Bretz 1925b, 1929; Bretz, Smith, and Neff 1956; Hewitt 1964, 1968, 1982; Baker 1973; Burbank 1983; Baker and Bunker 1985; Smith 1993; Smith and Fisher 1993; Cornwell 1998). Although each location exhibits successions that can vary from the theme noted above, many successions are composed of stacks of sequences that are repeated examples of the theme, punctuated by additional sequences selected from the six or are truncated examples of the theme with sequences missing. The characteristics of the sequences are detailed in the following sections.

11.4.1 Coarse Parallel-Bedded Units

11.4.1.1 Description

The base of sequences frequently can be characterized by thick beds (e.g. 5–10 m) of very coarse gravel (Baker 1973; Fraser and Bleuer 1988; Moody 1988; Fisher and Smith 1993; O'Connor 1993; Maizels 1993, 1997b; Smith 1993; Russell and Marren 1999; Carling *et al.* 2009b). These thick units may be *apparently* homogeneous and chaotic (Lord and Kehew 1987; Maizels 1993; Rushmer 2006, 2007; Duller *et al.* 2008; Carling *et al.* 2009b; Duller, Mountney, and Russell 2010), or sheet-like, internally unstratified and crudely graded (Mutti *et al.* 1996; Marren, Russell, and Rushmer 2009). Units may coarsen upwards (Smith 1993), or may exhibit large-scale low-angle clinoforms dipping down system (e.g. Carling *et al.* 2009b) (Figure 11.3). Sequences are unarmoured at bedding planes.

However, if grain-size is little differentiated in the vertical then any bedding may be crude and a degree of bed amalgamation is common in thicker units. Bedding is crudely horizontal and clast-supported and rarely is imbrication evident (Lord and Kehew 1987; Rushmer 2006; Smith 2006) and so clast-scale palaeocurrent indicators are rare. Where particle orientation is evident it can be either *a*-parallel or *a*-transverse to the inferred flow direction (e.g. Duller, Mountney, and Russell 2010). Transverse orientations usually are related to slow flow speeds and parallel orientations with high-speed flows (Johansson 1963; Rust 1972a; Carling and Glaister 1987). The interstices between larger gravel clasts are normally infilled by sands and granules and only local lenses of openwork gravels occur. These coarse units usually are devoid of silt and clays. The boulder concentrations are not matrix-supported and rarely do large intraclasts occur.

Variations on this general theme include: (i) concentrations of very-large semi-rounded boulders as stringers in otherwise finer sediments, possibly due to melt-out of boulder-rich ice floes in the water above the accreting bar; (ii) thick chaotic units of angular boulders which are sandwiched within finer bar sediments, probably due to large-scale collapse of rock walls adjacent to the bar (Figure 11.4); and (iii) concentrations of angular boulders that diminish down-flow from rock promontories as rock was torn from the bluffs (Carling *et al.* 2009b) and deposited within the accreting proximal portion of bars within the embayments downstream of the promontories (Carrivick *et al.* 2007). Thus basal coarse sequences are not found far into embayments and tributaries. Substantial deposits within megaflood successions that might be due to flood-induced large-scale collapse of valley rock-walls do not seem to have been reported in any detail (Carling *et al.* 2009b) but if suspect deposits are to be identified then analogy with the sedimentology of other large rock-fall deposits might prove fruitful (Shugar and Clague 2012).

11.4.1.2 Interpretation

In part, the thickness and coarseness is owing to the flux of coarse debris that may occur at the beginning of a large powerful flood (Waitt 1980; Maizels 1993, 1997a; Mutti *et al.* 1996), but usually have been related to rapid bedload deposition in tranquil embayments adjacent to a highly energetic

Figure 11.3 Large-scale, planar to slightly convex-up clinoforms near the base of: (a) Komdodj bar, Altai, the author (1.9 m tall) provides scale – flow left to right; (b) the Kezek-Jala bar, Altai, Laura Hayes (1.8 m tall) provides scale – flow left to right.

floodway transporting very coarse material at the base of the flow. Flow separation enables the boulders to be deposited efficiently in the basal shear zone between main flow and the flow separation cell (Baker 1973; Carling 1989; O'Connor 1993; Carrivick *et al.* 2007). Where *a*-parallel clast orientations occur these have be interpreted as evidence for deposition from bedload in a strong unidirectional flow, whereas *a*-transverse orientations have been interpreted as deposition in weaker flows

Figure 11.4 (a) A thick sequence of crudely stratified angular coarse gravels (including <4 m boulders centre-right) overlie flood gravels shown in B; (b) finer well-sorted beds of flood gravel in the Komdodj bar (Altai). Figure (1.8 m tall) in foreground for scale.

(e.g. O'Connor 1993; Benvenuti and Martini 2002). In those cases where good sorting and clear particle orientation are not evident it is probable that rapid deposition occurred (within macroturbulence of the shear zone – Craig 1987) from a relatively congested bedload layer (Smith 1993) at the base of a deep flood; the converse, in shallow prolonged floods, may be a stronger fabric due to a propensity for clasts to become orientated (Maizels 1993; Lønne 1997; Russell and Marren 1999; Cutler *et al.* 2002). Thick amalgamated gravel beds also point to rapid deposition from a sustained sediment supply and sustained flow (Kneller 1995; Kneller and Branney 1995). Progressive coarsening-up has been interpreted as indicative of a change in sediment supply (Russell and Marren 1999; Kataoka 2011) or more probably, given the basal location, as deposition on the rising stage of the flood (Maizels 1991, 1993, 1995). As the deposit builds in height, deposition of coarser material is subsequently precluded on the evolving depositional surface and finer sediments are laid-down conformably above. Thus the upper surfaces of these basal coarse gravel beds are not subject to extended steady flows and so do not armour (Rushmer 2006). Beds are not armoured except in shallow flows (Lord and Kehew 1987; Maizels 1997b).

11.4.2 Large-Scale Clinoforms

11.4.2.1 Description

Large-scale clinoforms of coarse gravel may be encountered in the main-channel flood deposits (Fahnestock and Bradley 1973; Dawson 1989; O'Connor 1993; Russell and Knudsen 1999a,b, 2002; Carling *et al.* 2002, 2009a,b; Benito and O'Connor 2003) and in more flood-marginal environments (Figure 11.5). Typically, clinoform sequences occur above a coarse basal sequence (e.g. Fraser and Bleuer 1988; Winsemann, Asprion, and Meyer 2007; Carrivick *et al.* 2007) and a single clinoform may be several metres high (Carling *et al.* 2002), with dips of up to 10–20°. Highly diagnostic are examples of angle-of-repose sand or gravel cross-bedding within giant bars that dip away from the main channel (Bretz 1925b; Bretz, Smith, and Neff 1956; Allison 1978; Maizels 1997a; Carrivick *et al.* 2007; Carling *et al.* 2002, 2009b) into embayments and tributary valleys. Although there are no detailed studies of large clinoforms, the larger scale examples tend to occur in or close to the main channel whereas smaller scale examples tend to occur within embayments. The stratigraphy and sedimentology of these smaller clinoforms within bars seems similar to the structure of large gravel dunes found in other locations (Carling 1996; Russell, Sharpe, and Bajc 2007). The large-scale clinoforms tend to consist of a single sequence of tabular sub-parallel beds rather than sigmoidal cross-beds. Stacked sequences are uncommon (Russell, Sharpe, and Bajc 2007) and usually are separated by high-angle and sub-horizontal disconformities. High-angle disconformities and cut-and-fill structures within

Figure 11.5 Granule and coarse sand cross-beds dipping into a tributary valley, Little Jaloman, Altai. Note the white silt bed at the unconformity. (a) Tabular cross-bedding with dips of 15°–20°, Alexey Rudoy (1.9 m tall) for scale. (b) Slightly convex-up cross-bedding dipping at c. 20°, Juergen Herget (1.9 m tall) for scale. Palaeoflow left to right, modern flow right to left.

clinoform sequences are more common in the positions proximal to the main flow. At these locations the interference flow patterns between the backwater and main floodway was complex and highly unsteady. Away from the proximal locations, cross-bedded strata are thinner and uncommon. Gravel cross-strata usually consist of decimetre-thick (rarely 0.75–1.5 m thick) poorly sorted pebble and cobble layers (rarely boulders) that are normally graded, interspersed by layers of poorly sorted granules and coarse sand (Russell and Marren 1999; Russell, Sharpe, and Bajc 2007), although thinner cross-beds in granule and coarse sand sizes also occur (Figure 11.6). Boulders occur only rarely within cobble or pebble strata. Openwork is more common in pebble strata, in which case the pebbles seem to be well sorted, whereas cobble strata tend to be poorly sorted with matrix-fill. In the case of the coarser cross-strata, there can be a weak tendency for coarser particles to be found towards the lee toe such that the tangential grading is fine-to-coarse with respect to the inferred flow direction. Coarse-gravel cross-strata may approach the angle-of-repose (O'Connor 1993) but usually tend to dip at angles below that of repose (e.g. 10–25°). Such large-scale coarse-grained cross-bedding has been reported widely (Bretz 1959; Malde 1968; Fraser and Bleuer 1988; Russell 1992; O'Connor 1993; Fisher and Smith 1993; Maizels 1997a; Russell and Marren 1999; Evans 2000; Rushmer 2006; Carrivick and Rushmer 2006; Smith 2006. Marren and Schuh (2009) suggest such individual units, if thick enough (5–10 m), can be indicative of deposition from high-magnitude floods. This supposition is reasonable provided the depositional context is accordant, as large-scale coarse-grained cross-beds can also form at bar fronts prograding into deep scours in larger braided rivers (Collinson 1970; Rust 1972b; Siegenthaler and Huggenberger 1993; Lunt, Bridge, and Tye 2004) as well as at delta fronts (Shaw 1975; Clemmensen and Houmark-Nielsen 1981). Large-scale cross beds that grade up into near-surface layers of horizontally-bedded gravel that exhibits armouring by preferentially orientated boulders at the subaerial surface (O'Connor 1993) reflect deposition by a single flood in relatively shallow water. Large-scale festoon cross-bedding has been reported by O'Connor (1993) and Herget (2005).

11.4.2.2 Interpretation

Large gravel clinoforms may be the product of (i) delta, (ii) bar, or (iii) dune progression. The distinction is difficult due to similar leeside sorting in all three cases (Kleinhans 2005) but context may be of assistance in making interpretations. For example, basal clinoforms in the main floodway have been interpreted as owing to downflow progression of a large-scale 'delta-form' bar front at the base of a catastrophic flood (Fisher and Smith 1993; Russell and Marren 1999; Carling *et al.* 2009b). The progradation of thick basal units from the mainstem floodway into backwaters, including back-flooded tributaries, is highly diagnostic of megaflooding. These clinoforms almost certainly represent initial progradation of a barform into an embayment during the early stages of flood deposition. Although reactivation surfaces sometimes are recorded (Russell and Marren 1999), the usual lack of cross-cutting beds indicates steady progradation in a steadily strengthening current, rather than periods of reactivation in fluctuating currents. Smaller scale cross-bedding and cross-bedding found high, or distally, within the succession is more likely related to dune migration (Russell, Sharpe, and Bajc 2007).

Considering gravel dunes, Carling (1996) ascribed the low leeside angles and the coarse–fine couplets of cross-beds, not to avalanching processes but to the bedform progressing by deposition from bedload sheets running-out down the lee sides. Bedload gravel sheets tend to be coarse at their fronts and fine up-flow such that in the lee of a dune the coarse-gravel fractions deposit first to be overrun by the finer-gravel tail, giving the repetitive coarse-fine structure in the vertical profile. Fine sediment readily infiltrates the depositing cobble layers but the finer-gravel layers tend to resist infilling. Silt and clay may occur as 'fines-coatings' on individual grains (due to later groundwater movements) but otherwise are absent in these cross-bedded sequences. The absence of fines may be ascribed to

Figure 11.6 (a) Upwards coarsening horizontally bedded to slightly wavy rhythmic laminations. Beds commence with fine sands (light colour) and coarsen upwards to fine pebbles. Thickness of beds is around 0.10–0.20 m with individual laminations >5 mm. Location: Little Jaloman bar, Altai. Palaeoflow left to right of the image. (b) Proximal portion of tributary-blocking bar downstream of glacial lake Vitim. Sequence of horizontally bedded coarse-grained beds. Usually, beds consist of clast-supported pebble or cobbles which locally can be matrix supported and which lack imbrication. Matrix consists of well-sorted coarse sand and granules. Locally beds are devoid of gravel but comprise coarse well-sorted sands in rhythmic succession. Coarser beds tend to grade upwards from large cobbles through sand. Sharp erosion unconformities occur locally and tend to have a concentration of 200–300-mm blocks at the base of the bed. John Jansen (1.8 m tall) for scale. (See also Margold *et al.* 2011). (c) Normally graded coarse sand and fine gravel rhythmic bedding. Note erosional base to some beds. Dip of beds is 10–15° down flow. Sequence at Skeiðarársandur, Iceland in embayment setting. Lense cap (55 mm) for scale. (See also Russell & Knudsen 1999).

the high power of the floods keeping fines in suspension until further down the system. The absence of these finer components also explains the presence of openwork pebble beds more generally. Large-scale festoon cross-bedding has been associated with changing flow patterns in recirculating flow cells (Baker 1973; O'Connor 1993) and as a falling-stage reworking of outer bar margins (Herget 2005).

11.4.3 Horizontally Bedded Laminations

11.4.3.1 Description

Parallel-planar to slightly wavy horizontal or very low-angle bedding is typical of sands and granule beds forming sequences that are laterally extensive (Cornwell 1998; Carling *et al.* 2009b). A distinctive feature of such beds in backwaters is that they can thin and fine away from the main flood channel (Maizels 1997a; Carling *et al.* 2009b). These sequences often occur above large-scale clinoform and coarse basal sequences. Upward fining (normal grading) and upward coarsening (reverse grading) beds can be repeated in the vertical or can intersperse each other in highly distinctive lithofacies (see Figure 11.9). The lower contact of each bed is usually conformable and sharp (e.g. Cornwell 1998), although examples of erosional contacts occur (Benito and O'Connor 2003; Hanson *et al.* 2012). Even thin beds usually are laminated although amalgamation (especially in coarser grainstocks) may make this obscure (Duller, Mountney, and Russell 2010). Individual beds typically range in thickness from <10 to 150 mm. Thinner beds tend to be more discontinuous (e.g. Russell and Knudsen 1999b; their figure 5), although thicker beds can be traced with little variation in thickness for decametres. Evidence of rippling or the development of low-amplitude very long-wavelength bedforms is slight but more prevalent in the upper parts of these sequences (Atwater 1984, 1986; Carling *et al.* 2009b). If thick massive beds occur, which are relatively rare in this sequence, they are usually in the lower portions of the stack of beds. Occasional pebble stringers may occur (e.g. Russell and Knudsen 1999b) together with isolated dropstones. Sets of thinner beds may form thicker bed-sets which also fine or coarsen upwards and these thicker beds themselves may form sets within larger-scale fining up sequences. The repetition of the depositional style in the vertical both at the thin-bed scale and at the thicker bed-sets scale provides strong evidence of rhythmicity in depositional process. Despite variation, normal graded beds tend to predominate (e.g. Spencer and Jaffe 2002). Although similar stacked sequences form a substantial portion of the succession, generally there is an overall fining upwards and a thinning of beds from base to top of this portion of the flood succession (Maizels 1997a; Carling *et al.* 2009b). Although small-scale bedding is usually conformable within sequences, contacts between rhythmic sequences often are unconformable.

11.4.3.2 Interpretation

Thick units of diffuse stratification showing no vertical trends in grain size are best explained as due to sustained depositing flow conditions in a steady current (Maizels 1993, 1997a; Kneller 1995; Kneller and Branney 1995). Occasional erosional contacts, often lined by pebble stringers in otherwise uniform finer sediments, may indicate short periods of strengthening current (Plink-Björklund and Ronnert 1999; Benito and O'Connor 2003). Despite differences in interpretation, a commonality is that these rhythms are deposited from high concentration flow onto traction carpets (e.g. Nemec and Muszyński 1982). In the case of massive beds, the bedload transport rate is weak in comparison with the 'rain-out' of suspension load. In the case of laminated beds, bedload transport is more vigorous, flow pulsing leads to unsteady rain-out of grains from suspension and near-bed high suspended sediment concentrations may lead to periodic density flows. Thus the latter style of deposition is more akin to 'turbidites' rather than the traction-dominated deposits of turbulent, high-energy shallow fluvial floods. Significant support for a turbidite interpretation is gained from video-recordings from which observations of jökulhlaup flow pulses (with periods of seconds to minutes) were matched qualitatively with stacks of normally graded rhythmic bedding exposed in section following flood recession (Russell and Knudsen 1999b). The suggestion that these coarse-grained laminated beds are akin to turbidites is controversial as turbidites are usually reported from fine sediments. This matter is considered further.

Normal-grade bedding has been ascribed to suspension-deposition from slowly waning low-density turbidity flows (Winsemann, Asprion, and Meyer 2007) but also has been related to waxing (Russell 2005) or waning high-density turbidity flows (Russell and Knudsen 1999b). A predominance of such normal-graded beds would imply deposition on the waning stage of the flood, which is consistent with the superior position of these beds in successions. Fining–coarsening–fining sets are also common. Inverse grading classically has been ascribed to waxing high-density flows (Costa 1974; Fisk 1974; Allen 1981; Fraser and Bleuer 1988; Maizels 1991, 1993) although, in these cases, Maizels (1997a) invoked grain-flow mechanisms to explain lamination – related to late-stage shearing of the sediment body (Vrolijk and Southard 1997). Although the waxing grain-flow explanation might be pertinent in some cases, it is unlikely to explain the ubiquity of coarsening-up lamination in some deposits and indeed steady stresses can induce upward-coarsening thin (<100 mm) beds (Sallenger 1979; Hiscott and Middleton 1980; Lowe 1982). Rather, the mechanism of inverse grading is better explained not as grain-flow or 'debris-flow', but as due to en-mass settling from high-concentration turbid currents into thin traction carpets (Talling *et al.* 2012). At the base of turbidity currents, thin layers of sediment can commonly form 'traction carpets' (Kuenen 1966; Hiscott 1994). A sequence of laminae and thin beds quickly accumulates (Kuenen 1966; Bannerjee 1977; Vrolijk and Southard 1997; Sumner, Amy, and Talling 2008), which may exhibit grading from base to top although individual laminae and thin beds are ungraded. The repetition of similar sets may represent rapid variations in sediment supply (Rust and Romanelli 1975; Osterkamp and Costa 1987) but by analogy to turbidite deposits may be related to flow pulsing (Postma 1986; Postma and Cruickshank 1988; Todd 1989; Sohn 1997; Cheel 1990). Thus the depositional mechanism is controlled by the overlying flow, which is an important contrast to debris flows that propagate downslope due to their own weight. Further consideration of the analogy of horizontally bedded megaflood laminations with turbidity-current deposits is provided by Carling *et al.* (2009b) and in section 11.5.

11.4.4 Ripple and Dune Cross-Lamination

11.4.4.1 Description

Ripple lamination and low-amplitude-dune cross-lamination usually occur towards the top of a sequence of horizontally bedded laminations. These superior beds tend to be of finer grain-size (sand and occasionally granules) than inferior horizontally bedded laminations (Figure 11.7). Tabular cross-bedding and trough cross-bedding are apparent in beds typically 0.10–0.15 m thick, but exceptionally up to 1.5 m thick (Maizels 1993; Russell and Knudsen 1999b; Russell, Sharpe, and Bajc 2007). Climbing ripple sequences are usually limited in height due to the small thickness of individual beds. Festoon cross-bedding in coarse sands and gravels, at any scale, is rather rare (Figure 11.8). Small-scale festoon cross-bedding has been reported by Maizels (1997a) in locations atop laminated bed sequences.

11.4.4.2 Interpretation

Ripples cannot form in fully turbulent flow (i.e. not in grain sizes >0.7 mm; Southard 1991), which stricture, together with the finer grain sizes, are strong indicators of slackening dilute flows. High accumulation rates (>0.4–0.6 mm s^{-1}; Talling *et al.* 2012) from suspension fallout suppress ripple

Figure 11.7 Laminated basalt coarse-sands, locally showing ripple or low-amplitude dune bedding. Sediments in gravel pit near the Rio Laja, Tucapel, Chile (Thiele *et al.* 1998). Horizontal field of view ~ 1.5 m.

development (Bannerjee 1977; Arnott and Hand 1989; Sumner, Amy, and Talling 2008), and so the presence of ripples is an important indicator of the upper limit to the accumulation rate from which suspended sediment concentrations might be deduced. The presence of ripples and ripple cross-lamination implies developing turbulent low-density flow as, although ripples have been induced in highly constrained laboratory laminar flows, ripples do not occur in 'natural' laminar flows and are significantly suppressed by high suspensions. Dunes can occur in fully turbulent flow in coarse-grained bedstocks (Carling 1999) but the muted amplitude of the bedforms in bar sections has resulted in a reticence to term low-amplitude bedforms in coarser sediments as 'dunes'. The presence of low-amplitude bedforms in beds lying above horizontal bedding has been interpreted as due to slackening subcritical flow (Carling *et al.* 2009b). Fully turbulent near-bed conditions can also occur in low-density turbidity currents that result in ripple cross-lamination (Baas, Best, and Peakall 2011), and in turbidity flows cross-bedding is more likely to be evident in coarser sediments than in finer sediments (Allen 1970). In contrast, in high-concentration flows turbulence is damped, reducing bedform amplitude (Best and Bridge 1992) deposition rates increase and hence aggregation is rapid. Thin beds exhibiting festoon cross-bedding have been associated with late-stage reworking (Maizels 1997a), which also implies variable flows.

11.4.5 Silt Beds

11.4.5.1 Description

Thin laminated silt beds are the least common lithofacies but may cap an individual gravel bed or a sequence, usually conformably but occasionally unconformably (Figure 11.9). Although silt is the most commonly reported grain-size for this sequence, fine sand or clay sometimes characterize these beds. Occasionally the beds show ripple cross-bedding and the silt layer can be convoluted with either the overlying or underlying beds or both (e.g. O'Connor 1993). Massive silt beds and draped lamination occur in some successions, although the lack of structure in the former has been related to bioturbation (Smith 1993). The laminated silt bed in Figure 11.9 has an erosive contact with the

Figure 11.8 (a) Festoon bedding in sand due to dune progression. Interpreted by Winsemann, Asprion, and Meyer (2009) as high-velocity jet efflux deposits (rather than open-water) on the Porta subaqueous ice-contact fan, Germany. Christian Brandes (1.75 m tall) provides scale. (b) Festoon bedding in the Walters Butte Bonneville flood bar, Snake River, Idaho. Main section is c. 3.5 m high. (*Source*: Digital Atlas of Idaho: http://imnh.isu.edu/digitalatlas/hydr/lkbflood/tour/lbfvt5.htm)

clinoforms below whereas the contact with the flood granules above is largely conformable. This thick silt layer in the Kezek-Jala bar (Figure 11.9) can be correlated with a thinner silt layer in the Little Jaloman valley 5 km distant (see silt layers in Figure 11.5a and b), so this particular example represents a system-scale depositional event (following erosion) rather than a highly localized sedimentary signature. Exceptions to the presence of only thin silt beds do exist, however, with Glenn (1965) and O'Connor *et al.* (2001) reporting silt–clay deposits up to 30 m thick in slackwater areas.

Figure 11.9 Fine sand and silt bed deposited on erosive upper boundary to large-scale clinoforms, Kezek-Jala bar, Altai. Road sign is 2.5 m high. Flow left to right.

11.4.5.2 Interpretation

In the case of both small-scale jökulhlaups (Maizels 1993; Russell and Marren 1999), the Altai megaflood (Carling *et al.* 2009b), and Missoula megaflood (Waitt 1980, 1985; Smith 1993; Shaw *et al.* 1999) the caps to sequences are interpreted as waning stage deposition of fine-grained material from suspension following an overall aggrading event. Marren (2002b) also reports silt drapes during high-stage deposition onto the bar surfaces as well as during waning flow. However, in lower sequences, including the coarse basal sequences, occasional silt beds capping gravel beds can occur which seem to be associated with waning stage deposition following erosion-only events. Frequent occurrence of laminated clay–silt beds capping flood sandy-gravel sequences can represent deposition into lakes (Baker *et al.* 1991) but the lack of such frequent intercalation supports the waning flow origin for beds of fines (Rigby 1982; Baker *et al.* 1991). Massive and draped deposits likely indicate deposition in unsheared flows, possibly ponded waters (Smith 1993; Mutti *et al.* 1996) of low turbulence intensity (Bretz 1929). Convolution often is due to rapid deposition of a superior unit onto a recently deposited inferior unit.

The presence of a distinct thin silt bed between otherwise very similar fine-gravel rhythms implies very rapid cessation of the preceding flood and a short period of time for silt deposition before rapid onset of the next gravel-depositing event. This interpretation would require strong but infrequent large-scale pulsing within an individual flood to enter the depositional embayment or a series of individual floods in quick succession that preserved the silt before subaerial erosion could have removed it.

Neither of the above explanations is completely satisfactory, but all reports of this silty megaflood sequence lack detailed assessment of the lamination style and the grain-size grading which would permit more thorough interpretation of depositional processes. Talling *et al.* (2012) explore in more detail the implications of variation in mud facies for depositional processes of turbidites. Druitt (1995) suggested that thin beds of fines can occur atop of beds of coarser sediments due to elutriation during settling of dense suspensions, which can also explain silt-rich gravel layers (Benito and O'Connor 2003). Sometimes silt has been interpreted to be of aeolian origin and thus to indicate a hiatus between floods (Waitt 1980, 1985; Benito and O'Connor 2003).

11.4.6 Debris Flow

11.4.6.1 Description

Occasionally coarse units deposited on the basal bedrock or on an erosional contact with underlying sediments have the characteristics of debris-flow deposits (Lord and Kehew 1987; Maizels 1993) or, if not definitely debris flow, these units can consist of coarse, angular clasts of local derivation channelled into underlying beds (Waitt 1980). Reported debris-flow intervals are predominately ungraded, coarse with angular clasts set in a matrix, but otherwise most literature provides little justification for the debris flow interpretation. By definition these deposits are due to non-Newtonian flows.

11.4.6.2 Interpretation

Inverse-graded gravel beds are sometimes interpreted being formed due to debris flow (Winsemann, Asprion, and Meyer 2007), as are genuinely massive and thick gravel beds (Shanmugam 2000). These particular interpretations are contrary to the majority opinion of such beds in a fluvial megaflood context (e.g. Russell and Knudsen (2002) for an alternative view; Duller, Mountney, and Russell 2010). Basal occurrence of debris flow sediments has been interpreted as pre-flood deposits, or debris flow deposits due to the mobilization of pre-flood deposits by the flood waters (Russell and Knudsen 1999b). Often basal units are interpreted as relatively dilute 'debris flow' transitional to a more fluid phase (e.g. Mutti *et al.* 1996). Mid-succession planar parallel-stratified pebble beds sandwiched in sandy rhythmite sequences have been interpreted to represent thin debris flows and turbulent density flows triggered by slope failures (Winsemann, Asprion, and Meyer 2007) or ascribed to the distal ends of thicker but dilute debris flows.

Sequence-capping debris-flow beds also occur that, unlike basal examples, tend to be thin and matrix rich (Figure 11.10). Bjornstad (1980) argued that poorly sorted thin-beds, resembling debris flow, capping individual rhythms might represent debris fallout from ice floes. Presumably during strong flood flow any ice floe meltout-sediment is intimately mixed with fluvial deposits but becomes evident in the sequences during the relatively calm conditions at the end of a flood or flood pulse. Carling *et al.* (2009b) ascribe thin units of distinctively red-coloured angular clasts to the top of depositional rhythms, as they sit on silt or thin coarse-gravel layers atop sets of rhythmically bedded laminations (Figure 11.10b). In this specific case the debris flow units are ascribed to falling-stage flooding and the emergent phase when the embayment of deposition was relatively tranquil. At this time, small failures of the regolith (induced by the flood under-cutting neighbouring slopes) would spread across the recently deposited flood sediment surface and be preserved. In contrast, during the main flood these small flows of colluvium and scree would be intimately mixed into the turbulent flow and 'lost' within the depositing sediment body (Carling *et al.* 2009b). In other cases a succession-capping gravel has been viewed as colluvium deposited between floods (Waitt 1980; Benito and O'Connor 2003).

Figure 11.10 (a) Coarse angular debris flow-bed capping silt-rich gravel (white colour) and coarse-gravel rhythmic horizontally-bedded laminations below. Lobes of silt extending downwards are superficial reworking due to downwash and do not represent the bedding boundary. Little Jaloman bar, Altai. Notebook is 220 mm in length. (b) Fine-grained debris flow (c. 2 m thick) including convoluted silt rip-up clast (arrowed) overlie coarsening-up series of coarse-gravel horizontal laminations. Little Jaloman bar, Altai. Camera bag is 220 mm in length.

11.5 Discussion

Many authors have noted the comparison of catastrophic flood units with Bouma turbidite sequences (Bretz *et al.* 1956; Baker 1973; Bjornstad 1980; Bunker 1980; Waitt 1980; Rigby 1982; Baker and Bunker, 1985; Baker *et al.* 1991; Russell and Knudsen 1999b, 1999; Maizels 1997a). Specifically, Maizels (1993) was able to relate a 12+ m thick succession to the observed waxing and waning flow characteristics of a jökulhlaup. This analogy between flood deposits and turbidites is important as it implies distinctive transport and depositional processes could be at work within megafloods which are akin to density driven processes most commonly seen in the marine environment. However, subsequent authors (e.g. Smith 1993; Carling *et al.* 2002, 2009b) have been reluctant to label coarse-grained megaflood deposits as turbidites. From the perspective of the author this reluctance is because: some authors are reluctant to accept the transport of coarse sediment (especially coarse gravel) in suspension; turbidites are associated with marine and lake environments and 'not' from fluvial deposits; there is limited direct evidence for the turbidity current process in relation to megaflood deposition.

Despite the three points noted above, megaflood deposits are often well-interpreted by reference to turbidity current literature (Carling *et al.* 2009b) and turbidity currents are *often* invoked in relation to large flood deposition within subglacial tunnel valleys (e.g. Bennett *et al.* 2007). Thus a fining upwards megaflood sequence that is similar to the Bouma sequence (Figure 11.11) might be readily interpreted as

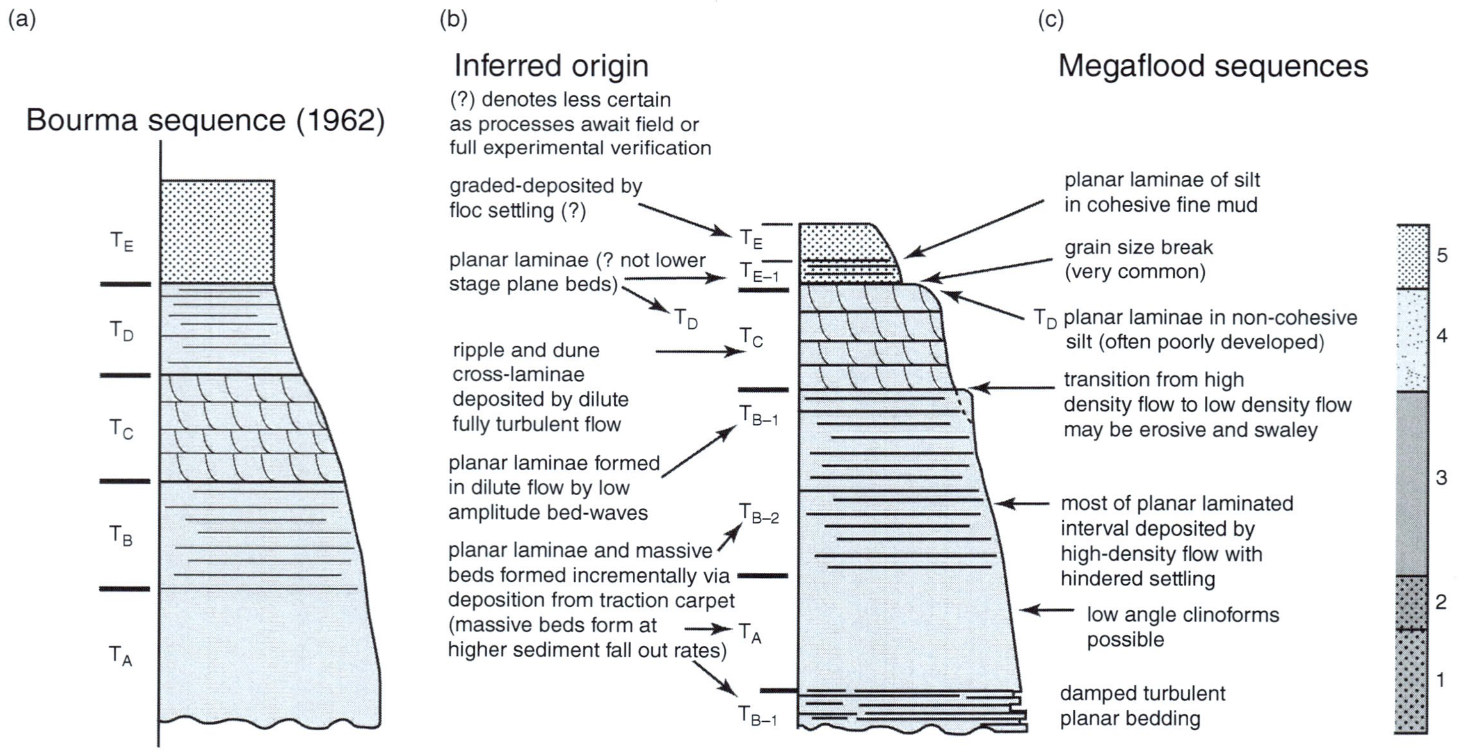

Figure 11.11 Generalized commonly found sequences within turbidites. (a) The sequence of Bouma (1962); (b) The sequence of Talling *et al*. 2012, modified to accord with the interpretation of this chapter as applied to megaflood sequences. (c) The inferred depositional processes are shown and the sequences 1 through 5 common to megaflood deposits are noted (right). Debris-flow intervals are not shown due to insufficient detail of the characteristics of megaflood debris-flow facies in published literature. (Modified and reproduced from Talling I. (2012) with permission.)

owing to continuous deposition from suspension within waning high-density turbid currents (Baker *et al.* 1991). Descriptions of depositional processes from turbidity currents (e.g. Lowe 1982; Kneller 1995) seem to apply aptly to megaflood 'Bouma-like' sequences (Winsemann, Asprion, and Meyer *et al.* 2007). Thus it is not unreasonable to view deeply back-flooded tributaries and embayments adjacent to main floodways as analogous to depositional basins in which turbidity currents might occur (Shaw 1975).

Despite the attractiveness of a Bouma-type model, the literature is replete with unresolved argument as to the nature of gravity flow deposition and the reader is referred to the detailed considerations of the complex mix of the possible processes and products of gravity flows provided by Nemec (1990), Benvenuti and Martini (2002) and by Talling *et al.* (2012). Herein it would be futile to speculate further as to the specific depositional mechanisms for any example pertaining to a megaflood. Rather, specific studies are required of the potential for gravity current development in megaflood depositional environments and these will require a detailed consideration of the rheology of the sediment-laden flows (Fan 1985). Nevertheless a simple example suffices to demonstrate that turbidity flows might readily evolve along depositional gradients in flood backwaters. The Knapp–Bagnold criterion (Parker, Fukushima, and Pantin 1986) is a basic necessary condition for self-sustaining turbidity currents to evolve:

$$\frac{US}{v_s} > 1 \qquad (11.1)$$

where U is the velocity of the flow, S is the gradient of the depositional surface, and v_s is the settling velocity of a particle from dense suspension. A consideration of Equation 11.1 for a 1 mm grain settling on gentle slopes of up to 10° shows that the criterion is more than satisfied for velocities as low as 0.5–0.7 m s^{-1}, and laminated turbidity-current deposits occur during experimental conditions (Toniolo, Parker, and Voller 2007).

Thus a basic vertical model of deposition can be derived that is common to many individual megaflood depositional successions in embayments:

Sequence 1 basal coarse parallel-bedded units
Sequence 2 large-scale clinoforms (often absent)
Sequence 3 horizontally-bedded laminations
Sequence 4 ripple and dune cross-lamination
Sequence 5 silt bed
Sequence 6 succession-capping debris flow (often absent or placed at the base of each succession).

The sedimentology of each constituent unit, when considered in temporal sequence, implies that the bulk of the sediment load is deposited more-or-less continuously as the nature of the flow is transformed during the hydrograph from initially traction-dominated, through traction with substantive fallout to suspension fallout alone (Sequences 1–5). Thus, this succession can be interpreted as a single cycle of waxing and waning flow deposits and, in this sense, it is similar to other temporal models for deposition proposed for small-scale freshwater floods (Mulder and Syvitski 1995), small-scale jökhulhlaups (Maizels 1997b), turbidites, ignimbrites, and lahars (Manville *et al.* 1999; Choux, Druitt, and Thomas 2004; Duller *et al.* 2008; Kataoka 2011). The model sequences together represent sustained deposition from high concentrations of sediment in flows that are generally initially accelerating and then decelerating at the scale of the hydrograph, but with significant shorter period pulsing due to pulsing of the hydrograph or the development of coherent turbulent flow structures, including highly-turbid currents. The prominent development of 'structureless', diffusive and amalgamated stratigraphy, and the presence of 'floating' boulders without flow separation structures around them, indicate that bedload transport often was ineffective in sorting sediment at the depositional interface due to the very high rate of suspended sediment deposition onto a diffuse accretion surface (*vis.* Arnott and Hand 1989). Stacks of such similar model successions probably represent deposition from several flood events, and the similarity in the successions then reflects similarity in the grain-size and hydrodynamics of the sequent floods. Sets that repeat within some of the sequences, most commonly sequences 3 and 4, are usually referred to as rhythms in the megaflood literature and, at this scale, represent fluctuations in individual flood hydrographs and/or pulsing of the flow related to the development of turbidity currents.

11.6 Conclusions

Megaflood beds often appear to be massive but on close inspection are found to be amalgamated, consisting of layers of similar sized grains. Bed sets commonly are near-horizontal, planar, and laterally extensive. Successions are not channelized but rather are laterally extensive, indicating coherent deposition over large areas. Large-scale cross-bedding occurs commonly as a discrete sequence, and the scale indicated by the thickness of the sets can be an important indicator of the exceptional depth of the depositing flow. In contrast to the extent of depositional sequences, major erosional contacts, notably steeply dipping examples, are uncommon.

A common stack of sequences is a repetitive motif both in the vertical at any one location and also when comparing geographically distinct outcrops. This succession consists of basal coarse parallel-bedded units, large-scale clinoforms, horizontally bedded laminations, ripple and dune cross-lamination, silt-beds and, in some cases, succession-capping debris flow. The succession can be interpreted as resulting from a single cycle of waxing and waning flood flow, generally from high-concentration suspensions onto plane beds. The high energy of the floods suspends fine gravel and coarse sand, which grains later are deposited as relatively coarse beds. Fines are rare, being transported further down system. Each stack of sequences reflects the acceleration and deceleration of a flood wave. Bedload transport often appears to have contributed little to parallel-bedded layer sequences, being suppressed by high-rates of suspension fallout and consequent rapid accretion.

Seeming rhythmic repetition of similar sequences is common, and may indicate deposition from serial floods. Although often coarse-grained, the sequences resemble turbidity flow deposits; the Bouma sequence in particular being a good analogue. Better attention needs to be given to the detailed sedimentology of individual beds, rather than the gross stratigraphy, such that improved process interpretation can lead to improved models of megaflood depositional events. A key issue to be tested using physical and numerical models is the potential for gravity current development in megaflood depositional environments and to relate the resultant coarse-grained deposits to the better known fine-grained deposits of deep-water turbidites.

Acknowledgements

The reviewer, Jim O'Connor, is thanked for his comments, which led to the provision of a better introductory focus to the paper. I should like to thank the organizers of GBR 8 Japan for the invitation to prepare this chapter, which is a condensation (with some additional new text) of a prior published paper: Carling, PA 2013, Freshwater megaflood sedimentation: What can we learn about generic processes? *Earth-Science Reviews* 125, 87–113. Copyright permission to reproduce portions of the original text was granted by Elsevier.

References

Alho, P, Baker, VR, and Smith, LN 2010. Paleohydraulic reconstruction of the largest Glacial Lake Missoula drainings. *Quaternary Science Reviews* 29, 3067–3078.

Allen, JRL 1970. The sequence of sedimentary structures in turbidites, with special reference to dunes. *Scottish Journal of Geology* 6, 146–161.

Allen, JRL 1981. Sediments and processes on a small alluvial stream-dominated Devonian alluvial fan, Shetland Islands. *Sedimentary Geology* 29, 31–66.

Allison, IS 1978. Late Pleistocene sediments and floods in the Willamette valley. *The Ore Bin* 40, 177–191 and 193–202.

Arnott, RWC and Hand, BM 1989. Bedforms, primary structures and grain fabric in the presence of suspended sediment rain. *Journal of Sedimentary Research* 59, 1062–1069.

Atwater, BF 1984. Periodic floods from glacial Lake Missoula into the Sanpoil Arm of glacial Lake Columbia, northeastern Washington. *Geology* 12, 464–467.

Atwater, BF 1986. Pleistocene glacial-lake deposits of the Sanpoil River Valley, northeastern Washington. *US Geological Survey Bulletin*, 1661.

Baas, JH, Best, JL, and Peakall, J 2011. Depositional processes, bedform development and hybrid flows in rapidly decelerated cohesive (mud–sand) sediment flows. *Sedimentology* 58, 1953–1987.

Baker, VR 1973. *Paleohydrology and Sedimentology of Lake Missoula Flooding in Eastern Washington.* Special Paper 144, Geological Society of America; 73 pp.

Baker, VR 1997. Megafloods and glaciation. In *Late Glacial and Postglacial Environmental Changes: Quaternary, Carboniferous–Permian and Proterozoic* (ed. IP Martini). Oxford University Press: Oxford; 98–108.

Baker, VR 2002. High-energy megafloods: Planetary settings and sedimentary dynamics. In *Flood and Megaflood Processes and Deposits: Recent and Ancient Examples* (eds IP Martini, VR Baker, and G. Garzon). IAS Special Publication 32, Blackwell Science: Oxford; 3–15.

Baker, VR 2007. Greatest floods – largest rivers. In, *Large Rivers: Geomorphology and Management* (ed. A Gupta). John Wiley & Sons: New York; 65–74.

Baker, VR 2009. Overview of megaflooding: Earth and Mars., In *Megaflooding on Earth and Mars* (eds DM Burr, PA Carling, and VR Baker). Cambridge University Press; 1–12.

Baker, VR 2013. Global late Quaternary fluvial paleohydrology with special emphasis on paleofloods and megafloods. In *Treatise in Fluvial Geomorphology* (ed. EE Wohl). Elsevier.

Baker, VR and Bunker, RC 1985. Cataclysmic Late Pleistocene flooding from glacial Lake Missoula: a review. *Quaternary Science Reviews* 4, 1–41.

Baker, VR, Bjornstad, BN, Busacca, AJ *et al.* 1991. The Quaternary Geology of the Columbia Plateau. In *Quaternary Nonglacial Geology: Conterminous U.S.* (ed. RB Morrision). The Geology of North America, Vol. K-2, Geological Society of America: Golden, CO; 215–250.

Bannerjee, I 1977. Experimental study on the effect of deceleration on the vertical sequence of sedimentary structures in silty sediments. *Journal of Sedimentary Petrology* 47, 771–783.

Benito, G and O'Connor, JE 2003. Number and size of last-glacial Missoula floods in the Columbia River valley between the Pesco Basin, Washington, and Portland, Oregon. *Geological Society of America Bulletin* 115, 624–638.

Bennett, MR, Huddart, D, and Thomas, GSP 2007. The Newbigging esker system, Lanarkshire, Southern Scotland: a model for composite tunnel, subaqueous fan and supraglacial sedimentation. In *Glacial Sedimentary Processes and Products* (eds M Hambrey, P Christoffersen, NF Glasser, and B Hubbard). Blackwell Science: Oxford.

Benvenuti, M and Martini, IP 2002. Analysis of terrestrial hyperconcentrated flows and their deposits. In *Flood and Megaflood Processes and Deposits: Recent and Ancient Examples* (eds IP Martini, VR Baker, and G. Garzon). IAS Special Publication 32, Blackwell Science: Oxford; 167–193.

Best, JL and Bridge, J 1992. The morphology and dynamics of low amplitude bedwaves upon upper stage plane beds and the preservation of planar laminae. *Sedimentology* 39, 737–752.

Bjornstad, BN 1980. *Sedimentology and Depositional Environment of the Touchet Beds, Walla Walla River basin, Washington.* Rockwell Hanford Operations, RHO-BW1-SA-44: Richland, Washington; 104 pp.

Bouma, A 1962. *Sedimentology of Some Flysch Deposits: a Graphic Approach to Facies Interpretation.* Elsevier: Amsterdam; 168 pp.

Bretz, JH 1925a. The channeled scablands of the Columbia Plateau. *Journal of Geology* 31, 617–649.

Bretz, JH 1925b. The Spokane flood beyond the channelled scablands. *Geological Society of America, Bulletin* 33, 97–115.

Bretz, JH 1929. Valley deposits immediately west of the channelled scabland of Washington: II. *Journal of Geology* 37, 505–541.

Bretz, JH 1959. Washington's channeled scabland. *Washing Division of Mines and Geology Bulletin* 45, 57 pp.

Bretz, JH, Smith, HT, and Neff, GE 1956. Channeled scabland of Washington: new data and interpretations. *Geological Society of America, Bulletin* 67, 957–1049.

Bunker, RC 1980. *Catastrophic flooding in the Badger Coulee area, south-central Washington.* MA thesis, University of Texas: Austin, TX.

Burbank, DW 1983. Multiple episodes of catastrophic flooding in the Peshawar basin during the past 700,000 years. *Geological Bulletin of the University of Peshawar* 16, 43–49.

Carling, PA 1989. Hydrodynamic models of boulder-berm deposition. *Geomorphology* 2, 319–340.

Carling, PA 1996. Morphology, sedimentology and palaeo-hydraulic significance of large gravel dunes: Altai Mountains, Siberia. *Sedimentology* 43, 647–664.

Carling, PA 1999. Subaqueous gravel dunes. *Journal of Sedimentary Research* 69, 534–545.

Carling, PA 2013. Freshwater megaflood sedimentation: what can we learn about generic processes? *Earth-Science Reviews* 125, 87–113.

Carling, PA and Glaister, MS 1987. Rapid deposition of sand and gravel mixtures downstream of a negative step: the role of matrix-infilling and particle-overpassing in the process of bar-front accretion. *Journal of the Geological Society* 144, 543–551.

Carling, PA, Kirkbride, AD, Parnachov, S*et al.* 2002. Late-glacial catastrophic flooding in the Altai Mountains of south-central Siberia: a synoptic overview and introduction to flood deposit sedimentology. In *Flood and Megaflood Processes and Deposits: Recent and Ancient Examples* (eds IP Martini, VR Baker, and G. Garzon). IAS Special Publication 32, Blackwell Science: Oxford; 17–35.

Carling, PA, Burr, DM, Johnson, TF, and Brennand, TA 2009a. A review of open-channel megaflood depositional landforms on Earth and Mars. In *Megaflooding on Earth and Mars* (eds DM Burr, PA Carling, and VR Baker). Cambridge University Press; 33–49.

Carling, PA, Martini, IP, Herget, J *et al.* 2009b, Megaflood sedimentary fill: Altai Mountains, Siberia. In *Megaflooding on Earth and Mars* (eds DM Burr, PA Carling, and VR Baker). Cambridge University Press; 243–264.

Carling, PA, Villanueva, I, Herget, J *et al.* 2010. Unsteady 1D and 2D hydraulic models with ice-dam break for Quaternary megaflood, Altai Mountains, southern Siberia. *Global and Planetary Change* 70, 24–34.

Carravick, JL and Rushmer, EL 2006. Understanding high-magnitude outburst floods. *Geology Today* 22, 60–65.

Carrivick, JL, Pringle, JK, Russell, AJ, and Cassidy, NJ 2007. GPR-derived architecture and stratigraphy of outburst flood sedimentation within a bedrock valley system, Hraundalur, Iceland. *Journal of Environmental and Engineering Geophysics* 12, 127–143.

Cheel, RJ 1990. Horizontal lamination and the sequence of bed phases and stratification under upper-flow-regime conditions. *Sedimentology* 37, 517–529.
Choux, C, Druitt, T, and Thomas, N 2004. Stratification and particle segregation in flowing polydisperse suspensions, with applications to the transport and sedimentation of pyroclastic density currents. *Journal of Volcanology and Geothermal Research* 138, 223– 241.
Clarke, GKC, Leverington, DW, Teller, JT, and Dyke, AS 2004. Paleohydraulics of the last outburst flood from Glacial Lake Agassiz and the 8200 BP cold event. *Quaternary Science Reviews* 23, 389–407.
Clemmensen, LB and Houmark-Nielsen, M 1981. Sedimentary features of a Weichselian glaciolacustrine delta. *Boreas* 10, 229–245.
Collinson, JD 1970. Bedforms of the Tana River, Norway. *Geografiska Annaler* 52A, 31–56.
Cornwell, K 1998. Quaternary break-out flood sediments in the Peshawar basin of northern Pakistan. *Geomorphology* 25, 225–248.
Costa, JE 1974. Stratigraphic, morphologic, and pedologic evidence of large floods in humid environments. *Geology* 2, 301–303.
Craig, RG 1987. Dynamics of a Missoula flood. In *Catastrophic Flooding* (eds L Mayer and D Nash). Allen and Unwin, Boston; 305–332.
Cutler, PM, Colgan, PM, and Mickelson, DM 2002. Sedimentologic evidence for outburst floods from the Laurentide Ice Sheet margin in Wisconsin, USA: implications for tunnel-channel formation. *Quaternary International* 90: 23–40.
Dawson, M 1989. Flood deposits within the Severn main terrace. In *Floods: Hydrological, Sedimentological and Geomorphological Implications* (eds K Beven and PA Carling). John Wiley and Sons: Chicester, UK; 253–265.
Druitt, TH 1995. Settling behaviour of concentrated dispersions and some volcanological applications. *Journal of Volcanology and Geothermal Research* 65, 27–39.
Duller, RA, Mountney, NP, and Russell, AJ 2010. Particle fabric and sedimentation of structureless sand, Southern Iceland. *Journal of Sedimentary Research* 80, 562–577.
Duller, RA, Mountney, NP, Russell, AJ, and Cassidy, NC 2008. Architectural analysis of a volcaniclastic jökulhlaup deposit, southern Iceland: sedimentary evidence for supercritical flow. *Sedimentology* 55, 939–964.
Evans, DJA 2000. Quaternary geology and geomorphology of the Dinosaur Provincial Park area and surrounding plains, Alberta, Canada: identification of former glacial lobes, drainage diversions and meltwater flood tracks. *Quaternary Science Reviews* 19, 931–958.
Fahnestock, RK and Bradley, WC 1973. Knik and Matanuska rivers, Alaska: a contrast in braiding. In *Fluvial Geomorphology* (ed. M Morisawa). Allen and Unwin: London; 220–250.
Fan, J 1985. Experimental studies on the density currents in settling basins. *Scientia Sinica A* 28, 319–336.
Fisher, TG and Smith, DG 1993. Exploration for Pleistocene aggregate resources using process-depositional models in the Fort McMurray region, NE Alberta, Canada. *Quaternary International* 20, 71–80.
Fisk, LH 1974. Inverse grading as stratigraphic evidence of large floods: comment. *Geology* 2, 613–615.
Fraser, GS and Bleuer, NK 1988. *Sedimentological Consequences of Two Floods of Extreme Magnitude in the Late Wisconsinan Wabash Valley*. Special Paper 229, Geological Society of America: Boulder, CO; 111–125.
Glenn, JL 1965. *Late Quaternary sedimentation and geologic history of the North Willamette Valley, Oregon*. Unpublished PhD thesis, Oregon State University: Corvallis; 231 pp.
Hanson, MA, Lian, OB, and Clague, JJ 2012. The sequence and timing of large late Pleistocene floods from glacial Lake Missoula. *Quaternary Science Reviews* 31, 67–81.
Herget, J 2005. *Reconstruction of Ice-dammed Lake Outburst Floods in the Altai Mountains, Siberia*. Special Paper 386, Geological Society of America: Boulder, CO.

Hewitt, K 1964. The great ice-dam. *Indus* 5, 18–30.
Hewitt, K 1968. Records of natural damming and related events in the upper Indus basin. *Indus* 10, 17–24.
Hewitt, K 1982. Natural dams and outburst floods of the Karakoram Himalaya. In *Hydrological Aspects of Alpine and High Mountain Areas* (ed. JW Glen). Publication 138, IAHS: Wallingford, UK; 259–269.
Hiscott, RN 1994. Traction-carpet stratification in turbidites: fact or fiction? *Journal of Sedimentary Research* A64, 204–208.
Hiscott, RN and Middleton, GV 1980. Fabric of coarse deep-water sandstones, Tourelle Formation, Quebec, Canada. *Journal of Sedimentary Petrology* 50, 703–722.
Johansson, CE 1963. Orientation of pebbles in running water. A laboratory study. *Geografiska Annaler* 45, 85–111.
Kataoka, K 2011. Geomorphic and sedimentary evidence of a gigantic outburst flood from Towada caldera after the 15 ka Towada-Hachinohe ignimbrite eruption, northeast Japan. *Geomorphology* 125, 11–26.
Kleinhans, MG 2005. Grain-size sorting in grain flows at the lee side of deltas. *Sedimentology* 52, 291–311.
Kneller, BC 1995. Beyond the turbidite paradigm: physical models for deposition of turbidites and their implications for reservoir prediction. In *Characterization of Deep-marine Clastic Systems* (eds AJ Hartley and DJ Prosser). Special Publication 94, Geological Society: London; 31–49.
Kneller, BC and Branney, MJ 1995. Sustained high-density turbidity currents and the deposition of thick massive sands. *Sedimentology* 42, 607–616.
Kuenen, PH 1966. Experimental turbidite lamination in a circular flume. *Journal of Geology* 74, 523–545.
Lønne, I 1997. Facies characteristics of a proglacial turbiditic sand-lobe at Svalbard. *Sedimentary Geology* 109, 13–35.
Lord, ML and Kehew, AE 1987. Sedimentology and paleohydrology of glacial-lake outburst deposits in southeastern Saskatchewan and northwestern North Dakota. *Geological Society of America, Bulletin* 99, 663–673.
Lowe, DR 1982. Sediment gravity flows II: deposition models with special references to the deposits of high-density turbidity currents. *Journal of Sedimentary Petrology* 52, 279–297.
Lunt, IA, Bridge, JS, and Tye, RS 2004. A quantitative, three-dimensional depositional model of gravelly braided rivers. *Sedimentology* 51, 377–414.
Maizels, JK 1991. Origin and evolution of Holocene sandurs in areas of jökulhlaup drainage, south Iceland. In *Environmental Change in Iceland: Past and Present* (eds. JK Maizels and C Caseldine). Kluwer: Dordrecht; 267–300.
Maizels, JK 1993. Lithofacies variations within sandur deposits: the role of runoff regime, flow dynamics and sediment supply characteristics. *Sedimentary Geology* 85, 299–325.
Maizels, JK 1995. Sediments and landforms of modern proglacial terrestrial enviroments. In *Modern Glacial Environments; Processes, Dynamics and Sediments* (ed. J Menzies). Butterworth-Heinemann: Oxford; 365–416.
Maizels, JK 1997a. Jökulhlaup deposits in proglacial areas. *Quaternary Science Reviews* 16, 793–819.
Maizels, JK 1997b. Palaeohydrology from fluvioglacial sediments: potential for inferring glacier variations. In *Glacial Fluctuations During the Holocene* (ed. B Frenzel). ESF project 'European Palaeoclimate and Man' Special Issue. *Paläoklimaforschung* 24(16), 182 pp.
Malde, HE 1968. *The catastrophic Late Pleistocene Bonneville Flood in the Snake River Plain, Idaho.* Professional Paper 596, US Geological Survey; 52 pp.
Mangerud, J, Jakobsson, M, Alexanderson, H *et al.* 2004. Ice-dammed lakes and rerouting of the drainage of northern Eurasia during the Last Glaciation. *Quaternary Science Reviews* 23, 1313–1332.
Manville, V, White, JDL, Houghton, BF, and Wilson, CJN 1999. Paleohydrology and sedimentology of a post-1.8 ka breakout flood from intracaldera Lake Taupo, North Island, New Zealand. *Geological Society of America, Bulletin* 111, 1435–1447.

Margold, M, Jansson, KN, Stroeven, AP, and Jansen, JD 2011. Glacial Lake Vitim, a 3000 km^3 outburst flood from Siberia to the Arctic Ocean. *Quaternary Research* 76, 393–396.

Marren, PM 2002a. Fluvial–lacustrine interaction on Skeiðarársandur, Iceland: implications for sandur evolution. *Sedimentary Geology* 149, 43–58.

Marren, PM 2002b. Glacier margin fluctuations, Skaftafellsjökull, Iceland: implications for sandur evolution. *Boreas* 31, 75–81.

Marren, PM and Schuh, M 2009. Criteria for identifying jökulhlaup deposits in the sedimentary record. In *Megaflooding on Earth and Mars* (eds DM Burr, PA Carling, and VR Baker). Cambridge University Press: Cambridge; 225–242.

Marren, PM, Russell, AJ, and Rushmer, L 2009. Sedimentology of a sandur formed by multiple jökulhlaups, Kverkfjöll, Iceland. *Sedimentary Geology* 213, 77–88.

Miall, AD 1977. A review of the braided river depositional environment. *Earth Science Reviews* 13, 1–62.

Miall, AD 1978. Lithofacies types and vertical profile models in braided river deposits: a summary. In *Fluvial Sedimentology* (ed. AD Miall). Memoir 5, Canadian Society of Petroleum Geologists; 597–604.

Miyamoto, H, Itoh, K, Komatsu, G *et al.* 2006a. Numerical simulations of large-scale cataclysmic floodwater: a simple depth-averaged model and an illustrative application. *Geomorphology* 76, 179–192.

Miyamoto, H, Komatsu, G, Baker, VR *et al.* 2006b. Cataclysmic Scabland flooding: insights from a simple depth-averaged numerical model. *Environmental Modelling and Software* 22, 1400–1408.

Moody, UL 1988. *Late Quaternary stratigraphy of the Channeled Scabland and adjacent areas.* PhD thesis, University of Idaho: Moscow, ID.

Mulder, T and Syvitski, JPM 1995. Turbidity currents generated at river mouths during exceptional discharges to the world oceans. *Journal of Geology* 103, 285–299.

Mutti, E, Davoli, G, Tinterri, R, and Zavala, C 1996. The importance of ancient fluvio-deltaic systems dominated by catastrophic flooding in tectonically active basins. *Memoire di Scienze Geologiche* 48, 233–291.

Nemec, W 1990. Aspects of sediment movement on steep delta slopes. In *Coarse-Grained Deltas* (eds A Colella and DB Prior). Special Publication 10, International Association of Sedimentologists, Blackwell Scientific: Oxford; 29–73.

Nemec, W and Muszyński, A 1982. Volcaniclastic alluvial aprons in Tertiary of Sophia District (Bulgaria). *Annales Societas Geologorum Polaniae* 52, 239–303.

O'Connor, JE 1993. *Hydrology, Hydraulics, and Geomorphology of the Bonneville Flood.* Special Paper 274, Geological Society of America: Boulder, CO; 83 pp.

O'Connor, JE and Baker, VR 1992. Magnitudes and implications of peak discharges from glacial Lake Missoula. *Geological Society of America, Bulletin* 104, 267–279.

O'Connor, JE, Sarna-Wojcicki, A, Wozniak, KC *et al.* 2001. *Origin, Extent and Thickness of Quaternary Geologic Units in the Willamette Valley, Oregon.* Professional Paper 1620, US Geological Survey; 52 pp.

Osterkamp, WR and Costa, JE 1987. Changes accompanying an extraordinary flood on a sand-bed stream, In *Catastrophic Flooding* (eds *L Mayer and D Nash).* Allen and Unwin: London; 201–224.

Parker, G, Fukushima, Y, and Pantin, HM 1986. Self-accelerating turbidity currents. *Journal of Fluid Mechanics* vol. 171, 145–181.

Plink-Björklund, P and Ronnert, L 1999. Depositional processes and internal architecture of late Weichselian ice-margin submarine fan and delta settings, Swedish west coast. *Sedimentology* 46, 215–234.

Postma, G 1986. Classification of sediment gravity flow deposits based on flow conditions during sedimentation. *Geology* 14, 291–294.

Postma, G and Cruickshank, C 1988. Sedimentology of a late Weichselian to Holocene terraced fan delta, Varangerfjord, northern Norway. In *Fan Deltas: Sedimentology and Tectonic Settings* (eds W Nemec and RJ Steel). Blackie and Son: London; 144–157.

Rigby, JG 1982. *The Sedimentology, mineralogy, and depositional environment of a sequence of Quaternary catastrophic flood-derived lacustrine turbidites near Spokane, Washington*. MSc thesis, University of Idaho: Moscow, ID.

Rushmer, EL 2006. Sedimentological and geomorphological impacts of the jökulhlaup (glacial outburst flood) in January 2002 at Kverkfjöll, northern Iceland. *Geografiska Annaler* 88A, 43–53.

Rushmer, EL 2007. Physical-scale modelling of jökulhlaups (glacial outburst floods) with contrasting hydrograph shapes. *Earth Surface Processes and Landforms* 32, 954–963.

Russell, AJ 1992. *Geomorphological effects of jökulhlaups, west Greenland*. PhD thesis, University of Aberdeen.

Russell, AJ 2005. Catastrophic floods. In *Encyclopedia of Geology* (eds RC Selby, LRM Cocks, and IR Primer). Elsevier: Oxford; 628–641.

Russell, AJ and Knudsen, Ó 1999a. Controls on the sedimentology of the November 1996 jökulhlaup deposits, Skeiðarársandur, Iceland. In *Fluvial Sedimentology VI* (eds ND Smith and J Rogers). Special Publication 28, International Association of Sedimentologists, Blackwell Science: Oxford; 315–329.

Russell, AJ and Knudsen, Ó 1999b. An ice-contact rhythmite (turbidite) succession deposited during the November 1996 catastrophic outburst flood (jökulhlaup), Skeiðarárjökull, Iceland. *Sedimentary Geology* 127, 1–10.

Russell, AJ and Knudsen, Ó 2002. Jökulhlaup deposits at the Ásbyrgi Canyon, northern Iceland: sedimentology and implications for flow type, In *The Extreme of the Extremes: Extraordinary Floods* (eds A Snorasson, HP Finnsdóttir, and M Moss). Proceedings Symposium at Reykjavik, Iceland, July 2000, Publication 271, IAHS: Wallingford,UK; 107–112.

Russell, AJ and Marren, PM 1999. Proglacial fluvial sedimentary sequences in Greenland and Iceland: a case study from active proglacial environments subject to jökulhlaups, In *QRA Technical Guide Number 7* (eds AP Jones, ME Tucker, and JK Hart). Quaternary Research Association: London; 171–120.

Russell, HAJ, Sharpe, DR, and Bajc, AF 2007. Sedimentary signatures of the Waterloo Moraine, Ontario, Canada. In *Glacial Sedimentary Processes and Products* (eds MJ Hambrey, P Christoffersen, NF Glasser, and B Hubbard). Blackwell Science: Oxford; 85–108.

Rust, BR 1972a. Pebble orientation in fluviatile sediments. *Journal of Sedimentary Petrology* 42, 384–8.

Rust, BR 1972b. Structure and process in a braided river. *Sedimentology* 18, 221–245.

Rust, BR and Romanelli, R 1975. Late Quaternary subaqueous outwash deposits bear Ottawa, Canada. In *Glaciofluvial and Glaciolacustrine Sedimentation* (eds AV Jopling and BC McDonald). Special Publication 23, Society of Economic Paleontologists and Mineralogists: Tulsa, OK; 177–192.

Sallenger, AH 1979. Inverse grading and hydraulic equivalence in grain-flow deposits. *Journal of Sedimentary Petrology* 49, 553–562.

Shanmugam, G 1997. The Bouma Sequence and the turbidite mind set. *Earth Science Reviews* 42, 201–229.

Shaw, J 1975. Sedimentary successions in Pleistocene ice-marginal lakes. In *Glaciofluvial and Glaciolacustrine Sedimentation* (eds AV Jopling and BC McDonald). Special Publication 23, Society of Economic Paleontologists and Mineralogists: Tulsa, OK; 281–303.

Shaw, J, Munro-Stasiuk, M, Sawyer, B *et al.* 1999. The Channeled Scabland: back to Bretz? *Geology* 27, 605–608.

Shugar, DH and Clague, JJ 2012. The sedimentology and geomorphology of rock avalanche deposits on glaciers. *Sedimentology* 58, 1762–1783.

Siegenthaler, C and Huggenberger, P 1993. Pleistocene Rhine gravel: deposits of a braided river system with dominant pool preservation. In, *Braided Rivers* (eds JL Best and CS Bristow). Special Publication 75, Geological Society: London; 147–162.

Smith, GA 1993. Missoula flood dynamics and magnitudes inferred from sedimentology of slack-water deposits on the Columbia Plateau, Washington. *Geological Society of America, Bulletin* 105, 77–100.

Smith, DG and Fisher, TG 1993. Glacial Lake Agassiz: the northwestern outlet and paleoflood. *Geology* 21, 9–12.
Smith, LN 2006. Stratigraphic evidence for multiple drainings of glacial Lake Missoula along the Clark Fork River, Montana, USA. *Quaternary Research* 66, 311–322.
Sohn, YK 1997. On traction-carpet sedimentation. *Journal of Sedimentary Research* 67A, 502–509.
Southard, JB 1991. Experimental determination of bed-form stability. *Annual Review of Earth Planetary Sciences* 19, 423–455.
Spencer, PK and Jaffee, MA 2002. Pre-Late Wisconsinian glacial outburst floods in Southeastern Washington – the indirect record. *Washington Geology* 30, 9–16.
Sumner, E, Amy, L, and Talling, PJ 2008. Deposit structure and processes of sand deposition from a decelerating sediment suspension. *Journal of Sedimentary Research* 78, 529–547.
Talling, PJ, Masson, DG, Sumner, EJ, and Malgesini, G 2012. Subaqueous sediment density flows: depositional processes and deposit types. *Sedimentology* 59, 1937–2003.
Todd, SP 1989. Stream-driven, high-density gravelly traction carpets: possible deposits in the Trabeg Conglomerate Formation, SW Ireland and some theoretical considerations of their origin. *Sedimentology* 36, 513–530.
Tómasson, H 1973. Hamfarahlaup í Jöklsá á Fjöllum, *Náttúruf-reaðingurinn* 43, 2–34.
Toniolo, H, Parker, G, and Voller, V 2007. Role of ponded turbidity currents in reservoir trap efficiency. *Journal of Hydraulic Engineering* 136, 579–595.
Vrolijk, PJ and Southard, JB 1997. Experiments on rapid deposition of sand from high-velocity flows. *Geoscience Canada* 24, 45–54.
Waitt, RB 1980. About forty last-glacial Lake Missoula jökulhlaups through southern Washington. *Journal of Geology* 88: 653–679.
Waitt, RB 1985. Case for periodic, colossal jökulhlaups from Pleistocene glacial Lake Missoula. *Geological Society of America, Bulletin* 96, 1271–1286.
Winsemann, J, Asprion, U, and Meyer, T 2007. Lake-level control on ice-margin subaqueous fans, glacial Lake Rinteln, Northwest Germany, In *Glacial Sedimentary Processes and Products* (eds MJ Hambrey, P Christoffersen, NF Glasser, and B Hubbard). Blackwell Science: Oxford; 121–148.

Discussion

Discussion by Hervé Piégay

Do we have studies questioning the effects of such events on the valley settings and the present fluvial patterns? Can we expect the observed channel patterns in such valleys to be significantly different from those observed in similar adjacent valleys not having experienced such events? This question may feed the debate on theories of catastrophism versus uniformitarism.

Discussion by James Bathurst

The general applicability of models can be maximized by testing against, and incorporating, data from as wide a range of environments and from as extreme a set of conditions as possible. Space exploration is increasingly providing data from other planets within the Solar System which can be used to extend models that were previously limited to Earth-based geomorphologies. To what extent, therefore, does the geomorphological evidence of past very large floods on Mars provide helpful constraints or insights for outbreak flood models on Earth?

Discussion by Gordon Grant of both Paul Carling and Mikael Attal

I am wondering how to reconcile the two perspectives offered by Carling and Attal (Chapter 12). On the one hand, Carling argues that the stratigraphy left behind by very large floods is interpretable in light of specific mechanisms of entrainment and deposition that occur during those events, and suggests a 'standard model' for catastrophic flood deposits, if only as a heuristic tool for interpreting flood sequences. On the other hand, Attal argues convincingly that very large floods are greatly influenced by patterns of sediment supply that play out in both spatial and temporal realms, which are conditioned by sequences of floods of different competences and capacities. One consequence of this should be that large flood deposits should vary dramatically in their structure and stratigraphy. Yet despite these complex controls, basin sediment yields tend to be rather systematic over time. How do we make sense of these seemingly disparate observations?

Response by the Author

It is important to identify any influence of glaciation post flood on fluvial patterns as well as the influence of megafloods. Phillips and Desloges (2014) termed this influence on river systems 'glacial conditioning' and Marren and Toomath (2014 have argued that moraines significantly influence many proglacial river courses and patterns. Megafloods *per se* have two major impacts on post-flood river courses (Holocene in most examples). (i) Megafloods often infill mountain valleys with sediment, including bars blocking tributaries. As a result, post-megaflood fluvial systems flowed as braided systems over the megaflood fill within the valleys. However, rapid incision follows (Carling 2013), such that the modern tributaries often breach the valley-side bars and the main stems are incised into the megaflood valley fill and are flanked by braidplain terraces. On bar tops the sediments can be well sorted and highly permeable, inhibiting surface-water drainage and surface incision. (ii) Landforms, such as giant bars, can be so large that prior drainage patterns are disrupted and consequently Holocene drainage has had limited impact on the megaflood landforms (Bretz *et al.* 1956), having to circumnavigate them.

To date knowledge of megafloods on Mars largely has been conditioned by our understanding of such processes on Earth. Carling *et al.* (2009a,b) exemplify some of these analogies. However, on Earth megaflood landforms have often been subject to significant degradation by subaerial processes, or are masked by vegetation. In contrast, on Mars there is an opportunity to view flood landscapes with minimal subaerial alteration and no masking by vegetation. The significant improvement in resolution and temporal frequency of imaging by both satellite and rover-mounted sensors means that some Martian surfaces (e.g. Gale crater) in just a few years could be better investigated than many terrestrial surfaces. The rover 'Curiosity' has only been in operation for around a year (Newsom *et al.* 2015) but already images of pebbles are being used to infer fluvial processes (Szabo *et al.* 2015) and this knowledge could be applied to Earth-based problems.

The observations of Carling and Attal are not actually disparate. The typical variability in sediment supply within megafloods due to dam failure was highlighted by Carling (2013, his section 2.2). Stratigraphic sequence structure is driven primarily by changes in the power of the flood as it waxes and wanes. In contrast, sediment supply tends to condition quantity and composition of deposits rather than the overall structure and stratigraphy. In my presentation, I noted that I was largely considering valley-confined high-power flood tracts. Different sediment compositions and stratigraphies might be anticipated farther down the depositional gradient where low powers occurred. For example, although in the gravel-rich units (Carling 2013) silt and clay is largely absent (due to high-power), silt and clay can be found in abundance in distal parts of depositional tracts. For example, Glenn (1965) and O'Connor *et al.* (2001) reported silt–clay deposits up to 30 m thick in distal slackwater areas of the Missoula tract.

References

Newsom, HE, Mangold, N, Kah, LC *et al.* 2015. Gale crater and impact processes – Curiosity's first 364 Sols on Mars. *Icarus* 249, 108–128.

Marren, PM and Toomath, SC 2014. Channel pattern of proglacial rivers: topographic forcing due to glacier retreat. *Earth Surface Processes and Landforms* 39, 943–951.

Phillips, RTJ and Desloges, JR 2014. Glacially conditioned specific stream powers in low-relief catchments of the southern Laurentian Great Lakes. *Geomorphology* 206, 271–287.

Szabo, T, Domokos, G, Grotzinger, JP, and Jerolmack, DJ 2015. Reconstructing the transport history of pebbles on Mars. *Nature Communications* 6: 8366. DOI: 10.1038/ncomms9366.

12

Linkage Between Sediment Transport and Supply in Mountain Rivers

Mikaël Attal

12.1 Introduction

Sediment transport in rivers occurs episodically during floods. In an ideal 'transport limited' world (Willgoose, Bras, and Rodriguez-Iturbe 1991) where the whole landscape would be made of loose sediment, sediment would always be available so the magnitude of a given flood would control sediment calibre (as a function of flow competence) and sediment flux (as a function of transport capacity), where sediment flux represents the amount of sediment transported by the river through a given reach per unit time. Such a situation may arise in alluvial rivers but is not necessarily applicable to mountainous areas, where rivers tend to start their journey and where most of the clastic sediment is produced. Sediment in mountain rivers is supplied from hillslopes and rivers typically exhibit a mixed bedrock–alluvial morphology (Bathurst, Leeks, and Newson 1986a, b; Howard, Dietrich, and Seidl 1994; Turowski *et al.* 2008a). In such circumstances, sediment availability becomes crucial in controlling the amount of sediment that is transported during a given flow event, in particular for the most extreme events during which transport capacity may exceed sediment flux due to limited sediment availability (Howard, Dietrich, and Seidl 1994; Wohl 2008; Morche and Schmidt 2012; Yager *et al.* 2012; Bennett *et al.* 2014). In this chapter, the issue of sediment supply and availability and its impact on sediment transport will be discussed, in particular with respect to high-magnitude, low-frequency events.

First, I review the evidence for the influence of sediment sources on the characteristics of the sediment available for transport in mountain rivers, in terms of grain size and lithological content. Field and modelling studies are used to illustrate the effect of spatial changes in source characteristics as well as episodic sediment supply to channels on the characteristics of the sediment available for transport. In section 12.3, I discuss the impact of variable grain size and amount of sediment available in the channel (at the reach scale rather than at the bedform scale) on the characteristics of the sediment mobilized during floods of varying magnitude, with particular emphasis on the effect of boulders on sediment mobility and on the impact of extreme flood events.

Gravel-Bed Rivers: Processes and Disasters, First Edition. Edited by Daizo Tsutsumi and Jonathan B. Laronne.

12.2 Sediment Supply to Mountain Rivers and its Influence on the Characteristics of the Sediment Available for Fluvial Transport

Shear stress applied on the river bed during a given flood event will control the transport capacity (e.g. Meyer-Peter and Müller 1948) and flow competence (e.g. Buffington and Montgomery 1997) at a given location. Total or partial sediment entrainment will occur depending on the grain size of the sediment available in the river at the considered location; the amount of transportable sediment available will control whether the river will be 'at capacity' or 'under capacity' (Figure 12.1). The grain size and volume of sediment available within the channel thus exert a key control on the grain size and volume of sediment exported during a flood event of a given magnitude (e.g. Yager *et al.* 2012; Bennett *et al.* 2014). These characteristics will vary both in space and time under the effect of abrasion and transport processes, but also in relation to the distribution of sources of sediment along the river. The relative importance of source characteristics (distribution of sources along the river, flux of sediment from sources, grain size, lithological content), abrasion, and selective transport and deposition depends on the location along the river. In particular, a distinction can be made between upland areas ('mountains') and sedimentary basins (Figure 12.2). In sedimentary basins, the supply of new fresh material to rivers is restricted to tributaries that may bring less 'mature' sediment to the main stem river. The importance of sediment sources thus becomes negligible compared with the abrasion of sediment particles during transport events (floods) and selective transport and deposition that the predominantly depositional environment will promote (Paola *et al.* 1992; Fedele and Paola 2007). In the mountains, rivers primarily incise into bedrock and lower the base-level for hillslope erosion processes (Burbank *et al.* 1996). In such context, selective transport and deposition in the river are inoperative in the long term, since all the sediment supplied to the river is eventually evacuated from the mountain range (Attal and Lavé 2006). The characteristics and distribution of the sources of sediment and the abrasion of sediment particles during transport are thus the main control on fluvial sediment characteristics in upland areas. In the following sections, I review theoretical and field evidence for the influence of sources of sediment on spatial and temporal variations in the characteristics of the sediment available for transport in mountain (upland) rivers.

12.2.1 Spatial Variations

Sources of sediment tend to be distributed along the course of rivers actively incising into bedrock (Figure 12.2). Grains transported by rivers are reduced in size by abrasion, yet downstream fining is not necessarily observed, due to the sustained supply of fresh material (Heller *et al.* 2001; Attal and Lavé 2006; Sklar *et al.* 2006; Chatanantavet *et al.* 2010). For example, Attal and Lavé (2006) and Sklar *et al.* (2006) modelled the evolution of the grain size of fluvial sediment along a river flowing through a uniformly uplifted/eroded landscape in which hillslopes supply the same type of sediment: sediment supply was considered spatially uniform and the models assumed no net deposition. Both studies showed that the fluvial sediment grain size initially decreased before reaching an asymptote when the quantity lost by abrasion was counterbalanced by the supply of fresh material from the hillslopes. This phenomenon was documented along the Hoh River, Olympic Peninsula, Washington, where no downstream fining was observed despite the high erodibility of the clasts, which tend to develop thick, easily abraded, weathering rinds (Heller *et al.* 2001). Heller *et al.* (2001) concluded that 'the continuous resupply of grains strongly attenuates the rate of downstream fining, despite the fact that these weathered grains abrade relatively rapidly'.

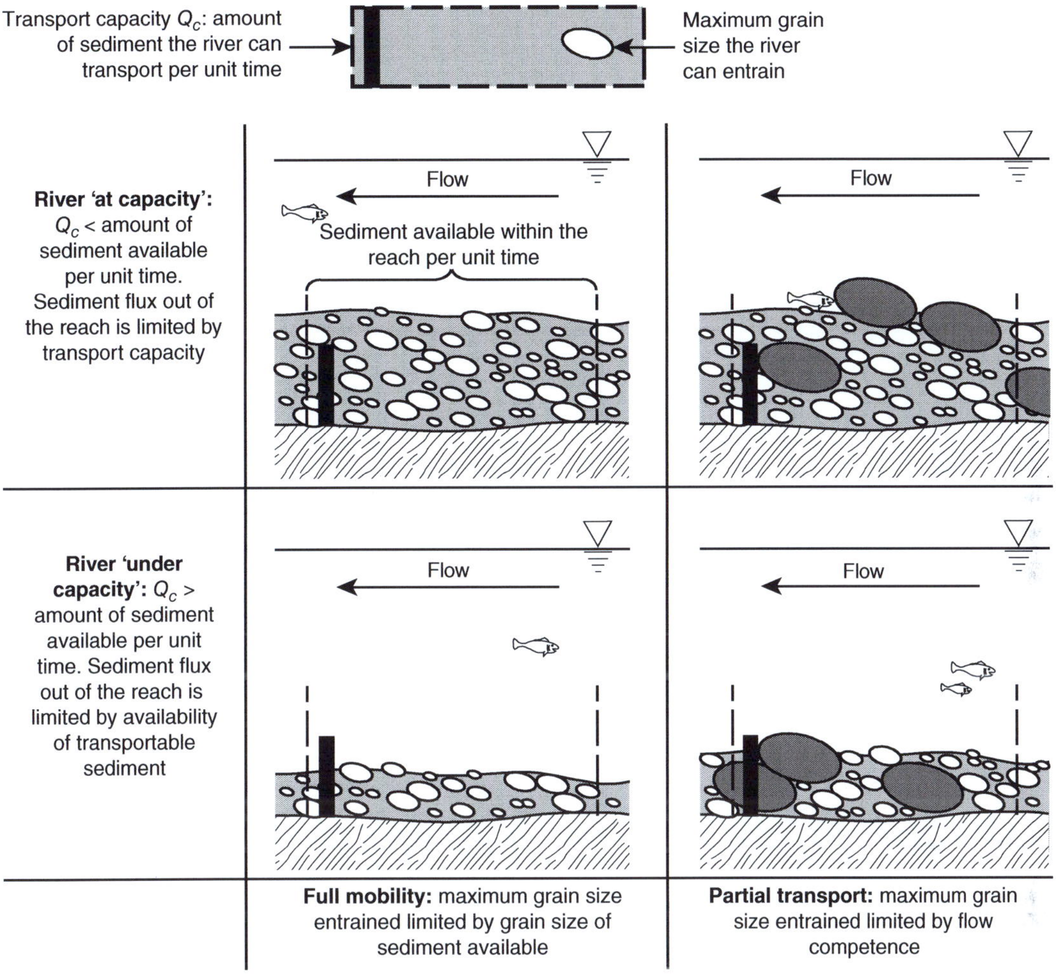

Figure 12.1 Schematic illustration of the interplay between flow competence, transport capacity, amount, and grain size of sediment available for transport, and the properties of the sediment exported from a given mixed bedrock–alluvial reach during a flood event. Grey box at the top is a schematic representation of the amount of sediment the river can transport out of the reach at the time considered (transport capacity – black rectangle is a size guide in all diagrams); the maximum grain size entrained at that time is shown in the box (flow competence). The four diagrams show the reach considered (delineated by the bold dashed lines) in different situations. In the left-hand diagrams, the sediment available is finer than what the river can transport, leading to full mobility. In the right-hand diagrams, some grains are larger than what the river can transport (dark grains), leading to partial transport (e.g. Wilcock and McArdell 1997). Top and bottom panels illustrate 'at capacity' and 'under capacity' scenarios, respectively, resulting from (im)balance between the amount of sediment available for transport and transport capacity. Note that this figure is based on a very simplistic model where the river transports all the grains that are smaller than a given size; it ignores factors that can decrease the mobility of the sediment mixture as a whole, such as the armouring effect, clast interlocking and/or a large proportion of large clasts (e.g. Lisle and Church 2002).

Wolcott (1988) was one of the first to demonstrate the importance of the grain-size distribution of sediment sources in setting the grain-size distribution of the sediment transported by rivers. Wolcott compared the grain size of fluvial sediment in two catchments in Oregon with the grain size of the sources of sediment for the rivers (hillslopes, bars, and channel banks). He showed that the grain size

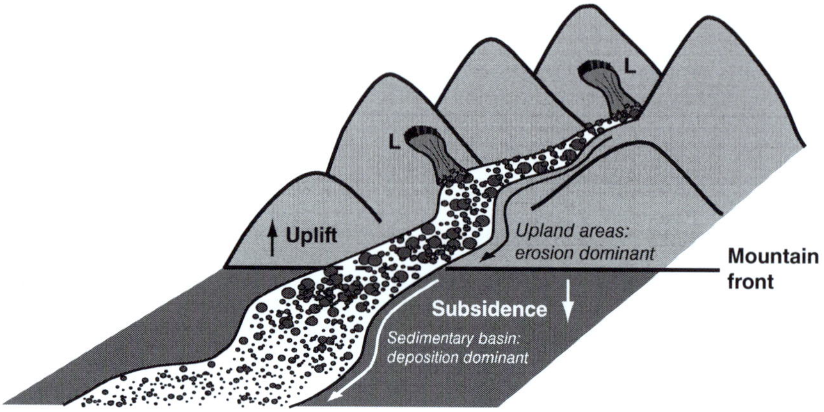

Figure 12.2 Schematic depicting the fate of sediment particles from upland areas to sedimentary basins. Note that this diagram presents a situation that typically arises in tectonically active areas, with a sharp transition from (uplifted) mountainous area to (subsiding) sedimentary basins; the transition will be more subdued in tectonically quiescent areas. Upstream of the mountain front, sediment particles are progressively reduced in size by abrasion as they are transported downstream; selective transport and deposition are inoperative on the long-term since all sediment supplied to the river is eventually evacuated from the mountain range; supply of fresh material from hillslopes (e.g. landslides 'L') and tributaries all along the river course can prevent downstream fining. Downstream of the mountain front, the depositional environment and the absence of fresh sediment supply leads to downstream fining under the combined effects of selective transport and deposition and sediment particle abrasion during fluvial transport.

of the sources of sediment was influenced by the lithology of the underlying rocks: the weathering of sandstones and siltstones produced sediment with a bimodal grain-size distribution, in contrast to basalt-derived sediment, which exhibited a unimodal grain-size distribution. The grain-size distribution of the sediment transported by rivers was found to be bimodal in the basin underlain by sandstones and siltstones and unimodal in the basin underlain by basalt (Wolcott 1988). Attal and Lavé (2006) also found that rock type influences the characteristics of the sediment at the source along the Marsyandi River (Himalayas): landslides in quartzite and gneiss supply coarser sediment than landslides in schist, with median grain size, D_{50}, in the ranges 18–74 mm and 3–23 mm, respectively. Marshall and Sklar (2012) further found a relationship between abundance of rock fragment in soils and climate. In particular, higher precipitation rate was found to correlate with lower rock fragment abundance and reduced occurrence of bimodal size distributions across sites spanning a range of environmental conditions in the Hawaiian islands, the Sierra Nevada (California), and Cascade Mountains (Washington) (Marshall and Sklar 2012).

The type of rock supplied to the fluvial system also influences fluvial sediment characteristics through differences in rock resistance to abrasion, which govern how quickly each sediment particle is reduced in size. In areas where rocks of different lithologies are supplied to rivers, the most resistant lithologies (e.g. granite, quartzite, and volcanics) end up being overrepresented in the bedload compared with the least resistant lithologies (e.g. schist and sandstone), which are rapidly reduced in size (e.g. Mezaki and Yabiku 1984; Parker 1991; Attal and Lavé 2006, 2009). The grain size and amount of sediment available for transport is therefore strongly controlled by the relative contribution of the different rock types within the catchment, with the most resistant lithologies such as quartzite likely to dominate the coarser fraction (cobble, boulders), whereas the least resistant lithologies will contribute essentially to the finest fractions (gravel, sand, and suspended load) (Attal and Lavé 2006, 2009).

The introduction in the river of sediment with distinct characteristics from a point source tends to have a local effect on fluvial sediment that dissipates downstream (Sklar *et al.* 2006). On the other hand, when the change in sediment supplied to the river is sustained, theoretical (Sklar *et al.* 2006) and field studies (Attal and Lavé 2006; Whittaker *et al.* 2008; Whittaker, Attal, and Allen 2010; Attal *et al.* 2015) have shown that the characteristics of the fluvial sediment can be significantly affected. The field studies showed that the median grain size, D_{50}, and 84th percentile, D_{84}, of fluvial sediment increased significantly both on the surface and in the subsurface in response to a coarsening of sediment sources (Figure 12.3). In these cases, change in the grain size of sediment sources was not due to a change in lithology but to a change in the process that generates sediment: the coarsening was associated with a change either from predominantly glacially derived to landslide-derived sediment (Attal and Lavé 2006) or from soil-derived to landslide-derived sediment (Whittaker *et al.* 2008; Whittaker, Attal, and Allen 2010; Attal *et al.* 2015). The increase in fluvial sediment grain size was spatially coincident with the change in sediment source, suggesting an immediate response of fluvial sediment characteristics to changes in source characteristics.

12.2.2 Temporal Variations

Sediment supply to rivers tends to be an episodic process due to the episodic nature of the triggers: sediment will be supplied to rivers via landslides, debris flows, dry ravel, or bank collapse following storms, earthquakes, wildfires, snowmelt, and/or rising river stage (e.g. Bathurst, Leeks, and Newson 1986a, b; Benda and Dunne 1997; Gabet 2003; Dadson *et al.* 2003, 2004; Lancaster 2008; Yanites *et al.* 2010; Singer *et al.* 2013; Bennett *et al.* 2014). As a result, the availability of sediment in a given river reach can significantly change through time, depending on the magnitude of the disturbance (the trigger) and its recurrence interval. Point sources (e.g. isolated landslides) tend to cause pulses of sediment with a signature that reflects the type and magnitude of the supply process; these pulses tend to dissipate downstream due to the diffusive nature of the sediment transport process (Benda and Dunne 1997; Cui *et al.* 2003a, b; Sklar *et al.* 2006). In addition to sediment transport, temporary storage in channels (e.g. in floodplains) can also buffer the pulse-like nature of sediment supply (e.g. Métivier and Gaudemer 1999; Phillips 2003; Lancaster 2008; Blöthe and Korup 2013). The influence of sources on temporal variations in sediment availability and characteristics (e.g. grain size) will therefore be greatest in low-order streams or when the supply events are of high magnitude. I present below a few examples to illustrate these points.

Death Valley (California) is bordered by numerous small catchments with well-developed alluvial fans. Channels tend to be entrenched in fan surfaces dating from the Pleistocene (Dorn 1996). Mosaic Canyon drains one of these small catchments (~11 km^2) on the western side of Death Valley (Figure 12.4). The main lithologies exposed in the catchment are white marble and black schist. As the river incises into its alluvial fan, it exposes strata with very different characteristics. In particular, the succession includes beds with up to 95% very angular marble clasts with median grain size ~3–4 cm alternating with beds comprising up to 95% subrounded schist clasts with median grain size ~1 cm (Figure 12.4). The most logical explanation for this observation is a source effect, whereby sediment availability in the basin is dominated by point sources such as landslides: fluvial transport events that led to the formation of the schist-dominated or marble-dominated beds would have occurred following supply events that led to the availability of either easily abraded schist or comparatively resistant marble, respectively.

Attal and Lavé (2006) made a similar observation in a much larger basin. At Chame, around 50 km downstream of the source of the Marsyandi River (Himalayas), the contributing drainage area is 680 km^2. The rocks in the contributing area are essentially schists and limestones from the Tethyan

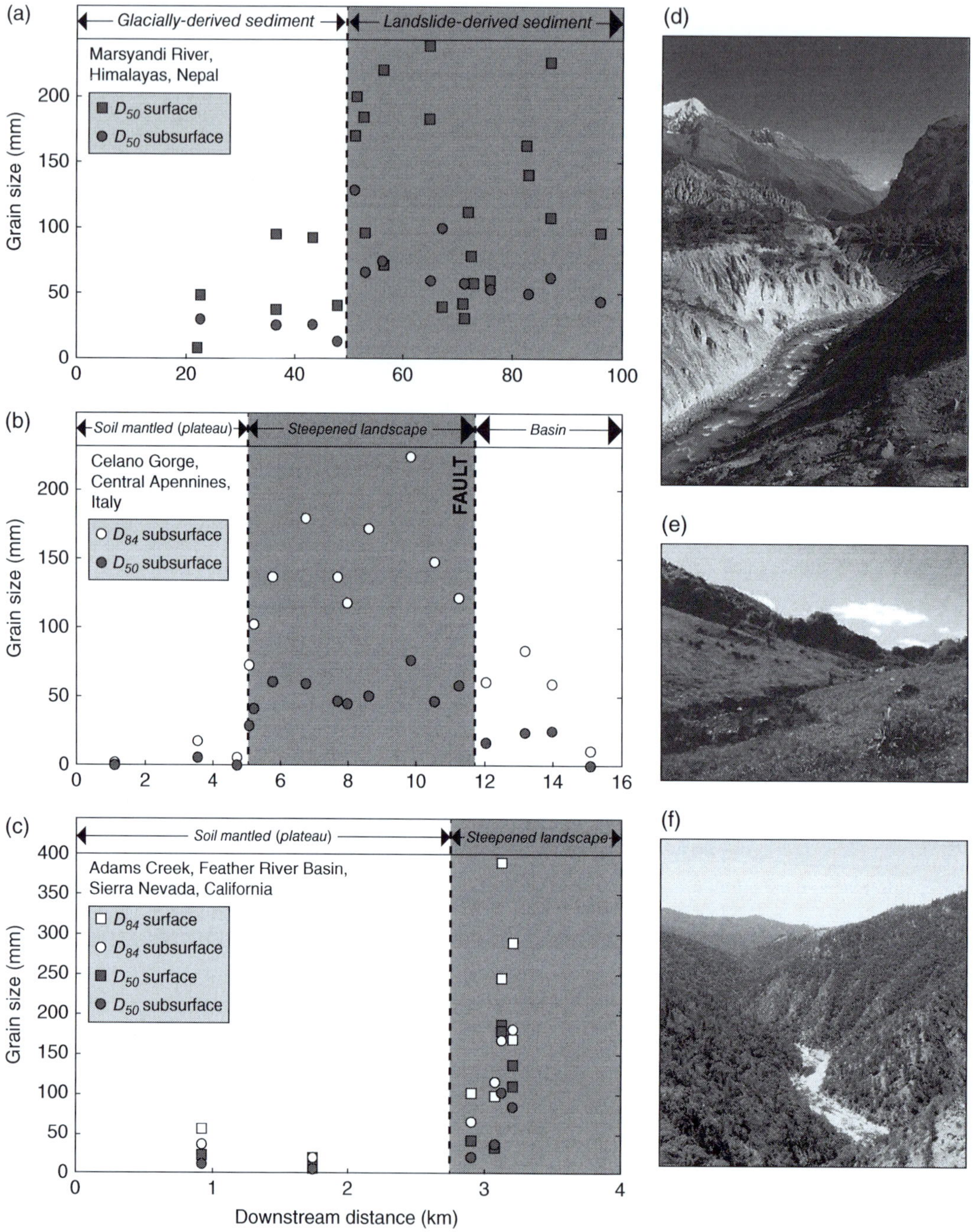

Figure 12.3 Fluvial sediment grain size as a function of distance downstream along three rivers exhibiting prominent changes in source type (note change in scale on *x* and *y* axes). (a) The increase in grain size along the Marsyandi River coincides with a change in source type from glacial till (D_{50} = 6–24 mm) to landslide-derived sediment (D_{50} = 3–74 mm). (Adapted from Attal and Lavé (2006).) (b) In response to an increase in fault throw rate, the landscape steepened upstream of the Fucino fault (Celano Gorge). Increase in fluvial sediment grain size coincides with the transition from low-relief landscape (source of sediment = soil) to steepened landscape (source of sediment = scree cones and landslides). (Adapted from Whittaker, Attal, and Allen (2010).) (c) The situation in Adams Creek is very similar to (b), except that the steepening is caused by rapid incision of the Feather River, of which Adams Creek is a tributary. Fluvial sediment coarsening is observed at the transition from low-relief landscape (source of sediment = soil, D_{50} = 0.1–1.2 mm) to steepened landscape (source of sediment = scree cones and landslides, D_{50} = 35–66 mm). (Adapted from Attal *et al.* (2015).) Photographs illustrating the different domains. (d) Marsyandi River tapping into glacial sediment. (e) Soil-mantled hillslopes in low-relief landscape in the Apennines. (f) Steepened landscape along the Feather River, Sierra Nevada.

Figure 12.4 Mosaic Canyon, Death Valley, California. (a) Location of Mosaic Canyon (Imagery © 2015 DigitalGlobe, Landsat, USDA Farm Service Agency, Map data © 2015 Google). (b and c) Quaternary fan sedimentary succession exposed near the mouth of the catchment following entrenchment, looking upstream and downstream, respectively. The succession includes beds with contrasted grain size, grain angularity, and lithological content. (d) Detail of two successive beds: the lower bed has around 95% subrounded schist clasts with median grain size ~1 cm, whereas the upper bed has around 95% very angular marble clasts with median grain size ~3–4 cm.

Series, with the gneiss from the Higher Himalayan Crystalline representing a few per cent of the contributing area (Attal and Lavé 2006). Yet, the fraction coarser than 4 cm in modern gravel bars is made of nearly 40% gneiss; in the vicinity, terrace deposits contain nearly 100% limestone clasts (Attal and Lavé 2006). This observation suggests a significant recent supply of gneiss upstream of Chame, possibly associated with landslides. Similar observations were made in the tributaries Naur and Dudh Khola, which join the Marsyandi River downstream of Chame.

Most sources of sediment for rivers tend to be point sources rather than diffuse. However, external factors can cause generalized supply by point sources at a basin scale. Lane *et al.* (2008) showed that coarse sediment supply from hillslopes and tributaries to the Upper Wharfe, an upland river in the Yorkshire Dales (UK), tends to happen following powerful storms, irrelevant of the antecedent conditions (dry or wet); these events typically happen as a result of convective storms in summer, demonstrating a seasonal control on sediment supply to the channel in this context. Major external disturbances can also trigger widespread sediment supply over a given area. Dadson *et al.* (2003, 2004) and Yanites *et al.* (2010) show a strong link between sediment supply to rivers in Taiwan and both storm intensity and ground shaking during earthquakes. Yanites *et al.* (2010) estimated that the magnitude 7.6 Chi-Chi earthquake in Central Taiwan and subsequent Typhoon Toraji in 2001 caused 50 000 landslides within a 3000 km^2 region. Within this region, Yanites *et al.* (2010) made a rough estimate of the amount of debris produced of $\sim$0.5–1.3 × 10^4 Mt, 'equivalent to stripping $\sim$0.6–1.7 m of material off the landscape' (Yanites *et al.* 2010). This extraordinary 'pulse' of sediment caused widespread aggradation in the rivers in the area, with up to 18 m vertical aggradation in places. On a larger scale, Pratt-Sitaula *et al.* (2004) document widespread aggradation in excess of 100 m in the Marsyandi Valley, Himalayas. They date two aggradation phases at 50–35 kyr and $\sim$8 kyr BP, that they interpret as resulting from monsoon intensification. The shift to wetter conditions would have caused widespread hillslope destabilization, supplying a considerable amount of sediment that takes thousands of years to be evacuated.

12.3 Influence of Varying Sediment Availability on Sediment Transport and Export During Floods

In section 12.2 I reviewed evidence for the influence of sources on the availability and characteristics (grain size, amount, and lithological content) of sediment in mountain rivers. In this section, I discuss the influence of spatial and temporal variations in sediment availability and characteristics on sediment transport during flood events of varied magnitude, including extreme events.

12.3.1 Geomorphic Work as a Result of the Interplay Between Flood Magnitude and Sediment Supply

Because sediment availability can be limited in mixed bedrock–alluvial mountain rivers, the largest floods are not necessarily associated with the greatest amount of sediment transported. Instead, floods of different magnitudes will do different types of geomorphic work depending on whether sediment is readily available and whether the threshold for entrainment is overcome (Bathurst, Leeks, and Newson 1986a, b; Costa and O'Connor 1995; Wohl 2008; Yager *et al.* 2012; Bennett *et al.* 2014). Lane *et al.* (2008) observed, for example, a dichotomy in the effects of short-lived, high-intensity storm events and longer-lived, moderate storm events on sediment dynamics in the Upper Wharfe catchment (Yorkshire Dales, UK). They found that sediment supply to tributaries and main channel tends to occur as a result of short periods of intense rainfall, typically in summer. During such events, they documented limited sediment transport in the main channel that they link to the short duration of the events; sediment transport in the main channel was essentially happening in winter, following moderate events that were able to sustain sediment transport for longer.

In mountain rivers, sediment aggradation can lead to the development of alluvial planform patterns (such as braiding or meandering) within a system constrained by bedrock. This 'alluvial overprint' (Carling 2009; Meshkova and Carling 2012; Meshkova, Carling, and Buffin-Bélanger 2012; Turowski *et al.* 2013) can influence sediment dynamics in response to floods of different magnitude. Turowski *et al.* (2013) document two contrasting behaviours that they call 'flood-depositing' and 'flood-cleaning', which may be conceptualized by a model in which sediment supply to a given reach Q_s is a power function of water discharge Q (Turowski 2012):

$$Q_s = cQ^{\lambda} \tag{12.1}$$

where c and λ are constants. The 'flood-depositing' case corresponds to $\lambda > 1$; in this case, low-frequency, high-magnitude floods supply more sediment than they can evacuate, leading to significant sediment deposition within the reach in question. Sediment is subsequently progressively evacuated through time during small and moderate floods. In the 'flood-cleaning' case ($\lambda < 1$), sediment progressively builds up in the channel following high-frequency, low-magnitude events; an extreme event then flushes a significant part or all of the sediment out of the reach. Turowski *et al.* (2013) review evidence from the literature and report that both behaviours are observed in both bedrock and alluvial rivers. They also show that the dominant behaviour can change spatially and potentially in time within a single basin, with some large basins (Clyde River, Australia (Nanson 1986) and Sabie River, South Africa (Heritage *et al.* 2004)) exhibiting alternating flood-cleaning and flood-depositing reaches. In fact, they link the two behaviours to channel–hillslope coupling, inferring that sediment must be supplied from hillslopes during aggrading phases, that is, during high-magnitude events in flood-depositing streams and during small and moderate floods in flood-cleaning streams. The relative timing of floods and sediment supply to the channel is therefore crucial in controlling the amount of sediment transported in a given reach and exported from a basin during floods of varying magnitudes (Bathurst, Leeks, and Newson 1986a, b).

It is notable that following major disturbances such as the ones documented in Taiwan (Dadson *et al.* 2003, 2004; Yanites *et al.* 2010) or along the Marsyandi River in the Himalayas (Pratt-Sitaula *et al.* 2004) (section 12.2.2), sediment availability will not limit sediment transport even for extreme events for a significant amount of time, due to the colossal nature of vertical sediment aggradation within the valleys, from tens of metres to in excess of 100 m. Pratt-Sitaula *et al.* (2004) and Yanites *et al.* (2010) estimate the time for evacuation of sediment to be in the order of hundreds to thousands of years. During this period, rivers are likely to be 'at capacity' during floods, the main constraint on sediment entrainment being the need to overcome the threshold for incipient sediment motion. These basin-scale alluviation events are typically associated with generalized slope instability and landsliding (Dadson *et al.* 2003, 2004; Pratt-Sitaula *et al.* 2004; Yanites *et al.* 2010), leading to a wide range of sediment grain sizes being supplied to the river, up to boulders a few metres in size (Attal and Lavé 2006; Whittaker *et al.* 2008; Whittaker, Attal, and Allen 2010; Attal *et al.* 2015) that may be mobilized only during the most extreme events. The coarseness of sediment present in the channel may therefore act as the main inhibitor of sediment transport (e.g. Wohl 1992; Jansen 2006), which will be discussed in the following section. The generalized destabilization of hillslopes may temporarily exhaust the sources of sediment; regeneration of sediment on hillslopes (e.g. soil development) would be possible during the hundreds to thousands of years that the river will need to evacuate the valley fill.

12.3.2 Grain Size and Sediment Mobility: the Boulder Issue

As mentioned in section 12.2 (Figure 12.1), the maximum grain size of the sediment entrained during a flood will be limited either by the grain size of the sediment available for transport (if sediment available is relatively fine-grained with respect to flow competence) or by flow competence (if a wide range of grain sizes is available, up to boulder size). The coarseness of the sediment supplied to rivers can therefore exert an important control on the grain size of the sediment transported during a given flood: a very powerful flood will not be able to transport coarse sediment if no coarse sediment is available for transport. Such a situation may arise in low relief, soil-mantled landscapes (Figure 12.3b, c, e) (Whittaker *et al.* 2008; Whittaker, Attal, and Allen 2010; Attal *et al.* 2015) or where large amounts of sediment lacking large clasts is available within a catchment (e.g. badlands or conglomeratic bedrock, e.g. Mather and Hartley (2005)). In other types of mountainous landscapes, sediment available for transport is likely to include large clasts up to boulder size supplied by landslides, debris flows or rock falls, or left behind by glaciers (e.g. Chatanantavet *et al.* 2010; Attal *et al.* 2015).

In mountain rivers, boulders can contribute to the development of bedforms that tend to limit the mobility of the sediment as a whole. Many studies have documented the formation of step-pool morphology in the presence of boulders, in particular where the size of bed materials is large relative to the channel size (see reviews on the topic by Chin and Wohl 2005; Church and Zimmermann 2007; Comiti and Mao 2012). The formation of steps is associated with the largest clasts acting as anchor points against which other clasts become imbricated and stop moving. The steps are therefore made of a variety of clast sizes, with the boulders acting as keystones. Moderate floods tend to maintain this morphology by contributing to the scouring of the pools upstream of steps, whereas high-magnitude, low-frequency events are able to destroy the steps and mobilize the sediment 'frozen' in the steps and between them (e.g. Gintz, Hassan, and Schmidt 1996; Lenzi, Mao, and Comiti 2004, 2006; Church and Zimmermann 2007; Wohl 2008; Turowski *et al.* 2009, 2013; Morche and Schmidt 2012; Yager *et al.* 2012). In the case of isolated boulders, a series of approaches have been proposed to estimate the magnitude of the flood capable of setting a clast of a given size in motion (e.g. Clarke 1996; Carling and Tinkler 1998; Carling, Hoffmann, and Blatter 2002; Carling *et al.* 2002; Stokes, Griffiths, and Mather 2012). The different methods have been shown to produce significantly different results (see discussion in Stokes, Griffiths, and Mather 2012), nevertheless they can be used to give a rough estimate of the flow depth required to mobilize a boulder of a given size. Using the equation described in Stokes, Griffiths, and Mather (2012) with the same values of fluid and boulder density and the same values of lift, drag, and sliding coefficients, the following flow depths for incipient motion were found: in a channel with a slope ranging between 0.1 and 0.01 (which is typical of mountain rivers), the flood that would entrain 0.5 and 1 m diameter boulders would have a minimum depth ranging between 0.9 and 1.8 and between 1.5 and 3.0 m, respectively. Abundant and/or large boulders may therefore be considered as a potentially significant hindrance to sediment entrainment during floods (e.g. Wohl 1992; Yager, Dietrich, and Kirchner 2007), in particular in the least steep reaches of mountain rivers.

Beyond the bedform scale, a few studies have provided evidence for the influence of boulders on channel geometry that was linked to reduced sediment mobility and bed armouring. Howard, Dietrich, and Seidl (1994) argue that boulders sourced locally from the walls of the Grand Canyon (Colorado) reduce alluvium mobility and therefore river incision, causing a negative feedback: rapid incision generates deep narrow gorges from the walls of which boulders will be produced, which will then inhibit river incision. Brocard and van der Beek (2006) and Johnson *et al.* (2009) both document significant channel steepening in response to local supply of boulders along actively incising rivers. Johnson *et al.* (2009) document differences in the geometry of channels incising into similar sandstones in the Henry Mountains (Utah): channels that have abundant large diorite clasts (cobbles and boulders) sourced from localized exposures in the headwaters are less incised and steeper than channels with no diorite clasts. They propose that the cover effect provided by the not very mobile coarse sediment inhibits incision and that the steepening would represent a long-term response of the channel to transport the sediment load (evidence for the transport of the diorite clasts exists in the form of imbrication and boulder jams). Brocard and van der Beek (2006) show that the local supply of large boulders from limestone cliffs causes a steepening of rivers incising into marls in the Western Alps (southeast France) (their figure 5). They argue that the boulders are too large to be transported by the river and that the armouring caused by abundant boulders would effectively be equivalent to limestone bedrock. Clarke and Hansen (1996) made similar observation of 'extremely stable boulders' that are unlikely to be entrained under the current climatic regime in small basins in an arid environment in California (Anza Borrego Desert). This effect could literally be called 'bedrock aggradation': the river will need to abrade or break the boulders into fragments of smaller size before entrainment could happen and the river could incise into the underlying bedrock. As in Johnson *et al.* (2009), Brocard and van der Beek interpret the local steepening as a long-term response of the channel to provide the extra work required to deal with the boulders. 'Bedrock aggradation' in the form of very large boulders in channels is widespread in steep landscapes (Figure 12.5) and tends to cause channel adjustment because only infrequent extreme flood events would be able to mobilize them, as discussed in the next section.

12.3.3 Sediment Transport During Extreme Flood Events

Extreme floods are exceptional events with recurrence intervals typically exceeding a century, that are capable of transporting very coarse sediment in colossal quantities, cause significant bedrock erosion, and therefore produce major landscape change (e.g. Baker 1978; Wohl 1992; Baker and Kale 1998; Montgomery *et al.* 2004; Gupta *et al.* 2007; Garcia-Castellanos *et al.* 2009; Warner *et al.* 2010; Baynes *et al.* 2015). Such events can be caused by powerful storms or typhoons, landslide dam break, glacial lake outburst, or subglacial volcanic eruptions; whereas some of these causes will be restricted to some area (e.g. subglacial volcanic eruptions), others will be common over geological time scales in mountainous landscapes (e.g. landslide dam breaks).

Extreme events may have the ability of mobilizing the entire grain population, leading to full mobility even in the presence of very large boulders. Evidence for transport of boulders up to a few metres in size during extreme floods abounds in the literature (e.g. Williams 1983; Beaty 1989; Baker and Kale 1998; Wohl 1992; Carrivick, Russell, and Tweed 2004; Mather and Hartley 2005; Lamb *et al.* 2008; Carling *et al.* 2009; Marren, Russell, and Rushmer 2009; Russell, Duller, and Mountney 2010; Stokes, Griffiths, and Mather 2012; Carling 2013), although hyperconcentrated or debris-flow processes rather than fluvial processes may have been involved in some of these cases (e.g. Mather and Hartley 2005). Extreme floods therefore play a crucial role in transporting sediment that could be considered stable over centennial to millennial time scales, due to their coarseness and/or the development of resistant bedforms such as armoured layers or pools and riffles (e.g. Wohl 1992, 2008; Lisle and Church 2002; Jansen 2006; Yager, Dietrich, and Kirchner 2007; Turowski *et al.* 2009, 2013; Baggs Sargood *et al.* 2015). The size of the largest clasts transported during such events may be limited by the coarseness of sediment available for transport (Mather and Hartley 2005).

In terms of sediment flux, such extreme events are likely to result in net sediment export from the mountain area to areas downstream where transport capacity is significantly reduced, typically as a result of slope reduction and/or channel/valley widening (e.g. sedimentary basin, Figure 12.2; Carling 2013). At the reach scale, the flood will not necessarily have a 'cleansing' effect in the mountainous reaches in the sense that bedrock will not necessarily be subsequently exposed. Figure 12.6

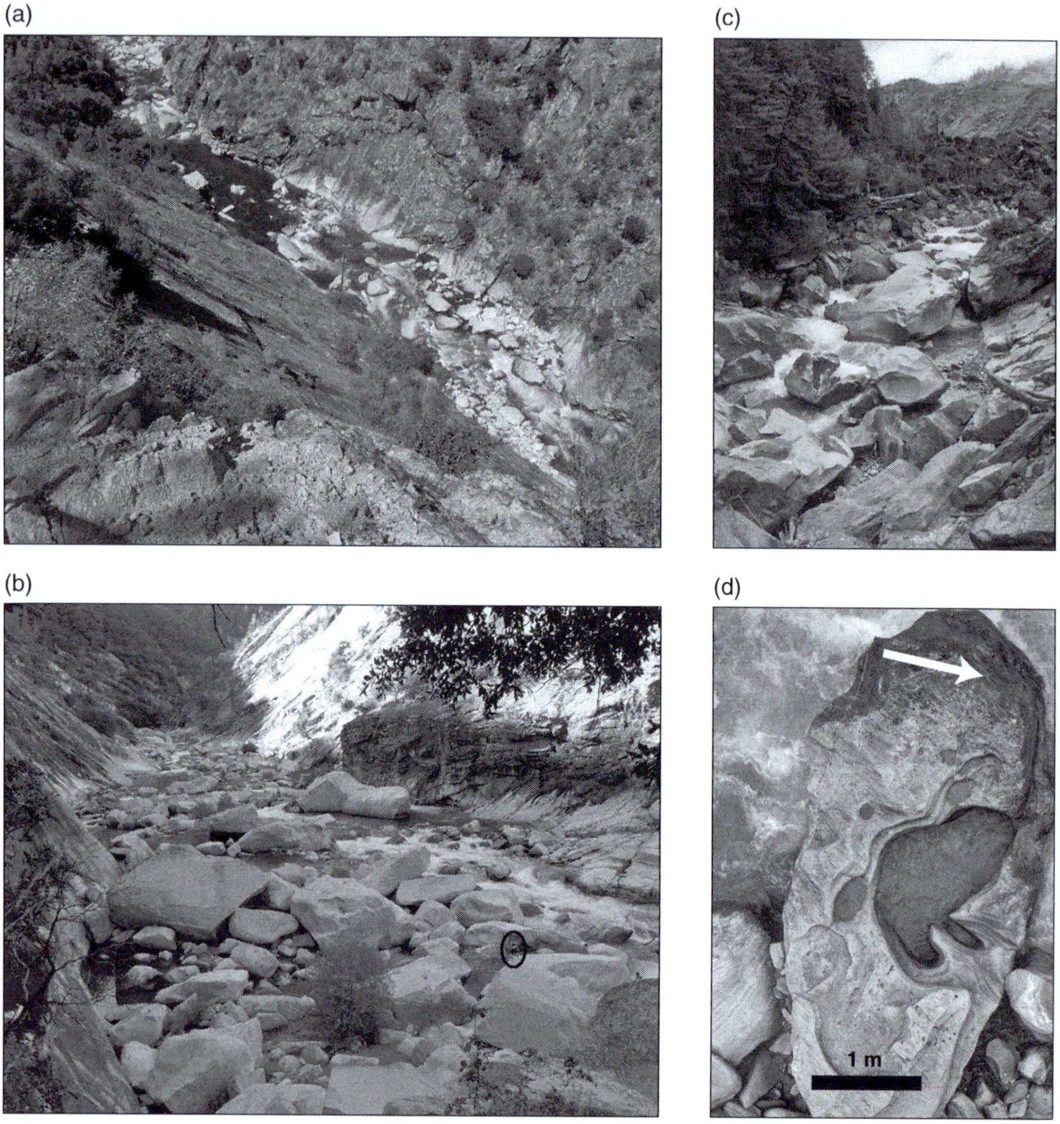

Figure 12.5 Examples of large boulder concentrations inhibiting bedrock erosion along actively incising rivers. Where concentration of boulders a few metres in size is high, their effect may be compared with 'bedrock aggradation'. Boulders typically exhibit fluting and potholing, testifying to their low mobility; only extreme floods may move them. (a) General view and (b) close up of the Feather River, California, where it incises into the Bald Rock granitic pluton. (Adapted from Attal *et al.* (2015).) River is ~30 m wide; it flows towards bottom right-hand corner of picture in (a) and towards camera in (b). Person is circled in (b) for scale. (c) Marsyandi River, Himalayas, where it incises into the silicified Annapurna limestone downstream of Pisang (Nepal); channel is ~20 m wide and river flows towards camera. (d) Large gneiss boulder in Marsyandi River showing intense fluting and potholing on its lee side (picture taken from above, arrow indicates flow direction). Abrasion by suspended- and bedload is effective during floods, when the boulder is submerged. Over time, abrasion can efficiently reduce the size of boulders, potentially to a size that may be mobilized during a moderate flood.

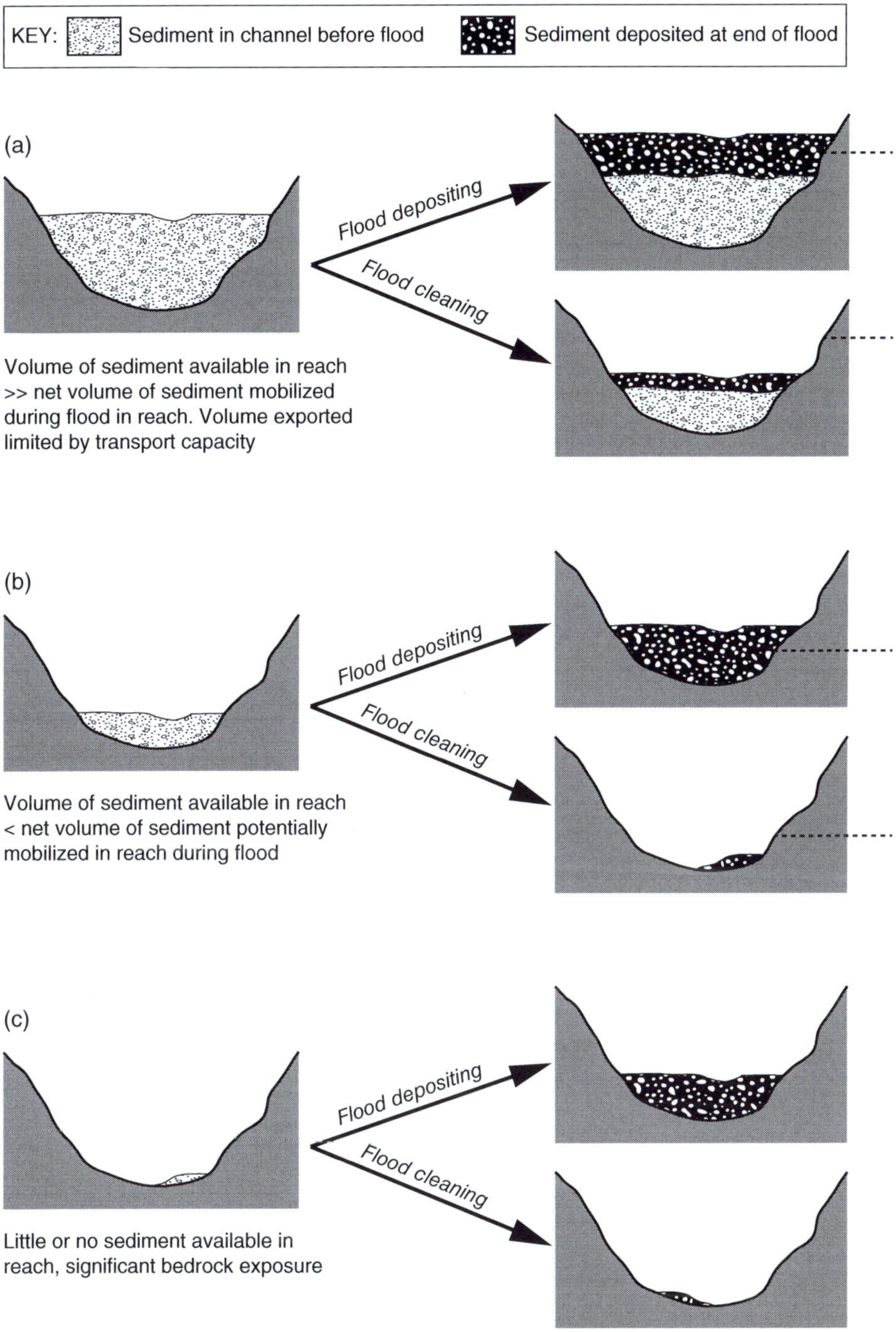

Figure 12.6 Schematic evolution of valley fill following an extreme flood, considering three different initial conditions and depending on whether the reach is 'flood-depositing' or 'flood-cleaning' according to Turowski *et al.*'s (2013) model. In (a) the flood is not powerful enough to mobilize all sediment in the reach and bedrock is not exposed. In (b) and (c) all sediment is mobilized in the reach; little or no sediment is preserved in the channel following the flood in the flood-cleaning scenario. Note that bedrock erosion is likely to happen in all (b) and (c) scenarios and may therefore contribute to the sediment load; in the flood-depositing case in (b), bedrock is not exposed before or after the flood but sediment moves and interacts with the bedrock during the flood ('dynamic cover effect'; Turowski, Lague, and Hovius 2007), likely leading to significant bedrock erosion. Dashed line on the diagrams to the right represents pre-flood sediment level. Changes in channel shape due to bedrock erosion are not represented.

schematically represents different scenarios based on the initial amount of sediment present in the channel and whether the reach considered behaves in a 'flood-depositing' or 'flood-cleaning' way (Turowski *et al.* 2013). The volume of sediment exported from the reach will be limited by transport capacity if considerable amounts of sediment are present in the channel before the flood (e.g. following a phase of generalized aggradation – Figure 12.6a) (e.g. Pratt-Sitaula *et al.* 2004) and/or if the reach is behaving in a flood-depositing way (Figure 12.6). Turowski *et al.* (2013) give examples of flood-depositing reaches where low-frequency, high-magnitude floods lead to aggradation, which they interpret as a result of synchronous excess sediment supply from hillslopes; the Taiwan example discussed in previous sections exemplifies this behaviour. During a single event a flood may be depositing at low discharge, eroding at high discharge, and depositing again at low discharge. Whether this results in net aggradation or erosion depends on the dependence of the transport capacity on discharge, sediment availability, and the shape of the hydrograph (e.g. Rickenmann 1997).

Sediment flux will be limited by sediment availability if the reach is behaving in a flood-cleaning way and the initial volume of sediment in the channel is limited (Figure 12.6b, c). In such situations, the extreme flood will have a cleansing effect, flushing out sediment and exposing bedrock (see example in Figure 12.7; e.g. Wohl 1992; Jansen 2006; Turowski *et al.* 2013; Baggs Sargood *et al.* 2015). Subsequent floods of any magnitude are likely to have limited sediment flux for the time needed to replenish the stock of sediment available for transport within the channel. The time between floods is therefore crucial in controlling the geomorphic response and amount of sediment that will be exported by a flood of a given magnitude (Harvey 1984; Cenderelli and Wohl 2003; Wohl 2008). Regarding extreme floods, the effect of this 'replenishment time' may be considerable both for the amount and type of sediment that will be exported from a catchment, due to the fact that extreme flood events can also lead to significant bedrock erosion. I present a simple model in Figure 12.8 that illustrates the potential effect of sediment exhaustion and recurrence time of extreme floods on the flux and type of sediment exported. The model describes the origin and volume of sediment exported from a given river section during extreme events as a function of the time between events (relative to the time needed for sediment replenishment in the channel). The model stresses that sediment exported during two extreme events of similar magnitude may represent very different volumes and include very different proportions of material sourced from bedrock, depending on the time between events, bedrock resistance to erosion, and the necessity for tools to detach bedrock. As a result, the two events of similar magnitude may export sediment with very contrasting characteristics (volume, lithological content, roundness of clasts, and grain-size distribution), with bedrock constituting a local source of likely angular sediment lacking fines (< sand size). This behaviour exemplifies the need to understand the internal dynamics of river basins before interpreting changes in sedimentary successions in terms of tectonics or climatic signals, as autogenic processes and bed reorganization can lead to the shredding of environmental signals (e.g. van de Wiel and Coulthard 2010; Jerolmack and Paola 2010).

Hartshorn *et al.* (2002) document a possible feedback mechanism between extreme events and sediment replenishment. They show that an extreme flood event along the Liwu River in Taiwan during the powerful typhoon Billis in 2000 (maximum flow depth 12 m) caused considerable bedrock erosion. However, erosion peaked on the banks of the channel, a few metres above the mean low-flow level. They concluded that in the studied channel, bed lowering is driven by moderate flood events, whereas extreme events tend to widen the channel and therefore transmit the erosion signal to the hillslopes. The undercutting of the hillslopes by an extreme flood may cause widespread destabilization and generalized mass movement and therefore increased sediment supply and replenishment to the channel (see also Hovius and Stark 2006; Turowski *et al.* 2008b; Turowski 2012). Morche and Schmidt (2012) provide evidence for such an effect: following a landslide dam break along the

(a)

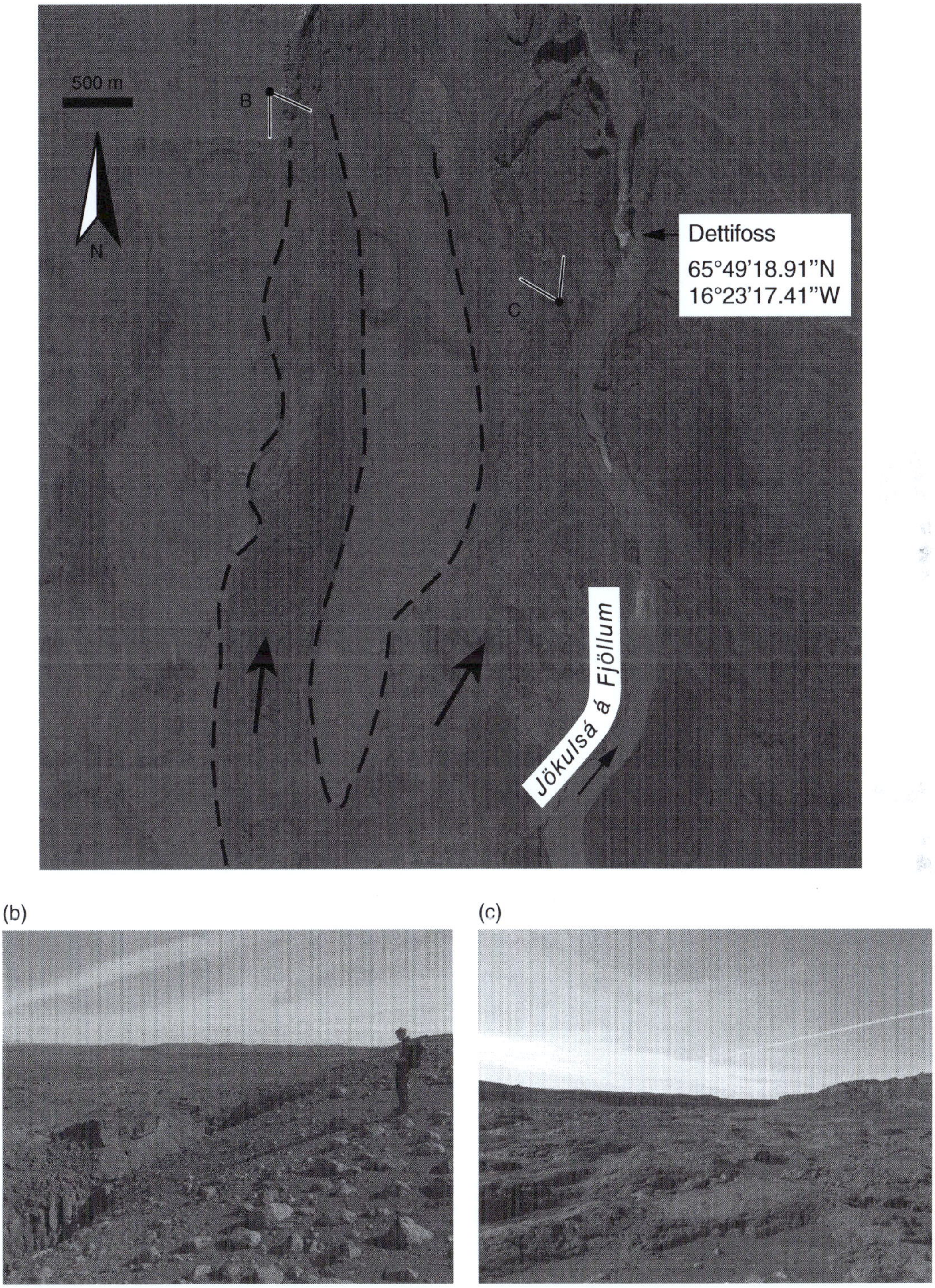

(b)

(c)

Figure 12.7 Evidence for the impact of extreme floods along the Jökulsá a Fjöllum, northeast Iceland. (a) Satellite image (Imagery © 2015 Cnes/Spot Image, DigitalGlobe, Map data © 2015 Google) showing the scablands around the Dettifoss waterfall. B and C show the location of the photographs (b) and (c), respectively. Dashed lines delineate thick glacial sediment deposits. One or more extreme flood event(s) cleared the glacial sediment along the path illustrated by the large arrows during the Holocene, exposing basalt bedrock (Baynes *et al.* 2015). (b) Glacial sediment, palaeoflood channel and sediment 'island' in the background. (c) Exposed bedrock in the path of the palaeoflood. If a new extreme flood follows the same flow path, sediment in this area will be essentially sourced from bedrock erosion.

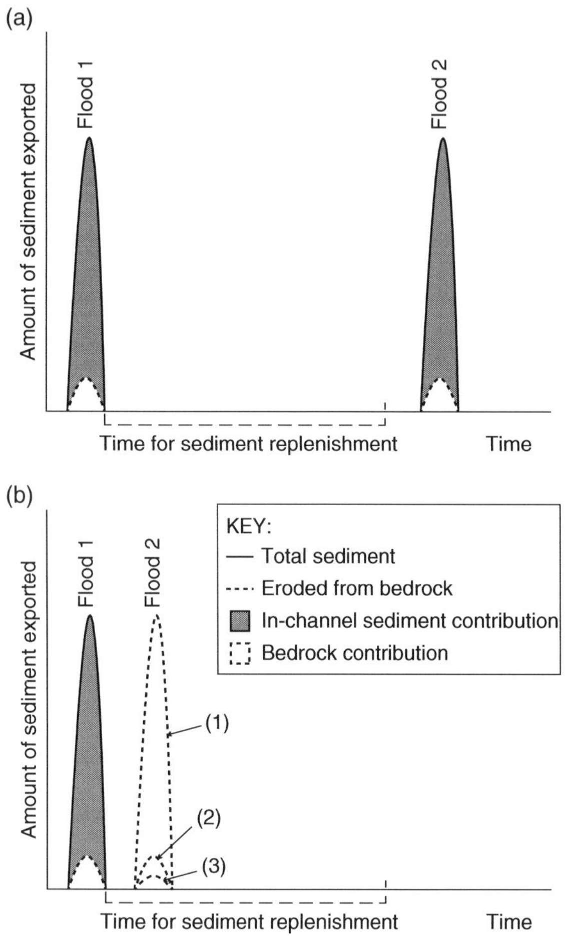

Figure 12.8 Simple model of sediment transport and bedrock erosion during extreme events. Here, two large single events of similar magnitude occur during the time considered. These events have the capacity to flush sediment out of the considered river section and cause bedrock erosion. In (a) the second flood occurs after sediment has replenished in the channel and the amount and type of sediment exported is similar to during the first flood. In (b) the second flood occurs very shortly after the first one and the sediment is essentially sourced from bedrock erosion. Different scenarios can be envisaged: (1) bedrock is highly erodible and the amount eroded is limited by transport capacity; (2) bedrock erosion is limited by the availability of material that can be detached, for example blocks that can be plucked from bedrock due to open joints and cracks in the near subsurface (e.g. Whipple, Hancock, and Anderson 2000); (3) bedrock erosion is inhibited by the lack of tools for erosion by abrasion and/or plucking (Whipple, Hancock, and Anderson 2000; Sklar and Dietrich 2001). Note that this model does not make any assumption regarding the mechanisms that lead to sediment replenishment and whether replenishment is progressive or episodic. It just assumes that a given amount of time is needed for sediment availability in the channel to equate sediment availability before the first flood.

Partnach River (Bavarian Alps), field observations and monitoring of water and sediment transport along the river show a sustained increase in the bedload contribution that the authors interpret as partly resulting from the recoupling of the channel with hillslopes downstream of the dam. They observed that the outburst flood cut the feet of several talus cones and measured elevated transport rates, in particular for bedload, that were sustained for 4 years after the event (see also Gintz, Hassan, and Schmidt 1996; Lenzi, Mao, and Comiti 2004; Turowski *et al.* 2009).

An extreme flood is likely to cause considerable sediment transport and therefore considerable sediment deposition at the end of the flood (e.g. Carling 2013). Sediment will be deposited where the flow loses power, for example where the valley widens such as at the transition from the confined channel to the sedimentary basin (Figure 12.2) or at the coast (e.g. Russell, Duller, and Mountney 2010). Widespread deposition will also happen at the end of the flood. At a local scale, flow constrictions, valley widening, slope reduction, plunge pools at cataracts, and bends may reduce stream power (e.g. backwater effect) and cause sediment deposition, even within the mountainous realm (e.g. Wohl 1992; Meshkova and Carling 2012). The extreme event may have opposing effects on the mobility of in-channel sediment for subsequent, potentially smaller floods. Mobility may be increased through the destruction of resistant bedforms and displacement or removal of boulders (Lenzi, Mao, and Comiti 2004, 2006; Wohl 2008; Turowski *et al.* 2009, 2013; Morche and Schmidt 2012); for example,

Lenzi, Mao, and Comiti (2004) document increased bedload transport for floods of a given magnitude following a large flood in an Alpine catchment. However, the event may also result in the formation of boulder bars that will be stable until the next extreme event or until the blocks are reduced in size by abrasion (e.g. Wohl 1992; Jansen 2006; Baggs Sargood *et al.* 2015).

12.4 Concluding Remarks

In mountain rivers, the amount and grain size of sediment transported by a flood of a given magnitude depend on flow competence and transport capacity, and also on the characteristics of the sediment available for transport. The latter vary in space and time due to the episodic nature of the sediment supply processes and to variations in the nature of the sources (e.g. glacigenic sediment, soil or landslide; lithology of source rock). Change in the nature of sources can be driven by tectonics and climate through landscape steepness, the impact of glaciation and deglaciation, precipitation and storm intensity, and/or the stabilizing effect of vegetation. A spatial or temporal change in tectonics or climate may therefore influence the type and amount of sediment supplied to mountain rivers, which should be reflected in the characteristics of the sediment transported by rivers and exported to sedimentary basins.

In steep landscapes, boulders up to a few metres in size are commonly supplied to channels and can inhibit sediment transport and bedrock erosion when abundant. Extreme flood events are crucial in mobilizing the whole grain population, including boulders that can be considered stable over centennial to millennial time scales. Extreme floods can have either a depositing or a cleaning effect (Turowski *et al.* 2013): the former occurs when substantial amounts of excess sediment are supplied simultaneously to the event, leading to net aggradation; in the latter, the flood may flush all sediment available out of a given reach, leading to bedrock exposure and a sediment flux limited by sediment availability. The nature of the sediment exported from a reach or basin during extreme events will also depend on the time between events and on the time needed for in-channel sediment replenishment. Because extreme events can flush sediment out of a reach, transport (and deposit) large boulders, and erode bedrock, they can have a long-lasting legacy on sediment storage and mobility in a given reach. These observations stress the need for integrating the sources of sediment and the episodic processes that lead to sediment supply to channels and sediment transport in channels when modelling sediment dynamics or bedrock erosion in mountain catchments, or the development of stratigraphy in sedimentary basins (Walling 1983; Yager *et al.* 2012; Turowski *et al.* 2013; Bennett *et al.* 2014).

Acknowledgement

I am very grateful to Jens Turowski and Paul Carling for constructive comments that helped improve the manuscript.

References

Attal, M and Lavé, J 2006. Changes of bedload characteristics along the Marsyandi River (central Nepal): implications for understanding hillslope sediment supply, sediment load evolution along fluvial networks, and denudation in active orogenic belts. *Geological Society of America Special Paper* 398, 143–171. DOI:10.1130/2006.2398(09).

Attal, M and Lavé, J 2009. Pebble abrasion during fluvial transport: Experimental results and implications for the evolution of the sediment load along rivers. *Journal of Geophysical Research* 114, F04023. DOI:10.1029/2009JF001328.

Attal, M, Mudd, SM, Hurst, MD *et al.* 2015. Impact of change in erosion rate and landscape steepness on hillslope and fluvial sediments grain size in the Feather River basin (Sierra Nevada, California). *Earth Surface Dynamics* 3, 201–222. DOI:10.5194/esurf-3-201-2015.

Baggs Sargood, M, Cohen, TJ, Thompson, CJ, and Croke, J 2015. Hitting rock bottom: morphological responses of bedrock-confined streams to a catastrophic flood. *Earth Surface Dynamics* 3, 265–279. DOI:10.5194/esurf-3-265-2015.

Baker, VR 1978. Large-scale erosional and depositional features of the Channeled Scabland. In *The Channeled Scabland* (eds Baker, VR and Nummedal, D). NASA: Washington, DC; 81–115.

Baker, VR and Kale, VS 1998. The role of extreme floods in shaping bedrock channels. In *Rivers Over Rock: Fluvial Processes in Bedrock Channels* (eds Tinkler, KJ and Wohl, EE). Geophysical Monograph 107, American Geophysical Union Press: Washington, DC; 153–165,

Bathurst, JC, Leeks, GJL, and Newson, MD 1986a. Field measurements for hydraulic and geomorphological studies of sediment transport-the special problems of mountain streams. In *Measuring Techniques in Hydraulic Research* (ed. Wessels, ACE). AA Balkema: Rotterdam, Netherlands; 137–151.

Bathurst, JC, Leeks, GJL, and Newson, MD 1986b. Relationship between sediment supply and sediment transport for the Roaring River, Colorado, USA. In *Drainage Basin Sediment Delivery* (ed. Hadley, RF). IAHS Publication 159, Centre for Ecology and Hydrology: Wallingford, UK; 105–117.

Baynes, ERC, Attal, M, Niedermann, S *et al.* 2015. Erosion during extreme flood events dominates Holocene canyon evolution in north-east Iceland. *Proceedings of the National Academy of Sciences* 112(8), 2355–2360. DOI:10.1073/pnas.1415443112.

Beaty, CB 1989. Great big boulders I have known. *Geology* 17, 349–352. DOI:10.1130/0091-7613(1989)017<0349:GBBIHK>2.3.CO;2.

Benda, L and Dunne, T 1997. Stochastic forcing of sediment supply to channel networks from landsliding and debris flow. *Water Resources Research* 33(12), 2849–2863.

Bennett, GL, Molnar, P, McArdell, BW, and Burlando, P 2014. A probabilistic sediment cascade model of sediment transfer in the Illgraben. *Water Resources Research* 50. DOI:10.1002/2013WR013806.

Blöthe, JH and Korup, O 2013. Millennial lag times in the Himalayan sediment routing system. *Earth and Planetary Science Letters* 382, 38–46. DOI:10.1016/j.epsl.2013.08.044.

Brocard, GY and van der Beek, PA 2006. Influence of incision rate, rock strength, and bedload supply on bedrock river gradients and valley-flat widths: field-based evidence and calibrations from western Alpine rivers (southeast France). *Geological Society of America Special Paper* 398, 101–126. DOI:10.1130/2006.2398(07).

Buffington, JM and Montgomery, DR 1997. A systematic analysis of eight decades of incipient motion studies, with special reference to gravel-bedded rivers. *Water Resources Research* 33, 1993–2029.

Burbank, DW, Leland, J, Fielding, E *et al.* 1996. Bedrock incision, rock uplift and threshold hillslopes in the northwestern Himalayas. *Nature* 379, 505–510.

Carling, PA 2009. Geomorphology and sedimentology of the Lower Mekong River. In *The Mekong: Biophysical Environment of an International River Basin* (ed. Campbell, IC). Academic Press: New York; 77–111.

Carling, PA 2013. Freshwater megaflood sedimentation: what can we learn about generic processes? *Earth Science Reviews* 125, 87–113. DOI:10.1016/j.earscirev.2013.06.002.

Carling, PA and Tinkler, KJ 1998. Conditions for entrainment of cuboid boulders in bedrock streams: an historical review of literature with respect to recent investigations. In *Rivers Over Rock: Fluvial Processes in Bedrock Channels* (eds Tinkler, KJ and Wohl, EE). Geophysical Monograph 107, American Geophysical Union Press: Washington, DC; 19–34.

Carling, PA, Hoffmann, M, and Blatter, AS 2002. Initial motion of boulders in bedrock channels. In *Ancient Floods, Modern Hazards* (eds House, PK, Webb, RH, Baker, VR, and Levish, DR). American Geophysical Union, Washington, DC; 147–160. DOI:10.1029/WS005p0147Carling.

Carling, PA, Hoffmann, M, Blatter, AS, and Dittrich, A 2002. Drag of emergent and submerged regular obstacles in turbulent flow above bedrock surface. In *Rock Scour due to Falling High-Velocity Jets* (eds Schleiss, AJ and Bollaert, E). Swets and Zeitlinger/Balkema: Lausanne, Switzerland; 83–94.

Carling, PA, Herget, J, Lanz, JK *et al.* 2009. Channel-scale erosional bedforms in bedrock and in loose granular material: character, processes and implications. In *Megaflooding on Earth and Mars* (eds Burr, DM, Carling, PA, and Baker, VR). Cambridge University Press; Cambridge; 13–32.

Carrivick, JL, Russell, AJ, and Tweed, FS 2004. Geomorphological evidence for jökulhlaups from Kverkfjöll volcano, Iceland. *Geomorphology* 63, 81–102.

Cenderelli, DA and Wohl, EE 2003. Flow hydraulics and geomorphic effects of glacial-lake outburst floods in the Mount Everest region, Nepal. *Earth Surface Processes and Landforms* 28, 385–407.

Chatanantavet, P, Lajeunesse, E, Parker, G *et al.* 2010. Physically based model of downstream fining in bedrock streams with lateral input. *Water Resources Research* 46, W02518. DOI:10.1029/2008WR007208.

Chin, A and Wohl, E 2005. Towards a theory for step pools in stream channels. *Progress in Physical Geography* 29, 275–296. DOI:10.1191/pp449ra.

Church, M and Zimmermann, A 2007. Form and stability of step-pool channels: research progress. *Water Resources Research* 43. DOI:10.1029/2006WR005037.

Clarke, AO 1996. Estimating probable maximum floods in the Upper Santa Ana Basin, Southern California, from stream boulder size. *Environmental and Engineering Geoscience* 2, 165–182.

Clarke, AO and Hansen, CL 1996. The recurrence of large boulder movement in small watersheds of the Anza Borrego Desert, California. *Yearbook of the Association of Pacific Coast Geographers* 58, 28–61. DOI:10.1353/pcg.1996.0007.

Comiti, F and Mao, L 2012. Recent advances in the dynamics of steep channels. In *Gravel Bed Rivers: Processes, Tools, Environments* (eds Church, M, Biron, P, and Roy, A). John Wiley & Sons, Chichester, UK; 353–374. DOI:10.1002/9781119952497.ch26.

Costa, JE and O'Connor, JE 1995. Geomorphically effective floods. In *Natural and Anthropogenic Influences in Fluvial Geomorphology* (eds Costa, JE, Miller, AJ, Potter, KW, and Wilcock, PR). Geophysical Monograph 89, American Geophysical Union Press: Washington, DC; 45–56,

Cui, Y, Parker, G, Lisle, TE *et al.* 2003a. Sediment pulses in mountain rivers: 1. Experiments. *Water Resources Research* 39(9), 1239. DOI:10.1029/2002WR001803.

Cui, Y, Parker, G, Pizzuto, J, and Lisle, TE 2003b. Sediment pulses in mountain rivers: 2. Comparison between experiments and numerical predictions. *Water Resources Research* 39(9), 1240. DOI:10.1029/2002WR001805.

Dadson, SJ, Hovius, N, Chen, HG *et al.* 2003. Links between erosion, runoff variability and seismicity in the Taiwan orogen. *Nature* 426, 648–651. DOI:10.1038/nature02150.

Dadson, SJ, Hovius, N, Chen, H *et al.* 2004. Earthquake-triggered increase in sediment delivery from an active mountain belt. *Geology* 32, 733–736. DOI:10.1130/G20639.1.

Dorn, RI 1996. Climatic hypotheses of alluvial-fan evolution in Death Valley are not testable. In *The Scientific Nature of Geomorphology* (eds Rhoads, BL and Thorn CE). John Wiley & Sons: Chichester, UK; 191–220.

Fedele, JJ and Paola, C 2007. Similarity solutions for fluvial sediment fining by selective deposition. *Journal of Geophysical Research* 112, F02038. DOI:10.1029/2005JF000409.

Gabet, EJ 2003. Sediment transport by dry ravel. *Journal of Geophysical Research* 108(B1), 2049. DOI:10.1029/2001JB001686.

Garcia-Castellanos, D, Estrada, F, Jimenez-Munt, I *et al.* 2009. Catastrophic flood of the Mediterranean after the Messinian salinity crisis. *Nature* 462, 778–781. DOI:10.1038/nature08555.

Gintz, D, Hassan, MA, and Schmidt, K-H 1996. Frequency and magnitude of bedload transport in a mountain river. *Earth Surface Processes and Landforms* 21, 433–445. DOI:10.1002/(SICI)1096-9837(199605)21:5<433::AID-ESP580>3.0.CO;2-P.

Gupta, S, Collier, JS, Palmer-Felgate, A, and Potter, G 2007. Catastrophic flooding origin of shelf valley systems in the English Channel. *Nature* 448, 342–345. DOI:10.1038/nature06018.

Hartshorn, K, Hovius, N, Dade, WB, and Slingerland, RL 2002. Climate-driven bedrock incision in an active mountain belt. *Science* 297, 2036–2038. DOI:10.1126/science.1075078.

Harvey, AM 1984. Geomorphological response to an extreme flood: a case from southeast Spain. *Earth Surface Processes and Landforms* 9, 267–279.

Heller, PL, Beland, PE, Humphrey, NF *et al.* 2001. Paradox of downstream fining and weathering-rind formation in the lower Hoh River, Olympic Peninsula, Washington. *Geology* 29, 971–974.

Heritage, GL, Large, ARG, Moon, BP, and Jewitt, G 2004. Channel hydraulics and geomorphic effects of an extreme flood event on the Sabie River, South Africa. *Catena* 58, 151–181. DOI.10.1016/j.catena.2004.03.004.

Hovius, N and Stark, CP 2006. Landslide-driven erosion and topographic evolution of active mountain belts. In *Landslides from Massive Rock Slope Failure* (eds Evans, SG, Scarascia Mugnozza G, Strom, A, and Hermanns, RL). Springer: Dordrecht, Netherlands; 573–590.

Howard, AD, Dietrich, WE, and Seidl, M 1994. Modeling fluvial erosion on regional to continental scales. *Journal of Geophysical Research* 99 (B7), 13,971–13,986.

Jansen, JD 2006. Flood magnitude–frequency and lithologic controls on bedrock river incision in post-orogenic terrain. *Geomorphology* 82, 39–57. DOI:10.1016/j.geomorph. 2005.08.018.

Jerolmack, DJ and Paola, C 2010. Shredding of environmental signals by sediment transport. *Geophysical Research Letters* 37, L19401. DOI:10.1029/2010GL044638.

Johnson, JPL, Whipple, KX, Sklar, LS, and Hanks, TC 2009. Transport slopes, sediment cover, and bedrock channel incision in the Henry Mountains, Utah. *Journal of Geophysical Research* 114, F02014. DOI:10.1029/2007JF000862.

Lamb, MP, Dietrich, WE, Aciego, SM *et al.* 2008. Formation of Box Canyon, Idaho, by Megaflood: Implications for Seepage Erosion on Earth and Mars. *Science* 320, 1067–1070. DOI: 10.1126/science.1156630.

Lancaster, ST 2008. Evolution of sediment accommodation space in steady state bedrock-incising valleys subject to episodic aggradation. *Journal of Geophysical Research* 113, F04002. DOI:10.1029/2007JF000938.

Lane, SN, Reid, SC, Tayefi, V *et al.* 2008. Reconceptualising coarse sediment delivery problems in rivers as catchment-scale and diffuse. *Geomorphology* 98, 227–249. DOI.10.1016/j.geomorph.2006.12.028.

Lenzi, MA, Mao, L, and Comiti, F 2004. Magnitude–frequency analysis of bed load data in an Alpine boulder bed stream. *Water Resources Research* 40, WR002961.

Lenzi, MA, Mao, L, and Comiti, F 2006. Effective discharge for sediment transport in a mountain river: Computational approaches and geomorphic effectiveness. *Journal of Hydrology* 326, 257–276. DOI:10.1016/j.hydrol.2005.10.031.

Lisle, TE and Church, M 2002. Sediment transport–storage relations for degrading, gravel bed channels. *Water Resources Research* 38(11), 1219. DOI:10.1029/2001WR001086.

Marren, PM, Russell, AJ, and Rushmer, EL, 2009. Sedimentology of a sandur formed by multiple jökulhlaups, Kverkfjöll, Iceland. *Sedimentary Geology* 213, 77–88.

Marshall, JA and Sklar, LS 2012. Mining soil databases for landscape-scale patterns in the abundance and size distribution of hillslope rock fragments. *Earth Surface Processes and Landforms* 37, 287–300. DOI:10.1002/esp.2241.

Mather, AE and Hartley, A 2005. Flow events on a hyper-arid alluvial fan: Quebrada Tambores, Salar de Atacama, northern Chile. In *Alluvial Fans: Geomorphology, Sedimentology, Dynamics* (eds Harvey, AM, Mather, AE, and Stokes, M). Special Publication 251, Geological Society: London; 9–24.

Meshkova, LV and Carling, P 2012. The geomorphological characteristics of the Mekong River in northern Cambodia: a mixed bedrock–alluvial multi-channel network. *Geomorphology* 147–148, 2–17. DOI.10.1016/j.geomorph.2011.06.041.

Meshkova, LV, Carling, P, and Buffin-Bélanger, T 2012. Nomenclature, complexity, semi-alluvial channels and sediment-flux-driven bedrock erosion. In *Gravel Bed Rivers: Processes, Tools, Environments* (eds Church, M, Biron, P, and Roy, A). John Wiley & Sons, Chichester, UK; 424–431.

Métivier, F and Gaudemer, Y 1999. Stability of output fluxes of large rivers in south and east Asia during the last 2 million years: implications on floodplain processes. *Basin Research* 11, 293–303.

Meyer-Peter, E and Müller, R 1948. Formulas for bed-load transport. In *Proceedings of the 2nd Meeting of the International Association for Hydraulic Structures Research*, Delft, Netherlands; 39–64,

Mezaki, S and Yabiku, M 1984. Channel morphology of the Kali Gandaki and the Narayani Rivers in central Nepal. *Journal of Nepal Geological Society* 4, 161–176.

Montgomery, DR, Hallet, B, Yuping, L *et al.* 2004. Evidence for Holocene megafloods down the Tsangpo River gorge, southeastern Tibet. *Quaternary Research* 62, 201–207. DOI:10.1016/j.yqres.2004.06.008.

Morche, D and Schmidt, K-H 2012. Sediment transport in an alpine river before and after a dambreak flood event. *Earth Surface Processes and Landforms* 37, 347–353. DOI:10.1002/esp.2263.

Nanson, GC 1986. Episodes of vertical accretion and catastrophic stripping – a model of disequilibrium flood-plain development. *Geological Society of America Bulletin* 97, 1467–1475.

Paola, C, Parker, G, Seal, G *et al.* 1992. Downstream fining by selective deposition in a laboratory flume. *Science* 258, 1757–1760.

Parker, G 1991. Selective sorting and abrasion of river gravel. II: applications. *Journal of Hydraulic Engineering–ASCE* 117(2), 150–171.

Phillips, J 2003. Alluvial storage and the long-term stability of sediment yields. *Basin Research* 15, 153–163.

Pratt-Sitaula, B, Burbank, DW, Heimsath, A, and Ojha, T 2004. Landscape disequilibrium on 1,000–10,000 year scales, Marsyandi River, Nepal, central Himalaya. *Geomorphology* 58, 223–241. DOI:10.1016/j.geomorph.2003.07.002.

Rickenmann, D 1997. Sediment transport in Swiss torrents. *Earth Surface Processes and Landforms* 22, 937–951. DOI:10.1002/(SICI)1096-9837(199710)22:10<937::AID-ESP786>3.0.CO;2-R.

Russell, AJ, Duller, R, and Mountney, NP 2010. Volcanogenic jökulhlaups (glacier outburst floods) from Mýrdalsjökull: impacts on proglacial environments. In *The Mýrdalsjökull Ice Cap, Iceland: Glacial Processes, Sediments and Landforms on an Active Volcano* (eds Krüger, J and Kjær, KH). Elsevier: Amsterdam; 181–207.

Singer, MB, James, LA, Aalto, R *et al.* 2013. Enduring legacy of toxic fans via episodic redistribution of California gold mining debris. *Proceedings of the National Academy of Sciences* 110(46), 18436–18441. DOI:10.1073/pnas.1302295110.

Sklar, LS and Dietrich, WE 2001. Sediment and rock strength controls on river incision into bedrock. *Geological Society of America Bulletin* 29, 1087–1090.

Sklar, LS, Dietrich, WE, Foufoula-Georgiou, E *et al.* 2006. Do gravel bed river size distributions record channel network structure? *Water Resources Research* 42, W06D18. DOI:10.1029/2006WR005035, 2006.

Stokes, M, Griffiths, JS, and Mather, A 2012. Palaeoflood estimates of Pleistocene coarse grained river terrace landforms (Río Almanzora, SE Spain). *Geomorphology* 149–150, 11–26. DOI:10.1016/j.geomorph.2012.01.007.

Turowski, JM 2012. Semi-alluvial channels and sediment-flux-driven bedrock erosion. In *Gravel Bed Rivers: Processes, Tools, Environments* (eds Church, M, Biron, P, and Roy, A). John Wiley & Sons, Chichester, UK; 401–416. DOI:10.1002/9781119952497.ch29.

Turowski, JM, Lague, D, and Hovius, N 2007. Cover effect in bedrock abrasion: a new derivation and its implications for the modeling of bedrock channel morphology. *Journal of Geophysical Research* 112, F04006. DOI:10.1029/2006JF000697.

Turowski, JM, Hovius, N, Wilson, A, and Horng, M-J 2008a. Hydraulic geometry, river sediment and the definition of bedrock channels. *Geomorphology* 99, 26–38. DOI:10.1016/j.geomorph.2007.10.001.

Turowski, JM, Hovius, N, Hsieh, M *et al.* 2008b. Distribution of erosion across bedrock channels. *Earth Surface Processes and Landforms* 33, 353–363. DOI:10.1002/esp.1559.

Turowski, JM, Yager, EW, Badoux, A *et al.* 2009. The impact of exceptional events on erosion, bedload transport and channel stability in a step-pool channel. *Earth Surface Processes and Landforms* 34, 1661–1673. DOI:10.1002/esp.1855.

Turowski, JM, Badoux, A, Leuzinger, J, and Hegglin, R 2013. Large floods, alluvial overprint, and bedrock erosion. *Earth Surface Processes and Landforms* 38(9), 947–958. DOI:10.1002/esp.3341.

Van de Wiel, MJ and Coulthard, TJ 2010. Self-organized criticality in river basins: Challenging sedimentary records of environmental change. *Geology* 38, 87–90. DOI:10.1130/G30490.1.

Walling, DE 1983. The sediment delivery problem. *Journal of Hydrology* 65, 209–237.

Warner, NH, Gupta, S, Kim, J-R *et al.* 2010. Retreat of a giant cataract in a long-lived (3.7–2.6 Ga) Martian outflow channel. *Geology* 38 (9), 791–794. DOI:10.1130/G31268.1.

Whipple, KX, Hancock, GS, and Anderson, RS 2000. River incision into bedrock: mechanics and relative efficacy of plucking, abrasion and cavitation. *Geological Society of America Bulletin* 112, 490–503.

Whittaker, AC, Attal, M, Cowie, PA *et al.* 2008. Decoding temporal and spatial patterns of fault uplift using transient river long profiles. *Geomorphology* 100, 506–526. DOI:10.1016/j.geomorph.2008.01.018.

Whittaker, AC, Attal, M, and Allen, PA 2010. Characterizing the origin, nature and fate of sediment exported from catchments perturbed by active tectonics. *Basin Research* 22, 809–828. DOI:10.1111/j.1365–2117.2009.00447.x.

Wilcock, PR and McArdell, BW 1997. Partial transport of a sand/gravel sediment. *Water Resources Research* 33(1), 235–245. DOI:10.1029/96WR02672.

Willgoose, G, Bras, RL, and Rodriguez-Iturbe, J 1991. A coupled channel network growth and hillslope evolution model, 1, Theory. *Water Resources Research* 27, 1671–1684.

Williams, GP 1983. Paleohydrological methods and some examples from Swedish fluvial environments. I. Cobble and boulder deposits. Geografiska Annaler 65A (3–4): 227–243.

Wohl, EE 1992. Gradient irregularity in the Herbert Gorge of northeastern Australia. *Earth Surface Processes and Landforms* 17, 69–84.

Wohl, EE 2008. Review of effects of large floods in resistant-boundary channels. In *Gravel-Bed Rivers VI: From Process Understanding to River Restoration* (eds Habersack, HP, and Massimo, R). Developments in Earth Surface Processes, Vol. 11, Elsevier: Amsterdam; 181–211. DOI:10.1016/S0928–2025(07)11125–1.

Wolcott, J 1988. Nonfluvial control of bimodal grain-size distributions in river-bed gravels. *Journal of Sedimentary Petrology*, 58, 979–984.

Yager, EM, Dietrich, WE, and Kirchner, JW 2007. Calculating bed load transport in steep boulder bed channels. *Water Resources Research* 43, W07418. DOI.10.1029/2006WR005432.

Yager, EM, Turowski, JM, Rickenmann, D, and McArdell, BW 2012. Sediment supply, grain protrusion, and bedload transport in mountain streams. *Geophysical Research Letters* 39, L10402. DOI:10.1029/2012GL051654.

Yanites, BJ, Tucker, GE, Mueller, KJ, and Chen, Y-G 2010. How rivers react to large earthquakes: evidence from central Taiwan. *Geology* 38(7), 639–642. DOI:10.1130/G30883.1.

Discussion

Discussion by Michael Church

Particle abrasion is cited in the paper as a significant mechanism for grain size diminution downstream in gravel-bed rivers. Another, much neglected mechanism is particle disintegration by mechanical weathering *in situ*. The wetting–drying, possibly freeze–thaw environment of stream beds represents one of the more severe environments on the planet for physical weathering of rock. In mountain rivers, freeze–thaw is apt to be particularly important. However, it obviously is a difficult phenomenon to study and most rock abrasion likely results from a combination of weathering and destruction during movement (Bradley 1970; Jones and Humphrey 1997).

References

Bradley, WC 1970. Effect of weathering on abrasion of granitic gravel, Colorado River (Texas). *Geological Society of America Bulletin* 81, 61–80.

Jones, LS and Humphrey, NF 1997. Weathering-controlled abrasion in a coarse-grained, meandering reach of the Rio Grande: implications for the rock record. *Geological Society of America Bulletin* 109, 1080–1088.

Discussion by Hervé Piégay

In your presentation you did not speak about connectivity, which is presently a critical issue when exploring sediment supply within a catchment and considering management scenarios. Sediment supply at a given point of the network is strongly linked to the upstream longitudinal pattern of transport capacity. Can you tell us more about this?

Discussion by David Milan

In the summer of 2007, an exceptional flood took place on Thinhope Burn, South Tyne catchment, UK (Milan, 2012). This event resulted in substantial reworking of the valley floor, mobilizing boulders of up to 1 m in diameter. New boulder berms were formed, and some of the large boulders were deposited on parts of the floodplain, without the floodplain surface being stripped of vegetation. In addition to the processes mentioned by Attal, it is suggested that non-Newtonian or hyperconcentrated flow processes can operate during extreme events that permit extremely high sediment transport rates and mobilization of very coarse clasts.

Reference

Milan, DJ 2012. Geomorphic impact and system recovery following an extreme flood in an upland stream: Thinhope Burn, northern England, UK. *Geomorphology* 138, 319–328.

Discussion by Gordon Grant of both Mikael Attal and Paul Carling

I am wondering how to reconcile the two perspectives offered by Carling and Attal. On the one hand, Carling argues that the stratigraphy left behind by very large floods is interpretable in light of specific mechanisms of entrainment and deposition that occur during those events, and suggests a 'standard model' for catastrophic flood deposits, if only as a heuristic tool for interpreting flood sequences. On the other hand, Attal argues convincingly that very large floods are greatly influenced by patterns of sediment supply that play out in both spatial and temporal realms, which are conditioned by sequences of floods of different competences and capacities. One consequence of this should be that large flood deposits should vary dramatically in their structure and stratigraphy. Yet despite these complex controls, basin sediment yields tend to be rather systematic over time. How do we make sense of these seemingly disparate observations?

Response by the Author

I would like to thank M. Church, H. Piégay, D. Milan and G. Grant for their thoughtful comments.

M. Church discusses the potential for 'weathering in place' to significantly contribute to gravel size reduction in rivers. Based on my field observations, I would say that this contribution is probably low in places where bedload transport is frequent and very energetic (e.g. powerful Himalayan rivers with yearly monsoon). However, I agree that it is a possibly overlooked mechanism where (i) sediment remains stored on gravel bars for significant periods of times; (ii) physical weathering processes are very efficient; and/or (iii) clasts are made of material sensitive to weathering (e.g. porous, grainy, and/or jointed). Wetting–drying and freeze-thaw in mountain rivers, as well as thermal expansion in arid environments, are likely to weaken particles, reducing their size and making them more susceptible to abrasion. This is illustrated by the studies mentioned in the discussion and also by Heller *et al.* (2001), and is certainly a phenomenon that requires more attention.

H. Piégay mentions connectivity as a key control on sediment availability at a given point in a river. I agree this is a very crucial point but chose to focus my attention to the reach scale due to the complexities associated with defining connectivity at the basin scale. Indeed, connectivity is strongly controlled by climate, tectonics, vegetation, hydrological regime, and lithologies exposed, to cite only a few controlling parameters, leading to potentially very different river responses to an extreme flood event. I tried to emphasize in the chapter that the availability of sediment upstream will control the sediment flux and excess transport capacity in a given reach during a flood. I believe that defining sediment connectivity at the basin scale (from hillslopes to channels) to forecast the sediment response to a given flood is one of the key challenges in geomorphology. I make reference to Bennett *et al.*'s (2014) approach which considers different 'reservoirs' with different degrees of connectivity, which I think is a step in the right direction.

Regarding the influence of hyperconcentrated flow processes on the export of large quantities of potentially coarse sediment, I completely agree with D. Milan's point. Extreme flood events may be associated to extreme precipitation, which increases the likelihood of slope failure and, therefore, the potential for hyperconcentrated flows. For example, extreme flooding and sediment transport affected a large part of Venezuela in December 1999 following a period of very intense rainfall, causing widespread landsliding and debris flows (e.g. Lopez *et al.* 2003). I also mention the flood deposits documented by Mather and Hartley (2005) in Chile, for which a hyperconcentrated flow origin is suggested.

G. Grant wonders how to reconcile my perspective with P. Carling's (Chapter 11). I believe the two approaches are not contradictory. Yes, the amount and coarseness of sediment exported will depend on sediment availability. Yet, as discussed in the chapter, it is likely that, due to their very high magnitude, extreme floods will result in net sediment export from the mountain realm to the 'basin', where basin could represent a sedimentary basin at the front of a mountain range, an intermontane basin or a locally wide valley. In these locations, I believe that P. Carling's model (in terms of bedforms and sediment structure for example) will apply, whatever the source of sediment (i.e., sediment already available for transport or bedrock erosion).

Reference

Lopez, JL, Perez, D, and Garcia, R 2003. Hydrologic and geomorphologic evaluation of the 1999 debris-flow event in Venezuela. In *Debris-Flow Hazards Mitigation: Mechanics, Prediction, and Assessment* (eds Rickenmann, D and Chen, C). Millpress: Rotterdam, Netherlands; 989–1000.

13

Geomorphic Responses to Dam Removal in the United States – a Two-Decade Perspective

Jon J. Major, Amy E. East, Jim E. O'Connor, Gordon E. Grant, Andrew C. Wilcox, Christopher S. Magirl, Mathias J. Collins, and Desiree D. Tullos

13.1 Introduction

In the past few decades there has been increasing popular and political momentum in the United States to remove dams. What began with removal of small dams impounding modest sediment volumes has progressed to removal of large dams impounding millions of cubic meters of sediment. The reasons for removal are varied, but commonly relate to diminished function, costly maintenance, concerns over safety, and to acknowledgment that dams impair fish passage and other ecosystem functions and services (e.g., Babbitt 2002; Poff and Hart 2002; O'Connor, Major, and Grant 2008; Service 2011; Lovett 2014). However, decisions to remove dams are clouded by uncertainties over potential environmental benefits and detriments of removal. Concerns and uncertainties surround sediment release especially (e.g., Downs *et al.* 2009). Although a number of small (<10 m tall) and a few tall (≥10 m) dams had been removed by the early 2000s (e.g., Bednarek 2001; Hart *et al.* 2002; Poff and Hart 2002; Pohl 2002, 2003; Bellmore *et al.* 2015, 2016), there were few detailed studies of geomorphic responses to those removals from which to draw conclusive insights. As a result, hypotheses regarding potential geomorphic responses of river systems to dam removal in the 1990s and early 2000s were drawn largely from analogies and concepts of fluvial geomorphology (Doyle, Stanley, and Harbor 2002; Pizzuto 2002). Since then, newer physical and numerical models that simulate channel evolution in reservoirs and downstream distributions of eroded sediment have sharpened insights on possible river responses (e.g., Cantelli *et al.* 2007; Cui and Wilcox 2008; Grant *et al.* 2008; Downs *et al.* 2009; Greimann 2013; Podolak and Wilcock 2013; Ferrer-Boix, Martín-Vide, and Parker 2014; Gartner, Magilligan, and Renshaw 2015). Furthermore, a number of detailed field studies have now examined responses to dam removal (Bellmore *et al.* 2015; Table 13.1). As a result of these more recent modeling and field studies, we are positioned to test the veracity of posited hypotheses regarding geomorphic responses to dam removals. And as the number of post-removal field studies has grown, we can assess whether common responses have emerged. Understanding outcomes of dam removals has a broader context than simply helping dam owners and stakeholders make decisions. Better knowledge of how river systems respond to sediment pulses introduced by dam removals informs assessments of hazards from natural dam failures; sediment injections from natural processes such as landslides (Hoffman and Gabet 2007), wildfires (Shakesby and Doerr 2006), and volcanic eruptions

Gravel-Bed Rivers: Processes and Disasters, First Edition. Edited by Daizo Tsutsumi and Jonathan B. Laronne.

Table 13.1 Dam removals and failures studied in detail between 1990 and 2015.

Dam (State)	River	Height (m)	Reservoir sediment volume (m^3)	Sediment composition*	Removal strategy[†]	Sources
Brownsville (OR)	Calapooia	2–3	10 000–15 000	Gravel	Sudden	Walter and Tullos (2010); Kibler, Tullos, and Kondolf (2011); Tullos, Finn, and Walter (2014)
La Valle (WI)	Baraboo	~2	140 000	45% sand, 40% silt, 15% clay	Sudden	Doyle, Stanley, and Harbor (2003); Greene, Krause, and Knox (2013)
St. John (OH)	Sandusky	2.2	200 000	Gravel, sand	Sudden	Cheng and Granata (2007)
Brewster (IL)	Brewster Creek	2.4	18 000	70%–99% silt and clay	Phased	Straub (2007)
Secor (OH)	Ottawa	2.5	5000–9000	Sand	Sudden	Harris (2008); Harris and Evans (2014)
Manatawny (PA)	Manatawny Creek	2.5	Uncertain; mostly dredged	Sand, gravel	Phased	Bushaw-Newton *et al.* (2002); Skalak *et al.* (2011)
Simkins (MD)	Patapsco	3	67 000	Sand, fine gravel	Sudden	MJ Collins (unpublished data)
Rockdale (WI)	Koshkonong	3.3	287 000	35% sand, 45% silt, 20% clay	Sudden	Doyle, Stanley, and Harbor (2003)
Anaconda (CT)	Naugatuck	3.3	11 900	Sand, gravel	Sudden	Wildman and MacBroom (2005)
Stronach (MI)	Pine	3.6–5	800 000	70% sand, 30% gravel	Phased	Burroughs *et al.* (2009)
Munroe Falls (OH)	Cuyahoga	3.7	Uncertain	70% sand, 20% mud, 10% gravel	Phased	Rumschlag and Peck (2007); Peck and Kasper (2013)
Homestead (NH)	Ashuelot	4	Uncertain	Sand	Sudden	Gartner, Magilligan, and Renshaw (2015)
Merrimack Village (NH)	Souhegan	4	62 000	95% sand	Sudden	Pearson, Snyder, and Collins (2011); Santaniello, Snyder, and Gontz (2013)
IVEX (OH)	Chagrin	7.4	236 000	Mud	Sudden	Evans *et al.* (2000); Evans (2007)
Savage Rapids (OR)	Rogue	12	150 000	70% sand, 30% gravel	Sudden	Bountry, Lai, and Randle (2013)
Milltown (MT)	Clark Fork	12.8	5.5 million[‡]	Sand	Sudden[$]	Evans and Wilcox (2014)
Marmot (OR)	Sandy	15	750 000	~50% gravel, 50% sand	Sudden	Major *et al.* (2012)
Elwha (WA)	Elwha	32	4.9 million	47% mud, 53% sand and gravel	Phased	Randle *et al.* (2015); Warrick *et al.* (2015)
Condit (WA)	White Salmon	38	1.8 million	60% sand, 35% mud, 5% gravel	Sudden	Wilcox, O'Connor, and Major (2014); Colaiacomo (2014)
Barlin (Taiwan)	Dahan	38	10.5 million	Gravel, sand	Sudden	Tullos and Wang (2014)
Glines Canyon (WA)	Elwha	64	16.1 million	44% mud, 56% sand and gravel	Phased	Randle *et al.* (2015); Warrick *et al.* (2015

* Mud is combined silt and clay (< 0.063 mm).
[†] Full base-level fall was achieved within minutes to days in sudden removals, and weeks to years in phased removals.
[‡] Approximately 40% of this sediment was excavated before breaching as part of Superfund remediation efforts.
[$] Part of the infrastructure was removed in phases while contaminated sediment was excavated. However, sediment release owed to sudden breaching of the remaining remnants of the dam.

(Pierson and Major 2014); behavior of legacy sediment (James 2013; Merritts *et al.* 2013); and restoration activities such as gravel augmentation (Sklar *et al.* 2009).

More than 87 000 dams taller than 1.8 m have been catalogued in the United States (Figure 13.1; US Army Corps of Engineers 2013), and more than 1100 have been removed intentionally – most since 2000 (American Rivers 2014). Since the early 2000s, intentional dam removals and some unintentional failures have released varying magnitudes of diversely composed sediment to river systems. Removed dams have been as tall as 64 m (Figure 13.2a) and have impounded sediment volumes ranging from a few thousands to several millions of cubic meters (Figure 13.2b). Stored sediment compositions have ranged from predominantly mud (≤0.063 mm) to predominantly gravel (Table 13.1). During some removals, stored sediment was released abruptly when the full barrier height of a dam was breached suddenly; in others, sediment was released more slowly as base-level changed gradually during phased removal of a dam.

Generally, effects of dam removals have been site specific, but an analysis of about 20 well-studied removals (Table 13.1) reveals common patterns and controls on geomorphic response. The removals examined, however, are geographically limited. They are distributed unevenly, mostly in the northern and eastern United States, and all but one large-dam removals are in the Pacific Northwest (Figure 13.1). They are mainly run-of-river structures that passed water freely over their crests (Csiki and Rhoads 2010) or impounded small to moderate reservoir pools (≤30 million m^3; Randle *et al.* 2015). Absent are removals of dams that stored very large volumes of water, that markedly altered flow, or that impounded large volumes of mostly mud. Here, we discuss patterns of and controls on geomorphic response within a contextual framework of volume and composition of stored reservoir sediment, style and rate of dam removal, processes eroding sediment from the reservoir upon dam breaching, and river hydrology during and after dam removal. Our analysis encompasses sandy alluvial and bedrock river systems in addition to those that are gravel-bedded. We conclude by summarizing common findings that have emerged from these studies.

13.2 Reservoir and Downstream Channel Responses to Dam Removal

Variations among physical settings, dam characteristics, and removal strategies have led to widely varying channel responses after removals. Common to most removals, however, is a coupled-system response of reservoir erosion and downstream sediment transport and deposition. But the rates and processes by which this coupled system operates vary among settings.

13.2.1 Reservoir Erosion

Trajectories of reservoir erosion and channel development following dam removal have been posited to follow conceptual channel-evolution models describing response to base-level fall. Doyle, Stanley, and Harbor (2002) proposed a multistage model that describes a spatial and temporal evolutionary sequence in which channel initiation begins at the dam site and migrates upstream through the reservoir. Initially, a narrow channel starts incising after reservoir water has drained. Channel-incision owing to headcut (a near-vertical step change in channel profile), knickpoint (a width-limited abrupt inflection in channel profile), or knickzone (a nearly channel-spanning abrupt inflection in channel profile) migration proceeds upstream. Channel widening owing to mass-wasting of channel banks follows incision. An aggradational stage of development follows this degradational stage as upstream erosion delivers and deposits abundant sediment to the downstream reach of reservoir. The aggradational stage promotes continued channel widening, but at a diminishing rate. As rates of sediment

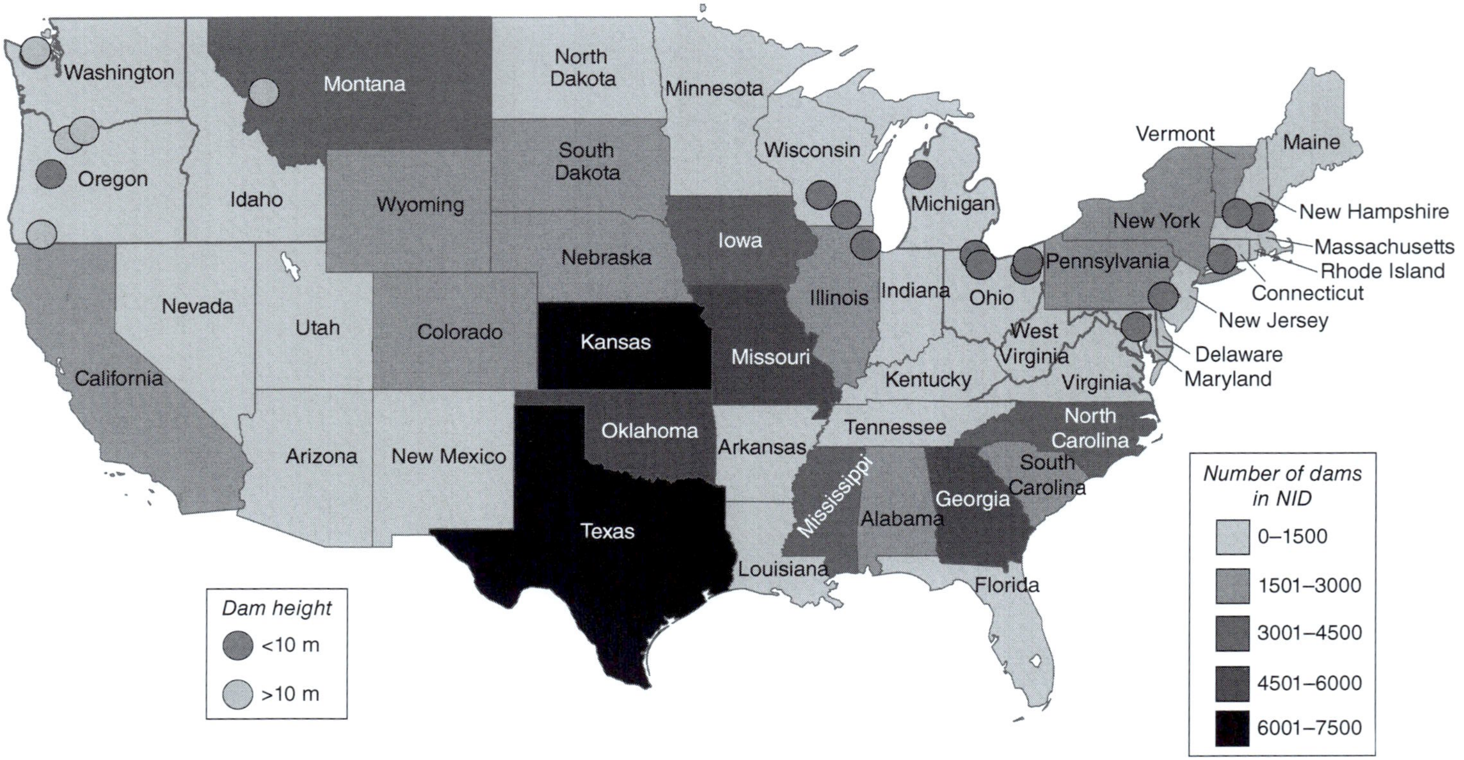

Figure 13.1 Map of United States showing the number of dams in each state catalogued by the US Army Corps of Engineers (2013) National Inventory of Dams (NID). Circles (color coded by threshold of dam height) show the geographic distribution of dam removals that inform this chapter (see Table 13.1).

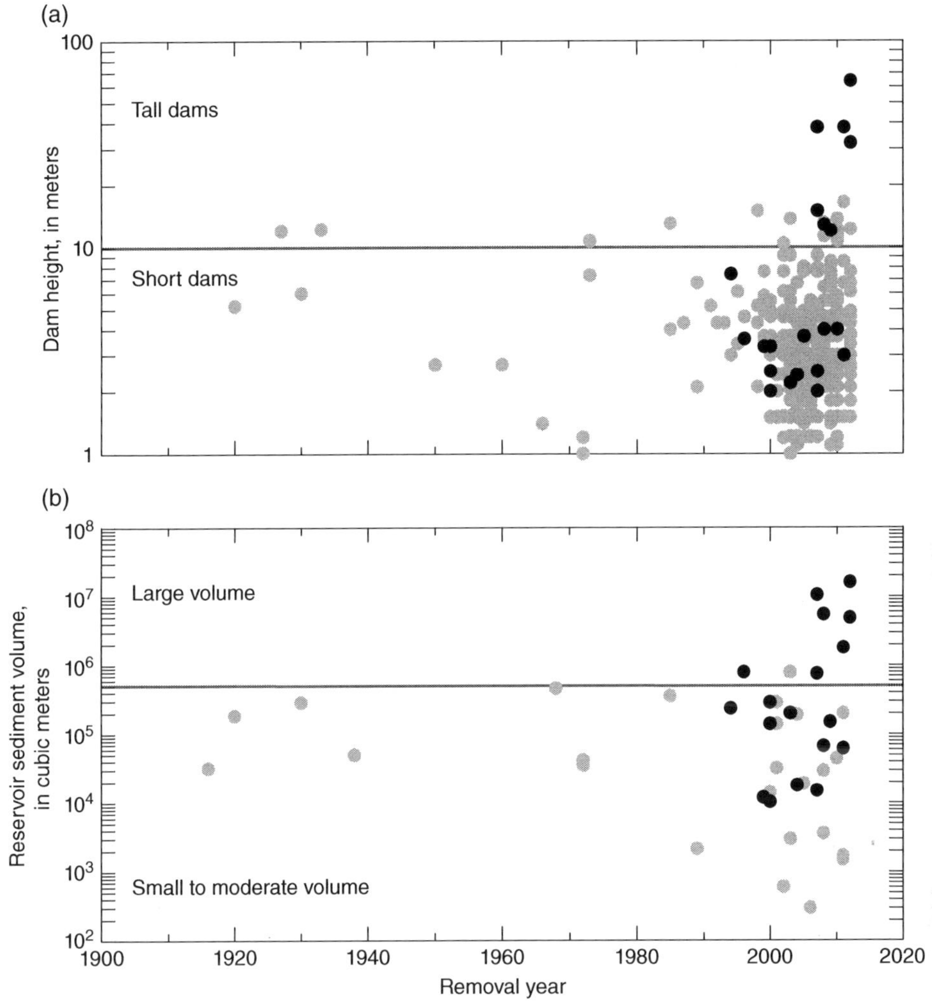

Figure 13.2 Dam height and impounded sediment volumes of documented dam removals (American Rivers 2014; Bellmore *et al.* 2015). Black circles show dam removals that inform this chapter (see Table 13.1).

delivery and aggradation decline, vegetation colonizes a newly developed floodplain, the channel stabilizes, and a new equilibrium condition establishes.

The proposed channel-evolution model generally describes conditions observed in reservoirs filled with sediment following sudden removal of a dam. In many instances, particularly when reservoir sediment was coarse grained (sand and gravel), erosion trajectories and channel development after dam removal progressed through several of the proposed stages very rapidly, especially when full base-level fall from reservoir elevation to downstream channel was sudden (e.g., Pearson, Snyder, and Collins 2011; Major *et al.* 2012; Collins *et al.* 2014; Wilcox, O'Connor, and Major 2014). Channel widening may involve mass wasting, but in coarse, cohesionless fill nearly continuous raveling rather than discrete mass wasting may also feed bank sediment directly to the channel. Erosion of filled reservoirs composed of abundant fine sediment (sand and mud) can proceed through the proposed stages rapidly, or it can proceed slowly owing to tempered headcut or knickpoint migration, which limits the rate of stage progression. Incision through fine sediment commonly leads to mass wasting (e.g., Evans *et al.* 2000; Doyle, Stanley, and Harbor 2003; Wilcox, O'Connor, and Major 2014). In some instances, aggradation of the lower reservoir reach owing to input of sediment from upstream may have contributed to channel widening (e.g., Bushaw-Newton *et al.* 2002; Doyle, Stanley, and Harbor 2003; Cheng and Granata, 2007), but mostly it did not.

Erosion and channel development in partly filled reservoirs can proceed as proposed if there is little longitudinal variation in sediment transport. Otherwise, sediment eroded from the upstream part of the reservoir can be mostly redeposited in the downstream part without advancing past the dam site (e.g., Cheng and Granata 2007; Randle *et al.* 2015). Redistribution of sediment can occur when a reservoir contains little sediment or is influenced by grade control (e.g., Bushaw-Newton *et al.* 2002; Cheng and Granata 2007; Skalak *et al.* 2011; Greene, Krause, and Knox 2013; Gartner, Magilligan, and Renshaw 2015), or when a dam is removed in phases and retains a gradually dwindling pool. For example, phased removals of Elwha and Glines Canyon dams, Washington, triggered upstream erosion and delta progradation across reservoir widths before pool water was fully released (Randle *et al.*, 2015). Delta progradation led to wide, braided reaches within the reservoirs. Only after bedload sediment began passing the vestige of Glines Canyon Dam did the channel through its reservoir become narrowly incised and begin delivering substantial amounts of sediment downstream (Magirl *et al.* 2015; Randle *et al.* 2015).

Trajectories of reservoir-channel evolution can be modified by unexpected changes in boundary conditions. New boundary conditions after dam removal can result from sediment coarsening with depth, especially in gravel-rich reservoirs, or from exhumed bedrock, wood, or man-made structures. These roughness elements can limit bed erosion and affect channel development. Coarsening of the reservoir channel bed following breaching of Marmot Dam (Oregon) – from bed winnowing and exposure of coarser gravel at depth – rapidly slowed the rate of incision (Major *et al.* 2012). After breaching of Condit Dam (Washington), erosion exhumed a 4.5-m-tall timber-crib cofferdam that halted local channel incision until it was removed manually several months later (Wilcox, O'Connor, and Major 2014). Similarly, a timber-crib structure exhumed behind Brownsville Dam (Oregon) limited the rate and extent of channel incision (Zunka, Tullos, and Lancaster 2015). Reservoir erosion behind Secor Dam (Ohio) exposed a buried beaver dam that temporarily stalled knickzone migration (Harris and Evans 2014). Large wood as well as bedrock exhumed behind Merrimack Village Dam (New Hampshire) affected reservoir erosion and channel evolution (Pearson, Snyder, and Collins 2011). And an erosion-resistant reach of boulders exhumed behind Homestead Dam (New Hampshire) acted as a grade control that prohibited knickpoint migration, and thus limited channel incision (Gartner, Magilligan, and Renshaw 2015). There, altered boundary conditions also largely decoupled bed incision and bank erosion during evacuation of reservoir sediment.

Reservoir geometry can affect rates and magnitudes of sediment evacuation. In several reservoirs having small aspect ratios (width/length) or in those where lateral erosion could swing unobstructed, erosion rapidly evacuated large percentages (≥40%) of impounded sediment after full base-level fall was achieved (Figure 13.3) (e.g., Wildman and MacBroom 2005; Epstein 2009; Pearson, Snyder, and Collins 2011; Major *et al.* 2012; Sawaske and Freyberg 2012; Bountry, Lai, and Randle 2013; Wilcox, O'Connor, and Major 2014; Collins *et al.* 2014; Warrick *et al.* 2015). By contrast, for reservoirs in which lateral erosion and landslides were limited, transport capacity diminished longitudinally, or the rate of base-level fall was strongly moderated, the rate and quantity of sediment evacuation were also limited (e.g., Evans *et al.* 2000; Doyle, Stanley, and Harbor 2003; Cheng and Granata 2007; Straub 2007; Burroughs *et al.* 2009).

In nearly all removals, erosion increased channel gradient through the impoundment reach. Gradually that gradient approximated the pre-dam gradient (e.g., Burroughs *et al.* 2009; Pearson, Snyder, and Collins 2011; Major *et al.* 2012; Wilcox, O'Connor, and Major 2014). In the absence of lateral

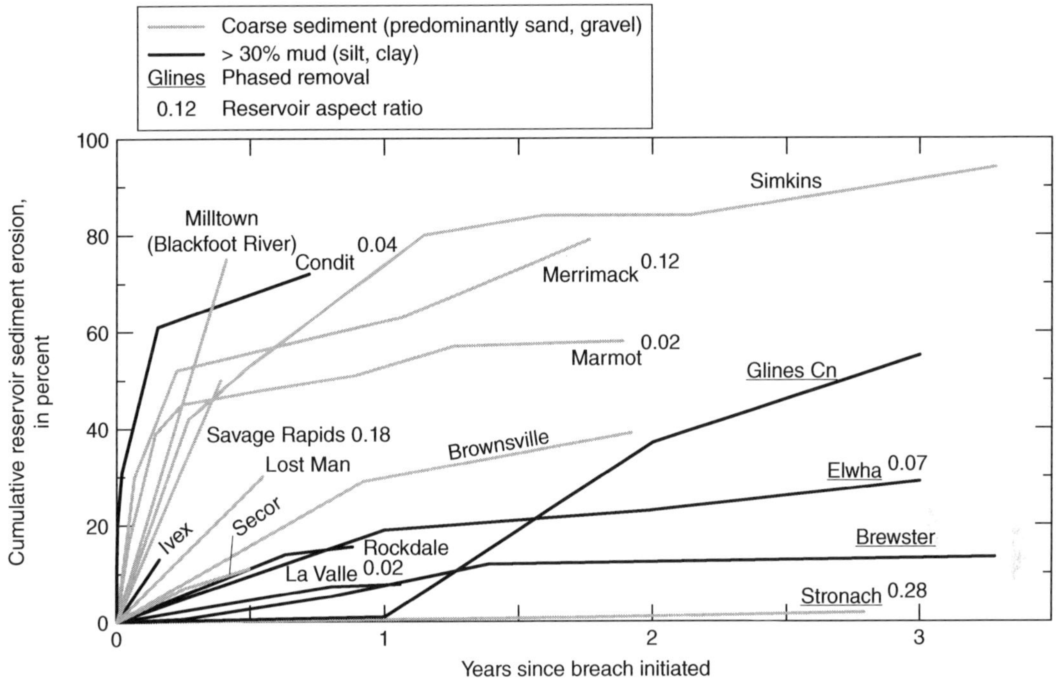

Figure 13.3 Percentage of reservoir sediment eroded with time since dam removal. Numbers near dam names indicate reservoir aspect ratio (width-to-length). The great increase in erosion at Glines Canyon Dam during the second year after removal began largely reflects final loss of pool water and release of bedload past the remnant structure. Through the first year after removal began, eroded sediment was mostly redeposited in a delta that prograded across the partly filled reservoir. (Original plot by Sawaske and Freyburg (2012); this version adapted from Grant and Lewis (2015).)

erosion, landslides, or large floods that can access remnant sediment, approximation of the pre-dam gradient limits the amount of reservoir sediment that can be evacuated. In cases where a reservoir reach contains little sediment, where dam removal is incomplete or has been replaced with grade-control, where negative transport gradients exist, or where conditions cause sediment redistribution rather than evacuation, reservoir profile change may be limited and renewed deposition may be induced (e.g., Cheng and Granata 2007; Skalak *et al.* 2011; Greene, Krause, and Knox 2013; Gartner, Magilligan, and Renshaw 2015).

13.2.2 Downstream Deposition

Abrupt increases in sediment delivery, even during phased removals, produce downstream effects of varying spatial and temporal extent. Downstream effects typically include channel aggradation; changes in channel gradient, width, and bed texture; filling of pools; and formation or modification of channel bars; some of these channel modifications are persistent, others are transient. In many cases, the greatest sediment impacts were close to dam sites, but varying channel morphology, volume and caliber of released sediment, and sediment delivery processes can extend or suppress sediment effects in the downstream direction.

13.2.2.1 Spatial Extents of Sediment Impacts

Spatial gradients in transport capacity owing to variations in channel slope, size and size-distribution of released sediment, and river hydrology affect sediment deposition and thus the spatial extent of sediment impact (e.g., Cui and Wilcox 2008; Gartner, Magilligan, and Renshaw 2015). Commonly, sand and finer sediment travel farther downstream than does gravel (Figure 13.4). In general, the greater and coarser the gravel content of released sediment, the more limited the initial spatial extent of downstream impact. In some cases downstream sediment sinks, such as deep pools (Bountry, Lai, and Randle 2013), other reservoirs (Evans *et al.* 2000; Peck and Kasper 2013; Collins *et al.* 2014), confluences with larger rivers (Pearson, Snyder, and Collins 2011; Wilcox, O'Connor, and Major 2014), or coastal zones (Warrick *et al.* 2015), have limited the spatial extent of sediment impact regardless of transport capacity. Where downstream transport was not interrupted, sediment released from removals of tall (≥ 10 m) dams has generally traveled farther than sediment released from removals of short (< 10 m) dams (Bellmore *et al.* 2015; Grant and Lewis 2015). Principal impacts of released sediment containing more than 20% gravel typically have extended no more than about 5 km downstream – a reach over which most gravel accumulated (e.g., Cheng and Granata 2007; Burroughs *et al.* 2009; Kibler, Tullos, and Kondolf 2011; Major *et al.* 2012; Bountry, Lai, and Randle 2013; Tullos and Wang 2014). Although released sand has traveled farther downstream than gravel, it has sometimes spread so thinly along channels that accumulations are difficult to detect (e.g., Major *et al.* 2012; Collins *et al.* 2014). By contrast, large-volume pulses of gravel, sand, and mud released by removals of Milltown Dam (Montana) and Glines Canyon Dam (Washington) traveled at least 15–20 km downstream (Evans and Wilcox 2014; East *et al.* 2015; Warrick *et al.* 2015).

13.2.2.2 Changes in Channel Gradient

Sediment erosion and downstream delivery after dam removal can alter channel gradient. The trajectory of change in channel gradient through the reservoir reach and downstream is toward a smooth profile. This new profile is mostly similar to the pre-dam profile, but not always. Exhumed bedrock, man-made structures, and coarse alluvium may inhibit channel incision in the reservoir reach and prevent development of a smooth profile. Downstream, valley shape and morphology influence flow and stress distributions, which can alter sediment accumulation and affect channel gradient.

Large releases of sand or sand and gravel generally aggraded river channels within a few kilometers downstream of dam sites and increased channel gradient (e.g., Burroughs *et al.* 2009; Pearson, Snyder, and Collins 2011; Major *et al.* 2012; Wilcox, O'Connor, and Major 2014). By contrast, sand and mud, whether released abruptly or during phased removals, have had relatively little effect on channel gradient (e.g., Evans *et al.* 2000; Doyle, Stanley, and Harbor 2003; Rumschlag and Peck 2007; Skalak *et al.* 2011).

Downstream changes in channel gradient are not always correlated positively with the volume and caliber of released sediment. Despite release of a large amount of sand and gravel following catastrophic failure of Barlin Dam, Taiwan, which induced as much as 7 to 10 m of aggradation, channel gradient within 1 km of the dam site remained relatively uniform owing to the narrow valley configuration. However, sediment accumulation in a wider reach farther downstream decreased channel gradient (Tullos and Wang 2014). Release of muddy sand and gravel to the Elwha River caused little change in downstream channel gradient, except in the lowermost 1.5 km of channel affected by tidal variation, because most of the released sediment passed through (East *et al.* 2015; Warrick *et al.* 2015). Similarly, release of tens of thousands of cubic meters of sand and gravel by removal of Savage Rapids Dam (Oregon) caused little gradient change downstream because it mostly filled deep pools (Bountry, Lai, and Randle 2013; Tullos, Finn, and Walter 2014).

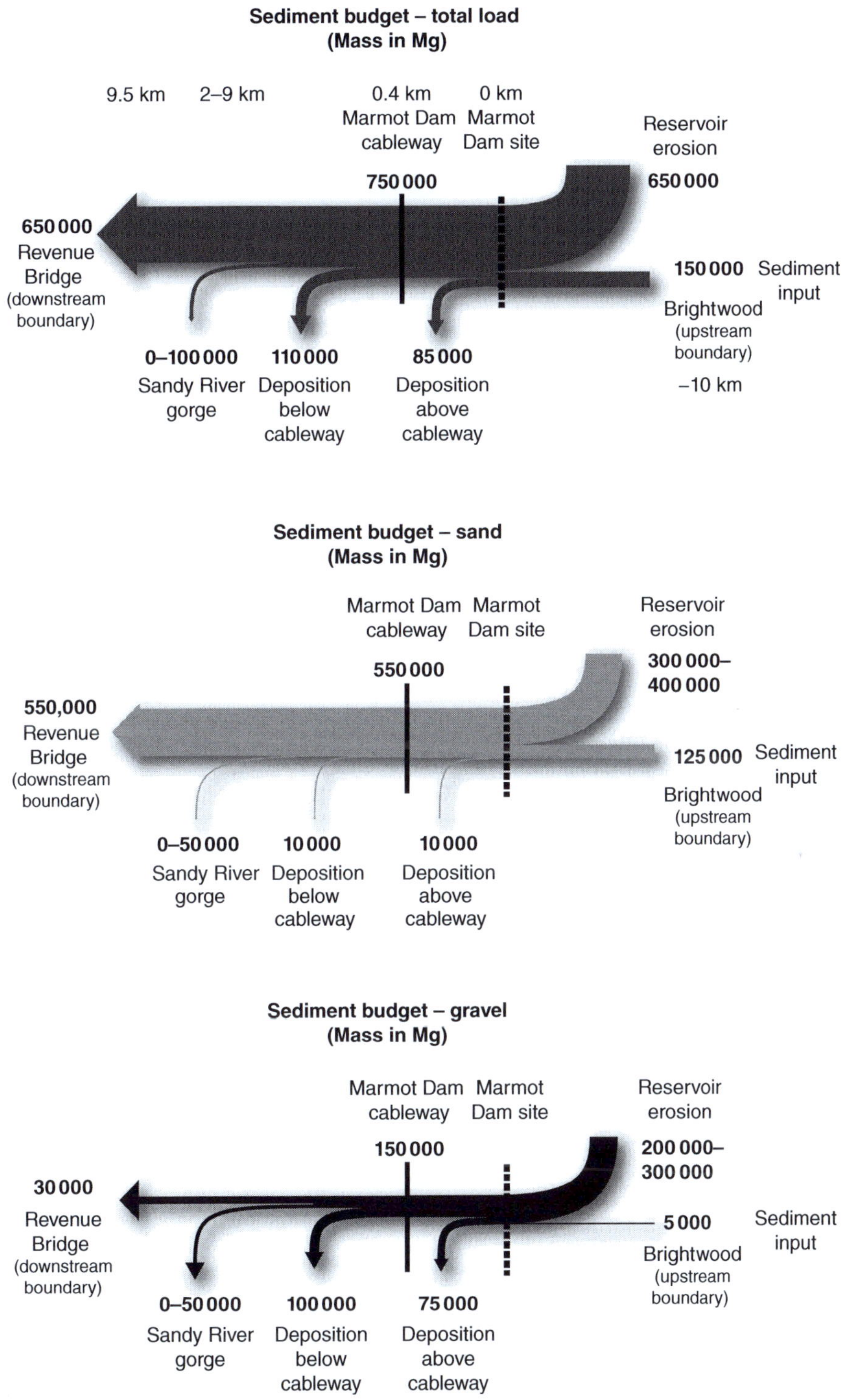

Figure 13.4 First-year compositional budgets for sediment released by removal of Marmot Dam (Oregon). Figure shows estimated upstream input to reservoir, erosion from reservoir, loss to depositional storage, and output at the exit of Sandy River gorge, ~10 km downstream of the dam site. (From Major *et al.* 2012.)

Most documented gradient changes have been in low-gradient (≤ 0.0006) reaches, even in steep mountain rivers. At Marmot Dam, the channel gradient in the first kilometer below the dam site was exceptionally shallow (~0.0001) compared to that farther upstream and downstream, and in this reach a large amount of sediment accumulated initially. Likewise, below Stronach (Michigan) and Merrimack Village dams, sediment accumulated in low-gradient (0.0006) reaches. Downstream of several other dam removals (or following sediment flushing from extant dams), sediment commonly filled pools (Wohl and Cenderelli 2000; Bountry, Lai, and Randle 2013; Tullos, Finn, and Walter 2014; Wilcox, O'Connor, and Major 2014) or uniformly aggraded already steep channel reaches (Tullos and Wang 2014), and therefore caused little change in overall channel gradient. Thus, morphologic variations in valley width, depth, and locations of pools, and other downstream influences that can affect base level (such as larger rivers or reservoirs) can affect sediment accumulation and gradient changes.

13.2.2.3 Changes in Channel Width

After dam removal, some downstream channels widened considerably whereas others changed little. Channels subject to substantial aggradation commonly widened because reaches became braided and the river spread across its floodplain at nearly all flows (e.g., Pearson, Snyder, and Collins 2011; Major *et al.* 2012; East *et al.* 2015) (Figure 13.5). By contrast, widths of channels that accumulated little sediment or that were morphologically constrained changed little (e.g., Bushaw-Newton *et al.* 2002; Doyle, Stanley, and Harbor 2003; Cheng and Granata 2007; Straub 2007; Peck and Kasper 2013).

Figure 13.5 Photographs showing temporal channel changes immediately downstream of Marmot Dam. The dam was breached on October 19, 2007. Horizontal bars and dot indicate identical locations among photographs.

Sometimes, valley morphology affected active channel width more than did sediment deposition (e.g., Skalak, Pizzuto, and Hart 2009; Skalak *et al.* 2011; Tullos and Wang 2014; Wilcox, O'Connor, and Major 2014).

13.2.2.4 Changes in Channel Bed Texture

Sediment released by dam removal can shift grain-size distribution of the downstream channel bed toward that of the reservoir deposit, but that shift can vary spatially and temporally. Dam emplacement commonly induces coarsening (winnowing) of the downstream bed as sediment supply is restricted. As a result, dam removal commonly induces fining of the channel bed until upstream input declines and the river reworks deposited sediment. Though the bed may coarsen after initial sediment input is reworked, it may remain finer than the pre-removal bed under new equilibrium conditions. Typically, most fining is restricted to within a few kilometers of the dam site (e.g., Cheng and Granata 2007; Rumschlag and Peck 2007; Skalak *et al.* 2011; Major *et al.* 2012; Colaiacomo 2014; Harris and Evans 2014), but following some removals changes in bed texture have extended at least a few tens of kilometers downstream (Draut and Ritchie 2015; Evans and Wilcox 2014; East *et al.* 2015). Bed fining is not a universal response to dam removal, however. Along some gravel-bed rivers, released sediment coarsened the channel bed (Burroughs *et al.* 2009; Kibler, Tullos, and Kondolf 2011; Tullos and Wang 2014); in others, bed texture changed little. Muddy sand released by removals of low-head dams on rivers in the Midwestern United States had negligible impact on downstream channel texture (e.g., Doyle, Stanley, and Harbor 2003).

Even where downstream fining in a gravel-bed river is negligible, deposition of fine sediment may be ecologically significant. After removals of Elwha River and Milltown dams, fine sediment infiltrated and clogged interstitial pore spaces in gravel beds for months (Draut and Ritchie 2015; Evans and Wilcox 2014). Although these fine deposits were at most a few centimeters thick, they likely degraded habitat quality for benthic macroinvertebrates and fish redds (Wood and Armitage 1997; Jones *et al.* 2012). On Elwha River, habitat degradation occurred before arrival of the main sediment pulse (East *et al.* 2015).

13.2.2.5 Changes in Channel Patterns and Bedforms

Sediment release by dam removal can have other short-lived effects on channel morphology besides altering channel gradient, width, or bed texture. These effects include changes in channel pattern, filling of channel pools, and creation and modification of channel bars. Glide morphology can replace pool–riffle morphology when hydraulic controls are buried (Kibler, Tullos, and Kondolf 2011; East *et al.* 2015). Aggradation ≥ 1 m can change a channel from a single-thread to multithread pattern (Figure 13.5) (e.g., Major *et al.* 2012; Tullos and Wang 2014; Wilcox, O'Connor, and Major 2014), and new bar formation can increase channel sinuosity and braiding index (East *et al.* 2015). Multithread patterns commonly persist during high rates of bedload transport. After sediment-transport rates diminish and aggradation ceases, channels incise and revert to single-thread patterns. Reversion to a single-thread channel can happen swiftly, within days to months (e.g., Pearson, Snyder, and Collins 2011; Major *et al.* 2012; Wilcox, O'Connor, and Major 2014). Even where gross channel patterns were not altered, channel pools commonly filled (e.g., Wohl and Cenderelli 2000; Bountry, Lai, and Randle 2013; Peck and Kasper 2013; Harris and Evans 2014; Wilcox, O'Connor, and Major 2014) and channel relief was modified (Zunka, Tullos, and Lancaster 2015).

Sediment release by dam removal may create new channel bars or modify existing bars. New bars may grow or existing bars may enlarge as released sediment moves downstream (Wohl and Cenderelli 2000; Doyle, Stanley, and Harbor 2003; Kibler, Tullos, and Kondolf 2011; Major *et al.* 2012; East *et al.* 2015; Zunka, Tullos, and Lancaster 2015). But even these geomorphic changes can be short-lived. Within weeks of breaching Marmot Dam, bars flanking the channel within 1 km of the dam site were finer grained and positioned in locations independent of bars that existed prior to breaching. Within 2 months, however, bar surfaces had coarsened, and within a year bar locations, forms, and compositions were similar to those before breaching (Major *et al.* 2012). After removing Brownsville Dam, downstream bar area and volume within 1 km of the dam site increased several fold and new riffles formed – changes that persisted for at least 2 years (Kibler, Tullos, and Kondolf 2011). But valley morphology and river dynamism may influence formation and modification of channel bars more than does sediment release by dam removal. After failure of Barlin Dam, forms and distributions of downstream bars remained unchanged even after passage of 10^7 m^3 of gravel and sand (Tullos and Wang 2014).

13.3 Factors Influencing Responses to Dam Removals

The varying responses to dam removal – both in the reservoir and downstream – are affected by several controlling factors. Studies so far suggest important controls include dam height, reservoir sediment volume and composition, removal strategy, and downstream channel and valley morphology. Hydrology and watershed setting appear to be subordinate influences on the initial geomorphic response downstream, but may have great influence on longer-term response. Sawaske and Freyburg (2012) noted similar influential factors in an analysis of responses to 12 small-dam removals. O'Connor, Duda, and Grant (2015) briefly summarize more recent findings.

13.3.1 Dam Height and Removal Strategy

Dam height influences geomorphic response to removal in two ways. First, it affects the amount of sediment and water in storage. Tall dams typically store large volumes (>500 000 m^3) of sediment and may impound large volumes of water. By contrast, short dams are typically run-of-river structures that store small to moderate volumes of sediment and limited water, and they typically have passed sediment for many years. Second, dam height controls the overall base-level fall induced by dam removal.

Removal strategy affects the rate at which full base-level fall is achieved, which influences the rate of reservoir erosion and downstream sediment delivery. Full base-level fall is achieved rapidly during sudden removal of a dam, or gradually during phased removal. Sudden breaching of tall dams, including Savage Rapids Dam (12 m), Milltown Dam (12.8 m), Marmot Dam (15 m), Condit Dam (38 m), and failure of Barlin Dam (38 m) induced rapid and extensive erosion in the respective reservoirs (Figure 13.3). Sudden breaching led to rapid but varying degrees of channel incision, channel widening, knickpoint and knickzone advance, and, at Condit Dam, highly mobile landslides (Epstein 2009; Major *et al.* 2012; Bountry, Lai, and Randle 2013; Tullos and Wang 2014; Wilcox, O'Connor, and Major 2014). By contrast, phased removals of 64-m-tall Glines Canyon Dam and 32-m-tall Elwha Dam, neither of which impounded a reservoir filled with sediment, resulted in comparatively slower sediment release because rates of base-level fall were modulated and because deltas had to prograde through dwindling pools before bedload passed the dam sites (Randle *et al.* 2015). Although reservoir erosion was mostly by channel incision and fluvial lateral widening, many small-scale slumps and bank collapses (1–5 m tall) also contributed to sediment delivery (Randle *et al.* 2015). However, neither extensive landslides nor pronounced knickpoint migration occurred.

Erosional processes similar to those observed after breaching of tall dams have been observed after sudden and phased removals of smaller dams. In many instances, those processes have removed sediment more slowly from reservoirs behind small dams (Figure 13.3) (e.g., Doyle, Stanley, and Harbor 2003; Harris and Evans 2014). However, following removals of 4-m-tall Merrimack Village Dam (New Hampshire; Pearson, Snyder, and Collins 2011) and 3-m-tall Simkins Dam (Maryland;

Collins *et al.* 2014), rapid incision and channel widening in sandy reservoir sediment removed more than 40% of the stored volumes within months (Figure 13.3). By contrast, phased removals of small dams (e.g., Brewster Creek and Stronach dams; Table 13.1) strongly modulated erosion of reservoir sediment (Straub 2007; Burroughs *et al.* 2009; Figure 13.3). Although similar processes erode and evacuate sediment from reservoirs impounded by both small and tall dams, these examples show the rate and magnitude of base-level fall induced by dam-removal strategy has a strong influence on the rate and magnitude of reservoir-sediment erosion.

13.3.2 Relations Among Reservoir Sediment Volume, Rate of Sediment Release, and Background Sediment Flux

Two dimensionless ratios, both incorporating a river's background sediment flux, may provide predictive insights on the extent and duration of geomorphic impact upon dam removal. One ratio, V^*, provides a relation between the volume of stored reservoir sediment and volume of average annual sediment load at the dam site (e.g., Burroughs *et al.* 2009; Major *et al.* 2012; Tullos and Wang 2014; Wilcox, O'Connor, and Major 2014). The other ratio, E^*, provides a relation between the volume of sediment eroded in the first year after removal (or the first year after sediment began passing a dam site if removal was phased) and background sediment flux (Grant and Lewis 2015).

The ratio V^* provides a measure of the potential magnitude of perturbation to the downstream channel. It may also provide some insight on potential duration of the downstream response. For example, if V^* is relatively small (≤20), the overall downstream impact may be modest (though locally impressive) because the downstream channel receives a sediment input not substantially greater than that to which it is equilibrated. Also, the primary downstream response to sediment release may be of relatively short duration (months to years) because the river will likely redistribute released sediment swiftly and reequilibrate. By contrast, if V^* is large (several tens to hundreds), the river system may receive a sediment input much greater than that to which it is equilibrated and may take many years, and possibly decades, to effectively redistribute released sediment and reequilibrate.

Two removals highlight the potential utility of V^* as a gross predictor of order-of-magnitude duration of principal downstream geomorphic response. Marmot Dam stored about 750 000 m^3 of sediment. At its location, the average annual sediment flux of Sandy River is about 100 000 to 200 000 m^3. Hence, removal of Marmot Dam released an effectively small volume of sediment to Sandy River – probably less than a decade's worth ($V^* < 10$; Figure 13.6). Despite swift erosion of a large amount of stored sediment (>40% within 8 weeks; Figure 13.3; Major *et al.* 2012), suspended-sediment loads returned to background level within months of removal and pre-removal channel form and texture just downstream of the dam site – the reach most affected by released sediment – began to reestablish (Figure 13.5; Major *et al.* 2012). Within a year of removal, channel form and texture in that reach had largely reestablished though bed elevation and relief remained significantly changed from pre-removal conditions (Major *et al.* 2012; Zunka, Tullos, and Lancaster 2015). Within about 5 years after removal, morphology and longitudinal profile of the channel extending 2 km downstream of the dam site had mostly stabilized (J.J. Major unpublished data; Zunka, Tullos, and Lancaster 2015). By contrast, two dams on Elwha River stored much greater relative volumes of sediment (Figure 13.6). The Glines Canyon (~16 million m^3) and Elwha (~5 million m^3) dams impounded a combined ~21 million m^3 of sediment (Randle *et al.* 2015). The average annual sediment load of Elwha River entering Lake Mills behind Glines Canyon Dam is about 200 000–300 000 m^3 (Curran *et al.* 2009; Magirl *et al.* 2015). Thus, Glines Canyon Dam impounded about 50–80 years of average annual sediment load (V^* ~50–80), and Elwha Dam impounded about 15–25 years of (unregulated) average sediment load (V^* ~15–25). Over the first 2 years of phased removal of these dams, about 7 million m^3 of

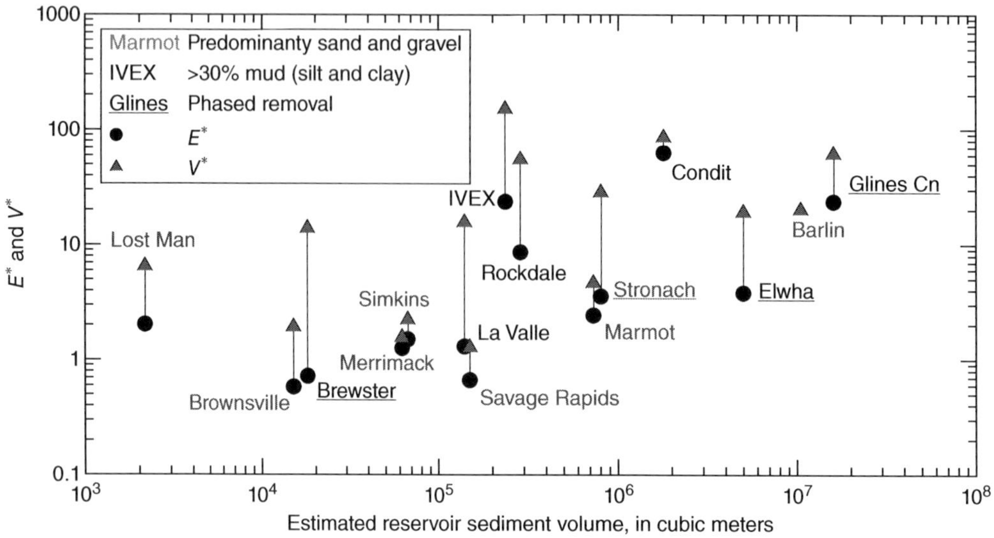

Figure 13.6 Volumes of stored sediment (V^*) (triangles) and volumes of sediment eroded during first year following dam removal (E^*) (black dots) relative to background sediment fluxes as a function of stored sediment volumes for recent dam removals. Vertical lines added for clarity to identify data points for given dams. Data for Glines Canyon Dam are for the second year following initiation of breaching because that is when most sediment, especially bedload sediment, began passing the dam site (see Figure 13.3 for additional information).

sediment was released (Randle *et al.* 2015), an amount equivalent to about 20–40 years of average annual load. Most of that released sediment passed through the 22-km-long channel corridor below Glines Canyon Dam and was delivered to a coastal zone (Warrick *et al.* 2015). However, the sediment deposited along the channel induced changes in elevation, pattern, and texture (East *et al.* 2015) to which the river is still responding some 4 years (at the time of this writing) after removals began. Sediment release from the former Lake Mills reservoir will continue over the coming years, especially during high-discharge events. The duration of major downstream response remains to be determined. These two examples show the relative magnitude of stored sediment to background sediment flux can influence the duration of downstream geomorphic response, especially if the river is able to access and erode large amounts of stored reservoir sediment.

Though the relative magnitude of the amount of sediment in storage broadly influences the nature and duration of downstream geomorphic response, the rate at which sediment is released is clearly an influential factor. Grant and Lewis (2015) approached the influence of rate of sediment release using the dimensionless ratio E^*. They examined the utility of using this ratio for predicting patterns of downstream sediment transport. For a limited number of removals in the US Pacific Northwest, they found that E^* did a better job of predicting the transport distance of coarse sediment (sand and gravel) than fine sediment (mud and fine sand). In general, fine sediment was transported downstream as suspended load and its transport distance correlated poorly with E^*. By contrast, coarse sediment is transported as bedload and its transport distance – when not interrupted by downstream sediment sinks such as other reservoirs or confluences with larger rivers – appears to correlate positively with E^*. That is, the greater the initial sediment delivery relative to background sediment flux, the farther downstream coarse sediment (mostly sand) is transported. If this simple empirical relation holds for other regions (especially for small dams on low-gradient rivers), it may prove a useful tool for quickly estimating possible spatial consequences of dam removals.

13.3.3 Reservoir-Sediment Grain Size

Sediment grain size (caliber) influences reservoir erosion processes and downstream effects of released sediment. In conjunction with the style of dam removal and degree to which sediment fills a reservoir, sediment grain size influences the nature and pace of reservoir erosion. It also affects consequent transport and storage of sediment downstream, and the distribution and residence time of the transported sediment.

Upon dam removal, reservoirs impounding coarse sediment commonly erode more completely and more rapidly than do those impounding fine sediment (Figure 13.3). The principal exception to this is Condit Dam, where highly mobile landslides swiftly evacuated a great amount of fine sediment. Differences in sediment characteristics, such as cohesion and consolidation of fine sediment, sediment stratification, and degree of saturation affect these granulometric differences in erosion rates. But the apparent differences may be biased by the limited number of removals and removal strategies examined to date, especially among large-dam removals. Few large dams impounding mostly fine sediment have been removed. And among those dams which we examined that impounded fine sediment, nearly half were removed incrementally, which modulated the rate of base-level fall (Figure 13.3). Nevertheless, differences among grain-size distributions of reservoir sediment appear to influence erosion rates upon dam removal.

Grain size of released sediment affects downstream impacts and responses regardless of removal strategy. Released fine sediment commonly has less downstream impact than coarse sediment. Apart from small-volume fine-sediment deposition as interstitial fill, most fine sediment is transported downstream as suspended load that temporarily increases turbidity and suspended-solids concentration (Doyle, Stanley, and Harbor 2003; Granata, Cheng, and Nechvatal 2008; Major *et al.* 2012; Magirl *et al.* 2015). Following some removals, fine sediment accumulated thinly on floodplains or along channel margins when flows were either deep enough to spill overbank or had relatively little transport capacity relative to fine-sediment supply (e.g., Evans *et al.* 2000; Draut and Ritchie 2015). In instances where downstream reservoirs interrupted sediment transport, substantial amounts of fine sediment accumulated (Evans *et al.* 2000).

Even large-volume releases containing abundant fine sediment have had relatively little downstream impact. Although abundant fine sediment at Condit Dam promoted conditions for hyperconcentrated flow upon breaching, that flow selectively draped the downstream channel corridor with sand and largely passed mud through the system (Wilcox, O'Connor, and Major 2014). Most of the fine sediment released by the concurrent removals of the Elwha River dams passed through the 22-km-long downstream corridor (Warrick *et al.* 2015). Fine sediment retained along the channel corridor initially infiltrated the pre-removal gravel bed, formed small bars along the channel margin, or draped floodplain channels (Draut and Ritchie 2015; East *et al.* 2015), but its deposition prompted little geomorphic response.

Coarse sediment, by contrast, can prompt substantial downstream geomorphic response. Coarse sediment release can increase bedload transport many fold, aggrade channel beds, coarsen or fine downstream bed texture, and modify channel pattern and morphology. These effects commonly are pronounced when large amounts of coarse sediment are released rapidly, especially when the amount released overwhelms transport capacity.

Coarse-sediment released by removals of small dams, mostly sand, has had modest impact relative to that released by removals of tall dams. Although sediment released by some small-dam removals caused substantial (1–3 m) channel aggradation, most releases induced only minor changes in bed elevation within a few kilometers downstream of a dam site. Commonly, those releases had little to modest effect on channel pattern owing mainly to pool filling, modification of local channel relief, or minor modifications to channel bars and riffles. They have also had varying effects on channel-bed texture. Typically, the effects of coarse sediment released by small-dam removals have been short-lived (months to perhaps a few years). By contrast, downstream impacts of coarse sediment released by removals of tall dams have been more profound regardless of removal strategy. Voluminous releases that have delivered sediment pulses that rival substantial natural events in scale have aggraded channel beds by as much as 10 m (Tullos and Wang 2014) and substantially modified channel relief, channel patterns, and bed texture (Major *et al.* 2012; Colaiacomo 2014; Tullos and Wang 2014; Wilcox, O'Connor, and Major 2014; East *et al.* 2015; Zunka, Tullos, and Lancaster 2015).

The effects of releasing large volumes of coarse sediment appear to last longer if the sediment contains a substantial proportion of gravel. After breaching Marmot Dam, gravel that accumulated in the 2-km-long reach downstream of the dam site (Figure 13.4) has remained largely in storage for nearly a decade after removal (J.J. Major unpublished data). Released sand that moved much farther downstream had limited and apparently short-lived effects (Major *et al.* 2012). Sand and minor gravel released by removal of Condit Dam swiftly aggraded the channel bed several kilometers downstream of the dam site by as much as 1 m after passage of the initial hyperconcentrated flow, but within days of attaining maximum aggradation that fill was incised nearly to the pre-removal level of channel riffles (Wilcox, O'Connor, and Major 2014). Such rapid incision shortly after maximum aggradation was due, in part, to a lack of gravel capable of armoring the bed surface and to rapid reduction of sediment supply.

13.3.4 Breach and Post-Breach Hydrology

There is little correlation between the proportion of sediment evacuated from the reservoir and magnitude of river flow during and after dam breaching. Several dams have been breached during low to moderate flows – flows much smaller than the 2-year-return-interval flow. Yet, during breaching and in the ensuing days to months these moderate flows in many cases rapidly evacuated large fractions of impounded sediment (Figure 13.7; Pearson, Snyder, and Collins 2011; Major *et al.* 2012; Bountry, Lai, and Randle 2013; Collins *et al.* 2014; Tullos and Wang 2014; Wilcox, O'Connor, and Major 2014; Magirl *et al.* 2015; Zunka, Tullos, and Lancaster 2015). Even during phased removals, such as those on Elwha River, moderate flows eroded large fractions of reservoir sediment if the rate of base-level fall induced sufficient change in slope to mobilize a broad range of sediment sizes (e.g., Magirl *et al.* 2015; Randle *et al.* 2015). By contrast, large flows have been most important for accessing and evacuating harder-to-reach remnant sediment that becomes isolated along reservoir margins or stranded on terraces, and in redistributing sediment deposited downstream (Pearson, Snyder, and Collins 2011; Major *et al.* 2012; Harris and Evans 2014; Tullos and Wang 2014). In gravel-rich rivers, large flows become especially important for evacuating remnant reservoir sediment and redistributing it as the channel within the reservoir and downstream armors.

Breaches that released substantial water volumes generated brief peak flows equivalent to low-frequency, high-magnitude floods. But in two well-documented examples, those water releases did not erode large amounts of sediment from reservoirs. Evans *et al.* (2000) documented a release of about 38 000 m^3 of water within minutes during failure of IVEX Dam (Ohio), producing a 466 m^3/s peak flow (>100-year flood?), yet only about 25 000 m^3 (10%) of impounded sediment had eroded after 2 months. Breaching of Condit Dam released 1.6 million m^3 of water in a little more than an hour, generating a 420 m^3/s peak flow (>100-year flood). However, substantial evacuation of reservoir sediment did not occur until landslides became prevalent after the water had drained (Wilcox, O'Connor, and Major 2014).

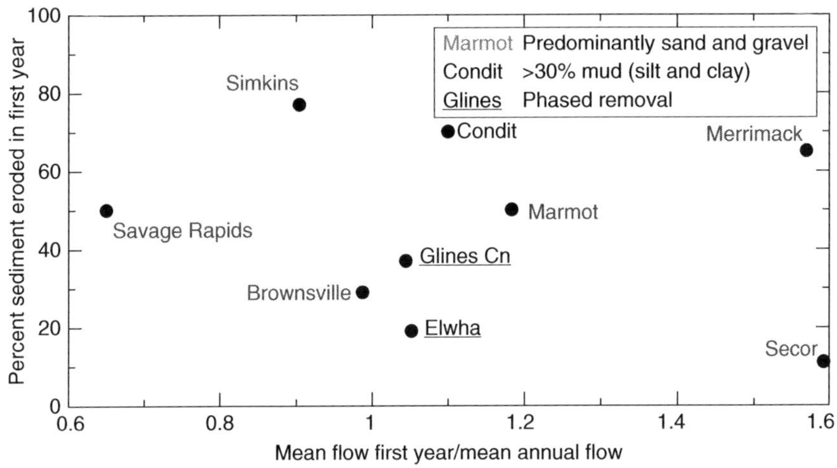

Figure 13.7 Fraction of reservoir sediment volume eroded as function of relative mean-flow magnitude during first year following dam removal. Data for Glines Canyon Dam are for the second year following initiation of breaching because that is when most sediment, especially bedload sediment, began passing the dam site (see Figure 13.3 for additional information). Data for Secor Dam represent fraction of sediment eroded within 6 months of breaching.

Lack of correlation between rate of sediment erosion and flow magnitude reflects the relative importance of process-based erosion versus event-based erosion, a conceptual model proposed by Pizzuto (2002) and extended by Pearson, Snyder, and Collins (2011). Process-based erosion, which is largely independent of flow magnitude, occurs when there is a great change in reservoir equilibrium upon dam removal (sudden channel initiation, rapid migration of headcuts and knickpoints, etc.). Event-based erosion follows when high flows are needed to access and exceed transport thresholds for remnant sediment (Pearson, Snyder, and Collins 2011). Removals to date highlight the general validity of this conceptual model.

Under disequilibrium conditions caused by reservoir-sediment release, modest flows can also transport abundant sediment downstream. For example, during the first year after removal of Marmot Dam, peak flows were no larger than 60% of the discharge of a 2-year flood and mean annual flow was only 18% greater than long-term mean flow. Under these conditions, unit-width bedload fluxes within 10 km downstream of the dam site approached some of the highest values measured for gravel-bed rivers (3–4 kg/s/m; Major *et al.* 2012). During the phased removals of the Elwha River dams, 2 years of muted flows having peak discharges about 70% of the discharge of a 2-year flood transported 90% of the released sediment (~9 million tonnes) to the river mouth (Magirl *et al.* 2015; Warrick *et al.* 2015). Release of a small volume of gravel (~15 000 m^3) by removal of Brownsville Dam and subsequent modest flows during the year after removal led to substrate coarsening, growth of channel bars, and an increase in habitat heterogeneity within a few hundred meters of the dam site, but little change farther downstream (Kibler, Tullos, and Kondolf 2011).

To date, there has yet to be an intentional dam removal contemporaneous with, or followed closely by, a large, naturally occurring flood. Thus, there is uncertainty regarding exactly how a river system might respond to such a scenario. Major *et al.* (2012) speculated that the sequencing of flow magnitudes at Marmot Dam influenced reservoir-sediment erosion at the time scale of individual events, but had little effect on total erosion 2 years after removal. They further speculated that had large-magnitude flows occurred at or shortly after breaching, a similar proportion of reservoir erosion would have been achieved after 2 years, but likely by a different erosion trajectory. Comparison of erosion trajectories at Merrimack Village Dam (MVD) and Simkins Dam (Figure 13.3) support this speculation. Both dams were of similar height and stored similar sediment volumes and compositions (Table 13.1), had similar upstream contributing areas, similar hydroclimate regimes, and were removed suddenly. No large flows (≥10-year event) occurred within 2 years of removal of MVD (Pearson, Snyder, and Collins 2011) whereas a 10-year event occurred within months of removal of Simkins Dam (Collins *et al.* 2014). Similar proportions of reservoir sediment eroded after 2 years from each dam site, but by slightly different trajectories.

13.3.5 Downstream Valley Morphology

Downstream valley morphology can affect overall river system response to sediment release from dam removal. Valley morphology can act as a filter that affects sediment storage and transport and influences the fate of differential sediment distribution. After removal of Marmot Dam, released gravel accumulated largely along a narrow, low-gradient reach immediately downstream of the dam site. By contrast, nearly all released sand passed much farther downstream (Figure 13.4; Major *et al.* 2012). Following failure of Barlin Dam, resistant valley walls and valley-forced meandering affected the locations, sizes, and geometries of channel bars, channel width, and lateral migration (Tullos and Wang 2014). Following breaching of Condit Dam, a narrow, bedrock-confined reach immediately below the dam passed much of the released sediment several kilometers downstream to a broader, lower-gradient reach where it accumulated thickly (Colaiacomo 2014; Wilcox, O'Connor, and Major 2014). In the weeks following breaching, bed elevation within the bedrock reach adjusted toward the pre-removal level of riffles but pools remained filled (Colaiacomo 2014). After removal of Savage Rapids Dam, the overall bed elevation in the 3-km reach downstream of the dam site changed little, but spatially progressive changes in residual pool depths reflected downstream advance of sediment for at least 2 years after removal (Tullos, Finn, and Walter 2014).

Existing morphology can also affect bed relief, and thus channel complexity, that develops from deposition of released sediment. Sediment released to low-relief channels can increase local relief by building channel bars, but it can decrease relief in high-relief channels by filling pools, by replacing pool–riffle morphology with glides, and by forming short-lived mid-channel bars (Wohl and Cenderelli 2000; East *et al.* 2015; Zunka, Tullos, and Lancaster 2015). The style, timing, and duration of downstream sediment transport can also be affected by channel morphology. Channel bars and pools can serve as sites of temporary, or even long-lived, sediment accumulation (Wohl and Cenderelli 2000; Doyle, Stanley, and Harbor 2003; Kibler, Tullos, and Kondolf 2011; Harris and Evans, 2014), and floodplains can trap sediment even during low flows if channel aggradation is sufficient (e.g., Evans *et al.* 2000; Burroughs *et al.* 2009; East *et al.* 2015). These sediment-storage reservoirs can limit impact farther downstream by disrupting or buffering sediment transport and can extend the duration of impact of released sediment.

13.3.6 Watershed Geologic Setting and Geographic Context

Watershed geologic setting and geographic context can strongly affect response to dam removal. Where a dam sits in a watershed broadly influences the impact its removal has on the downstream channel. Its position influences the caliber and quantity of reservoir sediment, its hydrological regime, and the potential for the downstream channel to store or transmit sediment – all of which affect the spatial extent of downstream impact. Geographic context also includes the presence of dams upstream or downstream of the removed dam, which influences sediment supply and downstream transport and storage.

Though geologic and geographic context can affect short-term response to dam removal, they likely have greater influence on long-term response. Skalak, Pizzuto, and Hart (2009) argue that many rivers in the US Mid-Atlantic region are bedrock channels mantled with a thin gravel cover. This morphology is related to low rates of sediment supply, transport, and storage typical of the region. Thus, stability of exposed bedrock and boulders exerts a very strong control on overall channel morphology, which influenced long-term erosion, transport, and storage of sediment. Removals of small dams in this region typically release short-lived pulses of sediment commonly finer grained than the composition of the downstream bed. Although these pulses may briefly alter channel morphology and bed composition, sediment supplies eventually approach watershed delivery rates. Thus, long-term response to sediment delivery is more likely to be controlled by the configuration and stability of exposed bedrock and immobile boulders than by modest sediment release from dam removal. As a result, gross geomorphic characteristics of channels may remain relatively unaffected by small-dam removals (Skalak, Pizzuto, and Hart 2009; Skalak *et al.* 2011).

Responses to removals of tall dams in the US Pacific Northwest and to a dam failure in Taiwan also exemplify influences of geology and geography. Sandy River, Oregon, drains Mount Hood, an active volcano in the Cascade Range. Large volumes of volcanic sediment are stored along the length of Sandy River, and provide locally dynamic sediment sources. Upon removal of Marmot Dam, released sand passed far downstream. However, river dynamism and local sediment sources masked the distribution of that sand. The configuration of confined bedrock and wider alluvial reaches along White Salmon River affected the transport and storage of sediment released by breaching of Condit Dam (Colaiacomo 2014; Wilcox, O'Connor, and Major 2014). Similarly, valley-forced morphology as well as basin-wide variations in caliber and quantity of sediment transport owing to lithological influences affected rates and types of channel adjustments to the sediment pulse delivered by failure of Barlin Dam (Tullos and Wang 2014).

13.4 Time Scales of Channel Responses to Dam Removals

Channel perturbation involves a reaction time over which channel conditions diverge from a state of stability, and a relaxation time during which perturbed channel conditions converge toward a state of stability (e.g., Graf 1977; Pierson and Major 2014). In the context of dam removal, reaction times may be short or prolonged – hours to days following sudden removal, or months to years during phased removal. Relaxation times can be highly variable, and systems may converge toward an equilibrium state that differs from the pre-dam state. Following dam removal, reaction and relaxation timescales typically overlap. Although large amounts of reservoir sediment can be evacuated swiftly following initial breaching of a dam, erosion commonly continues to remove and deliver sediment downstream for protracted periods – months to years, perhaps decades. Thus, the channel downstream can simultaneously be responding to and relaxing from an initial perturbation while being perturbed further by additional sediment delivery. Nevertheless, relaxation times toward a state of relative stability following dam removals have been shorter than commonly anticipated – years, not decades. Pizzuto (2002) hypothesized that evolution of an equilibrium channel through reservoir sediment might take at least a decade depending on mass and grain size of the sediment. Yet in many cases more than a third, and sometimes two-thirds, of stored sediment has been evacuated within weeks to months of dam removal (e.g., Sawaske and Freyburg 2012; Figure 13.3), and relatively stable channels within reservoirs have established swiftly. Even during phased removals of Elwha River dams, approximately a third of the total sediment stored behind two dams was evacuated in the first 2 years of the removal project (Randle *et al.* 2015; Warrick *et al.* 2015), and about half evacuated after 3 years (C.S. Magirl unpublished data). But owing to continued erosion, reservoir-reach channels on Elwha River have yet to stabilize after 4 years.

Major downstream channel adjustments following dam removal commonly achieved apparent stability within months to years. In several instances, bed textures, channel patterns and gradients, bar morphologies, and sediment transport rates approached pre-removal or pre-dam conditions within a year or two, although pools commonly remained filled (Bushaw-Newton *et al.* 2002; Doyle, Stanley, and Harbor 2003; Pearson, Snyder, and Collins 2011; Major *et al.* 2012; Peck and Kasper 2013; Coloaiacomo 2014; Tullos, Finn, and Walter 2014; Tullos and Wang 2014; Kim, Toda, and Tsujimoto 2015; Zunka, Tullos, and Lancaster 2015). The principal exception is Elwha River owing to continued delivery of large amounts of sediment. But now that the dams have been removed completely and at least half the sediment has been evacuated, and mostly passed to the coastal zone, stability may be achieved within a few years. Although major channel adjustments commonly occur swiftly, minor adjustments, such as decimeter-scale fluctuations of bed elevation, migration of small secondary bars, and adjustments of particle sorting, may persist for several years (Skalak *et al.* 2011; Greene, Krause, and Knox 2013; Peck and Kasper 2013).

Although rivers appear to recover rapidly after perturbations induced by dam removals, long-term assessments of responses to removals are rare. As of 2015, the longest lasting responses to dam removals have resulted from accumulation of gravel, local aggradation and filling of deep pools with sand, or substantial deposition in downstream reservoirs that interrupted sediment transport (e.g., Evans *et al.* 2000; Evans 2007; Skalak *et al.* 2011; Major *et al.* 2012; Peck and Kasper 2013: Zunka, Tullos, and Lancaster 2015). Seven years after removal of Marmot Dam, the profile of Sandy River in the first kilometer below the dam site still reflects substantive gravel accumulation, and minor fluctuations of bed elevation and local redistribution of sediment among channel bars persist (J.J. Major unpublished data). Five years after removal of Munroe Falls Dam (Ohio) sand deposited on the channel bed 2–3 km downstream of the dam site continued to influence activation of meander-bend chutes at all but the lowest flow, particle sorting persisted, and sandy deltaic deposits remained at the head of a reservoir 5 km downstream (Peck and Kasper 2013). Four years after removal of the low-head Manatawny Dam (Pennsylvania), which impounded sand and gravel, modest channel adjustments continued both upstream and downstream of the dam site (Skalak *et al.* 2011). And 12 years after failure of IVEX Dam, sand released more slowly from a delta at the upper end of the reservoir-reach had delivered twice the volume to a downstream reservoir as did the initial flush of fine sediment released by failure (Evans 2007). The overall findings from recent dam removal studies of rapid relaxation of sediment perturbations present a contrast with the hypothesis of much longer-term adjustments to breaching of low-head, colonial-age mill dams in the US Mid-Atlantic region (Merritts *et al.* 2013). That hypothesis proposes that erosion of sediment impounded by these structures has persisted for decades to centuries after breaching, and largely accounts for most of the present-day, fine-grained load delivered downstream.

Construction or exposure of grade-control features after dam removal can affect long-term sediment accumulation and erosion in reservoirs and consequent downstream response. An analysis of geomorphic response 10 years after removal of La Valle Dam (Wisconsin) showed that a man-made riffle constructed at the dam site altered flow hydraulics and sediment transport (Greene, Krause, and Knox 2013). That riffle induced renewed sedimentation in the reservoir reach, which subsequent floods eroded and redistributed downstream. Similar hydraulic and sediment-transport conditions established behind Homestead Dam (New Hampshire) after resistant boulders were exposed just upstream of the dam site (Gartner, Magilligan, and Renshaw 2015).

Local and system-wide relaxation times also depend upon the manner in which a sediment pulse moves downstream. A pulse of sediment introduced to a channel may disperse in place and diffuse downstream, translate downstream as a discrete sediment wave, or exhibit hybrid behavior (Lisle *et al.* 2001; Lisle 2008; Sklar *et al.* 2009). The manner of transport affects the magnitude and duration of downstream impact – a sediment pulse that disperses downstream may induce lower-magnitude changes, but have longer-lasting impact than does a pulse translating as a discrete wave. Theoretical, experimental, and field studies indicate sediment pulses on gravel-bed rivers propagate by translation rather than dispersion when pulse volume is small relative to channel dimensions, grain size is finer than the downstream river bed, and Froude number is low (<0.4) (Lisle *et al.* 2001; Lisle 2008; Sklar *et al.* 2009; Madej *et al.* 2009; Pryor *et al.* 2011). By contrast, sediment dispersion is favored when sediment volume is large relative to channel dimensions, grain-size distribution is broad and contains particles coarser than the channel bed, and Froude number is high (>0.4). There are few studies of dam removals that specifically address the manner of downstream sediment propagation (e.g., Evans, Huxley, and Vincent 2007; Thomas *et al.* 2014), but some provide provocative, and sometimes contradictory, observations. Within a week of breaching Condit Dam, Wilcox, O'Connor, and Major (2014) documented a distinct rise and fall of the channel bed a few kilometers downstream of the dam site as large amounts of sand and some gravel passed. This rapid change of channel-bed elevation, even though sediment volume was large relative to channel dimensions and Froude number was high (~1), suggests passage of a translating sediment wave. Similar cycles of downstream aggradation and degradation were observed following release of gravelly sediment by failure of Barlin Dam (Tullos and Wang 2014). By contrast, East *et al.* (2015) suggest that progressive but sustained low-magnitude changes in bed elevation on Elwha River during phased dam removals reflect sediment-pulse dispersion – a result that owes largely to the great volume of sediment released. Subsequent to removal of Marmot Dam, proximally deposited gravel was redistributed, but by 7 years after removal there was no evidence that it had translated downstream as a discrete sediment wave (Major *et al.* 2012; J.J Major unpublished data).

Regardless of how sediment moves downstream, when its supply exceeds a river's transport capacity, deposition occurs and bed texture and morphology are modified. Experiments show introduced sediment pulses obliterate much of the downstream channel morphology at the reach scale and change both mean and spatial patterns of bed elevation and texture (Lisle *et al.* 2001; Cui *et al.* 2003; Sklar *et al.* 2009; Podolak and Wilcock 2013). Experiments also show bed topography and textural patterns adjust over different time scales (Podolak and Wilcock 2013). But even after channels have achieved largely stable topographic and textural patterns similar to those extant before passage of a sediment pulse, spatially variable secondary textural and topographic adjustments may persist, including changes in particle sorting, minor bed-elevation fluctuations, and migration of secondary bars much smaller in scale than major channel bars (Podolak and Wilcock 2013). Although these persistent secondary adjustments may induce only modest geomorphic changes to the downstream channel, they may have important ecological ramifications.

13.5 Common Findings from Analyses of Responses to Dam Removals

Despite the variety of dam heights, reservoir sediment volumes and compositions, and removal strategies, a number of common findings have emerged with regard to geomorphic responses following dam removals.

1) Rivers are resilient. After dam removals, rivers have largely achieved apparent stability – at least within their range of natural variability – within a few years rather than decades. Despite concerns about rapidly releasing possibly large volumes of sediment to river systems, most removals to date have shown adverse geomorphic impacts (such as complete bed-texture transformation, enhanced flood risk, or damage to fish habitat) are short-lived and have been neither as damaging nor as long-lasting as feared. Although locally minor steps in reservoir profiles sometimes developed owing to exposure of resistant bedrock, boulders, or buried structures, concerns about development of stepped profiles capable of impeding upstream fish passage have been unfounded thus far.
2) Modest streamflows can effectively evacuate large percentages of reservoir sediment and move it downstream. Rapid evacuation and transport is associated with two key factors: (i) sediment composition; and (ii) establishment of extraordinary river capacity and competence owing to abrupt change in base level during dam breaching. In general, reservoir sediment composed dominantly of sand and gravel erodes more swiftly and more completely than does reservoir sediment dominated by mud. The nature and pace of erosion can be strongly influenced, however, by exposure of natural and man-made features that can act as grade-controls and impose new boundary conditions.
3) Sediment volume and caliber exert a dominant control on downstream response to dam removal. Large amounts of released gravel have had more substantial and longer-lasting downstream effects than has released sand and mud. But even the most substantial channel changes caused by accumulation of coarse sediment (such as large changes in bed elevation, changes in bed texture, and loss of pre-removal channel pattern) have not had long-lasting adverse geomorphic effects.
4) Location of a dam within a watershed, existing valley morphology, and geologic and geographic settings can strongly influence channel response to dam removal. Locations and forms of channel bars may be governed largely by existing valley morphology. Rivers attain equilibrium more quickly within confined bedrock reaches than within wider, lower-gradient alluvial reaches. Regional lithology and bedrock exposure may exert a greater influence on channel morphology than either dam emplacement or sediment release by dam removal.
5) Predicting time to recovery from sediment release following dam removal remains challenging. Although a river will converge toward a state of equilibrium with the supplied sediment load, it may not return to its pre-dam state. Assessment of metrics such as V^* (ratio of impounded sediment volume to background sediment flux) or E^* (ratio of sediment volume eroded within 1 year of removal to background sediment flux) can provide qualitative insights regarding an order-of-magnitude time it may take for a river to converge to a state of stability. In general, rivers converged toward stability conditions within a few years when both V^* and E^* had values of ≤ 20.

These findings have been extracted from a limited number of field studies across geographically restricted regions of the United States and a single dam failure abroad. Hence, we have no information on geomorphic responses to dam removals from a vast region of the United States having great topographic, climatic, and ecological diversity. Detailed studies of responses to removals in underrepresented regions would bolster our ability to test the general applicability of these findings. Furthermore, there has yet to be clear integration of physical and biological responses to dam removals. Tighter integration of the nature and pace of geomorphic and ecological responses – and their interactions – is needed to develop more thorough assessment of the impacts of sediment release by dam removal.

Consistent and thorough documentation of physical and ecological responses to dam removals is needed to improve our understanding of environmental benefits and detriments of dam removal. We recognize that all studies are subject to logistical and financial constraints. Nevertheless, we encourage the community to document as thoroughly as possible the following conditions: (i) pre-removal, and if possible pre-emplacement, morphological and ecological conditions of a river system; (ii) reservoir-sediment volume, grain size, and spatial distribution of grain size; (iii) regional background sediment flux; (iv) rates and processes of reservoir-sediment erosion; (v) rates and magnitudes of downstream sediment transport; (vi) styles, rates, and magnitudes of post-removal channel changes downstream; and (vii) long-term geomorphic adjustments to sediment release. Such measurements will improve both our understanding of the factors that influence responses to removals and quantification of sediment budgets. Accurate sediment budgets would not only improve our understanding of the fate of released sediment, but could also improve our ability to evaluate potential physical and ecological impacts below dam sites. They would also provide critical data for evaluating numerical models that predict physical responses to dam removals.

Acknowledgments

We are grateful to many colleagues who helped us collect data following dam removals we investigated, and to colleagues who generously shared data with us. Ryan Bellmore and Jeff Duda compiled an extensive database of dam-removal publications, which we leveraged. We thank Jeff for his assistance creating Figure 13.1. The US Geological Survey Powell Center for Analysis and Synthesis provided us the opportunity and facilities to gather to discuss what has been learned from dam removals. We thank Jim Pizzuto, Noah Snyder, and Katie Skalak for their reviews.

References

American Rivers 2014. *List of Dams Removed 1998–2014*. Available from www.americanrivers.org (last accessed April 17, 2015).

Babbitt, B 2002. What goes up, may come down. *BioScience* 52, 656–658.

Bednarek, AT 2001. Undamming rivers: a review of the ecological impacts of dam removal. *Environmental Management* 27, 803–814.

Bellmore, J, Vittum, K, Duda, J, and Greene, S 2015. *USGS Dam Removal Science Database*. DOI:10.5066/F7K935KT.

Bellmore, JR, Duda, JJ, Craig, LS *et al.* 2016. Status and trends of dam removal research in the United States. *WIREs Water*. DOI:10.1002/wat2.1164, 13 p.

Bountry, JA, Lai, YG, and Randle, TJ 2013. Sediment impacts from the Savage Rapids Dam removal, Rogue River, Oregon. In *The Challenges of Dam Removal and River Restoration* (eds JV De Graff and JE Evans). *Geological Society of America Reviews in Engineering Geology* 21, 93–104.

Burroughs, BA, Hayes, DB, Klomp, KD *et al.* 2009. Effects of Stronach Dam removal on fluvial geomorphology in the Pine River, Michigan, United States. *Geomorphology* 119, 96–107.

Bushaw-Newton, KL, Hart, DD, Pizzuto, JE *et al.* 2002. An integrative approach towards understanding ecological responses to dam removal: the Manatawny Creek study. *Journal of the American Water Resources Association* 38, 1581–1599.

Cantelli, A, Wong, M, Parker, G, and Paola, C 2007. Numerical model linking bed and bank evolution of incisional channel created by dam removal. *Water Resources Research* 43, W07436. DOI:10.1029/2006WR005621.

Cheng, F and Granata, T 2007. Sediment transport and channel adjustments associated with dam removal: field observations. *Water Resources Research* 43, W03444. DOI:10.1029/2005WR004271.

Colaiacomo, EJ 2014. *Downstream spatial and temporal response to dam removal, White Salmon River, WA*. MS thesis, University of Montana.

Collins, MJ, Boardman, G, Banks, W *et al.* 2014. Dam breach release of non-cohesive sediments: channel response and recovery rates. Abstract EP31E–07 presented at *2014 Fall Meeting, American Geophysical Union*, San Francisco, CA, December 15–19.

Csiki, S and Rhoads, BL 2010. Hydraulic and geomorphological effects of run-of-river dams. *Progress in Physical Geography* 34, 755–780.

Cui, Y and Wilcox, A, 2008. Development and application of numerical models of sediment transport associated with dam removal. In *Sedimentation Engineering: Processes, Measurements, Modeling, and Practice* (ed. MH Garcia). *ASCE Manuals and Reports on Engineering Practice* 110, 995–1020.

Cui, Y, Parker, G, Lisle, TE *et al.* 2003. Sediment pulses in mountain rivers: 1. Experiments. *Water Resources Research* 39. DOI:10.1029/2002WR001803.

Curran, CA, Konrad, CP, Higgins, JL, and Bryant, MK 2009. Estimates of sediment load prior to dam removal in the Elwha River, Clallam County, Washington. *US Geological Survey Scientific Investigations Report* 2009–5221, 18 pp.

Downs, PW, Cui, Y, Wooster, JK *et al.* 2009. Managing reservoir sediment release in dam removal projects: An approach informed by physical and numerical modelling of non-cohesive sediment. *International Journal of River Basin Management* 7, 433–452.

Doyle, MW, Stanley, EH, and Harbor, JM 2002. Geomorphic analogies for assessing probable channel response to dam removal. *Journal of the American Water Resources Association* 38, 1567–1579.

Doyle, MW, Stanley, EH, and Harbor, JM 2003. Channel adjustments following two dam removals in Wisconsin. *Water Resources Research* 39, 1011. DOI:10.1029/2002WR001714.

Draut, AE and Ritchie, AC 2015. Sedimentology of new fluvial deposits on the Elwha River, Washington, USA, formed during large-scale dam removal. *River Research and Applications* 31, 42–61. DOI:10.1002/rra.2724.

East, AE, Pess, GR, Bountry, JA *et al.* 2015. Large-scale dam removal on the Elwha River, Washington, USA: River channel and floodplain geomorphic change. *Geomorphology* 228, 765–786.

Epstein, JA 2009. *Upstream geomorphic response to dam removal: The Blackfoot River, Montana*. MS thesis, University of Montana.

Evans, E and Wilcox, AC 2014. Fine sediment infiltration dynamics in a gravel-bed river following a sediment pulse. *River Research and Applications* 30, 372–384.

Evans, JE 2007. Sediment impacts of the 1994 failure of IVEX Dam (Chagrin River, NE Ohio): a test of channel evolution models. *Journal of Great Lakes Research* 33(Special Issue 2), 90–102.

Evans, JE, Huxley, JM, and Vincent, RK 2007. Upstream channel changes following dam construction and removal using a GIS/remote sensing approach. *Journal of the American Water Resources Association* 43, 1–15.

Evans, JE, Mackey, SD, Gottgens, JF, and Gill, WM 2000. Lessons from a dam failure. *Ohio Journal of Science* 100(5), 121–131.

Ferrer-Boix, C, Martín-Vide, JP, and Parker, G 2014. Channel evolution after dam removal in a poorly sorted sediment mixture: Experiments and numerical model. *Water Resources Research* 50. DOI:10.1002/2014WR015550.

Gartner, JD, Magilligan, FJ, and Renshaw, CE 2015. Predicting the type, location and magnitude of geomorphic responses to dam removal: role of hydrologic and geomorphic constraints. *Geomorphology* 251, 20–30.

Graf, WL 1977. The rate law in fluvial geomorphology. *American Journal of Science* 277, 178–191.

Granata, T, Cheng, F, and Nechvatal, M 2008. Discharge and suspended sediment transport during deconstruction of a low-head dam. *Journal of Hydraulic Engineering* 134, 652–657.

Grant, GE and Lewis, SL 2015. The remains of the dam: What have we learned from 15 years of US dam removals?. In *Engineering Geology for Society and Territory 3 – River Basins, Reservoir Sedimentation, and Water Resources* (eds G Lollino, M Arattano, M Rinaldi *et al.*). Springer International: Switzerland; 31–35. DOI:10.1007/978-3-319-09054-2_7.

Grant, G, Marr, JDG, Hill, C *et al.* 2008. Experimental and field observations of breach dynamics accompanying erosion of Marmot cofferdam, Sandy River, Oregon. In *ASCE Proceedings of the 2008 World Environmental and Water Resources Congress* (eds RW Babcock and R Walton), Honolulu, HI, May 12–16; 10 pp.

Greene, SL, Krause, AJ, and Knox, JC 2013. A decade of geomorphic and hydraulic response to the La Valle Dam project, Baraboo River, Wisconsin. *Journal of the American Water Resources Association* 49, 1473–1484.

Greimann, B 2013. Prediction of sediment erosion after dam removal using a one-dimensional model. In *The Challenges of Dam Removal and River Restoration* (eds JV De Graff and JE Evans). *Geological Society of America Reviews in Engineering Geology* 21, 59–66.

Harris, N 2008. *Sedimentological response of the 2007 removal of a low-head dam, Ottawa River, Toledo, Ohio*. MS thesis, Bowling Green State University.

Harris, N and Evans, JE 2014. Channel evolution of sandy reservoir sediments following low-head dam removal, Ottawa River, northwestern Ohio, U.S.A. *Open Journal of Modern Hydrology* 4, 44–56.

Hart, DD, Johnson, TE, Bushaw-Newton, KL *et al.* 2002. Dam removal: challenges and opportunities for ecological research and river restoration. *BioScience* 52, 669–681.

Hoffman, DE and Gabet, EJ 2007. Effects of sediment pulses on channel morphology in a gravel-bed river. *Geological Society of America Bulletin* 119, 116–125.

James, LA 2013. Legacy sediment: definitions and processes of episodically produced anthropogenic sediment. *Anthropocene* 2, 16–26.

Jones, JI, Murphy, JF, Collins, AL *et al.* 2012. The impact of fine sediment on macro-invertebrates. *River Research and Applications* 28, 1055–1071.

Kibler, K, Tullos, D, and Kondolf, M 2011. Evolving expectations of dam removal outcomes: downstream geomorphic effects following removal of a small, gravel-filled dam. *Journal of the American Water Resources Association* 47, 408–423.

Kim, S, Toda, Y, and Tsujimoto, T 2015. Geomorphological and riparian vegetation responses following a low-head dam removal: a study based on literature review. *International Journal of River Basin Management* 13, 315–324.

Lisle, TE 2008. The evolution of sediment waves influenced by varying transport capacity in heterogeneous rivers. In *Gravel Bed Rivers VI: From Process Understanding to River Restoration* (eds H Habersack, H Piegay, and M. Rinaldi). Elsevier: Amsterdam; 443–472. DOI:10.1016/S0928-2025(07)11136-6.

Lisle, TE, Cui, Y, Parker, G *et al.* 2001. The dominance of dispersion in the evolution of bed material waves in gravel-bed rivers. *Earth Surface Processes and Landforms* 26, 1409–1420.

Lovett, RA 2014. Rivers on the run. *Nature* 511, 521–523.

Madej, MA, Sutherland, DG, Lisle, TE, and Prior, B 2009. Channel responses to varying sediment input: a flume experiment modeled after Redwood Creek, California. *Geomorphology* 103, 507–519.

Magirl, CS, Hilldale, RC, Curran, CA *et al.* 2015. Large-scale dam removal on the Elwha River, Washington, USA: Fluvial sediment load. *Geomorphology* 246, 669–686.

Major, JJ, O'Connor, JE, Podolak, CJ *et al.* 2012. Geomorphic response of the Sandy River, Oregon, to removal of Marmot Dam. *US Geological Survey Professional Paper* 1792, 64 pp.

Merritts, D, Walter, R, Rahnis, M *et al.* 2013. The rise and fall of Mid-Atlantic streams: millpond sedimentation, milldam breaching, channel incision, and streambank erosion. In *The Challenges of Dam*

Removal and River Restoration (eds JV De Graff and JE Evans). *Geological Society of America Reviews in Engineering Geology* 21, 183–203.

O'Connor, JE, Duda, JJ, and Grant, GE 2015. 1000 dams down and counting. *Science* 348(6234), 496–497.

O'Connor, J, Major, J, and Grant, G 2008. The dams come down – unchaining U.S. rivers. *Geotimes* 53(3), 22–28.

Pearson, AJ, Snyder, NP, and Collins, MJ 2011. Rates and processes of channel response to dam removal with a sand-filled impoundment. *Water Resources Research* 47, W08504. DOI:10.1029/2010WR009733.

Peck, JA and Kasper, NR 2013. Multiyear assessment of the sedimentological impacts of the removal of the Munroe Falls Dam on the middle Cuyahoga River, Ohio. In *The Challenges of Dam Removal and River Restoration* (eds JV De Graff and JE Evans). *Geological Society of America Reviews in Engineering Geology* 21, 81–92.

Pierson, TC and Major, JJ 2014. Hydrogeomorphic effects of explosive volcanic eruptions on drainage basins. *Annual Review of Earth and Planetary Science* 42, 469–507.

Pizzuto, JE 2002. Effects of dam removal on river form and process. *BioScience* 52, 683–691.

Podolak, CJP and Wilcock, PR 2013. Experimental study of the response of a gravel streambed to increased sediment supply. *Earth Surface Processes and Landforms* 38, 1748–1764.

Poff, NL and Hart, DD 2002. How dams vary and why it matters for the emerging science of dam removal. *BioScience* 52, 659–668.

Pohl, MM 2002. Bringing down our dams – trends in American dam removal rationales. *Journal of American Water Resources Association* 38, 1511–1519.

Pohl, MM 2003. American dam removal census – available data and data needs. In *Dam Removal Research – Status and Prospects* (ed. WL Graf). H.J. Heinz Center for Science, Economics, and Environment: Washington, DC; 29–39.

Pryor, BS, Lisle, TE, Montoya, DS, and Hilton, S 2011. Transport and storage of bed material in a gravel-bed channel during episodes of aggradation and degradation: a field and flume study. *Earth Surface Processes and Landforms* 36, 2028–2041.

Randle, TJ, Bountry, JA, Ricthie, A, and Wille, K 2015. Large-scale dam removal on the Elwha River, Washington, USA: Erosion of reservoir sediment. *Geomorphology* 246, 709–728.

Rumschlag, JH and Peck, JA 2007. Short-term sediment and morphologic response of the middle Cuyahoga River to the removal of the Munroe Falls Dam, Summit County, Ohio. *Journal of Great Lakes Research* 33, 142–153.

Santaniello, DJ, Snyder, NP, and Gontz, AM 2013. Using ground-penetrating radar to determine the quantity of sediment stored behind the Merrimack Village Dam, Souhegan River, New Hampshire. In *The Challenges of Dam Removal and River Restoration* (eds JV De Graff and JE Evans). *Geological Society of America Reviews in Engineering Geology* 21, 45–57.

Sawaske, S and Freyburg, DL 2012. A comparison of past small dam removals in highly sediment-impacted systems in the U.S. *Geomorphology* 151, 50–58.

Service, RF 2011. Will busting dams boost salmon?. *Science* 334(6058), 888–892.

Shakesby, RA and Doerr, SH 2006. Wildfire as a hydrological and geomorphological agent. *Earth-Science Reviews* 74, 269–307.

Skalak, K, Pizzuto, J, and Hart, DD 2009. Influence of small dams on downstream channel characteristics in Pennsylvania and Maryland – implications for the long-term geomorphic effects of dam removal. *Journal of the American Water Resources Association* 45, 97–109.

Skalak, K, Pizzuto, J, Egan, J, and Allmendinger, N 2011. The geomorphic effects of existing dams and historic dam removals in the U.S. Mid-Atlantic region. In *Sediment Dynamics upon Dam Removal* (eds AN Papanicolaou and BD Barkdoll). American Society of Civil Engineers: Reston, VA; 83–96.

Sklar, L, Fadde, J, Venditti, JG *et al.* 2009. Translation and dispersion of sediment pulses in flume experiments simulating gravel augmentation below dams. *Water Resources Research* 45, W08439. DOI:10.1029/2008WR007346.

Straub, TD 2007. *Erosion dynamics of a stepwise small dam removal, Brewster Creek Dam near St. Charles, Illinois*. PhD dissertation, Colorado State University.

Thomas, RJ, Constantine, JA, Gough, P, and Fussell, B 2014. Rapid channel widening following weir removal due to bed-material wave dispersion on the River Monnow, Wales. *River Research and Applications* 31, 1017–1027.

Tullos, D and Wang, H.-W 2014. Morphological responses and sediment processes following a typhoon-induced dam failure, Dahan River, Taiwan. *Earth Surface Processes and Landforms* 39, 245–258.

Tullos, DD, Finn, DS, and Walter, C. 2014. Geomorphic and ecological disturbance and recovery from two small dams and their removal. *PLoS One* 9(9), e108091. DOI:10.1371/journal.pone.0108091, 18 p.

US Army Corps of Engineers 2013. *National Inventory of Dams*. Available from http://nid.usace.army.mil (last accessed April 20, 2015).

Walter, C and Tullos, DD 2010. Downstream channel changes after a small dam removal – using aerial photos and measurement error for context; Calapooia River, Oregon. *River Research and Applications* 26, 1220–1245.

Warrick, JA, Bountry, JA, East, AE *et al.* 2015. Large-scale removal on the Elwha River, Washington, USA: Source-to-sink sediment budget and synthesis. *Geomorphology* 246, 729–750.

Wilcox, AC, O'Connor, JE, and Major, JJ 2014. Rapid reservoir erosion, hyperconcentrated flow, and downstream deposition triggered by breaching of 38 m tall Condit Dam, White Salmon River, Washington. *Journal of Geophysical Research – Earth Surface* 119, 1376–1394.

Wildman, LAS and MacBroom, JG 2005. The evolution of gravel bed channels after dam removal: Case study of the Anaconda and Union City Dam removals. *Geomorphology* 71, 245–262.

Wohl, EE and Cenderelli, DA 2000. Sediment deposition and transport patterns following a reservoir sediment release. *Water Resources Research* 36, 319–333.

Wood, PJ and Armitage, PD 1997. Biological effects of fine sediment in the lotic environment. *Environmental Management* 21, 203–217.

Zunka, JPP, Tullos, DD, and Lancaster, ST 2015. Effects of sediment pulses on bed relief in bar-pool channels. *Earth Surface Processes and Landforms* 40, 1017–1028.

Discussion

Discussion by Timothy J. Randle and Jennifer A. Bountry

This is a good comprehensive review and analysis. A comparison of numerical and physical model predictions with the results from actual dam removal would be helpful in identifying where models can provide useful and accurate results and the areas where models need to be improved.

Dam height is likely just a surrogate for reservoir sediment volume, especially if the reservoirs are full of sediment when the dams are removed. A V^* ratio of about 10 may be a more appropriate index for conditions when the overall downstream impact may be modest and not imply too much precision.

From a practical prospective, E^* may not be a "useful tool for quickly estimating possible spatial consequences of dam removals" because it is difficult to predict ahead of time, except for relatively small sediment volumes. V^* may be the more useful predictive index.

The mud portion of reservoir sediment and the width of the reservoir relative to the alluvial river width may be useful indices for predicting the percentage of reservoir sediment that erodes.

Discussion by Kimberly Hill

In the presentation of dam removal monitoring, the topic of size dependence of stored and transported sediment was discussed along with the geomorphic response of the river to the dam removal. There is both recent and historic evidence that the addition of fine particles to a gravel bed can affect the mobility of the bed and therefore might influence the geomorphic response. It would be interesting to know how and with what frequency: the smaller particles in transport and storage were monitored and geomorphic response was determined. Do you see any short-term or long-term evidence of ecological effects such as fines fouling or sand capping?

Discussion by Mike Miles

The authors indicate that the downstream response to dam removal is strongly influenced by the volume and caliber of the stored sediment. However, the volume of stored organic material is also potentially important.

Cleveland Dam is located on Coquitlam River near Vancouver (British Columbia) on the west (wet) coast of Canada. The dam was constructed in 1954 and the Coquitlam reservoir was drained in 1992 to facilitate seismic up-grading. Channel incision through the reservoir 'sediments' exposed substantial quantities of buried large wood, smaller wood pieces, and particulate organic matter as in the following figure; in places these materials comprised > 50% of the total thickness of reservoir deposits (which generally exceeded 2 m).

This observation has implications for how we estimate the volume of "sediment" deposited in a reservoir, as comparative surveys over time will overestimate the mineral constituent. The deposited organic material will affect biological oxygen demand (BOD) as it is flushed downstream and residual organic materials can create water quality issues. The organic materials also have the potential to adversely affect downstream engineering works, such as water infiltration plants (e.g., following dam removal on Elwha River in Washington State, USA; Tim Randle and Brian Cluer, pers. comm. 2015).

These observations prompt three specific questions:

1) Do the authors have any data on regional variations in the organic content of reservoir deposits?
2) Can you suggest how to determine the volume, caliber, and distribution of deposited organic material?
3) Do you have examples of how the erosion of deposited organic materials following dam removal has influenced downstream physical and biological processes?

Sediment, wood, and other organic material deposited in the reservoir behind Cleveland Dam on Coquitlam River near Vancouver (British Columbia).

Reply by the Authors

Randle and Bountry comment on four points: (i) Models: not all dam removals need to be modeled, and relatively few removals have been modeled, but we agree that comparisons of physical- and numerical-model predictions with outcomes from dam removals are needed for removals having the potential to advance scientific knowledge. Such comparisons have been done for some removals (e.g., Marmot Dam). The few model comparisons that have been published show promise for predicting the nature and pace of reservoir erosion and distribution of downstream sediment. Space limitation prevented us from delving into this topic. (ii) Dam height: there are two ways dam height influences geomorphic responses – it affects the amount of water and sediment in storage and it controls overall base-level fall. Sediment volume, however, does not necessarily correlate directly with dam height. Variations in upstream sediment delivery, poor trapping efficiency, diverse valley geometries, and episodic scour of reservoir sediment can complicate relations between dam height and sediment volume. (iii) V^* and E^*: given the small data set, an order-of-magnitude V^* threshold is perhaps more appropriate; E^* may be a useful predictor *if* first-year erosion *could* be roughly estimated, and *if* relations between E^* and sediment travel distance are robust beyond the small data set for which the metric was developed. More data are needed. (iv) Reservoir sediment composition and aspect ratio: Figure 13.3 shows relations among reservoir sediment composition, aspect ratio, and erosion. More data are needed before specific indices can be developed with confidence.

Regarding Hill's comments, several studies monitored post-removal transport of fine suspended sediment by directly measuring concentrations or by monitoring turbidity. However, such monitoring efforts typically last for relatively short time periods (a few months to a few years). As we noted (section 13.2.2.4) fine sand and mud can infiltrate gravels after dam removal and adversely affect ecological conditions despite negligible geomorphic impacts. Gravel mobility is enhanced when finer sediment, especially sand, is released by dam removal.

Miles points to a significant issue that needs study – the role of organics in dam removal. There are two types of organics of concern: large woody debris, which can affect reservoir erosion and downstream geomorphic response, and particulate organic matter, even small amounts of which can affect water quality. In section 13.2.1 we note that exhumed large woody debris can hinder reservoir erosion; it can also control the location of new bar formation downstream from a former dam site (e.g., East *et al.* 2015). We are unaware of any data that highlight regional variations of organic content in reservoir deposits. Without extensive coring or geophysical data from reservoir deposits, we do not know of any efficient way to estimate the volume, caliber, and distribution of organic material, and dam owners may not wish to invest in such surveys. However, this represents an important area of research.

14

Reservoir Sediment Flushing and Replenishment Below Dams: Insights from Japanese Case Studies

Tetsuya Sumi, Sameh Kantoush, Taymaz Esmaeili, and Giyoung Ock

14.1 Introduction

Dam construction disrupts the longitudinal continuity of the river system and interrupts the action of the belt conveyor of sediment transport (Kondolf 1997). Reservoirs can trap and permanently store virtually the entire sediment load delivered from the upstream basin (Petts 1979; Williams and Wolman 1984). Upstream of the dam, all bedload sediment and all or part of the suspended load are deposited in the stagnant water of the reservoir that results in reduction of reservoir capacity and its upstream reaches influenced by backwater. Current gross storage capacity in the world is 6000 km^3 with 45 000 large dams, and total storage loss and annual sedimentation rate are about 570 km^3 (12%) and 31 km^3/yr (0.52%/yr), respectively. If additional new development projects are not considered, total capacity will be decreased to less than half by 2100.

The rapid reservoir sedimentation not only decreases the storage capacity, but also increases the probability of flood inundation due to heightening of the bed elevations at the upstream end of the reservoir and the confluences of the tributaries (Liu *et al.* 2004b). Immediately downstream sediment load is greatly reduced. In addition, typical downstream changes in the flow regime include a reduction in the magnitude of peak flows and a possible increase in the magnitude of low flows (Williams and Wolman 1984). As a result, the downstream river may adjust in an attempt to re-establish an asymptotic equilibrium between the channel and the discharge with sediment load being transported. Possible adjustments include channel-bed erosion or deposition, channel widening or narrowing, and changes in channel pattern. Downstream impacts develop through discontinuity in downstream gradients, e.g., sediment supply, water quality, temperature, flow, and sediment regimes. Morphological effects on the river channel (e.g., Kondolf and Matthews 1993; Kantoush, Sumi, and Kubota 2010) include riverbed incision, riverbank instability, groundwater overdrafting, damage to bridges, embankments, and levees (e.g., Kondolf 1997; Batalla 2003), and changes in channel width. In order to solve these problems, an integrated approach is necessary for sediment management through flow and sediment regimes (Sumi and Kantoush 2010).

In many countries, various countermeasures have been implemented to decrease sediment accumulation and loss of storage capacity. These are: (i) reducing the sediment inflow by erosion control and upstream sediment trapping; (ii) routing sediments by sediment bypass facilities, off stream reservoirs,

Gravel-Bed Rivers: Processes and Disasters, First Edition. Edited by Daizo Tsutsumi and Jonathan B. Laronne.

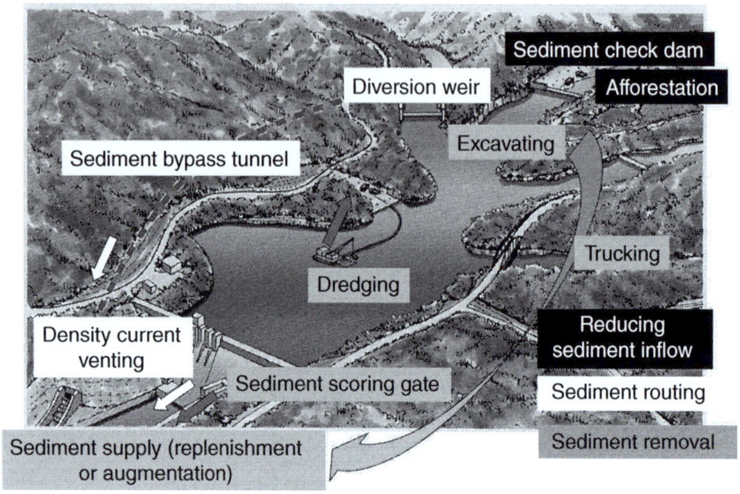

Figure 14.1 Reservoir sediment management measures in Japan.

sediment sluicing (drawdown routing), and venting of turbid density currents; and (iii) removing sediment by mechanical dry excavation, dredging, drawdown/pressure flushing and Hydrosuction Sediment Removal System (i.e., HSRS). Kantoush, Sumi, and Takemon (2011) reviewed several case studies in Japan and Kondolf *et al.* (2014) collected experiences from five continents. Current reservoir sediment management measures in Japan are illustrated in Figure 14.1. Among these methodologies, sediment bypass (Sumi *et al.* 2005) and drawdown flushing are considered to be permanent remedial measures for Japanese reservoirs. Feasibility conditions on drawdown flushing have been discussed by Atkinson (1996), White (2001), and Palmieri *et al.* (2003). Sumi (2008) reviewed sediment-flushing efficiency in the Kurobe River, Japan. In addition, sediment replenishment or augmentation after mechanical excavation has been intensively implemented in order to mitigate the adverse effects below dams by supplying mainly coarse sediments. Gaeuman (2014) reported coarse sediment augmentation on the Trinity River, and Bunte (2004) summarized gravel mitigation and augmentation below hydroelectric dams in salmon-bearing Pacific Coast gravel-bed rivers in the United States from a geomorphological perspective. Sakurai and Hakoishi (2013) have Japanese case studies and a numerical model to formulate the erosion and transport process of sediment replenishment.

In this chapter, current overviews of the mentioned measures regarding the reservoir sediment management as well as detailed planning and operational procedures associated with drawdown flushing and sediment replenishment are presented.

14.2 Present State of Reservoir Sedimentation in Japan

14.2.1 Reservoir Storage Loss

As of 2013, from 971 dams accounting for approximately one third of 3000 Japanese dams with height over 15 m, annual changes in sedimentation volume and the shape of accumulated sediment were reported (Sumi 2013). Probably, only Japan has established such a nationwide survey system, and has accumulated such amount of data as a considerably valuable record on a global basis.

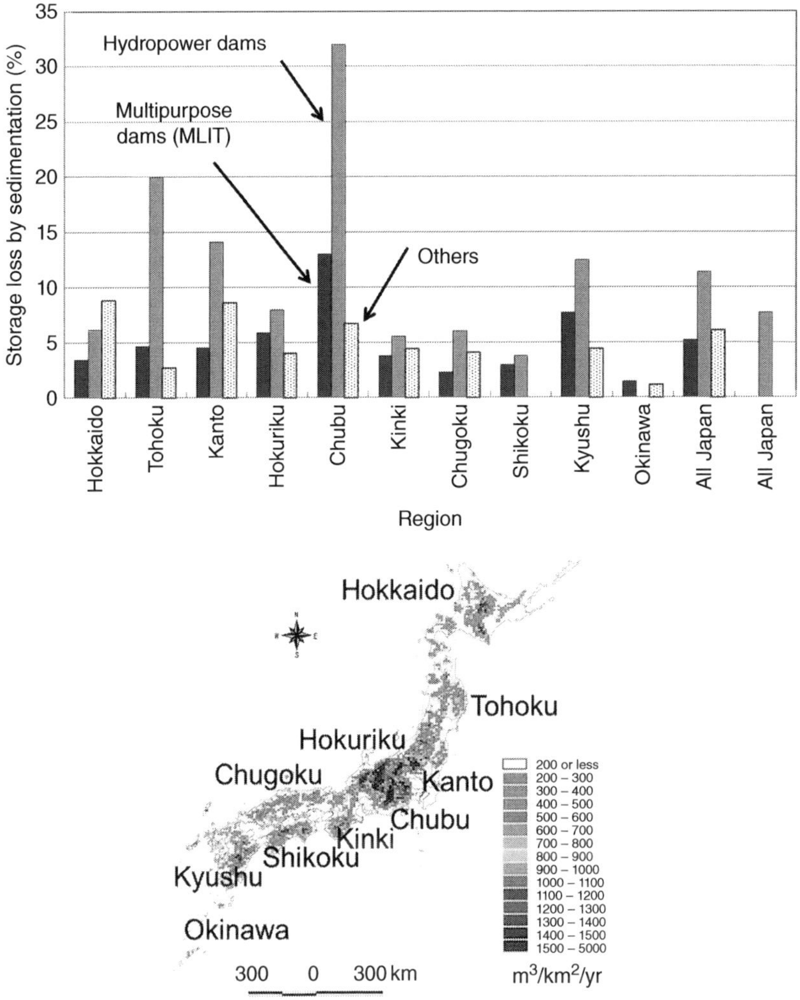

Figure 14.2 Reservoir storage loss by sedimentation and sediment yield potential map in Japan (After Sumi (2013) and Okano *et al.* (2004)).

Figure 14.2 shows reservoir storage losses by sedimentation depending on regions and categories based on purpose: three columns show multipurpose, hydropower, and other-purpose dams. Figure 14.2 also shows a "Sediment yield potential map of Japan" that has been extracted utilizing the GIS technique, which employs input data such as reservoir sedimentation records, existing geographical features, and geological data. (Okano *et al.* 2004). This map is currently used to check sedimentation planning for newly constructed dams and to estimate future sedimentation amount for existing dams.

14.2.2 Comprehensive Sediment Management in the Sediment Routing System

Nowadays, a new concept of sediment management is considered in Japan. River fluvial systems are composed of six segments: headwater, mountain, valley, alluvial fan, floodplain, and delta. The amount

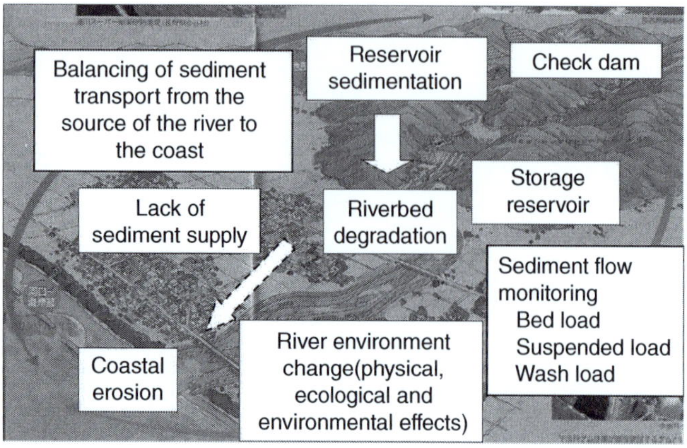

Figure 14.3 Comprehensive sediment management in the sediment routing system (Kantoush, Sumi, and Takemon 2011).

of sediment supplied from rivers to coasts was drastically reduced due to the construction of many check dams and storage reservoirs in mountainous areas and also because of the growth in gravel mining from riverbeds in Japan after World War II. As a result, various problems emerged such as riverbed degradation at downstream channel locations, fixing of river channels, erosion of coastal areas, and severe environmental changes along rivers and coastal areas. Environmental changes are largely dependent on armoring of the riverbed and lack of sediment transport (Williams and Wolman 1984). Such changes cause too much vegetation growth in the river channel and loss of suitable habitats for native aquatic species (Sumi and Kantoush 2010). Following a recommendation by the River Council of Japan in 1997, comprehensive sediment management has been proposed in order to recover sound sediment transport with regard to quantity and quality in the sediment routing system, as shown in Figure 14.3 (Japanese Ministry of Land, Infrastructure, Transport and Tourism 2005). The sediment routing system includes six segments and a coastal segment to which sediment has been transported there. For storage dams, sediment supply to the downstream river is strongly required in order to reduce the storage loss for reservoir sustainability and to mitigate adverse environmental impacts as much as possible for river restoration purposes.

14.3 Selecting Suitable Sediment Management Options

14.3.1 Classification of Sediment Management Measures

Sediment management in reservoirs is generally classified into the three approaches: (i) to reduce sediment inflow to reservoirs; (ii) to route sediment inflow to prevent accumulation in reservoirs; and (iii) to remove sediment accumulated in reservoirs. Sumi and Kantoush (2011) have classified selected examples in Japan and Europe according to these methodologies (Figure 14.4).

14.3.2 Promotion Strategy of Reservoir Sedimentation Management

In order to increase the number of good examples for reservoir sediment management, it is important to establish a guideline to select appropriate sediment management measures. ICOLD (2009) released

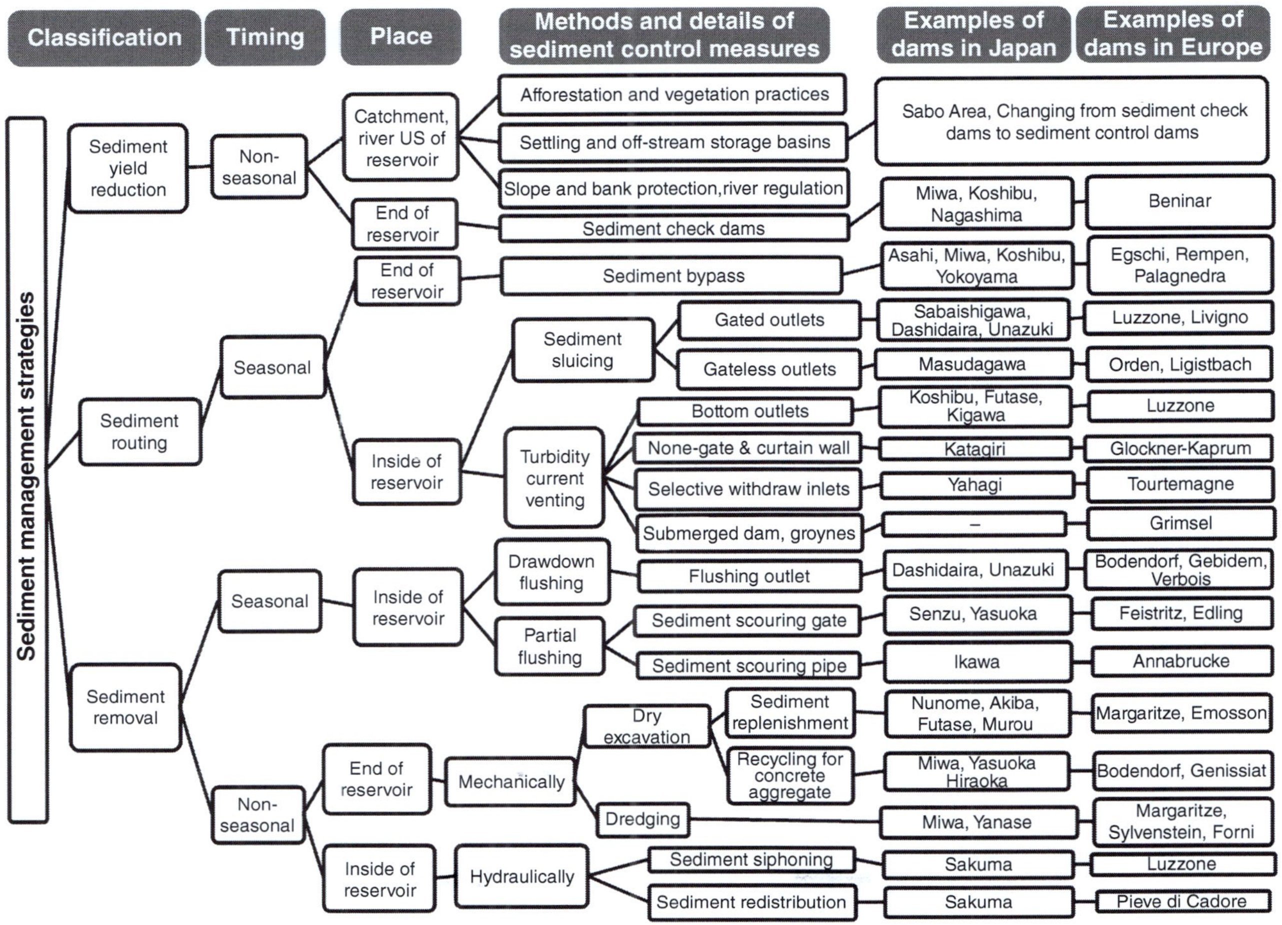

Figure 14.4 Classification of sediment control measures (Sumi and Kantoush 2011).

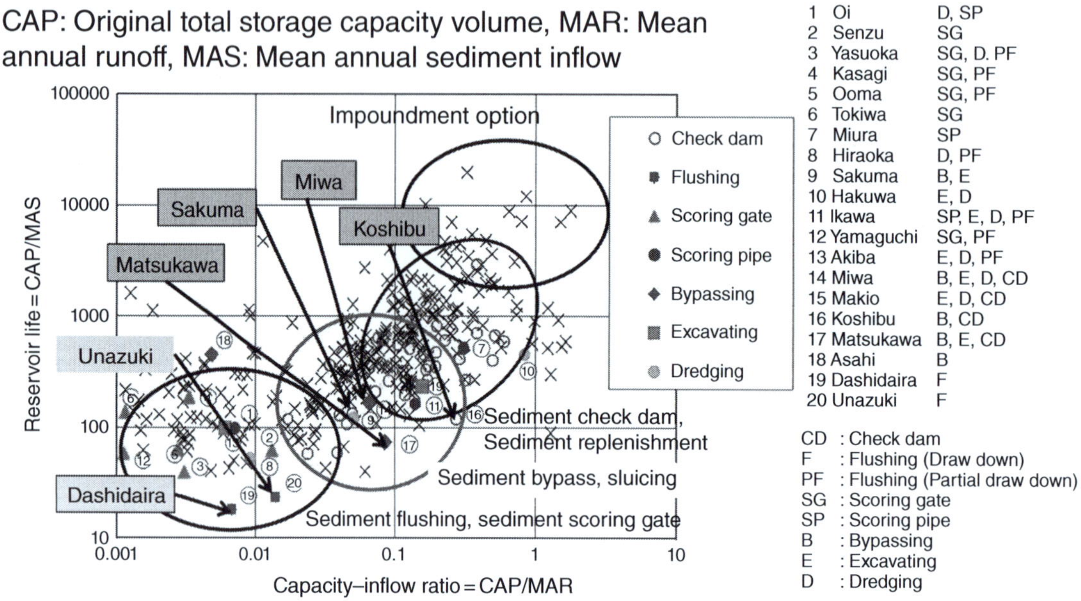

Figure 14.5 Representative sediment management measures in Japanese reservoirs (Sumi 2008).

Bulletin 147 entitled "Sedimentation and Sustainable Use of Reservoirs and River Systems." This bulletin discusses the upstream and downstream fluvial and morphological impacts of reservoir sedimentation and possible mitigation measures. In this bulletin, the current state and possible future sediment deposition in reservoirs have been investigated globally with the aid of the ICOLD Register on Dams. This bulletin also investigates the impacts of dams on the ecological features related to fluvial and morphological changes, and guidelines are proposed to try and mitigate the impacts on the downstream river morphology. Finally, the RESCON model (REServoir CONservation; Palmieri *et al.* 2003), which considers the life-cycle approach and reservoir conservation, is presented. This model, developed by the World Bank, is a prototype model to guide suitable options based on the conditions of the reservoirs. Sumi (2008) has analyzed Japanese dams based on the parameter of the turnover rate of water (CAP/MAR = total capacity/mean annual runoff) and sediment (CAP/MAS = total capacity/mean annual inflow sediment), as shown in Figure 14.5. It is thought that the selected sediment management measures can be classified by these two parameters. It is understood that selected measures actually change in the order of the sediment flushing, sediment bypass, sediment check dam, and excavating and dredging, as CAP/MAR increases (decrease in the turnover rate). This occurs owing to the considerable dependence of a selected sediment management measure on the amount of available water.

14.4 Sediment Flushing

14.4.1 Classification of Sediment Flushing

Among these approaches, flushing is considered as an economic approach to restore the storage capacity of reservoirs with severe deposition. "Sediment flushing" refers to an operation which

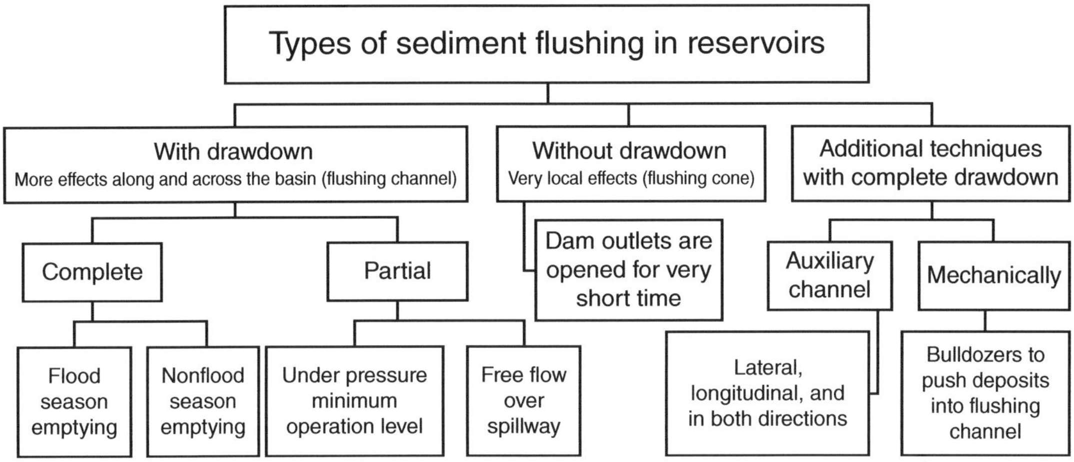

Figure 14.6 Classification of sediment flushing techniques (Kantoush *et al.* 2010a).

flushes sediment accumulated in the reservoir downstream through sediment flushing facilities using the tractive forces of the accelerated river flow. Basically, there are two types of flushing operations, with and without drawdown, and optional techniques can be also used with the complete drawdown flushing as shown in Figure 14.6 (Kantoush *et al.* 2010). Sediment flushing is performed at many dams all over the world, as shown in Table 14.1 (Sumi 2008). The necessary conditions for adopting sediment flushing in dam reservoirs are: 1-they are equipped with bottom outlets (sediment flushing outlets) through which the reservoir level is drawndown and flowing water can be discharged in an open channel during sediment flushing: 2-sufficient amount of water is secured for a series of operations of reservoir level drawdown, open channel discharge, and reservoir refill. Reservoir drawdown by opening the bottom outlet generates and accelerates unsteady flow towards the outlet (Morris and Fan 1998). This accelerated flow possesses an increased stream power, which consequently erodes a channel through the deposits and flushes the fine and coarse sediments through the outlet (Lai and Shen 1995, 1996; Shen 1999). During this process a progressive and a retrogressive erosion pattern can occur in the tail and delta reaches of the reservoir, respectively (Batuca and Jordaan 2000).

Sediment flushing is considered to be an extremely effective technique for discharging out sediment in terms of harnessing the tractive force in a natural river channel. However, when this technique is introduced, an extensive study is required in the planning stages, considering such conditions as inflow, sediment inflow, storage capacity, grain-size distribution, and reservoir operation. It should be also considered that information about flushing experience in the real scaled reservoirs with respect to flushing channel formation is scarce. One of the phenomena in reservoirs that is not well investigated and theoretically explained is the formation of flushing channels in the delta of the reservoir (Sloff, Jagers, and Kitamura 2004). These channels can be found in the deltaic deposition mainly in wide reservoirs. Another important aspect is to consider measures concerning environmental impacts under sediment flushing processes. In Japan, intensive studies on sediment flushing operations have been implemented in the Dashidaira and the Unazuki reservoirs in the Kurobe River.

Table 14.1 Global distribution of sediment-flushing in dam reservoirs.

Name of dam	Country	Year dam completed	Dam height (m)	Initial storage capacity (*CAP*) (million m^3)	Mean annual sediment inflow (*MAS*) (million m^3)*	1/(mean turnover ratio) (= *CAP/MAR*)	Reservoir life (= *CAP/MAS*)	Average flushing discharge (m^3/s)	Flushing duration (hours)	Flushing frequency (per year)
Dashidaira	Japan	1985	76.7	9.01	0.62	0.00674	14.5	200	12	1
Unazuki	Japan	2001	97	24.7	0.96	0.014	25.7	300	12	1
Gebidem	Switzerland	1968	113	9	0.5	0.021	18.0	15	70	1
Verbois	Switzerland	1943	32	15	0.33	0.00144	45.5	600	30	3
Barenburg	Switzerland	1960	64	1.7	0.02	0.000473	85.0	90	20	5
Innerferrera	Switzerland	1961	28	0.23	0.008	0.00018	28.8	80	12	5
Genissiat	France	1948	104	53	0.73	0.00467	72.6	600	36	3
Baira	India	1981	51	9.6	0.3	0.00489	32.0	90	40	1
Gmund	Austria	1945	37	0.93	0.07	0.00465	13.3	6	168	NA
Hengshan†	China	1966	65	13.3	1.18	0.842	11.3	2	672	2–3
Santo Domingo	Venezuela	1974	47	3	0.08	0.00667	37.5	5	72	NA
Jen-shan-pei†	Taiwan	1938	30	7	0.23	NA	30.4	12.2	1272	1
Guanting	China	1953	43	2270	60	1.5	37.8	80	120	NA
Guernsey	United States	1927	28.6	91	1.7	0.0433	53.5	125	120	NA
Heisonglin	China	1959	30	8.6	0.7	0.6	12.3	0.8	72	NA
Ichari	India	1975	36.8	11.6	5.7	0.00218	2.0	2.16	24	NA
Ouchi-Kurgan†	Former USSR	1961	35	56	13	0.00376	4.3	1000	2400	NA
Sanmenxia†	China	1960	45	9640	1600	0.224	6.0	2000	2900	NA
Sefid-Rud†	Iran	1962	82	1760	50	0.352	35.2	100	2900	NA
Shuicaozi	China	1958	28	9.6	0.63	0.0186	15.2	50	36	NA

*Average after dam completion.
†Sluicing dams.

14.4.2 Case Study in the Kurobe River, Japan

14.4.2.1 Outline of the Kurobe River and Planning of Coordinated Sediment Flushing

The Kurobe River in the eastern region of Toyama Prefecture is a representative steep river in Japan that stretches over 85 km in a 682 km^2 drainage basin (Figure 14.7). The Unazuki Dam (completed in 2001, height 97 m, gross capacity of reservoir 24 700 000 m^3), which is under the control of Ministry of Land, Infrastructure, Transport and Tourism, is located at the farthest point downstream of the Kurobe River. The Dashidaira Dam (completed in 1985, height 76.7 m, gross capacity of reservoir 9 010 000 m^3) owned by Kansai Electric Power Co. Ltd., is located upstream of the Unazuki Dam. These two dams have an extremely large amounts of sediment inflow compared to their gross storage capacity; therefore, they were the first cases in Japan that were built with full-scale sediment flushing facilities (sediment flushing gates). Sediment flushing has been conducted in the Dashidaira reservoir since 1991. Since 2001, when the Unazuki Dam was completed, *sediment flushing* and *sediment sluicing* have been conducted coordinately for the two dams (Figure 14.7). Here, *sediment flushing* refers to a drawing down operation just after the peak water flow in the first flood event of the year. *Sediment*

(a)

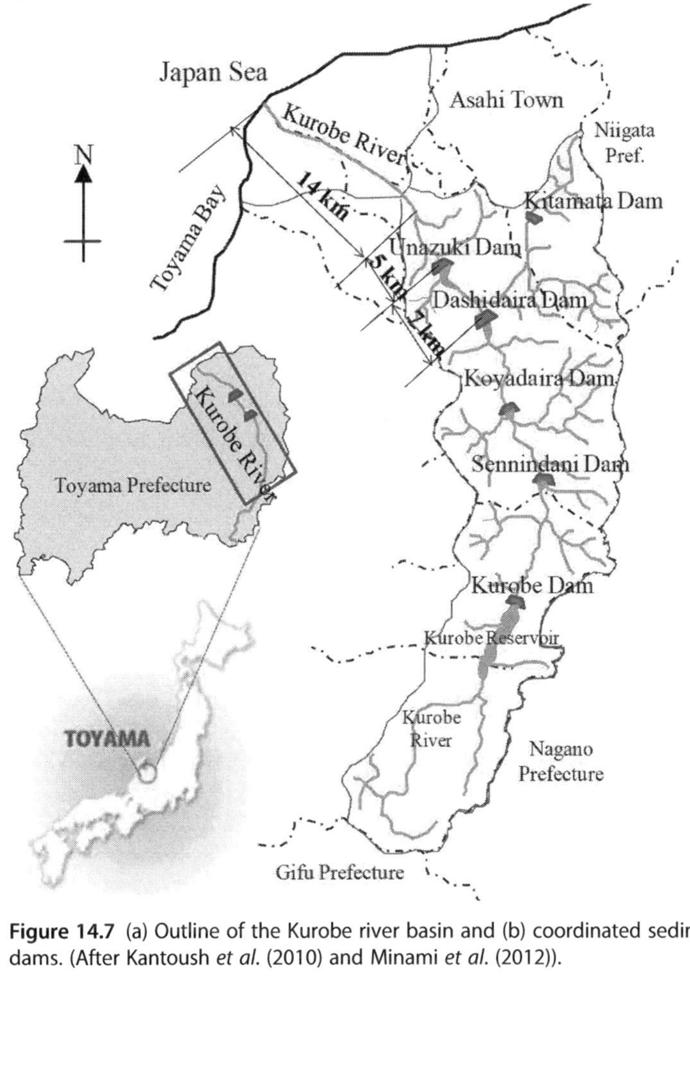

Figure 14.7 (a) Outline of the Kurobe river basin and (b) coordinated sediment flushing of Dashidaira and Unazuki dams. (After Kantoush *et al.* (2010) and Minami *et al.* (2012)).

(b)

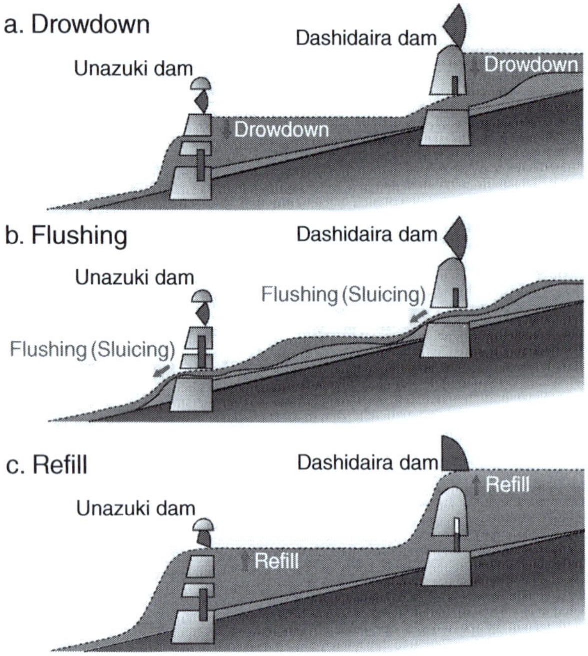

Figure 14.7 (*Continued*)

sluicing refers to an operation to sluice the additional sediment flowing into a reservoir after sediment flushing, using an operation similar to the sediment flushing.

Kokubo, Itakura, and Harada (1997), Liu *et al.* (2004a, b), Sumi and Kanazawa (2006), Sumi (2008), Sumi, Nakamura, and Hayashi (2009) and Minami *et al.* (2012) presented an overview of coordinated sediment flushing operations at the Dashidaira and the Unazuki Dams, which are typical sediment flushing sites in Japan. While it is important to maintain high sediment flushing efficiency in a sediment flushing operation, it is also necessary to minimize the environmental impacts of flushed out sediment in downstream areas. For this reason, the recent sediment flushing results from the Dashidaira Dam have been analyzed to evaluate the sediment flushing efficiency and the quality of flushed out water using discharged suspended sediment concentrations (i.e., SS).

14.4.2.2 Results of Coordinated Sediment Flushing

Since the completion of the Unazuki Dam in 2001, sediment flushing and sediment sluicing operations have been conducted almost annually (Figure 14.8). When a flood inflow discharge exceeds 300 m^3/s (250 m^3/s in some special cases) at the Dashidaira Dam for the first time of the year between June and July, a coordinated sediment flushing is performed. When a flood inflow discharge exceeds 480 m^3/s at the Dashidaira Dam after sediment flushing, a sediment sluicing is performed. These sediment flushing and sluicing practices have been conducting in coordination with the Kurobe River Sediment Flushing Evaluation Committee and the Kurobe River Sediment Management Council, monitoring the natural flow regime in the Kurobe River as well as the impacts of sediment discharge downstream.

(a)

(b)

Figure 14.8 Photographs of (a) Unazuki and (b) Dashidaira dams during coordinated sediment flushing operation (Sumi *et al.* 2009).

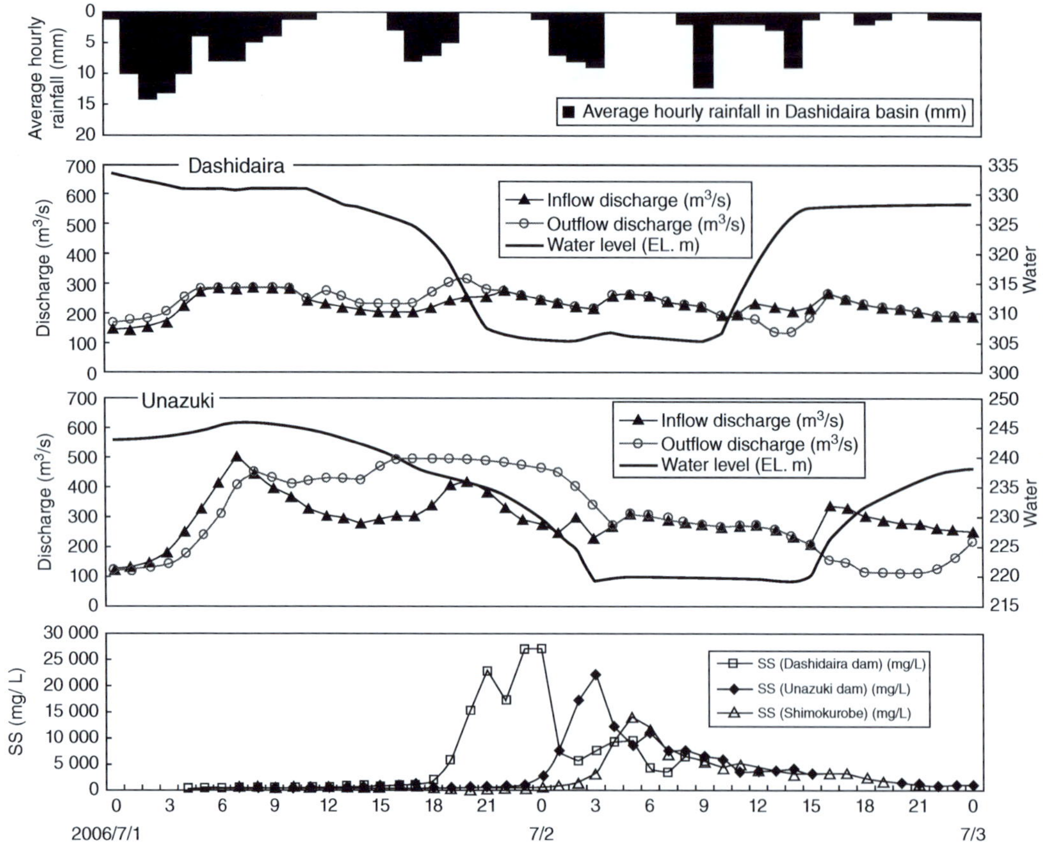

Figure 14.9 Recorded available data during the coordinated sediment flushing operation in 2006 (Sumi *et al.* 2009).

Sediment flushing operation includes the following procedures: drawing down the reservoir water level, keeping free-flow state for several hours, and recovering water level. The amount of time for the free-flow sediment flushing operation depends largely on the target amount of sediment to be flushed out, which is planned before the sediment flushing operation. Figure 14.9 exemplifies a sediment flushing operation performed in July 2006. A free-flow state was continued for 12 h to flush 240 000 m^3 of deposited sediment.

Table 14.2 shows the amount of sediment flushed out of Dashidaira Dam from 2001 to 2007. Sediment flushing is performed during the first big flood in the rainy season (i.e., between June and July) with a peak discharge about 500 m^3/s and sediment sluicing is conducted with a peak discharge about 700 m^3/s. The average drawdown period is 12 h for both operations. The flushing volume data was obtained by measuring the cumulative amount of sediment in the dam reservoir after the previous year's sediment flushing operation till end of May. This quantity fluctuates depending on flooding events outside the sediment-flushing season from autumn to spring.

The major outcomes of sediment flushing operations up to now are as follows.

1) At the Kurobe River, both sediment flushing and sluicing are performed approximately once a year without major environmental impacts (Sumi and Kanazawa 2006) by restricting sediment flushing operations to appropriate sediment-flushing seasons (June to August), and times when the natural

Table 14.2 Sediment flushing and sluicing operations at the Dashidaira Dam (Sumi, Nakamura, and Hayashi 2009).

Year and operation	Maximum discharge inflow (m^3/s)	Average discharge inflow (m^3/s)	Flushing volume (10^3 m^3)	Maximum SS (mg/L)	Average SS (mg/L)
2001 Flushing	333	277	590	90 000	15 000
2001 Sluicing	491	273		29 000	6700
2002 Flushing	362	215	60	22 000	4500
2003 Flushing	777	217	90	69 000	7100
2004 Flushing	356	229	280	42 000	10 000
2004 Sluicing	1152	281		16 000	7300
2005 Flushing	958	290	510	47 000	17 000
2005 Sluicing 1	835	275		90 000	16 000
2005 Sluicing 2	790	250		40 000	7300
2006 Flushing	308	246	240	27 000	6500
2006 Sluicing 1	378	203		12 000	2500
2006 Sluicing 2	685	264		27 000	5200
2006 Sluicing 3	529	196		7400	1800
2007 Flushing	418	245	120	25 000	3500
Average of flushing	502	246	270	46 000	9100
Average of sluicing	694	249		31 600	6700
Average of all data	598	247	270	38 800	7900

water flow rate exceeds a certain level, 300 m^3/s (250 m^3/s in some special cases) or more for sediment flushing, and 480 m^3/s or more for sediment sluicing.

2) The sediment-flushing efficiency at the Dashidaira Dam, which is calculated by using the quantity of flushed sediment and the quantity of water used during the free-flow phase, was about 0.01–0.03.
3) Discharged SS at the Dashidaira Dam during a sediment flushing operation was 8000 mg/L on average (40 000 mg/L at the peak), averaged over 14 sediment flushing operations. The average SS values are closely related to the quantity of flushed sediment as well as the quantity of water used during a flushing operation. Furthermore, peak SS depended on the rate of water level drawdown in the reservoir.
4) The Dashidaira Dam is currently at an equilibrium state in terms of its sediment, and the quantity of sediment passing through is approximately one million cubic meters annually. In contrast, sediment is still being accumulated at the Unazuki Dam since it is not in a conditions of equilibrium. While the majority of sediments of grain sizes larger than 2 mm are trapped in the reservoir, about 70% of the sediment, which is mostly of grain size smaller than 2 mm, is sluiced.
5) Active sand bars are observed in the river channels downstream, which are considered as a positive effect of sediment transport process during a coordinated sediment flushing operation in two successive dams. In particular, the supply of sandy materials has caused riverbed aggradation in some sections and has reversed armoring at all sections.

6) Sediment flushing operations ensure that the surface layer of accumulated sediment on the bottom of the reservoir is continually replaced with fresh sediments, decreasing the organic materials and the eutrophication indices.
7) In order to prevent accumulation of fine sediment on the sand bars in the downstream river channel after sediment flushing operations, a rinsing discharge from both dams is practiced. This is particularly effective to reduce the environmental impact after sediment flushing operations when a large volume of fine sediments has been discharged.
8) Evacuation channels have been prepared as shelters for many species of fish, such as Ayu (*Plecoglossus altivelis*), during the high turbidity in the main river due to a sediment flushing operation. They are showing a high performance by securing clear water sources to these channels during the flushing operation.

14.4.3 Improvement of Sediment Flushing

14.4.3.1 Relevant Key Factors for Sediment Flushing

In order to improve the sediment flushing operation, the following three relevant key factors should be studied.

1) Flushing efficiency: scoured sediment volume/water used.
2) Flushing effect: scoured sediment volume/total deposited sediment volume before flushing.
3) Environmental impacts: dissolved oxygen (DO) decreases rapidly with increase in SS concentration and fine sediment deposition on the downstream riverbed

14.4.3.2 Flushing Efficiency

There is a high interest to measure the efficiency of sediment flushing operations. There is a limited number of dams, shown in Table 14.1, surveyed in detail for collecting the sediment flushing data. Sumi (2008) evaluated efficiency of reservoir sediment flushing in the Kurobe River and compared the results with other dams in the world, such as the Gebidem and the Verbois dams in Switzerland, and the Baira dam in India, where the flushing data were recorded. In this case, the water consumption for sediment flushing has been calculated only during the full drawdown period, although discharge of fine sediment may start during the reservoir drawdown period.

Figure 14.10 shows sediment flushing efficiency ($F_e = S/W$) calculated by flushed out sediment volume (S) and the water consumption volume (W). Among these, the sediment flushing efficiency in the Gebidem dam is comparatively high since sediment flushing is performed with a low flow discharge for a long time. Moreover, in the Baira dam, flushing efficiency is also comparatively high although there is some variation. On the other hand, since the Verbois dam is located at the mainstream of the Rhone River and the sediment flushing is executed with a large amount of water from Lac Leman, the sediment flushing efficiency is not high. In the Dashidaira dam, the sediment flushing efficiency is also not so high similar to the Verbois dam.

In these four dams, sediment flushing is strictly managed in the Dashidaira and the Verbois dams through preventing any remarkable water quality change to the downstream river by maintaining a sufficient amount of water relative to sediment. Considering the downstream river environment implies the application of sufficient volume of water that suppresses the flushing efficiency. In the Kurobe River, the sediment flushing is executed by securing enough river discharge just after the natural floods. Also, the additional discharge has been recently introduced to wash out the fine sediment silted in the downstream river channel after sediment flushing operations; thus sediment flushing is using a larger volume of water.

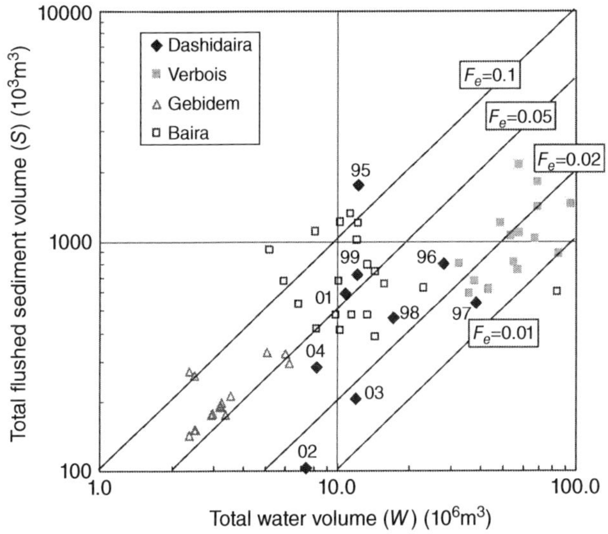

Figure 14.10 Total used water volume and flushed out sediment volume in sediment flushing (Sumi 2008).

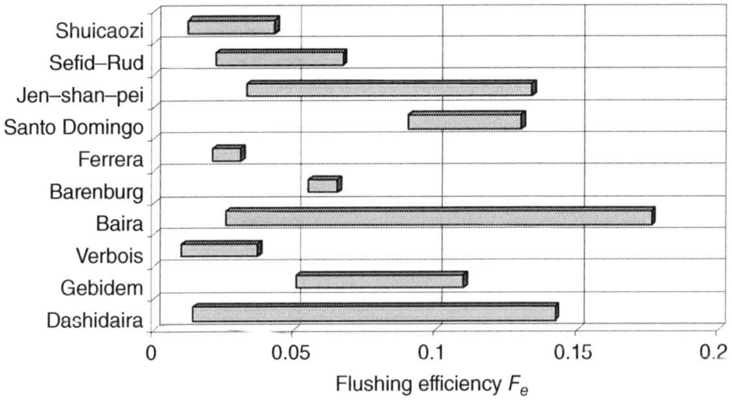

Figure 14.11 Variation range of sediment flushing efficiency in dams with sediment flushing operation (Sumi 2008).

The sediment flushing efficiency F_e of other dams is shown in Figure 14.11. The variation range of flushing efficiency is between 0.01 and 0.15 for most of the cases. There are also some cases with the flushing efficiency less than 0.05, which relates to flushing operations with less harmful environmental impacts. According to research on the feasibility evaluation of the sediment flushing, a feasible range of the sediment flushing can be obtained by the following equation using the parameter shown in Figure 14.5 (Sumi 2008). Here, the sediment flushing efficiency and the proportion of water consumption by the sediment flushing to the mean annual runoff (*MAR*) volume are shown by F_e and β, respectively. Here, *CAP* and *MAS* are the total capacity and the mean annual inflow sediment, respectively

$$\frac{CAP}{MAS} > \frac{\dfrac{CAP}{MAR}}{F_e\left(\beta - \dfrac{CAP}{MAR}\right)} \tag{14.1}$$

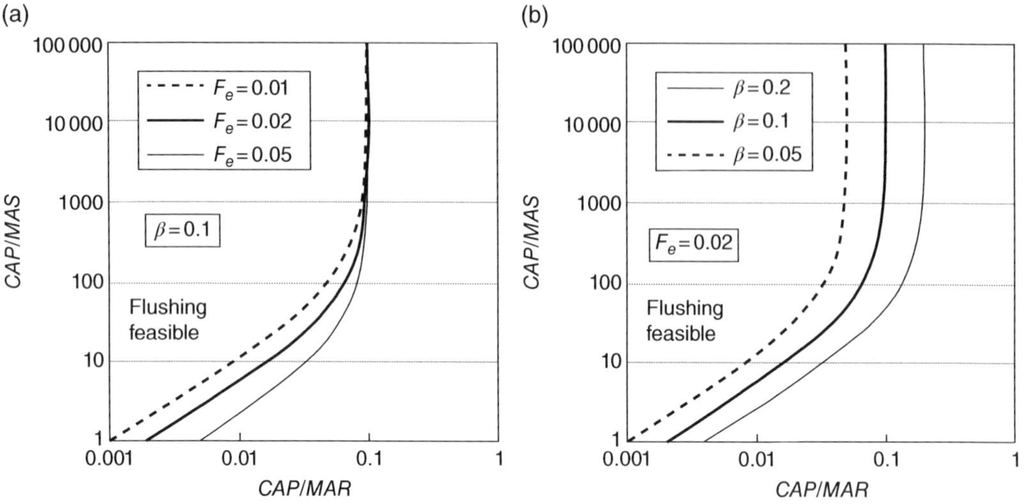

Figure 14.12 Extracted diagrams using equation 14.1 when (a) the proportion of water consumption to the mean annual runoff volume β is fixed; (b) sediment flushing efficiency F_e is fixed.

In Figure 14.12(a) and (b), a feasible range of the sediment flushing in the case that F_e changes to 0.01, 0.02, and 0.05 under fixed $\beta = 0.1$, and also in the case that β changes to 0.05, 0.1, and 0.2 under fixed $F_e = 0.2$ are shown, respectively. Feasible ranges are shown on the left-hand side of each line. According to which, change in F_e mainly influences within a small range of *CAP*/*MAS* and even a small turnover rate of the reservoir, e.g. large *CAP*/*MAR* introduces a possible rise of F_e under constant β. If river environmental conservation is considered, the possible range of the sediment flushing becomes narrower because it should estimate a low F_e. However, if β can be increased, the possibility of sediment flushing will increase under the same F_e since the water volume ratio that can be used for sediment flushing increases.

14.4.3.3 Flushing Channel Formation in the Deposited Reservoir Sediment

Kantoush and Schleiss (2009) conducted an experimental study on channel formation during the flushing of shallow reservoirs with different geometries. The flushing efficiency changes widely by various factors such as reservoir configuration, elevation of sediment flushing gates, volume and grain size of deposited sediment, discharge rate during a sediment flushing, and duration from the start of drawdown flushing. During the flushing processes two types of erosion patterns can occur; a progressive erosion in the tail reach of the reservoir, and a retrogressive erosion in the delta reach of the reservoirs. In order to predict the flushing channel width and location, it is important to understand the erosion process of sediment inside reservoirs.

Figure 14.13 shows erosion processes of deposited sediment in the Dashidaira reservoir in 2004. Longitudinal and lateral erosion were formed by riverbed degradation and side-bank erosion. A meandering flushing channel in the accumulated sediment was also formed. Not all the sediment is flushed out during the free flow duration. However, the sediment on both banks gradually falls by stream bank erosion. During very high discharges in 2004 the channel was able to break through the sand bar with bend cut-off and forming a wide channel section. Nevertheless, these cut-off events can occur only when water level at the dam in not completely drawn down. Simultaneously, the flow

Figure 14.13 Flushing channel formed in the Dashidaira reservoir. This photograph was taken when the water level had completely drawn down (Kantoush *et al.* 2010).

becomes concentrated in the existing deep channel. The flushing operation has a significant contribution to maintain the storage capacity, which extends reservoir lifetime.

The key parameters of the flushing channel when the reservoir is fully drawn down are location, width, side and longitudinal slopes, as well as shape. Some of them have been investigated experimentally in shallow reservoir geometries with different shapes (Kantoush and Schleiss 2009). The width of the flushing channel W_f is estimated as proposed by the RESCON model (Palmieri *et al.* 2003) and Atkinson (1996). To predict W_f, Atkinson (1996) suggests the empirical relationship ($W_f = KQ_f^{0.5}$) based on prototype measurements in four reservoirs of China, United States and India. In this equation, W_f is the flushing-channel width, K is a coefficient dependent on bed material type (in this case K is 12.8) and Q_f is the discharge that will contribute to flushing channel formation.

Figure 14.14 shows the relationship and a comparison of observed flushing-channel widths in the Dashidaira reservoir with four other reservoirs. The measured flushing-channel width in the Dashidaira reservoir was about 170 m at a distance of 640 m from the dam under a discharge of 250 m^3/s. Therefore, Kantoush *et al.* (2010a) suggested that value of K should be reduced to half for Japanese reservoirs (i.e., 6.4 for K).

14.4.3.4 Modelling of Sediment Flushing

There are studies on sediment flushing simulation using one-, and two-dimensional models. One-dimensional numerical modeling usually uses some assumptions and simplifications that could not represent the real nature of the complex interacting flow–sediment field during the flushing process. More advanced two-dimensional numerical models were used by Olsen (1999) and Badura, Knoblauch, and Schneider (2008) for simulating the flushing in both a physical model and prototype scale studies. Two-dimensional depth-averaged numerical models can not directly simulate a complex three-dimensional flow field, including secondary flows in channel bends, although they have made a strong contribution in natural sediment transportation processes.

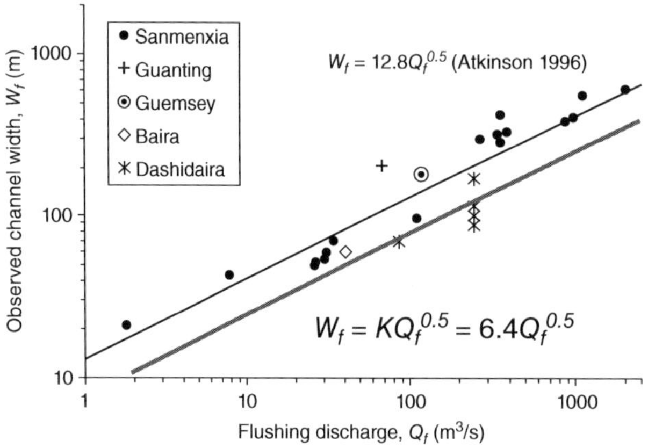

Figure 14.14 Comparison of Dashidaira dam's flushing-channel width with Atkinson's (1996) relationship (Kantoush *et al.* 2010).

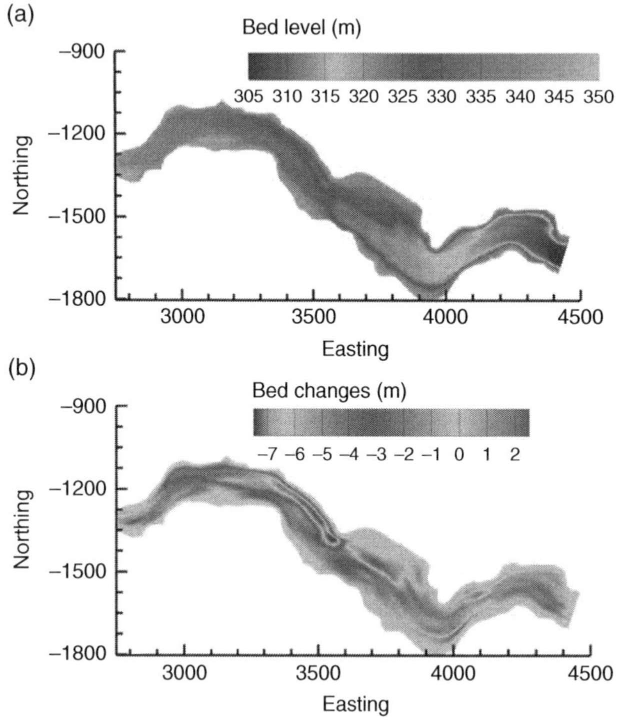

Figure 14.15 (a) Measured bed topography of Dashidaira reservoir after the flushing operation in June 2012; (b) corresponding measured bed changes after the flushing operation (Esmaeili *et al.* 2015). (See insert for color representation of this figure.)

Three-dimensional numerical models are still under development for application in this field. Khosronejad *et al.* (2008) and Haun and Olsen (2012a,b) used three-dimensional models for simulating the flushing process in the physical model and prototype scale, respectively. The complexity of three-dimensional flow patterns is better highlighted over the existing bedforms, on the deformed bed, during the free-flow state (Haun and Olsen 2012a; Esmaeili *et al.* 2014). Thus, application of three-dimensional numerical models is essential when the velocity variation over the flow depth plays a major role (e.g. in channel bends). Recently, application of three-dimensional numerical models became feasible for simulating the sediment transport process owing to the increase in calculation power of computers. Examples are the sediment transport simulation in Three Gorges project (Fang and Rodi 2003), simulation of the flushing process in rectangular reservoirs (Esmaeili, Sumi, and Kantoush 2014), Bodendorf reservoir (Haun *et al.* 2012), Angostura reservoir (Haun and Olsen 2012b; Haun *et al.* 2013), and also in an Alpine reservoir (Harb *et al.* 2014).

In order to develop structural or nonstructural methodologies to increase the flushing effect by controlling flushing channel formation, Esmaeili *et al.* (2015) studied sediment flushing processes in the Dashidaira reservoir during 2012. In the study, a computational fluid dynamic (CFD) code, SSIIM 2, developed by Olsen (2014) was used to perform the numerical simulations. The measured bed levels before and after the flushing operation were used to set up and validate the calculated morphological bed changes. Figure 14.15 demonstrates the measured bed levels after the flushing operation, together with the bed changes that were extracted by comparing the measured

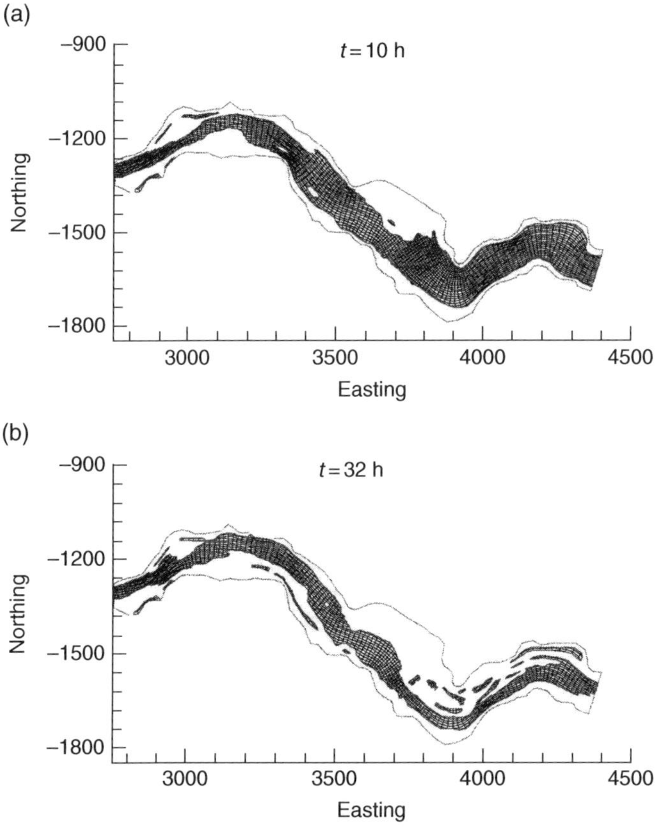

Figure 14.16 Computational grid at (a) the beginning of preliminary drawdown stage ($t = 10$ h) and (b) during the free-flow condition ($t = 32$ h) (Esmaeili *et al.* 2015).

bed levels before and after flushing. As can be seen, there is a significant bed erosion along the left embankment of lower area in the upstream half of the reservoir. Similar to the observations in the prototype, a distinctive flushing channel appeared in the numerical simulations during the free-flow condition. Figure 14.16 shows the computational grid adjustment at the beginning of preliminary drawdown stage, and also within the free-flow condition with a low water head in the reservoir. Cells with a lower water head than a specified value are removed from the computational domain due to the employed wetting/drying algorithm. Outputs reveal that flow deflects to the right-hand side in the middle part of the reservoir during the free-flow condition and, therefore, the flushing channel location is close to the right bank. Using this model, the effect of additional artificial discharge that can be supplied from upstream reservoirs during the free flow state was tested. It showed considerable effects on the bed changes and flushed-out sediment volume by increasing the flushing efficiency.

14.4.4 Environmental Impacts

Historically, adverse effects of the fine sediment on aquatic ecosystems have been largely reviewed (e.g. Newcombe and MacDonald 1991; Newcombe and Jensen 1996; Bilotta and Brazier 2008). Particularly, environmental impacts of sediment flushing have been studied in terms of water quality, benthic communities, fish mortality, and the modification of riverbed substrate (e.g., Gray and Ward 1982; Rambaud *et al.* 1988; Garric, Migeon, and Vindimian 1990; Wohl and Cenderelli 2000; Crosa *et al.* 2010). Adverse environmental impacts by sediment flushing can be summarized as rapid DO decrease with the increase in SS concentration during the free-flow discharge. Fine sediment deposition on the downstream riverbed has been recognized as one of the major environmental impacts.

The fine sediment can jam fish gills, thus greatly degrade their capability to absorb oxygen from water. Effects of the sediment flow on fish generally depend on comprehensive factors such as SS, DO, sediment particle grading, exposure duration, and fish category. Garric, Migeon, and Vindimian (1990) reported lethal effects of water quality degradation on brown trout caused by dam sediment flushing. Gerster and Rey (1994) proposed some implementing guidelines based on experiences of sediment flushing in Switzerland and France, and Staub (2000) pointed out that hypoxia would immediately become a critical factor for living fish when DO was below 2 mg/L or the SS exceeded 30 000 mg/L. Merle (2000) concluded that hypoxia (DO < 2 mg/L) showed more significant effects than high SS (> 30 000 mg/L) on the ecosystem. Crosa *et al.* (2010) proposed that the highest SS level should be controlled below 10 000 mg/L during sediment flushing periods in order to prevent significant negative impacts on the ecosystem of an alpine stream.

Bouchard (2000) proposed a one-dimensional sediment transport model (COURLIS) to predict the impact of SS during a reservoir emptying and flushing considering erosion of deposited cohesive sediment in reservoirs. Valette and Jodeau (2012) illustrated case studies of emptying and flushing of reservoirs using this model. Sumi and Kanazawa (2006) assessed environmental influences of the SS based on observations during sediment flushing periods in the Kurobe River, Japan with the stress index method (Newcombe and MacDonald 1991; Newcombe and Jensen 1996). In order to minimize environmental impacts, Sumi, Nakamura, and Hayashi (2009) discussed possible mitigation measures that provide evacuation channels for fish during the sediment flushing operation near downstream riverbanks, and also proposed rinsing discharge with clear water to minimize the long-term effect of fine sediment deposition on the riverbed after sediment flushing.

14.5 Sediment Replenishment

14.5.1 Definition and Objectives

Sediment replenishment basically consists of dredging or excavating the sediments accumulated in reservoirs, transporting them to the reaches just below the dam, and distributing the sediments along downstream channels by the following natural or artificial peak flows. The sediment sorting process occurs within a reservoir according to particle size. Coarse particles mainly deposit in the upstream area and finer particles can reach the hydraulic structure (Okano 2004; Knoblauch 2006). This size distribution process within a reservoir can be applicable to prepare selectively for downstream replenishment, optimal material, because coarser sediments (gravel and sand) are more beneficial for river systems than silt, which makes substratum clogging and causes high turbidity (Hartmann 2009). Check-dams, built upstream of main dams, can trap coarser sediments before entering a reservoir and facilitate their removal by land-based excavation. Thus, it does not require any water-level modification in the larger reservoir (Okano 2004). Currently, Japan is a leading country in the field of sediment replenishment application, with nearly 25% of dams resorting to sediment excavation for downstream replenishment.

Bunte (2004) reviewed gravel augmentation projects which have been conducted below hydroelectric dams in salmon-bearing Pacific Coast gravel-bed rivers in the United States from a geomorphological perspective. Because in the region, salmon habitats became severely degraded since large dams were constructed, resulting in the dramatic decline of salmon populations over the past several decades. In order to mitigate the poor quantity and quality of spawning habitat, spawning-sized gravels are artificially added to channels downstream of dams.

Gravel augmentation for enhancing spawning riffles has been undertaken downstream from at least 18 dams in California (Kondolf and Matthews, 1993; Bunte, 2004; Kondolf *et al.*, 2014). Among them, the Trinity River in Northern California is the focus of one of the most comprehensive programs to restore salmonid habitat by a combination of ecological flow and coarse sediment augmentation which began in 1976 and continues to date. By virtue of the combination approaches, the Trinity River Restoration Program (TRRP) has been at the forefront of efforts to restore bedload supply and transport in the regulated reaches below dams in the United States (Gaeuman 2012), developing systematic sediment management techniques and accumulating empirical data that cover various aspects of physical and ecological researches.

Ock, Sumi, and Takemon (2013) compiled sediment replenishment activities comparing cases of Japan and the United States. Though they share a similar concept, the implementation techniques have been developed based on both site-specific requirements and restrictions, for instances, mainly for reservoir sedimentation management or more for focusing on downstream river restoration. Compared with other sediment management strategies in reservoirs such as sediment flushing, bypassing, and sluicing options, sediment volume for downstream replenishment is very limited and insufficient to make up for the sediment deficit in rivers caused by dams. This is because of many restrictions in trapping volume in the upstream of reservoirs, trucking capacity, acceptable dumping sites, and transport capacity in the downstream river channel. Nevertheless, this approach has been widely applied in Japanese and US rivers because of immediate benefits, particularly to restore the downstream habitat primarily with regards to improved fish spawning habitat (Wheaton, Pasternack, and Merz 2004).

14.5.2 Implementation of Sediment Replenishment

Regarding implementation techniques of sediment replenishment, a number of restoration projects described the difficulty of implementation activity (Harvey *et al.* 2005). Several implementation techniques have evolved in response to available spaces, release flow regimes, costs and accessibility as well as primary objectives.

However, these applications still require a systematic development in the stages of planning and implementation, both of which are in concert with specific objectives and local hydrophysical restrictions in the basin, reservoir, and river. In particular, incorporating with flushing flows (magnitude, frequency, and timing), determining quantity (amount added) and quality (grain size and source materials) of coarse sediment, and selecting an effective implementation technique for adding and transporting sediment in the field are key factors in achieving a successful result in ecological river management and restoration.

14.5.2.1 Scale of Sediment Replenishment

Bunte (2004) classified the sediment augmentation approach into three spatial scales: local, multireach, and river-segment wide. Based on these scales, target sediment volume and expected effects will differ considerably:

- direct creation of spawning habitat by one-time gravel additions in stream areas with low erosion potential;
- passive gravel augmentation at a logistically convenient location to create spawning habitat where the added gravel deposits naturally;
- intensive multireach stream restoration (covering several reaches, perhaps 20–50 stream widths long);
- river segment-wide gravel augmentation and sediment management plan to restore geomorphological and biological functioning of the stream (10–100 km).

14.5.2.2 Types of Sediment Input

Regarding sediment input methodologies, Ock, Sumi, and Takemon (2013) summarized these types as shown in Figure 14.17.

In-channel bed stockpile is a relatively old method that started from 1970, but even now it is widely used for supplementation of riffles or pool tails. This approach places directly spawning gravels within the low-flow channel to provide immediate usable habitat features. This method, however, involves in-channel work at low flows, which may increase turbidity downstream. Sediment that is largely immobile under low flow would start to transport downstream when the flow exceeds the bed-mobility threshold.

High-flow stockpile method places gravel along the channel bank margin to be distributed downstream by high flows. This method assumes that the river will transport sediment and reshape the channel during high-flow events. Where peak flows in stream channel have a short duration and high magnitude, it would be applicable as an efficient method. It is a common method currently applied in Japan. This approach can add relatively large amounts of gravel at relatively low cost, but is limited to the volume piled and the number of suitable sites.

Point-bar stockpile method introduces coarse sediment to augment or create a point bar. The augmentation is accomplished using site-specific low-flow and bankfull channel dimensions of the reaches. The volume of coarse sediment introduced can be exaggerated because some of it may be routed downstream during high-flow events. However, the coarse sediment stockpiled at the inside bend in a meandering channel would be limited to erosion and transport downstream during high flows.

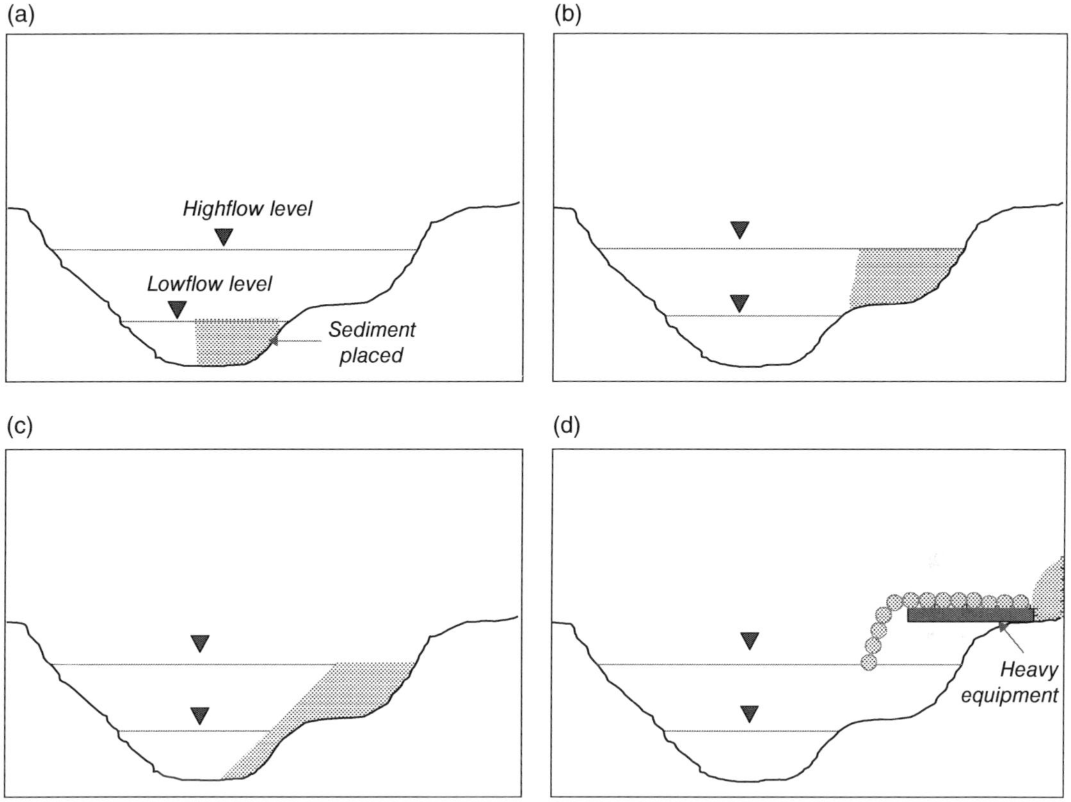

Figure 14.17 Sediment replenishment methods according to sediment placement or injection types: (a) in-channel bed stockpile, (b) high-flow stockpile, (c) point-bar stockpile, and (d) high-flow direct injection (Ock *et al.* 2013).

Gaeuman (2012) proposed the process of *high-flow direct injection*, which directly introduces gravel to the river channel during a high flow event using heavy equipment such as a conveyor belt. While the high-flow stockpile can be quickly exhausted in a large flow, this approach can allow new sediment to replace the material already transported. Also, an advantage of this method over a high-flow stockpile is that a larger volume and coarser gravel can be introduced during the course of a flood.

14.5.3 Environmental Effects and Monitoring

14.5.3.1 Creating Habitat for Spawning and Other Salmon Life-Stages

Bunte (2004) summarized environmental effects of gravel augmentation mainly on creating habitat for spawning and other salmon life-stages. Kondolf and Wolman (1993) reported that suitable spawning gravel must be well-rounded, the right size distribution for spawning fish, which means D_{50} particle size should be between 1 and 10% of the fish length, and without containing fine sediments (very fine gravel smaller than approximately 6.3 mm, sand, silt, and clay), which reduces intragravel flow and the dissolved oxygen content. Based on this information, spawning gravel quality should be carefully designed and monitored in the field after implementation, such as size distribution of bed material on spawning riffles, amount of fine sediment in spawning areas, and intragravel flows and dissolved oxygen content.

14.5.3.2 Suspended POM Retention Capacity and Hyporheic Flow Effect

Ock *et al.* (2013) showed that geomorphological changes in downstream reaches (e.g. lengthened riffles and restored bars) increased particulate organic matter (POM) retention, which is an ability of the channel to reduce the POM concentration and provide aquatic ecosystems with a primary energy resource (Tockner *et al.* 1999). However, in regulated rivers below dams, input of large amount of plankton from dam outflows influences POM quantity and quality, and subsequently could disturb the downstream ecosystem foodweb (Akopian *et al.* 1999). Thus, the retention efficiency of reservoir-derived plankton can be important component for recovering a normal state of trophic structure. In this sense, this "natural filtering" function, is significant for enhancing self-purification along channels and restoring the downstream ecosystem foodweb.

Ock *et al.* (2015) investigated water temperature alongside a gravel bar created by sediment augmentation and demonstrated that hyporheic flow between mainstream and subsurface waters beneath the bar provides thermal heterogeneity along the perimeter of gravel bars by cooling, buffering and lagging temperatures. For maximum hyporheic cooling benefit in summer, it is a strategy to increase the hyporheic flow rate, which is a function of hydraulic gradient and substrate permeability (Tonina and Buffington 2007). Coarse-sized gravels from gravel augmentation, allow a high permeability and increase hydraulic gradients within a gravel bar (Boulton 2007).

14.5.4 Case Studies in Japan

In order to reduce the sediment inflow, sediment trapping with check dams above reservoirs has been commonly implemented in Japan. In this technique, a low check dam is constructed at the end of a reservoir and regularly excavated mechanically to remove the trapped sediment, which mainly consists of coarse bedload sediments. (Figures 14.1, 14.4, and 14.5). Traditionally, the trapped coarse sediments have been used for construction materials. Recently, sediment replenishment has been carried out at least in 15 dams in Japan (Okano *et al.* 2004) including the Kizu River basin system (Kantoush, Sumi, and Kubota 2010; Kantoush and Sumi 2011).

In Japan, a high-flow stockpile is commonly used in order to avoid artificial turbid flow that is released through the side bank erosion at low flows. Sakurai and Hakoishi (2013) showed the location of typical case studies in (Figure 14.18). In the sediment replenishment measure, sediment replenishment volume and grain size are recognized as key factors for a successful management in the river basin to create and maintain physical habitats, and aquatic and riparian ecosystems. Figure 14.19 illustrates the relationship between the annual excavated sediment volumes from reservoirs and annual reservoir sedimentation volumes within reservoirs (Sumi and Kantoush 2011). Although sediment replenishment can be considered as a sediment management approach in reservoirs, its quantity compared to annual reservoir sedimentation is still very small (i.e., between 0.1% to 10%). Figure 14.20 shows the grain size distributions of replenished sediment, which are very different for each case because of difference in the available location to take the sediment out from each reservoir. In most cases in Japan, sediment replenishment is focusing both on reducing the reservoir sedimentation and on enhancing the river channel improvement, such as by detaching algae on the riverbed material (Okano *et al.* 2004). Creating new habitats for spawning and other fish life-stages is a new challenging topic in Japan. Further studies, and sharing the outcomes obtained from advanced case studies, are necessary when the scale and volume of sediment replenishment increases.

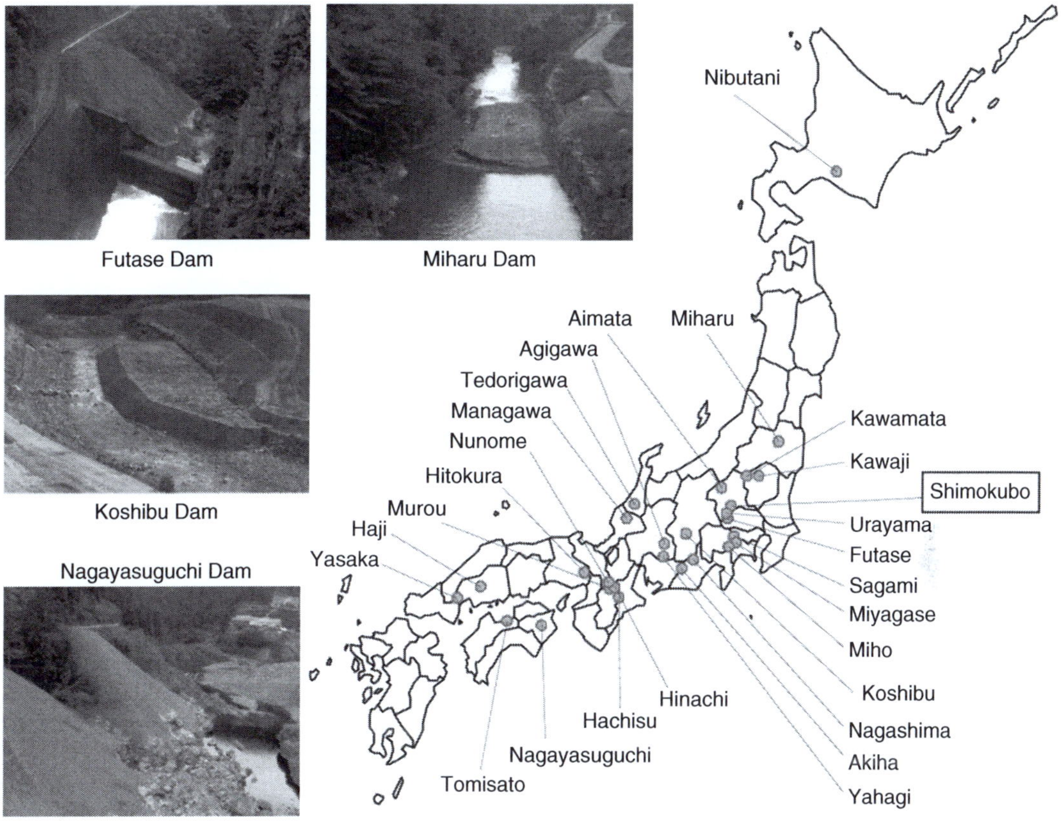

Figure 14.18 Location map of dams with sediment replenishment in Japan (Sakurai and Hakoishi 2013).

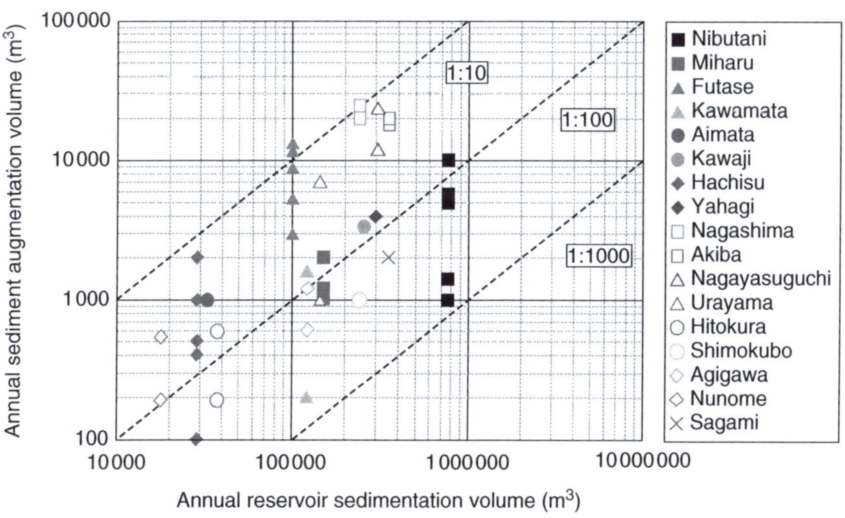

Figure 14.19 Relationship between annual reservoir sedimentation and replenishment volumes (Sumi and Kantoush 2011).

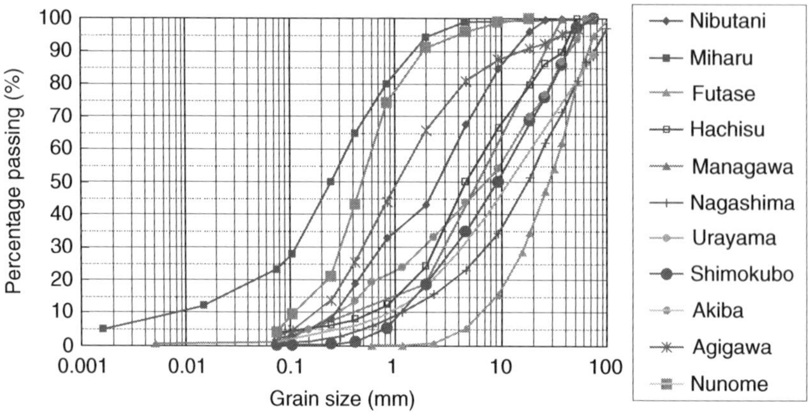

Figure 14.20 Grain-size distributions of sediment replenishment (Sumi and Kantoush 2011).

14.6 Conclusions

In the case of reservoir sediment management, it is necessary to assess the current situation and possible future scenarios from a reservoir sustainability point of view. In order to increase the number of best practices for reservoir sediment management, it is important to establish a suitable guideline to select appropriate sediment management measures based on both river basin and reservoir conditions. On this point, the turnover rate of water (*CAP*/*MAR*) and sediment (*CAP*/*MAS*) can be the key parameters to design reservoir sediment management strategy. For the sediment management plan, a combination of flow and sediment release should be appropriately designed to meet the demands of various functions based on hydrology, water quality, river morphology, and ecosystem data, etc. Furthermore, the integrated sediment management approach should be considered in sediment routing systems that cover not only the river basin but also coastal areas. Among several updated methodologies, effective and eco-friendly sediment flushing and replenishment techniques have been intensively applied in Japan. Even though the target volume of sediment is very different between these approaches, positive influences should be clarified from the point of view of both reservoir sustainability and downstream environmental improvement. Sediment flushing is now a common approach worldwide but problems still exist, such as how to increase flushing efficiency by minimizing environmentally adverse impacts. Sediment replenishment or augmentation can be more easily applied as a start-up option without installation of any large facilities, both for reducing reservoir sedimentation and for improving the downstream river environment. In order to maximize the benefit, incorporating flushing flows (magnitude, frequency, and timing), determining quantity (amount added) and quality (grain size and source materials) of coarse sediment, and selecting an effective implementation technique for adding and transporting sediment in the fields, are key factors in achieving a successful result in ecological river management and restoration. For the future, it is important to increase the scale and volume of sediment replenishment, and to establish sustainable management in order to maintain long-term benefits.

Acknowledgment

The authors gratefully acknowledge insightful and constructive comments and suggestions by two anonymous reviewers.

References

Akopian, M, Garnier, J, and Pourriot, R 1999. A large reservoir as a source of zooplankton for the river: structure of the populations and influence of fish predation. *Journal of Plankton Research* 21, 285–297.

Atkinson, E 1996. *The Feasibility of Flushing Sediment from Reservoir.* Report OD137, HR Wallingford: UK.

Badura, H, Knoblauch, H, and Schneider, J 2008. *Pilot Project Bodendorf, Work Package 8.* Report of the EU Interreg IIIB Project ALPRESREV: Austria.

Batalla, RJ 2003. Sediment deficit in rivers caused by dams and instream gravel mining. A review with examples from NE Spain. *Cuaternario y Geomorfología* 17(3–4), 79–91.

Batuca, DG and Jordaan, JM 2000. *Silting and Desilting of Reservoirs.* A. A. Balkema: Rotterdam

Bilotta, GS and Brazier RE 2008. Understanding the influence of suspended solids on water quality and aquatic biota. *Water Resources* 42. DOI:10.1016/j.watres.2008.03.018, 2849–2861.

Bouchard, JP 2000. Optimization of flushing conditions; forecasting of efficiency and impact. *International Workshop and Symposium on Reservoir Sedimentation Management,* Toyama City, Japan; 169–176.

Boulton, AJ 2007. Hyporheic rehabilitation in rivers: restoring vertical connectivity. *Freshwater Biology* 52, 632–650.

Bunte, K 2004. *Gravel Mitigation and Augmentation Below Hydroelectricdams: a Geomorphological Perspective.* Stream Systems Technology Center, Rocky Moumtains Research Station, Forestry Services, US Department of Agriculture: Fort Collins, CO; 144 pp.

Crosa, G, Castelli, E Gentili, G, and Espa, P 2010. Effects of suspended sediments from reservoir flushing on fish and macroinvertebrates in an alpine stream. *Aquatic Science* 72(1), 85–95. DOI:10.1007/s00027-009-0117-z.

Esmaeili, T, Sumi, T, and Kantoush, SA 2014. Experimental and numerical study of flushing channel formation in shallow reservoirs. *Journal of Japan Society of Civil Engineers. Series B1 (Hydraulic Engineering)* 70-4, I-19–I-24 .

Esmaeili, T, Sumi, T, Kantoush, SA, and Schleiss, AJ 2014. Three dimensional numerical modelling of flow field in rectangular shallow reservoirs. *Proceedings of 7th River Flow Conference Reservoir Sedimentation* (eds Shleiss *et al.*). Lausanne, Switzerland; 11–19.

Esmaeili, T, Sumi, T Kantoush, SA *et al.* 2015. Numerical study on flushing channel evolution, Case study of Dashidaira Reservoir, Kurobe River. Journal of Japan Society of Civil Engineers, Series B1 (Hydraulic Engineering) 71-4, I-115–I-120.

Fang, HW and Rodi, W 2003. Three-dimensional calculations of flow and suspended sediment transport in the neighborhood of the dam for the Three Gorges Project (TGP) reservoir in the Yangtze River. *Journal of Hydraulic Research,* 41(4), 379–394.

Gaeuman, D 2012. Mitigating downstream effects of dams. In *Gravel-Bed Rivers: Processes, Tools, Environments* (eds Church M, Roy AG, and Biron PM). John Wiley & Sons: Chichester, UK; 182–189.

Garric, J, Migeon, B, and Vindimian, E. 1990. Lethal effects of draining on Brown Trout: a predictive model based on field and laboratory studies. *Water Resources* 24(1), 59–65. DOI:10.1016/0043-1354(90)90065-E.

Gerster, S and Rey, P 1994. Ökologische Folgen von Stauraumspülungen, Bundesamt für Umwelt. *Wald und Landschaft (BUWAL), Schriftenreihe Umwelt* 219.

Gray, LJ and Ward, JV 1982. Effects of sediment releases from a reservoir on stream macroinvertebrates. *Hydrobiologia* 96, 177–184. DOI:10.1007/BF02185433.

Hartmann, S 2009. *Sustainable Sediment Management of Alpine Reservoirs Considering Ecological and Economical Aspects.* ALPRESERV 7, Institut für Wasserwesen: Neubiberg, Germany.

Harb, G, Haun, S, Schneider, J, and Olsen, NRB 2014. Numerical analysis of synthetic granulate deposition in a physical model study. *International Journal of Sediment Research* 29(1), 110–117.

Harvey, B, McBain, S, Reiser, D *et al.* 2005. *Key Uncertainties in Gravel Augmentation: Geomorphological And Biological Research Needs for Effective River Restoration.* CALFED Science Program and Ecosystem Restoration Program Gravel Augmentation Panel Report.

Haun, S and Olsen, NRB 2012a. Three-dimensional numerical modelling of the flushing process of the Kali Gandaki hydropower reservoir. *Lakes and Reservoirs: Research and Management* 17(1), 25–33.

Haun, S and Olsen, NRB 2012b. Three-dimensional numerical modeling of reservoir flushing in a prototype scale. *International Journal of River Basin Management* 10(4), 341–349.

Haun, S, Dorfman, C, Harb, G, and Olsen, NRB 2012. 3D numerical modeling of the reservoir flushing of the Bodendorf reservoir, Austeria. *Proceedings of 2nd IAHR European Conference*, Munich, Germany.

Haun, S, Kjærås, H, Løvfall, S, and Olsen, NRB 2013. Three-dimensional measurements and numerical modelling of suspended sediments in a hydropower reservoir. *Journal of Hydrology* 479, 180–188.

ICOLD 2009. *Sedimentation and Sustainable Use of Reservoirs and River Systems* (ed Basson, G). Bulletin 147, International Commission on Large Dams: Paris.

Japanese Ministry of Land, Infrastructure, Transport and Tourism 2005, '. *Technical Criteria for River Works: Practical Guide for Planning*',. http://www.japanriver.or.jp/publish/ book/ksgk_plan.pdf; , pp.121-1–123.

Kantoush, SA and Schleiss, AJ 2009. Channel formation during flushing of large shallow reservoirs with different geometries. *Environmental Technology* 30(8), 855–863.

Kantoush, SA and Sumi, T 2011. Sediment replenishing measures for revitalization of japanese rivers below dams. *Proceedings of the 34th IAHR World Congress*; 2838–2846.

Kantoush, SA, Sumi, T, and Kubota, A 2010. Geomorphic response of rivers below dams by sediment replenishment technique. *River Flow 2010*, Braunschweig, Germany; 1155–1163.

Kantoush, SA, Sumi, T, and Takemon, Y 2011. Lighten the load. *International Water Power and Dam Construction*, May; 38–45.

Kantoush, SA Sumi, T, Suzuki, T, and Murasaki, M 2010. Impacts of sediment flushing on channel evolution and morphological processes: case study of the Kurobe River, Japan. *River Flow 2010*, Braunschweig, Germany; 1165–1173.

Khosronejad, A, Rennie, CD, Salehi Neyshabouri, AA, and Gholami, I 2008. Three dimensional numerical modeling of sediment release. *Journal of Hydraulic Research* 46(2), 209–223.

Kokubo, T, Itakura, M, and Harada, M 1997. Prediction methods and actual results on flushing of accumulated deposits from Dashidaira Reservoir. *19th ICOLD Congress*, Florence; Q. 74, R.47, 761–791.

Kondolf, GM 1997. Hungry water: effects of dams and gravel mining on river channels. *Environmental Management* 21(4), 533–551.

Kondolf, GM and Matthews, WVG 1993. *Management of Coarse Sediment in Regulated Rivers of California*. Report No. 80, University of California Water Resources Center: Davis, CA.

Kondolf, GM and Wolman MG 1993. The sizes of salmonid spawning gravels. *Water Resources Research* 29(7), 2275–2285.

Kondolf, GM, Gao, Y, Annandale, GW *et al.* 2014. Sustainable sediment management in reservoirs and regulated rivers: experiences from five continents. *Earth's Future* 2, 256–280. DOI:10.1002/ 2013EF000184.

Knoblauch, H 2006. *Sustainable Sediment Management of Alpine Reservoirs Considering Ecological and Economical Aspects*. ALPRESERV, Vol. 4.

Lai, JS and Shen, HW 1995. Degradation flushing processes in reservoir. *HYDRO 2000*, Thomas Telford: London; 4.

Lai, JS and Shen, HW 1996. Flushing sediment through reservoirs. *Journal of Hydraulic Research* 34(2), 237–255.

Liu J, Minami S, Otsuki H *et al.* 2004a. Prediction of concerted sediment flushing. *Journal of Hydraulic Engineering ASCE* 130(11), 1089–1096.

Liu J, Minami S, Otsuki H *et al.* 2004b. Environmental impacts of coordinated sediment flushing. *Journal of Hydraulic Research* 42, 461–472.

McBain and Trush 2004. *Coarse Sediment Management Plan for the Lower Tuolumne River.* Final Report, Prepared for the Tuolumne River Advisory Committee by Stillwater Sciences: Davis, CA.

Merle, G 2000. Some environmental aspects of flushing. *International Workshop and Symposium on Reservoir Sedimentation Management,* Toyama City, Japan; 195–202.

Minami S, Noguchi K, Otsuki H *et al.* 2012. Coordinated sediment flushing and effect verification of fine sediment discharge operation in Kurobe River. *Proceedings of the International Symposium on Dams for a Changing World, The 80th Annual Meeting of ICOLD.*

Morris, GL and Fan, J 1998. *Reservoir Sedimentation Handbook: Design and Management of Dams, Reservoirs, and Watersheds for Sustainable Use.* McGraw-Hill.

Newcombe, CP and Macdonald, DD 1991. Effects of suspended sediments on aquatic ecosystems. *North American Journal of Fishery Management* 11(2), 72–82. DOI:10.1577/1548-8675 0112.3.CO.

Newcombe, CP and Jensen, JOT 1996. Channel suspended sediment and fisheries: a synthesis for quantitative assessment of risk and impact. *North American Journal of Fishery Management* 16. DOI:10.1577/1548-8675(1996)0162.3.CO;2, 913–918.

Ock, G, Sumi, T and Takemon, Y 2013. Sediment replenishment to downstream reaches below dams: implementation perspectives. *Hydrological Research Letters* 7(3), 54–59. DOI:http://dx.doi.org/10.3178/hrl.7.54.

Ock, G, Kondolf, GM, Takemon, Y, and Sumi, T 2013. Missing link of coarse sediment augmentation to ecological functions in regulated rivers below dams: comparative approach in Nunome River, Japan and Trinity River, California, US. *Advances in River Sediment Research* (eds Fukuoka, S, Nakagawa, H, Sumi, T, and Zhang, H). Taylor & Francis Group: London; 1531–1538.

Ock, G, Gaeuman, D, McSloy, J and Kondolf, GM 2015. Ecological functions of restored gravel bars, the Trinity River, California. *Ecological Engineering,* 83, 49–60.

Okano, M, Kikui, H, Ishida, H, and Sumi, T 2004. Reservoir sedimentation management by coarse sediment replenishment below dams. *9th International Symposium on River Sedimentation* II; 1070–1078.

Olsen, NRB 1999. Two-dimensional numerical modeling of flushing processes in water reservoirs. *Journal of Hydraulic Research* 3(1), 3–16.

Olsen NRB 2014. *A Three Dimensional Numerical Model for Simulation of Sediment Movement in Water Intakes with Multiblock Option.* Department of Hydraulic and Environmental Engineering, The Norwegian University of Science and Technology; 215.

Palmieri, A, Shah, F, Annandale, GW, and Dinar, A 2003. *Reservoir Conservation Volume 1: the RESCON Approach.* World Bank.

Petts, GE 1979. Complex response of river channel morphology subsequent to reservoir construction. *Progress in Physical Geography* 3, 329–362.

Rambaud, J, Clair, M, Sandret, J *et al.* 1988. Experience acquise dans les vidanges de retenues par Electricite de France et la Compagnie Nationale du Rhone. *The 16th ICOLD Congress,* San Francisco; Q.60.

Sakurai T and Hakoishi N 2013. Numerical simulation of sediment supply from dam reservoirs to downstream by the placed sediment. *Advances in River Sediment Research* – Fukuoka *et al.* (eds), Taylor & Francis Group, ISBN 978-1-138-00062-9, 1193–1199.

Shen HW 1999. Flushing sediment through reservoirs. *Journal of Hydraulic Research,* 37(6), 743–757.

Sloff, CJ, Jagers, HRA, and Kitamura, Y 2004. Study on the channel development in a wide reservoir. *River Flow 2004,* Napoli, Italy; 811–819.

Staub, E 2000. Effects of sediment flushing on fish and invertebrates in Swiss Alpine rivers. *International Workshop and Symposium on Reservoir Sediment Management*, Toyama City, Japan; 185–193.

Sumi, T 2008. Evaluation of efficiency of reservoir sediment flushing in Kurobe River. *4th International Conference on Scour and Erosion*, Tokyo; 608–613.

Sumi, T 2013. Reservoir sedimentation database and selecting suitable sediment management options in Japan. *Proceedings of the International Symposium on Changing Times: Infrastructure Development to Infrastructure Management, 81st Annual Meeting of ICOLD*; 1717–1727.

Sumi, T and Kanazawa, H 2006. Environmental study on sediment flushing in the Kurobe River. *22nd Congress of ICOLD*, Barcelona; Q85-R16, 219–242.

Sumi, T and Kantoush, SA 2010. Integrated management of reservoir sediment routing by flushing, replenishing, and bypassing sediments in Japan river basins. *8th International Symposium on EcoHydraulics (ISE 2010)*, COEX, Seoul; 12–16.

Sumi, T and Kantoush, SA 2011. Sediment management strategies for sustainable reservoir. *Proceedings of the International Symposium on Dams and Reservoirs under Changing Challenges, 79th Annual Meeting of ICOLD*; 353–362.

Sumi, T, Nakamura, S, and Hayashi, K 2009. The effect of sediment flushing and environmental mitigation measures in the Kurobe River. *23rd Congress of ICOLD*, Brasilia, Q89-R6.

Sumi, T, Okano, M, and Takata, Y 2005. Reservoir sedimentation management with bypass tunnels in Japan. *The 9th International Symposium on River Sedimentation*, II; 1036–1043.

Tockner, K, Pennetzdorfer, D, Reiner, N *et al.* 1999. Hydrological connectivity, and the exchange of organic matter and nutrients in a dynamic river-floodplain system (Danube, Austria). *Freshwater Biology* 41, 521–535.

Tonina, D and Buffington, JM 2007. Hyporheic exchange in gravel bed rivers with pool-riffle morphology: laboratory experiments and three-dimensional modeling. *Water Resources Research* 43, W01421.

Valette, E and Jodeau M 2012. How to predict the sedimentological impacts of reservoir operations?. *ICSE 6*, Paris; 1329–1336.

Wheaton, JM, Pasternack, GB, and Merz, JE 2004. Spawning habitat rehabilitation – I. *Conceptual approach and methods. International Journal of River Basin Management* 2, 3–20. DOI:10.1080/15715124.2004.9635218.

White R 2001. *Evacuation of Sediments from Reservoirs*. Thomas Telford Publishing: London.

Williams, GP and Wolman, MG 1984. Downstream Effects of Dams on Alluvial Rivers. Professional Paper 1286, US Geological Survey.

Wohl, EE and Cenderelli, DA 2000. Sediment deposition and transport patterns following a reservoir sediment release. *Water Resources Research*, 36, 319–333. DOI:10.1029/1999WR900272.

15

Bedload Transport in Laboratory Rivers: The Erosion–Deposition Model

Eric Lajeunesse, Olivier Devauchelle, Florent Lachaussée, and Philippe Claudin

15.1 Introduction

When the shear stress exerted by a river on its bed exceeds a threshold, superficial sediments are entrained by the flow. For moderate values of the shear stress, they move by rolling, sliding, and bouncing while gravity maintains them close to the bed surface (Bagnold 1973). This so-called 'bedload transport' deforms the bed, and the resulting interaction between flow and sediment transport generates a beautiful variety of river shapes and coastal morphologies (Gomez 1991; Church 2006; Seminara 2010). Understanding this process requires a sound theory of bedload transport.

Since the pioneering work of Shields (1936), much field, experimental, and theoretical work has been devoted to bedload (e.g. Meyer-Peter and Müller 1948; Einstein 1950; Bagnold 1956; Ashida and Michiue 1973; Engelund and Fredsoe 1976; Fernandez-Luque and Van Beek 1976; Bridge and Dominic 1984; Gomez and Church 1989; Gomez 1991; Nino and Garcia 1994; Charru, Mouilleron, and Eiff 2004; Ancey *et al.* 2008; Ancey 2010; Lajeunesse, Malverti, and Charru 2010; Durán, Andreotti, and Claudin 2012, 2014; Furbish, Ball, and Schmeeckle 2012; Furbish, Roseberry, and Schmeeckle 2012; Furbish *et al.* 2012; Roseberry, Schmeeckle, and Furbish 2012; Ancey and Heyman 2014). Most studies focus on the prediction of the sediment transport rate at a specific time and location, $q_s(x, t)$. The net exchange of sediment between the bed and the moving layer is then deduced from mass conservation, sometimes referred to as the 'Exner equation' in the context of fluvial geomorphology (Seminara 2010),

$$\frac{\partial h}{\partial t} + \delta v \frac{\partial q_s}{\partial x} = 0 \tag{15.1}$$

where we introduce the streamwise coordinate x, the time t, the volume of a sediment particle, δv, and the bed elevation $h(x, t)$. The sediment flux, $q_s(x, t)$, is expressed in number of particles per unit river width and time $[L^{-1}T^{-1}]$.

To investigate bedload transport, we can use a laboratory channel filled with a bed of plastic sediments (Figure 15.1). A pump located at the upstream end of the channel sets a steady, uniform and laminar flow entraining the bed sediments. A sediment feeder imposes a constant sediment flux at the channel inlet. After a few hours, the sediment bed reaches a steady-state characterized by a constant slope. We then deposit a plume of dyed sediments at the bed surface and monitor its evolution by means of a camera placed above the channel. We observe that the plume spreads as it is transported

Gravel-Bed Rivers: Processes and Disasters, First Edition. Edited by Daizo Tsutsumi and Jonathan B. Laronne.

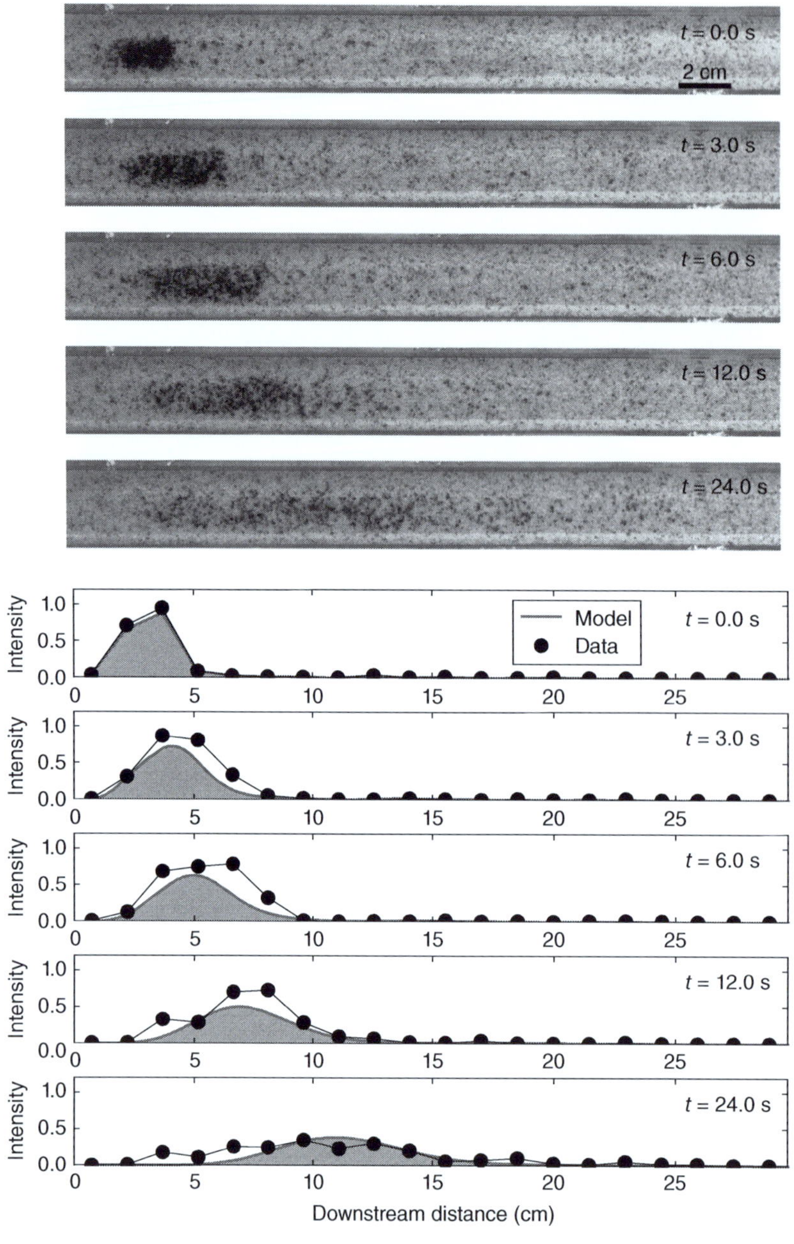

Figure 15.1 A plume of dyed sediments in an experimental channel (1 m long and 3 cm wide). Top: snapshots of the sediment bed. Bottom: intensity profiles (dots) averaged along the channel width (levels of grey) used as a proxy for the average tracer concentration. The bed of sediments (plastic grains of size 720 ± 120 μm and density 1520 ± 50 kg m^{-3}) is sheared by a mixture of water and glycerol (viscosity 10^{-2}Pas and density 1155 kgm^{-3}). The flow and sediment discharges, respectively 0.82 L min^{-1} and 0.14 g min^{-1}, are constant during the experiment. The ratio of the concentration of moving particles to the concentration of static particles is $\alpha \sim 8 \times 10^{-2}$. The plume of tracers (bed sediments dyed in black) is deposited in the channel once the sediment bed has reached steady state. Solid line: numerical solution of Equations (15.23) and (15.24) for $\alpha = 0.08$. The parameters $V = 0.046$ m s^{-1} and $t_s = 0.081$ s are adjusted to the experimental data.

downstream (Figure 15.1). This indicates that the dyed particles move with varied velocities. In particular, some tracer particles initially in motion settle on the bed where they remain trapped. These trapped particles are replaced by sediments freshly entrained from the bed to keep the sediment discharge constant.

These observations suggest a continuous exchange of particles between the sediment bed and the layer of moving grains, referred to as the 'bedload layer'. Equation (15.1) accounts for the evolution of the bed topography. However, it does not explicitly describe the exchange of particles between the bed and the bedload layer, which occurs even when the sediment flux is uniform. Yet, this exchange is at the heart of several processes such as the emplacement of alluvial deposits, the determination of the residence time of sediments in a drainage basin, and the propagation of sediment tracers in a river bed (Carretier, Regard, and Soual 2007; Gayer, Mukhopadhyay, and Meade 2008; Willenbring and von Blanckenburg 2010).

Sayre and Hubbell (1965) were the first to experiment sediment tracing in a natural stream. In November 1960, they monitored the propagation of about 18 kg of radioactive sand deposited on the bed of a sandy river in Nebraska (United States). Their data exhibit the same qualitative behaviour as our laboratory experiment: the plume of radioactive sand spread as it was entrained downstream by the flow (Figure 15.2).

Since this pioneering work, sediment tracing has become a popular tool to investigate the properties of bedload transport in natural streams (Reid, Frostick, and Layman 1985; Wilcock 1997; Ferguson and Wathen 1998; Haschenburger and Church 1998; Bradley, Tucker, and Benson 2010;). The recent development of radio-frequency identification (RFID) allows the insertion of passive integrated transponders in pebbles (Lamarre *et al.* 2005; Nichols 2004). These small electronic tags provide the means to track individual pebbles (Figure 15.3). RFID tracking is used to study bedload threshold at a fixed location, particle vertical mixing rates, sediment travel lengths, bed and bedform mobility, and particle storage in the sediment bed (Habersack 2001; Ferguson *et al.* 2002; Nikora *et al.* 2002; Haschenburger and Wilcock 2003; Hodge, Hoey, and Sklar 2011; Nathan Bradley and Tucker 2012; Hassan *et al.* 2013; Phillips, Martin, and Jerolmack 2013; Phillips and Jerolmack 2014). The interpretation of these observations requires that we understand the exchange of particles between the bed and the bedload layer.

The erosion–deposition model of bedload transport, first proposed by Charru, Mouilleron, and Eiff (2004), is based on the description of the exchange of particles between the bed and the bedload layer. This heuristic model, inspired by laboratory experiments, describes how this exchange sets the number of moving particles and determines the sediment transport rate (Charru, Mouilleron, and Eiff 2004; Lajeunesse, Malverti, and Charru 2010; Seizilles *et al.* 2014). It is versatile enough to describe the formation of various bedforms, the transport of mixed grain sizes, and the propagation of tracers (Charru 2006; Charru and Hinch 2006; Devauchelle *et al.* 2010a,b; Houssais and Lajeunesse 2012; Lajeunesse *et al.* 2013).

This chapter is a review of the erosion–deposition model. We begin with a derivation of the model equations based on recent laboratory experiments. We discuss the role of turbulence and the predictions of the model in terms of bedforms development. The chapter ends with a preliminary investigation of the propagation of a plume of tracer sediments.

15.2 The Erosion–Deposition Model

15.2.1 Laboratory Observations

Particles entrained by the flow move by rolling, sliding, and bouncing until they settle back on the bed. In this chapter we restrict ourselves to the case where the shear stress is close to the entrainment

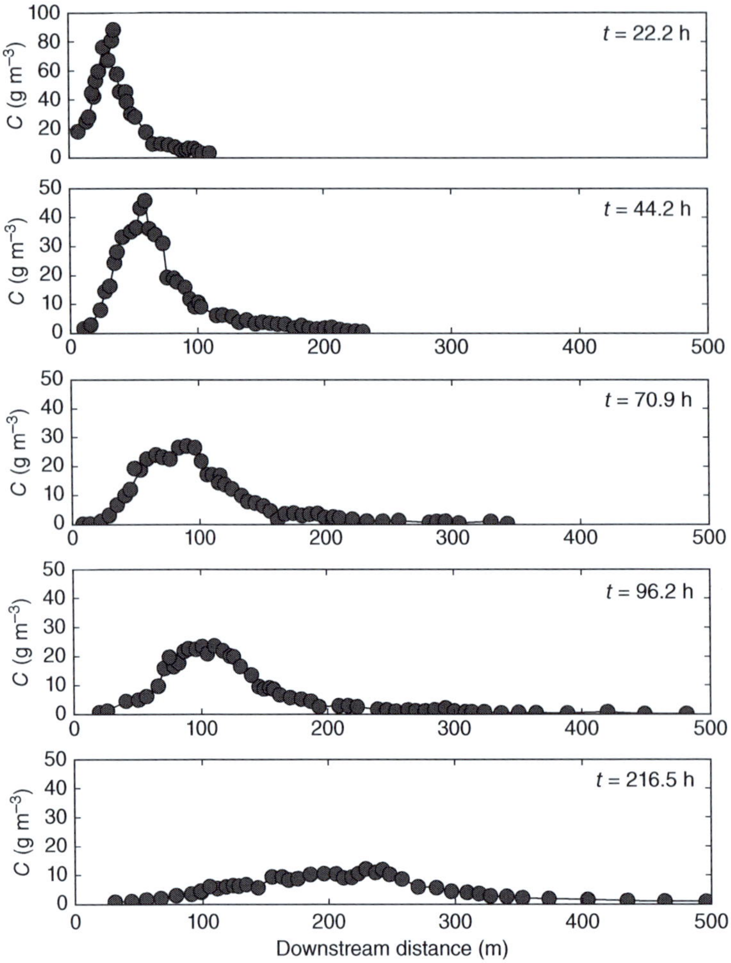

Figure 15.2 Evolution of the concentration of radioactive sand injected into the North Loup River (Nebraska, United States) in November 1960. (After Sayre and Hubbell (1965) and Bradley, Tucker, and Benson (2010).)

threshold. Image analysis shows that, in this regime, the bedload layer is approximately one grain-diameter thick (Figure 15.4). This confinement implies that bedload transport is controlled by the characteristics of the flow near the sediment bed. In particular, experiments show that bedload transport is primarily controlled by the ratio of the flow-induced shear stress to the weight of a particle, usually referred to as the Shields parameter (Shields 1936; Meyer-Peter and Müller 1948):

$$\theta = \frac{\tau}{(\rho_s - \rho) g d_s} \tag{15.2}$$

where we introduce the shear stress exerted on the bed τ, the grain size d_s, the sediment density ρ_s, the fluid density ρ, and the gravity g. When the Shields parameter exceeds a threshold value, θ_t, the shear stress is strong enough to overcome the weight of a particle and to entrain it in the bedload layer. As the Shields stress increases, bedload transport intensifies (Meyer-Peter and Müller 1948).

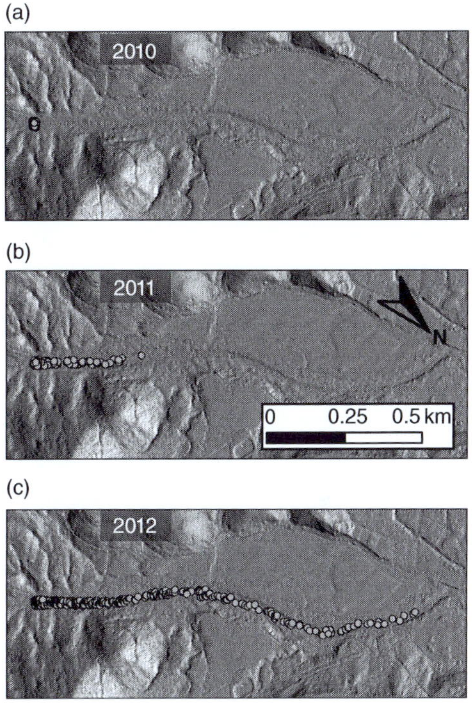

Figure 15.3 RFID-tagged pebbles in the Mameyes River (Luquillo Critical Zone Observatory, northeast Puerto Rico). Distribution of tracers in (a) May 2010, (b) summer 2011, and (c) summer 2012. (*Source*: Phillips, http://www.earth-surf-dynam.net/2/513/2014/. Used under CC-BY 3.0 https://creativecommons.org/licenses/by/3.0/)

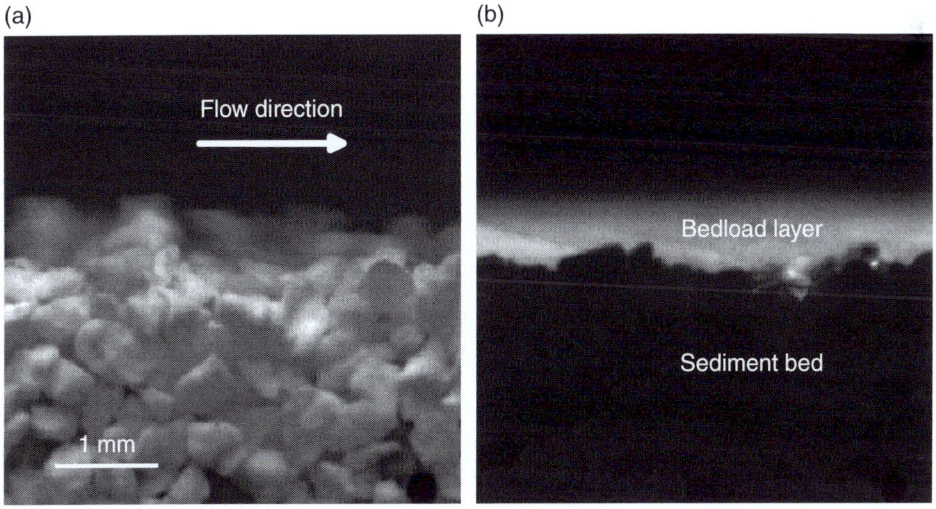

Figure 15.4 (a) Side-view of bedload transport in the experimental flume of Figure 15.1. The sediment bed, made of plastic grains (size $d_s \approx 720 \pm 120\,\mu$m, density $\rho_s = 1520 \pm 50$ kg m^{-3}) is sheared by a viscous flow of glycerol (viscosity $\eta = 10^{-2}$Pas, density $\rho = 1155$ kgm^{-3}, $Re \approx 50$). (b) Average amplitude of the difference between two successive pictures. The bedload layer appears in white.

The ensemble-averaged sediment transport rate, q_s [$L^{-1}T^{-1}$], can be expressed in terms of the particles' velocity distribution,

$$q_s = n \int_0^\infty F(v)\,v\,\mathrm{d}v, \tag{15.3}$$

where $F(v)$ is the probability distribution function (PDF) of the particle velocities, v, and $n \ll 1$ is the surface concentration of the bedload layer, i.e. the number of moving particles per unit bed area (Furbish *et al.* 2012; Ballio *et al.* 2014).

Using a camera placed a few centimetres above our experimental channel, we track the grain trajectories and compute their streamwise velocities (Seizilles *et al.* 2014). The resulting velocity distribution exhibits a peak centred around $v = 0$ with an asymmetric tail on its right-hand side (Figure 15.5a). Since the camera is placed above the channel, the free surface waves cause a spurious motion which affects even resting particles. We calibrate this noise by plotting the distribution of apparent velocities for static particles (dark grey curve on Figure 15.5a). The result suggests that we may distinguish two populations of particles: (i) particles at rest on the bed corresponding to the central peak in Figure 15.5a; (ii) particles belonging to the bedload layer corresponding to the distribution tail. Discarding the central peak, an exponential function approximates reasonably the velocity distribution in the bedload layer (Figure 15.5a), in accordance with previous experiments involving either laminar or turbulent flows (Figure 15.5b; Charru, Mouilleron, and Eiff 2004; Lajeunesse, Malverti, and Charru 2010; Furbish *et al.* 2012; Seizilles *et al.* 2014).

This observation suggests that the streamwise velocity distribution of particles sheared by a flow is the sum of two contributions (Figure 15.5a),

$$F(v) = \begin{cases} \dfrac{1}{V} e^{-v/V} & \text{if } v > 0 \\ \delta(v) & \text{otherwise} \end{cases} \tag{15.4}$$

where $\delta(v)$ is the Dirac function and V is the average velocity of bedload particles. Equation (15.4) distinguishes between two reservoirs : (i) the particles at rest represented by a Dirac distribution and (ii) the bedload layer represented by an exponential distribution.

A complete description of sediment transport requires a theoretical formulation of the velocity distribution of particles $F(v)$ and its dependency on the flow conditions (Furbish *et al.* 2012). The erosion–deposition model adopts a simplified approach by treating the bedload layer as a uniform reservoir of independent particles moving at velocity V (Charru, Mouilleron, and Eiff 2004). Accordingly, the bedload flux Equation (15.3) becomes

$$q_s = nV, \tag{15.5}$$

To calculate the bedload transport rate, we need to relate the average particle velocity, V, and the surface concentration, n, to the flow conditions.

15.2.2 Average Particle Velocity

The expression of the average particle velocity depends on the flow regime. In a viscous flow we expect the particles to move at about the velocity of the fluid that entrains them. The average particle velocity should therefore scale with the fluid velocity in the bedload layer, i.e. about one grain diameter above the bed surface,

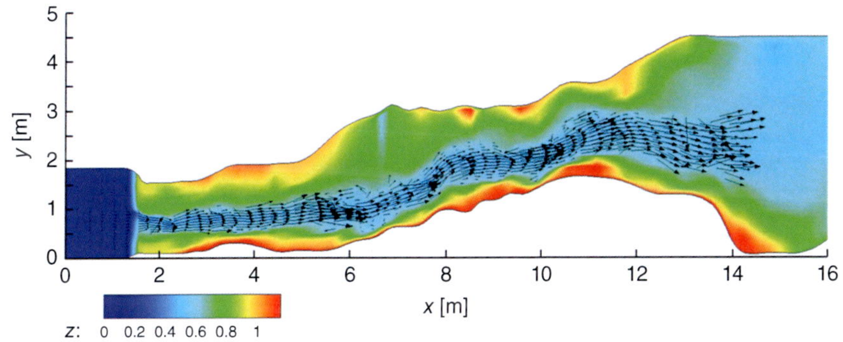

Figure 1.4 Snapshot of the velocity field during the propagation of a dam-break flow over a complex bathymetry using the SWE model of Guerra *et al.* (2014). The two-dimensional model yields the flooded area and instantaneous depth-averaged velocities at every point, incorporating the inundation of previously dry zones of the terrain. (*Source*: Guerra 2014. Reproduced with permission of American Society of Civil Engineers.)

Gravel-Bed Rivers: Processes and Disasters, First Edition. Edited by Daizo Tsutsumi and Jonathan B. Laronne.

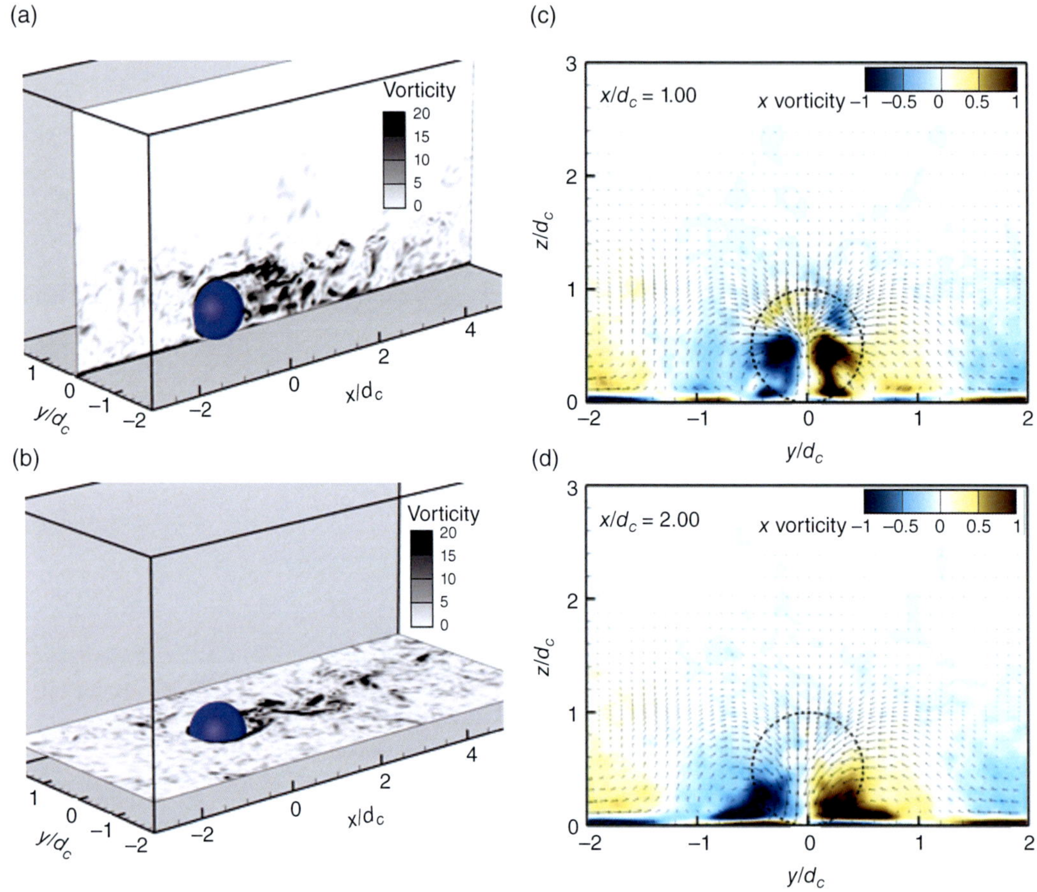

Figure 2.9 (a) Instantaneous lateral vorticity, ω_y, distribution within the streamwise-vertical, x–z, plane through the boulder centerline; (b) instantaneous vertical vorticity, ω_z, distribution within the streamwise-transverse, x–y, plane through the boulder center; (c) time-averaged streamwise vorticity, $\bar{\omega}_x$, distribution within the transverse-vertical, y–z, plane at $x/d_c = 1.00$; (d) time-averaged streamwise vorticity, $\bar{\omega}_x$, distribution within the transverse-vertical, y–z, plane at $x/d_c = 2.00$. (From Papanicolaou *et al.* 2012.)

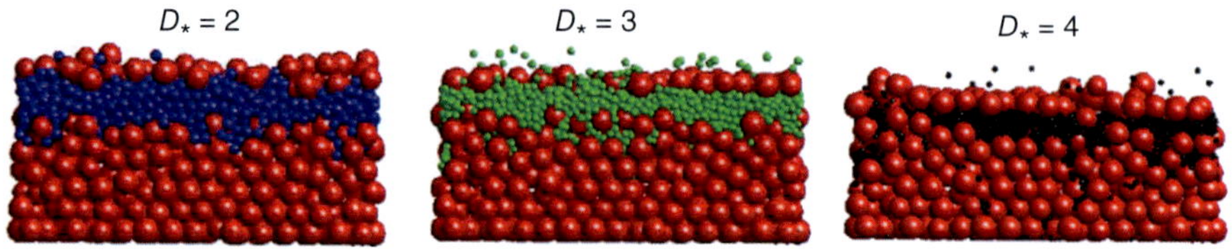

Figure 3.4 Snapshots from the near steady-state conditions from the end of each simulation. The grain size ratio is provided above.

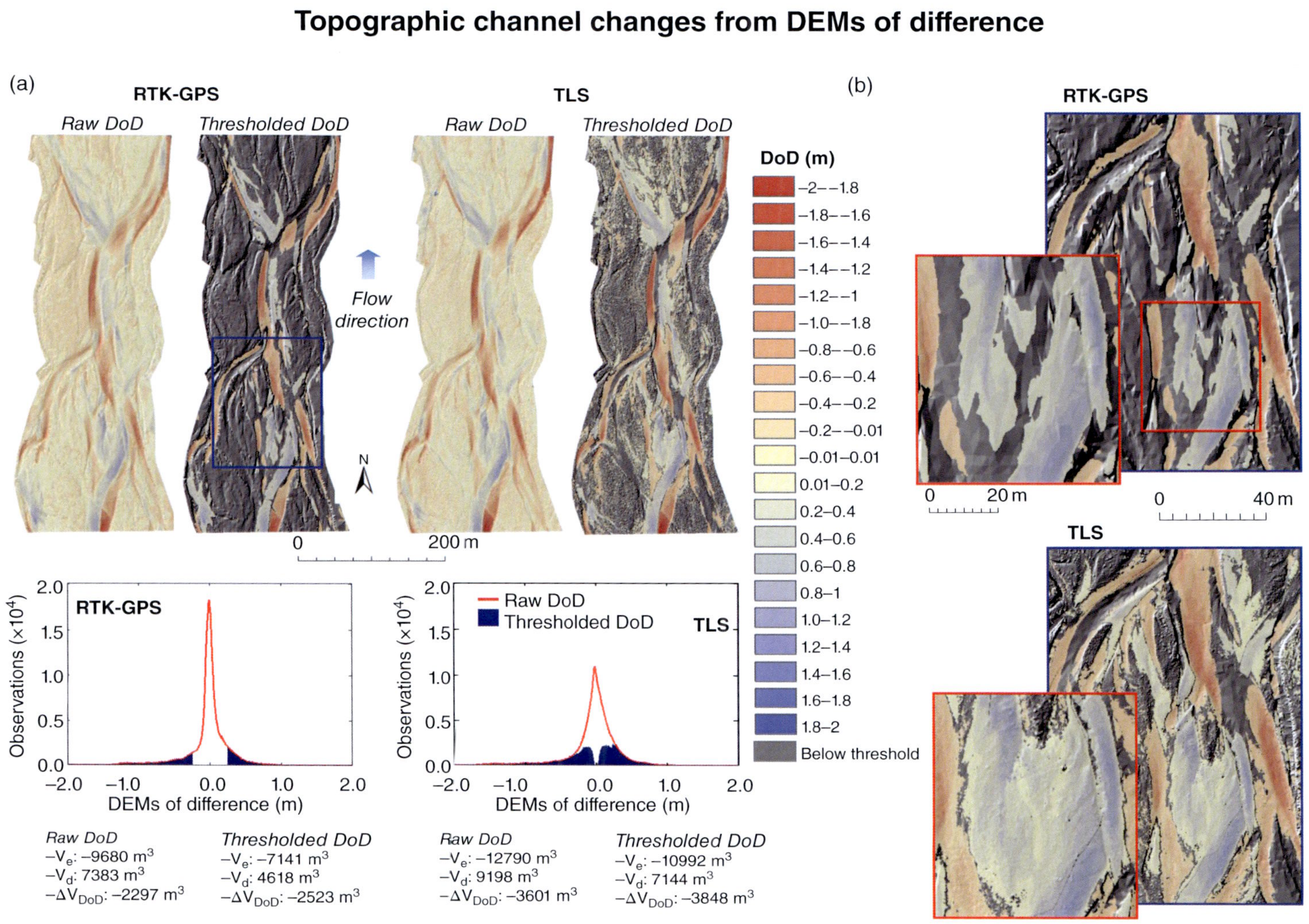

Figure 5.8 Illustration of additional erosion and deposition mechanisms identified with higher precision, hyperscale TLS HRT data versus traditional RTK-GPS HRT. (a) Comparison of reach-scale DoDs derived from RTK-GPS versus TLS DEMs for same time periods (2006–2007) from the Feshie River (Scotland). (b) Detailed close-up of changes on a gravel bar.

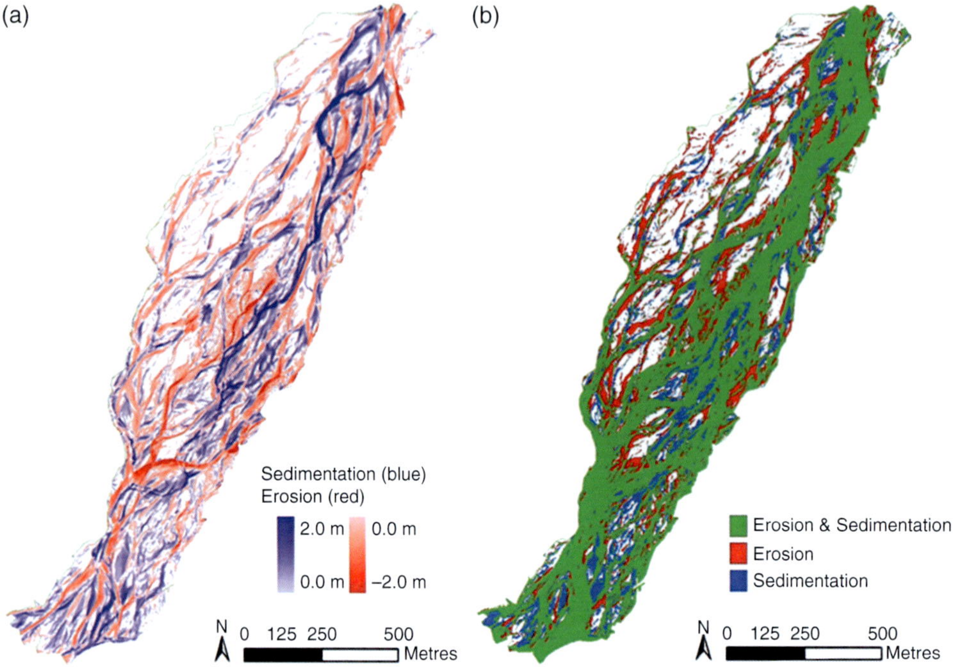

Figure 5.10 (a) Net annual thresholded DoD from the River Rees and (b) classification of regions that experienced just erosion (red), just sedimentation (blue), versus a more complicated, event-transgressive mix of erosion and sedimentation during a flood season (2009–2010).

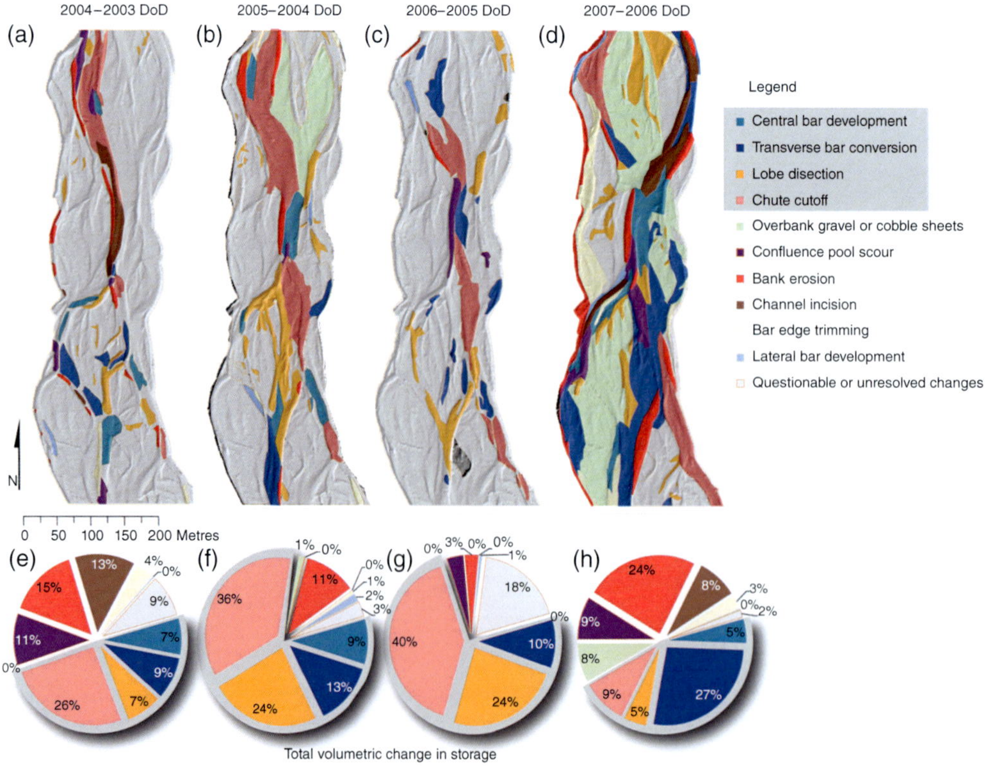

Figure 5.13 Example of using budget segregation of DoD to quantify morphodynamic signatures. (Figure reproduced from Wheaton *et al.* 2013.)

Figure 8.7 Interpolated spatially distributed apparent bedload velocity (m/s) in Rees River braided river anabranch. Data collected January 27–29, 2011. Locations with statistically significant bedload velocity marked in yellow. Grid coordinates (m) in New Zealand Transverse Mercator. The underlying aerial photograph was taken on February 27, 2011, after a high-flow event that caused morphological change within the study area. Maximum v_a was 0.55 m/s. Conversion of v_a to g_b depends on which Rees River calibration relation is utilized (see Table 8.2). Maximum g_b calculated as follows: the sand with gravel relation ($g_b = 1.3v_a - 0.063$), the gravel with sand relation ($g_b = 1.8v_a + 0.041$), and the gravel relation ($g_b = 7.0v_a + 0.022$) are 0.65, 1.0, and 3.9 kg/m/s, respectively. (*Source*: Williams *et al.* 2015. Reproduced with permission of John Wiley & Sons.)

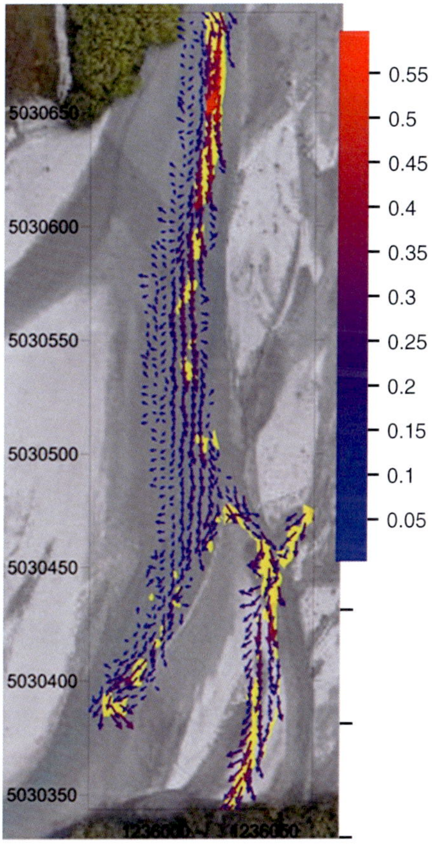

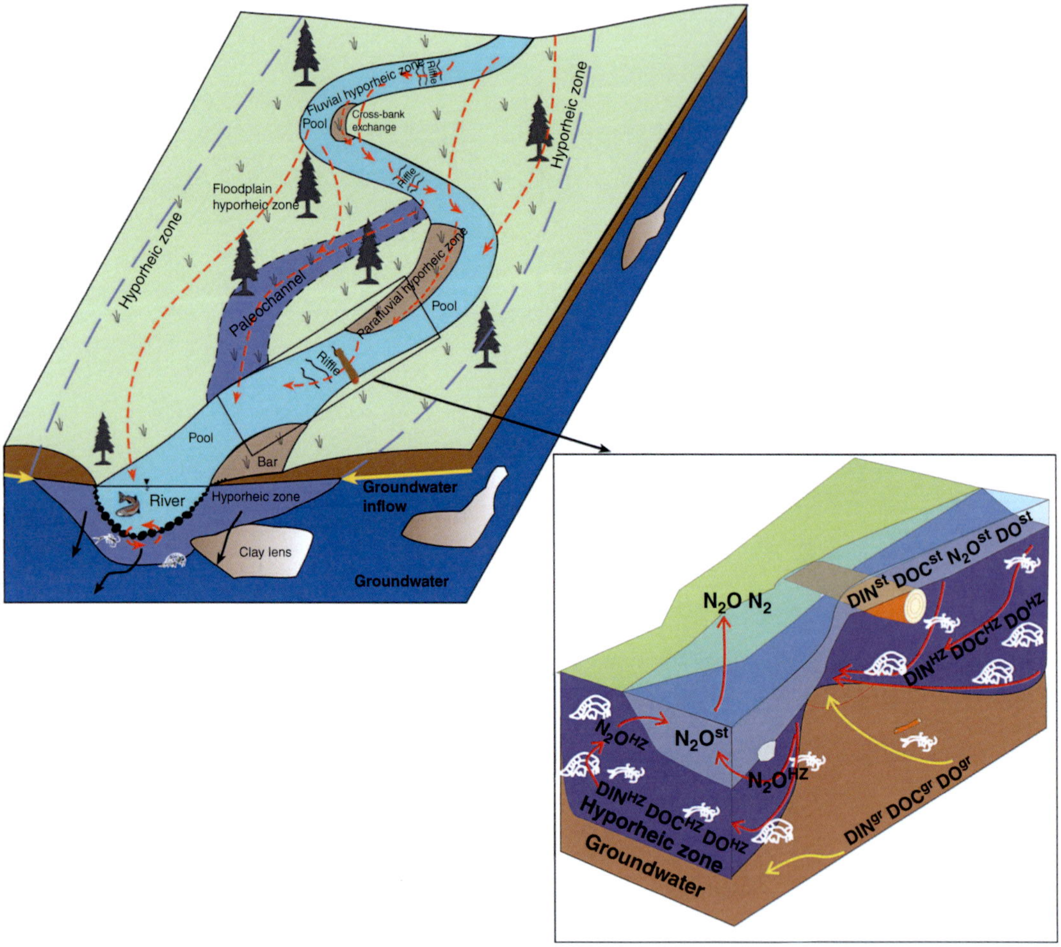

Figure 9.1 Fluvial, parafluvial, and floodplain hyporheic zones are depicted at the channel-unit scale around riffles and at channel-reach scale between meanders. The close-up shows the microscale fluvial hyporheic zone delineated by the biological method, shown by the presence of hyporheos, macroinvertebrates leaving in the hyporheic zone shown as white and green icons, geochemical method, shown by the dark purple color, which indicates more than 10% stream water, and hydraulic method, shown by the deepest red lines. Pathways of reactive solutes, dissolved organic carbon (DOC), dissolved inorganic nitrogen (DIN), and dissolved oxygen (DO), are also shown, with superscript *gr* for groundwater constituents, superscript *st* for stream and HZ for hyporheic zone (Modified from Tonina and Buffington 2009a).

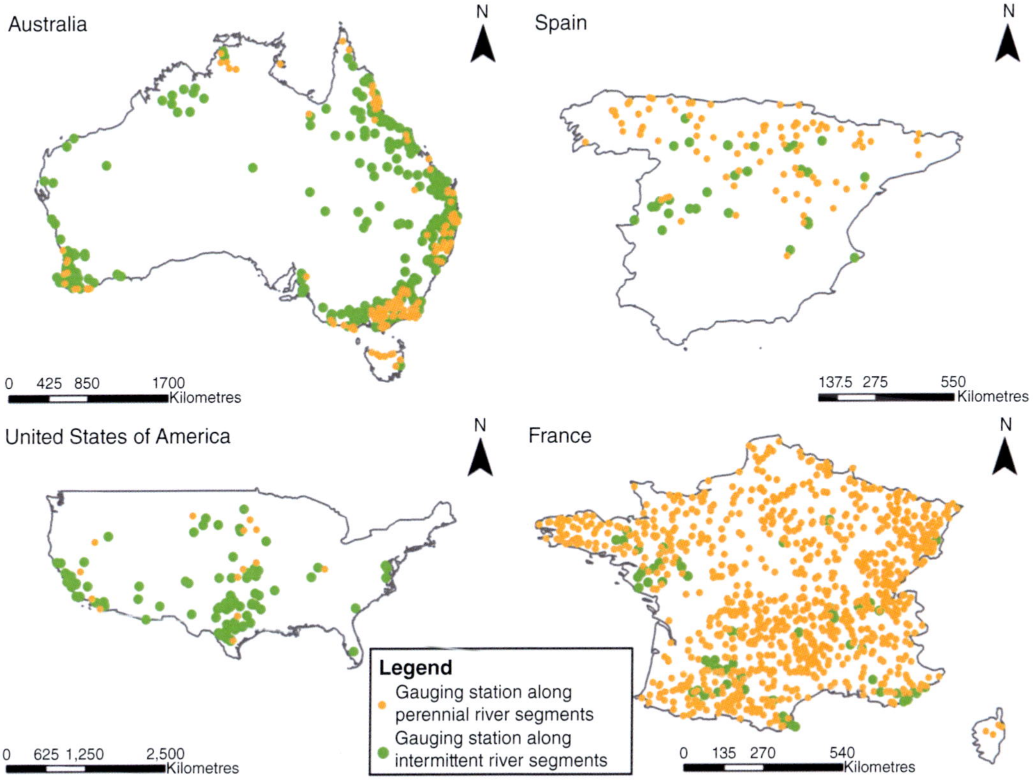

Figure 10.3 Classification of selected gauging stations across Australia, France, United States, and Spain. (Modified after Hoeve and De Vries (2014).)

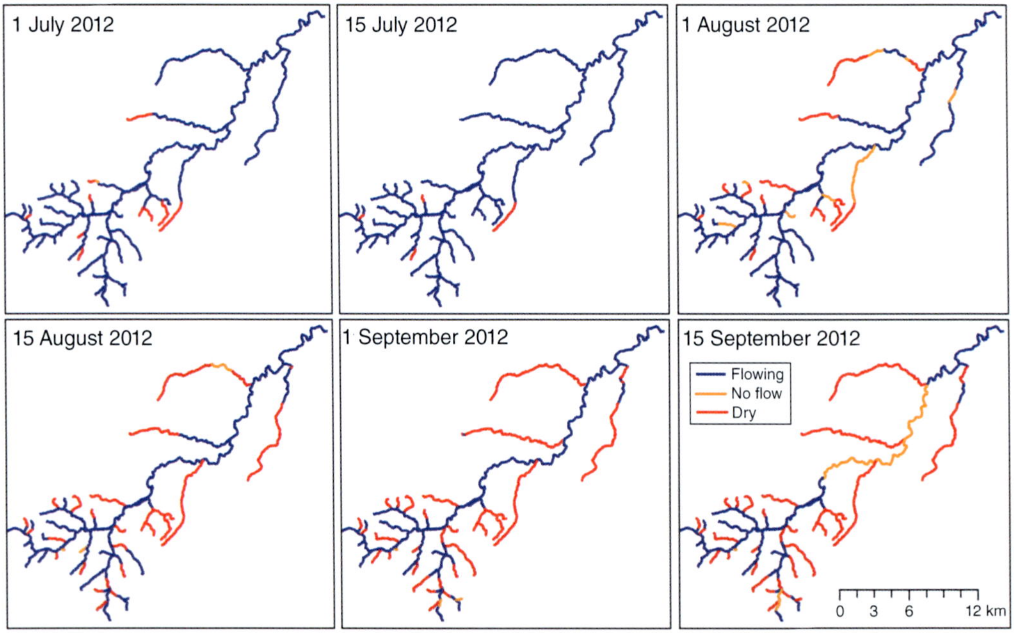

Figure 10.12 Temporal dynamics in patch composition and configuration of the Thouaret catchment during summer 2012: blue, lotic patches; orange, lentic patches; red, terrestrial patches. (Modified after Datry *et al.* (2016).)

Figure 14.15 (a) Measured bed topography of Dashidaira reservoir after the flushing operation in June 2012; (b) corresponding measured bed changes after the flushing operation (Esmaeili *et al.* 2015).

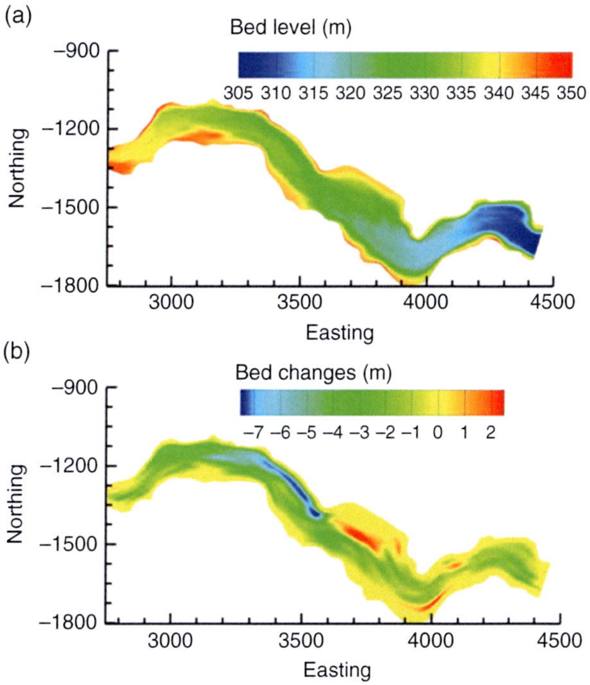

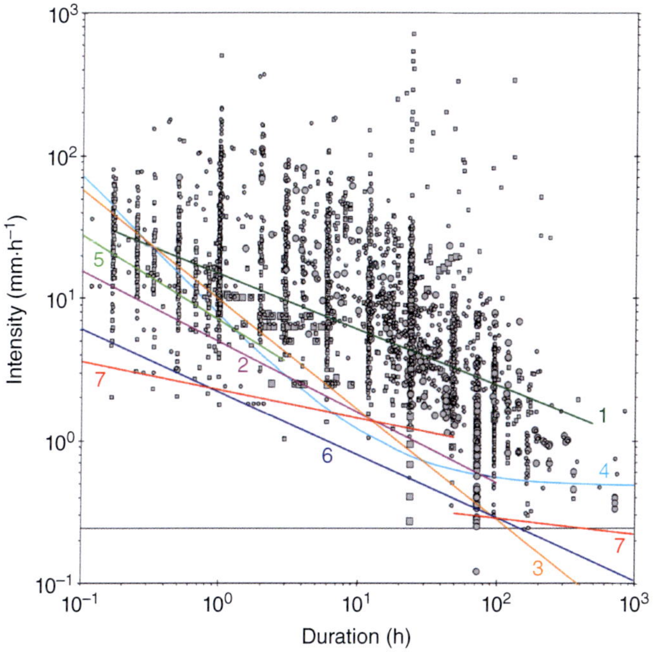

Figure 17.10 Plot of mean rainfall intensity versus rainstorm duration for a global (worldwide) dataset, and critical rainfall thresholds for landslide triggering: square, debris flow; dot, shallow failure; small symbol, single triggered landslide; large symbol, multiple triggered landslides; 1–6, rainfall thresholds proposed by various authors; 7, thresholds inferred from the probability estimates of the rainfall conditions for two different rainfall periods ($D < 48$ h, and $D \geq 48$ h). The grey horizontal line shows a rainfall intensity of 0.25 mm h^{-1}. (*Source*: Guzzetti *et al.* 2008. Reproduced with permission of Springer.)

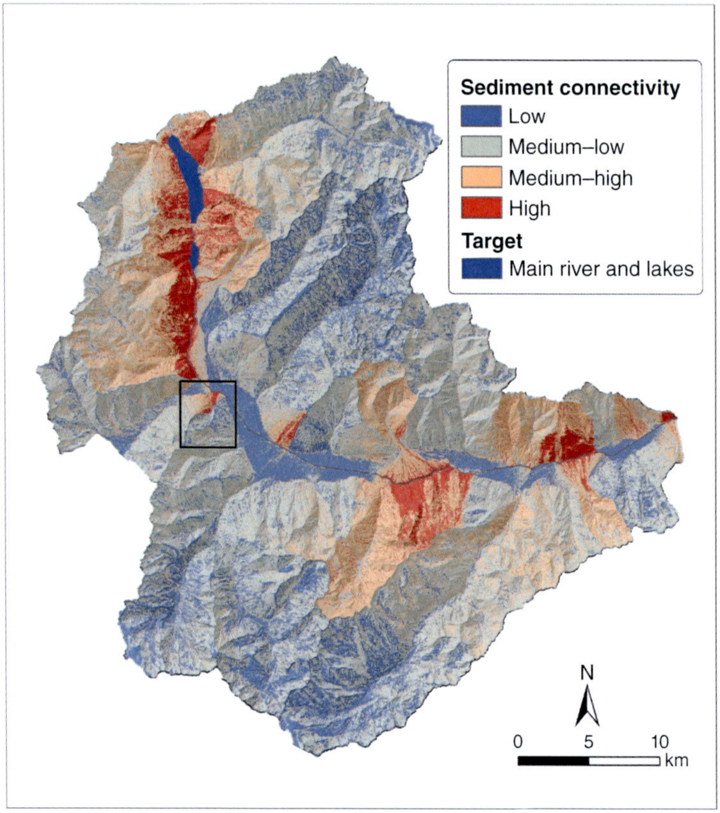

Figure 17.12 Map of the sediment connectivity index *IC* (equation 17.5) for the upper Venosta Valley showing the linkage of different sectors of the river basin with the main river. Four classes have been identified according to the Jenks natural breaks method; the rectangle identifies the enlarged portion shown in Figure 17.13.

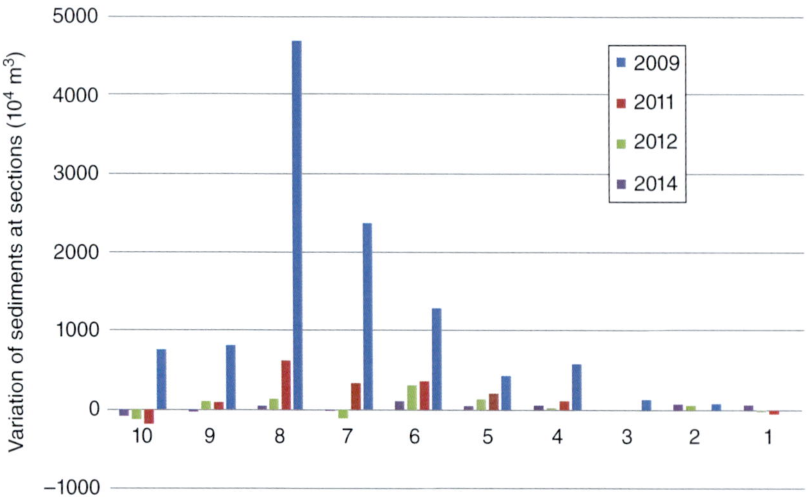

Figure 18.20 Variation in volume of sediment along the main Taimali River channel during 2009–2014.

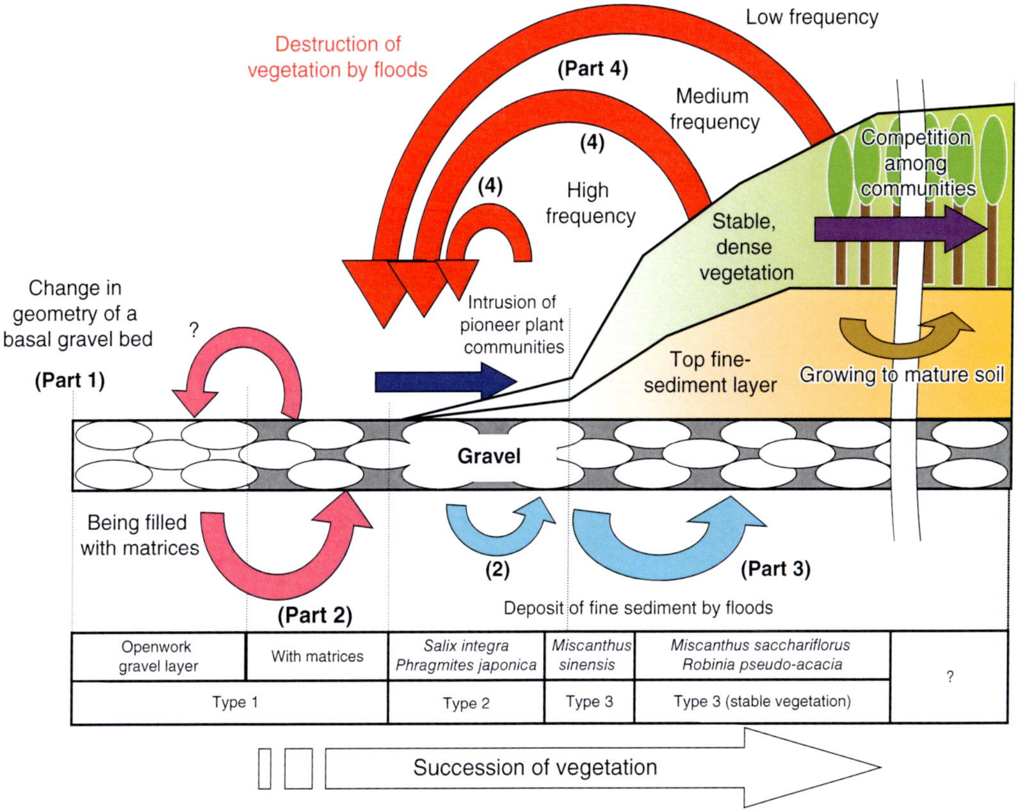

Figure 19.10 Framework of the simulation model to calculate surface state changes.

Figure 20.11 Intensive sand mining at the upper Gendol River, 2009, with Mount Merapi in the background.

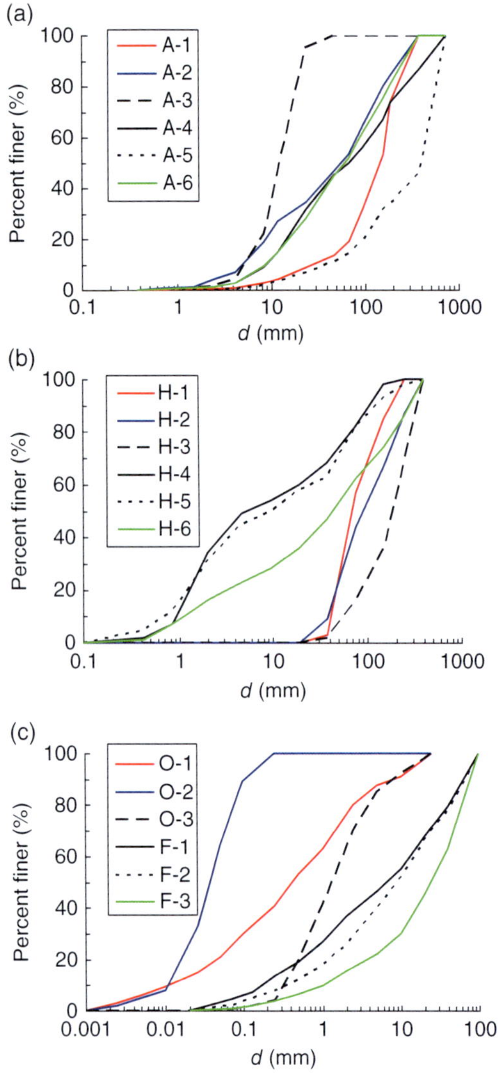

Figure 21.7 Grain-size distributions of natural sediment mixtures sampled in (a) Ai River, (b) Hino River, and (c) Ohtaki River and Fukadani bare slope (Sulaiman, Tsutsumi and Fujita 2007).

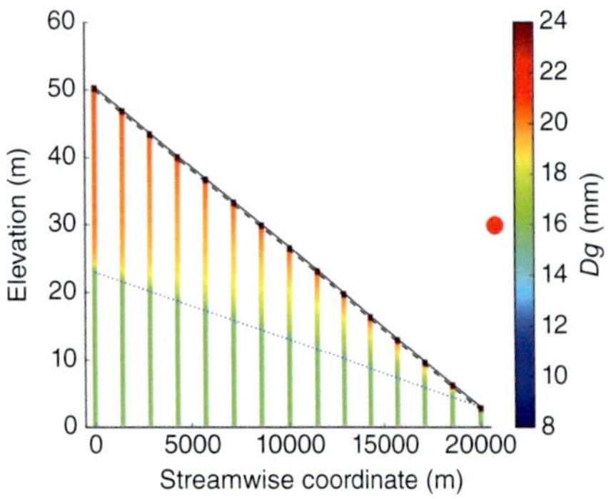

Figure 23.5 Grain size stratigraphy of the channel bed deposit. The continuous black line represents the bed elevation, the dashed line is the elevation of the active layer – substrate interface, the dotted line is the initial bed elevation, and the color scale represents the geometric mean diameter of the sediment stored in the channel bed. The dot next to the scale indicates the reach-averaged measured geometric mean diameter equal to 16 mm.

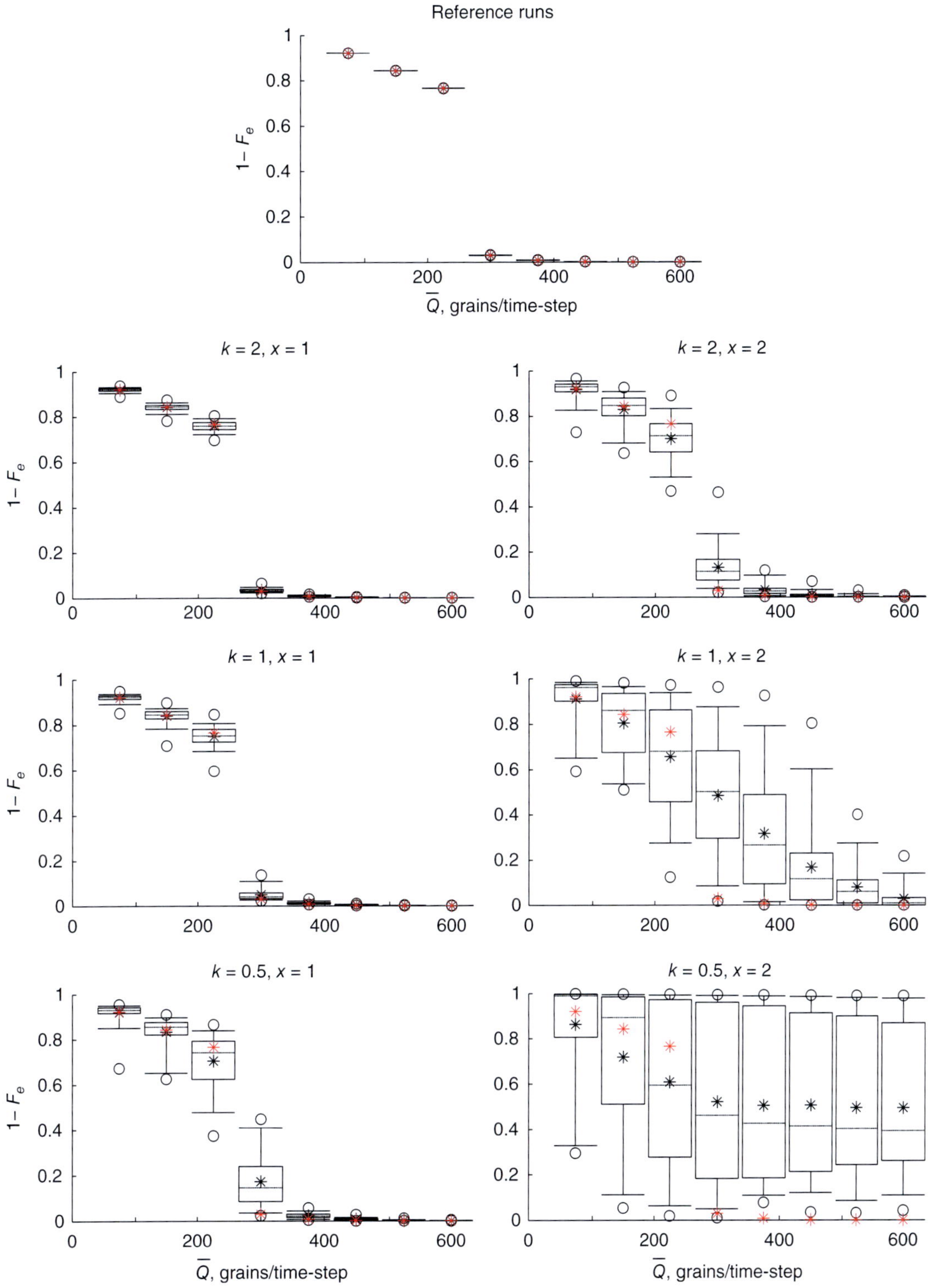

Figure 24.5 Box plots of sediment cover variation during the model runs. Lowest variability in sediment supply is given by $k = 2$, $x = 1$ and highest variability by $k = 0.5$, $x = 2$. Data are taken from time-steps 5000 to 20 000: 5000 is used because all runs with uniform sediment input show steady-state behaviour after this time. For each time-step, the sediment cover is calculated from the entire model domain. Boxplots show maximum and minimum (o), 5th and 95th percentiles (whiskers), Inter Quartile Range (IQR) and median (box and line), and mean (∗). Red means are values for the reference runs with constant sediment input.

(a)

(b)

Figure 26.1 Examples of mixed-sediment morphodynamics. (a) Intervention of sediment nourishment in the Rhine River (Netherlands). Courtesy Rijkswaterstraat, Netherlands. (b) Outlet of a sediment bypass tunnel near Alvaschein, Canton of Grisons, Switzerland (downstream view). (Courtesy M. Facchini, ETH Zurich, Switzerland.)

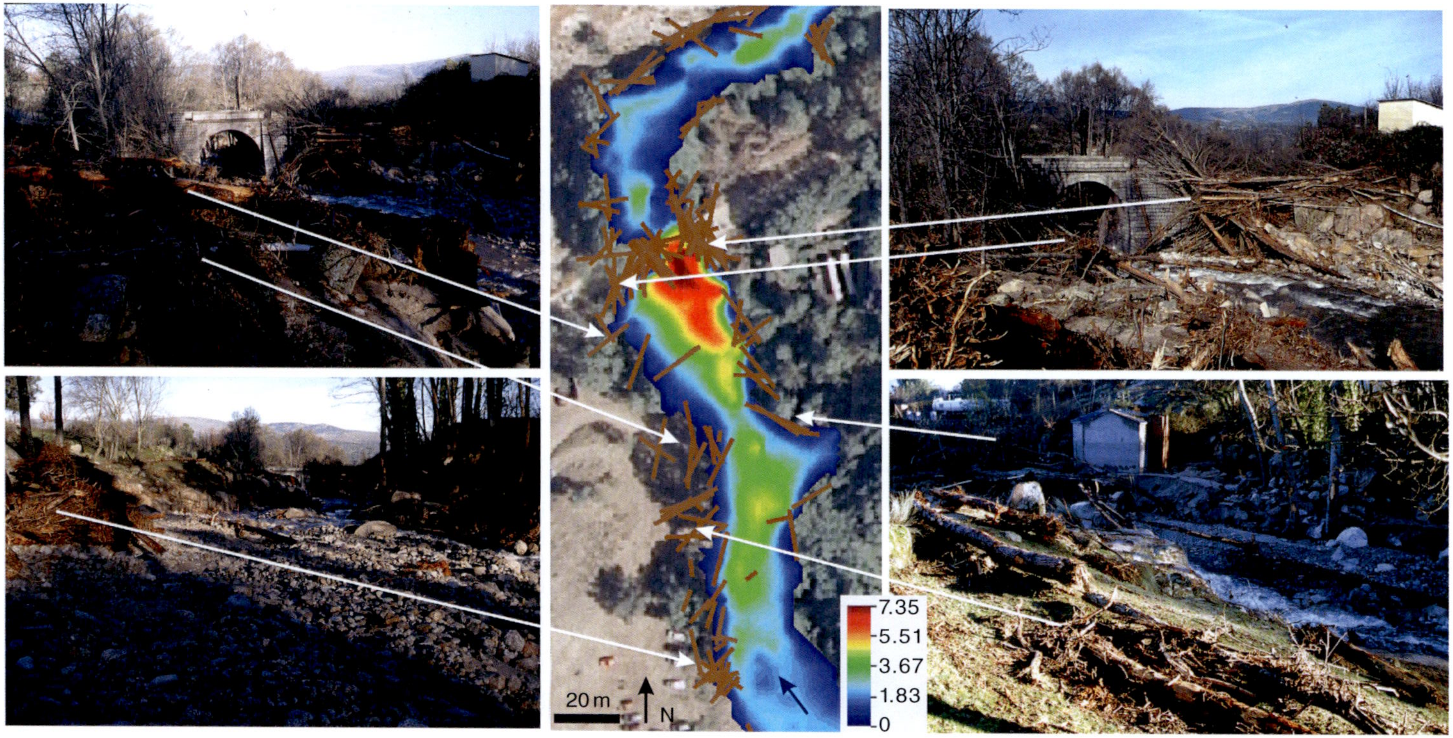

Figure 27.7 Field observations and model results for the simulation of the bridge clogging process assuming a semi-congested wood transport scenario. (Modified from Ruiz-Villanueva *et al.* (2014b).)

(b)

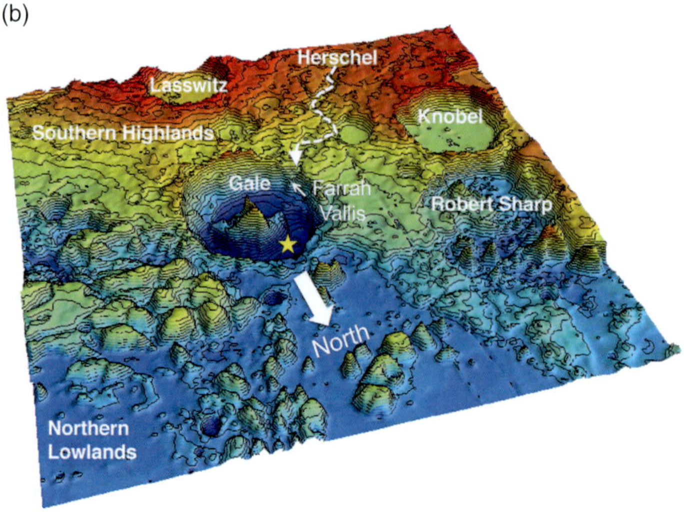

Figure 28.1 (b) A closer view of Gale Crater and its surrounding terrain. The lowest elevations in Gale Crater occur on the northwest portion of the crater floor, near the MSL landing site. Gale's impact disrupted a pre-existing valley network originating near Herschel Crater; this was likely one of the main early sources of water into Gale Crater. (Based on Irwin *et al.* (2005).)

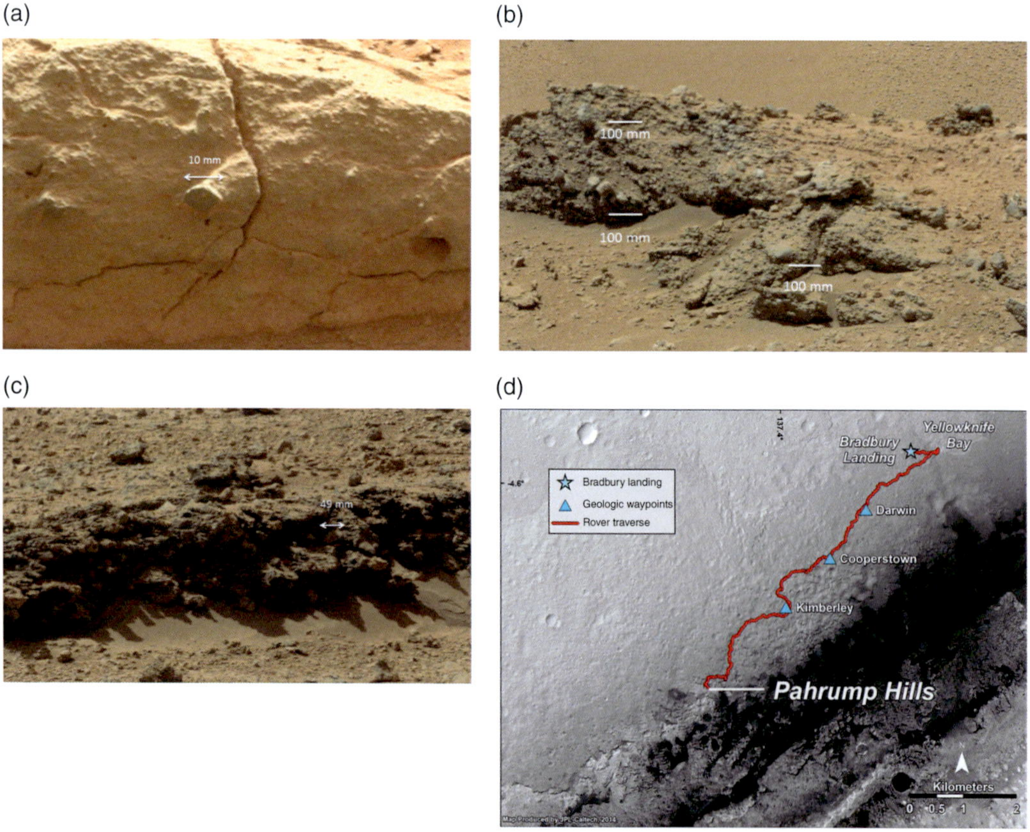

Figure 28.4 Three examples of gravel exposures as observed by the Mastcam cameras on the *Curiosity* rover. (a) An image from Yellowknife Bay, taken on Sol 153 (Mastcam 100 MR0850001), showing isolated stones in poorly sorted sandstones. (b) Mastcam image (MR002747) taken on Sol 646 from Occum Pond location showing a coarse conglomerate. (c) Mastcam image (MR002462) taken on Sol 588 from Point Coulomb location showing another example of a coarse gravel conglomerate. (d) Drive path of the *Curiosity* rover. Image is available at NASA web page as Mars-MSL-curiosity-rover-location-map-pahrump-hills-pia18607-br2.

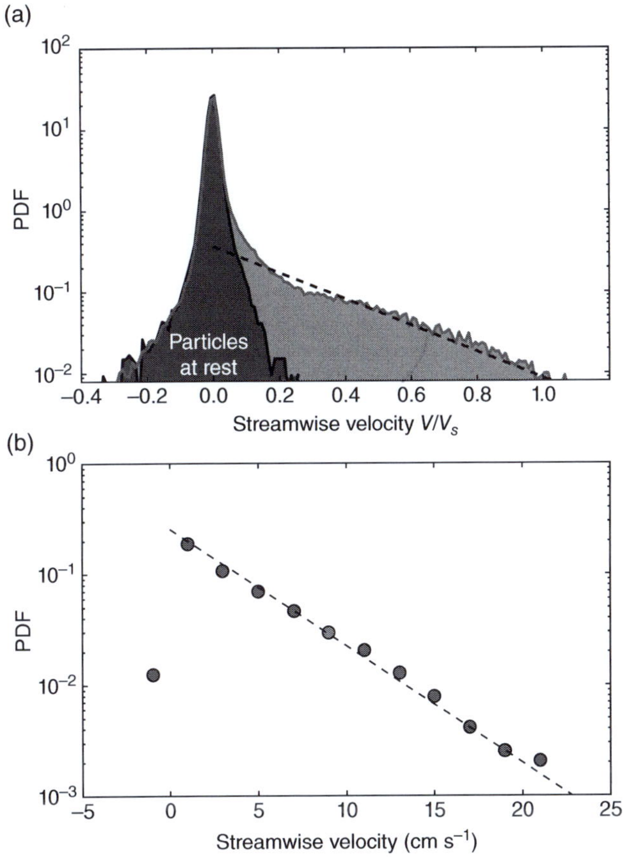

Figure 15.5 (a) Streamwise velocity distribution of plastic sediments (density $\rho_s = 1520 \pm 50$ kg m^{-3}, size $d_s = 344 \pm 90\,\mu$m) entrained by a laminar flow ($Re \approx 500$). The velocity distribution (light grey) is obtained by tracking dyed particles, including those resting on the bed. The spurious noise induced by free surface waves is characterized by the distribution of apparent velocities computed for static particles (dark grey). The tail of the velocity distribution is reasonably well fitted by an exponential function (dashed line). (After Seizilles *et al.* (2014).) (b) Streamwise velocity distribution of sand particles (density $\rho_s = 2500$ kg m^{-3}, size $d_s = 500 \pm 50$ µm) sheared by a turbulent flow ($Re \approx 3000$). Particles are discarded when slower than a small cut-off velocity. The velocity distribution is fitted by an exponential function (dashed line). (After Roseberry, Schmeeckle, and Furbish (2012).)

$$V \propto d_s \frac{\partial u}{\partial z} \tag{15.6}$$

where we introduce the flow velocity u and the vertical coordinate z. In a viscous flow, the shear rate $\partial u/\partial z$ is proportional to the shear stress, τ, so that

$$V \propto \frac{d_s}{\eta} \tau \propto V_s\, \theta \tag{15.7}$$

where η is the fluid viscosity and V_s is the settling velocity of the particle. Equation (15.7) accords with experiments (Charru, Mouilleron, and Eiff 2004; Lajeunesse, Malverti, and Charru 2010; Seizilles *et al.*, 2014).

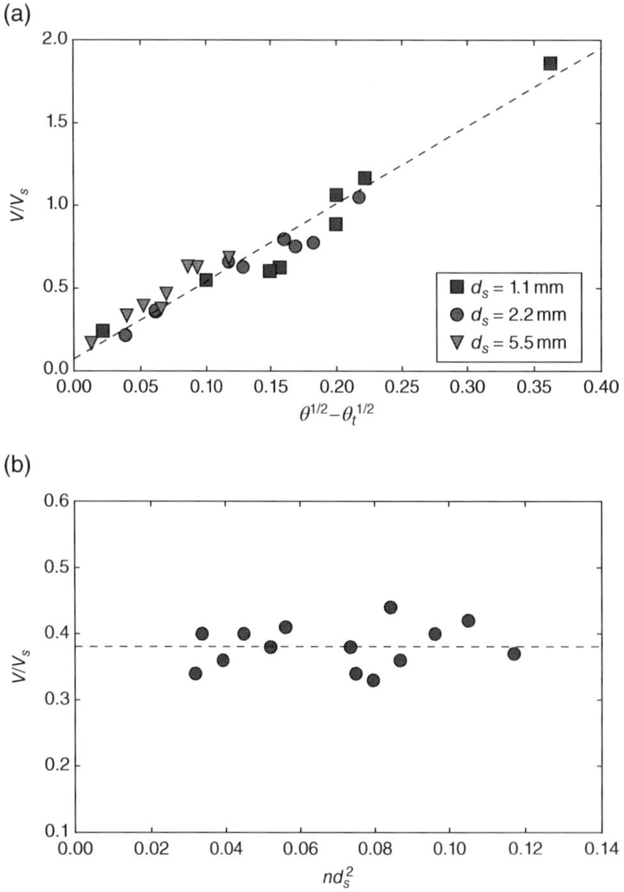

Figure 15.6 (a) Average particle velocity, V/V_s, as a function of $\left(\theta^{1/2}-\theta_t^{1/2}\right)$ for quartz grains in a turbulent flow. A fit of the data by Equation (15.8) (dashed line) leads to $a = 4.4 \pm 0.2$ and $b = 0.11 \pm 0.03$. (After Lajeunesse, Malverti, and Charru (2010).) (b) Average particle velocity, V/V_s, as a function of the concentration of moving sediments, nd_s^2, measured near the threshold. Plastic grains (density $\rho_s = 1520 \pm 50$ kg m^{-3}, size $d_s = 344 \pm 90$ μm) are entrained by a laminar flow ($Re \approx 500$). (After Seizilles *et al.* (2014).)

In a turbulent flow, the particle velocity is expected to depend on the shear velocity, $u_\star = (\tau/\rho)^{1/2} = V_s\theta^{1/2}$. Laboratory experiments and numerical simulations show that

$$V = V_s\left[a\left(\theta^{1/2}-\theta_t^{1/2}\right)+b\right]. \tag{15.8}$$

where a and b are dimensionless coefficients (Figure 15.6a) (Lajeunesse, Malverti, and Charru 2010; Durán, Andreotti, and Claudin 2012).

In both laminar and turbulent flows, the average particle velocity, V, scales like the particle settling velocity, V_s, which depends on the flow regime. In a viscous flow, it is the Stoke velocity $V_s = (\rho_s-\rho)gd_s^2/18\eta$, whereas it reads $V_s = ((\rho_s-\rho)gd_s/\rho)^{1/2}$ in a turbulent flow.

In a laminar flow, the particle velocity tends towards a finite value near threshold (Figure 15.6b; Seizilles *et al.* 2014). This question is still a matter of debate for turbulent flows (Lajeunesse, Malverti, and Charru 2010; Durán *et al.* 2014).

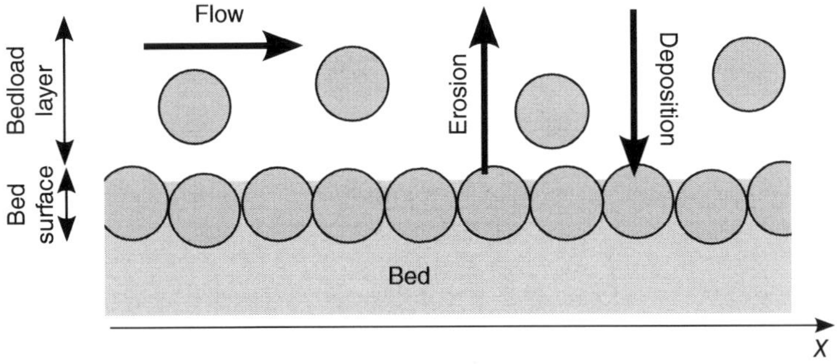

Figure 15.7 Granular bed sheared by a laminar flow. Moving grains settle on the bed, while static grains are entrained by the flow.

15.2.3 Surface Concentration of Moving Particles

The bedload layer results from a constant exchange of particles with the bed. The associated balance reads (Figure 15.7):

$$\frac{\partial n}{\partial t} + V\frac{\partial n}{\partial x} = \dot{n}_e - \dot{n}_d \tag{15.9}$$

where we define the erosion rate, $\dot{n}_e$, as the number of bed particles set in motion per unit time and surface. Similarly, we introduce the deposition rate $\dot{n}_d$ defined as the number of bedload particles settling on the bed per unit time and surface. The mass balance needs to be complemented with expressions for the erosion and deposition rates.

Experimental observations suggest that a particle entrained by the flow bounces on the bed at a rate proportional to d_s/V_s (Charru *et al.* 2007; Lajeunesse, Malverti, and Charru 2010; Seizilles *et al.* 2014). Each time the particle hits the bed, it can be trapped with a finite probability. Assuming a uniform probability, the deposition rate scales like the concentration of moving particles times the bouncing frequency:

$$\dot{n}_d \propto n\frac{V_s}{d_s}. \tag{15.10}$$

Similarly, laboratory experiments suggest that the erosion rate increases with the excess Shields stress, $\theta-\theta_t$ (Charru, Mouilleron, and Eiff 2004; Lajeunesse, Malverti, and Charru 2010; Seizilles *et al.* 2014). At first order, we also expect it to be proportional to the concentration of particles at the bed surface, $\sigma \approx 1/d_s^2$:

$$\dot{n}_e \propto \frac{f(\theta-\theta_t)}{d_s^2}\frac{1}{t_s} \tag{15.11}$$

where $t_s = d_s/V_s$ and f is a growing function.

The form of f can be deduced from steady-state experiments. In this configuration, the mass balance (15.9) reduces to $f(\theta-\theta_t) \propto n d_s^2$. Measurements of the concentration of bedload particles (see Figure 15.8) show that f is about linear in both laminar and turbulent flows (Charru, Mouilleron, and Eiff 2004; Lajeunesse, Malverti, and Charru 2010; Seizilles *et al.* 2014). The erosion rate is therefore:

$$\dot{n}_e \propto \frac{1}{d_s^2}\frac{V_s}{d_s}(\theta-\theta_t). \tag{15.12}$$

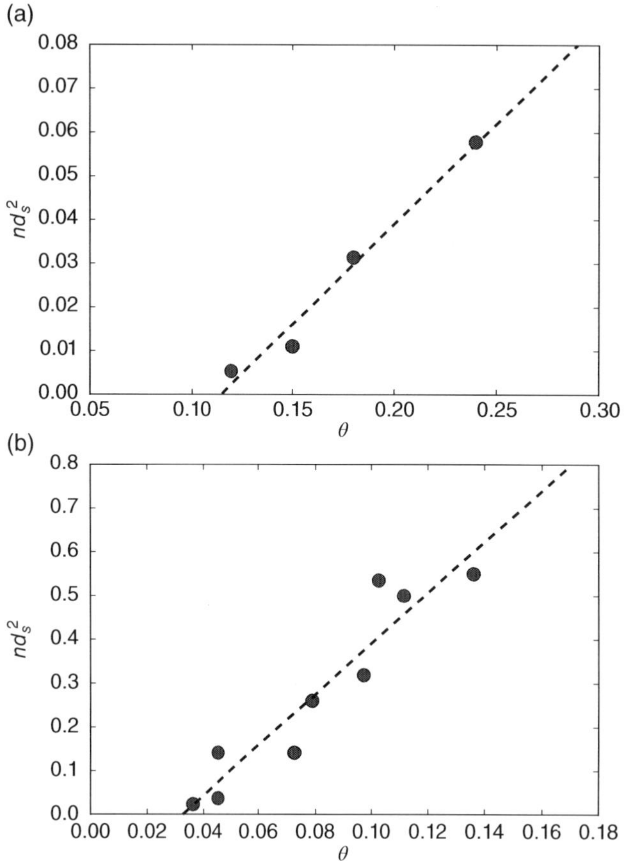

Figure 15.8 Surface concentration of moving grains, nd_s^2, versus Shields number, θ. (a) Spherical acrylic beads (density $\rho_s = 1180\,\mathrm{kg\,m^{-3}}$, size $d_s = 580 \pm 100\,\mu\mathrm{m}$) entrained by a viscous flow of silicon oil ($Re \approx 40$). (After Charru, Mouilleron, and Eiff (2004).) (b) Quartz grains (size $d_s = 2.2$ mm) entrained by a turbulent flow. (After Lajeunesse, Malverti, and Charru (2010).)

15.2.4 Bedload Transport Rate

Equations (15.5), (15.6), (15.9), (15.10) and (15.12) describe bedload transport according to the erosion–deposition model. The resolution of these equations for given flow conditions provides the erosion, deposition, and bedload transport rates.

In the case of a steady and spatially uniform flow above a flat topography, the erosion and deposition rates balance each other. In this equilibrium regime, the sediment flux reads

$$q_{s,\mathrm{eq}} \propto V d_s (\theta - \theta_t) \qquad (15.13)$$

where the index 'eq' denotes equilibrium. The average particle velocity V is given by either (15.7) or (15.8) depending on the nature of the flow.

Near the threshold of entrainment, the particle velocity scales as the settling velocity so that the sediment transport rate becomes a linear function of the excess Shields stress,

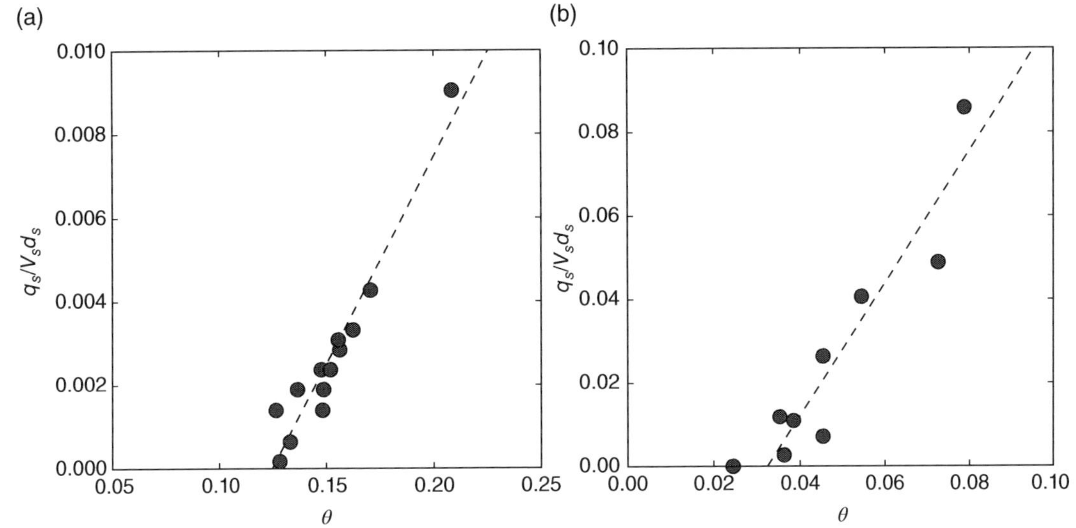

Figure 15.9 Sediment transport rate versus Shields number close to the entrainment threshold. (a) plastic sediments (density $\rho_s = 1520 \pm 50$ kg m^{-3}, size $d_s = 344 \pm 90$ μm) entrained by a laminar flow ($Re \approx 500$). (After Seizilles *et al.* (2014).) (b) Quartz grains (size $d_s = 2.2$ mm) entrained by a turbulent flow. (After Lajeunesse, Malverti, and Charru (2010).)

$$q_s \propto V_s d_s(\theta - \theta_t), \tag{15.14}$$

in agreement with measurements (Figure 15.9).

The erosion–deposition model proves reasonable over a broad range of shear stress (Charru, Mouilleron, and Eiff 2004; Ancey 2010; Lajeunesse, Malverti, and Charru 2010). At high shear stress, however, the bedload layer becomes vertically stratified (Recking *et al.* 2009; Durán *et al.* 2012). Equation (15.5) then breaks down, and must be replaced with more detailed diphasic models (e.g. Chiodi, Claudin, and Andreotti 2014; Revil-Baudard and Chauchat 2013).

15.3 Deposition Length and Bedforms

Rivers exhibit a beautiful variety of bedforms such as ripples, dunes, and bars (Charru, Andreotti, and Claudin 2013; Seminara 2010). These alluvial morphologies result from the feedback between the flow and the bed geometry: (i) the shape of the sediment bed sets the fluid velocity and, therefore, the sediment transport rate; (ii) sediment transport modifies in turn the shape of the bed.

The mathematical description of this feedback requires an expression for the sediment flux above a wavy bed. When the bed evolves slowly enough, we may consider that the flow is stationary (quasi-static approximation). The shear stress then depends on the bed's shape only, and so does the sediment flux. Assuming that particles have negligible inertia, their velocity V adapts instantaneously to the shear stress. Equations (15.5), (15.6), (15.9), (15.10) and (15.12) thus reduce to

$$\ell_d \frac{\partial q_s}{\partial x} = q_{s,\text{eq}} - q_s, \tag{15.15}$$

where $q_{s,\text{eq}}$ is the equilibrium sediment flux predicted from Equation (15.13) and ℓ_d is the average length travelled by a particle before it bounces back on the bed ($\ell_d \sim Vt_s \sim Vd_s/V_s$). According to

Equation (15.15), the sediment flux adjusts to the local shear stress over the typical distance ℓ_d. This non-locality of sediment discharge is consistent with other models proposed in the literature (Bell and Sutherland 1983; Phillips and Sutherland 1989, 1990). In most of these studies, however, non-locality is introduced heuristically. In the erosion–deposition model, it results spontaneously from the exchange of particles between the bed and the bedload layer: entrainment and deposition depend on local parameters through Equations (15.10) and (15.12), and the sediment flux adapts with a delay described by the relaxation Equation (15.15).

The length ℓ_d is analogous to the saturation length introduced in the context of aeolian sand transport, where it results from grain inertia and scales like $d_s\rho_s/\rho$, regardless of the wind velocity (Sauermann, Kroy, and Herrmann 2001; Andreotti, Claudin, and Douady 2002; Hersen 2005; Claudin and Andreotti 2006; Narteau *et al.* 2009; Andreotti, Claudin, and Pouliquen 2010; Charru, Andreotti, and Claudin 2013; Pähtz *et al.* 2013).

Bedload transport generates instabilities: a small perturbation on a flat bed can cause the sediment flux to further increase its size. Charru (2006) successfully describes the early stages of ripple growth in laboratory experiments with a linear stability analysis. The relaxation mechanism embedded in Equation (15.15) introduces a cut-off length that stabilizes the short wavelengths (Lagrée 2003; Kouakou and Lagrée 2005; Lagrée *et al.* 2009; Fourriere, Claudin, and Andreotti 2010). As a result, the size of the most unstable ripples scales with the deposition length ℓ_d, in agreement with experiments (Figure 15.10a). Similarly, the deposition length seems to control the size of the rhomboid pattern and of alternate bars (Figure 15.10b; Devauchelle *et al.* 2010a,b; Andreotti *et al.* 2012).

Laminar free-surface flows are certainly distinct from turbulent flows in many regards. For instance, the shape of the vertical velocity profile differs between the two, and so does the friction law. However, these quantitative differences do not affect the structure of the associated shallow-water equations (Devauchelle *et al.* 2007). Similarly, the nature of the flow changes the velocity of bedload particles (Equations (15.7) and (15.8)), but does not qualitatively alter the mechanism by which the bed

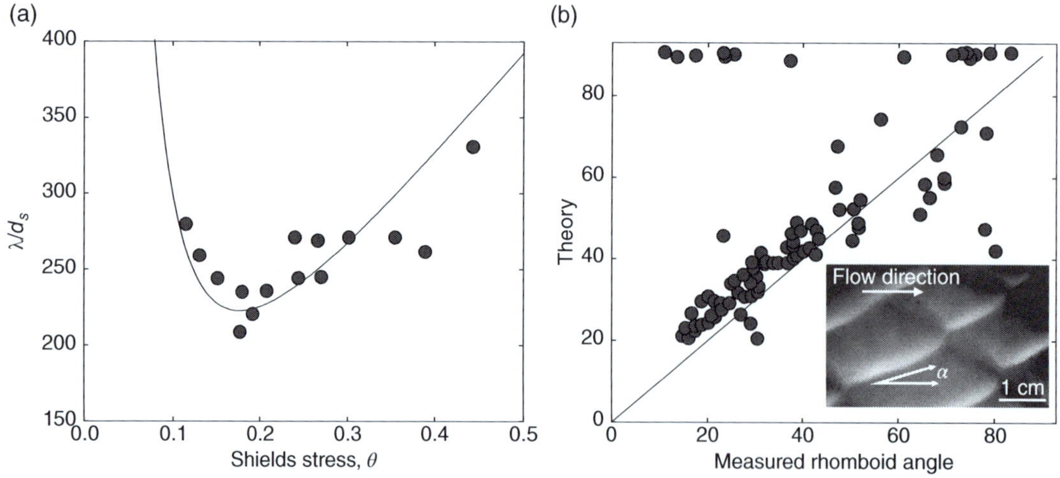

Figure 15.10 (a) Ripples wavelength λ as a function of Shields stress for sand grains in water ($d_s = 200\ \mu m$). Dots: experimental measurements by Coleman and Melville (1996). Solid line: prediction of the erosion–deposition model (Charru 2006). (b) Opening angle α of a rhomboid pattern formed by a viscous flow shearing a bed of glass beads. Dots: comparison between a linear stability analysis and observations. Inset: example of rhomboid pattern. (After Devauchelle *et al.* (2010b).)

exchanges particles with the bedload layer. This analogy explains why many alluvial structures found in nature also form in small laboratory experiments (Coleman and Eling 2000; Paola *et al.* 2001, 2009; Metivier and Meunier 2003; Malverti, Lajeunesse, and Metivier 2008; Devauchelle *et al.* 2010a; Lajeunesse *et al.* 2010; Seizilles *et al.* 2013; Guerit *et al.* 2014).

15.4 Spreading of a Plume of Tracers

Even in steady state, the bedload layer constantly exchanges particles with the static bed. To illustrate this mechanism, we now consider a plume of tracer particles entrained by bedload transport. We separate the particles into two groups with the same physical properties, hereafter referred to as 'marked' and 'unmarked'. For the sake of simplicity, we furthermore assume that sediment transport is uniform and steady. In this configuration, Equations (15.5), (15.6), (15.9), (15.10) and (15.12) predict that the Shields stress (and consequently the mean particle velocity), the erosion and deposition rates, and the surface concentration of moving particles are constant. Equation (15.13) gives the resulting transport rate.

Let us define ϕ as the proportion of marked grains in the moving layer. Similarly, ψ is the proportion of marked grains on the bed surface. The concentration of marked grains, defined as the ratio of the number of marked particles to the surface concentration of particles, is

$$c = \frac{n\phi + \sigma\psi}{n+\sigma} = \frac{\alpha}{\alpha+1}\phi + \frac{1}{\alpha+1}\psi \tag{15.16}$$

where we introduce $\alpha = n/\sigma \sim nd_s^2$, the ratio of the surface concentration of moving particles, n, to the concentration of static particles, σ (this ratio is <1). The total amount of tracers,

$$M = \int_{-\infty}^{\infty} c\mathrm{d}x, \tag{15.17}$$

is conserved. The mass balance for marked grains in the moving layer reads

$$\frac{\partial}{\partial t}(\phi n) + \frac{\partial}{\partial x}(\phi n\ V) = \dot{n}_e\psi - \dot{n}_d\phi. \tag{15.18}$$

In steady-state, erosion compensates deposition:

$$\dot{n}_e = \dot{n}_d = \frac{n}{t_s}, \tag{15.19}$$

where t_s is the characteristic settling time ($t_s \sim d_s/V_s$, see Equation 15.10). Combining Equations (15.18) and (15.19), we find the evolution equation for the proportion of marked grains in the moving layer:

$$\frac{\partial\phi}{\partial t} + V\frac{\partial\phi}{\partial x} = \frac{1}{t_s}(\psi - \phi). \tag{15.20}$$

We further assume that the flow can only entrain grains from the bed surface. In other words, the active layer is one-grain-diameter thick. This hypothesis is consistent with a small departure from threshold. Balancing the fluxes of marked grains, we find

$$\sigma\frac{\partial\psi}{\partial t} = \dot{n}_d\phi - \dot{n}_e\psi \tag{15.21}$$

which, when combined with Equation (15.19), leads to the evolution equation for the proportion of marked grains on the bed surface:

$$\frac{\partial \psi}{\partial t} = -\frac{\alpha}{t_s}(\psi - \phi). \tag{15.22}$$

Complemented with initial and boundary conditions, Equations (15.20) and (15.22) describe the evolution of the plume of marked particles.

In dimensionless form, Equations (15.20) and (15.22) read

$$\frac{\partial \phi}{\partial \hat{t}} + \frac{\partial \phi}{\partial \hat{x}} = (\psi - \phi) \tag{15.23}$$

$$\frac{\partial \psi}{\partial \hat{t}} = -\alpha(\psi - \phi). \tag{15.24}$$

where $\hat{t} = t/t_s$ and $\hat{x} = x/(V t_s)$ are dimensionless variables. A single parameter controls this system of equations: the ratio of surface densities α, which characterizes the average distance between grains in the bedload layer. Since the erosion–deposition model assumes that the behaviour of an individual particle is independent from other particles, we can only expect it to be valid when moving particles are sufficiently far away from each other, that is when α is small or, equivalently, when the Shields parameter is near threshold.

We did not measure the ratio of surface densities for the experiment of Figure 15.1. However, its value can be estimated from the sediment flux and the particle settling velocity: $\alpha \sim 8 \times 10^{-2}$. We solved Equations (15.23) and (15.24) for $\alpha = 8 \times 10^{-2}$ with finite volumes (Lajeunesse *et al.* 2013). The agreement with experimental data suggests that the erosion–deposition model reasonably represents the plume dynamics.

We now turn our attention to the behaviour of the sediment plume at long durations. To do so, we solve numerically Equations (15.23) and (15.24), starting from an arbitrary configuration, and let it evolve for a long period (Figure 15.11). As expected, the plume spreads as it is transported downstream by the flow. Initially, its shape is skewed towards the direction of propagation. This skewness later disappears, and the plume resembles a normal distribution at long durations.

The average position of the plume, defined as

$$\langle X \rangle = \frac{1}{M}\int_{-\infty}^{\infty} c x \, \mathrm{d}x, \tag{15.25}$$

increases with time and quickly reaches a constant velocity (Figure 15.12). Similarly, the variance of the plume, defined as

$$\sigma_X^2 = \frac{1}{M}\int_{-\infty}^{\infty} c(x - \langle X \rangle)^2 \, \mathrm{d}x \tag{15.26}$$

increases linearly with time after a short transient, thus mimicking diffusion. We observe this 'effective diffusion' behaviour regardless of α. This suggests that, at long durations, a plume of tracers advances at constant velocity and diffuses linearly, a behaviour reminiscent of the linear advection-diffusion equation. This observation is consistent with the diffusive 'intermediate range' regime introduced within the conceptual model of Nikora *et al.* (2002).

In the model presented here, the spread of the plume results only from the continuous exchange of particles between the bedload layer and the immobile bed. In reality, the distribution of particle velocities in bedload transport is continuous (Lajeunesse, Malverti, and Charru 2010; Furbish, Ball, and Schmeeckle

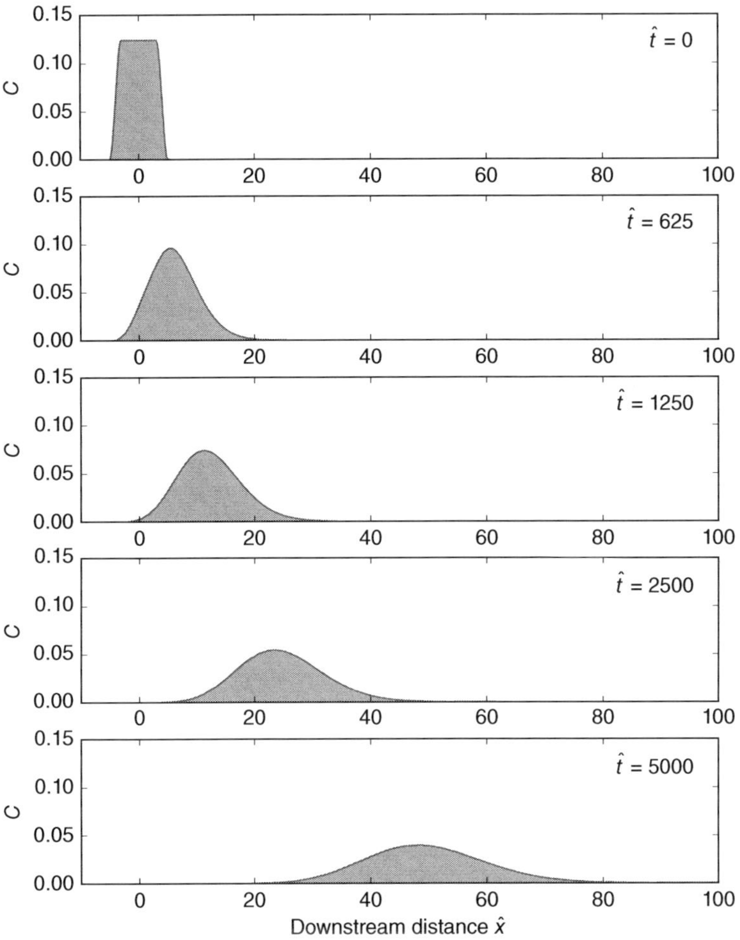

Figure 15.11 Numerical computation of the concentration of tracer particles as a function of the dimensionless distance *X*, for $\alpha = 0.01$. The initial length of the plume is $L = 10$. The concentration is normalized so that the total amount of tracers is 1.

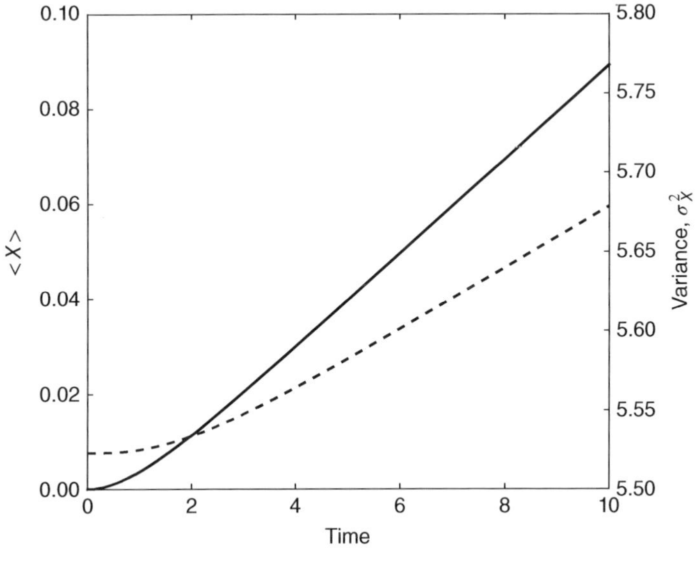

Figure 15.12 Mean position $<X>$ (solid line, left axis) and variance σ_X^2 (dashed line, right axis) of a plume of tracers as a function of time for the numerical run of Figure 15.11. After a short transient, the average position and the variance of the plume increase linearly with time.

2012a; Furbish, Roseberry, and Schmeeckle 2012c). Therefore, moving grains in the bedload layer also disperse due to velocity fluctuations. This effect, in addition to the continuous exchange of particles between the bedload layer and the immobile bed, is likely to enhance the dispersion of the plume of tracers.

Although derived in the context of bedload transport, Equations (15.23) and (15.24) apply to any steady uniform system in which some passive tracer is exchanged between two layers moving at different velocities. Among such systems are the Taylor dispersion, distillation, and chromatography (Taylor 1953; James, Postel, and Sepúlveda 2000).

15.5 Conclusions

Laboratory experiments reveal that, for moderate values of the shear stress, bedload transport involves a continuous exchange of particles between the sediment bed and a thin bedload layer of entrained grains (Fernandez-Luque and Van Beek 1976; Van Rijn 1984; Nino and Garcia 1994; Charru, Mouilleron, and Eiff 2004; Lajeunesse, Malverti, and Charru 2010). Taking advantage of this observation and using experimental particle velocity distributions, the erosion–deposition model provides a simple description of bedload transport. The bedload layer is treated as a uniform reservoir of independent particles moving at velocity V (Charru, Mouilleron, and Eiff 2004). The exchange of particles between the sediment bed and the bedload layer sets the surface concentration of moving particles, n, and therefore the sediment transport rate, $q_s = nV$.

For a steady flow over a varying topography, the sediment flux adjusts to a change of shear stress over a characteristic deposition length. This relaxation plays a crucial role in selecting the wavelength of bedforms such as ripples, rhomboids, and bars (Charru 2006; Charru and Hinch 2006; Devauchelle *et al.* 2010b). The equations describing bedload transport and the development of bedforms exhibit a strong mathematical analogy in laminar and turbulent flows, thus explaining the 'unreasonable effectiveness of stratigraphic and geomorphic experiments' (Paola *et al.* 2009).

A preliminary investigation indicates that the erosion–deposition model also accounts for the propagation of a plume of tracer sediments entrained by bedload transport. It suggests that, after a short transient, the plume reaches asymptotically an advection-diffusion regime, in which it advances with a constant velocity while its variance increases linearly with time. A more quantitative assessment of the model through systematic comparison between theory and experiments is currently under progress.

The results discussed in this chapter are based on idealized laboratory experiments (uniform grain size, constant flow, and sediment discharges). However, the discharge of most rivers fluctuates significantly throughout the year, and the sediments they transport are often broadly distributed in size. This is specially true for gravel-bed rivers. Recent progress in extending the erosion–deposition model to the transport of mixed grain sizes should allow us to investigate more realistic configurations (Houssais and Lajeunesse 2012; Houssais *et al.* 2015).

Notation

c	Concentration of tracer particles
d_s	Sediment size
g	Gravity
$F(v)$	Probability distribution function of bedload particle velocities
$\ell_d \sim Vt_s \sim Vd_s/V_s$	Average flight length travelled by a particle before it settles back on the bed

n	Surface concentration of bedload particles in number of particles per unit bed area
$\dot{n}_e$	Erosion rate in number of particles per unit time and surface
$\dot{n}_d$	Deposition rate in number of particles per unit time and surface
q_s	Sediment flux in number of particles per unit river width and time
$q_{s,\mathrm{eq}}$	Sediment flux in the case of steady and uniform transport
t	Time
$t_s \sim d_s/V_s$	Characteristic settling time
$u_\star$	Shear velocity of the flow
V	Average velocity of bedload particles
V_s	Sediment settling velocity
x	Downstream distance
$\langle X \rangle$	Average position of the plume of tracers
$\alpha = n/\sigma$	Ratio of the surface concentrations of the bedload layer to the bed surface
η	Fluid viscosity
ϕ	Proportion of tracers in the moving layer
ψ	Proportion of tracers on the bed surface
ρ	Fluid density
ρ_s	Sediment density
$\sigma \sim 1/d_s^2$	Surface concentration of static particles at the bed surface
σ_X^2	Variance of the plume of tracers
τ	Shear stress exerted by the flow on the bed
θ	Shields number
θ_t	Threshold Shields number

Acknowledgements

We are grateful to Colin Phillips and Douglas Jerolmack for the data of Figure 15.3 and to Francesco Ballio for his insightful review of the manuscript. We also thank F. Métivier and F. James for fruitful discussions. Part of the experimental work was funded by the EC2CO program of the INSU CNRS.

References

Ancey, C 2010. Stochastic modeling in sediment dynamics: Exner equation for planar bed incipient bed load transport conditions. *Journal of Geophysical Research* 115, F00A11. DOI:10.1029/2009JF001 260.

Ancey, C and Heyman, J 2014. A microstructural approach to bedload transport: mean behavior and fluctuations of particles transport rates. *Journal of Fluid Mechanics.* 744, 129–168.

Ancey, C, Davison, A, Bohm, T *et al.* 2008. Entrainment and motion of coarse particles in a shallow water stream down a steep slope. *Journal of Fluid Mechanics* 595, 83–114. DOI: 10.1017/S0022112007008 774.

Andreotti, B, Claudin, P, and Douady, S 2002. Selection of dune shapes and velocities part 2: a two-dimensional modelling. *The European Physical Journal B* 28(3), 341–352.

Andreotti, B, Claudin, P, and Pouliquen, O 2010. Measurements of the aeolian sand transport saturation length. *Geomorphology* 123(3), 343–348.

Andreotti, B, Claudin, P, Devauchelle, O *et al.* 2012. Bedforms in a turbulent stream: ripples, chevrons and antidunes. *Journal of Fluid Mechanics* 690, 94–128.

Ashida, K and Michiue, M 1973. Studies on bed-load transport rate in open channel flows. In *IAHR International Symposium on River Mechanics*, Bangkok; Vol. 1, 407–417.

Bagnold, R 1956. The flow of cohesionless grains in fluids. *Philosophical Transactions of the Royal Society of London, Series A, Mathematical and Physical Sciences* 249(964), 235–297.

Bagnold, R 1973. The nature of saltation and of bedload transport in water. *Philosophical Transactions of the Royal Society of London, Series A, Mathematical and Physical Sciences* 332, 473–504.

Ballio, F, Nikora, V, and Coleman, SE 2014. On the definition of solid discharge in hydro-environment research and applications. *Journal of Hydraulic Research* 52(2), 173–184.

Bell, RG and Sutherland, A. 1983. Nonequilibrium bedload transport by steady flows. *Journal of Hydraulic Engineering* 109(3), 351–367.

Bradley, DN, Tucker, GE, and Benson, DA 2010. Fractional dispersion in a sand bed river. *Journal of Geophysical Research* 115, F00A09.

Bridge, J and Dominic, D 1984. Bed load grain velocity and sediment transport rates. *Water Resources Research* 20(4), 476–490.

Carretier, S, Regard, V, and Soual, C 2007. *In situ* cosmogenic nuclides in river bedload. implications for catchment-scale erosion rate and surface exposure dating. *Geochimica et Cosmochimica Acta* 71(15), A147–A147.

Charru, F 2006. Selection of the ripple length on a granular bed sheared by a liquid flow. *Physics of Fluids* 18, 121508. DOI: 0.1017/S002211200500 786X.

Charru, F and Hinch, E 2006. Ripple formation on a particle bed sheared by a viscous liquid. Part 1. Steady flow. *Journal of Fluid Mechanics* 550, 111–121.

Charru, F, Andreotti, B, and Claudin, P 2013. Sand ripples and dunes. *Annual Review of Fluid Mechanics* 45, 469–493.

Charru, F, Mouilleron, H, and Eiff, O 2004. Erosion and deposition of particles on a bed sheared by a viscous flow. *Journal of Fluid Mechanics* 519, 55–80.

Charru, F, Larrieu, E, Dupont, JB, and Zenit, R 2007. Motion of a particle near a rough wall in a viscous shear flow. *Journal of Fluid Mechanics* 570, 431–453.

Chiodi, F, Claudin, P, and Andreotti, B 2014. A two phase flow model of sediment transport: transition from bed-load to suspended-load. *Journal of Fluid Mechanics* 755, 561–581.

Church, M 2006. Bed material transport and the morphology of alluvial river channels. *Annual Reviews in Earth and Planetary Science* 34, 325–354.

Claudin, P and Andreotti, B 2006. A scaling law for aeolian dunes on mars, venus, earth, and for subaqueous ripples. *Earth and Planetary Science Letters* 252 (1), 30–44.

Coleman, S and Eling, B 2000. Sand wavelets in laminar open-channel flows. *Journal of Hydraulic Research* 38, 331–338.

Coleman, SE and Melville, BW. 1996. Initiation of bed forms on a flat sand bed. *Journal of Hydraulic Engineering* 122(6), 301–310.

Devauchelle, O, Josserand, C, Lagrée, PY, and Zaleski, S 2007. Morphodynamic modeling of erodible laminar channels. *Physical Review E* 76(5), 056 318.

Devauchelle, O, Malverti, L, Lajeunesse, E *et al.* 2010a. Rhomboid beach pattern: a laboratory investigation. *Journal of Geophysical Research* 115, F02 017.

Devauchelle, O, Malverti, L, Lajeunesse, E *et al.* 2010b Stability of bedforms in laminar flows with free surface: from bars to ripples. *Journal of Fluid Mechanics* 642, 329–348.

Durán, O, Andreotti, B, and Claudin, P 2012. Numerical simulation of tubulent sediment transport, from bed load to saltation. *Physics of Fluids* 24(103306).

Durán, O, Andreotti, B, and Claudin, P 2014. Turbulent and viscous sediment transport–a numerical study. *Advances in Geosciences* 37(37), 73–80.

Einstein, H 1950. *The Bed-Load Function for Sediment Transportation in Open Channel Flows*. Soil Conservation Service, US Department of Agriculture.

Engelund, F and Fredsoe, J 1976. A sediment transpot model for straight alluvial channels. *Nordic Hydrology* 7(5), 293–306.

Ferguson, R and Wathen, S 1998. Tracer-pebble movement along a concave river profile: virtual velocity in relation to grain size and shear stress. *Water Resources Research* 34(8), 2031–2038.

Ferguson, RI, Bloomer, DJ, Hoey, TB, and Werritty, A 2002. Mobility of river tracer pebbles over different timescales. *Water Resources Research* 38(5), 3–1.

Fernandez-Luque, R and Van Beek, R 1976. Erosion and transport of bed-load sediment. *Journal of Hydraulic Research* 14, 127–144.

Fourriere, A, Claudin, P, and Andreotti, B 2010. Bedforms in a turbulent stream: formation of ripples by primary linear instability and of dunes by nonlinear pattern coarsening. *Journal of Fluid Mechanics* 649, 287–328.

Furbish, D, Ball, A, and Schmeeckle, M 2012. A probabilistic description of the bed load sediment flux: 4. Fickian diffusion at low transport rates. *Journal of Geophysical Research* 117(F3), F03 034.

Furbish, D, Roseberry, J, and Schmeeckle, M 2012. A probabilistic description of the bed load sediment flux: 3. The particle velocity distribution and the diffusive flux. *Journal of Geophysical Research* 117(F3), F03 033.

Furbish, D, Haff, P, Roseberry, J, and Schmeeckle, M 2012. A probabilistic description of the bed load sediment flux: 1. Theory. *Journal of Geophysical Research* 117(F3), F03 031.

Gayer, E, Mukhopadhyay, S, and Meade, BJ. 2008. Spatial variability of erosion rates inferred from the frequency distribution of cosmogenic ^{3}He in olivines from Hawaiian river sediments. *Earth and Planetary Science Letters* 266(3), 303–315.

Gomez, B 1991. Bedload transport. *Earth-Science Reviews* 31(2), 89–132.

Gomez, B and Church, M 1989. An assessment of bed load sediment transport formulae for gravel bed rivers. *Water Resources Research* 25, 1161–1186.

Guerit, L, Métivier, F, Devauchelle, O *et al.* 2014. Laboratory alluvial fans in one dimension. *Physical Review E* 90(2), 022 203.

Habersack, H 2001. Radio-tracking gravel particles in a large braided river in new zealand: a field test of the stochastic theory of bed load transport proposed by Einstein. *Hydrological Processes* 15(3), 377–391.

Haschenburger, JK and Church, M 1998. Bed material transport estimated from the virtual velocity of sediment. *Earth Surface Processes and Landforms* 23(9), 791–808.

Haschenburger, JK and Wilcock, PR. 2003. Partial transport in a natural gravel bed channel. *Water Resources Research* 39(1).

Hassan, MA, Voepel, H, Schumer, R, Parker, G, and Fraccarollo, L 2013. Displacement characteristics of coarse fluvial bed sediment. *Journal of Geophysical Research: Earth Surface* 118(1), 155–165.

Hersen, P 2005. Flow effects on the morphology and dynamics of aeolian and subaqueous barchan dunes. *Journal of Geophysical Research* 110, F04S07. DOI:10.1029/2004JF000 185.

Hodge, RA, Hoey, TB, and Sklar, LS. 2011. Bed load transport in bedrock rivers: the role of sediment cover in grain entrainment, translation, and deposition. *Journal of Geophysical Research: Earth Surface* 116(F4).

Houssais, M and Lajeunesse, E 2012. Bedload transport of a bimodal sediment bed. *Journal of Geophysical Research* 117, F04 015.

Houssais, M, Ortiz, C, Durian, D, and Jerolmack, DJ 2015. Onset of sediment transport is a continuous transition driven by fluid shear and granular creep. *Nature Physics* 6(6527).

James, F, Postel, M, and Sepúlveda, M 2000. Numerical comparison between relaxation and nonlinear equilibrium models. application to chemical engineering. *Physica D: Nonlinear Phenomena* 138(3), 316–333.

Kouakou, KKJ and Lagrée, PY. 2005. Stability of an erodible bed in various shear flows. *The European Physical Journal B – Condensed Matter and Complex Systems* 47(1), 115–125.

Lagrée, PY. 2003. A triple deck model of ripple formation and evolution. *Physics of Fluids (1994-present)*, 15(8), 2355–2368.

Lagrée, PY, Devauchelle, O, Nguyen Thu-Lam, KD *et al.* 2009. Erosion structures in laminar flumes. In *AIP Conference Proceedings, Powders and Grains* (eds M Nakagawa and S Luding), Golden, CA; Vol. 1145, 963–966.

Lajeunesse, E, Malverti, L, and Charru, F 2010. Bedload transport in turbulent flow at the grain scale: experiments and modeling. *Journal of Geophysical Research* 115, F04 001. DOI:10.1029/2009JF001 628.

Lajeunesse, E, Malverti, L, Lancien, P *et al.* 2010. Fluvial and submarine morphodynamics of laminar and near-laminar flows: a synthesis. *Sedimentology* 57, 1–26.

Lajeunesse, E, Devauchelle, O, Houssais, M, and Seizilles, G 2013. Tracer dispersion in bedload transport. *Advances in Geoscience.* 37, 1–6.

Lamarre, H, MacVicar, B, and Roy, AG. 2005. Using passive integrated transponder (pit) tags to investigate sediment transport in gravel-bed rivers. *Journal of Sedimentary Research* 75(4), 736–741.

Malverti, L, Lajeunesse, E, and Metivier, F 2008. Small is beautifull: upscaling microscale experimental results to the size of natural rivers. *Journal of Geophysical Research* 113, F04004. DOI:10.1029/2007JF000 974.

Metivier, F and Meunier, P 2003. Input and output mass flux correlations in an experimental braided stream. implications on the dynamics of bed load transport. *Journal of Hydrology* 271(1–4), 22–38.

Meyer-Peter, E and Müller, R 1948. Formulas for bed-load transport. In *Proceedings, 2nd Congress, International Association of Hydraulic Research* (ed. S Stockholm); 39–64.

Narteau, C, Zhang, D, Rozier, O, and Claudin, P 2009. Setting the length and time scales of a cellular automaton dune model from the analysis of superimposed bedforms. *Journal of Geophysical Research Earth Surface* 114, F03 006.

Nathan Bradley, D and Tucker, GE. 2012. Measuring gravel transport and dispersion in a mountain river using passive radio tracers. *Earth Surface Processes and Landforms* 37(10), 1034–1045.

Nichols, M 2004. A radio frequency identification system for monitoring coarse sediment particle displacement. *Applied Engineering in Agriculture* 20(6), 783–787.

Nikora, V, Habersack, H, Huber, T, and McEwan, I 2002. On bed particle diffusion in gravel bed flows under weak bed load transport. *Water Resources Research* 38(6), 17–1.

Nino, Y and Garcia, M 1994. Gravel saltation. part i: Experiments. *Water Resources Research* 30(6), 1907–1914.

Pähtz, T, Kok, JF, Parteli, EJ, and Herrmann, HJ. 2013. Flux saturation length of sediment transport. *Physical Review Letters* 111(21), 218 002.

Paola, C, Mullin, J, Ellis, C *et al.* 2001. Experimental stratigraphy. *GSA Today* 11(7), 4–9.

Paola, C, Straub, K, Mohrig, D, and Reinhardt, L 2009. The 'unreasonable effectiveness' of stratigraphic and geomorphic experiments. *Earth-Science Reviews* 97(1), 1–43.

Phillips, BC and Sutherland, AJ. 1989. Spatial lag effects in bed load sediment transport. *Journal of Hydraulic Research* 27(1), 115–133.

Phillips, B and Sutherland, A 1990. Temporal lag effect in bed load sediment transport. *Journal of Hydraulic Research* 28(1), 5–23.

Phillips, C and Jerolmack, D 2014. Dynamics and mechanics of bed-load tracer particles. *Earth Surface Dynamics* 2, 513–530.

Phillips, CB, Martin, RL, and Jerolmack, DJ 2013. Impulse framework for unsteady flows reveals superdiffusive bed load transport. *Geophysical Research Letters* 40(7), 1328–1333.

Recking, A, Frey, P, Paquier, A, and Belleudy, P 2009. An experimental investigation of mechanisms involved in bed load sheet production and migration. *Journal of Geophysical Research* 114, F03 010. DOI:10.1029/2008JF000 990.

Reid, I, Frostick, LE, and Layman, JT 1985. The incidence and nature of bedload transport during flood flows in coarse-grained alluvial channels. *Earth Surface Processes and Landforms* 10(1), 33–44.

Revil-Baudard, T and Chauchat, J 2013. A two-phase model for sheet flow regime based on dense granular flow rheology. *Journal of Geophysical Research: Oceans* 118(2), 619–634.

Roseberry, J, Schmeeckle, M, and Furbish, D 2012. A probabilistic description of the bed load sediment flux: 2. Particle activity and motions. *Journal of Geophysical Research* 117(F3), F03 032.

Sauermann, G, Kroy, K, and Herrmann, H 2001. Continuum saltation model for sand dunes. *Physics Reviews E* 64(3; Part 1), 31 305–31 305.

Sayre, W and Hubbell, D 1965. *Transport and Dispersion of Labeled Bed Material, North Loup River, Nebraska*. Technical Report 433-C, Professional Paper, US Geological Survey.

Seizilles, G, Devauchelle, O, Lajeunesse, E, and Métivier, F 2013. Width of laminar laboratory rivers. *Physics Reviews E* 87, 052 204.

Seizilles, G, Lajeunesse, E, Devauchelle, O, and Bak, M 2014. Cross-stream diffusion in bedload transport. *Physics of Fluids* 26(1), 013 302.

Seminara, G 2010. Fluvial sedimentary patterns. *Annual Review of Fluid Mechanics* 42, 43–66.

Shields, AS 1936. Anwendung der aehnlichkeitsmechanik und der turbulenzforschung auf die geschiebebewegung. *Mitteilung der Preussischen Versuchsanstalt fur Wasserbau und Schiffbau* 26.

Taylor, G 1953. Dispersion of soluble matter in solvent flowing slowly through a tube. *Proceedings of the Royal Society of London. Series A. Mathematical and Physical Sciences* 219(1137), 186–203.

Van Rijn, L 1984. Sediment transport, part i: bed load transport. *Journal of Hydraulic Engineering* 110(10), 1431–1456.

Wilcock, PR. 1997. Entrainment, displacement and transport of tracer gravels. *Earth Surface Processes and Landforms* 22(12), 1125–1138.

Willenbring, JK and von Blanckenburg, F 2010. Meteoric cosmogenic beryllium-10 adsorbed to river sediment and soil: applications for earth-surface dynamics. *Earth-Science Reviews* 98(1), 105–122.

Discussion

Discussion by D.J. Furbish

I applaud your use of laminar flow experiments, as they have much to offer in terms of contributing to a deeper understanding of sediment motions in response to moving fluids, including turbulent flows. Regarding your reference to particle diffusion, the touchstone measure of particle spreading involves the mean squared displacement of a Brownian-like particle, where the calculation, involving all paired coordinate positions separated by a specified time interval τ, survives an arbitrary translation in time t. In contrast, the mean squared displacement of sediment particles involving start-and-stop motions, where the particles are treated as a virtual plume, reveals a different feature of particle spreading. Within this context, please comment on your calculations of the mean squared displacement in relation to your description of particle diffusion.

Discussion by Astrid Blom on Eric LaJeunesse *et al.* and David Furbish

Both the authors and David Furbish have worked on the modelling of the streamwise dispersion of grains. The models presented by both parties are very interesting yet seem to lack one physical mechanism inducing streamwise dispersion: the temporary burial of particles within the bed. Crickmore and Lean did pioneering work on this subject (Crickmore and Lean 1962a,b). Can the authors elaborate on the relevance of this mechanism in the field (and in the laboratory), and, if relevant, if and how they intend to incorporate the mechanism in their future modelling efforts of streamwise dispersion? Regarding temporary particle burial, what do the authors expect to gain from insights due to the recent work by Pelosi and co-authors (Pelosi *et al.* 2014)?

References

Crickmore, M and Lean, G 1962a. The measurement of sand transport by means of radioactive tracers. *Proceedings of the Royal Society of London Series A* 266, 402–421.

Crickmore, M and Lean, G. 1962b. The measurement of sand transport by the time-integration method with radioactive tracers. *Proceedings of the Royal Society of London Series A* 270, 27–47.

Pelosi, A, Parker, G, Schumer, R, and Ma, H. 2014. Exner-based master equation for transport and dispersion of river pebble tracers: derivation, asymptotic forms, and quantification of nonlocal vertical dispersion. *Journal of Geophysical Research: Earth Surface* 119(9), 1818–1832.

Discussion by David Milan

In contrast to the flume experiments of Lajeunesse *et al.*, natural gravel-bed rivers usually have bedforms composed of a mixture of gravels and sand, and have variable discharge and sediment supply. Furthermore, the sand fraction can move selectively over a stable gravel surface during less competent flows. Sand tracing experiments conducted on the gravel-bed river Rede, UK, demonstrate that pool–riffle bedforms control spatial patterns of dispersal and as a consequence the downstream location of the mean concentration of the tracer plume (Milan and Large 2014). In this study, the shape of the downstream sand tracer plume did not show the same form as that shown in the flume by Lajeunesse *et al.*, instead showing a lumpy dispersal pattern, with tracer peaks coinciding with the location of topographic highs on the bed long-profile.

Reference

Milan, DJ and Large, AR 2014. Magnetic tracing of fine-sediment over pool–riffle morphology. *Catena* 115, 134–149.

Response by the Authors

Sensu stricto, diffusion results from the motion of a large number of Brownian particles. This collective motion exhibits two characteristics (Einstein 1905):

- it generates a macroscopic flux of particles directed from regions of high concentration to regions of low concentration, with a magnitude proportional to the concentration gradient;
- the mean squared displacement of the particles (i.e. the variance) increases linearly with time.

As D.J. Furbish points out, the spreading mechanism discussed in this chapter does not rely on Brownian motion. Instead, it involves the continuous exchange of particles between two layers: the bedload layer, where particles travel at the constant velocity V, and the sediment bed, where particles are at rest. The difference of velocity between the two layers progressively smears out the concentration distribution in the direction of the flow, enhancing the rate at which it spreads in that direction. In the limit of long durations, the plume of tracers spreads linearly with time (i.e. its variance increases linearly with time), thus mimicking diffusion. This is why we use the word "diffusion" to describe the long-term behaviour of the plume. This is of course a misuse of language and an expression such as 'effective diffusion' would be more appropriate. Following D.J. Furbish's comment, we have slightly modified the end of section 15.4.

Although presented in the context of bedload transport, effective diffusion occurs in a variety of physical systems in which layers moving at different velocities exchange a passive tracer. A typical example is Taylor dispersion where a passive tracer diffuses in a Poiseuille flow through a circular pipe. The combination of the shear rate and transverse molecular diffusion creates an effective diffusion in the flow direction (Taylor 1953). Other examples of effective diffusion include flow in porous media and chromatography (Van Genuchten and Wierenga 1976; James *et al.* 2000).

As Astrid Blom points out, the streamwise dispersion of grains results from the combined action of at least three different processes: (i) the fluctuations of bedload particles velocities; (ii) the continuous exchange of particles between the sediment bed and the bedload layer; (iii) the temporary burial of particles within the bed. These processes are likely influenced by the grain-size distribution and the presence of bedforms.

Given the complexity of the problem, David Furbish and ourselves chose to focus on idealized laboratory configurations (uniform grain size, constant flow, and sediment discharges, etc.), allowing us to isolate and investigate separately a single mechanism at a time.

David studied the fluctuations of bedload particles velocities. He developed a description of bedload transport based on the tools of statistical physics which allows us to understand the shape of the distribution of particle velocities and its influence on sediment transport, including streamwise dispersion (Furbish, Ball, and Schmeeckle 2012; Furbish, Roseberry, and Schmeeckle 2012).

In this chapter, we focus on the description of the exchange of particles between the bedload layer and the immobile bed. Experiments and modelling reveal that this process strongly contributes to the streamwise dispersion of bedload tracers even in the simplistic configuration of a steady-state sediment bed that neither aggrades nor degrades.

As Astrid Blom suggests, temporary burial of particles within the bed plays an important role in natural streams. Incorporating this mechanism in future models requires two supplementary equations: (i) the first is the Exner equation, accounting for total sediment mass balance inside the bed; (ii) the second describes the mass balance of tracers inside the bed. Complementing the erosion–deposition model with two such equations is rather straightforward in the case of a sediment bed of homogeneous grain size.

The case of a sediment bed involving a wide range of grain sizes is more complicated (Blom and Parker 2004). Most models rely on the arbitrary splitting of the erodible bed into an active layer and a substrate, introduced first by Hirano (1971). In this context, the continuous formulation of mass balance within the bed developed by Parker *et al.* (2000) and its recent application to the dispersion of tracers by Pelosi *et al.* (2014) appears promising.

This chapter is a review of the erosion–deposition model of bedload transport. The section devoted to the propagation of a plume of bedload tracers is a preliminary study aimed at illustrating the model in an idealized laboratory configuration (uniform grain size, flat sediment bed, constant flow and sediment discharges). As David Milan points out, natural rivers, and in particular gravel-bed rivers, often transport strongly heterogeneous sediments over bedforms, and their flow and sediment discharges vary significantly through time. Future efforts should aim at incorporating all these effects in models of streamwise dispersion. Our answer to the discussion raised by Astrid Blom presents how this might be achieved.

References

Blom, A and Parker, G 2004. Vertical sorting and the morphodynamics of bed form-dominated rivers: a modeling framework. *Journal of Geophysical Research – Earth Surface* 109(F2).

Einstein, A 1905. Über die von der molekularkinetischen theorie der wärme geforderte bewegung von in ruhenden flüssigkeiten suspendierten teilchen. *Annalen der Physik* 322, 549–560. DOI:10.1002/andp.19053220806.

Hirano, M 1971. River bed degradation with armouring. *Transactions of the Japanese Society of Civil Engineers* 3(2), 194–195.

Parker, G, Paola, C, and Leclair, S 2000. Probabilistic exner sediment continuity equation for mixtures with no active layer. *Journal of Hydraulic Engineering* 126(11), 818–826.

Van Genuchten, MT and Wierenga, P 1976. Mass transfer studies in sorbing porous media. I: analytical solutions. *Soil Science Society of America Journal* 40(4), 473–480.

16

Bedforms, Structures, Patches, and Sediment Supply in Gravel-Bed Rivers

Jeremy G. Venditti, Peter A. Nelson, Ryan W. Bradley, Dan Haught, and Alessandro B. Gitto

16.1 Introduction

In sand-bedded rivers there is a well-known sequence of bedforms that develops with increasing flow strength. For a specific grain size, the sequence includes: lower stage plane bed, ripples, dunes, upper stage plane bed, antidunes, chutes-and-pools, and cyclic steps (see recent reviews in Venditti (2013) and Cartigny *et al.* (2014)). These features control flow resistance and bed-material sediment flux rates, and leave sedimentary structures in the rock record that are often used to infer paleoflow conditions from an understanding of the modern sedimentary processes responsible for bedform development (Venditti 2013). In gravel-bedded rivers, a comparable sequence of bedforms with increasing flow strength has not been identified. Nevertheless, these rivers develop a distinct suite of bed features that are larger than individual clasts and smaller than channel-scale features (e.g. bars or step-pool features). These bed features include (i) gravel dunes (e.g. Carling 1999), (ii) mobile, migrating patches typically referred to as bedload sheets (e.g. Whiting *et al.* 1988; Nelson *et al.* 2009), (iii) sand bedforms developed over an otherwise immobile gravel bed (sand ribbons, barchans and dunes; e.g. Kleinhans *et al.* 2002), (iv) pebble clusters (e.g. Dal Cin 1968; Brayshaw 1984), (v) stone lines (Larrone and Carson 1976) or transverse ribs (e.g. Koster 1978; Allen 1984), and (vi) reticulate stone cells (e.g. Church, Hassan, and Wolcott 1998; Hassan and Church 2000). There have been many important insights into how these bed features develop and how they influence flow dynamics and sediment transport rates, but there has been little insight into how or whether the features are interrelated, and the conditions required for the emergence of one type of bed feature as opposed to another. Here, we demonstrate the critical role that sediment supply and differential mobility of the grain-size distribution play in the types of bedforms that emerge. We end with a discussion about what controls the emergence of these various bedforms, bed-surface structures, and patches in a common conceptual framework.

16.2 Bedload Transport, Sediment Supply, and Bed Mobility

It is commonly assumed that the primary control on bedload transport rates is the flow strength, often characterized by the mean boundary shear stress. The idea has been embedded in a wide range

Gravel-Bed Rivers: Processes and Disasters, First Edition. Edited by Daizo Tsutsumi and Jonathan B. Laronne.

of bedload transport formulae (e.g. Meyer-Peter and Muller 1948; Parker, Klingman, and McLean 1982; Parker 1990). However, observations in rivers and experimental channels have revealed that bedload transport rates can vary by several orders of magnitude under steady uniform flow conditions due in part to the emergence of various bedforms, patches, and structures of an armored gravel surface. Nelson *et al.* (2009) provide an example of the interplay between bedload transport and patches of similarly sized bed material (Figure 16.1). They fed a unimodal, lognormally distributed gravel with a median size $D_{50} = 8$ mm and range from 2 to 32 mm to a 25 m long and 0.86 m wide channel. Flow rate in the channel was held constant and was above the threshold to move the gravel mixture. When the sediment output rate matched the supply rate, distinct patches were identified and mapped and the supply to the channel was reduced.

The patchiness that emerged on the bed in the experiments is one type of a wide variety of sediment patches of similar grain size and sorting that occur in gravel-bedded rivers. Following the terminology proposed for bar types (e.g. Seminara 1998), sediment patches can be identified as either "forced" (spatially persistent associated with topographic controls), "fixed" (spatially persistent due to coarsening), or "free" (migrating patches) (Nelson *et al.* 2009). The migrating free patches shown in Figure 16.1b have been referred to as bedload sheets, typically described as low-amplitude bedforms with heights of 1–2 coarse grain diameters that fine toward the tails (Whiting *et al.* 1988). The downstream migration of these patches cause a nearly two orders of magnitude change in the bedload transport rate (Figure 16.1a). Recking *et al.* (2009) document entirely similar behavior in the transport of a tri-modal gravel mixture due to the formation and migration of bedload sheets.

As the sediment supply to a channel is progressively reduced, the bed-surface grain size becomes substantially less heterogeneous (Figure 16.1b). This occurs because the active bedload transport corridor on the bed narrows (Lisle, Iseya, and Ikeda 1993; Nelson *et al.* 2009) and the patches become less distinct (Figure 16.1b). This feeds back on the sediment transport rates, which progressively exhibit less variability (Figure 16.1a). Once the sediment supply to the channel is removed, the bed becomes uniformly coarse (Figure 16.1b). A more subtle response that occurs due to the reduction of sediment supply is a progressive change in the size distribution of the particles transported as bedload. Under conditions of high sediment supply, the channel moves particles in proportion to their presence in the sediment supply, but as the supply is reduced, some of the coarser particles get left behind in the bed and coarse patches expand. Below, we explore how sediment supply and differential mobility of the grain size distribution impact the emergence of different types of bed features in gravel-bedded rivers. But first, we provide some quantitative measures of the sediment supply and differential mobility.

16.2.1 Sediment Supply

The rate at which bedload material is supplied to a channel exerts a strong control on the bed-surface grain size (Kuhnle and Southard 1988; Dietrich *et al.* 1989; Buffington and Montgomery 1999; Nelson *et al.* 2009). In channels where sediment supply is not limited, there is no difference between the surface and subsurface grain-size distributions (Laronne and Reid 1993; Laronne *et al.* 1994; Reid and Laronne 1995). This is commonly observed in unvegetated arid watersheds where there is ample supply of all sediment sizes and flows are typically flashfloods with a rapid decline in flow that prevents reworking of the surface after general bedload movement has ceased (Laronne *et al.* 1994). In contrast, seasonal and perennial streams can exhaust their sediment supply, forcing adjustments to the bed surface (Reid and Laronne 1995).

Channels where the sediment supply is limited with respect to the transport capacity are typically armored (Parker and Klingeman 1982) and have a bed surface slightly coarser than the bedload, which often has a grain-size distribution similar to the subsurface. When the sediment supply is nearly equivalent to the transport capacity of the channel, active bedload transport takes place across much of the

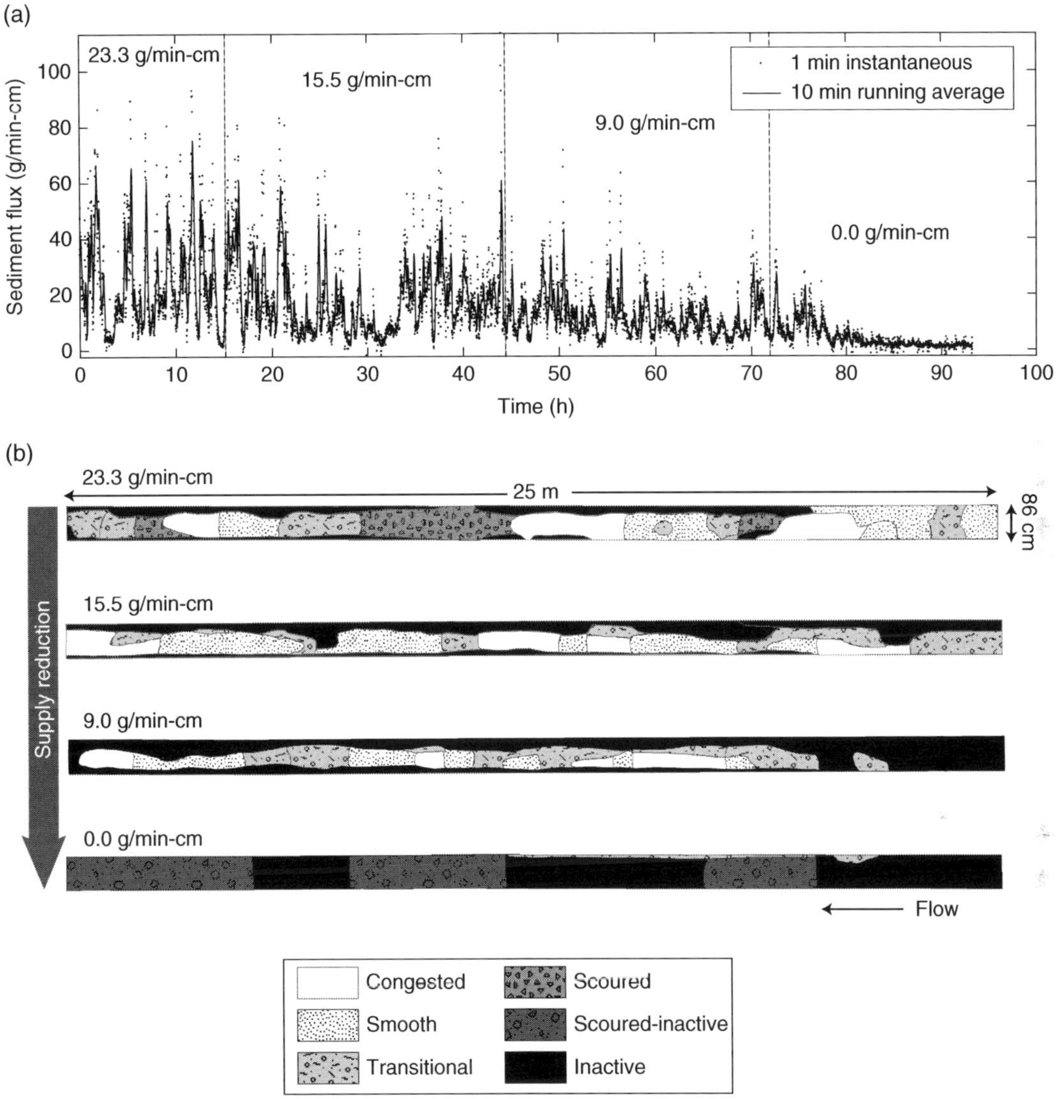

Figure 16.1 Response of (a) bedload transport and (b) bed-surface texture to reductions in sediment supply from 23.3 to 15.5 to 9.0 to 0.0 g/min-cm. At each supply rate, distinct patches were identified and classified as "congested" (coarse, $D_{50} = 4.68$ mm), "transitional" ($D_{50} = 3.63$ mm), "smooth" (fine, $D_{50} = 2.62$ mm), and "inactive" (coarse zones with no active bedload transport, $D_{50} = 5.49$ mm) following Iseya and Ikeda (1987). Bed areas that appeared to have a bimodal grain-size distribution were classified as "scoured," (because the intermediate-sized material was absent, exposing fine grains. See Nelson *et al.* (2009) for further details on the experimental procedure. (*Source*: Nelson *et al.* (2009). Reproduced with the permission of AGU.)

channel width and there is a small fraction of the bed surface grain-size distribution that can be mobilized only during extreme flood flows with higher transport capacity. Reduction of the sediment supply narrows the zone of active transport, the bedload fines and the mean bed surface grain-size becomes coarser (Dietrich *et al.* 1989; Lisle, Iseya, and Ikeda 1993; Nelson *et al.* 2009).

Dietrich *et al.* (1989) argued that the effects of sediment supply on the bed surface grain-size distribution can be quantified with respect to the transport capacity. Invoking a conventional functional form where the bedload transport rate is proportional to the excess boundary shear stress raised to a power (e.g., Meyer-Peter and Muller, 1948),

$$q_b = k(\tau - \tau_c)^{1.5}, \tag{16.1}$$

the transport rate over the coarse surface layer normalized by the transport rate over a surface as fine as the subsurface or load is

$$q_* = \left(\frac{\tau - \tau_{cs}}{\tau - \tau_{cl}}\right)^{1.5} \tag{16.2}$$

where q_b is the bedload transport rate per unit width, k is a constant, τ is the boundary shear stress, τ_c is the critical shear stress for particle entrainment and the subscripts s and l indicate values of τ_c for the bed surface and the load. Writing the shear stresses in nondimensional form

$$\tau^* = \frac{\tau}{(\rho_s - \rho)gD_{50}} \tag{16.3}$$

allows Equation 16.2 to be rewritten as

$$q_* = \left(\frac{\tau/\tau_{cl} - D_{50s}/D_{50l}}{\tau/\tau_{cl} - 1}\right)^{1.5} \tag{16.4}$$

where D_{50} is the median of the grain-size distribution. Dietrich *et al.* (1989) showed that when the sediment supply matches the river's ability to transport the load, q_* is unity. As the sediment supply decreases with respect to the transport capacity, the bed surface becomes increasingly armored, as in Figure 16.1. Figure 16.2 shows the variation in q_* with the ratio between the median size of the bed surface (D_{50s}) and load (D_{50l}) using a variety of data sets where a systematic, stepwise reduction of sediment supply was undertaken at a constant discharge. We assume that D_{50s}/D_{50l} is equivalent to the armor ratio under the assumption that the bedload and subsurface size distributions are the same. The data conform with the original observation from Dietrich *et al.* (1989) in that there is an asymptotic decline in q_* as the bed gets coarser and armors. Included in Figure 16.2 are data from Church, Hassan, and Wolcott (1998), whose experiments did not have a sediment feed, but instead started with a gravel–sand mixture and progressively armored at a constant flow. This data set does not show a clear asymptotic decline in q_* with the assumed armor ratio equivalent, but the authors do note that there was an asymptotic decline in sediment flux from the beginning of their experiment as the upstream bed armored and fed winnowed sediment to the downstream end of the channel where flux was measured.

Building on the underlying concept from Dietrich *et al.* (1989), Church and Hassan (2005) proposed that the value of q_* could be more easily estimated without assuming a specific functional form of the sediment transport equation by calculating $q_{*\,(C\,\&\,H)}$ as the ratio of the bedload transport rate for a bed surface that was armored after 96 hours of flow to the transport rate of the same bed in the first hour of their experiment, before it armored, at the same flow strength. It is assumed that the unarmored bed transport rate is approximately equal to the capacity for the grain-size distribution and hydraulic conditions (depth, slope, velocity). Extrapolating that concept, $q_{*\,(C\,\&\,H)}$ can be calculated for an experiment with sediment feed reductions as the bedload transport rate after the bed has adjusted to a new sediment feed divided by the transport rate for an unarmored bed, before any feed reductions.

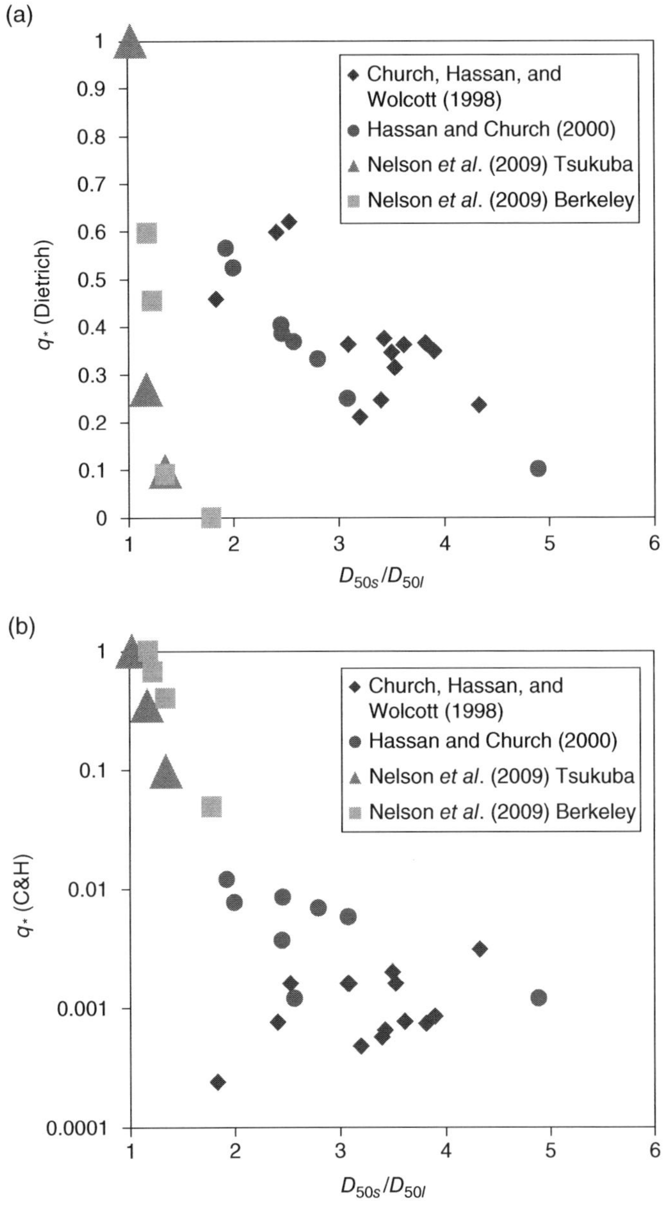

Figure 16.2 Variation in q_*, calculated using the (a) Dietrich *et al.* (1989) and (b) Church and Hassan (2005) methods, with the assumed armor ratio equivalent, D_{50s}/D_{50l}. Both experiments reported by Nelson *et al.* (2009) had bedload sheets with the exception of the lowest q_* value for the Berkeley experiments. Tsukuba data are the same data presented in Dietrich *et al.* (1989). The Church, Hassan, and Wolcott (1998) and Hassan and Church (2000) experiments both produced clusters, ribs, and reticulate cells.

Figure 16.2b shows the variation in $q_{*\,(C\,\&\,H)}$ with the assumed armor ratio equivalent, revealing the same asymptotic behavior as q_* with D_{50s}/D_{50l}. However, the q_* values are 2 orders of magnitude larger than the $q_{*\,(C\,\&\,H)}$ values for the Church *et al.* (1998) and Hassan and Church (2000) data. The difference between the two methods lies in the assumed functional form of the bedload transport equation in Dietrich *et al.* (1989). Church and Hassan (2005) show that transport rates in their experiments are overpredicted by ~2 orders of magnitude by the Meyer-Peter Muller Equation using a value of $\tau_c^* = 0.045$, so $q_* >> q_{*\,(C\,\&\,H)}$. They also suggest that the development of bed structures increases τ_c^* by ~50%. Examination of Figure 16.2 suggests that the experiments reported in Nelson *et al.* (2009), Church *et al.* (1998) and Hassan and Church (2000) all appear to form a continuous asymptotic decline in $q_{*\,(C\,\&\,H)}$ with increasing D_{50s}/D_{50l}. The Nelson *et al.* (2009) data reveal a sharp decline in $q_{*\,(C\,\&\,H)}$ and the Church *et al.* (1998) and Hassan and Church (2000) experiments form the stable asymptote. As discussed below, differences in the relative mobility of the grain size distributions in the Nelson *et al.* (2009), Church *et al.* (1998), and Hassan and Church (2000) experiments are probably responsible for the different behavior.

16.2.2 Mobility of the Grain-Size Distribution

The mobility of grain-size distributions affects the size of the material available for transport and therefore the bed's response to changes in sediment supply. Gravel mixtures can experience full, partial, selective, and equal mobility. There is some confusion in the literature over the precise definition of these terms, and universally accepted definitions have not been adopted. Here we follow the definitions of Parker (2008), which appear to capture all the possible behaviors. Partial transport occurs when the coarse tail of the bedload size distribution is finer than that of the bed surface (Figure 16.3a). Selective transport occurs when all grain sizes on the bed are found in the bedload, but the bedload size distribution is finer than the bed surface (Figure 16.3b). Equal mobility occurs when the size distributions of the bedload and the bed surface are the same (Figure 16.3c). Equal mobility of the bedload with respect to the subsurface sediment (rather than the surface) can occur in the presence of a coarse surface layer (e.g. Parker and Klingeman, 1982), which is a special case of selective mobility (Figure 16.3d).

Most gravel-bed streams exist in a condition of partial mobility during flows below bankfull, but have selective mobility during bankfull flows. True equal mobility, where the size distribution in transport is the same as the surface, occurs when the fraction of particle size i in the bedload p_i and the fraction of that size on the bed f_i are the same for all grain sizes. Essentially, for equal mobility the coarse surface layer must disappear. This condition is rare in gravel-bed rivers, a conjecture supported by the widespread use of hiding functions in sediment transport formulae to account for the fact that $p_i/f_i = 1$ is rare. Equal mobility with respect to the subsurface is more typically observed at high flows (Powell, Reid, and Laronne 2001) and can produce mobile armor layers (Wilcock and DeTemple 2005).

The true definitions of equal, selective, and partial mobility depend on the shape of the p_i/f_i curve with grain size (Figure 16.4). For equal mobility, the curve is flat across a range of grain sizes. For selective mobility, p_i/f_i declines for the large grain sizes, but p_i/f_i is greater than 0 for the largest grain size class, while for partial mobility, p_i/f_i is zero for larger grain sizes. It is not always practical to calculate p_i/f_i to determine the transport regime as this requires detailed measurements of both the bed surface and bedload grain-size distributions. Differential mobility conditions can be indexed by examining the ratio of the boundary shear stress to the shear stress required to move a particular grain size, τ/τ_{ci} (Wilcock and McArdell 1993). A practical choice for that grain size is the 84th percentile of the bed material (D_{84}) because it is diagnostic of whether the coarse tail of the grain-size distribution

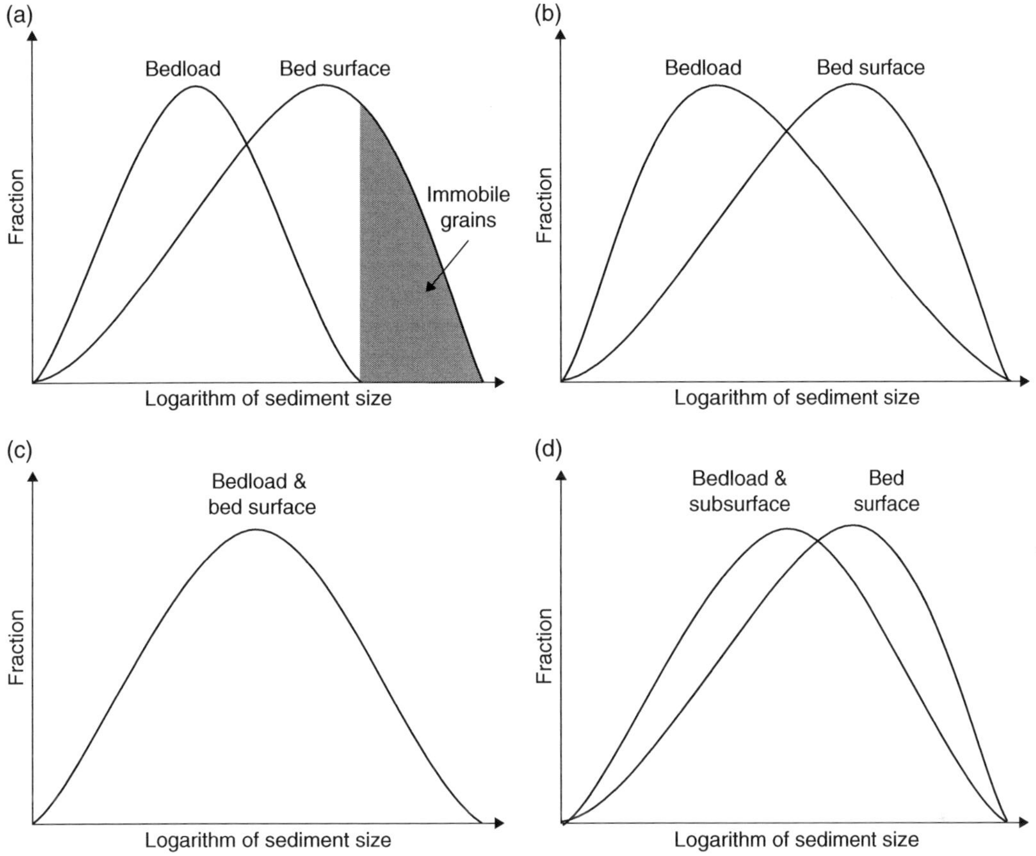

Figure 16.3 Hypothetical grain-size distributions for (a) partial, (b) selective and (c) equal mobility in gravel-bed rivers based on Parker (2008). Panel (d) depicts a special case of selective mobility where transport is equally mobile with respect to the subsurface sediment.

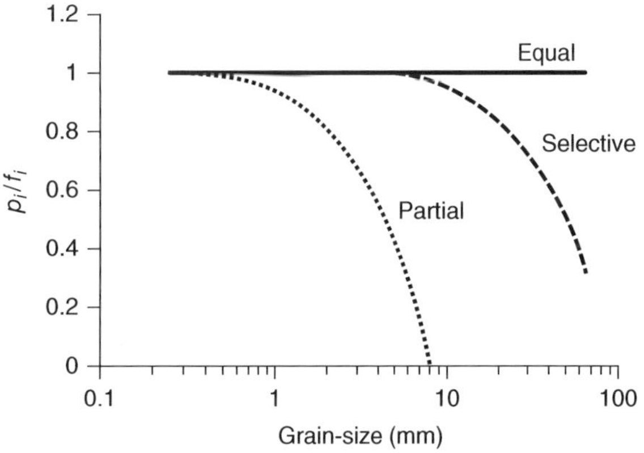

Figure 16.4 Conceptual illustration of how the ratio of fraction of grain size i in the bedload p_i and on the bed f_i varies with grain size for partial, selective, and equal mobility.

is moving. Partial mobility occurs when τ/τ_{c84} <1 because the shear stress is not competent to move the large grain sizes on the bed. For the same reason, selective mobility occurs when $\tau/\tau_{c84} > 1$. The arguments presented in Wilcock and McArdell (1993) suggest that when $\tau/\tau_{c84} > 2.1$, all grains are equally mobile. Some field studies have noted markedly higher values, which are thought to be linked to structuring of the bed surface material (cf. Powell *et al.* 2001).

While τ/τ_{c84} may be a useful criterion, it requires accurate characterization of shear stress applied to sediment grains on the bed, which is not easy because detailed measurements of the flow are needed as well as assumptions or measurements of the critical shear stress required to entrain a particular grain size. Furthermore, the differences reported for the critical values of τ/τ_{c84} for equal mobility make it impractical to use. An alternative is to cast τ/τ_{c84} into a ratio of the D_{84} for the load (or subsurface) and the bed surface. When $D_{84s}/D_{84l} = 1$, the grain size distribution is equally mobile. When the D_{84s}/D_{84l} >1, the grain-size distribution is selectively or partially mobile because large particles are underrepresented or absent in the load. To our knowledge, there is no theoretical D_{84s}/D_{84l} threshold at which conditions should change from selective to partial mobility. However, experimental conditions that exhibit strict partial mobility ($p_i/f_i = 0$ for the large grains), have values of D_{84s}/D_{84l} >2 (cf. Wilcock and McArdell 1993; Church, Hassan, and Wolcott 1998; Hassan and Church, 2000; Kleinhans *et al.* 2002), so we adopt this empirical threshold here to distinguish between selective and partial mobility.

16.3 Bed Features in Gravel-Bed Rivers

The types of bed features that emerge on a gravel bed are dictated by the sediment supply, relative to transport capacity, and the mobility regime. Figure 16.5 shows how the various types of bed features observed in gravel-bedded rivers plot in the phase space created by $q_{*\,(C\,\&\,H)}$ and D_{84s}/D_{84l}. Below, we

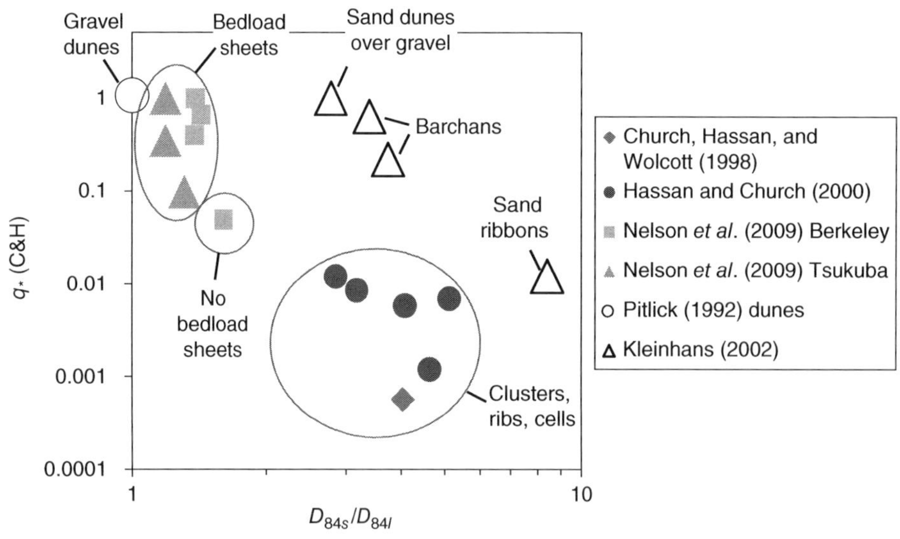

Figure 16.5 Bedforms developed under differing sediment supply and mobility regimes. Equal mobility occurs when $D_{84s}/D_{84l} = 1$, selective mobility conditions occur when $1 < D_{84s}/D_{84l} < 2$, and partial mobility conditions occur when $D_{84s}/D_{84l} > 2$.

provide some notes on the morphology and occurrence of each type of bed feature and attempt to elucidate the sediment supply and mobility conditions required for each bed feature.

16.3.1 Equal Mobility Regime

Figure 16.5 suggests that the bed features developed under equal mobility conditions in gravel-bedded rivers – where the supply equals the capacity – are gravel dunes. Here, we distinguish between gravel dunes and sand dunes developed in mixed size sediment where sand dunes may form over an otherwise immobile gravel bed (e.g. Kleinhans *et al.* 2002; Tuijnder, Ribberink, and Hulscher 2009) or sand dunes with a subsidiary gravel component (Blom, Ribberink, and De Vriend 2003), which we treat separately below. Early work on gravel transport argued that it was a flat-bed phenomenon and classical bedforms developed in sand-bedded channels were absent in gravel-sized sediments (see Carling (1999) and references therein). This idea is captured in many bedform phase diagrams showing that the dune field terminates at grain sizes larger than ~4 mm (e.g. Allen 1984). Yet at high shear stresses, gravel dunes have been observed to emerge from an otherwise flat bed during flood flows in the field (c.f. Dinehart 1989, 1992a,b; Pitlick 1992) and gravel dunes are commonly observed to develop over bars following flood flows (Figure 16.6; see also Baker 1984). Carling (1999) convincingly argued that this has led to misidentification of many features that are probably gravel dunes in both laboratory and field flows.

Gravel dunes have been identified in flows where the nondimensional Shields stress (τ_*; Equation 16.3) exceeds 0.1 (e.g. Hubbell *et al.* 1987; Kuhnle and Southard 1988; Dinehart 1989, 1992a,b; Carling 1999), which is approximately twice the typical entrainment threshold ($\tau_c^* = 0.045$) for gravel mixtures (Miller,

Figure 16.6 Gravel dunes formed on a bar in the Columbia River near Revelstoke, British Columbia. (Photograph courtesy of Rolf Kellerhals.)

McCave, and Komar1977; Yalin and Karahan 1979). Dinehart (1989, 1992a) described the growth of gravel dunes from incipient conditions that reach a maximum height at $\tau_* = 0.25$ and washed out to a flat bed at $\tau_* = 0.3$. This behavior is nearly identical to hydraulically-controlled sand dunes.

Carling, Richardson, and Ikeda (2005) showed that shear stress greatly in excess of the critical value for entrainment is not an absolutely necessary condition for fine gravel dune development, provided the flow persists for a long enough time. Similar observations have been made for dunes in sand-bedded rivers. Venditti, Church, and Bennett (2006) argued that bedforms are initiated instantaneously when the shear stress greatly exceeds the critical shear stress for entrainment, where the sand layer can be fluidized forming hydrodynamic instabilities between the flow and the bedload transport layer that are imprinted on the bed. They also noted that sand dunes can be initiated from defects in the bed when the flow is at or below the threshold of motion for the whole sediment bed, but the development process took much longer. Carling, Richardson, and Ikeda (2005) show a classic case of defect-type initiation of gravel dunes at shear stresses below where the bed can fluidize. However, gravel dunes would certainly be more common when formed during flood flows where the shear stresses are high enough for the transport layer to behave like a continuous fluid media so that they could form instantaneously.

Supply and Mobility Conditions

The sediment supply and mobility conditions for gravel dunes have never been described in the literature. The information provided in the classic references is insufficient to properly assess these parameters primarily because of the difficulty in measuring bedload and the bed-surface grain size simultaneously. However, data reported by Pitlick (1992) for the North Toutle River provides some insight into the conditions necessary to produce gravel dunes. We used Pitlick's (1992) reported bedload transport measurements and shear stress to calculate the bedload capacity of an unarmored bed from the Meyer-Peter and Muller (1948) equation to determine a value of $q_{*\,(C\,\&\,H)}$. We also used bed material D_{84}, measured ~6 months after the bedload measurements to estimate the ratio D_{84s}/D_{84l}, under the assumption that this value of D_{84} was characteristic of the channel-bed surface. The calculations show that, on average for the dune beds, $q_{*\,(C\,\&\,H)} = 1.08$ and $D_{84s}/D_{84l} = 0.99$ (Figure 16.5). We attempted the same calculations using a number of other data sets (notably Dinehart 1992a) and found the only reported bed-surface measurement (Dinehart 1992b) was much finer than the bedload, suggesting that unlike the Pitlick (1992) measurement, it was not characteristic of the channel at higher flows. This suggests that gravel dunes form under conditions where the supply equals or nearly equals the capacity and the transport condition is near equal mobility. Furthermore, gravel dunes develop under the same conditions as sandy bedforms, which are hydraulically controlled insofar as their emergence and dimensions can be predicted from the mean hydraulics in a channel (i.e., depth, slope, velocity, and shear stress).

16.3.2 Selective Mobility Regime

The bed features that plot in the selective mobility range of Figure 16.5 ($1 \leq D_{84s}/D_{84l} \leq 2$) are bedload sheets. First identified by Whiting *et al.* (1988), they appear in river channels as bands of the coarser fractions of the bed material that grade upstream to the finer fractions. These features are commonly aligned across the primary flow path in river channels and migrate downstream, but it is not uncommon to see bedload sheets shoaling onto bars as a result of secondary flow patterns (Wooldridge and Hickin, 2005). The formative mechanism for bedload sheets has not been fully elucidated, but the sorting patterns across bedload sheets suggest that they pose abrupt changes in roughness and turbulence structure (Best 1996; and references therein) and Seminara, Colombini, and Parker (1996) proposed that the stress perturbation due to this sorting structure allows for the growth of bedload sheets.

Bedload sheets are widely thought to be a precursor to gravel dune development (Bennett and Bridge 1995a,b) when flows are sufficiently large that grain inertia is overcome to begin the vertical stacking process required for dune growth. This can only happen quickly in flows well in excess of the critical threshold (Carling 1999). Some authors regard bedload sheets as analogous to dunes (cf. Bridge 1993), but the evidence for this conjecture is not yet well supported. The numerous attempts to include bedload sheets into the bedform phase diagrams used to predict bedform occurrence in sandy sediments (cf. Best 1996; Carling 1999; Kleinhans *et al.* 2002) have only been partially successful at distinguishing between bedform types in poorly sorted sediments. Indeed, bedload sheets commonly fall into the dune existence field on bedform phase diagrams. This has provided some evidence of the linkage between bedload sheets and dunes formed in coarse sediments (gravel and sand–gravel mixes).

Bedload sheets migrate downstream as a consequence of the "catch and mobilize" process, in which large grains are caught in the wakes of other large grains, followed by infilling of their interstices by smaller particles, which can in turn smooth out hydraulic wakes causing large particles to be remobilized (Whiting *et al.* 1988). The low-amplitude coarse front of a bedload sheet develops when a sufficient concentration of coarse particles accumulates (so-called "gravel jams"; Iseya and Ikeda 1987). Relative to the finer sheet tail, this coarse front is hydraulically rough and has relatively large friction angles (e.g., Buffington, Dietrich, and Kirchner 1992; Kirchner *et al.* 1990), which provide distrainment sites where other particles are likely to become temporarily deposited. As fine particles fill the interstices of this coarse front, the bed becomes smoother and near-bed flow accelerates, which disproportionately increases the drag on the coarse particles because they are exposed further into the flow than fine particles (Venditti *et al.* 2010). This causes the coarse particles to become remobilized, and once they are plucked from the bed, they tend to roll over finer bed material until they once again become caught in a coarse sheet front.

The remobilization of coarse particles through grain interactions with fine particles and consequent migration of bedload sheets in gravel–sand mixtures agrees conceptually with observations that the presence of sand can increase the mobility of gravel (e.g., Iseya and Ikeda 1987; Ferguson, Prestegaard, and Ashworth 1989; Wilcock 1998; Wilcock, Kenworthy, and Crowe *et al.* 2001; Wilcock and Kenworthy 2002; Wilcock and Crowe 2003; Curran and Wilcock, 2005). Indeed, most field and flume observations of bedload sheets have occurred under conditions with gravel–sand mixtures (e.g., Iseya and Ikeda 1987; Whiting *et al.* 1988; Kuhnle and Southard 1988; Dietrich *et al.* 1989; Wilcock 1992; Bennett and Bridge 1995a; Bunte *et al.* 2004; Kuhnle *et al.* 2006; Madej *et al.* 2009). However, Nelson *et al.* (2009) showed that bedload sheets, and the catch-and-mobilize process, can occur in a unimodal gravel without any sand present. Thus it seems likely that it is the ratio of coarse to fine particles in the distribution, rather than the presence of sand, that determines whether bedload sheets are able to develop and migrate.

The dynamics (wavelength and celerity) of bedload sheets are controlled in large part by the rate of sediment supply. Nelson *et al.* (2009) showed that under conditions of high sediment supply, bedload sheets tended to move downstream more quickly and sheets were spaced more closely together than those observed under low supply conditions. They suggested that these phenomena can be explained by simple mass balance arguments where, because the fronts of bedload sheets tend not to grow beyond a height of 1–2 coarse grain diameters, the bedform celerity is proportional to the sediment transport rate (i.e., the sediment supply), and the contributing bed area necessary to provide enough coarse particles to form the sheet front (that is, the space between consecutive bedload sheets) shrinks with increasing sediment supply.

Supply and Mobility Conditions

The migration process of bedload sheets precludes the possibility that they are formed under a partial mobility condition because the whole grain-size distribution must experience transport in order to pass the waveform downstream. This suggests that channels with bedload sheets should exhibit an equal mobility condition, integrated over the waveform, with respect to the sediment supply, along the corridor of active bedload transport. Because the zone of active transport narrows with decreasing sediment supply as immobile coarse patches expand on the channel margins (Lisle, Iseya, and Ikeda 1993; Nelson *et al.* 2009), under low to moderate sediment supply conditions the grain-size distribution of the whole bed surface will likely be coarser than the distribution of the load and the channel will therefore be in a state of selective mobility. If the sediment supply becomes so low that the coarse patches extend across the entire width of the channel, bedload sheets should stop moving and gradually disappear.

We see this behavior in the Nelson *et al.* (2009) experiments with bedload sheets formed in pure gravel (Figure 16.1). As sediment supply to a channel is reduced, the zone of active transport narrows, bedload sheet migration rate declines until, under conditions of zero sediment supply, the sheets eventually disappear and the bed becomes uniformly coarse (Nelson *et al.* 2009). Had sand been part of the bed material mixture, it is likely the bed at this point would have entered a partial mobility condition as the sand was winnowed from the bed. But because the bed was composed entirely of gravel, the bed surface grains became interlocked against one another, trapping the fine material below the surface (Wydgza *et al.* 2005). This highlights a degree of freedom that generally does not exist in a sand-bedded channel. If the supply to a sand-bedded channel is reduced, the bed will erode to compensate for the sediment flux divergence. In a gravel-bedded channel, erosion of the bed will also occur, but as the bed erodes, the bed can become coarser under selective mobility conditions reducing and stabilizing the erosion.

The forgoing argument and Figure 16.5 appear to suggest that bedload sheets form under selective mobility conditions when the ratio of sediment supply to transport capacity for the subsurface is $0.1 < q_* < 1$. What happens if the transport regime is equally mobile rather than selective? It would be expected that incipient dunes begin to form (e.g. sediment feed experiments reported in Kleinhans 2002).

16.3.3 Partial Mobility Regime

There are two distinct types of bedforms that plot in the partial mobility regime of Figure 16.5. Partial-mobility conditions can occur with relatively high sediment supply ($q_* > 0.1$) when sand is transported over an immobile armored gravel bed. Under low sediment supply ($q_* < 0.1$) conditions in gravel mixtures, bed structures develop.

16.3.3.1 High Sand Supply Bedforms (q_* >0.1)

At high values of q_*, supply-limited sandy bedforms can develop over an otherwise immobile gravel bed. This is a special condition of partial transport whereby the gravel-bed armor is developed, but finer material continues to be supplied to the channel bed. This can occur through a gravel–sand transition where sand is supplied to the bed from suspension (e.g. Venditti and Church 2014; Venditti *et al.* 2015), when coarse gravel supply to the channel is episodic but sand supply from the floodplain and hillslopes is persistent, or where there has been a permanent shift in sediment supply from gravel to sand, and thus the gravel bed is in the process of being buried. The bedforms that develop under conditions of a static coarse bed and a fine sediment supply are well documented and include sand-ribbons, barchans, and dunes with gaps in the troughs that progressively close as the sediment supply increases until there is a fully alluvial sand bed (Figure 16.7a; Kleinhans *et al.* 2002; Tuijnder,

Ribberink, and Hulscher 2009; Tuijnder and Ribberink 2012). Figure 16.7b–e shows the results of a simple phenomenological experiment where sand was fed to a 1-m wide Plexiglas channel, revealing that as the sand supplied to an otherwise sediment-starved flow increases, the sequence of bedforms arises even under the simplest experimental conditions, without additional roughness elements. This sequence of bedforms is not limited to mixed gravel–sand-bedded rivers, as it has been well-documented in eolian environments (Bagnold 1941; Lancaster 1995), tidal environments (Allen 1968; Ernstsen *et al.* 2005) and on the sea floor (Lonsdale and Malfait 1974; Lonsdale and Spiess 1977), suggesting that a definable suite of bedforms develops under sand supply-limited conditions in many flow environments at the Earth's surface.

There has now been extensive work in flumes on this sequence of bedforms over otherwise immobile gravel beds in laboratories (e.g. Kleinhans *et al.* 2002; Grams 2006; Tuijnder, Ribberink, and Hulscher 2009; Tuijnder and Ribberink 2012) and the field (Kleinhans *et al.* 2002; Venditti *et al.* 2009). There has also been some work on dune development in mixed size sediments where there is a subsidiary gravel component in the bed mixture that forms an armor layer in the dune troughs, but also plays a role in active transport over the dunes (e.g. Blom, Ribberink, and De Vriend 2003). Generally, the work has shown that sand supply to the bed controls the bedform morphology (Figure 16.7a) and, in particular, it is the thickness of the sand coverage that sets the types of bedforms that emerge (Kleinhans *et al.* 2002; Venditti *et al.* 2009). Bedforms also grow with transport stage in addition to sediment supply (Figure 16.7a). Kleinhans *et al.* (2002) recognized this and cast their results in the form of a phase diagram defined by a bed mobility parameter (transport stage) and a ratio of the predicted to measured bedload, which markedly improved the distinction between bed features where sand was being transported over immobile gravel.

Supply and Mobility Conditions

It is clear from the existing literature that this sequence of sandy bedforms developed over an otherwise immobile bed is characteristic of the partial mobility regime because the gravel bed is generally immobile and sandy bedforms are migrating over it. Unfortunately, there have not been any systematic experiments exploring these features where the sediment supply and transport rate are independent, as occurs in sediment feed flumes. Most work on these types of bedforms in mixed gravel–sand mixtures has been undertaken in sediment recirculating flumes (e.g., Kleinhans *et al.* 2002; Blom, Ribberink, and De Vriend 2003; Kuhnle *et al.* 2006) where the sediment supply is controlled by the transport rate in the flume, allowing a partial mobility condition to persist in equilibrium (Wilcock and McArdell 1993). If a gravel–sand mixture is fed into a flume, and the flow is only competent to carry the sand, the gravel will cause aggradation until the slope of the channel can pass the incoming supply of all grain sizes, eventually shifting the transport condition to a selective mobility condition where bedload sheets are more likely to form, at least along a narrow active transport corridor (Figure 16.1). So whether this distinctive suite of bedforms can form with a mixed sand–gravel sediment supply is unclear. Data presented in Kleinhans (2002) from a sediment feed flume did not reveal the sequence of bedforms observed in his recirculating flume experiments; instead, he found sand ribbons and bedload sheets coexisted in the selective mobility regime and mixed sand–gravel dunes developed as the transport conditions approached equal mobility. Hence, we speculate that sandy bedforms developed over an otherwise immobile gravel bed can be sustained only if the gravel supply is terminated.

Within this context, our placement of the Kleinhans *et al.* (2002) data in Figure 16.5 should be viewed with some caution. In order to increase the sand supply, they increased the shear stress to disturb the armor layer and release more sand from the bed until sand dunes developed. In our calculation of $q_{*(C\&H)}$, we regarded the fully sand-bed transport rate as the unarmored transport rate and

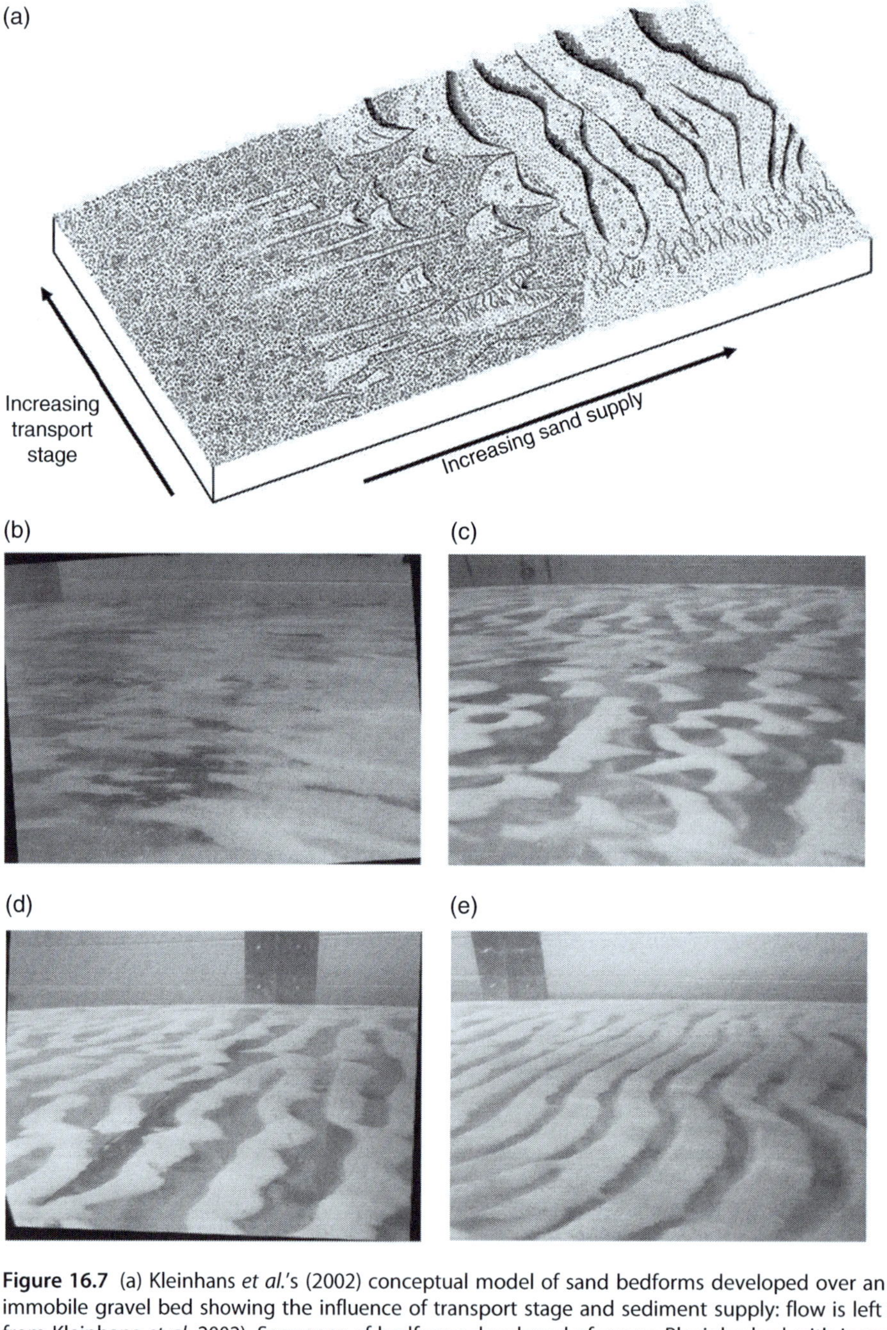

Figure 16.7 (a) Kleinhans *et al.*'s (2002) conceptual model of sand bedforms developed over an otherwise immobile gravel bed showing the influence of transport stage and sediment supply: flow is left to right. (Modified from Kleinhans *et al.* 2002). Sequence of bedforms developed of over a Plexiglas bed with increasing sand supply showing (b) patches of sand akin to sand ribbons, (c) barchan dunes, (d) interconnected barchan dunes, and (e) laterally continuous dunes: flow is right to left.

subsequent states (with lower shear stresses) as the transport rate after the bed had armored. Nevertheless, the data conform to the other patterns observed in Figure 16.5. At low values of $q_{*\,(C\,\&H)}$, sand ribbons form. Similar patches of sand have been described as beds are winnowed of finer particles (e.g. Hassan and Church 2000). As $q_{*\,(C\,\&\,H)}$ increases, the barchans form, and then when $q_{*\,(C\,\&\,H)} \approx 1$, sand dunes form with an underlying gravel armor layer. The bed material is still composed of a gravel–sand mixture; however, the experiment remained in the partial mobility regime. If transport stage increases in a channel, and the armor layer can be entrained, a shift to selective mobility dunes could occur (e.g. Blom, Ribberink, and De Vriend 2003), but these would be hydraulically controlled features.

16.3.3.2 Low Gravel Supply Bed Structures (q_* <0.1)

The bed features that plot in the partial mobility range when $q_{*\,(C\,\&\,H)}$ <0.1 are structural features of armor layers that include "pebble" clusters (Dal Cin 1968; Brayshaw 1984,1985; Strom and Papanicolaou 2009), transverse ribs (McDonald and Banerjee 1971; McDonald and Day 1978; Koster 1978) or stone lines (Laronne and Carson 1976), and reticulate stone cells described by Church, Hassan, and Wolcott (1998) and others (e.g. Gustavson 1974; Laronne and Carson 1976; Hassan and Church 2000). A clear taxonomy for these structural features of a gravel bed has not been presented in the literature, so for simplicity, we adopt the terms clusters, ribs, and reticulate cells.

Cluster bedforms are closely nested groups of clasts aligned roughly parallel with the flow and are widely considered the dominant microform roughness in gravel-bedded rivers (Dal Cin 1968; Brayshaw 1984). They are thought to form on the falling limb of hydrographs when the largest stones in the bed material, keystones, making up the D_{95} of the grain size distribution, stop moving. These keystones lie above the mean bed elevation of the surface armor and accumulate particles larger than the D_{75} on the upstream stoss side and particles smaller than the D_{50} in the lee of the keystone (Brayshaw 1984) forming the characteristic morphology shown in Figure 16.8a, d, and e. Brayshaw *et al.* (1983) shows that the lee deposit forms as particles are captured in the keystone wake and interacting flow fields can lead to particle capture on the stoss side. However, it has also been shown that clusters can be formed by particle interactions, without consideration of the fluid dynamics (Tribe and Church 1999). The features tend to be longer in the streamwise direction than in the cross-stream direction, are not regularly spaced, and have been observed to occupy 5–10% of the total bed area. Cluster size varies with the caliber of the sediment, but clusters up to 1.2 m in length have been documented (Brayshaw 1984). The stoss-side deposit forms as shear stress declines below the threshold for movement of the larger stones, and the lee-side deposit develops when the conditions permit deposition of the D_{50}. The flow field developed around a keystone is critical to the interaction of the large stones in the bed material because once a moving clast is caught in the upstream perturbed flow field of the keystone or its wake, it is essentially removed from the sediment load (Brayshaw, Frostick, and Reid 1983; Malmaeus and Hassan 2002).

The bed material comprising a cluster can typically be reentrained only by movement of the keystone. Billi (1988) describes a process whereby as flows increase, the lee particles are entrained, followed by the stoss particles, followed by dislodgement of the keystones at high flow. As such, cluster bedforms wash out at high flows and are reestablished on the declining limb of hydrographs (Billi 1988; De Jong 1991; Strom and Papanicolaou 2009), although continued disaggregation has been documented on the falling limb of hydrographs (Reid, Brayshaw, and Frostick 1984). Aggregation and disaggregation of groups of particle clasts is a fundamental component of gravel transport processes. Strom *et al.* (2007) show that clusters can form in uniform sediments, and it is common to see temporary clusters form as part of the transport process in mixed sized sediments. These temporary clusters formed during the transport process are distinguished from the more persistent cluster bedforms created under partial mobility conditions where the largest particles on the bed are immobile.

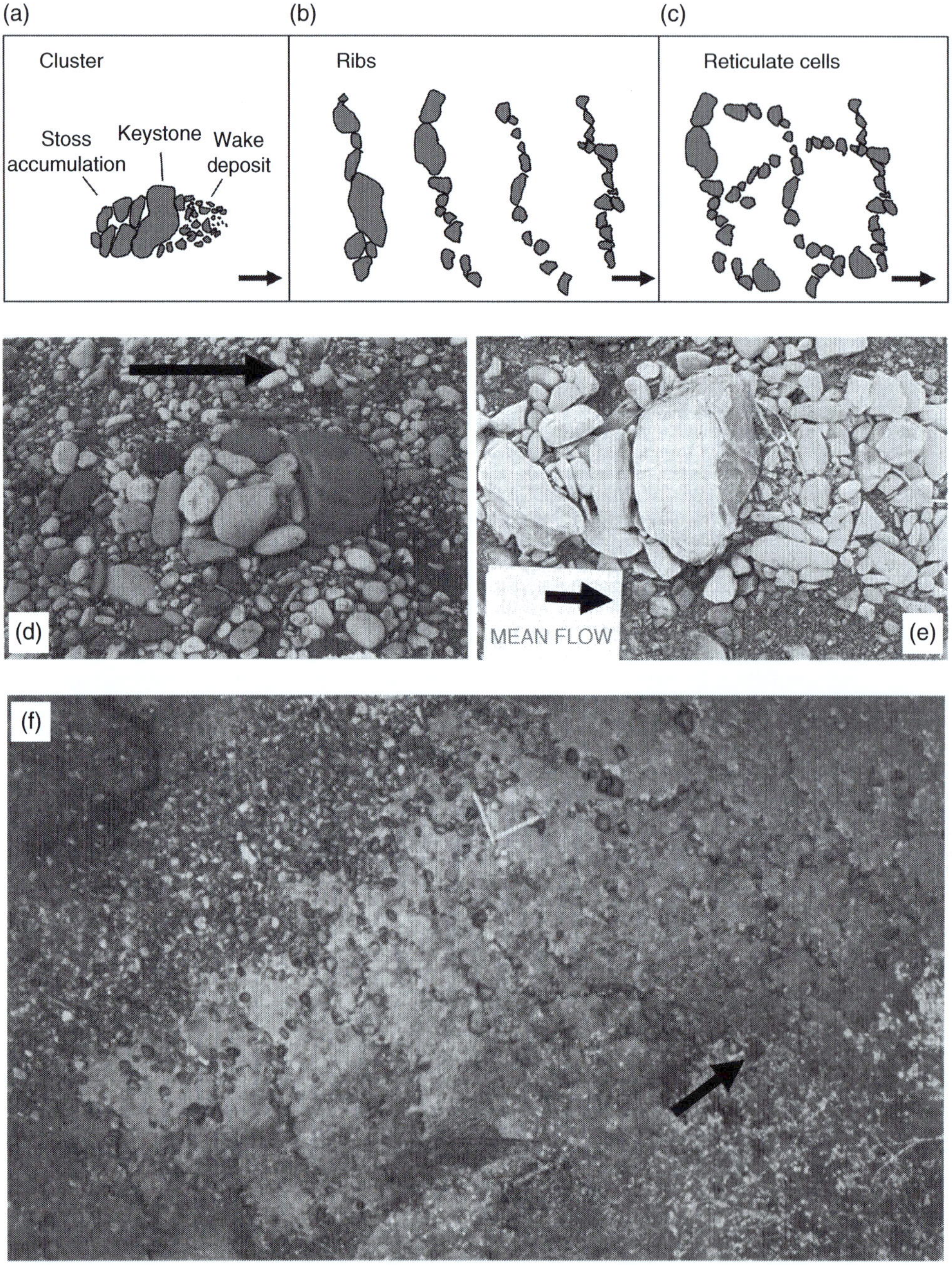

Figure 16.8 (a–c) Planview cartoons of the structural features of an armored surface, including (a) clusters, (b) ribs, and (c) reticulate cells. In panels b and c, only the large grains on the bed (> D_{84}) are depicted. Planview photographs of (d) a gravel cluster from Piave River, Italy, (e) a gravel cluster from Afon Elan, Wales, and (f) ribs and reticulate cells on the bed of Harris Creek, British Columbia, Canada. Flow direction is highlighted in all panels by the black arrows. ((d) *Source*: Cin 1968. Reproduced with permission of Elsevier; (e) from Brayshaw 1984, and (f) photograph courtesy of Michael Church.)

After clusters develop, intermediate features known as stone lines or transverse ribs may form (Figure 16.8b and f). There are considerable differences in the sedimentology and genetic relation among the various features that have been described as transverse ribs. Morphologically, they have been described as regularly spaced rows of pebbles, cobbles, or boulders lying transversely across the primary flow direction (McDonald and Banerjee 1971; Gustavson 1974; McDonald and Day 1978; Koster 1978; Allen 1983, 1984; Church, Hassan, and Wolcott 1998). Like clusters, transverse ribs are typically not more than a few grain diameters high and scale with the clast sizes. They typically extend across the flow for lengths greater than their alongstream spacing. The formative process for transverse ribs has been linked to supercritical flow, where the ribs develop under stationary hydraulic jumps (Koster 1978; McDonald and Day 1978; Allen 1983, 1984). This explanation may be plausible, but it requires hydraulic jumps to be stationary in space and time long enough for large particles to organize, which seems unlikely in a natural channel. Furthermore, some transverse ribs described in the literature look like the reticulate cells described by Church, Hassan, and Wolcott (1998), others may be bedload sheets, and still others are obviously step-pool morphologies (see examples provided by Macdonald and Banerjee (1971) and McDonald and Day (1978)), none of which are linked to supercritical flows. Step-pool morphologies appear to be linked to the jamming of and interlocking of large stones, on steep slopes where the flow plays a subsidiary role in setting the length and hydraulic roughness of the step-pool morphology (Zimmerman and Church 2001; Church and Zimmerman 2007), but their stability is argued to be related to the stochastic arrangement of clasts in the channel (Zimmerman, Church, and Hassan 2010). Our use of the term ribs is to describe the class of transverse features that form in gravel-bedded rivers under supply-limited conditions and excludes more exotic features such as transverse lines of pebbles formed over a silt bed (Macdonald and Banerjee 1971) or transverse lines of vegetation formed on concrete (Allen 1984), and channel-scale, step-pool morphology (Macdonald and Banerjee 1971).

Church, Hassan, and Wolcott (1998) described cluster bedforms as an intermediate stage in the development of stable reticulate cells, which form as a mixed-size bed develops an armor layer (Figure 16.8c). They consider these structures a stable self-organized product of the armoring process whereby particles move from less stable positions to more stable positions. Church, Hassan, and Wolcott (1998) argue that reticulate cells are derived from ribs formed under subcritical flow conditions. The cluster growth between the ribs interconnects them to form a web of cellular structures. Laronne and Carson (1976) describe similar features as closed structures that can trap finer-grained material. The features described by Church, Hassan, and Wolcott (1998) are on average 10 D_{84} long and 6 D_{84} wide. Structural development varies with the sediment transport, which is reasonable because higher transport rates require more frequent exchanges between mobile sediment and the bed. If larger particles are moving, the clasts forming the reticulate structures will be entrained. Hassan and Church (2000) reported experiments where they fed sediment to an otherwise immobile bed and found that greater sediment supply rates resulted in more weakly developed cells.

Some insight into the controlling mechanisms of bed structures has been acquired by simplified numerical modeling. Tribe and Church (1999) simulated the development of clusters, ribs, and reticulate cells using a simple rules-based approach and showed that these features may be formed by grain-to-grain interactions, without any consideration of the flow. In their model, particles stop on contact with other grains and may cause those grains to move. The density of large stones in the simulation domain controlled the development of different structures. When the domain had a coverage of 5–10% of large stones, linear along-stream cluster-like features developed. When the coverage was 20–30% of the domain, a loosely transverse series of interconnected ridges developed that were similar to reticulate cells, and when the coverage was > 40%, rib features emerged. When grains were allowed to rotate, transverse lines became more tightly packed if an entrainment probability rule (smaller grains are more likely to be entrained) and a neighbor rule (stone is entrained only if it has few close or touching neighbors) were implemented. Tribe and Church (1999) concluded that grain rotation and domain coverage were the primary controls on the emergence of reticulate cells observed by Church, Hassan, and Wolcott (1998), with the entrainment probability and the neighbor rule reinforcing the tight packing. A more probabilistic approach that includes entrainment thresholds based on shear stresses, resistance fields about particles for distrainment, a vertical protrusion-dependent entrainment rule, and a particle dropping rule to prevent vertical accretion, has produced more realistic simulations (Malmaeus and Hassan 2002).

Supply and Mobility Conditions

Clusters are formed by particles that have stopped moving, thereby allowing the accumulation of other clasts on the stoss side and the formation of a fine wake deposit. This would suggest they are a feature associated with partial mobility conditions. Similarly, ribs and reticulate cells are built from these clustered particles during the armoring process, so they too are characteristic of the partial mobility regime. In fact, the development of bed structures enhances partial mobility conditions by making it more difficult to entrain coarse particles. Field and flume measurements of interlocked coarse grains have shown that the vertical plucking force required to remove a grain from a static, dry bed can be 3–5 times the weight of the grain (Wyzdga *et al.* 2005). Feeding sediment to the bed at even nominal amounts leads to less well-defined cellular structures (Hassan and Church 2000), indicating they should be formed when sediment supply is low relative to the capacity.

16.4 A Phase Diagram for Bed Features in a Gravel-Bedded River

In sand-bedded rivers, the emergence of various bedforms have been effectively linked to measures of flow strength (e.g. velocity, bed shear stress, or nondimensional shear stress) and grain size (see Southard and Boguchwal 1990; and reviews in Allen (1984), Southard (1991) and Venditti (2013)). Bedform phase diagrams based on the mean flow velocity and grain size are particularly good discriminators of bedform types in sand-bedded rivers where there is no sediment supply limitation. Other combinations of flow strength and grain size are less effective. For example, bedform phase diagrams based on the shear stress and grain size often exhibit overlap of several bedform types in the phase space (Venditti 2013). There have been attempts to place bed features of gravel-bedded rivers on bedform phase diagrams developed for sand-bedded rivers with varying success. Carling (1999) added gravel dunes and bedload sheets to the mean velocity–grain size phase diagrams from Southard and Boguchwal (1990) and bed shear stress–grain size phase diagram of Allen (1984). The exercise is important in the present context, because it shows gravel dunes are hydraulically controlled. Indeed, if depth is taken into account in the Southard and Boguchwal (1990) diagram, gravel dunes plot in the same phase space (extended for larger grain sizes) as sand dunes. However, there is concerning overlap between plane-bed, dunes, and bedload sheets across the phase space on both diagrams, which prevents their use as a predictive tool in gravel-bedded rivers.

Placing the bedforms, structures, and patches that develop in gravel-bedded rivers on the flow-strength-based diagrams developed for sand-bedded rivers explicitly assumes flow strength exerts first-order control on these bed features. While we cannot discount the effect of flow strength, the discussion above and Figure 16.5 suggest that a more useful phase diagram may be formed by variables representing the sediment supply (relative to the transport capacity) and the differential mobility of

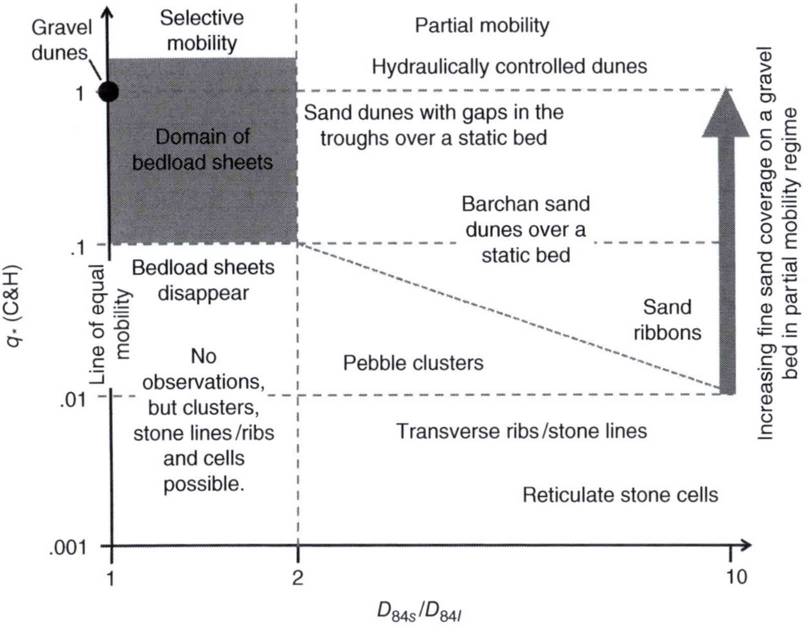

Figure 16.9 Conceptual bedform phase diagram for gravel beds, partly based on Figure 16.5.

particles in a mixture. Both of these are dependent on flow strength, but we assert that the flow strength itself plays a secondary role in the types of bed features that emerge.

Figure 16.9 shows a conceptual phase diagram for bed features developed under sediment supply-limited conditions in gravel-bedded rivers. The placement of the various features is based partly on Figure 16.5, but is also constrained by the above discussion of the sediment supply and mobility conditions required for the formation and stability of each bed feature. The phase diagram shows our proposed line of equal mobility at $D_{84s}/D_{84l} = 1$, selective mobility conditions between $1 < D_{84s}/D_{84l} < 2$ and partial mobility condition when $D_{84s}/D_{84l} > 2$. The bed is aggradational when $q_* > 1$. The tradeoff between surface armoring and erosion in response to the sediment flux divergence created when $q_* < 1$ is difficult to assess qualitatively, but a slow reduction in sediment supply, as is typical on the waning phase of hydrographs, would favor surface coarsening rather than bed incision and this is borne out by experimental work (Dietrich *et al.* 1989; Church, Hassan, and Wolcott 1998; Hassan and Church 2000; Nelson *et al.* 2009). It is of course possible to have $D_{84s}/D_{84l} < 1$, but this should only occur under aggradational conditions where the bed surface is being buried by coarser particles, so conditions to the left of $D_{84s}/D_{84l} = 1$ and $q_* < 1$ should not be physically possible.

These conditions define unique existence fields for bedforms, mobile patches, and structures in gravel-bedded rivers. We show gravel dunes occurring under equal mobility conditions when the supply is closely matched by the transport capacity. The dune area we identify on the phase diagram is a single point, which is too restrictive. For example, dunes observed in Kleinhans' (2002) sediment feed experiments would plot in the selective mobility region near the line of equal mobility, but uncertainties regarding the nature of the bed surface prevent us from including them in Figure 16.5. Furthermore, there is no reason why gravel dunes cannot form under aggradational conditions. In fact, their presence in the rock record (Carling 1999) dictates this. Nevertheless, our fundamental argument here is that gravel dunes are hydraulic phenomena and not developed by sorting processes, and our placement of them on this diagram is commensurate with that idea. Figure 16.9 is therefore best suited to highlighting bed features developed under supply-limited conditions.

The domain of bedload sheets lies within the selective mobility regime and we have indicated that they occur when $q_* > 0.1$, based on the experimental data presented in Nelson *et al.* (2009). They show that as sediment supply is reduced and the bed coarsens, the patchiness responsible for bedload sheet migration disappears. What happens to bedload sheets when $q_* << 0.1$ is not clear as there are no experimental data in that range. The bed may simply become static, but it is more likely that after bedload sheets disappear, the bed will shift to a partial mobility regime where the coarsest stones on the surface become immobile and the sequence of clusters, ribs, and reticulate cells emerge. The transport conditions described by Church, Hassan, and Wolcott (1998) as producing these bed features persisted during the final, zero-feed phase of the Nelson *et al.* (2009) Berkeley experiments when the coarse particles became interlocked and generally immobile (Wydzga *et al.* 2005). A logical question is whether bedload sheets can occur under aggradational conditions. Bennett and Bridge (1995a,b) and Madej *et al.* (2009) both report archetypical bedload sheets under aggradational conditions. This extends the domain of bedload sheets beyond our nominal placement based on Figure 16.5 to values of q_* up to ~1.5.

The clusters, ribs, and reticulate cells described by Church, Hassan, and Wolcott (1998) and Hassan and Church (2000) dominate when $q_* << 0.1$. Our placement of these bedforms in a sequence from higher to lower q_* and offset diagonally is purposeful. Hassan and Church (2000) describe the reticulate structures as being best developed when there is essentially no sediment supply and the bed has fully armored. Therefore, it is reasonable to expect that clusters will form first as the bed armors. As the bed continues to armor and the transport becomes increasingly partially mobile, transverse ribs will form. Finally, once the sediment transport rate has reached a stable asymptote, the development process is complete and the reticulate cells emerge as the dominant feature of the bed structure. Nevertheless, the available data do not distinguish between these features and we have taken some artistic liberty in their placement on the phase diagram.

At higher values of q_* and when D_{84s}/D_{84l} is large, Kleinhans *et al.* (2002) report sand ribbons emerging as the dominant bedform on the coarse surface layer. In a partial mobility regime, an increase in q_* represents increasing fine sand coverage on a gravel bed, which leads to the formation of barchans. Sand dunes with gaps in the troughs over a static bed form as q_* is further increased until the bed becomes aggradational with fine sediment and transitions to a fully alluviated sand bed with hydraulically controlled bedforms. Our placement of the sequence of sand ribbons, barchans, and dunes with gaps in the troughs diagonally is purposeful. Our expectation is that as q_* approaches 1, the grain-size distribution will become less partially mobile (more of the larger grains will move) and may transition to a stable condition of hydraulically controlled, mixed-size sand dunes with coarse troughs and finer stoss slopes and crests (Blom, Ribberink, and De Vriend 2003).

16.5 Perspective and Conclusions

We have reviewed the impact of sediment supply on bedload transport rates and the emergence of various bedforms, structures, and patches in gravel-bedded channels. Sediment supply, relative to capacity, appears to play a critical role in the emergence and disappearance of bed features and a distinct suite of features develops under equal, selective, and partial mobility conditions. Dunes appear to be the dominant bedform under equal mobility conditions where the sediment supply is closely matched by the transport capacity and they can form under aggradational conditions. Bedload sheets occur under conditions of selective transport when the grain-size distribution is wide enough to allow coarse particles to be mobilized by the fine fraction. Under the special case of a fine sediment supply to an otherwise coarse immobile gravel bed (partial mobility), a sequence of sandy bedforms (sand ribbons, barchans, and dunes with gaps in the troughs) will emerge until the fine sediment supply overwhelms the bed and it becomes sand-bedded. When the sediment supply to a gravel bed is low compared to the capacity in the partial mobility regime, the bed evolves a sequence of clusters, transverse ribs, and reticulate cells as the bed coarsens. These represent the end product of the coarsening process and the organization of the armor layer in a gravel-bedded channel.

Our framework provides an opportunity to explore controls on local bed features regardless of the source of the sediment supply. In the laboratory experiments we examined, the sediment feed provided the sediment supply. In the field, sediment supply is typically derived from the upstream reach, where it may be controlled by hydraulics. However, sediment supply may also be controlled by upstream or within-reach hillslope processes (landslides, debris flows, direct particle delivery) or bank erosion. Our framework does not distinguish between these sources because local bed features are responding to whatever is provided from immediately upstream.

Our framework also simplifies a complex process in gravel-bedded channels: variable discharge. Because our framework focuses on grain-size distributions and relative sediment supply, it is insensitive to variable flow. The bed features record the last flow capable of moving particles. As such, we could use the dominant bed features and grain-size distributions of the bed surface and subsurface (assumed to be equivalent to the load) to estimate the relative sediment supply conditions in a channel without knowing the flow. This could then be inverted to estimate the flow if supply could be estimated from downstream sediment deposits that capture all of the bedload. Our framework could therefore be developed into a paleoflow estimator. However, some consideration needs to be given to determine whether the predicted flow is the flow of interest. As with sandy bedforms, it is not always clear what aspects of variable flow (peak magnitude, duration of flow, steepness of rising and falling limbs of the flow) are responsible for the development of the bed features we observe in the field.

There are some impediments to applying the framework we developed. There are timescales of bed adjustment to consider when sediment supply is changed or changing, especially when the upstream sediment supply is controlled hydraulically. For sandy bedforms over a gravel bed, the timescale of adjustment is dependent on their size. For gravel clusters, the timescale of adjustment can be very short; clustering is an inherent process of gravel transport, so a sharp recessional hydrograph would leave clusters on a bed. Better developed stoss and lee deposits and interlocking caused by particle vibration would result from longer recessional hydrographs with long periods of partial mobility. Ribs and reticulate cells also become better developed through time as particles jostle into position on the recessional limb of hydrographs.

The greatest impediment to using the framework we developed here is the absence of data to verify the phase diagram. Our attempt to form a phase diagram from observations has been hampered by the availability of data to populate a diagram. Many experimental arrangements have used sediment-recirculating flumes (or partial-recirculating flumes where sand is recirculated and gravel is not) where the sediment supply is intricately intertwined with the channel hydraulics. Essentially, the sediment supply is always equivalent to the transport capacity, so q_* cannot be calculated reliably. While much can be learned about the transport process from these sediment recirculating experiments, these data are difficult to capture within our framework. Further hampering our ability to populate a phase diagram is that in order to quantify equal, selective, and partial mobility conditions, we need to know the behavior of the coarsest fractions of the bedload with respect to their presence on the bed. Few existing publications report sufficient information to do so. This is in part because many of the early publications on phenomena in gravel-bedded rivers documented features not previously reported and pre-date the pioneering work of Wilcock and collaborators on the differential mobility of grain-size distributions. However, we examined over 75 publications and found just four with the necessary information, a few with enough information for us to estimate parameters, and another few with enough information to intuit where they might lie on a phase diagram. This points to the need for a community data repository to facilitate storage and dissemination of existing and new sediment experimental data in a common format, which has recently been discussed extensively at workshops and scientific meetings (e.g., Hsu *et al.* 2015).

A better understanding of sediment supply controls on bed features in gravel-bedded rivers and verification of our phase diagram can ultimately be achieved with new field observations and laboratory experiments specifically designed to simultaneously measure bedload flux, the grain-size distributions of the bed surface and bedload, and bed morphology under different sediment supply rates for a variety of gravel and gravel–sand mixtures. We expect such work will result in refinement of the ideas presented herein and bring our understanding of bed features in gravel-bedded rivers to the same level that we understand bedforms developed in sand-bedded rivers.

Acknowledgments

This work was supported by an NSERC Discovery Grant to JGV. Michael Church and Rolf Kellerhals kindly provided previously unpublished photographs from their work. The authors thank Ian Reid and one anonymous reviewer for insightful comments that helped us sharpen the chapter and better understand how our ideas, developed largely from laboratory experiments, apply to gravel-bed rivers.

References

Allen, JRL 1968. *Current Ripples.* North Holland Publishing Company: Amsterdam.

Allen, JRL 1983. A simplified cascade model for transverse stone-ribs in gravelly streams. *Proceedings Royal Society of London Series A* 385(1789), 253–266.

Allen, JRL 1984. *Sedimentary Structures: Their Character and Physical Basis.* Elsevier Science Publishers: New York.

Bagnold, RA 1941. *The Physics of Blown Sand and Desert Dunes.* Methuen: London.

Baker, VR 1984. Flood sedimentation in bedrock fluvial systems. In *Sedimentology of Gravels and Conglomerates* (eds EH Koster and RJ Steel). Memoir 10, Canadian Society of Petroleum Geologists: Calgary; 87–98.

Bennett, SJ and Bridge, JS 1995a. The geometry and dynamics of low-relief bed forms in heterogeneous sediment in a laboratory channel, and their relationship to water flow and sediment transport. *Journal of Sedimentary Research* A65(1), 29–39.

Bennett, SJ and Bridge, JS 1995b. An experimental study of flow, bedload transport and bed topography under conditions of erosion and deposition and comparison with theoretical models. *Sedimentology* 42(1), 117–146.

Best, J 1996. The fluid dynamics of small-scale alluvial bedforms. In *Advances in Fluvial Dynamics and Stratigraphy* (eds PA Carling and MD Dawson). John Wiley & Sons: Chichester, UK; 67–125.

Billi P 1988. A note on cluster bedform behavior in a gravel-bed river. *Catena* 15(5), 473–481.

Blom, A, Ribberink, JS, and De Vriend, HJ 2003. Vertical sorting in bed forms: flume experiments with a natural and a trimodal sediment mixture. *Water Resources Research* 39(1025), 1–13.

Brayshaw, AC 1984. Characteristics and origin of cluster bedforms in coarse-grained alluvial channels. In *Sedimentology of Gravels and Conglomerates* (eds EH Koster and RJ Steel). Memoir 10, Canadian Society of Petroleum Geologists: Calgary; 77–85.

Brayshaw, AC 1985. Bed microtopography and entrainment thresholds in gravel-bed rivers. *Geological Society of America Bulletin* 96(2), 218–223.

Brayshaw, AC, Frostick, LE, and Reid, I 1983. The hydrodynamics of particle clusters and sediment entrainment in coarse alluvial channels. *Sedimentology* 30(1), 137–143.

Bridge, JS 1993. The interaction between channel geometry, water flow, sediment transport and deposition in braided rivers. In *Braided Rivers* (eds JL Best and CS Bristow). Special Publication 75, Geological Society of London; 13–71.

Buffington, JM and Montgomery DR 1999. Effects of sediment supply on surface textures of gravel bed rivers. *Water Resources Research* 35(11), 3523–3530.

Buffington, JM, Dietrich, WE, and Kirchner JW 1992. Friction angle measurements on a naturally formed gravel streambed: implications for critical boundary shear stress. *Water Resources Research* 28(2), 411–425.

Bunte, K, Abt, SR, Potyondy, JP, and Ryan, SE 2004. Measurement of coarse gravel and cobble transport using portable bedload traps. *Journal of Hydraulic Engineering* 130(9), 879–893.

Carling, PA 1999. Subaqueous gravel dunes. *Journal of Sedimentary Research* 69(3), 534–545.

Carling, PA, Richardson, K, and Ikeda, H 2005. A flume experiment on the development of subaqueous fine-gravel dunes from a lower-stage plane bed. *Journal of Geophysical Research-Earth Surface* 110 (F04S05), 1–15.

Cartigny, MJB, Ventra, D, Postma, G, and van Den Berg JH 2014. Morphodynamics and sedimentary structures of bedforms under supercritical-flow conditions: new insights from flume experiments. *Sedimentology* 61(3), 712–748.

Church, M and Hassan, MA 2005. Estimating the transport of bed material at low rate in gravel armoured channels. In *Geomorphological Processes and Human Impacts on River Basins* (eds RJ Batalla and C Garcia). Publication 299, IAHS: Wallingford, UK; 141–153.

Church, M and Zimmermann, A 2007. Form and stability of step-pool channels: Research progress. *Water Resources Research* 43(W03415), 1–21.

Church, M, Hassan, MA, and Wolcott, JF 1998. Stabilizing self-organized structures in gravel-bed stream channels: field and experimental observations. *Water Resources Research* 34(11), 3169–3179.

Curran, JC and Wilcock PR 2005. Effect of sand supply on transport rates in a gravel-bed channel. *Journal of Hydraulic Engineering* 131(11), 961–967.

Dal Cin, R 1968. Pebble clusters: their origin and utilization in the study of paleocurrents. *Sedimentary Geology* 2(4), 233–241.

De Jong, C 1991. A reappraisal of the significance of obstacle clasts in cluster bedform dispersal. *Earth Surface Processes and Landforms* 16(8), 737–744.

Dietrich, WE, Kirchner, JW, Ikeda, H, and Iseya, F 1989. Sediment supply and the development of the coarse surface layer in gravel-bedded rivers. *Nature* 340, 215–217.

Dinehart, RL 1989. Dune migration in a steep, coarse-bedded stream. *Water Resources Research* 25(5), 911–923.

Dinehart, RL 1992a. Evolution of coarse gravel bedforms: field measurements at flood stage. *Water Resources Research* 28(10), 2667–2689.

Dinehart, RL 1992b. Gravel-bed deposition and erosion by bedform migration observed ultrasonically during storm flow, North Toutle River, Washington. *Journal of Hydrology* 136(1–4), 51–71.

Ernstsen, VB, Noormets, R, Winter, C *et al.* 2005. Development of subaqueous barchanoid-shaped dunes due to lateral grain size variability in a tidal inlet channel of the Danish Wadden Sea. *Journal of Geophysical Research-Earth Surface* 110(F04S08), 1–13.

Ferguson, RI, Prestegaard, KL, and Ashworth PJ 1989. Influence of sand on hydraulics and gravel transport in a braided gravel bed river. *Water Resources Research* 25(4), 635–643.

Grams, PE 2006. *Sand transport over a coarse and immobile bed.* PhD thesis, Johns Hopkins University.

Gustavson, TA 1974. Sedimentation on gravel outwash fans, Malaspina Glacier Foreland, Alaska. *Journal of Sedimentary Petrology* 44(2), 374–389.

Hassan, MA and Church M 2000. Experiments on surface structure and partial sediment transport. *Water Resources Research* 36(7), 1885–1895.

Hubbell, DW, Stevens, HH, Skinner, JV, and Beverage, JP 1987. *Laboratory Data on Coarse-Sediment Transport for Bedload-Sampler Calibrations.* Water-Supply Paper 2299, US Geological Survey; 31 pp.

Hsu, L, Martin, RL, McElroy, B *et al.* 2015. Data management, sharing, and reuse in experimental geomorphology: challenges, strategies, and scientific opportunities. *Geomorphology* 244, 180–189.

Iseya, S and Ikeda, H 1987. Pulsation in bedload transport rates induced by a longitudinal sediment sorting: a flume study of sand and gravel mixtures. *Geografiska Annaler* 69(1), 15–27.

Kirchner, JW, Dietrich, WE, Iseya, F, and Ikeda H 1990. The variability of critical shear stress, friction angle, and grain protrusion in water-worked sediments. *Sedimentology* 37(4), 647–672.

Kleinhans, MG 2002, *Sorting out sand and gravel; sediment transport and deposition in sand-gravel bed rivers.* PhD thesis, Universiteit Utrecht.

Kleinhans, MG, Wilbers, AWE, De Swaff, A, and Van Den Berg JH 2002. Sediment supply limited bedforms in sand-gravel bed rivers. *Journal of Sedimentary Research* 72(5), 629–640.

Koster, EH 1978. Transverse ribs: their characteristics, origin and paleohydraulic significance. In *Fluvial Sedimentology* (ed. AD Miall). Memoir 5, Canadian Society of Petroleum Geologists: Calgary; 161–186.

Kuhnle, RA and Southard JB 1988. Bed load transport fluctuations in a gravel bed laboratory channel. *Water Resources Research* 24(2), 247–260.

Kuhnle, RA, Horton, JK, Bennett, SJ, and Best JL 2006. Bed forms in bimodal sand–gravel sediments: Laboratory and field analysis. *Sedimentology* 53(3), 631–654.

Lancaster, N 1995. *Geomorphology of Desert Dunes.* Routledge: London.

Laronne, JB and Carson MA 1976. Interrelationships between bed morphology and bed-material transport for a small, gravel-bed channel. *Sedimentology* 23(1), 67–85.

Laronne, JB and Reid, I 1993. Very high bedload sediment transport in desert ephemeral rivers. *Nature* 366, 148–150.

Laronne, JB, Reid, I, Yitshak, Y, and Frostick, LE 1994. The nonlayering of gravel streambeds under ephemeral flood regimes. *Journal of Hydrology* 159(1–4), 353–363.

Lisle, TE, Iseya, F, and Ikeda H 1993. Response of a channel with alternate bars to a decrease in supply of mixed-size bed load: a flume experiment. *Water Resources Research* 29(11), 3623–3629.

Lonsdale, P and Malfait B 1974. Abyssal dunes of foraminiferal sand on the Carnegie Ridge. *Geological Society of American Bulletin* 85(11), 1697–1712.

Lonsdale, P and Spiess FN 1977. Abyssal bedforms explored with a deeply towed instrument package. *Marine Geology* 23, 57–75.

Madej, MA, Sutherland, DG, Lisle, TE, and Pryor B 2009. Channel responses to varying sediment input: A flume experiment modeled after Redwood Creek, California. *Geomorphology* 103(4), 507–519.

Malmaeus, JM and Hassan MA 2002. Simulation of individual particle movement in a gravel streambed. *Earth Surface Processes and Landforms* 27(1), 81–97.

McDonald, BC and Banerjee, I 1971. Sediments and bedforms on a braided outwash plain. *Canadian Journal of Earth Sciences* 8(10), 1282–1301.

McDonald, BC and Day, TJ 1978. An experimental flume study on the formation of transverse ribs. *Geological Survey of Canada Current Research Part A* 78-1A, 441–451.

Meyer-Peter, E and Muller, R 1948. Formulas for bed-load transport. *Proceedings of International Association for Hydraulic Structures Research 2nd Meeting*, Stockholm, 7–9 July 1948; 39–61.

Miller, MC, McCave, IN, and Komar, PD 1977. Threshold of sediment motion under unindirectional currents. *Sedimentology* 41(4), 883–903.

Nelson, PA, Venditti, JG, Dietrich, WE *et al.* 2009. Response of bed surface patchiness to reductions in sediment supply. *Journal of Geophysical Research-Earth Surface* 114(F02005), 1–18.

Parker, G 1990. Surface-based bedload transport relation for gravel rivers. *Journal of Hydraulic Research* 28(4), 417–436.

Parker, G 2008. Transport of gravel and sediment mixtures. In *Sedimentation Engineering: Theories, Measurements, Modeling and Practice* (ed. MH Garcia). Manual and Reports on Engineering Practice No. 110, American Society of Civil Engineers: Reston, VA; 165–264.

Parker, G, and Klingeman PC 1982. On why gravel bed streams are paved. *Water Resources Research* 18(5), 1409–1423.

Parker, G, Klingman, PC, and McLean DG 1982. Bedload and size distribution in paved gravel-bed streams. *ASCE Journal of the Hydraulics Division* 108, 544–571.

Pitlick, J 1992. Flow resistance under conditions of intense gravel transport. *Water Resources Research* 28(3), 891–903.

Powell, DM, Reid, I, and Laronne, JB 2001. Evolution of bedload grain-size distribution with increasing flow strength and the effect of flow duration on the caliber of bedload sediment yield in ephemeral gravel-bed rivers. *Water Resources Research* 37(5), 1463–1474.

Recking, A, Frey, P, Paquier, A, and Belleudy P 2009. An experimental investigation of mechanisms involved in bed load sheet production and migration. *Journal of Geophysical Research-Earth Surface* 114(F03010), 1–13.

Reid, I and Laronne, JB 1995. Bedload sediment transport in an ephemeral stream and a comparison with seasonal and perennial counterparts. *Water Resources Research* 31(3), 773–781.

Reid, I, Brayshaw, AC, and Frostick, LE 1984. An electromagnetic device for automatic detection of bedload motion and its field applications. *Sedimentology* 31(2), 269–276.

Seminara, G 1998. Stability and morphodynamics. *Meccanica* 33(1), 59–99.

Seminara, G, Colombini, M, and Parker, G 1996. Nearly pure sorting waves and the formation of bedload sheets. *Journal of Fluid Mechanics* 312, 253–278.

Southard, JB 1991. Experimental determination of bedform stability. *Annual Reviews of Earth Science* 19, 423–455.

Southard, JB and Boguchwal, LA 1990. Bed configurations in steady unidirectional water flow. Part 2. Synthesis of flume data. *Journal of Sedimentary Petrology* 60(5), 658–679.

Strom, KB and Papanicolaou, AN 2009. Occurrence of cluster microforms in mountain rivers. *Earth Surface Processes and Landforms* 34(1), 88–98.

Strom, KB, Papanicolaou, AN, Evangelopoulos, N, and Odeh, M 2007. Microforms in gravel bed rivers: Formation, disintegration and effects on bedload transport. *Journal of Hydraulic Engineering* 130(6), 554–567.

Tribe, S and Church, M 1999. Simulation of cobble structure on a gravel streambed. *Water Resources Research* 35(1), 311–318.

Tuijnder, AP and Ribberink, JS 2012. Experimental observation and modelling of roughness variation due to supply-limited sediment transport in uni-directional flow. *Journal of Hydraulic Research* 50(5), 506–520.

Tuijnder, AP, Ribberink, JS, and Hulscher, SJMH 2009. An experimental study into the geometry of supply-limited dunes. *Sedimentology* 56(6), 1713–1727.

Venditti, JG 2013. Bedforms in sand-bedded rivers. in *Treatise on Geomorphology, Volume 9, Fluvial Geomorphology*, ed. Shroder, J. (Editor in Chief), Wohl, E. (Volume Editor), Academic Press, San Diego, USA), 137–162.
Venditti, JG and Church, M 2014. Morphology and controls on the position of a gravel-sand transition: Fraser River, British Columbia. *Journal of Geophysical Research-Earth Surface* 119(9), 1959–1976.
Venditti, JG, Church, M, and Bennett, SJ 2006. On interfacial instability as a cause of transverse subcritical bed forms. *Water Resources Research* 42(W07423), 1–10.
Venditti, JG, Nittrouer, JA, Humphries, RP, and Allison, MA 2009. Supply-limited bedforms in a gravel-sand transition. *Eos, Transactions of the AGU 90(52), Fall Meeting Supplement*, Abstract EP21B-0586.
Venditti, JG, Dietrich, WE, Nelson, PA *et al.* 2010. Mobilization of coarse surface layers in gravel-bedded rivers by finer gravel bed load. *Water Resources Research* 46(W07506), 1–10.
Venditti, JG, Domarard, N, Church, M, and Rennie, CD 2015. The gravel–sand transition: Sediment dynamics in a diffuse extension. *Journal of Geophysical Research – Earth Surface* 120(6), 943–963.
Whiting, PJ, Dietrich, WE, Leopold *et al.* 1988. Bedload sheets in heterogeneous sediment. *Geology* 16(2), 105–108.
Wilcock, PR 1992. Experimental investigation of the effect of mixture properties on transport dynamics. In *Dynamics of Gravel-Bed Rivers* (ed. P Billi). John Wiley & Sons: Chichester, UK; 109–139.
Wilcock, PR 1998. Two-fraction model of initial sediment motion in gravel-bed rivers. *Science* 280, 410–412.
Wilcock, PR and Crowe, JC 2003. Surface-based transport model for mixed size sediment. *Journal of Hydraulic Engineering* 129(2), 120–128.
Wilcock, PR and DeTemple, BT 2005. Persistence of armor layers in gravel bed streams. *Geophysical Research Letters* 32(L08402), 1–4.
Wilcock, PR and Kenworthy, ST 2002. A two-fraction model for the transport of sand/gravel mixtures. *Water Resources Research* 38(10), 1–12.
Wilcock, PR and McArdell, BW 1993. Surface-based fractional transport rates: mobilization thresholds and partial transport of a sand–gravel sediment. *Water Resources Research* 29(4), 1297–1312.
Wilcock, PR, Kenworthy, ST, and Crowe, JC 2001. Experimental study of the transport of mixed sand and gravel. *Water Resources Research* 37(12), 3349–3358.
Wooldridge, CP and Hickin, EJ 2005. Radar architecture and evolution of channel bars in wandering gravel-bed rivers: Fraser and Squamish Rivers, British Columbia, Canada. *Journal of Sedimentary Research* 75, 844–860.
Wydzga, MA, Hassan, M, Venditti, JG, and Dunne, T 2005. Can interlocked grains reduce the mobility of gravel bed rivers? *Eos, Transactions of the AGU 86(52), Fall Meeting Suppliment*, Abstract H53B-0462.
Yalin, MS and Karahan E 1979. Inception of sediment transport. *ASCE Journal of the Hydraulics Division* 105, 1433–1443.
Zimmermann, A and Church M 2001. Channel morphology, gradient profiles and bed stresses during flood in a step-pool channel. *Geomorphology* 40(3–4), 311–327.
Zimmermann, A, Church, M, and Hassan, MA 2010. Step-pool stability: testing the jammed state hypothesis. *Journal of Geophysical Research - Earth Surface* 115(F02008), 1–16.

Discussion

Discussion by Jie Qin

The author presents an impressive phase diagram which represents a promising framework to investigate bed features emerged on gravel surfaces. This diagram is an improved version of another diagram presented by the author (Venditti, Nelson, and Dietrich 2008). In both diagrams, the authors argue that the emergence of the bed features is controlled by sediment conditions (e.g. sediment supply and grain-size distribution), while hydraulic conditions play a subsidiary role in their development. I agree that sediment conditions play an important role in developing the bed structures as demonstrated in previous literature, but the relative importance of sediment conditions and hydraulic conditions may need to be elaborated. Supposing that there is a simple recirculating flume experiment filled with heterogeneous gravel materials, the flow discharge increased gradually from zero until reaching an equal mobility situation. During this experiment, different bed features can be readily observed, while, in my opinion, these features are mainly determined by hydraulic conditions or by both hydraulic and sediment conditions.

Reference

Venditti, JG, Nelson, PA, and Dietrich, WE 2008. The domain of bedload sheets. In *Marine and River Dune Dynamics III* (eds DP Parsons, T Garlan, and JL Best). University of Leeds.

Discussion by Jonathan Laronne

The attempt to construct phase diagrams for bedforms smaller than the channel scale but larger than individual grains excepting dunes in gravel-bedded rivers is very interesting. The main argument is that "the emergence of these features is controlled by sediment supply and the differential mobility of grain-size distributions (equal, selective, and partial mobility)." If supply is the controlling mechanism, ideally a given river reach or large bedform such as a bar would be characterized by one bedform. Nonetheless, except for starved gravel-bedded rivers, all the rest are characterized by a wealth of bedforms of the type examined in these flume runs. They form intricate patches of more mobile and less mobile bed materials. As such, sediment supply must be very locally controlled and not reach-controlled. This also attests to the fact that facies of fluvial conglomerates include a variety of fabrics and grain size distributions within a given unit.

Response by the Authors

Most gravel-bed rivers exhibit discrete patches of similar grain size that may be free (migrating sorting features), fixed (spatially persistent sorting features caused by weak, grain-scale, topographic controls and local coarsening) or forced (spatially persistent features due to strong topographic controls such as bar morphology and channel obstructions) (Nelson *et al.*, 2009, 2010). The occurrence of these patches is related to reach-scale sediment supply, insofar as channels without a supply will eventually become uniformly coarse and immobile. Our phase diagram considers the features developed due to local sediment supply conditions. It does not consider the variability in bed features that would emerge across channel-scale topographic features such as bars, steps, and pools due to spatial gradients in local sediment supply. Nevertheless, it is likely that there are characteristic sequences or patterns of bed features that are related to these channel-scale topographic features, which directly reflect reach-scale sediment supply conditions. Identification of these reach-scale patterns remains a problem for further research.

The relative role of hydraulics and sediment supply in creating bed features is not fully resolved. In uniform sediment, bedforms that emerge are controlled by the interaction of a turbulent flow field with the labile bed and sediment supply is not an important consideration so long as it is high enough to keep the bed fully alluviated. In non-uniform sediment, the bed surface grain-size distribution adjusts to sediment supply (Dietrich *et al.* 1989). As such, the quantity and caliber of sediment supplied from upstream exerts a dominant role in what types of bed features emerge in response to a particular flow. If flow is held constant and upstream sediment supply is varied, the type of bed feature that emerges depends on the sediment supply (see Figure 16.1). The flow is only important insofar as it is responsible for the type of transport condition it generates (partial, selective, or equal) for a particular size distribution. Sediment recirculating flumes exhibit fundamentally different behavior than sediment feed flumes in that the quantity and caliber of sediment supplied is dependent on flow strength because the sediment that exists the flume is what is supplied to the upstream end (cf. Parker and Wilcock, 1993; Wilcock and DeTemple, 2005). In this case, the types of bed features that emerge may superficially appear to be controlled by hydraulic conditions, but ultimately hydraulics is setting the sediment supply, which in our framework, sets the bed feature.

References

Nelson, PA, Dietrich, WE, and Venditti, JG 2010. Bed topography and the development of forced bed surface patches. *Journal of Geophysical Research – Earth Surface* 115, F04024. DOI:10.1029/2010JF001747.

Parker, G and Wilcock, P 1993. Sediment feed and recirculating flumes: fundamental difference. *Journal of Hydraulic Engineering* 119, 1192–1204.

17

Linking Debris Flows and Landslides to Large Floods in Gravel-Bed Rivers

Lorenzo Marchi

17.1 Introduction

The interactions between landslides (including debris flows) and river systems are a topic of great relevance for both the understanding of the evolution of mountain basins and channel networks and the implications for watershed management and flood-hazard control (Costa 1984; Benda and Dunne 1997; Korup 2005, 2013). Landslides and debris flows cause sudden and spatially concentrated pulses of sediment to travel from hillslopes and small lateral channels to main rivers. One of the most dramatic outcomes is the blockage of the receiving stream by landslide dams (Costa and Schuster 1988). Even when channel blockage does not occur, the impacts on the river system are remarkable, influencing geomorphological and sedimentological settings of valley floors (Webb *et al.* 1988; Benda 1990). Channel aggradation, which is one of the most common outcomes of sediment pulses delivered from landslides to river systems, favours channel widening and lateral instability, including avulsion (Korup 2004; Korup, McSaveney, and Davies 2004).

The interactions of landslides and debris flows with rivers during major floods are similar to those occurring during normal flow conditions, but they are characterized by faster evolution and higher intensity of the involved processes. Due to the generally high transport capacity of the receiving stream during floods, sediment pulses from landslides lead to increased sediment transport (Wohl and Pearthree 1991). However, channel instability due to the aggradation of channel beds and floodplains, which occurs if sediment input exceeds the transport capacity, may severely increase the flood hazard. During floods, the probability of the receiving stream becoming blocked can be lowered because high discharge can erode and entrain landslide accumulations and debris-flow deposits, but there is a higher probability of paroxysmal evolution (fast filling of the landslide lake and erosion of the landslide dam) if channel damming does occur.

The spatial and temporal scales of hydrometeorological triggering events (Figure 17.1) control the interactions of landslides and debris flows with floods in gravel-bed rivers. Debris flows and shallow landslides can be triggered by convective cells that, because of their limited extent and short duration, do not cause relevant floods in gravel-bed rivers, which drain larger basins than debris-flow channels. Longer rainstorms involving larger areas, ascribed to mesoscale storm systems, may trigger both debris flows in small, steep channels and floods in larger rivers. In the latter scenario, in addition to the occurrence of multiple hazards (landslides, debris flows, and floods), the overlapping of the flood response in catchments of different extents enhances the possible impacts of individual

Gravel-Bed Rivers: Processes and Disasters, First Edition. Edited by Daizo Tsutsumi and Jonathan B. Laronne.

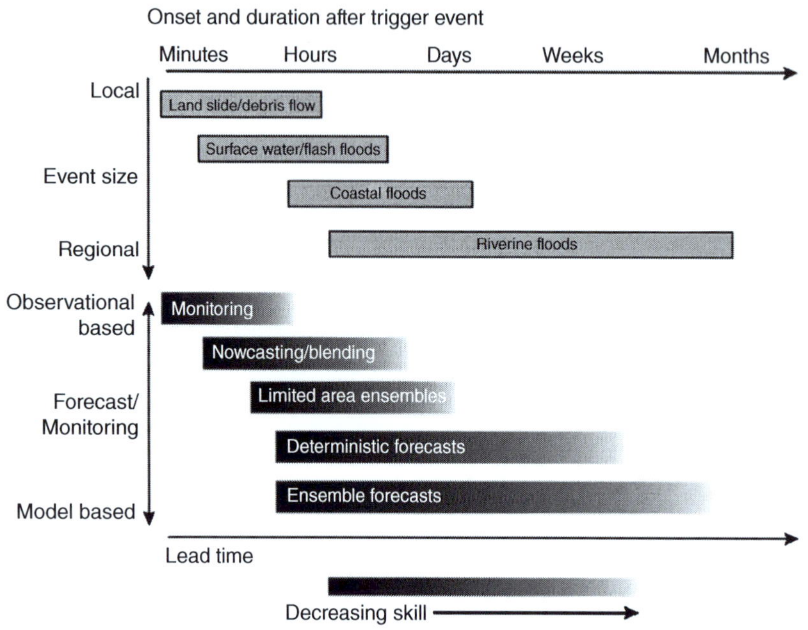

Figure 17.1 Spatial and temporal scales of (upper part) water-related hazards and (lower part) forecasting/monitoring strategies. Floods in gravel-bed rivers mostly lie in the domain of surface-water/flash floods. (*Source*: Alfieri *et al.* 2012. Reproduced with permission of Elsevier.)

processes (Borga *et al.* 2014). The probability of concurrence with floods in gravel-bed rivers is particularly high for debris flows and shallow rainfall-induced landslides, which are caused by intense rainfall that also induce floods in gravel-bed rivers draining basins of small to medium size. In addition, large deep-seated landslides, although less closely linked to flood hydrology, may occur during major floods, resulting in dramatic impacts on the fluvial system.

This chapter aims to outline, with examples from the literature and the experience of the author, different types of interactions of landslides and debris flows with floods in mountain gravel-bed rivers (section 17.2), and present an overview on the approaches suitable to predict the occurrence of landslides and debris flows and their linkage with river systems, with attention to both the temporal and spatial scales of the processes (section 17.3). Section 17.4 summarizes the space scale at which different types of instability influence the river system, and discusses hazard prediction and management issues. Finally, section 17.5 presents some conclusions and delineates aspects that could deserve attention in future research.

17.2 Interactions Between Mass Wasting and Floods in Gravel-Bed Rivers

Examples of different types of interactions between landslides and gravel-bed rivers are presented hereafter:

- changes in the type of flow process;
- increase in sediment transport and catchment sediment yield (both during and following the event – months to years);

- formation and collapse of landslide dams;
- erosion and transport of large wood debris.

17.2.1 Changes in Flow Process Type

In steep gravel-bed streams that drain rugged catchments, landslides can change the type of flow processes occurring during large floods, triggering debris flows in channels that are normally affected by floods with bedload. The Chieppena Torrent (Eastern Alps, Italy) provides an example of such processes. On 4 November 1966, intense and long-lasting rainfall caused major flooding in several basins of northeastern Italy. In mountain catchments, the melting of relevant snowpacks already present enhanced the influence of rainfall on runoff formation and slope instability. The Chieppena Torrent lies in one of the sectors most severely hit by this regional-scale flood event. Widespread shallow landslides on the forested slopes of the catchment (Figure 17.2) blocked the main channel with sediment and wood-debris dams. The sudden collapse of the landslide dams caused two major debris-flow surges that flowed through the channel and entrained more sediment, which also occurred due to the destruction of stone-masonry check dams, leading to the deposition of approximately 10^6 m^3 of sediment on the alluvial fan. The catchment area (27 km^2 at the fan apex) exceeds the area of most catchments affected by debris flows in the Eastern Italian Alps (Marchi and D'Agostino 2004), while the average channel slope (19%), combined with the channel confinement, is compatible with debris-flow propagation. The alteration to the normal sediment supply process and the formation and failure of landslide dams thus caused a change in the flow processes of the Chieppena Torrent during the flood of November 1966, leading to the formation of an extreme debris flow. An analysis of historical records regarding the floods of the Chieppena Torrent also shows that for at least one previous event (September 1882) the failure of landslide dams triggered a debris-flow (Cerato 1999), whereas no landslide activity was reported for more frequent floods with bedload (Marchi and Cavalli 2007).

Figure 17.2 Shallow landslides on the side slopes and destroyed check dams in the upper part of the Chieppena Torrent: mf, landslide deposit; fq, phyllites. The remnants of the destroyed check dams are visible on the right bank. (*Source*: Venzo and Largaiolli 1968. Reproduced with permission of Giulio Antonio Venzo.)

Other remarkable examples of the evolution of water flows into debris flows due to mass-wasting events have been documented for lake outbursts in proglacial environments (Clague and Evans 2000; O'Connor, Hardison, and Costa 2001; Chiarle *et al.* 2007). O'Connor, Hardison, and Costa (2001) describe the formation of debris flows due to the incorporation of large amounts of loose material downstream of the breaches of moraine-dammed lakes in the Oregon Coast Range (United States). The transformation of water flows into debris flows occurred within 500 m for all the cases studied; the debris flows travelled several kilometres; farther downstream, after a large volume of sediment had been deposited, several of them evolved into sediment-laden streamflow. Sediment entrainment in glacial-lake outburst floods may alter the flow processes in gravel-bed rivers without attaining the high solid concentration of debris flows. Cenderelli and Wohl (1998) have referred to water floods or to flow processes intermediate between hyperconcentrated flow and water flood to describe the deposits left by two large outburst floods in the Mount Everest region.

17.2.2 Increase in Sediment Transport

An increase in sediment transport in receiving rivers is one of the most important and common impacts of the delivery of sediment due to debris flows and landslides. This process, which represents an expected outcome of sudden and spatially concentrated inputs of sediments concomitant with the large transport capacity of flood flows, is commonly recognized but seldom assessed quantitatively.

A sequence of landslides and floods in Taiwan since 1999 has provided an ideal natural experiment to assess the interactions between landslides and sediment transport in gravel-bed rivers. The Chi-Chi earthquake (M_w 7.6, 21 September 1999) triggered widespread soil and bedrock landslides in central Taiwan. In a landslide inventory implemented after Typhoon Toraji (July 2001) the number of landslides was higher and the total landslide area was more than three times larger than after the earthquake. Most post-typhoon landslides were located in areas where the earthquake had not triggered any. Moreover, rainfall-triggered landslides were more closely connected to the river channel than the landslides caused by the earthquake, which had remained mostly confined to hillslopes (Dadson *et al.* 2004). Dadson *et al.* (2004) suggested that widespread landslide occurrence after Typhoon Toraji was favoured by the Chi-Chi earthquake through a loss of cohesion and frictional strength in the rock masses. The impact of the landslides triggered by the Chi-Chi earthquake and Typhoon Toraji and other smaller post-earthquake typhoons on river sediment transport has been assessed by Dadson *et al.* (2004), who analysed records of suspended sediment loads: rating curves relating water discharge to the suspended sediment concentration show an increase up to approximately four times from pre- to post-earthquake conditions (Figure 17.3). In addition to the increase of sediment transport, the landslides triggered in central Taiwan by the Chi-Chi earthquake and subsequent typhoons resulted in dramatic channel aggradation (Chen 2008)

A study by Chen *et al.* (2011) in the mountain catchments of southeastern Taiwan analysed the combined influence of earthquakes and typhoons on landslides and the subsequent increase in fluvial sediment transport over a longer time span (from 1996 to 2007). The role of earthquakes in triggering slope failures and weakening bedrock and the increased occurrence of landslides during typhoons were confirmed. However, Chen *et al.* (2011) noted that the influence of widespread landslides on sediment transport in the studied catchments began declining after 1–3 years.

17.2.3 Channel Damming and Failures of Landslide Dams

Although the processes of river damming due to landslides are similar during normal flow conditions (see e.g. Becker *et al.* 2007) and during floods, the latter situation is particularly challenging for flood hazards and risk, especially due to the fast evolution of the phenomena. Lake filling is particularly rapid

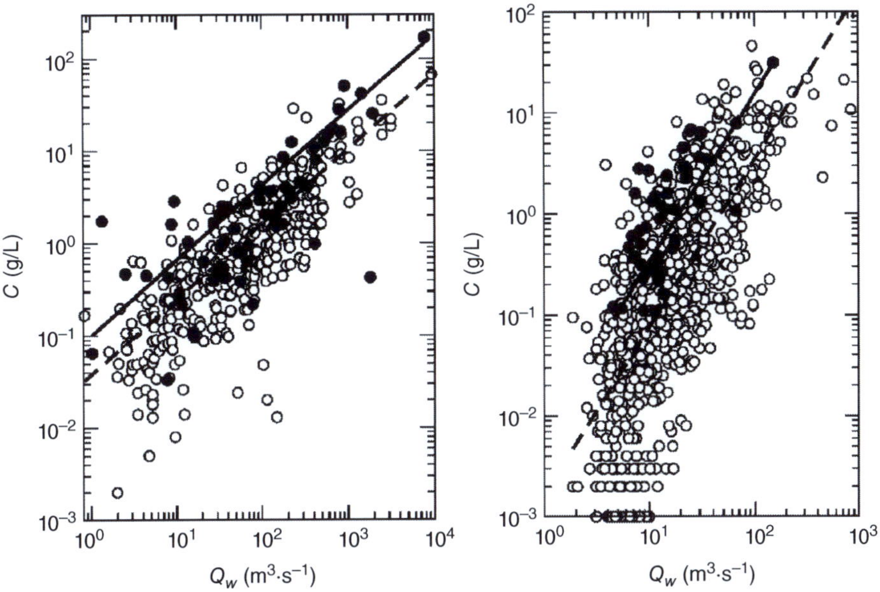

Figure 17.3 Suspended-sediment rating curves for two rivers in central Taiwan. The open circles show measurements made before the Chi-Chi earthquake (September 1999); the filled symbols show measurements after the earthquake. (*Source*: Dadson *et al.* 2004. Reproduced with permission of The Geological Society of America.)

due to large inflow from the upstream river basin, and the overtopping (and possible breaching) of the landslide dam takes place a short time after dam formation, shortening the warning time. Moreover, although river obstruction substantially reduces discharge downstream from the dam, the runoff from the downstream tributaries is not intercepted by the landslide dam: if dam breaching occurs, the flood surge from the temporary lake sums with the rainfall-generated flood in the downstream channel.

The problem of channel blockage and the failure of temporary dams is linked to the relationships between the magnitudes of landslides and debris flows and the size of the affected fluvial system. Landslides of moderate magnitude may prove sufficient to block small gravel-bed rivers that are confined to narrow valleys (Costa and Schuster 1998), whereas very large landslides are required to occupy the wide valley floors of major gravel-bed rivers. Moreover, the intensity of the flood influences the possible occurrence of channel blockage. While large deep-seated landslides may completely block channels even during large floods, accumulations of smaller landslides and debris flows, which would be able to obstruct the receiving channel in normal flow conditions or during moderate floods, are easily mobilized when they are synchronous with major floods. In such cases, it can be argued that the concurrence of landslides with large floods may avoid the most dramatic geomorphic impacts of the former, i.e. the formation and possible collapse of a landslide dam.

A recent, well-documented example of the blockage of a gravel-bed river by a deep-seated landslide is the Shiaolin landslide in Taiwan. On 9 August 2009, Typhon Morakot, one of the most severe typhoons to hit Taiwan in recent years (Chien and Kuo 2011; Shieh and Chen, Chapter 18 this volume), triggered a landslide that destroyed Shiaolin village, causing more than 400 casualties: the total landslide volume amounted to 2×10^{6} m^{3}; the landslide accumulation formed a 60 m high dam in the 80 m wide channel of the Chishan River (Tsou, Feng, and Chigira 2011; Feng 2012); breaching of the dam, which occurred due to erosion by overtopping water 1 h and 24 min after formation, resulted in a severe flood surge (Feng 2012). Despite the attenuation of the dam-break surge along the

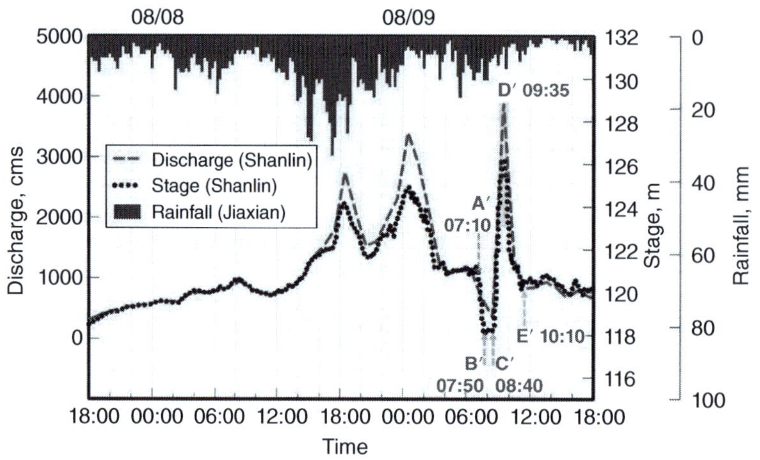

Figure 17.4 Measured river stage at a gauge station located 30.5 km downstream from the Shiaolin landslide dam; rainfall was measured at a rain gauge located between the landslide and the river-stage measuring station. (From Feng 2012.)

30.5 km channel reach between the landslide site and the closest river gauging station, the peak discharge exceeded the maximum values associated with the rainfall induced flood caused by the most intense phase of Typhoon Morakot (Figure 17.4).

River-stage measurement stations, if present (as in the Chishan River), can record flood surges that originated from dam breaches. However, their suitability for determining the discharge of dam-break flood depends on their distance from the landslide dam, and extreme dam-break floods may damage or disrupt gauging stations. Other sensor networks may prove of utmost importance to integrate river-stage measurements: in the case of the Shiaolin landslide, Feng (2012) demonstrated the suitability of broadband seismic recordings for the interpretation of the processes in landslide dam formation and breaching.

The Val Pola landslide in the Italian Alps provides an example of the interactions between debris flows, deep-seated landslides, and river channels during a flood. Between 15 and 20 July 1987, intense rainfall hit the Adda River basin in Lombardy (northern Italy), causing a major flood, debris floods and debris flows in the tributaries, and widespread shallow landslides. On 18 July, a debris fan with an estimated volume of 0.6×10^6 m^3 (Govi, Gullà, and Nicoletti 2002) that formed from a debris flow in a small tributary (Val Pola Creek) dammed the Adda River, forming a lake that attained an extension up to 0.266×10^6 m^2 (Azzoni *et al.* 1992) and a volume of approximately 0.5×10^6 m^3 on 28 July (Govi, Gullà, and Nicoletti 2002). On that day, a large rock avalanche affected the valley flank adjacent to the Val Pola Creek (Azzoni *et al.* 1992; Govi, Gullà, and Nicoletti 2002; Crosta, Chen, and Lee 2004). The volume of the original rock mass was estimated to be approximately 34×10^6 m^3, and more material was entrained along the rock avalanche path, resulting in an accumulation of approximately 41×10^6 m^3 (Govi, Gullà, and Nicoletti 2002). The rock avalanche filled the valley floor, rose on the opposite side of the valley and displaced the water in the lake that formed on 18 July due to debris fan blockage. The wave travelled upstream and destroyed three villages. Two of them had been evacuated in the previous days, but 27 people were killed in the third village upstream. The landslide dam (Figure 17.5) formed a lake that had a volume of 22×10^6 m^3 when completely full (Costa 1991).

Different to most landslide dams (Costa and Schuster 1988; Ermini and Casagli 2003), the Val Pola dam did not fail in the days immediately after its formation. However, a serious threat of dam failure

Figure 17.5 Val Pola landslide and lake. (*Source*: Photo by Ceccato, reproduced with permission of CNR IRPI Torino.)

remained, urging control measures to be taken, including the construction of a bypass tunnel, reduction and armouring of the downstream slope of the landslide dam, installation of pumping stations, and controlled overtopping of the landslide dam (Costa 1991). This last measure, which was decided after the acceleration of lake infilling caused by a flood on the night of 23–24 August, is probably the most unusual of the control measures adopted due to the inherent risk of producing a failure in the landslide dam. The remedial measures were finally successful, and a substantial reduction in the lake level and volume was achieved in September and October 1987. As noted by Costa (1991), 'The Val Pola landslide was a disaster that could have been worse'. This event can be used as an example of the concurrence of different extreme hydrologic and geomorphic processes involving slopes and an important mountain river, and as a case in which complex and expensive control measures permitted the effective management of flood risk caused by a large landslide dam.

In addition to these recent channel-damming events caused by large rockslides, two cases of channel blockage caused by debris flows in the European Alps, identified from historical documents and maps, are described below.

The Moscardo Torrent (catchment area 4.1 km^2) is one of the most active debris-flow channels in the Eastern Italian Alps. Documents from historical archives and maps show that its deposits have dammed the receiving stream (the But River) several times forming a lake (Figure 17.6). No continuous records regarding the formation and extinction of the lake exist, but historical documents support the presence of the lake since the 15th century (Marinelli 1898). An analysis of a core sample from a borehole drilled in an area previously occupied by the lake has confirmed the presence of lacustrine deposits at different depths, which are intercalated with fluvial deposits from the But River and debris-flow deposits from the Moscardo Torrent (CNR IRPI, unpublished data). The concurrent activity of the debris-flow catchment (obstruction of the main valley caused by large debris flows and lake

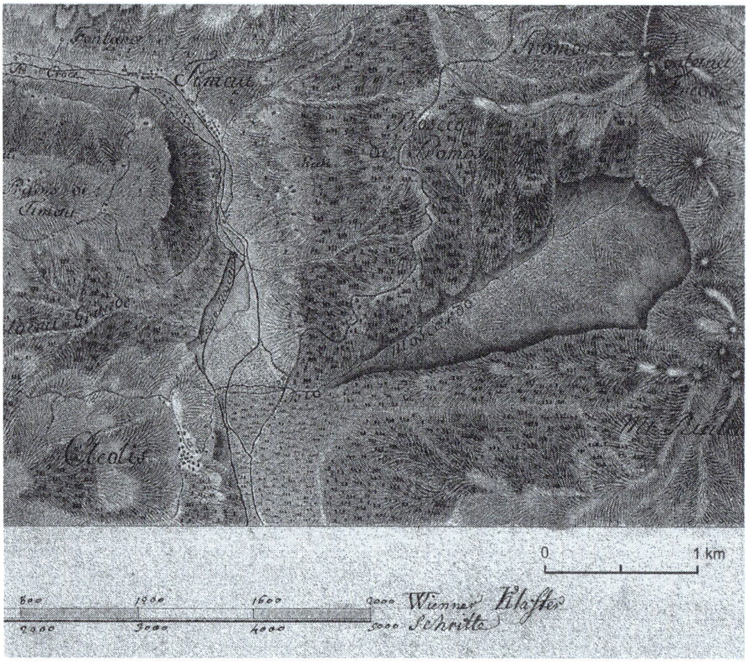

Figure 17.6 The lake formed from the debris-flow deposits of the Moscardo Torrent in the Austrian map of 1805.

formation) and gravel-bed river (erosion of debris-flow deposits during large floods and the extinction of the lake) has dominated fluvial processes in the sector of the But River valley over centuries. In recent decades, the debris-flow magnitude has been relatively moderate (Marchi, Arattano, and Deganutti 2002; Blasone *et al.* 2014), and only small, short-duration obstructions that were quickly eroded by ordinary floods in the main stream have occurred. This is mainly ascribed to the absence of extreme debris flows and, to a lesser extent, the effect of torrent control projects implemented in the Moscardo Torrent catchment. The large availability of loose debris and the presence of deep-seated landslides in the Moscardo catchment (Marcato *et al.* 2012) suggest that the future occurrence of extreme debris flows potentially able to dam the receiving stream cannot be excluded.

In the Poucet Torrent (Maurienne Valley, Western Alps, France), the presence of an important road that is endangered or damaged when debris flows clog the receiving stream (Arc River) has favoured the collection of a long time series of debris flows. Debris-flow records from historical documents (Figure 17.7) show that complete clogging of the receiving stream is quite common when sufficient debris-flow magnitudes (greater than approximately 20 000 m^3) occur. Event descriptions from historical records provide unequivocal evidence of the damming of the Arc River for eight of the 39 debris flows that were recorded from 1740 to 1996. Eight more debris flows have partially clogged the main stream. The first events are reported at time intervals spaced by several decades, whereas a number of debris flows have been reported in the most recent years. The descriptions of the oldest events depict extreme debris flows that dammed the main river and formed rather large, although ephemeral, lakes (Mougin 1914). For most debris flows that occurred in recent decades, the lack of reported damage or descriptions such as 'the debris flow reaches the Arc River' (i.e., without damming it) or 'debris-flow deposits at the outlet of the gorge' (Brochot, Marchi, and Lang 2002) suggest small or moderate magnitudes for these debris flows. The observed increase in debris-flow frequency in recent decades can be

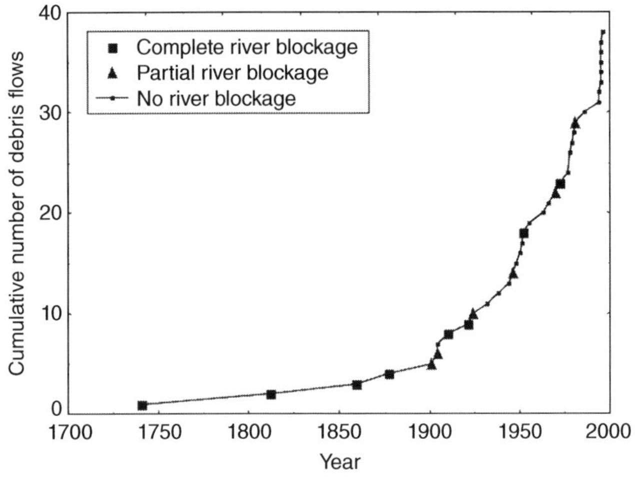

Figure 17.7 Time series of debris flows in the Poucet Torrent; the events that dammed the Arc River are shown. (Figure based on table 3 of Brochot, Marchi, and Lang (2002), with data from SOGREAH (1979) and Bonnet-Staub (1998).)

ascribed, as in other alpine catchments (e.g. Marchi and Tecca 2006), to the more careful collection and conservation of documents reporting debris-flow occurrences. This permitted the recording of small debris flows that had minor geomorphic impacts. Stationarity in the occurrence of debris flows in the Poucet Torrent since 1860 can be demonstrated (Brochot, Marchi, and Lang 2002) if only the largest debris flows, which include those that completely or partially dammed the Arc River, are considered.

Cases of channel obstruction caused by debris flows and landslides that do not produce persistent effects and do not undergo a paroxysmal evolution are rather common. A recent debris flow in the Chilean Andes (Sepulveda *et al.* 2015) outlines typical features of such events. On 21 January 2013, a debris flow in the catchment of Estero San Alfonso (central Chile) partially obstructed the receiving stream (Maipo River) (Figure 17.8). The mostly fine-grained debris-flow deposit moderately increased the river level upstream from the confluence and was eroded by the Maipo River without major impacts on downstream discharge. However, the strong increase in sediment transport resulting from the delivery of debris-flow deposits to the main stream disrupted the drinkable water supply to a sector of the city of Santiago de Chile for 24 h (Bonelli *et al.* 2013). No quantitative data on the impact of the Estero San Alfonso debris flow on the discharge and sediment transport of Maipo River downstream of the fan dam are available; other cases of channel obstruction due to lateral debris flows, followed by a non-catastrophic erosion of the debris-flow deposits, show a sudden decrease of water stage in the receiving stream, followed by a fast recovery of the previous level and a spike of turbidity (Boultbee *et al.* 2006).

17.2.4 Erosion and Delivery of Large Wood Debris

Large wood debris influences the morphological settings and biotic conditions in river channels and may enhance inundation hazards during floods (Lancaster, Hayes, and Grant 2001; Gurnell *et al.* 2002; May and Gresswell 2003a, b; Mazzorana *et al.* 2009; Wohl, Ogden, and Goode 2009; Rigon, Comiti, and Lenzi 2012; Wohl and Ogden 2013; Ruiz-Villanueva *et al.* 2014). In forested catchments, landslides and debris flows may deliver significant amounts of large wood debris to river systems.

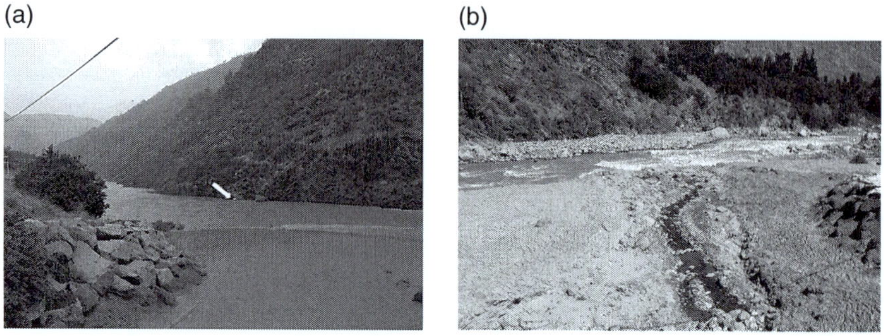

Figure 17.8 (a) Partial obstruction of Rio Maipo (central Chile) caused by the debris-flow deposit of a tributary (Estero San Alfonso) on 21 January 2013; the arrow indicates a level gauging station that was almost submerged by the flow. (b) The channel after the erosion of the debris-flow deposits. (*Source*: Reproduced with permission of Sebastian Vicente Perez and Alfonso Labra Holzapfel (Dirección General de Aguas, Chile).)

Wood-debris recruitment and transfer processes are greatly influenced by the size and characteristics of landslides. Although large volumes of wood debris can be recruited and delivered to main channels by deep-seated landslides, the large masses of rock and debris predominate over wood. At the opposite end of the magnitude range of instability phenomena, the limited width of small debris-flow channels (<5–10 m) can prove insufficient to deliver trees to the main streams. Large wood debris can also alter the dynamics of debris flows: according to Lancaster, Hayes, and Grant (2003), when wood debris represents a non-negligible part of the debris-flow mass, it can reduce runout velocity and length.

A sound evaluation of large-wood-debris entrainment and transport during floods should thus include the estimation of the volume of wood debris removed from the slopes and debris-flow channels and the effectiveness of its transfer to river channels. Assessing the budget of large wood debris during a major flood is of utmost interest to recognize the relative contributions of different sources (landslides, debris flows, and erosion of channel banks and islands) and as a preliminary step toward the recognition of the factors that control wood debris inputs, delivery through the river system, and export downstream from the basin outlet.

A study on two catchments in northwestern Italy (Pogliaschina and Gravegnola, with drainage areas of 25 and 34 km^2, respectively) is proposed as an example of such budget assessment (Lucía *et al.* 2015). These two catchments lie in the sector of the Magra River basin (1700 km^2) most severely affected (return period exceeding 200 years) by a major flash flood that occurred on 25 October 2011 (Nardi and Rinaldi 2015). Field surveys assessed the large-wood-debris volume deposited in the channel and evaluated the geomorphic effects of the flood. Landslides and channel widening were mapped through interpretation of pre- and post-event aerial photographs in a GIS, complemented by field observations. The eroded wood-debris volumes were computed by multiplying the areas involved in the landslides or widening channels by unit stand volumes reported in forest inventories. The amounts of wood debris recruited, expressed per kilometre of channel length, were high in both catchments: 4800 in Pogliaschina and 9800 in Gravegnola (Lucía *et al.* 2015). The much larger wood-debris erosion in the Gravegnola is ascribed to more intense channel widening (the median of pre-event width/post-event width was 8.00 in Gravegnola versus 2.85 in Pogliaschina) (Lucía *et al.* 2015) and to larger landslide areas coupled to the main channels (0.105 km^2 versus 0.014 km^2) (Figure 17.9). The coupling of hillslopes with a river increases when the widening channel reaches the foot of the slopes: this occurred in several semi-confined sectors of the channel network in Gravegnola.

The recruitment from floodplains accounted for 79% of total wood-debris erosion in Gravegnola and 68% in Pogliaschina, whereas landslides (mostly earth flows) in forest-covered slopes played a

Figure 17.9 Shallow landslide on a forested slope coupled to the Gravegnola Creek.

relatively minor but not a negligible role. Large wood debris delivered to the channel system by the landslides, even if less relevant than wood-debris erosion from the floodplain, would have been sufficient to clog bridges, thus increasing flood damage.

17.3 Approaches to Prediction

The impact of landslides and debris flows on flood dynamics in gravel-bed rivers has urged the development of methods to predict these processes and assess their possible interactions.

Recognition of incipient slope instability that is susceptible to paroxysmal evolution into deep-seated landslides and flood forecasting are important sectors of engineering geology and hydrology, respectively. A review of methods for the prediction of large landslides and flood forecasting would require an extensive treatment, which is outside the scope of this work. The analysis presented hereafter will focus on shallow landslides and debris flows, which typically occur at multiple locations during major floods, thus requiring a spatially distributed assessment of their occurrence, and are usually triggered by intense precipitation, similar to floods in gravel-bed rivers.

Methods for the prediction of shallow landslides and debris flows need to be considered at both temporal and spatial scales. The temporal scale considers the time occurrence of shallow landslides and debris flows during a rainstorm. Due to the different durations of the events (Figure 17.1) and different mechanisms that cause shallow landslides, debris flows, and floods in gravel-bed rivers, the predictive methods are different and only an integrated analysis enables recognizing the interactions between the different processes. The spatial scale includes the identification of slopes and catchments susceptible to shallow landslides and debris flows and the assessment of their possible linkage with gravel-bed rivers.

17.3.1 Temporal Scale

A well-established approach to the recognition of rainfall that may produce landslides and debris flows, which is also suitable in real-time operational warning systems (Alfieri *et al.* 2012), is based on the exceedance of rainfall thresholds. Rainfall thresholds for landslides and debris flows identify the minimum values of event cumulative rainfall or rainfall intensity above which instability phenomena are likely to occur. Thresholds may include antecedent precipitation (Glade, Crozier, and Smith 2000) or other hydrological variables, e.g. discharge values used to synthesize the combined effects of rain, rain on snow, snowmelt, and antecedent moisture (Jakob and Weatherly 2003). These critical thresholds can be established through process-based approaches or through empirical or statistical procedures (Wieczorek and Glade 2005; and references therein). Following early studies by Caine (1980) and Innes (1983), a number of rainfall thresholds have been developed in different geographical areas and at different (local, regional, global) scales. Guzzetti *et al.* (2007, 2008) have reviewed a number of rainfall thresholds for landslides and debris flows and proposed new ones based on statistical procedures (Figure 17.10).

The large variability of rainfall in space and time gives rise to uncertainties in the thresholds for landslides and debris flows. A recent study, carried out in an alpine region (7400 km^2) where both hourly rainfall data from a dense rain-gauge network (one every 74 km^2) and a comprehensive database of debris flows (405 debris-flow initiation points identified from 2000 to 2010) are available, has aimed to assess the impact of the spatial representativeness of rainfall data on the critical thresholds of debris flows (Nikolopoulos *et al.* 2014). The results showed that uncertainty in rainfall estimation has a major impact on the estimated intensity–duration thresholds, with rainfall at the measurement sites predicting the exceedance of the critical thresholds for rainfall rates lower that those actually required for rainfall triggering. The main concern is thus the frequent occurrence of false alarms caused by the underestimation of the critical rainfall thresholds required to trigger debris flows.

17.3.2 Spatial Scale

17.3.2.1 Shallow Landslide Modelling

Distributed models have been developed to analyse the control of topography and the hydraulic and mechanical characteristics of soils on shallow landslides (Montgomery and Dietrich 1994; Pack, Tarboton, and Godwin 1998). These models couple a saturated subsurface flow model with the infinite slope stability approach, representing topography through digital terrain models. The model SHALSTAB (Montgomery and Dietrich 1994) includes a module that simulates landslide routing, which assesses the possibility that material mobilized by landslides reaches the channel network. Steady-state stability models (Montgomery and Dietrich 1994; Pack, Tarboton, and Goodwin 1998) assess slope stability/instability for constant rainfall rates, thus basically representing the control of topography on shallow landslides. Quasi-dynamic models (Iida 1999, 2004; Borga, Dalla Fontana, and Cazorzi 2002, Lanni *et al.* 2012) take into account the intensity and duration of rainfall, thus representing a closer link between the topographic, hydrologic, and mechanical characteristics of landslide-prone hillslopes and landslide-triggering rainfall. The possibility of application in broad areas is a valuable feature of these models, which have provided satisfactory performances in ranking hillslopes according to their susceptibility to shallow landsliding. The difficulty in predicting the location of

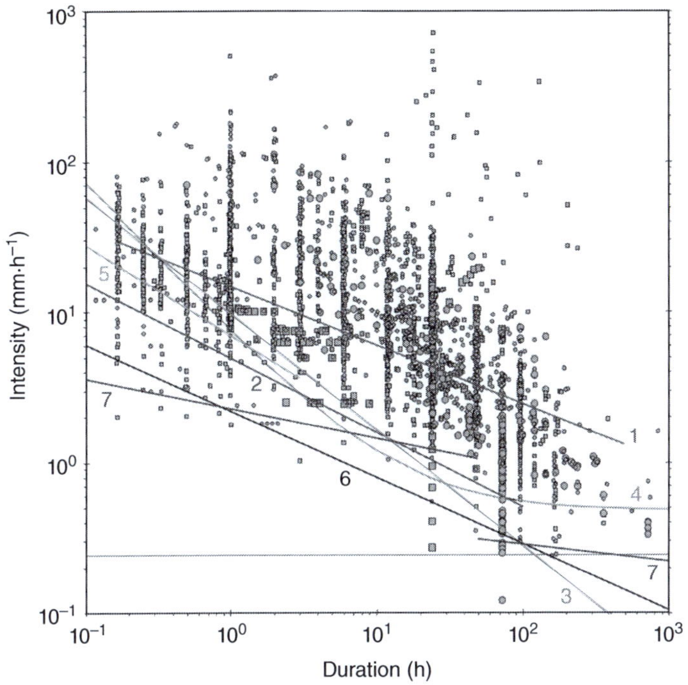

Figure 17.10 Plot of mean rainfall intensity versus rainstorm duration for a global (worldwide) dataset, and critical rainfall thresholds for landslide triggering: square, debris flow; dot, shallow failure; small symbol, single triggered landslide; large symbol, multiple triggered landslides; 1–6, rainfall thresholds proposed by various authors; 7, thresholds inferred from the probability estimates of the rainfall conditions for two different rainfall periods ($D < 48$ h, and $D \geq 48$ h). The grey horizontal line shows a rainfall intensity of 0.25 mm h^{-1}. (*Source*: Guzzetti *et al.* 2008. Reproduced with permission of Springer.) (See insert for color representation of this figure.)

individual slope failures, which can be referred to the simplified representation of the processes, as well as to the difficulty of accounting for the spatial variability of soils, still limits their systematic use for landslide hazard assessment. Modelling the contribution of landslides to sediment yield (Burton and Bathurst 1998; Bathurst *et al.* 2006) is a further step in the analysis of the interactions between slope instability and the river system.

17.3.2.2 Morphometric Assessment of Debris-Flow Susceptibility

The conditions for the occurrence of debris flows are, in addition to sudden water inflows, the availability of loose debris prone to mobilization and steep topography. These conditions are frequently met in many mountainous regions, but geological, topographic, and geomorphological conditions that control susceptibility to debris flows show a great variability in space. Thus, methods are required to rapidly identify debris-flow prone slopes and catchments at river basin or regional scales, and their possible interactions with flooding on the main streams. Such analyses can be subdivided into two steps:

1) identification of catchments potentially prone to debris flows (requiring sediment and morphological conditions favourable for its erosion and transfer downstream);
2) assessment of runout distances and inundation areas.

The presence of sediment available for erosion and transport can be assessed through field observations and the interpretation of aerial photographs (Brardinoni, Slaymaker, and Hassan 2003). In partly glaciated or recently deglaciated basins, the geomorphological recognition permits also identifying moraine-dammed or ice-dammed lakes in basin headwaters, the failure of which could produce outburst floods capable of evolving into debris flows or hyperconcentrated flows in the downstream channels (Cenderelli and Wohl 1998; O'Connor, Hardison, and Costa 2001). The multi-temporal analysis of aerial photographs enables the detection of variations in the presence and activity of sediment source areas that can be associated with debris-flow occurrences. This is of particular relevance in catchments where debris-flow occurrences are controlled by the presence of sufficiently mobilizable material (weathering-limited catchments *sensu* Bovis and Jakob 1999), which may require a long period for debris recharge after sediment evacuation caused by a previous debris flow (Jakob, Bovis, and Oden 2005).

The recognition of the susceptibility to debris flows implies a basic differentiation among debris-flow catchments and bedload flood catchments. This discrimination is relevant when dealing with the linkage between sediment inputs from tributaries and floods in gravel-bed rivers. Although both bedload floods and debris flows can deliver large amounts of sediment from tributaries to the channels that drain the main valley, debris flows, due to their higher concentrations of solids, may prove more effective than water floods with bedload in altering valley morphology at the confluence, thus inducing aggradation and instability in the receiving stream or even damming it.

Morphometric descriptors of the drainage basin and the alluvial fan have proven effective in recognizing catchments where debris flows can reach the lower parts of a channel, dominating the depositional processes at the fan. Jackson, Kostaschuck, and MacDonald (1987), in a study in the Canadian Rocky Mountains, showed that debris-flow catchments and 'fluvial' (floods with bedload) catchments lie in separate sectors of a plot of basin ruggedness versus fan slope. The catchment ruggedness was expressed by the Melton number (Melton 1965), which is computed as the ratio of catchment relief to the square root of the catchment area. This approach was successfully applied in other geographical regions, mostly located under humid climates at middle latitudes (Marchi, Pasuto, and Tecca 1993; Marchi and Brochot 2000; De Scally and Owens 2004). Amongst the most recent studies, which based the differentiation between debris-flow catchments and bedload catchments on statistical procedures instead of empirically defined thresholds, the work by Bertrand, Liébault, and Piégay (2013) deserves to be cited. In that study, the application of linear discriminant analysis (Figure 17.11) to a large dataset of catchments and fans resulted in the definition of robust morphometric thresholds for defining the flow response at the catchment outlet.

Although effective at rapidly recognizing debris-flow catchments, these methods based on simple morphometric indices of catchments and alluvial fans show several limitations. As acknowledged by the authors that developed and proposed these methods (e.g. Jackson, Kostaschuck, and MacDonald 1987; Marchi, Pasuto, and Tecca 1993), a simple descriptor of basin ruggedness can prove inadequate in representing sediment transport processes at the catchment outlet: despite the high basin ruggedness indicated by the Melton number, a debris flow may fail to reach the alluvial fan if low-slope regions along the main channel induce sediment deposition and the transformation of debris flows into less concentrated types of sediment transport. Moreover, different processes, encompassing bedload transport, hyperconcentrated flow, and debris flow, may occur in the same catchment and even during the same event. Indeed, the recognition of dominating long-term processes, witnessed by the geomorphic parameters of drainage basins and alluvial fans, do not provide information regarding the occurrence of a specific type of sediment transport during a given event.

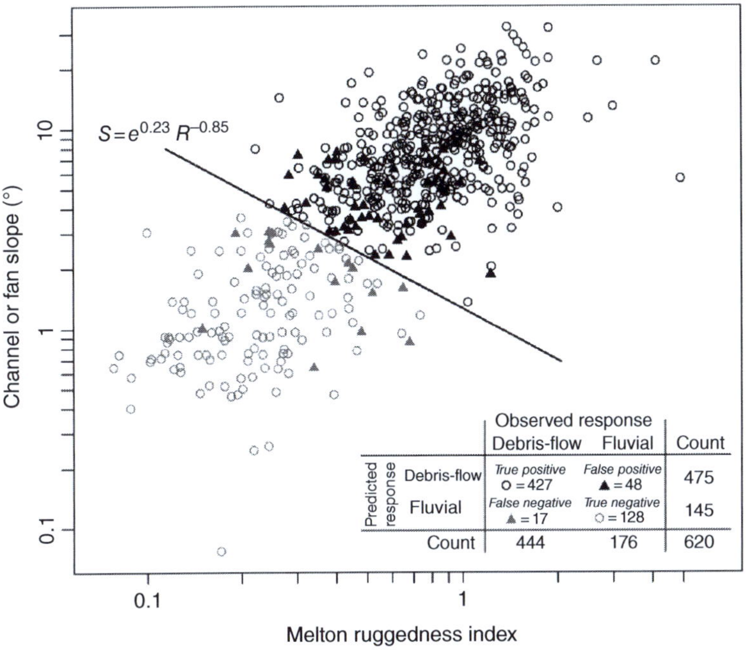

Figure 17.11 Plot of fan or fluvial-channel slope versus the Melton ruggedness number, with a differentiation between fluvial and debris-flow catchments according to a linear discriminant analysis model. (*Source*: Bertrand, Liébault, and Piégay 2013. Reproduced with permission of Springer.)

17.3.2.3 Models for Debris-Flow Runout and Inundation Areas

The recognition of channels susceptible to debris flows must be followed by the assessment of the possible interactions of debris flows with the river in the main valley. This involves the problem of evaluating the total distance travelled by debris flows (entire flow path) or their runout length in the depositional part of the flow path.

There are several methods that predict these variables (Rickenmann 2005; and references therein). We recall here some of the simplest methods to compute the distance travelled by debris flows, which can be applied to a fast preliminary assessment of the possible interactions between debris flows in steep tributaries and the rivers in main valleys, and some more sophisticated two-dimensional numerical models suitable for a more detailed analysis of debris-flow propagation and deposition.

Burton and Bathurst (1998) incorporated a module for debris-flow runout distance, which includes the assessment of the fraction of material that is delivered to the channel system, into the physically based model SHETRAN (Ewen, Parkin, and O'Connell 2000). The rules for debris-flow propagation and stopping defined by Burton and Bathurst (1998) have been successfully integrated into procedures for debris-flow susceptibility assessment at regional scales (Cavalli and Grisotto 2006).

Based on regression analysis using mainly data from debris flows in Switzerland, Rickenmann (1999) developed an equation for the total travel distance L, i.e. the distance travelled by a debris flow from the initiation point to the lowest point of deposition:

$$L = 1.9M^{0.16}{H_e}^{0.83} \tag{17.1}$$

where M is the debris-flow volume and H_e is the elevation difference between the initiation point and the lowest deposition point of a debris flow. The same author also proposed an equation for runout distance on fan L_f:

$$L_f = 15M^{\frac{1}{3}} \tag{17.2}$$

In the framework of empirical-statistical methods, the model proposed by Iverson, Schilling, and Vallance (1998) for lahars represents a step toward a more comprehensive assessment of debris-flow deposition, as it computes both the runout length and inundation areas. This model is based on scaling relations between the debris-flow volume V, inundation area B, and cross-sectional flow area A:

$$A = 0.05\,V^{\frac{2}{3}} \tag{17.3}$$

$$B = 200\,V^{\frac{2}{3}} \tag{17.4}$$

The model takes into account the topography of the deposition area, possibly represented by a digital terrain model. Other researchers (Crosta, Cucchiaro, and Frattini 2003; Berti and Simoni 2007, 2014) followed this approach and extended it to other types of debris flows. Propagation extent of debris flows is also computed by models devised for application at the regional scale with minimum data requirements, such as the model Flow-R (Horton *et al.* 2013), which essentially needs a digital terrain model as input.

At a higher degree of complexity, two-dimensional physically based models enable an enhanced representation of the propagation and deposition of debris flows. These models assess flow velocity and depth at various stages of the flow process and the extent and thickness of the deposits at the end of the event. Notable models include: the well-known quasi-two-dimensional commercial code FLO-2D (O'Brien, Julien, and Fullerton 1993); the software package RAMMS (Christen *et al.* 2012), which also includes tools for the simulation of other types of flows such as wet snow avalanches; and the fully two-dimensional, mobile-bed, two-phase TRENT2D (Armanini, Fraccarollo, and Rosatti 2009; Rosatti and Begnudelli 2013). The use of these models is recommended for specific cases that require a thorough analysis of the interactions between debris flows in steep lateral channels and water floods in the gravel-bed rivers of main valleys, whereas their systematic application to a large number of debris-flow catchments could be hampered by large data requirements.

17.3.2.4 A Metric for Sediment Connectivity

The concept of sediment connectivity, i.e. the linkage between sediment fluxes in different parts of the landscape, is implicit in the application of the methods for debris-flow runout and inundation areas in order to determine possible interactions with the receiving stream. Korup (2013) differentiates three coupling interfaces between hillslopes and channels, i.e. axial, when at least half the landslide runout occurs along the channel; point source, which occurs when the interaction between the landslide/debris-flow accumulation enters the channel following a direction normal or oblique to the receiving channel; and buffered, when the land surface or water bodies intercept the landslide runout and prevent or delay interactions with the channel network. The axial coupling interface is likely to occur in mountainous, steeper reaches of gravel-bed rivers, whereas point-source coupling and buffered interactions prevail in downstream channels, where the bedslope is usually lower and the channel is usually less confined, with both factors reducing the length of landslide runout along the receiving channel.

Geomorphometric indices that identify the coupling/decoupling of hillslopes with the channel network help to evaluate the supply of sediment from landslides to rivers. In the case of debris-flow channels, indices depicting connectivity patterns emphasize the relative contributions of different sectors of the catchment to the sediment supply and help evaluate the role of alluvial fans in coupling/decoupling the upstream catchment with the receiving river.

The application of a distributed, topography-based index of sediment connectivity to an Alpine basin drained by a gravel-bed river is proposed below. The index of connectivity *IC*, which was originally developed by Borselli, Cassi, and Torri (2008), is a topography-based index that consists of an upslope and a downslope component following this formula:

$$IC = log_{10}\left(\frac{D_{up}}{D_{dn}}\right) \tag{17.5}$$

in which D_{up} expresses the potential for the downward routing of the sediment produced upslope, and D_{dn} takes into account the flow path length that a particle has to travel to arrive to the nearest target of interest (e.g. a river channel):

$$D_{up} = \bar{W} \cdot \bar{S} \cdot \sqrt{A} \tag{17.6}$$

where A is the upslope contributing area, $\bar{W}$ is the average weighting factor of the upslope area, and $\bar{S}$ is the average slope gradient of the upslope area.

$$D_{dn} = \Sigma_i \frac{d_i}{W_i \cdot S_i} \tag{17.7}$$

where d_i is the length of the flow path along the ith cell according to the steepest downslope direction, and W_i and S_i are the weighting factor and slope gradient of the ith cell, respectively.

Cavalli *et al.* (2013) proposed a modified version of the sediment connectivity index *IC*. The variations with respect to the original index of connectivity included the use of the *D*-infinity multiple flow algorithm (Tarboton 1997) for computation of upslope area, and a new approach to the computation of the weighting factor *W*. In the original model, the weighting factor, which is related to the impedance to the flow, was derived from the *C*-factor of the USLE and RUSLE models (Wischmeier and Smith 1978; Renard *et al.* 1997), thus including land-use and soil-surface properties. Two alternatives have been proposed for the impedance parameter *W*: the Manning roughness coefficient (Goldin 2015) and a local measure of topographic surface roughness (Cavalli and Marchi 2008). This second option requires a high-resolution digital terrain model.

A sector of the Venosta Valley (Eastern Alps, Italy), corresponding to the upper Adige river basin at a drainage area of 1209 km^2, was chosen to evaluate the capability of the *IC* index to recognize the variability in the sediment connectivity of hillslopes and debris-flow channels with a gravel-bed river in an Alpine landscape with complex morphology (Goldin 2015). The test area has rugged relief, with elevation ranging from 550 to 3905 m; debris flows dominate sediment transfer from most of the tributaries, with fluvial processes prevailing only in a few channels draining larger areas. The Adige River and two reservoirs in the upper part of the main valley were chosen as targets for sediment delivered from hillslopes and tributaries, and Manning *n* was used to compute the impedance parameter in equations 17.6 and 17.7.

Figure 17.12 shows that effective coupling of hillslopes and debris-flow catchments with the main river and lakes occurs only where the main channel is confined between the valley flanks or where the reservoirs occupy the whole valley floor. In other sectors of the Venosta Valley, the valley floor acts as a buffer for sediment transfer from hillslopes and debris-flow channels, resulting in their decoupling from the main stream. The Venosta Valley is characterized by the presence of a cluster of very large debris-flow fans, which formed in the post-glacial period. Some fans occupy the whole floodplain, forcing the main river to flow close to the opposite valley flank, causing the opposite valley flank

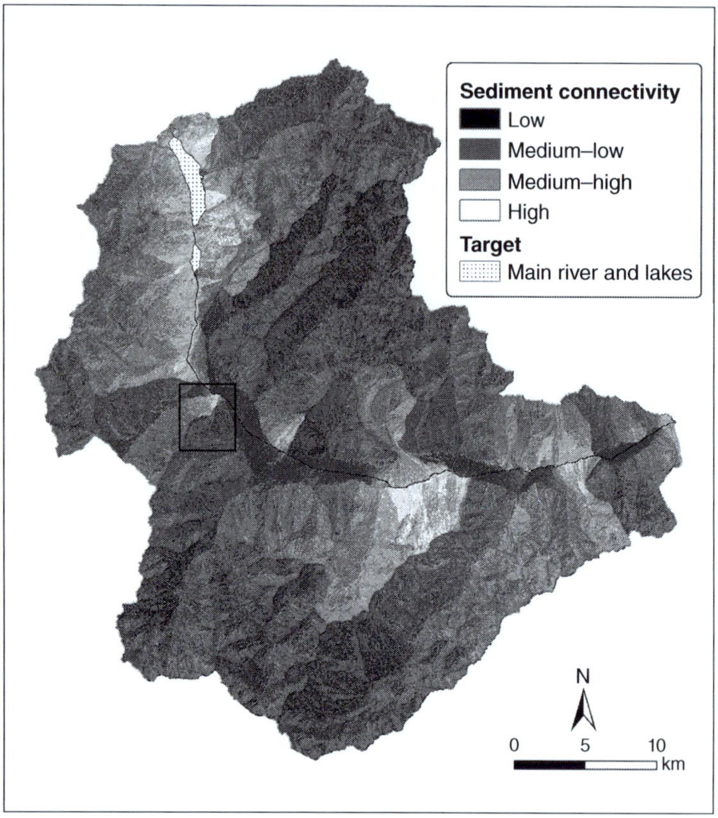

Figure 17.12 Map of the sediment connectivity index *IC* (equation 17.5) for the upper Venosta Valley showing the linkage of different sectors of the river basin with the main river. Four classes have been identified according to the Jenks natural breaks method; the rectangle identifies the enlarged portion shown in Figure 17.13. (See insert for color representation of this figure.)

to be closely connected with the Adige River (Figure 17.12). The surfaces of the debris-flow fans show great variability in sediment connectivity, with the highest *IC* values occurring along the preferred directions of debris-flow propagation and in areas linked to the main stream (Figure 17.13). While the topography-based *IC* index depicts the variability in the topographic control on the sediment connectivity of the debris-flow fans, a closer analysis of sediment supply processes at different time scales would be necessary for assessing if the alluvial fans act as buffers or link the sediment source areas to the receiving river (Harvey 2001).

This index and other topography-based metrics for sediment connectivity should thus complement field observations and the interpretation of aerial photographs, not replace or act as a surrogate for them. The objective assessment of a topography-based metric of sediment connectivity has to be combined with expert judgment that values knowledge of the sediment entrainment and delivery. In the framework of a comprehensive assessment of sediment connectivity, topography-based metrics, such as the *IC* sediment connectivity index, show some valuable features, i.e.: (i) they enable the fast recognition of sediment transfer patterns at river basin or regional scales; (ii) topographic parameters are easily derived from digital terrain models; (iii) they define scenarios that are suitable for outlining sediment management alternatives.

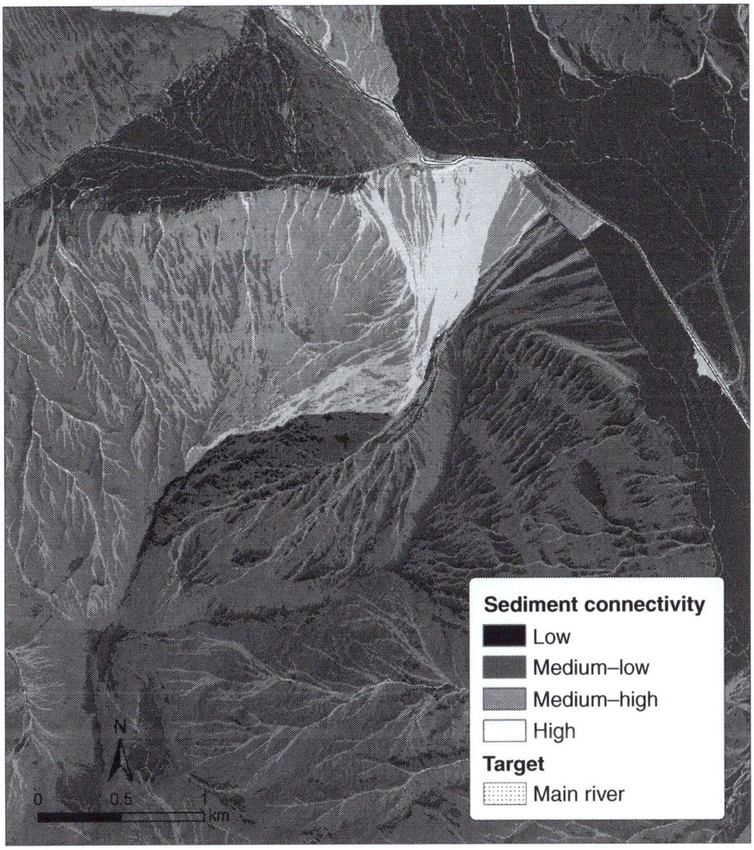

Figure 17.13 Variability in the sediment connectivity index *IC* of a debris-flow fan and its upstream catchment (see Figure 17.12 for location).

17.4 Discussion

The time scale at which landslides and debris flows interact with floods in gravel-bed rivers shows relatively small differences (from minutes to a few hours) between different classes of instability phenomena, which are triggered by the same factors (usually intense rainfall) that cause floods in gravel-bed rivers (Figure 17.1). In contrast, the long-term impact on the fluvial systems shows great differences depending on the type and size of landslides. The studies on widespread landslides triggered by typhoons on slopes weakened by an earthquake in Taiwan show a strong increase in sediment transport due to debris supply from the landslides in the years immediately after the typhoon (Dadson *et al.* 2004), followed by an attenuation in the subsequent period (Chen *et al.* 2011). The impact of debris flows of moderate magnitude on gravel-bed rivers can be limited to temporary damming, with the sediment deposited easily removed by flows of the receiving stream, as was documented for the Poucet Creek and the Estero San Alfonso (Figures 17.7 and 17.8). A stronger and less ephemeral influence on the morphology of the valley floor of the main stream is observed when large and frequent debris flows deliver huge amounts of sediment that alter the valley planform floor and the longitudinal profile of the main stream, and form long-lasting lakes upstream of the debris fan dam, as was documented for the Moscardo Torrent (Figure 17.6). Large landslides, in addition to the immediate dramatic consequences in case of landslide dam failure, may influence the morphology and the evolution of gravel-bed rivers for hundreds or thousands of years, altering their morphology and even transforming a bedrock channel into an alluvial channel. An outstanding example can be found in Pratt-Sitaula *et al.* (2004), who ascribe to a single, very large landslide the most recent (approximately 4000 years ago) of the three main alluviation events that have filled a river reach in Nepal Himalaya in Quaternary times. Persistent disturbances in gravel-bed rivers can be produced by landslide dams and lakes, especially if several landslides occur, even at different times, in the same river system (Hewitt 2006; Hewitt, Clague, and Orwin 2008). If the landslide dam does not fail in the days or weeks after formation, sedimentation occurs in the lake upstream of the barrier, which becomes a local base-level for the river, whereas downstream of the landslide dam sediment-depleted water may incise valley deposits or the bedrock. Numerical modelling of fluvial response, which includes dam scenarios and non-dam scenarios (van Gorp *et al.* 2014), has indicated for dam scenarios lower long-term sediment yields due to the preservation of part of the sediment trapped by the barrage. Other remarkable effects of landslide damming, which can impact the fluvial system on time scales of 10^3 years, are drainage rearrangement and epigenetic gorge formation (van Gorp *et al.* 2014).

The prediction of the interactions between landslides and floods in gravel-bed rivers has been considered in this chapter from the point of view of the spatial location and temporal occurrence of the instability processes. The spatial recognition of landslides and debris flows that could deliver sediment and large wood debris and potentially dam the rivers is relevant for hazard areas mapping and preparation of emergency plans, whereas the temporal assessment of landslides and debris flows plays a central role in emergency management during the floods.

Potential and limitations of several methods for the assessment of sediment and wood-debris supply associated with debris flows and shallow landslides, i.e. morphometric recognition of debris-flow catchments, modelling of the topographic control on shallow landslides, evaluation of mobilizable large wood-debris volumes, and assessment of debris-flow runout, have been discussed in the previous section. The prediction of large deep-seated landslides that could dam gravel-bed rivers in mountain valleys is especially difficult due to their rare occurrence, which hamper the collection of data suitable for the assessment of predisposing factors. In most cases, moreover, large slope instabilities are difficult to detect because of their slow evolution before the catastrophic failure. Useful elements for a preliminary screening of valley flanks potentially susceptible to large landslides may derive from regional inventories of rock-slope failures (Jarman *et al.* 2014) and deep-seated gravitational slope deformation (Crosta, Frattini, and Agliardi 2013), which can be associated with the occurrence of large fast-moving instabilities. Once a large deep-seated landslide has been detected, its evolution can be monitored (e.g. Dixon *et al.* 2012) and modelled for assessing possible failure and interactions with the river system: see Hungr (1995) for a runout analysis model and Bossi *et al.* (2013) for an example of application to a large landslide in the Alps.

The management of hazards and risks associated with the concurrence of landslides, debris flows, and floods in gravel-bed rivers requires coupling real-time flood forecasting models with landslide prediction based on critical rainfall thresholds. The threats due to formation of landslide lakes are especially challenging because of the magnitude and complexity of the processes involved, and require skills in soil and rock mechanics as well as in hydraulics; the experience arising from positive and negative outcomes of past events, such as the Val Pola landslide lake, may provide useful elements for the management of such emergencies.

17.5 Conclusions

The review presented on the interaction of landslides and debris flows with river floods has highlighted some of the most relevant, and potentially most hazardous, features of the relations between these processes. Landslide and debris flows may increase the flood hazard in the mountain sectors of gravel-bed rivers through different mechanisms, ranging from channel instability induced by higher sediment supply to paroxysmal phenomena related to the formation and failure of landslide dams. Several approaches requiring skills in geomorphology, engineering geology, hydrology, and hydraulics exist for the analysis of these processes, which are usually assessed separately. Indeed, their proper integration is still a challenging task.

Korup (2013) summarized some topics that appear to be relevant for future research on the influence of landslides on fluvial systems, including: (i) landslide erosion in zero-order basins, and debris flows or colluvial channels; (ii) key process thresholds on the formation, failure, and consequences of landslide dams; (iii) time-dependent linkage between landslide erosion and delivery to the channel network; (iv) residence times of landslide deposits; and (v) improvement of numerical forecasting of landslide-induced sediment pulses.

With particular regard to landslides and debris flows that occur during major floods, and with attention to hazards management, the following issues will deserve attention in future studies: (i) improved assessment of the spatial and temporal scales of precipitation that can trigger landslides, debris flows, and floods in gravel-bed rivers under different climatic and geomorphological conditions, including assessment of uncertainties in landslide forecasts based on rainfall thresholds; (ii) recognition of catchments susceptible to debris flows for specific rainfall events, by combining morphological channel characteristics with runoff-related variables; (iii) refinement in the numerical modelling of the interactions between sediment pulses from debris flows and landslides and floods in the receiving streams; (iv) modelling large wood-debris recruitment and delivery to gravel-bed rivers by landslides and debris flows; (v) development of semi-dynamic metrics of sediment connectivity to include hydrometeorological variables in the representation of coupling/decoupling between hillslopes, debris-flow channels, and gravel-bed rivers.

Acknowledgements

The author wishes to thank the colleagues Alessandro Pasuto and Marco Cavalli for their valuable suggestions. Francesco Comiti and an anonymous reviewer provided helpful reviews on the first version of this paper.

References

Alfieri, L, Salamon, P, Pappenberger, F *et al.* 2012. Operational early warning systems for water-related hazards in Europe. *Environmental Science and Policy* 21, 35–49.

Armanini, A, Fraccarollo, L, and Rosatti, G 2009. Two-dimensional simulation of debris flows in erodible channels. *Computers and Geosciences* 35, 993–1006.

Azzoni, A, Chiesa, S, Frassoni, A, and Govi, M 1992. The Valpola landslide. *Engineering Geology* 33, 59–70.

Bathurst, JC, Burton, A, Clarke, BG, and Gallart, F 2006. Application of the SHETRAN basin-scale, landslide sediment yield model to the Llobregat basin, Spanish Pyrenees. *Hydrological Processes* 20, 3119–3138.

Becker, JS, Johnston, DM, Paton, D *et al.* 2007. Response to landslide dam failure emergencies: issues resulting from the October 1999 Mount Adams landslide and dam-break flood in the Poerua River, Westland, New Zealand. *Natural Hazards Review* 8, 35–42.

Benda, L 1990. The influence of debris flows on channels and valley floors in the Oregon Coast Range, U.S.A. *Earth Surface Processes and Landforms* 15, 457–466.

Benda, L and Dunne, T 1997. Stochastic forcing of sediment supply to channel networks from landsliding and debris flow. *Water Resources Research* 33, 2849–2863.

Berti, M and Simoni, A 2007. Prediction of debris flow inundation areas using empirical mobility relationships. *Geomorphology* 90, 144–161.

Berti, M and Simoni, A 2014. DFLOWZ: A free program to evaluate the area potentially inundated by a debris flow. *Computers and Geosciences* 67, 14–23.

Bertrand, M, Liébault, F, and Piégay, H 2013. Debris-flow susceptibility of upland catchments. *Natural Hazards* 67, 497–511.

Blasone, G, Cavalli, M, Marchi, L, and Cazorzi, F 2014. Monitoring sediment source areas in a debris-flow catchment using terrestrial laser scanning. *Catena* 123, 23–36.

Bonelli, S, Poblete, D, Bustos, E, and Maureira, F 2013. *Reporte visita Estero San Alfonso - Cajón del Maipo.* Unpublished report, Pontificia Universidad Católica de Chile, Centro de Cambio Global, Santiago de Chile. (In Spanish.)

Bonnet-Staub, I 1998. *Mécanismes d'initiation des laves torrentielles dans les Alpes Françaises - contribution à la maîtrise du risque.* PhD thesis, École des Mines, Paris. (In French.)

Borga, M, Dalla Fontana, G, and Cazorzi, F 2002. Analysis of topographic and climatic control on rainfall-triggered shallow landsliding using a quasi-dynamic wetness index. *Journal of Hydrology* 268, 56–71.

Borga, M, Stoffel, M, Marchi, L *et al.* 2014. Hydrogeomorphic response to extreme rainfall in headwater systems: flash floods and debris flows. *Journal of hydrology* 518, 194–205.

Borselli, L, Cassi, P, and Torri, D 2008. Prolegomena to sediment and flow connectivity in the landscape: a GIS and field numerical assessment. *Catena* 75, 268–277.

Bossi, G, Frigerio, S, Mantovani, M *et al.* 2013. Hazard assessment of a potential rock avalanche in South Tyrol, Italy: 3D modeling and risk scenarios. *Italian Journal of Engineering Geology and Environment* 6, 221–227.

Boultbee, N, Stead, D, Schwab, J, and Geertsema, M 2006. The Zymoetz River rock avalanche, June 2002, British Columbia, Canada. *Engineering Geology* 83, 76–93.

Bovis, MJ and Jakob, M 1999. The role of debris supply conditions in predicting debris flow activity. *Earth Surface Processes and Landforms* 24, 1039–1054.

Brardinoni, F, Slaymaker, O, and Hassan, MA 2003. Landslide inventory in a rugged forested watershed: a comparison between air-photo and field survey data. *Geomorphology* 54, 179–196.

Brochot, S, Marchi, L, and Lang, M 2002. L'estimation des volumes des laves torrentielles: méthodes disponibles et application au torrent du Poucet (Savoie). *Bulletin of Engineering Geology and the Environment* 61, 389–402. (In French.)

Burton, A and Bathurst, JC 1998. Physically based modelling of shallow landslide sediment yield at a catchment scale. *Environmental Geology* 35, 89–99.

Caine, N 1980. The rainfall intensity–duration control of shallow landslides and debris flows. *Geografiska Annaler* 62, 23–27.

Cavalli, M and Grisotto, S 2006. Individuazione con metodi GIS delle aste torrentizie soggette a collate detritiche: applicazione al bacino dell'alto Avisio (Trento). *Quaderni di Idronomia Montana* 26, 83–94. (In Italian.)

Cavalli, M and Marchi, L 2008. Characterisation of the surface morphology of an alpine alluvial fan using airborne LiDAR. *Natural Hazards and Earth System Sciences* 8, 323–333.

Cavalli, M, Trevisani, S, Comiti, F, and Marchi, L 2013. Geomorphometric assessment of spatial sediment connectivity in small Alpine catchments. *Geomorphology* 188, 31–41.

Cenderelli, DA and Wohl, EE 1998. Sedimentology and clast orientation of deposits produced by glacial-lake outburst floods in the Mount Everest region, Nepal. In *Geomorphological Hazards in High Mountain Areas* (eds J Kalvoda and CL Rosenfeld). Kluwer Academic Publishers: Dordrecht; 1–26.

Cerato, M 1999, *Contributi per una storia delle sistemazioni idrauliche e forestali in Valsugana*. Provincia Autonoma di Trento, Azienda Speciale di Sistemazione Montana, Trento, Italy. (In Italian.)

Chen, C-Y 2008. Sedimentary impacts from landslides in the Tachia River Basin, Taiwan. *Geomorphology* 105, 355–365.

Chen, H, Len, G-W, Lu, M-H *et al.* 2011. Effects of topography, lithology, rainfall and earthquake on landslide and sediment discharge in mountain catchment of southeastern Taiwan. *Geomorphology* 133, 132–142.

Chiarle, M, Iannotti, S, Mortara, G, and Deline, Ph 2007. Recent debris flow occurrences associated with glaciers in the Alps. *Global and Planetary Change* 56, 123–136.

Chien, F-C, and Kuo, C 2011. On the extreme rainfall of Typhoon Morakot (2009). *Journal of Geophysical Research: Atmospheres* 116, D05104.

Christen, M, Bühler, Y, Bartelt, P *et al.* 2012. Integral hazard management using a unified software environment: numerical simulation tool 'RAMMS' for gravitational natural hazards. In *Proceedings of the 12th Congress INTERPRAEVENT* (eds G Koboltschnig, J Hübl, and J Braun). International Research Society INTERPRAEVENT: Klagenfurt; Vol. 1, 77–86.

Clague, JJ and Evans, SG 2000. A review of catastrophic drainage of moraine-dammed lakes in British Columbia. *Quaternary Science Reviews* 19, 1763–1783.

Costa, JE 1984. Physical geomorphology of debris flows. In *Developments and Applications of Geomorphology* (eds JE Costa and PJ Fleisher). Springer-Verlag: Berlin, Heidelberg; 268–317.

Costa, JE 1991. Nature, mechanics, and mitigation of the Val Pola landslide, Valtellina, Italy, 1987–1988. *Zeitschrift für Geomorphologie* 35, 15–38.

Costa, JE and Schuster, RL 1988. The formation and failure of natural dams. *Geological Society of America Bulletin* 100, 1054–1068.

Crosta, GB, Chen, H, and Lee, CF 2004. Replay of the 1987 Val Pola landslide, Italian Alps. *Geomorphology* 60, 127–146.

Crosta, GB, Cucchiaro, S, and Frattini, P 2003. Validation of semiempirical relationships for the definition of debris-flow behavior in granular materials. In *Debris Flows Hazards Mitigation: Mechanics, Prediction and Assessment* (ed D Rickenmann and C. Chen). Millpress: Rotterdam; 821–831.

Crosta, GB, Frattini, P, and Agliardi, F 2013. Deep seated gravitational slope deformations in the European Alps. *Tectonophysics* 605, 13–33.

Dadson, SJ, Hovius, N, Chen, H *et al.* 2004. Earthquake triggered increase in sediment delivery from an active mountain belt. *Geology* 32, 733–736.

De Scally, F and Owens, IF 2004. Morphometric control and geomorphic response on fans in the southern Alps, New Zealand. *Earth Surface Processes and Landforms* 29, 311–322.

Dixon N, Spriggs MP, G. Marcato G, and Pasuto, A 2012. Landslide hazard evaluation by means of several monitoring techniques, including an acoustic emission sensor. In *Landslides and Engineered Slopes: Protecting Society through Improved Understanding* (eds E Eberhardt, C Froese, K Turner, and S Leroueil). Taylor and Francis: London; 1405–1411.

Ermini, L and Casagli, N 2003. Prediction of the behaviour of landslide dams using a geomorphological dimensionlex index. *Earth Surface Processes and Landforms* 28, 31–47.

Ewen, J, Parkin, G, and O'Connell, PE 2000. SHETRAN: distributed river basin flow and transport modeling system. *Journal of Hydrologic Engineering ASCE* 5, 250–258.

Feng, Z-Y 2012. The seismic signatures of the surge wave from the 2009 Xiaolin landslide-dam breach in Taiwan. *Hydrological Processes* 26, 1342–1351.

Glade, T, Crozier, M, and Smith, P 2000. Applying probability determination to refine landslide-triggering rainfall thresholds using an empirical 'Antecedent Daily Rainfall Model'. *Pure and Applied Geophysics* 157, 1059–1079.

Goldin, B 2015. *Geomorphometric analysis and sediment dynamics in mountainous basins: spatial and temporal scales*. PhD thesis, University of Padova, Italy.

Govi, M, Gullà, G, and Nicoletti, PG 2002. Val Pola rock avalanche of July 28, 1987, in Valtellina (Central Italian Alps). *Reviews in Engineering Geology* 15, 71–89.

Gurnell, AM, Piégay, H, Swanson, FJ, and Gregory, SV 2002. Large wood and fluvial processes. *Freshwater Biology* 47, 601–619.

Guzzetti, F, Peruccacci, S, Rossi, M, and Stark, CP 2007. Rainfall thresholds for the initiation of landslides in central and southern Europe. *Meteorology and Atmospheric Physics* 98, 239–267.

Guzzetti, F, Peruccacci, S, Rossi, M, and Stark, CP 2008. The rainfall intensity–duration control of shallow landslides and debris flows: an update. *Landslides* 5, 3–17.

Harvey, AM 2001. Coupling between hillslopes and channels in upland fluvial systems: implications for landscape sensitivity, illustrated from the Howgill Fells, northwest England. *Catena* 42, 225–250.

Hewitt, K 2006. Catastrophic landslides and their effects on the Upper Indus streams, Karakoram Himalaya, northern Pakistan. *Geomorphology* 26, 47–80.

Hewitt, K, Clague, JJ, and Orwin, JF 2008. Legacies of catastrophic rock slope failures in mountain landscapes. *Earth-Science Reviews* 87, 1–38.

Horton, P, Jaboyedoff, M, Rudaz, B, and Zimmermann, M 2013. Flow-R, a model for susceptibility mapping of debris flows and other gravitational hazards at a regional scale. *Natural Hazards and Earth System Sciences* 13, 869–885.

Hungr, O 1995. A model for the runout analysis of rapid flow slides, debris flows, and avalanches. *Canadian Geotechnical Journal* 32, 610–623.

Iida, T 1999. A stochastic hydro-geomorphological model for shallow landsliding due to rainstorm. *Catena* 34, 293–313.

Iida, T 2004. Theoretical research on the relationship between return period of rainfall and shallow landslides. *Hydrological Processes* 18, 739–756.

Innes, JL 1983. Debris flows. *Progress in Physical Geography* 7, 469–501.

Iverson, RM, Schilling, SP, and Vallance, JW 1998. Objective delineation of lahar-inundation hazard zones. *Geological Society of America Bulletin* 110, 972–984.

Jackson, LE, Kostaschuck, RA, and MacDonald, GM 1987. Identification of debris flow hazard on alluvial fans in the Canadian Rocky Mountains. *Reviews in Engineering Geology* 7, 115–124.

Jakob, M and Weatherly, H 2003. A hydroclimatic threshold for landslide initiation on the North Shore Mountains of Vancouver, British Columbia. *Geomorphology* 54, 137–156.

Jakob, M, Bovis, M, and Oden, M 2005. The significance of channel recharge rates for estimating debris-flow magnitude and frequency. *Earth Surface Processes and Landforms* 30, 75–766.

Jarman, D, Calvet, M, Corominas, J *et al.* 2014. Large-scale rock slope failures in the Eastern Pyrenees: identifying a sparse but significant population in paraglacial and parafluvial contexts. *Geografiska Annaler: Series A, Physical Geography* 96, 357–391.

Korup, O 2004. Landslide-induced river channel avulsions in mountain catchments of southwest New Zealand. *Geomorphology* 63, 57–80.

Korup, O 2005. Geomorphic imprint of landslides on alpine river systems, southwest New Zealand. *Earth Surface Processes and Landforms* 30, 783–800.

Korup, O 2013. Landslides in the fluvial system. In *Fluvial Geomorphology* (ed. JF Shroder). Vol. 9, *Treatise on Geomorphology* (ed. E Wohl). Elsevier: Amsterdam; 244–259.

Korup, O, McSaveney, MJ, and Davies, TRH 2004. Sediment generation and delivery from large historic landslides in the Southern Alps, New Zealand. *Geomorphology* 61, 189–207.

Lancaster, ST, Hayes, SK, and Grant, GE 2001. Modeling sediment and wood storage and dynamics in small mountainous watersheds. In *Geomorphic Processes and Riverine Habitat, Water Science and Application*, Vol. 4 (eds JM Dorava, DR Montgomery, BB Palcsak and FA Fitzpatrick). American Geophysical Union: Washington, DC; 85–102.

Lancaster, ST, Hayes, SK, and Grant, GE 2003. Effects of wood on debris flow runout in small mountain watersheds. *Water Resources Research.* 39, 1168.

Lanni, C, Borga, M, Rigon, R, and Tarolli, P 2012. Modelling shallow landslide susceptibility by means of a subsurface flow path connectivity index and estimates of soil depth spatial distribution. *Hydrology and Earth System Sciences* 16, 3959–3971.

Lucía, A, Comiti, F, Borga, M *et al.* 2015. Dynamics of large wood during a flash flood in two mountain catchments. *Natural Hazards and Earth System Sciences* 15, 1741–1755.

Marcato, G, Mantovani, M, Pasuto, A *et al.* 2012. Monitoring, numerical modelling and hazard mitigation of the Moscardo landslide (Eastern Italian Alps). *Engineering Geology* 128, 95–107.

Marchi, L and Brochot, S 2000. Les cônes de déjection torrentiels dans les Alpes françaises : morphométrie et processus de transport solide torrentiel. *Revue de Géographie Alpine* 88, 23–38. (In French.)

Marchi, L and Cavalli, M 2007. Procedures for the documentation of historical debris flows: application to the Chieppena Torrent (Italian Alps). *Environmental Management* 40, 493–503.

Marchi, L and D'Agostino, V 2004, Estimation of debris-flow magnitude in the Eastern Italian Alps. *Earth Surface Processes and Landforms* 29, 207–220.

Marchi, L and Tecca, PR 2006. Some observations on the use of data from historical documents in debris-flow studies. *Natural Hazards* 38, 301–320.

Marchi, L, Arattano, M, and Deganutti, AM 2002. Ten years of debris-flow monitoring in the Moscardo Torrent (Italian Alps). *Geomorphology* 46, 1–17.

Marchi, L, Pasuto, A, and Tecca, PR 1993. Flow processes on alluvial fans in the Eastern Italian Alps. *Zeitschrift für Geomorphologie* 37, 447–458.

Marinelli, G 1898. *Guida del Friuli III - Guida della Carnia*, reprint 1981. Società Alpina Friulana: Udine. (In Italian.)

May, CL and Gresswell, RE 2003a. Large wood recruitment and redistribution in headwater streams in the southern Oregon Coast Range, U.S.A. *Canadian Journal of Forest Research* 33, 1352–1362.

May, CL and Gresswell, RE 2003b. Processes and rates of sediment and wood accumulation in headwater streams of the Oregon Coast Range, USA. *Earth Surface Processes and Landforms* 28, 409–424.

Mazzorana, B, Zischg, A, Largiader, A, and Hübl, J 2009. Hazard index maps for woody material recruitment and transport in alpine catchments. *Natural Hazards and Earth System Sciences* 9, 197–209.

Melton, MA 1965. The geomorphic and paleoclimatic significance of alluvial deposits in Southern Arizona. *Journal of Geology* 73, 1–38.

Montgomery, DR and Dietrich, WE 1994. A physically based model for the topographic control on shallow landsliding. *Water Resources Research* 30, 1153–1171.

Mougin, P 1914. *Les torrents de la Savoie.* Grands établissements de l'imprimerie générale: Grenoble. (In French.)

Nardi, L, and Rinaldi, M 2015. Spatio-temporal patterns of channel changes in response to a major flood event: the case of the Magra River (central-northern Italy). *Earth Surface Processes and Landforms* 40, 326–339.

Nikolopoulos, EI, Crema, S, Marchi, L *et al.* 2014. Impact of uncertainty in rainfall estimation on the identification of rainfall thresholds for debris flow occurrence. *Geomorphology* 221, 286–297.

O'Brien, JS, Julien, PY, and Fullerton, WT 1993. Two-dimensional water flood and mudflow simulation. *Journal of Hydraulic Engineering ASCE* 119, 244–259.

O'Connor, JE, Hardison, JH 3rd, and Costa, JE 2001. *Debris flows from failures of Neoglacial-age moraine dams in the Three Sisters and Mount Jefferson wilderness areas, Oregon, 2001.* Professional Paper 1606, US Geological Survey: Washington, DC.

Pack, RT, Tarboton, DG, and Goodwin, CN 1998. The SINMAP approach to terrain stability mapping. *8th Congress of the International Association of Engineering Geology*, Vancouver, Canada.

Pratt-Sitaula, B, Burbank DW, Heimsath, A, and Ojha, T 2004. Landscape disequilibrium on 1000–10,000 year scales Marsyandi River, Nepal, central Himalaya. *Geomorphology* 58, 223–241.

Renard, K, Foster, GR, Weesies, GA *et al.* 1997, *Predicting soil erosion by water: a guide to conservation planning with the Revised Universal Soil Loss Equation (RUSLE).* Agriculture Handbook 703, US Department of Agriculture: Washington, DC.

Rickenmann, D 1999. Empirical relationships for debris flows. *Natural Hazards* 19, 47–77.

Rickenmann, D 2005. Runout prediction methods. In *Debris-flow Hazards and Related Phenomena* (eds M Jakob and O Hungr). Praxis Springer: Berlin, Heidelberg; 305–324.

Rigon, E, Comiti, F, and Lenzi, MA 2012. Large wood storage in streams of the eastern Italian Alps and the relevance of hillslope processes. *Water Resources Research* 48, W01518.

Rosatti, G and Begnudelli, L 2013. Two-dimensional simulation of debris flows over mobile bed: enhancing the TRENT2D model by using a well-balanced Generalized Roe-type solver. *Computers and Fluids* 71, 179–195.

Ruiz-Villanueva V, Díez-Herrero A, Ballesteros JA, and Bodoque JM 2014. Potential large woody debris recruitment due to landslides, bank erosion and floods in mountain basins: a quantitative estimation approach. *River Research and Applications* 30, 81–97.

Sepúlveda, SA, Moreiras, SM, Lara, M, and Alfaro, A 2015. Debris flows in the Andean ranges of central Chile and Argentina triggered by 2013 summer storms: characteristics and consequences. *Landslides* 12, 115–133.

SOGREAH 1979. *Etude hydraulique de la protection de la RN 6 (Route du Frejus Lyon-Turin) au confluent de l'Arc et du Torrent du Pousset.* Rapport d'ingénierie: Grenoble. (In French.)

Tarboton, DG 1997. A new method for the determination of flow directions and contributing areas in grid digital elevation models. *Water Resources Research* 33, 309–319.

Tsou, C-Y, Feng, Z-Y, and Chigira, M 2011. Catastrophic landslide induced by the Typhoon Morakot, Shiaolin, Taiwan. *Geomorphology* 127, 166–178.

Van Gorp, W, Temme, AJAM, Baartman, JEM, and Schoorl, JM 2014. Landscape evolution modelling of naturally dammed rivers. *Earth Surface Processes and Landforms* 39, 1587–1600.

Venzo, GA and Largaiolli, T 1968. Il bacino del Chieppena (Trentino) - Stratigrafia - Tettonica - Geomorfologia e Idrologia - Fenomeni Franosi. *Memorie del Museo Tridentino di Scienze Naturali* 17, 1–103; two maps at 1:10000 scale. (In Italian.)

Webb, RH, Pringle, PT, Reneau, SL, and Rink, GL 1988. Monument Creek debris flow, 1984: Implications for formations of rapids on the Colorado River in Grand Canyon National Park. *Geology* 16, 50–54.

Wieczorek, GF and Glade, T 2005. Climatic factors influencing occurrence of debris flows. In *Debris-flow Hazards and Related Phenomena* (eds M Jakob and O Hungr). Praxis Springer: Berlin, Heidelberg; 325–362

Wischmeier, WH and Smith, DD 1978, *Predicting Rainfall Erosion Losses – a Guide to Conservation Planning.* Agriculture Handbook 537, US Department of Agriculture: Washington, DC.

Wohl, E and Ogden, FL 2013. Organic carbon export in the form of wood during an extreme tropical storm, Upper Rio Chagres, Panama. *Earth Surface Processes and Landforms* 38, 1407–1416.
Wohl, EE and Pearthree, PP 1991. Debris flows as geomorphic agents in the Huachuca Mountains of southeastern Arizona. *Geomorphology* 4, 273–292.
Wohl, E, Ogden, FL, and Goode, J 2009. Episodic wood loading in a mountainous neotropical watershed. *Geomorphology* 111, 149–159.

Discussion

Discussion by James Bathurst

The SHETRAN model for shallow landslide sediment yield at the catchment scale uses a rule-based approach to determine if material released in a shallow landslide will travel downhill and potentially reach the channel network (Burton and Bathurst, 1998). Under forest cover, slope failures are assumed to release such energy and mass (through overcoming both root strength and soil cohesion) that a debris flow will develop and transport the landslide material downhill. For grass cover, root binding is relatively weak and landslide occurrence is more closely controlled by the soil cohesion. Debris flows are therefore assumed not to develop, at least on planar slopes. In the author's experience, are these assumptions realistic and to what extent can debris-flow occurrence be enhanced or inhibited by slope vegetation cover?

Discussion by Gordon Grant of Both Lorenzo Marchi and Chjeng-Lun Shieh

It is my impression that while we currently have a fairly large set of case studies for where landslide dams have occurred in the past and the geomorphic consequences of both their persistence and failure, our ability to predict future events and their persistence is still rather immature. In particular, we are only just beginning to develop empirical, much less theoretical models of how and where landslide dams occur in relation to geology and both rainfall triggering and seismic events. What is your sense of the state-of-the-art with respect to predicting the occurrence of landslide dams in different landscapes, and their persistence over time?

Discussion by Philippe Belleudy

Considering risk assessment by managers and especially the predetermination of risk, is the practical use (option 1) separate from flood and additional consideration of landslide occurrences, or (option 2) an integrated consideration of the flood and debris hazards in the case of favourable morphological watersheds?

Discussion by Kimberly Hill

This chapter concluded with a compelling list of topics for future study to improve hazard management concerning interactions between river floods, landslides, and debris flows. The first includes a consideration of the scales of precipitation that can trigger these events in and around gravel-bed rivers under different climatic and geomorphological conditions. Others include consideration of channel-scale to hillslope-scale effects such as rainfall event predictions, and modelling of system response to sediment pulses. Could you comment on whether a complementary body of research considering particle-scale effects might be useful, or if the events appear overwhelmed by the enormity of the failures? An example of particle-scale influences could include how climate history has determined typical local particle-size distributions and how the particle-size distributions subsequently may control flow behaviours, bed material entrainment, and other system dynamics.

Discussion by Mike Miles and Colin Thorne of Lorenzo Marchi, Chjeng-Lun Shieh and Yu-Shiu Chen, and Koichi Fujita

It is common practice to design river-related infrastructure and regulate floodplain management on the basis of a design event or 'hazard', such as the flood with a return period of 1:100 years. This criterion, together with an estimate of the infrastructure or building's design life, allows determination of the cumulative risk that the design event will be exceeded during the lifetime of the project as follows:

$$P = 1 - \left(1 - \frac{1}{T}\right)^{L_p}$$

where P, is probability of exceeding a design criterion (%), L_p, is anticipated project lifespan (years), and T is average return period in years as a design criterion.

Recent experience in Canada suggests that land managers, and even senior engineers, commonly fail to appreciate either the cumulative risk of exceeding the design event for long-lived infrastructure (such as levees) and structures (such as residences), or the importance of selecting a suitably conservative design standard for communities vulnerable to life-threatening hazards, such as debris flows. Recent studies, such as those by Jakob (2014), recommend that land-use zoning or engineering design be based on an acceptable level of risk, which varies with the severity of the consequence associated with failure.

In this context, please could you explain the basis for design standards/events commonly employed in your country and what factors are considered when determining what cumulative probability of failure is considered to be acceptable?

Reference

Jakob, M 2014, *Debris Flood Hazards and Risks at Cougar Creek.* http://www.canmore.ca/index.php?option=com_docmanandtask=doc_downloadandgid=4954 (accessed 21 October 2015).

Reply by Lorenzo Marchi

The assumptions recalled by James Bathurst about debris flows on forested slopes and grass-covered slopes are undoubtedly realistic. Nevertheless, if shallow landslides in unforested slopes occur close to the stream network they can reach it, thus contributing to debris-flow initiation in steep channels or increasing the magnitude of debris flows originated in upstream channel reaches. In my experience, vegetation cover has generally a positive effect in controlling debris-flow occurrence. However, given the complexity of the relations between vegetation and slope stability (see Marston (2010) for a review), specific conditions may result in a less positive role of vegetation. For instance, poor forest health may give rise to factors favourable to slope instability, such as increase in soil saturation or loss of soil strength as roots decay (Johnson and Wilcock 2002). It also deserves to be reminded that debris flows in forested catchments deliver to the receiving stream not only sediment, but also large wood debris, which can greatly enhance flood hazards.

The comment by Gordon Grant underlines the difficulty to progress from the description of specific cases of large-scale landslides that dammed gravel-bed rivers to more general models suitable for predicting such events. Data on formation of landslide dams have been collected and empirical-statistical models suitable to predict their persistency have been proposed (Costa and Schuster 1988; Ermini and Casagli 2003). The recognition of the conditions leading to large slope failures that could result in landslide dam formation is still in an early phase. The issue implies the assessment of geostructural, topographic, and climate conditions that may determine large-scale landslides. Some studies (e.g., Crosta, Frattini, and Agliardi 2013; Jarman *et al.* 2014) provide elements suitable for recognition of sites potentially susceptible to large-scale landslides. These studies at regional scale need to be integrated at local scale, with the recognition of clues of possible occurrence of large landslides and the evaluation of topographic settings relevant for the formation of landslide dams.

I agree that, as suggested by Kimberly Hill, particle-scale influences could have a role in future research on the interactions between landslides and floods in gravel-bed rivers. When focusing on the triggering and runout of debris flows and landslides, the particle-size of material involved is one of the factors that deserve to be taken into account in the geotechnical and rheological analyses of these processes. A closer link with flood processes in gravel-bed rivers arises when considering the interactions of landslide deposits with the fluvial system. For instance, particle size of landslide accumulations may influence the mechanical and hydraulic behaviour of landslide dams, thus affecting their persistence. In the case of debris flows, the influence of particle-size on sediment transport in the receiving stream and landform of the main valley also deserve further investigations.

The comments by Philippe Belleudy, Mike Miles, and Colin Thorne deal with flood hazard and risk assessment, which represents one of the main practical implications of the interactions of landslides and debris flows with gravel-bed rivers. According to the national regulation, the floodplain in Italy is mapped into three classes depending on inundation probability: high frequency, corresponding to floods with return period of 20–50 years, moderate frequency, with return period of 100–200 years, and low frequency, with return period of 300–500 years. The design of structures exposed to floods or aimed at the protection from floods and the evaluation of the cumulative probability of failure depends on the type of infrastructure. For hydraulic works, the design return period is minimum (<10 years) for sewerage systems and maximum (≥1000 years) for large dams. With regard to the topic of the chapter, it can be noted that the design of structures and floodplain mapping are usually performed for water floods only, i.e. without considering possible interactions of landslides with the river system. Landslide hazard and risk are the objects of separate analyses, carried out using methods different from those applied for floods. The integrated consideration of joint occurrence of floods and landslides (i.e., the option 2 of the comment by Philippe Belleudy) represents a real progress toward better management of natural hazards in mountain watersheds. Such integrated analysis has been implemented in particular cases, in which active landslides show potential interactions with rivers, but is still a challenging task and has not yet entered the practice of hazard assessment.

References

Johnson, AC, and Wilcock, P 2002. Association between cedar decline and hillslope stability in mountainous regions of southeast Alaska. *Geomorphology* 46, 129–142.

Marston, RA 2010. Geomorphology and vegetation on hillslopes: interactions, dependencies, and feedback loops. *Geomorphology*116, 206–217.

18

Gravel Riverbed Processes Resulting from Large-Scale Landslides

Chjeng-Lun Shieh and Yu-Shiu Chen

18.1 Introduction

Taiwan is located at the junction between the Eurasian Plate and the Philippine Sea Plate. This location gives Taiwan the characteristics of a fragile geology, and steep slopes that result in rapid river flow. During the summer and autumn seasons Taiwan has frequent typhoons and storm events that cause both surface erosion and sediment transport, resulting in landslides and debris flows, which may further result in significant property damage and potential loss of life. This compounding effect can also propagate downstream damaging larger areas (Shieh *et al.* 2010; Chen *et al.* 2011).

In Taiwan, storm events and typhoons have resulted in large-scale disasters. For example, a tropical depression on 7 August 1959 caused the '87 flooding' and resulted in 1046 deaths. The debris flow alone from the '87 flooding' caused more than 70 deaths, and 852 injuries at Shanhu Village, Shirtou Township, and Changhua County. In addition to loss of life, 27 466 buildings were completely destroyed and 18 303 buildings damaged. Typhoons can be equally damaging: typhoons Gloria, Daina, and Elsie in the 1960s; typhoons Nora and Betty in the 1970s; typhoons Cecil and Wayne in the 1980s; typhoons Ofelia and Herb in the 1990s; typhoons Toraji, Nari, Mindulle, Haitang, Kalmaegi, Sinlaku, and Morakot occurring after 2000. In 2009, Typhoon Morakot caused the '88 flooding' and is considered the largest disaster in Taiwanese history.

From the '87 flooding' in 1959 to the '88 flooding' in 2009, 20 large-scale events occurred, with an average occurrence every 2.5 years across this 50-year span. However, this average is misleading because eight among the 20 typhoons occurred between 2004 and 2013. Using this time frame, the average occurrence is every 1.25 years in 'the recent time'. In addition to an increase in the frequency of large-scale events, there has also been an increase in their scale. Global warming and climate change is one potential cause for this trend of high intensity, long duration, and large cumulative amounts of rainfall. In addition to these, the type of disaster changed from a single event to large-scale, compound disasters (Chen *et al.* 2011). For example the largest rainfall event in Taiwan's history occurred in August 2009 during Typhoon Morakot. This event yielded a rainfall intensity of 126 mm/h during 100 h resulting in an accumulated rainfall of 3000 mm. A 10 000 km^2 area experienced a cumulative rainfall of more than 1000 mm. This event caused many sediment disasters in central, southern, and eastern Taiwan (Shieh *et al.* 2009). One of the disaster sites was the Shaolin Village in Jiasian Township, Kaohsiung County. This village was completely destroyed from the combination of a deeply

Gravel-Bed Rivers: Processes and Disasters, First Edition. Edited by Daizo Tsutsumi and Jonathan B. Laronne.

seated landslide, natural dam formation, natural dam breach, and intense flooding (Chen *et al.* 2011; Tsai *et al.* 2012; Kuo *et al.* 2012, 2013).

Typically, strong rainfall events result in transport of sediments deposited in downstream channels or at cliff faces. This deposition changes the terrain, becoming a sediment source. The boundary of the deposition is termed the 'front wave'. Sediment movement and deposition processes are not only large-scale in space but also long in time.

In contrast to the 'small-scale', such as a sand wave, and the 'medium-scale', such as sand bar, the 'large-scale' refers to river catchment in terms of space and time scales in the order of decades. A large-scale landslide event usually includes a subsequent sediment transport process, which moves the sediment to downstream areas. This process influences the terrain, river flow conditions and sediment transport characteristics due to changes of riverbed slope, width, and depth.

This chapter discusses cases of two Taiwanese rivers: the Shoufeng River (Case I) in Hualien County and the Taimali River (Case II) in Taitung County (Figure 18.1). The discussion focuses on riverbed processes related to large-scale landslides based on measurements of river cross-sections, aerial photographs, satellite images, digital terrain models, topographic data, and records from hydrological stations. In Case I a large-scale landslide (1.5×10^7 m^3) occurred in 1984 affecting an area of about 270 ha. In Case II a large landslide (1.1×10^8 m^3; 200 m maximum landslide depth) occurred in 1973 affecting an area of about 250 ha. This landslide resulted in the formation of a landslide dam that gave rise to an upstream lake. This lake had a length exceeding 1000 m, a depth of 10 m and an area of 70 ha, storing a total of 5 330 000 m^3 of water.

18.2 Case Study: Shoufeng River

18.2.1 Shoufeng River

The Shoufeng River catchment area is located in the Hualien County, eastern Taiwan. The downstream villages of the Shoufeng River are mostly composed of aboriginal people at the Wenland north bank and Zilin south bank villages. The upstream catchment area is covered by natural vegetation with no population.

The Shoufeng River has a 21 373 ha catchment. The river originates in the Achan Mountain located in the central mountain range of Taiwan. The catchment has a diverse set of terrain characteristics such as mountain, valley, meandering river, and an alluvial plain. Due to a hard lithology, irregular plumes and dendritic terrain are common. The palmate type steep river in the upstream area indicates that erosion is considerable in this catchment, which results in large sediment movements on hillslopes. According to satellite landslide surveillance images, the landslide area was 1397 ha (Figure 18.2, December 2008). With a catchment elevation between 62 m and 3096 m, the area above 1000 m comprises 73.3% of the Shoufeng River catchment. Hillslopes are very steep: 46.2% (30.12%) of the catchment area has slopes in the respective range 30–45° (>45°). With a southeast flow direction, the river enters the Hualien River, thereafter to the Pacific Ocean.

18.2.2 Disaster History

The Shoufeng River catchment has a fragile geology, discontinuous strata, and steep slopes. These characteristics tend to cause slope collapse and high sediment yields. During a typhoon or storm event, rivers discharge sediment downstream resulting in dramatic riverbed changes.

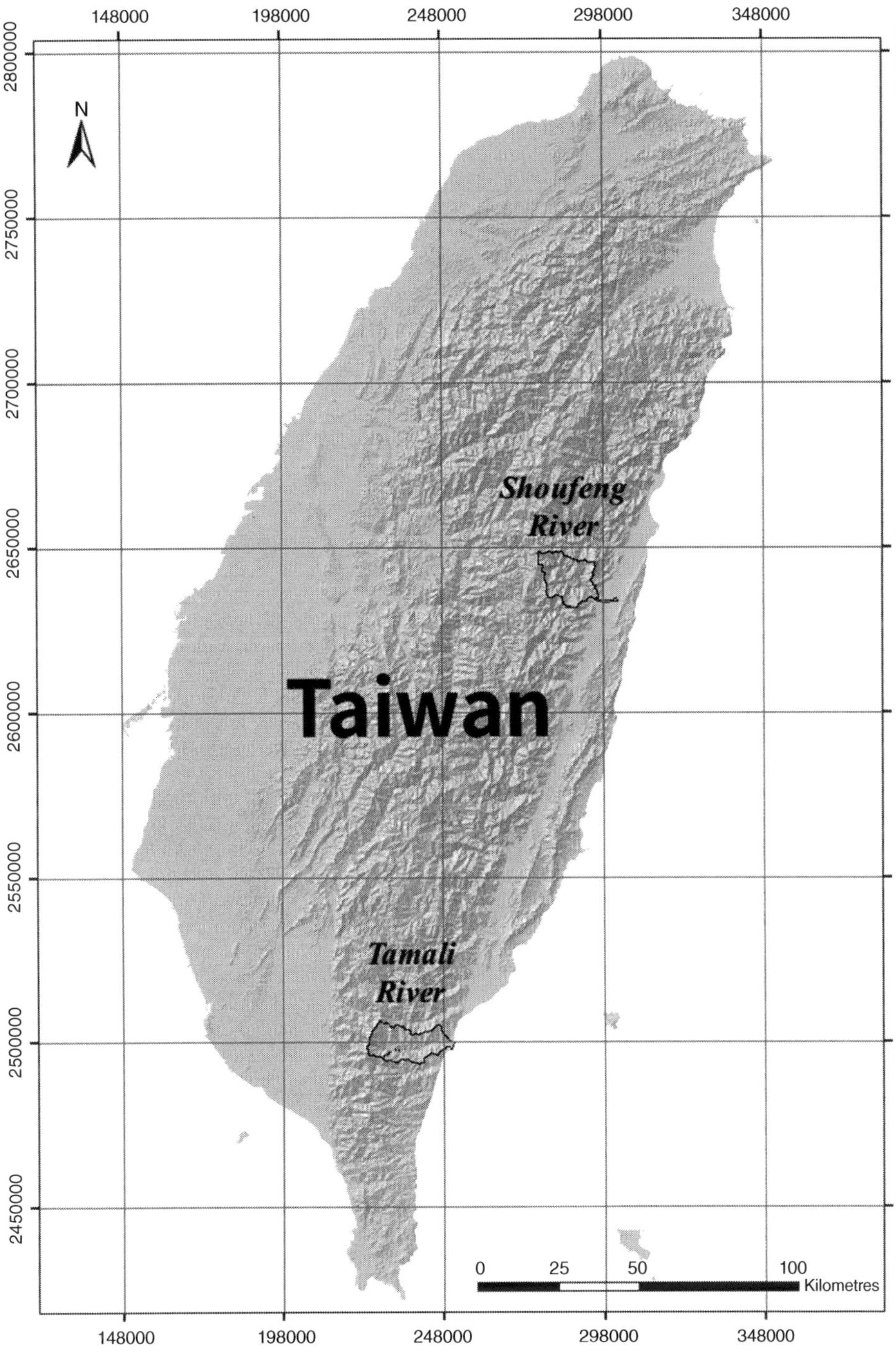

Figure 18.1 Locations of the two study cases in Taiwan.

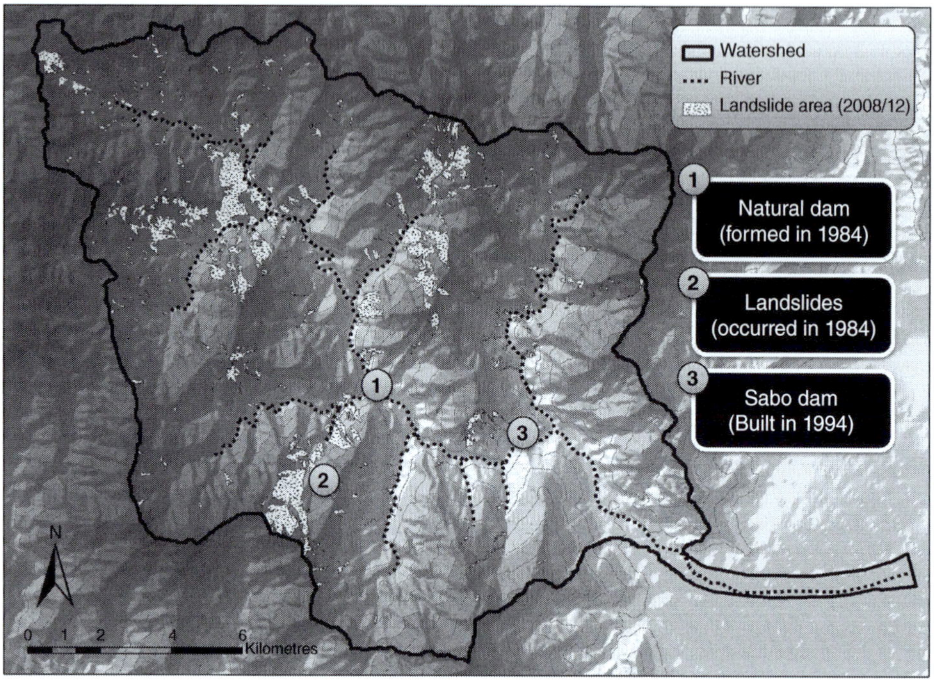

Figure 18.2 The topography and landslide area in the Shoufeng River catchment.

In 1984, a natural dam was formed at the branch river location at 20 K + 000 (distance from outlet) in the upstream area. The height of the dam was 50–60 m. The volume of the dam body was about 15×10^6 m^3 (Figure 18.3). Due to sediment movement downstream, multiple damage occurred to the embankment at the Xilin Village area between 1984 and 1994. The most serious disaster occurred in 1990 during Typhoon Dot. The Wangzon embankment was breached along 500 m. The resulting flood entered a railway tunnel located below the river and resulted in multiday disruption in transportation. In 1994, the Forest Service Office had to construct a check dam and relocate the across-river railway on a bridge into an underground rail tunnel for protection. After the check dam was built the only embankment damage occurred in 1998 during Typhoon Zeb and Typhoon Babs, with 170 m and 80 m embankment damages, respectively. The peak water discharges were as follows during Typhoon Nora (1490 m^3/s) in 1973, Typhoon Ike (1160 m^3/s) in 1981, Typhoon Clara (1160 m^3/s) in 1981, Typhoon Andy (1230 m^3/s) in 1983, and Typhoon Bilis (1010 m^3/s) in 2000.

18.2.3 Transformation of the Catchment

River cross-sectional measurements between 1984 and 2013 allowed calculating the variation of sediment changes and discussion of the transformation process. Figure 18.4 shows the location of each river section. Figures 18.5–18.7 present sediment transformation information where each line represents the accumulated erosion/deposition volume for each section of the river. A positive value represents deposition and a negative value represents erosion after 1984. The slope represents the speed of erosion/deposition.

Figure 18.3 Aerial photographs (left: 11/04/1980) before and (right: 08/03/1986) after the landslide.

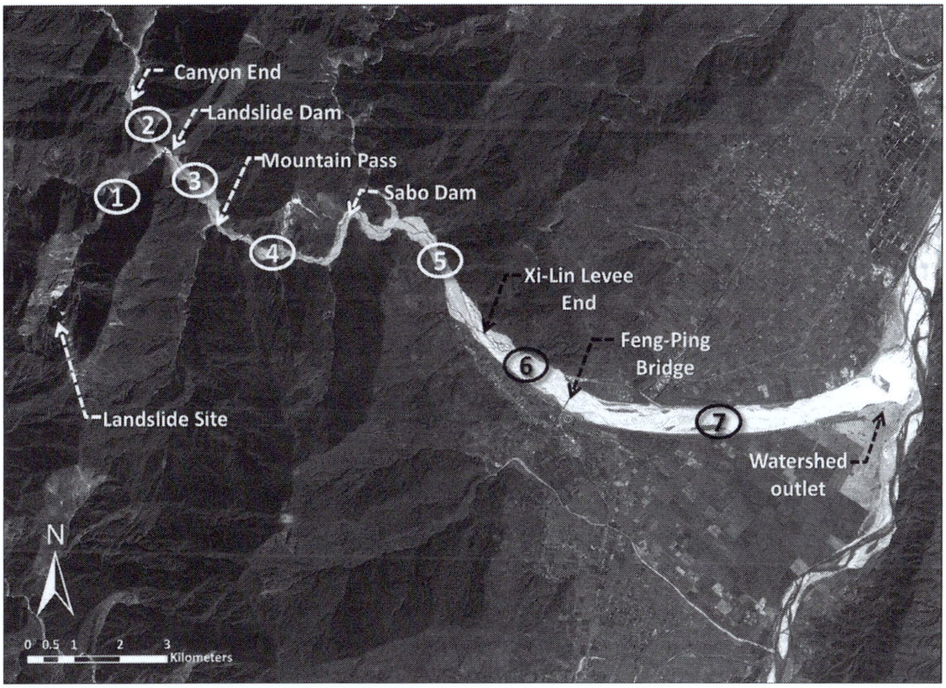

Figure 18.4 Illustration of analysis of transformation of the catchment.

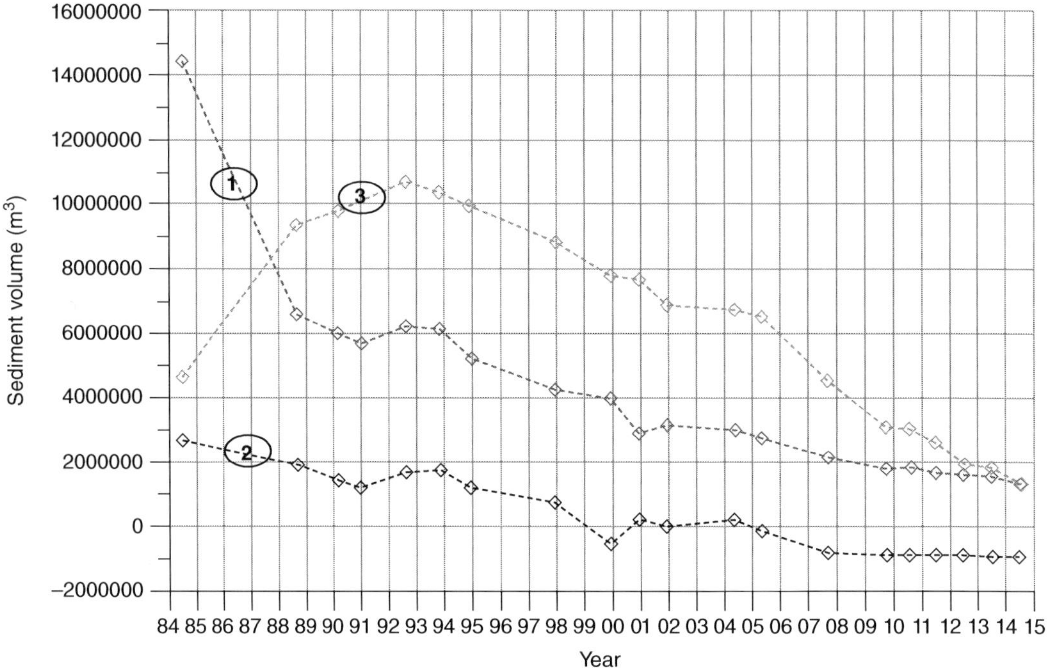

Figure 18.5 Cumulative sediment volume for sections 1–3 of the Shoufeng River.

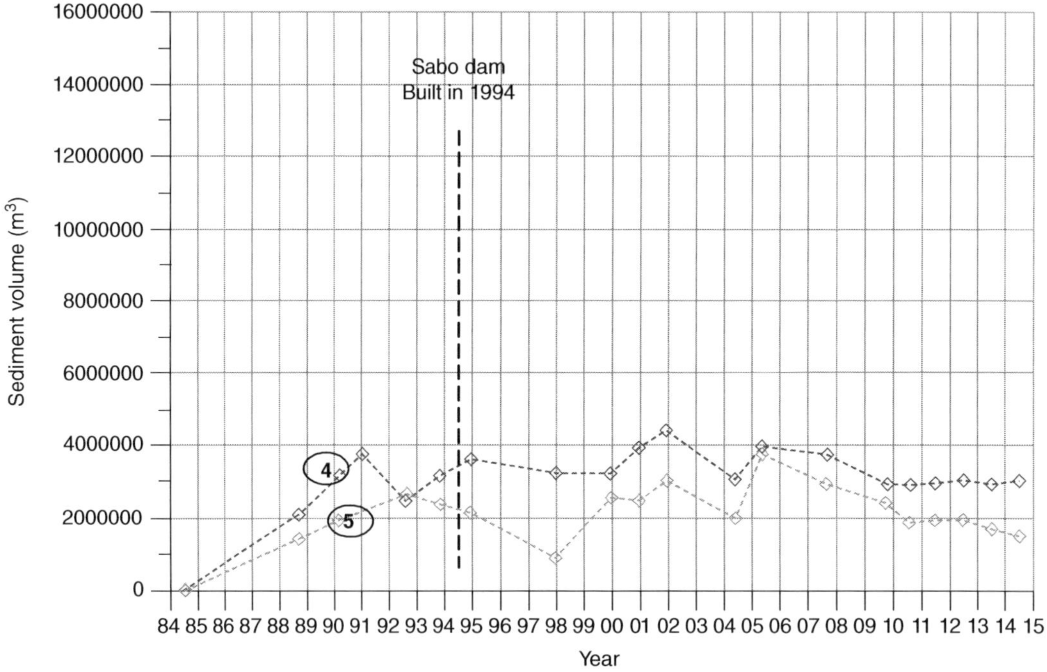

Figure 18.6 Cumulative sediment volumes for section 4–5 of the Shoufeng River.

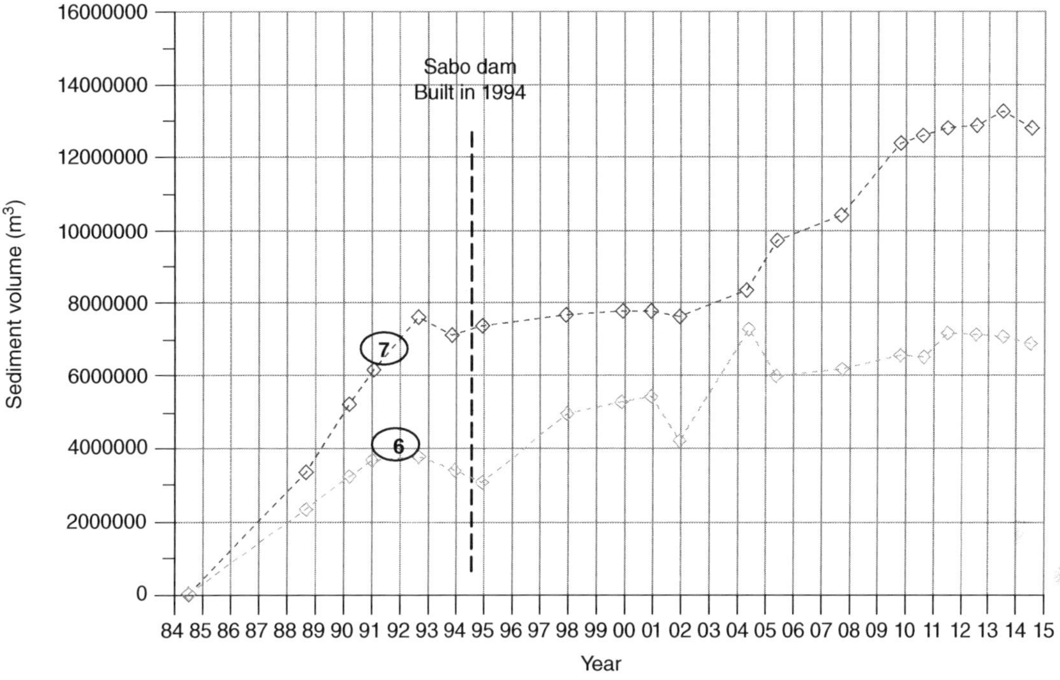

Figure 18.7 Cumulative sediment volume for section 6–7 of the Shoufeng River.

As shown in Figure 18.7, the deposited sediment in 1984 moved downstream. The sediment volume dramatically increased between the landslide location (section 1 in Figure 18.4) and the location of the landslide dam (section 2 in Figure 18.4). Section 3 shows that between 1984 and 1993, the first 10 years after the occurrence of the large-scale landslide, deposition occurred at the upstream area of the natural throat (see Figure 18.5). The peak deposition at section 3 occurred in 1993. After 1993 erosion of sections 1 and 2 continued, and erosion at section 3 started. The Main River No. 1 sabo dam was completed in 1994, and started to block sediment in the period 1994–2003, the second decade after the large-scale landslide (Figure 18.6). After 2003, the check dam reached maximum sediment deposition volume. Deposition began at sections 6 and 7 between 2004 and 2013 (Figure 18.6), the third decade after the large-scale landslide.

A set of three cross-sectional profiles and the average elevation of the riverbed are presented in Figures 18.8 and 18.9, showing cross-sectional profiles for the branch that occurred from the landslide (section 1 in Figure 18.4). The elevation increased 50 m after the landslide in 1984 and gradually decreased after the event. In 2014 the elevation of the riverbed approached that of 1970 before the landslide occurred. The elevational changes are parabolic-like.

In the subcatchment of the landslide area, the riverbed elevation increased from 475 m to 525 m immediately after the event and then gradually decreased to 490 m in 1994, which is a decreasing rate

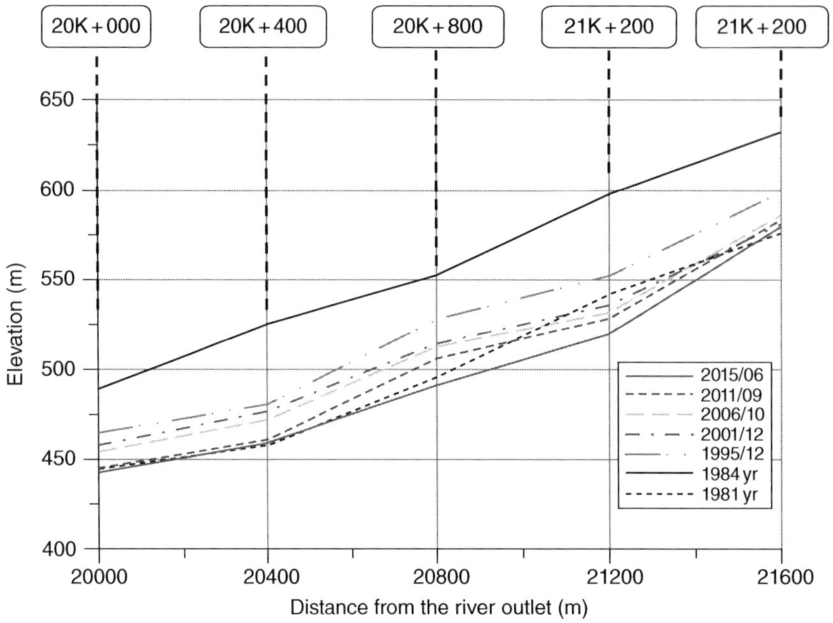

Figure 18.8 Temporal changes in riverbed elevation at the branch river channel.

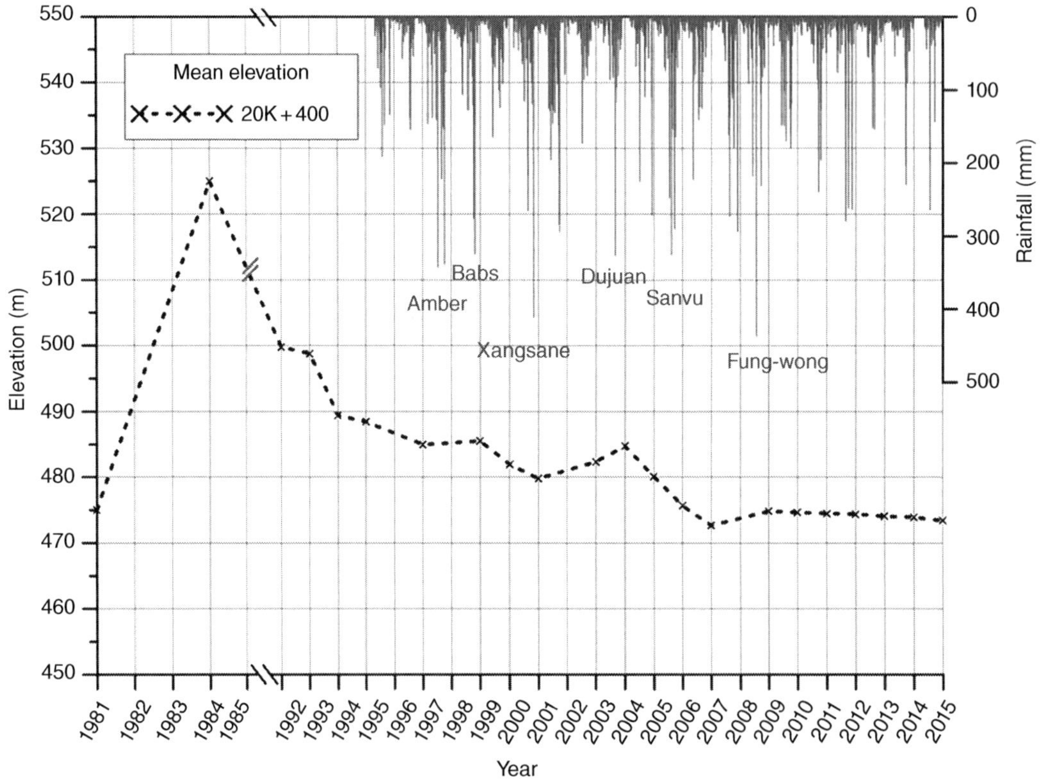

Figure 18.9 Change in riverbed elevation for upstream and downstream areas of the mountain pass.

of about 3.5 m/yr over a 10-year period. Between 1994 and 2004 the elevation reached 485 m with a decreasing rate of 0.5 m/yr, and between 2004 and 2014 the riverbed elevation tended toward the elevation before the landslide event, with a decreasing rate of about 1 m/yr.

Figures 18.10–18.12 show river profiles and cross-sections upstream and downstream of the natural throat (18 K + 400). The upstream area of the natural throat is the main deposition area within the first decade after the landslides event (1984–1993). In 1993 the riverbed elevation was 20 m above the elevation before the landslide event. After 1993, the elevation continuously decreased 12 m at 19 K + 200 and 5 m at 18 K + 400. Since the mountain pass restricted the transport of sediment, the downstream area had milder changes in elevation. The largest elevation difference was about 3 m in the downstream area of the mountain pass.

Figures 18.13–18.15 show river profiles and the cross-section for the upstream area and riverbed elevation hydrograph at the main river check dam. The check dam was completed in 1994. The river section near the check dam was the main deposit area between 1994 and 2003. Riverbed elevation in 1994 was 7 m higher than before the landslide event occurred in 1981. After the check was completed, the variation of riverbed elevation was about 1 m. This indicates that the check dam has the capability to stabilize the riverbed. Downstream of the check dam the elevation decreased slightly after the dam was completed and thereafter slightly increased to an average elevation of 275 m.

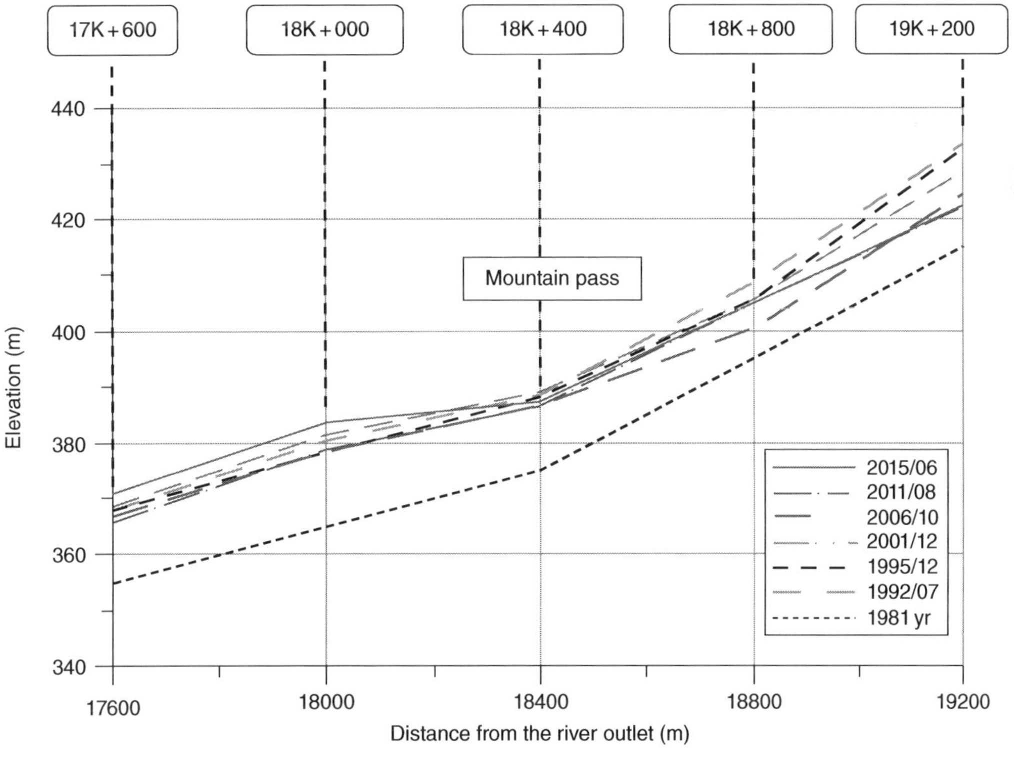

Figure 18.10 Changes in riverbed elevation upstream and downstream of the mountain pass.

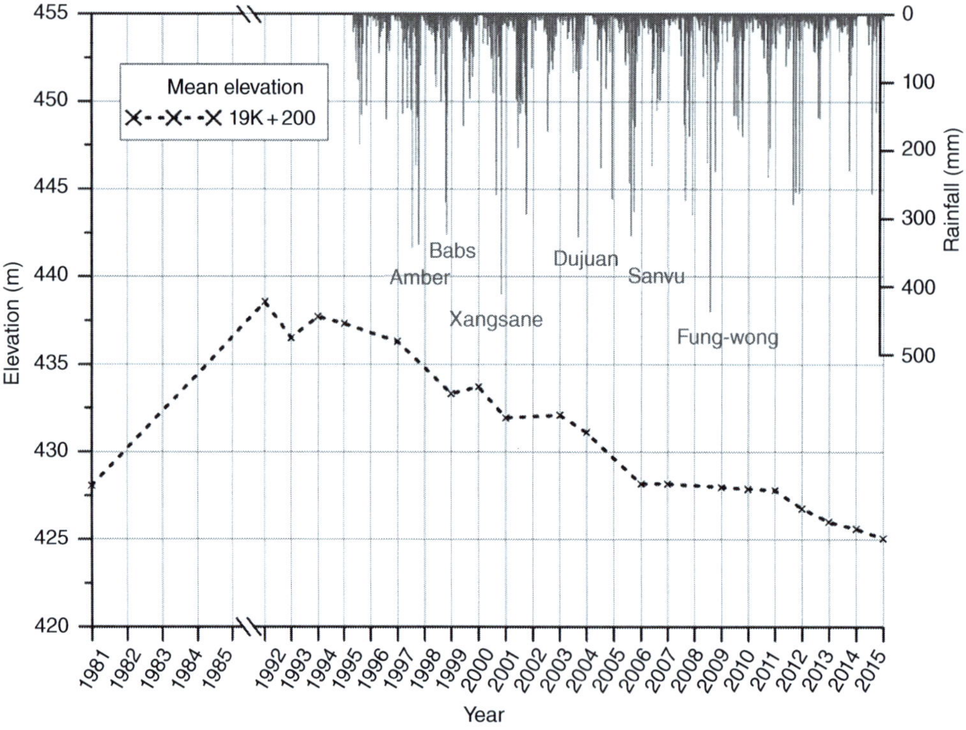

Figure 18.11 The change in riverbed elevation upstream of the mountain pass.

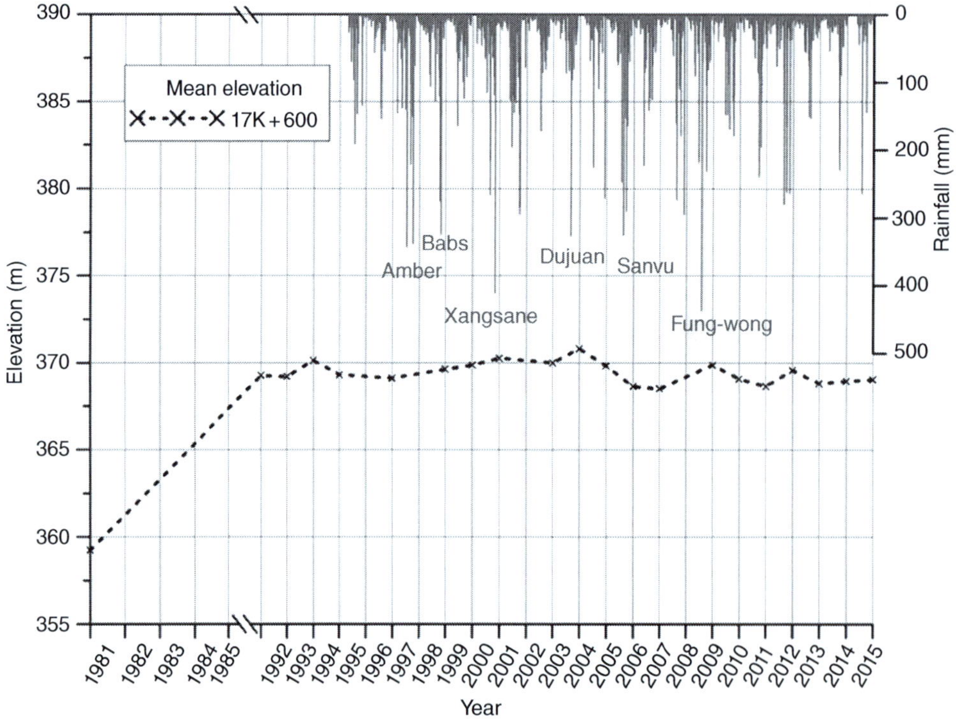

Figure 18.12 The change in riverbed elevation downstream of the mountain pass.

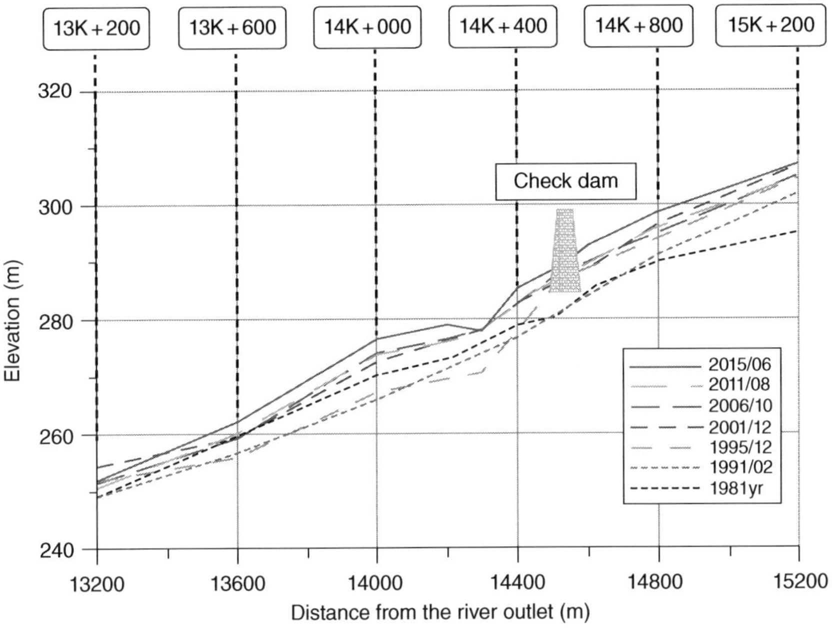

Figure 18.13 Changes in riverbed elevation upstream and downstream of the check dam.

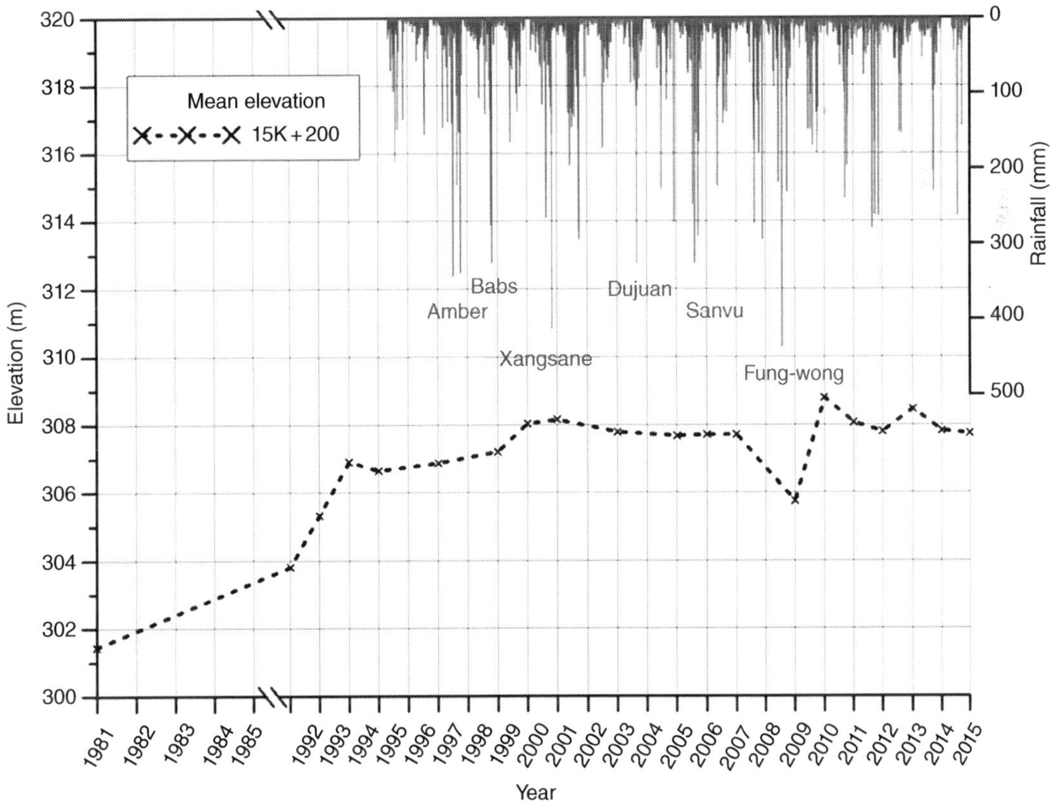

Figure 18.14 Change in riverbed elevation upstream of the check dam.

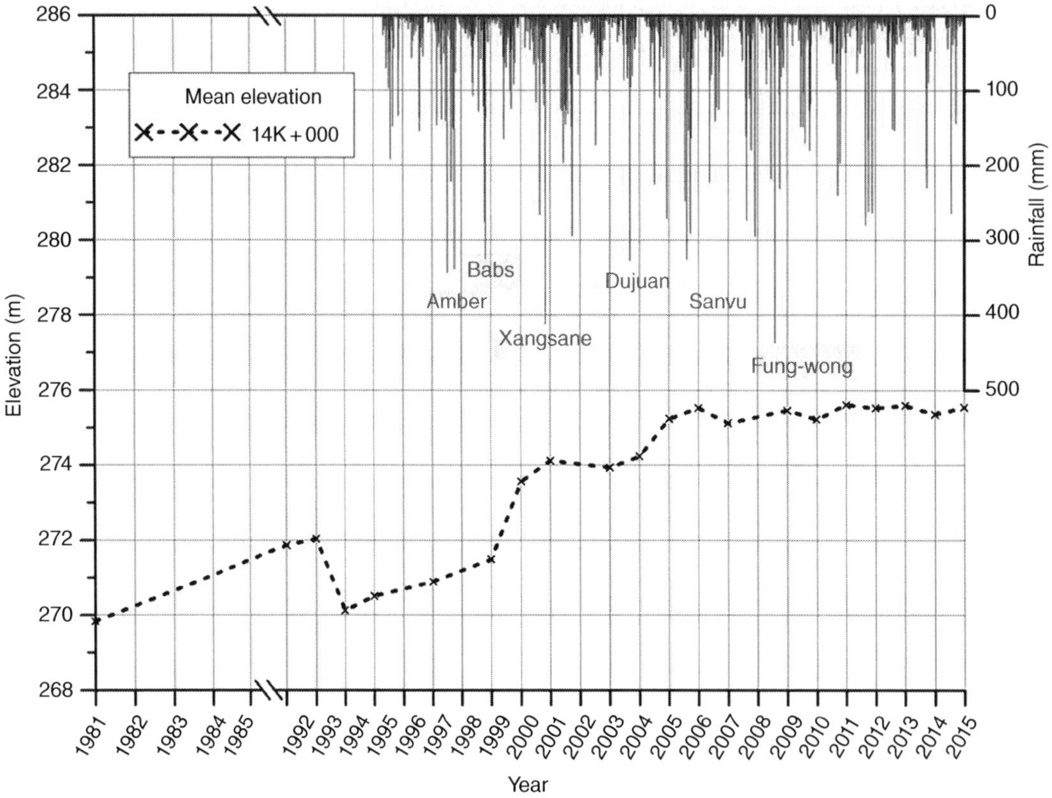

Figure 18.15 Change in riverbed elevation downstream of the check dam.

18.3 Case Study: Taimaili River

18.3.1 Background

The Taimaili River, flowing eastwards towards the Pacific Ocean, is located in Taitung County in eastern Taiwan (Figure 18.16). The catchment area is 21 753 ha, the river length is 36.3 km, and the average slope is 1/15. The majority of the catchment area is mountainous with an average elevation of 860 m, a maximum elevation of 3088 m and an elevation range of 62–3096 m. The area with elevation between 500 m and 1000 m comprises 43.2% of the catchment area. In terms of slope, 50.4% of the catchment area is between 30° and 45° and 14.4% of the catchment area is steeper than 45°.

Tectonically, the structures are oriented north–south or northeast–southwest. The outcrops comprise hard shale, slate, phyllite, metamorphic sand, and schist. The local aquifer includes quaternary alluvium, alluvial sedimentary layers, terrace-stacked layers, laterite plateau, and coastal sedimentary layers.

The main river channel of the Taimali River flows from west to east. The main tributaries are located northwards from the trunk stream and include the Dolidoli, Malidu (Hei River), Manuer (Bai River), Maliwu, and Cuno Creeks. The main southern tributary is Pulami Creek. Two natural mountain passes are located at the confluence of Pulami and Maliwu Creeks. Four villages (Buchachila, Galan, Shitou, and Taiho) are located downstream of Taimali River. Remote sensing images show that development areas are located in the valley close to the river and the alluvial fan. The middle and upstream areas have remained an original forest.

18.3.2 Disaster History

The fragile geologic situation brings rise to frequent collapse and weathering of the land surface. The hydrological data and remote sensing images show that rainfall increases from east to west and the landslide areas are mainly located in the upstream areas.

The most serious disaster in history was Typhoon Morakot in 2009. The landslide covered 15.2% of the upstream area. The large-scale landslide at Baoshanshe was 200 m deep, and 2.8 km long, with a volume of 110×10^6 m^3. The landslide material piled at the river channel forming a landslide dam with a dam height of 100 m and 4 km length. The flood transported sediment resulting in the destruction of the levees at Galan and Shitou villages, damage of South-Taimali railway, freeways, and sediment deposition at the outlet of the catchment. The flood and the transported sediment respectively destroyed 47 and 35 residential buildings in Granville and Sitou villages. The flood caused traffic disruption of railway and bridges, sediment deposition over large areas, and inundation of 400 ha of farmland (Kuo *et al.* 2012).

The 12 satellite images collected between 1997 and 2012 show that the terrain collapse rate increased since 1998. In 2005 (Figure 18.17), Typhoon Haitang increased the rate of collapse from 2.1% to 3.4%, representing 700 ha of collapsed terrain. In 2009 the collapse rate was further increased to 15.2%, which represents 3000 ha of collapsed terrain. In the upper catchment collapse rates were very high, 33.1% and 28.2% of the Dolidoli and Baoshanshe basins respectively.

In the Taimali catchment, the collapse rate increased from 3.86 to 15.2% during Typhoon Morakot. The collapse rate gradually increased to 11.84% in December 2013. However, large-scale sediment movement is still ongoing in the catchment. In 2010, the flowing waters brought by Typhoon Fanapi moved the sediment to fill up the landslide lake created by Typhoon Morakot. The deposited sediment is still being transported downstream.

18.3.3 Transformation of the Taimali River Catchment

Figure 18.18 shows ten segments of the Taimali River catchment based on the terrain characteristics, such as confluence, mountain pass, natural dam, and village. Typhoon Morakot caused large-scale landslides and resulted in elevation increments upstream of Pulami creek (Figure 18.19). Between 2005 and 2013 the aggradation was about 5 m at 15 K + 000 and about 80 m at 22 K + 000 (natural dam location). The longitudinal river slope thereby increased from 1.8% to 2.9%.

After Baoshanshe a landslide lake was filled by sediment in 2009, the largest scour located at 22 K + 000, with the landslide dam 20 m deep. In 2012, Typhoon Tembin and '0610' storm event flushed sediment downstream. The riverbed elevation at the Dolidoli creek confluence aggraded about 10 m. As of 2014 this phenomenon continues. Aggradation was about 3–6 m between Hei and Pulami creek. This phenomenon shows that the deposition area moves downstream.

Since 2011, two years after Typhoon Morakot, segment 10 began degrading (Figure 18.20). This phenomenon continuously expanded downstream and reached segments 9 and 7 in 2014. The main deposition area moved from the landslide lake location toward the mountain pass at Pulami Creek.

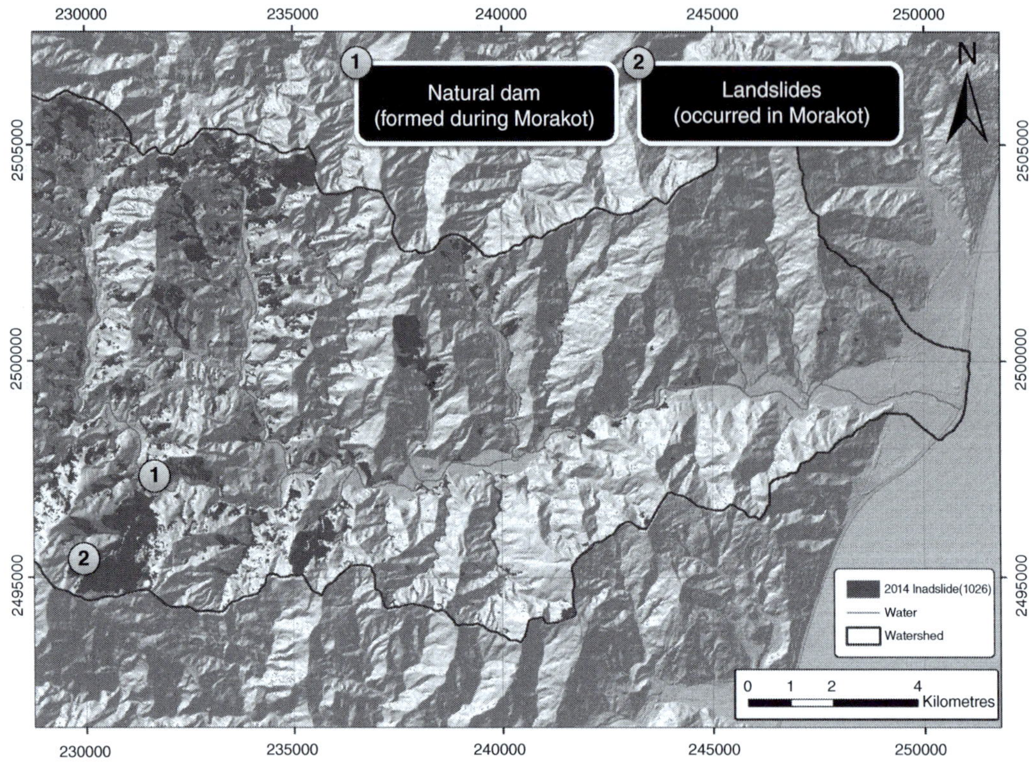

Figure 18.16 The topography and landslide area in the Taimali River catchment.

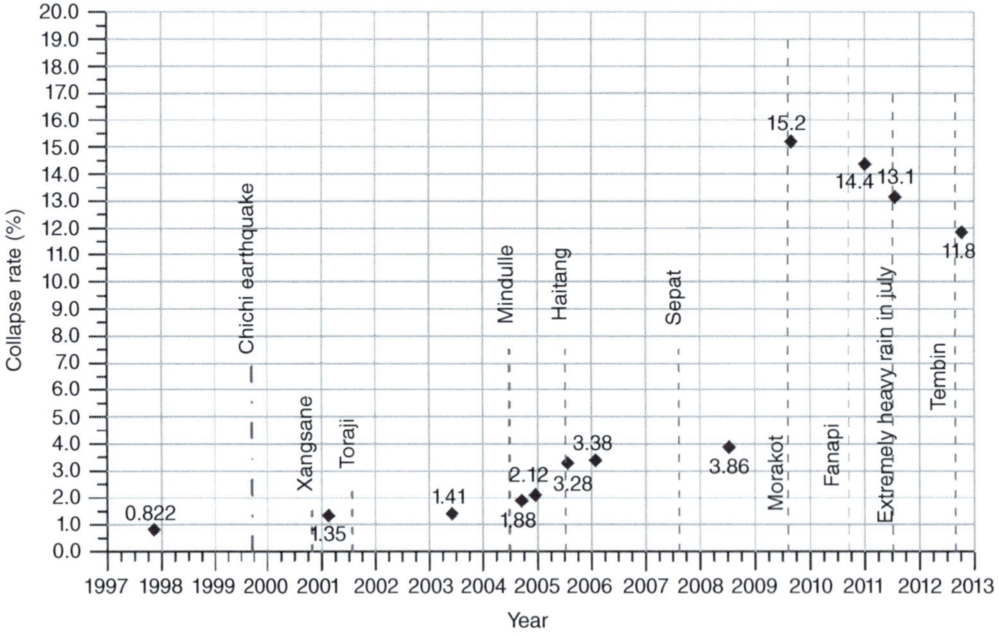

Figure 18.17 Terrain collapse rate in the Taimali River catchment.

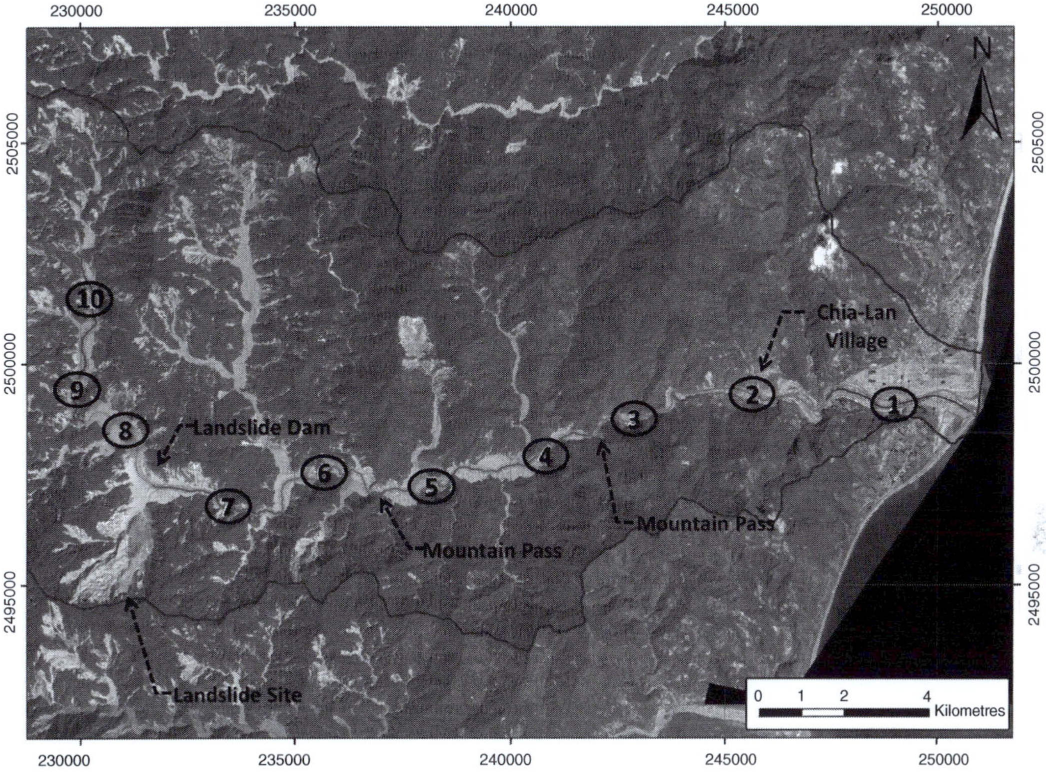

Figure 18.18 Location of the ten segments of the Taimali River catchment.

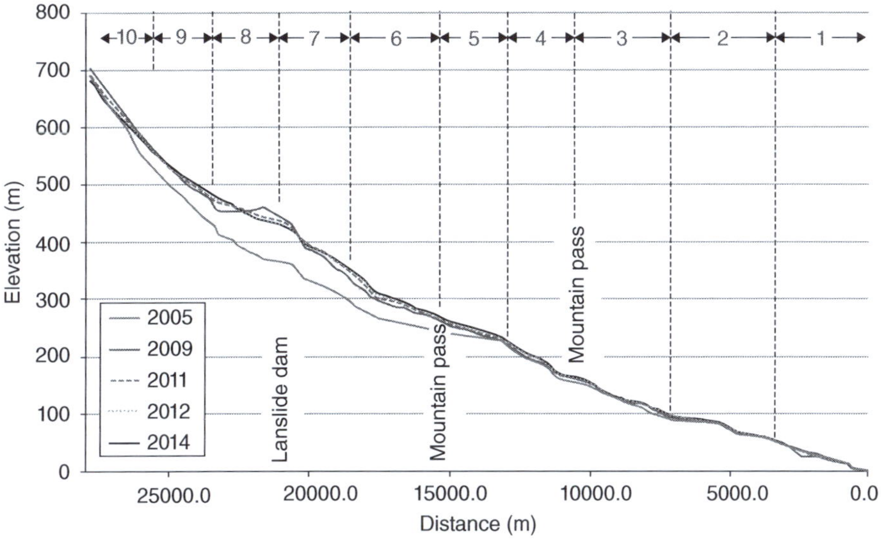

Figure 18.19 Elevation changes of the Taimali River bed along the ten segments.

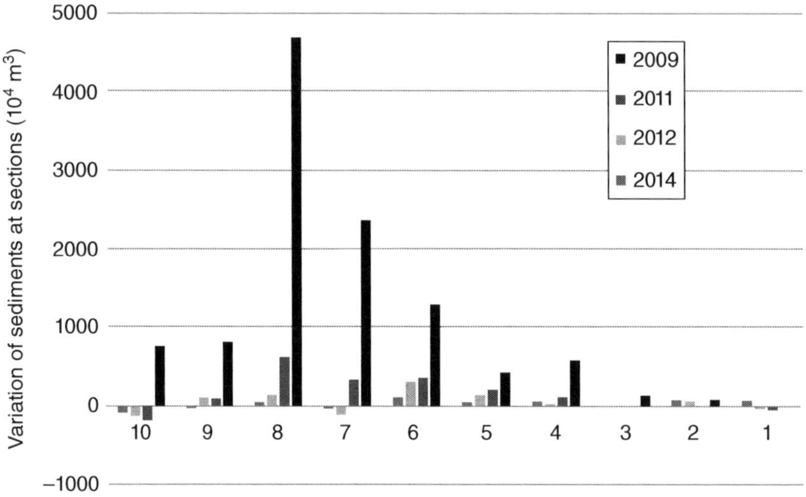

Figure 18.20 Variation in volume of sediment along the main Taimali River channel during 2009–2014. (See insert for color representation of this figure.)

18.4 Conclusion

This study exemplifies data from the Shoufeng and Taimali rivers in order to better understand the process of sediment movement from large landslide areas through rivers, using erosion and subsequent deposition data based on a 30-year observation period. Since the movement is large-scale both in area and time, large-scale riverbed processes occur. During the early stage of a large-scale landslide event the sediment yield is larger than the equilibrium sediment transport. Especially, the riverbed aggrades near the depositional area of landslides. With time, erosion of the landslide sediment generates aggradation downstream. The variation in riverbed change is parabolic, with the rate of change in riverbed elevation decreasing with time.

The character and amount of sediment moved is affected by natural mountain passes and check dams. According to the Shoufeng River case, the changes in riverbed elevation are linear in the upstream area of the mountain pass. Below the mountain pass the change in riverbed elevation is small. A check dam is an artificial structure to control the sediment, which can stabilize the riverbed for a long period. For the Shoufeng River, the upstream and downstream areas of a check dam had 3 m variation in elevation and remained relatively stable.

In 2009 Typhoon Morakot caused the most serious sediment disaster in Taiwanese history due to record-breaking rainfall intensity, cumulative rainfall depth, and duration, as well as the large area affected. Large-scale landslides occurred in many river catchments. These landslides caused sediment deposition in rivers and massive fluvial aggradation.

In the past check dams were used to control sediment yield. In this study natural mountain passes are shown to have the capability for large-scale sediment control. Therefore, this study suggests that natural mountain passes with artificial check dams can serve as efficient locations to control sediment movement and provide better sediment management results.

Acknowledgements

We would like to express our heartfelt thanks to the organizers of the GBR8 Workshop for the invitation to submit an invited chapter, and for editorial comments that improved the manuscript. We appreciate the Hualien and Taitung Forest District Offices, Forestry Bureau for their unwavering support for monitoring and research.

References

Chen, YS, Kuo, YS, Lai, WC *et al.* 2011. Reflection of typhoon morakot – the challenge of compound disaster simulation. *Journal of Mountain Science* 8(4), 571–581.

Kuo, YS, Chiu, SY, Chen, KT, and Shieh, CL 2012. Geometry parameters for earthquake-induced and rainfall-induced landslide dams in Taiwan. *Journal of the Taiwan Disaster Prevention Society* 4(2), 143–152. (In Chinese.)

Kuo, YS, Tsai, YJ, Chen, YS *et al.* 2013. Movement of deep-seated rainfall-induced landslide at Hsiaolin Village during Typhoon Morakot. *Landslides* 10(2), 191–202.

Shieh, CL, Wang, CM, Lai, WC *et al.* 2009. The composite hazard resulted from Typhoon Morakot in Taiwan. *Journal of the Japan Society of Erosion Control Engineering* 62(4), 61–65.

Shieh, CL, Wang, CM, Chen, YS *et al.* 2010. An overview of disasters resulted from Typhoon Morakot in Taiwan. *Journal of Disaster Research* 5(3), 236–244.

Tsai, YJ, Kuo, YS, Chen, YS *et al.* 2012. The landslide movement process at Shiao Lin Village during Typhoon Morakot. *International Journal of Erosion Control Engineering* 5(1), 46–51.

Discussion

Discussion by Mike Miles and Colin Thorne of Lorenzo Marchi, Chjeng-Lun Shieh and Yu-Shiu Chen, and Koichi Fujita

It is common practice to design river-related infrastructure and regulate floodplain management on the basis of a design event or 'hazard', such as the flood with a return period of 1:100 years. This criterion, together with an estimate of the infrastructure or building's design life, allows determination of the cumulative risk that the design event will be exceeded during the lifetime of the project as follows:

$$P = 1 - \left(1 - \frac{1}{T}\right)^{L_p}$$

where P, is probability of exceeding a design criterion (%), L_p, is anticipated project lifespan (years), and T is average return period in years as a design criterion.

Recent experience in Canada suggests that land managers, and even senior engineers, commonly fail to appreciate either the cumulative risk of exceeding the design event for long-lived infrastructure (such as levees) and structures (such as residences), or the importance of selecting a suitably conservative design standard for communities vulnerable to life-threatening hazards, such as debris flows. Recent studies, such as those by Jakob (2014), recommend that land-use zoning or engineering design be based on an acceptable level of risk, which varies with the severity of the consequence associated with failure.

In this context, please could you explain the basis for design standards/events commonly employed in your country and what factors are considered when determining what cumulative probability of failure is considered to be acceptable?

Reference

Jakob, M 2014. *Debris Flood Hazards and Risks at Cougar Creek.* http://www.canmore.ca/index.php?option=com_docmanandtask=doc_downloadandgid=4954 (accessed 21 October 21 2015).

Discussion by Shinji Egashira

It is very well known that a large amount of sediment released by a huge landslide influences sediment transport characteristics and associated channel change for many years. This study provides characteristics of riverbed variation and a suggestion for planning countermeasures, assuming that the longitudinal profile of the reach, even if it changes largely due to a catastrophic event, tends to approach its original configuration. This is very suggestive and constructive for deciding a suitable plan of countermeasures.

However, I introduce one event that is not subject to your assumption and your idea. A catastrophic event including many deep-seated landslides and debris flows took place in 1989 along a reach of the Kumano River in Japan. Since then, flood inundations have occurred very often in the Hongu district located along a middle reach where the very famous 1000 year old Kumano-hongu-taisya Shrine was considerably damaged by the 1989 event. This fact means that the river configuration changed very much due to the event but it has not recovered. Hence, when dealing with catastrophic events it should be considered that there are events that cause quite a different channel configuration from the original one.

Discussion by Gordon Grant of Both Lorenzo Marchi and Chjeng-Lun Shieh

It is my impression that while we currently have a fairly large set of case studies for where landslide dams have occurred in the past and the geomorphic consequences of both their persistence and failure, our ability to predict future events and their persistence is still rather immature. In particular, we are only just beginning to develop empirical, much less theoretical models of how and where landslide dams occur in relation to geology and both rainfall triggering and seismic events. What is your sense of the state-of-the-art with respect to predicting the occurrence of landslide dams in different landscapes, and their persistence over time?

Reply by the Authors

The most common practical experience to design river-related or sediment-related infrastructures (such as levees, dykes, check dams and debris-flow control works) in Taiwan has been based on the natural occurrence hazard events such as floods and debris flows with a return period of 1:50 to 1:100 years for regulating river management, except for the capital river flooding control, which based on a period of 1:200 years.

However, the disaster software items of forecasting, evacuation, and emergency operating system should be taken into consideration seriously to meet the insufficiency of engineering construction for the cumulative risk of disaster reduction. This can be an acceptable level of residence property damages but would not be allowable for losing a person's life.

Replying to Egashira, there is one significant example of the Shoufeng River in eastern Taiwan where a total sediment volume of 1.6×10^7 m^3 occurred in 1983. Also, a landslide dam resulted from these sediments, induced by many large-scale landslides in the upper reach of Shoufeng River. Recently, the reach profiles of the river tend to recover the original river configuration due to the huge sediment loads transported by floods during the past 30 years.

A different situation is that where almost 80% of total sediment has been trapped in the upstream of Taimali River in eastern Taiwan since a large-scale landslide took place in 2009. In summary, the recovery of original river configuration should be seriously considered by the following parameters:

- rainfall conditions
- amount of total sediment yield
- distribution of sediment particle sizes
- available of engineering counter measures.

The Kumano River (Japan) with a sediment yield volume exceeding 1×10^9 m^3 caused by catastrophic events that took place in 1889 will likely require more years to be transported by floods until it more completely recovers to its original configuration.

As for the discussion of Grant, the prediction of the potential area of landslide dams in Taiwan has just begun to develop an empirical model based on field investigations and laboratory simulation studies. Our ability to predict the potential area of landslide dams is still immature with reference to landslide dam formation, but we have already established the potential large-scale landslide database that can be used to calculate or estimate the volume and runout distance of sediment and whether it may block river flow. This database supplies a reference to evaluate the possibility of large-scale landslides to develop further to the occurrence of a landslide dam. More information is available in Chen (2015).

Reference

Chen, K-T 2015. *Study on interpretation of potential locations for landslide dam*. PhD dissertation, National Cheng Kung University; 157 pp. (In Chinese.)

19

Gravel-Bed River Management Focusing on Finer Sediment Behaviour

Koichi Fujita

19.1 Introduction

Channel capacity, defined in this chapter as the area bounded by the riverbed, the levee slopes, and the levee design stage of a cross section, is a major factor in governing flood-flow capacity (the maximum discharge that a river reach can flow without flooding over levees). Since flood flow capacity is, in turn, directly related to the probability of flooding occurrence, the appropriate maintenance of channel capacity is crucial to flood disaster prevention and mitigation for alluvial rivers.

Particularly in Japan, river administrative agencies have made great efforts toward increasing channel capacity by excavation and dredging as well as levee raising and withdrawing, while the construction of flood control facilities such as dam reservoirs and retarding basins has been also urged. Consequently, a key to river-channel management is how much work will be needed to maintain channel capacity and how the amount of work can be minimized by wise management strategy.

While the channel capacity over a river reach under examination is a primary target of channel management, the degradation of flood-flow capacity attributable to hydraulic retarding effects of in-stream luxuriant vegetation should also be considered. Vegetation has a larger role in fluvial systems where the height of vegetation growth seasonally coincides with flood occurrence, as is the case with Japanese rivers.

The importance of comprehensive policy implementation has been increasingly emphasized. In this context, flood-risk management (FRM) approaches have been examined and adopted particularly in countries with large areas of high residual risk. For example, a government expert group from the following four countries summarizes such approaches being practiced in Japan, Netherlands, United Kingdom, and United States, covering a wide variety of measures developed from flood-disaster risk assessments (Alphen *et al.* 2011). In addition, many experts share the perception that FRM approaches are favourable for addressing adaptation to climate change. In 2008 a policy report on climate-change adaptation strategies was submitted to the Japanese Ministry of Land, Infrastructure, Transport and Tourism (MLIT) (Subcommittee on Climate Change Adaptation for Flood Control, the Panel on Infrastructure Development 2008), and it emphasizes the need of conducting investigations not only from a contracted viewpoint of conventional flood control but from various angles based on risk assessments. The policy report was updated in August 2015 in a far wider perspective.

Gravel-Bed Rivers: Processes and Disasters, First Edition. Edited by Daizo Tsutsumi and Jonathan B. Laronne.

In flood-risk management, flooding probability reduction through structural measures does not necessarily play a dictating role as it once did. In light of high and immediate effectiveness of channel-capacity augmentation, however, this does not mean that efforts to advance engineering of flood-flow capacity management have been undervalued. An engineering framework of river-channel management needs to be improved further, so that river administrators can appropriately realign channel-capacity augmentation and maintenance among various FRM approaches on the basis of how far and how they should rely on that approach from a long-term viewpoint.

Thorne and Evans provide a negative view as to whether channel excavation for flood disaster protection can coexist with sustainable channel maintenance and river environment conservation (Evans *et al.* 2008, p. 146–148). Examination and implementation of 'room for river' projects in the Netherlands called for the idea of 'cyclic rejuvenation' that introduced channel revitalization for enhancing the fluvial environment to aims of ordinary maintenance work (Duel, Baptist, and Penning 2001). Moro, Kazama, and Fukuoka (2011) stress the necessity for planning of channel dredging that will not be accompanied by significant rebounds, showing effectiveness of large-scale dredging made in the Lower Tone River, Japan since 1936. These indicate that prediction of post-improvement channel responses is still a core technology at the stage of comprehensive policy formation.

As its title indicates, this chapter focuses on the role of sediment finer than gravel in determining channel capacity, and accordingly flood-flow capacity, though the channel management discussed in the following sections is still targeted at gravel-bed rivers. There are two reasons for adopting this standpoint.

First, sediment resources generally supply gravel much less than finer sediment such as sand and silt. This does not necessarily mean that the reduction rate of channel capacity due to gravel transport is less dominant, because most of the finer sediment passes through gravel-bed reaches without being significantly deposited under ordinary conditions. However, once a gravel-bed river acquires conditions such that finer sediment can be even partly deposited, the rate of consequent channel-capacity reduction can exceed permissible levels. In this sense, the potential contribution of fine sediment to causing channel-capacity reduction deserves more attention in terms of the channel management for gravel-bed rivers.

Second, finer sediment is capable of vitalizing vegetation growth by forming a surface layer on a gravel bed. Vegetation can grow to have much greater biomass on fine sediment than on gravel.

The linkage between flood-flow capacity management in gravel-bed rivers and finer sediment behaviour has not been sufficiently examined particularly from a practical viewpoint, and the above standpoint is worth taking up as the core subject of this chapter.

The next section provides background information to help understand the following sections. Section 19.3 identifies several important points for framing the channel management of gravel-bed rivers on the basis of observed facts and some principles, which underlie this introduction. Requirements from management sides are emphasized over science-driven motivations. Sections 19.4 and 19.5 deal with representative phenomena concerning fine-sediment behaviour in gravel-bed rivers that resulted in channel-capacity reduction. As concluding remarks, section 19.6 discusses an engineering framework for gravel-bed management, based on the knowledge provided in the previous sections.

19.2 Background Information

This section explains the nomenclature used in the chapter and basic characteristics of Japanese alluvial rivers necessary for understanding the following sections. While these contents were basically

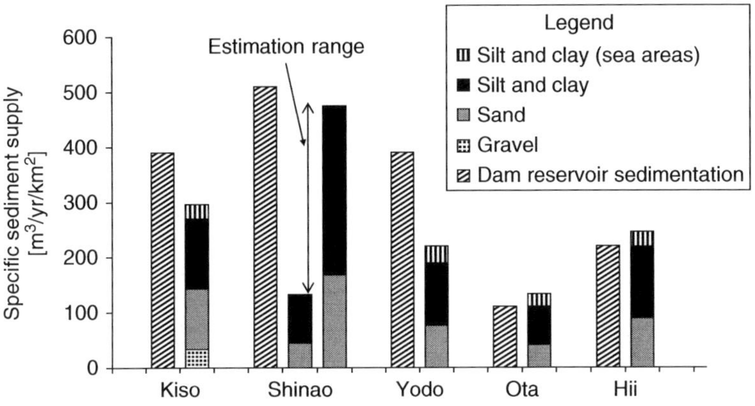

Figure 19.1 Quantity of sediment supplied to alluvial rivers for each grain-size range investigated, based on the bulk volume and grain sizes of their alluvia.

derived from observations and analyses of alluvial rivers in Japan (Fujita 2008), they should be universal for alluvial rivers in the sense that they are beneficial to discussion of a channel management framework.

19.2.1 Grain-size Distribution of Sediment Supplied to Alluvial Rivers

The grain-size distribution shown below is based on bulk volume in sedimentary condition. From analyses of alluvia volumes and grain sizes (Fujita, Yamamoto, and Akabori 1998), the grain-size distribution of sediment supplied to alluvial rivers from sediment sources over approximately the past 10 000 years is estimated to be 50–60% silt and clay, 30–40% sand, and < 10% gravel and larger (Figure 19.1). The grain-size distribution of sediment deposited in dam reservoirs (Figure 19.2), which was calculated from strata along a longitudinal section of each dam reservoir estimated by use of borehole sampling data, typically corresponds well to these percentages (Sakurai, Kashiwai, and Ooguro 2002).

19.2.2 Material *m*, *s*, and *t* for Classification of Riverbed Material

Large differences in grain size will be noticed when walking across a river looking down at its bed. This simple observation leads to the classification of riverbed material into *m* (main), *s* (sub) and *t* (transient) types (Figure 19.3).

Material *m* forms the bottom of the main channel, playing a primary role in governing, for example, the formation of longitudinal bed profiles, riverbed configurations, roughness coefficients of the main channel, scouring depth, and habitat structures composed of riffles and pools.

Material *s*, much finer than material *m*, is formed by the deposition of suspended fine sediment onto material *m* beside the main channel. Its thickness greatly varies longitudinally. Material *t*, much finer than material *m*, is temporary and unstable, being flushed out or reduced sharply by even small floods.

19.2.3 Classification of River Channels Based on Longitudinal Segment Types

The longitudinal profile of major alluvial rivers in Japan can be modelled rather well by the aggregate of several segments, each of which has almost the same bed-material characteristics, longitudinal

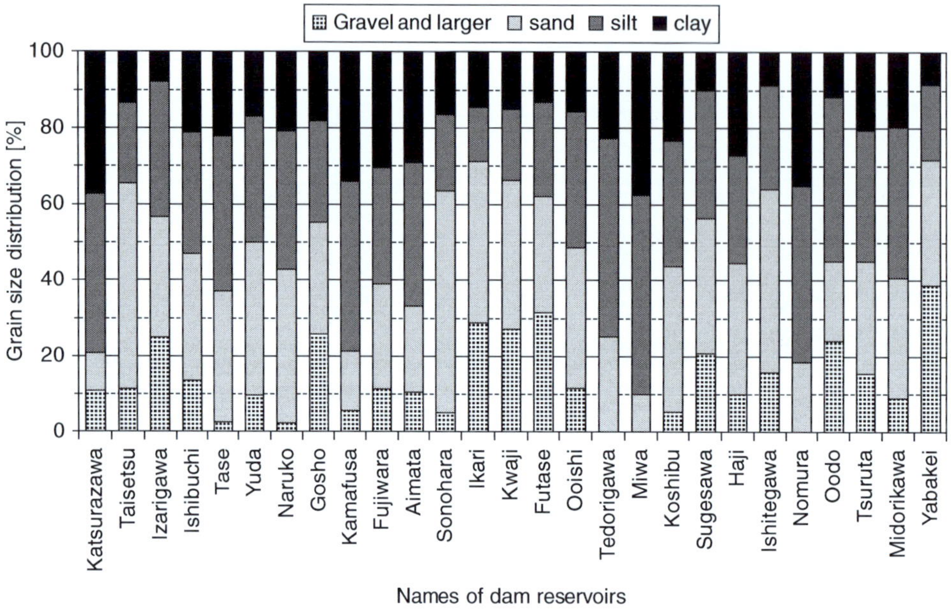

Figure 19.2 Quantity of sediment supplied to alluvial rivers for each grain size range investigated, based on the bulk volume and grain sizes of deposits in dam reservoirs.

gradient, channel dimensions, and hydraulic properties. Divided segments can be classified into several types according to the average grain size of their material *m* or according to the geomorphologic categories of the alluvial fan zone, the natural levee zone, and the delta zone. When emphasizing the former, *G* segments, where material *m* is gravel, and *S* segments, where material *m* is sand, are used. For the latter, segment types of 1 (alluvial fan zones), 2 (natural levee zones), and 3 (delta zones) are used. This segment-based categorization and resulting framework of river engineering were proposed by Yamamoto (1986).

Figure 19.3 illustrates a combination of the two above classifications, which helps systematically grasp the overall image of the spatial distribution of bed material in the segment types. While material *s* layers can become as large as floodplains in natural levee zones, they remain relatively thin in alluvial fan zones.

19.2.4 Exchange Type and the Pass-Through Type for Grain-Size Group Movement

The behaviour of sediment with a wide range of grain sizes can be effectively understood by tracking the movement of each of grain-size group. The manner in which sediment is transported, exchanges with riverbed material, and influences river topography all vary greatly according to grain-size groups.

In addition to the approach of tracking sediment movement for each grain-size group, it is also important to categorize macroscopic movement as either exchange type or pass-through type (Fujita, Yamamoto, and Akabori 1998; Cui and Parker 1998; Egashira *et al.* 2000).

Exchange-type movement is defined as the macroscopic transport of a grain-size group in which the group is transported with significant exchange with material *m*. The flux of sediment transport

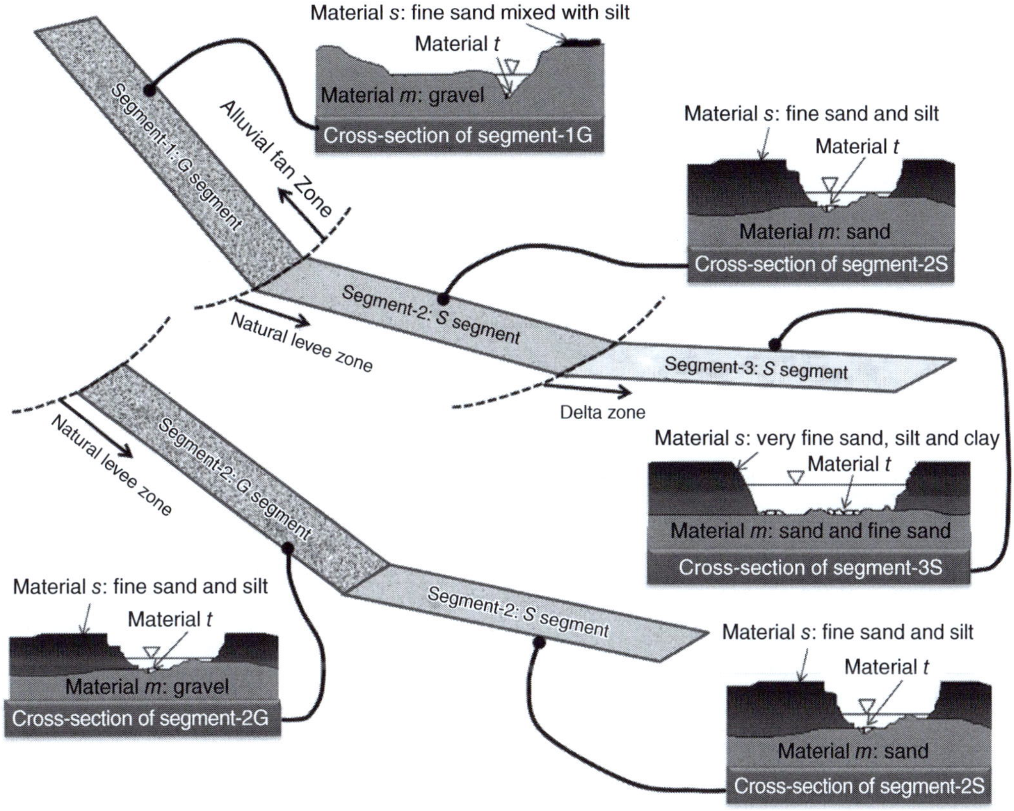

Figure 19.3 Illustration of the macroscopic classification of river channels and bed material.

belonging to this type varies longitudinally, and the transport of material *m* is of the exchange type, which is explained by its definition.

Pass-through type movement is defined as the macroscopic transport in which a particular grain-size group exchanges little with material *m* (at the most, to a degree that fills gaps in the main riverbed material on the surface). Unlike exchange-type movement, sediment transport of this type cannot contribute to riverbed variation. The flux of sediment transport of this type varies longitudinally only a little. Some silt, clay, and finer sand become material *s* and material *t*, but their quantities are much smaller than the total flux, or their deposition is temporary.

19.2.5 Class-A and Class-B River Systems

A river system in Japan that is especially important for the national economy and land conservation is designated as a Class-A river system by the Japanese River Act. It is administered by MLIT (the central government). The others are designated as Class-B river systems, which are administered by prefectural governors (local governments).

19.3 Vital Points to Advance Channel Management Strategy

19.3.1 Sensitivity of Flooding Probability to Riverbed Variation

As mentioned in section 19.1, the primary target in channel-capacity augmentation is to lower the probability of flooding and to maintain this lowered probability. In this context, the importance of river-channel management is governed by the degree to which riverbed variation influences the probability. The essence of this degree is examined below by using a simple method.

When longitudinal variation in flood-flow capacity is considered, local channel excavation may result in flood-disaster risk transfer to a downstream reach. This indicates that strategic excavation can be a tool for longitudinal control of flooding risk, though consensus building becomes a big issue. This chapter, however, does not extend its discussion toward that dimension, focusing instead on the fundamental local relationship between channel capacity and flood conveyance. In Japan channel excavation is designed very carefully so that the overall balance of flooding risk in a river system will be kept in an appropriate manner.

Equation 19.1 expresses a precipitation rate r (mm/h) for a certain duration corresponding to a return period T (years) on the basis of the Fair formula (Fair, Geyer, and Okun 1970, pp. 58–59), where m is a parameter of the formula. Equation 19.2 gives the peak flood discharge Q (m^3/s) caused by the precipitation rate r, where the parameter p is introduced to take into account non-linearity in the rainfall–runoff relationship. Equation 19.3 gives the relation between water depth h and flood discharge Q, which can be obtained by applying the Manning formula to uniform flow conditions in a rectangular and prismatic channel. In these three equations, a quantity with subscript 0 denotes a baseline.

$$\frac{r}{r_0} = \left(\frac{T}{T_0}\right)^m. \tag{19.1}$$

$$\frac{Q}{Q_0} = \left(\frac{r}{r_0} - 1\right) \cdot p + 1. \tag{19.2}$$

$$\frac{h}{h_0} = \left(\frac{Q}{Q_0}\right)^{\frac{3}{5}}. \tag{19.3}$$

By solving the above equations simultaneously,

$$\frac{\Delta T}{T_0} = \left\{\frac{1}{p}\left(\frac{\Delta h}{h_0} + 1\right)^{\frac{5}{3}} + 1 - \frac{1}{p}\right\}^{1/m} - 1 \tag{19.4}$$

is obtained, where $\Delta h = h - h_0$ and $\Delta T = T - T_0$. Furthermore, the first-order Taylor expansion of Equation 19.4 with regard to $\Delta h / h_0$ leads to the following.

$$\left.\frac{\partial\left(\frac{\Delta T}{T_0}\right)}{\partial\left(\frac{\Delta h}{h_0}\right)}\right|_{\frac{\Delta h}{h_0}=0} = \frac{5}{3pm}. \tag{19.5}$$

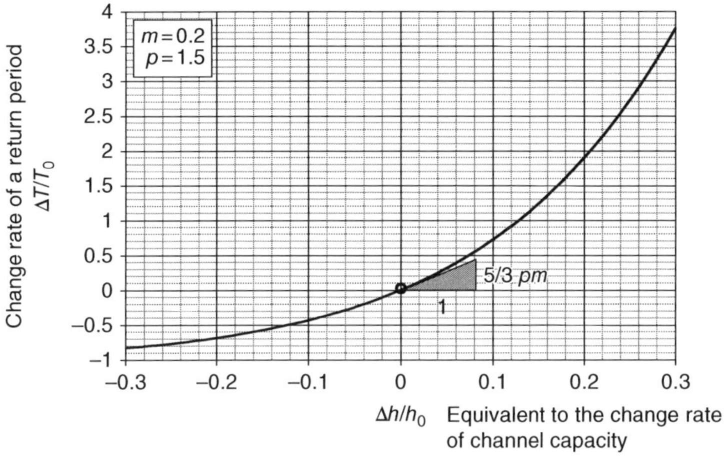

Figure 19.4 Relation between the change rates of channel capacity and a return period of flooding occurrence.

Assuming that the difference in elevation between a levee design stage and a mean riverbed height is h and that Δh is fully attributable to riverbed variation, a fundamental sense concerning the sensitivity of flooding-occurrence probability to channel-capacity change (represented by riverbed variation) can be obtained from Equation 19.4 or 19.5. For example, Figure 19.4 can be drawn with 0.2 and 1.5 given to m and p, respectively: an m value of 0.2 is in the middle of the range of m values calculated from hydrological statistical analyses using data from the Japanese Automated Meteorological Data Acquisition System (Suzuki and Terakawa 1996): a p value of 1.5 is also representative of rainfall–runoff relations calculated for flood-disaster prevention planning of major rivers in Japan. This figure indicates that a $\Delta h/h_0$ change of 10% can result in a $\Delta T/T_0$ change as high as 50%. Hence, it can be concluded that flooding probability is likely to be highly sensitive to changes in channel capacity.

19.3.2 Transition in Total Channel Capacity of Major Reaches of Japanese Rivers

The total channel capacity was calculated by longitudinally integrating the channel capacity for cross sections along all river reaches directly administered by the MLIT. This channel capacity analysis was conducted by the River Division of The National Institute for Land and Infrastructure Management (NILIM), covering data since 1970. The longitudinal intervals of cross-sectional survey data used for the analysis ranged from about 100 m to about 500 m. The channel capacity for each cross section was defined as the area bounded by the riverbed, the levee slopes, and the high-water level. Since a single value (the most recent) was adopted for the high-water level at each cross section, a factor in affecting the total capacity calculation was riverbed variation alone except levee alignment modifications, which were less dominant in the analysis period. Interpolation and extrapolation with respect to time were used to cope with the lack of data.

The upper line in Figure 19.5 shows the 40-year transition in total channel capacity of all G segments (see section 19.2.3) administered by the MLIT. The figure indicates that the total capacity has increased by about 20%. The increase rates in the 1970s and 1980s are larger than those in the subsequent two decades. Considering the relation shown in Figure 19.4, it can be understood that this capacity increase has contributed to the reduction of flooding probability, though $\Delta h/h_0$ in Figure 19.4 cannot be simply replaced by increase rates of the total channel capacity. It should also

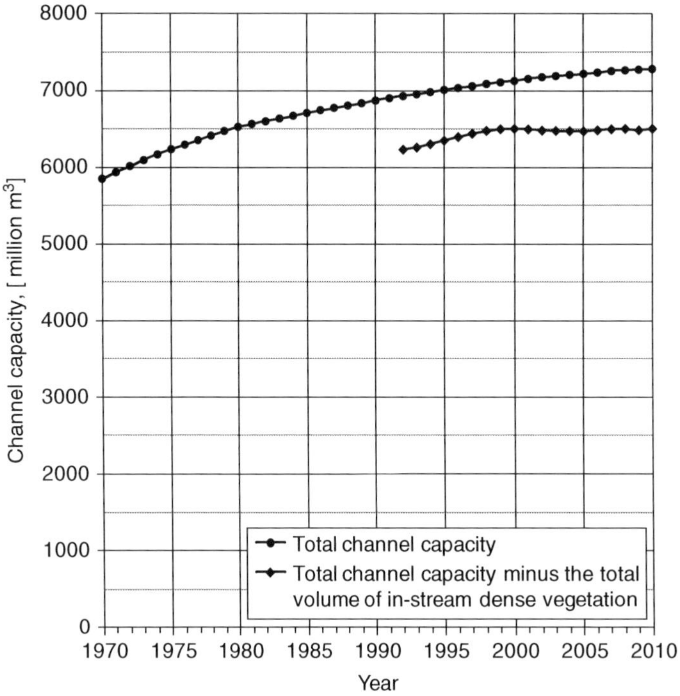

Figure 19.5 Transition in total channel capacity for all G segments directly administered by the MLIT.

be noted that the risk of flood disaster has been lowered in Japan by a great effort through a combination of channel-capacity enlargement, peak-flood-discharge regulation by dam reservoirs, retarding basins, and diversion and cut-off channels, and various non-structural measures. Nevertheless, Figure 19.5 delivers a clear message: wise river channel management for maintaining channel capacity enlarged is a key to flood-disaster prevention and mitigation.

The lower line in Figure 19.5 shows the total channel capacity minus the total volume of in-stream dense vegetation (hereinafter, the total channel capacity with vegetation considered). In the calculation of the latter, the spatial area and height of each luxuriant woodland was measured and summed by using data from the national censuses on river environments that the MLIT has conducted for each river system in a 5 year cycle since the early 1990s. Woodlands are considered dead-water areas in hydraulic calculation for obtaining flood-flow capacity and flood water stages, and the lower line comes to provide additional information on flood-flow capacity.

While the total channel capacity with vegetation considered had increased at the same rate as the total channel capacity until the late 1990s, since then it has remained nearly constant. This suggests that woodland expansion, presumably becoming noticeable around the late 1990s, retarded the increase in total channel capacity with vegetation considered to some extent. As areas with dense vegetation are usually on higher riverbeds, meaning shallower water depth during a flood, the influence of channel capacity reduction due to vegetation growth on flood-flow capacity is probably not as large as that due to riverbed variation.

The above tendency emphasizes the importance of regarding channel capacities both with and without vegetation considered as targets of channel management for flood-disaster reduction. Accordingly, woodland expansion should be examined along with geomorphic changes leading to channel-capacity reduction, though the difference in their influence on flood-flow capacity should be taken into account.

19.3.3 Potential Impacts of Sediment Supply on Channel Capacity

In order to obtain a primary sense of the impacts, rough estimates have been made. Table 19.1 summarizes the steps of this estimation, which focuses on all the *G* segments of every Class-A river system. There are 109 such rivers. In the estimation, the calculations for the targeted river systems were summed and treated as a whole. Discussion of the estimation results is therefore limited to a macroscopic perspective based on gross estimates.

Complementary explanation on the table is as follows. The total river basin area of the Class-A rivers (the first row) occupies about 65% of the total national land area. The assumption of overall sediment-yield intensity (the second row) was based on analyses of sediment volume deposited in dam reservoirs (Okano *et al.* 2004; Fujita *et al.* 2009, pp. 3/11–16). In the calculation of r_{dc} (the fourth row), duplication of dam catchments on each river basin was eliminated. Values of r_{gc} and r_{sf} (the fifth and sixth rows) were assumed, with reference to the results introduced in section 19.2.1. A r_{ies} value of 0.7 (the seventh row) was derived from the assumption that half the silt and finer material supplied from sediment sources passes entirely through without causing any substantial geomorphic change in reaches of alluvial fan zones and natural levee zones. This assumption is based on the observation that significant deposits finer than 0.01 mm are not likely to be found except in delta zones, while corresponding sediment ratios of wash-load usually range from 30% to 60% according to recent observations for Japanese rivers (Yamamoto 1986; Shigeta *et al.* 2008; Mochizuki, Amano, and Fujita 2011). The total channel capacity as of 2010 (the eighth row) is from Figure 19.5. Since several assumed values were used to produce the estimates presented in the bottom two rows (δ_{gc} and δ_{sf}'), and since they were calculated on a mean annual basis, δ_{gc} and δ_{sf}' are of order-of-magnitude estimations.

Taking these limitations into account, we proceed to examine the estimate results. The estimated δ_{gc} value is of the order of 0.001 or 0.1%. According to this estimate, the decrease ratio to the total channel capacity will be around 10% in 100 years. However, the estimated δ_{sf}' value is 10 times larger, giving a ratio of 100% after 100 years.

Before discussing what these two estimates mean in the context of channel management, we need to understand the difference in 'possibility' between δ_{gc} and δ_{sf}'. While most gravel and coarser material supplied to a *G* segment remain in the segment unless it flows directly into the sea (the minority case, seen in just 21 of the 109 Class-A river systems), most of the sand and finer material will pass through the segment as mentioned in section 19.2.4. To make this difference clear, a description of the 'Theoretically possible maximum' is included in the explanation of δ_{sf}' in Table 19.1.

19.3.4 Strategies for Channel Capacity Management

In light of the above, several points are important for furthering strategic channel management.

First, gravel and coarser sediment supply is not likely to cause highly serious problems in maintaining the gross channel capacity. Therefore, preparing countermeasures against abrupt increases in sediment supply induced by huge impacts on sediment sources are more beneficial than long-term gross gravel budget control. This point is founded on the fact that Japanese river administrators had already achieved the channel capacity enlargement by a mean annual increase rate of about 0.5% for the 40 years (Figure 19.5), which is much larger than the estimated δ_{gc} value.

Second, river administrators need to pay attention to the possibility of scenarios in which sand or finer sediment deposition and a resulting substantial channel capacity reduction occur. As the estimated δ_{sf}' value is larger than even the achieved increase rate, they will be confronted with significant

Table 19.1 Estimates of potential impacts of sediment supply on total channel capacity.

	Notation	Unit		Source	Expression
Total basin area of the Class-A rivers	A_{rb}	km^2	2.4E + 05	Data	
Overall sediment yield intensity	I_{sy}	mm/yr	0.8	Assumed	
Total sediment supply per year	S_{ts}	m^3/yr	1.9E + 08	Calculated	$A_{rb} \cdot I_{sy}$
Ratio of the total dam catchment area to the total basin area of the Class-A rivers	r_{dc}		0.23	Data	
Ratio of gravel and coarser sediment supply to the total	r_{gc}		0.05	Assumed	
Ratio of sand and finer sediment supply to the total	r_{sf}		0.9	Assumed	
Proportion of effective sediment in causing a significant geomorphic change in the total sand and finer sediment supply	r_{ies}		0.7	Assumed	
Total channel capacity as of 2010	V_{tcc}	m^3	7.3E + 09	Data	
Overall decrease rate of the total channel capacity due to gravel and coarser sediment supply	δ_{gc}	1/yr	0.10%	Calculated	$S_{ts} \cdot r_{gc} \cdot (1 - r_{dc})/V_{tcc}$
Theoretically possible maximum rate of overall decrease in the total channel capacity due to sand and finer sediment supply	δ_{sf}'	1/yr	1.3%	Calculated	$S_{ts} \cdot r_{sf} \cdot r_{ies} \cdot (1 - r_{dc})/V_{tcc}$

difficulties in maintaining the enlarged channel capacity or further enlargement if such a scenario is realized by chance.

Finally, administrators also need to pay attention to sediment-related scenarios allowing dense vegetation to grow and expand actively in gravel-bed rivers, because channel capacity with vegetation considered should be a target of channel management, as noted in section 19.3.2.

There are two notable scenarios related to the second and the third points that deserve attention to substantiate the above discussion. One is the role of finer sediment behaviour in the expansion of dense vegetation, and the other is accretion of floodplains by finer sediment deposition. The next two sections introduce these.

19.4 Role of Finer Sediment in the Expansion of Dense Vegetation Areas in Segment-1G Reaches

19.4.1 Deposition of a Top Fine-Sediment Layer as an Essential Process

Significant and monotonous expansion of dense vegetation areas has been prevalent in gravel-bed rivers in Japan since around the mid-1980s. This has called for studies, because such expansion is a factor in degrading flood discharge capacity and can impair in-stream ecosystems.

To tackle this problem, various studies have been carried out. The following are results from a field study on the Nagata Bridge district of the Tama River located in Tokyo Metropolis, Japan (Fujita *et al.*

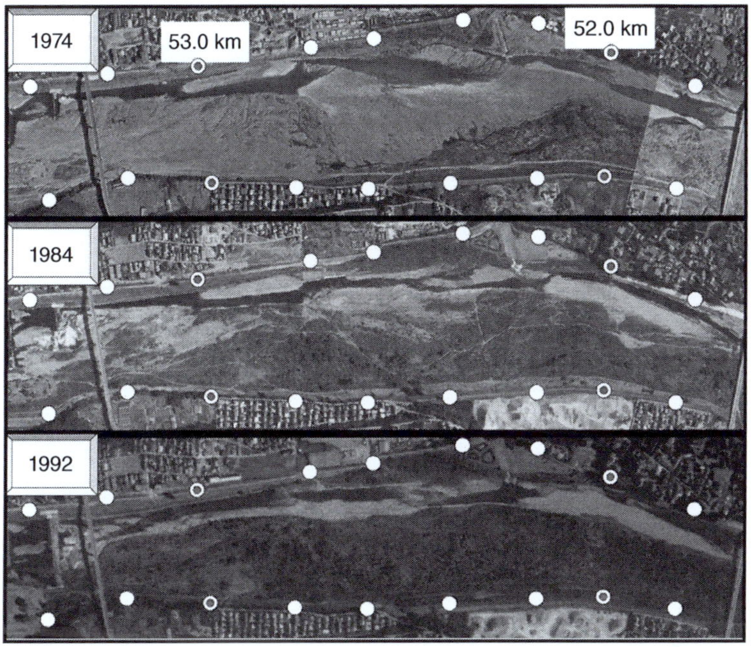

Figure 19.6 Expansion of dense vegetation areas in the Nagata Bridge district of the Tama River (upstream is to the left).

2003). The river reach of the district, classified as Segment-1G (Figure 19.3), has a slope of 1/220, an average bed material size of 35 mm, and a width of 275 m at a distance of about 52 km from the river mouth. Bare gravel areas, which were dominant as of 1974 (after the 1974 flood), were replaced by stable vegetation areas by the early 1990s (Figure 19.6). In the meantime the river reach changed to a compound cross-section type, forming narrow low-water and wide high-water channels. After the 1974 flood, only the 1981, 1982, and 1983 floods made flows on the high-water channel, which is illustrated in Figure 19.7. Figure 19.8 indicates that the expansion of stable vegetation areas was caused by the shift from a Type 1 to a Type 3 surface. Type 1 surfaces are bare or very sparse herbaceous communities; Type 2 surfaces have sparse plants such as *Phragmites japonica* or *Salix integra*; and Type 3 surfaces have dense vegetation cover such as *Miscanthus sinensis, Miscanthus sacchariflorus, Ambrosia trifida* or *Robinia pseudo-acacia*. It is obvious from Figure 19.8 that the three floods in the early 1980s triggered the expansion, instead of reviving bare gravel areas.

The scenario of the Nagata Bridge district case was identified as follows: according to the scenario, the expansion would not have occurred if any of the following four processes had been missing:

1) The vast bare gravel land (Type 1) formed by the 1974 flood had surface conditions allowing sparse vegetation to slowly grow. Consequently, by 1981 the gravel surface had been thinly covered with 'pioneer plant communities' of Type 2.
2) The geomorphic change to the compound cross-section type shown in Figure 19.7 played an important role in ensuring that the floods that occurred after 1974 did not disturb the higher bed.
3) The 1981, 1982, and 1983 floods were large enough to supply sufficient fine sediment all over the channel. The three floods were therefore able to widely form top fine-sediment layers on the higher

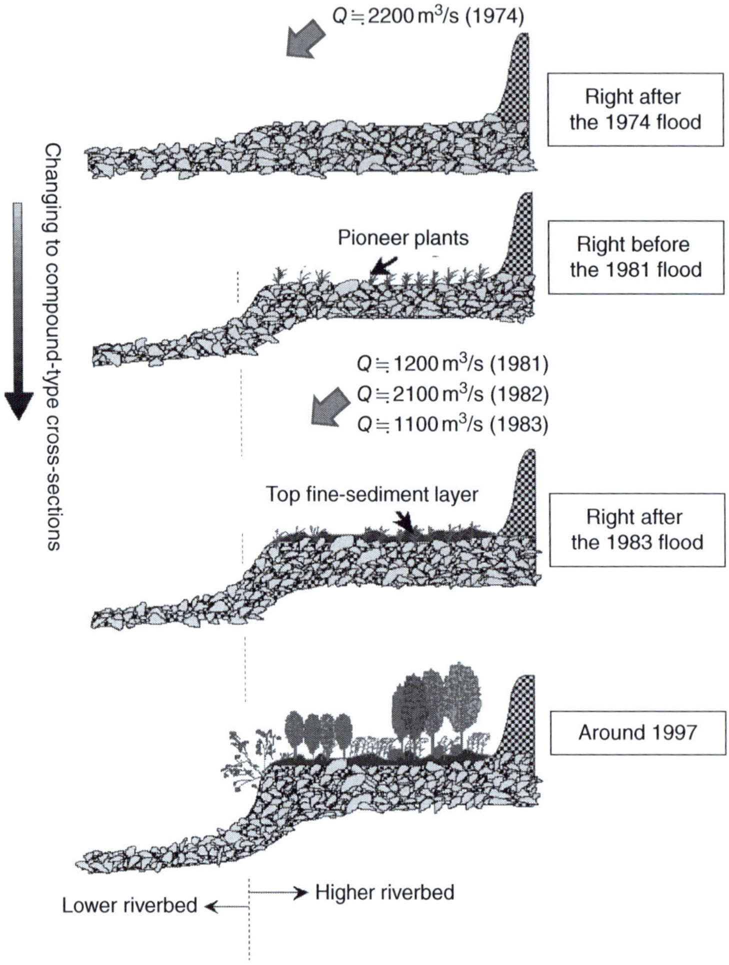

Figure 19.7 Schematic illustration of the processes of rapid expansion of dense vegetation areas in the Nagata Bridge district of the Tama River (cross-sectional views downstream).

gravel-bed, with the help of the flow retarding effect of the pioneer plants. The illustration in Figure 19.9 defines the layer. The grain sizes mainly range from silt to fine sand.

4) Since then Type 3 vegetation has been the primary contributor to the vegetation area expansion, rapidly intruding because the top fine-sediment layers were far better for vegetation growth than bare-gravel beds.

It is worth noting that a physical process played an indispensable role in the expansion of dense vegetation areas. In general, a top fine-sediment layer formed on a gravel-bed reach of Segment-1G is only a few tens of centimetres thick at the most, and does not substantially raise the riverbed. However, it has an accelerating effect on the prevalence of vegetation. Since physical processes have closer relations with channel design and management (including sediment transport control) than do biological processes, it is important to identify their roles in targeted phenomena even when they are seemingly related to biological processes such as vegetation growth.

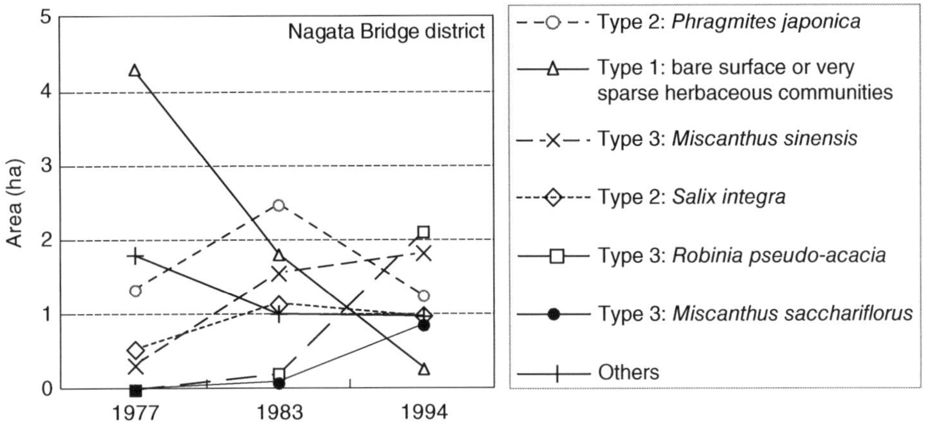

Figure 19.8 Transition in the total area of each type of surface state.

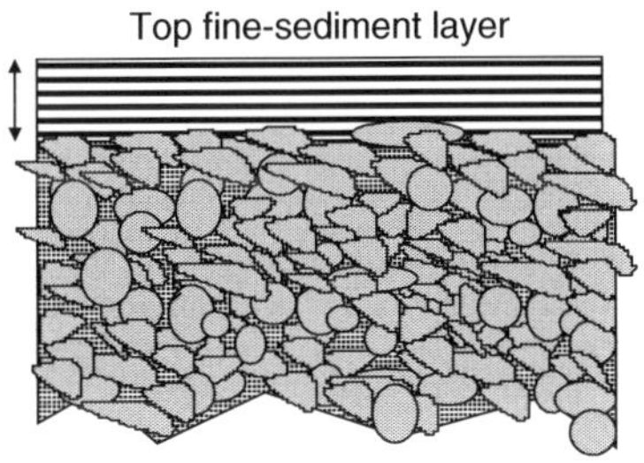

Figure 19.9 Top fine-sediment layer.

19.4.2 Simulation of the Expansion and Extinction of Stable Vegetation Areas

To make good use of the foregoing scenario for channel management practice, river administrators need as an engineering tool a simulation method for calculating change in dense vegetation areas. A simple, but effective, model for such simulation was developed with the following four parts (Fujita *et al.* 2003) (Figure 19.10):

1) The geometry of a basal gravel-bed at the time of interest is given.
2) Pioneer plant communities have grown on the bare gravel surface when a calm (non-disturbed) period exceeds T_{pr} (years).
3) With the hydraulic retarding effect of the pioneer plants, floods form top fine-sediment layers originating from pass-through type sediment supplied as wash load. When their total thickness D_{ts} exceeds a threshold D_c, the bed turns into a stable, dense vegetation area.
4) A flood with τ_* (non-dimensional shear stress for material m) exceeding a threshold τ_{*ce} restores the area to a bare gravel surface by disturbing the riverbed.

Thus, the overall vigour of the vegetation is determined through the balance between Parts 2 and 3 versus Part 4.

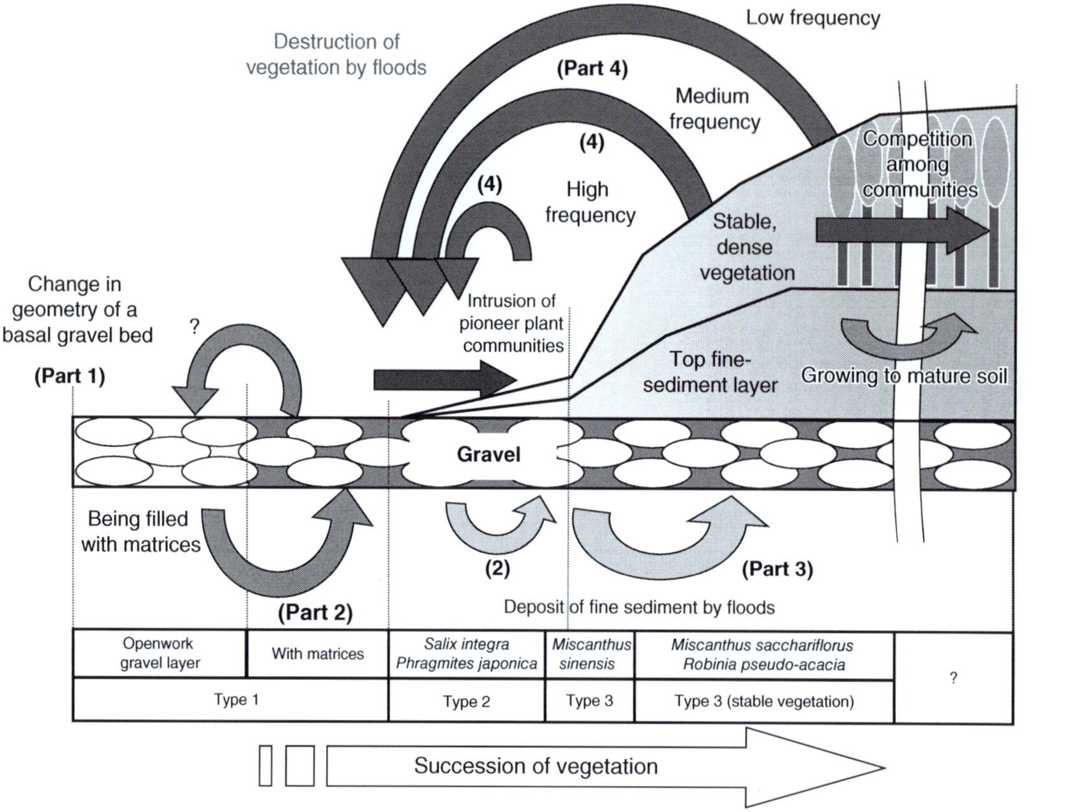

Figure 19.10 Framework of the simulation model to calculate surface state changes. (See insert for color representation of this figure.)

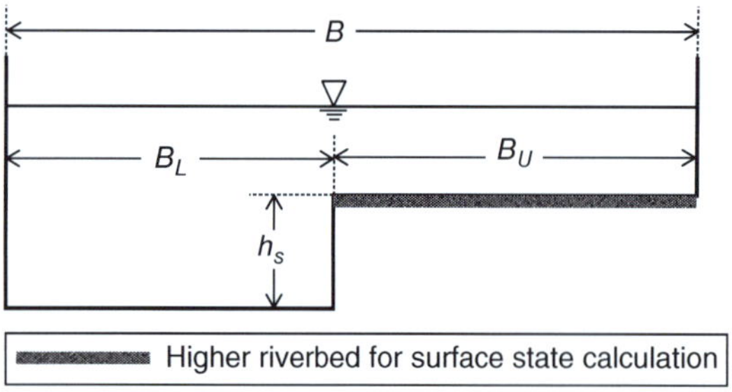

Figure 19.11 Channel cross-section for the prototype simulation model.

For simplicity, a prismatic, compound-type channel composed of two rectangular cross-sections (Figure 19.11) was adopted for a prototype simulation model so that uniform flow calculations would be applicable. Given the geometry of a target river (B, B_U, B_L, h_s), channel slope I_b, time series of flood discharges Q, Manning's roughness coefficients for the two beds with different elevation n_U and n_L, the load of fine sediment supply Q_s expressed as proportional to Q to the power of p_s, the representative grain sizes of material m D_R and of top fine-sediment layers d_{ts}, and other parameters (T_{pr}, D_c, τ_{*ce}), the simulation model provides information on the surface states of the higher bed. The validity of the model was successfully confirmed for the Nagata Bridge district case.

19.4.3 Use of the Simulation Model for Channel Management Strategy

The diagram shown in Figure 19.12 was devised to make practical use of the simulation. In the diagram, the vertical axis is the stable vegetation duration ratio (defined as the ratio of the total duration of stable, dense vegetation to the calculation period – 1000 years), and the horizontal axis is the surface-state alternation cycle (defined as the average interval of stable vegetation appearance for the calculation period). Plotting calculated results as a diagram shows overall riverbed characteristics in terms of the Types 1, 2, and 3 surface states.

An instructive example of how to use the diagram is as follows. The large circle in Figure 19.12 indicates the calculation result for the baseline case with the following conditions given: the time series of flood discharges was generated by the use of the same stochastic parameters as the flood occurrence in the Nagata Bridge district: $B_L = 70$ m, $B_U = 200$ m, $h_s = 2.5$ m, $T_{pr} = 3$ yr, and $d_{ts} = 0.15$ mm: $Q_s = 2 \times 10^{-7} \cdot Q^2$, where the unit of Q and Q_s is m^3/s: $D_c = 0.03$ m, $\tau_{*ce} = 0.05$, $D_R = 0.06$ m, and the given geometry of the basal gravel-bed was kept unchanged. These conditions correspond roughly to those of the Nagata Bridge district. Each line spreading out from the large circle shows a series of calculated results for the cases where the parameter corresponding to each line alone deviates from the value of the baseline case.

Figure 19.12 shows how each channel condition influences overall characteristics of a riverbed. Conversely, this information can be used for finding the way to regulate those conditions so that desired characteristics can be obtained.

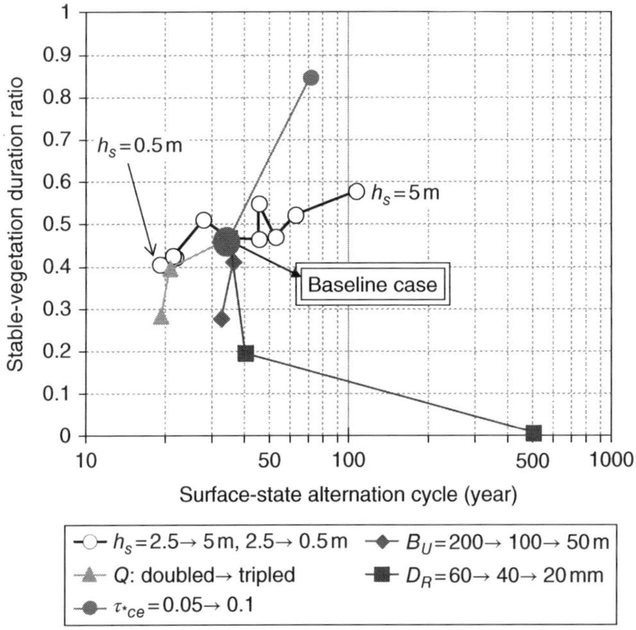

Figure 19.12 Diagram for characterizing a riverbed with respect to stable vegetation prevalence.

Let us examine this possibility further. With regard to the long-term reduction of efforts for vegetation control, it is desirable to establish or revive a sound disturbance system under which alternation of bare gravel-beds and vegetation areas is frequent enough to prevent Type 3 vegetation from growing large and dense. Such a system is also favourable for the conservation of ecosystems indigenous to a gravel-bed river because it is likely to give riverbeds spatial non-uniformity to maintain sufficient bare gravel surfaces with different ages. Thus, the surface-state alteration cycle shown in Figure 19.12 is regarded as a key index.

That figure shows that h_s (relative height; see Figure 19.11) slightly increases the stable-vegetation-duration ratio and sharply increases the surface-state alternation cycle. This means that the riverbed surface is altered much less frequently. Although there are other factors capable of changing the cycle, in general they cannot be practically modified by human intervention. The relative height h_s is therefore considered to be the most important factor.

As is demonstrated above, the simulation model has the potential to provide instructive information on how to design a river channel that does not require much dependence on laborious maintenance work, including the direct and frequent removal of vegetation.

After examination of the prototype simulation model, it was extended to a two-dimensional calculation-based model (Hattori, Fukushima, and Takeuchi 2013). In the extended model the riverbed variation caused by the transport of material m can be calculated as well as the surface states. Moreover, Type 4 is added as another surface state that represents a more progressed succession than Type 3, and the growth curve of woods is introduced for Type 4 calculation. This extension has enabled us to make more detailed and precise examination about river channel modifications and subsequent management plans.

19.5 Floodplain Accretion by Finer Sediment Deposition and Resulting Channel Narrowing in Segment-2G Reaches

19.5.1 Channel Narrowing in the Sendai and the Powder Rivers

In this chapter 'channel narrowing' is defined as a reduction of channel width that occurs over a reach with no sharp bend more than ten times as long as the original width. The author raises two cases of typical channel narrowing (Fujita *et al.* 1996): in the Sendai River located in Kagoshima Prefecture, Japan and in the Powder River in Montana and Wyoming in the United States. The latter is based on the long-term surveys of cross-sectional changes in a 93 km long river reach by Moody and Meade (1990, 2014). The target reaches of both rivers belong to Segment-2G (Figure 19.3). Outlines of the two cases are described below.

19.5.1.1 The Sendai River Case

With cut-off construction completed in a reach of the Sendai River in 1979, the channel was widened to about twice its original width. The longitudinal slope of the reach became about 1/1300. The mean grain size of the bed material (material m) is 2–10 mm (gravel).

Figure 19.13 shows the observed cross-sectional changes of a representative section (68.8 km from the river mouth) in the reach. The figure shows that the new bank and floodplain grew on the right bank, and that it reached the height of the remaining floodplain only 10 years after the cut-off construction. The surfaces of the new banks and floodplains were covered with luxuriant vegetation. Figure 19.14 shows strata along a longitudinal trench crossing the section at the highest point of the new floodplain as of 1990. While the new floodplain deposits were mostly fine sand, very fine sand,

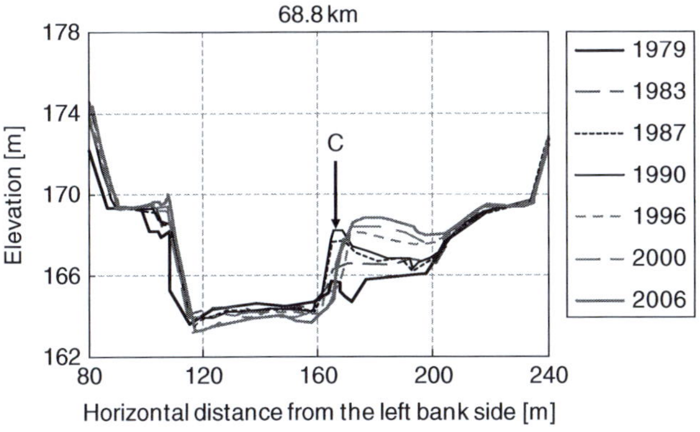

Figure 19.13 Observed cross-section changes in the 68.8-km section of the Sendai River.

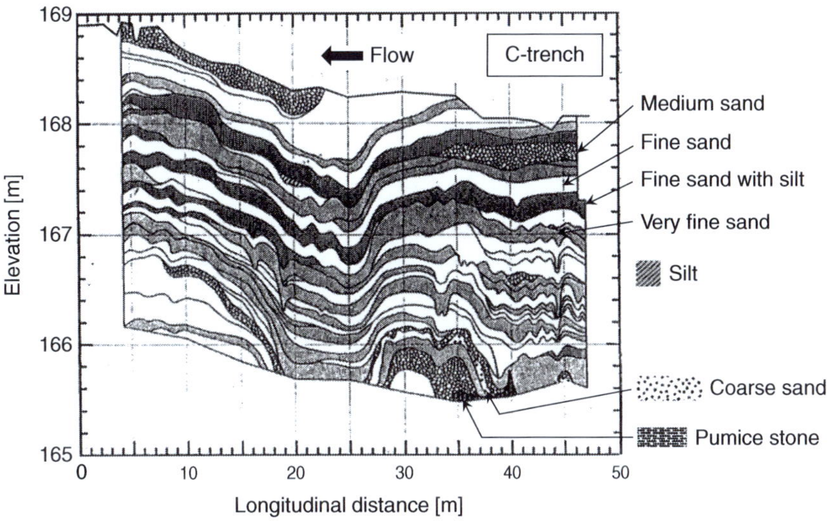

Figure 19.14 Strata of the new floodplain along a longitudinal trench starting downstream from point C in the 68.8-km section shown in Figure 19.13.

and fine sand mixed with silt, the bed material (material m) was fine gravel. This indicates that material for the new floodplain was material s carried by pass-through-type transport. The elevation of the foot of the new floodplain was within the normal water-stage range for the section.

19.5.1.2 The Powder River Case

In 1978 the Powder River, a right-bank tributary of the Yellowstone River, experienced an extremely large flood, and its banks were eroded severely in many reaches, where their channel widths were consequently substantially widened. The channel had been subject to normal discharges since the event. Therefore, channel changes after the 1978 flood are considered as a response to forced channel widening, similar to the Sendai River case. The overall longitudinal slope in the surveyed reach is

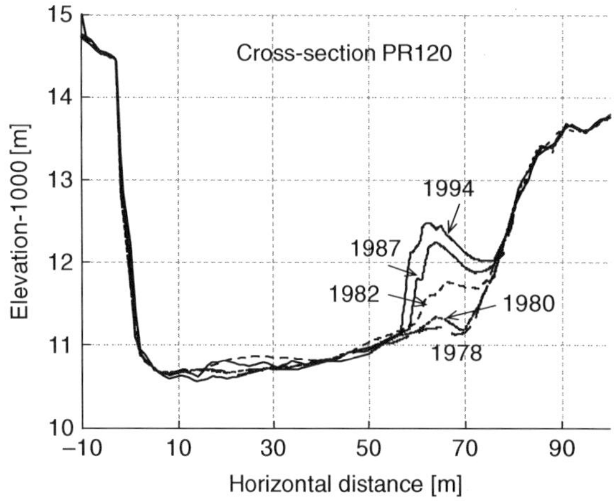

Figure 19.15 Observed cross-section changes in the PR120 section of the Powder River.

around 1/900. Its stable channel width is 45–55 m on average. No levee had been constructed in the reach. After the large flood, new banks and floodplains developed in several reaches, which led to width reduction. The author now focuses on the approximately 800 m long reach including Cross Section PR120, 120 km from the upstream end of the survey.

Figure 19.15 shows changes in the cross section after the 1978 flood. Notably, the new bank and floodplain had grown on the right-hand bank and consequently the channel width reduced remarkably. The horizontal position of the feet of the new bank had not moved significantly since the very early stage of the floodplain formation. The figure also indicates that the change in the riverbed height at the bottom of the cross section is negligible compared to that of the new bank and floodplain zone, as is the case of the Sendai River in Figure 19.13.

Figure 19.16 shows strata and material of the new bank and floodplain with the expression of vegetation cover. The figure indicates that most strata were made of fine or very fine sand mixed with silt. Since the average grain size of the bed material (material *m*) was 20–30 mm (gravel), the new floodplain deposits were much finer than the bed material and the floodplain deposits corresponded to material *s*. The new floodplain above the normal water stages was fully covered with vegetation, and there were dense roots inside the bank and floodplain.

19.5.1.3 Commonalities Between the Two Cases

There are many commonalities between the channel-narrowing cases described above. Especially important, with a direct relation to the second point raised in section 19.3.4, is that the narrowing was caused by the rapid growth of new floodplains and banks and was almost completed within 10 years. Also, they were formed through deposition of sediment as fine as the wash load (i.e., categorized as pass-through type; see section 19.2.4).

19.5.2 Mechanism of the Channel Narrowing by Fine Sediment Deposition

This section demonstrates that the growth of vegetation in part of the original riverbeds was an essential condition for the subsequent floodplain formation and the resultant channel narrowing (Fujita

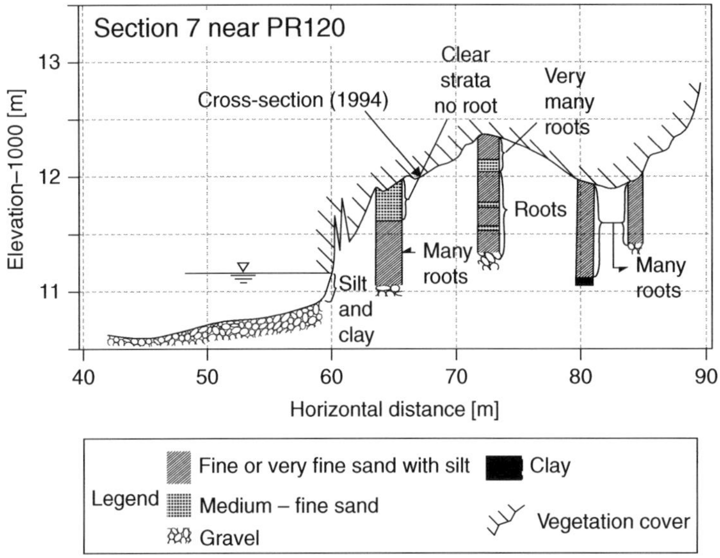

Figure 19.16 Strata of the new floodplain in the PR120 section of the Powder River.

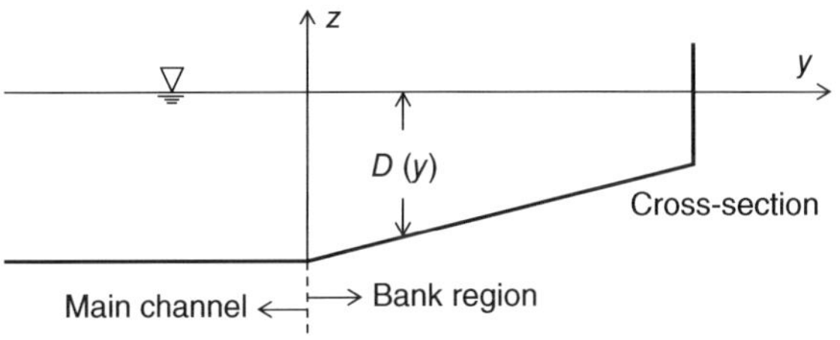

Figure 19.17 Model river channel for the calculation of silt and very fine sand deposition.

et al. 1996). First, to show the importance of vegetation for the entire process, the author examines whether fine sediment deposition can take place without any vegetation on a riverbed.

19.5.2.1 Hydraulic Conditions for Fine-Sediment Deposition

The prismatic model channel shown in Figure 19.17 was created, being composed of a rectangular cross section with an added gently sloped river bank with a transverse slope of 1:35. The added bank enables lateral transport of suspended load toward the right bank due to lateral diffusion induced by the change in water depth. This channel roughly modelled the cross-sectional shape of PR120 prior to the floodplain formation. Hereinafter, the added bank and the flow area above it are referred to collectively as 'bank region'.

The following equations were used to calculate sediment concentration and deposit rates:

$$\bar{c} = \frac{1}{D}\int_{-D}^{0} c \cdot dz, \tag{19.6}$$

$$\varepsilon_y \cdot \frac{d^2}{dy^2}(\bar{c}D) + \frac{d\varepsilon_y}{dy}\frac{d}{dy}(\bar{c}D) - \varepsilon_y \frac{dD}{dy}\frac{dc_b}{dy} - \left(\frac{d\varepsilon_y}{dy}\frac{dD}{dy} + \varepsilon_y \frac{d^2 D}{dy^2}\right) c_b + \left[q_{wsub}\sqrt{1 + \left(\frac{dD}{dy}\right)^2} - c_b w_0 \right] = 0, \tag{19.7}$$

$$\bar{c} = c_b \frac{1}{6Z}\{1 - \exp(-6Z)\}, \tag{19.8}$$

$$Z = \frac{w_0}{\kappa u_*}. \tag{19.9}$$

Figure 19.17 shows the coordinates, where the water surface is set as $z = 0$: c is sediment concentration at an arbitrary point; D is water depth; $\bar{c}$ is vertically averaged sediment concentration; ε_y is a diffusion coefficient with respect to the transverse direction; c_b is sediment concentration at the bottom surface; w_0 is settling velocity of fine sediment; κ is Karman's constant; and u_* is friction velocity for a riverbed. In this, q_{wsub} is defined as the pick-up rate, which is the vertical upward (volumetric) flux of suspended sediment rising from unit riverbed area per unit time; the value is generally a function of w_0, u_*, and d_{ws} (grain size of fine sediment to be deposited).

Equation 19.7 is obtained by vertically integrating simplified diffusion equations with regard to the sediment concentration in the y–z surface (Parker 1978; Izumi and Ikeda 1991). This simplification is based on the hypothesis that the flow and sediment concentration do not change in the longitudinal direction.

The following equations were used for ε_y and u_*:

$$\varepsilon_y = \frac{\beta' \kappa u_* D}{6}, \tag{19.10}$$

$$u_* = \sqrt{gDI_b}. \tag{19.11}$$

Here, I_b is a longitudinal slope and β' was set to 2.

The handling of the term in brackets in Equation 19.7 distinguishes this calculation from bed-material-load-based calculation. The value of the term multiplied by −1 corresponds to a rate of deposition. Since pass-through-type fine material carried through the main channel and its rate of deposition are now being focused on, in cases where the value of the term in brackets would be greater than 0, this term is considered to be equal to 0. Thus, while q_{wsub} was calculated by substituting u_* in the Itakura–Kishi formula (Itakura and Kishi 1980), this conditional expression was also adopted.

Solving Equation 19.7 by substitution of Equations 19.8 and 19.9 gives $\bar{c}$ and the rate of deposition. The boundary condition on the right-hand side vertical wall can be derived; since the flux of suspended sediment in the y-direction must be zero there, the following differential equation must be satisfied:

$$-\varepsilon_y \cdot D\frac{d\bar{c}}{dy} + \varepsilon_y \frac{dD}{dy}(c_b - \bar{c}) = 0. \tag{19.12}$$

Equation 19.12 must also be satisfied where the sediment concentration falls to zero before it reaches the wall.

The boundary condition at $y = 0$ (the boundary between the main channel and the bank region) was given as follows. First, a discharge Q flowing through the main channel with a width of 40 m and a velocity coefficient of 12 was computed by using the uniform-flow calculation. Q_s (fine sediment load) was computed from the discharge Q and the rating curve calibrated with data for the Powder River. Then, Q_s was transformed into $\bar{c}$ to be given as the boundary condition. In this way, the behaviour of

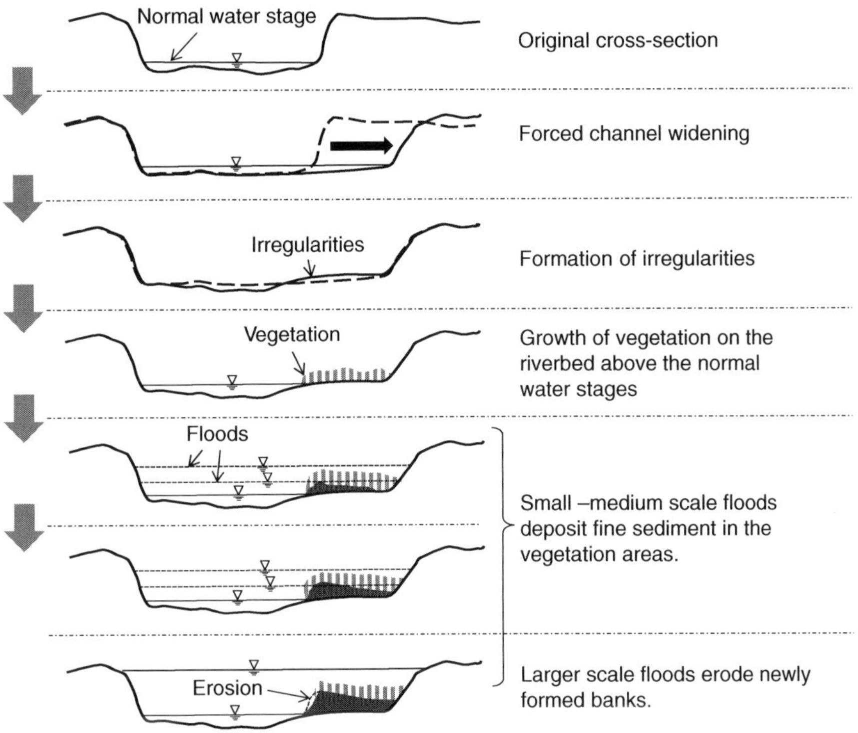

Figure 19.18 Scenario of the evolution of channel narrowing.

fine sediment in the bank region can be related not only to local hydraulic conditions but also to the total transport rate of fine sediment carried directly from sediment sources.

Values for d_{ws} were set as 0.1 mm and 0.05 mm. Given the water depths and the duration that were able to contribute to the floodplain formation at its early stages, the calculation by the foregoing steps revealed that at any water depth the calculated deposit rates were far lower than the volumes surveyed in the Powder River. This occurred because the sediment grain sizes were too small for the bracket term of Equation 19.7 to become negative, even at locations adjacent to the right-hand side water edge.

After the failure in identifying any target hydraulic condition from the model channel, vegetation was introduced to the bank region. It then eventually became possible to reproduce the initial stages of the floodplain formation. The explanation for this is that the introduced vegetation depresses the friction velocity at its area to low enough for fine sediment deposition to overcome the pick-up rate, whether the vegetation is submerged or non-submerged.

19.5.2.2 Scenario of the Channel Narrowing

By integrating all the observations and results presented above into a probable scenario, the following can be obtained (see also Figure 19.18).

1) The widened riverbed was much flatter than the slopes of the riverbanks, but it had irregularity to the extent that part of the bed was exposed above the normal water stages.
2) For several years until floodplain formation began, there had been no floods capable of removing pioneer vegetation on the bed. In the meantime, the riverbed above the normal water stages had

been covered with vegetation. This vegetation had a permeability value low enough to deposit fine sediment in both standing and fallen vegetation states.

3) Small-scale floods took place under these circumstances, and fine sediment that would not have been deposited without vegetation was deposited on the vegetation areas due to their retarding effect.
4) The fine sediment deposits included mud material. As the vegetation was replaced by other kinds more suited to the new site environment, highly dense vegetation continued to grow on the floodplain during the formation process.
5) The new floodplain formation was accelerated by the difference in velocity between the original riverbed and the new floodplain due to the elevation gap and the retarding effect of the vegetation.
6) While the floodplain formation continued, floods came to occasionally erode the bank such that the channel width was gradually adjusted to an equilibrium state with regard to a representative hydraulic power of floods. The cohesiveness of the new floodplain played a considerable role in the erosion process as a resistance factor.

19.5.3 Development into Planar Two-Dimensional Calculation of Floodplain Accretion

This section introduces an outline of the calculation founded on the mechanism and the scenario demonstrated in the previous section (Oonuma *et al.* 2007).

The central idea is as follows. While the transport of fine sediment is incorporated into the ordinary two-dimensional suspended-load calculation, the bracketed term in Equation 19.7 with regard to the fine sediment is set to 0 outside vegetation areas. Inside the vegetation areas, q_{wsub} in the term is expressed by Equation 19.13 to model the entrapment effect of vegetation hindering fine sediment pickup at the bed surface.

$$q_{wsub} = w_0 c_b (1-\sigma). \tag{19.13}$$

Here, σ is a variable ranging from 0 to 1 to quantify the efficiency of the entrapment effect. Its value is determined according to the density of vegetation cover.

Figure 19.19 illustrates the framework of the planar two-dimensional calculation with the above idea implemented. According to the channel-narrowing scenario, calculations of floodplain accretion need to include biological processes in such a manner that where, when, and how densely vegetation will grow on a riverbed are incorporated. In this context, the calculation process for vegetation conditions presented inside the thick solid rectangle in the figure is an important difference from the ordinary planar two-dimensional calculation.

The calculation method was applied to a reach of the Onga River in Fukuoka Prefecture, Japan. In addition to geomorphic changes, changes in vegetation area accompanying floodplain accretion for the 11 years beginning in 1994 were calculated for the 4-km reach, the downstream end of which is 28.8 km from the river mouth. The year 1994 was immediately after the completion of low-water channel widening. The 11 years can therefore be considered as the period of channel response to the widening. Figure 19.20 shows the results. Comparing Figures 19.20a and 19.20b clearly shows the ontogeny and expansion of new floodplains in response to the widening. The calculation result shown in Figure 19.20c has successfully reproduced the response. The floodplain formation downstream of the 30.8-km section was much less active. This difference is attributable to the most downstream weir, which raises low-water stages and hinders vegetation growth. This shows the importance of incorporating the thick solid rectangle part in Figure 19.19 to the ordinary two-dimensional calculation.

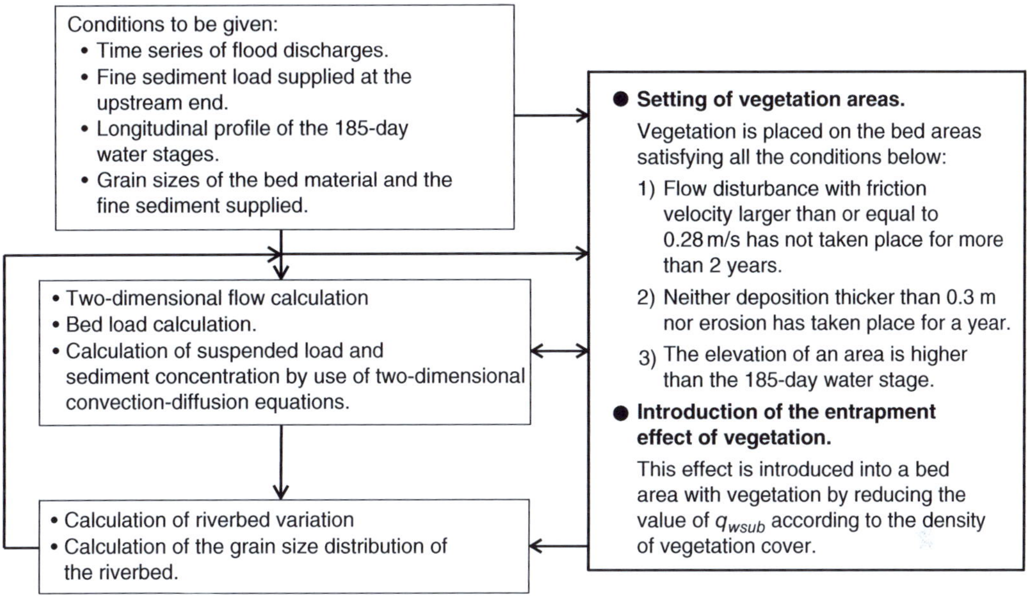

Figure 19.19 Framework of the planar two-dimensional calculation of riverbed variation incorporating the behaviour and the hydraulic effects of vegetation.

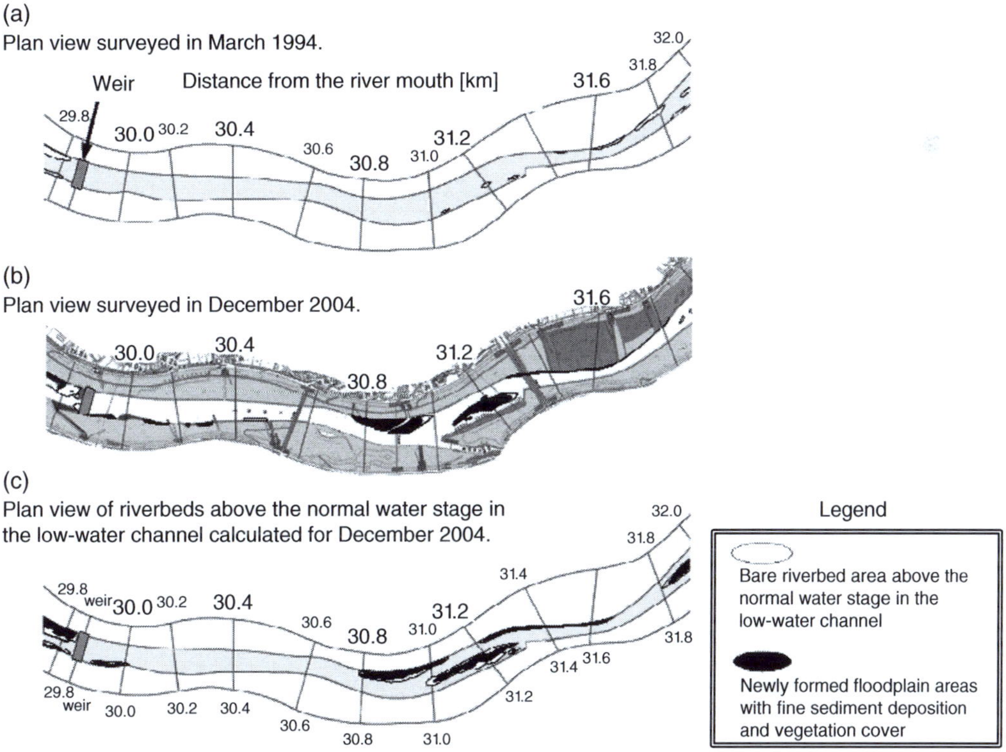

Figure 19.20 Accretion of the floodplains observed between 1993 and 2004 in the Onga River and comparison with the calculated changes.

19.6 Engineering Framework for Gravel-Bed River Management

One aim of this chapter was to make it clear through the examination presented that fine sediment is capable of playing a substantial role in geomorphic changes and vegetation growth, which may result in channel-capacity reduction below an allowable level. The behaviour of fine sediment, therefore, should be a major target for gravel-bed river management, as well as gravel of material m. In light of the comparison of the channel changes examined in the previous two sections, in 1G segments, the effects of vegetation growth are presumably more significant, whereas in 2G segments the main effect is fine sediment deposition. Interestingly, this difference is likely to reflect an essential distinction between alluvial fan zones and natural levee zones around a fluvial system.

An important commonality among the channel changes previously introduced is the existence of physical and direct impacts that trigger subsequent changes. For instance, channel narrowing began immediately after forced widening by man-made modifications in the Sendai River case and an extremely large flood in the Powder River case, and expansion of dense vegetation areas in the Tama River accompanied the basal geomorphic change to the compound cross-section type. This commonality suggests that control of channel changes is best effected by focusing on triggering events. Namely, appropriate design or regulation of initial impacts should make subsequent channel changes more favourable for maintenance work.

This idea is particularly important in Japan. Although great efforts have been made for river improvement works including channel-capacity enlargement like that shown in Figure 19.5, the present safety levels in terms of flood-disaster prevention are still considerably lower than the national standards for many rivers in Japan. This means that channel modification to increase capacity will continue to be a reliable and practical measure. Accordingly, well-designed modifications are expected to result in the further reduction of subsequent maintenance work.

To design channel modifications that are suitable both for achieving immediate goals and for minimizing subsequent maintenance work, it is crucial to predict the channel changes likely to accompany the modifications. To that end, the following two stages seem necessary: first, a scoping stage, in which scenarios of channel change to be examined should be widely screened or prioritized; second, prediction of channel changes for the selected scenarios.

Since it is likely that some future channel changes will be beyond what administrators and researchers can predict, continual observations and analyses are needed to find novel phenomena that might develop into obstacles to channel maintenance and to record these as working knowledge. Furthermore, a cycle should be established in such a way that the knowledge will be systematically incorporated into the framework of channel-change prediction methods in the cycle.

In this context, this chapter contributes to building and adding scenarios concerning fine sediment behaviour to potentially influence channel management, and to providing the prediction methods for them. In addition, a common fundamental approach behind the contents in this chapter may become a new important element of the framework: namely, treating the behaviour of gravel and finer sediment separately with the understanding of its large difference in response time and spatial scale. This difference seems analogous to a frequency spectrum with two distinct major peaks. A rational engineering approach may be taken, beginning with an examination of how to control phenomena corresponding to such peaks individually.

After prediction methods have been prepared, it is necessary to handle the uncertainty of the predicted results. In addition to the lack of data for calibration and the inherent limitations on accuracy that originate from the prediction models themselves, the setting of miscellaneous conditions induces uncertainty. One important example is providing future flood discharges that cannot be known exactly, but that have a strong influence on the calculation results. The diagram in Figure 19.12 avoids this problem by averaging extremely long-term variation. This approach, however, has its own limitations: channel changes with time cannot be traced while overall characteristics of a riverbed are provided.

Considering such inevitable uncertainty, the author emphasizes that calculating future channel changes should not ncessarily be seen as 'prediction', but rather as a powerful scenario analysis tool for making wise judgements. In this context, for example, 'worst' and 'most favourable' cases are calculated, and these are used for discussing management strategy.

This approach is compatible with adaptive management. Indeed, in the Onga River case, using multiple calculation results for possible channel narrowing, river administrators with the Kyushu Regional Development Bureau of the MLIT have monitored channel changes as well as the difference between actual and calculated changes, so that they can continuously revise the original management plan.

The remaining point to raise now is the perspective of integration. Channel modifications and subsequent maintenance work are used not only in flood disaster prevention but also to promote the restoration and conservation of natural fluvial environments. River-channel management therefore needs to encompass flood control, channel maintenance, and environmental conservation. Taking a more comprehensive approach, researchers and engineers engaged in river management are strongly encouraged to establish situations that benefit all stakeholders, even in cases with many constraints. Since geomorphic changes including in-stream vegetation can be connected closely with environmental evaluation, the best use of outcomes of research and development on channel changes must lead to such goals.

Notation

Material *m*, *s*, *t*	Classifications of riverbed material defined in section 19.2.2
r	Precipitation rate for a certain duration (mm/h)
T	Return period for r (years)
m	Parameter in Equation 19.1 based on the Fair formula
Q	Flood discharge (m^3/s)
p	Parameter to take into account non-linearity in the rain-runoff relationship in Equation 19.2
h	Water depth in a rectangular and prismatic channel
Δh	Variation of h from baseline h_0
ΔT	Variation of T from baseline T_0
Type 1	A gravel-bed surface state of bare ground or with very sparse herbaceous communities
Type 2	A gravel-bed surface state of ground with sparse plants
Type 3	A gravel-bed surface state of ground covered with dense vegetation
T_{pr}	Threshold in terms of a calm (non-disturbed) period at which pioneer plant communities start to grow
D_{ts}	Thickness of a top fine-sediment layer
D_c	Threshold in terms of D_{ts} at which a stable, dense vegetation area starts to form

τ_*	Non-dimensional shear stress for material m
τ_{*ce}	Threshold in terms of τ_* at which a flood flow restores a riverbed with vegetation cover to a bare gravel surface
B	River width
B_U	Width of the higher riverbed of a compound-type channel
B_L	Width of the lower riverbed of a compound-type channel
h_s	Relative height of the higher riverbed to the lower one in a compound-type channel
n_U	Manning's roughness coefficient for the higher riverbed of a compound-type channel
n_L	Manning's roughness coefficient for the lower riverbed of a compound-type channel
D_R	Representative grain size of material m
d_{ts}	Representative grain size of a top fine-sediment layer
Q_s	Load of fine sediment supply
p_s	The power to which Q is raised to obtain a proportional relation to Q_s
I_b	Channel slope
c	Sediment concentration at an arbitrary point
$\bar{c}$	Vertically averaged sediment concentration
c_b	Sediment concentration at a bottom surface
w_0	Settling velocity of fine sediment
κ	Karman's constant
u_*	Friction velocity for a riverbed
q_{wsub}	Vertical upward (volumetric) flux of suspended sediment rising from unit riverbed area per unit time (pick-up rate)
D	Local water depth
ε_y	Diffusion coefficient with respect to the transverse direction
d_{ws}	Grain size of fine sediment to be deposited
β'	Ratio of ε_y to a diffusion coefficient with respect to the vertical direction
σ	Variable to quantify the efficiency of the fine-sediment entrapment by vegetation in Equation 19.13

Acknowledgements

Section 19.4 is based on research under a project conducted by the River Ecology Research Group of Japan. The author owes large part of this chapter to collaboration and discussions with people belonging to the PWRI, MLIT, USGS, KICT, consulting firms, and NILIM, including Koichi Yamamoto, John A. Moody, Samhee Lee, Satoshi Watanabe, Toshihiro Tanoue, Atsushi Hattori, among others. The author would like to express appreciations to the reviewers for their important and stimulating comments.

References

Alphen, JV, Bourget, L, Elliott, C *et al.* 2011. *Flood Risk Management Approaches – as Being Practiced in Japan, Netherlands, United Kingdom, and United States.* IWR Report No: 2011-R-08. http://www.iwr.usace.army.mil/Portals/70/docs/iwrreports/2011-R-08.pdf (accessed on 3 February 2016).

Cui, Y and Parker, G 1998. The arrested gravel front: stable gravel–sand transitions in rivers. Part 2: general numerical solution. *Journal of Hydraulic Research* 36(2), 159–182.

Duel, H, Baptist, MJ, and Penning, WE 2001. *Cyclic Floodplain Rejuvenation: a New Strategy Based on Floodplain Measures for Both Flood Risk Management and Enhancement of the Biodiversity of the River Rhine.* CFR Project, Main Report, NCR Report 14-2001, Delft Cluster and IRMA-SPONGE: Delft.

Egashira, S, Jin, H, Takebayashi, H *et al.* 2000. Bed variation and sediment budget in the downstream reach of Kizu River. *Annual Journal of Hydraulic Engineering, JSCE* 44, 777–782. (In Japanese.)

Evans, EP, Simm, JD, Thorne, CR *et al.* 2008. *An Update of the Foresight Future Flooding 2004 Qualitative Risk Analysis.* Cabinet Office: London.

Fair, GM, Geyer, JC, and Okun, DA 1970. *Elements of Water Supply and Wastewater Disposal,* 2nd edn. John Wiley & Sons, Toppan Company.

Fujita, K 2008. How is the gap between the concept and practice of integrated sediment management bridged?. *Proceedings of the Fourth International Conference on Scour and Erosion 2008,* Keynote-3; 41–62.

Fujita, K, Yamamoto, K, and Akabori, Y 1998. Evolution mechanisms of the longitudinal bed profiles of major alluvial rivers in Japan and their implications for profile change prediction. *Journal of Hydraulic, Coastal and Environmental Engineering, JSCE* 600(2-44), 37–50. (In Japanese.)

Fujita, K, Moody, JA, Uda, T, and Fujii, M 1996. Accretion of flood-plains by wash load deposit and resulting channel narrowing. *Journal of Hydraulic, Coastal and Environmental Engineering, JSCE* 551(2-37), 47–62. (In Japanese.)

Fujita, K, Lee, S, Watanabe, S *et al.* 2003. Mechanism and simulation of the expansion and extinction of stable vegetation areas in a gravel-bed alluvial fan river. *Journal of Hydraulic, Coastal and Environmental Engineering, JSCE* 747(2-65), 41–60. (In Japanese.)

Fujita, K, Tomita, Y, Oonuma, K *et al.* 2009. *A Fundamental Approach for Assessing the Downstream Effects of a High Dam on River Characteristics in Order to Recognize the Change in Aquatic Organisms and Stream Ecosystem.* Technical Note 521 of NILIM and Technical Note 4140 of PWRI. (In Japanese.)

Hattori, A, Fukushima, M, and Takeuchi, Y 2013. *Development of the Engineering Tool for the Examination of Channel Modification and Maintenance Incorporating the Environmental Impacts on Aquatic Organisms and Stream Ecosystem.* Comprehensive Research Report 3/1-22, Gokase River Research Group of the River Ecology Research Group of Japan. (In Japanese.)

Itakura, T and Kishi, T 1980. Open channel flow with suspended sediments. *Proceedings of the ASCE, Journal of the Hydraulics Division* 106(HY8), 1325–1343.

Izumi, N and Ikeda, S 1991. Stable channel cross-section of straight sand–silt rivers with active sediment transport. *Journal of Hydraulic, Coastal and Environmental Engineering* 429(2-15), 57–66. (In Japanese.)

Mochizuki, T, Amano, K, and Fujita, K 2011. Grasp of sediment movement focusing on capturing function of fine sediment in Lake Jusan. *Annual Journal of Hydraulic Engineering, JSCE* 55, s1513–1518. (In Japanese.)

Moody, JA and Meade, RH 1990, *Channel Changes at Cross Sections of the Powder River between Moorhead and Broadus, Montana, 1975–1988.* Open-File Report 89-407, US Geological Survey.

Moody, JA and Meade, RH 2014. Ontogeny of point bars on a river in a cold semi-arid climate. *Geological Society of America Bulletin* 126(9/10), 1301–1316.

Moro, Y, Kazama, S, and Fukuoka, S 2011. Change of river improvement works of the Lower Tone River and effectiveness of channel dredging. *Advances in River Engineering, Committee on Hydroscience and Hydraulic Engineering, JSCE* 17, 101–106. (In Japanese.)
Okano, M, Takayanagi, J, Fujii, T, and Ando, A 2004. Study of reservoir sedimentation management based on actual sediment inflow: the case of development of the sediment yield potential map. *Journal of Japan Society of Dam Engineers* 14(3), 167–176. (In Japanese.)
Oonuma, K, Fujita, K, Satou, Y, Nishimoto, N, Matsuki H, and Inoue, Y 2007. Case study on the integrated evaluation of channel modifications in a Segment–2 reach by calculating subsequent channel changes and their influence on the function of flood control and environmental conservation. *Advances in River Engineering, Committee on Hydroscience and Hydraulic Engineering, JSCE* 13, 375–380. (In Japanese.)
Parker, G 1978. Self-formed straight rivers with equilibrium banks and mobile bed. Part 1. The sand–silt river. *Journal of Fluid Mechanics* 89, 109–125.
Sakurai, T, Kashiwai, J, and Oguro, M 2002. Capture and accumulation of fine sediment in multi-purpose dam reservoirs in Japan. *International Symposium On Reservoir Management In Tropical And Sub-Tropical Regions Proceedings*, Vol. 1. International Commission on Large Dams; 772–786.
Shigeta, K, Nihei, Y, Sakai, A, and Ootsuka, S 2008. Fundamental study on suspended-sediment transport of main rivers flowing into Tokyo Bay. *Annual Journal of Hydraulic Engineering, JSCE* 52, 913–918. (In Japanese.)
Subcommittee on Climate Change Adaptation for Flood Control, the Panel on Infrastructure Development 2008. *Climate Change Adaptation Strategies to Cope with Water-Related Disasters due to Global Warming.* Policy Report to MLIT: Japan. http://www.mlit.go.jp/river/basic_info/jigyo_keikaku/gaiyou/kikouhendou/pdf/draftpolicyreporttext.pdf (accessed on 3 February 2016).
Suzuki, T and Terakawa, A 1996. Probable rainfall map based on rainfall-duration–frequency relationship using AMEDAS data. *Civil Engineering Journal* 38(12), 20–25. (In Japanese.)
Yamamoto, K 1986. *Study on the Characteristics of Alluvial River Channels.* Report 168 of PWRI; 1–66. (In Japanese.)

Discussion

Discussion by Mike Miles and Colin Thorne

It is common practice to design river-related infrastructure and regulate flood plain management on the basis of a design event or 'hazard', such as the flood with a return period of 1:100 years. This criterion, together with an estimate of the infrastructure or building's design life, allows determination of the cumulative risk that the design event will be exceeded during the lifetime of the project as follows:

$$P = 1 - \left(1 - \frac{1}{T}\right)^{L_p}$$

where P is probability of exceeding a design criterion (%), L_p is anticipated project lifespan (years), and T is average return period in years as a design criterion.

Recent experience in Canada suggests that land managers, and even senior engineers, commonly fail to appreciate either the cumulative risk of exceeding the design event for long-lived infrastructure (such as levees) and structures (such as residences), or the importance of selecting a suitably conservative design standard for communities vulnerable to life-threatening hazards, such as debris flows. Recent studies, such as those by Jakob (2014), recommend that land-use zoning or engineering design be based on an acceptable level of risk, which varies with the severity of the consequence associated with failure.

In this context, please could you explain the basis for design standards/events commonly employed in your country and what factors are considered when determining what cumulative probability of failure is considered to be acceptable?

Reference

Jakob, M 2014, *Debris flood hazards and risks at Cougar Creek.* http://www.canmore.ca/index.php?option=com_docman&task=doc_download&gid=4954 (accessed on 21 October 2015).

Discussion by Mike Miles, Colin Thorne and Dave Polster

This chapter describes the efforts currently being undertaken in Japan to maintain flood conveyance capacity in aggrading rivers, principally through ongoing maintenance (repeated dredging) and capital works (excavation to enlarge flood control channels).

Recent work in western Canada has used soil bio-engineering techniques to capture mobile sediment upstream of flood sensitive infrastructure. Techniques such as live gravel bar staking employs cuttings of willows (*Salix*) and cottonwoods (*Populus*) that are placed into gravel bars using an excavator (Polster 2006; Figure 19.21). These materials root vegetatively, increase channel roughness, trap fine sediment, and continue growing through the deposited sediment. Over time a seral vegetation community is developed, which provides a variety of ecosystem services, as well as reducing the volume of sediment (especially fine sediment) input to flood control channels downstream.

Where sediment source control is not feasible, managing fine sediment transfer through the fluvial system by promoting beneficial deposition in selected reaches may represent a viable approach to reducing the frequency with which it is necessary to dredge or enlarge flood control channels downstream. Hence, our question is whether the potential for using bio-engineering to manage fine sediment dynamics (and so improve the efficiency with which flood control channels are maintained) has been evaluated in Japan?

(a) (b)

Figure 19.21 Using vegetation to trap sediment. (a) Example of live gravel bar staking. (b) Vegetation development and sediment deposition 8 years after construction.

Reference

Polster, DF 2006. Soil bioengineering for riparian restoration. Paper presented at the *Canadian Land Reclamation Association 2006 Conference*, Ottawa, Ontario, August 20-23; 313–322.

Discussion by James Bathurst

The author shows that fine sediment deposits on gravel bars provide a foundation for vegetation growth and is concerned that the resulting increase in flow resistance may significantly reduce channel flood capacity. However, in the UK, Natural Flood Management deliberately increases channel and riparian flow resistance (e.g. using large woody debris and log structures) to hold back floodwaters in headwater streams so as to reduce potential flooding in more vulnerable areas downstream. In this process, encouraging the deposition of finer sediments in the headwaters would have beneficial consequences for the downstream areas, albeit at the expense of localised flooding in the headwaters themselves. Is there scope to adopt this approach also in Japanese mountain rivers?

Discussion by Colin Thorne

I commend the author for his thoughtful account of issues raised by sediment management for flood risk reduction in Japan. He correctly identifies that deposition of fine sediment and colonization of flood control channels by vegetation present challenging problems for river management. His scenario-based model and management framework provide the basis for informed management of sediment, vegetation, and flood risk.

Where I beg to differ from the author is regarding *how* sediment and vegetation should be managed in order to keep flood risk at acceptable levels *sustainably* in a future characterized by climate change that is likely to increase the magnitude, frequency, and duration of flood events. Frankly, I do not believe it will be possible to excavate channels sufficiently large to contain all future floods and would stress the need to manage sediment sources and pathways, with dredging conducted only as a last resort.

Would the author agree that river management should optimise operation of the river system as a whole, with targets set not only for flood-risk management but also water quality, drought resilience, and continuous provision of ecological services that are not just economically valuable, but intrinsic to the quality of life for the people who live and work along the river?

Reply by Koichi Fujita

The MLIT technical criteria for river works provide for approximate references on five scale ranges corresponding to river-system importance. Design floods with return periods from 100 to 200 years have been set for main reaches of most Class-A rivers, accordingly.

While Japan is deeply dependent on flood-prone areas, the overall progress level of river improvement works even for the Class-A rivers is still far below their planning scales. The above design floods are, therefore, regarded as long-term goals. In these situations the revision of flood-prevention planning scales has not been a central issue.

On the other hand, coping with consequences by events over protection levels has been becoming an important target particularly for the last decade. A recent development is a new policy 'Toward rebuilding FRM-conscious society' issued by the MLIT in December 2015. It states that non-structural measures to effectively enable people to save their own lives should be enhanced and that structural measures to ease hazards by events exceeding design levels of existing structures should be widely introduced. This policy typically shows that, while improvement works will be made consistently toward long-term goals, Japanese policy has been extended to consequences of mitigation, considering occurrence of rainfalls exceeding present and future protection levels.

In the above development the author does not have a positive view of adopting the cumulative risk as it is, because it links protection levels to project life spans. The latter is not relevant to protection of people, and the determination of protection levels should have closer relation to life spans of people and sustainability of societies and nations.

Using bio-engineering methods is an interesting idea to the author as well, though such methods have not been actively examined for Japanese rivers. Their feasibility is governed by whether the necessary riparian room can be secured. Relations between floodplain growth rates and proportionality coefficient α relating Q_s to Q^{p_s} are linear on the whole (Fujita *et al.* 1996), indicating that sediment concentration needs to be considerably decreased. In light of a large amount of fine-sediment supply, cases with sufficient room available seem to remain exceptional in Japan.

The author understands that the approach is underlain by the fundamental idea: a river should respond to events exceeding a design level in such a mild manner that people can resiliently manage. When its adoption is examined, however, the sharp division between steep mountains and alluvial plains should be considered unfavourable as a Japanese geographical feature. The feature makes it difficult to find sufficient headwater areas that can substantially reduce flood discharge downstream with diminished suffering due to flooding. This may lead to an alternative of allowing floods to flow to particular parts within large alluvial plains. When moving toward this idea, we cannot avoid serious discussion on the equity issue and the practical approach of consensus building.

The author definitely agrees that river management should be through a holistic approach. Where I may differ from the discussant is on emphasis. The author would rather deepen the discussion on how we can have a good command of excavation as a tool than draw a universal boundary of its applicability. For example, in some cases excavation that will not cause rebounds could be devised; laborious maintenance work could be rationalized if protected areas are highly populated with huge assets; and wise maintenance could restore an ecosystem triggering fluvial-process revitalization. Researchers and engineers need to continuously improve the engineering framework for optimizing operation of each individual river with wide scope and thorough consideration of local situations.

20

Lahar Flow Disaster, Human Activities, and Risk Mitigation on Volcanic Rivers: Case Study of Rivers on Mount Merapi Slopes, Indonesia

Djoko Legono and Adam Pamudji Rahardjo

20.1 Introduction

The dynamics of flow (water and sediment) of rivers on Mount Merapi slopes are mainly affected by volcanic activity and also by hydroclimatological behaviour in the surrounding area. Volcanic activity of Mount Merapi includes eruptions followed by pyroclastic flows, while heavy rainfall may trigger debris or lahar flows. Through anthropogenic processes, mining activities along these rivers also affect the flow behaviour, contributing to significant changes in riverbed configuration. The presence along rivers of such huge sediment volumes of potential economic value is utilized as job opportunity alternatives by government authorities. However, uncontrolled and excessive mining activity may produce negative impacts that contribute to failure of important infrastructure, such as severe erosion downstream of sediment control structures and scouring around bridge piers, as well as instability of riverbanks. Although local government authorities have issued recommendations and regulations regarding the allowable location and volume of extractable sediment, excessive sand and gravel mining still take place. Local rainfall intensity–cumulative relationships, with results compared against the global rainfall intensity–duration relationships, are utilized to develop a warning criterion against the occurrence of lahar flows.

Mount Merapi, which is located in Java Island at 7°32′15.49″ S 110°27′04.03″ E (Figure 20.1), is the most active volcano in Indonesia; it erupts frequently and produces huge amounts of material ejected from the summit of the mountain. Climatic conditions of the Mount Merapi area are tropical monsoon with high-intensity rainfall occurring in the wet season (November–April) and low-intensity rainfall in the dry season (May–October). The average annual rainfall is in the range 2500–3000 mm, temperature range is 20°–33 °C and relative humidity 80–90%. Small-scale Mount Merapi eruptions often occur every 2 years, whereas relatively large-scale eruptions occur every 10 years, such as in 1994 and 2006 (Directorate General of Water Resources 2009).

Mount Merapi is located on the border of two provinces: Central Java Province and Yogyakarta Province (Figure 20.2). The area affected by Mount Merapi activity is very random, although the more frequently affected areas are found on the west, southwest, south, and the southeast. Most of rivers that originate from Mount Merapi flow towards the Pacific Ocean (south coast of Yogyakarta) through two trunk rivers, the Progo and Oyo Rivers. There are at least 13 rivers originating from the summit of Mount Merapi. Several of them flow across national roads and important urban areas such as Yogyakarta City, the runway of the Yogyakarta International Airport, and the world cultural heritage site of Prambanan Temple (Figure 20.3).

Gravel-Bed Rivers: Processes and Disasters, First Edition. Edited by Daizo Tsutsumi and Jonathan B. Laronne.

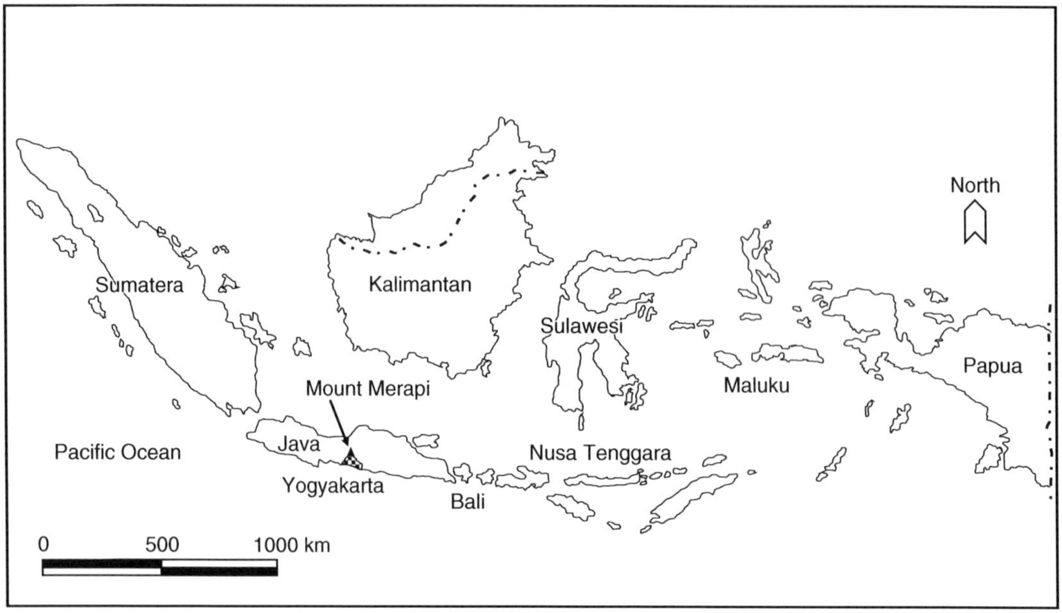

Figure 20.1 Map of Indonesia and location of Mount Merapi.

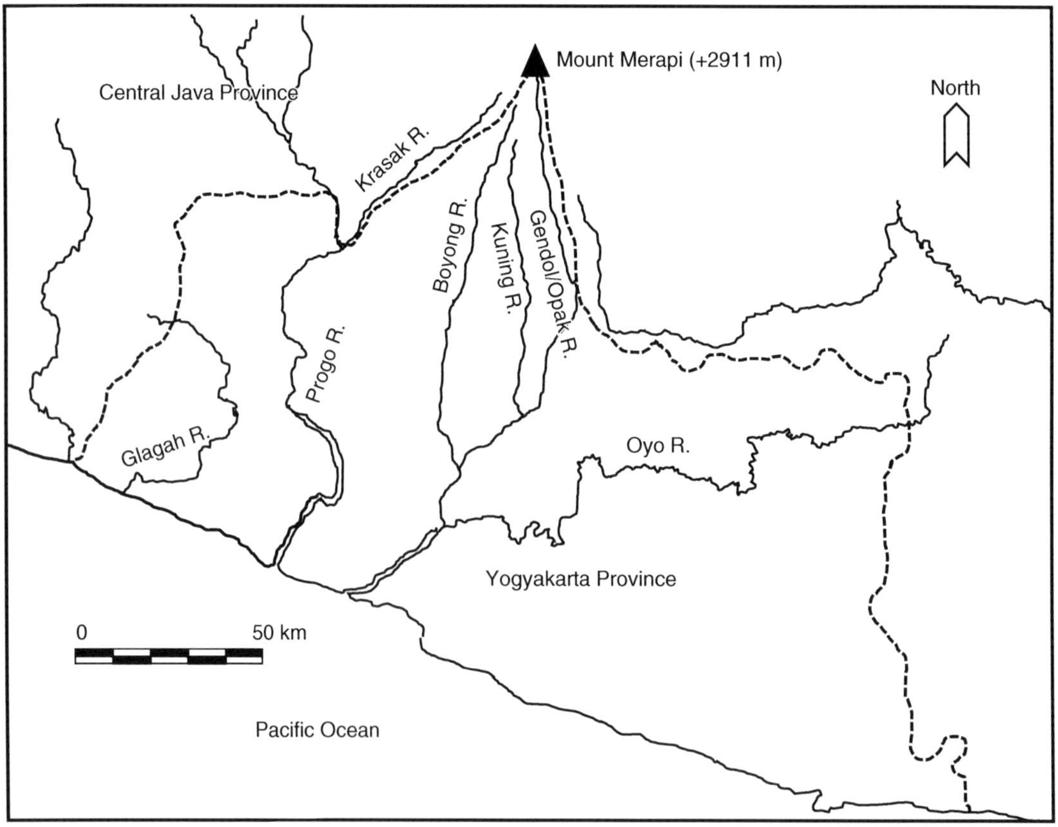

Figure 20.2 Province boundary and location of Mount Merapi.

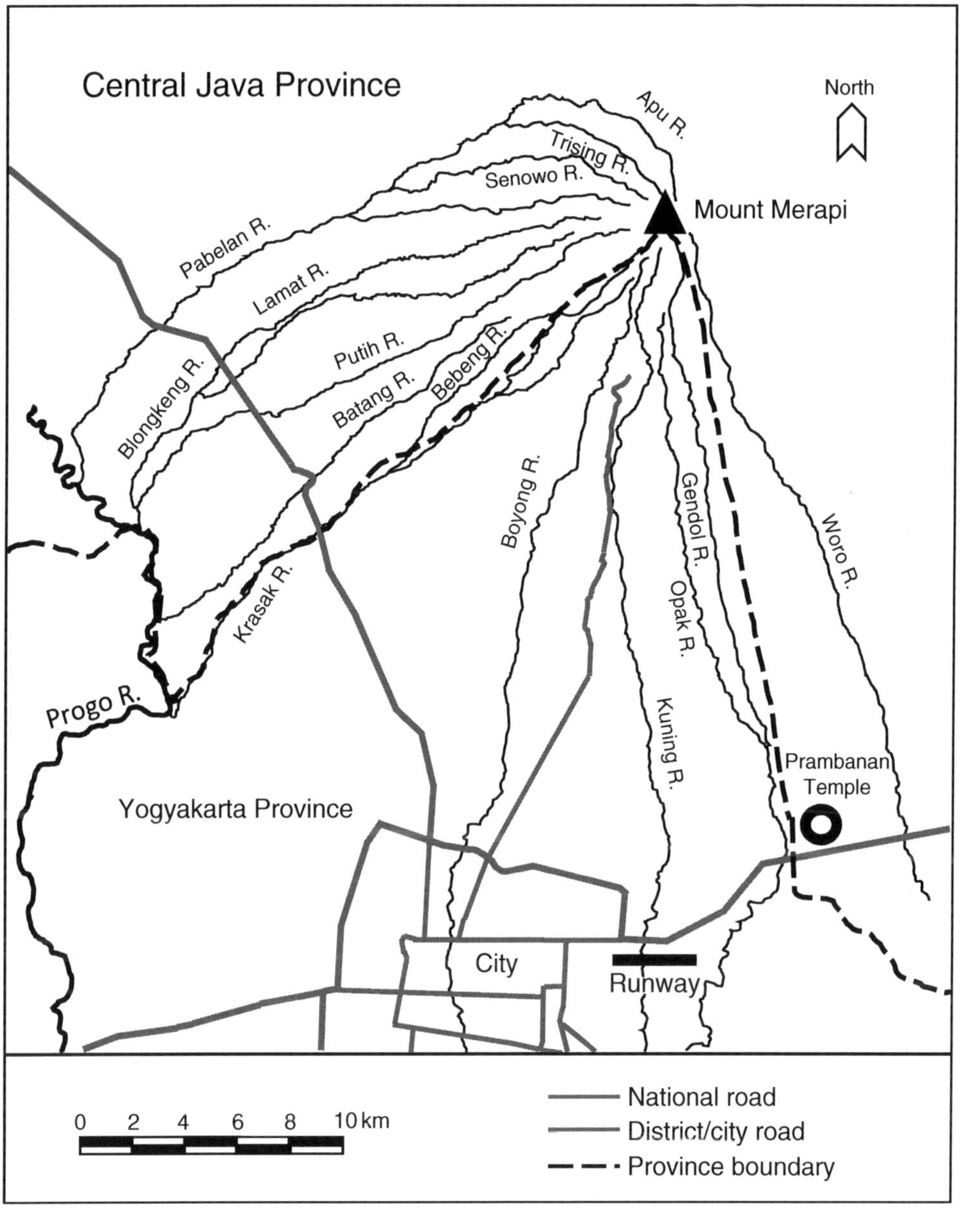

Figure 20.3 Rivers that originate from Mount Merapi.

Mount Merapi lies along two important regional faults: the transversal fault that separates East Java and Central Java, and the longitudinal fault along Java Island. The summit of Mount Merapi has a caldera that frequently forms one or two lava domes. Figure 20.4 shows the lava dome formation several days after the 3 November 2010 eruption. The formation (i.e. position, size, and height) of the lava dome takes place randomly. The size of the lava dome varies from 2 to 4 million m^3 at an elevation 2900 m with a height of 100 m. The lava dome collapses through two mechanisms: high pressure from

Figure 20.4 Lava dome at the summit of Mount Merapi following the 2010 eruption.

volcanic ejection, or instability of the dome. The collapse may produce pyroclastic flow that carries coarser particles along the deepest parts of river valleys. At the same time, very fine particles with high temperature form a cloud that moves downstream with very high velocity. This may take place because the slope around the summit is very steep. Mount Merapi produces both primary disasters (pyroclastic flow, lava flow, ash fall, and rock fall) and secondary disasters such as lahar flows. The 2010 Mount Merapi eruption produced more than 130 million m^3 of sediment. Additional to the lava flow disaster, these huge amounts of sediment moved 16 km along the Gendol River. The lahar flow disasters, which caused damage downstream, were induced by tropical and orographic rainfall on the Merapi slope, with intensity > 50 mm/h in the wet season. In addition to negative impacts, the eruption also contributes a sediment source that benefits human habitation. The average sediment production from Mount Merapi activity is estimated to be 3 million m^3 per year.

The 2010 Mount Merapi eruption had an impact on almost all rivers, with varying degrees of damage in each. The lava flow in the Gendol River was the longest historic flow (Directorate General of Water Resources 2009). Indeed all 13 rivers originating from Mount Merapi have already experienced lava flows (see Figure 20.5): from the Apu River in the northwest to the Woro River in the southeast (Lavigne *et al.* 2000). In general, the western rivers and southern rivers can be viewed separately.

On the west, the channels between the Pabelan River and the Krasak River have a length of 20–30 km, and catchment areas 22–45 km^2. However, the Pabelan River is 46 km long and has a watershed area of 103 km^2.

This chapter demonstrates that in order to cope with such problems of huge sediment supply and lahar flow disaster risk can be reduced while taking benefit of sediment material by mining activity, which has been successfully implemented in the Mount Merapi area.

20.2 Riverbed Characteristics

The bed configuration of rivers originating in Mount Merapi varies considerably and is affected by hydraulics as well as by human interference. Riverbeds in the upper reaches of these rivers consist of a majority of boulders of 200–5000 mm diameter, but also gravel, sand, and relatively small amounts

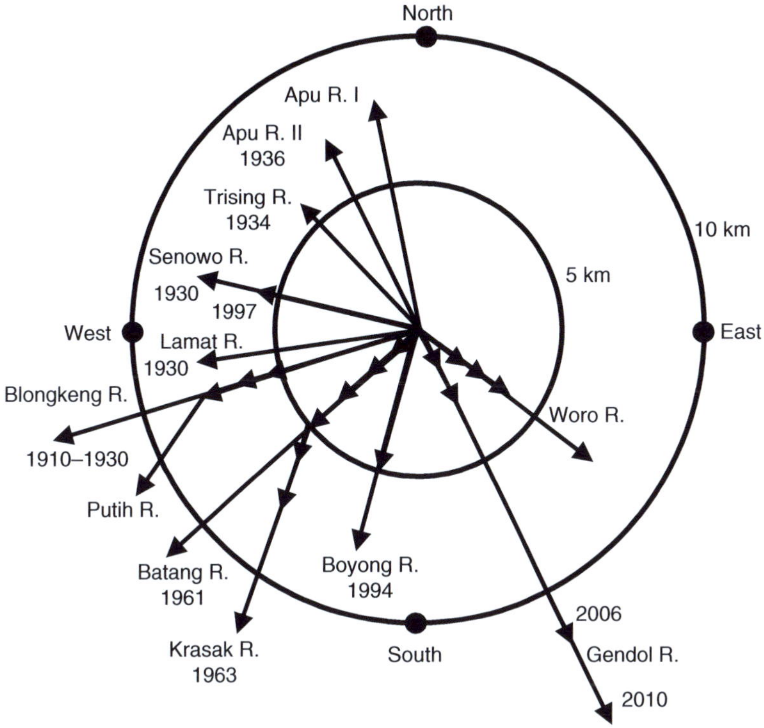

Figure 20.5 Lava flow at rivers due to Mount Merapi eruptions.

of fine material. The undisturbed bulk density of the riverbed (C_*) is 0.6–0.7, whereas the specific density is 2.69–2.90 ton/m^3 (Directorate General of Water Resources 2009). During prolonged periods, sand and gravel mining activities change the bed roughness since the coarse bed material is left behind. These activities may also change the wetted perimeter, since mining on the bank areas widens the riverbed cross section and changes the bank slope. The most valuable material is sand, followed by gravel and boulders. Figures 20.6–20.8 show the respective riverbed condition of the Pabelan, Putih, and Gendol Rivers.

Surveys to identify sediment characteristics were conducted in several rivers in the Mount Merapi area (Bebeng, Trising, Apu, Putih, Blongkeng, Woro, Krasak, and Kuning rivers; Directorate General of Water Resources 2009). A specific technique of sampling (combination between sieving and direct measurement) riverbed material was carried out to determine the grain-size distribution of the Gendol River. Three adjacent cross-sections A, B, C of a certain reach were selected, three bed material samples were taken in each cross-section and nine samples were analyzed. The results showed that the median diameter (D_{50}) lower truncated at 0.063 mm was 60 mm (see Figure 20.9), whereas D_{16} and D_{84} were 18 mm and 90 mm respectively. These values show a significant difference compared with those observed in global volcanic rivers, especially the Paez River of Nevado del Huila Volcano in Columbia, where the D_{16}, D_{50}, and D_{84} were reported to be 0.018, 0.042, and 0.081 mm respectively (Worni *et al.* 2008).

Riverbed characteristics are very dynamic depending upon the lava formation being ejected from Mount Merapi eruption, heavy rainfall triggering lahar flow occurrence, as well as sand mining activities.

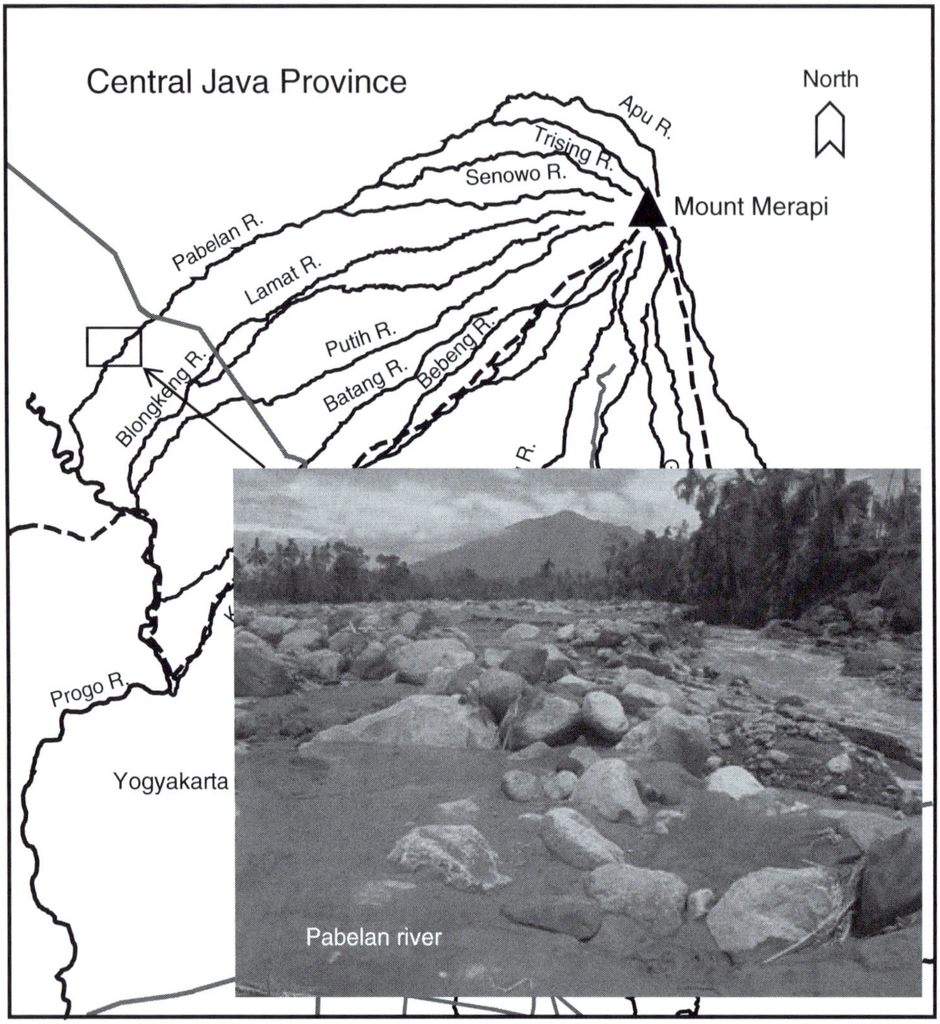

Figure 20.6 Riverbed condition of the Pabelan River.

20.3 Human Activities

The sediment of rivers in the Mount Merapi area has become a potential source that supports livelihood locally. The most valuable source that may be found from rivers in the Mount Merapi area is sand. Surveys on properties of riverbed material showed that only relatively few trucks carrying material out from the river were loading gravel and coarser sediment, with the bulk being sand. This indicates that the exploitation of building materials (gravels and stones) is minor to that of sand (Directorate General of Water Resources, 2009). Moreover, it was also found that the consumption of sediment coarser than sand from the Mount Merapi varies considerably.

The intensity of sand mining fluctuates annually depending on sand availability and the market need. Other factors such as government projects (river normalization) and opening of new private quarry sites also contribute to small fluctuations. Data on the volume of sand mining activity is very

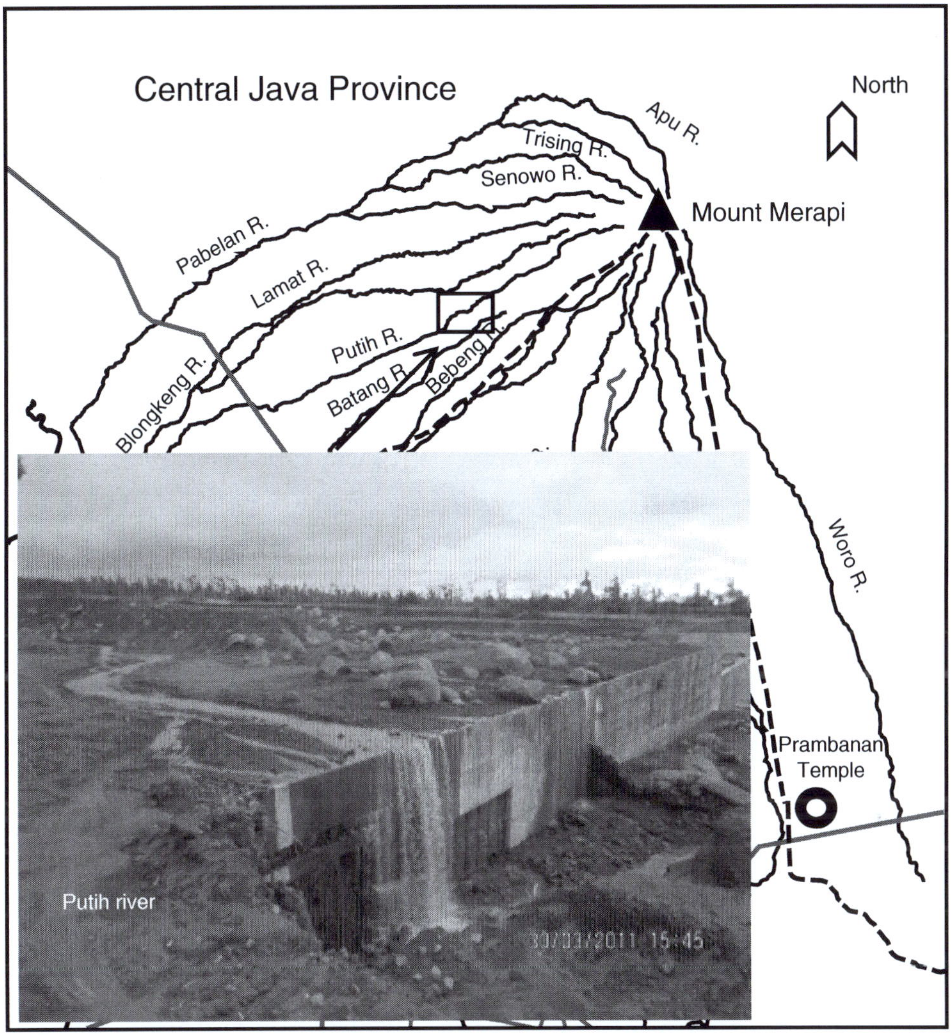

Figure 20.7 Riverbed condition of the Putih River.

difficult to obtain even though the activity is well recorded by the government. There is resistance by miners to report since they try to avoid paying higher taxes. The average yearly sediment production within the period of 2007 through 2012 is estimated at 2 million m^3. The rapid decline of sand mining from 2001 until 2006 suggests that sand stock around Mount Merapi had been reduced due mining activities while no eruption occurred. Hence, it is obvious that the availability of sand in the surroundings of Mount Merapi is the primary limitation to sand mining. The balance between input (pyroclastic material) and output (sand mining production) is important to prevent environmental degradation. Since 1997 the average sand mining volume has been relatively high (3.55 million m^3). Practically, the maximum sand mining volume should be equal to the average pyroclastic sediment from the average annual Mount Merapi eruption. Sand mining is helpful in availing sufficient space upstream of sediment control structures. The sufficient space may avoid huge amounts of sediment to flow

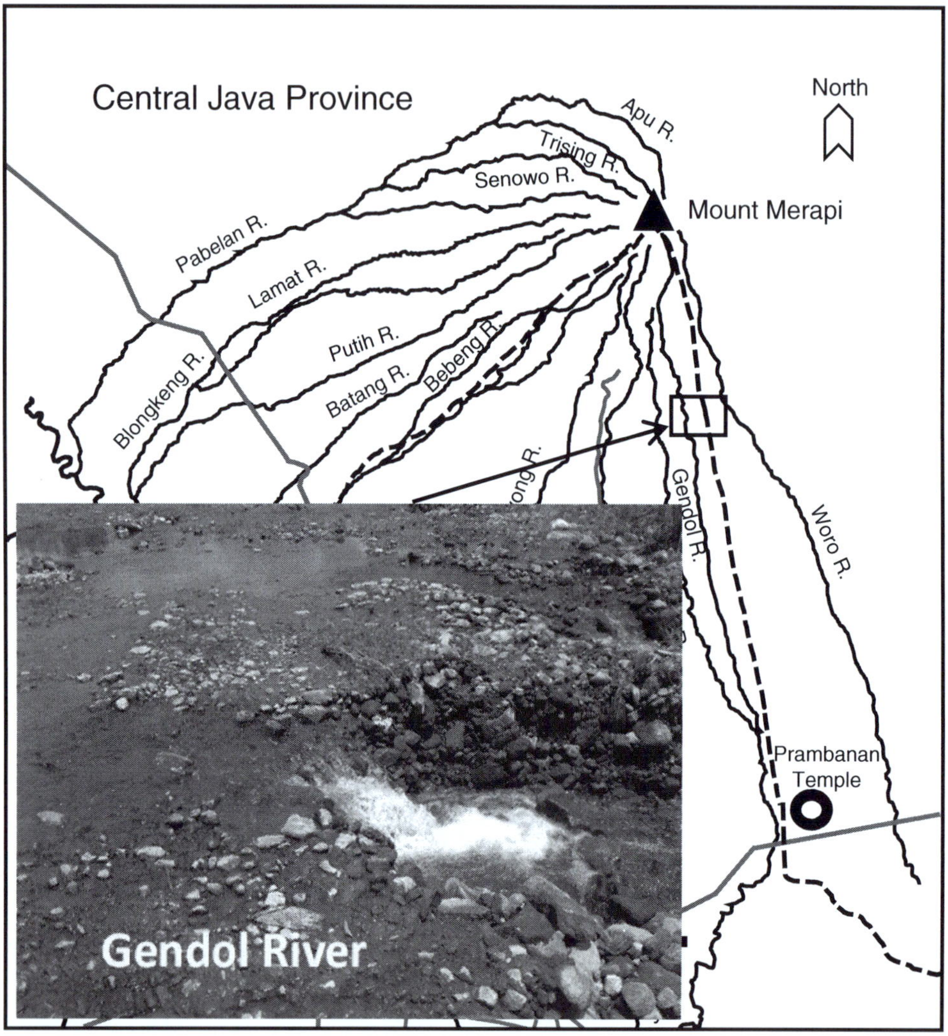

Figure 20.8 Riverbed condition of the Gendol River.

downstream, thereby damaging important facilities (houses, commercial areas, public facilities, etc.). Sand mining may be negative when it is excessive and contributes to riverbed degradation.

In addition to the lack of regulation being applied to sand mining, job opportunities provided by sand mining operations (sand collectors, truck drivers, truck owners, building contractors, etc.) was another reason for instigating law enforcement regulations; a survey indicated that jobs better than sand miners were not available.

To avoid riverbed degradation due to excessive withdrawal of sand, thus also avoid damage to sabo dam facilities, an allowable yearly sand mining volume of 2 million m^3 has been proposed. The 2010 Mount Merapi eruption produced more than 130 million m^3 of sediment. Between 2010 and 2013 there was no limitation of sand extraction from the Mount Merapi area.

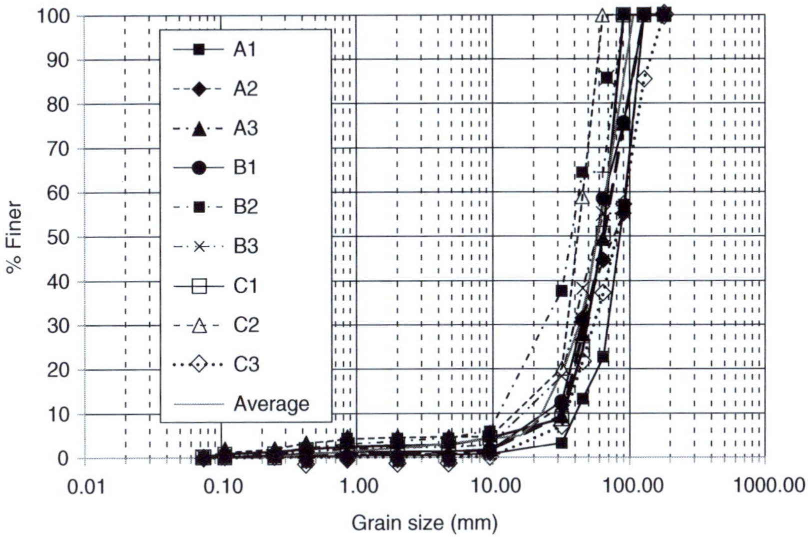

Figure 20.9 Grain-size distribution of bed material from the Gendol River.

20.3.1 Sand Mining in the Upper Pabelan River

Sediment production by the Upper Pabelan River (i.e. tributary of the Apu River) before the 2010 Mount Merapi eruption was considered small and not many sand miners were present. Most of the miners come from the same village of the dam. The situation of the quarry is shown in Figure 20.10. At the Apu River, some coarse gravel was selected and left aside but most of it was not used. Some of the coarser components were covered by finer-grained materials to pave roads for the trucks to enter further upstream for sand mining. The road to the quarry is quite steep, hence trucks leave the quarry with less than full capacity.

20.3.2 Sand Mining in the Gendol River

Sand mining activities in the Gendol River (Figure 20.11) are carried out either by utilizing heavy equipment or manually. The heavy equipment operated near (downstream of) the dam, while the conventional sand miners carried out activities further downstream. This causes not only riverbank instability but also structure instability.

20.3.3 Sand Mining in the Woro River

The Woro River underwent very intensive sand mining (more than five in-stream quarry sites). The high intensity of sand mining activities on the Woro River took place due to very weak enforcement of regulation.

After the pyroclastic flow in 2006, sediment filled the sediment-control basins. Sand was quickly mined and within 2 months the land surface level was reduced by approximately 3 m. Figure 20.12 shows the level of land surface 3 m below the crest of the dam. Further downstream the river was also heavily mined. As a consequence, the toe of the dam was undermined and damaged.

Figure 20.10 Non-intensive mining at the upper Pabelan River, 2009.

Figure 20.11 Intensive sand mining at the upper Gendol River, 2009, with Mount Merapi in the background. (See insert for color representation of this figure.)

Figure 20.12 Overintensive mining at the Woro River, 2009.

20.4 Sediment Management and Risk Mitigation

20.4.1 Damage Due to the 2010 Mount Merapi Eruption

It was a very unfortunate situation that the 2010 Mount Merapi eruption occurred at the beginning of the rainy season, i.e. 26 October 2010, when high rainfall intensity occurred while the sediment source was large and unstable (Gonda, Legono, and Santosa 2013). The first rain of the rainy season after the eruption caused damage to structures and the sediment control mechanism of the Putih River did not function properly. The increase of riverbank erosion was considerable and was affected by the huge source of sediment from upstream.

The rainfall triggered high-intensity lahar flows along the Putih, Pabelan, and Gendol Rivers, causing damage to important infrastructures such as national roads and bridges, and to villages. When such lahar flows pass through bridge structures, potential danger may arise in the form of scouring around bridge piers as well as bridge abutments. Compared with flow without sediment content, the destructive power of a high-density lahar flow is double in terms of scouring around bridge abutments and bridge piers (Hübl *et al.* 2009; Legono, Wignyosukarto, and Saputra 2011; Wardoyo *et al.* 2013). Flow with a high concentration of sediment may cause structural damage through several mechanisms, such as the scour around bridge abutments and scour around bridge piers, as well as the impact force on bridge piers.

Figure 20.13 shows damage that occurred to a bridge over the Pabelan River through the combined mechanism scouring around bridge abutment and bridge piers, with concomitant impact of the lahar flow on the bridge piers, which in turn reduces stability. The gradual decrease in structural integrity of the bridge piers was obvious prior to the failure of the bridge due to the frequent lahar flows that had occurred during the rainy season.

The sediment control of rivers originating from Mount Merapi is designed to reduce the destructive power of possible lahar flows, particularly those triggered by heavy rainfall. Sediment control structures are designed to resist a certain degree or level of lahar flow, below which they are not supposed to fail. However, determining the degree of lahar flow to be used as criteria for designing sediment control in such rivers still requires more understanding. During the 1994 and 2006 Mount Merapi eruptions, only about 3 and 8 million m^3 of material, respectively, was produced by the volcano (Directorate General of Water Resources 2009). It has also been reported that an eruption similar to that in 2010 occurred more than 1000 years ago. This interval indicates why the 2010 Mount Merapi eruption should be considered an extraordinary disaster, for which locals had little awareness and preparedness.

20.4.2 The Human Dimension

Proper shelter planning is critical to minimize the potential for additional victims. Poor planning of emergency shelters and camps can create new disaster victims due to disease outbreaks and malnutrition if shelter is inadequate and supplies of food, clean water, and medicine are not timely. Shelter planning should also take into account the quality of life and livelihoods for displaced populations. For example, 50–70% of people displaced by the 2010 eruption of Mount Merapi ignored evacuation orders and consistently returned (in some cases daily) to danger zones during the crisis because of the need to care for livestock and to check their possessions.

The lack of activities and work programmes in the evacuation camps also can result in people leaving the shelters. In addition, if schools are used as shelters, public education suffers because evacuees occupy the school buildings. In countries with limited relief resources, people may be better served if extended families can temporarily house impacted relatives during emergencies. Community leaders, with assistance from scientists, can encourage residents to develop their own evacuation and relocation strategies.

20.4.3 Integrated Sediment Management

It is important to know that unless sand mining is implemented in an integrated way, negative impacts may persist. This integrated way should incorporate participation of both parties: the local government authority and the community. The sediment supply from Mount Merapi is a natural blessing for the utilization of sand as construction material. However, the intensive sand mining activities can cause severe riverbed degradation, which in turn can contribute to environmental degradation in the form of lowering groundwater level, unstable sediment control structures due to local scouring, etc.

In terms of economic development, the approximate 2 million m^3 per year of sediment deposit that is produced by Mount Merapi volcanic activity is a volume that contributes income not only for locals, but also those from outside areas. Therefore, human interference on sediment migration in the form of extracting the sediment becomes an attractive activity, unless sufficient job alternatives are made available. Ideally, sediment withdrawal from rivers of the Mount Merapi area should not be too much so as to cause river degradation and damage to infrastructure, nor not too little so as to cause river aggradation with the potential for future lahar flows to destroy highly populated areas surrounding the rivers. In order to sustain the sand mining activity, a proper sand mining management needs to be established as the follows:

1) develop the community institutions as part of the proposed institutional framework of sediment management;
2) improve the manuals and plans of proper sediment management and disaster management;

Figure 20.13 Bridge pier damage due to lahar flow (the Pabelan River).

3) formulate the action plans based on the improved manuals on sediment management and disaster management;
4) implement pilot field activities of proper sediment management practice, including its evaluation and recommendation for sustainable activity.

In 2009, a systematic framework for sediment management institution was established to control sand mining activities, particularly in the involvement of the community as the important players in sand mining. Through the long-term approach in familiarizing the aspiration and perception of stakeholders involved in sand mining activity in the Mount Merapi area, an institutional framework of sand mining management was introduced, three community institutions were established, and several actions were conducted as a pilot project. The goal would be a promising community development that will enable support of proper sand and gravel mining management, and sustainable disaster management of Mount Merapi as a whole. In order to conduct and implement proper sand mining management provided with community participation, establishment of community-based institutions is important. To accelerate the implementation of proper sand and gravel mining management, community-based institutions were established in three model villages. Since these institutions were newly established, it was necessary to develop their capacity to operate the organization and to implement their activities in coordination with government agencies and local communities.

20.4.4 Minimizing Risk by Establishing Lahar-Flow Warning Criteria

The role of scientists for risk assessment is minimizing loss of life due to disasters (Pierson, Wood, and Driedger 2014). Mass movements such as landslides and lahar flows are the most common sediment-related disasters. The occurrence of mass movements is mostly dependent on rainfall intensity (Chen,

Saito, and Oguchi 2015). According to Chen *et al.*, mass movements in Taiwan mostly occur in high-intensity periods, but some events occur before and after the rainfall peak. Both antecedent and peak rainfall play important roles in triggering landslides, whereas lahar flow occurrence is more related to peak rainfall rather than antecedent rainfall. Since rivers in the Mount Merapi area have a high potential as sediment sources, lahar flow occurrence may present danger to sand miners. Efforts have to be introduced to allow sand miners to evacuate when lahar flows occur. Several studies have been conducted to establish warning criteria as the first step of providing information to the sand miners (Directorate General of Water Resources 2009). This was carried out based on the guidelines recommended by the Japan Ministry of Land and Infrastructure and Transportation (MLIT 2004). Warning criteria for lahar flow disaster after the 2010 Mount Merapi eruption were studied for the Putih River, thereafter the Critical Line (CL), the Warning Line (WL), and the Evacuation Line (EL) were introduced (Yulinsa, Legono, and Rahardjo 2014). The determination of CL at Gunung Maron rainfall station (see Figure 20.14) was carried out by connecting the two points of the lahar flow occurrence.

The values of RH1M (maximum 1-h rainfall) of 54 mm and RH2M (maximum 2-h rainfall) of 81 mm were used to determine the WL and EL as follows. A horizontal line was drawn from the RH1M in the vertical axis. As it crosses the critical line, the vertical EL line can be drawn from the cross point. This line gives a R2 of 54 mm. The WL line is then determined by shifting the EL line to the left

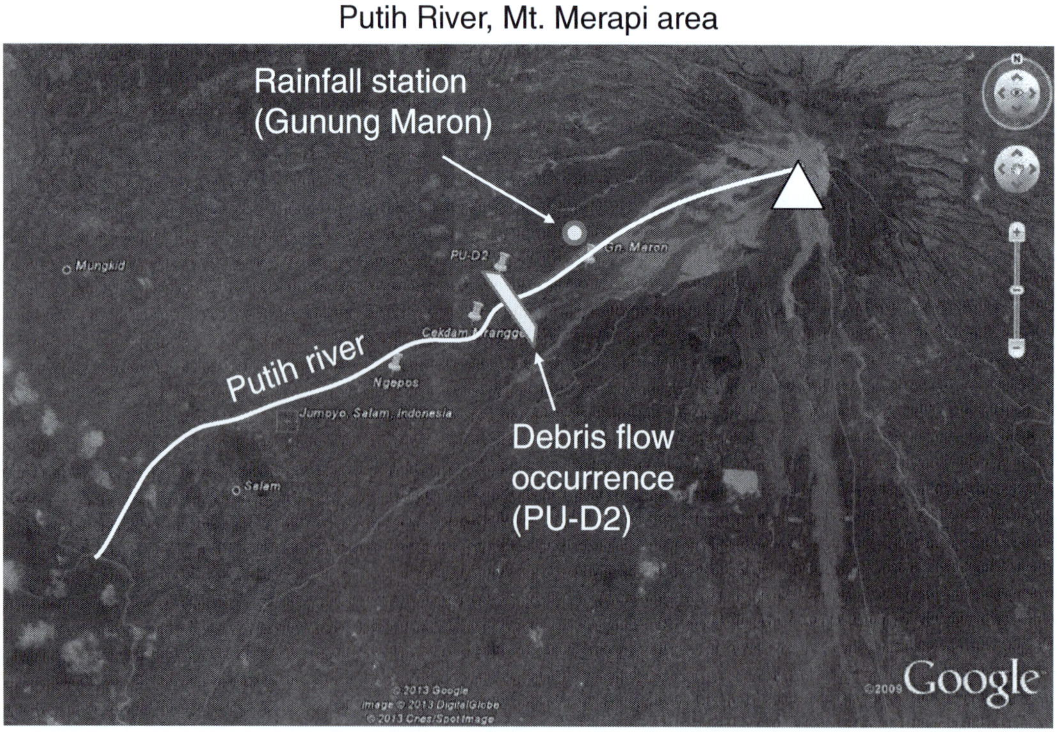

Figure 20.14 Gunung Maron rainfall station of the Putih River. (*Source*: Google Maps.)

with RH2M – RH1M (=27 mm). This WL gives a R1 of 27 mm. Both R1 and R2 are the standards of warning summarized as follows.

- standard cumulative rainfall for warning (R1 = 27 mm);
- standard cumulative rainfall for evacuation (R2 = 54 mm).

Further utilization of the warning criteria persists in the form of plotting the actual rainfall occurrence over the above graphs. Some examples were adopted to draw the 'snake' (undulating) line developed from the rainfall that occurred on 27– 30 November 2011 as shown in Figures 20.15 and 20.16 for the rainfall condition and 'snake' line respectively.

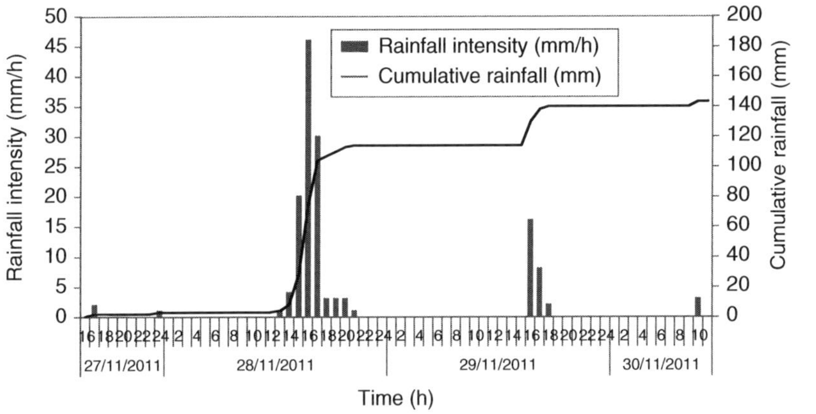

Figure 20.15 Gunung Maron rainfall data of the Putih River.

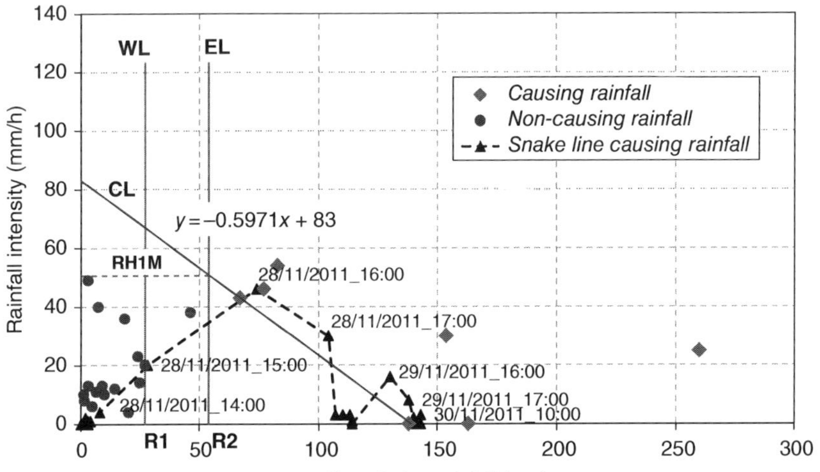

Figure 20.16 Gunung Maron-based 'snake' line of the Putih River.

It is often found that when plotting the 'snake' line, there were rainfall data in which the series was still within the safe zone, even when a lahar flow occurred. This occurs because in plotting the 'snake line' the antecedent rainfall was not taken into account, while in determining the warning criteria it was included. In fact, this is a matter of practicality, but eventually it is much better in terms of providing safer warning criteria to the community. The rainfall duration D, rainfall intensity I, cumulative rainfall E, and antecedent rainfall (AD) are the most common variables considered in developing warning criteria of lahar flow occurrence (Giannecchini, Galanti, and Avanzi 2012). In particular, landslide initiation is frequently related to rainfall intensity and duration (Guzzetti *et al.* 2008). In the case of lahar flow occurrence in the Mount Merapi area, particularly during the rainy season, the rainfall intensity may reach 150 mm/h, but only for a very short duration (< 30 min). By considering the critical line as shown in Figure 20.16, and replacing the notation of x into E and y into I, the rainfall thresholds can be written as follows (see Eq. 20.1).

$$I = -0.5971E + 83 \tag{20.1}$$

where I is rainfall intensity (mm/h) and E is cumulative rainfall (mm).

As an example, from Eq. 20.1 it can be estimated that at a rainfall intensity of 40 mm/h, a cumulative rainfall of 72 mm will initiate lahar flow. This means that the rainfall duration takes place in 1.80 h. The critical line is a rainfall threshold for lahar flow initiation. Guzzetti *et al.* (2008) further discussed the utilization of rainfall threshold as the basis of warning criteria for lahar flow occurrence and proposed

$$I = 2.28 \times D^{-0.20} \tag{20.2}$$

where I is rainfall intensity (mm/h) and D is rainfall duration (h) ($0.1 < D < 48$). The calculation of rainfall intensity by applying Eq. 20.2 with 1.80 h rainfall provides a 2.08 mm/h rainfall intensity. This indicates that the rainfall intensity estimation as performed by Eq. 20.1 and Eq. 20.2 has significant discrepancy. Further investigations regarding such warning criteria for the Mount Merapi area remain open, until sufficient lahar flow data are collected.

20.5 Conclusions

Mount Merapi is an active volcano that produces huge amounts of sediment, which may contribute potential disaster but at the same time also provides valuable natural resources. Among of the sediment sources are boulders, gravel, and particularly sand.

In order to sustain the development of sediment control structures in the Mount Merapi area, an integrated approach on sediment management taking into account the society, economy, and environment, including locality of community is highly essential.

Risk mitigation on potential lahar flow disasters should include the development of warning criteria of lahar flow occurrence. In the case of the Putih River, the Gunung Maron-based warning criteria with the 27 mm and 54 mm continuous cumulative rainfall for warning line and evacuation line respectively may be applied. A necessary update needs to be introduced considering additional data regarding rainfall and lahar flows.

Acknowledgements

The authors would like to express profound gratitude to the Institute for Research and Community Services, Universitas Gadjah Mada for provision of funds to carry out this study, and to two reviewers who contributed comments and suggestions to improve this chapter.

References

Chen, C, Saito, H, and Oguchi, T 2015. Rainfall intensity–duration condition for mass movements in Taiwan. *Progress in Earth and Planetary Science* 2, 14. DOI: 10.1007/s10346-007-0112-1

Directorate General of Water Resources 2009. *Community and Institution Development at Mt. Merapi Area*. Final Report, YEC, Ministry of Public Works, Government of Indonesia.

Giannecchini, R, Galanti, Y, and Avanzi, GD 2012. Critical rainfall thresholds for triggering shallow landslides in the Serchio River Valley (Tuscany) Italy. *Natural Hazards and Earth System Sciences* 12, 829–842.

Gonda, Y, Legono, D, and Santosa, UB 2013. Lahar flows and flash floods after the 2010 Eruption of Mt. Merapi. *Proceeding of 4th International Seminar of HATHI*, 6–8 September, Yogyakarta, Indonesia.

Guzzetti, F, Peruccacci, S, Rossi, M, and Stark, CP 2008. The rainfall intensity–duration control of shallow landslides and lahar flows: an update. *Landslides* 5, 3–17.

Hübl, J, Suda, J, Proske, D *et al.*2009. Lahar flow impact estimation. *International Symposium on Water Management and Hydraulic Engineering*, Ohrid, Macedonia.

Lavigne, F, Thouret, J, Voight, B *et al* 2000. Lahars at Merapi Volcano, Central Java: an overview. *Journal of Volcanology and Geothermal Research* 100, 423–456.

Legono, D, Wignyosukarto, B, and Saputra, A 2011. The influence of lahar flow on stability of river structures due to scour and impact forces. *Proceedings of International Seminar, HATHI*, Jakarta, 15–16 July 2011.

MLIT 2004. *Development of Warning and Evacuation System against Sediment Disaster in Developing Countries*. Ministry of Land, Infrastructure and Transport: Japan.

Pierson, TC, Wood, NJ, and Driedger, CL 2014. Reducing risk from lahar hazards: concepts, case studies, and role of scientists. *Journal of Applied Volcanology* 3, 16. DOI: 10.1186/s13617-014-00164.

Wardoyo, W, Legono, D, Jayadi, R, and Fathani, TF 2013. Analyzing sediment transport mechanism and related hydraulic structure damage after Mt. Merapi eruption in Gendol River. *Journal of Basic and Applied Scientific Research* 3(1), 849–857.

Worni, R, Huggel, C, Stoffel, M, and Pulgarin, B. 2008. Challenges of modeling current very large lahars at Nevado del Huila Volcano, Columbia. *Bulletin of Volcanology, Journal of the International Association of Volcanology and Chemistry of the Earth Interior (IAVCEI)* 70(5). DOI 10.1007/s00445-011-0522-8.

Yulinsa, N, Legono, D, and Rahardjo, AP 2014. Assessment of lahar flow warning criteria of Putih River after 2010 Mt. Merapi Eruption. *Proceeding of IAHR-APD*, Hanoi, 21–25 September.

21

A Method for Estimating the Porosity of Sediment Mixtures and Application to a Bed-Porosity Variation Model

Masaharu Fujita, Muhammad Sulaiman, and Daizo Tsutsumi

21.1 Introduction

Issues related to void structure of riverbed materials have both ecological and engineering aspects. From an ecological point of view, the interstitial spaces of bed materials are used as the habitat of aquatic biota such as fish and aquatic insects. A different structure of the interstitial spaces provides a habitat for different aquatic insects. A flood causing bed variation often changes the void structure of the bed material. As a result, an existing habitat could be lost, but a new habitat may be formed. If finer sediments come to occupy the interstitial spaces of coarser bed materials during a flood, the porosity and the permeability of the bed material decrease. Such a change in habitat condition ordinarily occurs in the bed deformation process. If finer sediments are supplied to a riverbed covered by a completely developed armour coat at a small rate, the fine sediments could infiltrate into the interstitial spaces of the coarser bed material and move into the void spaces. As a result, the bed level will not be easily changed. In this case, the habitat condition changes without a variation in bed level. Also, the change in porosity of bed materials affects the infiltration process of river flow. This process, one of the habitat conditions for aquatic biota, is an important issue for habitat conservation.

From an engineering point of view, a problem in the analysis of bed variation is evident because the porosity is not always constant. So far many bed-variation models have been proposed, and those models have been made under an assumption of a constant porosity. In a sense this assumption makes the calculation much easier. However, the models are not available for assessing the changes in interstitial spaces and porosity of bed material. Also, those models create an unreasonable result in case of fine sediment supply to an armoured riverbed because the riverbed deformation could be absorbed by the change of porosity.

For this problem on bed variation, it is essential to consider the sediment infiltration process into immobile gravels and sediment movement in the void of bed materials. Michiue *et al.* (1994) developed a model of the infiltration process of fine sediments into a gravel bed with large porosity, and applied the model to the calculation of non-equilibrium suspended-load concentration. Fujita *et al.* (2000) proposed a method for simulating bedload transport of finer sediment on immobile gravel beds and the process by which the transported sediment fills the void of the bed material. They have shown that the depth of fine sediment deposition in the void is dependent on the sediment supply rate. Wooster *et al.* (2008) performed a similar study on fine sediment infiltration.

Gravel-Bed Rivers: Processes and Disasters, First Edition. Edited by Daizo Tsutsumi and Jonathan B. Laronne.

So far the influence of the porosity change due to sedimentation on the riverbed environment has been widely investigated (e.g. Wood and Armitage 1997; Erman and Ligon 1988; Yamada and Nakamura 2002). Boulton *et al.* (1998) studied the void space as a habitat of aquatic organisms and the role of void space in controlling conductivity. Regarding the initial porosity of sediment deposits, Komura (1963) proposed a relation between the porosity and the median diameter. Wu and Wang (2006) modified this relation using a large number of measured data. However, these are results for uniform sediment. The porosity of particle mixtures has been studied in the research field of powder technology (Standish and Borger 1979; Yu and Standish 1993; Desmond and Weeks 2014). A computer random-packing model of spherical particles is a very useful tool to obtain the porosity. Suzuki and Oshima (1983) employed a random-packing model to find the positions of particles in a vessel by dropping and rolling. Nolan and Kavanagh (1993) improved the random-packing model for a lognormal distribution. Kenneth and Weeks (2014) developed a different random packing model to obtain the packing fraction. Tsutsumi, Fujita, and Sulaiman (2006) proposed a new idea for a random-packing model. They developed a method to find the lowest position where a spherical particle randomly selected from a grain-size distribution can be located stably in a vessel. They discussed a relation between porosity and geometrical parameters of grain-size distribution using this method. Additionally, there are many previous empirical porosity predictors and theoretical porosity predictors (Carling and Reader 1982; Peronius and Sweeting 1985; Yu and Standish 1993; Wu and Wang 2006; Wooster *et al.* 2008). Frings, Schüttrumpf, and Vollmer (2011) reviewed these predictors and summarized their accuracy.

Flushing of sediment from a reservoir is an effective method to keep the function of the reservoir and also to improve the sediment-poor condition downstream. For such a management, a method for assessing the change in interstitial spaces and porosity of bed materials is required under the above-mentioned background. Sulaiman, Fujita, and Tsutsumi (2007) presented a framework of a bed variation model to simulate porosity change as well as bed variation. A primitive numerical model for sediment mixture of two particle groups with very different grain-size distributions has already been presented under this framework. It cannot be used, however, for continuous grain-size distributions. Additional to the basic equations, new equations on the change in porosity are required for continuous grain-size distributions. Compaction is one of the very essential factors. However, an equation governing compaction has not yet been studied. Grain-size distribution is another factor affecting porosity. A different type of grain-size distribution has a different porosity. If the bed material is assumed to be always mixed at a given degree of compaction, the porosity is dependent only on the grain-size distribution. The porosity can be empirically obtained with geometric parameters. Tsutsumi, Fujita, and Sulaiman (2006) presented a numerical method for packing particles in a vessel to obtain the porosity of the sediment mixture. This method was applied to obtain a relation between porosity and the geometric parameters of grain-size distributions.

This study aims to first develop a submodel to estimate the porosity of bed material in a bed-porosity variation model. Identifying the type of grain-size distribution and relating the porosity to the geometric parameters of grain-size distribution are important to establish the submodel. First, the grain-size distributions of bed material are classified into three typical types and those geometric parameters are identified. Then a method for geometrically identifying the grain-size distribution type is presented. Second, a relation between the geometric parameters of a typical grain-size distribution and the porosity is obtained by means of a particle-packing simulation and experiment. Third, an algorithm for identifying the type of grain-size distribution and estimating the porosity is indicated. Finally, a bed-porosity variation model is developed and the numerical simulation results are introduced to show the performance of the model.

Natural bed materials often have probability-density distributions of grain size with more than two peaks. We, however, discuss mainly sediment mixtures with a single peak, namely unimodal grain-size distributions.

21.2 Identification of Grain-Size Distribution

21.2.1 Classification

Unimodal grain-size distributions of bed materials are classified into three typical distributions: namely, lognormal, Talbot, and anti-Talbot. These typical distributions and their probability-density distributions are shown in Figure 21.1, where f is the percentage of the finer grain size and $p = \mathrm{d}f/\mathrm{d}d$ is the probability density of grain size (Sulaiman, Tsutsumi, and Fujita 2007).

21.2.1.1 Lognormal Distribution

The most typical probability-density distribution of grain size is a lognormal distribution. Grain-size distribution of the lognormal type and the probability-density distribution are shown in Figure 21.1a. The probability-density distribution has a single peak and is symmetric around the mean value. The probability density function of the lognormal distribution is written as

$$p\left(\frac{d}{d_{mg}}\right) = \frac{1}{\sqrt{2\pi}\sigma_L}\frac{d_{mg}}{d}\exp\left[-\frac{\{\ln(d/d_{mg})\}^2}{2(\sigma_L)^2}\right] \tag{21.1}$$

where d_{mg} is the geometric average of grain size and σ_L is the standard deviation of $\ln d$. The geometric parameter of the lognormal distribution is only $\Pi_1 = \sigma_L$ if the diameter d is normalized with d_{mg}.

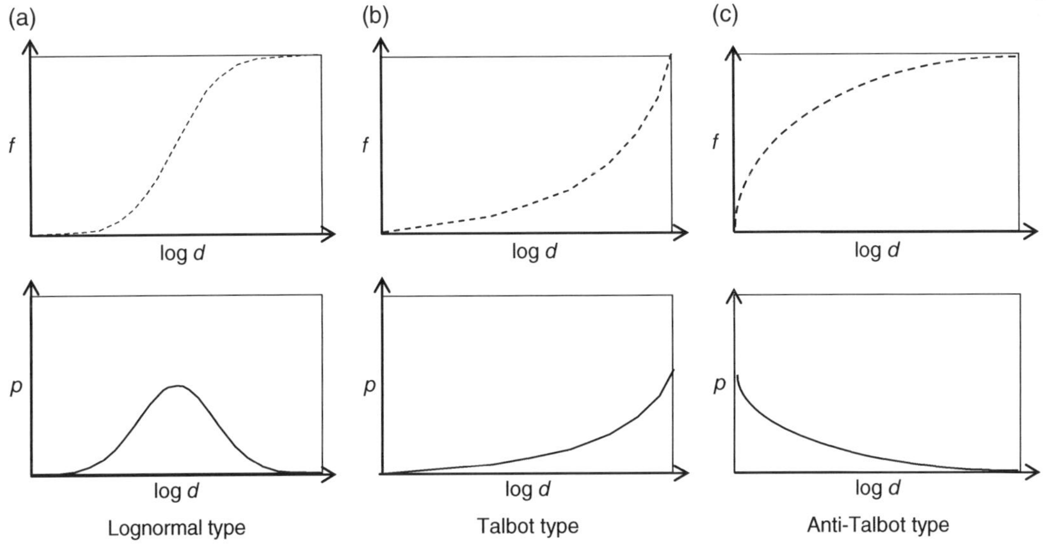

Figure 21.1 Typical types of unimodal grain-size distributions and their probability-density distributions (Sulaiman, Tsutsumi and Fujita 2007).

21.2.1.2 Talbot Distribution

It is well-known that the surface bed material of mountainous rivers covered by an armour has a grain-size distribution of Talbot type as shown in Figure 21.1b (Tatsuzawa, Hayashi, and Hasegawa 1998). The shape of the grain-size distribution is concave and the probability-density distribution is skewed towards the high-end tail. The original function of the Talbot distribution is written as

$$f(d) = \left(\frac{d}{d_{\max}}\right)^{n} \tag{21.2}$$

where $d_{\max}$ is the maximum grain size and n is a coefficient > 1 termed the Talbot number. Sulaiman, Tsutsumi, and Fujita (2007) pointed out that Eq. (21.2) does not always fit with the grain-size distributions of natural Talbot-type materials, and thereby they have modified the original Talbot function with consideration of the minimum grain size as shown in Eq. (21.3).

$$f(d) = \left(\frac{\log d - \log d_{\min}}{\log d_{\max} - \log d_{\min}}\right)^{n_T} \tag{21.3}$$

where n_T is a coefficient > 1 and in this chapter we use Eq. (21.3) as a Talbot distribution instead of Eq. (21.2). The geometric parameters of the modified Talbot distribution are the Talbot number $\Pi_1 = n_T$ and the ratio of maximum and minimum grain sizes, $\Pi_2 = d_{\max}/d_{\min}$.

21.2.1.3 Anti-Talbot Distribution

The anti-Talbot type of grain-size distribution and its probability-density distribution are shown in Figure 21.1c. The shape of the grain-size distribution is convex and the probability-density distribution is skewed towards the low-end tail of the distribution. The grain-size distribution is similar to the modified Talbot distribution, but the Talbot number n_T is < 1.

$$f(d) = \left(\frac{\log d - \log d_{\min}}{\log d_{\max} - \log d_{\min}}\right)^{n_T} \tag{21.4}$$

The geometric parameters of the anti-Talbot distribution are $\Pi_1 = n_T$ and the ratio of the maximum to the minimum diameter of a particle $\Pi_2 = d_{\max}/d_{\min}$.

21.2.2 Identification

21.2.2.1 Geometric Indices of Grain-Size Distribution

A grain-size distribution type can be visually identified by the shape of the distribution and the probability-density distribution. However, this visual identification is not available for bed-variation models. Thus, Sulaiman, Tsutsumi, and Fujita (2007) introduced the geometric indices β and γ to identify the distribution type. The indices β and γ are defined in Eqs (21.5) and (21.6), designating the relative location of the grain size d_{peak} for the peak probability density and the median grain size d_{50} between the minimum size $d_{\min}$ and the maximum size $d_{\max}$.

$$\beta = \frac{\log d_{\max} - \log d_{\text{peak}}}{\log d_{\max} - \log d_{\min}} \tag{21.5}$$

$$\gamma = \frac{\log d_{\max} - \log d_{50}}{\log d_{\max} - \log d_{\min}} \tag{21.6}$$

21.2.2.2 Indices of Typical Grain-Size Distributions

Figure 21.2a shows the lognormal distribution (LN) curves with the same mean value. Both β and γ of the lognormal distribution are equal to 0.5 as shown in Figure 21.3. The indices β and γ of the lognormal distribution are plotted on the centre point (0.5, 0.5). Figure 21.2b shows the Talbot (T) distribution curves with a same d_{max}/d_{min} and different Talbot numbers. Figure 21.3 shows

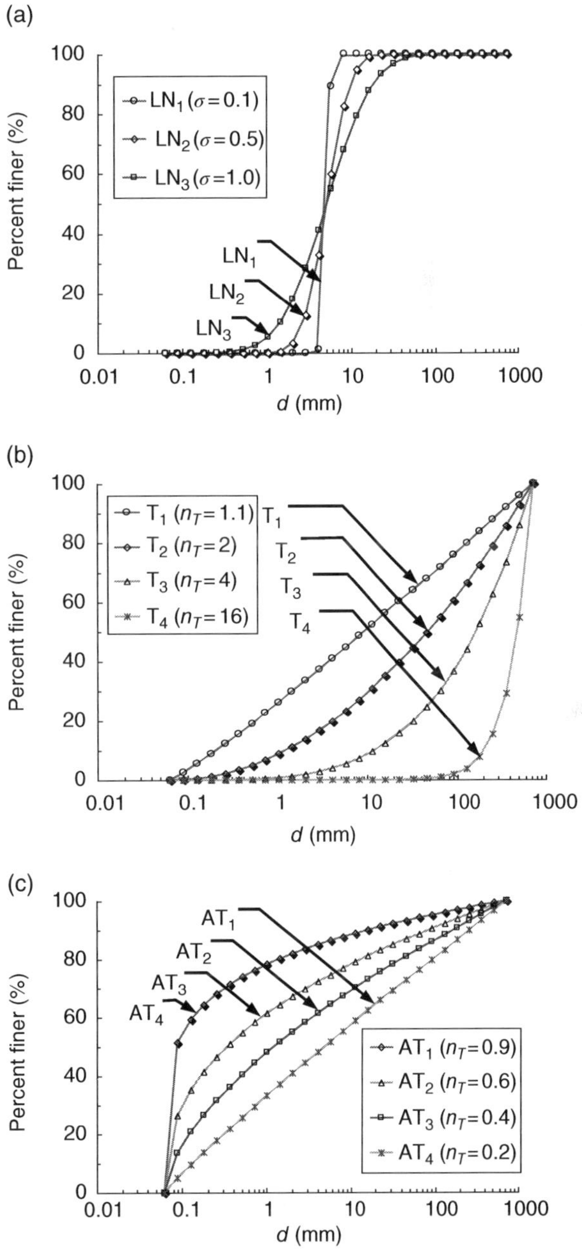

Figure 21.2 (a) Lognormal distributions with different standard deviations, (b) Talbot and (c) anti-Talbot distributions with different Talbot numbers.

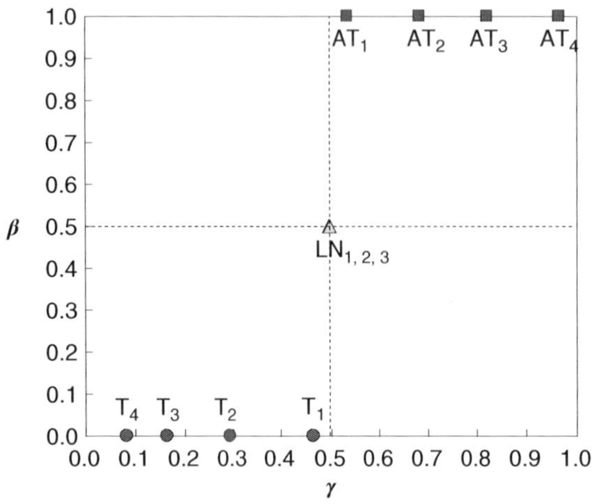

Figure 21.3 The relation between β and γ of lognormal, Talbot, and anti-Talbot distributions shown in Figure 21.2.

the geometric indices for each distribution shown in Figure 21.2b. The index β of the Talbot distribution is 0.0. The index γ changes from 0.0 to 0.5 depending on the Talbot number n_T and γ decreases with increase in the Talbot number. If $n_T = 1.0$, the distribution is linear and γ is 0.5. The indices β and γ are plotted on the line expressed by $\beta = 0.0$ $(0 < \gamma < 0.5)$. Figure 21.2c shows the anti-Talbot (AT) distribution curves with a same $d_{\max}/d_{\min}$ and different Talbot numbers. The indices β and γ for each distribution are shown in Figure 21.3. The index β is 1.0. The value of γ varies from 0.5 to 1.0 depending on the value of n_T and the value of γ increases with decrease in n_T. If $n_T = 1.0$, the distribution is linear and $\gamma = 0.5$. The indices β and γ are plotted on the line expressed by $\beta = 1.0$ $(0 .5 < \gamma < 1.0)$.

21.2.2.3 Indices of Other Grain-Size Distributions

Almost all of the natural grain-size distributions have transitional shapes between the typical grain-size distributions. Figure 21.4a shows a grain-size distribution between Talbot distribution and lognormal distribution. M_1 looks like a Talbot distribution and M_3 is visually close to a lognormal distribution. When the distribution changes from M_1 to M_3, the plot of β and γ approaches the centre point (0.5, 0.5) from the origin (0,0) along the line expressed by $\beta = \gamma$ as shown in Figure 21.5. Figure 21.4b shows the transition shape from lognormal distribution to anti-Talbot distribution. N_1, N_2, and N_3 gradually change from lognormal type to anti-Talbot type and the plot of β and γ approaches the point (1, 1) from the centre point along the line expressed by $\beta = \gamma$ as shown in Figure 21.5.

The distributions as shown in Figure 21.4c and d have transition shapes between linear and lognormal distributions. The indices β and γ are plotted in Figure 21.5. All of those distributions have the same median size. If β is close to 0.5, the distribution becomes lognormal. If β is close to 0.0 or 1.0, the distribution becomes linear.

Five distributions shown in Figure 21.4e have a different mean size, but the same peak grain size d_{peak}, midway between the minimum and maximum sizes. The indices β and γ are plotted in Figure 21.5. P_1 and O_1 are a unimodal distribution, but P_2 and O_2 are a bimodal distribution. When γ decreases to be < 0.25 or increases to be > 0.75 for $\beta = 0.5$, it is impossible to draw any unimodal distributions like P_2 and O_2 and the grain-size distribution must be bimodal.

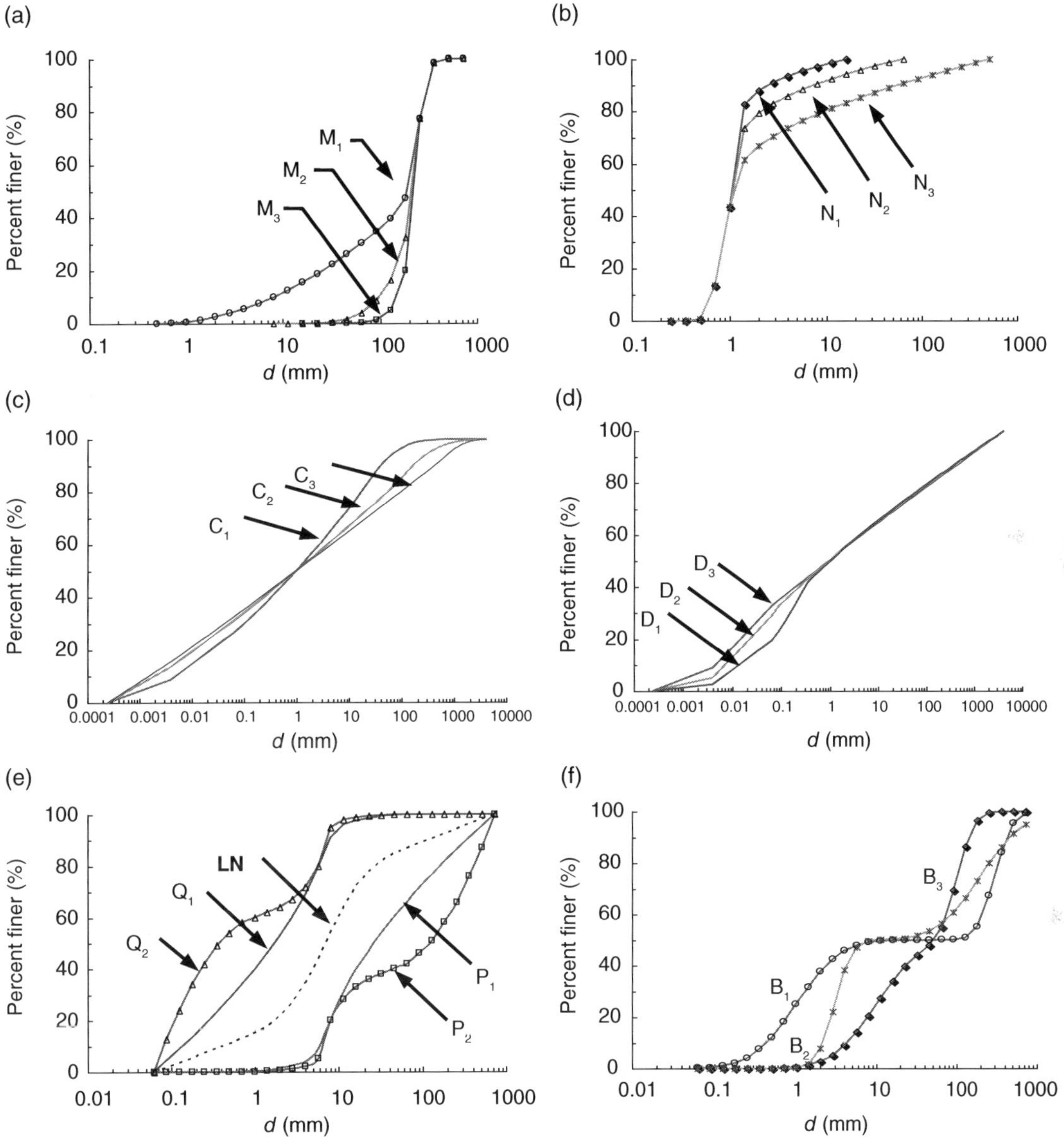

Figure 21.4 Transition type: (a) Talbot-lognormal, (b) lognormal–anti-Talbot, (c) lognormal–linear (Talbot), (d) lognormal–linear (anti-Talbot), (e) other distributions with $\beta = 0.5$, and (f) bimodal distributions.

Figure 21.4f shows three bimodal grain-size distributions. The geometric indices β and γ are obtained in the same way. The indices β and γ of bimodal distributions can be plotted in all regions ($0 < \beta < 1$ and $0 < \gamma < 1$).

21.2.2.4 Borderlines for Identification

The indices of Talbot and anti-Talbot distributions are on Line-1 ($\beta = 0$ and $0 < \gamma < 0.5$) and Line-2 ($\beta = 1.0$ and $0.5 < \gamma < 1.0$) in Figure 21.6. The indices of the lognormal distribution are plotted just

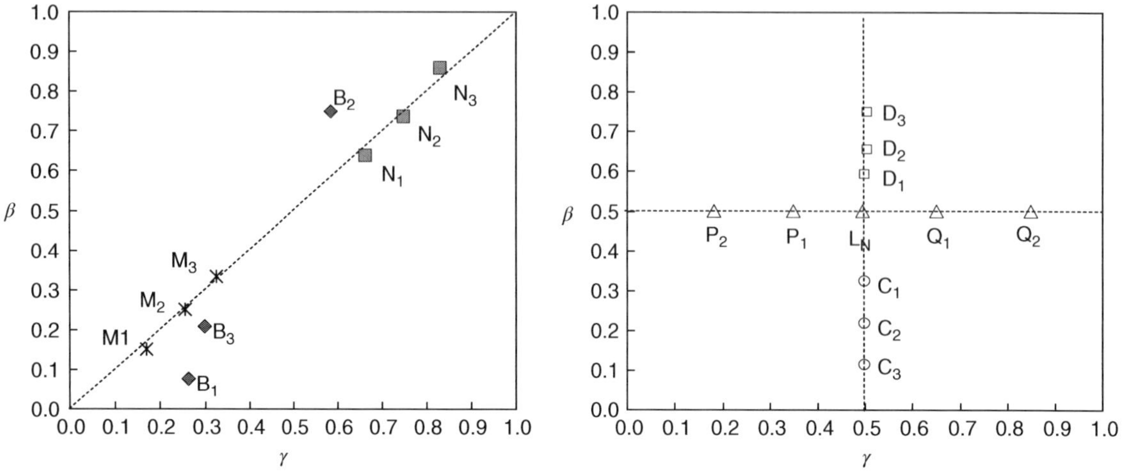

Figure 21.5 The relation between β and γ of the grain-size distributions shown in Figure 21.4.

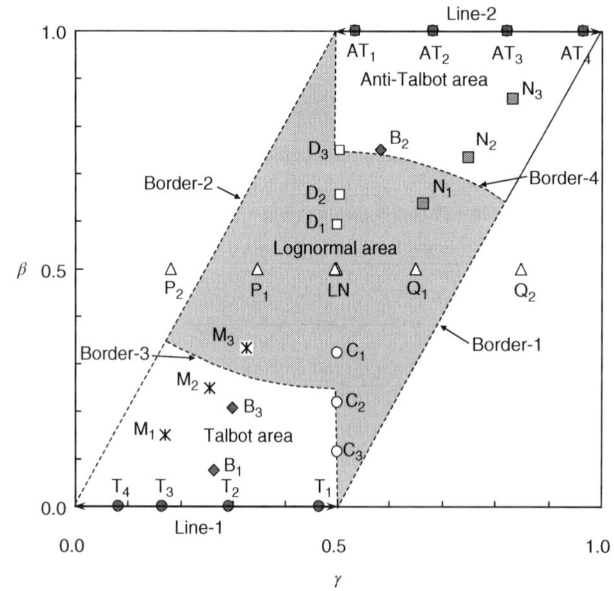

Figure 21.6 Diagram indicating Talbot, anti-Talbot, and lognormal region.

on the centre point (0.5, 0.5). The indices of the other distribution are plotted on the area of $0 < \beta < 1$ and $0 < \gamma < 1$, apart from Line-1, Line-2, and the centre point. However, as previously mentioned, there is an area where no unimodal distribution exists. From a geometric analysis, an area where unimodal distribution exists is surrounded by Border-1, Border-2, $\beta = 0$, and $\beta = 1.0$, as shown in Figure 21.6.

It seems reasonable and best that grain-size distribution types are identified with the distance to the point (γ, β) from Line-1 (corresponding to Talbot), Line-2 (corresponding to anti-Talbot), or the centre point (corresponding to lognormal). According to this criterion, the border line between the Talbot distribution and the lognormal distribution (Border-3) is written as Eq. (21.7) and the border line between anti-Talbot and lognormal distribution (Border-4) is expressed as Eq. (21.8).

$$\text{Border}-3: \beta = (0.5-\gamma)^2 + 0.25 \tag{21.7}$$

$$\text{Border}-4: \beta = -(0.5-\gamma)^2 + 0.75 \tag{21.8}$$

The regions of the Talbot, anti-Talbot, and lognormal distributions are indicated in Figure 21.6. The indices of the grain-size distributions shown in Figure 21.4 were plotted on the classification diagram of grain-size distribution type (Figure 21.6). According to this criterion: M_1 and M_2 are identified as a Talbot distribution; N_2 and N_3 are classified into anti-Talbot; M_3, N_1, P_1, Q_1, C_1, D_2, and D_2 are identified as lognormal. These results seem to be appropriate visually. P_2 and Q_2 are bimodal distributions and outside the unimodal area.

21.2.2.5 Application to Natural Bed Materials

The identification method presented above is applied to the following data and the validity of the method is verified. The samples were taken from the Ai River (Fujita 2004), Hino River (MLIT 2004), Ohtaki River (Ashida and Fujita 1987), and a Fukadani bare slope. The grain-size distributions are shown in Figure 21.7 (Sulaiman, Tsutsumi, and Fujita 2007). A, H, O, and F denote the samples from Ai River, Hino River, Ohtaki River, and a Fukadani bare slope, respectively. The samples are classified by the characteristics of sampling points as follows:

1) surface bed material (A-1, 5, H-1, 2, 3);
2) subsurface bed material (A-2, 6, H-4, 5, 6);
3) sand bar (A-3);
4) sediment deposition in a sabo dam (A-4);
5) sediment deposition in a valley immediately after Ontake Lanslide 1984 (O-1);
6) sediment deposition in a reservoir (O-2, 3);
7) sediment produced on a bare slope (F-1, 2, 3).

The armour sediment was collected to the depth of the maximum grain size. In field measurement, coarser materials are sometimes truncated because of handling difficulty. The coarser materials were, however, not cut off at samples A, O, and F. The report for H did not mention whether the coarser materials were truncated.

The indices β and γ of the distributions are plotted on the classification diagram of grain-size distribution type (Figure 21.8). The plots are categorized into five types by visual identification to Talbot, lognormal, Talbot–lognormal transition, anti-Talbot–lognormal transition, and bimodal. The indices of the grain-size distributions visually identified as Talbot or lognormal distributions are plotted in Talbot area and lognormal area, respectively. The indices of the grain-size distributions that were visually identified as transition were plotted near Border-3 or Border-4. This result indicates that this classification method is available for the identification of the grain-size distribution type.

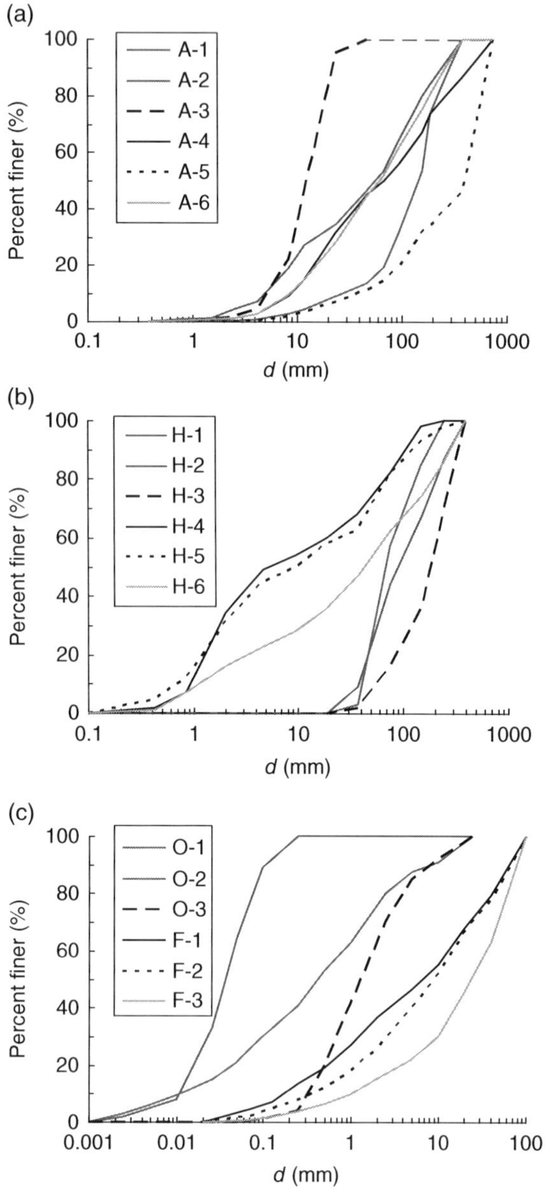

Figure 21.7 Grain-size distributions of natural sediment mixtures sampled in (a) Ai River, (b) Hino River, and (c) Ohtaki River and Fukadani bare slope (Sulaiman, Tsutsumi and Fujita 2007). (See insert for color representation of this figure.)

21.3 Relationship Between the Geometric Parameters of Grain-Size Distributions and Porosity

After the identification of distribution type, the relation between the geometric parameters of the three typical distributions and the porosity is necessary in a bed-porosity variation model. To obtain the relation, systematic data on the porosity for the geometric parameters are necessary. In this chapter we employed a particle-packing numerical model and a measurement method to collect the data.

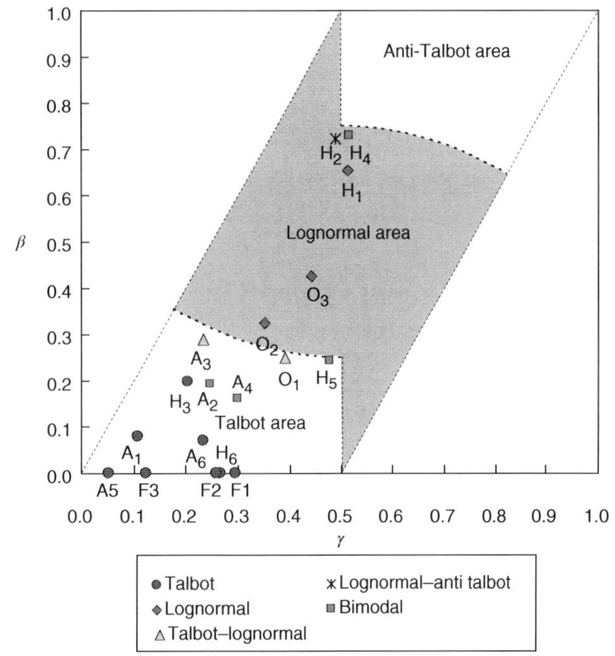

Figure 21.8 The relation between β and γ of natural grain-size distributions shown in Figure 21.7.

21.3.1 Particle-Packing Simulation

Tsutsumi, Fujita, and Sulaiman (2006) developed a numerical method for packing particles in a vessel to obtain the porosity of a mixture of spherical particles. The vessel size should be several times larger than the maximum grain size, however, it takes much execution time for wide grain-size distributions. By this limitation, the vessel sizes were from two to five times the maximum grain size in the simulation. The sizes of packed particles and the packing order are determined from the grain-size distribution using random numbers. The particles are placed in the lowest stable position available in the vessel. The porosity of the particle mixture depends on not only the grain-size distribution but also the degree of compaction. However, the compaction degree cannot be intentionally controlled in the model. As a result of simulation, the compaction degree is given. The defect of this method is that it takes too much time to pack the particles with wide grain-size distributions in a vessel. For such wide distribution, it is better to use a measurement method according to the actual sediment mixture.

21.3.2 Measurement Method

21.3.2.1 Previous Studies

Methods for the measurement of gravel-bed porosity are described by Carling and Reader (1982), Milhous (2002), and Gayraud and Philippe (2003). Carling and Reader (1982) explained a frozen-sediment core sampling method. Milhous (2002) developed a simple method for measuring porosity below the armour coat using a wooden frame and a plastic sheet. Gayraud and Philippe (2003) analysed 99 frozen cores collected from 15 French streams to investigate the influence of grain-size distribution and vertical bed packing on the interstitial habitat. However, the relation between the geometric parameters of typical grain-size distributions and the porosity was not clearly described.

21.3.2.2 A Measurement Method

Porosity is defined as the ratio of the pore volume to the bulk volume of a given sample. To obtain the porosity, it is necessary to measure the bulk volume of the material and the volume of sediment. As a technique to measure the bulk volume, a water displacement method (ASTM 1994) is easily used in field surveying. The measuring method is as follows (Figure 21.9):

1) Place and set the PVC ring vessel stable and horizontal on a flat ground surface. Then, place a plastic sheet sticking firmly to the ring and the ground surface.
2) Fill water up to a level in the ring vessel and record the volume of the water V_1 (Figure 21.9a).
3) Remove the surface sediment under the ring vessel by a given depth (generally the maximum grain size).
4) Place the plastic sheet back inside of the ring vessel. Fill the water again to the same level and measure the volume of the water V_2 (Figure 21.9b).
5) Calculate the bulk volume of the sample (V_t) by subtracting V_1 from V_2.
6) Measure the weight and the grain-size distribution of the removed sediment.
7) Measure the specific gravity of the sediment and convert the volume of the sediment V_s.
8) Calculate the porosity.

This method is very simple and applicable for field surveying. However, the accuracy of the porosity is reduced if the roughness of the ground surface is different before and after sediment removal (Figure 21.9a and b). When the porosity of the armour coat is measured, the surface of the armour coat is rough, but a smooth surface of the lower layer appears after the removal of the armour coat. If the standard method is used, the void in the vicinity of the original surface is neglected (Figure 21.9). To improve this technical problem, we used a method to fill the void of the surface layer with salt up to approximately cover the coarse material before setting the plastic sheet in step **1** as shown in Figure 21.9c.

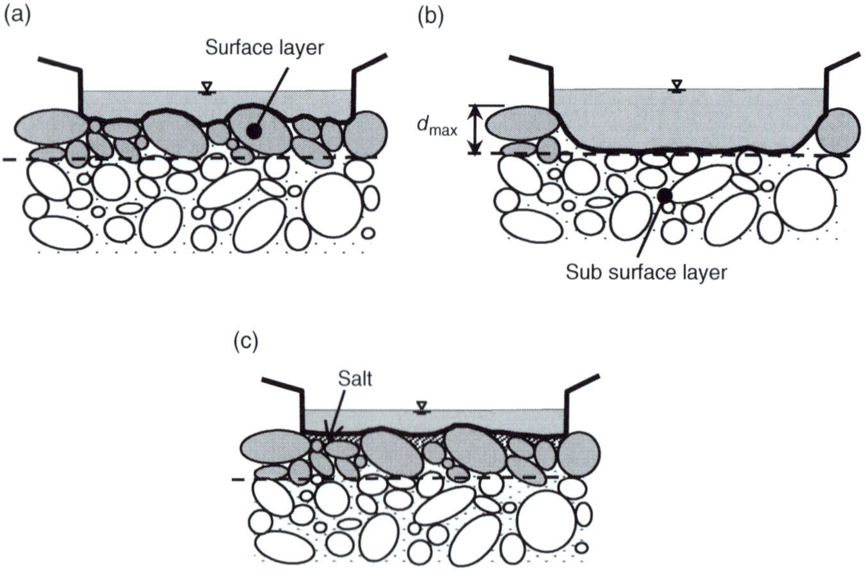

Figure 21.9 Schematic instruction of a porosity measuring method.

21.3.3 Simulation and Measurement Results

21.3.3.1 Method and Conditions

Porosity of sediment mixtures was measured by simulation and laboratory experiment. The simulation was conducted for sediment mixtures of lognormal and Talbot types. The standard deviation of the lognormal distribution σ_L ranged from 0.01 to 1.5. The ratios of d_{max} to d_{min} of Talbot type were 10 and 100, and the Talbot number ranged from 1 to 12. The simulation was conducted five times for each sample.

In the laboratory experiments, the method described in section 21.3.2.2 was employed for two types of grain-size distribution. Several sediments were mixed well in a plastic box 60 cm in length, 40 cm in width, and 17 cm in height, to make 20 sediment mixtures with different types of grain-size distribution; consisting of five lognormal distributions with different standard deviation (i.e., 0.02, 0.253, 0.29, 0.303, and 1.6) and 10 Talbot distributions. The mixed sediment was scraped flat and compacted evenly by repeatedly dropping the wood cylinder 5 cm. A wood cylinder with a diameter of 15 cm and a height of 10 cm and with a steel handle (total weight = 2.38 kg) was used to compact the material. The grain-size distribution and the porosity were measured five times for each sample.

21.3.3.2 Porosity of Lognormal Distributions

Figure 21.10 shows the mean value of the simulated porosity for each lognormal distribution (Sulaiman, Tsutsumi, and Fujita 2007). In this figure, the measured porosity is also plotted. The error bar of each plot indicates the range between maximum and minimum values of five repetitions.

There is no great difference in the trend of porosity change with standard deviation between the measurement and the simulation. However, the measurement result is somewhat greater than the simulated one. The discrepancy may be due to differences in the shape of particles, the degree of compaction, and the size grading of the sediment mixture in the box. Particularly in the measurement, it is very difficult to mix the sediment evenly. Consequently, the coarser particles lie at higher positions than the finer particles. This grading process makes the porosity greater, while in the simulation the particles were mixed evenly. Thus, the particle packing in the simulation might be denser than the packing of particles in the measurement.

Considering that the porosity of natural uniform sediment ($\sigma_L = 0$) is about 0.4, the simulated porosity is underestimated. If the simulated porosity shifts to the natural porosity, the relation between standard deviation and porosity is modified as indicated by the dashed line in Figure 21.10.

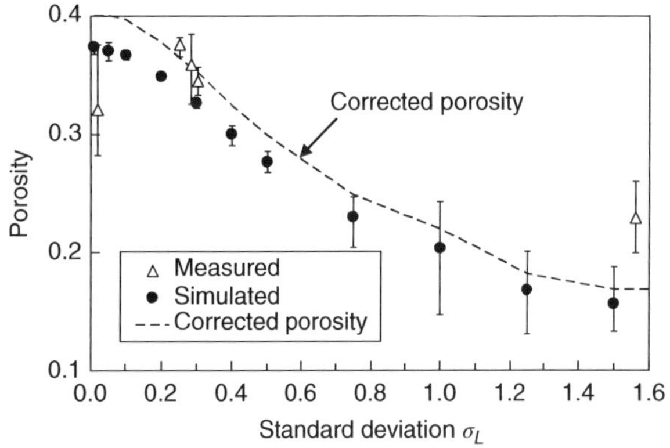

Figure 21.10 The relation between porosity and standard deviation of lognormal distribution.

21.3.3.3 Porosity of Talbot Distributions

The mean porosities of five repetitions of the measured porosity for 10 different Talbot distributions are shown in Figure 21.11 (Sulaiman, Tsutsumi, and Fujita 2007). The relationship between the Talbot number n_T and the simulated porosity is shown in Figure 21.11. The simulation result shows that the porosity increases with Talbot number and decreases with the ratio of d_{max} to d_{min}. The measured porosity has a similar tendency, but both are rather different quantitatively for the same Talbot number and the same ratio of d_{max} to d_{min}. It seems that the reason for this discrepancy is also due to the differences in particle shape, degree of compaction, and the size grading of the sediment mixture.

Considering two relations between the simulated porosity and Talbot number n_T are roughly parallel in the simulation as shown in Figure 21.11, the simulated porosity may be adjusted to the measurement – the corrected relations are drawn as shown in Figure 21.11. More simulations and experiments are necessary to improve this porosity predictor.

21.3.3.4 Porosity of Bimodal Distributions

Sediment with a bimodal grain-size distribution, a binary sediment mixture, is a mixture of two materials with quite different mean diameter. If each material is composed of a uniform sediment and the porosity of the two materials is λ_s, the porosity of the binary sediment mixture is calculated by a simple analysis of a two-size mixture according to the following equations

$$\lambda = \lambda_s - \frac{p}{1-p}(1-\lambda_s) \quad \text{for} \quad p \le \frac{\lambda_s}{(1+\lambda_s)} \tag{21.9}$$

$$\lambda = \frac{p\lambda_s}{1-\lambda_s(1-p)} \quad \text{for} \quad p > \frac{\lambda_s}{(1+\lambda_s)} \tag{21.10}$$

where p is the proportion of the finer material. The relation between λ and p for $\lambda_s = 0.35$ is shown in Figure 21.12. It is found that the porosity has a minimum value (0.1225) at $p = 0.259$ (Sulaiman 2008).

Sulaiman (2008) performed a packing simulation to obtain the porosities of binary sediment mixtures shown in Table 21.1. Material A and B have a lognormal distribution with standard deviation $\sigma_L = 0.2$, with d_{50} equal to 1 mm and 4 mm, respectively. Simulation was carried out for seven different mixtures of A and B. The proportion of finer material A, p, is 0, 0.2, 0.25, 0.3, 0.5, 0.75, and 1.0. The relationship between the proportion, p, and the porosity for the binary sediment mixture is shown in Figure 21.12. The porosity of a mixture of two materials depends on the fractional concentrations of

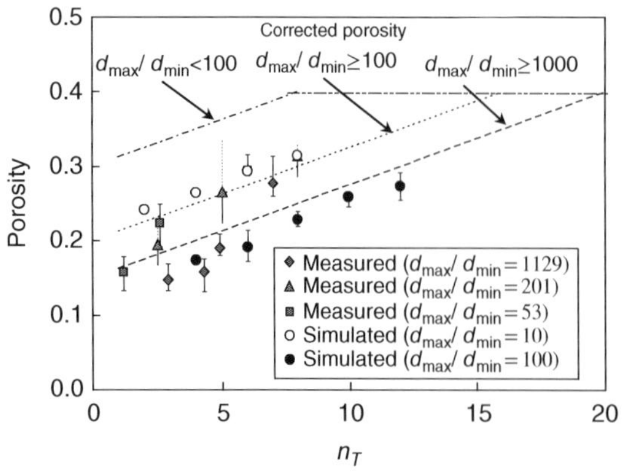

Figure 21.11 The relation between porosity and Talbot number for Talbot distribution.

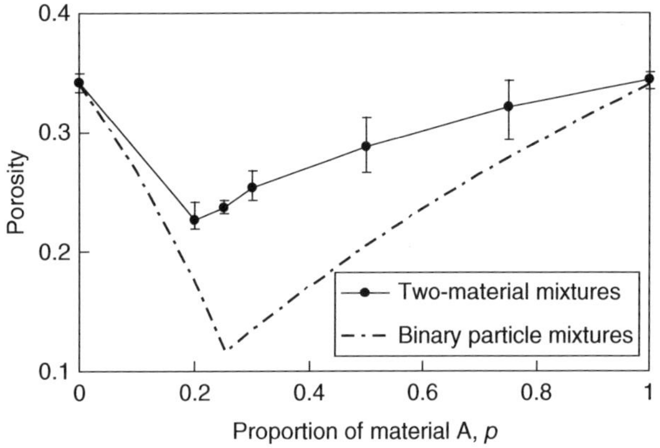

Figure 21.12 Relationship between the proportion of material *A*, *p*, and the porosity of the mixtures.

Table 21.1 Characteristic parameters of sediment mixtures for the bimodal distribution.

Fraction	σ_L	d_{min} (mm)	d_{50} (mm)	d_{max} (mm)
A	0.2	0.5	1.0	2.0
B	0.2	2.0	4.0	8.0

each material. Comparing with the porosity of binary sediment mixtures, the changes of porosity with p for binary particle mixtures and binary mixtures are similar and a minimum porosity exists near $p = 0.2$.

21.4 An Algorithm for Estimating the Porosity

It is very important include the results from sections 21.2 and 21.3 in the establishment of a model of the variation of bed porosity. Therefore, a subprogram for identifying the grain-size distribution type and obtaining the porosity is explained in this section. The subprogram involves four stages as shown in Figure 21.13.

21.4.1 Input of Grain-Size Distribution Data

The grain size of bed materials is divided into several classes in the bed-variation model. The representative grain sizes $dd(i)$ of class-i and the proportion $f_s(i)$ are generally used in calculation. However, the minimum grain size and the maximum grain size are necessary to obtain the geometric indices β and γ. Also, to obtain the Talbot number by fitting the grain-size distribution with the Talbot distribution, the ith grain size $d(i)$ and the percentage finer $f(i)$ are necessary. Therefore, $d(i)$ and $f(i)$ as well as $dd(i)$ and $f_s(i)$ are used in the calculation program. As the representative grain size $dd(i)$ is the geometric average of the uppermost size and the lowest size of class i, $d(i)$ and $f(i)$ are calculated by Eqs (21.11) and (21.12) if the minimum grain size $d(1)$ or the maximum grain size $d(n)$ is given.

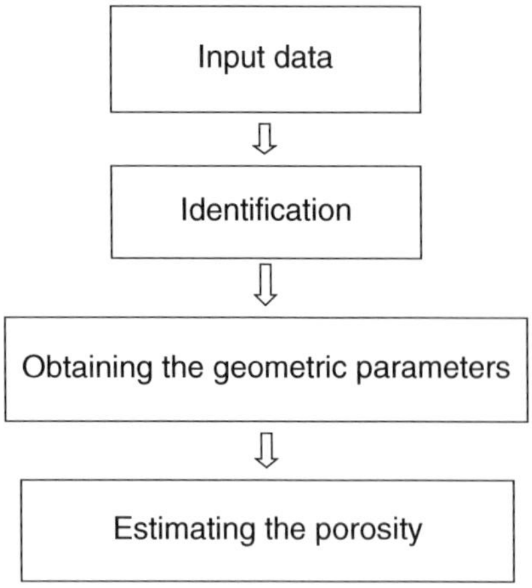

Figure 21.13 Calculation procedure for classifying the grain-size distribution and estimating the porosity.

$$d(i+1) = \frac{dd(i)^2}{d(i)} \tag{21.11}$$

$$f(i) = \sum_{j=1}^{i-1} f_s(j) \quad (i > 1), f(1) = 0.0 \tag{21.12}$$

21.4.2 Identification and Classification

Next, the grain-size distribution is classified into Talbot and lognormal types. The geometric indices β and γ are calculated with input data on grain-size distribution. Then, the geometric type of the grain-size distribution is identified using the diagram on classification of grain-size distribution type shown in Figure 21.6.

21.4.3 Obtaining the Geometric Parameters of Grain-Size Distribution

The geometric parameters of grain-size distribution are obtained by a curve-fitting. The geometric parameter of the lognormal distribution is only the standard deviation σ_L. The standard deviation for the lognormal distribution is calculated from the grain-size distribution data by Eq. (21.13).

$$\sigma_L^2 = \sum_{i=1}^{N} \left(\ln d_i - \overline{\ln d}\right)^2 . fs_i \tag{21.13}$$

The geometric parameter of the Talbot distribution is the Talbot number n_T and the ratio of the maximum and minimum diameter, d_{max}/d_{min}. The minimum error method is best to obtain the Talbot number, but an easier method is the individual five-point fitting method, as proposed in this paper.

The Talbot number by one-point fitting for percent finer of x% is calculated as

$$n_{T(x\%)} = \frac{\ln\left(f\left(d_{(x\%)}\right)\right)}{\ln\left(\frac{\log d_{(x\%)} - \log d_{min}}{\log d_{max} - \log d_{min}}\right)} \tag{21.14}$$

After the Talbot numbers are obtained for five points (16%, 25%, 50%, 75%, and 84%), the Talbot number is finally obtained from the average of n_T (x %) as follows:

$$n_T = \frac{n_{T(16\%)} + n_{T(25\%)} + n_{T(50\%)} + n_{T(75\%)} + n_{T(84\%)}}{5} \tag{21.15}$$

21.4.4 Estimating the Porosity

The porosity is estimated by using the relation between the geometric parameters of grain-size distribution and the porosity as respectively shown in Figures 21.10 and 21.11 for the lognormal and Talbot distributions. If this algorithm is installed in a bed-variation model able to calculate the change of grain-size distribution of bed material, the porosity of bed material is also obtained at each time-step.

21.5 Application to Bed-Porosity Variation Model

21.5.1 Basic Equations

Considering the change of porosity, the continuity equation of sediment is expressed as

$$\frac{\partial}{\partial t}\int_{z_o}^{z}(1-\lambda)dz + \frac{1}{B}\frac{\partial Q_s}{\partial x} = 0 \tag{21.16}$$

where λ is the porosity, z is the bed level, z_o is the datum of bed level, Q_s is the total sediment transport rate, B is the channel width, t is time, and x is an axis along the flow.

The continuity equation of each grain size fraction is calculated by

$$\frac{\partial}{\partial t}\int_{z_o}^{z}(1-\lambda)p_j dz + \frac{1}{B}\frac{\partial Q_{sj}}{\partial x} = 0 \tag{21.17}$$

where j is the grain size class, p_j is the existing ratio of size class j, and Q_{sj} is the total sediment transport of class j.

The porosity is a function of geometric parameters, Π_1, Π_2, Π_3....

$$\lambda = f_n(\Pi_1, \Pi_2, \Pi_3, \ldots\ldots) \tag{21.18}$$

For the lognormal distribution $\Pi_1 = \sigma_L$ and for the Talbot distribution $\Pi_1 = d_{max}/d_{min}$ and $\Pi_2 = n_T$. The relations between λ and Π are shown in Figures 21.10 and 21.11.

21.5.2 Conditions of Simulation

The bed-porosity variation model is applied to the following conditions. The channel length is 15 m and the channel slope is 1/100. The downstream end is fixed. Two sediment mixtures of different types are initially lain. In Case C-1, a sediment mixture with a lognormal distribution is shown in Figure 21.14. Under the condition of no sediment supply, water is supplied at a rate of 0.02 m^3/s. In Case C-2, a sediment mixture with a Talbot distribution shown in Figure 21.14 is lain. The sediment with median sizes from 2.8 to 5.9 mm is supplied at a rate of 8×10^{-6} m^2/s, and water dischrge is 0.02 m^3/s.

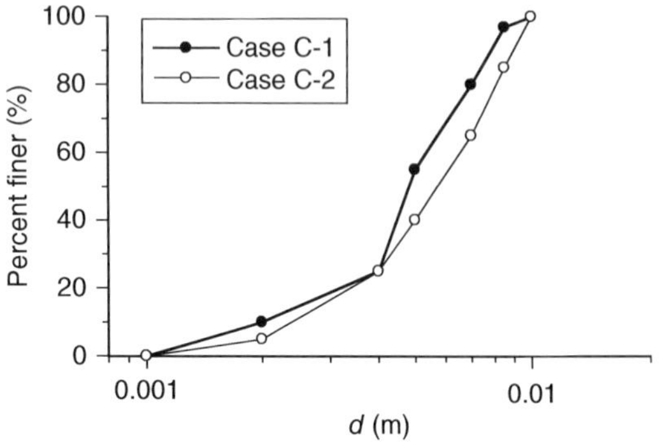

Figure 21.14 Initial grain-size distributions of bed materials.

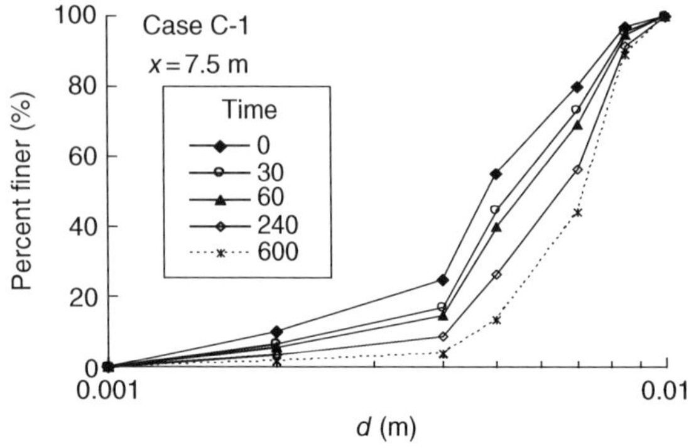

Figure 21.15 Change in grain-size distributions of bed materials at middle section.

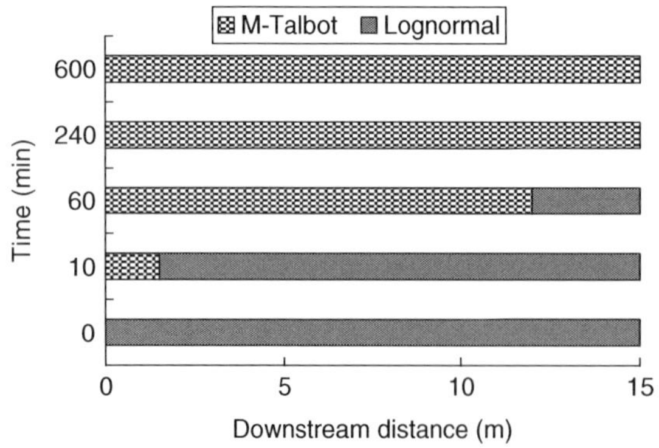

Figure 21.16 Change in type of grain-size distribution.

21.5.3 Results

In Case C-1 the initial grain-size distribution is classified as a lognormal distribution by the presented model. Under the no sediment supply condition, armouring develops. Figure 21.15 shows the change of the grain-size distribution at the middle part of the channel. Figure 21.16 shows the change of the grain-size distribution type. These figures show the initial lognormal type distribution as it changes to a Talbot distribution with armouring process from the upstream side. Figure 21.17 shows the change

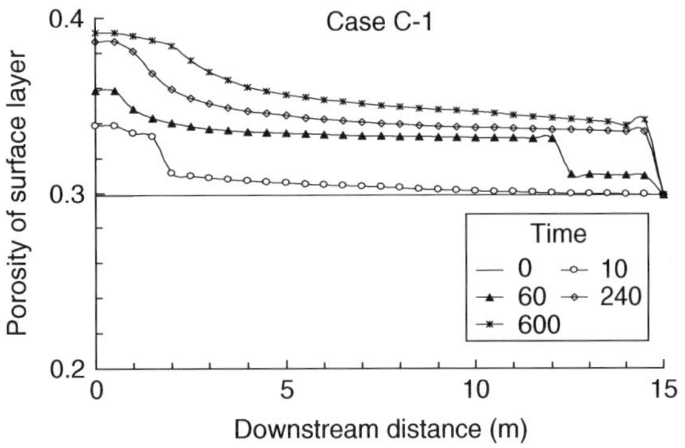

Figure 21.17 Change in porosity of bed materials.

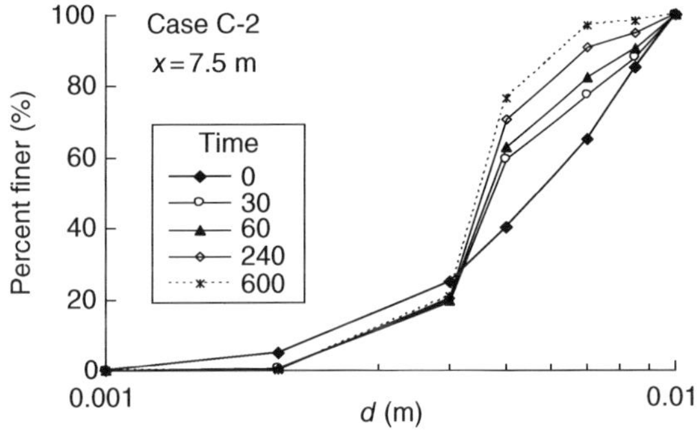

Figure 21.18 Change in grain-size distribution of bed materials at middle section.

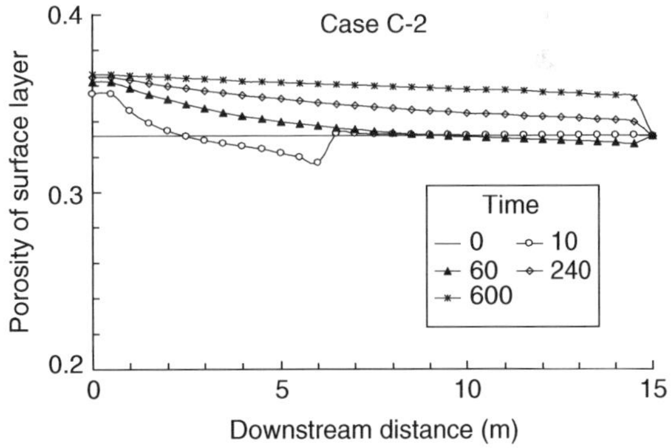

Figure 21.19 Change in porosity of bed materials.

in porosity of the bed material. The porosity increases to 0.4 from the upstream side. This increase in porosity with armouring process looks reasonable because this feature is observed in natural rivers.

In Case C-2, sediment supply is low resulting in bed degradation. Figure 21.18 shows the change of the grain-size distribution at the middle part of the channel. This figure shows the initial Talbot type distribution changing to a lognormal type distribution with fine sediment deposition. Figure 21.19 shows the change of the porosity. The porosity increases almost evenly over the whole flow area because the sediment became more uniform.

We introduced the absorption of bed deformation by porosity change as one of the characteristics of sediment transport in rivers with armour. Case C-2 was a condition where we expected that the model would simulate this feature, but instead bed degradation occurred. If we carefully select the hydraulic and sediment conditions the model can show this performance.

21.6 Conclusions

A method for identifying the grain-size distribution type of sediment mixtures and estimating the porosity has been proposed and a bed-porosity variation model has been developed. The results are as follows:

1) The grain-size distribution type is classified into Talbot, lognormal, and anti-Talbot types by the geometric indices designating the relative locations of the grain size for the peak probability density and d_{50} between the minimum and maximum sizes.
2) A classification diagram of grain-size distribution types has been presented. The validity and applicability of this diagram were examined using natural grain-size distributions.
3) For a lognormal distribution there is no large difference in the tendency of the change of porosity with the standard deviation in the simulation and measurement. However, the measurement result is somewhat larger than the simulation result because of the grain size grading due to packing the sediment mixture in the box. The relation between the porosity and the standard deviation in the simulation was modified using the measurement result.
4) For the Talbot distribution it is possible to roughly estimate the porosity using the simulated relation between porosity and the geometric parameter of grain-size distribution. However, as there is a rather large difference between the simulation and the measurement, more investigation is necessary to obtain a better relation.
5) As a technique to measure the porosity in the field, a water displacement method that is an ASTM standard was introduced. A technical problem of this method was pointed out, and an improvement has been proposed.
6) A bed-porosity variation model is developed and applied to two simple cases of sorting and fine sediment deposition. Some aspects remain to be improved in the model, but additional information on grain-size distribution type and porosity could be shown in the bed-variation model.

Acknowledgements

We thank the reviewers of this chapter for providing some good suggestions and recommendations.

References

Ashida, K and Fujita, M 1987. Simulation of reservoir sedimentation. *Annuals, DPRI, Kyoto University* 30 (B-2), 407–419. (In Japanese.)

ASTM 1994. *Standard Test Method for Density of Soil and Rock in Place by the Water Replacement Method in a Test Pit.* American Society for Testing and Materials Standard D 5030–89.

Boulton, AJ, Findlay, S, Marmonier, P *et al.* 1998. The functional significance of the hyporheic zone in streams and rivers. *Annual Review of Ecology, Evolution, and Systematics* 29, 59–81.

Carling, PA and Reader, NA 1982. Structure, composition and bulk properties of upland stream gravels. *Earth Surface Processes and Landforms* 7, 349–365.

Desmond, KW and Weeks, ER 2014. Influence of particle size distribution on random close packing of spheres. *Physical Review* E90, 022204.

Erman, DC and Ligon, FK 1988. Effects of discharge fluctuation and the addition of fine sediment on stream fish and macroinvertebrates below a water-filtration facility. *Environmental Management* 12, 85–97.

Frings, RM, Schüttrumpf, H, and Vollmer, S 2011. Verification of porosity predictors for fluvial sand–gravel deposits. *Water Resources Research* 47, W07525.

Fujita, M 2004. Study on Riverbed Morphology as Habitat for Aquatic Living Things. *Report of River Environment Fund.* (In Japanese.)

Fujita, M, Sawada, T, Mizuyama, T, and Kinoshita, A 2000. Movement of sediment removed at Sabo Dam and its impact on river environment. *Journal of Hydraulic Engineering, JSCE* 44, 1215–1220. (In Japanese.)

Gayraud, S and Philippe, M 2003. Influence of bed-sediment features on the interstitial habitat available for macroinvertebrates in 15 French streams. *International Reviews in Hydrobiology* 88, 77–93.

Kenneth, WD and Weeks, ER 2014. Influence of particle size distribution on random close packing of spheres. *Physics Review Series E* 90, 022204. DOI:10.1103/PhysRevE.90.022204.

Komura, S 1963. Discussion of sediment transport mechanics: introduction and properties of sediment. *Journal of Hydraulics Division, ASCE* 89(1), 262–266.

Michiue, M, Fujita, M, Kitagawa, T, and Mimura, K 1994. Deposition and transport process of suspended load on a channel bed with a large porosity. *Journal of Hydraulic Engineering, JSCE* 38, 609–614. (In Japanese.)

Milhous RT 2002. Measurement of the bed material of gravel-bed rivers. *Hydraulics Measurement and Experimental Methods 2002*, Estes Park, CO; CD –ROM.

MLIT 2004. *Technical Report on Integrated Sediment Management of Hino River Basin 2003.* Hinogawa River Office, MLIT. (In Japanese.)

Nolan, GT and Kavanagh, PE 1993. Computer simulation of random packings of spheres with log-normal distributions. *Powder Technology* 76, 309–316.

Peronius, N and Sweeting, TJ 1985. On the correlation of minimum porosity with particle size distribution. *Powder Technology* 42, 113–121.

Standish, N and Borger, DE 1979. The porosity of particulate mixtures. *Powder Technology* 22, 121–125.

Sulaiman, M 2008. Study on porosity of sediment mixtures and a bed-porosity variation model. *PhD thesis*, Kyoto University.

Sulaiman, M, Fujita, M, and Tsutsumi, D 2007. Bed variation model considering porosity change in riverbed material. *Journal of the Japan Society of Erosion Control Engineering* 60(1), 11–18.

Sulaiman, M, Tsutsumi, D, and Fujita, M 2007. Porosity of sediment mixtures with different type of grain size distribution. *Journal of Hydraulic Engineering, JSCE* 51, 133–138.

Suzuki, M and Oshima, T 1983. Estimation of the co-ordination number in a multi-component mixture of spheres. *Powder Technology* 35, 159–166.

Tatsuzawa, H, Hayashi, H, and Hasegawa, K 1998. Roles of heterogeneous property of bed materials for the formation of step-pool system in mountain streams. *Journal of Hydraulic Engineering, JSCE* 42, 1075–1080. (In Japanese.)

Tsutsumi, D, Fujita, M, and Sulaiman, M 2006. Changes in the void ratio and void structure of riverbed material with particle size distribution. In *River, Coastal and Estuarine Morphodynamics*, Vol. 2 (eds G Parker and MH Garcia). Taylor & Francis: London; 1059–1065.

Wood, PJ and Armitage, PD 1997. Biological effect of fine sediment in the lotic environment. *Environmental Management* 21, 203–217.
Wooster, JK, Dusterhoff, SD, Cui, Y *et al.* 2008. Sediment supply and relative size distribution effects on fine sediment infiltration into immobile gravels. *Water Resources Research* 44, WO3424.
Wu, WW and Wang, SSY 2006. Formulas for sediment porosity and settling velocity. *Journal of Hydraulic Engineering, ASCE* 132(8), 858–862.
Yamada, H and Nakamura, F 2002. Effect of fine sediment deposition and channel works on periphyton biomass in the Makomanai river, northern Japan. *River Research and Applications* 18, 481–493.
Yu, AB and Standish, N 1993. A study of packing particles with mixture size distribution. *Powder Technology* 76, 113–124.

Discussion

Discussion by Astrid Blom

The authors discuss the existing 'analytical method for a two-sizes mixture' by Yu and Standish (1987). How well does the authors' complex porosity model reproduce the result of this relatively simple analytical method by Yu and Standish (1987)? A parameterized porosity model (for instance, porosity as a function of the volume fraction contents of grain size classes in the bed, and the relative grain sizes of the size classes) would be ideal for large-scale mixed-sediment (numerical) models. How can the authors' complex porosity model be used to derive such a parameterized porosity model? Do the authors have the ambition to develop such a parameterized porosity model based on their complex porosity model?

Reference

Yu, AB and Standish, N 1987. Porosity calculations of multi-component mixtures of spherical particles. *Powder Technology* 52, 233–241.

Discussion by Kimberly Hill

This presentation involved new methods for estimating the porosity of bed sediments as it varied with grain-size distributions. I was wondering if the authors had yet validated their new method using previously published work such as that on packing efficiencies of mixtures historically published in the powder technology literature (e.g. review by Cumberland and Crawford 1987). Often, the work in the powder technology community has focused on the question of optimal packing efficiencies for particular industrial applications, so the solid fraction (or porosity) may not be exactly equal to that expected in river beds. On the other hand, since the previous work on optimal packing has considered a variety of different size distributions, from binary mixtures to lognormal size distributions, it may provide useful data for baseline comparisons with the size distributions of concern for riverbeds.

Reference

Cumberland, DJ and Crawford, RJ 1987. *The Packing of Particles*. Elsevier: New York.

Discussion by Richard Measures and Colin Thorne

In the presentation of this chapter a simple analytical model was presented for the porosity of a binary mixture made up of small and large spherical particles with very different diameters. This model is fully described by Sulaiman (2008). A feature of this porosity relationship for such binary mixtures is the well-defined minimum porosity as the mixture moves from 'coarse packing', where the relatively low proportion of fine sediment can be contained within the void spaces between the large particles, to 'fine packing', where the fine sediment can no longer fit in the voids between the large particles without increasing the overall mixture volume (Figure 21.12). The proportion of fine particles required for minimum porosity is a function of the porosity of the individual sediment fractions. For the case shown in Figure 21.12, where both fractions have porosity 0.35, minimum porosity (and hence maximum bulk density) occurs with 25.9% fines. For fractions with lower porosities (like those typical of non-spherical particles), minimum porosity occurs with a lower proportion of fines.

During laboratory experiments to investigate the effect on gravel mobility of adding sand to a gravel-bed in a flume, Wilcock, Kenworthy, and Crowe (2001) observed that increasing the proportion of sand from 15 to 27% caused the most rapid increase in gravel transport rate. Further increasing the proportion of sand to 34% did not produce a proportionate increase in gravel mobility. It is notable that in Wilcock, Kenworthy, and Crowe's experiments the sand percentage associated with maximum gravel mobility (27%) corresponds closely with the proportion of fines required to achieve minimum porosity (25.9%). Is this a coincidence or is it associated with the switch from 'coarse packing' to 'fine packing' which corresponds to the large particles (i.e. the gravel) becoming matrix-supported rather than framework-supported?

Do the authors agree that the similarities between the thresholds identified for particle packing relationships in this paper and gravel mobility (Wilcock, Kenworthy, and Crowe 2001) suggests that research into particle packing and porosity could be valuable for informing transport relations for widely graded sediments, by identifying whether the coarsest fractions are framework or matrix-supported?

Reference

Wilcock, PR, Kenworthy, ST, and Crowe, JC 2001. Experimental study of the transport of mixed sand and gravel. *Water Resources Research* 37(12), 3349–3358. http://dx.doi.org/10.1029/2001WR000683

Reply by the authors

Regarding Astid Blom's query, the presented particle packing simulation model can analyse the porosities of any grain-size distribution. As described in our chapter, the simulation obtains the relation between the calculated porosity and the geometric parameters of the size distribution. This is an important point in our chapter. A disadvantage of this method is that the calculation time for wide-range size distributions is long. Hence, for practical purposes, a simple method such as a parameterized porosity model is necessary for wide-range size distributions, but it is difficult to develop such a simple method based on the presented particle-packing model. However, as the porosity of a sediment mixture of two quite different sizes has been analysed by a parameterized porosity model (Yu and Standish 1987), we may develop a method for binary mixtures of quite different mean grain sizes based on this parameterized porosity model. We simulated the porosity of binary sediment mixtures by the presented particle-packing model. Figure 21.12 shows the relation between the simulated porosity and the proportion of the finer fraction. The relation is similar to that of two-sizes mixture. Thus, this may be a promising idea for wide-ranging binary size distributions.

Kimberly Hill suggests referring to the powder technology literature. We have validated the presented particle-packing model using our experimental result, but have not compared with results published in the powder technology literature. Even though the objectives of both studies may differ, there are many common points. It seems that results in the field of powder technology provide many suggestions for this study.

It is very interesting to discuss the mobility of gravel in bed material composed of a gravel–sand mixture from an aspect of the change in porosity of the bed material as proposed by Measures and Thorne. If we mix sand in a gravel bed, the threshold condition of gravel movement may be reduced. This is also explained by Egiazaroff's Theory (Egiazaroff 1965). The position of gravel on the bed surface is relatively high when sand is mixed in the gravel and larger velocity acts on the gravel. This is the reason why gravel mobility increases. However, in a situation of fine packing (the proportion of sand is larger than 25.9% when the porosity of gravels is 0.35), the gravel particles are buried in the sand matrix. As a result the mobility of gravel is reduced. As mentioned by Measures and Thorne, the mobility is dependent on whether the coarsest fractions are framework or matrix-supported. In the coarser packing, where the gravel is framework-supported, the mobility decreases with the proportion of sand fraction and the highest mobility of gravel appears at minimum porosity.

References

Cumberland, DJ and Crawford, RJ 1987. *The Packing of Particles.* Elsevier: New York.

Egiazaroff, IV 1965. Calculation of nonuniform sediment concentration. *Proceedings of the American Society of Civil Engineers* HY4, 225–247.

Wilcock, PR, Kenworthy, ST, and Crowe, JC 2001. Experimental study of the transport of mixed sand and gravel. *Water Resources Research* 37(12), 3349–3358. http://dx.doi.org/10.1029/2001WR000683

Yu, AB and Standish, N 1987. Porosity calculations of multi-component mixtures of spherical particles. *Powder Technology* 52, 233–241.

22

Gravel Sorting and Variation of Riverbeds Containing Gravel, Sand, Silt and Clay

Masato Sekine and Yuki Hiramatsu

22.1 Introduction

In the reach of a river just downstream of a dam the riverbed may progressively undergo degradation and armouring. The primary reason for this is that the sediment supply from upstream is decreased to almost zero by the presence of the dam. According to field observations in such a river reach, the bed is typically found to be composed of sediment with an extremely wide size distribution. In other words, the framework of the riverbed consists of larger sediment sizes such as boulders or cobbles, which cannot move even during floods. Cobble sizes are typically exposed on the bed surface. Sand, gravel, and silt, on the other hand, fill the void space of the cobble framework. In the case of such a riverbed, it is clear that the mechanisms of sediment transport and riverbed deformation processes are strongly influenced by these exposed cobbles. Vertical sorting of sediment is found to be present in the void spaces of the cobble framework. It is difficult to explain this structure clearly based on current knowledge, and appropriate numerical prediction is still beyond reach. Therefore, further fundamental research is necessary in order to elucidate the relevant mechanisms. One purpose of this study is to obtain a clear understanding of the vertical sorting of riverbed sediment occurring under the condition of dynamic equilibrium. A series of experiments under systematically varied conditions were performed to realize this understanding. Half this chapter is devoted to 'Experimental Set 1', the experiments of which were specifically designed to address equilibrium vertical sorting. As to the phenomenon discussed in this chapter, several studies have been conducted on the process of sand infiltration into the framework material (Blaschke *et al.* 2003; Gibson *et al.* 2009; Datry *et al.* 2014).

In addition to the above, another sediment transport phenomenon is also addressed herein: riverbed variation in a reach of tidal river. Mud in the clay size range flows into a sand or gravel river reach under the influence of tidal action from the coastal area, and deposits on the bed. The sediment transport that occurs under this condition is much more complicated than that on a bed composed of sand or gravel only. This is because the mechanism by which clay is eroded from a riverbed by water flow is only poorly understood at present. More specifically, our understanding of the mechanisms by which sand and gravel moving as bedload erode a clay bed is insufficient. It is important to realize that a clay bed can be eroded by the action of water flow alone, but erosion can also be accomplished by abrasion as sand or gravel is transported over the bed. In addition, the deposition of sand or gravel can prevent erosion of a clay bed. For example, when sand or gravel is supplied from upstream by an amount above

Gravel-Bed Rivers: Processes and Disasters, First Edition. Edited by Daizo Tsutsumi and Jonathan B. Laronne.

a certain critical value, sand or gravel can be emplaced beneath the surface of a clay riverbed, so that a mixed layer of clay and sand (or gravel) is formed. Furthermore, when clay and sand (or gravel) are mixed sufficiently in this layer, the bed can no longer be eroded, and the riverbed reaches an approximate dynamic equilibrium state. The latter half of this chapter is devoted to presentation and discussion of the results of this experiment. This experimental set is referred to as 'Experimental Set 2' herein. It is worth noting that Nishimori and Sekine (2009) have studied the erosion process of a clay riverbed by the action of flowing water only, including presentation of an erosion rate formula. On the other hand very little research has been performed in regard to the influence of transported sand or gravel on the erosion of a clay riverbed (but see Harsanto *et al.* (2012) and Sekine, Oka, and Nomura (2013)).

Based on the results of the above-mentioned two experimental sets, we attempt to obtain a better understanding of influence of the amount of supplied sand or gravel on the vertical structure of riverbeds.

22.2 Summary of Experiments

22.2.1 Experiment on a Bed with an Extremely Wide Range of Sediment Grain Size (Experimental Set 1)

A schematic view of the experimental apparatus used in this investigation is illustrated in Figure 22.1a. The flume is an open channel with a rectangular cross-section. The length, width, and gradient of the channel are 16 m, 0.2 m, and 1/250 respectively. A reach with a movable bed is located between 8.5 and 14 m from the upstream end of the channel. In this reach, the bottom of the channel is 0.05 m lower than that of the channel upstream or downstream. The erodible-bed reach was filled with sediment up to this elevation difference. The nature of the sediment in question is explained below.

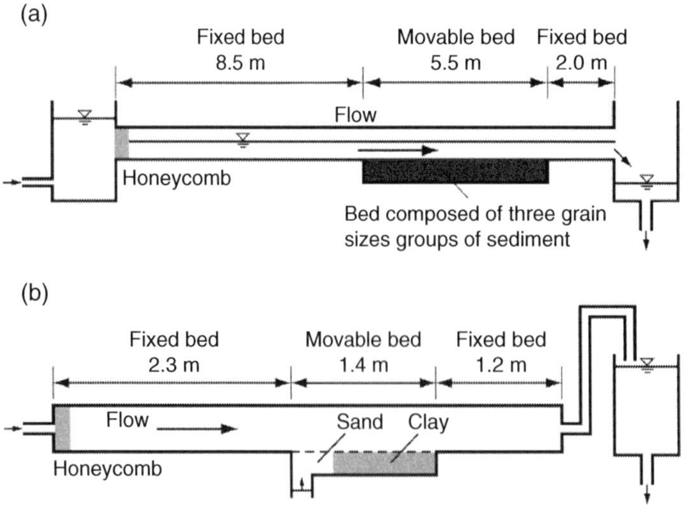

Figure 22.1 Schematic view of experimental apparatus: (a) Experimental Set 1 for a bed with an extremely wide range of sediment sizes; (b) Experimental Set 2 for sand transport over a clay bed.

The basic structure of the framework of the bed consisted of the largest of the three sediment sizes, referred to as L-particles in this chapter. Aluminium balls of size $D_L = 50$ mm diameter with a specific weight of 3.98 were used as L-particles. L-particles do not move at all under the range of conditions of this study. Glass beads with a specific weight of 2.5 were used as the intermediate size of sediment. These are referred to as M-particles; their diameter is $D_M = 2$ mm. M-particles move as bedload. Silica sand No. 7 was used as the smallest size of sediment, i.e. the S-particles, which have a specific weight of 2.65. This material consists of well-sorted sand with a characteristic grain size D_S = 0.21 mm. S-particles can move as suspended load. It should be noted that each size differs from its nearest alternative size(s) by one order of magnitude, and that the following relation is satisfied $D_L \gg D_M \gg D_S$. For reference, the dimensionless critical tractive force (critical Shields number) of M-particles is τ^*_{MC} = 0.042, and the settling velocity of S-particles is w_{oS} = 0.025 m/s.

As mentioned above, the framework of the bed consisted of L-particles; these were positioned to form an orthogonal lattice. Figure 22.2a illustrates this basic bed structure. The void space between the L-particles is filled with uniformly mixed M- and S-particles: the state of the bed is seen in

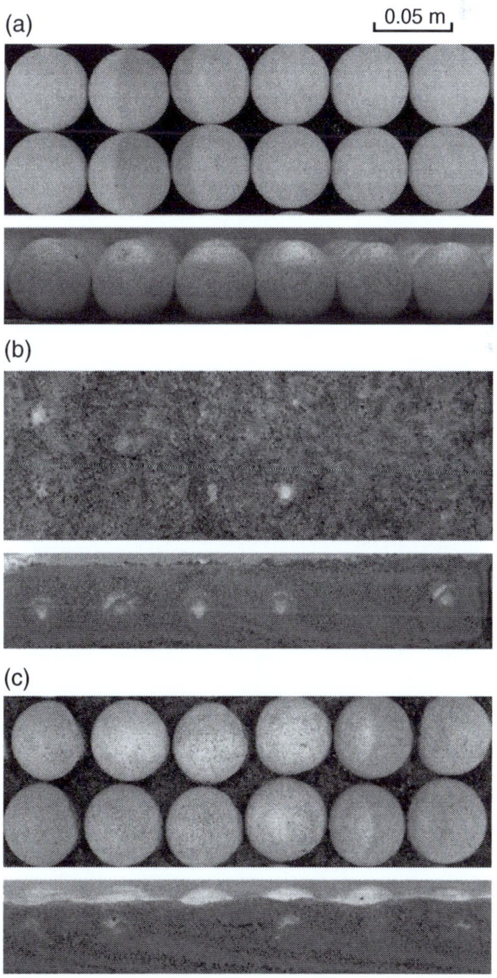

Figure 22.2 Typical state of bed (Experimental Set 1): (a) basic structure of the bed, which constituted by L-particles (white); (b) state in which the elevation of the top of an L-particle is equivalent to that of an M- or S-particle; (c) state of the bed at static equilibrium.

Figure 22.2b. It has been set so as to satisfy the condition that the elevation of the top of an L-particle is equivalent to that of the M- and S-particles between the L-particles.

The experimental conditions are indicated in Table 22.1. The supply rate of M- and S-particles per unit time from the upstream end of a movable bed reach is defined respectively as Q_M and Q_S, and the flow discharge of water is Q_w. These were kept constant during each experiment. As seen from Table 22.1, three sets of experiments were conducted. In the series of experiments named Case 1A and 1B, Q_w is set 5×10^{-3} m^3/s, and the average friction velocity $\bar{u}^*$ on the bed is 0.045 m/s. In Case 1C, on the other hand, $Q_w = 9 \times 10^{-3}$ m^3/s and $\bar{u}^* = 0.050$ m/s. In these experiments, Q_M was set to 1.2×10^{-6} m^3/s in Case 1A, and was set to 0.4×10^{-6} m^3/s in Cases 1B and 1C. Under the conditions $Q_w = 5 \times 10^{-3}$ m^3/s and $\bar{u}^* = 0.045$ m/s, the value $Q_M = 1.2 \times 10^{-6}$ m^3/s has the following physical meaning. By applying the Meyer-Peter and Muller (1948) formula, the equilibrium bedload transport rate Q_{Be}

Table 22.1 Experimental Set 1: conditions and results.

Case	Q_w (m^3/s)	$\bar{u}^*$(m/s)	u_M^*(m/s)	Q_M ×10^6 (m^3/s)	Qs ×10^6 (m^3/s)	Δ_M/D_L	Δ_S/D_L	R_M	Stage
1A-0	0.0052	0.047	0.033	0.0	0.0	0.30	0.36	–	0
1A-1	0.0052	0.042	0.042	1.2	0.0	0.03	0.38	1.00	4
1A-2	0.0052	0.043	0.043	1.2	0.4	0.03	0.07	0.80	4
1A-3	0.0051	0.040	0.040	1.2	1.4	0.00	0.00	0.56	4
1A-4	0.0052	0.041	0.041	1.2	2.8	0.02	0.02	0.37	4
1A-5	0.0053	0.042	0.042	1.2	3.1	0.02	0.02	0.28	4
1B-0	0.0052	0.047	0.033	0.0	0.0	0.30	0.36	–	0
1B-1	0.0051	0.046	0.041	0.4	0.0	0.15	0.36	1.00	1
1B-2	0.0051	0.046	0.042	0.4	0.2	0.15	0.20	0.82	1
1B-3	0.0053	0.047	0.041	0.4	0.4	0.15	0.16	0.72	2
1B-4	0.0052	0.045	0.042	0.4	2.5	0.11	0.11	0.24	3
1B-5	0.0049	0.039	0.039	0.4	3.8	0.00	0.00	0.11	3
1B-6	0.0050	0.038	0.038	0.4	4.5	0.00	0.00	0.08	3
1C-0	0.0092	0.053	0.035	0.0	0.0	0.34	0.40	–	0
1C-1	0.0090	0.051	0.042	0.4	0.0	0.20	0.42	1.00	1
1C-2	0.0093	0.052	0.042	0.4	0.4	0.22	0.26	0.79	1
1C-3	0.0090	0.052	0.043	0.4	1.4	0.20	0.21	0.64	1
1C-4	0.0091	0.052	0.042	0.4	2.5	0.21	0.22	0.74	2
1C-5	0.0091	0.052	0.043	0.4	3.4	0.20	0.20	0.34	3
1C-6	0.0091	0.052	0.045	0.4	4.5	0.19	0.19	0.30	3
1C-7	0.0088	0.048	0.042	0.4	5.0	0.17	0.17	0.24	3
1C-8	0.0091	0.051	0.044	0.4	5.7	0.17	0.17	0.15	3

Q_w, flow discharge of water; $\bar{u}^*$, average friction velocity acting on bed; u_M^*, actual friction velocity acting on M-particle; Q_M, supply rate of M-particles; Q_S, supply rate of S-particles; R_M, volume ratio of M-particles to total volume of M- and S-particles in the void space.

for M-particles was evaluated to be 1.2×10^{-6} m^3/s under these conditions. In Case 1A, Q_M is set equal to Q_{Be}. Therefore, the bed attains the state shown in Figure 22.2b under the condition of dynamic equilibrium. Cases 1B and 1C correspond to experiments in which Q_M is much less than the value of Q_{Be}. Under this condition, it can be expected that the top of the L-particles would be exposed to some extent on the bed surface at the dynamic equilibrium state, even though S-particles are supplied at some rate from the upstream end. The sediment supply rate of S-particles Q_S was set to values in the range between 0 and about 5.7×10^{-6} m^3/s. This value was set to be less than that of the equilibrium suspended sediment transport rate of S-particles. In this study, we investigated the process by which the bed evolved toward dynamic equilibrium.

In these experiments, which are summarized in Table 22.1, Cases 1A-0, 1B-0, and 1C-0 correspond to experiments under the condition $Q_M = Q_S = 0$. Under this condition, the bed attains a static equilibrium state, which proves to be clearly different from the dynamic equilibrium state. In this study, the static equilibrium state of the bed was also investigated.

The initial bed condition was set as follows. In principle, the initial state of the bed was set to be the same as the static equilibrium state, but some experiments were performed under a different initial bed condition such that the upper surface elevation of M- and S-particles exposed on the bed surface was equivalent to the top of L-particles. This state is shown in Figure 22.2b. As a result, it was verified that the bed attains a similar state of vertical structure as when the initial state corresponds to static equilibrium, so that there appears to be no influence from the initial condition. The volume ratio of M-particles to total volume of both M- and S-particles in the void space between L-particles is defined as R_M in this chapter. For the initial condition of the bed, R_M was set equal to 0.5 in the experiments.

The measurements in this study can be summarized as follows. With regard to the water flow field, both the flow discharge and water surface profile were measured. The water surface profile was measured every 0.5 m in the reach between 9.5 and 13 m from the upstream end of the channel. The average bed shear stress was evaluated based on the gradient of the water surface. The effective bed shear stress τ' was obtained after eliminating side-wall effects. It should be noted that the value τ' corresponds to the average shear stress exerted over the entire bed. The actual shear stress acting on M- and S-particles exposed on the bed surface between L-particles is equal to or smaller than the value of τ'. This effect is called 'hiding', such that M- and S-particles are sheltered by L-particles. The friction velocity $\bar{u}^*$ was evaluated by means of the definition $\bar{u}^* = \sqrt{\tau'/\rho}$; the value in each case is shown in Table 22.1. The bedload transport rate of M-particles and the suspended transport rate of S-particles were measured at the downstream end of the channel.

Actual measurements of the vertical structure of bed were carried out in the reach between 11.2 and 11.5 m. In this investigation, the thickness of the exchange layer (or alternatively surface layer) was measured directly along with the thickness of the mixed layer in almost all runs corresponding to Cases 1B and 1C. The particles in this exchange layer are gradually replaced with the moving particles as bedload or suspended load. The thickness of this layer is defined as δ_E in this chapter. The way in which this thickness was measured is as follows. M-particles painted red were supplied from the beginning of the experiment. The supply rate of these particles was kept constant during this experiment until the bed attained dynamic equilibrium, at which time the bed no longer showed variation. After verifying equilibrium, we then replaced red M-particles in the feed with blue M-particles, and continued the run at equilibrium. After confirming that no red M-particles were found in the bedload exiting the downstream end of the channel, the flow of running water was stopped. The thickness δ_E of this exchange layer was evaluated by measuring the height of the upper and lower bounding surfaces of blue M-particles.

22.2.2 Experiments on a Clay Bed (Experimental Set 2)

A schematic view of the second experimental apparatus is illustrated in Figure 22.1b. The experiment was conducted in a circulating closed conduit of 4.9 m in length. The cross-section of this conduit is a square with 0.1 m width and 0.1 m height. The bed and walls are made of acrylic plates. The bottom of the conduit was set as follows. In the reach between 2.3 and 3.7 m from the upstream end of the conduit, the bottom is 0.05 m lower than that of the bottom upstream or downstream. This depression allows the insertion of sediment to make a movable bed. In this reach, there is much deeper depression, i.e. with a depth of 0.2 m located at the upstream end. Sand was placed in this zone. Sand could then be supplied for transport downstream by raising the bottom plate in this portion so as to supply a constant amount of sand. In the upstream and downstream reaches of the movable bed, the bottom of the conduit was fixed, and a rubber board with dense pyramid-like roughness elements with a height of 3 mm was attached on the bottom. This promoted the development of the flow boundary layer upstream of the erodible-bed reach. In addition, the ceiling plate of the conduit in this movable bed reach is removable. By removing this plate it is possible to set an initial clay bed, and also measure the local height of the bed surface before and after the experiment.

In this second experimental set, kaolin with a grain size of 0.007 mm and a specific weight of 2.65 was used as bed material. The clay content ratio R_{cc}, here defined as the weight ratio of clay to the sum of clay and sand, was set to $R_{cc} = 1.0$ in the initial bed. In addition, silica sand No. 3 was used to represent sand or gravel. It was supplied from the upstream end and it moved as bedload on the clay bed. The grain size of this silica sand is 1.56 mm, and its specific weight is 2.65.

The state of the movable bed can be explained with reference to Figure 22.1b. The upstream 0.4 m reach of the depression, which included the deeper depression, was covered with silica sand No. 3. In the downstream 1.0 m reach of the depression, the clay was placed and allowed to consolidate sufficiently under still water with a depth of 0.15 m. These two reaches are referred to as the sand-bed reach and the clay-bed reach. The clay bed was prepared by the same technique as that developed by Nishimori and Sekine (2009). In addition, the water content ratio was set to be $R_{wc} = 0.8$, where R_{wc} is defined as the weight ratio of water to the clay in the test sample.

The measurement procedure is summarized as follows. The elevation of the bed surface was measured using a laser displacement sensor made by the Keyence Corporation. These measurements were conducted in a reach of 0.4 m in length corresponding to between 0.7 and 1.1 m from the upstream end of the movable bed. Measurement lines were set in the streamwise direction at equal lateral intervals of 5 mm, for 17 measurement lines in total. Therefore, the total number of measuring points of bed elevation was 1360 (80 points in every measurement line times 17 lines). In this experiment, silica sand was supplied by gradually raising a plate at the bottom of the deep depression of the sand-bed reach. The water discharge, the pressure difference in the conduit, and the supply rate of sand transported onto the reach with a clay bed were measured every 3 min. The supply rate of sand per unit time was evaluated by measuring the volume of the sand that flowed out from this reach. This was achieved by means of analysis of images taken from the side during the experiment with a video camera. In addition, the bedload transport rate of sand was also measured at the downstream end of the channel. The pressure difference between the upstream and downstream ends of the movable bed was measured using a pressure sensor made by the Keyence Corporation, which was connected to holes along the sidewall. The vertical structure of the bed in the clay-bed reach was investigated by taking samples as follows. Individual grains of sand were carefully extracted with tweezers. The mixed-layer thickness, as defined below, and the volume ratio of clay and sand in it were then measured.

The experimental conditions are shown in Table 22.2. These experiments can be classified into three cases depending upon the flow discharge of water, the water temperature, and the supply rate

Table 22.2 Experimental Set 2: conditions.

Case	Q_w (m^3/s)	τ^*	$Q_{sg} \times 10^4$ (m^3/s/m)	T_w (°C)
2A-0	0.0045	0.079	0.00	23.2–25.6
2A-1	0.0045	0.079	0.02	23.2–25.6
2A-2	0.0045	0.079	0.04	23.2–25.6
2A-3	0.0045	0.079	0.06	23.2–25.6
2A-4	0.0045	0.079	0.10	23.2–25.6
2B-0	0.0050	0.097	0.00	23.8–25.0
2B-1	0.0050	0.097	0.02	23.8–25.0
2B-2	0.0050	0.097	0.06	23.8–25.0
2B-3	0.0050	0.097	0.10	23.8–25.0
2B-4	0.0050	0.097	0.12	23.8–25.0
2B-5	0.0050	0.097	0.16	23.8–25.0
2C-0	0.0050	0.097	0.00	10.5–15.5
2C-1	0.0050	0.097	0.02	10.5–15.5
2C-2	0.0050	0.097	0.04	10.5–15.5
2C-3	0.0050	0.097	0.05	10.5-15.5
2C-4	0.0050	0.097	0.08	10.5-15.5
2C-5	0.0050	0.097	0.12	10.5-15.5
2C-6	0.0050	0.097	0.14	10.5-15.5
2C-7	0.0050	0.097	0.18	10.5-15.5
2C-8	0.0050	0.097	0.22	10.5-15.5

Q_w, flow discharge of water; τ^*, dimensionless tractive force; Q_{sg}, supply rate of silica sand; T_w, water temperature.

of sand per unit time to the clay-bed reach. The average water temperature was 24.6, 24.1, and 13.5 °C in Cases 2A, 2B, and 2C, respectively. Cases 2A and 2B were conducted using nearly the same water temperature, but the flow discharge and tractive force were different. Cases 2B and 2C correspond to experiments under the same flow discharge, but with different water temperature. The dimensionless tractive force (Shields number) τ^* for silica sand used here was 0.079 in Case 2A and 0.097 in Cases 2B and 2C. The dimensionless critical tractive force τ_c^* was equal to 0.046. In order to verify the accuracy of the measured results, experiments were conducted several times under the same conditions. Other control experiments were also conducted under the same conditions except that sand was not supplied. Cases 2A-0, 2B-0, and 2C-0 correspond to experiments under such conditions.

22.3 Vertical Sorting and Variation of the Riverbed with Extremely Wide Range of Sediment Sizes

This chapter is focused on the dynamic equilibrium state of a gravel bed with an extremely wide range of sediment grain sizes. The vertical structure of the bed is discussed first. The characteristic state of the bed in dynamic equilibrium can be classified into several stages, as illustrated in Figure 22.3. In this figure, the large arcs drawn in broken lines represent L-particles, and black circles represent M-particles. The grey portion corresponds to S-particles (including the water in the void space). The mean vertical distance from the top of an L-particle to the upper surface of an M- or S-particle exposed on the bed surface between L-particles is defined as Δ_M and Δ_S, respectively.

Let us consider first the static equilibrium state of such a bed, which is termed Stage-0 in Figure 22.3a. Cases 1A-0, 1B-0, and 1C-0 correspond to the experiments that were conducted without any supply of either M- and S-particles. In these cases, the bed attains the equilibrium state shown in Figure 22.3a. A layer consisting of M-particles only was formed at the top of the bed between the L-particles. It was found that the thickness of this layer takes a roughly constant value corresponding to 1–2 times the diameter of an M-particle (Sekine, Hiramatsu and Kadoi 2013). This result is consistent with the experimental results of Ashida and Fujita (1986), obtained for a bed composed of only two grain sizes. In their experiment, sediment of one size could not move at all. This grain

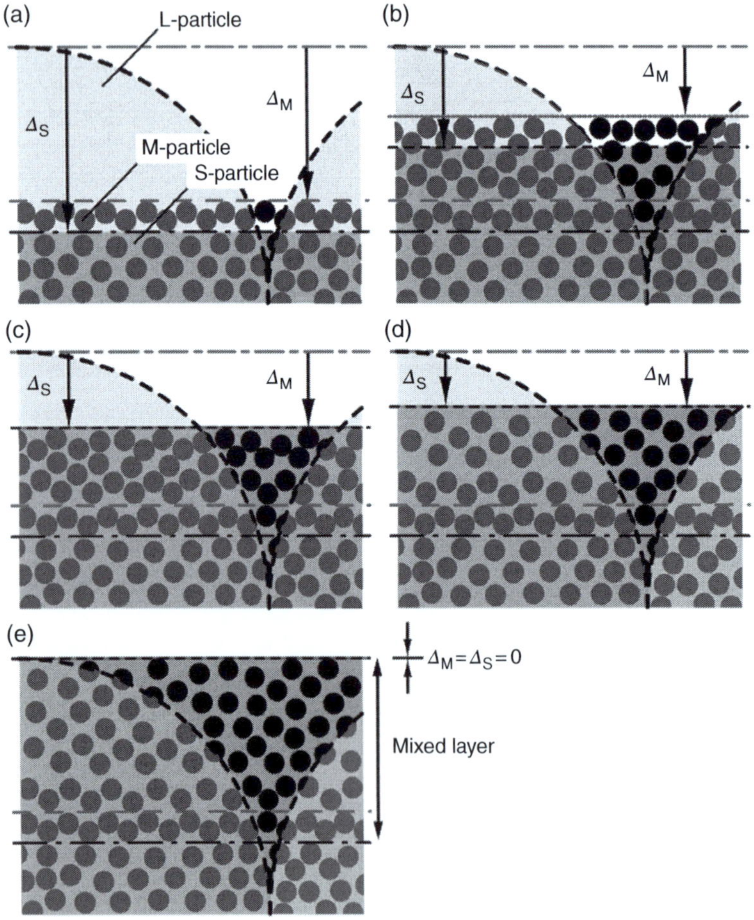

Figure 22.3 Schematic diagram of the bed at equilibrium (Experimental Set 1): (a) the static equilibrium state and (b)–(e) the dynamic equilibrium state – (b) Stage-1, (c) Stage-2, (d) Stage-3, (e) Stage-4.

size corresponded to the M-particle size in this study. The other size in their experiments moved as suspended load: it corresponds to the S-particle size of the present study.

In the static equilibrium state, the tractive force exerted on the upper surface of M-particles can be thought to be the critical tractive force $\tau_{\rm mc}$ for motion of an M-particle itself. The tractive force on the upper surface of an S-particle, on the other hand, becomes equivalent to the critical value of sediment suspension for that size. The hiding effect of L-particles on the two smaller sizes (M and S) was investigated on the basis of a series of experiments at the static equilibrium condition. Experiments on beds composed of two grain-size groups, i.e. either L–M or L–S, were conducted without supply of M- and S-particles (Sekine, Hiramatsu and Kadoi 2013). The hiding function H_i is defined as the ratio of the friction velocity u_e^* that is actually exerted on the upper surface of an M- or S-particle exposed on the bed surface to the average value $\bar{u}^*$ acting on the entire bed surface. The relation between H_i and $\Delta_{\rm M}$ / D_L or $\Delta_{\rm S}$ /D_L is illustrated in Figure 22.4. In this figure, the circles that are labelled 'L–M bed' or 'L–S bed' denote the experimental results in the case where the bed is composed of L- and M-particles, or L- and S-particles, respectively. The data can be fit to the solid line in Figure 22.4, which represents the following equation:

$$H_i \equiv u_e^* / \bar{u}^* = \left[1 + \frac{1}{k_e^2}\left(\frac{\Delta}{D_L}\right)^2\right]^{-1/2} \tag{22.1}$$

in which k_e is a constant of proportionality that takes a value of 0.3. In this equation, Δ denotes $\Delta_{\rm M}$ or $\Delta_{\rm S}$ appropriately. This relation enables us to evaluate the actual tractive force exerted on M- and S-particles. Ashida and Fujita (1986) proposed a similar function that expresses the hiding effect. In Table 22.1, the value of u_M^* corresponds to the estimated values of u_e^* for M-particles obtained from this equation when $\Delta = \Delta_{\rm M}$.

We discuss here the results of experiments conducted under some kind of critical state. This is termed Stage-4 in Figure 22.3e. In Cases 1A-1–1A-5 in Table 22.1, the bed reaches the state $\Delta_{\rm M}$ = 0, for which L-particles are almost buried and the elevation of upper surface of M-particle is equivalent to that of the top of L-particle. In these experiments, the flow discharge of water was $Q_w = 5 \times 10^{-3}$ m^3/s, and supply rate of M-particles per unit time was $Q_M = 1.2 \times 10^{-6}$ m^3/s. The dimensionless tractive force $\tau_{\rm M}^*$ for the M-particle size was calculated by considering the hiding function as follows;

Figure 22.4 Relation between $\Delta_{\rm M}$ /D_L, $\Delta_{\rm S}$ /D_L and hiding coefficient H_i (Experimental Set 1). The parameters $\Delta_{\rm M}$ and $\Delta_{\rm S}$ are defined in Figure 22.3.

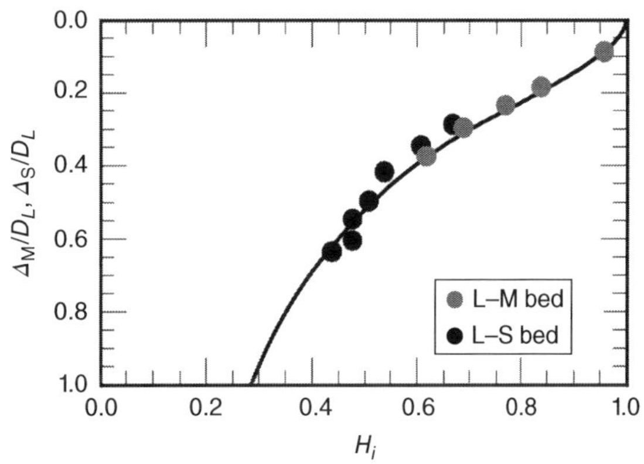

$$\tau_{\mathrm{M}}^{*} = \frac{(H_i\,\bar{u}^{*})^2}{R_s\, g\, D_M} \tag{22.2}$$

in which R_s is the submerged specific gravity of an M-particle (=1.5) and g is the gravitational acceleration. Calculated values of τ_{M}^{*} are almost constant at 0.059. By using this value, that of the equilibrium bedload transport rate Q_{Be} was estimated by the Meyer-Peter and Muller (1948) formula

$$Q_{Be} = B \times q_B = 8.0 \times B \times \sqrt{R_S\, g\, D_M^3} \times \left(\tau_{\mathrm{M}}^{*} - \tau_{\mathrm{Mc}}^{*}\right)^{3/2} \tag{22.3}$$

in which B is the channel width (= 0.2 m). The result of the calculation is $Q_{Be} = 1.2 \times 10^{-6}\ \mathrm{m^3/s}$. In Case 1A, the supply rate Q_M of M-particles was equivalent to the equilibrium value of Q_{Be}. It can be easily expected that the bed surface attains the dynamic equilibrium state $\Delta_{\mathrm{M}} = 0$ even if S-particles are supplied at some small amount. From Table 22.1, it can be confirmed that Δ_{M} was almost zero regardless of the supply rate of S-particles, at least within the range of Q_S studied here.

If Q_M is set to be greater than Q_{Be}, or if Q_S is above the range in given in Table 22.1, L-particles can be buried by M-particles and S-particles, and the bed becomes composed of M- and S-particles only. In such a case, we can apply the 'gravel–silt-packed bed model' of Sekine (2010) to predict the variation in time and space of such a bed.

In this chapter, the vertical sorting of the bed was discussed primarily under the condition that L-particle sizes are not buried, but instead are exposed on the bed surface at a state of dynamic equilibrium. Such experiments are termed Cases 1B and 1C herein. In these experiments, the supply rate of M-particle was set to be much smaller than Q_{Be}. And the supply rate of S-particles is changed in Cases 1B-1–1B-6 and 1C-1–1C-8 in Table 22.1. It should be noted that the values of Q_S used in the experiments were sufficiently small that the state of the bed corresponded to one those of Figure 22.3b–d.

In order to classify the results of the experiments, we divide them into Stages 0, 1, 2, and 3, as illustrated in Figure 22.3. Stage 0 described in Figure 22.3a corresponds to static equilibrium, i.e. Cases 1A-0, 1B-0, and 1C-0 in Table 22.1. Stages 1, 2 and 3, described in Figure 22.3b–d respectively, correspond to dynamic equilibrium, at which M- and S-particle were mobile. Consider cases for which the flow discharge Q_w and supply rate of M-particles Q_M are set constant. For each combination of Q_w and Q_M, Stages 1, 2, and 3 are realized for progressively higher supply rates of S-particles, as described below.

0) At static equilibrium, i.e. Stage 0 shown in Figure 22.3a, there is a layer of only M-particles between L-particles that is free of S-particles. The black dash-dot line in Figure 22.3a denotes the location of the interface above which there are no S-particles. This same black dash-dot line is also shown in Figure 22.3b–e for reference.
1) When the supply rate Q_S of S-particles is kept sufficiently small, a layer consisting solely of M-particles still forms below the bed surface. This state, which is called Stage 1, is illustrated in Figure 22.3b. The bed structure was similar to that of the static equilibrium state, i.e. Stage 0 described in Figure 22.3a, but a deposition layer forms in this stage below the bottom of a layer of M-particles. The layer corresponds to that between the black dashed line and the black dash-dot line in Figure 22.3b. And, in the range of these experiments, the thickness of a layer of M-particles takes about the same value as that of the static equilibrium state.
2) When Q_S is set to be sufficiently large, a layer consisting only of M-particles does not appear, and instead a uniform mixed layer of M- and S-particles forms on the bed. Here we call this an M–S particle mixed layer. The upper surface of the mixed layer moves upward as shown in Figure 22.3d corresponding to Stage 3. The upper surface eventually approaches the top of the L-particles when Q_S is sufficiently large. In addition, the volume ratio R_M of M-particles to the sum of M- and

S-particles decreases. This means that the distance between individual M-particles becomes larger than that in the mixed layer at Stage 2 explained later.

3) Stage 2 corresponds to the intermediate state between Stages 1 and 3, which is illustrated in Figure 22.3c. In this stage, the vertical structure in a mixed layer is uniform, and Δ_S is completely equal to Δ_M in this Stage. The volume ratio R_M in this layer is almost the same as that in the layer from the black dash-dot line to the black dashed line at Stage 1 in Figure 22.3b. The value of R_M is in the range 0.70–0.80. And the distance between individual M-particles does not change in both layers. The porosity, which is defined as the ratio of water volume in this layer to the total volume, was about 0.4.
4) When Q_S or Q_M is set greater than the value of equilibrium sediment transport rate, L-particles are buried in the dynamic equilibrium state. This is illustrated in Figure 22.3e corresponding to Stage 4.

The experimental results shown in Figure 22.5 serve to quantify the above behaviour based on actual measured data. The horizontal axis corresponds to Q_S, and the vertical axis corresponds to the two ratios Δ_M / D_L and Δ_S / D_L, where Δ_M and Δ_S are defined in Figure 22.3 and D_L is the diameter of the L-particle size. For reference, in Stages 2 and 3 Δ_M was precisely equal to Δ_S, and therefore two circle symbols for Δ_M and Δ_S overlap at the higher values of Q_S.

The formation of this mixed layer of M- and S-particles merits more discussion. From a dynamic point of view, it is natural to assume that this mixed layer consists of two layers, that is, an exchange layer (surface layer) and a deposition layer. There exists an exchange layer of M–S particles just beneath the bed surface, and below that, there is a deposition layer. M–S particles in the former layer are continually and repeatedly exchanged with particles transported as bedload or suspended load. In contrast, any particles in the deposition layer are not replaced by transported particles.

The photograph in Figure 22.6 was taken at the end of the experiment of Case 1C-4, which was at Stage 2. It reveals the vertical structure of the bed along a section corresponding to a side view. In this experiment, the colour of M-particles supplied was changed from red to blue after dynamic equilibrium was reached, as mentioned above. A gradual change in colour serves to verify that an exchange layer formed in the upper part of M–S particle mixed layer. Eventually, this exchange layer became composed of blue M-particles only. It was found that the thickness of this exchange layer δ_E for this experiment was about twice the diameter of M-particles D_M. The results of measurements of δ_E are summarized in Figure 22.7. The vertical axis is δ_E / D_M in this figure, and the horizontal axis is the supply rate Q_S of S-particles. This figure indicates a tendency for the thickness of the exchange layer δ_E to increase as the supply rate of S-particles increases. At Stage 3, however, the value of δ_E / D_M approaches a constant of about 3, that is, approximately three times the grain diameter D_M. In general, it was already known that the thickness of the exchange layer in gravel-bed rivers ranges from about one to three times the value of D_{90} in this layer (Parker 1986; Ashida, Egashira, and Nishino 1988). Here, D_{90} is the grain size such that 90% of the sediment is finer. However, in the case for which some of the bed material is so large that it cannot move, the thickness of the exchange layer should depend only on some size that characterizes larger moving particles, i.e. D_M in the present experiments. It is possible that this thickness is nearly independent of the diameter of large particles that do not move.

22.4 Variation of a Clay Bed Caused by Sand or Gravel Transport Over It

Now we consider the deformation process of a clay riverbed when sand or gravel are transported as bedload over it. One of the authors has already conducted fundamental experiments on the erosion process of a clay riverbed by the action of running water (Nishimori and Sekine 2009). The volume of

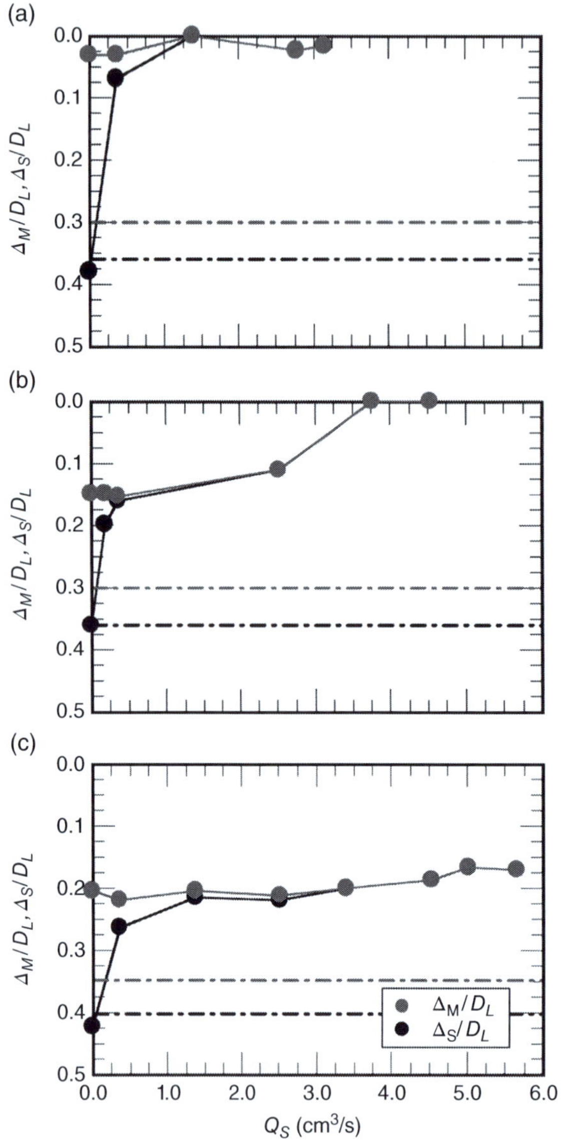

Figure 22.5 Relation between Δ_M/D_L, Δ_S/D_L and Q_S under dynamic equilibrium (Experimental Set 1): (a) Case 1A, (b) Case1B, and (c) Case 1C. The grey and black dashed lines represent the elevation of M- and S-particle at the static equilibrium state.

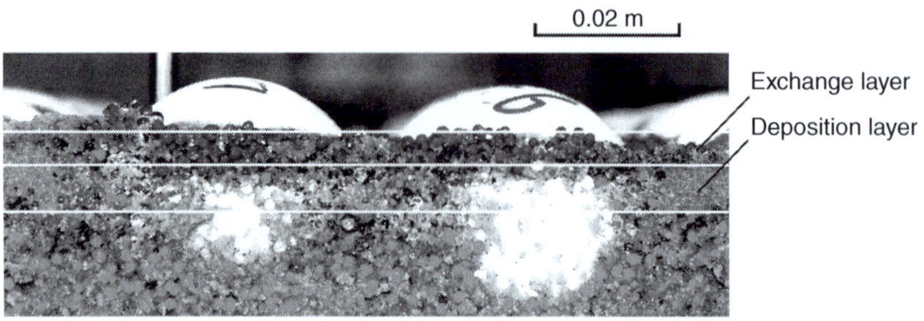

Figure 22.6 Side view of the bed at dynamic equilibrium (Experimental Set 1: Case 1C-4).

Figure 22.7 Relation between the thickness of exchange layer divided by particle diameter of M-particle size δ_E/D_M and the supply rate Q_S of S-particles per unit time (Experimental Set 1).

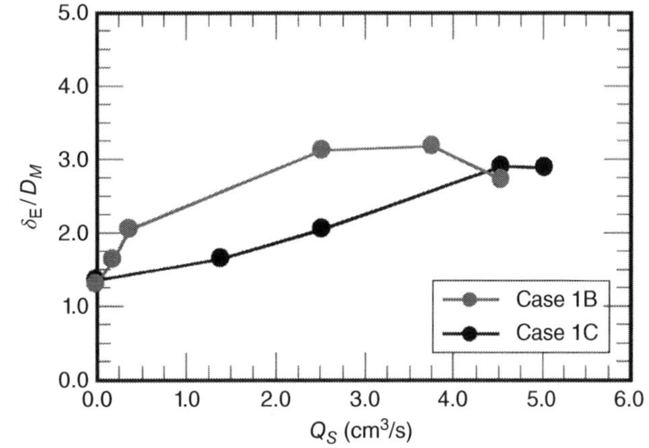

sediment eroded per unit time and unit area is defined as the erosion rate E_S. It was experimentally determined that the erosion rate E_S satisfies the following relation in the absence of a supply of sand or gravel:

$$E_S = \alpha\, R_{wc}^{2.5}\, u^{*3} \tag{22.4}$$

where α is a coefficient that takes a different value depending upon water temperature as well as clay mineral, R_{wc} is the water content ratio, i.e. the weight of water divided by the total weight of clay and u^* is the friction velocity. It should be noted that α is not a dimensionless coefficient.

The conditions of experiments conducted in this investigation are summarized in Table 22.2. Three series of experiments, from Cases 2A to 2C were carried out for a total of 20 experiments. In each case, tractive force and water temperature were kept constant, but the supply rate of the silica sand No. 3 varies from run to run.

Cases 2A and 2B correspond to experiments performed under different tractive forces and approximately constant water temperature. In Cases 2A and 2C, the tractive force was the same, but water temperature varies from experiment to experiment. It should be noted that more erosion occurs as the tractive force is increased, and also as water temperature increases. Each experiment was continued until the local elevation of the riverbed hardly changed along the entire reach of the movable bed. It is thought that this condition corresponds approximately to the dynamic equilibrium.

In Figure 22.8, the states of the riverbed surface under four different conditions are compared. Each photograph was taken after the end of the experiment in question (2A-1, 2A-2, 2A-3, and 2A-4). White zones consist of kaolin clay, and zones that are not white consist of silica sand. Case 2A-1 is the experiment for which the supply rate of sand is smallest in the series of Case 2A, and Case 2A-4 is the experiment for which the supply rate is largest. It can be seen that the volume of depositing sand, or the bed area covered with sand becomes larger as the supply rate of sand increases. Figure 22.8 illustrates the change of bed state: the areal fraction cover of the clay bed decreases with increasing supply of silica sand.

Quantitative results from these experiments are summarized in Figure 22.9. The horizontal axis is the value of the supply rate of silica sand Q_{sg} divided by the equilibrium bedload transport rate Q_{Be} calculated by the relation of Meyer-Peter and Muller (1948). The ratio of Q_{sg}/Q_{Be} is called herein the normalized supply rate of silica sand. The vertical axis of Figure 22.9a is the thickness of the mixed layer δ_{mix} divided by grain size of the silica sand D_{sg}. This mixed layer consists of clay and sand; and the

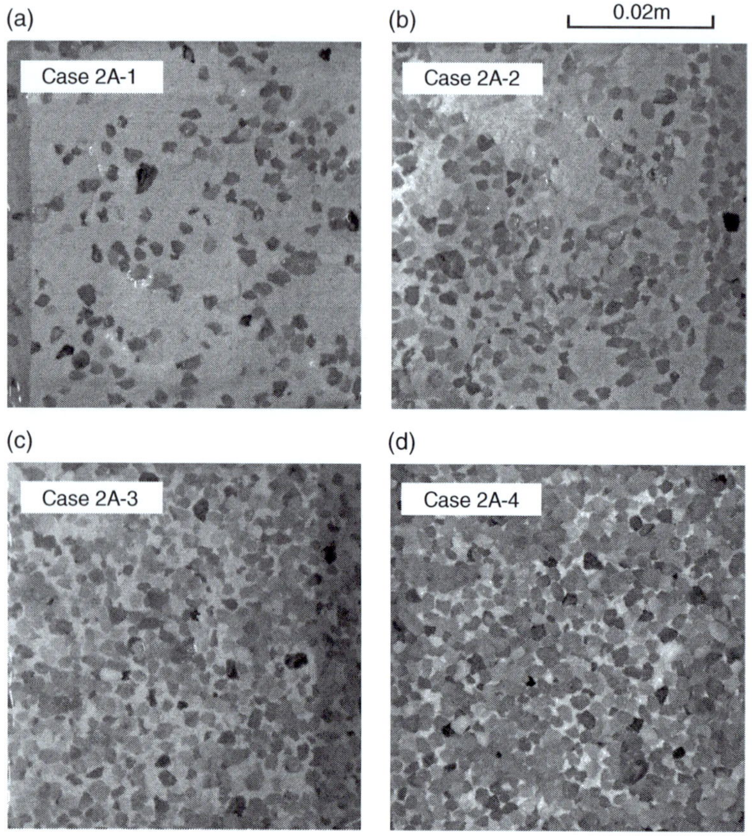

Figure 22.8 Plan view of the state of the bed surface for Cases 2A-1, 2A-2, 2A-3, and 2A-4: white zones are clay and grey zones are sand (Experimental Set 2).

sand transported as bedload can exchange to some extent with the sand in this layer. We were unable to confirm, however, whether or not this layer corresponds to a true exchange layer as described in the context of the experiments of Table 22.1. It is seen from Figure 22.9a that the thickness of this mixed layer is roughly equal to the grain size of the silica sand, and independent of the sand supply rate.

Figure 22.9b shows the relation between clay content ratio R_{cc} in the mixed layer and the normalized supply rate Q_{sg}/Q_{Be}. The clay content ratio R_{cc} in this mixed layer decreases as the supply rate becomes higher. This means that the degree to which the bed surface is covered with sand increases. But it can also be seen that R_{cc} no longer decreases as sediment supply rate Q_{sg} increases if Q_{sg} is larger than a certain critical value. Instead, R_{cc} tends to converge approximately to 0.3.

The relation between erosion rate of clay E_S^* and the normalized supply rate Q_{sg}/Q_{Be} is illustrated in Figure 22.9c. We define the erosion rate of clay per unit bed area E_S^* as follows. In this study, E_S^* is the net volume V_c of eroded clay during each experiment divided by the bed area A_c and the total time period T of this experiment. That is, $E_S^* = V_c/A_c/T$. It should be noted that the erosion rate E_S^* is not the instantaneous value of erosion rate under the dynamic equilibrium condition, but is the average

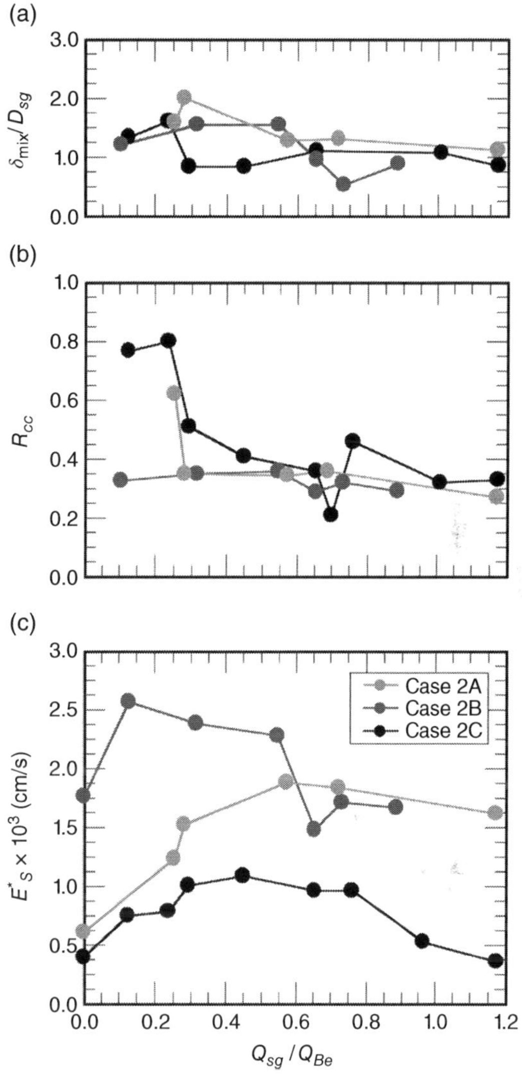

Figure 22.9 Characteristics of vertical structure of clay bed and erosion rate of clay (Experimental Set 2): (a) the thickness of mixed layer δ_{mix} divided by grain size D_{sg} of sand; (b) clay content ratio R_{cc} in mixed layer; (c) erosion rate of clay E_S^*.

value over the period of 10 min corresponding to each experiment. With regard to the process of erosion of clay, Nishimori and Sekine (2009) reported that wave-like undulations formed as a bed of pure clay was eroded by sediment-free water. They also showed that such bed undulations did not form when even a small amount of sand or gravel was contained in the bed. In the case of the present experiments, such bed waves also formed in the early stages of erosion of a clay bed (see Figure 22.10). Subsequently, however, sand deposited in the hollows of the undulations, and then worked in beneath the clay surface. We speculate that it is this process that led to the formation of a mixed layer beneath the bed surface. As can be seen from Figure 22.9c, we found that for the experiments of Cases 2A, 2B, and 2C, the relation between E_S^* and the normalized supply rate Q_{sg}/Q_{Be} had a maximum value in E_S^* possibly because a mixed layer is formed.

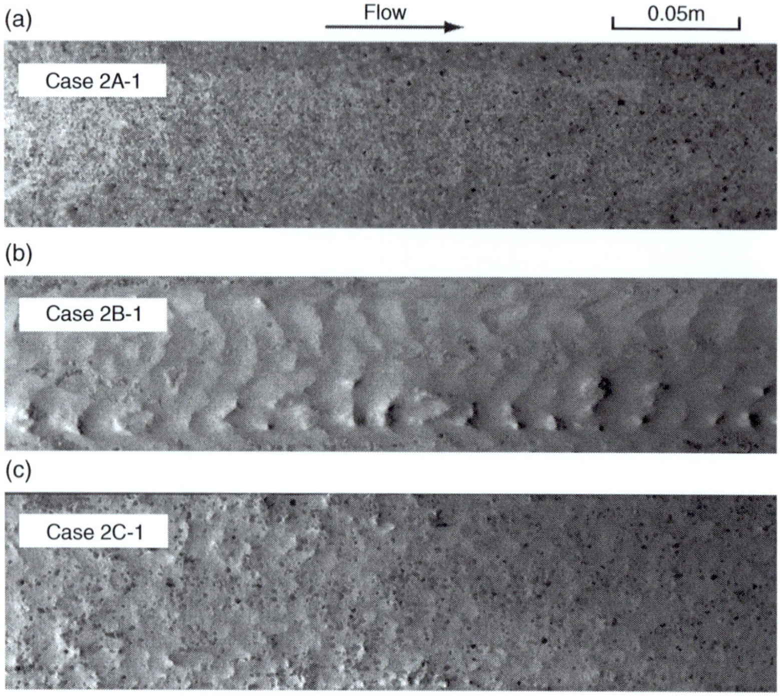

Figure 22.10 Plan view of the state of the bed surface in Cases 2A-1, 2B-1, and 2C-1 (Experimental Set 2).

Some further explanation in regard to Case 2A may be useful.

1) In the case where the supply rate Q_{sg} of silica sand is approximately equal to 60% of the equilibrium bedload transport rate Q_{Be}, the influence of abrasion by silica sand becomes greatest.
2) The volume of the eroded clay tends to decrease as the supply rate of silica sand increases more than the critical value. The reason for this is that the bed surface is partially coated by silica sand and the effect of inhibiting erosion is stronger. It should be noted that beyond some limit, there appears to be almost no influence on the net erosion volume of clay or E_S^* with increase in the rate of supply of silica sand.

22.5 Conclusions

We report here on two sets of experiments concerning riverbeds with wide ranges of grain sizes.

One set of experiments (22 experiments in total) is about the dynamic equilibrium state of a riverbed with an extremely wide range of grain sizes. The framework of the bed consisted of L-particles, which correspond to boulders or cobbles at the field scale. The void space between L-particles was composed of M- and S- particles, which correspond to gravel that can move as bedload and sand that can move as suspended load, respectively. The tractive force and the supply rates of M- and S-particles were varied from experiment to experiment. For a specified supply rate of M-particles, the state of the bed varied with increasing supply rate of S-particles. When the supply rate of S-particles was relatively small, a layer of only M-particles appeared immediately below the bed surface. When the supply rate of S-particles was increased, a mixed layer consisting of both M- and S-particle sizes formed just below the bed surface. It was found that the upper part of this layer is an exchange layer, with M-particles and S-particles exchanging with the bedload and suspended load at dynamic equilibrium. The rest of the mixed layer below this exchange layer is a deposition layer, so that the particles in it are not affected by the moving particles after the bed attains dynamic equilibrium.

The second set of experiments (20 in total) was conducted to investigate the process by which a clay bed is eroded by moving sand as bedload. The supply rate of sand was set to be different in each individual experiment. It was found that a mixed layer consisting of clay and sand was formed beneath the bed surface. The fraction of clay in this mixed layer decreased with increasing supply rate of sand, down to an asymptotic value of about 0.3. The formation of this mixed layer has significant implications for the process by which clay is eroded by water alone, and by a mixture of water and moving sand. The non-cohesive material of sand or gravel can increase the erosion of clay, but increasing cover of the clay by sand inhibits further erosion.

Notation

D_L, D_M, D_S	Sediment grain size of L, M, and S-particle, respectively
D_{90}	Grain size such that 90% of the sediment is finer
E_S^*	Erosion rate of clay
g	Gravitational acceleration
H_i	Hiding function
Q_{Be}	Equilibrium bedload transport rate
Q_M, Q_S	Supply rate of M- and S-particles, respectively
Q_{sg}	Supply rate of silica sand
Q_w	Flow discharge of water
R_M	Volume ratio of M-particles to total volume of both M- and S-particles in the void space between L-particles
R_s	Submerged specific gravity of sediment
R_{cc}	Clay content ratio
R_{wc}	Water content ratio
$\bar{u}^*$	Average friction velocity on the bed
u_e^*	Friction velocity that is actually exerted on the upper surface of an M- or S-particle exposed on the bed surface
u_M^*	Actual friction velocity acting on M-particle
w_{oS}	Settling velocity of S-particles
Δ_M, Δ_S	The vertical distance from the top of an L-particle to the upper surface of an M- or S-particle exposed on the bed surface between L-particles
Δ	Δ_M or Δ_S appropriately
δ_E	Thickness of exchange layer
δ_{mix}	Thickness of the mixed layer
τ'	Effective bed shear stress
τ_{MC}	Critical tractive force for motion of M-particle

τ_M^*	Dimensionless tractive force of M-particle
τ_{MC}^*	Dimensionless critical tractive force for motion of M-particle
τ_c^*	Dimensionless critical tractive force of silica sand No. 3.

Acknowledgement

This study was conducted with the financial support of a Grant-in-Aid for Scientific Research from the Ministry of Education, Science, and Culture of Japan (No. 21560546 and No. 24560629). The authors would like to thank T. Nomoto, N. Kenmotsu, Y. Kadoi, and H. Takano for their assistance with movable bed experiments. We are deeply grateful for Professor Gary Parker for his advice in writing this manuscript.

References

Ashida, K and Fujita, M 1986. Mechanism of particle suspension in an open channel. *Proceedings of JSCE* 375(II-6), 79–88. (In Japanese.)

Ashida, K, Egashira, S, and Nishino, T 1988. Mechanism of armoring phenomena in movable beds. *Annuals, Disaster Prevention Research Institute, Kyoto University* 31(B-2), 423–441. (In Japanese.)

Blaschke, AP, Steiner, KH, Schmalfuss, R *et al.* 2003. Clogging processes in hyporheic interstices of an impounded river, the Danube at Vienna, Austria. *International Review of Hydrobiology* 88(3–4), 397–413.

Datry, T, Lamouroux, N, Thivin, G *et al.* 2014. Estimation of sediment hydraulic conductivity in river reaches and its potential use to evaluate streambed clogging. *River Research and Applications.* DOI: 10.1002/rra.2784.

Gibson, S, Abraham, D, Heath, R and Schoellhamer D 2009. Vertical gradational variability of fines deposited in a gravel framework. *Sedimentology* 56(3), 661–676.

Harsanto, P, Toan, NMM, Takebayashi, H, and Fujita, M 2012. Effect of bedload transport on erosion rate of cohesive sediment. *Journal of JSCE, B1 (Hydraulic Engineering)* 68(4), I_19–I_24.

Meyer-Peter, E, and Müller, R 1948. Formulas for bed-load transport. *Proceedings of the 2nd Meeting, IAHR*, Stockholm; 39–64.

Nishimori, K and Sekine, M 2009. Erosion rate formula of cohesive sediment. *Journal of JSCE, B1 (Hydraulic Engineering)* 65(2), 127–140. (In Japanese.)

Parker, G 1986. On armoring. *Proceedings of JSCE* 375(II-6), 17–27. (In Japanese.)

Sekine, M 2010. Gravel–silt-packed bed model and its application to the river channel deformation with vegetation. *International Journal of Erosion Control Engineering* 3(2), 117–125.

Sekine, M, Hiramatsu, Y, and Kadoi, Y 2013. Equilibrium state of river bed with three grain groups of sediment. *The 12th International Symposium on River Sedimentation*; 393–400.

Sekine, M, Oka, Y, and Nomura, M 2013. Effect of sand transport on deformation of bed with cohesive sediment. *Journal of JSCE, B1 (Hydraulic Engineering)* 69(4), I_1027–I_1032. (In Japanese.)

23

Modeling Stratigraphy-Based Gravel-Bed River Morphodynamics

Enrica Viparelli, Astrid Blom, and Ricardo R. Hernandez Moreira

23.1 Introduction

In recent decades numerical models of gravel-bed river morphodynamics have been formulated and implemented to study fundamental problems such as downstream fining, i.e. the gradual decrease in size of the sediment in the channel bed (e.g. Parker 1991a,b; Hoey and Ferguson 1994, 1997; Cui, Parker, and Paola 1996; Wright and Parker 2005a,b), the long-term evolution of a gravel–sand transition (e.g. Cui and Parker 1998; Ferguson 2003), river response to changes in relative base level (e.g. Robinson and Slingerland 1998a), the effects of a cycled hydrograph (Parker, Hassan, and Wilcock 2008), and, more recently, applied to problems such as dam removal (Cui *et al.* 2006 – DREAM), sedimentation in reservoirs (Cui 2007), river response to changes in flow and sediment supply (e.g. Parker *et al.* 2003; Nelson *et al.* 2009), and the response of a river reach to river straightening (e.g. Talbot and Lapointe 2002).

Similar models were also developed to study variation in the vertical and in the streamwise direction of the sediment grain size distribution within channel bed deposits, i.e. the grain size stratigraphy or size stratification. Notable examples are the work of Ribberink (1987), to reconstruct the grain size stratigraphy emplaced by migrating dunes, of Spasojevich and Holly (1990), who presented a two-dimensional model that not only accounts for sediment sorting in the streamwise direction but also within the channel cross sections, and of Van Niekerk *et al.* (1992), who accounted for density and size sorting (see also Robinson and Slingerland 1998b). More recently, Viparelli *et al.* (2010) validated a procedure for the bookkeeping of grain size stratigraphy against laboratory experiments for the case of the lower plane bed bedload transport regime, i.e. a transport regime characterized by the absence of small scale bedforms. This procedure was then implemented in a one-dimensional morphodynamic model to study the impact of coarse gravel augmentation combined with high flow releases on the regulated Trinity River, California, United States (Viparelli *et al.* 2011), and in the MAST-1D model to describe the morphodynamic evolution of a channel–floodplain complex on the gravel bed Ain River, France (Lauer *et al.* 2014).

These numerical models have two elements in common that are relevant for the present study: (i) they account for the nonuniformity of the bed material, generally multiple grain sizes, possibly including coarse sand as well as gravel; and (ii) the interaction between the channel bed and the bedload transport is modeled with the active layer approximation or an adapted form of it.

Gravel-Bed Rivers: Processes and Disasters, First Edition. Edited by Daizo Tsutsumi and Jonathan B. Laronne.

In an active layer based model the channel bed deposit is divided into two sediment layers: the active layer, in which particles can interact with bedload transport; and the substrate, i.e. the deposit underneath the active layer, in which particles cannot directly interact with sediment transport (see e.g. Parker (2004) and references therein). The active layer is generally modeled as a thin mixed layer, i.e. no vertical variation of sediment grain size distribution, and the topmost part of the channel bed, while the grain size distribution of the sediment stored in the substrate can be allowed to vary in the vertical and in the streamwise direction.

In an active layer formulation, the bed material is transferred from the active layer to the substrate during channel bed aggradation, and from the substrate to the active layer during channel bed degradation. In other words, the exchange of sediment within the channel bed occurs for channel bed aggradation or degradation only. Thus, vertical sediment fluxes associated, for example, with bedform migration (Blom, Ribberink, and Parker 2008), and problems such as the infiltration of fine sediment in the channel bed (Viparelli *et al.* 2011), the dispersal of tracer stones and contaminants, or the prediction of the internal structure of alluvial deposits, cannot be adequately studied (e.g. Parker, Paola, and Leclaire 2000; Pelosi *et al.* 2014). Further, as in the active layer approach sediment entrainment and deposition are limited to the topmost part of the river bed, entrainment and deposition over a wider range of elevations cannot be accounted for.

To overcome the limitations related to the discrete nature of the active layer approximation, Parker, Paola, and Leclaire. (2000) proposed a continuous formulation to model sediment fluxes from the bedload transport to the channel bed and within the bed deposit. This formulation is based on the assumption that the particles within the alluvial portion of the bed deposit can interact with the sediment transport, i.e. there is always a finite probability that at the generic elevation z within the alluvial deposit a sediment particle is entrained in transport or deposited in the channel bed.

The difference between the continuous formulation, i.e. the Parker–Paola–Leclair (PPL) framework, and the active layer approximation is schematically represented in Figure 23.1 (adapted from Parker 2004), where the continuous black line denotes the channel bed elevation, and the black dashed line is the elevation of the active layer – substrate interface with respect to a specified datum. The probability that a sediment particle is entrained in transport or deposited in the channel bed at

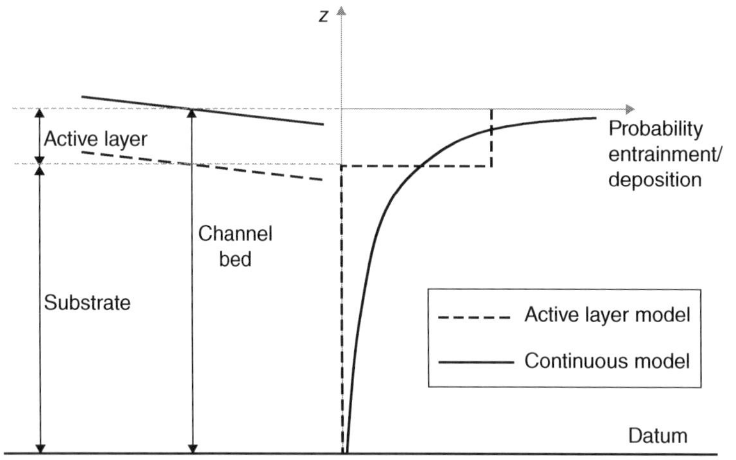

Figure 23.1 Schematic representation of the probability of entrainment and deposition in an active layer and in the continuous models. (Adapted from Parker 2004).

elevation z is modeled with a continuous function (continuous black line in Figure 23.1) in the PPL framework, and with a step function (dashed black line in Figure 23.1) in active layer based models.

We are not aware of any other implementation of the PPL framework to predict the grain size stratigraphy than that of Blom and Parker (2004), Blom *et al.* (2006), and Blom, Ribberink, and Parker (2008), who modeled the grain size stratigraphy emplaced in the dune regime at laboratory scale. Blom *et al.* (2006) and Blom, Ribberink, and Parker (2008) constrained the probability functions of sediment entrainment and deposition by applying a simple wave analysis to dune migration. In particular, they derived and implemented submodels to compute vertical sediment fluxes due to (i) dune migration and related sediment sorting over the lee faces, (ii) changes in bedform size, and (iii) channel bed aggradation and degradation. The validation of the continuous model against laboratory data (Blom *et al.* 2006; Blom, Ribberink, and Parker 2008) shows that the PPL framework is able to reproduce the stratigraphic details of an alluvial deposit emplaced in the dune regime. However, the Blom *et al.* (2006) and Blom, Ribberink, and Parker (2008) implementation of the continuous framework has a serious drawback: it rapidly becomes too computationally expensive as the spatial scale of the problem increases (Blom 2008).

Here we present a first attempt to use the PPL framework at field scale to predict the temporal evolution of a gravel bed river and simultaneously store information related to the emplacement of the grain size stratigraphy. While theoretically superior to the discrete layer modeling approach, the PPL framework requires the specification of the probability distribution functions of entrainment and of the local variation of bed elevation, very little information is available to constrain these functions at laboratory or field scale. As previously undertaken by other researchers (Parker and Perg 2005; Pelosi *et al.* 2014), in our numerical runs we hypothesize the shape of the probability functions based on the information available in the literature.

Due to the lack of information on how the probability density functions (PDFs) change with flow characteristics, the sediment transport regime, and the channel geometry, we introduce rather severe assumptions that we hope to relax in the near future. In particular, we assume that (i) the time scale of the variations in channel bed elevation due to aggradation and degradation is long compared to the time scale for the adjustment of the bedload transport and thus the PDFs can be assumed steady and uniform; and (ii) the sediment transport regime does not change during the simulations, and the portion of the channel bed that preferentially interacts with the sediment transport remains constant in time.

We acknowledge that these assumptions do not account for the spatial and temporal variability in channel geometry or the interannual variability of flow rates and sediment loads, for example. However, laboratory experiments currently performed by the authors with very slow aggradation rates in a sediment feed flume reveal that in the case of constant flow discharge, sediment feed rate, and channel width the assumption of steady and uniform PDFs is a reasonable approximation.

Our objective is to investigate if the PPL framework is able to characterize an approximately equilibrium state for a gravel bed river after a phase of aggradation, which is the first and necessary step to test the applicability of a modeling framework of river morphodynamics. We call this approach in which we run the numerical model towards the equilibrium state of the considered reach a 'zeroing model run'. In other words, we present and discuss the results of 'zeroing model runs' in terms of channel slope and spatial variation of grain size on the bed surface, in the substrate, and in bedload transport. Although maybe not exciting, zeroing runs are fundamental for reliable field scale applications of models of river morphodynamics. In general, zeroing runs are performed to (i) estimate parameters that cannot be estimated from field data, (ii) verify that they can reasonably reproduce conditions of mobile bed equilibrium, or (iii) determine relatively reasonable initial conditions for runs that describe contemporary and/or future scenarios (e.g. Cui 2007, Lauer and Parker 2008a,b;

Ferguson and Church 2009, Viparelli, Nittrouer, and Parker 2015). The runs presented in this chapter pertain to the second group, i.e. they are performed to determine if the models are able to reasonably reproduce conditions of mobile bed equilibrium. We perform our numerical runs on a simplified river reach of which the parameter values are based on those of the Trinity River, California, United States (Viparelli *et al.* 2011). It is important to note that we do not attempt to simulate the geological history of the Trinity River.

Notwithstanding the model simplifications, the comparison between model results and field observations shows that (i) the continuous model is able to reasonably reproduce the mobile bed equilibrium of a gravel bed river in terms of channel slope and mean annual bedload transport rates and grain size distributions of the channel bed surface, and (ii) 10 000 year-long simulations can be performed reasonably fast – from 20 min to a few hours – on a PC. In other words, the implementation of the PPL framework at field scale presented in this chapter demonstrates that the discrete layer based approach, first introduced by Hirano in 1971 (Hirano 1971; Ribberink 1987; Parker 1991a,b, 2004) and used in the vast majority of the morphodynamic models, can be substituted by a theoretically superior continuous approach, which requires comparable computational costs and times.

23.2 Model Formulation

The two numerical models discussed herein are an active layer model (e.g. Parker 1991a,b) and a vertically continuous model based on the PPL probabilistic framework for sediment continuity (Parker, Paola, and Leclaire 2000). These models are not site-specific; their formulation is general and allows for applications at field and laboratory scales.

The model governing equations are the shallow water equations for the flow and the grain size specific equations of conservation of bed material. Notwithstanding the differences related to the derivation of the equations of conservation of channel bed material, the two models have the following basic assumptions and approximations in common:

1) They describe the long-term evolution of a river channel and do not account for the exchange of sediment between the channels and their floodplains, thus the wash load is not accounted for in the calculations (e.g. Viparelli *et al.* 2011).
2) The bed material is a mixture of sand and gravel. It is modeled in terms of M characteristic grain sizes and is assumed to be transported exclusively as bedload (e.g. Parker 1991a,b).
3) The volumetric bed material load is assumed to be orders of magnitude smaller than the flow rate, so that the quasi-steady approximation (de Vries 1965) holds for the flow. To further simplify the problem, the flow is assumed to be locally uniform (e.g. Viparelli *et al.* 2011).
4) The channel bed is fully alluviated, i.e. mixed bedrock–alluvial reaches cannot be modeled with the present formulations (Viparelli, Nittrouer, and Parker 2015).
5) The simplified channel cross section is assumed to be rectangular with specified effective width B (e.g. Lauer and Parker 2008a,b).
6) The interannual variability of flow discharge is modeled in terms of a characteristic flood flow, Q_w, and a flow intermittency, I_f, denoting the fraction of time that the river is morphologically active (Paola, Heller, and Angevine 1992).
7) The river reach is modeled as a sediment feed flume analog, whereby the total amount of sediment stored in the channel bed is allowed to change in time because we are interested in modeling the long-term evolution of the system (e.g. Viparelli *et al.* 2011).

23.2.1 Governing Equations for the Flow

The governing equations are the shallow water equations of mass and momentum balance

$$\frac{\partial H}{\partial t} + \frac{\partial UH}{\partial x} = 0 \tag{23.1}$$

$$\frac{\partial UH}{\partial t} + \frac{\partial U^2 H}{\partial x} = -gH\frac{\partial H}{\partial x} + gHS - C_f U^2 \tag{23.2}$$

where x and t respectively denote the downchannel and the temporal coordinates, H is the flow depth, U is the mean flow velocity, g is the acceleration of gravity, S represents the channel bed slope, and C_f is a nondimensional friction coefficient. In the simulations presented herein the nondimensional friction coefficient is assumed to be constant, which is appropriate to model the long-term evolution of fluvial systems (Parker 2004).

Equations (23.1) and (23.2) are simplified with the quasi-steady approximation (de Vries 1965), and further assuming that the flow is locally uniform. Dropping the time dependence in equation (23.1), the continuity equation for the flow simplifies to

$$q_w = UH \tag{23.3}$$

where q_w denotes the water discharge per unit channel width, equal to Q_w/B.

Dropping the time dependence in equation (23.2) and further assuming that the flow is locally uniform, the momentum equation for the flow reduces to

$$S = S_f \tag{23.4}$$

where S_f is the friction slope defined as $C_f Fr^2$, with Fr^2 denoting the Froude number squared defined as $U^2/(gH)$.

23.2.2 Governing Equations for the Bed Material

The divergence form of the Exner equation of conservation of total channel bed material, i.e. summed over the M grain characteristic grain sizes, takes the form

$$c_b \frac{\partial \eta}{\partial t} = -I_f \frac{\partial q_{bT}}{\partial x} \tag{23.5}$$

where c_b denotes the volume fraction of the channel bed that is bed material and thus c_b is equal to (1 – bulk bed porosity), η is the elevation of the channel bed, and q_{bT} represents the total volumetric bed material load per unit channel width.

The volumetric bed material load is computed with the surface based bedload relation of Wilcock and Crowe (2003) for sediment mixtures. The total volumetric bedload transport rate per unit channel width, q_{bT}, is equal to the sum of the M volumetric bedload transport rates per unit width of sediment in each characteristic grain size range, q_{bi}, computed as

$$q_{bi} = \frac{u_*^3}{Rg} F_{si} W_i^* \tag{23.6}$$

where R is the submerged specific gravity of the bed material defined as $(\rho_s - \rho)/\rho$, with ρ_s and ρ respectively denoting the density of the sediment and of the water; u_* is the shear velocity; the subscript i refers to the generic grain size range; F_{si} represents the fraction of sediment in the bed surface;

and W_i^* is a nondimensional parameter representing the grain size specific bedload transport rate, i.e. the bedload transport rate of sediment in the different grain size ranges. The shear velocity, u_*, is defined as $(\tau_b/\rho)^{0.5}$, where τ_b is the bed shear stress defined as ρgHS.

The nondimensional parameter W_i^* is a function of the parameter ϕ_i, which is defined in equation (23.7a) as

$$\phi_i = \frac{\tau_{sg}^*}{\tau_{ssrg}^*}\left(\frac{d_i}{D_{sg}}\right)^{-b} \tag{23.7a}$$

$$W_i^* = G(\phi_i) \tag{23.7b}$$

where D_{sg} denotes the geometric mean diameter of the bed surface; d_i is the generic characteristic grain size; τ^*_{sg} and τ^*_{ssrg} respectively are the nondimensional bed shear stress, i.e. the Shields number and its reference value; and b is an exponent equal to

$$b = \frac{0.67}{1 + \exp\left(1.5 - D_i/D_{sg}\right)} \tag{23.8}$$

The Shields number is defined as $\tau_b/(\rho RgD_{sg})$, and its reference value is

$$\tau_{ssrg}^* = 0.021 + 0.015\exp(-20F_{ss}) \tag{23.9}$$

with F_{ss} denoting the fraction of sand in the bed surface.

Finally, the function G in equation (23.7b) is equal to

$$G(\phi_i) = \begin{cases} 0.002\phi_i^{7.5}; & \phi_i < 1.35 \\ 14\left(1 - \dfrac{0.894}{\phi_i^{0.5}}\right)^{4.5}; & \phi_i \geq 1.35 \end{cases} \tag{23.10}$$

Equations (23.1)–(23.10) are solved in both the active layer and the continuous model. The main differences between the two models are (i) the form of the grain size specific equations of conservation of bed material, and (ii) the definition of the grain size distribution of the bed surface. These differences are discussed below.

23.2.3 Grain Size Specific Equations of Conservation of Bed Material in the Active Layer Model

The active layer is the mixed region in the topmost part of the bed deposit, in which the particles constantly interact with the bedload. Depending on the transport regime and on the preferential mode of sediment transport, the active layer thickness scales with the characteristic grain size of the channel bed surface or with the size of the bedforms, whichever is larger.

The divergence form of the grain size based equation of conservation of sediment in the active layer is (e.g. Parker 2004)

$$c_b\left[\frac{\partial F_{ai}L_a}{\partial t} + f_{Ii}\frac{\partial(\eta - L_a)}{\partial t}\right] = -I_f\frac{\partial q_{bi}}{\partial x} \tag{23.11}$$

where F_{ai} denotes the volume fraction of active layer sediment, L_a is the active layer thickness, and f_{Ii} denotes the volume fraction of sediment at the interface between the active layer and the substrate. When equation (23.11) is summed over all the grain sizes, the Exner equation, i.e. equation (23.5), is obtained.

Hoey and Ferguson (1994) and Toro-Escobar, Parker, and Paola (1996), among others, showed that the volume fraction of bed material at the active layer – substrate interface, f_{Ii}, during channel bed aggradation is best described with a weighted average of the grain size distributions of the active layer and of the load. During channel bed degradation the f_{Ii} are equal to the fractions of substrate material, $f_{sub,i}$. In particular, Hoey and Ferguson (1994) proposed the following closure, which is implemented in the active layer model presented herein

$$f_{Ii} = \begin{cases} f_{sub,i}; & \dfrac{\partial \eta}{\partial t} < 0 \\ \alpha F_{ai} + (1-\alpha) p_i; & \dfrac{\partial \eta}{\partial t} > 0 \end{cases} \tag{23.12}$$

where p_i represents the fraction of sediment in the generic grain size range in the bedload, i.e. q_{bi}/q_{bT}.

The implementation of a surface based bedload relation, e.g. the Wilcock and Crowe (2003) model discussed above, is straightforward. It suffices to assume that the grain size distribution of the bed surface in equations (23.6)–(23.10) is equal to the grain size distribution of the active layer. Equation (23.11) can then be solved to compute the time rate of change of the fractions of active layer sediment with the aid of equation (23.12), and thus update the grain size distribution of the active layer.

To update the grain size distribution of the substrate during channel bed aggradation or degradation, we implemented the Viparelli *et al.* (2010) procedure to access and store grain size stratigraphy.

The substrate is divided into N mixed layers of specified thickness, L_s. The thickness of the substrate topmost layer is smaller or equal to L_s and it is allowed to change in time. During channel bed degradation, substrate sediment is directly transferred to the active layer. The thickness of the topmost layer decreases in time, until the entire layer is eroded and the channel bed degrades into the next layer.

During channel bed aggradation, sediment with grain size distribution given by equation (23.12) is transferred to the topmost mixed layer of the substrate. The grain size distribution of the sediment stored in this layer is updated by averaging the grain size distributions of the newly deposited sediment and of the substrate. As the channel bed aggrades, the thickness of the topmost layer increases in time, until it becomes equal to L_s. At this point, a new mixed layer is added to the substrate. The procedure is formulated to handle relatively rapidly aggradation or degradation, i.e. one or more substrate layers may be deposited or eroded.

23.2.4 Grain Size Specific Equations of Conservation of Bed Material in the Continuous Model

The implementation of the PPL probabilistic framework for sediment continuity (Parker, Paola, and Leclaire 2000) is not as simple as the implementation of an active layer model because of the assumption that a particle at elevation z within the channel bed deposit can always be entrained in transport. Consequently, the grain size distribution of the channel bed can vary in time within the entire deposit and not only in the upper part of the deposit, i.e. the active layer and the topmost substrate layer.

23.2.4.1 The Elevation Specific Equation of Conservation of Bed Material

In the probabilistic framework the interaction between the channel bed and the sediment transport is modeled in terms of a probability distribution of instantaneous bed elevations (Parker, Paola, and Leclaire 2000). The geometry represented in Figure 23.2 is considered in the derivation of the PPL framework.

The continuous black line represents an instantaneous realization of bed profile at elevation η_{inst} above the datum. The dashed black line is the mean bed elevation at elevation η above the datum. If we draw a line at elevation z, a fraction of the line will fall within the bed deposit (thick grey segments), and the rest will be in flowing water (dashed grey segments). The average fraction of the line that falls within the bed deposit at elevation z is denoted with $P_e(z)$. It can be interpreted as the probability that the instantaneous bed elevation $\eta_{inst} > z$. Thus, P_e tends to 1 deep inside the bed deposit ($z \rightarrow -\infty$), and tends to 0 well above the channel bed in the water column ($z \rightarrow +\infty$). The probability that $\eta_{inst} \leq z$ is thus equal to $(1 - P_e)$, and the probability density that the instantaneous bed elevation is at elevation z is

$$p_e = -\frac{\partial P_e}{\partial z} \tag{23.13}$$

By definition, the integral of p_e over the entire deposit has to be equal to 1 and the average bed elevation, η, is equal to the mean of the probability density of the bed elevation (Parker, Paola, and Leclaire 2000).

We further assume that the instantaneous realizations of bed elevation, η_{inst}, are statistically uniform in space and time over scales that are large compared to those associated with each instantaneous realization, but small compared to the scales associated with channel bed aggradation and degradation.

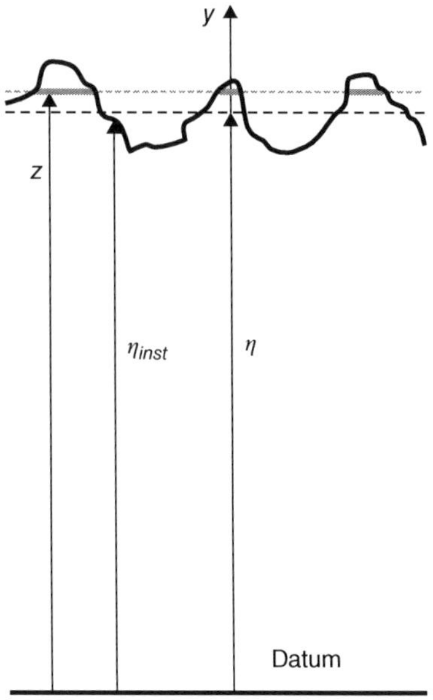

Figure 23.2 Schematic representation of the instantaneous realization of bed elevation, η_{inst}, the average bed elevation, η, and of the vertical coordinates z and y.

In general, P_e and p_e are functions of space – x and z – and time, while η is a function of the streamwise coordinate x and t. To simplify the problem, Parker, Paola, and Leclaire (2000) introduced the vertical coordinate y, which is boundary-attached to the mean bed elevation (Figure 23.2) and defined as $y = z - \eta$.

In the y coordinate system P_e and p_e respectively represent the complementary cumulative distribution function and the PDF of the instantaneous bed elevation fluctuations around its mean value, $\eta_{inst} - \eta$. The mean of the probability density of bed elevation in the y reference system is thus constant in time and equal to zero.

Here we consider the following form of the equation of conservation of channel bed material in the generic grain size range at elevation y in the bed deposit (Parker, Paola, and Leclaire 2000; Pelosi *et al.* 2014):

$$c_b\left[P_e\left(\frac{\partial f_i}{\partial t} - \frac{\partial f_i}{\partial y}\frac{\partial \eta}{\partial t}\right) + f_i\left(\frac{\partial P_e}{\partial t} + p_e\frac{\partial \eta}{\partial t}\right)\right] = I_f\left(D_{iy} - E_{iy}\right) \tag{23.14}$$

where f_i is the fraction of bed material in the generic grain size range at elevation y, and D_{iy} and E_{iy} respectively denote the deposition and entrainment rates of bed material in the generic grain size range at elevation y.

Equation (23.14) is solved to compute the time rate of change of the volume fractions of bed material within the bed deposit, and to then update the grain size stratigraphy of the bed deposit. Thus, in the continuous model a bookkeeping procedure for the storage of the grain size stratigraphy, such as that proposed by Viparelli *et al.* (2010), is not required, because the grain size distribution of the entire substrate is numerically computed at each time step.

The integration of equation (23.14) over the entire bed deposit expresses the grain size based equation of conservation of bed material in the channel bed, i.e. the time rate of change of sediment in the generic grain size range in the bed deposit is equal to the net deposition rate of sediment within the control volume:

$$\int_{-\infty}^{+\infty} c_b\left[P_e\left(\frac{\partial f_i}{\partial t}-\frac{\partial f_i}{\partial y}\frac{\partial \eta}{\partial t}\right)+f_i\left(\frac{\partial P_e}{\partial t}+p_e\frac{\partial \eta}{\partial t}\right)\right]dy = I_f(D_i-E_i) \quad (23.15)$$

where D_i and E_i respectively denote the grain size specific volume rates of particle deposition and entrainment per unit bed area (Wong *et al.* 2007), defined as

$$D_i=\int_{-\infty}^{+\infty} D_{iy}dy \quad (23.16a)$$

$$E_i=\int_{-\infty}^{+\infty} E_{iy}dy \quad (23.16b)$$

We reduce equation (23.15) to equation (23.11) under the assumption that the active layer thickness, L_a, represents the average thickness of the deposit above the maximum depth of scour $-L_{na}$, i.e. $P_e=1$ for $y<-L_{na}$ (Parker, Paola, and Leclaire 2000; Parker 2004), thus

$$L_a=\int_{-L_{na}}^{+\infty} P_e dy=\int_{\eta-L_{na}}^{+\infty} P_e dz \quad (23.17)$$

For simplicity, we integrate the left-hand side of equation (23.15) with respect to the fixed, i.e. not boundary-attached, z coordinate. In the z coordinate system the maximum depth of scour is at elevation $z=\eta-L_{na}$ and the apparent convective flux of sediment due to the temporal change in mean bed elevation $(\partial P_e f_i/\partial y)(\partial\eta/\partial t)$ is equal to zero (Parker 2004). Thus the left-hand side of equation (23.15) simplifies to

$$\begin{aligned} c_b\int_{-\infty}^{+\infty}\frac{\partial P_e f_i}{\partial t}dz &= c_b\frac{\partial}{\partial t}\int_{-\infty}^{+\infty} P_e f_i dz = c_b\left(\frac{\partial}{\partial t}\int_{-\infty}^{\eta-L_{na}} P_e f_i dz+\frac{\partial}{\partial t}\int_{\eta-L_{na}}^{+\infty} P_e f_i dz\right)\\ &= c_b\left[P_e f_i|_{\eta-L_{na}}\frac{\partial(\eta-L_{na})}{\partial t}+\frac{\partial}{\partial t}\int_{\eta-L_{na}}^{+\infty} P_e f_i dz\right] = c_b\left[f_{Ii}\frac{\partial(\eta-L_{na})}{\partial t}+\frac{\partial F_{ai}L_a}{\partial t}\right]\end{aligned} \quad (23.18)$$

where we define the fractions of sediment at the active layer – substrate interface as

$$f_{Ii}=f_i|_{\eta-L_{na}} \quad (23.19)$$

and the volume of sediment in each grain size range in the active layer is a weighted average over the average amount of sediment in the bed deposit above the elevation $\eta-L_{na}$,

$$F_{ai}L_a=\int_{\eta-L_{na}}^{+\infty} P_e f_i dz \quad (23.20)$$

Noting that the left-hand sides of the divergence and the entrainment forms of the grain size specific equation of channel bed material, i.e. equations (23.11) and (23.14), (i) express the grain size specific conservation of channel bed material, and (ii) can be reduced to the same form – equations (23.15)–(23.20), the right-hand sides of these equations should also be equal, i.e.

$$E_i - D_i = \frac{\partial q_{bi}}{\partial x} \tag{23.21}$$

Equation (23.21) expresses the equivalence between the grain size based divergence and entrainment forms of the Exner equation of conservation of channel bed material. This result is in agreement with the work of Parker, Paola, and Leclaire (2000), Blom *et al.* (2006) and Pelosi and Parker (2014). Starting from the probabilistic formulation of bedload transport (Einstein 1937) and using the definition of particle step length, i.e. the average distance travelled by a particle once entrained in transport until a period of rest begins Λ, it was demonstrated that the entrainment and the divergence forms of the Exner equation are equivalent if the average particle step length is small compared to the length of the modeled domain, L, i.e. $\Lambda/L < 0.01$ (Pelosi and Parker 2014).

Since the average nondimensional particle step length λ, defined as the ratio between the average step length Λ and the characteristic grain diameter D, is generally assumed to vary between 80 and 250 in the case of uniform bed material, and in the case of poorly sorted bed material λ is smaller than in the case of uniform sediment and equal to 10–30 (Einstein 1942, 1950; Nakagawa, Tsujumoto, and Nakano 1982; Tsujimoto and Motohashi 1990), it is reasonable to assume that the divergence and the entrainment form of the Exner equation are equivalent for field scale applications.

23.2.4.2 Calculation of the Entrainment and Deposition Rates

The solution of equation (23.14) requires specific closure relations to compute the grain size specific entrainment and deposition rates at elevation y in the bed deposit, i.e. E_{iy} and D_{iy}. As in Wong *et al.* (2007) and Pelosi *et al.* (2014), we express E_{iy} and D_{iy} respectively as the products of the volume rates per unit bed area, E_i and D_i, and the PDFs of entrainment and deposition at elevation y, p_E and p_D. Since we model the bed material as a mixture of different grain sizes, we further assume that the probability densities of entrainment and deposition are grain size specific, thus

$$E_{iy} = E_i p_{Ei} \tag{23.22a}$$

$$D_{iy} = D_i p_{Di} \tag{23.22b}$$

We relate the grain size specific volume rate of deposition per unit bed area, D_i, to the grain size specific entrainment volume rate of entrainment, E_i, via the particle step length, i.e. the particles deposited at the downchannel coordinate x were entrained in transport upstream. This is mathematically expressed as (Nakagawa and Tsujimoto 1980; Parker, Paola, and Leclaire 2000; Blom and Parker 2004):

$$D_i(x) = \int_0^\infty E_i(x-r) f_{si}(r)\,dr \tag{23.23}$$

where $f_{si}(r)$ denotes the PDF that a particle in the generic grain size range, once entrained into transport, travels a distance r before a period of rest begins. Under the assumption of constant nondimensional particle step length $\lambda = \Lambda_i/D_i$ equation (23.23) simplifies to

$$D_i(x) = E_i(x - \Lambda_i) \tag{23.24}$$

Regardless of how we model the particle step length, the calculation of D_i reduces to the calculation of the grain size specific volume rates of entrainment per unit bed area, E_i. To our knowledge, grain size specific formulations to compute E_i in gravel bed rivers are not available in the literature (Parker 2008b; Wilcock 2001), while several grain size specific bedload relations applicable at the field scale are readily available to estimate q_{bi}. An example of such bedload relations is the Wilcock and Crowe (2003) model illustrated at beginning of this section and implemented in our numerical models.

In a step length formulation, the grain size specific bedload transport rate in a cross section is equal to the sum of all the particles that are entrained in transport upstream, and that are deposited downstream of the cross section. Thus, we can first compute q_{bi} with a bedload transport model, and then we estimate E_i from the step length formulation. If we denote the downchannel coordinate of our cross section with x, this is mathematically expressed as follows (Nakagawa and Tsujimoto 1980; Parker, Paola, and Leclaire 2000):

$$q_{bi} = \int_0^{\infty} E_i(x-r)\int_r^{\infty} f_s(r')dr'dr \tag{23.25}$$

In the case of constant particle step length, Λ_i, equation (23.25) simplifies to

$$q_{bi} = \int_0^{\Lambda_i} E_i(x-r)dr \tag{23.26}$$

At mobile bed equilibrium, or when the time scales of channel bed aggradation and degradation are long compared to those for the adjustment of the bedload transport regime, $q_{bi} \approx E_i\Lambda_i$ (Einstein 1950; Blom and Parker 2004).

For simplicity, in the numerical model presented herein we assume that E_i is equal to q_{bi}/Λ_i, and D_i is computed with equation (23.24). The implementation of a more rigorous formulation is a necessary future step in the development of stratigraphy models of gravel bed river morphodynamics.

23.2.4.3 Definition of the Grain Size Distribution of the Bed Surface

The last problem that needs to be addressed for the implementation of the continuous model is the definition of the grain size distribution of the bed surface, which is needed in the surface based calculation of bed material load.

Noting that the PDF of bed elevation fluctuations, p_e, represents the probability that the instantaneous realization of the bed elevation, η_{inst}, is at elevation y, we define the volume fraction of sediment on the bed surface as the average of the volume fraction of sediment at elevation y in the bed deposit weighted over p_e:

$$F_{si} = \int_{-\infty}^{+\infty} f_i p_e dy \tag{23.27}$$

23.2.4.4 Implementation of the Grain Size Specific Equation of Mass Conservation

To compute the time rate of change of the generic fraction of sediment with equation (23.14), the time rates of change of channel bed elevation, η, and the complementary cumulative distribution function of bed elevation fluctuations, P_e, have to be known. We compute the time rate of change of channel-bed elevation with equation (23.5), and the time rate of change P_e by summing equation (23.14) over all the grain sizes as

$$c_b\left(\frac{\partial P_e}{\partial t} + p_e\frac{\partial \eta}{\partial t}\right) = I_f\left(D_y - E_y\right) \tag{23.28}$$

where D_y and E_y respectively denote the total volume deposition and entrainment rate per unit bed area at elevation y, and are defined as

$$D_y = \sum_{i=1}^{M} D_{iy} = \sum_{i=1}^{M} D_i p_{Di} \tag{23.29a}$$

$$E_y = \sum_{i=1}^{M} E_{iy} = \sum_{i=1}^{M} E_i p_{Ei} \tag{23.29b}$$

Unfortunately we are not aware of any dataset to reasonably constrain the PDFs of bed elevation fluctuations, p_e, entrainment, p_{Ei}, and deposition, p_{Di}, for nonuniform bed material at the field scale. In addition, it is evident from equations (23.14), (23.31), (23.29a) and (23.29b) that these functions should not be arbitrarily specified.

Here, in analogy with the simplifications carried out to estimate E_i, we assume that the temporal scales that characterize the changes in channel bed elevation are long compared to those that characterize the bedload transport regime. We further assume that the characteristics of the bedload transport regime do not significantly change during the calculations, and thus P_e is assumed to be constant in time. Under this assumption and with the aid of equation (23.5), equation (23.28) simplifies to

$$\sum_{i=1}^{M} D_i p_{Di} - \sum_{i=1}^{M} E_i p_{Ei} + p_e \frac{\partial q_{bT}}{\partial x} = 0 \tag{23.30}$$

which is a relation linking the PDFs of bed elevation fluctuations, entrainment, and deposition. In particular, equation (23.30) shows that given a bedload transport regime, i.e. p_e, the probabilities of entrainment and deposition cannot be arbitrarily specified. They are functions of the time rate of change of average channel bed elevation and of the grain size specific entrainment and deposition rates.

The grain size specific form of equation (23.30) takes the form

$$D_i p_{Di} - E_i p_{Ei} + p_e \frac{\partial q_{bi}}{\partial x} = 0 \tag{23.31}$$

Equation (23.31) can be solved with the aid of equation (23.21), for example, to compute the grain size based PDF of deposition, given the PDFs of bed surface elevation and entrainment as

$$p_{Di} = \frac{E_i}{D_i} p_{Ei} - p_e \left(1 - \frac{E_i}{D_i}\right) \tag{23.32}$$

Equation (23.32) shows that under equilibrium conditions, i.e. when $E_i = D_i$, the grain size specific PDF of deposition is equal to the PDFs of entrainment of particles in the same grain size range. However, under disequilibrium conditions, the PDF of deposition depends on both the probabilities of entrainment and bed elevation fluctuation, and it changes in time depending on the relative magnitude of the volumetric entrainment and deposition rates of particles in each grain size range.

Finally, under the assumption of a steady PDF of bed surface elevation equation (23.14) reduces to

$$c_b \left[P_e \frac{\partial f_i}{\partial t} + \frac{\partial \eta}{\partial t} \left(p_e f_i - P_e \frac{\partial f_i}{\partial y} \right) \right] = I_f \left(D_{iy} - E_{iy} \right) \tag{23.33}$$

With the aid of equations (23.5), (23.13), (23.22a), and (23.22b), and recalling that $y = z - \eta$, equation (23.33) further simplifies to

$$c_b P_e \frac{\partial f_i}{\partial t} = I_f \left[D_i p_{Di} - E_i p_{Ei} - \frac{\partial q_{bT}}{\partial x} \frac{\partial f_i P_e}{\partial y} \right] \tag{23.34}$$

Substituting equation (23.31) in equation (23.34), we find the equation that describes the time rate of change of the grain size distribution of the channel bed deposit for the simplified case considered herein:

$$c_b P_e \frac{\partial f_i}{\partial t} = -I_f \left[p_e \frac{\partial q_{bi}}{\partial x} + \frac{\partial q_{bT}}{\partial x} \frac{\partial f_i P_e}{\partial y} \right] \tag{23.35}$$

It is important to note here that due to the assumption of steady probability density of bed elevation fluctuations and to the definition of the grain size distribution of the bed surface the grain size specific equation of conservation of channel bed material at elevation y does not depend on the PDFs of entrainment and deposition.

23.2.5 The Calculation Procedure

The model initial conditions are the longitudinal profile of channel bed elevation, $\eta_{t=0}$, and the grain-size distribution of the initial channel bed, $F_{a,t=0}$ and $f_{sub,t=0}$ in the active layer model, and $f_{t=0}$ in the continuous model. The upstream boundary conditions are a characteristic flood flow, and a mean annual bed material load with its grain size distribution. The channel width, B, is assumed to be constant in the downchannel direction and in time. Since the flow is modeled as quasi-normal, i.e. steady and locally uniform, the downstream boundary condition is a specified channel bed elevation at the downstream end of the modeled reach (Parker 2004).

Given the initial and boundary conditions, the model computes the water depth and the mean flow velocity with equations (23.3) and (23.4); and the Shields number and the shear velocity from their definitions. The relevant parameters of the bedload transport relation are computed with equations (23.7a), (23.8) and (23.9), and the M nondimensional parameters W_i^* are estimated with equations (23.7b) and (23.10).

The grain size specific volumetric bedload transport rates per unit channel width, q_{bi}, are computed with equation (23.6). The volumetric total bedload transport rate per unit channel width, q_{bT}, is computed by summing the M values of q_{bi}. Equation (23.5) is finally solved to compute the time rate of change of channel bed elevation, and thus the new longitudinal channel bed profile.

In the active layer model, once the time rate of change of channel bed elevation is known, equation (23.11) is numerically integrated to compute the time rate of change of each fraction of bed material on the bed surface. The grain size distributions of the active layer and of the substrate are then updated and the flow equations are solved again.

In the continuous model, the grain size based rates of entrainment and deposition are computed as a function of the volumetric bed material load per unit channel width, q_{bi}, and of the average particle step length, Λ_i. The grain size specific PDFs of entrainment are model input parameters, and the model computes the PDFs of deposition with equation (23.32). The grain size distribution of the channel bed is updated with equation (23.35) and the flow equations are solved again.

23.3 Application to a Case Inspired by the Trinity River, California, United States

We applied the active layer model and the continuous model to study how the numerical predictions of the mobile bed equilibrium and the grain size stratigraphy of a channel bed change using either an active layer or a probabilistic formulation. In agreement with previous studies (e.g. Parker and Wilcock

1993), we define mobile bed equilibrium as a condition in which (i) the channel bed elevation does not change in time; (ii) the grain size specific volumetric bedload transport rates per unit channel width, q_{bi}, are constant in space and time and equal to the mean annual bed material load imposed at the upstream end of the modeled reach; and (iii) the grain size distribution of the bed surface is constant in space and time. Whether the definition of grain size distribution of the bed surface given in equation (23.27) will converge to a steady and uniform distribution of the channel bed surface at equilibrium is one of the questions that we explore in the continuous model runs presented below.

The modeled reach is the ~ 20 km long reach that is inspired by the Trinity River, California, United States, downstream of Lewiston Dam where long and detailed field observations allow for the characterization of the unregulated (pre-dam) flow and sediment transport regime studied in Viparelli *et al.* (2011). The upstream end of the modeled domain is located at the New Bridge, and the historical (pre-dam) series of flow discharges used to constrain the pre-dam flood regime was measured at the US Geological Survey Lewiston station located at the Old Bridge. The grain size distribution of the present substrate is represented in Figure 23.3. The number of characteristic grain size ranges is $M = 9$, the geometric mean diameter, D_g, of the present substrate is 16 mm and the geometric standard deviation is 5.13.

As in Viparelli *et al.* (2011), due to the lack of information on the unregulated Trinity River, we assume that the modeled reach was in equilibrium and thus we invoke the strong form of equal mobility (Parker and Klingeman, 1982; Parker and Toro-Escobar 2002) to justify the assumption that the grain size distribution of the mean annual bed material load is equal to the grain size distribution of the substrate. The input parameters of the model runs are those reported by Viparelli *et al.* (2011), and are summarized in Table 23.1.

Noticeable differences between the Viparelli *et al.* modeling effort and the numerical runs presented herein are: (i) Viparelli *et al.* modeled the flow in terms of a cycled hydrograph and the channel width as a function of the flow discharge, here we use a constant characteristic flood flow and channel width equal to the reference values of figure 7 in Viparelli *et al.* (2011), i.e. $Q_w = 368$ m^3/s, $B = 75$ m; and (ii) the friction coefficient C_f in Viparelli *et al.* is computed with a Manning–Strickler formulation as a function of D_{a90}, i.e. the diameter such that 90% of the active layer sediment is finer, here we use a nondimensional Chezy friction coefficient, $Cz = C_f^{-0.5}$, of 12 to match the measured channel

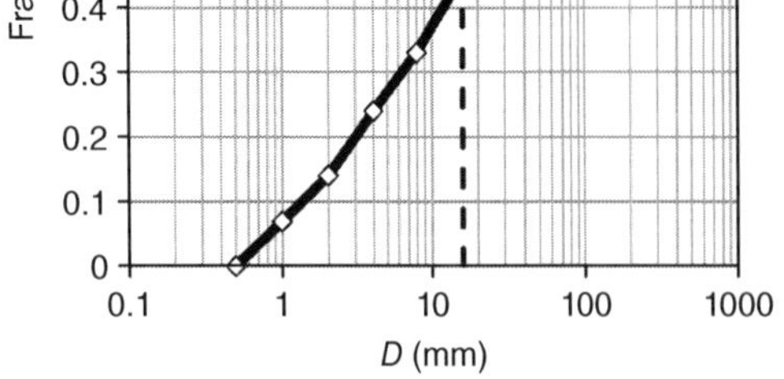

Figure 23.3 Grain size distribution of the present and of the modeled reach of the Trinity River, California, United States.

Table 23.1 Model input parameters for the numerical runs.

Parameter	Value
Q_w characteristic flood flow	368 m^3/s
I_f flow intermittency	0.01
G_s mean annual bed material load	24 000 t/yr
B channel width	75 m
L length of the modeled reach	20 km
λ_p bulk porosity of the channel bed	0.4
ρ_s density of the bed material	2650 kg/m^3
ρ density of the water	1000 kg/m^3
Cz nondimensional Chezy friction coefficient	12
Λ/D nondimensional particle step length	25
L_s maximum thickness of the layers for the storage of the stratigraphy	0.5 m
Δy vertical distance between computational nodes in the continuous model	0.2–0.4 m
Number of computational nodes in the downchannel direction	15
Δt temporal interval	0.002–0.2 yr^{-1}

bed slope, $S = 0.0024$ m/m. Finally, as in Viparelli *et al.* (2011), we multiply the reference Shields number for significant bedload transport of the original Wilcock and Crowe (2003) model, i.e. equation (23.9), times 0.5 to obtain a reasonably good agreement between active layer model predictions and field observations.

The initial condition for all the numerical runs is given in terms of a longitudinal profile of channel-bed elevation with constant initial slope $S_I = 0.001$ and a downstream channel bed elevation $\eta_d = 3$ m. The initial grain size distribution of the bed deposit is assumed to be constant in space and equal to the grain size distribution of the present substrate represented in Figure 23.3. Since we are interested in modeling the grain size stratigraphy of the bed deposit, the initial condition is chosen to allow for significant channel bed aggradation during the runs. In particular, the simulations are specifically designed to study how the grain size stratigraphy of the channel bed deposit differs between the active layer and the continuous model, and from one continuous simulation to the other as a function of the PDFs of bed elevation fluctuation.

23.3.1 The Active Layer Model Run

In the runs presented herein it is assumed that small scale bedforms are absent and the active layer thickness, L_a, scales with D_{a90} (Parker 2008b). In particular, as in Viparelli *et al.* (2011), L_a is set equal to $2D_{a90}$, which shows a reasonably good agreement between field measurements and numerical predictions of the grain size distribution of the armour layer in the application (post-dam) runs. In particular, field estimates of D_{a50} and D_{a90} are ~ 90 mm and ~ 198 mm respectively, while the numerical model predicted $D_{a50} = 85$ mm and $D_{a90} = 202$ mm. Similarly, Ferguson and Church (2009) assumed an active layer thickness of $2D_{a84}$ to model a reach of the Fraser River, Canada. The maximum thickness of the mixed layers for the storage of stratigraphy is equal to 0.5 m. Although this is a very fine grid for field scale applications, it allows for comparison with the continuous model results.

The model is written in Python and the run time of the simulation presented here is 20 min on a Mac OS X laptop.

The grain size distribution of the sediment transferred to the substrate during channel bed aggradation is computed with equation (23.12). As in Viparelli *et al.* (2011) the α parameter in equation (23.12) is set equal to 0.2, a value that returns a numerical grain size distribution of the pre-dam substrate very similar to the grain size distribution measured on the present Trinity River. This value of the parameter α is dependent on the definition of the active layer thickness, since the choice of a thicker active layer would have resulted in a smaller amount of bedload material transferred to the substrate during channel bed aggradation.

Results of the active layer model are presented in Figure 23.4 in terms of temporal changes in channel bed elevation (Figure 23.4a), total volumetric bed material load (Figure 23.4b), and geometric mean diameter of the bedload and of the bed surface (Figure 23.4c and d). In Figure 23.4b–d the initial values

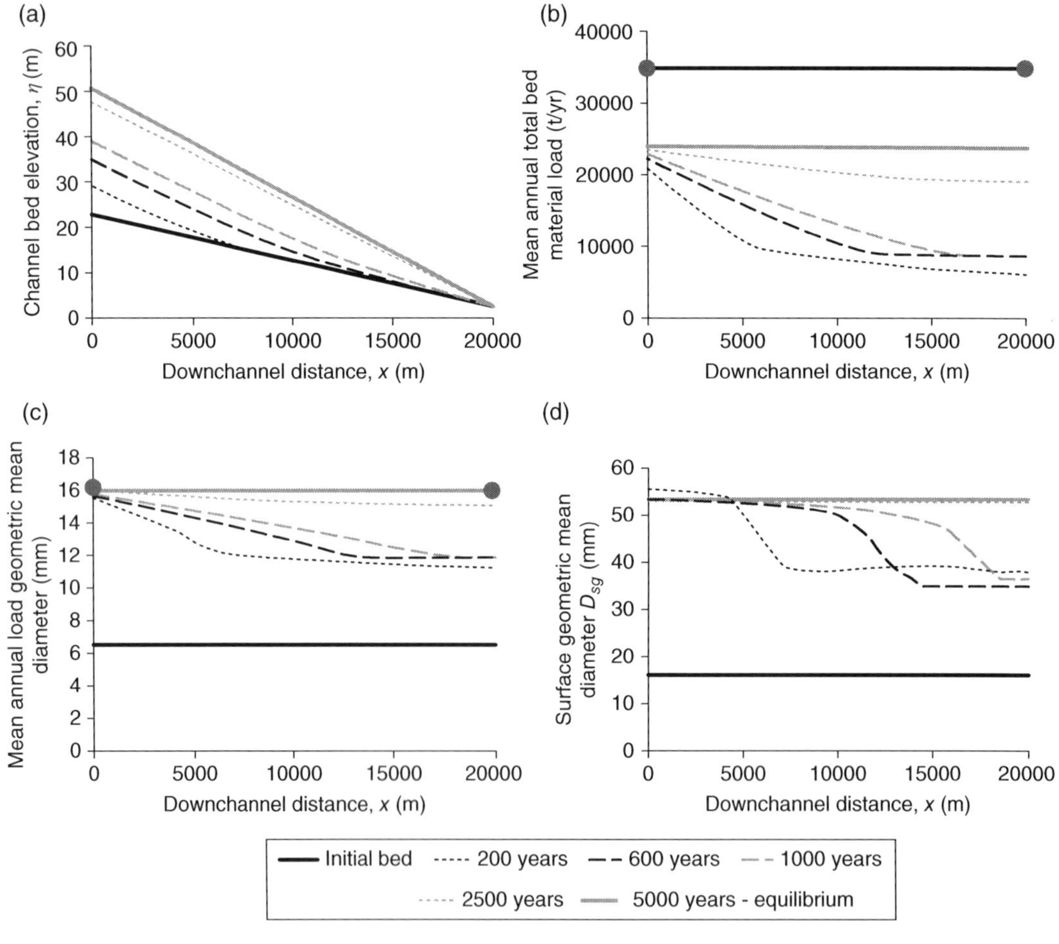

Figure 23.4 Results of the active layer model run. Spatial and temporal evolution of (a) channel bed elevation, (b) mean annual bed material load, (c) geometric mean diameter of the bedload, and (d) geometric mean diameter of the bed surface. The dots indicate the available field data.

of the bed material load and of the geometric mean diameters are affected by the initial condition of an unarmored bed, i.e. the coarse surface layer typical of many gravel bed rivers has not developed yet.

The evolution of the channel bed profile of Figure 23.4a is typical of morphodynamic models in which the relative base level is constant and the flow is assumed to be locally steady and uniform. Under these conditions, the downstream channel bed elevation is fixed in time and, as soon as the wave of aggradation generated at the upstream end of the domain reaches the downstream end, ~ 1000 years, the transition between the initial and the final equilibrium profiles can be approximated as a rotation around the downstream end of the domain. In the numerical run of Figure 23.4 the flow and the sediment transport reach conditions of mobile bed equilibrium in approximately 5000 years.

The spatial and temporal evolution of the total bed material load of Figure 23.4b shows that as the bed aggrades, the bedload transport capacity increases in the streamwise direction, with a consequent increase in the grain size of the bedload transport and of the active layer (Figure 23.4c and d). In other words, in the first stages of the aggradation, coarse sediment is preferentially deposited in the upstream portion of the modeled domain, while the fine material is transported downstream. As the channel bed aggrades, the channel slope increases, coarse particles are transported further downstream and eventually reach the end of the modeled domain. Unfortunately field measurements of the grain size distribution of the unregulated Trinity River bed surface are not available to validate the numerical predictions of the grain size distribution of the active layer.

At mobile bed equilibrium the channel bed slope, the total bed material transport rate, and its grain size distribution, as well as the grain size distribution of the active layer, are steady and uniform, i.e. the longitudinal profile does not show any upward concavity or the active layer surface any downstream fining. This is not only a consequence of the relatively short length of the modeled reach, ~ 20 km, but also and more importantly of the model simplifications. In particular, we neglect particle abrasion, changes in relative base level, and tributaries that contribute sediment and water to the main channel, i.e. some of the physical mechanisms that are responsible for the concavity of the longitudinal channel profiles (Sinha and Parker 1996).

The stratigraphic signal of the preferential deposition of bedload sediment is the pattern of downstream fining and upward coarsening represented in Figure 23.5 in terms of geometric mean diameter of the sediment stored in the bed deposit. In Figure 23.5 the channel bed at mobile bed equilibrium is represented by the continuous black line, the elevation of the active layer at mobile bed equilibrium, i.e. at the end of the numerical run, is represented by the dashed black line, and each coloured marker in the bed deposit denotes the elevation of the layers for the storage of the grain size stratigraphy. The colour scale associated with the layers for the storage of stratigraphy represents the geometric mean diameter of the sediment stored in the channel bed deposit in millimeters.

It is important to note here that: (i) the grain size distribution of the sediment transferred to the substrate during channel bed aggradation is a weighted average between the grain size distribution of the active layer and of the bedload, i.e. equation (23.12) with $\alpha = 0.2$, thus it is considerably finer than the coarse surface represented in Figure 23.4d; and (ii) the difference between the geometric mean diameter of the present Trinity River substrate, 16 mm, and of the numerical substrate, D_g = 17 – 22 mm, is always smaller than 6 mm, which is a more than acceptable result given the model simplifications and assumptions.

23.3.2 The Continuous Model Runs

In the absence of detailed bathymetric surveys of the unregulated Trinity River main channel to estimate the PDF of bed elevation fluctuations for the continuous model runs, we refer to the experimental work of Nakagawa and Tsujimoto (1979) and Wong *et al.* (2007) to constrain the shape of the

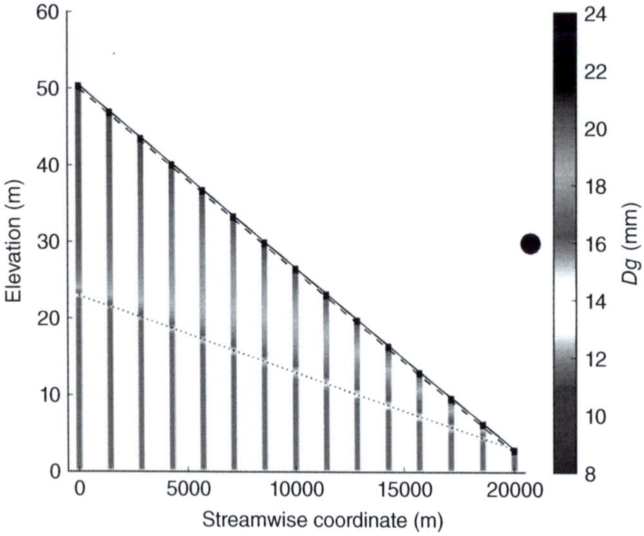

Figure 23.5 Grain size stratigraphy of the channel bed deposit. The continuous black line represents the bed elevation, the dashed line is the elevation of the active layer – substrate interface, the dotted line is the initial bed elevation, and the color scale represents the geometric mean diameter of the sediment stored in the channel bed. The dot next to the scale indicates the reach averaged measured geometric mean diameter equal to 16 mm. (See insert for color representation of this figure.)

probability distribution of bed elevation fluctuations. In particular, we assume that the PDF of bed elevation fluctuations can be modeled with a normal distribution function. However, due to the simplifications of the PPL modeling framework illustrated in section 23.2, and to the uncertainties related to the role of the nonuniform bed material on the amplitude of the bed elevation fluctuations, we further assume that the standard deviation of the bed elevation fluctuations is constant during the numerical runs presented herein. We also consider the ideal case of p_e lognormally distributed to investigate the effects of a hypothetical asymmetry of the PDF of bed elevation fluctuations that may arise, for example, from the spatial variability of channel topography.

We perform a total of four numerical runs reported in Table 23.2 to study how the morphodynamic evolution of the unregulated Trinity River changes with the PDF of bed elevation fluctuations. In particular, we consider the cases of bed elevation fluctuations normally and lognormally distributed to compare the cases of a symmetrical and an asymmetrical distribution.

Noting that fluctuations of the standard deviation of bed elevation should increase as the characteristic diameter of the bed surface increases (Wong *et al.* 2007), we consider two limiting cases: σ_e equal to three times the geometric mean diameter of the mean annual bed material load, 48 mm, and σ_e equal to 50 cm, ~ 30 times the geometric mean diameter of the active layer at mobile bed equilibrium.

In all the numerical runs the grain size specific equation of conservation of channel bed material, equation (23.35), is solved with a distance between the computational nodes in the y direction, Δy, of 20 cm. Each run simulates 10 000 years of channel bed evolution because the time required for the flow and the sediment transport to reach conditions of mobile bed equilibrium in the continuous model is longer than in the active layer model. The model run time ranged between 30 min and 5 h and 30 min on a MAC OS X computer, depending on the characteristics of the PDF shape and parameters.

Table 23.2 The continuous model runs. Here σ_e is the standard deviation of bed elevation fluctuations and p_e is the probability density function.

Run	σ_e (mm)	p_e	Run time (h)
1	48	Normal	12
2	500	Normal	2
3	48	Lognormal	2.5
4	500	Lognormal	0.5

The implementation of the lognormal distribution of bed elevation requires some care because for a variable to be lognormally distributed, the variable has to be greater than 0. We thus define an auxiliary y' vertical coordinate that is downward oriented, i.e. $y' \rightarrow \infty$ deep inside the bed deposit, and y' = mean of the lognormal distribution function for $y = 0$. For simplicity, we set the location parameter of the lognormal distribution equal to 0.

The continuous model results are very similar from one run to the other. For this reason we refer to Run 2, i.e. p_e normally distributed with $\sigma_e = 50$ cm, to describe the continuous model runs. We summarize the results in Figure 23.6, which is analogous to Figure 23.4, i.e. the temporal evolution of the long profile is shown in Figure 23.6a, and the spatial and temporal variability of the total mean annual bed material load, the geometric mean diameter of both the bedload transport and of the bed surface are respectively shown in Figure 23.6b–d.

As in the active layer model run, the temporal evolution of the channel bed profile is characterized by the formation of a downstream migrating wave of aggradation that reaches the downstream end of the modeled domain in ~2500 years, i.e. much later than in the active layer model runs.

A possible explanation for the different time scale of the channel bed evolution predicted between the active layer model and the continuous model is that the response of the initial unarmoured channel bed to the imposed flow and bedload transport rates in the two modeling frameworks is significantly different. This is evident in Figure 23.6b–d, i.e. not only is the change in bedload transport rate predicted with the continuous model more gradual than in the active layer model case, but also the geometric mean diameter of the bedload transport more gradually decreases in the streamwise direction in the continuous model runs. These relatively gradual changes in bedload transport rate and grain size in the streamwise direction result in the slower formation of a coarse bed surface in the continuous runs compared to the active layer run.

It could be argued that the differences between the active layer and the continuous model runs are related to the definition of the part of the bed deposit that exchanges sediment with the bedload, i.e. the active layer or equation (23.27). In other words, it could be argued that in the continuous model the entire deposit interacts with the bedload, and thus it takes longer for the system to adjust to the imposed sediment transport regime. Below we demonstrate that this is not the case.

The grain size stratigraphy predicted in the continuous model runs does not noticeably change from one run to the other, thus we report the grain size stratigraphy of Run 2 and we compare it with the stratigraphy predicted in the active layer model run. The comparison is presented in Figure 23.7, where the grain size stratigraphy is represented in terms of spatial variation of the geometric mean diameter and the continuous line refers to the average elevation of the channel bed and the dashed black line is the active layer – substrate interface in Figures 23.7a and c. The dashed line in Figure 23.7b represents the initial channel bed elevation. In particular, Figure 23.7a refers to the active layer run, i.e. Figure 23.5 but with a different color scale to represent the geometric mean diameter,

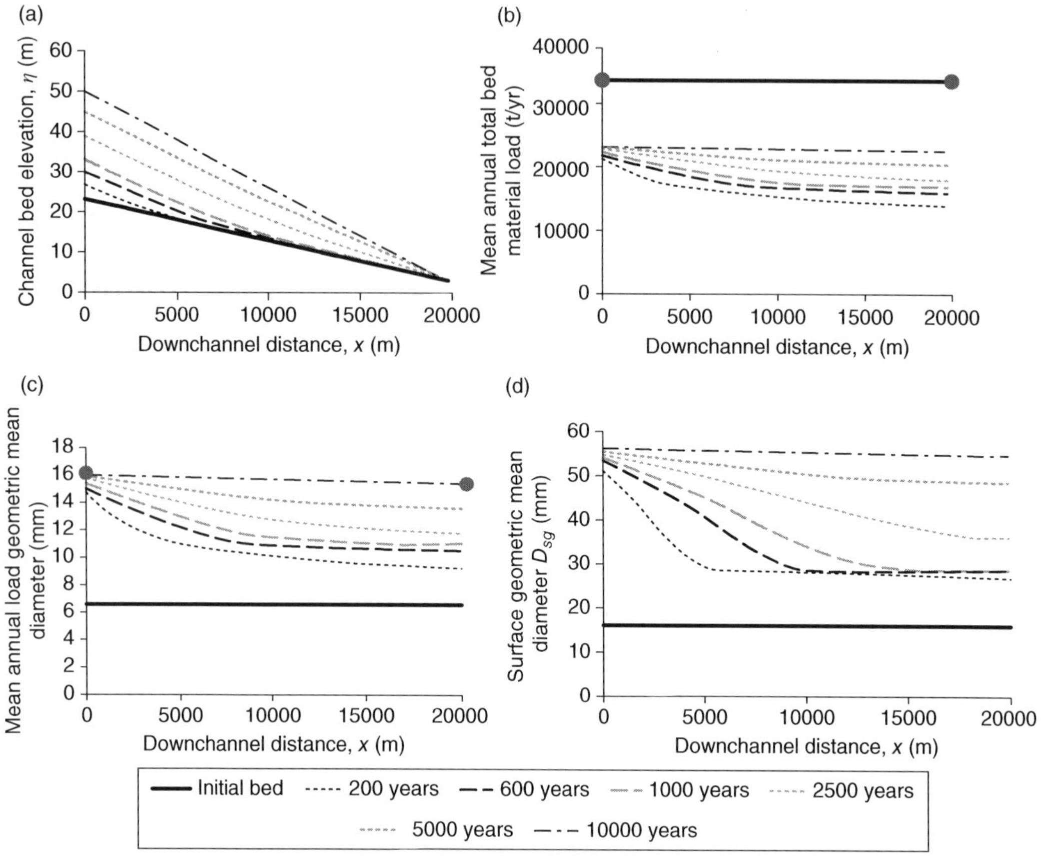

Figure 23.6 Results of the continuous model Run 2. Spatial and temporal evolution of (a) channel bed elevation, (b) mean annual bed material load, (c) geometric mean diameter of the bedload, and (d) geometric mean diameter of the bed surface. The dots indicate the available field data.

Figure 23.7b summarizes the stratigraphy predicted in the continuous model runs, and Figure 23.7c is the stratigraphy emplaced in a new active layer model run performed assuming that only the active layer sediment is transferred to the substrate during channel bed aggradation, i.e. $\alpha = 1$ in equation (23.12).

Although a clear pattern of downstream fining is visible within the bed deposit, the grain size stratigraphy predicted with the continuous model runs (Figure 23.7b) is significantly coarser than the present Trinity River substrate and than the stratigraphy predicted in the active layer model run with $\alpha = 0.2$ (Figure 23.7a). In addition, the spatial variation of the geometric mean diameter within the bed deposit in the continuous model runs closely resembles the stratigraphy predicted with an active layer model when $\alpha = 1$ (Figure 23.7c).

The spatial and temporal evolution of the channel bed profile, the bedload transport, and the geometric mean diameters of the bedload transport and of the bed surface, as well as the time to reach equilibrium condition are almost identical between the active layer model run with $\alpha = 1$ and the continuous model runs. We thus speculate that the different results of the active layer model run with $\alpha = 0.2$ and of the continuous model runs are dependent on how the vertical fluxes of sediment from the bedload to the channel bed are modeled.

In particular, we suspect that the assumption of a steady cumulative probability function of the bed elevation, and the consequent simplification of equation (23.14), i.e. the grain size specific conservation of sediment at elevation y in the bed deposit does not depend on the PDFs of entrainment and deposition, has to be relaxed to properly account for the interaction between

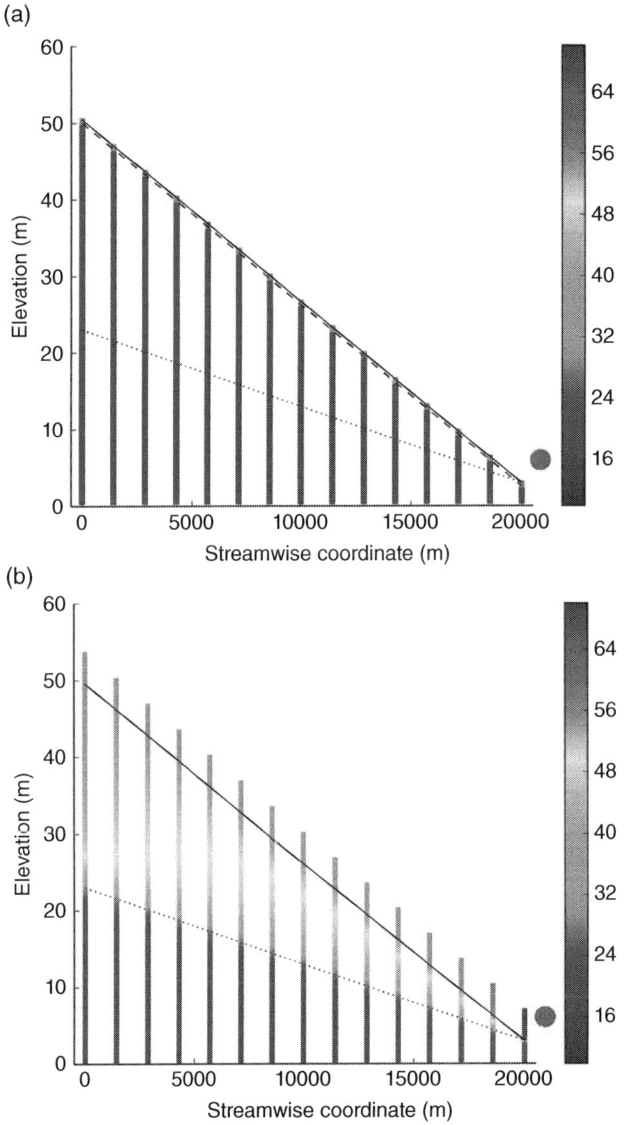

Figure 23.7 Grain size stratigraphy of the channel bed deposit (a) in the active layer model run with $\alpha = 0.2$, (b) in the continuous model Run 2, and (c) in the active layer model run with $\alpha = 1$. The continuous black line represents the bed elevation, the dotted lines are the initial bed elevations, the dashed line in (a) and (c) denote the elevation of the active layer – substrate interface. The scale represents the geometric mean diameter of the sediment stored in the channel bed. The dot next to the scale indicates the reach-averaged measured geometric mean diameter equal to 16 mm.

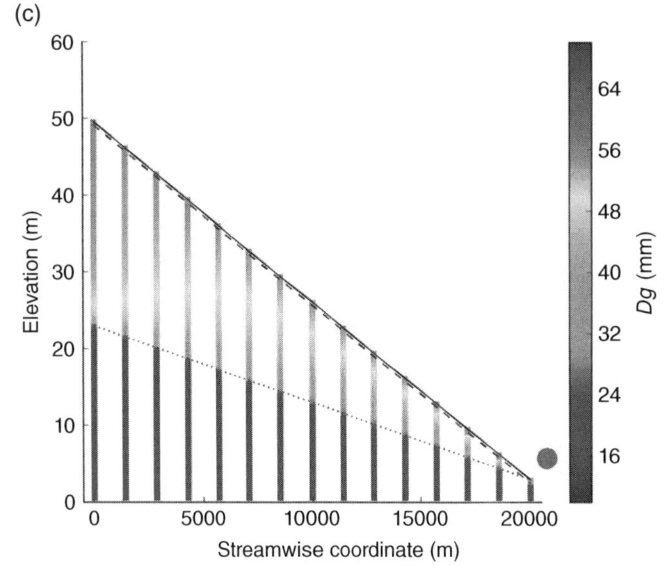

Figure 23.7 (*Continued*)

the bedload and the channel bed, and reasonably predict the grain size stratigraphy of the emplaced deposit.

23.4 Conclusions

The findings of our study are summarized as follows:

1) Under the assumption of steady and uniform probability density of bed elevation fluctuations, the elevation specific equation of conservation of bed material in the generic grain size range does not depend on the PDFs of entrainment and deposition. It depends only on the probability functions of bed elevation fluctuations.
2) The PPL framework for sediment continuity is not too computationally expensive and can be implemented at field scale when the sediment fluxes are modeled in terms of continuous probability functions of entrainment and deposition.
3) The active layer and the continuous framework predict the same conditions of mobile bed equilibrium in terms of channel bed slope, and geometric mean diameter of the bed surface.
4) The temporal evolution of the modeled domain predicted in the continuous model runs:
   - is not significantly dependent on the shape of the probability functions considered herein, i.e. normal and lognormal;
   - is almost identical to the temporal evolution of the modeled domain predicted with an active layer model in which only active layer sediment is transferred to the substrate during channel bed aggradation, i.e. $\alpha = 1$;
   - occurs on a temporal scale that is longer than the one that characterizes the results of an active layer model in which the grain size distribution of the sediment transferred to the substrate is

computed as a weighted average between the grain size distributions of the active layer and of the bedload, i.e. $\alpha = 0.2$.

5) The grain size stratigraphy predicted by the simplified continuous model is coarser that the grain size stratigraphy measured in the field. In particular, it is practically equal to the grain size stratigraphy predicted by an active layer model in which the active layer sediment is directly transferred to the substrate. This result is probably related to the assumption of steady probability distribution of bed elevation fluctuations, which will be relaxed in the near future.

Notation

b	Parameter of the Wilcock and Crowe (2003) bedload relation (–)
B	Effective channel width (m)
c_b	Fraction of channel bed that is bed material (–)
C_f	Friction coefficient (–)
D_{a50}	Median diameter of the active layer (mm)
D_{a84}	Diameter of the active layer such that 84% of the sediment is finer (mm)
D_{a90}	Diameter of the active layer such that 90% of the sediment is finer (mm)
d_i	Characteristic grain size (m)
D_i	Grain size specific volumetric deposition rate of sediment per unit bed area (m/s)
D_{iy}	Grain size specific deposition rate of sediment at elevation y (m/s)
D_{sg}	Geometric mean diameter of the bed surface (m)
D_y	Total deposition rate of sediment at elevation y (m/s)
E_i	Grain size specific volumetric entrainment rate of sediment per unit bed area (m/s)
E_{iy}	Grain size specific entrainment rate of sediment at elevation y (m/s)
E_y	Total entrainment rate of sediment at elevation y (m/s)
f_i	Volume fraction of sediment in the bed deposit (–)
f_{si}	Grain size specific probability density function of step length (1/m)
f_{Ii}	Volume fraction of sediment at the active layer – substrate interface (–)
$f_{sub,i}$	Volume fraction of sediment in the substrate (–)
Fr	Froude number (–)
F_{ai}	Volume fraction of sediment in the active layer (–)
F_{si}	Volume fraction of sediment in the bed surface (–)
F_{ss}	Volume fraction od sand in the bed surface (–)
g	Acceleration of gravity (m/s^2)
i	Subscript to denote the generic grain size range
I_f	Flood intermittency (–)

L	Length of the modeled domain (m)
L_a	Active layer thickness (m)
L_{na}	Maximum depth of scour (m)
L_s	Thickness of the grid for the storage of the stratigraphy (m)
M	Number of characteristic grain sizes (–)
N	Number of layers for the storage of the grain size stratigraphy (–)
p_{Di}	Grain size specific probability density of deposition (1/m)
p_e	Probability density function of bed elevation fluctuations (1/m)
p_{Ei}	Grain size specific probability density of entrainment (1/m)
p_i	Volume fraction of sediment in the bedload (–)
P_e	Complementary cumulative distribution function of bed elevation fluctuations (–)
q_{bi}	Grain size specific volumetric bedload transport rate per unit channel width (m^2/s)
q_{bT}	Total volumetric bedload transport rate per unit channel width (m^2/s)
q_w	Characteristic flood flow per unit channel width (m^2/s)
Q_w	Characteristic flood flow (m^3/s)
R	Submerged specific gravity of the sediment (–)
S	Channel slope (–)
S_f	Friction slope (–)
t	Temporal coordinate (s)
U	Mean flow velocity (m/s)
u_*	Shear velocity (m/s)
W^*_i	Nondimensional parameter representing the grain size specific bedload transport rate (–)
x	Downchannel coordinate (m)
y	Boundary attached vertical coordinate (m)
z	Upward oriented vertical coordinate (m)
α	Parameter controlling the grain size distribution of the sediment at the active layer – substrate interface (–)
Δy	Distance between the computational nodes in the y direction (m)
λ	Nondimensional particle step length (–)
Λ_i	Particle step length (m)
η	Average channel bed elevation (m)
η_{inst}	Instantaneous realization of channel bed elevation (m)
ρ	Water density (kg/m^3)
ρ_s	Sediment density (kg/m^3)
σ_e	Standard deviation of bed elevation fluctuations (m)

τ_b	Bed shear stress (Pa)
τ^*	Shields number (–)
τ^*_{sg}	Shields number defined with the geometric mean diameter of the bed surface (–)
τ^*_{ssrg}	Reference Shields number (–)
ϕ_i	Wilcock and Crowe (2003) bedload equation parameter (–) equation (23.7a)

Acknowledgments

The authors thank Gary Parker for the discussions on the implementation of the continuous framework, which motivated our study, and the reviewers C. Paola and R. Ferguson whose reviews improved the quality of the manuscript. Ricardo R. Hernandez Moreira is supported through the NSF grant EAR-1250641.

References

Blom, A 2008. Different approaches to handling vertical and streamwise sorting in modeling river morphodynamics. *Water Resources Research* 44, W03415. DOI:10.1029/2006WR005474.

Blom, A and Parker, G 2004. Vertical sorting and the morphodynamis of bed form-dominated rivers: a modeling framework. *Journal of Geophysical Research* 109, F02007. DOI:10.1029/2003JF000069.

Blom, A, Ribberink, JS, and Parker, G 2008. Vertical sorting and the morphodynamics of bed form-dominated rivers: a sorting evolution model. *Journal of Geophysical Research* 113, F01019. DOI:10.1029/2006JF000618.

Blom, A, Parker, G, Ribberink, JS, and de Vriend, HJ 2006. Vertical sorting and the morphodynamics of bed-form-dominated rivers: an equilibrium sorting model. *Journal of Geophysical Research* 111, F01006. DOI:10.1029/2004JF000175.

Cui, Y 2007. The unified gravel–sand (TUGS) model: simulating sediment transport and gravel/sand grain size distributions in gravel-bedded rivers. *Water Resources Research* 43, W10436. DOI:10.1029/2006WR005330.

Cui, Y and Parker, G 1998. The arrested gravel front: stable gravel-sand transitions in rivers. Part 2: general numerical solution. *Journal of Hydraulic Research* 36(2), 159–182.

Cui, Y, Parker, G, and Paola, C 1996. Numerical simulation of aggradation and downstream fining. *Journal of Hydraulic Research* 34(2) 185–204.

Cui, Y, Parker, G, Braudrick, C, and Dietrich, W.E 2006. Dam removal express assessment models (DREAM). Part 1: model development and validation. *Journal of Hydraulic Research* 44(3), 291–307.

De Vries, M 1965. Considerations about non-steady bed-load transport in open channels. *Proceedings, 11th Congress, International Association for Hydraulic Research*, Leningrad; 381–388.

Einstein, HA 1937. Der Geschiebetrieb als Wahrscheinlichkeitsproblem. In *Mitteilung der Versuchsanstalt fur Wasserbau an der Eidgenossische Technische Hochschule Zurich*. Rascher: Zurich, Switzerland. (English translation, *Sedimentation*, edited by H. W. Shen, C1–C105, Colo. State Univ, Fort Collins.)

Einstein, HA 1942. Formulas for the transportation of bed load. *Transactions of the American Society of Civil Engineering* 107, 561–597.

Einstein, HA 1950. *The Bed-Load Function For Sediment Transportation In Open Channel Flows*. Technical Bulletin 1026, Soil Conservation Service, US Department of Agriculture: Washington, DC; 78 pp.

Ferguson, R 2003. Emergence of abrupt gravel to sand transitions along rivers through sorting processes. *Geology* 31(2), 159–162.

Ferguson, R. and Church, M 2009. A critical perspective on 1-D modeling of river processes: Gravel load and aggradation in the lower Fraser River. *Water Resources Research* 45, W11424. DOI:10.1029/2009WR007740.

Hirano, M 1971. River bed degradation with armouring. *Transactions of the Japanese Society of Civil Engineers* 3(2), 194–195.

Hoey, T.B. and Ferguson, R 1994. Numerical simulation of downstream fining by selective transport in gravel bed rivers: Model development and illustration. *Water Resources Research* 30(7), 2251–2260.

Hoey, T.B. and Ferguson, R 1997. Controls of strength and rate of downstream fining above a river base level. *Water Resources Research* 33(11), 2601–2608.

Lauer, J.W. and Parker, G 2008a. Modeling framework for sediment deposition, storage, and evacuation in the floodplain of a meandering river: theory. *Water Resources Research* 44, W04425. DOI:10.1029/2006WR005528.

Lauer, J.W. and Parker, G 2008b. Modeling framework for sediment deposition, storage, and evacuation in the floodplain of a meandering river: application to the Clark Fork River, Montana. *Water Resources Research* 44, W08404. DOI:10.1029/2006WR005529.

Lauer, JW, Li, C, Viparelli, E., and Piegay, H 2014. MAST-1D: a size-specific sediment transport and tracer model with off-channel storage. *Proceedings World Environmental and Water Resources Congress*, Portland, Oregon, June 1–5.

Nakagawa, H and Tsujumoto, T 1979. Characteristics of sediment transport processes on duned beds analyzed by stochastic approach. *Bulletin of the Disaster Prevention Research Institute* 29(1), 45–63.

Nakagawa, H and Tsujumoto, T 1980. Sand bed instability die to bedload motion. *Journal of the Hydraulic Division* 106(12), 2029–2051.

Nakagawa, H, Tsujumoto, T, and Nakano, S 1982. Characteristics of sediment motion for respective grain sizes of sand mixtures. *Bulletin of the Disaster Prevention Research Institute* 32, 1–32.

Nelson, PA, Venditti, J, Dietrich, WE *et al.* 2009. Response of bed surface patchiness to reductions in sediment supply. *Journal of Geophysical Research* 114, F02005. DOI:10.1029/2008JF00 1144.

Paola, C, Heller, PL, and Angevine, CL 1992. The large-scale dynamics of grain-size variation in alluvial basins, 1: theory. *Basin Research* 4, 73–90.

Parker, G 1991a. Selective sorting and abrasion of river gravel. *I: theory. Journal of Hydraulic Engineering* 117(2), 131–149.

Parker, G 1991b. Selective sorting and abrasion of river gravel. II: applications. *Journal of Hydraulic Engineering* 117(2), 150–171.

Parker, G 2004, *1D Morphodynamics of Rivers and Turbidity Currents*. http://hydrolab.illinois.edu/people/parkerg//?q=people/parkerg/ (accessed January 2016]).

Parker, G 2008. Transport of gravel and sediment mixtures. In *Sedimentation Engineering Processes: Measurements, Modeling and Practice* (eds Garcia, MH). ASCE: Reston, VA; 165–251.

Parker, G and Klingeman, PC 1982. On why gravel bed rivers are paved. *Water Resources Research* 18(5), 1409–1423.

Parker, G and Perg, L 2005. Probabilistic formulation of conservation of cosmogenic nuclides: effect of surface elevation fluctuations on approach to steady state. *Earth Surface Processes and Landforms* 30, 1127–1144.

Parker, G and Toro-Escobar, CM 2002. Equal mobility of gravel in streams: the remains of the day. *Water Resources Research* 38(11), 46-1–46-8.

Parker, G and Wilcock, PR 1993. Sediment feed and recirculating flumes: fundamental difference. *Journal of Hydraulic Engineering* 119(11), 1192–1204.

Parker, G, Hassan, M, and Wilcock, PR 2008. Adjustment of the bed surface size distribution of gravel-bed rivers in response to cycled hydro-graphs. In *Gravel-Bed Rivers VI: From Process Understanding to River Restoration* (eds Habersack, H, Piegay, H, and Rinaldi, M). Elsevier: New York; 241–285.

Parker, G, Paola, C, and Leclaire, S 2000. Probabilistic Exner sediment continuity equation for mixtures with no active layer. *Journal of Hydraulic Engineering* 126(11), 818–826.

Parker, G, Toro-Escobar, CM, Ramey, M, and Beck, S 2003. Effect of floodwater extraction on mountain stream morphology. *Journal of Hydraulic Engineering* 129(11), 885–895.

Pelosi, A. and Parker, G 2014. Morphodynamics of river bed variation with variable bedload step length. *Earth Surface Dynamics* 2, 243–253.

Pelosi, A, Parker, G, Shumer, R, and Ma, H-B 2014. Exner based master equation for transport and dispersal of river pebble tracers: derivation, asymptotic forms, and quantification of nonlocal vertical dispersion. *Journal of Geophysical Research Earth Surface* 119, 1818–1832. DOI:10.1002/2014JF003130.

Ribberink, JS 1987. *Mathematical modelling of one- dimensional morphological changes in rivers with non- uniform sediment.* PhD thesis, Delft University, The Netherlands.

Robinson, RAJ, and Slingerland, RL 1998a. Grain-size trends, basin subsidence and sediment supply in the Campanian Castlegate Sandstone and equivalent conglomerates of central Utah. *Basin Research* 10, 109–127.

Robinson, RAJ and Slingerland, RL 1998b. Origin of fluvial grain-size trends in a foreland basin: the Pocono formation of the central Appalachian basin. *Journal of Sedimentary Research* 68(3), 473–486.

Sinha, SK and Parker, G 1996. Causes of concavity in longitudinal profiles of rivers. *Water Resources Research* 32(5), 1417–1428.

Spasojevic, M and Holly, FM Jr 1990. 2D bed evolution in natural watercourses: new simulation approach. *Journal of Waterways, Port, Coastal, and Ocean Engineering* 116(4), 425–443.

Talbot, T and Lapointe, M 2002. Numerical modeling of gravel bed river response to meander straightening: The coupling between the evolution of bed pavement and long profile. *Water Resources Research* 38(6). DOI: 10.1029/2011WR000330.

Toro-Escobar, CM, Parker, G, and Paola, C 1996. Transfer function for the deposition of poorly sorted gravel in response to streambed aggradation. *Journal of Hydraulic Research* 34(1), 35–53.

Tsujimoto, T, and Motohashi K 1990. Static armoring and dynamic pavement. *Journal of Hydroscience and Hydraulic Engineering* 8(1), 55–67.

Van Niekerk, A, Vogel, KR, Slingerland, RL, and Bridge, JS 1992. Routing of heterogeneous sediments over movable bed: model development. *Journal of Hydraulic Engineering* 118(2), 246–262.

Viparelli, E, Nittrouer, JA, and Parker, G 2015. Modeling flow and sediment transport dynamics in the lowermost Mississippi River, Louisiana, USA, with an upstream alluvial-bedrock transition and a downstream bedrock-alluvial transition: implications for land-building using engineered diversions. *Journal of Geophysical Research – Earth Surface* 120, 534–563. DOI: 10.1002/2014JF003257.

Viparelli, E, Sequeiros, OE, Cantelli, A *et al.* 2010. River morphodynamics with creation/consumption of grain seize stratigraphy 2: numerical model. *Journal of Hydraulic Research* 48(6), 727–741.

Viparelli, E, Gaeuman, D, Wilcock, PR, and Parker, G 2011. A model to predict the evolution of a gravel bed river under an imposed cyclic hydrograph and its application to the Trinity River. *Water Resources Research* 47, W02533. DOI: 10.1029/2010WR009164.

Wilcock, PR 2001. Toward a practical method for estimating sediment-tranasport rates in gravel bed rivers. *Earth Surface Processes and Landforms* 26, 1395–1408.

Wilcock, PR and Crowe, JC 2003. Surface-based transport model for mixed-size sediment. *Journal of Hydraulic Engineering*129(2), 120–128.

Wong, M, Parker, G, De Vries, P *et al.* 2007. Experiments on dispersion of tracer stones under lower-regime plane-bed equilibrium bedload transport. *Water Resources Research* 34, W03440. DOI: 10.1029/2006WR0517.

Wright, S and Parker, G 2005a. Modeling downstream fining in sand-bed rivers. I: formulation. *Journal of Hydraulic Research* 43(6), 612–619.

Wright, S and Parker, G 2005b. Modeling downstream fining in sand-bed rivers. II: application. *Journal of Hydraulic Research* 43(6), 620–630.

Discussion

Discussion by Philippe Belleudy

In your modeling, parameter Λ fixes the adaptation length of the effective transport to the transport capacity. Do you have a special adaptation of the length of the computational reaches in relation to Λ, in the case of an upstream sediment boundary condition, which is not the sediment transport capacity?

Discussion by Gordon Grant of Both Enrica Viparelli and Rebecca Hodge

Hodge has given us an interesting perspective on the importance of physical grain-grain interactions and what happens when those interactions are not present, for example in bedrock rivers. Viparelli has shown us how a probabilistic representation with an implicit grain interaction (e.g., the Wilcock–Crowe formulation) in the entrainment and active layer exchange can reveal aspects of bed evolution. I am interested in how these contrasting perspectives (i.e. grain physics versus probabilities) can inform each other. Specifically, does the Viparelli approach lose some key behavior in only using a bulk probabilistic function for entrainment? Conversely, what would be gained in the Hodge approach by incorporating a more probabilistic approach to entrainment?

Discussion by Kimberly Hill

While this model does not have the capability of explicit particle scale interactions, it has the promising capability to incorporate a spatially variable rule for particle scale effects, like segregation. This is particularly noteworthy in the case where the segregation rule may change with depth, e.g. by following winnowing behavior at the very top of the bed, and by following behavior similar to dense, creeping granular flows slightly below the surface. It is less obvious how longitudinal variability could be incorporated in segregation rules (such as different behaviors that are associated with the lee and stoss sides of bedforms). Can you comment on whether this longitudinal variability may be incorporated in your morphodynamic model, possibly through a probabilistic approach to the sorting rules?

Reply by the Authors

In our model the average nondimensional particle step length Λ is assumed to be constant and equal to 20. This means that the dimensional particle step length is equal to 20 times the characteristic particle grain diameter.

The sediment upstream boundary condition is not the sediment transport capacity. It is an imposed sediment input rate with a specified grain size distribution. This sediment input rate and size distributions are equal to the sediment transport capacity when the system reaches mobile-bed equilibrium conditions. The entrainment rate at the upstream end of the modeled reach and everywhere in the computational domain is, for simplicity, computed as a ratio between the imposed sediment input rate and the particle step length. The implementation of a more complex mathematical formulation based, for example, on a probabilistic description of the particle step length is relatively straightforward.

The assumption of a constant particle step length and the use of a surface-based bedload relation are both responsible for loss of details in the estimate of the entrainment rates. The problem, however, is that we are not aware of any field-based and grain size specific formulation to compute entrainment rates as a function of the flow properties and of the grain size distribution of the bed surface. Further, the bed surface is generally defined as the portion of the channel bed that actively exchanges sediment with the bedload transport and this can be described mathematically in several different ways. In summary, the formulation implemented in the model cannot capture all the details of the entrainment into bedload transport but it is probably one of the best formulations that can be implemented at the field scale.

How to incorporate particle segregation in the probabilistic modeling framework is an interesting problem. The implementation of physically based and grain size specific probability density functions of particle entrainment, associated with detailed relations to compute the entrainment rate, may reproduce segregation patterns averaged over a large number of bedforms. The mathematical form of these probability functions should depend (at least) on the characteristics of the flow, of the bed elevation fluctuations and of the sediment that is transported. These probability functions have not been derived or measured yet. Wong *et al.* (2007) proposed a functional relation for the case of uniform sediment and lower plane bed bedload transport, however the extension to sediment mixtures and other bedload transport conditions is not straightforward. We think that detailed DEM – discrete element method – modeling may help in the identification of mathematical expressions of the probability density functions for entrainment.

24

Sediment Processes in Bedrock–Alluvial Rivers: Research Since 2010 and Modelling the Impact of Fluctuating Sediment Supply on Sediment Cover

Rebecca A. Hodge

24.1 Introduction

Bedrock–alluvial rivers (rivers with a bedrock bed, but also a degree of alluvial cover), share many similarities with gravel-bed rivers, but also a number of important differences. This chapter starts with a brief overview of differences in sediment processes between bedrock and alluvial rivers. It then considers the advances that have been made in sediment processes in bedrock–alluvial rivers since the previous Gravel-Bed Rivers meeting in 2010. Field-based research is included alongside modelling as the two types of research do not occur in isolation. The chapter then identifies some cross-cutting themes and outlines some of the outstanding research questions that emerge from the literature review. It ends with an application of a numerical model to address the identified research question of how sediment cover responds to temporal variation in sediment flux.

24.2 Differences Between Sediment Processes in Alluvial and Bedrock–Alluvial Channels

Though there are many similarities between alluvial and bedrock–alluvial channels, the presence of a resistant boundary in the latter means that there are also substantial differences in the way in which sediment processes operate in these two systems. Within an alluvial channel, sediment processes are linked through a series of feedbacks to both the channel morphology and the flow (Figure 24.1). In a bedrock–alluvial system, feedbacks between sediment processes and other components are more complex (Figure 24.1). The channel boundary is composed of both bedrock and sediment, and so feedbacks between sediment transport processes and each of these components have to be considered. Furthermore, these two sets of feedbacks typically operate over very different timescales. Three key differences relating to sediment processes in alluvial and bedrock–alluvial channels are outlined below: sediment cover, sediment transport, and morphological change.

Sediment within a bedrock–alluvial channel typically forms a discontinuous cover, which can be characterized by both its extent and its grain-size distribution. The formation of sediment cover is in some ways analogous to the formation of bedforms such as bars in alluvial channels. However,

Gravel-Bed Rivers: Processes and Disasters, First Edition. Edited by Daizo Tsutsumi and Jonathan B. Laronne.

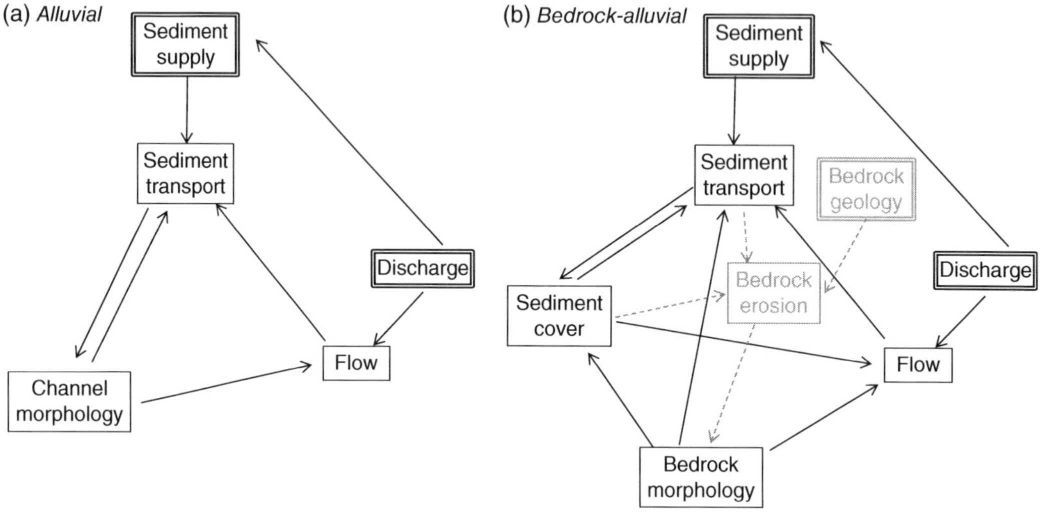

Figure 24.1 Relationships between flow, morphology, and sediment transport in (a) alluvial and (b) bedrock–alluvial rivers. Boxes with double outlines represent external controls. In (b), boxes and arrows in grey represent feedbacks that occur over the timescale of bedrock erosion.

whereas sediment bedforms in alluvial channels emerge from internal processes, the location of sediment patches in bedrock–alluvial channels has significant external controls (i.e. the bedrock morphology). Sediment transport processes are also affected by the two-component morphology. Grains on a smooth bedrock surface are entrained at a lower critical shear stress than grains within an alluvial bed (Hodge, Hoey, and Sklar 2011). Once in motion, sediment grains are unlikely to be deposited on a bedrock surface, and so sediment travel paths are primarily between sediment patches (Hodge, Hoey, and Sklar 2011). Consequently, sediment travel paths are determined by the spatial distribution of sediment cover.

Morphological change in an alluvial channel is determined by the erosion and deposition of sediment by the flow. In a bedrock–alluvial channel, morphological change is caused in two related ways (Figure 24.1); erosion and deposition of the sediment cover and erosion of the bedrock. The link between these two components is that sediment transport provides tools that abrade and pluck bedrock, while sediment cover protects bedrock from these erosive processes. Morphological change of sediment patches is typically much faster than that of bedrock. The growing research literature specifically on bedrock–alluvial channels is reviewed in the following section.

24.3 Review of Sediment Processes in Bedrock–Alluvial Rivers Since 2010

Since the start of 2010, over 40 published papers have addressed the role of sediment processes in bedrock–alluvial rivers. This review concentrates solely on papers with an explicit focus on sediment processes; for this reason papers such as those on pothole evolution (Ortega *et al.* 2014), plucking (Dubinski and Wohl 2013) and longer term channel evolution (e.g. Ferrier, Huppert, and Perron 2013) are not included. Although the example application in this chapter is of modelling sediment processes, it is instructive to consider the field data interpretations that have been reported in parallel with the modelling advances. This is because there is often a direct synergy between the use of field data to test modelling predictions and the use of modelling to test interpretations of field data. Modelling is taken in the broadest sense to include analytical, numerical, and physical models.

A summary of the range of research that has been undertaken in bedrock–alluvial rivers is presented in Table 24.1. The papers have been separated along two different dimensions. The first dimension is the primary approach that was used: physical model, numerical or analytical model, or field data. The second dimension is the scale of the research, and how it maps onto the relationships in Figure 24.1:

1) Grain-scale dynamics, i.e. research into the movement of individual grains and how they are affected by and/or affect bedrock morphology. This is indicated in Figure 24.1 by the arrows between sediment transport, bedrock erosion, and bedrock morphology.
2) The development of sediment cover over a static bedrock morphology. This scale integrates the movement of multiple individual grains over a larger spatial scale, and in Figure 24.1 incorporates the arrows between sediment cover, sediment transport, and bedrock morphology.
3) The same set of interactions as group 2, but with the addition that bedrock morphology can evolve. This group focuses at the temporal scale of individual flow events to annual observations.
4) The fourth group cover the same interactions as the previous group, but over geological (10^2–10^6 years) timescales.
5) Research that explicitly addresses flow characteristics in a bedrock–alluvial channel. Although this is not of itself a sediment process, it has direct implications for understanding of sediment processes in these systems.

24.3.1 Grain-Scale Dynamics

Much of the research on grain-scale dynamics since 2010 has been on saltation-driven abrasion, i.e. erosion of bedrock by saltating bedload. Sklar and Dietrich's (2004) seminal paper formulated a series of equations to predict the rate of saltation-driven abrasion, but made a number of necessary simplifications; these included assuming a flat bed topography and using saltation data from alluvial channels for parameterization. Recent research has gone some way towards addressing these assumptions. Chatanantavet *et al.* (2013) produced scaling relationships for saltation dynamics over a bedrock surface scale that use either Froude number or shear stress, and demonstrated that saltation velocities are independent of grain size.

The impact of convex, upstream-facing rock bedforms on saltation and erosion rates has been addressed using a range of approaches. The bedform topography alters the impact angle of the saltating grain, producing incision rates 10 to 100 times faster than with a planar bed (Huda and Small 2014). Flume work, field data, and numerical modelling have demonstrated that these convex upstream bedforms are themselves produced through the impacts of saltating grains (Sipos *et al.* 2011; Wilson, Hovius, and Turowski 2013; Wilson and Lave 2014), albeit starting with a surface that has some initial relief.

Recent work has measured components of the saltation-abrasion model in a small number of instrumented streams, with concurrent measurements of bedload transport, energy delivery to the bed, hydraulics, and bedrock erosion (Turowski *et al.* 2013; Beer *et al.* 2014). These data have potential to advance significantly our understanding of saltation-abrasion, although the instrumentation has so far been deployed in only a small number of settings.

Bedload tracers have been used to understand how grains are transported through bedrock–alluvial channels (Goode and Wohl 2010; Siddiqui and Robert 2010; Hodge, Hoey, and Sklar 2011). These

Table 24.1 A summary of published research into sediment processes in bedrock–alluvial rivers since the start of 2010. Papers with more than one approach are listed under their primary method. Any omissions are unintentional.

Scale	Physical models	Numerical or analytical models	Field data
Grain-scale dynamics	Saltation over a bedrock surface (Chatanantavet *et al.* 2013) Impact of convex bedforms on saltation and abrasion rates (Wilson and Lave 2014) Suspended sediment caused abrasion rates (Scheingross *et al.* 2014)	Influence of convex bedforms on abrasion rates (Sipos *et al.* 2011; Huda and Small 2014)	Bedload tracer data (Goode and Wohl 2010; Siddiqui and Robert 2010; Hodge, Hoey, and Sklar 2011) Field measurements of energy delivery and abrasion rates (Turowski *et al.* 2013; Beer *et al.* 2014) Evolution of convex bedforms (Wilson, Hovius, and Turowski 2013)
Sediment cover		Incision models incorporating the impact of grain-scale roughness on sediment cover (Johnson 2014; Inoue *et al.* 2014) Cellular automaton model of sediment cover (Hodge and Hoey 2012) Theoretical formulations for sediment cover (Nelson and Seminara 2012; Nelson, Pittaluga, and Seminara 2014)	
Morphology and sediment supply/ transport/cover (event to annual timescales)	Interactions between morphology, sediment cover and erosion (Johnson and Whipple 2010)	Bedrock erosion model including flow and sediment transport (Lai, Greimann, and Wu 2011) Physically-based model of downstream fining (Chatanantavet *et al.* 2010)	Fast changes in bedrock systems (Lamb and Fonstad 2010; Huang, Pan, and Liao 2013; Lague, Brodu, and Leroux 2013; Cook, Turowski, and Hovius 2014) Role of sediment in bedrock incision (Cook, Turowski, and Hovius 2013) Incision under different hydraulic regimes (Johnson, Whipple, and Sklar 2010) Theoretical relationships for formative events (Turowski *et al.* 2012) Downstream changes in channel properties and processes (Bekaddour *et al.* 2013; O'Connor *et al.* 2014; Toone, Rice, and Piégay *et al.* 2014) Bedload driven incision (in alluvial channels) (Hobley *et al.* 2011)
Morphology and sediment supply/ transport/cover (geological timescales)		Impacts of sediment cover relationships on channel evolution (Lague 2010; Valla, van der Beek, and Lague 2010) Landscape scale models incorporating sediment processes (Attal *et al.* 2011) Consideration of the stream power law (Lague 2014) Theoretical model of channel geometry (Yanites and Tucker 2010)	Channel morphology (Allen *et al.* 2013; Finnegan and Balco 2013) Influence of sediment supply and cover on long-term incision (Yanites *et al.* 2010, 2011; Jansen *et al.* 2011; Finnegan, Schumer, and Finnegan 2014)
Flow		Application of one-dimensional flow model (Van, Carling, and Atkinson 2012)	Flow in bedrock canyons (Venditti *et al.* 2014)

datasets show that bedload transport over bedrock surfaces tends to be size-independent, with size dependent distances when there is more extensive sediment cover (Siddiqui and Robert 2010; Hodge, Hoey, and Sklar 2011). Channel morphology has also been demonstrated to affect grain mobility and size selectivity (Goode and Wohl 2010). Grains primarily move between sediment patches, meaning that the spatial distribution of patches controls transport lengths (Hodge, Hoey, and Sklar 2011).

24.3.2 Sediment Cover

Whereas a range of approaches have been used to analyse grain-scale dynamics, research addressing the formation of sediment cover has primarily used numerical and analytical modelling. The one exception is the experiments of Inoue *et al.* (2014) in a flume cut into natural bedrock. These experiments measured the extent of sediment cover and the roughness of sections with and without cover. Inoue *et al.* (2014) then created a model of sediment cover formation, which has similarities to those developed by Nelson and Seminara (2012) and Johnson (2014). Nelson and Seminara (2012) demonstrated using a stability analysis how sediment cover will tend to form discrete patches. Inoue *et al.* (2014) and Johnson (2014) independently developed models that incorporate the effect of bed roughness on sediment cover extent, and the consequent changing roughness on the flow and transport capacity; bed roughness is defined as the ratio of bedrock hydraulic roughness to grain size. These models incorporate many of the interactions in Figure 24.1, but with a static channel morphology.

In the models of Inoue *et al.* (2014) and Johnson (2014), sediment cover is proposed to be a function of the ratio of alluvial layer thickness to bedrock macroroughness height. Model runs demonstrated that increased roughness increased critical shear stress for sediment entrainment and produced greater sediment cover. If sediment roughness is greater than bedrock roughness, runaway alluviation and sediment stripping can result. Hodge and Hoey (2012) used a simpler approach to the issue of sediment cover, developing a cellular automaton (CA) model in which the probability of entrainment of each individual grain was reduced if a grain had neighbouring grains. This simple rule implicitly incorporates the impact of alluvial cover in increasing critical entrainment shear stresses and/or reducing applied shear stresses. The model also reproduced a range of relations between cover fraction, sediment supply, and transport capacity observed by Johnson (2014) and Inoue *et al.* (2014). These numerical models start to explain the range of behaviours observed in flume experiments such as those of Chatanantavet and Parker (2008), and which have not been explained by previous models in which sediment cover was purely a function of the ratio of sediment supply to sediment transport capacity (Sklar and Dietrich 2004; Turowski *et al.* 2007).

24.3.3 Morphology and Sediment Supply/Transport/Cover: Event Timescales

In contrast to research with an explicit focus on sediment cover formation, research with a focus on interactions between changing channel morphology and sediment processes (i.e. the full suite of relationships in Figure 24.1) has used a broad range of approaches. Field data have shown that some bedrock–alluvial channels have the capacity for rapid change over short timescales from individual flow events to a year (Johnson, Whipple, and Sklar 2010; Lamb and Fonstad 2010; Cook, Turowski, and Hovius 2013, 2014; Huang, Pan, and Lia 2013; Lague, Brodu, and Leroux 2013). These studies demonstrate that there is not a single set of characteristics that can produce rapid erosion. The susceptibility of the bedrock to erosion is one factor; Lamb and Fonstad (2010) and Huang, Pan, and Lia's (2013) field sites had bedrock with a tendency for plucking processes, whereas Lague, Brodu, and Leroux (2013) had a field site with weakly consolidated mudstones. The occurrence of driving processes is also important; Cook, Turowski, and Hovius (2013) describe how erosion only occurred once sediment started to be transported along their study reach and abraded the bedrock.

Johnson, Whipple, and Sklar (2010) describe how the amount of erosion in Swett Creek under different sized events was also a function of sediment supply. Larger flash-flood events were net depositors of sediment and hence protected the bed, whereas smaller snowmelt events stripped sediment and caused erosion. In the conceptual model of Turowski *et al.* (2012), Swett Creek is a flood-depositing system. Turowski *et al.* (2012) divided systems into being flood-depositing or flood-clearing, and considered how the relationship between flow event size and sediment supply could affect the morphological evolution of the channel. Bekaddour *et al.* (2013) demonstrate how along a single channel there can be variation in sediment supply and the magnitude of events required to mobilize the bed sediment. Toone, Rice, and Piégay (2014) highlight the importance of bedrock–alluvial reaches on controlling channel width and sediment connectivity. Chatanantavet *et al.* (2010) used a one-dimensional model of a bedrock–alluvial river to demonstrate how spatial variation in grain size, sediment mobility, and alluvial cover can be produced by downstream variation in sediment supply, coupled with both abrasion and size-selective transport. O'Connor *et al.* (2014) analysed a number of rivers, and found that in alluvial channels there was a relationship between bed material transport, local transport capacity and bar area. In contrast, bedrock–alluvial channels in the same area did not show these relationships, with local conditions appearing to exert a greater influence.

Lai, Greimann, and Wu (2011) and Johnson and Whipple (2010) used numerical and physical modelling approaches to reproduce the suite of relationships in Figure 24.1 along a reach. Lai, Greimann, and Wu (2011) incorporated both bedrock channel bed and hydraulic and abrasive scour processes into a two-dimensional hydraulic and sediment transport model. The model was found to be able to reproduce the general pattern and depth of erosion observed in the field. Such a model could be used to try and isolate the different controls on the pattern and depth of erosion, similar to the approach taken by Johnson and Whipple (2010). Johnson and Whipple's (2010) physical model consisted of a weak concrete bed, which was eroded by flows of water and sediment. Systematic variation of flow and sediment discharges demonstrated the importance of sediment load and sediment cover in controlling abrasion rates, whereas shear stress was not found to be important. They also hypothesized that feedbacks between bed topography and roughness, flow, sediment cover formation and the spatial pattern of erosion may stabilize the bed roughness; some of these feedbacks are expressed in the previously discussed model of Johnson (2014).

Finally, some of the research on bedrock–alluvial rivers may also have applications to other alluvial channels; Hobley *et al.* (2011) demonstrated how the incision of rivers into poorly sorted glacial sediment can be described as a detachment-limited process, and constrained a function that describes how incision varies as a function of sediment flux in a similar manner to the model of Sklar and Dietrich (2004).

24.3.4 Morphology and Sediment Supply/Transport/Cover: Geological Timescales

A range of studies since 2010 have demonstrated the impact of sediment supply and consequent sediment cover on the longer-term evolution of bedrock–alluvial channels. Over the longest timescales, Finnegan, Schumer, and Finnegan (2014) demonstrated how cosmogenic-derived measures of incision rate decrease as the time period that they are measured over increases from 10^4 to 10^7 years. They attribute this decrease to heavy-tailed hiatuses in incision caused by alluvial deposition, i.e. a long-term cover effect. Yanites *et al.* (2010) also addressed cover effects by using field data to derive a conceptual model for how a bedrock–alluvial river might respond to an earthquake and consequent sediment input. Downstream changes in long-term sediment cover were found by Yanites *et al.* (2011) to produce spatial patterns in incision rate along the Peikang River. Valla, van der Beek, and Lague (2010) found that cover effects were important when modelling post-glacial channel incision; their model would only reproduce the field long profiles if gorge deepening in the model triggered significant lateral sediment fluxes. In contrast, Jansen *et al.* (2011) identify a long-term tools effect; they used cosmogenic dating to identify a long-term reduction in incision rate in a number of Scottish rivers, which they attribute to decreasing sediment supply following deglaciation. Attal *et al.* (2011) found that in a landscape evolution model a combined tools-and-cover effect was important; their model was best able to reproduce field data when the incision model incorporated a shear stress threshold that was equivalent to the shear stress required to entrain the median grain size. They describe this threshold as being the point at which tools are set in motion and bedrock is exposed.

The role of sediment supply in constraining other aspects of channel morphology has also been addressed. Finnegan and Balco (2013) described how strath terrace formation and abandonment could be linked to aggradation/degradation in a downstream alluvial reach; such changes are linked to, but not exclusively caused by, changes in sediment supply. Allan *et al.* (2013) identified how high sediment supply can cause rivers to respond to changes in bedrock strength by adjusting channel width rather than slope. Yanites and Tucker (2010) produced a conceptual model to explain the impact of sediment cover on channel geometry relationships, which was tested against data from a range of channels.

Finally, Lague (2010, 2014) has highlighted the importance of the sediment supply regime for long-term channel incision. Lague (2010) developed the model SSTRIM, which incorporates daily changes in sediment cover but can be run to model channel evolution over thousands of years. Within this model, the parameters that are able to adjust are channel slope and width. Lague found that over geological timescales, the amount of sediment cover was controlled by the stochasticity of the sediment supply and discharge, rather than the within-channel dynamics. However, this finding may also reflect the relatively simple way in which channel processes were represented. The incorporation of morphological feedbacks such as evolving bed roughness could make the within-channel dynamics relatively more important. Lague (2014) further highlights the importance of sediment supply by attributing the failure of stream power incision models to their lack of specific representation of the impact of sediment fluxes.

24.3.5 Flow

Although the focus of this chapter is on sediment processes, these cannot be understood in isolation from the flow. In Table 24.1 there is a lack of flow data from bedrock–alluvial systems since 2010. Prior to 2010, the literature on bedrock–alluvial flow typically consists of depth–discharge rating curves that were used to constrain one-dimensional models and to calculate roughness (e.g. Broadhurst and Heritage 1998; Heritage *et al.* 2004; Kidson, Richards, and Carling 2006). Since 2010, Van, Carling, and Atkinson (2012) have used stage and width data and a one-dimensional model for a bedrock–alluvial river with multiple channels and Lai, Greimann, and Wu (2011) have applied a two-dimensional flow model to a bedrock–alluvial channel.

Before 2010, there were relatively few attempts to measure directly flow within bedrock–alluvial channels. One exception is Tinkler (1997), who collected velocity data from a cross-section at a range of discharges. Another is Richardson and Carling (2006), who collected dye-dilution data from a straight bedrock channel across a wide range of discharges. Together, these data go towards understanding the hydraulic structure of the flow and how it changes with discharge. One development since 2010 is the dataset of Venditti *et al.* (2014), who collected acoustic Doppler current profiler (ADCP) data from a series of bedrock canyons along the Fraser River, Canada, resolving three-dimensional flow structures in two of these canyons. These data showed that the canyon morphology induced a type of flow circulation that has not been previously observed in bedrock–alluvial channels, and which may cause undercutting of the channel walls. A couple of previously mentioned papers have also included an amount of measured flow data; Inoue *et al.* (2014) measured flow data at a number of points along their bedrock flume, from which they back-calculated roughness parameters. Siddiqui and Robert (2010) measured velocity profiles and used them to calculate spatial variation in shear stress. However, the hydraulics of bedrock–alluvial channels remains a relatively under-researched area.

24.4 Literature Review Findings and Cross-Cutting Themes

This literature review has demonstrated the range and breadth of ongoing research into sediment processes in bedrock–alluvial rivers. Physical and numerical models and field data have been used to address research questions across a wide range of spatial and temporal scales. Furthermore, many studies combine multiple approaches. Inevitably, research is focusing more on some areas than on others. This section highlights some key areas of research focus and cross-cutting themes that emerge from the recent literature.

24.4.1 Saltation-Abrasion Model

One area that has particularly been developed since 2010 is the quantification of grain-scale dynamics, which has largely been motivated by the desire to extend and test the saltation-abrasion model of Sklar and Dietrich (2004). Much of this research has tended to focus on how a particular factor affects grain dynamics, such as the impact of surface topography (Wilson and Lave 2014; Huda and Small 2014), saltation trajectories on a bedrock surface (Chatanantavet *et al.* 2013) and the role of suspended load (Scheingross *et al.* 2014). At larger scales, the saltation-abrasion model has been applied in both a qualitative and a descriptive sense in order to explain observed behaviour. These applications include the derivation of a sediment-flux–sediment-cover function (Hobley *et al.* 2011) and observations of incision patterns in relation to differing sediment-supply regimes (Johnson, Whipple, and Sklar 2010; Cook, Turowski, and Hovius 2013). The saltation-abrasion model was also incorporated into the coupled flow-sediment-incision model of Lai, Greimann, and Wu (2011).

The use of the saltation-abrasion model at a wide range of spatial and temporal scales indicates its importance in this body of literature. Part of its attraction is that it produces a process-based prediction of incision rate from a relatively simple set of parameters, albeit with a number of (clearly stated) assumptions. There is, however, a potential tension in the current literature between the different uses of this model. On the one hand, research into grain-dynamics is testing model assumptions and identifying additional factors that could be added to the model. This is potentially at odds with larger-scale and longer-term applications, which often find that the current model parameters (e.g. sediment supply, flow regime) cannot be predicted with any certainty over the timescales of interest: in this scenario further complicating the model may be undesirable.

24.4.2 Channel Evolution

Another area of research focus since 2010 is field observations of channel evolution over both event to annual, and to geological timescales. This has in part been aided by the application of relatively new techniques such as cosmogenic dating (e.g. Jansen *et al.* 2011; Finnegan, Schumer, and Finnegan 2014) and terrestrial laser scanning (e.g. Wilson, Hovius, and Turowski 2013; Cook, Turowski, and Hovius

2014). Such data have furthered our understanding of the key factors affecting the morphological development of bedrock–alluvial systems across a range of scales.

24.4.3 Sediment Supply

One common theme of much of the research in Table 24.1 is the importance of sediment supply. Studies such as Cook, Turowski, and Hovius (2013) and Johnson, Whipple, and Sklar (2010) have demonstrated how sediment supply can affect channel evolution over a small number of flow events, whereas Lague (2010) and Finnegan, Schumer, and Finnegan (2014) highlighted the importance of the sediment supply regime in controlling sediment cover over millennial timescales. Valla, van der Beek, and Lague (2010), Bekaddour *et al.* (2013) and Chatanantavet *et al.* (2010) all identify the importance of lateral sediment supply for sediment properties and channel evolution. Both Lague (2010) and Turowski *et al.* (2012) explore possible relationships between discharge and sediment supply, and the impact that this might have on sediment cover. However, despite the recurrence of this theme, there are few data that can actually be used to quantify sediment supply regimes in bedrock–alluvial rivers.

24.5 Outstanding Research Questions

Table 24.1 demonstrates that advances are being made on a broad range of research areas related to sediment in bedrock–alluvial rivers. However, there are still notable omissions. This section outlines four research areas that would address some of these gaps.

24.5.1 Flow Research Questions

To reiterate, there are few flow datasets from bedrock–alluvial channels. There are obvious problems with collecting such data, i.e. the need to access channels under a range of flow conditions. Numerical and physical modelling approaches are affected by the absence of data specifying boundary conditions and channel morphology, and for these reasons the majority of flow modelling in these systems to date has been limited to one-dimensional approaches (e.g. Van, Carling, and Atkinson 2012). However, flow datasets comparable to those collected from alluvial channels in the field and flume (e.g. Buffin-Belanger, Roy, and Kirkbride 2000; Rice, Buffin-Bélanger, and Reid 2014) would help to develop our understanding of the hydraulics of bedrock–alluvial systems and how they feedback into the relationships outlined in Figure 24.1. Some outstanding research questions are posed below.

24.5.1.1 What are the Hydraulic Characteristics of Flow in Bedrock–Alluvial Systems at Different Spatial Scales and Discharges?

Some authors have suggested that flow in bedrock–alluvial systems tends to maintain a Froude number of around 1 across a range of discharges (Grant 1997; Tinkler 1997). Hypotheses have been put forward for how this occurs (Grant 1997; Tinkler 1997), but there are few datasets with which to test them. More datasets of high-resolution, three-dimensional, flow measurements (e.g. Venditti *et al.* 2014) from a range of channel geometries at multiple discharges would not only address reach-averaged issues such as Froude number, but would also help to quantify the impact that particular bedrock morphologies had on the local flow properties and consequent sediment dynamics. Such datasets could also be used to validate numerical flow models; the models could then be applied to different discharges and channel morphologies.

24.5.1.2 What Controls the Flow Resistance of a Bedrock–Alluvial Channel, and How Should it be Quantified?

Hydraulic roughness values for bedrock–alluvial rivers calculated from stage–discharge data are typically higher than for alluvial rivers (Manning's n of 0.03 to 2; Heritage *et al.* 2004). However, there is currently no robust mechanism for determining *a priori* what hydraulic roughness value is appropriate. Furthermore, it is not apparent which flow resistance law is more reliable over a range of discharges in bedrock–alluvial channels. An additional complication is that the variation between different bedrock–alluvial channel topographies means that any roughness value will vary significantly between different channels. High-resolution flow data would enable hydraulic roughness to be quantified at a range of discharges and spatial scales. Comparison between channel topography, flow structures, and hydraulic roughness may enable the key length scales that control the hydraulic roughness to be identified.

24.5.1.3 What is the Impact of Sediment Cover on Hydraulic Roughness?

Sediment cover will change the hydraulic roughness of a bedrock–alluvial channel, as observed by Inoue *et al.* (2014). Both Inoue *et al.* (2014) and Johnson (2014) hypothesized that sediment cover could either increase or decrease channel roughness, depending on grain size and bedrock morphology. Again, flow data from channels with a range of sediment cover are needed to quantify these effects.

24.5.2 Sediment Supply Research Questions

Sediment supply has been identified as being an important component of bedrock–alluvial rivers at a range of spatial and temporal scales. Over long timescales, it has been suggested that variation in sediment supply is a more important control on channel evolution than the details of the sediment processes within the channels (Lague 2010; Finnegan, Schumer, and Finnegan 2014). There is therefore a need to understand what the sediment supply regime of a bedrock–alluvial river is like, and what processes control it.

24.5.2.1 Is There a Typical Sediment Supply Regime of a Bedrock–Alluvial River?

All bedrock–alluvial rivers will have a sediment supply regime whereby the long-term sediment supply is less than the potential transport capacity of the channel. Beyond this similarity, the details of the regime have potential to vary greatly between different channels; these variations will be caused by different sediment sources, flow regimes, and degrees of coupling between the flow and sediment supply. While there is unlikely to be a typical regime, it is important to understand the similarities and differences between the regimes of different rivers.

Quantifying sediment supply regimes requires estimates of sediment fluxes, which are not straightforward to measure. Hard infrastructure and intensive monitoring approaches such at the Erlenbach (Rickenmann *et al.* 2012) can provide such data, but only at a single location. Suspended sediment data have been used to estimate bedload, but there are significant uncertainties (Turowski, Rickenmann, and Dadson 2010). Analysis of seismic noise from rivers has potential as a way of measuring bedload flux remotely (Burtin *et al.* 2011; Hsu, Finnegan, and Brodsky 2011). There is therefore a need to develop appropriate techniques and to collect concurrent flow and sediment transport data from a range of bedrock–alluvial rivers.

24.5.2.2 What Controls the Magnitude, Frequency, and Grain-size Distribution of Sediment Supply?

One of the primary causes of difference in sediment supply regime is the different potential sediment sources. Sediment could be sourced from the upstream channel, which could be alluvial or bedrock–alluvial, from the adjacent hillslopes through processes such as landslides and debris flows, or released from storage within the channel and floodplain. Each of these different sources will have its own magnitude and frequency characteristics and grain-size distribution. Understanding sediment supply regimes will also require an understanding of how these different sources operate. This suggests that collection of flow and bedload transport data should be carried out at multiple points along a channel network, and in conjunction with monitoring of hillslope activity.

24.5.3 Sediment Cover Research Questions

In Table 24.1, the scale of much of the field research that has addressed sediment cover in one way or another has either been focused on grain-scale dynamics, or larger-scale trends in reach-averaged sediment cover. The numerical models listed under sediment cover in Table 24.1 tend to consider reach-average properties, such as a single roughness value (Inoue *et al.* 2014; Johnson 2014). Consequently, little research has focused on the way in which sediment cover is distributed within a reach.

The pattern of sediment cover within a reach is important for a number of reasons. Grains primarily move between sediment patches, and therefore the interpatch spacing will control sediment-transport lengths (Hodge, Hoey, and Sklar 2011). Sediment cover determines the locations in the channel where bedrock is exposed and can be eroded (Turowski *et al.* 2008; Johnson, Whipple, and Sklar 2010). In Figure 24.1, sediment cover plays a key role in moderating the feedbacks between sediment transport, flow, erosion, and consequently bedrock morphology. Johnson and Whipple (2010) began to quantify these feedbacks using a physical model, and hypothesized that over several erosive events the feedbacks can determine the channel morphology and limit the channel roughness; the way in which these feedbacks operate in other channels and under different sediment supply conditions is currently unknown. Some outstanding research questions are therefore as follows.

24.5.3.1 What Controls the Extent and Spatial Pattern of Sediment Cover?

A type of field data that is largely absent in Table 24.1 is mapping of sediment cover within bedrock–alluvial channels. Various factors could control the location and extent of sediment cover, e.g. bedrock morphology, local hydraulics, sediment calibre, the sediment supply regime, and the history of previous sediment-supply events. Datasets of channel morphology and the locations and extent of sediment cover from a range of channels with varying morphologies and sediment supply regimes would enable the relative importance of these various factors to be assessed.

24.5.3.2 How Does Sediment Cover Respond to Changes in Controlling Factors?

Given the stochastic nature of sediment supply, it is unlikely that sediment cover is ever entirely in equilibrium with current sediment supply. The storage and release of sediment stored in a bedrock–alluvial channel will buffer against a certain amount of variation in sediment supply. Sediment storage can also mean that the impact of a large sediment supply event can persist in the channel for years. Long-term datasets of flow, sediment supply, and sediment cover from channels with different flow and sediment regimes would help to address this question. Physical and numerical models would provide another approach to answering this question.

24.5.3.3 To What Extent Does the Pattern of Sediment Cover Control the Morphological Development of the Channel?

Sediment cover development can impact erosion in ways including causing erosion to be focused on particular areas of the bed, and by changing the hydraulic roughness and hence flow and grain dynamics. Studies using a modelling approach (e.g. the physical model of Johnson and Whipple 2010, and the coupled flow, sediment and erosion model of Lai, Greimann, and Wu 2011) could be used to tackle this question by systematically varying flow, sediment supply, and initial channel morphology and measuring the resulting patterns of sediment cover, erosion, and morphological change. If there is a clear link between the sediment supply regime and consequent channel morphology, it is possible that analysis of channel morphology in the field may enable the long-term sediment supply regime to be estimated.

24.5.4 Upscaling Research Questions

The final research area is that of how to upscale between the different spatial and temporal scales in the rows of Table 24.1. Much of the research in the grain dynamics row is motivated by the aim of improving the saltation-abrasion model of Sklar and Dietrich (2004), which was originally designed to model long-term incision. However, research such as Lague (2010) and Finnegan, Schumer, and Finnegan (2014) has indicated that the details of grain dynamics may be unimportant at geological timescales, and that other factors such as the sediment supply regime may be dominant.

24.5.4.1 How Does the Importance of Factors Controlling Bedrock Erosion Change Over Different Timescales?

Addressing the previous research questions will help to determine the factors that control the interactions in Figure 24.1 over relatively short timescales. Assessing whether the same factors are important over longer timescales will require a combination of different approaches. One approach would be to take long-term models such as Lague's (2010) SSTRIM, and to develop the representation of grain dynamics within them; this could involve a two-dimensional representation of the channel morphology and flow dynamics (e.g. Lai, Greimann, and Wu 2011), improved roughness representation (Inoue *et al.* 2014; Johnson 2014), or improved grain dynamics (Huda and Small 2014). This would help to determine how much difference the details of grain dynamics make to geological scale evolution.

24.6 Implications for Modelling Sediment Processes in Bedrock–Alluvial Rivers

The above research questions make it clear that physical, numerical, and analytical modelling could play a significant role in furthering understanding of bedrock–alluvial rivers. Field data are complicated by the multiple controlling factors that affect the system evolution; furthermore, factors such as sediment supply will have varied spatially and temporally in a way that is hard to reconstruct. Modelling provides a complementary approach in which the different factors can be isolated and controlled. Some issues to be considered around modelling sediment processes in bedrock–alluvial rivers are as follows.

24.6.1 Appropriate Level of Process Representation

As outlined earlier, there is a potential divergence between the research on the saltation-abrasion model at different scales, with increasing complexity at grain-scales offset by the difficulty of applying even the simplest form of the model at longer time-scales. In any model of sediment processes, there is therefore a need to identify the appropriate level of process representation. This identification could be achieved through consideration of field data, or through contrasting model runs in which processes are or are not included, or are parameterized in different ways.

24.6.2 Stochastic Versus Deterministic Approaches

Morphological modelling of alluvial rivers often uses a deterministic continuum mechanics approach based around the Exner equation and a bedload transport relationship (e.g. Hoey and Ferguson 1994). Similar approaches have also been used for bedrock–alluvial rivers (e.g. Lague 2010; Lai, Greimann, and Wu 2011). However, aspects of the behaviour of bedrock–alluvial rivers suggest that a stochastic approach may be more representative. For example, the CA model of Hodge and Hoey (2012; see section 24.7) used a stochastic approach and simple rules to represent the non-linear impact that exposed bedrock has on sediment dynamics and hence sediment-cover formation.

24.6.3 Constraining Boundary Conditions and Input Parameters

Stochastic modelling approaches may also offer a solution to the problem of poorly constrained boundary conditions over geological timescales. With increased understanding of sediment supply regimes it may be possible to specify the distribution of flow and sediment supply events over long timescales, even if the exact time series cannot be determined. The model of Lague (2010) is one example of using a stochastic input of discharges to a deterministic model. Another approach to uncertain boundary conditions and input parameters is to use Monte-Carlo-type approaches.

24.6.4 Developing Physical Models

The previous issues have primarily been in relation to developing numerical and analytical models of bedrock–alluvial rivers. However, physical models also offer significant potential for research. Much of the physical models have been concentrated on grain-scale dynamics (e.g. Chatanantavet *et al.* 2013), possibly because of the difficulties of creating self-formed bedrock–alluvial channels in the flume. Those that have been created (e.g. Johnson and Whipple 2010) have been relatively small and thus quantifying the flow within them is not straightforward. New techniques such as terrestrial laser scanning and three-dimensional printing offer the possibility of creating Froude-scaled models of bedrock–alluvial rivers, in which flow and sediment dynamics can be quantified (Hodge and Hoey, 2016a, 2016b).

24.7 An Application of a Numerical Model of Sediment Processes

This application of a numerical model addresses the research question *How does sediment cover respond to changes in controlling factors?* Specifically, what is the response of sediment cover to temporally varying sediment supply? The model is the CA model of Hodge and Hoey (2012), which models the development of sediment cover in a bedrock channel under a specified sediment supply and constant discharge. Individual sediment grains are represented explicitly, and their behaviour is governed by empirical rules that mimic the differential entrainment of grains on bedrock and alluvial surfaces (Hodge, Hoey, and Sklar 2011). The aim of the CA model is not to reproduce all of the processes and feedbacks that affect sediment dynamics within a bedrock channel. Rather, it is to demonstrate that relatively complex behaviours, which replicate those observed in other settings, can be demonstrated through application of simple, plausible, rules. Consequently, the focus is on sediment grains, with no explicit representation of flow within the model.

Hodge and Hoey (2012) used this CA model to evaluate the form of the relationship between the input sediment flux and the resulting amount of sediment cover, under conditions of constant sediment supply. They found that the form of the relationship between sediment flux and sediment cover varied according to the way in which grain entrainment was parameterized. The resulting forms of the relationship included linear, sigmoidal, negative exponential, and runaway alluviation. This wide range of forms is consistent with those reported from experiments and theory (e.g. Sklar and Dietrich 2004; Turowski *et al.* 2007; Chatanantavet and Parker 2008), and suggests that the difference in grain mobility between bedrock and alluvial areas is a key parameter in determining the behaviour of sediment cover (see also Johnson (2014) and Inoue *et al.* (2014)).

24.7.1 Methods

24.7.1.1 CA Model Structure

In the CA model, the bedrock surface is represented by an array of 1000 cells long by 100 cells wide (Figure 24.2a). Each cell contains zero, one, or more grains; in the latter case the grains form a vertical stack. Grains in the model are assigned a probability of movement (P) per time-step, which is a function of the number of neighbouring cells containing grains (Figure 24.2b). Grains with five or more neighbouring cells with grains are considered to be in an alluvial cluster and have a lower probability of entrainment (P_c) than isolated grains (P_i), which have four or fewer neighbouring cells with grains. This parameterization reproduces the effects of alluvial cover in both increasing the entrainment shear stress of grains and increasing local roughness and so decreasing the flow shear stress. However, note that flow is not represented in the model beyond this parameterization.

In each time-step, random numbers and the values of P_i and P_c are used to determine which grains move. This is subject to the constraint that a maximum of four grains can be entrained from each cell, representing the presence of an active layer. The entrained grains are moved a fixed step-length downstream equivalent to 10 grain diameters. Previous work has shown that model results are not sensitive to this value (Hodge and Hoey 2012). A specified number of grains are also supplied to the upstream end of the model domain, which are evenly distributed across the model width. These model processes are outlined in Figure 24.2a. A smoothing algorithm is applied to the model to redistribute grains from cells with excessive stacks of grains compared to their downstream neighbour. Sediment cover is defined as the proportion of cells with immobile grains (i.e. grains that are not entrained) in a time-step. Over time, with a constant sediment input, the model normally equilibrates to a constant sediment cover (Figure 24.2c). Further details of the CA model are available in Hodge and Hoey (2012).

Previous work focused on the impact that the values of P_i and P_c had on the relationship between sediment supply and sediment cover under conditions of constant sediment supply. In each model run, sediment supply had a constant value between 12.5% and 100% of the model transport capacity (Q_t). Transport capacity is defined as the sediment flux that would be produced by a fully alluviated bed, i.e.:

$$Q_t = L P_c L_A w \tag{24.1}$$

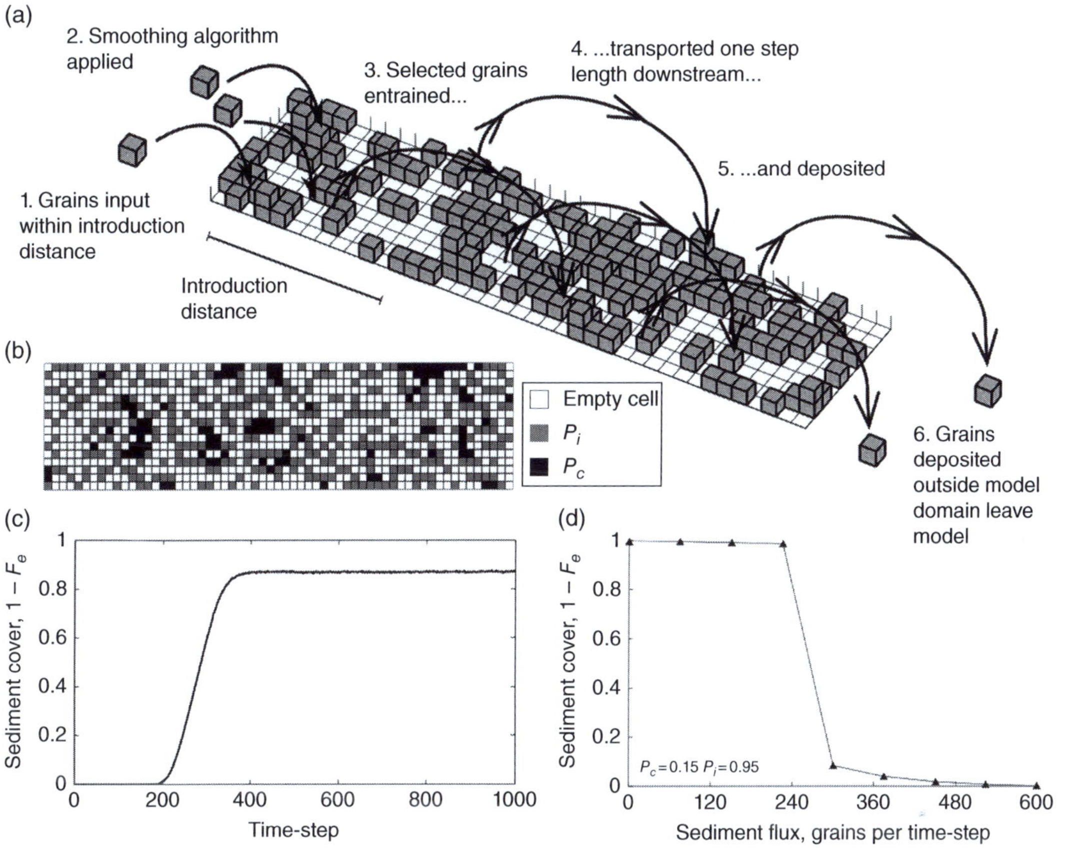

Figure 24.2 (a) A schematic of the CA model, illustrating the different operations that are applied within each model time-step. The full model domain is 100 cells wide by 1000 cells long. (b) A section of model domain showing the parameterization of grain entrainment. Grains where five or more neighbouring cells contain grains have entrainment probability P_c, whereas those with fewer neighbouring grains have probability of entrainment P_i. (c) Evolution of bedrock exposure ($1 - F_e$) in a downstream section of the model domain. (d) The relationship between sediment flux and $1 - F_e$ when $P_i = 0.95$ and $P_c = 0.15$.(Adapted from Hodge and Hoey (2012).)

where L is the grain step length (10 cells), L_A is the active layer depth (4 grains) and w is the model width (100 cells). As previously stated, Hodge and Hoey (2012) found that the form of the relationship between sediment supply and sediment cover varied depending on the values of P_i and P_c. Many combinations of P_i and P_c produced a sigmoidal relationship, with a rapid transition between high bedrock exposure at low sediment supply, and an alluvial bed at high sediment supply (e.g. Figure 24.2d). At low supply, most grains have few neighbours, probability P_i, and are relatively mobile. As supply increases, grains have more neighbours, and consequently more grains have probability P_c. This decreased probability of entrainment makes grains less mobile, and so alluvial cover rapidly develops. The model runs presented here use $P_i = 0.95$ and $P_c = 0.15$, which produces a sigmoidal distribution between sediment flux and cover (Figure 24.2d), and a maximum transport capacity of 600 grains per time-step.

24.7.1.2 Fluctuating Sediment Supply

In these new model runs, a temporally varying sediment input was used to evaluate the impact that variation in sediment flux has on long-term sediment cover. During each run, P_i and P_c and the active-layer depth were held constant, which is equivalent to applying a constant discharge. Such decoupling of flow and sediment supply is admittedly unlikely in a natural channel, although it could result from earthquake-triggered landslides, for example. Discharge was held constant in these runs in order to isolate the impact of sediment supply on sediment cover, and to avoid adding additional model complexity that could make the results harder to interpret.

For each model run, the sediment supply in each time-step was drawn from a specified distribution. A set of eight model runs was completed for each of six different sediment supply distributions. A seventh set of runs used a constant sediment input for comparison. Each of the eight different runs in each set used one of eight time-averaged sediment input rates $(\overline{Q_s})$. These rates are from 75 to 600 grains per time-step and correspond to 12.5% to 100% of the model transport capacity when the bed is fully alluviated (see Hodge and Hoey 2012). Each model run comprised 20000 time-steps; with a constant sediment input the model typically reached steady state sediment cover after 500 time-steps (Hodge and Hoey 2012).

The distributions of sediment supply are produced following the approach of Lague (2010). Sediment input into a river can be related to discharge as:

$$Q_s \propto Q^x \tag{24.2}$$

where Q_s is the sediment flux, Q is discharge, and x is a rating exponent. This relationship is useful, because formulations for distributions of Q have been produced, which can be used in turn to predict distribution of Q_s. The distribution of Q_s is used to drive the fluctuating sediment supply. The constant discharge in the CA model means that the distribution of Q is not used in the modelling, apart from as a way of generating a distribution of Q_s. The distribution of Q used here was developed for the landscape evolution model €ros (Davy and Crave 2000; Lague *et al.* 2005), but has since been shown to replicate distributions of Q across locations including mountain rivers (Lague *et al.* 2005). The distribution has a power law tail, and is given by:

$$pdf(Q^*) = \frac{k^{k+1}}{\Gamma(k+1)} exp\left(-\frac{k}{Q^*}\right) Q^{*-(2+k)} dQ^* \tag{24.3}$$

In the original application of equation (24.3), Q^* is the daily discharge normalized by the mean annual discharge. Q^* is therefore a weighting parameter, with a time-average value of 1. The variability of the discharges is set by k, with smaller values of k giving higher variability (Figure 24.3). Discharge data from rivers have fitted k values of between 0.1 and 4. Three values of k are used in these simulations: 0.5, 1, and 2.

Equation (24.3) was used to produce three time series of 20 000 values of Q^*, one for each of the three values of k. There is no temporal correlation within the series. Each of the three series has a mean value of 1 ± 0.006. For each model run, each time series of Q^* was used to produce a time series of Q_s using:

$$Q_s(t) = Q^*(t)\overline{Q_s} \tag{24.4}$$

where $\overline{Q_s}$ is the time-averaged sediment input for that model run (between 75 and 600 grains per time-step) and $Q_s(t)$ and $Q^*(t)$ are the values at time t. This application of equation (24.4) is equivalent to using $x = 1$ in equation (24.2), i.e. the variability in sediment supply is linearly related to the variability in discharge predicted by equation (24.3). A second set of three time series of Q^* were produced by

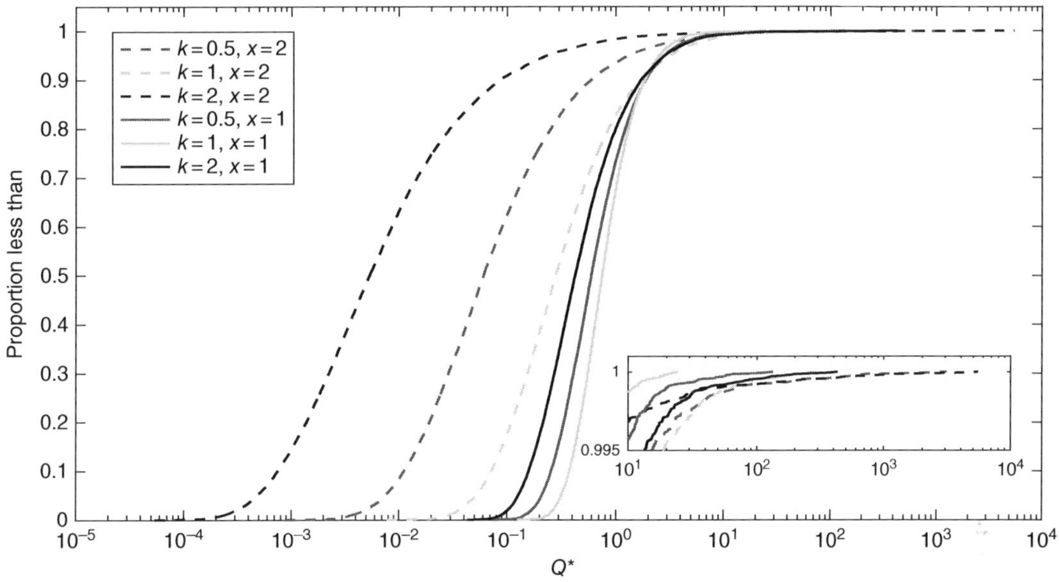

Figure 24.3 Distributions of Q^* for $k = 0.5$, 1, and 2 and $x = 1$ and 2. Q^* is a weighting parameter that determines the variability of sediment supply to the model. k represents the discharge (Q) variability, with lower values of k giving higher variability. $x = 1$ means that sediment supply is proportional to Q, whereas $x = 2$ means that sediment supply is proportional to Q^2. The inset shows the upper tail of the distributions.

Table 24.2 Properties of the time series of Q^* under different values of k and x. Each time series contains 20000 values.

Values of k and x	Median	Maximum	Maximum / median
$k = 2, \quad x = 1$	0.75	25	33
$k = 1, \quad x = 1$	0.59	134	227
$k = 0.5, x = 1$	0.42	428	1019
$k = 2, \quad x = 2$	0.28	308	1100
$k = 1, \quad x = 2$	0.060	3076	51267
$k = 0.5, x = 2$	0.0054	5469	1012778

squaring the values of Q^* and then rescaling so that the series mean was still 1; new time series of Q_s were then calculated using equation (24.4). These new values of Q^* are equivalent to using $x = 2$ in equation (24.2), and produce greater variability in sediment supply (Figure 24.3).

To recap, the CA model was run with six different time series of Q^* (Figure 24.3), drawn from each of six distributions that were created using combinations of three values of k (0.5, 1, and 2) and two values of x (1 and 2). The medians and maximum values of Q^* for each of 20 000-long time series drawn from the six distributions are shown in Table 24.2. For each combination of k and x, the model was run with eight different values of $\overline{Q_s}$ from 75 to 600 grains per time-step. For comparison, a seventh set of runs used a constant sediment input of $\overline{Q_s}$.

24.7.2 Results

Figure 24.4 shows an example of the model domain at different time-steps during model runs with different values of x. In the model run with less variable sediment input ($k = 2$ and $x = 1$), sediment cover develops across the model domain early on, and is persistent throughout the rest of the run. When sediment input is more variable, patches of sediment cover develop, but are transported downstream through the model domain. Consequently, the model domain alternates between almost full sediment cover and bedrock with a patchy cover of grains.

Distributions of the amount of bedrock exposure in each of the model runs are shown in Figure 24.5. For consistency with previous work, bedrock exposure (rather than sediment cover) is reported: bedrock exposure is $1 - F_e$, where F_e is the proportion of the model domain that is covered with sediment. $1 - F_e = 1$ or 0 indicates full bedrock exposure or sediment cover respectively. $1 - F_e$ is important in the saltation-abrasion model (Sklar and Dietrich 2004) because it is the fraction of the bed surface that can be eroded.

In the reference runs (no temporal variation in sediment input), the amount of bedrock exposure over the entire model domain is constant once the sediment cover has reached a steady state. Comparison with the other model runs indicates that the greater the variability in sediment input, the greater the variability in bedrock exposure throughout the model run. For most parameterizations, the amount of variability tended to be highest at intermediate values of $\overline{Q_s}$. In model runs where $x = 1$ and $k = 1$ and 2, mean values of $1 - F_e$ are very close to those produced by the reference runs (Figure 24.5). When $x = 1$ and $k = 0.5$, and $x = 2$ and $k = 2$, mean values of $1 - F_e$ are close to the reference runs except for $\overline{Q_s} = 225$ and 300 grains per time-step (37.5 and 50% of capacity transport). When $x = 2$ and $k = 1$ and 0.5, mean values of $1 - F_e$ are different to the reference runs at most values of $\overline{Q_s}$. For the reference runs, the mean values of $1 - F_e$ display a sigmoidal pattern with increasing $\overline{Q_s}$. The mean values produced by more variable sediment supplies instead show a more linear decrease with increasing $\overline{Q_s}$.

Figure 24.5 shows average bedrock exposure over the entire model domain, however there are spatial variations within this area (Figure 24.4). Temporal variations between bedrock exposure at the upstream and downstream ends of the model domain are shown in Figure 24.6. When sediment input has the smallest temporal variation ($k = 2$, $x = 1$), the time series of $1 - F_e$ are very similar to the time series from the reference runs, particularly at the downstream end of the model. At intermediate levels of sediment supply variability ($k = 0.5$, $x = 1$ and $k = 2$, $x = 2$), there is an increased difference between

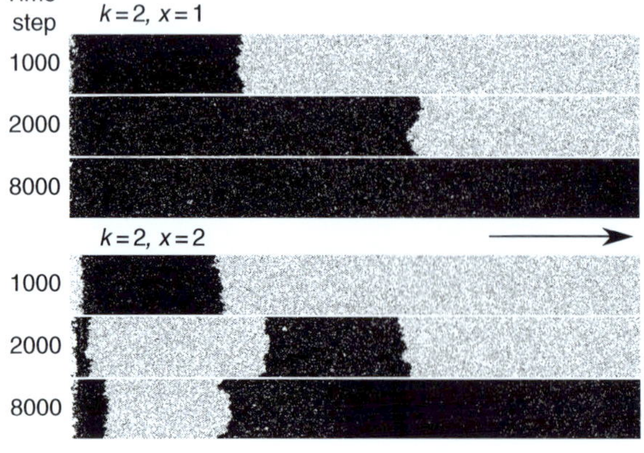

Figure 24.4 Illustrations of sediment cover in two model runs with $\overline{Q_s} = 300$. Sediment grains in a cluster (>4 neighbours) are plotted in black with isolated grains (≤4 neighbours) in grey. Grains move from left to right. Time-steps are selected to show how one run has uniform sediment cover by time-step 8000 whereas the other one does not.

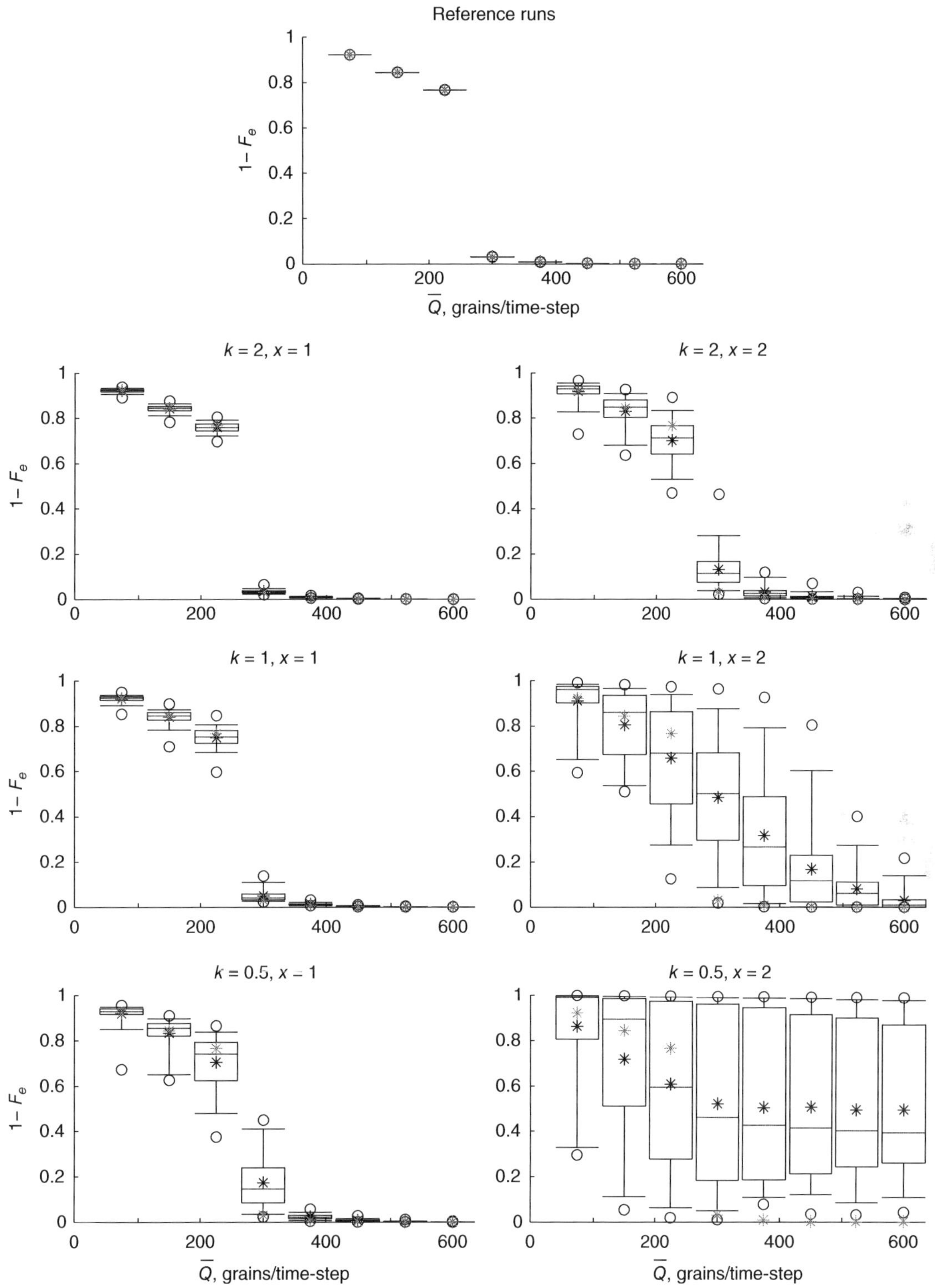

Figure 24.5 Box plots of sediment cover variation during the model runs. Lowest variability in sediment supply is given by $k = 2$, $x = 1$ and highest variability by $k = 0.5$, $x = 2$. Data are taken from time-steps 5000 to 20 000: 5000 is used because all runs with uniform sediment input show steady-state behaviour after this time. For each time-step, the sediment cover is calculated from the entire model domain. Boxplots show maximum and minimum (o), 5th and 95th percentiles (whiskers), Inter Quartile Range (IQR) and median (box and line), and mean (∗). Gray means are values for the reference runs with constant sediment input. (See insert for color representation of this figure.)

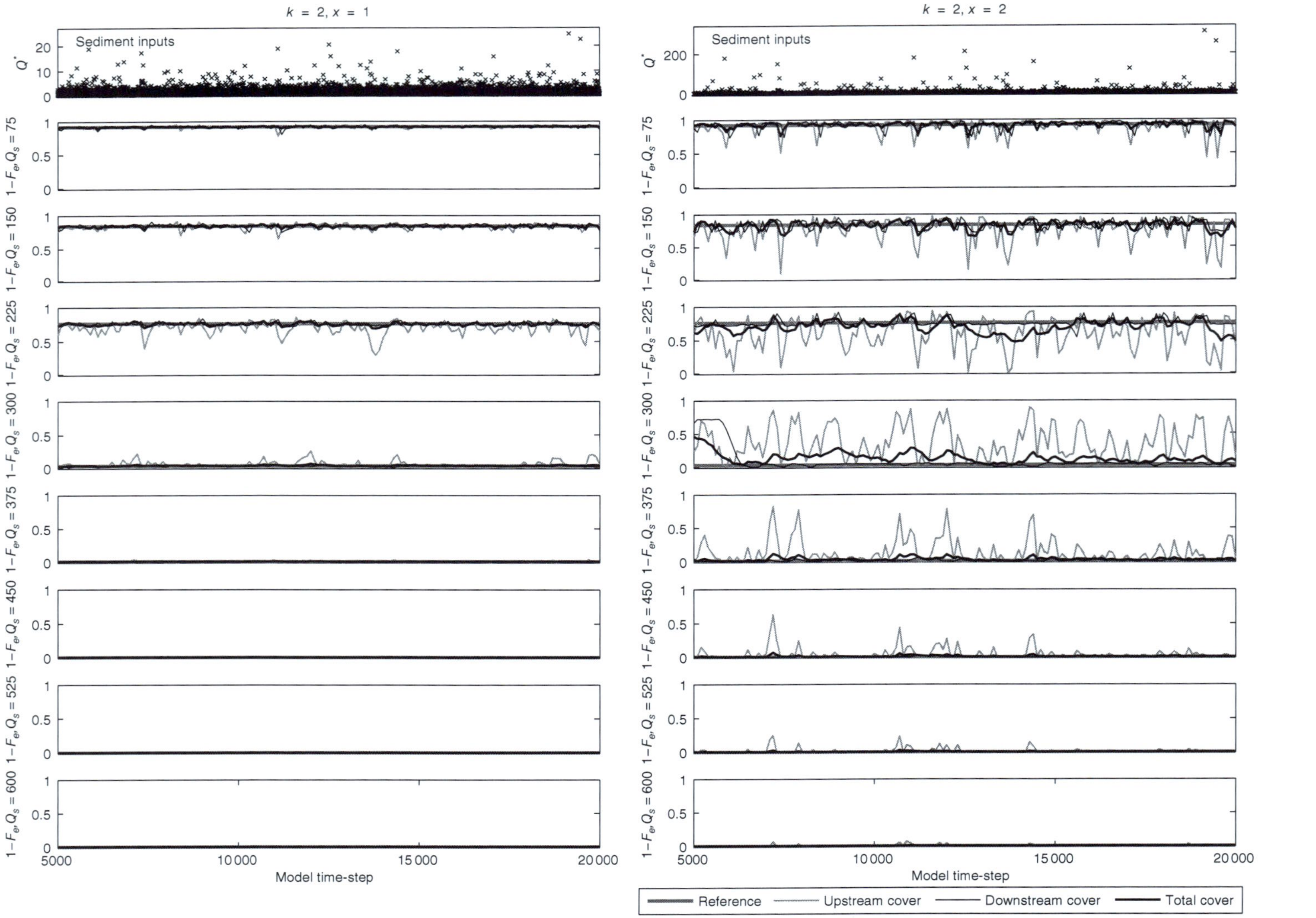

Figure 24.6 Time series of sediment cover in model runs with $k = 0.5$ and 2 and $x = 1$ and 2. Sediment cover is plotted as $1 - F_e$, i.e. 1 indicates complete bedrock exposure and 0 complete sediment cover. Upstream cover refers to the first 100 rows of the model domain, and downstream cover refers to the last 100 rows. In each set, the model domain is 1000 rows long. Reference is sediment cover in runs with constant sediment input. The top panel shows the sediment input, in terms of Q^*, which is the weighting applied to the mean sediment input.

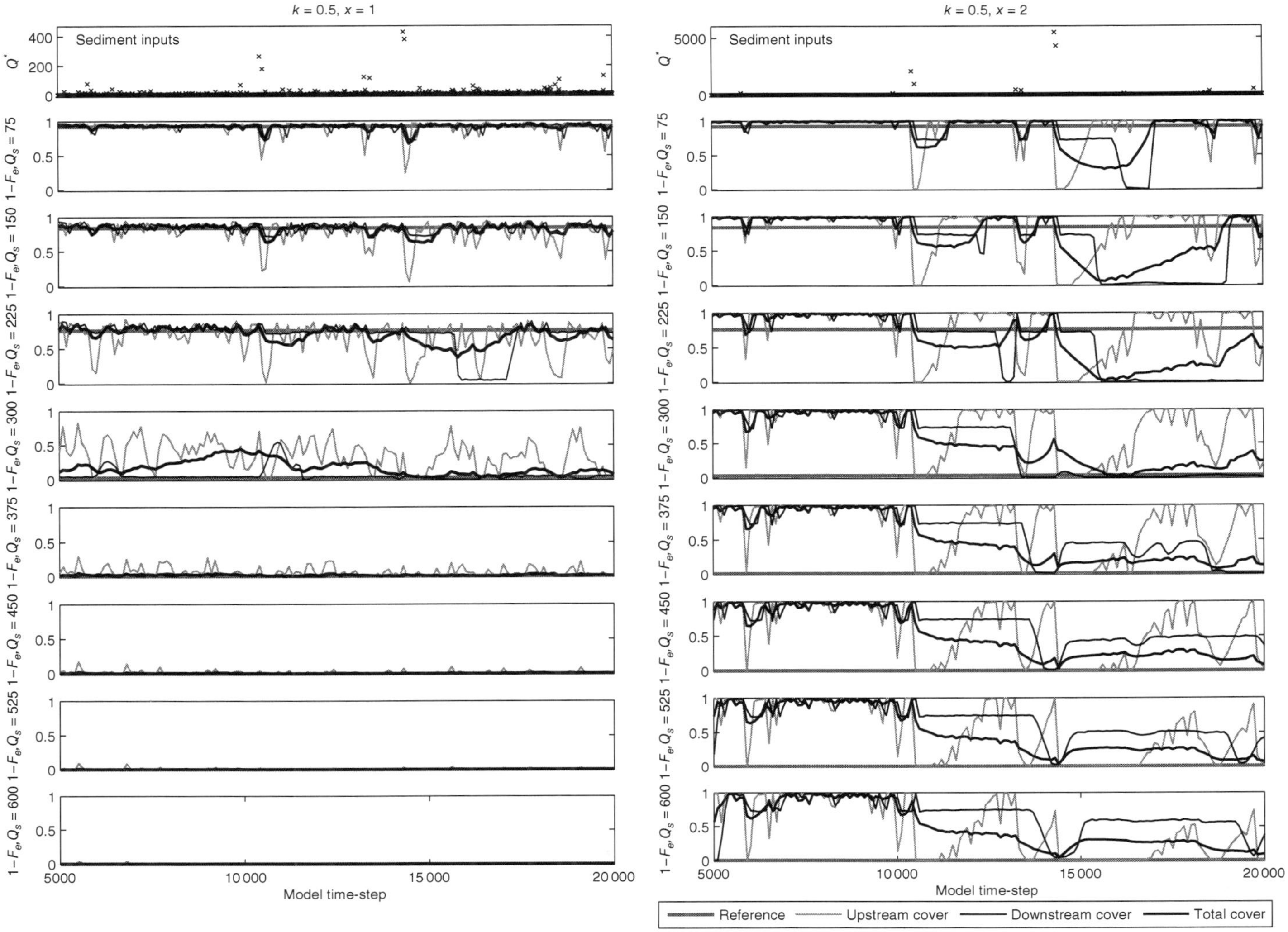

Figure 24.6 (*Continued*)

the upstream and downstream time series of $1 - F_e$. The downstream time series generally follows a similar trend to the reference data, whereas the upstream time series has greater temporal variation. As with Figure 24.5, the variability in the upstream time series is greatest when $\overline{Q_s} = 225$ or 300 grains per time-step.

At the most variable sediment input ($k = 0.5$, $x = 2$), there is little similarity between values of $1 - F_e$ for the reference runs and the model results, with values of $1 - F_e$ instead primarily responding to the stochastic sediment input. At all values of $\overline{Q_s}$ there are large decreases in $1 - F_e$ caused by the large sediment inputs at time-steps 10 450 and 13 290. The low values of $1 - F_e$ (i.e. sediment cover) persist for longer at larger values of $\overline{Q_s}$. As before, there is more variability at the upstream end of the model domain compared to the downstream end.

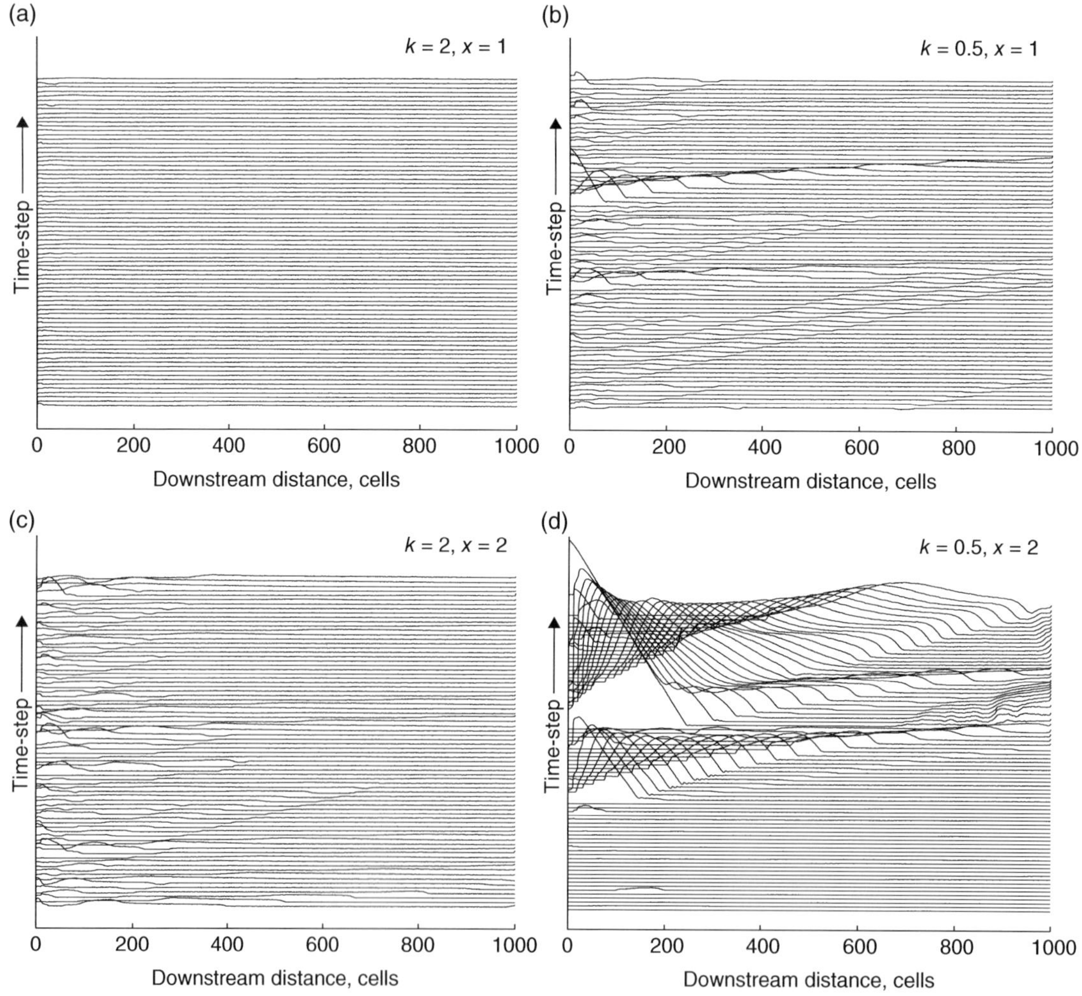

Figure 24.7 Stacked plots of the number of sediment grains in each row across the model domain between time-steps 5000 and 20 000. All runs have $\overline{Q_s} = 300$. All plots have the same vertical scale. These plots demonstrate how some sediment pulses are transmitted to the end of the model, whereas others are dispersed and not present at the downstream end.

Some of the differences in behaviour between the upstream and downstream ends of the model (Figure 24.6) can be explained by the transmission of sediment pulses along the model domain (Figure 24.7). In model runs with highly variable sediment input (k = 0.5, x = 2), Figure 24.7d shows how sediment pulses form at the upstream end of the model domain, and are transmitted downstream. The amplitude of the pulse decreases as it moves downstream, but it is still present at the downstream end. In model runs with less variable sediment input (Figure 24.7b and c), sediment pulses at the upstream end have been smoothed out before reaching the downstream end, meaning that there is less temporal variation in sediment cover.

24.7.3 Discussion

24.7.3.1 Key Findings

The aim of this set of experiments was to evaluate the impact that variation in sediment flux has on long-term sediment cover. In these model runs with variable sediment input, most of the time the number of grains entering the model domain is less that the model transport capacity, with occasional large pulses that are far greater than the transport capacity (Figure 24.3). The model can therefore be considered to be a flood-depositing system (Turowski *et al.* 2012), in which sediment is mainly supplied to the system in occasional large events and sediment deposits are eroded during the intervening periods.

The main finding is that the model shows two different regimes, depending on the variability of the sediment input. Regime one: smaller temporal variation in sediment input produces sediment cover that is comparable to that in model runs with a constant sediment input (Figure 24.6). Note that smaller is relative here; runs defined as having smaller variation still include sediment input distributions in which the maximum sediment input is 1000 times the median input (Table 24.2). Regime two: when sediment input is extremely variable (k = 0.5 or 1, x = 2, maximum : median = 51 267 or > 10^6), the occurrence of sediment cover is instead dominated by the timing of the largest sediment pulses (Figure 24.6).

There are two further trends to highlight. The first is that the pattern of sediment cover varies downstream over the model domain. The domain length of 1000 cells is equivalent to 100 grain-step lengths, so is actually relatively short. Despite this, time series of $1 - F_e$ are more variable at the upstream end than at the downstream end of the model. The second trend is that under regime one, the deviation from the reference runs varies with the value of $\overline{Q_s}$; intermediate values of $\overline{Q_s}$ = 225 or 300 grains per time-step show the greatest difference to the reference run value of $1 - F_e$.

These findings from the CA model are important because they demonstrate that a simple model can produce complex, non-linear patterns of sediment cover and consequently have implications for the development of sediment cover in bedrock–alluvial rivers. Model results are explained with reference to grain dynamics, and then the model implications are discussed.

24.7.3.2 Explanation of Model Behaviour

The two different model regimes can be interpreted as follows. In regime one, the development of sediment cover is determined by the interactions between sediment grains, whereas in regime two, the sediment cover is dominated by the variable sediment supply. In order to understand regime one, it is helpful first to consider how the model operates under constant sediment flux. At steady state, the sediment cover evolves to an equilibrium condition whereby the number of grains entrained from an area in a time-step is equal to the number of incoming grains. When sediment input is low, grains tend to have few neighbouring grains and are easily entrained (probability of movement P_i). Higher sediment input increases the number of grains in the model domain; consequently grains have more neighbours and are less mobile (probability of movement P_c). P_c is less than P_i (in this case, 0.15 compared to 0.95), and so when grains are predominantly P_c, a larger store of grains on the bed is required to produce a specified sediment flux. This switch from P_i to P_c can happen very rapidly, particularly when P_c is much smaller than P_i (Hodge and Hoey 2012). Near the threshold, additional sediment input will produce more grains with P_c. These more immobile grains become neighbours for grains deposited in the next time-step and so the new grains are also more likely to have entrainment probability P_c. This process leads to rapid development of sediment cover, known as runaway alluviation (Chatanantavet and Parker 2008). The switch from P_i to P_c accounts for the sigmoidal shape of the relationship between $1 - F_e$ and $\overline{Q_s}$ for the reference runs (Figure 24.5), with the drop in $1 - F_e$ at $\overline{Q_s}$ = 300 grains per time-step reflecting this transition from P_i to P_c.

When the sediment input is not extremely variable, these grain interactions buffer the system against perturbations (Figure 24.6). The occasional large inputs of sediment are not sufficient to cause a significant switch between P_i and P_c grains over most of the model domain, and so the value of $1 - F_e$ remains constant. Under regime one, temporal variations in $1 - F_e$ are most likely to occur when $\overline{Q_s}$ is 225 or 300 grains per time-step, i.e. when $\overline{Q_s}$ is around the values associated with a switch from mainly P_i to mainly P_c under constant sediment supply conditions. At these values of $\overline{Q_s}$, an increase or decrease in sediment input is more likely to change the balance between P_i and P_c, causing a non-linear change in $1 - F_e$. However, these changes are relatively short lived, with the system returning to its original value of $1 - F_e$ once the additional/depleted sediment has been entrained/replaced. Variations in sediment input are less important at high and low sediment fluxes. At high sediment fluxes, there is complete sediment cover, and only a very prolonged run of low sediment input will remove it. Under low flux conditions, grains are more mobile and the sediment pulses are relatively smaller, and quickly removed in subsequent time-steps.

Under both regimes the storage and release of sediment in the upstream section of the model modulates the supply of sediment to the rest of the model. Sediment pulses are lengthened and reduced in amplitude as they are transmitted downstream because of the constraint that a maximum of four grains can be entrained in a time-step from any one cell. Stored sediment at the upstream end continues to be released during periods with low sediment inputs. Downstream sections of the model consequently receive a smoothed sediment supply and hence have less temporal variation in $1 - F_e$ (Figure 24.5).

Under regime two (e.g. when $k = 0.5$ or 1 and $x = 2$), the variation in sediment supply overrides the grain interactions. The fluctuations between long periods with very low sediment inputs and the rare very large sediment pulses mean that the model is not able to form a steady state sediment cover. For example, in Figure 24.6d there is very little sediment input prior to the sediment pulse at time-step 10 450; consequently, all runs, regardless of the value of $\overline{Q_s}$ have high bedrock exposure. After the sediment pulse, the extent of the sediment cover that forms downstream is constrained by the release of stored sediment from upstream, rather than the long-term rates of sediment supply. This results in runs at all values of $\overline{Q_s}$ having a similar downstream value of $1 - F_e = 0.73$ after this initial sediment pulse.

24.7.3.3 Model Comparison and Implications

The identification of two different sediment cover regimes shows the importance of the interaction between grain dynamics and sediment supply on sediment cover in bedrock–alluvial systems. Under regime one, grain dynamics play a significant role in modulating the variability of sediment supply, even with relatively variable sediment inputs (e.g. maximum : median flux = 1000) and over short distances (100 grain-transport lengths). Sediment variability only has a significant impact on sediment cover in regime two, which occurs when variability is very large (e.g. maximum : median flux > 50 000).

These findings suggest that variability of sediment input is mainly important in rivers that are subject to extreme fluctuations, e.g. monsoonal climates, and/or in sections of rivers that are close to the sources of sediment pulses. Over geological time, the impact of variability in sediment supply on sediment cover is maybe not important unless sediment supply is extremely variable. However, the findings of Finnegan, Schumer, and Finnegan (2014), who observed that channel incision rate decreased over increasing time periods, suggest that heavy-tailed sediment supply events are not uncommon over geological timescales.

These results show similarities to those produced by Lague (2010) using the model SSTRIM, despite the differences between the two models. SSTRIM is a more complex, two-dimensional model that reproduces flow, sediment transport, and channel geometry at a number of nodes along a channel, over a daily time-step and can be run for thousands of years. Channel erosion is driven with the saltation-abrasion model of Sklar and Dietrich (2004). The extent of sediment cover is determined by spreading the volume of sediment stored at each node across the channel area. Lague (2010) ran SSTRIM using equation (24.3) to produce a distribution of flow discharges. Sediment input fluxes were calculated from the water discharge using $x = 1$ and 2.

Lague (2010) found that changing the value of x produced different model behaviours; $x = 1$ produced a permanent partial cover regime in which the amount of sediment cover fluctuated around its mean long-term value, whereas $x = 2$ produced an intermittent cover regime where the bed alternated between full and no sediment cover. These regimes are comparable to those produced by the CA model under lower, and high, sediment variability respectively. The similar behaviour of the two models can be explained by the fact that they both have the capacity to store and release sediment within the channel, suggesting that this is an important control on long-term sediment cover.

It is also instructive to consider the differences between the two models. In most runs of SSTRIM, discharge was covaried with sediment input, whereas discharge is effectively constant in the CA model. The impact of increasing discharge could be replicated in the CA model by increasing P_i and P_c, and/or increasing the number of grains that can be entrained from a single cell. Implementing these changes would mean that sediment waves could be more readily transmitted down the model domain, producing more variability in sediment cover at the downstream end of the model. The change may also shift the occurrence of regime two to conditions with less sediment variability than those observed here. Another difference is that SSTRIM does not incorporate any non-linear grain dynamics. The internal relationships between the extent of sediment cover and the relative proportions of grains with P_i and P_c may account for why SSTRIM displays a wider range of sediment cover values when $x = 1$ than is seen in the CA model.

These findings show the importance of grain dynamics in determining sediment cover in bedrock–alluvial rivers. Variable sediment supply seems to be important only when the range of sediment inputs is extremely large. However, in order to apply these findings to actual rivers, more information is required on the typical range of sediment inputs that are received by bedrock–alluvial rivers in different environments. Data on the downstream variation in sediment cover from a range of systems is also needed to compare against the downstream variation predicted by the model.

24.8 Conclusions

This chapter has addressed the issue of sediment processes in bedrock–alluvial channels through two different approaches: (i) a review of the literature since 2010, which identifies the main themes and outstanding research questions; and (ii) a modelling application that demonstrates how one of the identified research questions can be addressed. The review demonstrated that research into sediment processes in bedrock–alluvial rivers has covered a broad range of approaches across a range of spatial and temporal scales. Despite the significant advances that have been made since 2010, there are still many outstanding research questions, particularly around flow, sediment supply, sediment cover, and upscaling between dynamics at the grain-scale and over geological time. In the second part of the chapter, the application of the CA model illustrates how complex sediment-cover relationships can emerge from simple interactions between sediment grains. The model results indicate that grain dynamics are able to modulate a reasonable amount of variation in sediment supply, producing sediment cover that is temporally invariant. Variations in sediment supply override the impact of grain dynamics only when sediment supply is extremely variable (e.g. maximum : median flux > 50000). Variability in sediment supply therefore needs to be quantified in natural channels, and considered when modelling channel incision.

Acknowledgements

Thanks to Trevor Hoey, Rob Ferguson, and Dimitri Lague for useful discussions and comments. Thanks also to Joel Johnson and Liran Goren for useful reviews that helped to improve the clarity of the chapter.

References

Allen, GH, Barnes, JB, Pavelsky, TM, and Kirby, E 2013. Lithologic and tectonic controls on bedrock channel form at the northwest Himalayan front. *Journal of Geophysical Research – Earth Surface* 1806–1825. DOI:10.1002/jgrf.20113.

Attal, M, Cowie, PA, Whittaker, AC *et al.* 2011. Testing fluvial erosion models using the transient response of bedrock rivers to tectonic forcing in the Apennines, Italy. *Journal of Geophysical Research – Earth Surface* 116(F2), F02005. DOI:10.1029/2010JF001875.

Beer, AR, Turowski, JM, Fritschi, B, and Rieke-Zapp, DH 2014. Field instrumentation for high-resolution parallel monitoring of bedrock erosion and bedload transport. *Earth Surface Processes and Landforms.* DOI:10.1002/esp.3652.

Bekaddour, T, Schlunegger, F, Attal, M, and Norton, KP 2013. Lateral sediment sources and knickzones as controls on spatio-temporal variations of sediment transport in an Apine river. *Sedimentology* 60(1), 342–357. DOI:10.1111/sed.12009.

Broadhurst, LJ and Heritage, GL 1998. Modelling stage–discharge relationships in anastomosed bedrock-influenced sections of the Sabie River system. *Earth Surface Processes and Landforms* 23(5), 455–465. DOI:10.1002/(SICI)1096–9837(199805)23:5<455::AID-ESP860>3.0.CO;2–2.

Buffin-Bélanger, T, Roy, AG, and Kirkbride, AD 2000. On large-scale flow structures in a gravel-bed river. *Geomorphology* 32(3–4), 417–435. DOI:10.1016/S0169–555X(99)00106–3.

Burtin, A, Cattin, R, Bollinger, L *et al.* 2011. Towards the hydrologic and bed load monitoring from high-frequency seismic noise in a braided river: The “torrent de St Pierre”, French Alps. *Journal of Hydrology* 408(1–2), 43–53. DOI:10.1016/j.jhydrol.2011.07.014.

Chatanantavet, P. and Parker, G. 2008. Experimental study of bedrock channel alluviation under varied sediment supply and hydraulic conditions. *Water Resources Research* 44, W12446. DOI:10.1029/2007WR006581.

Chatanantavet, P, Lajeunesse, E, Parker, P *et al.* 2010. Physically based model of downstream fining in bedrock streams with lateral input. *Water Resources Research*, 46, W02518. DOI:10.1029/2008WR007208.

Chatanantavet, P, Whipple, KX, Adams, M, and Lamb, MP 2013. Experimental study on coarse-grain saltation dynamics in bedrock channels. *Journal of Geophysical Research – Earth Surface* 1161–1176. DOI:10.1002/jgrf.20053.

Cook, KL, Turowski, JM, and Hovius, N, 2013. A demonstration of the importance of bedload transport for fluvial bedrock erosion and knickpoint propagation. *Earth Surface Processes and Landforms* 38(7), 683–695. DOI:10.1002/esp.3313.

Cook, KL, Turowski, JM, and Hovius, N, 2014. River gorge eradication by downstream sweep erosion. *Nature Geoscience* 7(9), 682–686. DOI:10.1038/ngeo2224.

Davy, P and Crave, A 2000. Upscaling local-scale transport processes in large-scale relief dynamics. *Physics and Chemistry of Earth Part A* 25(6–7), 533–541. DOI:10.1016/S1464-1895(00)00082-X.

Dubinski, IM, and Wohl, E 2013. Relationships between block quarrying, bed shear stress, and stream power: *A physical model of block quarrying of a jointed bedrock channel Geomorphology* 180–181(0), 66–81. DOI:10.1016/j.geomorph.2012.09.007.

Ferrier, KL, Huppert, KL, and Perron, JT 2013. Climatic control of bedrock river incision. *Nature* 496(7444), 206–209. DOI:10.1038/nature11982.

Finnegan, NJ and Balco, G 2013. Sediment supply, base level, braiding, and bedrock river terrace formation: Arroyo Seco, California, USA. *Geological Society of America Bulletin*. DOI:10.1130/B30727.1.

Finnegan, NJ, Schumer, R, and Finnegan, S 2014. A signature of transience in bedrock river incision rates over timescales of 104–107 years. *Nature* 505(7483), 391–394. DOI:10.1038/nature12913.

Goode, JR and Wohl, E 2010. Coarse sediment transport in a bedrock channel with complex bed topography. *Water Resources Research* 46, W11532. DOI:10.1029/2009WR008135.

Grant, GE 1997. Critical flow constrains flow hydraulics in mobile-bed streams: A new hypothesis. *Water Resources Research* 33(2), 349–358. DOI:10.1029/96WR03134.

Heritage, GL, Moon, BP, Broadhurst, LJ, and James, CS 2004. The frictional resistance characteristics of a bedrock-influenced river channel. *Earth Surface Processes and Landforms* 29(5), 611–627. DOI:10.1002/esp.1057.

Hobley, DEJ, H. D. Sinclair, HD, Mudd, SM, and Cowie, PA, 2011. *Field calibration of sediment flux dependent river incision Journal of Geophysical Research*, 116, F04017. DOI:10.1029/2010JF001935.

Hodge, RA, and Hoey, TB 2012. Upscaling from grain-scale processes to alluviation in bedrock channels using a cellular automaton model. *Journal of Geophysical Research – Earth Surface* 117, F01017. DOI:10.1029/2011JF002145.

Hodge, RA, Hoey, TB, and Sklar, LS 2011. Bedload transport in bedrock rivers: the role of sediment cover in grain entrainment, translation and deposition *Journal of Geophysical Research*. DOI:10.1029/2011JF002032.

Hodge, RA and Hoey, TB 2016a. A Froude-scaled model of a bedrock-alluvial channel reach: 1. Hydraulics. *Journal of Geophysical Research: Earth Surface* 121, 1578–1596.

Hodge, RA and Hoey, TB 2016b. A Froude-scaled model of a bedrock-alluvial channel reach: 2. Sediment cover. *Journal of Geophysical Research: Earth Surface* 121, 1597–1618.

Hoey, TB and Ferguson, RI 1994. Numerical-simulation of downstream fining by selective transport in gravel-bed rivers – model development and illustration. *Water Resources Research* 30(7), 2251–2260.

Hsu, L, Finnegan, NJ, and Brodsky, EE 2011. A seismic signature of river bedload transport during storm events. *Geophysical Research Letters* 38. DOI:10.1029/2011GL047759.

Huang, M-W, Pan, Y-W, and Liao, J-J 2013. A case of rapid rock riverbed incision in a coseismic uplift reach and its implications. *Geomorphology* 184, 98–110. DOI:10.1016/j.geomorph.2012.11.022.

Huda, SA, and Small, EE 2014. Modeling the effects of bed topography on fluvial bedrock erosion by saltating bed load *Journal of Geophysical Research – Earth Surface* 119(6), 1222–1239. DOI:10.1002/2013JF002872.

Inoue, T, Izumi, N, Shimizu, Y, and Parker, G 2014. Interaction among alluvial cover, bed roughness and incision rate in purely bedrock and alluvial-bedrock channel. *Journal of Geophysical Research – Earth Surface* 2014JF003133. DOI:10.1002/2014JF003133.

Jansen, JD, Fabel, D, Bishop, P. *et al.* 2011. Does decreasing paraglacial sediment supply slow knickpoint retreat? *Geology* 39(6), 543–546. DOI:10.1130/G32018.1.

Johnson, JPL 2014. A surface roughness model for predicting alluvial cover and bed load transport rate in bedrock channels. *Journal of Geophysical Research – Earth Surface* 2013JF003000. DOI:10.1002/2013JF003000.

Johnson, JPL, and Whipple, KX, 2010. Evaluating the controls of shear stress, sediment supply, alluvial cover, and channel morphology on experimental bedrock incision rate. *Journal of Geophysical Research*, 115, F02018. DOI:201010.1029/2009JF001335.

Johnson, JPL, Whipple, KX, and Sklar, LS 2010. Contrasting bedrock incision rates from snowmelt and flash floods in the Henry Mountains, Utah. *Geological Society of America Bulletin* 122(9–10), 1600 –1615. DOI:10.1130/B30126.1.

Kidson, RL, Richards, KS, and Carling, PA, 2006. Hydraulic model calibration for extreme floods in bedrock-confined channels: case study from northern Thailand. *Hydrological Processes* 20(2), 329–344. DOI:10.1002/hyp.6086.

Lague, D 2010. Reduction of long-term bedrock incision efficiency by short-term alluvial cover intermittency. *Journal of Geophysical Research* 115. DOI:10.1029/2008JF001210.

Lague, D 2014. The stream power river incision model: evidence, theory and beyond. *Earth Surface Processes and Landforms* 39(1), 38–61. DOI:10.1002/esp.3462.

Lague, D, Hovius, N and Davy, P 2005. Discharge, discharge variability, and the bedrock channel profile. *Journal of Geophysical Research* 110, F04006. DOI:10.1029/2004JF000259.

Lague, D, Brodu, N, and Leroux, J 2013. Accurate 3D comparison of complex topography with terrestrial laser scanner: application to the Rangitikei canyon (N-Z). *ISPRS Journal of Photogrammetry and Remote Sensing* 82, 10–26. DOI:10.1016/j.isprsjprs.2013.04.009.

Lai, YG, Greimann, BP, and Wu, K 2011. Soft bedrock erosion modeling with a two-dimensional depth-averaged model. *Journal of Hydraulic Engineering – ASCE* 137(8), 804–814. DOI:10.1061/(ASCE)HY.1943–7900.0000363.

Lamb, MP and Fonstad, MA 2010. Rapid formation of a modern bedrock canyon by a single flood event. *Nature Geoscience* 3(7), 477–481. DOI:10.1038/NGEO894.

Nelson, PA and Seminara, G 2012. A theoretical framework for the morphodynamics of bedrock channels. *Geophysical Research Letters* 39, L06408. DOI:201210.1029/2011GL050806.

Nelson, PA, Pittaluga, MB, and Seminara, G, 2014. Finite amplitude bars in mixed bedrock-alluvial Channels. *Journal of Geophysical Research – Earth Surface* 119, 566–58. DOI:10.1002/2013JF002957.

O'Connor, JE, Mangano, JF, Anderson, SW *et al.* 2014. Geologic and physiographic controls on bed-material yield, transport, and channel morphology for alluvial and bedrock rivers, western Oregon. *Geological Society of America Bulletin* B30831.1. DOI:10.1130/B30831.1.

Ortega, JA, Gómez-Heras, M, Perez-López, R, and Wohl, E, 2014. Multiscale structural and lithologic controls in the development of stream potholes on granite bedrock rivers. *Geomorphology* 204, 588–598. DOI:10.1016/j.geomorph.2013.09.005.

Rice, SP, Buffin-Bélanger, T, and Reid, I 2014. Sensitivity of interfacial hydraulics to the microtopographic roughness of water-lain gravels. *Earth Surface Processes and Landforms* 39(2), 184–199. DOI:10.1002/esp.3438.

Richardson, K and Carling, PA, 2006. The hydraulics of a straight bedrock channel: Insights from solute dispersion studies. *Geomorphology* 82(1–2), 98–125. DOI:10.1016/j.geomorph.2005.09.022.

Rickenmann, D, Turowski, JM, Fritschi, B *et al.* 2012. Bedload transport measurements at the Erlenbach stream with geophones and automated basket samplers. *Earth Surface Processes and Landforms* 37(9), 1000–1011. DOI:10.1002/esp.3225.

Scheingross, JS, Brun, F, Lo, DY *et al.* 2014. Experimental evidence for fluvial bedrock incision by suspended and bedload sediment. *Geology* G35432.1. DOI:10.1130/G35432.1.

Siddiqui, A, and A. Robert, 2010. Thresholds of erosion and sediment movement in bedrock channels *Geomorphology* 118(3–4), 301–313. DOI:10.1016/j.geomorph.2010.01.011.

Sipos, AA, Domokos, G, Wilson, A, and Hovius, N, 2011. A discrete random model describing bedrock profile abrasion. *Mathematical Geosciences* 43(5), 583–591. DOI:10.1007/s11004-011–9343–8.

Sklar, LS and Dietrich, WE 2004. A mechanistic model for river incision into bedrock by saltating bed load. *Water Resources Research* 40, W06301. DOI:10.1029/2003WR002496.

Tinkler, KJ 1997. Critical flow in rockbed streams with estimated values for *Manning's n Geomorphology* 20(1–2), 147–164. DOI:10.1016/S0169–555X(97)00011–1.

Toone, J, Rice, SP, and Piégay, H 2014. Spatial discontinuity and temporal evolution of channel morphology along a mixed bedrock-alluvial river, upper Drôme River, southeast France: contingent responses to external and internal controls *Geomorphology* 205, 5–16. DOI:10.1016/j.geomorph.2012.05.033.

Turowski, JM, Lague, D, and Hovius, N 2007. Cover effect in bedrock abrasion: a new derivation and its implications for the modeling of bedrock channel morphology. *Journal of Geophysical Research* 112, F04006. DOI:10.1029/2006JF000697.

Turowski, JM, Rickenmann, D, and Dadson, SJ 2010. The partitioning of the total sediment load of a river into suspended load and bedload: a review of empirical data. *Sedimentology* 57(4), 1126–1146. DOI:10.1111/j.1365-3091.2009.01140.x.

Turowski, JM, Hovius, N, Wilson, A, and Horng, M-J, 2008. Hydraulic geometry, river sediment and the definition of bedrock channels. *Geomorphology* 99(1–4), 26–38. DOI:10.1016/j.geomorph.2007.10.001.

Turowski, JM, Badoux, A, Leuzinger, J, and Hegglin, R 2012. Large floods, alluvial overprint, and bedrock erosion. *Earth Surface Processes and Landforms* 38, 947–958. DOI:10.1002/esp.3341.

Turowski, JM, Böckli, M, Rickenmann, D, and Beer, AR 2013. Field measurements of the energy delivered to the channel bed by moving bed load and links to bedrock erosion. *Journal of Geophysical Research – Earth Surface* 118(4), 2438–2450. DOI:10.1002/2013JF002765.

Valla, PG, van der Beek, PA, and Lague, D 2010. Fluvial incision into bedrock: Insights from morphometric analysis and numerical modeling of gorges incising glacial hanging valleys (Western Alps, France). *Journal of Geophysical Research* 115(F2), F02010. DOI:10.1029/2008JF001079.

Van, TPD, Carling, PA, and Atkinson, PM, 2012. Modelling the bulk flow of a bedrock-constrained, multi-channel reach of the Mekong River, Siphandone, southern Laos. *Earth Surface Processes and Landforms* 37(5), 533–545. DOI:10.1002/esp.2270.

Venditti, JG, Rennie, CD, Bomhof, J *et al.* 2014. Flow in bedrock canyons. *Nature* 513(7519), 534–537. DOI:10.1038/nature13779.

Wilson, A and Lave, J 2014. Convergent evolution of abrading flow obstacles: Insights from analogue modelling of fluvial bedrock abrasion by coarse bedload. *Geomorphology* 208, 207–224. DOI:10.1016/j.geomorph.2013.11.024.

Wilson, A, Hovius, N, and Turowski, JM, 2013. Upstream-facing convex surfaces: Bedrock bedforms produced by fluvial bedload abrasion. *Geomorphology* 180–181(0), 187–204. DOI:10.1016/j.geomorph.2012.10.010.

Yanites, BJ and Tucker, GE, 2010. Controls and limits on bedrock channel geometry. *Journal of Geophysical Research* 115(F4), F04019. DOI:10.1029/2009JF001601.

Yanites, BJ, Tucker, GE, Mueller, KJ, and Chen, J-G 2010. How rivers react to large earthquakes: Evidence from central Taiwan. *Geology* 38(7), 639–642. DOI:10.1130/G30883.1.

Yanites, BJ, Tucker, GE, Hsu, H *et al.* 2011. The influence of sediment cover variability on long-term river incision rates: An example from the Peikang River, central Taiwan. *Journal of Geophysical Research* 116, F03016. DOI:10.1029/2010JF001933.

Discussion

Discussion by David Milan

Many bedrock–alluvial rivers in semi-arid areas have vegetation in the channel. For example in the Kruger Park Rivers, South Africa, *Breonadia salacina* can become established by growing in cracks and fissures in the bedrock during periods of quiescent flow, whilst *Fragmites mauritianus* may grow rapidly in areas of sediment deposition. Once established, vegetation can have a significant influence upon flow and sediment transport dynamics, until finally it is stripped out by the rare catastrophic flows that are the dominant geomorphic agents in these systems (Entwistle *et al.* 2014).

Reference

Entwistle, N, Heritage, G, Tooth, S, and Milan, D 2014. Anastomosing reach control on hydraulics and sediment distribution on the Sabie River, South Africa. In *Sediment Dynamics from the Summit to the Sea*, (Eds Jun Xu, Y Allison, MA Bentley *et al.*), Publication 367, IAHS: Wallingford, UK; 215–219.

Discussion by Gordon Grant

Hodge has given us an interesting perspective on the importance of physical grain–grain interactions and what happens when those interactions are not present, for example in bedrock rivers. Viparelli has shown us how a probabilistic representation with an implicit grain interaction (e.g. the Wilcock–Crowe formulation) in the entrainment and active-layer exchange can reveal aspects of bed evolution. I am interested in how these contrasting perspectives (i.e. grain physics versus probabilities) can inform each other. Specifically, does the Viparelli approach lose some key behaviour in using only a bulk probabilistic function for entrainment? Conversely, what would be gained in the Hodge approach by incorporating a more probabilistic approach to entrainment?

Discussion by Kimberly Hill

You mentioned a few details of different behaviour of mixtures in bedrock–alluvial rivers. In simpler granular systems (where the system is more constrained and particle–particle interactions dominate), it has been shown that the segregation rules can change dramatically with packing fraction, granular temperature, and other boundary conditions (e.g. Fan and Hill 2011). It could be imagined that changes in loose sediment cover of a bedrock channel may also lead to changes in granular temperature and packing fraction, which might provide a suitable test bed for the relevance of the observations in simpler granular systems to riverbeds. Can you comment on whether there have been any observations of differences in sorting behaviours as the sediment cover varies in bedrock rivers, either in terms of local segregation rules or more reach-scale sorting patterns?

Reference

Fan, Y and Hill, KM 2011. Phase transitions in shear-induced segregation of granular materials. *Physical Review Letters* 106(21), 218301.

Reply by Rebecca Hodge

I thank the discussants for their varied and interesting comments. I agree with Milan that vegetation is another possible influence on sediment dynamics in bedrock–alluvial systems. In my cellular automaton (CA) model, vegetation could be represented as a time-dependent decrease in the mobility of patch sediment. This would encourage further accumulation and vegetation establishment in a positive feedback. In an actual channel, vegetation growing in cracks and fissures, as described by Milan, could also provide the initial nucleus for sediment patch deposition. The dynamics of sediment–vegetation interaction will also depend on the sediment regime. Positive feedbacks are likely to operate under relatively consistent sediment inputs. Under very variable sediment inputs, the CA model demonstrates that this variability controls the sediment cover, and so vegetation is likely to have a lesser impact on sediment cover extent.

Grant asks about the potential of a more probabilistic approach to entrainment. The CA model does use a probabilistic approach, insofar as random numbers determine which grains move in a given time-step. However, a single probability is used for each of the cluster and the isolated grains (P_c and P_i), and only four grains can potentially be entrained from each cell in a time-step. The model of Viparelli *et al.* (Chapter 23, this volume) uses probability density functions to describe grain entrainment as a function of grain size and elevation within the bed. A comparable approach for the CA model would be to parameterize entrainment as a continuous function of the number of surrounding grains. This would probably straighten the strongly sigmoidal shape of the relationship between sediment flux and cover under constant sediment input, but might have little effect when sediment cover is driven by sediment supply. Furthermore, there are very few data available with which such a function could be parameterized.

Hill raises a good question about the comparability of behaviours observed in the granular physics literature and in bedrock–alluvial channels. There is clearly synergy between the fluvial community and granular physicists, albeit that rivers probably have more complex boundary conditions, grain sizes, and shapes. In bedrock–alluvial channels there do seem to be differences in sorting processes as a function of sediment cover extent; Hodge, Hoey, and Sklar (2011) showed that tracer transport distances were more size-dependent in a river with 80% cover than one with 20% cover. However, there are very few field data with which to generalize and extend this observation. There is, therefore, a real need for more data on sediment dynamics (e.g. using tracers) and spatial patterns of grain size (both within and between reaches, and including the locations and grain-size distribution of sediment sources).

25

Modelling Braided Channels Under Unsteady Flow and the Effect of Spatiotemporal Change of Vegetation on Bed and Channel Geometry

Hiroshi Takebayashi

25.1 Introduction

The formative mechanisms of bars and channels have been studied by several researchers. Early studies of the formative mechanism of bars and channels used theoretical analyses (Werner 1951; Reynolds 1963; Einstein and Shen 1964; Anderson 1967; Hansen 1967; Shen and Komura 1968; Callender 1969; Engelund, and Skovgaard 1973; Parker 1976; Fredsøe 1978; Colombini, Seminara and Tubino 1987; Watanabe *et al.* 2001), field surveys (Ashworth *et al.* 1992; Thorne, Russell, and Alam 1993; Wu *et al.* 2005), and flume experiments (Ashida and Shiomi 1966). Trials for reproducing bars and channels using numerical analysis have also been conducted (Murray and Paola 1994; Takebayashi, Egashira, and Kuroki 1997; Enggrob and Tjerry 1999; Takebayashi, Egashira, and Okabe 2004; Jang and Shimizu 2005a). These studies clarified the formative mechanism of bars and channels. However, the effect of vegetation and the unsteady flow characteristics on the formative mechanism of bars and channels has not been considered satisfactorily in these studies.

To establish a river regulation work plan for flood control, water utilization, and environmental conservation in a balanced manner, quantitative prediction of temporal changes of both water flow and bed geometry are needed. In recent years, the growth of vegetation in rivers has been remarkable all over the world, because disturbance by floods to vegetation on bars was reduced due to the decrease in flood peak discharge as a result of flood control. Vegetation growth has an impact on the geometry of bars and channels. Figure 25.1 shows the temporal change of the vegetated area on bars of the Yoshino River in Japan. Several reservoirs have been constructed in the upstream region of the Yoshino River, thereby suppressing the maximum water discharge of floods. For that reason, it is easy for vegetation to grow on bars. Figure 25.2 shows the temporal change of cross-sectional bed geometry in the Yoshino River. The bed is degradated in the main channel and aggraded on bars because sediment deposits around vegetation and water is forced to flow in the main channel. These phenomena are likely to suppress bifurcation of channels and the development of new channels on bars.

The effect of vegetation on the geometric characteristics of bars and channels is discussed by many researchers and reviewed by Corenblit *et al.* (2007), Gurnell, Bertoldi, and Corenblit (2012), Curran and Hession (2013), and Camporeale *et al.* (2013). Tsujimoto (1999) reproduced bed deformation characteristics in and around vegetation by use of horizontal two-dimensional analysis. Jang and

Gravel-Bed Rivers: Processes and Disasters, First Edition. Edited by Daizo Tsutsumi and Jonathan B. Laronne.

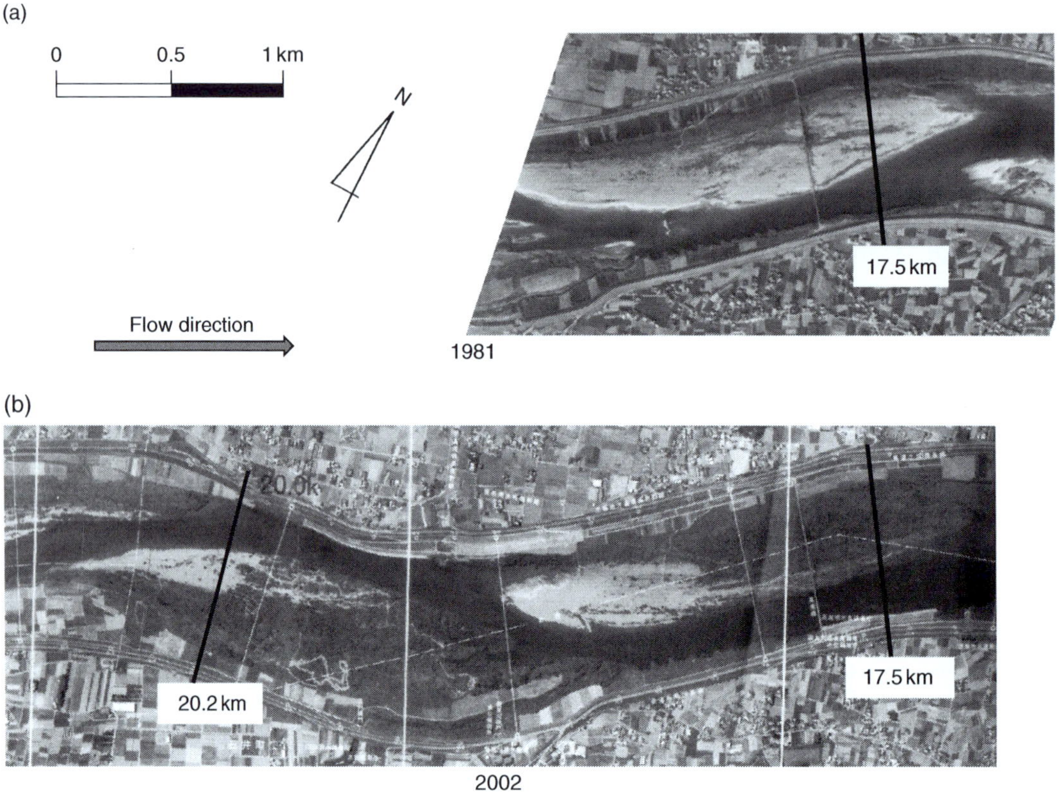

Figure 25.1 Temporal change of vegetated area on bars of the Yoshino River between (a) 1981 and (b) 2002.

Shimizu (2005b) performed both flume experiments and numerical analysis on the formation process of bars and channels considering vegetation. Perucca, Camporeale, and Ridolfi (2007) performed simple numerical analysis of meandering channels considering vegetation. Perona *et al.* (2009) applied the previous model (Perucca, Camporeale, and Ridolfi 2007) to both meandering and braided channels considering vegetation. Takebayashi and Okabe (2009) developed a horizontal two-dimensional bed deformation analysis model considering temporal change of vegetation density and seepage flows; the equations are written in a Cartesian coordinate system. However, the reproducibility of the model was not discussed in comparison to experimental flume data. Crosato and Samir Saleh (2011) and Nicholas (2013) also performed bed deformation analysis on the formation process of bars and channels considering vegetation as a binary classification (vegetation exists or not) and concluded that the existence of vegetation reduces the number of channels. Murray and Paola (2003), Hooke *et al.* (2005), and Coulthard, Hicks, and Van de Wiel (2007) performed numerical analyses on the formation process of braided channels considering vegetation by use of cellular modeling. Tal and Paola (2007, 2010) also performed flume experiments of braided channels considering vegetation and the effect of the vegetation on the decrease of the number of channels is observed. It is considered that to quantitatively predict spatiotemporal changes of water flow and bed geometry in a river, analyses in which the spatiotemporal distribution of growth of vegetation is introduced are necessary. However, findings obtained so far are insufficient to assess the influence of unsteady flow characteristics and spatiotemporal distribution characteristics of vegetation overgrowth upon flow and bed deformation.

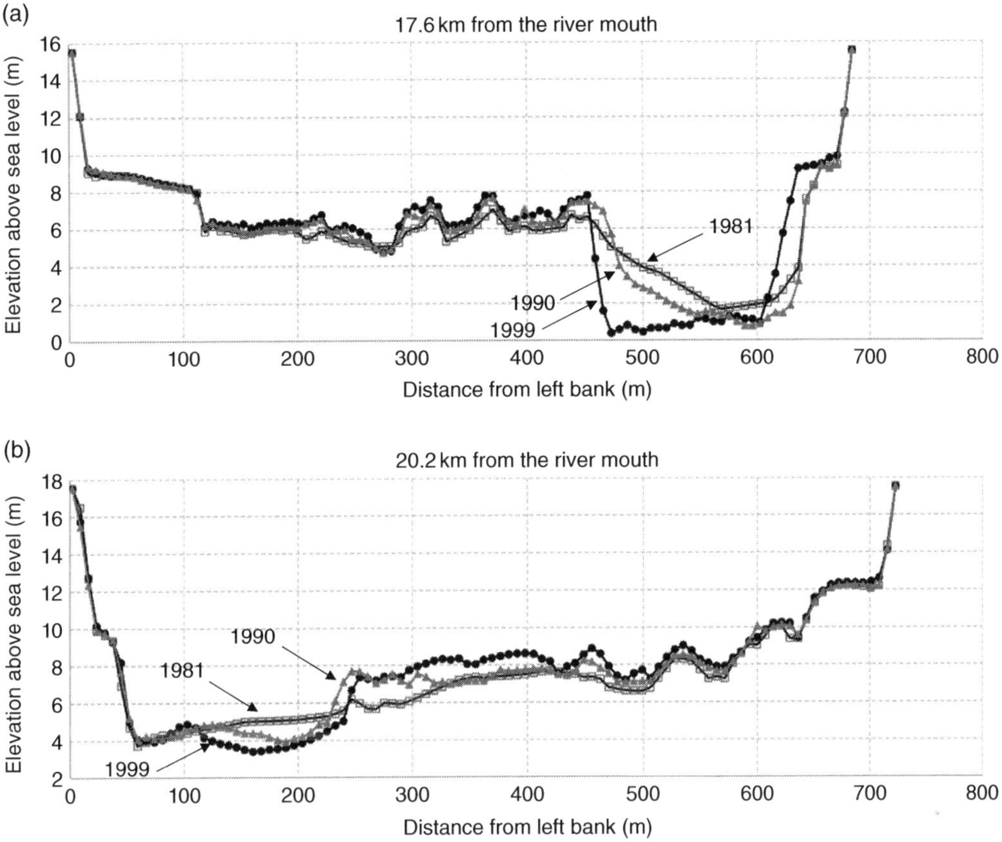

Figure 25.2 Temporal change of cross-sectional bed geometry at longitudinal locations (a) 17.5 km and (b) 20.2 km upstream from the mouth of the Yoshino River.

In this study, braided channels under unsteady flow conditions are reproduced by both flume experiments and a horizontal two-dimensional bed deformation numerical analysis. Furthermore, invasion, growth, and washing away processes of vegetation are introduced in the numerical analysis model and reproduction of braided channels with vegetation is attempted. Bed degradation causes the washing away process in the model, because most vegetation on bars is not washed away by water flow without bed degradation in the Yoshino, Kako, and Kizu rivers, Japan. Furthermore, the effects of both the unsteady characteristics of water supply and the presence of vegetation on geometric characteristics of braided channels are discussed based on quantitative data from the numerical analysis. Islands are formed in braided channels. There is no surface water flow on islands, but groundwater flows in the bed. Groundwater flow must be considered to reproduce island formation and deformation processes because groundwater affects shoreline formation. Hence, both surface water and groundwater flows are considered in the model. Additionally, depth integrated two-dimensional flow and bed deformation analyses considering the vegetation were performed for the Yoshino River, Japan, incorporating the spatiotemporal distribution of vegetation drag characteristics on the basis of the flooding that occurred in 2004. The governing equations are written in a generalized curvilinear coordinate system to apply the model to the Yoshino River. A discussion of the effect of the temporal change of the vegetation density on the bed deformation characteristics follows.

25.2 Numerical Analysis Method

25.2.1 Framework of the Numerical Model

The relationship between a Cartesian coordinate system and a generalized curvilinear coordinate system follows.

$$J = \frac{1}{\left(\frac{\partial x}{\partial \xi}\frac{\partial y}{\partial \eta} - \frac{\partial x}{\partial \eta}\frac{\partial y}{\partial \xi}\right)} \tag{25.1-1}$$

$$\frac{\partial \xi}{\partial x} = J\frac{\partial y}{\partial \eta} \tag{25.1-2}$$

$$\frac{\partial \eta}{\partial x} = -J\frac{\partial y}{\partial \xi} \tag{25.1-3}$$

$$\frac{\partial \xi}{\partial y} = -J\frac{\partial x}{\partial \eta} \tag{25.1-4}$$

$$\frac{\partial \eta}{\partial y} = J\frac{\partial x}{\partial \xi} \tag{25.1-5}$$

where ξ and η are the coordinates along the longitudinal and the transverse directions in a generalized curvilinear coordinate system, respectively, and x and y are the coordinates in a Cartesian coordinate system.

Computation of surface flow is carried out using the governing equation of the horizontal two-dimensional flow-averaged depth. Groundwater is treated as a horizontal two-dimensional saturation flow. Exchange of momentum between surface water and groundwater flow is not considered; only mass exchange between them is considered. The conservation of mass including inflow and outflow of mass by seepage flow is taken into consideration as shown in the following equation.

$$\Lambda\frac{\partial}{\partial t}\left(\frac{z}{J}\right) + \frac{\partial}{\partial \xi}\left(\frac{hU}{J}\right) + \frac{\partial}{\partial \eta}\left(\frac{hV}{J}\right) + \frac{\partial}{\partial \xi}\left(\frac{h_g U_g}{J}\right) + \frac{\partial}{\partial \eta}\left(\frac{h_g V_g}{J}\right) = 0 \tag{25.2}$$

where t is the time and z is the water surface level. Depth of surface flow is represented as h, seepage-flow depth is h_g, and U and V represent the contravariant depth-averaged flow velocity above the bed along ξ and η coordinates, respectively. These velocities are defined as:

$$U = \frac{\partial \xi}{\partial x}u + \frac{\partial \xi}{\partial y}v \tag{25.3}$$

$$V = \frac{\partial \eta}{\partial x}u + \frac{\partial \eta}{\partial y}v \tag{25.4}$$

where u and v represent depth-averaged flow velocity above the bed along x and y coordinates, respectively. U_g and V_g represent the contravariant depth-averaged seepage flow velocity along ξ and η coordinates, respectively. These velocities are defined as:

$$U_g = \frac{\partial \xi}{\partial x} u_g + \frac{\partial \xi}{\partial y} v_g \tag{25.5}$$

$$V_g = \frac{\partial \eta}{\partial x} u_g + \frac{\partial \eta}{\partial y} v_g \tag{25.6}$$

where depth averaged seepage flow velocities along x and y coordinates in a Cartesian coordinate system are shown as u_g and v_g, respectively, Λ is a parameter related to the porosity in the soil, wherein $\Lambda = 1$ as $z \geq z_b$, and $\Lambda = \lambda$ as $z < z_b$, where z_b is the bed level and λ is the porosity in the soil. Seepage flow is assumed to be horizontal two-dimensional saturation flow. Momentum equations of surface water are as follows:

$$\begin{aligned}
&\frac{\partial}{\partial t}\left(\frac{hU}{J}\right) + \frac{\partial}{\partial \xi}\left(\frac{hUU}{J}\right) + \frac{\partial}{\partial \eta}\left(\frac{hUV}{J}\right) \\
&-\frac{hu}{J}\left(U\frac{\partial}{\partial \xi}\left(\frac{\partial \xi}{\partial x}\right) + V\frac{\partial}{\partial \eta}\left(\frac{\partial \xi}{\partial x}\right)\right) - \frac{hv}{J}\left(U\frac{\partial}{\partial \xi}\left(\frac{\partial \xi}{\partial y}\right) + V\frac{\partial}{\partial \eta}\left(\frac{\partial \xi}{\partial y}\right)\right) \\
&= -gh\left(\frac{1}{J}\left(\left(\frac{\partial \xi}{\partial x}\right)^2 + \left(\frac{\partial \xi}{\partial y}\right)^2\right)\frac{\partial z}{\partial \xi} + \frac{1}{J}\left(\frac{\partial \xi}{\partial x}\frac{\partial \eta}{\partial x} + \frac{\partial \xi}{\partial y}\frac{\partial \eta}{\partial y}\right)\frac{\partial z}{\partial \eta}\right) - \frac{\tau_{b\xi}}{\rho J} - \frac{F_{v\xi}}{\rho J} \\
&+\frac{1}{J}\left(\frac{\partial \xi}{\partial x}\right)^2\frac{\partial}{\partial \xi}(h\sigma_{xx}) + \frac{1}{J}\frac{\partial \xi}{\partial x}\frac{\partial \eta}{\partial x}\frac{\partial}{\partial \eta}(h\sigma_{xx}) + \frac{1}{J}\frac{\partial \xi}{\partial y}\frac{\partial \eta}{\partial x}\frac{\partial}{\partial \eta}(h\tau_{yx}) + \frac{1}{J}\frac{\partial \xi}{\partial y}\frac{\partial \xi}{\partial x}\frac{\partial}{\partial \xi}(h\tau_{yx}) \\
&+\frac{1}{J}\frac{\partial \xi}{\partial x}\frac{\partial \eta}{\partial y}\frac{\partial}{\partial \eta}(h\tau_{xy}) + \frac{1}{J}\frac{\partial \xi}{\partial x}\frac{\partial \xi}{\partial y}\frac{\partial}{\partial \xi}(h\tau_{xy}) + \frac{1}{J}\left(\frac{\partial \xi}{\partial y}\right)^2\frac{\partial}{\partial \xi}(h\sigma_{yy}) + \frac{1}{J}\frac{\partial \xi}{\partial y}\frac{\partial \eta}{\partial y}\frac{\partial}{\partial \eta}(h\sigma_{yy})
\end{aligned} \tag{25.7}$$

$$\begin{aligned}
&\frac{\partial}{\partial t}\left(\frac{hV}{J}\right) + \frac{\partial}{\partial \xi}\left(\frac{hVU}{J}\right) + \frac{\partial}{\partial \eta}\left(\frac{hVV}{J}\right) \\
&-\frac{hu}{J}\left(U\frac{\partial}{\partial \xi}\left(\frac{\partial \eta}{\partial x}\right) + V\frac{\partial}{\partial \eta}\left(\frac{\partial \eta}{\partial x}\right)\right) - \frac{hv}{J}\left(U\frac{\partial}{\partial \xi}\left(\frac{\partial \eta}{\partial y}\right) + V\frac{\partial}{\partial \eta}\left(\frac{\partial \eta}{\partial y}\right)\right) \\
&= -gh\left(\frac{1}{J}\left(\frac{\partial \xi}{\partial x}\frac{\partial \eta}{\partial x} + \frac{\partial \xi}{\partial y}\frac{\partial \eta}{\partial y}\right)\frac{\partial z}{\partial \xi} + \frac{1}{J}\left(\left(\frac{\partial \eta}{\partial x}\right)^2 + \left(\frac{\partial \eta}{\partial y}\right)^2\right)\frac{\partial z}{\partial \eta}\right) - \frac{\tau_{b\eta}}{\rho J} - \frac{F_{v\eta}}{\rho J} \\
&+\frac{1}{J}\frac{\partial \eta}{\partial x}\frac{\partial \xi}{\partial x}\frac{\partial}{\partial \xi}(h\sigma_{xx}) + \frac{1}{J}\left(\frac{\partial \eta}{\partial x}\right)^2\frac{\partial}{\partial \eta}(h\sigma_{xx}) + \frac{1}{J}\frac{\partial \eta}{\partial y}\frac{\partial \xi}{\partial x}\frac{\partial}{\partial \xi}(h\tau_{yx}) + \frac{1}{J}\frac{\partial \eta}{\partial y}\frac{\partial \eta}{\partial x}\frac{\partial}{\partial \eta}(h\tau_{yx}) \\
&+\frac{1}{J}\frac{\partial \eta}{\partial x}\frac{\partial \xi}{\partial y}\frac{\partial}{\partial \xi}(h\tau_{xy}) + \frac{1}{J}\frac{\partial \eta}{\partial x}\frac{\partial \eta}{\partial y}\frac{\partial}{\partial \eta}(h\tau_{xy}) + \frac{1}{J}\frac{\partial \eta}{\partial y}\frac{\partial \xi}{\partial y}\frac{\partial}{\partial \xi}(h\sigma_{yy}) + \frac{1}{J}\left(\frac{\partial \eta}{\partial y}\right)^2\frac{\partial}{\partial \eta}(h\sigma_{yy})
\end{aligned} \tag{25.8}$$

where g is gravity, ρ is the water density, and $\tau_{b\xi}$ and $\tau_{b\eta}$ represent the contravariant shear stress along ξ and η coordinates, respectively. These shear stresses are defined as:

$$\tau_{b\xi} = \frac{\partial \xi}{\partial x}\tau_{bx} + \frac{\partial \xi}{\partial y}\tau_{by} \tag{25.9}$$

$$\tau_{b\eta} = \frac{\partial \eta}{\partial x}\tau_{bx} + \frac{\partial \eta}{\partial y}\tau_{by} \tag{25.10}$$

where τ_{bx} and τ_{by} are the shear stress along x and y coordinates, respectively. These are defined as;

$$\tau_{bx} = \tau_b \frac{u_b}{\sqrt{u_b^2 + v_b^2}} \tag{25.11}$$

$$\tau_{by} = \tau_b \frac{v_b}{\sqrt{u_b^2 + v_b^2}} \tag{25.12}$$

$$\frac{\tau_b}{\rho} = u_*^2 \tag{25.13}$$

$$u_*^2 = \frac{n_m^2 g}{R^{1/3}}\left(u^2 + v^2\right) \text{ or } u_*^2 = \frac{u^2 + v^2}{\left(6 + 2.5\ln\frac{h}{k_s}\right)^2} \tag{25.14}$$

where u_* is the friction velocity, n_m is the Manning's roughness coefficient, R is the hydraulic radius, k_s is the roughness height, and u_b and v_b represent velocity near the bed surface along the x and y coordinates, respectively. Velocities near the bed are evaluated using the curvature radius of streamlines as follows:

$$u_b = u_{bs}\cos\alpha_s - v_{bs}\sin\alpha_s \tag{25.15}$$

$$v_b = u_{bs}\sin\alpha_s + v_{bs}\cos\alpha_s \tag{25.16}$$

$$u_{bs} = 8.5u_* \tag{25.17}$$

$$v_{bs} = -N_* \frac{h}{r} u_{bs} \tag{25.18}$$

where $\alpha_s = \arctan(v/u)$, N_* is a dimensionless coefficient of secondary flow intensity and takes a value of 7.0 (Engelund 1974) and r is the curvature radius of streamlines obtained from the depth-integrated velocity field as follows (Shimizu and Itakura 1991)

$$\frac{1}{r} = \frac{1}{\left(u^2 + v^2\right)^{3/2}}\left\{u\left(u\frac{\partial v}{\partial x} - v\frac{\partial u}{\partial x}\right) + v\left(u\frac{\partial v}{\partial y} - v\frac{\partial u}{\partial y}\right)\right\} \tag{25.19}$$

σ_{xx}, σ_{yy}, τ_{xy} and τ_{yx} are turbulence stress as follows:

$$\sigma_{xx} = 2\nu\frac{\partial u}{\partial x} - \frac{2}{3}k_t \tag{25.20}$$

$$\sigma_{yy} = 2\nu\frac{\partial v}{\partial y} - \frac{2}{3}k_t \tag{25.21}$$

$$\tau_{xy} = \tau_{yx} = \nu \left(\frac{\partial v}{\partial x} + \frac{\partial u}{\partial y} \right) \tag{25.22}$$

$$\nu = \frac{\kappa}{6} u_* h \tag{25.23}$$

$$k_t = 2.07 u_*^2 \tag{25.24}$$

where ν is the coefficient of kinematic eddy viscosity, κ is the Karman constant, and k_t is the depth-averaged turbulence kinetic energy (Nezu and Nakagawa 1993). $F_{v\xi}$ and $F_{v\eta}$ are the contravariant drag forces due to vegetation along ξ and η coordinates, respectively. These drag forces are defined as

$$F_{v\xi} = \frac{\partial \xi}{\partial x} F_{vx} + \frac{\partial \xi}{\partial y} F_{vy} \tag{25.25}$$

$$F_{v\eta} = \frac{\partial \eta}{\partial x} F_{vx} + \frac{\partial \eta}{\partial y} F_{vy} \tag{25.26}$$

where F_{vx} and F_{vy} are the drag forces due to vegetation along the x and y coordinates, respectively, as follows (Tsujimoto 1999):

$$F_{vx} = F_v \frac{u}{\sqrt{u^2 + v^2}} \tag{25.27}$$

$$F_{vy} = F_v \frac{v}{\sqrt{u^2 + v^2}} \tag{25.28}$$

$$\frac{F_v}{\rho} = \frac{1}{2} C_{dv} \lambda_v \left(u^2 + v^2 \right) h_v \tag{25.29}$$

where C_{dv} is the coefficient of form drag varying due to vegetation shape. The order of C_{dv} is 0.1–1.0 and there is no information on the spatiotemporal change of the vegetation shape here. Hence, 1.0 is used for the simplicity. λ_v is vegetation density. The vegetation density is determined so that it is increased linearly from 0 to 0.05 1/m as follows:

$$\lambda_v = 0.05 \left(t_d / T_f \right) \tag{25.30}$$

where t_d is the accumulated dry bed time and T_f is the time that the relevant region is filled with vegetation. Figure 25.3 shows the model of the temporal change of vegetation. Bed degradation is assumed to cause vegetation extinction; vegetation is washed out when the bed level decreases below the level at which the vegetation was established, so that the density is set to 0. h_v is the water depth acting

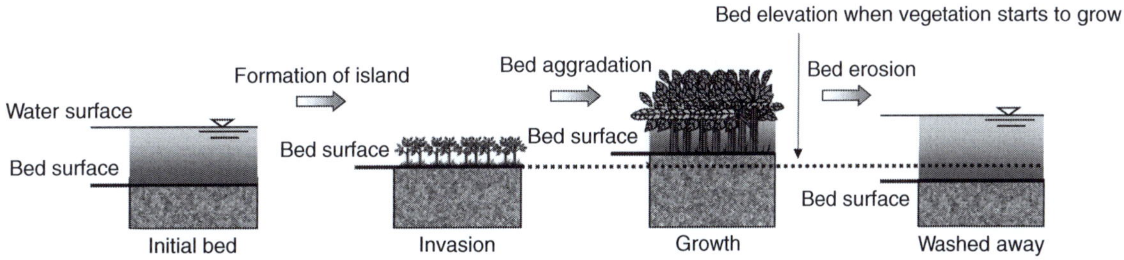

Figure 25.3 Schematic model of the temporal change in vegetation.

on vegetation. When vegetation height is larger than water depth, h_v is equal to water depth. When vegetation height is lower than water depth, h_v is equal to vegetation height.

u_g and v_g represent velocity of seepage flow along the x and y coordinates, respectively, as follows:

$$u_g = -k_{gx}\left(\frac{\partial \xi}{\partial x}\frac{\partial z_b}{\partial \xi} + \frac{\partial \eta}{\partial x}\frac{\partial z_b}{\partial \eta}\right) \tag{25.31}$$

$$v_g = -k_{gy}\left(\frac{\partial \xi}{\partial y}\frac{\partial z_b}{\partial \xi} + \frac{\partial \eta}{\partial y}\frac{\partial z_b}{\partial \eta}\right) \tag{25.32}$$

where k_{gx} and k_{gy} are the coefficients of permeability along the longitudinal and the transverse directions, respectively. When the water depth of surface flow becomes less than the mean diameter of the bed material, the surface flow is computed only in consideration of the pressure term and bed shear-stress term in the momentum equation of surface flow.

Grain-size distribution is evaluated using the sediment transport multilayer model (Takebayashi, Egashira, and Okabe 2003).

$$\frac{\partial}{\partial t}\left(\frac{c_b E_b f_{bk}}{J}\right) + (1-\lambda)F_{bk}\frac{\partial}{\partial t}\left(\frac{z_b}{J}\right) + \left(\frac{\partial}{\partial \xi}\left(\frac{q_{b\xi k}}{J}\right) + \frac{\partial}{\partial \eta}\left(\frac{q_{b\eta k}}{J}\right)\right) = 0 \quad \begin{cases} F_{bk} = f_{d1k}, \ \partial z_b/\partial t \le 0 \\ F_{bk} = f_{bk}, \ \partial z_b/\partial t \ge 0 \end{cases} \tag{25.33}$$

$$\frac{\partial}{\partial t}\left(\frac{E_{d1} f_{d1k}}{J}\right) - F_{dk}\frac{\partial}{\partial t}\left(\frac{E_{d1}}{J}\right) = 0 \quad \begin{cases} F_{dk} = f_{d1k}, \ \partial z_b/\partial t \le 0 \\ F_{dk} = f_{bk}, \ \partial z_b/\partial t \ge 0 \end{cases} \tag{25.34}$$

In the formulae above, f_{bk} is the concentration of bedload of size class k in the bedload layer, f_{dmk} is the sediment concentration of size class k in the mth bed layer, c_b is the depth-averaged concentration of bedload, E_{be} is the equilibrium bedload-layer thickness estimated by the following equation (Egashira and Ashida 1992):

$$\frac{E_{be}}{d_m} = \frac{1}{c_b \cos\theta(\tan\phi - \tan\theta)}\tau_{*m} \tag{25.35}$$

where d_m is the mean diameter of bedload, ϕ is the angle of repose, and τ_{*m} is the non-dimensional shear stress of the mean diameter. E_{sd} is the sediment-layer thickness on a non-erodible sediment bed. E_b is the bedload-layer thickness as follows:

$$E_b = E_{be} \qquad E_{sd} \ge E_{be}\frac{c_b}{1-\lambda} \tag{25.36}$$

$$E_b = E_{sd}\frac{1-\lambda}{c_b} \qquad E_{sd} \le E_{be}\frac{c_b}{1-\lambda} \tag{25.37}$$

$q_{b\xi k}$ and $q_{b\eta k}$ are the bedload components of size class k in ξ and η directions, respectively, as follows:

$$q_{b\xi k} = \frac{\partial \xi}{\partial x}q_{bxk} + \frac{\partial \xi}{\partial y}q_{byk} \tag{25.38}$$

$$q_{b\eta k} = \frac{\partial \eta}{\partial x}q_{bxk} + \frac{\partial \eta}{\partial y}q_{byk} \tag{25.39}$$

q_{bxk} and q_{byk} are the bedload components of size class k in the x and y directions, respectively, as follows (Ashida and Michiue 1971; Kovacs and Parker 1994; Liu 1991):

$$q_{bxk} = q_{bk}\cos\beta_k \tag{25.40}$$

$$q_{byk} = q_{bk}\sin\beta_k \tag{25.41}$$

$$q_{bk} = 17\frac{\rho\, u_{*e}^3}{(\rho_s-\rho)g}\left(1-\sqrt{K_c}\frac{u_{*ck}}{u_*}\right)\left(1-K_c\frac{u_{*ck}^2}{u_*^2}\right)f_{bk}r_b \tag{25.42}$$

Therein, ρ_s is the sediment density. u_{*e} is the effective shear velocity as follows:

$$u_{*e}^2 = \frac{u^2+v^2}{\left(6+2.5\ln\frac{h}{d_m(1+2\tau_{*m})}\right)^2} \tag{25.43}$$

The non-dimensional critical friction velocity of size class k is evaluated as follows (Ashida and Michiue 1971):

$$u_{*ck}^2 = u_{*cm}^2\left[\frac{\log_{10}19}{\log_{10}(19d_k/d_m)}\right]^2\frac{d_k}{d_m} \qquad d_k/d_m \ge 0.4 \tag{25.44-1}$$

$$u_{*ck}^2 = 0.85u_{*cm}^2 \qquad d_k/d_m \le 0.4 \tag{25.44-2}$$

Iwagaki's (1956) formula, which is formulated for uniform bed material, is used for evaluating u_{*cm} as follows:

$$u_{*cm}^2 = 80.9d_m \qquad d_m \ge 0.303 \tag{25.45-1}$$

$$u_{*cm}^2 = 134.6d_m^{31/22} \quad 0.118 \le d_m < 0.303 \tag{25.45-2}$$

$$u_{*cm}^2 = 55.0d_m \qquad 0.0565 \le d_m < 0.118 \tag{25.45-3}$$

$$u_{*cm}^2 = 8.41d_m^{11/32} \quad 0.0065 \le d_m < 0.0565 \tag{25.45-4}$$

$$u_{*cm}^2 = 226d_m \qquad d_m < 0.0065\ (\text{Unit: cm}) \tag{25.45-5}$$

r_b is the function related to thickness of bedload layer as follows:

$$r_b = \frac{E_b}{E_{be}} \tag{25.46}$$

K_c is a correction factor due to the influence of bed inclination on sediment motion (Kovacs and Parker 1994; Liu 1991).

$$K_c = 1+\frac{1}{\mu_s}\left[\left(\frac{\rho}{\rho_s-\rho}+1\right)\cos\alpha\tan\theta_x+\sin\alpha\tan\theta_y\right] \tag{25.47}$$

where α is the angle of deviation of near-bed flow from the x direction as follows.

$$\alpha = \arctan\left(\frac{v_b}{u_b}\right) \tag{25.48}$$

μ_s is the coefficient of static friction. θ_x and θ_y are bed inclinations in the x and y directions, respectively. These inclinations are evaluated as follows

$$\theta_x = \arctan\left(\frac{\partial \xi}{\partial x}\frac{\partial z_b}{\partial \xi} + \frac{\partial \eta}{\partial x}\frac{\partial z_b}{\partial \eta}\right) \tag{25.49}$$

$$\theta_y = \arctan\left(\frac{\partial \xi}{\partial y}\frac{\partial z_b}{\partial \xi} + \frac{\partial \eta}{\partial y}\frac{\partial z_b}{\partial \eta}\right) \tag{25.50}$$

Hence, the local bed slope along the direction of bedload for mean diameter (θ) is obtained as follows:

$$\sin\theta = \cos\beta_m \sin\theta_x + \sin\beta_m \sin\theta_y \tag{25.51}$$

where β_m is the deviation angle of bedload of the mean diameter to the depth-averaged flow direction. The deviation angle of bedload of size-class k to the depth-averaged flow direction (β_k), which depend on the near-bed flow and inclination of the bed, is calculated by the following relation

$$\tan\beta_k = \frac{\sin\alpha - \Pi\Theta_y\left(\dfrac{u_{*ck}^2}{u_*^2}\right)\tan\theta_y}{\cos\alpha - \Pi\Theta_x\left(\dfrac{u_{*ck}^2}{u_*^2}\right)\tan\theta_x} \tag{25.52}$$

$$\Pi = K_{ld} + 1/\mu_s \tag{25.53}$$

$$\Theta_y = \frac{1}{1 + \tan^2\theta_x + \tan^2\theta_y} \tag{25.54}$$

$$\Theta_x = \Theta_y + \frac{\rho}{\rho_s - \rho}\cos^2\theta_x \tag{25.55}$$

where K_{ld} (=0.85) is the ratio of lift force to drag force. Evolution of bed elevation is estimated by means of following formula.

$$\frac{\partial}{\partial t}\left(\frac{c_b E_b}{J}\right) + (1-\lambda)\frac{\partial}{\partial t}\left(\frac{z_b}{J}\right) + \left(\frac{\partial}{\partial \xi}\left(\sum_{k=1}^{n}\frac{q_{b\xi k}}{J}\right) + \frac{\partial}{\partial \eta}\left(\sum_{k=1}^{n}\frac{q_{b\eta k}}{J}\right)\right) = 0 \tag{25.56}$$

In equation (25.56), n represents the total number of size classes of sediment. Local bed slope was reset to the angle of repose at calculation points where the slope becomes steeper than the angle of repose.

The continuum equation and the momentum equations for surface flow are discretized by use of the total variation diminishing (TVD) MacCormack scheme (Causon 1989).

25.2.2 Hydraulic Conditions

Hydraulic conditions for the flume tests and the numerical computations to discuss the effect of the seepage flow are Case 0 shown in Table 25.1. These hydraulic conditions are classified into the formative condition of braided channels (Egashira and Takebayashi 1999). The other hydraulic conditions for the numerical analysis of the braided channels are shown in Table 25.1. Figure 25.4 shows these conditions on the mesoscale bed-configuration-regime criterion (Kuroki and Kishi 1982). The calculation domain is a straight channel with solid side banks and movable bed. The

Table 25.1 Hydraulic conditions.

Case	Type	Vegetation	Width (m)	Water discharge (m^3/s)	Non-dimensional shear stress	Width : depth ratio
0	Experiment and analysis	No	1	0.00190	0.041	155
1	Experiment and analysis	No	1.5	0.00250	0.039	239
2	Experiment and analysis	No	1.5	0.00125–0.00375	0.026, 0.050	363, 188
3	Analysis	Yes	1.5	0.00250	0.039	239
4	Analysis	Yes	1.5	0.00125–0.00375	0.026, 0.050	363, 188
5	Analysis	Yes	1.0	0.00125–0.00375	0.033, 0.064	190, 98
6	Analysis	Yes	0.55	0.00250	0.072	48
7	Analysis	Yes	0.8	0.00133	0.039	128
8	Analysis	Yes	0.26	0.00043	0.039	42
9	Analysis	No	0.55	0.00125–0.00375	0.047, 0.091	73, 38
10	Analysis	Yes	0.55	0.00125–0.00375	0.047, 0.091	73, 38

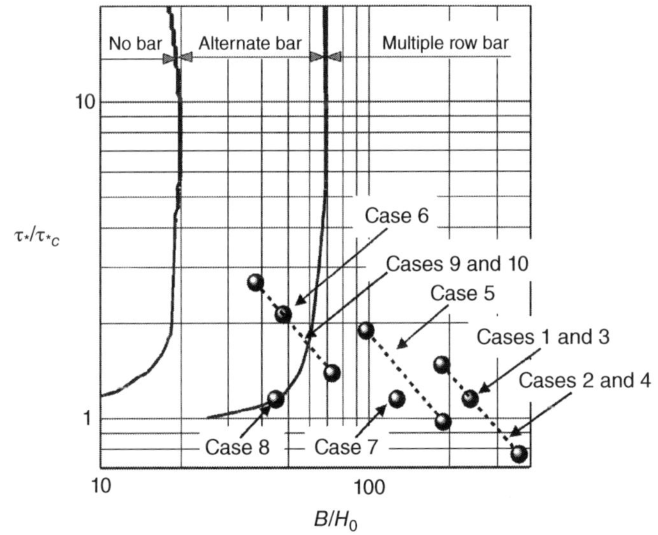

Figure 25.4 Hydraulic conditions of flume experiments and numerical analysis on the mesoscale bed configuration regime criterion of Kuroki and Kishi (1982). The non-dimensional critical shear stress is 0.034.

bed slope is 1/100. Initial bed geometry is flat with small disturbances as shown in Figure 25.5. The bed materials are treated as uniform sediment with a particle diameter of 1 mm. Cases 1 and 2 are the same conditions as used in flume experiments. Cases 3 and 4 are hydraulic conditions that are identical to those of Cases 1 and 2, but density of vegetation is taken into account. Water discharge under Case 5 is the same as that under Case 4, but the width is changed. Water discharge under Case 6 is the same as

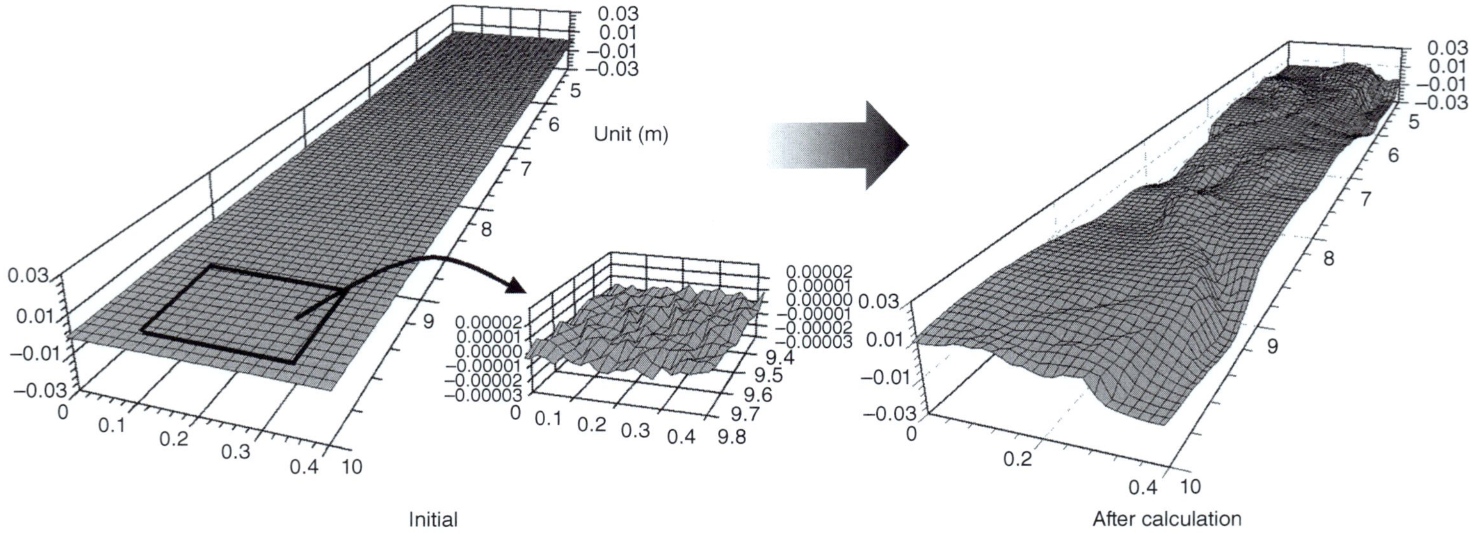

Figure 25.5 Initial bed geometry.

Table 25.2 Vegetation density and water discharge for the Yoshino River calculations.

Case	Vegetation density (1/m)	Water discharge (m^3/s)
11	0.05	7596
12	0.01	11 908
13	0.005	15 886
14	0.08	7596
15	0.02	11 908
16	0.08 and 0.02	Figure 25.6
17	0.05	Figure 25.6
18	0	Figure 25.6

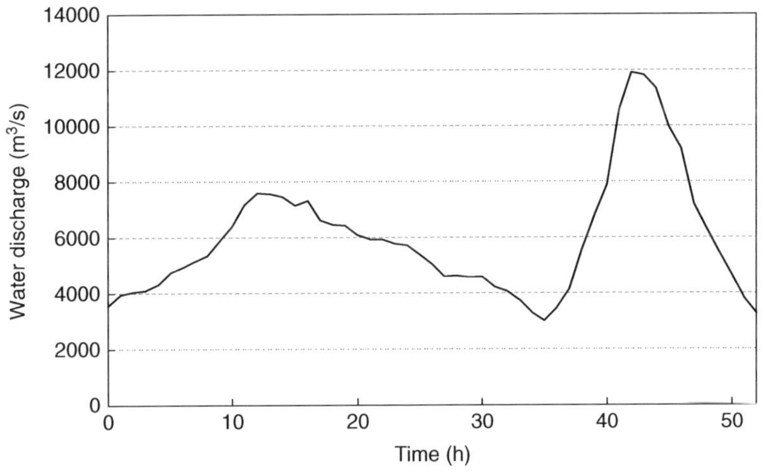

Figure 25.6 Hydrograph in Cases 16–18.

that under Case 3, but the width is changed and the formative regime is of alternate bars that are approximately in the formative region of multiple row bars. In Cases 7 and 8 the width is changed, with the same unit width water discharge conditions as those used in Case 3. In Cases 9 and 10, Case 6 is converted to unsteady conditions that are used as the formative hydraulic conditions of alternate bars at a large water discharge and as the formative hydraulic condition of multiple row bars at a small water discharge. Vegetation growth is not considered in Case 9.

Water discharge from the upstream end and vegetation density for the numerical analysis of the Yoshino River are shown in Table 25.2 and Figure 25.6. Measured water-surface elevation data are used as the downstream boundary condition. For the initial shape of the bed, results of laser surveys carried out in 2003 are used. The horizontal distribution of vegetation and the numerical analysis grids

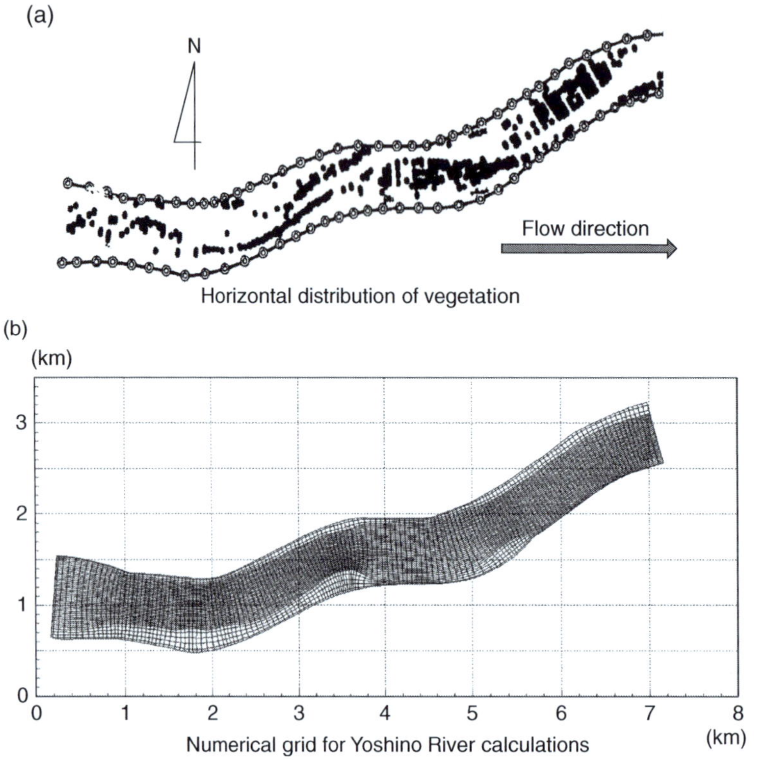

Figure 25.7 Horizontal distribution of (a) vegetation and (b) numerical grid for Yoshino River calculations.

are shown in Figure 25.7. The vegetation height is estimated from the height of the vegetation crest obtained from laser surveying performed in 2003 and from bed-elevation data. The temporal change model of vegetation as shown in Figure 25.3 is not used here, because of the lack of information on vegetation in the Yoshino River. Here, for the temporal change of vegetation density, a value that can reproduce the measured water level most precisely is calculated by the horizontal two-dimensional water flow model and is used for the bed deformation analysis, as described later. Numerical analyses are carried out under a total of eight conditions: Cases 11 through 18. In Cases 11 through 15, in which only the flow is analysed, the maximum water discharge of each flood is used as the analysis water discharge: 7600 m^3/s measured for Typhoons 10 and 11 (1 August 2004) for Cases 11 and 14; 11 900 m^3/s for Typhoon 16 (31 August 2004) for Cases 12 and 15; and 15900 m^3/s for Typhoon 23 (20 October 2004) for Case 13. Manning's roughness coefficient is set to 0.03. This is considered to be the value for non-vegetated land and a constant value is given. In Cases 11 through 13, all vegetation crests are higher than the water-surface level. Cases 14 and 15 have the same hydraulic conditions as those used in Cases 11 and 12, respectively, and the measured vegetation height is considered. The reduction of projecting area of the vegetation with regard to flows, resulting from submergence of the vegetation in the water, is examined in this condition. In Cases 16 through 18, for which bed deformation analysis is performed, hydrographs taken during Typhoons 10 and 11 and during Typhoon 16 are combined, as shown in Figure 25.6. The mean grain diameter of the bed materials is set to 3.2 cm from field measurements. In Case 16, values of the density are changed for every flood using values of density in Cases 14 and 15. For Case 17, a constant value of 0.05 1/m, the average of values of density of the vegetation obtained in Cases 14 and 15, is given. Case 18 is a condition without vegetation.

The river basin area of the Yoshino River is 3750 km^2 and the total length of the main channel is 198 km. For the present study, the research zone is from the Kakihara weir to the number 10 weir (both are fixed weirs), from 17 to 24 km upstream of the river mouth. The river width is approximately 800 m and is nearly constant throughout the zone; 3–4 bars with wavelengths of approximately 3 km exist. Figure 25.1 shows an overview of the river channel including the study zone. Bars are covered mostly by woody plant vegetation and grass plant vegetation such as vines and reeds. Woody plant vegetation, consisting primarily of willows, has been rooted since the late 1980s, when large-scale gravel mining was banned and dam reservoirs had been constructed. Since then, their distribution has expanded year by year.

25.3 Flume Experiments: Method and Hydraulic Conditions

The rectangular channels used in the experiments have lengths of 15 m in Case 0 and 17 m in Case 1 and Case 2. The bed slope is 1/100 and the initial bed geometry is made flat. The bed materials are nearly uniform sediments with average particle diameter of 1 mm. Table 25.1 summarizes the hydraulic conditions used in the experiments. An experiment was carried out for Case 0 to investigate the effects of seepage flow on the geometry of braided channels. Water was supplied from the upstream end under a constant rate. Hydraulic conditions in Case 0 were within the formative regime of multiple row bars (Kuroki and Kishi 1982). Experiments were carried out for Case 1 and Case 2 to investigate the effects of unsteady characteristics of water and sediment supply on the geometry of braided channels. The sole difference between Case 1 and Case 2 is the method of supplying water and sediments. Under the unsteady conditions (Case 2), water discharge was changed every 30 min between 1.25 l/s and 3.75 l/s. The water supply was short enough not to change bed configuration completely in a cycle. Consequently, the bed configurations differ from those corresponding to each water discharge. Sediment was supplied so that the initial bed elevation at the upstream end would not be lowered. Hydraulic conditions in Case 1 and Case 2 (both large and small water discharges) were all within the formative regime of multiple row bars (Kuroki and Kishi 1982). Although the non-dimensional shear stress at a small water discharge was less than the non-dimensional critical shear stress, sediment moved when the bed configuration was generated and flows concentrated in channels.

Photographs were taken every hour; measurements of the bed geometry were taken 6 hours later, when no change was noticed in channel configuration.

25.4 Results and Discussion

25.4.1 Effects of Unsteady Characteristics of Water Supply on Bed and Channel Configuration

Figure 25.8 shows the correlation between island area and the number of islands per square metre with an area larger than island areas. The spatial scale of islands without groundwater was smaller than that obtained by flume tests. Nonetheless, the distribution for larger islands in the flume data was well reproduced when the groundwater is considered. The inundation process on bars and formative process of puddles caused those differences. When water overflowed onto bars, water seeped fast and surface water appeared after filling the bed with water. Inundated water was, however, treated as surface water flow, when the groundwater was neglected. Hence, the area of islands decreased. Furthermore,

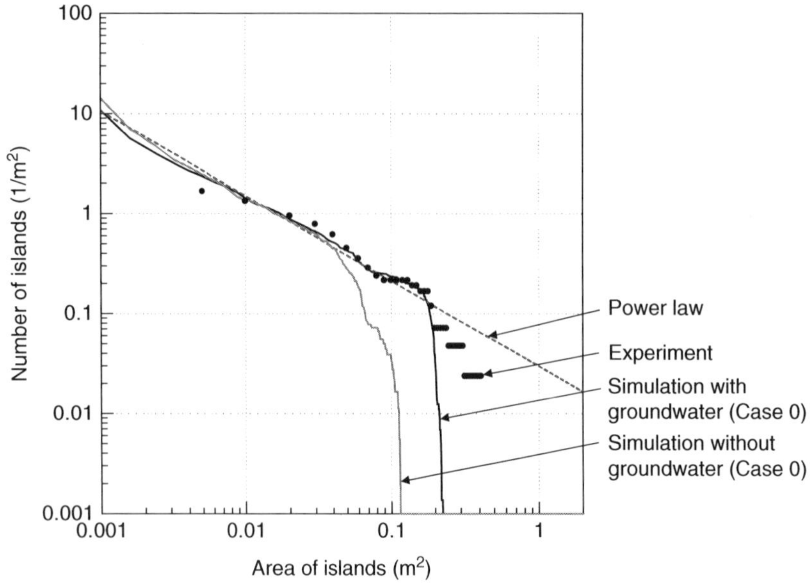

Figure 25.8 Variation of island area with the number of islands per square metre that have an area larger than island area.

many puddles were formed on bars on the decreasing process of water surface in flume tests because water flow on the braided rivers was unsteady.

Figure 25.9 shows bed geometries obtained through flume experiments and numerical analysis under steady (Case 1) and unsteady (Case 2) water and sediment supply conditions where vegetation growth was ignored. These hydraulic conditions are in the formative regime of multiple row bars. Some parts of the bed geometries of flume experiments are not reproduced well by numerical analyses, because braided channels have chaotic characteristics and the strict bed geometry cannot be reproduced. However, some identical characteristics exist in both flume experiments and numerical analyses; the bed elevation of the dry bed regions during low flow was higher and flattened under unsteady water and sediment supply conditions and fewer channels occurred under unsteady conditions than under steady conditions. In addition, the bed elevation in the channel was further lowered in unsteady conditions relative to steady conditions. This lowering is due to the fact that, at a large water discharge, sediment was transported to and deposited at higher bed-elevation areas, thereby choking smaller scale channels. At lower water discharge, flows were concentrated to larger scale channels with the deepest bed, thereby further developing channels.

Figure 25.10 shows bed geometries obtained through numerical analyses under steady water and sediment supply conditions where vegetation growth is neglected (Case 6), unsteady water and sediment supply conditions where vegetation growth is neglected (Case 9), and unsteady water and sediment supply conditions where vegetation growth is taken into account (Case 10). These hydraulic conditions are in the formative regime of alternate bars. In Case 6, vegetation does not grow because no dry bed regions are formed, even if vegetation growth is taken into account. The wavelength of bars in unsteady water and sediment supply conditions was longer than that in steady conditions as previously reported (Watanabe *et al.* 2001). Furthermore, in unsteady conditions the water-surface slope became steeper from a riffle towards a pool during low flow, and the riffles were eroded.

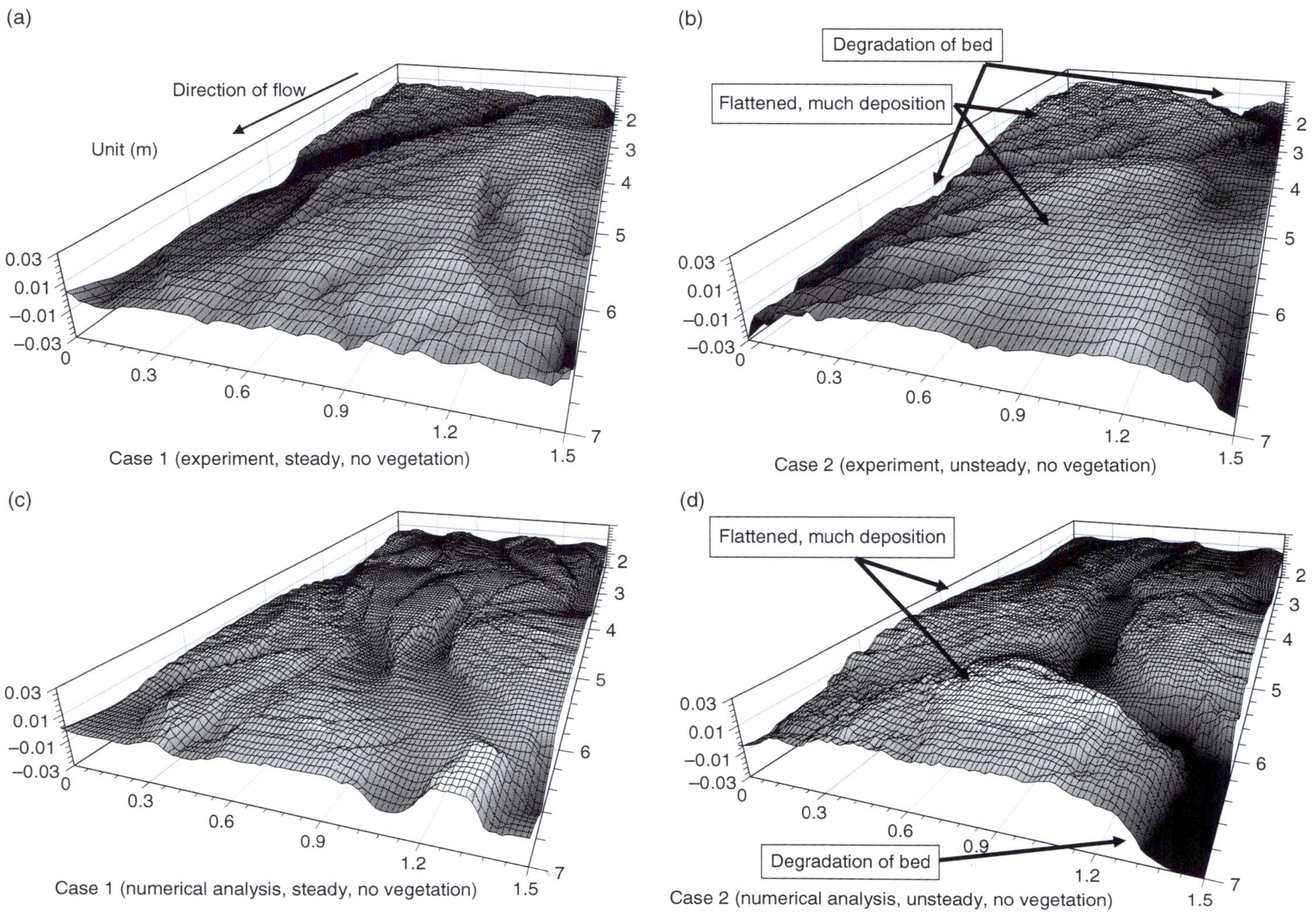

Figure 25.9 Bed geometry: flume experiments and numerical analysis in Cases 1 and 2.

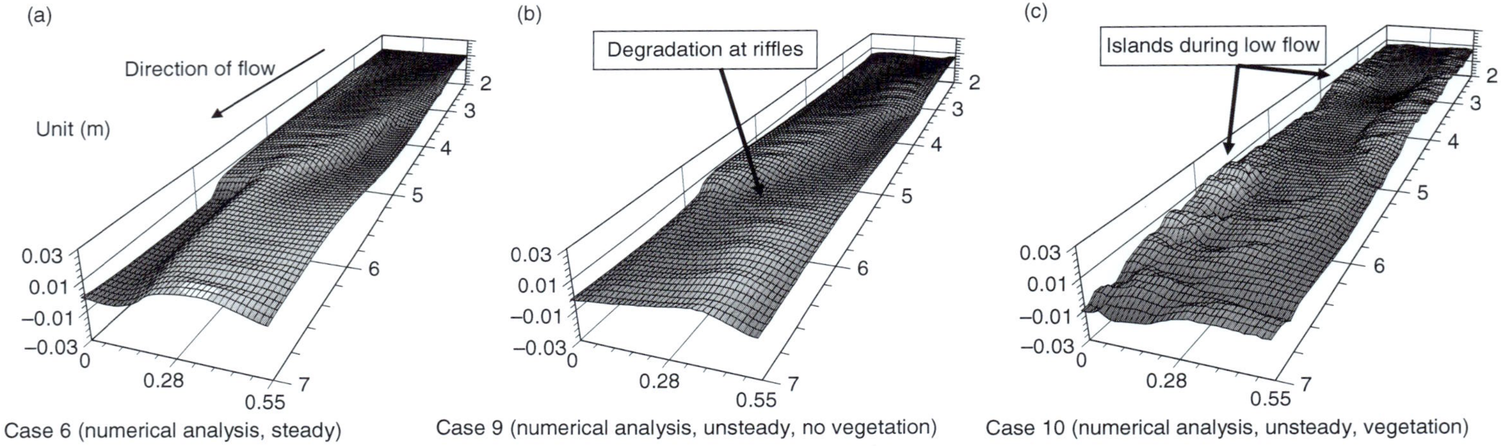

Figure 25.10 Bed geometry: numerical analysis in Cases 6, 9 and 10.

25.4.2 Effects of Vegetation Growth on Bed and Channel Configuration

Figure 25.11 shows bed geometries obtained from the numerical analyses under steady (Case 3) and unsteady (Case 4) water and sediment supply conditions. These hydraulic conditions are in the formative regime of multiple row bars. A comparison of Cases 1 and 2 is made with the same conditions, but ignoring the vegetation growth. When vegetation growth is considered, sediment deposition was accelerated in the area of higher bed elevation. Furthermore, the bed elevation in the channel was lowered in both steady and unsteady conditions. Simultaneously, the number of channels decreased as discussed by Crosato and Samir Saleh (2011). In the unsteady condition (Case 4) one meandering channel with smaller anabranches was formed. The formative region of a meandering channel appears with greater width : depth ratio in the field with rigid banks, if both vegetation growth and unsteady characteristics of water and sediment supply are considered. Care should be taken when defining width. Under formative hydraulic conditions of alternate bars, no dry bed region is formed with steady water and sediment supply conditions. Hence, the width of flow is equal to the width of the analysis domain. However, under formative hydraulic conditions of multiple row bars, the channel width does not have to be the same as the width of the analysis domain because a dry bed region can be formed even under steady water and sediment supply conditions. Hence, the width of the analysis domain becomes braiding width or floodplain width. With previous linear bed instability analyses, the treatment of dry bed regions within the analysis domain is difficult. For this reason, the formative mechanism of bars with dry bed regions has not been studied well. As a result, it is considered that the concept that 'formative hydraulic conditions of alternate bars are equivalent to formative hydraulic conditions of meandering channels' has been declared in many studies without discussion of the formative mechanism of floodplains.

Figure 25.12 shows bed geometries obtained through numerical analyses of Case 5 using a smaller width : depth ratio, in which unsteady water and sediment supply conditions are used and where effects of vegetation growth are considered. This condition is in the formative region of multiple row bars. Compared with Case 4 shown in Figure 25.11, despite a smaller width : depth ratio, the main channel in Case 5 is bifurcated and the number of channels with largest channel scale is greater than that of Case 4. These differences contradict the previous findings derived from linear bed instability analysis. It is considered that vegetation growth produces these differences. Figure 25.13 shows a horizontal distribution of the degree of vegetation growth obtained from Cases 4 and 5. The degree of vegetation growth 0 is for no vegetation and 1 is for perfect growth. Figure 25.13 shows that, compared

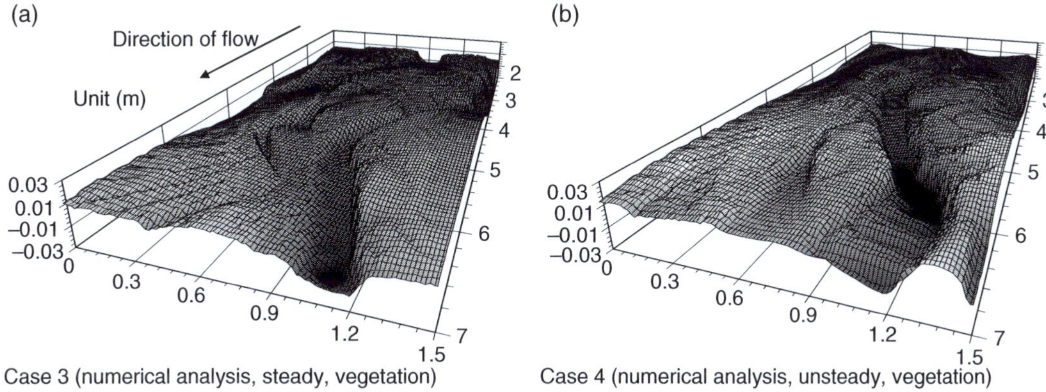

Figure 25.11 Bed geometry: numerical analysis in Cases 3 and 4.

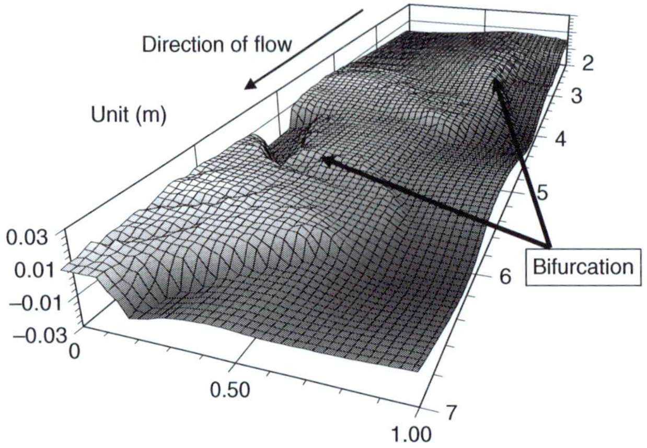

Figure 25.12 Bed geometry: numerical analysis in Case 5.

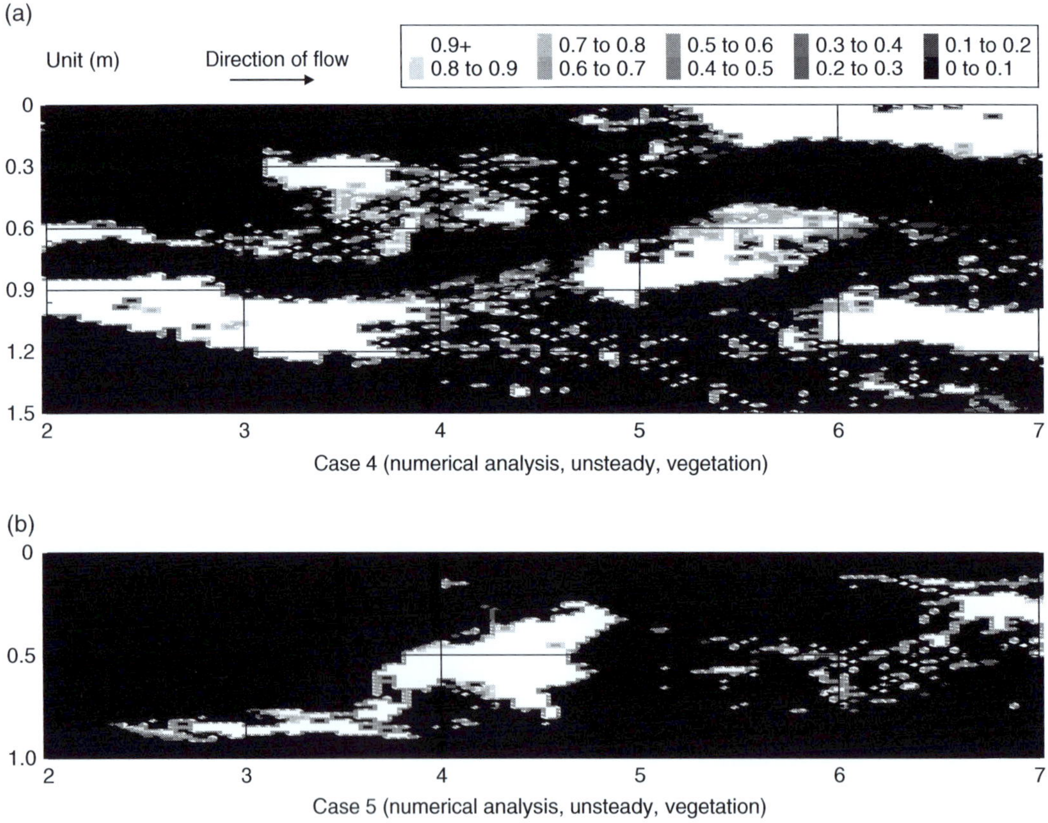

Figure 25.13 Vegetation growth rate: numerical analysis in Cases 4 and 5 (0, no vegetation; 1, perfect growth).

with Case 4, the vegetated area is narrower with regard to the total area of the analysis domain in Case 5 with a smaller width : depth ratio. This is due to narrowing the width. Disturbances to bars by flowing water are increased and vegetation growth is suppressed. It is considered that suppression of vegetation growth enhanced channel deformation and accelerated generation of channel bifurcations.

Figure 25.14 shows bed geometries obtained through numerical analyses of Cases 7 and 8 with a smaller width : depth ratio under steady water and sediment supply conditions, in which effects of vegetation growth are considered and the non-dimensional shear stress is identical with that of Case 3. Comparison of Case 3 and Case 7 shows that, despite a greater width : depth ratio, the number of main channels in Case 3 is only 1–2, which is nearly equal to that of Case 7. The same discussions as those used for Cases 4 and 5 are also applicable here. Bifurcation of channels is suppressed by vegetation growth, even with increased width : depth ratio and where the number of channels is not considerably increased. Furthermore, the bed geometry in Case 8 shows formation of nearly one channel flowing alternately from one bank to the other. In other words, under the formative condition for braided channels, the number of channels is determined by a balance between instability of the bed surface and resistance characteristics of vegetation. Under the condition in which resistance characteristics of the vegetation are low, the number of channels increases as the width : depth ratio increases, thereby forming braided channels. With greater resistance characteristics of vegetation, the number of channels remains below a maximum value at a certain level of width : depth ratio; a meandering channel is formed with a higher width : depth ratio.

Figure 25.10 shows bed geometries obtained through numerical analysis of Case 10 under unsteady water and sediment supply conditions in which the effects of vegetation growth are considered. The hydraulic conditions are in the formative region of alternate bars. Although deposition occurred alternately along the right and left banks, by considering the unsteady characteristics of water and sediment supply, as well as vegetation growth, a channel configuration is formed that

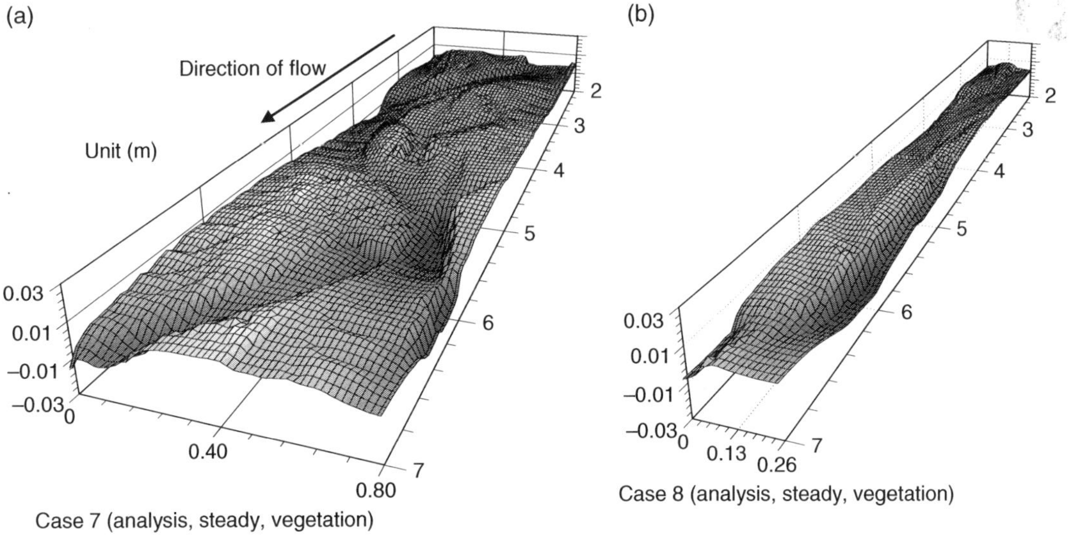

Figure 25.14 Bed geometry: numerical analysis in Cases 7 and 8.

has entirely different geometric characteristics from those of well-known alternate bars. This difference arises due to the fact that the riffles are eroded during low flow. In addition, vegetation grows on areas where the bed elevation is higher during low flow, sediment is deposited only on the vegetated area resulting in increased bed elevation, and flows are concentrated in the main channel with lower bed elevation. Concentration of flows to lower bed elevation areas because of vegetation growth is similar to such a field in which the dry bed regions are formed. It is therefore considered that under formative hydraulic conditions of alternate bars close to formative hydraulic conditions of multiple row bars, phenomena in natural rivers tend to shift to a greater width : depth ratio.

Cohesive alluvium in the floodplain reduces sediment transport rate. Therefore, it is expected that a meandering channel forms if cohesive sediments occur, even if vegetation does not grow (Ashida and Sawai 1977).

25.4.3 Effects of Spatiotemporal Change of Vegetation on the Yoshino River

Table 25.2 shows the vegetation density used in the analyses. In Figure 25.15 the measured water level and analysis water level along with right- and left-hand banks are compared. In Cases 11 through 13, values of the density are varied so that the measured water level was best reproduced. Here, vegetation density was used as the only calibration parameter. Observation of Cases 11 through 13 indicates that, for reproduction of the measured water level, the vegetation density should be changed for every flooding. The vegetation inclined to the downstream direction and the vegetation disappeared because of bank erosion. In addition, flushing-away of plants and leaf-fall occurred. It is then considered that the vegetation densities in Cases 12 and 13 were smaller than in Case 11. In other words, the vegetation density is expected to change considerably due to flooding and leaf-fall during one year. When vegetation height is considered (Cases 14 and 15), values of the density are greater than those of Cases 11 and 12. That may have been caused because some vegetation was submerged, thereby reducing the projection area of the vegetation with regard to flows. The resultant decrement of vegetation density in Cases 11 and 12 was supplemented by increased density.

Figure 25.16 shows the transverse section at 20.2 km and 17.5 km upstream of the river mouth after flooding by Typhoons 10 and 11 (1 August 2004) and 16 (31 August 2004). At the 20.2 km section, differences of bed deformation on the bars are remarkable; this section is located at the upstream region of the vegetation area on the bar formed along the right bank. For the section at 17.5 km, differences of bed deformation characteristics in the lower channel are remarkable; this section is located at the downstream region of the vegetation area on the bar formed along the left bank. It is apparent from Figure 25.16 that deposition at the 20.2 km section occurred at the vegetation region 300–400 m from the left bank for both Cases 16 (vegetation density changed by flooding) and 17 (constant vegetation density); in addition, the difference of sedimentation thickness is as large as approximately 60 cm. Comparison of Cases 16 and 18 reveals that in the latter, where no vegetation is provided, the bed in the lower channel was raised and the bed was lowered in the vegetation region of 300–400 m from the left bank, suggesting that the transverse bed slope decreased. At the 17.5 km section, vegetation grew at the upstream region, and sediment was not transported to the vegetation region on the bars. Hence, no bed deformation occurred on the bar in Cases 16 and 17. A difference of bed elevation (maximum 70 cm) dominates in the lower channel. In Case 18, bed deformation was observed not only in the lower channel but also on the bar. The bed elevation on the bar was lowered and bed elevation in the lower channel was raised, thereby the bed deformed as the transverse bed slope decreased.

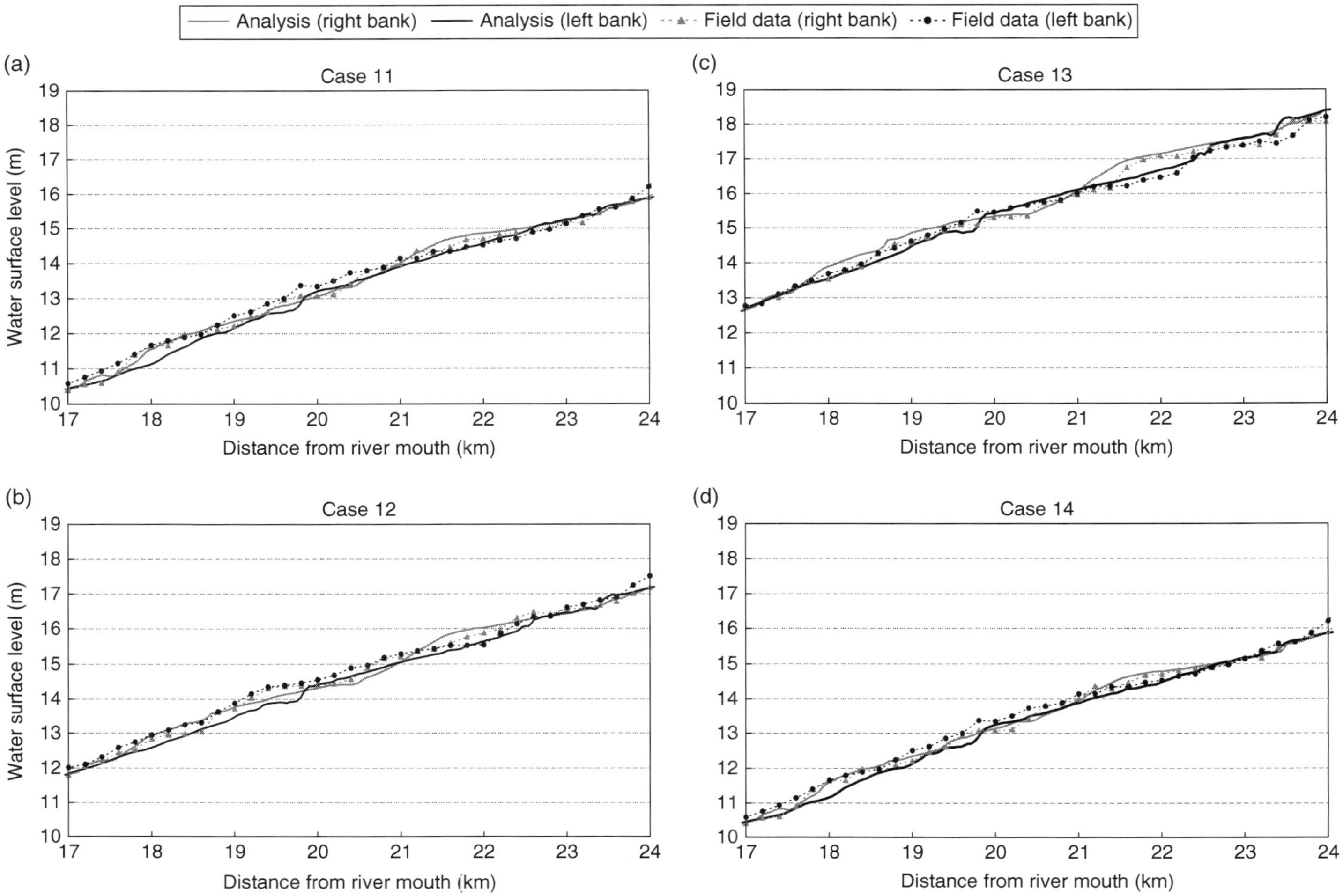

Figure 25.15 Comparison of water-surface elevation between numerical analyses and field data.

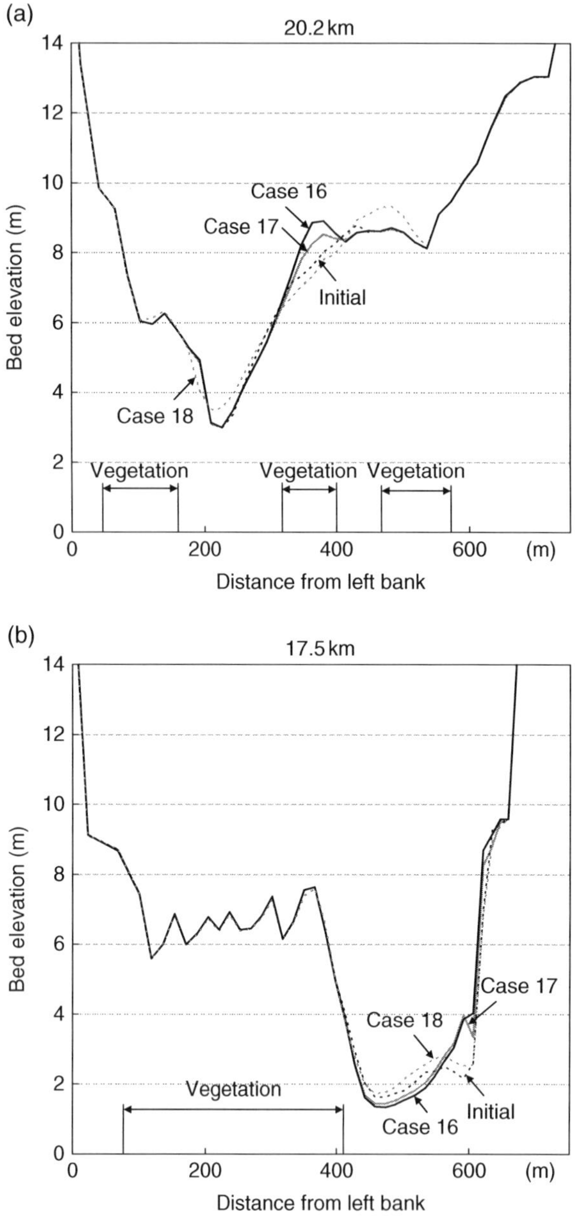

Figure 25.16 Bed geometries at 20.2 km and 17.5 km after the analysis.

25.5 Conclusions

Braided channels under unsteady flow conditions are reproduced by both flume experiments and a horizontal two-dimensional bed deformation numerical analysis. Furthermore, invasion, growth and washing away processes of vegetation are introduced in the numerical analysis model and the effects of both the unsteady characteristics of water supply and the presence of vegetation on the geometric characteristics of braided channels are discussed based on quantitative data from the numerical analysis. Both surface water and groundwater flows are considered in the model. Finally, the depth integrated two-dimensional flow and bed deformation analyses considering vegetation were performed for the Yoshino River, Japan, incorporating the spatiotemporal distribution of vegetation drag characteristics on the basis of the flooding that occurred in 2004. This allowed discussion of the effect of spatiotemporal change of the vegetation density on bed deformation characteristics. Results are summarized as follows.

1) The spatial scale of islands without groundwater is smaller than that obtained by flume tests. However, the distribution for larger islands in flume data is reproduced well when groundwater is considered. The inundation process on bars and formative process of puddles causes these differences.
2) Both flume experiments and numerical analyses showed that, compared with steady conditions, the bed elevation on dry bed regions at low water flow is both elevated and flattened under formative hydraulic conditions of multiple row bars and unsteady conditions. In addition, compared with steady conditions, the bed elevation in the channel is lowered under unsteady conditions. These phenomena are due to sediment brought to and deposited in such areas where the bed elevation is high during high flow, thereby choking smaller scale channels. In addition, flows are concentrated to a larger scale channel where bed elevation is low, thereby further developing the channel.
3) In the formative region of alternate bars compared with steady hydraulic conditions, the wavelength of bars is lengthened under unsteady conditions. Furthermore, under unsteady conditions the water-surface slope becomes steeper from a riffle towards a pool during low flow, thereby bed elevation becomes lowered in riffles.
4) Under both steady and unsteady conditions, if the vegetation growth is taken into account, deposition of sediment on a higher bed elevation area is accelerated and bed elevation in the channel is lowered. In addition, under unsteady conditions, the number of channels is reduced and nearly one meandering channel is formed with a large width : depth ratio. This suggests that in a field with rigid banks, the formative region of meandering channels appears in conditions with larger width : depth ratio if vegetation grows and water and sediment supply are unsteady.
5) Because opportunities for bifurcation of channels are suppressed by vegetation growth even if the width : depth ratio is increased, the number of channels does not increase remarkably. This fact suggests that under formative hydraulic conditions of braided channels, the number of channels is determined by the balance between the instability of the bed surface and the resistance of vegetation. The number of channels increases as the width : depth ratio increases with small vegetation resistance. Consequently, it forms braided channels, whereas with large vegetation resistance the number of channels remains below a maximum value, resulting in a meandering channel with a large width : depth ratio.
6) Even if mean hydraulic conditions are in the formative region of alternate bars, a channel configuration is formed that has entirely different geometric characteristics from those of well-known alternate bars because of the vegetation growth during low flow under unsteady conditions. It is therefore considered that under formative hydraulic conditions of alternate bars close to formative hydraulic conditions of multiple row bars, phenomena in natural rivers tend to shift to a larger width : depth ratio.
7) Vegetation density varies during the year in the Yoshino River and is considerably dependent on flooding and leaf-fall.

8) When the submergence area of trees is expressed using vegetation height data, the measured water level cannot be reproduced unless far greater values than those used in a case in which crests of all vegetation are higher than water surface level are used for vegetation density.
9) The difference of bed deformation between a case ignoring temporal change of vegetation density and a case considering temporal change is greater on a bar in the upstream vegetation region and is greater in the lower channel in the downstream vegetation region. Furthermore, the transverse bed slope becomes milder if vegetation is ignored.

Notation

c_b	Depth-averaged concentration of bedload
C_{dv}	Coefficient of form drag (=1.0)
d_m	Mean diameter of bed material
E_b	Bedload-layer thickness
E_{be}	Equilibrium bedload-layer thickness
E_{sd}	Sediment layer thickness on non-erodible sediment bed
F_v	Drag forces attributable to vegetation along the depth-averaged flow direction
F_{vx}	Drag forces attributable to vegetation along x coordinate
$F_{v\xi}$	Drag forces attributable to vegetation along ξ coordinate
F_{vy}	Drag forces attributable to vegetation along y coordinate
$F_{v\eta}$	Drag forces attributable to vegetation along η coordinate
f_{bk}	Concentration of bedload of size class k in the bedload layer
f_{dmk}	Sediment concentration of size class k in the mth bed layer
g	Gravity
h	Surface flow depth
h_g	Seepage flow depth
h_v	Water depth acting on vegetation
J	Jacobian
k	Number of size class of sediment
K_c	Correction factor attributable to the influence of bed inclination on sediment motion
k_{gx}	Coefficients of permeability along x coordinate
k_{gy}	Coefficients of permeability along y coordinate
K_{ld}	Ratio of the lift force to the drag force (=0.85)
k_s	Roughness height
k_t	Depth-averaged turbulence kinetic energy
n	Total number of size classes of sediment
n_m	Manning's roughness coefficient

N_*	Coefficient of secondary flow intensity (=7.0)
q_b	Sediment transport rates along the sediment transport direction
q_{bx}	Sediment transport rates along x coordinate
$q_{b\xi}$	Sediment transport rates along ξ coordinate
q_{by}	Sediment transport rates along y coordinate
$q_{b\eta}$	Sediment transport rates along η coordinate
R	Hydraulic radius
r	Curvature radius of streamlines obtained from the depth-integrated velocity field
r_b	Function related to thickness of bedload layer
t	Time
t_d	Accumulated dry bed time
T_f	Time during which the relevant region is overgrown with vegetation
u	Depth-averaged flow velocity along x coordinate
U	Contravariant depth-averaged flow velocity along ξ coordinate
u_b	Velocity near the bed surface along x coordinate
u_{bs}	Velocity near the bed surface along depth-averaged flow direction
u_g	Depth-averaged seepage flow velocity along x coordinate
U_g	Contravariant seepage flow velocity along ξ coordinate
u_*	Friction velocity
u_{*c}	Non-dimensional critical friction velocity
u_{*cm}	Non-dimensional critical friction velocity of mean diameter
u_{*e}	Effective friction velocity
v	Depth-averaged flow velocity along y coordinate
V	Contravariant depth-averaged flow velocity along η coordinate
v_b	Velocity near the bed surface along y coordinate
v_{bs}	Velocity near the bed surface perpendicular to depth-averaged flow direction
v_g	Depth-averaged seepage flow velocity along y coordinate
V_g	Contravariant seepage flow velocity along η coordinate
x	Coordinates along the longitudinal direction in Cartesian coordinate system
y	Coordinates along the transverse direction in Cartesian coordinate system
z	Water surface level
z_b	Bed surface level
α	$\arctan(v_b/u_b)$
α_s	$\arctan(v/u)$
β_m	Deviation angle of bedload of mean diameter to the depth-averaged flow direction

β_k	Deviation angle of bedload of size class k to the depth-averaged flow direction
ϕ	Angle of repose
κ	Karman constant
Λ	Porosity parameter related to the relative elevation of the bed surface and water surface in bed sediment, wherein $\Lambda = 1$ as $z > z_b$, and $\Lambda = \lambda$ as $z < z_b$
λ	Bed material porosity
λ_v	Vegetation density
$\lambda_{v\max}$	Maximum vegetation density
μ_s	Coefficient of static friction
ν	Coefficient of kinematic eddy viscosity
Π	Coefficient of sediment particle motion
θ	Local bed slope along the direction of bedload for mean diameter
θ_x	Bed inclination along x coordinate
Θ_x	$\Theta_y + \frac{\rho}{\rho_s - \rho}\cos^2\theta_x$
θ_y	Bed inclination along y coordinate
Θ_y	$\frac{1}{1 + \tan^2\theta_x + \tan^2\theta_y}$
ρ	Water mass density
ρ_s	Sediment mass density
σ_{xx}, σ_{yy}, τ_{xy} and τ_{yx}	Turbulence stress
τ_b	Bed shear stresses along the flow direction
τ_{bx}	Bed shear stresses along x coordinate
$\tau_{b\xi}$	Bed shear stresses along ξ coordinate
τ_{by}	Bed shear stresses along y coordinate
$\tau_{b\eta}$	Bed shear stresses along η coordinate
τ_{*m}	Non-dimensional shear stress of mean diameter
ξ	Coordinate along the longitudinal direction in the generalized curvilinear coordinate system
η	Coordinate along the transverse direction in the generalized curvilinear coordinate system

Acknowledgements

This work was funded by a Grant-in-Aid for Scientific Research (B) (Representative: Yasuyuki Shimizu) from the Japan Society for the Promotion of Science and Grant-in-Aid for Scientific Research for young researchers (B) (Representative: Hiroshi Takebayashi). I also wish to acknowledge the helpful comments of the reviewers.

References

Anderson, AG 1967. On the development of stream meanders. *Proceedings of the 12th IAHR Congress*, Vol. 1, Fort Collins, CO; 370–378.

Ashida, K and Michiue, M 1971. An investigation of river bed degradation downstream of a dam. *Proceedings of 14th IAHR Congress*, Vol. 3; 247–255.

Ashida, K and Sawai, K 1977. A study on the stream formation process on a bare slope (3) – three dimensional channel form. *Disaster Prevention Research Institute Annuals, Kyoto University* 20, B-2. (In Japanese with English abstract.)

Ashida, K and Shiomi, Y 1966. Study on the hydraulic behaviors of dunes (meander) in channels. *Disaster Prevention Research Institute Annuals, Kyoto University* 9, 457–477. (In Japanese with English abstract.)

Ashworth, PJ, Ferguson, RI, Ashmore, PE *et al.* 1992, Measurements in a braided river chute and lobe 2. Sorting of bed load during entrainment, transport, and deposition. *Water Resources Research* 28(7), 1887–1896.

Callender, RA 1969. Instability and river channels. *Journal of Fluid Mechanics* 36(3), 465–480.

Camporeale, C, Perucca, E, Ridolfi, L, and Gurnell, A. M 2013. Modeling the interactions between morphodynamics and riparian vegetation. *Reviews of Geophysics*, 51, 1–36.

Causon, DM 1989. High resolution finite volume schemes and computational aerodynamics, nonlinear hyperbolic equation theory, computation methods and applications. *Notes on Numerical Fluid Mechanics* 24, 63–74.

Colombini M, Seminara G, and Tubino, M 1987. Finite-amplitude alternate bars. *Journal of Fluid Mechanics* 181, 213–232.

Corenblit, D, Tabacchi, E, Steiger, J, and Gurnell, AM 2007. Reciprocal interactions and adjustments between fluvial landforms and vegetation dynamics in river corridors: a review of complementary approaches. *Earth-Science Reviews* 84, 56–86.

Coulthard, TJ, Hicks, DM, and Van de Wiel, MJ 2007. Cellular modelling of river catchments and reaches: advantages, limitations and prospects. *Geomorphology* 90, 192 – 207.

Crosato, A and Samir Saleh, M 2011. Numerical study on the effects of floodplain vegetation on river planform style. *Earth Surface Processes and Landforms* 36(6), 711–720.

Curran, JC and Hession, WC 2013. Vegetative impacts on hydraulics and sediment processes across the fluvial system. *Journal of Hydrology* 505, 364–376.

Egashira, S and Ashida, K 1992. Unified view of the mechanics of debris flow and bed-load. *Advances in Micromechanics of Granular Materials* (eds HW Shen, M Satake, M Mehrabadi *et al.*). Elsevier: Amsterdam; 391–400.

Egashira, S and Takebayashi, H 1999. Self-formed low water channel in straight channel. *Proceedings of IAHR Symposium on River, Coastal and Estuarine Morphodynamics*; 509–518.

Einstein, HA and Shen, HW 1964. A study on meandering in straight alluvial channels. *Journal of Geophysical Research* 69, 5239–5247.

Engelund, F 1974. Flow and bed topography in channel bends. *Journal of Hydraulic Division, American Society of Civil Engineering* 100(HY11), 1631–1648.

Engelund, F and Skovgaard, O 1973. On the origin of meandering and braiding in alluvial streams. *Journal of Fluid Mechanics* 57, 289–302.

Enggrob, HG and Tjerry, S 1999. Simulation of morphological characteristics of a braided river. *Proceedings IAHR Symposium on River, Coastal and Estuarine Morphodynamics*, 6 September; 585–594.

Fredsøe, J 1978. Meandering and braiding of rivers. *Journal of Fluid Mechanics* 84(4), 609–624.

Gurnell, AM, Bertoldi, W, and Corenblit, D 2012. Changing river channels: the roles of hydrological processes, plants and pioneer fluvial landforms in humid temperate, mixed load, gravel bed rivers. *Earth-Science Reviews* 111, 129–141.

Hansen, E 1967. *On the Formation of Meanders as a Stability Problem*. Basic Research Progress Report 13, Hydraulic Laboratory, Technical University of Denmark.

Hooke, JM, Brookes, CJ, Duane, W, and Mant, JM 2005. A simulation model of morphological, vegetation and sediment changes in ephemeral streams. *Earth Surface Processes and Landforms* 30, 845–866.

Iwagaki, Y 1956. Hydrodynamic study on critical shear stress. *Proceedings of Japan Society of Civil Engineering* 41, 1–21. (In Japanese with English abstract.)

Jang, CL and Shimizu, Y 2005a. Numerical simulation of relatively wide, shallow channels with erodible banks. *Journal of Hydraulic Engineering, American Society of Civil Engineering* 131(7), 565–575.

Jang, CL and Shimizu, Y 2005b. Numerical simulations of the behavior of alternate bars with different bank strengths. *Journal of Hydraulic Research* 43(6), 596–612.

Kovacs, A and Parker, G 1994. A new vectorial bedload formulation and its application to the time evolution of straight river channels. *Journal of Fluid Mechanics* 267, 153–183.

Kuroki, M and Kishi, T 1982. Study on regime criteria of river morphology. *Annual Journal of Hydraulic Engineering, Japan Society of Civil Engineering* 26, 51–56. (In Japanese with English abstract.)

Liu, BY 1991. *Study on sediment transport and bed evolution in compound channels*. Doctoral thesis, Kyoto University.

Murray, A.B. and Paola, C 1994. A cellular model of braided rivers. *Nature* 371, 54–57.

Murray, AB and Paola, C 2003. Modelling the effect of vegetation on channel pattern in bedload rivers. *Earth Surface Processes and Landforms* 28, 131–143.

Nezu, I and Nakagawa, H 1993. *Turbulence in Open Channel Flows*. IAHR Monograph, Balkema: Rotterdam, The Netherlands; 53–56.

Nicholas, AP 2013. Modelling the continuum of river channel patterns. *Earth Surface Processes and Landforms* 38, 1187–1196.

Parker, G 1976. On the cause and characteristic scales of meandering and braiding in rivers. *Journal of Fluid Mechanics* 76(3), 457–480.

Perona, P, Camporeale, C, Perucca, E *et al.* 2009. Modelling river and riparian vegetation interactions and related importance for sustainable ecosystem management. *Aquatic Sciences* 71, 266–278.

Perucca, E, Camporeale, C., and Ridolfi, L 2007. Significance of the riparian vegetation dynamics on meandering river morphodynamics. *Water Resources Research* 43, 1–10.

Reynolds, AJ 1963. Waves on the erodible bed of an open channel. *Journal of Fluid Mechanics* 16(4), 113–133.

Shen, HW and Komura, S 1968. Meandering tendencies in straight alluvial channels. *Journal of Hydraulic Division, Proceedings of American Society of Civil Engineering* 94(HY4), 997–1016.

Shimizu, Y and Itakura, T 1991. Calculation of flow and bed deformation with a general non-orthogonal coordinate system. *Proceedings of XXIV IAHR Congress*, Vol. C-2, Madrid; 41–48.

Takebayashi, H and Okabe, T 2009. Braided streams with vegetation presence under unsteady flow. *Water Management, Institution of Civil Engineers* 162(3), 189–198.

Takebayashi, H, Egashira, S, and Kuroki, M 1997. Prediction of geometric characteristics of alternate bars. *Proceedings of 28th Congress of International Association for Hydraulic Research* (CD-ROM).

Takebayashi, H, Egashira, S, and Okabe, T 2003. Braided streams formed on beds with non-uniform sediment. *Proceedings of 3nd IAHR Symposium on River, Coastal and Estuarine Morphodynamics*, 1 September; 787–798.

Takebayashi, H, Egashira, S, and Okabe, T 2004. Numerical analysis of braided streams formed on beds with non-uniform sediment. *Journal of Hydroscience and Hydraulic Engineering, Japan Society of Civil Engineering* 22(2), 37–46.

Tal, M and Paola, C 2007. Dynamic single-thread channels maintained by the interaction of flow and vegetation. *Geology*, 35(4), 347–350.
Tal, M and Paola, C 2010. Effects of vegetation on channel morphodynamics: results and insights from laboratory experiments. *Earth Surface Processes and Landforms* 35(9), 1014–1028.
Thorne, CR, Russell, APG, and Alam, MK 1993. Plan form pattern and channel evolution of the Brahmaputra River, Bangladesh. In *Braided Rivers*. The Geological Society: London; 257–276.
Tsujimoto, T 1999. Fluvial processes in streams with vegetation. *Journal of Hydraulic Research* 37(6), 789–803.
Watanabe, Y, Tubino, M, Zolezzi, G, and Hoshi, K 2001. Behavior of alternate bars under unsteady flow conditions. *Proceedings of 2nd IAHR Symposium on River, Coastal and Estuarine Morphodynamics*, 10 September; 575–584.
Werner, PG 1951. On the origin of river meanders. *Transactions, American Geophysical Union* 32(6), 898–902
Wu, B, Wang, G, Ma, J, and Zhang, R 2005. Case study: river training and its effects on fluvial processes in the Lower Yellow River, China. *Journal of Hydraulic Engineering, American Society of Civil Engineering* 131(2), 85–96.

Discussion

Discussion by Philippe Belleudy

In the case of flow through vegetation, how do you share the friction slope between a drag force on stems and shear stress on the bottom, to be used for the calculation of the sediment transport capacity?

Discussion by Mike Miles and Colin Thorne

Dr Takebayashi posits that 'vegetation growth in rivers is absolutely essential for formation of single thread meandering channels'. Experiments demonstrate that reinforcement of cohesionless bed materials by plant roots is necessary for meandering in laboratory channels (Braudrick *et al.* 2009; Tal and Paola 2010). However meandering can occur on unvegetated streams such as the Onyx River in the Antarctic (see http://antarctica.recollect.co.nz/). This might be analogous to the meandering Martian channels described in Chapter 28 of this volume (also see Levy 2015).

Bank strength is affected by many factors. For example, on Onyx River bank erosion resistance could be increased by ice-rich permafrost (Seybold, Balks, and Harms 2010), or carbonate-rich sediments (Hendy *et al.* 1979). Can the modelling approach you describe reproduce single-thread meandering channels if factors *other than* vegetation are represented as increasing bank strength and, if so, how would these effects be calibrated?

References

Braudrick, CA, Dietrich, WE, Leverich, GT, and Sklar. LS 2009. Experimental evidence for the conditions necessary to sustain meandering in coarse-bedded rivers. *Proceedings of the National Academy of Sciences of the United States of America* 106(40), 16936–16941.
Hendy, CH, Healy, TR, Rayner, EM *et al.* 1979. Late Pleistoceme glacial chronology of the Taylor Valley, Antarctica, and the global climate. *Quaternary Research* 11(2), 172–184.

Levy, J 2015. A hydrological continuum in permafrost environments: the morphological signatures of melt-driven hydrology on Earth and Mars. *Geomorphology* 240, 70–82.
Seybold, C, Balks, MR, and Harms, DS 2010. Characterization of active layer water contents in the McMurdo Sound region, Antarctica. *Antarctic Science* 22(6), 633–645.

Reply by Hiroshi Takebayashi

As to the question by Philippe Belleudy, the bed shear stress and the drag force on streams are calculated separately. Sediment transport rate is calculated only by use of bed shear stress.

With respect to the comment by Mike Miles and Colin Thorne, indeed the sentence must be modified to 'Sediment deposition on both the bars and the floodplain and the smaller bank erosion rate at the outer bank of the curvature are absolutely essential for formation of single thread meandering channels. Vegetation growth in rivers accelerates sedimentation on bars and floodplain during flood and suppresses erosion at the outer bank.'

I think that the present modelling approach I describe cannot reproduce single-thread meandering channels if factors other than vegetation are represented as increasing bank strength. However, if the bed deformation analysis model of a bed composed of both non-cohesive and cohesive materials is introduced, single-thread meandering channels in unvegetated areas can be reproduced. Calibration of the erosion rate of the bank material must be required. Single-thread meandering channels in unvegetated areas with cohesive material are described in the main text.

26

Modelling of Mixed-Sediment Morphodynamics in Gravel-Bed Rivers Using the Active-Layer Approach: Insights from Mathematical and Numerical Analysis

Annunziato Siviglia, Guglielmo Stecca, and Astrid Blom

26.1 Introduction

The presence of sediment of different sizes (mixed sediment) is a characteristic feature of the morphology of gravel-bed rivers and has a fundamental role in their morphodynamic evolution. In fact, as the mobility of sediment is strongly influenced by their diameter, the impact of mixed sediment on morphodynamics is non-negligible in many cases of practical relevance. Illustrative examples are: (i) the construction of a dam, which diminishes the incoming sediment supply, generally promoting bed armouring accompanied by degradation of the downstream reach (e.g. Williams and Wolman 1984; Grant 2012); (ii) the opposite case of dam removal, in which a high sediment load usually having finer mean diameter than the armoured river bed is suddenly released and sediment pulses in mountain streams (e.g. Sklar *et al.* 2009), in which coarser/finer material than that forming the river bed is supplied by landslides, and sediment nourishment, in which local injections of coarse sediment, having lower mobility, are applied to prevent further degradation of the bed profile (see Figure 26.1a); (iii) dredging, in which the armour layer is removed and finer sediment is exposed to the flow; and (iv) the construction of sediment bypass tunnels conveying sediment from upstream a reservoir to the downstream reach during flood events (e.g. Facchini *et al.* 2015; see Figure 26.1b).

In order to quantify the morphodynamic response of river reaches to changes in sediment supply, and to design practical engineering interventions, a quantitative approach based on the use of mathematical models is nowadays widespread. Despite new developments over the past two decades (see e.g. Parker, Paola, and Leclair 2000; Blom 2008; Viparelli, Blom, and Moreira, Chapter 23, this volume) the most widely used mathematical model for the prediction of mixed-sediment morphodynamics is the active-layer model of Hirano (1971, 1972). He postulated the existence of an exchange layer (termed active layer in the English literature) in the upper part of the river bed, which is vertically mixed and provides sediment to the flow, giving rise to sediment discharge, and exchanges sediment with the substrate (see e.g. Parker 1991a,b; Viparelli *et al.* 2010). Since the complexity of this model generally hinders the development of analytical solutions, in most applications the problem is solved by numerical approximation techniques.

As pointed out by Mosselman (2012), physics-based models in morphodynamics invariably constitute simplified representations of reality, which poses concern over their prediction capability and suggests the need of their careful mathematical assessment. In the case of the active-layer model, three distinct issues need to be explored. First, as the active-layer concept is highly schematized and

Gravel-Bed Rivers: Processes and Disasters, First Edition. Edited by Daizo Tsutsumi and Jonathan B. Laronne.

(a)

(b)

Figure 26.1 Examples of mixed-sediment morphodynamics. (a) Intervention of sediment nourishment in the Rhine River (Netherlands). Courtesy Rijkswaterstraat, Netherlands. (b) Outlet of a sediment bypass tunnel near Alvaschein, Canton of Grisons, Switzerland (downstream view). (Courtesy M. Facchini, ETH Zurich, Switzerland.) (See insert for color representation of this figure.)

therefore a strongly simplified representation of reality, we shall question whether this mathematical model can faithfully represent the essential physical processes. Second, as it has been reported that the model can become conceptually unsuitable for morphodynamic predictions due to a switch in its mathematical character under some parameter settings (see e.g. Ribberink 1987; Stecca, Siviglia, and Blom 2014), we shall investigate how this affects the numerical solutions, and how this issue can be solved. Finally, we shall analyse how the accuracy in the numerical approximation crucially affects the prediction capability of the mathematical model.

In this chapter we seek to explore these issues by adopting a mathematical approach, so as to unravel the practical outcomes of the inner mathematical structure of the problem. To this end, after recalling the mathematical model (section 26.2) we will first review the present knowledge of its mathematical behaviour (section 26.3). With this insight, we will then assess a numerical solution (section 26.4) and identify the strengths and weaknesses of the underlying mathematical model in reproducing the physics. Finally, the concluding section (section 26.5) presents some research perspectives to be developed in future work.

26.2 The Saint-Venant–Hirano Model

The model we consider for the free surface flow is one-dimensional, hydrostatic, unsteady, and it is described by the Saint-Venant equations. Friction exerted by flow over a cohesionless bottom composed of mixed sediment induces sediment transport, which is assumed to occur only as bedload. The active-layer approach of Hirano (1971, 1972) is employed to account for changes in the grain-size distribution of the bed surface and for the development of size stratification.

26.2.1 Governing Equations

The Saint-Venant–Hirano model includes equations for the conservation of water mass (continuity equation)

$$\frac{\partial h}{\partial t} + \frac{\partial q}{\partial x} = 0, \tag{26.1}$$

for the momentum principle applied to the water phase

$$\frac{\partial q}{\partial t} + \frac{\partial}{\partial x}\left(\frac{q^2}{h} + \frac{1}{2}gh^2\right) + gh\frac{\partial \eta}{\partial x} = -ghS_f, \tag{26.2}$$

and for the conservation of the total sediment mass (the Exner equation):

$$\left(1-\lambda_p\right)\frac{\partial \eta}{\partial t} + \frac{\partial q_b}{\partial x} = 0. \tag{26.3}$$

Equations (26.1)–(26.3) form the Saint-Venant–Exner model, which governs one-dimensional morphodynamics when a single sediment fraction is considered (unisize-sediment case). Here, $t\,[T]$ is time, $x\,[L]$ is the streamwise coordinate, $h\,[L]$ is the flow depth, $q\,[L^2T^{-1}]$ is the flow discharge per unit width, $\eta\,[L]$ is the mean bed elevation at x and t, $g = 9.8\,\mathrm{ms}^{-2}$ is the acceleration due to gravity, $q_b\,[L^2T^{-1}]$ is the total sediment discharge, $\lambda_p\,[-]$ is the bed porosity, and $S_f\,[-]$ is the friction slope, to be specified by an appropriate closure relation. The depth-averaged flow velocity is defined as $u = q/h\,[LT^{-1}]$.

The active-layer approach provides additional continuity equations for mixed sediment. The sediment mixture is discretized into N fractions, each one of which is characterized by a representative grain diameter d_k and by sediment discharge q_{bk}, where $1 \le k \le N$ is an integer index. The total sediment discharge is then given by

$$q_b = \sum_{k=1}^{N} q_{bk} . \tag{26.4}$$

A vertical discretization of the bed into two layers is considered: the active layer and the substrate (see Figure 26.2).

The active layer is a sediment volume located just underneath the bed surface and represents the sediment that interacts with the flow, giving rise to sediment discharge. It is adjoined to the bed surface, and thus its vertical displacement varies with the mean bed elevation. It is vertically mixed and thus admits a unique vertically averaged grain-size distribution $F_{ak}(x,t)\,[-]$ along the vertical. As commonly assumed for the purpose of mathematical analysis of the model (e.g. Ribberink 1987; Stecca, Siviglia, and Blom 2014), we set the active layer thickness L_a constant in time.

The interface between the active layer and the substrate is located at elevation $z = \eta - L_a$. The substrate is not vertically mixed and has grain-size distribution $f_{sk}(x, z, t)$. As it is not directly affected by bedload discharge, it only has vertical mass exchange with the active layer due to net increase or decrease in the elevation of the interface associated to variation in bed elevation. This mass flux is compensated by consumption or creation of new substrate on top of the layer.

Assuming the porosity λ_p to be constant, these settings yield the following conservation equations for the mass of the kth sediment fraction in the active layer,

$$\left(1-\lambda_p\right)\left(\frac{\partial}{\partial t}(F_{ak}L_a) + f_k^I \frac{\partial}{\partial t}(\eta - L_a)\right) + \frac{\partial q_{bk}}{\partial x} = 0 \quad \text{for } 1 \le k \le N-1 . \tag{26.5}$$

Analogous equations can be written for the substrate,

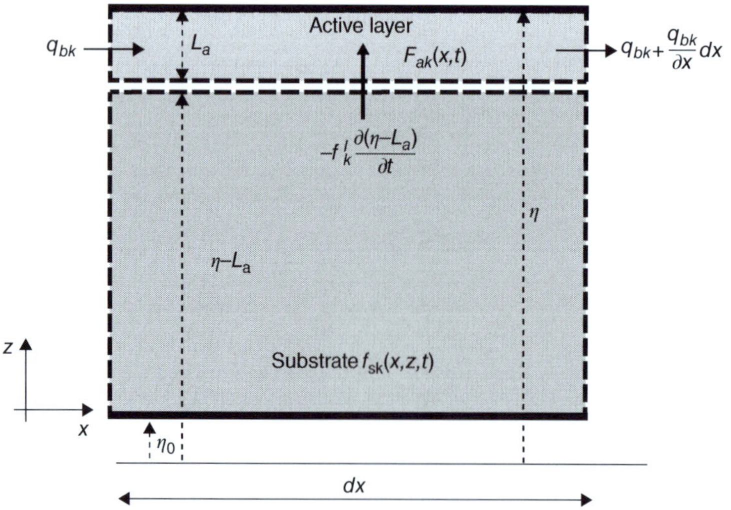

Figure 26.2 Schematic view and notation of the vertical discretization of the river bed and related sediment fluxes in the active-layer model. (*Source*: Stecca, Siviglia, and Blom (2014). Reproduced with permission of Elsevier.)

$$\frac{\partial}{\partial t}\int_{\eta_0}^{\eta-L_a} f_{sk}\mathrm{d}z - f_k^I\frac{\partial}{\partial t}(\eta-L_a)=0 \quad \text{for } 1\le k\le N-1. \tag{26.6}$$

Here, η_0 is a constant elevation datum and f_k^I is the grain-size distribution at the interface between the two layers (see Figure 26.2), given by (Hirano 1971, 1972)

$$f_k^I=\begin{cases}F_{ak} & \text{if } \partial(\eta-L_a)/\partial t>0\\ f_{sk}(z=\eta-L_a) & \text{if } \partial(\eta-L_a)/\partial t<0\end{cases}, \tag{26.7}$$

in agreement with the above-described vertical mass-exchange mechanism. Alternative relations to (26.7), whereby f_k^I in aggradational conditions is computed as a weighted average of the grain-size distribution of bedload and of that in the active layer (Parker 1991a; Hoey and Ferguson 1994; Toro-Escobar, Parker, and Paola 1996), are not considered in this analysis. By considering the constraint that the volumetric fraction content of all fractions must sum up to 1, only $N-1$ independent active-layer equations (26.5) and substrate equations (26.6) are required for describing the evolution of the grain-size distribution. Therefore, when the problem is written for N size fractions, the system includes $2N+1$ partial differential equations (PDEs), namely the three equations of the Saint-Venant–Exner model (26.1)–(26.3), $N-1$ active-layer equations (26.5), and $N-1$ substrate equations (26.6).

26.2.2 Closure Relations for Sediment Transport

The sediment discharge of each fraction q_{bk} results from the transport capacity Q_{bk} for the case that the bed would consist only of that fraction (yet, possibly including a hiding–exposure correction), weighted by the availability of that fraction in the active layer, given by the volumetric content F_{ak}, i.e.,

$$q_{bk}=F_{ak}Q_{bk}. \tag{26.8}$$

Q_{bk} is computed by suitable bedload transport relations (e.g. Meyer-Peter and Müller 1948), based on the quantification of the local shear stress exerted by the fluid on the bottom. This quantification essentially requires the knowledge of the friction coefficient, flow depth and velocity, sediment diameter d_k, and relative density of sediment.

The hiding factor reduces the mobility of fine grains in the mixture compared to the unisize-sediment case and increases that of coarse grains, thus promoting conditions closer to equal mobility with respect to the case without hiding correction.

Q_{bk} can be modified if a hiding–exposure correction (e.g. Egiazaroff 1965) is taken into account. The hiding factor reduces the mobility of fine grains in the mixture compared to the unisize-sediment case and increases that of coarse grains, thus promoting conditions closer to equal mobility with respect to the case without hiding correction. Therefore, the different sediment classes are characterized by different transport capacities. As grain-size selectivity of transport is the standard condition in rivers from low to relatively high transport conditions, it shall be retained in the analysis to properly appreciate the features of the mathematical model.

Noteworthy, the closure relation (26.8) is also supported by the results obtained in flume experiments by Houssais and Lajeunesse (2012). By investigating the bedload transport above a bed composed of a bimodal mixture, they observed that the average velocity and the surface density of moving grains obey the same equations as those reported for a bed of unisize sediment. The only evident difference concerns the critical Shields number for incipient sediment motion. Above a unisize-sediment bed, the latter depends on the particle Reynolds number through the Shields curve, whereas in the case of a bimodal mixture it varies with the fraction of the bed surface covered by fine grains. Thus, these observations also support the use of a hiding–exposure correction.

26.3 Mathematical Analysis

Let us consider open-channel flow over a mixed-sediment bed, in which the flow is sufficiently intense to transport sediment as bedload. If the channel slope and width are constant, and the grain-size distribution of the bed material does not vary spatially, the system allows for an equilibrium state consisting of uniform flow over a flat bed in which the sediment discharge is spatially constant. This condition results in no aggradation nor degradation in the channel.

Let us now impose a perturbation to this equilibrium, by locally modifying the bed topography or varying its grain-size distribution. A number of questions then arise. Will these initial perturbations propagate in the downstream or upstream direction, or both? What will be their propagation speed? What will be the influence of mixed sediment on the propagation of morphological perturbations, and how does such propagation change with respect to a case with unisize sediment?

The answer to these questions can be given through a mathematical analysis of the system of PDEs, and studying the characteristic speeds and the wave dynamics described by an appropriate linearization of the governing equations.

26.3.1 The System in Matrix–Vector Form for a Two-Fraction Mixture

To carry out the analysis we first conveniently recast the system (26.1)–(26.3) and (26.5) in a suitable matrix–vector formulation. Noticeably in the analysis the substrate equations (26.6) are discarded from the system, since they do not describe any propagating morphodynamic signal (Stecca, Siviglia, and Blom 2014). For the sake of simplicity, here we introduce and discuss the system written for a two-fraction mixture governed by five PDEs. The formulation for an indefinite number of fractions can be found in Stecca, Siviglia, and Blom (2014).

We assume inviscid flow, by which friction in the momentum equation (26.2) is neglected, thus restricting the outcomes to relatively short time and spatial scales.

With this assumption, the sought matrix–vector form is

$$\frac{\partial \mathbf{Q}}{\partial t} + \mathbf{A}(\mathbf{Q})\frac{\partial \mathbf{Q}}{\partial x} = \mathbf{0} \tag{26.9}$$

where $\mathbf{Q}$ is the vector of unknowns, given by

$$\mathbf{Q} = [h, q, \eta, L_a F_{a1}]^T \tag{26.10}$$

and $\mathbf{A} = \mathbf{A}(\mathbf{Q})$ is the coefficient matrix. The entries of the matrix can be conveniently expressed in terms of two non-dimensional numbers, which will then appear as control parameters in approximations to characteristic celerities. They are the well-known Froude number $F_r = q/\sqrt{gh^3}$ and the parameter ψ, defined as

$$\psi = \frac{1}{1-\lambda_p}\frac{\partial q_b}{\partial q}. \tag{26.11}$$

The latter is a measure of the intensity of total bedload in the flow (see e.g. de Vries 1965; Lyn and Altinakar 2002; Stecca, Siviglia, and Blom 2014), ranging between 0 (no sediment transport, fixed bed) and $\mathcal{O}(10^{-2})$ (high transport).

With these definitions, the coefficient matrix for the system considering a two-fraction mixture can be written as:

$$\mathbf{A} = \begin{bmatrix} 0 & 1 & 0 & 0 \\ u^2\left(\dfrac{1}{F_r^2}-1\right) & 2u & \dfrac{u^2}{F_r^2} & 0 \\ -u\psi & \psi & 0 & \dfrac{Q_{b1}-Q_{b2}}{(1-\lambda_p)L_a} \\ -u\gamma_1\psi & \gamma_1\psi & 0 & \dfrac{(1-f_1^I)Q_{b1}+f_1^I Q_{b2}}{(1-\lambda_p)L_a} \end{bmatrix}, \tag{26.12}$$

with

$$\gamma_1 = c_1 - f_1^I, \quad c_1 = \frac{1}{\psi}\frac{\partial q_{b1}}{\partial q}. \tag{26.13}$$

For our analytical studies, the system (26.9) is linearized, by assuming the entries of matrix **A** (26.12) as constant and related to the characteristics of undisturbed flow and sediment transport in the channel.

26.3.2 Analysis of Characteristics

The analysis of characteristics essentially consists of studying the propagation of hydromorphodynamic waves as described by the system of governing PDEs, with appropriate algebraic manipulations. Waves here represent carriers of information on hydromorphodynamic changes, which propagate across the domain at certain speeds, termed characteristic celerities.

We focus on the case of perturbations having infinitesimal amplitude. For this problem, a simple linear solution applies, and thus celerities are independent of the perturbation type and extent, and only to depend on the undisturbed flow and sediment transport. With reference to the linearized problem obtained by system (26.9), we analyse the eigenvalues of the system matrix (26.12), which will yield the mathematical relations for the celerities, for given values of the parameters F_r and ψ.

26.3.2.1 Unisize-Sediment Case

For the unisize-sediment case ($N = 1$ fraction), the analysis of Lyn (1987) and Lyn and Altinakar (2002), who generalized the previous study of de Vries (1965), gives approximations to three characteristic speeds. These are given in dimensionless form as

$$\lambda_{1,2}^* = 1 \pm \frac{1}{F_r}, \quad \lambda_3^* = \frac{\psi}{1-F_r^2}, \tag{26.14}$$

where $\lambda_i^* = \lambda_i/u$ ($i = 1, 2, 3$) are the nondimensional wave propagation celerities. The approximation (26.14) holds under well-developed subcritical or supercritical conditions, i.e if the Froude number F_r is sufficiently far from unity (say, $F_r < 0.8$ or $F_r > 1.2$).

Physically, the speed λ_3^* (26.14) yields the celerity of perturbations in bed elevation under unisize sediment conditions (de Vries 1965) and is termed the 'bed' celerity. It is positive under subcritical flow, denoting a wave which travels in the downstream direction, and negative under supercritical flow, indicating an upstream-travelling wave. The speeds $\lambda^*_{1,2}$ (26.14), which are equal to those of the fixed-bed Saint-Venant model, identify the celerity of waves mainly carrying perturbations of the flow variables at much faster pace (e.g. Lyn and Altinakar 2002).

Close to critical conditions (roughly for $0.8 < F_r < 1.2$), this decoupled interpretation of the role of waves does not apply, and an alternative approximation yields the speed of coupled hydromorphodynamic perturbations, for which we refer to Lyn (1987) and Lyn and Altinakar (2002).

26.3.2.2 Mixed-Sediment Case

In the case of mixed sediment, we restrict our attention to the case of high-transport, i.e., we assume that all fractions are transported by the flow giving rise to relatively high total sediment discharge. For this case analytical solutions to celerities are available. Following Stecca, Siviglia, and Blom (2014), when the model is set for N size fractions, the analysis reveals $N-1$ additional waves besides the three of the unisize-sediment model (26.14). Under the assumption of moderate grain-size selectivity, i.e. that the transport capacities of each fraction Q_{bk} have distinct (albeit not too different) magnitude, as appropriate under high-transport conditions, approximations to these new characteristic celerities are given by

$$\lambda^*_{3+k} = \frac{Q_{bk}\left(1 - f^I_k\right) + f^I_k Q_{bN}}{uL_a} \quad \text{for } 1 \le k \le N-1. \tag{26.15}$$

Equation (26.15) gives the speed of $N-1$ distinct 'sorting' waves, which describe the propagation of disturbances related to spatial gradients in the volumetric fraction content in the active layer. As the celerities are always positive regardless of the Froude regime, these waves invariably travel in the downstream direction, at slightly different pace, and progressively separate.

The 'sorting' waves have a prominent role over all the other waves in advecting changes in the grain-size distribution of the active layer (i.e. a streamwise sorting signal), which justifies their name. However, as we will show in section 26.3.4, these waves have a broader effect, as they also convey changes in bed elevation and hydraulic variables that are determined by the changes in grain-size distribution. For this reason, in their name we will always retain quotation marks in the name 'sorting' wave.

Solution (26.15) generalizes the earlier approximation of Ribberink (1987). The latter was obtained under the assumption of steady flow, using a simplified sorting model, where the individual active-layer equations are replaced by one equation for the evolution of the mean sediment diameter, and fully neglecting grain-size selectivity. Ribberink (1987) identified the two wave propagation celerities

$$\lambda^*_3 = \frac{\psi}{1 - F_r^2}, \quad \lambda^*_4 = \frac{q_b}{uL_a}. \tag{26.16}$$

Here, λ^*_3 is identical to the bed celerity in the unisize-sediment case (26.14) and has the same physical meaning, whereas λ^*_4 is a unique 'sorting' wave celerity resembling all the $N-1$ celerities in (26.15). Noteworthy, all the distinct 'sorting' wave celerities (26.15) reduce to λ^*_4 (26.16) when sediment transport is not grain-size selective.

In Figure 26.3a we assess these approximations against a reference solution obtained numerically, by plotting the dimensionless celerities as a function of the Froude parameter. For simplicity, only the subcritical Froude regime is considered. A five-fraction mixture is used, with a sharp unimodal Gaussian-shaped grain-size distribution, and high sediment transport is assumed ($\psi = 0.0075$). With reference to the case of bedform-dominated bed conditions of gently sloping, lowland rivers, the

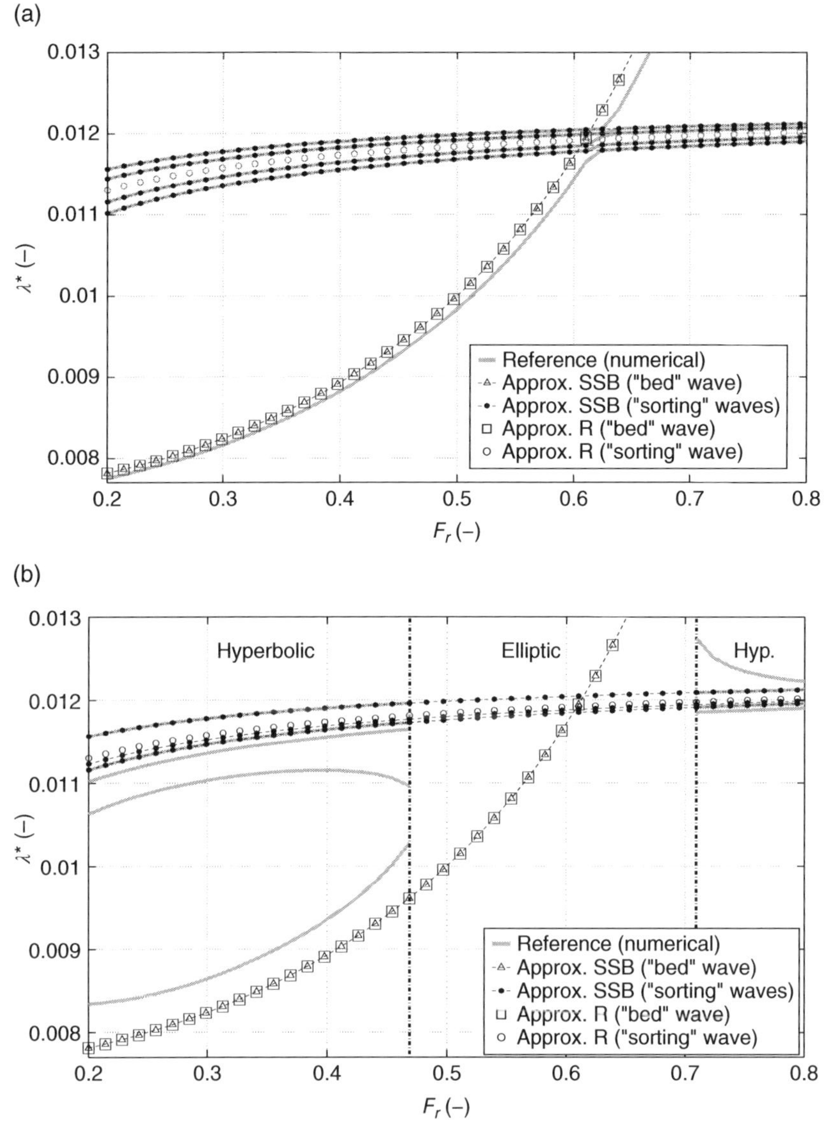

Figure 26.3 Plot of the characteristic celerities, for $N = 5$ fractions, under bedform-dominated and high-transport conditions, as a function of the Froude parameter. A sharp Gaussian grain-size distribution in the active layer is used: (a) aggradational conditions and (b) degradational conditions. Here, a sharp Gaussian distribution in the substrate is used, with moderately finer mean diameter than that in the active layer. The estimates of Stecca, Siviglia, and Blom (2014) and Ribberink (1987) are denoted with SSB and R, respectively. The view is focused on the waves having significant impact on morphodynamic changes. (Replotted from Stecca, Siviglia, and Blom (2014).)

active-layer thickness is set as a fraction of bedform height (Ribberink 1987), which scales with the flow depth, yielding $L_a = \mathcal{O}(10^{-1})h$.

The view in Figure 26.3a is restricted to the waves having significant impact in advecting morphodynamic changes (the four 'sorting' waves and the 'bed' wave), and leaves out the faster hydrodynamic waves $\lambda^*_{1,2}$ (26.14). The numerical solution (grey solid lines) is well matched by the approximation for the 'bed' celerity λ^*_3 (26.14) (solid circles) and the four 'sorting' celerities (26.15) (triangles). It is here confirmed that $N-1$ distinct 'sorting' speeds are described by the model set for N sediment fractions. The solution of Ribberink (1987), correctly describes the 'bed' celerity (squares), but is incapable of detecting the four distinct 'sorting' celerities. Yet, it provides a single speed estimate (open circles), which still well captures their average magnitude.

We finally concentrate on the relative speed of the 'sorting' and 'bed' waves. In general, their ratio λ^*_4/λ^*_3 in the approximation (26.16) is in the range $0.3 \div 4$ (Ribberink 1987) or $2 \div 8$ (Sloff and Mosselman 2012). Hence, the 'sorting' celerities introduced by the active-layer equations for mixed sediment are at least as fast as the 'bed' wave of the unisize-sediment model, and possibly several times faster. This is confirmed in Figure 26.3a, at least for Froude numbers lower than ~ 0.6.

In this analysis, the setup of the active layer is crucial. In fact, as 'sorting' celerities are proportional to the inverse of the active-layer thickness in (26.15) and (26.16), they can be in principle arbitrarily increased by setting thinner active layers. Therefore, at least a lower boundary to L_a shall apply, so as not to have sorting processes propagate at faster pace than hydrodynamic perturbations (Sieben 1997). Generally, in applications to plane beds, in which the active-layer thickness is usually set as a multiple of a characteristic diameter value, e.g. $L_a = 1.5 \div 3d_{90}$ (e.g. Parker 2004), the active layer is thinner, e.g. $L_a = \mathcal{O}(10^{-2})h$, than in bedform-dominated cases. Therefore higher 'sorting' speeds than those in Figure 26.3a are expected in plane-bed cases (Stecca, Siviglia, and Blom 2014). This marked dependence, which may be a symptom of an excessively rough and artificial schematization by the active-layer approach, theoretically explains the crucial role of L_a in controlling the celerity of streamwise sorting, as commonly observed in numerical applications.

In practical computations, the active-layer thickness is set as a calibration parameter, since the above scale relations for bedform-dominated and plane-bed cases do not provide precise values. Noteworthy, a direct field or laboratory measurement of L_a as the thickness of the sediment volume that interacts with the flow is impossible in the deterministic framework of the active-layer model. In fact, as the entrainment of sediment into the flow is better represented as a probabilistic function of elevation within the bed, a rigorous description of the flow–sediment interaction and quantification of its vertical extent would imply the adoption of a probabilistic framework associated with a vertical-continuous description (e.g. Parker, Paola, and Leclair 2000; Blom and Parker 2004). The discreteness of Hirano's active layer does not allow for inclusion of all physical mechanisms playing a role in sediment entrainment and deposition. For instance, it cannot account for the facts that (i) the bed surface is rather represented by a probability density function of bed elevation and (ii) the probability of entrainment increases with elevation in the bed and depends on grain-size. In light of these considerations, direct measurement as well as an optimal calibration of the active-layer thickness are therefore impossible.

26.3.3 Loss of Hyperbolic Character

Systems of PDEs can be be classified based on the behaviour of characteristic celerities. In particular, models describing wave propagation at real finite celerities are termed hyperbolic, whereas models having complex celerities are termed elliptic. The Saint-Venant–Hirano model is generally hyperbolic, as seen in the previous section. However, under some conditions, it may switch to elliptic behaviour (Ribberink 1987). In the case of the Saint-Venant–Hirano model, as well as for other models in classic mechanics that aim at describing the time evolution of physical systems, ellipticity represents a major conceptual drawback, a mathematical issue, a practical problem in numerical computations, and the symptom of oversimplification of some physical processes.

Conceptually, an elliptic model cannot represent a time-advancing problem, and is suitable only to model steady (equilibrium) problems in multidimensions. This can be explained by the way the boundary conditions are posed. While hyperbolic models require boundary conditions in space and only initial conditions in time, forming well-posed initial-value problems, elliptic problems require conditions over the full boundary in all independent variables. This, in the case of elliptic initial-value problems, leads to the meaningless requirement of a time boundary condition at final time, which makes the model unsuitable for morphodynamic predictions.

Mathematically, elliptic initial-value problems are unstable, since the amplitude of small oscillations grows at an unbounded rate as their wavelength tends to zero (e.g. Joseph and Saut 1990). In practice, their numerical solution by time-advancing algorithms may display nonphysical oscillations, which may amplify until simulation failure, as we will show in section 26.4.3.

Physically, ellipticity may indicate that some important regularising physics is neglected. It is not unusual for many different models in fluid mechanics (see e.g. Joseph and Saut 1990) to switch to an elliptic character when stretched beyond their range of validity. In section 26.5 we will discuss the multiple approaches that have been proposed to recover hyperbolicity, and sketch some perspectives.

26.3.3.1 Analytical Determination of the Hyperbolicity Domain

A first analysis of the hyperbolicity domain was put forward by Ribberink (1987), using the two-equation reduced model described in section 26.3.2. He found that the model is always hyperbolic under aggradational conditions, but hyperbolicity can be lost when the active layer degrades into a finer substrate.

Within degradational cases with a finer substrate, Sieben (1997) precisely identified one subcase characterized by unconditional ellipticity, which he termed 'crossing of eigenvalues'. This happens if the 'bed' celerity λ_3^* (26.16) is equal to (i.e., in a graphical representation, crosses) the 'sorting' celerity λ_4^* (26.16).

The analysis of Stecca, Siviglia, and Blom (2014) releases some of the simplifying assumptions in Ribberink (1987). Restricting their attention to the two-fraction setup, for which analytical manipulations are still feasible, they considered the fully-unsteady model, thus including any Froude regime, and accounted for grain-size selective conditions modelled by sediment transport relations of practical use. The latter feature allows not only consideration of high-transport cases as in Ribberink (1987), but the entire range of transport conditions including partial transport, for which the coarsest fraction is immobile.

If both fractions are mobile, they found the model to be always hyperbolic under aggradational conditions, and, under degradational conditions to be hyperbolic if the active layer degrades into a substrate that is coarser than a threshold grain-size distribution related to the bedload (see Stecca, Siviglia, and Blom 2014). Furthermore, they proved unconditional hyperbolicity under partial transport (i.e. with one immobile fraction).

Conversely, they found that hyperbolicity cannot be proven under conditions of degradation into finer substrate when both fractions are mobile. The system in this case may become elliptic, which confirms the findings of Ribberink (1987). Noteworthy, within this case, they found at least one subcase for which elliptic behaviour is straightforwardly demonstrated, which in practice resembles the 'crossing of eigenvalue' case of Sieben (1997).

26.3.3.2 Numerical Study of Hyperbolicity Under Degradation

To empirically extend the above results to multiple fractions, we refer to the the numerical hyperbolicity study in Stecca, Siviglia, and Blom (2014) for a five-fraction mixture. Results are reported in

Figure 26.3b by showing the non-dimensional wave celerities as a function of the Froude number. The setup is identical to that adopted in Figure 26.3a, except for bed degradation into a finer substrate in the present case. The grain-size distribution in the substrate resembles the Gaussian distribution in the active layer (see section 26.3.2), but with mean diameter shifted to a value 12.5 % smaller.

The numerical solution is not plotted when elliptic behaviour is detected in the reference numerical solution (solid grey lines). This happens around the crossing zone of the 'bed' celerity of the Saint-Venant–Exner model and the four 'sorting' celerities at $\mathrm{Fr} \sim 0.6$, which gives a visual explanation of the concept of 'crossing of eigenvalues'. Noteworthy, analytical solutions do not automatically detect this behaviour and mistakenly extend the hyperbolic region to the entire Froude domain. We finally recall that under aggradational conditions the system was found hyperbolic throughout the Froude domain (Figure 26.3a), which gives further empirical confirmation of the above findings obtained for the two-fraction case.

26.3.4 Linearized Wave Dynamics

We now investigate the role of the 'sorting' waves (26.15), introduced in addition to the three waves of the unisize-sediment model (26.14), in advecting morphodynamic changes. To this end, we take advantage of the linearized analytical solution developed by Stecca, Siviglia, and Blom (2014), who generalized a similar analysis conducted by Lyn and Altinakar (2002) for unisize sediment.

26.3.4.1 Mixed-Sediment Case

We consider an initial condition describing an equilibrium situation given by uniform flow over a movable bed throughout the domain, with a reference depth $h_R = 1\,\mathrm{m}$ and reference Froude number $F_{rR} = 0.65$. The flow is inviscid and uniform flow takes place over a horizontal bed profile ($\eta = \eta_R$). Sediment transport is intense ($\psi_R = 0.01$) and the bed is plane.

The grain-size distribution is discretized into $N = 5$ fractions, with diameters set on a linear scale for simplicity. All over the domain, the initial grain-size distribution in the active layer is equal for all fractions ($F_{ak} = F_{aR} = 1/N = 0.2$). A Gaussian-shaped perturbation in the grain-size distribution of the active layer, given by a small localized increase in the content of the second finest fraction up to $F_{a2} = 0.201$, and a corresponding decrease in the content of the other fractions is initially set at the domain centre ($x = 0$). This perturbation produces an imbalance in the bedload transport of each fraction, and triggers hydromorphodynamic changes that propagate along the characteristic directions at speeds (celerities) given by relations (26.14)–(26.15). Having adopted a plane-bed set-up of the active layer, under the assumed subcritical flow, 'sorting' waves are significantly faster than the 'bed' wave.

Figure 26.4a shows for the variation in flow depth ($h - h_R$), bed elevation ($\eta - \eta_R$), and fraction content in the active layer ($F_{ak} - F_{aR}$, for fractions from 1 to 4), with respect to the reference values, as functions of the dimensionless abscissa $x^* = x/(u_R t)$, u_R being the reference flow velocity. The view is restricted to the waves carrying significant morphodynamic changes, namely the 'bed' wave and the $N - 1 = 4$ 'sorting' waves, which are all downstream-travelling under the assumed subcritical conditions. The fastest-propagating waves, which travel at speed $\lambda^* = 1 \pm 1/F_{rR}$ in (26.14) in the downstream and upstream direction, carry only modest morphodynamic changes and are not shown.

The 'bed' wave is revealed by significant degradation of the bed profile at $x^* \approx 0.016$, which is not associated to any significant change of the grain-size distribution. Conversely, the four faster 'sorting' waves are clearly indicated by positive and negative peaks of the variation of the fraction content $F_{ak} - F_{aR}$ at $x^* \sim 0.04$, which confirms their role in advecting a streamwise sorting signal. In detail, each 'sorting' wave carries perturbations in the fraction content of all fractions. Besides these perturbations in the grain-size distribution, each 'sorting' wave also carries a local perturbation of the bed-elevation

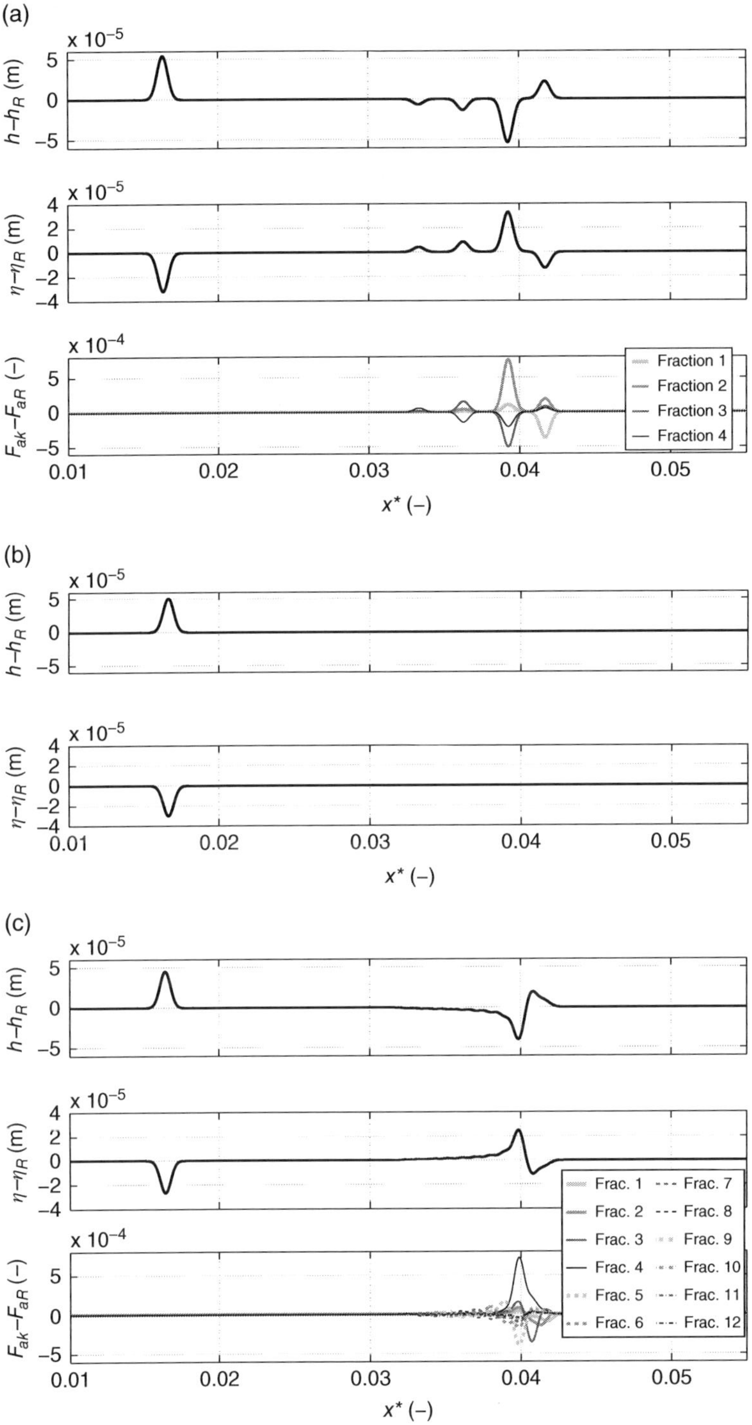

Figure 26.4 Wave propagation under subcritical flow and high transport conditions. The view focuses on the waves having significant impact on the propagation of morphodynamic changes. (a, c) Waves are triggered by a localized perturbation of the grain-size distribution in the active layer: (a) $N = 5$ fractions and (c) $N = 13$ fractions set-ups. (b) Wave propagation in the unisize-sediment case, induced by a localized initial scour. (*Source*: Stecca, Siviglia, and Blom (2014). Reproduced with permission of Elsevier.)

profile, which have been provoked by the imbalances in total bedload associated to the perturbations in volumetric fraction content in the active layer. 'Sorting' waves are thus shown here to be very effective in propagating changes in bed elevation, at faster pace than the 'bed' wave.

26.3.4.2 Comparison with the Unisize-Sediment Case

Sloff and Mosselman (2012), when summarizing two decades of numerical morphodynamic modelling of the Pannerden bifurcation in the Dutch Rhine, report that the propagation pace of morphological changes was predicted significantly better by a mixed-sediment model set-up, with respect to that observed with a unisize-sediment model. Therefore the two models predicted a completely different morphological response to floods.

Their observations can be explained theoretically comparing the wave dynamics of the two model set-ups. To this end, in Figure 26.4b we show some additional computations for a unisize-sediment case. To achieve the best comparison with the mixed-sediment computation in Figure 26.4b, here we set the sediment diameter and friction coefficient such that the same transport rate ($\psi_R = 0.01$) is achieved. Since, with unisize sediment, morphodynamic changes cannot be induced by perturbing the volume fraction content of the active layer, here we impose an initial localized scour of infinitesimal amplitude.

As 'sorting' waves are not present, the perturbation of the bed profile travels only along the slower 'bed' wave ($x^* \approx 0.016$). Hence, morphodynamic adjustment is predicted in this unisize-sediment case to happen at much slower pace than in the mixed-sediment cases of Figure 26.4a. These considerations visually explain the faster bed adaptation observed in their numerical computations for the mixed-sediment case.

26.3.4.3 Influence of the Discretization of the Granulometric Curve

Moving back to the mixed-sediment case, we finally aim at showing the influence of the discretization of the granulometric curve on the bed-elevation profiles, using the linearized solution. We carry out an additional computation, using similar settings as in the case of Figure 26.4a, but discretizing the granulometric curve into $N = 13$ fractions on a linear diameter scale.

The initial grain-size distribution in the active layer is equal for all fractions ($F_{ak} = F_{aR} = 1/N = 0.077$). The perturbation given by a localized increase by 0.001 is applied to the content of the fourth fraction, which has the same diameter as the second fraction of the five-fraction set-up in Figure 26.4a. The match between the two cases in terms of total transport ($\psi_R = 0.01$) is achieved by proper calibration of the friction coefficient.

Results for this 13-fraction case are presented in Figure 26.4c. It is shown here that the individual perturbations associated to sorting waves partially overlap, and a continuous pattern of sorting waves and associated bed changes is found for $0.032 < x^* < 0.043$. This pattern significantly differs from that of the five-fraction case in Figure 26.4a, which is characterized by individual bed humps and scours.

This exercise thus suggests that the discretization of the granulometric curve is crucial for determining the associated bed profiles, and is likely to play a role in more realistic computations. It would therefore be useful to derive general rules for such discretization. Unfortunately, these cannot be based on simple theoretical analyses as the present one, which can serve only to illustrate this issue. In fact, achieving (or not) wave separation not only depends on the number of fractions, but also on the longitudinal extent of perturbations. As this parameter is arbitrarily set in linearized solutions, it is always possible to recover wave separation by decreasing the standard deviation of Gaussian perturbations. Guidelines for optimal discretization of the granulometric curve shall instead be based on non-linear numerical modelling in more realistic cases, as we discuss in section 26.5.

26.4 Assessment of Numerical Solutions

The use of numerical solutions adds a full new set of uncertainties on the model prediction capability, which is related to the quality of the numerical approximation, to the uncertainties of a more fundamental nature which concern the possible oversimplification of the physics and the resulting mathematical shortcomings discussed in the previous section. Careful assessment of numerical solutions is therefore essential. In this section we present three numerical test cases, to show the importance of numerical accuracy and the practical consequences of the loss of hyperbolic character.

Concerning the assessment of numerical accuracy, we follow a two-step procedure. First, the benchmark given by analytical solutions in simplified and idealized cases is exploited, to check the model capability to reproduce some basic features. To this end, the analytical developments presented in section 26.3 represent a valuable asset. Here, we will assess the convergence of the numerical solution to the linearized wave dynamics described by the system of governing equations.

Afterwards, the model is applied to a real-world case. This step allows not only for further assessment of the numerics, but also provides insight on the capability of the mathematical model of describing the physics when numerical uncertainties are reduced. To this end, laboratory data collected in well-controlled conditions shall be used preferably, to reduce the uncertainties in the calibration of empirical submodels for friction and sediment transport. Here we will present the numerical simulation of laboratory experiment E8–E9 by Ribberink (1987), involving streamwise and vertical sorting (also see Blom 2008). Finally in section 26.4.3 we perform a simulation in the elliptic range showing the occurrence of related instability.

26.4.1 Numerical Modelling of Small-Amplitude Wave Dynamics

Here we consider the idealised case presented in section 26.3.4, for which the analytical linearized solution is known. We will use this linearized solution as a benchmark for the numerical solution, focusing on the impact of numerical accuracy on the model capability to capture the propagation of infinitesimal morphodynamic perturbations, in terms of wave speed and amplitude of travelling perturbations.

A remarkable advantage of this exercise for model assessment is its independence of calibration issues. In fact, although a different set-up of the sediment transport relation and active-layer thickness would alter the analytical profiles, still the numerical solution obtained with the same settings shall converge to these, as the mesh is progressively refined. Furthermore, despite the idealized case, this test still has physical significance, as 'sorting' wave propagation appears to have an impact on morphodynamic changes in real-world numerical computations (Sloff and Mosselman 2012) and thus capturing their theoretically predicted speed is crucial for a numerical model.

The settings of this test very closely resemble those of Figure 26.4a. The bed material is discretized with five sediment fractions and wave propagation is triggered by a small initial imbalance in the grain-size distribution of the active layer at the centre of the domain. The advection of hydro-morphodynamic waves creates travelling perturbation patterns essentially determined by streamwise sorting.

To assess the impact of the numerical discretization, we compare the outcomes of two numerical schemes. The DOT upwind scheme of Dumbser and Toro (2011) provides accurate solutions, by explicitly accounting for all waves in computations, whereas a simpler and less computationally expensive solution is achieved by the centred PRICE-C method of Canestrelli *et al.* (2009). Both methods are used here in second-order extension, which improves accuracy by returning sharper numerical profiles than those of basic first-order schemes.

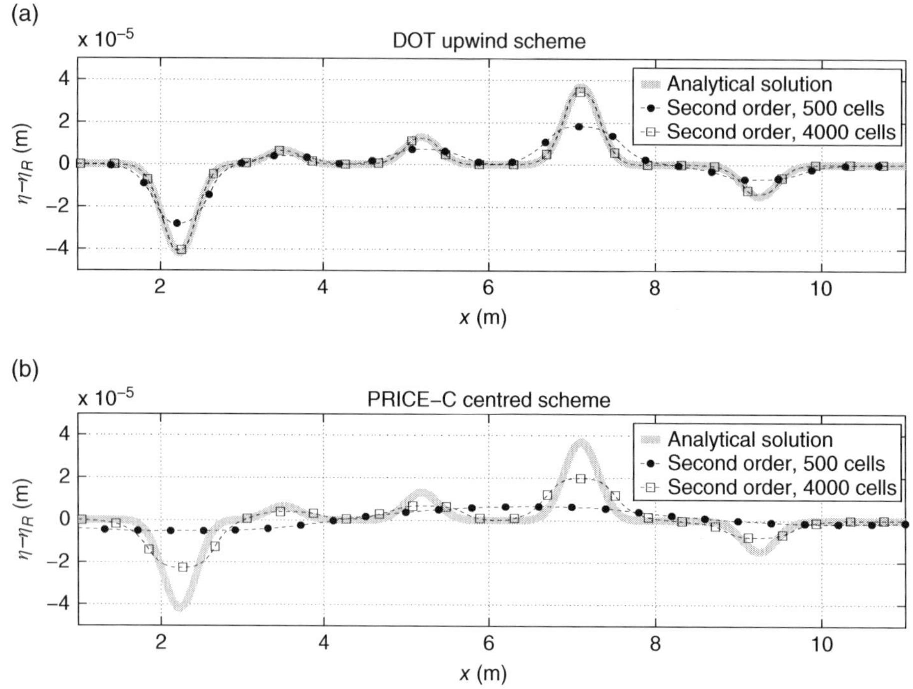

Figure 26.5 Convergence to the linearized wave propagation solution for $N = 5$ fractions under subcritical flow conditions: (a) DOT scheme and (b) PRICE-C scheme. (*Source*: Stecca, Siviglia, and Blom (2016). Reproduced with permission of Elsevier.)

In Figure 26.5 , the solution profiles of the two schemes obtained at different grid refinement, i.e. using a coarse mesh of 500 cells and a fine mesh of 4000 cells, are compared with the analytical solution, to observe their convergence. For brevity, we report only the bed-elevation profiles. Figure 26.5a refers to the DOT scheme and 5b to the PRICE-C scheme. The view is limited to the waves having significant impact on the bed profile, i.e., the 'bed' wave at $x \sim 2.3$ m and the four 'sorting' waves at $x \sim 3.5$ m, $x \sim 5.2$ m, $x \sim 7.1$ m, and $x \sim 9.3$ m.

In Figure 26.5a the solution of DOT overlaps the analytical solution when a fine mesh is employed, and thus the displacement and amplitude of perturbations is correctly captured. This is a consequence of using a refined upwind approach, which assigns the minimum amount of numerical diffusion needed for numerical stability to each wave, returning sharp profiles. In contrast, the solution by the centred PRICE-C scheme Figure 26.5b is affected by excessive spurious diffusion, which makes the individual perturbations almost indistinguishable unless high-mesh resolution is employed, and still in this case considerably dampens the amplitude of perturbation peaks.

This small-scale theoretical application thus shows how a poor numerical solution can hinder the observation of the basic physical features of the solution, and thus emphasizes the importance of accurate numerical techniques.

26.4.2 Streamwise and Vertical Sorting in a Flume Experiment

Here we investigate the suitability of the active-layer model in applications to a real-world case under controlled conditions. We consider the benchmark of flume experiment E8–E9 by Ribberink (1987), where morphodynamic changes due to streamwise and vertical sorting were measured.

The bed was gently sloping, covered by dunes, and composed of a bimodal mixture. The initial vertical profile displayed a downward coarsening trend, as reported in Figure 26.6a, where the volume fraction content of the fine fraction along the vertical is shown. Starting from initial equilibrium in the flow and sediment transport, morphodynamic changes were induced by altering the upstream sediment supply over time, as shown in Figure 26.6b, while keeping the hydrodynamic controls constant. A progressively increased amount of the coarse fraction was fed to the flume. After 30 h only the coarse fraction was supplied, while the total feed remained almost constant.

In the numerical solution, careful calibration of empirical submodels for friction, sediment transport, and active-layer thickness was enabled by the availability of accurate data and previous modelling experience (Ribberink 1987; Blom 2008). In detail, L_a is set equal to full bedform height to best match the observed speed of morphodynamic changes (Blom 2008).

The bed surface reacted to these changes in sediment feed by allowing the coarse material fed to prograde along the flume. This coarsening (sorting) wave, propagating in the downstream direction, is shown in Figure 26.6c and d, where experimental profiles of the fraction content of fines in the active layer F_{a1} at different times are reported by square markers. It is seen that the imposed coarsening at the upstream end, indicated by a progressive reduction in F_{a1}, progressively affects a more extended region of the flume.

The performance of the DOT and PRICE-C schemes in the reproduction of these profiles is assessed against the data in Figure 26.6c and d respectively. The profiles of DOT compare fairly well with the data, capturing the correct propagation speed of streamwise sorting processes, whereas the solution of PRICE-C, due to excessive diffusion, only provides trends of downstream fining, which show poor quantitative agreement with the experimental ones.

In Figures 26.6e and f we observe the measured feedback of streamwise sorting changes on bed elevation, by reporting the time series of bed elevation at three locations. The data (squares) show an initial phase of degradation, in which a degradational wave (scour) travels in the downstream direction. This is caused by the propagating coarsening signal described above, which determines a temporary reduction of incoming sediment load in correspondence of the coarsening front. Noteworthy, these observations confirm the interplay between streamwise sorting and bed-elevation changes, observed by the characteristic analysis in section 26.3. Eventually, the bed elevation increases, as the bed slope increases to transport the coarser material.

Once again, the two numerical schemes perform differently in this respect. In the solution of DOT in Figure 26.6e the initial degradation is correctly captured, whereas in the solution of PRICE-C in Figure 26.6f it is almost neglected, as a consequence of the incorrect reproduction of streamwise sorting profiles in Figure 26.6d.

This exercise confirms the validity of the active-layer schematization in the description of streamwise sorting and its interplay with bed elevation over relatively short time scales in this experiment, i.e. until vertical sorting comes also into play. A carefully calibrated numerical model based on accurate numerics is able to reproduce these changes as predicted by the mathematical model, providing good quantitative agreement with the data (Figure 26.6c and e), and can serve as a powerful tool for interpreting the physics. Conversely, an inaccurate approximation (Figure 26.6d and f) can hinder from reproducing these processes, although they are predicted by the underlying equations, possibly leading to misinterpretation of the physics.

However, the performance of the active-layer model with respect to vertical sorting is not satisfactory. Figure 26.6g shows the vertical sorting profile at final time (after 120 h), in terms of content of the fine fraction. The numerical profile of DOT agrees well with the experimental one in the active layer, where the model correctly predicts that the bed surface has significantly coarsened, but shows a marked mismatch in the substrate, as a consequence of the oversimplification of vertical fluxes by the active-layer schematization in dune-dominated cases.

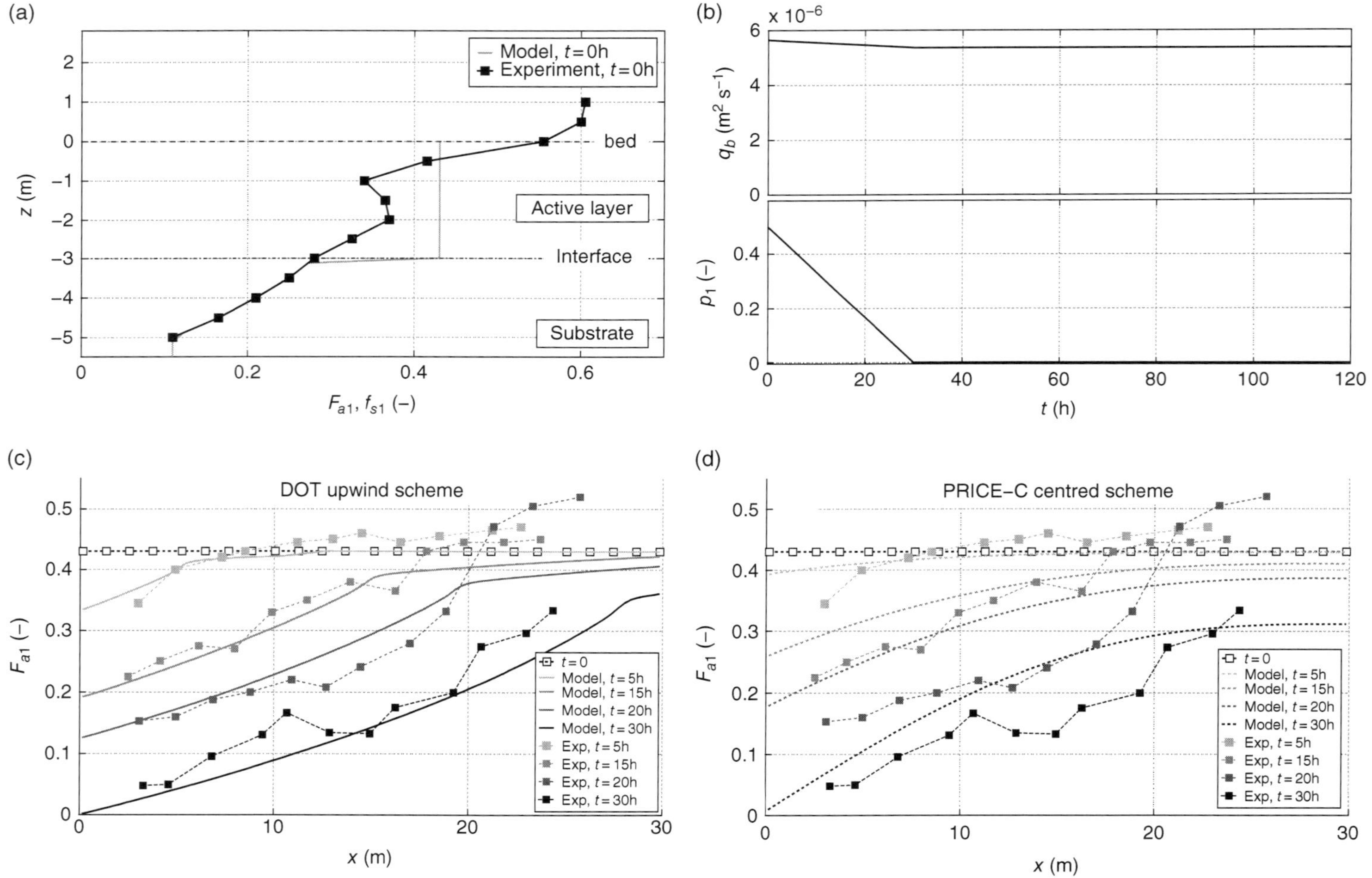

Figure 26.6 Numerical reproduction of experiment E8–E9 by Ribberink (1987). (a) Initial vertical sorting profile at the downstream end, given by the content of the fine fraction along the vertical. (b) Time evolution of sediment feed (total amount q_b and content of the fine fraction p_1). (c, d) Streamwise profiles of the content of the fine fraction F_{a1} in the active layer at different times. The model results of the DOT scheme (c) and PRICE-C scheme (d) are compared with the data. (e, f) Time evolution of bed elevation at three locations. The model results of DOT (e) and PRICE-C (f) are compared with the data. (g) Vertical sorting profile at final time, given by the content of the fine fraction. The results of DOT are compared with the data. (All figures except (d) and (f) are replotted from Stecca, Siviglia, and Blom (2016). All the laboratory data originate from Ribberink (1987).)

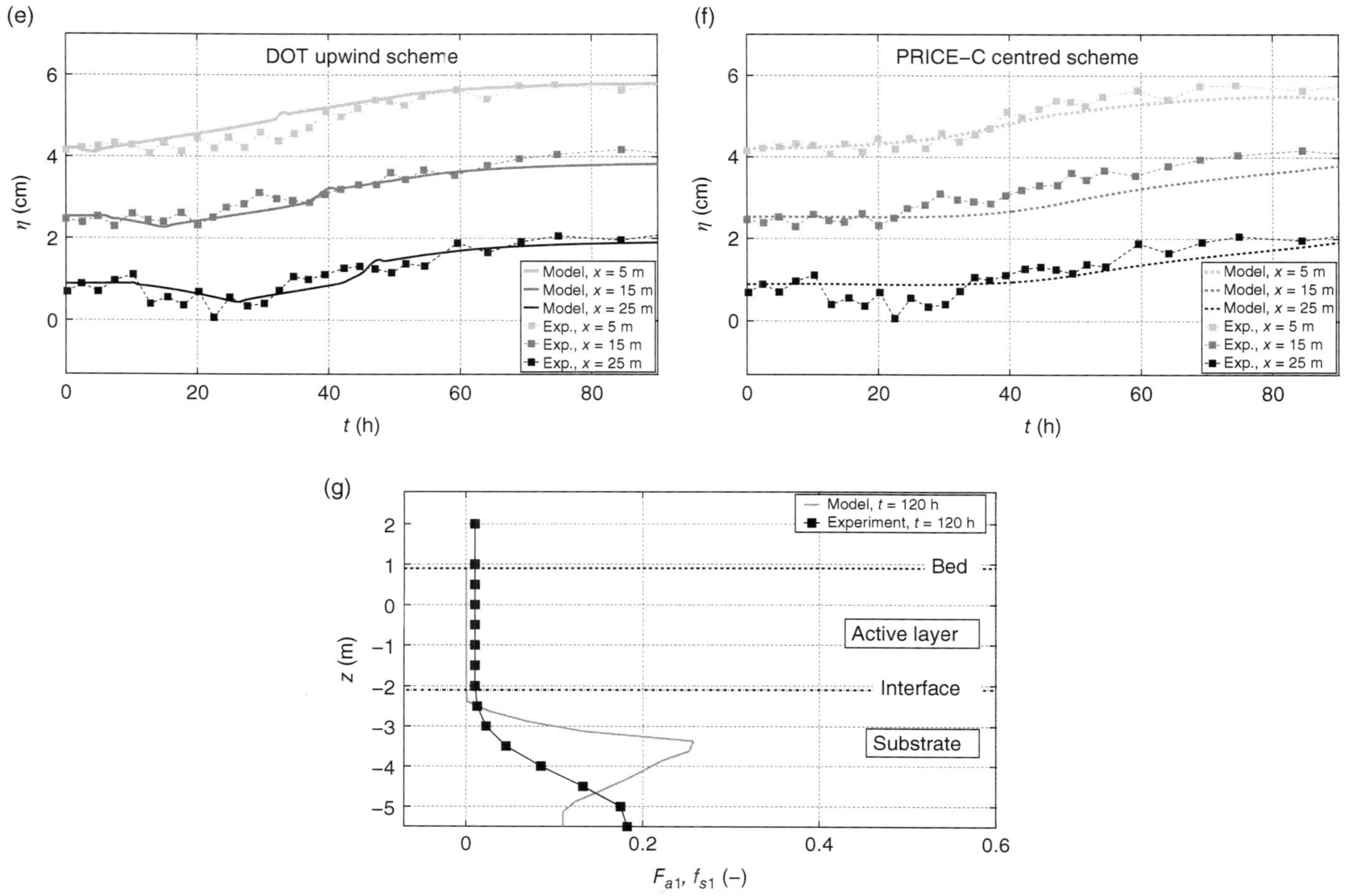

Figure 26.6 *(Continued)*

These physical shortcomings are potentially severe also for longitudinal sorting in longer-term computations, since the fraction content in the active layer can be altered by inaccurate representation of the vertical mass exchange (Stecca, Siviglia, and Blom 2016). This indicates the need for more sophisticated models of vertical sorting (see e.g. Blom, Ribberink, and de Vriend 2003; Blom 2008; Blom, Ribberink, and Parker 2008). The improvements of vertical-continuous models to prediction ability over longer spatial scales (e.g. the reach scale) shall be determined against real-world benchmarks, after a similar process of validation as that detailed in this section.

26.4.3 Failure of Numerical Solutions Under Elliptic Conditions

Unstable, or at least oscillatory behaviour of numerical simulations of the active-layer model conducted under elliptic conditions has been reported earlier (e.g. Ribberink 1987; Sieben 1997; Chavarrías *et al.* 2014).

Here in Figure 26.7 we present an original simulation conducted in the elliptic range, using the numerical model of Stecca, Siviglia, and Blom (2016). We set a mildly-sloping bed composed of a bimodal mixture (diameters $d_1 = 0.5$ mm and $d_2 = 10$ mm). The bed is initially armoured: the active layer, characterized by content of fines $F_{a1} = 0.1$, is coarser than the substrate, having content of fines $f_{s1} = 0.9$ (Figure 26.7a). An imposed M2 backwater profile (see, e.g., Chow 1959) induces bed degradation throughout the computation. These settings determine elliptic conditions, triggering instability. As a consequence, slight oscillations are detected in the profiles at $t = 0.5$ h in Figure 26.7b, which eventually grow towards finite amplitude in Figure 26.7c.

Noteworthy, these non-physical oscillations cannot be attributed to numerical instability, as they have nothing to do with the stability of the discretization algorithms, but rather with the stability of the underlying mathematical problem (e.g. Joseph and Saut 1990). It follows that, in the solution of an elliptic initial-value problem, all the reasonably accurate numerical solvers would display oscillations, regardless of the numerical techniques in use, and that an unstable numerical solution represents a faithful and correct approximation. Conversely, the solution of an inaccurate solver may look mistakenly stable and smooth, due to the added spurious diffusion which dampens oscillations. Such counterintuitive behaviour needs to be carefully considered when analysing the performance of numerical models. This example emphasizes the importance of analysing the mathematical structure of the system, which can unravel the reasons of failure in numerical solutions.

26.5 Conclusions and Research Perspectives

In this chapter we have reviewed the state of the art of theoretical and numerical developments in modelling mixed-sediment morphodynamics using the active-layer model.

First, we have presented the analysis of characteristics and the dynamics of small-amplitude waves, which theoretically yield the speed of hydromorphodynamic perturbations as described by the system of PDEs. By these means, we have shown the presence of distinct 'sorting' waves, which travel in the downstream direction with different velocities. These waves determine the adaptation of bed profiles at a faster pace than the 'bed' wave of the unisize-sediment model, in agreement with previous numerical observations.

The characteristic analysis reveals that, under degradation into a substrate which is finer than the active layer, the system may experience a switch in its mathematical character, from hyperbolic to elliptic. As only hyperbolic models yield well-posed initial-value problems, this behaviour represents a severe conceptual shortcoming in the description of the physics. Furthermore, as shown here by a

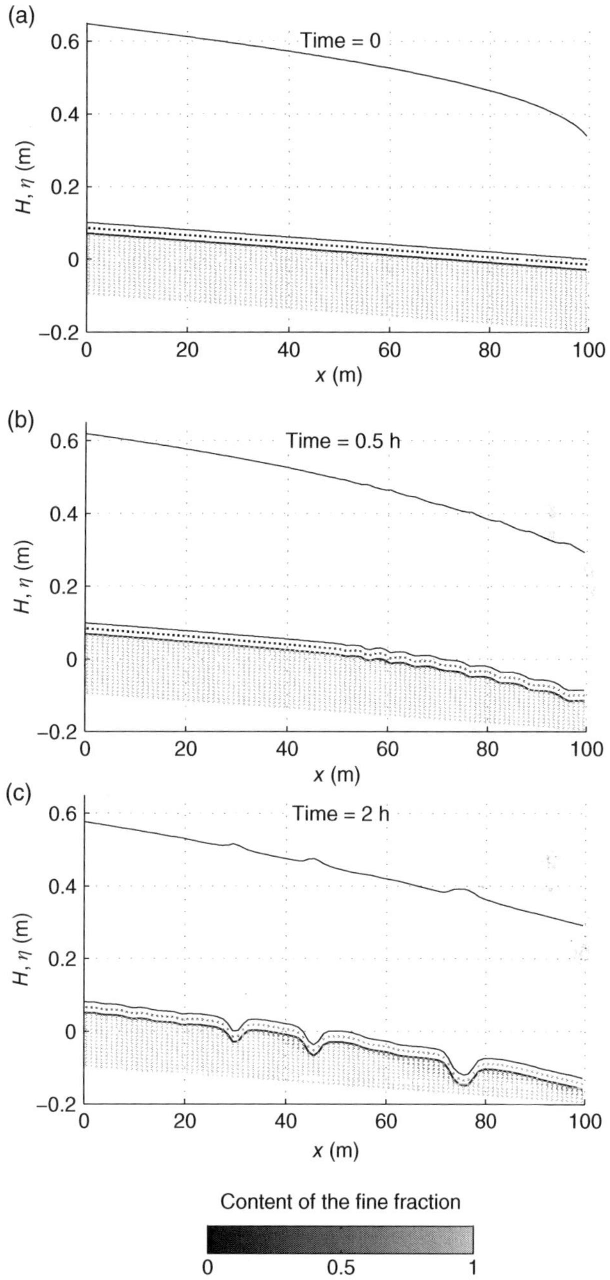

Figure 26.7 Numerical simulation conducted under elliptic conditions using the model of Stecca, Siviglia, and Blom (2016), concerning degradation of an armoured bed due to an M2 backwater profile: (a) initial condition, (b) initiation of instability, and (c) development of finite-amplitude nonphysical oscillations.

computational example, it is the source of instabilities of mathematical origin, and possibly the origin of the failure of numerical simulations.

In the spirit of Mosselman (2012), we have followed a two-step procedure for the assessment of numerical solutions, first using the theoretical benchmark of the linearized wave dynamics, and then against a real-world case in controlled conditions of a laboratory experiment. By comparing two numerical solutions, we have shown that numerical accuracy has a crucial impact on the ability of the model to correctly reproduce the speed and extent of morphodynamic changes predicted by the mathematical model.

With an accurate solution, the predictions of the mathematical model concerning the propagation of streamwise sorting and its interplay with the evolution of bed elevation have been found consistent with the experimental data. The active-layer model is thus adequate for representing streamwise sorting, as long as vertical sorting does not play an important role. However, the performance of the model with respect to vertical sorting has been found unsatisfactory, due to oversimplification in the description of vertical mass exchange.

Although the application of the active-layer model to laboratory data has produced somehow encouraging results, the analyses and results presented in this chapter show that there is need for improvement of the mathematical formulation, and for further assessment of the model. We thus identify the following four research areas.

26.5.1 Restoration of Hyperbolicity

Although the shortcomings connected to a possible switch to elliptic behaviour (sections 26.3.3 and 26.4.3) have been known since the pioneering work of Ribberink (1987), a unique well-established strategy to restore hyperbolicity does not exist, and various alternative approaches are documented.

A simple way, without model modifications, is to tweak the active-layer thickness, so as to move away from the region of 'crossing of eigenvalues' (Figure 26.4; Sieben and Sloff 1994). A closer look at the problem however suggests that ellipticity may be ultimately determined by oversimplification of the physics by the mathematical model. In detail, two alternative research lines focus (i) on the oversimplification in the description of sediment transport (streamwise processes) or (ii) of vertical mass exchange and size stratification (vertical processes). Following the first line, Sieben (1997) developed a stabilization strategy by considering an adaptation length in the transport formulation, which is in agreement with experimental evidence (e.g. Phillips and Sutherland 1989; Houssais and Lajeunesse 2012). Stecca, Siviglia, and Blom (2014) instead put forward the idea that the model simplifications that determine the occurrence of ellipticity could be connected to the neglect of diffusion terms in the bedload flux, and argued that introducing these terms could help restoring hyperbolicity. Alternatively, Ribberink (1987) focused on improving the description of vertical processes for dune-dominated beds by setting a three-layer model, in which an intermediate layer between the substrate and the active layer is located. This approach reduces the occurrence of ellipticity, but cannot fully prevent it (Sieben 1997). Finally, an ongoing PhD project (Chavarrías *et al.* 2014, 2015) seeks to explore a physics-based approach to restoration of hyperbolicity, which relies on modelling the direct entrainment of fine material from the substrate into transport under bed degradation.

26.5.2 Improved Description of Vertical Sorting

As shown in Figure 26.6g, the active-layer model produces poor predictions of the development of size stratification, due to oversimplification of vertical mass fluxes in its schematization. Alternative modelling concepts based on a vertical continuous framework (Parker, Paola, and Leclair 2000; Blom and Parker 2004; Blom *et al.* 2006; Blom, Ribberink, and Parker 2008), have instead been shown to perform much more accurately in this respect. Blom (2008) was the first to apply such a vertical-continuous model in long-term and large-scale computations (see Viparelli, Blom, and Moreira, Chapter 23, this volume). It is also worth exploring whether the advantage of vertical-continuous approaches can be replicated by properly parametrized, multiple active-layer formulations, and quantitatively assess the improvement.

26.5.3 Optimal Discretization of the Granulometric Curve

The discretization of the granulometric curve was found in section 26.3.4 to crucially affect the resulting bed-elevation profiles. It is therefore likely to also play a significant role in more realistic computations conducted with numerical models.

In field applications, the discretization of the granulometric curve is somehow arbitrary. In the case where the granulometric curve is characterized by two or more modes, the discrete fractions and their diameters can be set to match these modes. This does not apply to unimodal mixtures, for which a criterion for identification of the optimal number of fractions is not presently available.

In future work, we aim at exploring these issues through the numerical reproduction of laboratory data, for instance those concerning sediment pulses in mountain streams (e.g. Sklar *et al.* 2009). Our preliminary results with respect to this case show that a more detailed discretization determines faster propagating morphological changes and smoother bed profiles. The price that must be paid for a detailed discretization is a higher computational cost due to the large number of active-layer equations. In this respect the adoption of different modelling concepts based on a continuous description of the grain-size distribution is worth investigating.

26.5.4 Two-Dimensional Morphodynamics: Bar Instability

After assessing the validity and limitations of the active-layer formulation in one-dimensional cases, our goal is to move to two-dimensional applications in river morphodynamics.

Concerning the unisize-sediment case, two-dimensional morphodynamics has been intensively studied in the past decades, by analytical perturbative solutions of bar instability (e.g. Olesen 1984; Colombini, Seminara, and Tubino 1987; Seminara and Tubino 1992). In recent years, two-dimensional numerical models have been shown to be able to reproduce the dynamics of these individual morphological features in agreement with analytical theories (e.g. Siviglia *et al.* 2013), and thus proven to be reliable tools for morphodynamic prediction at the reach scale.

The question is thus whether this approach can be replicated in the modelling of two-dimensional morphodynamics with mixed sediment. Unfortunately in this field, the development of analytical morphodynamic theories has been more limited, and, as few experimental studies are available, there is a need for more laboratory and field data. The theory of Lanzoni and Tubino (1999) and the experiments of Olesen (1987) and Lanzoni (2000) concerning bar instability with mixed sediment could provide a fruitful morphodynamic benchmark for numerical models.

Acknowledgements

The authors thank the editors D. Tsutsumi and J. Laronne, and E. Lajeunesse (Institut de Physique du Globe de Paris) and one anonymous reviewer for their constructive comments and remarks, which definitely helped to improve the clarity and the content of this chapter.

The contribution of Guglielmo Stecca to this research was supported by a Marie Curie International Outgoing Fellowship within the 7th European Community Framework Programme (project 'BraidSideEarth', grant 2013-621886) and by Aspasia scholarship 015.007.051 of the Netherlands Organization for Scientific Research (NWO).

References

Blom, A 2008. Different approaches to handling vertical and streamwise sorting in modeling river morphodynamics. *Water Resources Research* 44, W03415.

Blom, A and Parker, G 2004. Vertical sorting and the morphodynamics of bedform-dominated rivers: A modeling framework. *Journal of Geophysical Research*, 109, F02007.

Blom, A, Ribberink, JS, and de Vriend, HJ 2003. Vertical sorting in bed forms: Flume experiments with a natural and a trimodal sediment mixture. *Water Resources Research* 39(2).

Blom, A, Ribberink, JS, and Parker, G 2008. Vertical sorting and the morphodynamics of bed form-dominated rivers: a sorting evolution model. *Journal of Geophysical Research* 113, F01019.

Blom, A, Parker, G, Ribberink, JS, and de Vriend, HJ 2006. Vertical sorting and the morphodynamics of bed-form-dominated rivers: an equilibrium sorting model. *Journal of Geophysical Research* 111, F01006.

Canestrelli, A, Siviglia, A, Dumbser, M, and Toro, E 2009. Well-balanced high-order centred schemes for non-conservative hyperbolic systems. Applications to shallow water equations with fixed and mobile bed. *Advances in Water Resources*, 32 (6), 834–844.

Chavarrías, V, Stecca, G, Viparelli, E, and Blom, A 2014. *Ellipticity in Modelling Mixed Sediment Morphodynamics*. Netherlands Centre for River Studies, University of Twente, Enschede, Netherlands.

Chavarrías, V, Stecca, G, Siviglia, A, and Blom, A 2015. Physically-based approach for the restoration of hyperbolicity in the Saint-Venant–Hirano model for mixed-sediment river morphodynamics. *RCEM Conference*, Iquitos, Peru.

Chow, VT 1959. *Open-Channel Hydraulics*, McGraw-Hill: New York; 680 pp.

Colombini, M, Seminara, G, and Tubino, M 1987. Finite-amplitude alternate bars. *Journal of Fluid Mechanics* 181, 213–232.

De Vries, M 1965. Considerations about non-steady bed-load transport in open channels. *11th Congress International Association for Hydraulics Research*, Leningrad, Russia.

Dumbser, M and Toro, EF 2011. A simple extension of the osher riemann solver to non-conservative hyperbolic systems. *Journal of Scientific Computing* 48(1–3).

Facchini, M, Siviglia, A, and Boes, R 2015. Downstream morphological impact of a sediment bypass tunnel – preliminary results and forthcoming actions. In *First International Workshop on Sediment BypassTunnels*, VAW Mitteilungen 232, ETH Zurich; 137–146.

Grant, GE 2012. The geomorphic response of gravel-bed rivers to dams: perspectives and prospect. In *Gravel Bed Rivers: Processes, Tools, Environments* (eds Church, M, Biron, P, and Roy, A). John Wiley & Sons: Chichester, UK.

Hirano, M 1971. River bed degradation with armoring. *Transactions of the Japanese Society of Civil Engineering* 3, 194–195.

Hirano, M 1972. Studies on variation and equilibrium state of a river bed composed of nonuniform material. *Transactions of the Japanese Society of Civil Engineering* 4, 128–129.

Hoey, TB and Ferguson, RI 1994. Numerical simulation of downstream fining by selective transport in gravel bed rivers: model development and illustration. *Water Resources Research* 30, 2251–2260.

Houssais, M and Lajeunesse, E 2012. Bedload transport of a bimodal sediment bed. *Journal of Geophysical Research* 117, F04 015.

Joseph, DD and Saut, JC 1990. Short-wave instabilities and ill-posed initial-value problems. *Theoretical and Computational Fluid Dynamics* 1, 191–227.

Lanzoni, S 2000. Experiments on bar formation in a straight flume: 2. Graded sediment. *Water Resources Research* 36, 3351–3363.

Lanzoni, S and Tubino, M 1999. Grain sorting and bar instability. *Journal of Fluid Mechanics* 393, 149–174.

Lyn, DA 1987. Unsteady sediment transport modeling. *Journal of Hydraulic Engineering, ASCE* 113(1), 1–15.

Lyn, DA and Altinakar, M 2002. St. Venant–Exner equations for near-critical and transcritical flows. *Journal of Hydraulic Engineering, ASCE* 128(6), 579–587.

Meyer-Peter, E and Müller, R 1948. Formulas for bed–load transport. In *2nd Meeting IAHR*, Stockholm.

Mosselman, E 2012. Modelling sediment transport and morphodynamics of gravel-bed rivers. In *Gravel Bed Rivers: Processes, Tools, Environments*. John Wiley & Sons, Chichester, UK; 101–115.

Olesen, KW 1987. *Bed topography in shallow river bends*. PhD thesis, Delft University of Technology, Delft, Netherlands.

Parker, G 1991a. Selective sorting and abrasion of river gravel. *I. Theory. ASCE Journal of Hydraulic Engineering* 117, 113–149.

Parker, G 1991b. Selective sorting and abrasion of river gravel. *II. Applications. ASCE Journal of Hydraulic Engineering* 117, 150–171.

Parker, G 2004. *E-Book on 1D Sediment Transport Morphodynamics with Applications to Rivers and Turbidity Currents*. https://csdms.colorado.edu/wiki/1D_Sediment_Transport_Morphodynamics_with_applications_to_Rivers_and_Turbidity_Currents

Parker, G, Paola, C, and Leclair, S 2000. Probabilistic exner sediment continuity equation for mixtures with no active layer. *ASCE Journal of Hydraulic Engineering* 126, 818–826.

Phillips, BC and Sutherland, AJ 1989. Spatial lag effects in bed load sediment transport. *Journal of Hydraulic Research* 27, 115–133.

Ribberink, JS 1987. *Mathematical modelling of one-dimensional morphological changes in rivers with non-uniform sediment*. PhD thesis, Delft University of Technology, Delft, Netherlands. http://repository.tudelft.nl/view/ir/uuid%3Abdfc1519-a71d-4752-83f7-3ebf1bb890e9/

Seminara, G and Tubino, M 1992. Weakly nonlinear-theory of regular meanders. *Journal of Fluid Mechanics* 244, 257–288.

Sieben, J 1997. *Modelling of hydraulics and morphology in mountain rivers*. PhD thesis, Delft University of Technology, Delft, Netherlands.

Sieben, J and Sloff, CJ 1994. Analysis of a 2D mathematical model for mountain rivers with graded sediments. In *International Workshop on Floods and Inundations and Related to Large Earth Movements*, Trento, Italy.

Sklar, LS, Fadde, J, Venditti, JG *et al.* 2009. Translation and dispersion of sediment pulses in flume experiments simulating gravel augmentation below dams. *Water Resources Research* 45.

Sloff, CJ and Mosselman, E 2012. Bifurcation modelling in a meandering gravel-sand bed rivers. *Earth Surface Processes and Landforms* 37, 1556–1566.

Stecca, G, Siviglia, A, and Blom, A 2014. Mathematical analysis of the Saint-Venant–Hirano model for mixed-sediment morphodynamics. *Water Resources Research* 50(10), 7563–7589.

Stecca, G, Siviglia, A, and Blom, A 2016. An accurate numerical solution to the Saint-Venant–Hirano model for mixed-sediment morphodynamics in rivers. *Advances in Water Resources* 93(A), 39–61. DOI:10.1016/j.advwatres.2015.05.022.

Toro-Escobar, CM, Parker, G, and Paola, C 1996. Transfer function for the deposition of poorly sorted gravel in response to streambed aggradation. *Journal of Hydraulic Research* 34(16), 35–53.

Viparelli, E, Sequeiros, OE, Cantelli, A *et al.* 2010. River morphodynamics with creation/consumption of grain size stratigraphy 2: numerical model. *Journal of Hydraulic Research* 46(6), 726–741.

Viparelli, E, Blom, A, and Moreira, R this volume. Modelling stratigraphy-based gravel-bed river morphodynamics, In Gravel-Bed Rivers: Processes and *Disasters* (eds D Tsutsumi and J Laronne). John Wiley & Sons: Chichester, UK;

Williams, GP and Wolman, MG 1984. *Downstream Effects of Dams on Alluvial Rivers*. Technical Report 1286, Professional Paper, US Geological Survey.

Discussion

Discussion by Philippe Belleudy

Can you exemplify the evidence of upstream bottom wave propagation in supercritical flow conditions?

Reply by the authors

A common example of upstream bottom wave propagation are the transportational cyclic steps created by flow over an erodible bed. Starting from a flat bed, if the Froude number of the approaching uniform water flow is sufficiently high, the flow and bed often co-evolve into a spatially varying pattern of deposition and erosion. This pattern initiates and subsequently maintains upstream migrating bottom steps (Taki and Parker 2005). The resulting spatially periodic bedforms can be successfully modelled using the unisize-sediment Saint-Venant–Exner model (Sun and Parker 2005).

References

Sun, T. and Parker, G. 2005. Transportational cyclic steps created by flow over an erodible bed. Part 2. Theory and numerical simulation. *Journal of Hydraulic Research* 43(5), 502–514.

Taki, K. and Parker, G. 2005. Transportational cyclic steps created by flow over an erodible bed. Part 1. Experiments. *Journal of Hydraulic Research* 43(5), 488–501.

27

Physical and Numerical Modelling of Large Wood and Vegetation in Rivers

Walter Bertoldi and Virginia Ruiz-Villanueva

27.1 Introduction

In the past decade several studies have shown the relevance of vegetation and wood as key elements controlling the ecological and physical evolution of rivers and their riparian environment. Hence, most river restoration projects focus on the management of vegetation and large woody elements to recover impacted river reaches (Bernhardt *et al.* 2007). It is recognized that evolutionary trajectories of rivers are strongly affected by the two-way interaction between fluvial morphology and riparian and aquatic plants, which act as 'ecosystem engineers' (Gurnell *et al.* 2012; Gurnell 2014). This indicates the need of bio-geomorphological succession models (Corenblit *et al.* 2007, 2011) as tools to describe, interpret, and predict the spatial and temporal dynamics of the geomorphological and the biological components of the river environment.

Fluvial processes determine water availability conditions, dispersion of seeds and fragments, removal and burial of vegetation, which in turn affect the flow field, the mechanical properties of the bed material, and the habitat structure. Relative riverbed elevation and proximity to the main river channel may be considered as the driving physical variables controlling these dynamics (Camporeale and Ridolfi 2006; Corenblit *et al.* 2014). However, a complete understanding and quantification of these processes is still lacking, hindered by the complexity of the feedbacks acting on different time-scales and by the occurrence of local effects. Only recently has there been an increasing consideration towards the quantification of these processes and collaboration through multidisciplinary teams to investigate river–vegetation interactions using physical and numerical models (Camporeale *et al.* 2013). Further investigation is needed to explore and single out the mechanistic rules controlling the eco-hydromorphological processes, and to express them through mathematical relationships. This is a necessary step towards a better assessment and prediction of the evolutionary trajectories of rivers subject to natural or human pressures.

In this chapter we discuss how physical and numerical modelling are crucial tools to explore these complex interactions. Field observations face the problem of disentangling the different processes. In a natural system, the chronological sequence of floods and the occurrence of local effects often limit the possibility to identify the role of the different parameters. Modelling has the advantage of controlling the initial and boundary conditions of the system and allows the investigation of the morphological evolution of river reaches subject to changes of the controlling variables. As indicated by Paola *et al.*

Gravel-Bed Rivers: Processes and Disasters, First Edition. Edited by Daizo Tsutsumi and Jonathan B. Laronne.

(2009), an integrated approach that takes advantage of physical and numerical modelling has a huge potential, as laboratory-scaled models may provide specific data for calibration of new numerical models, and together they can explore the interaction among the various eco-morphological components.

Vegetation is also a crucial element when addressing flood risk. In particular, large wood can drastically increase flooding when it interacts with structures such as weirs and bridges. Recent floods in forested mountain areas demonstrated that wood accumulations are one of the major problems that need to be addressed in future flood-protection schemes. Although still limited, wood modelling can provide new insight on these processes, as well as new tools for planning an integrated approach where the ecological quality and safety issues are equally addressed.

The inclusion of vegetation and large wood as active elements concurring in the evolution of rivers is therefore recognized as one of the frontiers in fluvial research and is one of the most active fields. Several reviews and special issues have been published in recent years (e.g. Camporeale *et al.* 2013; Gurnell 2013; Frostick *et al.* 2014). This chapter aims at reviewing recent improvements in vegetation and wood modelling, with particular focus on the reach-scale interactions and is organized as follows: in sections 27.2 and 27.3 we discuss physical and numerical modelling of vegetation, respectively; sections 27.4 and 27.5 address wood modelling in terms of both wood interaction with flow and river morphology; a final section is devoted to open issues and future challenges.

27.2 Physical Modelling of Vegetation

There is a wide range of laboratory experiments where vegetation has been included in order to investigate its effect on the flow field, sediment transport, or soil cohesion. Although in most cases hydraulics and fluid mechanics were the main topic addressed by the experiments, nowadays many studies aim at reproducing a more complex framework, where vegetation is one of the relevant variables in the three-way (flow, sediments, vegetation) fully coupled bio-geomorphological evolution of river reaches. The main challenges are related to correctly reproducing the complex shape and flexibility of real plants, and to accurately scale the relevant time scales in terms of vegetation growth and the (seasonal) interaction with floods. In the following, recent results of physical models involving vegetation are presented with reference to: (i) the effect of vegetation on the flow field; (ii) the characterization of vegetation growth (above and below ground); (iii) the effect of vegetation on river planform and bed topography. From a complementary point of view, the three sections explore vegetation as: (i) a passive element that modifies the flow field; (ii) a growing element that responds to environmental forcing; (iii) an 'ecosystem engineer'.

27.2.1 Flow–Vegetation Interaction

The impact of vegetation on flow has been investigated since many decades. Nepf (2012), Aberle and Järvelä (2013), and Vargas-Luna, Crosato, and Uijttewaal (2014) recently reviewed flow resistance, flow turbulence, and associated sediment transport. Here we report on the most recent results and challenges, with particular focus on the use and characterization of real vegetation. Laboratory experiments are the preferable approach in this case, as it is easy to control and measure in detail the flow field. A further advantage is the possibility to use plant surrogates, i.e. wooden or plastic elements, which may be rigid or flexible (for a recent review on experimental techniques, see Frostick *et al.* 2014). This allows a simplification of the problem, considering regular shapes and elements that are all equal, avoiding issues related to differences among individuals. Moreover, it is possible to better characterize vegetation parameters (dimension, spatial density, stiffness) and it is easier to ensure experiment reproducibility.

However, when quantifying the flow–vegetation interaction, scaling is a relevant issue (Nikora 2010). When adopting Froude similarity (as in most cases), or preserving the stem Reynolds number (relevant for turbulence studies), scaling has an impact on the computation of Young's module of elasticity (Frostick *et al.* 2014). Plants flexibility is particularly relevant for submerged vegetation, because it controls the interaction with canopy-scale vortices, which may induce regular oscillations called *monami* (Ackerman and Okubo 1993). This waving motion affects the flow field near and inside the canopy, enhancing the turbulent vertical transport of momentum, thus increasing the flow velocity in the canopy layer (Ghisalberti and Nepf 2006; Sukhodolov and Sukhodolova 2012).

Most existing models are designed for submerged vegetation that can be schematized with a simple element (Luhar and Nepf 2012). However, natural woody vegetation, i.e. shrubs and trees, show complex structures including heterogeneous plant parts. Västilä *et al.* (2013) performed a series of experiments measuring the drag force of several riparian plants, subject to flow of different magnitude and depth. They showed that the separate computation of the stem and the foliage component on drag force is preferable when considering emergent natural vegetation. The model proposed by Västilä and Järvelä (2014) suggests the leaf-area-to-stem-area ratio as a metric to quantify the different contribution of complex-shaped plants. This parameter is largely dependent on the species, but is also influenced by growing conditions (habitat, water availability), growth stage (with a decrease of almost one order of magnitude from twigs to mature trees), and seasonality (most typical riparian species in temperate climate, *Populus* and *Salix*, are deciduous). Advances in remote-sensing techniques, such as terrestrial laser scanner and structure-from-motion photogrammetric surveys, may help in the assessment of vegetation characteristics, in terms of leaf area and frontal-projected stem area per ground area (Antonarakis *et al.* 2009, 2010).

Stem configuration and specifically its streamwise orientation and density are the main parameters considered by Euler *et al.* (2014) to quantify local erosion and deposition induced by single plants (modelled in the laboratory through cylindrical obstacles). They showed that increased inclination and reduced leaf area are both likely to limit the total amount of erosion, the occurrence of an upstream horseshoe scour, and enhance downstream deposition (hence formation of new habitats).

27.2.2 Characterization of Plant Growth and Uprooting

The increased recognition of the role of vegetation resulted in a more extensive use of real plants in laboratory experiments, and in an improved attention of researchers when designing their experiments. Riparian vegetation is generally represented in bio-geomorphological physical experiments by smaller and fast growing species, such as *Medicago sativa* (alfalfa) (Gran and Paola 2001) or *Avena sativa* (Perona *et al.* 2012). In the past few years, several studies have been carried out with the aim to better characterize the growing phases of these plants (considering both the aerial and root systems) and to improve knowledge on their resistance to uprooting.

Clarke (2014) investigated the best practices for seeding and growing the two species, maximizing proportion of sprouted seeds and stem length. The two plants are different in terms of stem (and root) dimension, so they can be used to model different scenarios. Moreover, *Medicago sativa* germinates in about 3 days, whereas *Avena sativa* has a larger proportion of seeds germinating after a longer period of time. This is relevant in physical model design, because it may change the distribution of vegetation in space and time, and it may also reproduce a more complex vegetated patch, with individuals characterized by diverse growth stage.

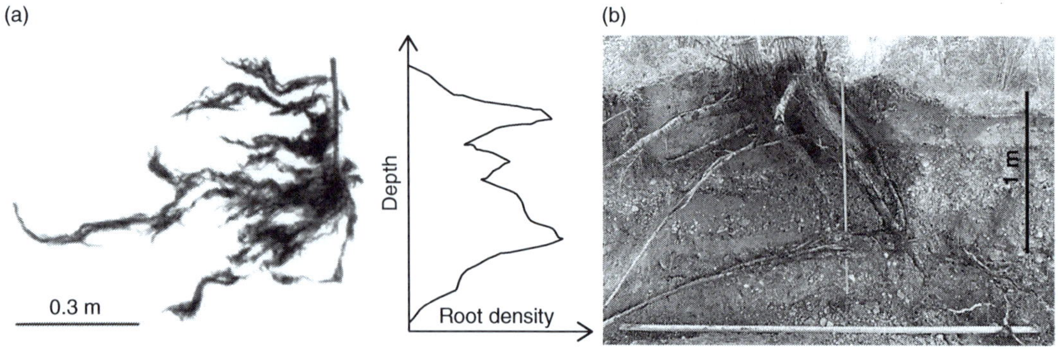

Figure 27.1 Characterization of root structures. (a) Root-density distribution of a willow cutting (modified from Pasquale *et al.* 2012). (b) Root structure of a poplar tree and sediment deposition layers on the Tagliamento River (Italy). (Photograph taken by James Holloway.)

Properties of roots and soil reinforcement are one of the new frontiers in laboratory modelling. Various sets of experiments have been designed with the aim to quantitatively describe root structure and to relate it to uprooting mechanisms. Erosion and uprooting of vegetation depends on the shear stress and drag exerted on the individual plant, but also on the type and form of its roots. Pasquale *et al.* (2012) performed detailed experiments using willow cuttings installed on a gravel bar. The development of below-ground biomass was monitored over three growing seasons, relating the vertical root density distribution to hydrological conditions (see Figure 27.1a). The authors demonstrated that it is possible to predict the depth of maximum root density as a function of bed topography and groundwater availability (see also Tron, Laio, and Ridolfi 2014), opening new opportunities for modelling the soil reinforcement of riparian plants and their survival to flooding conditions. Further understanding may come from site analysis of the relationship between root structure and sediment layer. Different morphological processes and floods with different magnitude may lead to grain-size sorting and therefore build sand/gravel layers where roots find different water availability conditions (e.g. the poplar root structure depicted in Figure 27.1b).

Root morphology and tensile strength of 14 riparian species were analysed by Polvi, Wohl, and Merritt (2014), who evaluated the added cohesion of trees, shrubs, and graminoids, as a function of sediment texture. These categories showed different root characteristics and soil reinforcement, providing a relevant dataset for predicting bank stabilization and the effect of vegetation on river morphological evolution. Edmaier *et al.* (2011, 2014) investigated uprooting mechanisms through an extensive experimental analysis of the properties of *Medicago sativa* and *Avena sativa*, considered as model riparian saplings. Two possible uprooting mechanisms were highlighted, when considering noncohesive sediments, with a slower mechanism related to sediment erosion and root exposure, suggesting that flood duration may also play a relevant role. Edmaier *et al.* (2014) showed that there is a difference between the role of the main root, which controls the maximum plant resistance, and the secondary roots, which are responsible for increasing the resistance to long-lasting disturbances.

Properties of uprooted plants versus those remaining unaffected by floods were investigated by Perona *et al.* (2012) and Crouzy *et al.* (2013) through a set of experiments with *Avena sativa*, and also analytically modelled by Crouzy and Perona (2012). Their statistical analysis indicated that floods contribute to the successional evolution of riparian vegetation by uprooting plants with specific features (in terms of root structure and growth stage) and possibly differentiating among species. Due to the fast growth rate of young plants, flood timing is therefore a crucial parameter that needs to be taken into account. Combining the previous works, Pasquale *et al.* (2014) reported on a dataset where willow cuttings growth and survival rates are associated with flood magnitude and morphological change. These field experiments likely represent the first attempt to fully couple vegetation–flow–sediment dynamics. They show that plant uprooting depends both on bed shear stress and on erosion and deposition processes. The resulting non-linear relationship can help explain the complex spatial and temporal distribution of vegetated patches in gravel-bed rivers.

27.2.3 Reach-Scale Models of River Evolution: a Two-Way Interaction

The impact of vegetation on reach-scale morphology in experimental channels has been investigated by several authors, starting from the seminal work by Gran and Paola (2001). In this case, no strict scaling is used, and vegetation (generally alfalfa seedlings) is employed as a way to increase sediment cohesion and bank stability. Paola *et al.* (2009) discussed how morphological experimental work is 'unreasonably effective', due to the inherent scale invariability of the main geomorphic processes. This is true also when considering vegetation, as long as the correct time scales are taken into account (Paola 2001). The use of uniformly distributed vegetation and reseeding during low-flow conditions causes the transition from a braided network to a single-thread, deeper channel, with a vegetated floodplain and characterized by a balance between bank erosion and accretion through point-bar deposition (Tal and Paola 2007, 2010). Braudrick *et al.* (2009) further demonstrated that the added cohesion provided by vegetation (along with fine suspended sediments) is the crucial ingredient to experimentally reproduce and maintain constant-width meandering channels. Van Dijk *et al.* (2013) quantified the stabilization effect of vegetation on bank erosion, with small-scale experiments forcing flow deflection against an erodible bank. Both seedling density and growth stage influenced the added cohesion and led to sharper and more irregular bank lines.

The effect of vegetation on reach-scale sediment transport was recently explored in two experimental studies on sand-bed rivers (Gorrick and Rodríguez 2012; Le Bouteiller and Venditti 2014). Both studies reported an increased sedimentation in and close to the vegetation patch, with the main flow diverting from the vegetated bank. Strongly non-linear processes are responsible for fast changes in the configuration, associated with slight changes in water level. Kim, Kimura, and Shimizu (2014) investigated the morphological impact of a vegetated quadrat, reproduced at laboratory scale through rigid cylindrical dowels. Changing the flow level below or above the average threshold for sediment transport induced large alterations in the resulting bed configuration. At lower flow magnitude, a deeper scour is formed on the side of the patch, whereas at high flow sediment deposition is more pronounced both upstream and inside the patch.

Most of the experiments involving vegetation considered a uniform distribution, representing seed dispersion by wind, or generally diffused seed availability on every dry surface. Van Dijk *et al.* (2013) revised this hypothesis and compared experiments with uniformly distributed vegetation against experiments where seeds are distributed hydrochorously. Results show that seed dispersion has a great impact on vegetation location, particularly in terms of its elevation and distance from the water surface. In turn, this determines the potential of uprooting by subsequent floods, as Pasquale *et al.* (2014) also observed. Moreover, different vegetation distributions are likely to form different planform morphologies, with uniformly distributed vegetation associated with single-thread meandering channels and spatially variable vegetation forming island braided systems, with seeds depositing at the upstream end of vegetated patches.

The availability of high resolution bed-topography surveys at both field and laboratory scale suggested representing and quantifying river morphology at the reach-scale through bed-elevation probability distribution functions. Figure 27.2 reports on different examples of such distributions,

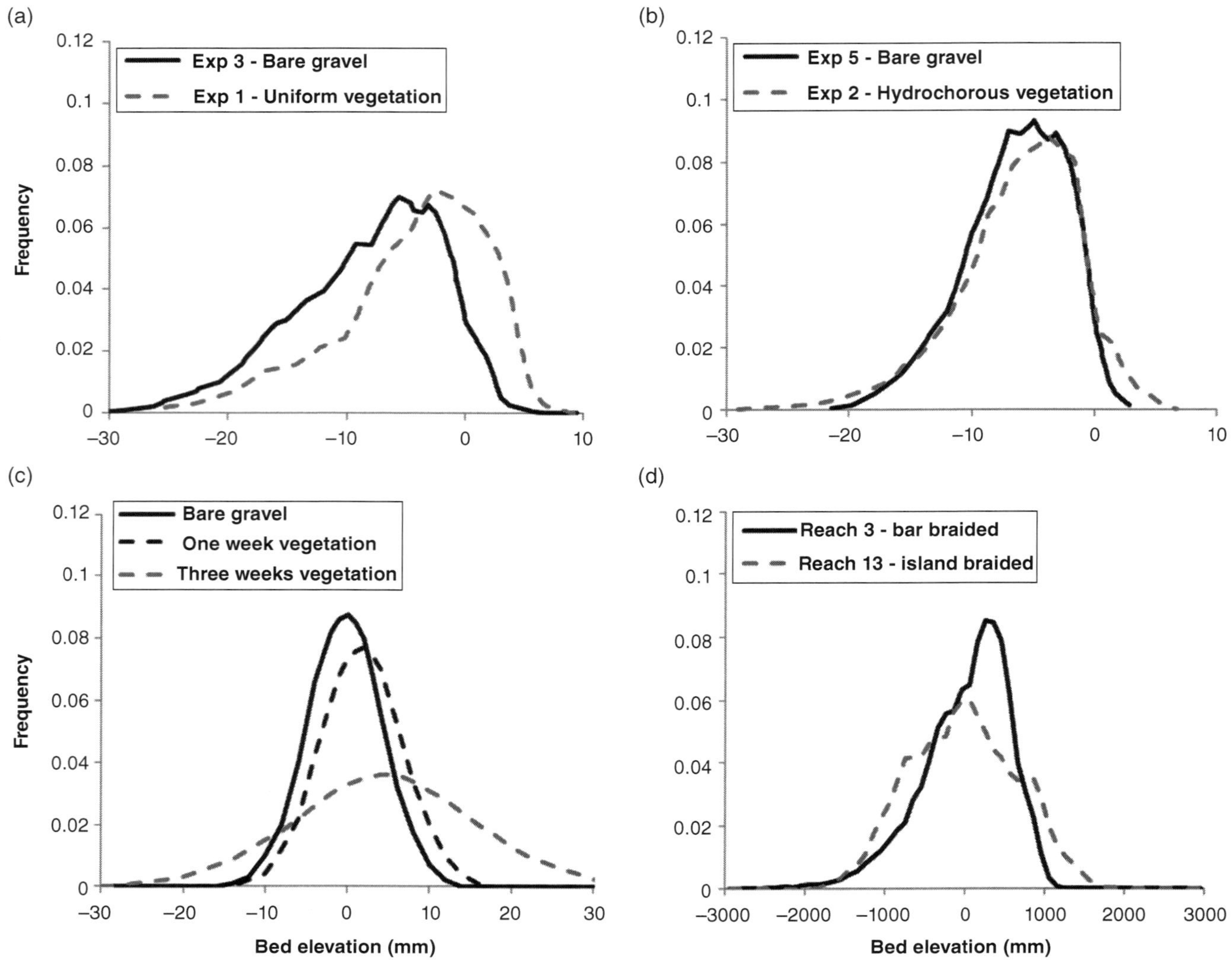

Figure 27.2 Comparison of bed-elevation frequency distribution with and without vegetation: (a, b)modified from van Dijk *et al*. (2013); (c) Hull experiments; (d) Tagliamento river, modified from Bertoldi *et al*. (2011). (*Source*: (a, b) van Dijk *et al*. (2013). Reproduced with permission of John Wiley & Sons.)

where it is possible to assess the effect of vegetation on river morphology. Three examples of laboratory experiments are compared to observations on the Tagliamento River, Italy (as reported by Bertoldi, Gurnell, and Drake 2011). Figure 27.2a and b show the result of the experiments by van Dijk *et al.* (2013). Even though the runs with and without vegetation were not characterized by the same parameters, the distributions indicate how vegetation is more likely to enhance deposition at higher elevation (by reduced transport capacity in the vegetated areas) and to scour deeper channels. The first process appears more pronounced in the case of uniformly distributed vegetation (Figure 27.2a) as opposed to hydrochorously distributed vegetation (Figure 27.2b). Figure 27.2c shows the results of experiments performed at the Total Environment Simulator, University of Hull, in the framework of the Hydralab EU network (see next section and Bertoldi *et al.* 2014b for further details). Vegetation colonization during a 4-week experiment is highlighted by a progressive broadening of the distribution and a shift of the peak towards higher elevations. These observations are in agreement to what was observed on several reaches of the Tagliamento River, where a gradient of vegetation cover is reflected by a range of different bed-elevation distributions (Figure 27.2d; Perona *et al.* 2012).

27.3 Numerical Modelling of Riparian Vegetation

The inclusion of vegetation in numerical models is recognized as a critical step towards a better prediction of the evolution of river morphology and its ecosystem. Camporeale *et al.* (2013) reviewed the progress in mathematical modelling of riparian and aquatic vegetation. Here we present the most recent developments, focusing on three aspects: (i) modelling the effect of flow on vegetation growth and distribution; (ii) modelling the effect of vegetation on bank processes; (iii) the inclusion of vegetation dynamics in two-dimensional shallow-water models. The main limitation is the scarcity of tested mathematical laws to evaluate the effect of vegetation on sediment transport and soil properties. Combined research is needed, where field (or laboratory) observations are designed in order to assess and quantify exactly the same parameters that are used in the numerical models. Otherwise, highly sophisticated models cannot be scientifically tested and validated.

27.3.1 The Effect of Flow and Climate Change on Vegetation Dynamics

Vegetation recruitment, growth, and survival depend strongly on the hydrologic regime, e.g. water availability and flood occurrence. Changes in base flow and high flow levels caused by water extraction and diversion, dam building, or climate change, may have an impact on vegetation distribution and composition. Several studies have addressed this issue, from a more ecological or hydrological point of view, providing different mathematical models able to describe this interaction (see Orellana *et al.* (2012) for a recent review).

River managers show increasing interest in the restoration of riparian and wetland areas and therefore need proper tools to evaluate the effectiveness of designed interventions at both the local and reach scale. For example, Hammersmark *et al.* (2010) used a coupled model that integrates a hydrological module (computing flow depth and groundwater levels), with individual plant distribution models. The result is a quantitative estimation of the habitat suitability for different species and groundwater availability, specifying the conditions more favourable for mesic or xeric vegetation.

Following the 'recruitment box' model proposed by Mahoney and Rood (1998), further researchers developed models able to assess the effect of modified flow dynamics on seedling recruitment (e.g. Dixon and Turner 2006). Benjankar *et al.* (2014) recently proposed a new spatially distributed model that includes five hydrophysical parameters for cottonwood recruitment (annual peak flow timing; shear stress; local stage recession; recruitment elevation above base flow; and duration of winter flooding). The model proved useful to predict cottonwood recruitment patterns and to guide reservoir flow releases in restoration programmes.

The succession of riparian vegetation from recruitment to mature forest (and back, in case of disturbances) was modelled with a rule-based vegetation model (CASiMiR-vegetation, Benjankar *et al.* 2011) that predicts dynamic succession and retrogression as a function of scour disturbance, shear stress, and flood duration on an annual time-step basis. The model was applied to different environments and proved to be a valid tool, thanks to its modelling approach that considers vegetation succession phases instead of specific local features (Rivaes *et al.* 2014). The main issues related to these types of model are: (i) the absence of a morphodynamic module that updates the bed topography during time and (ii) the lack of data on actual values of the most relevant hydraulic (and morphological) variables that drive the succession (or removal) of vegetation. However, they offer the great advantage of investigating the effect of flow and climate changes on the ecosystem structure. For example, Rivaes *et al.* (2014) showed that predicted changes in flow regimes at European level may cause a drastic adjustment of riparian environments, with reduction of the younger successional stages and potential disappearance of the more water-dependent species.

27.3.2 Effect on Bank Processes

The effect of roots on soil cohesion and the consequent stabilization effect on stream banks were modelled through a fibre bundle approach by Pollen and Simon (2005). Their RipRoot model accurately estimated the added reinforcement provided by riparian vegetation roots, thanks to the inclusion of progressive root breaking. The model was applied to different scenarios (Pollen-Bankhead and Simon 2010; Pollen-Bankhead *et al.* 2011; Polvi, Wohl, and Merritt 2014) and proved valid to assess plant pullout forces, including rhizome interconnectivity and seasonal cycle.

Root reinforcement is directly relevant in the case of low-energy streams, and in case of rivers with banks that do not exceed the rooting depth. For higher banks, riparian vegetation may decrease bank stability when hanging over the bank, thereby augmenting its weight. However, other factors may interfere and add stability, for example the deposition of fine organic cohesive soil, and the physical protection ensured by deposited trees and vegetated soil slumps (Micheli and Kirchner 2002).

Quantification of bank-retreat processes is a necessary step to improve the modelling of river planform evolution, particularly in the case of meandering rivers. Commonly, bank migration rate is linearly related to near-bank excess velocity through a very simple method that is useful for long-term theoretical analysis.

In order to improve accuracy of the prediction, a physically based bank-erosion formulation was proposed by Motta *et al.* (2012), which accounts for natural bank profiles and the presence of horizontal soil layers. Its application showed a good agreement with observed migration rates over a period of several decades, indicating the relevance of soil spatial heterogeneity, and of the different processes controlling bank erosion (as hydraulic erosion, cantilever, and planar failure).

Eke, Parker, and Shimizu (2014) proposed a further model where bank erosion and bank deposition processes are fully decoupled, therefore allowing for self-adjustment of channel width. The process of bank erosion is decelerated by the armouring effect of vegetated slump blocks, which may considerably reduce the erosion rate. Bank deposition is modelled considering the effect of vegetation encroachment, fine sediment deposition, and stabilization. Final rules of bank erosion can be expressed in a way similar to the common linear relationship to near-bank conditions, but all input parameters have specific physical meaning, and therefore have the potential to be measured in the field, improving the applicability of the model.

These improved representations of bank processes could be coupled with riparian vegetation dynamics, as proposed by Perucca *et al.* (2007), in order to estimate the mutual interaction between meander evolution and floodplain formation. Vegetation growth affects bank stability and thus modifies the meander shape, compared to the case of constant soil parameters. The model by Perucca, Camporeale, and Ridolfi (2007) also highlighted the relevant role played by the ratio between the temporal scale of vegetation establishment (and erosion) and the morphological timescale.

27.3.3 Two-Dimensional Shallow-Water Models

The first attempt to include vegetation dynamics in numerical models considered cellular automata, which are used to model braided river evolution. The effects of vegetation and the feedbacks on floodplain processes were included by Murray and Paola (2003) and Coulthard *et al.* (2007), who considered the root reinforcement as the main parameter influencing bed and bank erosion. These models showed the transition from a multi-thread (unvegetated) pattern to a single-thread (densely vegetated) system. A further cellular model proposed by Hooke *et al.* (2005) was used to simulate ephemeral streams, where vegetation has a major role in determining the impact of floods and the related morphological change.

Recently, two-dimensional models based on shallow-water equations (computational fluid dynamics, CFD) also approached the issue of taking into account vegetation-related processes. These models are largely used for studying rivers at reach-scale, and proved to accurately describe flow-field and bed-morphology changes.

In the first step, numerical models considered vegetation as an element affecting bed roughness and turbulence characteristics and thus suspended sediment transport. For example, Wu *et al.* (2005) developed a depth-averaged two-dimensional model where the effect of emergent and submerged rigid vegetation was taken into account by modifying bed shear stress and sediment transport capacity.

Crosato and Saleh (2011) and Li and Millar (2011) implemented more details into their shallow-water models in order to investigate the effect of riparian vegetation on the planform evolution of gravel-bed rivers. Li and Millar (2011) considered a vegetated floodplain, with different values of the bed roughness and an additional effect induced by the soil root reinforcement. This increases the friction angle, thus affecting bank erosion and migration, as well as the threshold for sediment motion. The new ingredient proposed by Crosato and Saleh (2011) is related to the spatial and temporal distribution of vegetation, which is not fixed, but plants colonize instantaneously all the newly formed exposed deposits. This model is therefore a first attempt to actually link morphological and vegetation dynamics. Both models confirmed the strong influence of vegetation on the planform configuration, with a transition from braided to wandering to single-thread with increasing vegetation density (or effect on roughness and sediment transport).

Nicholas *et al.* (2013) developed a model that also demonstrated the relevant effect of vegetation on the bed topography of large rivers. In this case, vegetation was included in a very simple way, in order to differentiate between 'channel' and 'floodplain' rules of evolution. Computational cells that do not experience flow deeper than a fixed threshold for a long enough period are transformed into floodplain cells, where sediment cohesion prevents vertical erosion. As a result, they may be reactivated only by lateral bank erosion. Nicholas *et al.* (2013) showed that changing vegetation-related parameters has an impact on bar formation and migration and consequently on island evolution and shape (Figure 27.3a). In particular, the vegetation growth time scale proved to be a crucial variable in order to reproduce observed morphologies. Moreover, the model highlights the role of the hydrologic

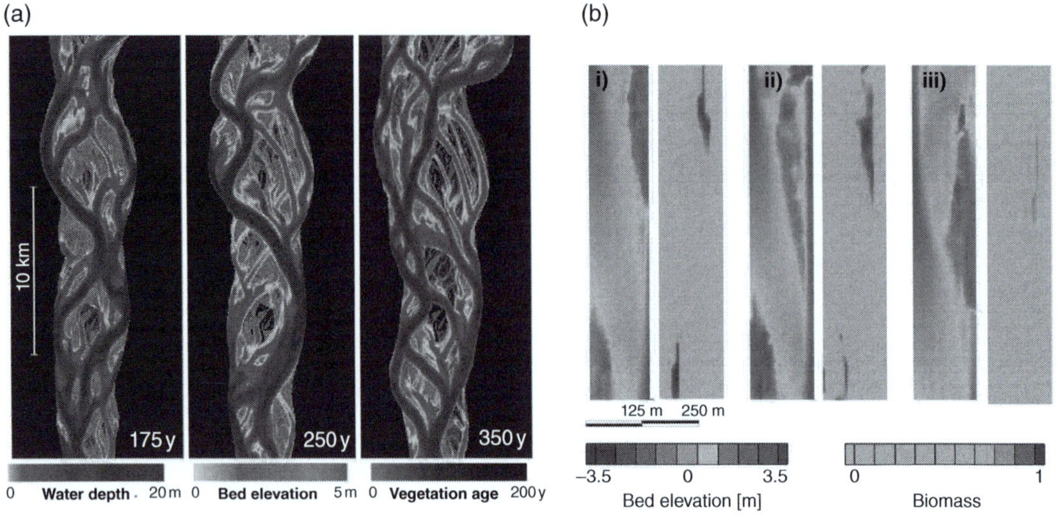

Figure 27.3 Examples of the numerical models results: (a) bed topography and vegetation age at three temporal steps for a large river (modified from Nicholas *et al.* 2013); (b) bed topography and vegetation biomass after three floods, for three different vegetation distributions (type 2, 7, 9), using the Bertoldi *et al.* (2014a) model. (*Source*: (b) Bertoldi *et al.* (2014a). Reproduced with permission of John Wiley & Sons.)

regime (in terms of flood magnitude variability) as the main driver of high bars formation. These bars are then more likely to become stable and be colonized by riparian vegetation, with additional increase of their stability.

All these models lacked a full coupling between hydraulics, sediment transport, and vegetation dynamics. This aspect was considered in the recent analytical model proposed by Perona *et al.* (2014) and in the numerical model proposed by Bertoldi *et al.* (2014a), who developed a new vegetation module for the two-dimensional shallow-water model BASEMENT (developed at VAW – ETH Zurich). In this case plant growth is described as a function of species characteristics and water availability, where the distance between riverbed and groundwater level is the main controlling parameter. The resulting vertical distribution is a bell-shaped function, in which width is a measure of the adaptability of the selected species. Vegetation biomass grows in time, reaching a maximum after a species-specific number of years. This approach allows investigation of the role of various relevant parameters such as flow discharge, vegetation roughness, root soil reinforcement, and groundwater depth. It also models how different plant species influence bed evolution, depending on diverse adaptation to water conditions, inundation tolerance, and growth rates. Figure 27.3b reports three examples of bed topography and vegetation biomass distribution obtained for the same morphological parameters, but different vegetation distributions. Comparison between the three panels shows that vegetation occurrence and location is strongly dependent on this parameter. The hypothesis of uniformly distributed vegetation is often not accurate enough to describe the actual river evolution.

This approach introduced some strong simplification when modelling the vegetation feedbacks, however, the main processes are taken into account and the relatively limited number of parameters may help in the assessment of the role of each component. Model results showed that the inclusion of dynamic vegetation is a necessary step when evaluating the evolutionary trajectories of rivers and their response to altered flow regimes.

27.4 Physical Modelling of Large Wood

In this section, recent results on laboratory experiments involving instream large wood are summarized and discussed. River scientists approach the presence of large wood from diverse points of view, ranging from flood hazard to their geomorphological and ecological roles in shaping the fluvial system. Logs or trees are generally modelled through simple elements, which lack their real complex physical and ecological properties. Uprooted trees often carry soil (and thus organic matter and fine sediments) and have the ability to sprout and form new trees. Despite these limitations, modelling of large wood has proven to be a useful tool for understanding transport and deposition dynamics. We propose classification into three main groups, where instream large wood is investigated for its impact on (i) flooding, (ii) river habitat and restoration, and (iii) river morphology.

27.4.1 Impact of Large Wood on Flooding

Entrainment by rising water is one of the fundamental issues when modelling the dynamics of large wood. Braudrick and Grant (2000) and Bocchiola *et al.* (2006) studied the entrainment of cylindrical logs in fixed-bed laboratory experiments, showing that there is a threshold flow depth for floating that depends on log orientation, diameter, and density. A new experiment by Crosato *et al.* (2013) pointed out the role of local accelerations of the flow field, regarding both local flow depth and streamline curvature. Detailed measurements of the flow field showed that an accurate hydrodynamic description is a necessary condition for developing numerical models simulating log entrainment. Moreover, preliminary results on experiments with a mobile bed showed that entrainment conditions may vary in the presence of bed scouring around the log, enhancing log drifting and further scouring.

When logs are entrained and start floating, they become a potential hazard for structures, particularly bridges and weirs, and may increase flooding probability and extent. Physical modelling was conducted at ETH Zurich to better understand and limit the hazard related to large wood. Schmocker and Hager (2011) investigated the blocking probability at bridge decks, exploring its dependence on log dimensions, bridge type, and flow characteristics, focusing on the conditions where the freeboard under the bridge is drastically reduced. The experimental runs demonstrated a strong random component, with a wide scatter of data and freeboard and flow Froude number as the main factors driving blockage probability. Scatter of the data is possibly due to local flow variability, but also because natural logs with irregular shapes and without branches were used (Figure 27.4a). Schmocker and Hager (2013) and Schmocker and Weitbrecht (2013) performed a series of laboratory experiments to evaluate the efficiency of a wood rack and to design a new structure to retain floating wood in a bypass channel located at outer river bends. Schmocker and Hager (2013) investigated the backwater effect and the wood accumulation in the presence of wood racks (Figure 27.4b). Their analysis showed that the estimate of the backwater depends mainly on the approach Froude number (and the wood volume) and it is practically not dependent on log size or shape. To avoid possible problems of rack overloading and flooding induced by the backwater, Schmocker and Weitbrecht (2013) proposed a different configuration, with a rack parallel to the main flow, where logs are stored in a side channel located at the outer bank of a bend. The physical modelling confirmed the effectiveness of this structure, which was then successfully implemented in a prototype.

27.4.2 Impact of Large Wood on Habitat

Large wood significantly contributes to the ecological functioning of the fluvial system (Abbe and Montgomery 1996). Many river-restoration projects include strategies for reintroduction of large

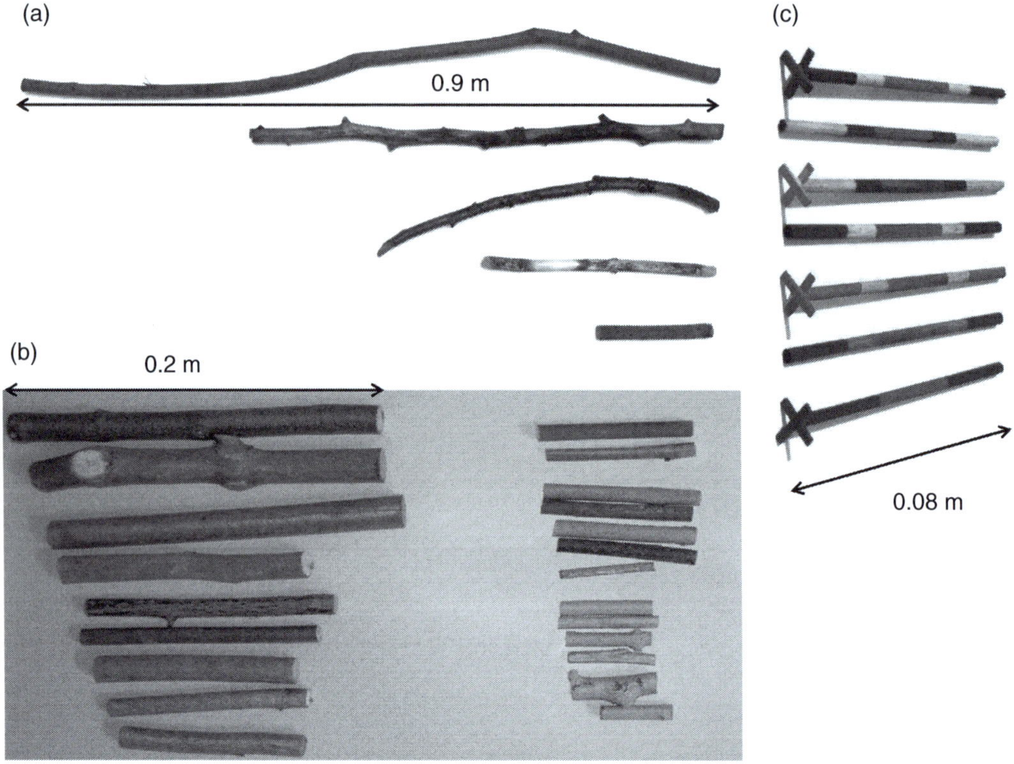

Figure 27.4 Example of the wood models used by (a) Schmocker and Hager (2011), (b) Schmocker and Hager (2013), and (c) Welber *et al.* (2013). (*Sources*: (a, b) Schmocker and Hager (2011, 2013). Reproduced with permission of American Society of Civil Engineers. (c) Welber *et al.* (2013). Reproduced with permission of John Wiley & Sons.)

wood to enhance habitat diversity (Nagayama and Nakamura 2010; Abbe and Brooks 2011). Scaled experiments may help with the design of these actions, assessing the potential implications of different restoration schemes. Large wood modifies the flow field, both at the local scale and, on average, at the reach scale, generally increasing the flow depth and decreasing flow velocity, thus creating more suitable habitats. Bocchiola (2011) explored the availability of fish habitats (measured through the wetted usable area index) by reproducing clustered and distributed wood inputs over a fixed, plane bed with rigid obstacles. These experiments confirmed the beneficial effect of logs on fish habitat, due to the more complex and variable flow conditions, and provided a tool to quantify this effect. Further experiments involving sediment transport and more complex morphologies may support a more detailed understanding of the ecological role of instream large wood and its sustainable and effective use in river restoration projects. Recently, Gallisdorfer *et al.* (2014) reported on preliminary outcomes of experiments with a combined fixed-bed and movable-bed aimed at exploring the impact of engineered log jams on river hydraulics and morphology. Results of the analysis show that these log jams have a great effect on the local flow field, decreasing the velocity near the structure, with a corresponding acceleration in the central part of the channel. The movable-bed runs highlighted the occurrence of local scour and deposition patterns, which extended markedly both downstream and across the channel width. In this case, log jams were fixed, and this probably enhanced their influence on bed morphology compared to the case of natural wood deposits.

27.4.3 Dynamics of Large Wood and River Morphology

Although wood is recognized as one of the crucial drivers for river morphology (Gurnell 2013), very few studies addressed this issue through experimental investigation. Braudrick *et al.* (1997) likely presented the first results on wood dispersion in a mobile bed channel, where large wood dynamics (transport, deposition, remobilization) are coupled with morphological evolution. Welber *et al.* (2013) and Bertoldi *et al.* (2014b) proposed a series of laboratory experiments showing that physical modelling is a suitable tool to quantify wood mobility and dispersion, provided that the key morphological processes are correctly scaled. In this case, the main variables are: (i) the ratio between log length and channel width, assuming application to large rivers, i.e. logs that do not span from one bank to the opposite; and (ii) the ratio between channel depth and log diameter, in order to scale floating conditions. The presence of root wad may also be reproduced in a simple way, adding cross-shaped elements to a cylindrical log (Figure 27.4c) in order to avoid log rolling. The coupling of instream wood and sediment dynamics demonstrated the strong relationship between these transport processes, with large wood deposition patterns mainly determined by the formation and shape of sediment bars. In the case of complex fluvial morphologies (such as braided networks), most wood is likely to deposit at bar-apex locations, and to travel a relatively short distance, with a probability of 30–50% to stop on the first bar (Welber *et al.* 2013). Also, resuspension patterns are strongly related to river morphological changes. At constant discharge and in a dynamic system, large wood turnover rate is fast, and comparable to the turnover of dry areas (Bertoldi *et al.* 2014b).

The analysis of deposition patterns in terms of downstream distribution and accumulation size showed that travel distance is primarily controlled by log diameter, whereas log length and presence of roots affect the tendency to form large jams. These experiments also highlighted the tendency of complex morphologies to create scattered distributions of logs, with jams generally including a limited number of logs (from 2 to 5). Single logs are also very common, counting for 30–40% of total (Figure 27.5a). This is a major difference compared to single thread, narrower rivers, where several reports cite the occurrence of log jams with tens or hundreds of elements.

Physical modelling has proven fundamental for providing the possibility to single out the effect of different parameters, which is difficult to perform with field data. Changes in log properties, log input rate, and discharge showed that the wood–morphology interaction is likely to exhibit a threshold behaviour, with relatively small changes affecting the spatial and temporal distribution

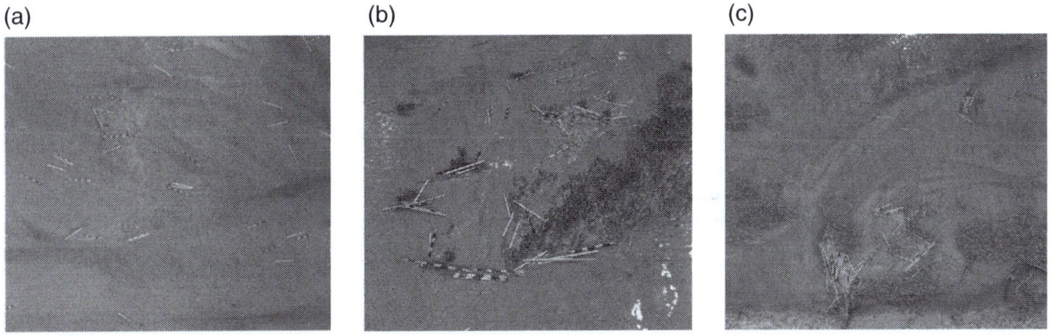

Figure 27.5 Images from the Hull experiments. (a) Vertical image of a run without vegetation; (b) lateral view of the vegetation–wood interaction; (c) vertical image showing the formation of a large wood jam. Flow is from left to right; wood dowels are 8 cm long.

of the wood deposits. For example, an increase of the wood input rate over 100 logs/h (in scaled conditions) resulted in a drastic increase of the log-jam size, suggesting that localized wood input (through forested bank erosion or tributary delivery) may alter wood dynamics in a relevant way.

However, experiments are still lacking some fundamental components, such as the role of fine sediments and the possibility of logs to sprout and act as nuclei of pioneer island formation. This may explain the absence of relevant effects of large wood on bed morphology in the experiments reported by Bertoldi *et al.* (2014b).

Bertoldi *et al.* (2015) tested the possibility of investigating the effects of living vegetation and large wood on river morphology in a coupled way. Figure 27.5 reports two examples of the wood deposits that were reproduced in these experiments, which have an impressive resemblance to many reported observations in the field (e.g. Abbe and Montgomery 1996). Experiments have confirmed that vegetation increases bank stability, reducing erosion and the number of active channels (Tal and Paola 2010; van Dijk *et al.* 2013). They have also shown how this affects large wood dynamics, promoting the formation of stable wood jams, where logs accumulated constantly during subsequent floods, further increasing the effect on river morphology. They confirmed the important joint impact of riparian woodland and large wood on river channel form and dynamics, illustrating their aggregate effects on the morphology.

27.5 Numerical Modelling of Instream Large Wood Transport

Wood mobility has been investigated using different approaches, from the earliest conceptual models for wood budgeting and transport rate, to the latest numerical techniques for wood hydrodynamics simulation. In this section we focus mainly on the latter. One of the most relevant steps when applying wood dynamics modelling is to establish wood initial and boundary conditions. A detailed understanding of the fluvial corridor and the riparian vegetation is needed to estimate large wood availability and dimensions in terms of length, diameter, and density. Detailed data at the level of individual logs are scarce. Possible approaches to overcome this limitation include a forest inventory using geographic information systems (Rigon, Comiti, and Lenzi 2012), as well as the application of fuzzy-logic principles (Ruiz-Villanueva *et al.* 2014d).

27.5.1 Modelling Large Wood Transport in Rivers Using a Computational Fluid-Dynamics Approach

Interactions of solid particles with flows are very common in nature. Therefore, over the years researchers have been working on modelling the dynamic behaviour of particle transport, and nowadays an extensive literature exists describing different methods for different purposes (Feldman 1966; Bangert and Sagdeo 1977; Zhang *et al.* 2001; Lakehal 2002; van Wachem *et al.* 2015). The use of CFDs is advantageous to better understand some complex river processes (Bates, Lane, and Ferguson 2005; Versteeg and Malalasekera 2007). Unlike sediment transport, modelling wood transport has been attempted only recently. Braudrick and Grant (2000) presented the first theoretical model addressing the hydraulic conditions for wood transport and provided the basic framework to approach wood transport. They considered wood entrainment with reference to a force-balance model acting on a cylinder, and they analysed the influence of different log characteristics (orientation, size, density, and presence of roots) on wood mobility, comparing the theoretical approach with flume experiments under steady and uniform flow. Many other studies on wood motion also used simplified continuity or Manning equations to estimate flow conditions and to analyse wood transport and deposition (Braudrick and Grant 2001; Wilcox and Wohl 2006; Bocchiola 2011).

Others used one-dimensional or two-dimensional CFD models, or a combination of both, first computing the hydraulics and then using the results to calculate wood mobilization and deposition (Merten *et al.* 2010; Comiti *et al.* 2012; Hafs *et al.* 2014). One of the first attempts to simulate wood transport was carried out by Mazzorana *et al.* (2010) on a cell-by-cell basis with unsteady flow conditions. They delineated the possible pathways for a given wood volume and computed the transport conditions using results from a hydrodynamic two-dimensional simulation at different time-steps.

Ruiz-Villanueva *et al.* (2014a) presented a deterministic hydrodynamic model to simulate the transport of multiple wood elements of different sizes under complex hydraulic conditions at short time-scales. This couples the two-dimensional hydrodynamic model IBER (Bladé *et al.* 2014; developed at FLUMEN – University of Catalonia) to a new module that allows the inclusion of wood in the simulations. The approach uses a mixed Eulerian–Lagrangian framework based on a modified version of the theoretical model from Braudrick and Grant (2000).

Therefore, when logs are placed in the modelling domain, the incipient motion is determined by the force balance acting on the centre mass of the log. Once the log is put in motion, translation and rotation are considered depending on the two-dimensional velocity field acting on the two ends of the log, and two possible transport mechanisms are implemented, sliding on the riverbed, or floating. The presence of wood reduces the available storage volume at every finite volume of the model domain, which is possible by adding a new shear stress (produced by the drag force of the logs) term in the Saint Venant equations. In addition, log velocity and movement is modified at every time-step when interactions with the river geometry or among logs occur.

The numerical model needs initial and boundary conditions for instream wood. The initial position of each log (x, y coordinates of mass centre and angle with respect to the flow), its length, diameter, and wood density for the initial time-step must be provided. Moreover, inlet boundary conditions (i.e. logs entering the simulation) can also be assigned to the simulation domain boundaries, specifying a number of wood pieces per minute and their characteristics. Stochastic variations of these parameters, together with position and angle, are used to characterize the wood rate entering the domain.

As usual in models, simplifications have to be adopted such as the geometry of the logs as cylinders, disregarding the effect of branches or roots. This simplification can result in a deviation when velocity and water level are calculated near to the log where complex flow is formed. Three-dimensional high resolution CFD modelling can then be used to represent more precisely the real shape and the local velocity field and turbulent flow conditions.

Allen and Smith (2012) reported one of the few three-dimensional modelling attempts to quantify the numerical effect of wood geometric simplifications on the surrounding flow field, comparing results of three-dimensional CFD modelling with experiments. Following the same approach Lai and Bandrowski (2014) presented a strategy for combining field observations and three-dimensional models. However, three-dimensional modelling of wood is still rarely carried out since mesh representation of complex wood shapes can be a daunting task and simulation of such flows is very challenging (in addition to the high computational requirements).

These models were proposed and applied to answer different research questions, helping to improve our knowledge about wood dynamics. Some of the analysed aspects are: (i) understanding control factors of wood transport and deposition; (ii) potential wood-related hazards; and (iii) influence of large wood on the fluvial ecosystem. In the following we present some examples of the first two topics.

27.5.2 Exploring Factors Controlling Large Wood Transport and Deposition

Linking direct field observations and numerical modelling allows us to analyse different factors controlling wood transport. The relationships between wood transport rate and discharge, or wood size and river morphology can be explored by setting different scenarios. An ongoing research project (FLORIST, funded by a Polish–Swiss Joint Research Programme; Kundewicz *et al.* 2014) focuses on this task, combining numerical modelling and field surveys to investigate large wood dynamics in the Czarny Dunajec River, southern Poland. This river was studied in detail regarding wood storage (Wyzga and Zawiesjka 2005, 2010), and therefore provides an excellent opportunity to verify and compare field observations and model results. Different controls on wood mobilization and transport can be analysed, including the effect of bed morphology, since the river presents contrasting geomorphic configurations. Numerical simulations provide the tool to investigate the effect of different flood regimes and different wood configurations, which are rather difficult to be observed in the field (for more details see Ruiz-Villanueva *et al.* 2015).

Preliminary results show that the wood transport ratio (i.e. the number of pieces in transport) strongly decreased with increasing wood volume, but this relation is not linear. The scatter of points in the diagrams (Figure 27.6) shows that different discharges resulted in different transport ratios for the same wood volume. Results highlight the role of bed morphology. Wood mobilization is significantly lower in multi-thread reaches (Figure 27.6b) than in single-thread configurations (Figure 27.6a), especially for larger wood volumes. Field observations monitoring tagged logs after several floods confirmed the modelling results.

In terms of wood shape, a decreasing linear relation between transport ratio and log diameter was found for different log lengths, when considering the single-thread reach. A change in the log length makes a significant difference in the transport ratio in this morphometric configuration (Figure 27.6c). However, this relationship is different for a multi-thread morphology where log length seems not to affect wood transport (Figure 27.6d). These observations are explained by the different flow configuration in the two morphologies. Water-depth and flow-velocity distributions are different in the two configurations, with a higher occurrence of deep flow in the single-thread reach, where thicker pieces can be transported more easily. On the other hand, in the wider multi-thread reach log length exerts a weaker control, as discussed also by Welber, Bertoldi, and Tubino (2013).

27.5.3 Potential Hazards Related to Instream Large Wood

Large wood may be entrapped at critical cross-sections (e.g. bridges) and therefore dramatically increase the destructive power of floods. This was reported after recent floods especially in forested mountain basins. One example is the Arroyo Cabrera (Tagus River Basin, central Spain), where a flash flood event with a high content of large wood transport took place in 1997. This small poorly gauged forested mountain catchment has since been monitored from a few days after this extreme flood happened, in addition several previous studies have provided important information with which to validate model results (Ruiz-Villanueva *et al.* 2012, 2014a,b).

Therefore, numerical modelling has allowed simulation of hazard processes related to large wood, such as bridge clogging and reconstructing wood deposition patterns. Ranges for the diameters, density, and length of wood have to be established based on knowledge of the riparian vegetation. When detailed information about wood is not available, stochastic variation of these parameters can be used to characterize each piece of wood entering the simulation. In addition a series of recruitment scenarios can be designed (congested, semi-congested, and steady, as defined by Braudrick *et al.* 1997) to define the temporal distribution of the recruited wood volume.

Modelling results showed the main depositional sites and areas prone to form wood jams (Figure 27.7). Moreover, the two-dimensional model quantified the probability of a log to be blocked at the bridge site, along with the effect of wood transport and deposition on the hydrodynamics.

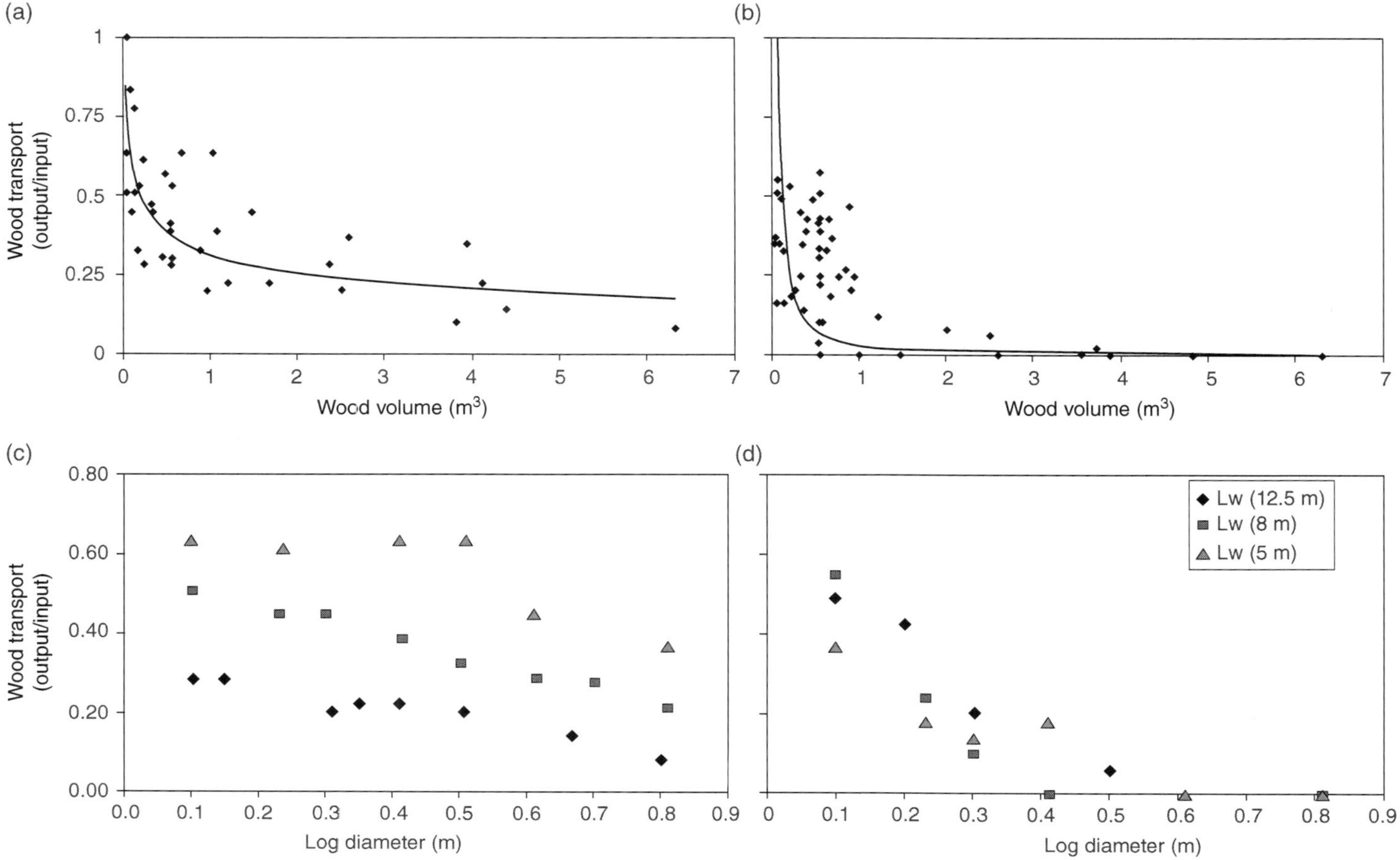

Figure 27.6 (a, b) Scatter plots and estimated regression relationships between large wood transport ratio and wood volume. (c, d) Relationship between wood transport ratio and log diameter for different log lengths. For (a, c) a single-thread reach and (b, d) a multi-thread reach. Lw is the wood length.

Figure 27.7 Field observations and model results for the simulation of the bridge clogging process assuming a semi-congested wood transport scenario. (Modified from Ruiz-Villanueva *et al.* (2014b).) (See insert for color representation of this figure.)

The main effect of the bridge clogging was a significant increase in the water depth and decrease in flow velocity, facilitating sediment and further wood deposition upstream of the bridge. Different wood recruitment scenarios provided different results. The model correctly reproduced that most of the wood was deposited in the bridge section, with further logs and wood jams deposited in slower and shallower flow areas along the river margins.

The presence of large wood had a significant impact on flood hazard. The reconstructed flood hydrograph produced a water level ~ 4.7 m at the bridge section in a hypothetical scenario without wood; this value increases to ~7.3 m in the case of a semi-congested transport scenario and > 8 m when considering a congested transport. A steady wood transport scenario resulted in a moderately affected case, where the bridge just partially blocked and the water depth did not reach the bridge deck, rising up to 6.8 m. The observed peak value was measured at 7.1 m, with the semi-congested scenario producing the best fit in this case.

This example demonstrated that models might be particularly important for practitioners who must assess and manage flooding risks associated with wood to identify large wood delivery sources and potentially critical sections and infrastructures (Ruiz-Villanueva *et al.* 2014c). However, systematic analysis still requires considerable research.

27.6 Future Challenges

Physical and numerical modelling are powerful tools that allow study of flow, sediment, and vegetation dynamics in a controlled environment and open new possibilities for understanding and disentangling the complex linkages in the bio-geomorphological evolution of the fluvial system. Many physical and numerical models have been proposed, with different details depending on the aim of the investigation. The spatial and temporal scales of the problem, as well as the research question, set the suitable type of model. As more detailed representations of complex processes require extended datasets and the availability of accurate measurements to assess the modelling parameters the combined use of physical and numerical models has not been little explored, but has great potential. Numerical models need detailed data for calibration and validation of the new modules including vegetation and large wood transport. In most cases, laboratory experiments are the only way to provide data with the necessary high spatial and temporal resolution, including details on sediment transport flux that rarely can be measured in the field. Physical and numerical models can investigate the role of different parameters under similar simplifying hypotheses, to understand the dynamics of the different components, and to highlight the existence of threshold behaviours and tipping points.

The advances in modelling techniques and the improvements in the overall understanding of large wood and vegetation dynamics suggest new challenges for the next-generation models.

1) Vegetation and wood have complex shapes, but are often schematized with very simple elements, such as cylindrical sticks. A better characterization in terms of stem/foliage distribution, root structure, and wood form may lead to a more accurate reproduction of the interaction with flow and sediments.
2) Inclusion of vegetation and large wood requires specific analysis on the relevant time scales. Comparable temporal evolution scales of both vegetation and riverbed morphology characterize dynamic riparian systems. This means that relatively small changes of one of the two may lead to large changes, where one of the two elements takes control of the other.
3) Discharge variability and flood occurrence timing are relevant parameters that should be included in future models. Most models dealing with hydraulic or morphological problems usually consider

constant flow or regular sequences of hydrographs. However, when considering living vegetation, flood timing and seasonality are crucial, as they may have a great impact on the recruitment dynamics, vegetation growth, and thus species composition (Mahoney and Rood 1998; Greet, Webb, and Cousens 2011).

4) There is a missing link between vegetation dynamics (particularly erosion) and wood production and transport. The two are obviously linked, however, no model (physical or numerical) has been proposed that addresses this issue.
5) Vegetation and wood are also linked when considering asexual reproduction of riparian species, such as *Populus* and *Salix*. Asexual reproduction benefits from physical damage to plants by fluvial processes and has therefore a different response to environmental conditions. Vegetation growth from deposited wood is a key process to reproduce the evolution of large rivers with exposed bars, such as the braided Tagliamento River.

Acknowledgements

The authors wish to thank Paolo Perona and Lukas Schmocker for the useful comments and suggestions that improved this manuscript.

References

Abbe, TB and Brooks, A 2011. Geomorphic, engineering, and ecological consideration when using wood in river restoration. In *Stream Restoration in Dynamic Fluvial Systems: Scientific Approaches, Analyses, and Tools* (eds A Simon, SJ Bennett and JM Castro). Monograph 194, American Geophysical Union: Washington, DC; 419–451.

Abbe, TB and Montgomery, DR 1996. Large woody debris jams, channel hydraulics and habitat formation in large rivers. *Regulated Rivers Research and Management* 12, 201–221.

Aberle, J and Järvelä, J 2013. Flow resistance of emergent rigid and flexible floodplain vegetation. *Journal of Hydraulic Research* 51, 33–45.

Ackerman, JD and Okubo, A 1993. Reduced mixing in a marine macrophyte canopy. *Functional Ecology* 7, 305–309.

Allen, JB and Smith, DL 2012. Characterizing the impact of geometric simplification on large woody debris using CFD. *International Journal of Hydraulic Engineering* 1, 1–14.

Antonarakis, AS, Richards, KS, Brasington, J, and Bithell, M 2009. Leafless roughness of complex tree morphology using terrestrial lidar. *Water Resources Research* 45, W10401.

Antonarakis, AS, Richards, KS, Brasington, J, and Muller, E 2010. Determining leaf area index and leafy tree roughness using terrestrial laser scanning. *Water Resources Research* 46, W06510.

Bangert, LH and Sagdeo, PM 1977. On fiber alignment using fluid-dynamic forces. *Textile Research Journal* 12, 773–780.

Bates, PD, Lane, SN, and Ferguson, RI 2005. *Computational Fluid Dynamics: Applications in Environmental Hydraulics*. John Wiley & Sons: Chichester, UK.

Benjankar, R, Egger, G, Jorde, K *et al.* 2011. Dynamic floodplain vegetation model development for the Kootenai River, USA. *Journal of Environmental Management* 92, 3058–3070.

Benjankar, R, Burke, M, Yager, E *et al.* 2014. Development of a spatially-distributed hydroecological model to simulate cottonwood seedling recruitment along rivers. *Journal of Environmental Management* 145, 277–288.

Bernhardt, E, Sudduth, E, Palmer, M *et al.* 2007. Restoring rivers one reach at a time: results from a survey of US river restoration practitioners. *Restoration Ecology* 15, 482–493.

Bertoldi, W, Gurnell, AM, and Drake, NA 2011. The topographic signature of vegetation development along a braided river: Results of a combined analysis of airborne lidar, color air photographs, and ground measurements. *Water Resources Research* 47, W06525.

Bertoldi, W, Siviglia, A, Tettamanti, S *et al.* 2014a. Modeling vegetation controls on fluvial morphological trajectories. *Geophysical Research Letters* 41, 7167–7175.

Bertoldi, W, Welber, M, Mao, L *et al.* 2014b. A flume experiment on wood storage and remobilization in braided river systems. *Earth Surface Processes and Landforms* 39, 804–813.

Bertoldi, W, Welber, M, Gurnell, AM *et al.* 2015. Physical modelling of the combined effect of vegetation and wood on river morphology. *Geomorphology* 246, 178–187.

Bladé, E, Cea, L, Corestein, G *et al.* 2014. Iber: river modelling simulation tool. *Revista Internacional de Métodos Numéricos para Cálculo y Diseño en Ingeniería* 30, 1–10.

Bocchiola, D 2011. Hydraulic characteristics and habitat suitability in presence of woody debris: a flume experiment. *Advances in Water Resources* 34, 1304–1319.

Bocchiola, D, Rulli, MC, and Rosso, R 2006. Flume experiments on wood entrainment in rivers. *Advances in Water Resources* 29, 1182–1195.

Braudrick, CA and Grant, GE 2000. When do logs move in rivers? *Water Resources Research* 36, 571–583.

Braudrick, CA and Grant, GE 2001. Transport and deposition of large woody debris in streams: a flume experiment. *Geomorphology* 41, 263–283.

Braudrick, CA, Grant, GE, Ishikawa, Y, and Ikeda, H 1997. Dynamics of wood transport in streams: a flume experiment. *Earth Surface Processes and Landforms* 22, 669–683.

Braudrick, CA, Dietrich, WE, Leverich, GT, and Sklar, LS 2009. Experimental evidence for the conditions necessary to sustain meandering in coarse-bedded rivers. *Proceedings of the National Academy of Science USA* 106, 16936–16941.

Camporeale, C and Ridolfi, L 2006. Riparian vegetation distribution induced by river flow variability: a stochastic approach. *Water Resources Research* 42, W10415.

Camporeale, C, Perucca, E, Ridolfi, L, and Gurnell, AM 2013. Modelling the interactions between river morphodynamics and riparian vegetation. *Reviews of Geophysics* 51, 1–36.

Clarke, L 2014. The use of live vegetation in geomorphological experiments: how to create optimal growing conditions. *Earth Surface Processes and Landforms* 39, 705–710.

Comiti, F, D'Agostino, V, Moser, M *et al.* 2012. Preventing wood-related hazards in mountain basins: from wood load estimation to designing retention structures. In *Proceedings of the 12th Congress INTERPRAEVENT* (eds G Koboltschnig, J Hübl and J Braun). International Research Society INTERPRAEVENT: Klagenfurt; Vol. 1, 651–662.

Corenblit, D, Tabacchi, E, Steiger, J, and Gurnell, AM 2007. Reciprocal interactions and adjustments between fluvial landforms and vegetation dynamics in river corridors: A review of complementary approaches. *Earth Science Reviews* 84, 56–86.

Corenblit, D, Baas, ACW, Bornette, G *et al.* 2011. Feedbacks between geomorphology and biota controlling Earth surface processes and landforms: a review of foundation concepts and current understandings. *Earth Science Reviews* 106, 307–331.

Corenblit, D, Steiger, J, Gonzalez, A *et al.* 2014. The biogeomorphological life cycle of poplars during the fluvial biogeomorphological succession: a special focus on *Populus nigra* L. *Earth Surface Processes and Landforms* 39, 546–563.

Coulthard, TJ, Hicks, DM, and Van de Wiel, MJ 2007. Cellular modelling of river catchments and reaches: advantages, limitations and prospects. *Geomorphology* 90, 192–207.

Crosato, A and Saleh, MS 2011. Numerical study on the effects of floodplain vegetation on river planform style. *Earth Surface Processes and Landforms* 36, 711–720.

Crosato, A, Rajbhandari, N, Comiti, F *et al.* 2013. Flume experiments on entrainment of large wood in lowland rivers. *Journal of Hydraulic Research* 51, 1–8.

Crouzy, B and Perona, P 2012. Biomass selection by floods and related timescales: Part 2. Stochastic modeling. *Advances in Water Resources* 39, 97–105.

Crouzy, B, Edmaier, K, Pasquale, N, and Perona, P 2013. Impact of floods on the statistical distribution of riverbed vegetation. *Geomorphology* 202, 51–58.

Dixon, MD and Turner, MG 2006. Simulated recruitment of riparian trees and shrubs under natural and regulated flow regimes on the Wisconsin River, USA. *River Research and Applications* 22, 1057–1083.

Edmaier, K, Burlando, P, and Perona, P 2011. Mechanisms of vegetation uprooting by flow in alluvial non-cohesive sediment. *Hydrology and Earth System Sciences* 15, 1615–1627.

Edmaier, K, Crouzy, B, Ennos, R *et al.* 2014. Influence of root characteristics and soil variables on the uprooting mechanics of *Avena sativa* and *Medicago sativa* seedlings. *Earth Surface Processes and Landforms* 39, 1354–1364.

Eke, E, Parker, G, and Shimizu, Y 2014. Numerical modeling of erosional and depositional bank processes in migrating river bends with self-formed width: Morphodynamics of bar push and bank pull. *Journal of Geophysical Research* 119, 1455–1483.

Euler, T, Zemke, J, Rodrigues, S, and Herget, J 2014. Influence of inclination and permeability of solitary woody riparian plants on local hydraulic and sedimentary processes. *Hydrological Processes* 28, 1358–1371.

Feldman, L 1966. Theoretical trajectory studies of light bodies in non-uniform two-dimensional flows. *Textile Research Journal* 9, 809–813.

Frostick, LE, Thomas, RE, Johnson, MF *et al.* 2014, *User Guide to Ecohydraulic Modelling and Experimentation: Experience of the Ecohydraulic Research Team (PISCES) of the HYDRALAB Network.* IAHR Design Manual, CRC Press/Balkema: Leiden, The Netherlands.

Gallisdorfer, MS, Bennett, SJ, Atkinson, JF *et al.* 2014. Physical-scale model designs for engineered log jams in rivers. *Journal of Hydro-environmental Research* 8, 115–128.

Ghisalberti, M and Nepf, H 2006. The structure of the shear layer over rigid and flexible canopies. *Environmental Fluid Mechanics* 6, 277–301.

Gorrick, S and Rodríguez, JF 2012. Sediment dynamics in a sand bed stream with riparian vegetation. *Water Resources Research* 48, W02505.

Gran, K and Paola, C 2001. Riparian vegetation controls on braided stream dynamics. *Water Resources Research* 37, 3275–3283.

Greet, JOE, Webb, JA, and Cousens, RD 2011. The importance of seasonal flow timing for riparian vegetation dynamics: a systematic review using causal criteria analysis. *Freshwater Biology* 56, 1231–1247.

Gurnell, AM 2013. Wood in fluvial systems. In *Treatise on Geomorphology 9, Fluvial Geomorphology* (eds J Shroder and E Wohl). Academic Press: San Diego, CA; 163–188.

Gurnell, AM 2014. Plants as river system engineers. *Earth Surface Processes and Landforms* 39, 4–25.

Gurnell, AM, Bertoldi, W, and Corenblit, D 2012. Changing river channels: The roles of hydrological processes, plants and pioneer fluvial landforms in humid temperate, mixed load, gravel bed rivers. *Earth Science Reviews* 111, 129–141.

Hafs, AW, Harrison, LR, Utz, RM, and Dunne, T 2014. Quantifying the role of woody debris in providing bioenergetically favorable habitat for juvenile salmon. *Ecological Modelling* 285, 30–38.

Hammersmark, CT, Dobrowski, SZ, Rains, MC, and Mount, JF 2010. Simulated effects of stream restoration on the distribution of wet-meadow vegetation. *Restoration Ecology* 18, 882–893.

Hooke, JM, Brookes, CJ, Duane, W, and Mant, JM 2005. A simulation model of the morphological, vegetation and sediment changes in ephemeral streams. *Earth Surface Processes and Landforms* 30, 845–866.

Kim, HS, Kimura, I, and Shimizu, Y 2015. Bed morphological changes around a finite patch of vegetation. *Earth Surface Processes and Landforms* 40, 375–388.

Kundzewicz, ZW, Stoffel, M, Kaczka, RJ *et al.* 2014. Floods at the northern foothills of the Tatra Mountains – a Polish–Swiss research project. *Acta Geophysica* 62, 620–641.

Lai, Y and Bandrowski, D 2014. Large wood flow hydraulics: a 3D modelling approach. In *Proceedings of the 7th International Congress on Environmental Modelling and Software Society (iEMSs)* (eds DP Ames, NWT Quinn, and AE Rizzoli), San Diego; 2163–2171.

Lakehal, D 2002. On the modelling of multiphase turbulent flows for environmental and hydrodynamic applications. *International Journal of Multiphase Flow* 28, 823–863.

Le Bouteiller, C and Venditti, JG 2014. Vegetation-driven morphodynamic adjustments of a sand bed. *Geophysical Research Letters* 41, 3876–3883.

Li, SS and Millar, RG 2011. A two-dimensional morphodynamic model of gravel-bed river with floodplain vegetation. *Earth Surface Processes and Landforms* 36, 190–202.

Luhar, M and Nepf, HM 2012. From the blade scale to the reach scale: A characterization of aquatic vegetative drag. *Advances in Water Resources* 51, 305–316.

Mahoney, JM and Rood, SB 1998. Streamflow requirements for cottonwood seedling recruitment–an integrative model. *Wetlands* 18, 634–645.

Mazzorana, B, Hübl, J, Zischg, A, and Largiader, A 2010. Modelling woody material transport and deposition in alpine rivers. *Natural Hazards* 56, 425–449.

Merten, E, Finlay, J, Johnson, L *et al.* 2010. Factors influencing wood mobilization in streams. *Water Resources Research* 46, W10514.

Micheli, ER and Kirchner, JW 2002. Effects of wet meadow riparian vegetation on streambank erosion: 2. Measurements of vegetated bank strength and consequences for failure mechanics. *Earth Surface Processes and Landforms* 27, 687–697.

Motta, D, Abad, JD, Langendoen, EJ, and Garcia, MH 2012. A simplified 2-D model for meander migration with physically based bank evolution. *Geomorphology* 163–164, 10–25.

Murray, AB and Paola, C 2003. Modelling the effect of vegetation on channel pattern in bedload rivers. *Earth Surface Processes and Landforms* 28, 131–143.

Nagayama, S and Nakamura, F 2010. Fish habitat rehabilitation using wood in the world. *Landscape and Ecological Engineering* 6, 289–305.

Nepf, HM 2012. Flow and transport in regions with aquatic vegetation. *Annual Review of Fluid Mechanics* 44, 123–142.

Nicholas, AP, Ashworth, PJ, Sambrook Smith GH, and Sandbach, SD 2013. Numerical simulation of bar and island morphodynamics in anabranching megarivers. *Journal of Geophysical Research* 118, 2019–2044.

Nikora, V 2010. Hydrodynamics of aquatic systems: An interface between ecology, biomechanics and environmental fluid mechanics. *River Research and Applications* 26, 367–384.

Orellana, F, Verma, P, Loheide, SP, and Daly, E 2012. Monitoring and modeling water-vegetation interactions in groundwater-dependent ecosystems. *Reviews of Geophysics* 50(RG3003.

Paola, C 2001. Modelling stream braiding over a range of scales. In *Gravel Bed Rivers V* (ed. MP Mosley). New Zealand Hydrological Society; 11–46.

Paola, C, Straub, K, Mohrig, D, and Reinhardt, L 2009. The unreasonable effectiveness of stratigraphic and geomorphic experiments. *Earth Science Reviews* 97, 1–43.

Pasquale, N, Perona, P, Francis, R, and Burlando, P 2012. Effects of streamflow variability on the vertical root density distribution of willow cutting experiments. *Ecological Engineering* 40, 167–172.

Pasquale, N, Perona, P, Francis, R, and Burlando, P 2014. Above-ground and below-ground Salix dynamics in response to river processes. *Hydrological Processes* 28, 5189–5203.

Perona, P, Molnar, P, Crouzy, B *et al.* 2012. Biomass selection by floods and related timescales: part 1. Experimental observations. *Advances in Water Resources* 39, 85–96.

Perona, P, Crouzy, B, McLelland, S *et al.* 2014. Ecomorphodynamics of rivers with converging boundaries. *Earth Surface Processes and Landforms* 39, 1651–1662.

Perucca, E, Camporeale, C, and Ridolfi, L 2007. Significance of the riparian vegetation dynamics on meandering river morphodynamics. *Water Resources Research* 43, W03430.

Pollen, N and Simon A 2005. Estimating the mechanical effects of riparian vegetation on stream bank stability using a fiber bundle model. *Water Resources Research* 41, W07025.

Pollen-Bankhead, N and Simon, A 2010. Hydrologic and hydraulic effects of riparian root networks on streambank stability: is mechanical root-reinforcement the whole story?. *Geomorphology* 116, 353–362.

Pollen-Bankhead, N, Thomas, RE, Gurnell, AM *et al.* 2011. Quantifying the potential for flow to remove the emergent aquatic macrophyte Sparganium erectum from the margins of low-energy rivers. *Ecological Engineering* 37, 1779–1788.

Polvi, L, Wohl, E, and Merritt, D 2014. Modeling the functional influence of vegetation type on streambank cohesion. *Earth Surface Processes and Landforms* 39, 1245–1258.

Rigon, E, Comiti, F, and Lenzi, MA 2012. Large wood storage in streams of the Eastern Italian Alps and the relevance of hillslope processes. *Water Resources Research* 48, W01518.

Rivaes, RP, Rodrìguez-Gonzàlez, PM, Ferreira, MT *et al.* 2014. Modeling the evolution of riparian woodlands facing climate change in three European rivers with contrasting flow regimes. *PLoS ONE* 9, e110200.

Ruiz-Villanueva, V, Bodoque, JM, Díez-Herrero, A *et al.* 2012. Reconstruction of an ungauged flash flood event with large wood transport and its influence on hazard patterns. *Hydrological Process* 27, 3424–3437.

Ruiz-Villanueva, V, Bladé-Castellet, E, Sánchez-Juny, M *et al.* 2014a. Two dimensional numerical modelling of wood transport. *Journal of Hydroinformatics* 16, 1077–1096.

Ruiz-Villanueva,V, Bladé-Castellet, E, Díez-Herrero, A *et al.* 2014b. Two-dimensional modelling of large wood transport during flash floods. *Earth Surface Processes and Landforms* 39, 438–449.

Ruiz-Villanueva, V, Bodoque, JM, Díez-Herrero, A, and Bladé, E 2014c. Large wood transport as significant influence on flood risk in a mountain village. *Natural Hazards* 74, 967–987.

Ruiz-Villanueva, V, Díez-Herrero, A, Ballesteros, JA, and Bodoque, JM 2014d. Potential large woody debris recruitment due to landslides, bank erosion and floods in mountain basins: a quantitative estimation approach. *River Research and Applications* 30, 81–97.

Ruiz-Villanueva, V, Wyżga, B, Zawiejska, J, Hajdukiewicz, M, and Stoffel, M 2015. Factors controlling large wood transport in a mountain river. *Geomorphology* in press.

Schmocker, L and Hager, WH 2011. Probability of drift blockage at bridge decks. *Journal of Hydraulic Engineering ASCE* 137, 269–274.

Schmocker, L and Hager, WH 2013. Scale modelling of wooden debris accumulation at a debris rack. *Journal of Hydraulic Engineering ASCE* 139, 827–836.

Schmocker, L and Weitbrecht, V 2013. Driftwood: risk analysis and engineering measures. *Journal of Hydraulic Engineering ASCE* 139, 683–695.

Sukhodolov, AN and Sukhodolova, TA 2012. Vegetated mixing layer around a finite-size patch of submerged plants: Part 2. Turbulence statistics and structures. *Water Resources Research* 48, W12506.

Tal, M and Paola, C 2007. Dynamic single-thread channels maintained by the interaction of flow and vegetation. *Geology* 35, 437–350.

Tal, M and Paola, C 2010. Effects of vegetation on channel morphodynamics: results and insights from laboratory experiments. *Earth Surface Processes and Landforms* 35, 1014–1028.

Tron, S, Laio, F, and Ridolfi, L 2014. Effect of water table fluctuations on phreatophytic root distribution. *Journal of Theoretical Biology* 360, 102–108.

Van Dijk, WM, Teske, R, van de Lagenweg, WI,and Kleinhans, MG 2013. Effects of vegetation distribution on experimental river channel dynamics. *Water Resources Research* 49, 1–17.

Van Wachem, B, Zastawny, M, Zhao, F, and Mallouppas, G 2015. Modelling of gas–solid turbulent channel flow with non-spherical particles with large Stokes numbers. *International Journal of Multiphase Flow* 68, 80–92.

Vargas-Luna, A, Crosato, A, and Uijttewaal, WSJ 2015. Effects of vegetation on flow and sediment transport: comparative analyses and validation of predicting models. *Earth Surface Processes and Landforms* 40, 157–176.

Västilä, K and Järvelä, J 2014. Modeling the flow resistance of woody vegetation using physically based properties of the foliage and stem. *Water Resources Research* 50, 229–245

Västilä, K, Järvelä, J, and Aberle, J 2013. Characteristic reference areas for estimating flow resistance of natural foliated vegetation. *Journal of Hydrology* 492, 49–60.

Versteeg, HK and Malalasekera, W 2007, *An introduction to computational fluid dynamics: the finite volume method*, Pearson Education Limited, Harlow, England.

Welber, M, Bertoldi, W, and Tubino, M 2013. Wood dispersal in braided streams: results from physical modeling. *Water Resources Research* 49, 7388–7400.

Wilcox, AC and Wohl, EE 2006. Flow resistance dynamics in step-pool stream channels: 1. Large woody debris and controls on total resistance. *Water Resources Research* 42, W05418.

Wu, WM, Shields, FD, Bennett, SJ, and Wang, SSY 2005. A depth-averaged two-dimensional model for flow, sediment transport, and bed topography in curved channels with riparian vegetation. *Water Resources Research* 41(W03015.

Wyżga, B and Zawiejska, J 2005. Wood storage in a wide mountain river: case study of the Czarny Dunajec, Polish Carpathians. *Earth Surface Processes and Landforms* 30, 1475–1494.

Wyżga, B and Zawiejska, J 2010. Large wood storage in channelized and unmanaged sections of the Czarny Dunajec River, Polish Carpathians: implications for the restoration of mountain rivers. *Folia Geographica* 41, 5–34.

Zhang, H, Ahmadi, G, Fan, F, and McLaughlin, JB 2001. Ellipsoidal particles transport and deposition in turbulent channel flows. *International Journal of Multiphase Flow* 27, 971–1009.

28

Fluvial Gravels on Mars: Analysis and Implications

William E. Dietrich, Marisa C. Palucis, Rebecca M. E. Williams, Kevin W. Lewis, Frances Rivera-Hernandez, and Dawn Y. Sumner

28.1 Introduction

Decades of satellite imagery of Mars have provided strong evidence of an extensive early history of fluvial processes (e.g. Carr and Clow 1981; Craddock and Howard 2002; Moore *et al.* 2003; Malin and Edgett 2003; Moore and Howard 2005; Irwin *et al.* 2008). Channel networks, giant outflow channels, meandering channels, deltas, and fans have been mapped across the planet, and the relative timing of fluvial erosion and deposition events has been estimated from the number and size of craters found on mapped units (e.g. Irwin *et al.* 2008; Carr 2012). These features make a strong case that the early Mars climate had an atmosphere capable of producing precipitation and lasting long enough for a sustained hydrologic cycle. The last period of widespread and extensive fluvial activity ended about 3.5 billions of years ago (e.g. Fassett and Head 2008), and yet while this is still fairly early in Mars' history, climate models suggest that a sustained atmosphere at this time was unlikely, and it remains difficult to predict for even wetter, earlier periods (e.g. Woodsworth *et al.* 2013; Carter *et al.* 2015). This interpretation, in part, has led to a reassessment of fluvial features. Some have proposed, based on modeling, that fans and deltas could form in just a few years, and thus might be the product of transient precipitation from a major impact (e.g. Kleinhans 2005; Kleinhans, van de Kasteele, and Hauber 2010; Mangold *et al.* 2012). Others have considered the possibility that features that appear to be of fluvial origin, may be gas driven, due to eruptive volcanism, or other processes involving little or no running water (e.g. Gulick 1998; Gulick and Baker 1990).

For the first time, as a result of the Mars Science Laboratory (MSL) mission and photographs obtained by the *Curiosity* rover, fluvial gravels have been directly observed on Mars. These observations strongly support previous satellite-based interpretations that fluvial processes have been active on Mars. They also allow us to examine shape, sorting, and grain size. Here we first describe initial observations and interpretations of Martian fluvial gravels. We then show some examples of exposed conglomerates that the rover has crossed during its first 10 km of travel. These observations present a fundamental challenge of great interest on Mars: how much paleohydrologic and climatic information can we extract from a gravel deposit. We review two approaches to reconstructing instantaneous flows and explore the problem of estimating flow when only the grain-size distribution is known. Finally, we

Gravel-Bed Rivers: Processes and Disasters, First Edition. Edited by Daizo Tsutsumi and Jonathan B. Laronne.

show how the total volume of sediment in a deposit (e.g. alluvial fan) has been used to estimate the runoff volume and thus place constraints on the climate history of Mars.

28.2 First Observations of Fluvial Conglomerates on Mars

Curiosity landed on Mars on August 6, 2012 (UTC) on the northern floor of Gale Crater (5.3°S 137.7°E), an impact structure that formed about 3.8–3.6 billion years ago (Thompson *et al.* 2011; Le Deit *et al.* 2013; Figure 28.1). This 157 km diameter crater has a 5 km central mountain of sedimentary rocks (Aeolis Mons) that towers by 2 km over the northern rim of the crater. Remarkably, it was the scour caused by the thrust from the descent engines during *Curiosity's* landing that revealed a lithified, almost concrete-like, appearance to the underlying deposit at the landing site. The exposure at the landing site and subsequently outcrops found in the initial traverse by the rover to the northeast are described in Williams *et al.* (2013) (Figure 28.2). The initial debate upon observing these deposits was about the transport mechanism, which is particularly significant in regards to the volume of water and hence climatic conditions during transport and deposition. Did these deposits record fluvial transport as bedload and suspended load or could these deposits record debris-flow processes (and consequently much less water involved)? Questions were initially raised as well as to whether the deposits did in fact record water-driven transport processes. The key properties of these deposits (Figure 28.2) are: (i) distinctly well-rounded shapes; (ii) local sorting such that like-sized particles were co-deposited; (iii) vertical variation in sorting (stratification); (iv) grain to grain contact (i.e. not matrix supported); (v) some locations of grain imbrication; and (vi) particle sizes up to 40 mm in diameter. Collectively, these properties strongly indicate fluvial transport. Perhaps most persuasive is the strong rounding of particles that are too large to be transported by wind. Not only does this suggest a collisional transport process (i.e. bedload transport) but also a travel distance, which Williams *et al.* (2013) suggest was at least a few kilometers. The median grain size from three outcrops found near the landing site ranged from 4.5 to 9.5 mm. Yingst *et al.* (2014) also found rounded pebbles lying on the surface at various locations over a distance of about 400 m from the initial landing site.

No sense of scale of the fluvial system was indicated in the outcrops; that is, we did not see channel-scale bedding features or preserved banks. These first observations were made in an elevated area (referred to as Bradbury Rise) within a closed basin that lies between Mt Sharp (Aeolis Mons) and the crater wall and into which the terminal end of the Peace Vallis fan enters (Figures 28.1c and 28.3). The Rise is on the shoulder of a ridge that connects the base of Mount Sharp to the crater wall. As Palucis *et al.* (2014) illustrate, this Rise lies downslope of the Peace Vallis fan (Figure 28.3) and it is possible that sheetfloods from the Peace Vallis fan swept across this area (see figure 15 in Palucis *et al.* 2014). Local topography, however, indicates that wind deflation has occurred, and the deposits may thus be exhumed and hence record an earlier period of deposition, perhaps associated with the processes that formed the ridge that crosses the crater floor. Malin and Edgett (2000) proposed that Gale was once entirely filled with sediment and then subsequently a ring of scour developed (presumably by wind deflation) that left a towering (~5 km) low-gradient mountain in the center of the crater in which remnants of the prior deposits are exposed (Figure 28.1). If this occurred, Palucis *et al.* (2014) argue that the Peace Vallis fan developed after the crater was subsequently exhumed. The origin of Mount Sharp sediments awaits direct observations by *Curiosity* of the basal sediments (Grotzinger *et al.* 2015)

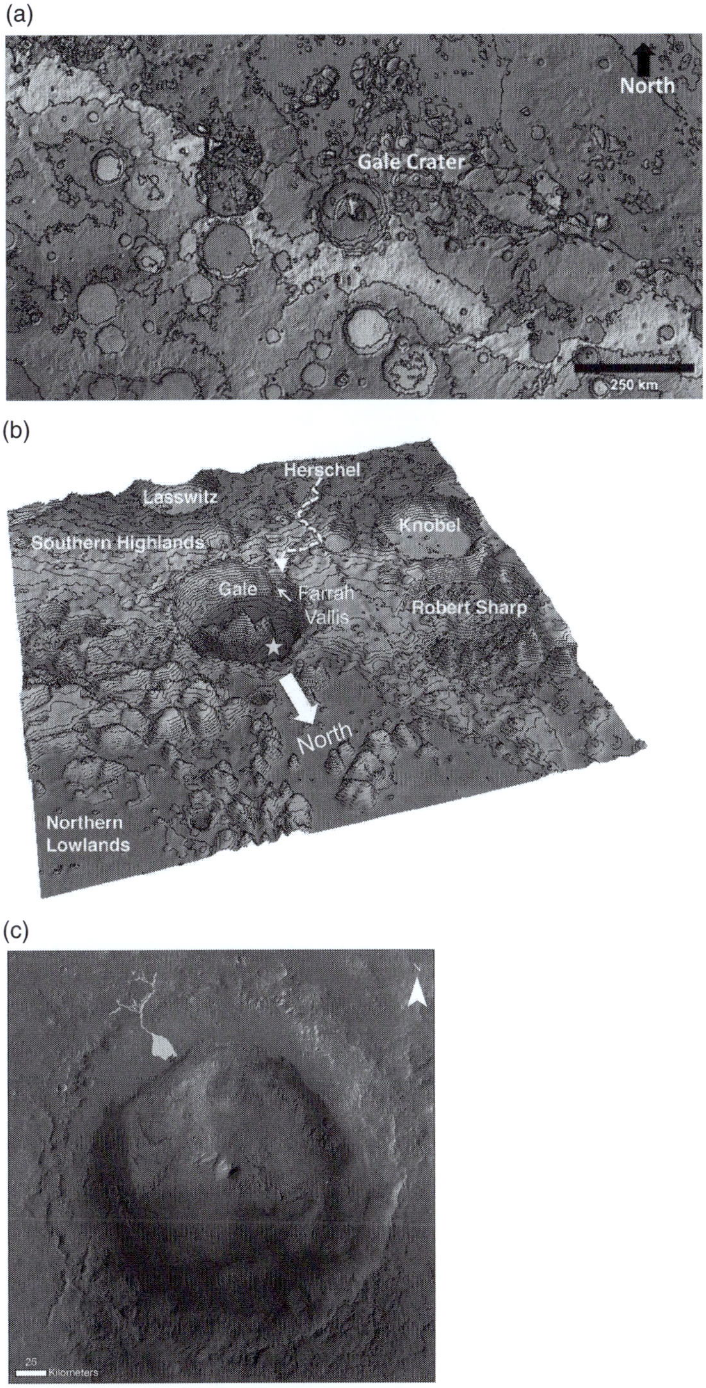

Figure 28.1 (a) Shaded relief map (with 1000 m contours) of Gale Crater and the surrounding region generated from MOLA (Mars Orbiter Laser Altimeter) data. Gale Crater is located on the crustal dichotomy between the cratered southern highlands and the smoother northern lowlands of Elysium Planitia. (b) A closer view of Gale Crater and its surrounding terrain. The lowest elevations in Gale Crater occur on the northwest portion of the crater floor, near the MSL landing site. Gale's impact disrupted a pre-existing valley network originating near Herschel Crater; this was likely one of the main early sources of water into Gale Crater. (Based on Irwin *et al.* (2005).) (c) Mosaic of satellite photographs of Gale Crater derived from Mars Reconnaissance Orbiter Context Camera CTX. Peace Vallis fan and the upstream incised channel network that provided sediment to the fan are shown in a light tone. The star is the *Curiosity* rover landing site. (See insert for color representation of Figure 28.1b.)

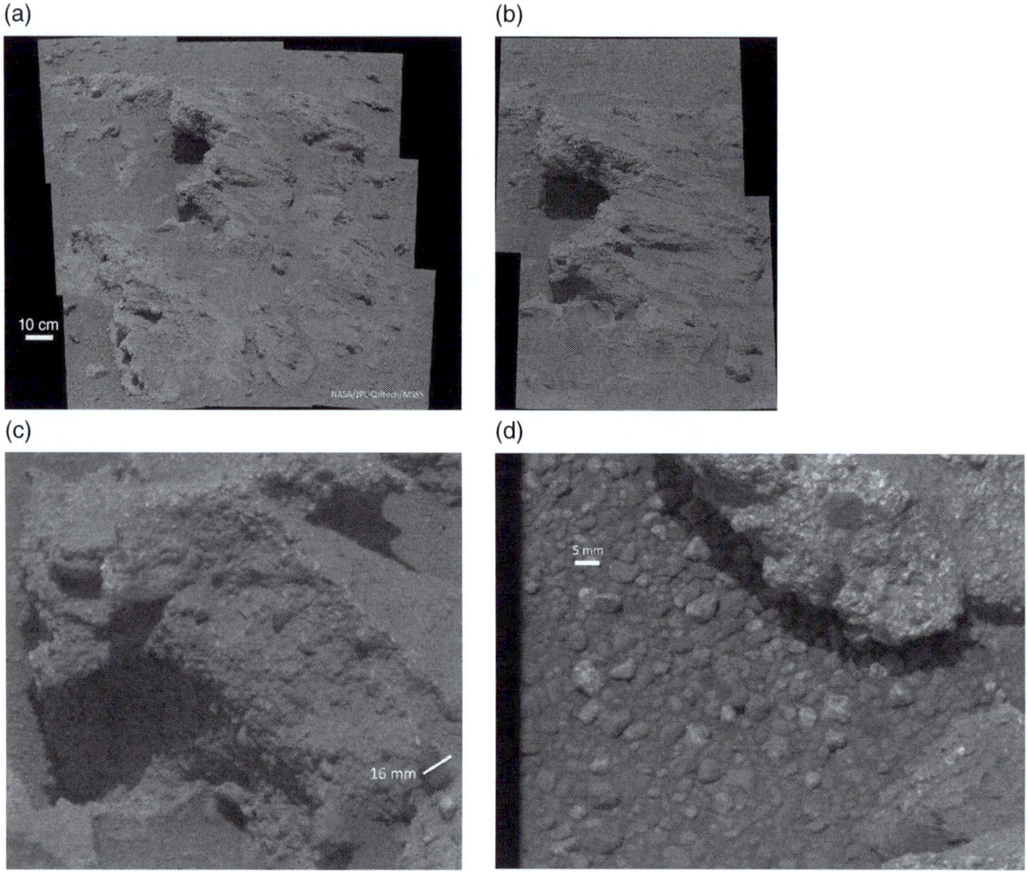

Figure 28.2 (a) Fluvial conglomerates exposed near the *Curiosity* rover landing site. The outcrop is referred to as "Hottah" and this mosaic was taken by the Mastcam camera on Sol 039 (Mastcam images 0039MR0177000000E, 0039MR0177001000E1, 0039MR0177002000E1, 0039MR0177003000E1, 0039MR0177004000E1, 0039MR0177005000E1, 0039MR0177006000E1, 0039MR0177007000E1, and 0039MR0177008000E1). (b) Close-up view of the Hottah conglomerates. (c) A 16 mm pebble within the conglomerate is highlighted on the right side of the image. (d) A close up view of the conglomerate as well as pebbles inferred to have eroded out from the outcrop. Only abrasion during water transport can produce grains of this size, shape, and rounding and therefore appears to be definitive evidence of fluvial gravel transport.

28.3 Some Fluvial Conglomerates on the Way to Mount Sharp

The rover has driven over 10 km since landing, and conglomerates have been noted periodically throughout this drive to Mount Sharp. Every photograph taken by the Mars Science Laboratory is available at http://mars.jpl.nasa.gov/msl/multimedia/raw/. Figure 28.4 shows some examples of exposures and Figure 28.4d shows the travel path of the rover. Figure 28.4a is from the Yellowknife Bay area at the distal end of the Peace Vallis fan. The pebbles range from 2 to 8 mm in diameter. These stones are big enough to have low friction angles relative to the much finer sand, but not so big as to be immobile. In terrestrial studies isolated pebbles in the sand beds are known to occur well downstream of the gravel–sand transitions in rivers (e.g. Venditti and Church 2014). Figure 28.4b and c show coarse

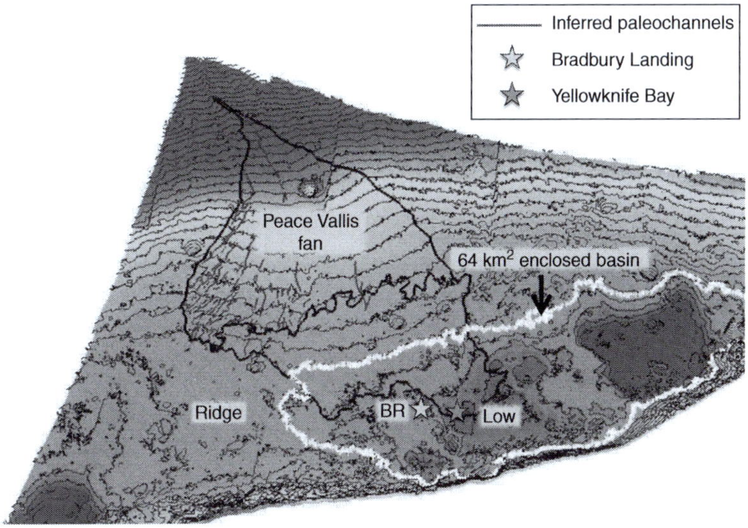

Figure 28.3 A three-dimensional view of the Peace Vallis fan system (vertical exaggeration a factor of 10) with a colorized elevation map and 10 m HiRISE contours. Adjacent to the "Ridge" and within the enclosed basin (denoted by the thick white contour) is Bradbury Rise ("BR") and to the east of Bradbury Rise is a local low ("Low") into which the lower fan enters. The location of the MSL *Curiosity* rover landing site (Bradbury Landing) is indicated by the light star and Yellowknife Bay is indicated by the shaded star.

poorly sorted conglomerates about 6 km southwest from the landing site (near Kimberly in Figure 28.4d). The deposits are weakly lithified and the grains are rounded. The largest grains exceed 50 mm in size. In Figure 28.4b the outcrop was about 19 m from the rover and in Figure 28.4c the outcrop was about 13 m from the rover.

Work is currently underway to place these gravel deposits in a stratigraphic and geomorphic setting (Grotzinger *et al.* 2015). Below we first review how channel flows can be estimated if channel slope, grain size, and width are known. Gravel outcrops observed by *Curiosity*, however, reveal no discernable channel-scale features, i.e. width, depth, or slope. We then discuss how from just grain size alone crude estimates of channel scale and discharge can be made.

28.4 Estimates of Stream Velocity, Channel Discharge, and Gravel Mobility on Mars

Before the *Curiosity* rover landed and revealed the first direct observations of fluvial gravels on Mars, many studies had relied on satellite imagery to infer fluvial processes on Mars, specifically gravel transport. These studies exploited morphologic data (e.g. channel widths and slopes) and inferences about grain size to estimate bankfull channel flows, sediment transport rates, and durations of the fluvial period. Such analyses have been important in indicating periods of free-flowing water on Mars, and consequently, climatic conditions favorable for such conditions. Here we first discuss the varying strategies used to infer flow rates from morphologic evidence and then review the contrasting challenge presented by the *Curiosity* imagery to date: gravel outcrops, but no strong morphologic context to constrain flow conditions.

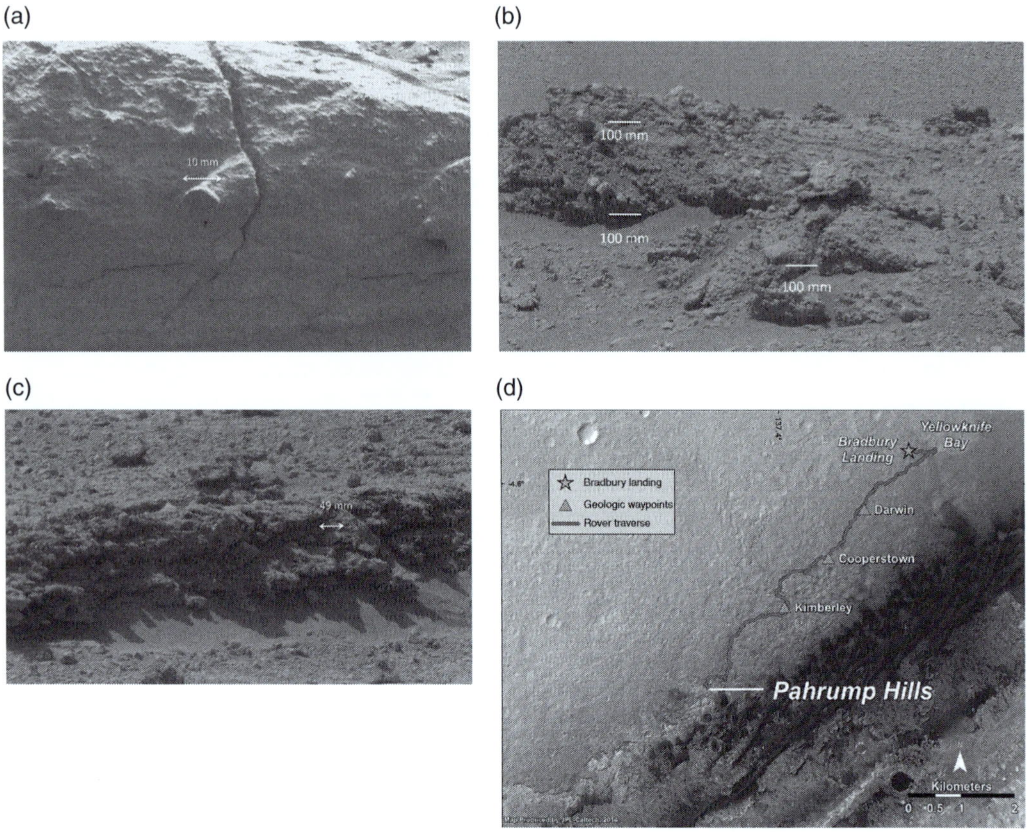

Figure 28.4 Three examples of gravel exposures as observed by the Mastcam cameras on the *Curiosity* rover. (a) An image from Yellowknife Bay, taken on Sol 153 (Mastcam 100 MR0850001), showing isolated stones in poorly sorted sandstones. (b) Mastcam image (MR002747) taken on Sol 646 from Occum Pond location showing a coarse conglomerate. (c) Mastcam image (MR002462) taken on Sol 588 from Point Coulomb location showing another example of a coarse gravel conglomerate. (d) Drive path of the *Curiosity* rover. Image is available at NASA web page as Mars-MSL-curiosity-rover-location-map-pahrump-hills-pia18607-br2. (See insert for color representation of this figure.)

Standard hydraulic and sediment transport relations applied to terrestrial systems are often applied to Mars to estimate stream discharge and runoff rates (see reviews in Wilson *et al.* 2004; Kleinhans 2005; Irwin *et al.* 2008; Williams, Irwin, and Zimbelman 2009; Grotzinger *et al.* 2013). Two general approaches have been used to estimate channel-flow depth and velocity, and channel discharge on Mars: (i) the threshold channel concept and (ii) hydraulic geometry and river morphology correlations. The threshold channel concept assumes: (i) that gravel-bed material only starts to move significantly once a critical boundary shear stress is reached and (ii) this flow condition roughly corresponds to bankfull discharge. There is theory and observation to support this approach (e.g. Parker 1978; Andrews 1984; Dade and Friend 1998; Ryan, Porth, and Troendle 2005; Parker *et al.* 2007). In contrast, hydraulic geometry (i.e. empirical power law relationships between hydraulic properties of channels, such as bankfull width and depth, and stream discharge), first defined by Leopold and Maddock (1953), have been used as alternative means of estimating flows without making any assumptions about sediment transport conditions. Empirical power law morphologic scaling (e.g. meander wavelength variation with channel width) has been used to infer morphologic properties that can then be used in hydraulic geometry relationships (e.g. Irwin *et al.* 2014). Although empirical studies under varying combinations of gravity, viscosity, and sediment and fluid density are not available, it is assumed that these simple scaling corrections can be applied to hydraulic geometry and morphologic correlations under Martian conditions (e.g. Irwin *et al.* 2008).

The system of equations most commonly used for estimating discharge in a channel with an observed channel width relies on the threshold channel concept, which is described in detail in section 28.4.1. By conservation of mass, discharge Q equals the mean velocity $\bar{u}$ times depth h times channel width w.

$$Q = \bar{u}wh \tag{28.1}$$

A widely used empiricism that relates mean velocity $\bar{u}$ to shear velocity u_*, where the shear velocity is defined as

$$u_* = \sqrt{\frac{\tau_b}{\rho}} = \sqrt{ghS} \tag{28.2}$$

(S is water surface slope, ρ is fluid density and g is gravity), is

$$\bar{u} = \left(\frac{8ghS}{f}\right)^{0.5} \tag{28.3}$$

In Equation (28.3), f is a representative frictional resistance, which in this case is the Darcy–Weisbach friction factor.

There are numerous empirical expressions for the friction factor (e.g. Ferguson 2013). One that is intended to work for a range of flow conditions from shallow, boulder-rich channels to deeper flows over gravel is that proposed by Ferguson (2007). Morgan *et al.* (2014) write the Ferguson (2007) equation using the Ferguson reported empirical coefficients fit to data as

$$\left(\frac{8}{f}\right)^{0.5} = 17.7\frac{h}{D_{84}}\left[56.3 + 5.57\left(\frac{h}{D_{84}}\right)^{\frac{5}{3}}\right]^{-0.5} \tag{28.4}$$

in which D_{84} is the grain size for which 84% of the particles sizes are smaller. This empirical fit was derived using data from a wide range of channel slopes (i.e. ~0.001 to ~0.2).

Flow depth, which appears in both Equations (28.1) and (28.4) to calculate velocity, is difficult to estimate on Mars. Many channels are canyons in which the flow during incision likely only occupied a portion of the total canyon depth. Width is more readily evident, although canyon widths can be misleading. Channels are commonly recognized by sinuous, relatively uniform width and flat-topped ridges (Figure 28.5; e.g. Williams, Irwin, and Zimbelman 2009; Burr *et al.* 2010). These inverted channels, in which cemented or armored beds are resistant to wind deflation compared to adjacent areas, may preserve channel width, but not directly reveal flow depth. Some theory or empiricism is typically needed to estimate depth and thus solve Equation (28.3).

28.4.1 Threshold Channel Concept

Terrestrial studies have found that in gravel-bedded channels significant bedload motion typically develops as flows approach bankfull discharge (e.g. Parker 2008). Furthermore, theory for gravel-bedded channels with gravel banks suggests that a stable channel cross-section, i.e. one that transmits

(a)

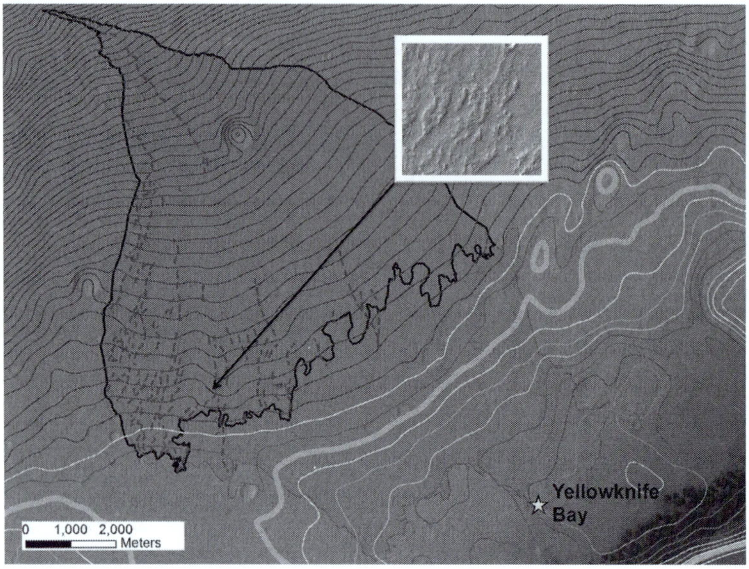

(b)

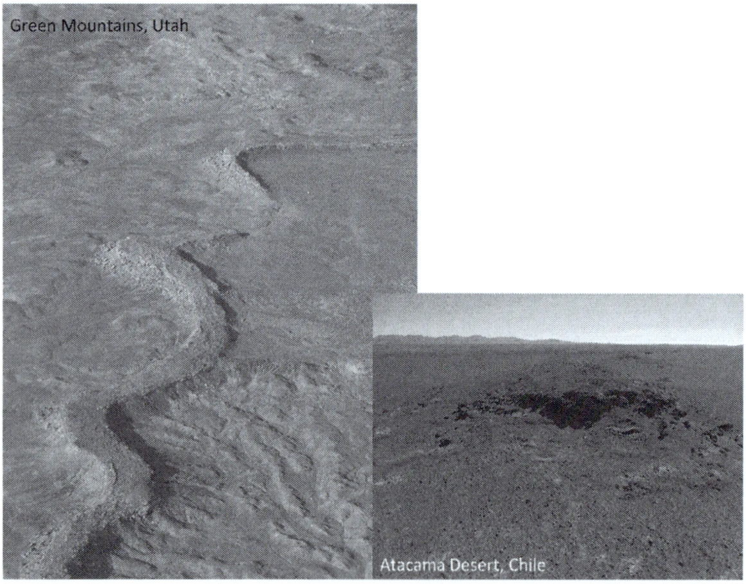

Figure 28.5 (a) From HiRISE imagery, 43 distinct ridges were mapped on the Peace Vallis fan system (shown here on the lower fan region as dashed lines; 5 m contour interval); they varied in width from 5 to 130 m, with an average width of 27 m (Palucis *et al.* 2014). (b) Images of an inverted channel segment from the Green Mountains, Utah where channel widths are about 54 m (Williams, Irwin, and Zimbelman 2009) and the Atacama Desert.

bedload without significant bank erosion, occurs when the boundary shear stress is just slightly above a representative critical boundary shear stress for the initiation of motion (e.g. Parker 1978; Ikeda, Parker, and Kimura 1988; Parker *et al.* 2007; Seizilles *et al.* 2013). Consequently, it is commonly assumed that bankfull flow in gravel-bedded channels generates a boundary shear stress equal to the critical boundary shear stress. This threshold channel concept provides a means to specify a flow depth if channel slope and grain size can be estimated, and this then enables calculation of mean flow velocity, $\bar{u}$, and discharge. The critical Shields number for initial motion τ_{*cr} can be written as

$$\tau_{*cr} = \frac{\rho g h S}{(\rho_s - \rho) g D_{50}} \tag{28.5}$$

in which ρ_s is the sediment particle density and D_{50} is the median grain size of the bed surface. If we write the grain roughness scale in Equation (28.4) as $D_{84} = aD_{50}$, we can solve for the flow depth,

$$h = \tau_{*cr} \frac{(\rho_s - \rho)}{\rho a S} D_{84} \tag{28.6}$$

Lamb, Dietrich, and Venditti (2008) found that the critical Shield's number is slope dependent,

$$\tau_{*cr} = 0.15 S^{0.25} \tag{28.7}$$

and several papers support this finding (e.g. Prancevic, Lamb, and Fuller 2014). Substitution into Equation (28.6) leads to

$$h = 0.15 S^{-0.75} (P/a) D_{84} \tag{28.8}$$

in which P is $\left(\frac{\rho_s - \rho}{\rho}\right)$.

Equation (28.8) implies that the flow depth at initial motion (which is the assumed bankfull flow depth) varies not quite inversely with slope and directly with the D_{84}. As Morgan *et al.* (2014) point out flow depth h also will vary with the concentration of fine sediment in the fluid, which influences fluid density ρ: the higher the concentration, the shallower the depth.

By combining Equations (28.2), (28.4), (28.7), and (28.8), the mean velocity $\bar{u}$ is

$$\bar{u} = 1.028 S^{-0.625} (P/a)^{1.5} (D_{84} g)^{0.5} \left\{56.3 + 0.236 S^{-1.25} (P/a)^{1.667}\right\}^{-0.5} \tag{28.9}$$

Substitution of various values of slope, holding all other parameters constant, shows that mean velocity increases with decreasing slope, a result opposite from Equation (28.2) alone (without the critical shear stress constraint). This reduction is non-linear with much greater velocity increases as slope decreases from 0.03 to 0.01 than from 0.003 to 0.0003 (approaching the lower limit at which gravel-bedded rivers are found on Earth). With decreasing slope, for critical shear stress flow conditions to be met, the depth will increase. Also, converse to expectations without the critical boundary shear stress constraint, the mean velocity is predicted to increase with the bed roughness length scale D_{84}, specifically to the square root of that grain size. As many have pointed out (e.g. Irwin *et al.* 2008), the reduced gravity of Mars relative to Earth leads to lower velocities, i.e. velocity varies as the square root of gravity, as shown in Equation (28.9).

Substituting Equations (28.8) and (28.9) into Equation (28.1) leads to

$$Q = 0.154 w S^{-1.375} (P/a)^{2.5} D_{84}{}^{1.5} g^{0.5} \left\{56.3 + 0.236 S^{-1.25} (P/a)^{1.667}\right\}^{-0.5} \tag{28.10}$$

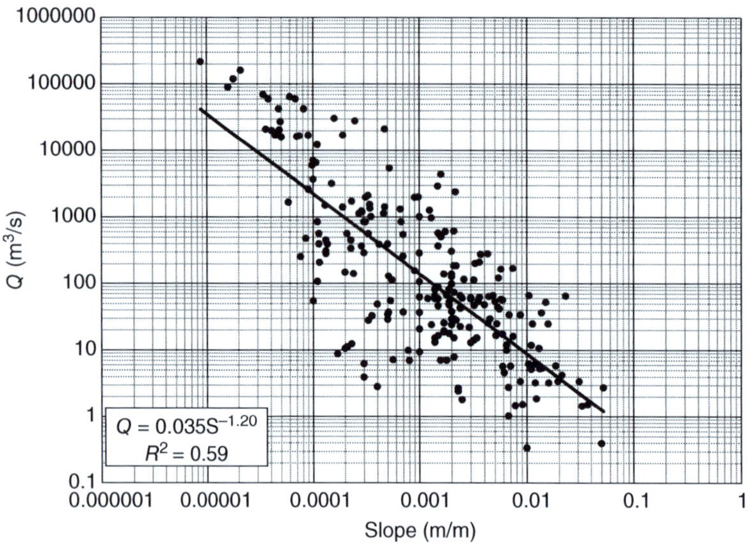

Figure 28.6 Observed bankfull discharge versus channel slope for 230 sites compiled by Li *et al.* (2014). Discharge is inversely proportional to slope with a power law exponent of −1.2 and an R^2 value of 0.59.

which illustrates the relative importance of estimates of slope S, relative density P, and grain size in the calculation of channel discharge. Performing various substitutions in Equation (28.10) and plotting the results shows that Q varies as an inverse power law of slope, with the power exponent ranging from −0.8 to −0.9. This is consistent with empirical correlations with terrestrial river systems. Figure 28.6 shows a plot of bankfull discharge versus channel slope from a recently published terrestrial river data compilation (i.e. Li *et al.* 2014). While the correlation is significant, the range of variation about the regression line for any particular slope value is two orders of magnitude. Substitutions into Equation (28.10) indicate that discharge depends on the relative density term P to an exponent of about 2.5, but the range of probable values of P is narrower than for S. Predicted discharge increases as the 1.5 power of D_{84}, thus an uncertainty of D_{84} of an order of magnitude equates to a 32-fold uncertainty in discharge. Further constraints on likely flow depths can be to use empirical width to depth ratios for terrestrial rivers (e.g. Palucis *et al.* 2014).

An alternative equation often used in terrestrial studies to Equations (28.3) and (28.4) is the vertically integrated law of the wall equation, using an empirical grain-size roughness length scale of $3.5D_{84}$ (e.g. Garcia 2008)

$$\bar{u} = \frac{u_*}{k}\ln\left(\frac{11h}{3.5D_{84}}\right) \tag{28.11}$$

in which k is von Karman's constant (0.4). Combining Equation (28.11) with (28.8) and setting the boundary shear stress equal to the critical boundary shear stress in calculating the shear velocity u_*, the velocity expression becomes,

$$\bar{u} = 0.968S^{0.125}(P/a)^{0.5}(D_{84}g)^{0.5}\ln\left(0.471S^{-0.75}(P/a)\right) \tag{28.12}$$

and the discharge is

$$Q = 0.145wS^{-0.625}(P/a)^{1.5}D_{84}{}^{1.5}g^{0.5}\ln\left(0.471S^{-0.75}(P/a)\right) \tag{28.13}$$

Equations (28.9) and (28.12) and (28.10) and (28.13) yield nearly identical results. Figure 28.7a shows the observed mean-velocity–flow-depth product for the rivers in the Li *et al.* (2014) compilation versus that predicted using Equation (28.13) divided by flow width. P was specified as 1.65 and a as 2.0 (D_{84} was calculated from reported D_{50} data). Only data for channels with a grain size of 10 mm or greater were used because the analysis by Li *et al.* (2014) showed that finer gravel-bedded rivers tend to have higher Shield's numbers τ_{*c}. The threshold flow depths (and discharges) are consistently low

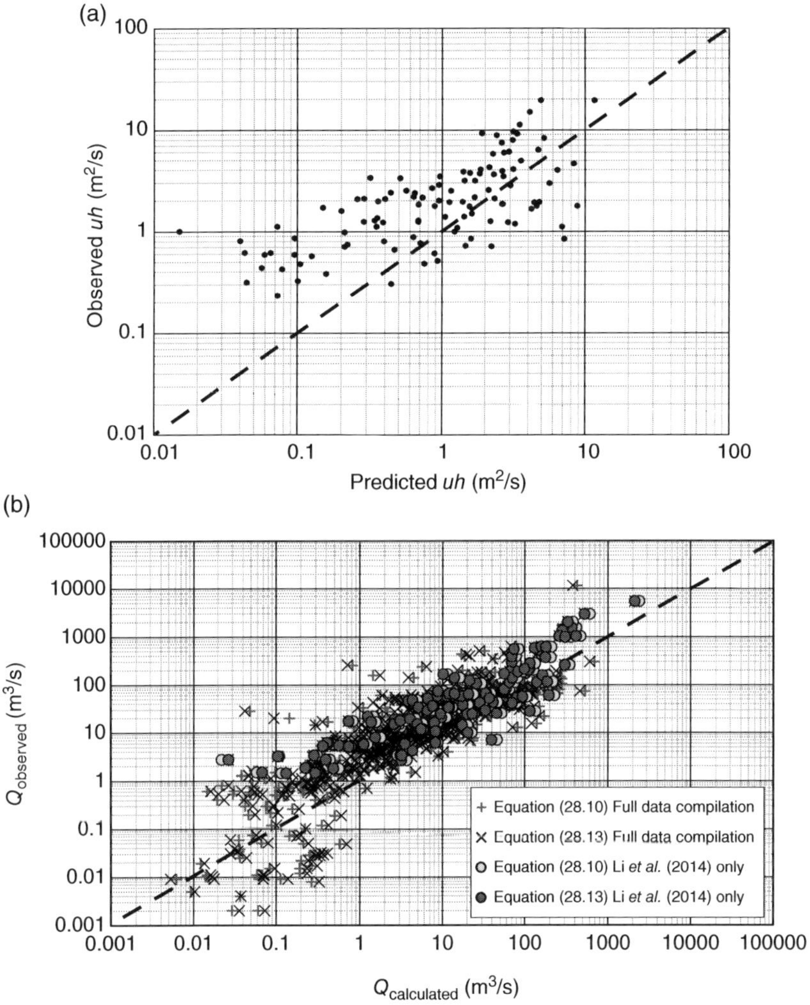

Figure 28.7 (a) Observed velocity–depth product (*uh*) plotted against predicted velocity–depth product in units of (m^2/s) using the Li *et al.* (2014) data compilation for grains larger than 10 mm. Prediction is based on Equation (28.13) (for calculating Q) divided by channel width for the threshold channel case. Width is not included here to eliminate the improvement obtained by plotting width on both axes. While there is considerable scatter, the data systematically deviate to the left of the 1:1 line at lower velocity–depth products, i.e. Equation (28.13) divided by width underpredicts the velocity–depth product. (b) Comparison of observed versus predicted discharge from Equations (28.10) and (28.13) for 709 rivers (from a compilation of Li *et al.* (2014; 230 points shown in dots), Trampush, Huzurbazar, and McElroy (2014; 196 rivers distinct from Li *et al.*), and a new compilation with an additional 283 rivers).

for shallower channels, suggesting that the boundary shear stress at bankfull was above critical or that additional roughness scales matter. Figure 28.7b shows a plot of observed and predicted discharge (from Equations (28.10) and (28.13)) for two sets of data: the Li *et al.* (2014) data and a compilation of data from Li *et al.* (2014), Trampush, Huzurbazar, and McElroy (2014) and additional rivers compiled by the authors (a supplement spreadsheet of all data is provided). Results of Equations (28.10) and (28.13) show only small differences. For most of the discharge range, the calculated discharge is less than observed.

28.4.2 Hydraulic Geometry Applications

An alternative to the threshold channel approximation for estimating channel flows is to use empirical bankfull hydraulic geometry relationships (e.g. Irwin *et al.* 2008, 2014). Using data from 230 terrestrial river reaches, Li *et al.* (2014) found bankfull shear velocity, u_{*bf}, to be

$$u_{*bf} = 35S^{0.26}(Pg\nu)^{0.333} \tag{28.14}$$

where ν is the kinematic viscosity. The data did not have a wide range of viscosities reported, and all of the reaches had the same gravity, so the actual dependence on these two terms was not tested.

Since $u_{*bf} = (ghS)^{0.5}$, the depth dependence is

$$h_{bf} = 1220(P\nu)^{0.667}g^{-0.333}S^{-0.47} \tag{28.15}$$

This relationship has the advantage over Equation (28.8) in that it is not grain-size dependent and it corresponds empirically to terrestrial bankfull depth dependence on slope. It also introduces a viscosity dependence that raises questions about water temperature (and possible salt content) and sediment concentration at times of flow on Mars. Figure 28.8 shows observed versus predicted depth using Equation (28.15). Although Equation (28.14) is derived using the dataset from Li *et al.* (2014), there

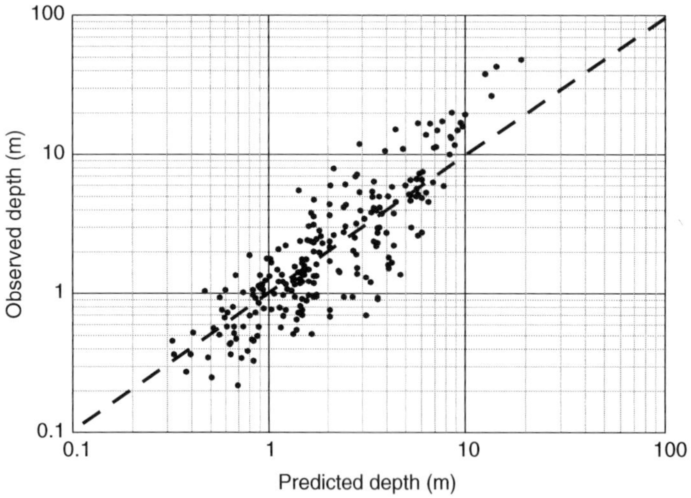

Figure 28.8 Observed bankfull depth versus predicted bankfull depth using the shear velocity relationship (Equation (28.14)) reported by Li *et al.* (2014). The data begin to systematically deviate to the left of the 1:1 line as depth increases, i.e. the relation suggested by Li *et al.* (2014) underpredicts the actual depth. A minor correction to equation (28.14) was published by Li *et al.*, (2016), after this chapter was written. This correction is sufficiently small that it makes imperceptible changes to this Figure and Figure 28.9, hence the equations and corresponding plots were not replaced.

is a systematic residual underprediction at deeper channel depths, which is also visible in the figures reported in Li *et al.* (2014).

Equations (28.15) and (28.8) predict the same depth (and consequently the same velocity) when

$$D_{84} = 8133 a\nu^{0.67} (Pg)^{-0.33} S^{0.28} \quad (28.16)$$

Equation (28.15) can also be used in roughness relationships to calculate velocity, as was shown above.

For the law of wall relationship (Equation (28.11)), the mean velocity is

$$\bar{u} = 87.3(Pg\nu)^{0.334} S^{0.265} \ln\left(3834(P\nu)^{0.667} g^{-0.33} S^{-0.47} D_{84}^{-1.0}\right) \quad (28.17)$$

Hence, if Equation (28.15) is used to estimate depth, a velocity calculation still requires a roughness estimate and consequently for gravel-bed channels a grain-size estimate. The discharge becomes

$$Q = 106506(P\nu)S^{-0.205} \ln\left(3834(P\nu)^{0.667} g^{-0.33} S^{-0.47} D_{84}^{-1.0}\right) w \quad (28.18)$$

There is a key distinction between the threshold channel approach and the use of empirical hydraulic geometry relationships. The bankfull hydraulic geometry, defined by Equations (28.17) and (28.18), predict opposite grain-size dependencies than the threshold channel model. All else held constant, mean velocity in (28.17) and discharge in (28.18) decreases with increasing grain size. Typically in terrestrial systems, however, the bed material coarsens with increasing slope (e.g. Ferguson *et al.* 1996; Montgomery and Buffington 1997; Dade and Friend 1998), such that flow velocity may vary little. For the recently reported Li *et al.* (2014) data, the median velocity value was about 1.52 m/s with a standard deviation of 0.73 m/s. This median value, corrected for reduced gravity (i.e. multiply by 0.61 (see Irwin *et al.* (2008)), may provide a useful first estimate of flow velocities.

Figure 28.9a shows observed and predicted velocity–depth products (uh) (Equation (28.18) divided by width). The scatter is large, reflecting both the systematic difference reported in Figure 28.8 and the poor predictability of velocity. A similar plot using a constant velocity of 1.52 m/s yields a better fit, but the range of observed values for any predicted uh remains two orders of magnitude.

To use the above expressions on Mars to estimate depth, velocity, and discharge, particle density relative to fluid density P, water surface slope S, fluid viscosity ν, and median or D_{84} grain sizes must be estimated. Particle density in most cases is likely close to 3.0 g/cm^3 (i.e. density of basalt), but fluid density is not known (e.g. Morgan *et al.* 2014). Surface slopes, a proxy for water surface slope, can be estimated from well-defined features (e.g. fans, deltas, inverted channel paths) if they exist and topographic data are available, but must be estimated for deposits out of geomorphic context (see next section). Grain-size estimates have tended to favor fairly coarse gravels. Wilson *et al.* (2004) cite surface rock sizes observed by the Viking and Pathfinder missions when proposing a D_{84} of 164 mm for calculations for outflow channel discharge analysis. Kleinhans, van de Kasteele, and Hauber (2010), in the absence of direct observations, have settled upon a D_{50} of 100 mm, regardless of the context. Morgan *et al.* (2014) interpret HiRISE 0.25 m/pixel imagery of likely paleochannel bed surfaces as indicating a median grain size of 125–250 mm. While such grain sizes may be appropriate for slopes on the order of 1% in an alluvial setting, it seems highly probable to be in error in canyon settings where deep high flows can scour the channel to bedrock or boulders (as discussed by Irwin *et al.* (2014)), or on low-gradient alluvial features (less than 0.1%). Collectively, these uncertainties mean that discharge calculations using the systems of equations above are at least an order magnitude out, but two orders of magnitude is probably common. As shown in Figures 28.7b and 28.9b, even when applied to terrestrial rivers, the uncertainty about the actual discharge remains large.

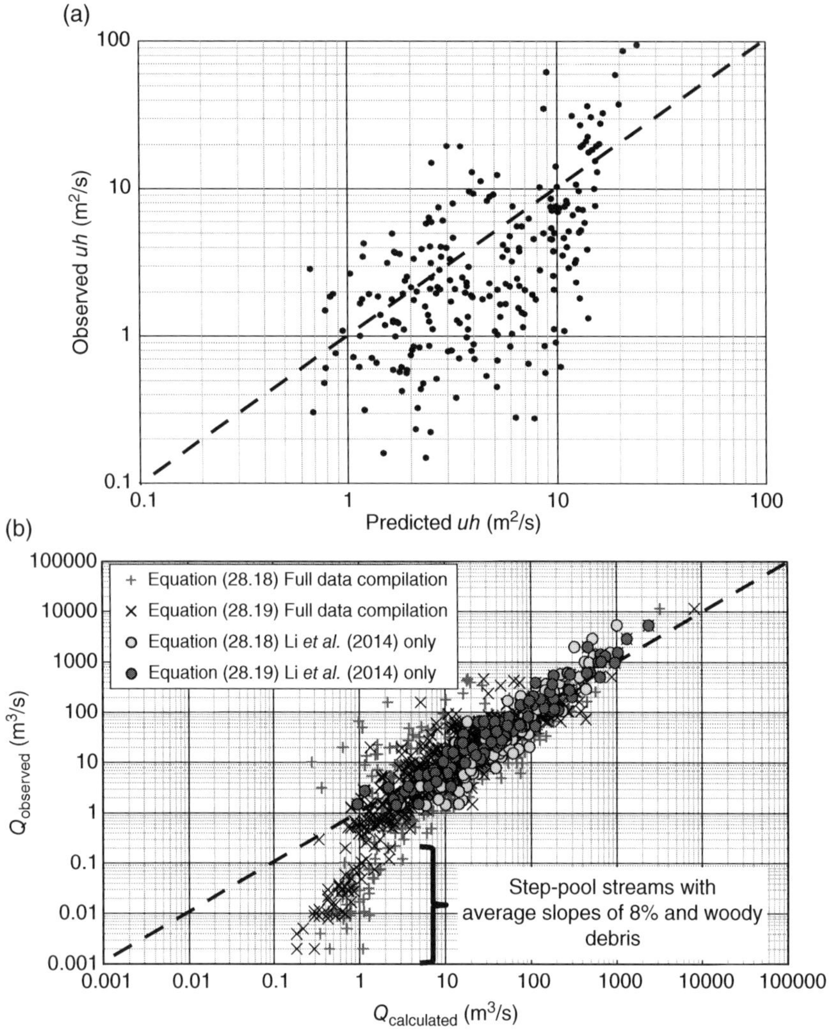

Figure 28.9 (a) Observed velocity–depth product (*uh*) plotted against predicted velocity–depth product in units of (m^2/s) using the Li *et al.* (2014) data compilation for all grain sizes. Prediction is based on Equation (28.18) (the hydraulic geometry case) divided by channel width. The scatter about the 1:1 line is large, due to both the differences reported in Figure 28.7 and the difficulty in predicting velocity. (b) Comparison of observed versus predicted discharge from Equations (28.18) and (28.19) for 709 rivers (from a compilation of Li *et al.* (2014; 230 points shown in colored dots), Trampush, Huzurbazar, and McElroy (2014; 196 rivers distinct from Li *et al.*), and a new compilation with an additional 283 rivers). The data points that obviously deviate from the 1:1 line are steep, step-pool streams located in the American west (namely Washington State and Montana) that have large amounts of woody debris (data from Wohl and Merritt 2005).

The most direct application of hydraulic geometry in Martian settings, one that avoids estimating parameters, is to use terrestrial correlates between bankfull discharge (corrected for Martian gravity) and bankfull width. For example, Irwin *et al.* (2014) measured meander geometry of the large channel bends in Eberswalde delta to estimate channel width and then used a correlation between the 2 year recurrence interval discharge event (corrected for Martian gravity) and bankfull width from the Missouri River Basin, a region inferred to have low bank strength, to estimate the discharge in the Eberswalde meanders. Two compilations of terrestrial data (with significant overlap in sources) provide data that yield these correlates (no gravity correction applied):

$$Q = 0.219w^{1.65} \tag{28.19}$$

from Li *et al.* (2014) representing 230 data points, and

$$Q = 0.134w^{1.72} \tag{28.20}$$

from Trampush, Huzurbazar, and McElroy (2014) representing 323 data points. The range of discharge values for a given width in these correlations is about an order of magnitude. Figure 28.9b shows the observed discharge plotted against the calculated discharge using Equations (28.18) and (28.19). The Li *et al.* (2014) data, which were used to parameterize Equations (28.18) and (28.19), are shown as circles, and the match is fairly close. The Trampush, Huzurbazar, and McElroy (2014) data plot similarly, but a subset of the data, collected on steep, step-pool channels, plot distinctly off the trend of the other data. These streams, which were originally studied by Wohl and Merritt (2005) and located in the western United States (Washington and Montana), were found to include abundant wood that could be altering channel morphologies. Figure 28.10 shows the observed discharge as a function of the channel width for combined data set in Li *et al.* (2014), Trampush, Huzurbazar, and McElroy (2014), and the compilation for this paper (a total of 654 unique data points, excluding the Wohl and Merritt (2005) data). The regression has an R^2 of 0.9,

$$Q = 0.20w^{1.66} \tag{28.21}$$

Geomorphic setting (canyon versus floodplain bordered), relative sediment supply and size, bank strength controls, vegetation, and runoff characteristics can cause specific locations to differ greatly from these broad empirical trends, hence application to studies on Mars, while perhaps useful, must be considered approximate. It is not obvious that using Equation (28.19), (28.20), or (28.21) based on

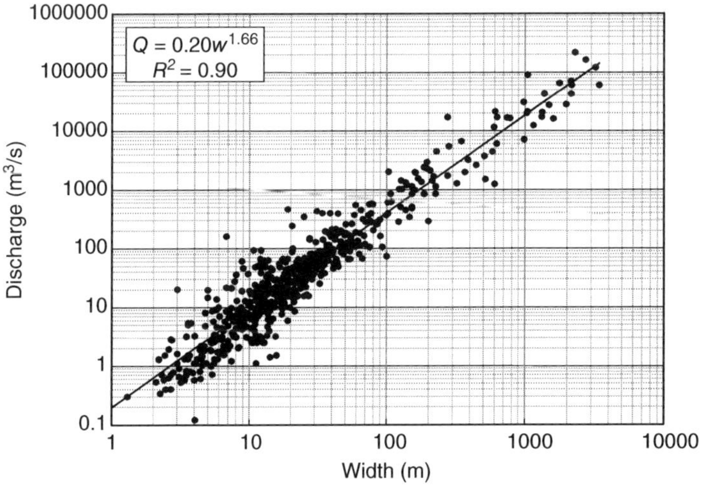

Figure 28.10 Bankfull discharge as a function of bankfull width for 654 river sites (excluding the data from Wohl and Merritt 2005).

width observations alone, yields a poorer estimate of discharge than the more mechanistic relationships described above.

28.4.3 Analysis of Martian Gravel Conglomerates

The gravel conglomerates directly observed on Mars with the *Curiosity* rover offer the opposite problem to that outlined in sections 28.4.1 and 28.4.2. In Gale Crater we can estimate grain sizes in the relatively thin exposures revealed in transit towards Mount Sharp, but we do not see channel widths and preserved channel slopes. There is no direct surface morphologic guide to estimating flows. This is the common circumstance for gravel deposits in the rock record on Earth as well, and strategies for estimating channel slope have been proposed for gravel (e.g. Paola and Mohrig 1995) and sand deposits (e.g. Lynds *et al.* 2014), although in both cited cases the flow depth must be determined from stratigraphic relationships.

Williams *et al.* (2013), in reporting on the first fluvial conglomerates on Mars, did grid by number counts on the thin exposed edges of cemented gravels to obtain grain-size distributions. The median grain size for three separate outcrops range from 4.5 to 9.5 mm and the corresponding D_{84} is 6.5 to 15 mm. They estimated the minimum flow to transport the gravels for the threshold channel for initial motion case, and used Equation (28.5) with two bracketing values of the critical Shields number (0.03 and 0047) and two estimated water surface slopes (0.01 and 0.001) to estimate flow depth (Equation (28.6) using the observed D_{50} and velocity (from Equation (28.11)). The slope estimates were based on the slope of the nearby Peace Vallis fan, the lower portion of which has a mean slope of about 0.01, and the empirical observation that gravel-bedded streams are typically steeper than 0.001. This led to bracketing depths and velocities between 0.27–0.89 m and 0.2–0.75 m/s, respectively.

Similar to the threshold versus hydraulic geometry approach to estimating flow properties described in the previous sections, empirical correlations provide another means of estimating flow properties when only grain size is known. The empirical compilations by Li *et al.* (2014) and Trampush, Huzurbazar, and McElroy (2014) show an approximate correlation between slope and median grain size in their selected terrestrial rivers (Figure 28.11a and b). For gravel sizes, especially 5–10 mm and coarser, there is essentially no relationship between slope and grain size. As many others have noted, rivers with grain sizes between 2 and 5 mm are rare (e.g., Venditti and Church 2014). According to Figure 28.11, for the grain sizes first observed at Gale Crater (i.e. 4.5–9.6 mm), a slope estimate of 0.001 to 0.01 appears to well represent the majority of the sites included in these terrestrial data sets. Notable exceptions to these slopes are the steeper, fine-grained fans studied by Stock *et al.* (2007) (and not included in either data compilation). Trampush, Huzurbazar, and McElroy (2014) report a correlation in which slope depends strongly on bankfull depth and weakly on grain size, but recommend the use of this correlation only for slopes less than 5×10^{-3}. They also note explicitly that their correlation should not be used for paleoslope analysis on Mars because it is strictly empirical and it is not established how fluid viscosity, density, and gravity may influence these relationships. In the absence of stratigraphic evidence of flow depth, however, only grain size can provide a guide to the scale of the flows that created the deposits imaged by *Curiosity*, and we can either guess based on terrestrial experience, use theory (which retains some level of empiricism in it), or use empirical relations with awareness for a need for caution.

If grain size can be used to estimate crudely the slope, then correlations with slope can point to channel dimensions. In the Li *et al.* (2014) data, bankfull channel depth varies inversely with slope (Figure 28.12),

$$h = 0.0567 S^{-0.52} \tag{28.22}$$

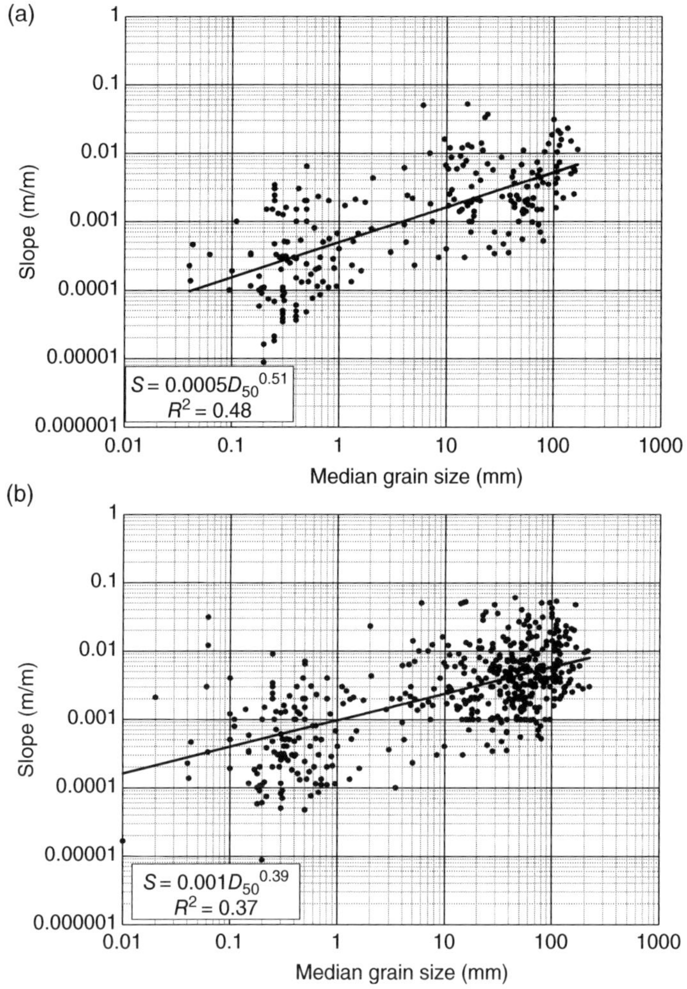

Figure 28.11 (a) Channel slope as a function of grain size using data compilation by Li *et al.* (2014). While there is considerable variability in the median grain size for a given slope, the data follow a general trend where slope is proportional to D_{50} raised to the one-half power. Reproduced with permission of Taylor & Francis.) (b) Channel slope as a function of grain size using data compilations by Trampush, Huzurbazar, and McElroy (2014). Again, while there is considerable variability in the median grain size for a given slope, the data follow a general trend where slope is proportional to D_{50}, though with this larger data set, the power law exponent is lowered to 0.4.

The R^2 is 0.73, but for slopes less than 6×10^{-5} the depths are systematically underpredicted. This slope corresponds to depths greater than about 10 m. A channel slope of 0.01 would predict a bankfull depth of 0.62 m, and a slope of 0.001 would predict a depth of 2.1 m. These depths are higher than those estimated from the limiting case of threshold of motion (i.e. Williams *et al.* (2013) report a depth range of 0.03–0.09 m for a slope of 0.01 and 0.27–0.89 m for a slope of 0.001). Because the deposits could record transport conditions well above threshold, and because of additional roughness in the channel other than skin friction scaled by D_{84}, these deeper flows – though strictly empirically

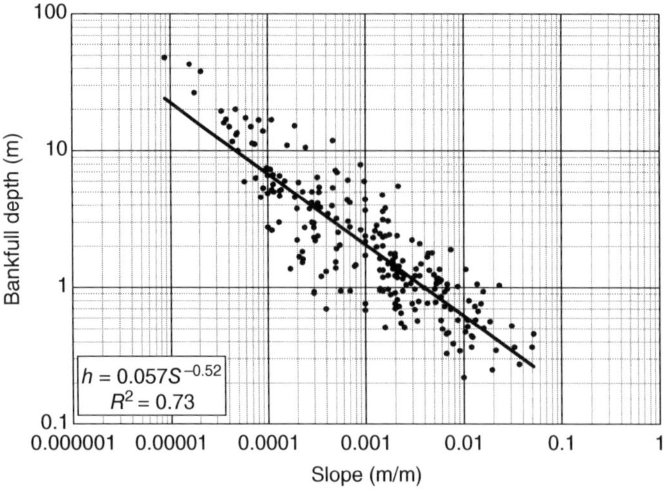

Figure 28.12 Bankfull depth as a function of slope using the Li *et al.* (2014) data compilation. Bankfull depth is inversely proportional to slope, with a power law exponent of −0.52.

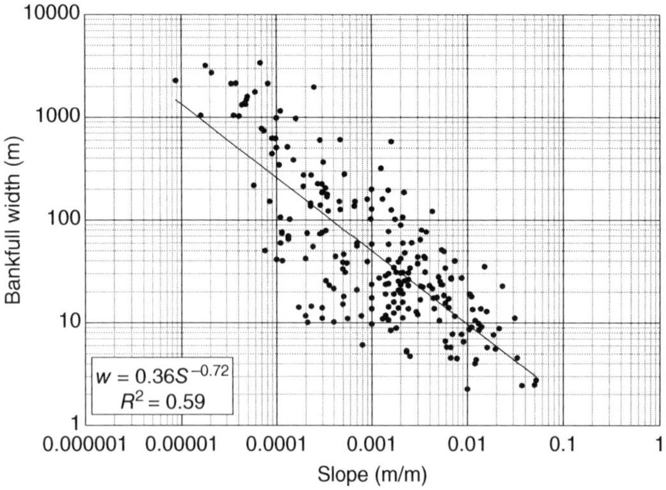

Figure 28.13 Bankfull width as a function of slope for the Li *et al.* (2014) data.

derived – may be better estimates. Similarly, the Li *et al.* (2014) data lead to a bankfull width to slope correlation of

$$w = 0.36S^{-0.72} \tag{28.23}$$

The R^2 is lower (0.59) and there is systematic underprediction of width below slopes of 10^{-4} (Figure 28.13). For any given slope, the range of observed slope is about an order of magnitude, so this correlation provides only crude guidance to flow scale. It suggests that slopes of 0.01 would typically have widths of about 10 m, and if the slope were reduced to 0.001, the width would be about 50 m.

Taking these empiricisms together, the fluvial conglomerates first documented by Williams *et al.* (2013) may have been deposited by a channel of about 0.62–2.1 m deep, 10–50 m wide, and if we assume the median flow velocity of terrestrial rivers reported in the Li *et al.* (2014) data set and correct for gravity (giving a mean velocity of 0.9 m/s) then the predicted discharge is about 5.6–94.5 m^3/s. Given the slope and depths, velocity could alternatively be calculated from Equation (28.9). If the estimated widths are used in the empirical discharge correlation with width for the Li *et al.* (2014) data set, the predicted bankfull discharges are 6 and 85.6 m^3/s for the 10 and 50 m channel widths, respectively. This falls in the range of 3.7–207 m^3/s estimated by Palucis *et al.* (2014) for channels on the nearby Peace Vallis fan. This analysis alone, cannot rule out, however, the possibility that the flows only slightly exceed threshold conditions and were relatively shallow and associated with poorly confined flows or sheet floods. Exposures that provide stratigraphic clues about flow depth will be particularly important in providing greater constraint on the character of the flows that transported the gravels observed in Gale.

28.5 Runoff Volume and Implications for Climate

The discharge analysis above provides only a sense of scale of the channel and the instantaneous bankfull flow. While peak flows from a given watershed area places constraints on possible runoff generation mechanisms, such as snow melt, rainfall-runoff, or impact melting of subsurface ice (e.g. Mangold *et al.* 2012; Morgan *et al.* 2014; Palucis *et al.* 2014), it does not reveal how much total runoff occurred to create a given fluvial feature.

Four recent examples illustrate the approaches taken to estimate total runoff volume on Mars. Mangold *et al.* (2012) employ a numerical model for delta construction by Kleinhans *et al.* (2010) to calculate a steady-state discharge of 8.7×10^4 m^3/s feeding the construction of Eberswalde delta, a site of many investigations directed at runoff and sediment transport (e.g. Malin and Edgett 2003; Moore *et al.* 2003; Jerolmack *et al.* 2004; Bhattacharya *et al.* 2005; Wood 2006; Lewis and Aharonson 2006; Pondrelli *et al.* 2008; Rice *et al.* 2011; Mangold *et al.* 2012; Irwin *et al.* 2014). Mangold *et al.* (2012) assume a constant feeder channel in a canyon 500 m wide, 25 m deep with a gradient of 0.034 and a surface grain size of only 100 mm. Assuming various sediment to water ratios (0.05–30%), Mangold *et al.* (2012) suggested that the 5 km^3 delta could be built in 2 weeks to two decades. The total water volume to build the delta for a 5000 km^2 source area would be as high as 2 km to as low as 3 m (for the exceptionally dense flows with high sediment to water ratios). The calculated steady-state discharge to the delta is about 40% of the mean daily discharge on the Amazon River even though the watershed area that drains to the Eberswalde delta is just 0.3% of that of the Amazon. The annual sediment load would also exceed that currently discharged from the Amazon River (about 1200×10^6 t/yr, Syvitski *et al.* (2003)). Irwin *et al.* (2014) offer a detailed analysis that disagrees with the Mangold *et al.* (2012) proposals, and instead argue for a relatively long-lived hydrologic period to build the meandering channels, avulsing delta system, and sustained lake level.

Irwin *et al.* (2014) calculate the runoff rate to Eberswalde delta needed to maintain a constant lake level despite evaporation losses. Their estimate ranged from 2.4–16 cm/yr of runoff production. To deposit the estimated 6 km^3 of delta sediment, they estimate it would take 1–10 km of runoff for a 5000 km^2 contributing area to 0.4–4 km for a 17 000 km^2 contribution area. This may have occurred over 10^4–10^5 years.

Morgan *et al.* (2014) estimate the volume of two intersecting fans in Saheki Crater to be 586 km^3. Assuming a sediment to water ratio in the runoff producing the fan of 0.2, they estimate that about 3.57 km of runoff was needed from the 793 km^2 watershed source area. This high sediment to water ratio is supported by the visibly coarse texture of the wind-exposed fan.

Palucis *et al.* (2014) estimate the volume of the Peace Vallis fan to be 0.93 km^3. They note that the sediment to water ratio in Chilean streams representing arid zones was reported to be about 0.0002 (Pepin *et al.* 2010). This leads to an estimate of 4650 km^3 of water runoff from the 730 km^2 watershed source area, or the equivalent of 6.4 km of runoff. Here we use a compilation of sediment yield and runoff data from 411 rivers (http://csdms.colorado.edu/wiki/Data:World_River_Sediment_Yields_-Database) to calculate the sediment to water ratio (assuming the bulk density of the deposited sediment is 1.9 t/m^3). The median value of the log-transformed data is −3.48, or a sediment to water ratio of 0.00033. Figure 28.14 shows the frequency distribution of log-transformed sediment to water volume ratios. When this sediment to water ratio is applied to the Peace Vallis fan system, it would suggest 3.9 km of runoff. Conservatively, Palucis *et al.* (2014) proposed that at least 0.6 km of runoff must have occurred to produce the fan.

In each of these cases, kilometers of total runoff from the watershed source areas are required to produce the observed depositional features. Note that evaporation and infiltration losses during runoff are not considered, so these numbers are likely minimum values. Because there cannot be kilometers of water equivalent stored as snow in these watersheds, this implies that the depositional features require a hydrologic cycle in which precipitation replenishes runoff. Morgan *et al.* (2014), for example, suggest that could be accomplished by an annual snow accumulation of about 3 m. Collectively, these analyses point to the need to have periods of 1000–100 000 year time scales of climatic conditions that support an active liquid hydrologic cycle on Mars. At each of the three sites, the timing of the fan/delta deposition may be similar (late Hesperian to early Amazon, i.e. 3.4–2.7 billion years ago), and thus may record the same climatic incident. Currently climate models have not been able to produce wet periods of this duration and magnitude.

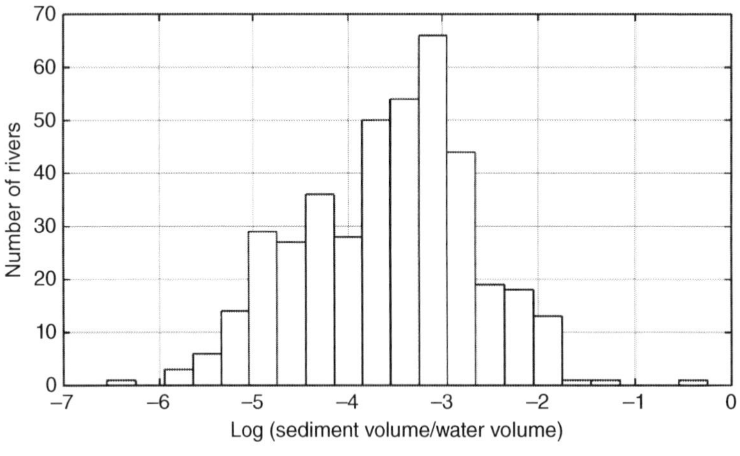

Figure 28.14 Distribution of log-transformed sediment volume to water volume ratio for 411 rivers of the world. Sediment yield was converted to sediment volume assuming a density of 2650 kg/m^3 and a solid fraction of 0.7 (i.e. porosity of 0.3 for gravel).

28.6 Conclusions

The paleocanyon channels, meandering rivers, fans, and deltas on Mars have typically been assumed to be gravel bedded. Satellite imagery enabled quantification of fluvial forms, giving estimates of channel widths and slopes. The threshold channel concept and hydraulic geometry relationships based on analysis of terrestrial rivers have been used to estimate channel-flow characteristics including depth, velocity, and bankfull discharge. The threshold channel approach is conservative in that it assumes that flows only just reach critical shear stress at bankfull, and terrestrial studies show that finer gravel beds may tend to have higher shear stresses. This approach has the advantage that it has a theoretical foundation and a minimum of parameters, but it requires an estimate of bed-surface grain size and currently satellite imagery cannot resolve particles smaller than boulders. Consequently, grain sizes are typically guessed. The hydraulic geometry employs empirical power law relationships derived from metadata analysis of terrestrial rivers. Channels are not assumed to be just at threshold of motion, but if mean flow velocity is to be estimated, bed-surface grain size still must be guessed. The simplest approach is to use coarse correlates, such as the power law relationship between bankfull discharge and bankfull width. While the correction for gravity differences can be computed in hydraulic geometry relationships, the influence of fluid density, viscosity, and possible changes in erosional mechanics due to velocity–depth–shear-stress differences have an unknown effect on the relevance of terrestrial river hydraulic geometry relationships to Mars. Uncertainty in discharge estimates, even when applying empirical relationships to terrestrial data, remains large. In application to Mars, uncertainty of plus or minus an order of magnitude may be common.

The images returned from the *Curiosity* rover provide the first direct observations of fluvial conglomerates. The deposits are finer (order 10 mm) than has been assumed in satellite-based studies (100 mm). The stones vary in rounding, but some are quite well rounded, providing strong evidence of their fluvial history. Given the fine pebble size of the gravels exposed, the fluvial system responsible for transport was likely in a relatively low-gradient system (below 1%) with small channels (<50 m wide). Although we can quantify grain size, we then have to estimate paleoslope and flow depth. Terrestrial slope dependency on grain size, and width and depth dependence on slope in rivers provide one way to estimate paleochannel characteristics in the conglomerates. Analysis of photographs taken by *Curiosity* is underway to seek further observations to constrain the paleoenvironment of these fluvial deposits. Unlike sandstones, which can be of fluvial or eolian origin, gravel conglomerates provide definitive evidence of free-flowing surface water on Mars.

Total runoff to produce a specific deposit is very important to estimate, as it provides significant constraints on the climate conditions required. We have not documented the volume of water responsible for depositional landforms on Earth. On Mars, this estimate relies on assumptions about the sediment concentration in runoff. For the Earth, the median concentration of all the major rivers of the world is $\sim 10^{-4}$. Very high concentrations (upwards of 30%) have been proposed, but would seem incompatible in systems with well-defined sinuous low-gradient channels. In studies of fan/delta features recently published, the runoff from source areas has been estimated to be on the order of 1 km, thus requiring a period with a sustained hydrologic cycle (annual precipitation in source areas and runoff to areas of deposition) lasting likely thousands of years. A large uncertainty is the intermittency of the climate conditions favorable for runoff. Channel morphology places constraints, however, on the scale of runoff, and coupled with the total runoff required to produce a deposit, forces fluvial deposition events to extend to periods of at least thousands of years. These sites may have been fluvially active in the same period, about 3 billion years ago, sufficiently late in Mars' history that climate models do not expect an active liquid-water-driven hydrologic cycle.

Acknowledgments

This research was conducted for the Jet Propulsion Laboratory, California Institute of Technology, under a contract to Malin Space Sciences Systems with NASA under the Mars Program Office. Martin Kleinhans and Aileen Yingst provided valuable reviews. Photographic images collected by the Mars Science Laboratory MAHLI/Mastcam/MARDI group led by the Malin Space Sciences Systems are revealing the fluvial history of Mars. We also thank the teams responsible for the orbital data, namely, the HiRISE, CTX, and HRSC teams.

References

Andrews, ED 1984. Bed-material entrainment and hydraulic geometry of gravel-bed rivers in Colorado. *Geol. Soc. Am. Bull.* 95, 371–378.

Bhattacharya, JP, Payenberg, THD, Lang, SC, and Bourke, M 2005. Dynamic river channels suggest a long-lived Noachian crater lake on Mars. *Geophys. Res. Lett.* 32, L10201. http://dx.doi.org/10.1029/2005GL022747.

Burr, DM, Williams, RME, Wendell, KD *et al.* 2010. Inverted fluvial features in the Aeolis/Zephyria Plana region, Mars: Formation mechanism and initial paleodischarge estimates. *J. Geophys. Res.* 115, E07011. DOI:10.1029/2009JE003496.

Carr, MH 2012. The fluvial history of Mars. *Philos. Trans. R. Soc. A.* 370, 2193–2215. DOI:10.1098/rsta.2011.0500.

Carr, MH and Clow, GD 1981. Martian channels and valleys: Their characteristics, distribution, and age. *Icarus* 48(1), 91–117.

Carter, J, Loizeau, D, Mangold, N *et al.* 2015. Widespread surface weathering on early Mars: a case for a warmer and wetter climate. *Icarus* 248, 373–382.

Craddock, RA and Howard, AD 2002. The case for rainfall on a warm, wet early Mars. *J. Geophys. Res.* 107. DOI:10.1029/2001JE001505.

Dade, W and Friend, PF 1998. Grain-size, sediment-transport regime, and channel slope in alluvial rivers. *J. Geol.* 106(6), 661–676.

Fasset, CI and Head, JW 2008. Valley network-fed, open-basin lakes on Mars: distribution and implications for Noachian surface and subsurface hydrology. *Icarus* 198, 37–56. DOI:10.1016/j.icarus.2008.06.016.

Ferguson, R 2007. Flow resistance equations for gravel- and boulder-bed streams. *Water Resour. Res.* 43, W05427. http://dx.doi.org/10.1029/2006WR005422

Ferguson, R 2013. Reach-scale flow resistance. In *Treatise on Geomorphology*, Vol. 9 (ed. Wohl, E). Academic Press: San Diego, CA; 50–68.

Ferguson, R, Hoey, T, Wathen, S, and Werritty, A 1996. Field evidence for rapid downstream fining of river gravels through selective transport. *Geology* 24(2), 179–182.

García, MH 2008. Sediment transport and morphodynamics. *Sedimentation Engineering: Processes, Measurements, Modeling, and Practice*, Vol. 110 (ed. García, MH). American Society of Civil Engineers: Reston, VA; 21–164.

Grotzinger, JP, Hayes, AG, Lamb, MP, and McLennan, SM 2013. Sedimentary processes on Earth, Mars, Titan, and Venus. *Comparative Climatology of Terrestrial Planets* (eds Mackwell, SJ *et al.*). University of Arizona: Tucson.

Grotzinger, JP, Gupta, S, Rubin, DM *et al.* 2015. Deposition, exhumation, and paleoclimate of an ancient lake deposit, Gale Crater, Mars. *Science* 350(6257), 1–12. DOI:10.1126/science.aac7575.

Gulick, VC 1998. Magmatic intrusions and a hydrothermal origin for fluvial valleys on Mars. *J. Geophys. Res. Planets* 103(E8), 19365–19387. DOI:10.1029/98JE01321.

Gulick, VC and Baker, VC 1990. Origin and evolution of valleys on Martian volcanoes. *J. Geophys. Res. Solid Earth* 95(B9), 14325–14344. DOI:10.1029/JB095iB09p14325.

Ikeda, S, Parker, G, and Kimura, Y 1988. Stable width and depth of straight gravel rivers with heterogeneous bed materials. *Water Resour. Res.* 24, 713–722.

Irwin III, RP, Craddock, RA, and Howard, AD 2005a. Interior channels in Martian valley networks: discharge and runoff production. *Geology* 33(6), 489–492. http://dx.doi.org/10.1130/G21333.1

Irwin III, RP, Howard, AD, Craddock, RA, and Moore, JM 2005b. An intense terminal epoch of widespread fluvial activity on early Mars: 2. Increased runoff and paleolake development. *J. Geophys. Res.* 110, E12S15. http://dx.doi.org/10.1029/2005JE002460

Irwin, RP, Lewis, KW, Howard, AD, and Grant, JA 2014. Paleohydrology of Eberswalde crater, Mars. *Geomorphology*, http://dx.doi.org/10.1016/ j.geomorph.2014.10.012.

Jerolmack, DJ, Mohrig, D, Zuber, MT, and Byrne, S 2004. A minimum time for the formation of Holden Northeast fan, Mars. *Geophys. Res. Lett.* 31, L21701. http://dx.doi.org/10.1029/2004GL021326

Kleinhans, MG 2005. Flow discharge and sediment transport models for estimating a minimum timescale of hydrological activity and channel and delta formation on Mars. *J. Geophys. Res, Planets* 110(E2), E12003.

Kleinhans, MG, van de Kasteele, HE, and Hauber, E 2010. Palaeoflow reconstruction from fan delta morphology on Mars. *Earth Planet. Sci. Lett.* 294, 378–392. http://dx.doi.org/10.1016/j.epsl.2009.11.025

Lamb, MP, Dietrich, WE, and Venditti, JG 2008. Is the critical Shields stress for incipient sediment motion dependent on channel-bed slope? *J. Geophys. Res. Earth Surface* 113, F02008. DOI:10.1029/2007JF000831.

Le Deit, L, Hauber, E, Fueten, F *et al.* 2013. Sequence of infilling events in Gale Crater, Mars: Results from morphology, stratigraphy, and mineralogy. *J. Geophys. Res. Planets.* DOI:10.1002/2012JE004322

Leopold, LB and Maddock,T 1953. *The Hydraulic Geometry Of Stream Channels And Some Physiographic Implications*. Professional Paper 252, US Geological Survey; Washington, DC.

Lewis, KW and Aharonson, O 2006. Stratigraphic analysis of the distributary fan in Eberswalde crater using stereo imagery. *J. Geophys. Res.* 111, E06001. http://dx.doi.org/10.1029/2005JE002558

Li, C, Czapiga, MJ, Eke, EC *et al.* 2014. Variable Shields number model for river bankfull geometry: bankfull shear velocity is viscosity-dependent but grain size-independent. *Journal of Hydraulic Research.* DOI:10.1080/00221686.2014.939113.

Li, C, Czapiga, MJ, Eke, EC *et al.* 2016. Closure to "Variable Shields number model for river bankfull geometry: bankfull shear velocity is viscosity-dependent but grain size-independent" *Journal of Hydraulic Research.* DOI:10.1080/00221686.2015.1137088

Lynds, RM, Mohrig, D, Hajek, EA, and Heller, PL 2014. Paleoslope reconstruction in sand suspended-load dominant rivers. *Journal of Sedimentary Research* 84, 825–836.

Malin, MC and Edgett, KS 2000. Sedimentary rocks on early Mars. *Science* 290, 1927–1937. DOI:10.1126/ science.290.5498.1927.

Malin, MC and Edgett, KS 2003. Evidence for persistent flow and aqueous sedimentation on early Mars. *Science* 302, 1931–1934. http://dx.doi.org/10.1126/science.1090544

Mangold, N, Kite, ES, Kleinhans, MG *et al.* 2012. The origin and timing of fluvial activity at Eberswalde Crater, Mars. *Icarus* 220, 530–551. http://dx.doi.org/10.1016/j.icarus.2012.05.026

Montgomery, DR and Buffington, JM 1997. Channel-reach morphology in mountain drainage basins. *GSA Bulletin* 109(5), 596–611.

Moore, JM and Howard, AD 2005. Large alluvial fans on Mars. *Journal of Geophysical Research* 110(E4), E04005. DOI:10.1029/2004JE002352.

Moore, JM, Howard, AD, Dietrich, WE, and Schenk, PM 2003. Martian layered fluvial deposits: implications for Noachian climate scenarios. *Geophys. Res. Lett.* 30(24), http://dx.doi.org/10.1029/ 2003GL019002.

Morgan, AM, Howard, AD, Hobley, DE *et al.* 2014. Sedimentology and climatic environment of alluvial fans in the Martian Saheki crater and a comparison with terrestrial fans in the Atacama Desert. *Icarus* 229, 131–156.

Palucis, MC, Dietrich, WE, Hayes, A *et al.* 2014. The origin and evolution of the Peace Vallis fan system that drains to the Curiosity landing area, Gale Crater, *Mars. J. Geophys. Res. Planets* 119, 705–728. DOI:10.1002/2013JE004583.

Paola, C and Mohrig, D 1996. Palaeohydraulics revisited: Paleoslope estimation in coarse-grained braided rivers. *Basin Res.* 8, 243–254.

Parker, G 1978. Self formed rivers with stable banks and mobile bed: Part II, the gravel river. *J. Fluid Mech.* 89(1), 27–148.

Parker, G 2008. Transport of gravel and sediment mixtures. In *Sedimentation Engineering: Processes, Measurements, Modeling, and Practice*, Vol. 110 (ed. MH García). American Society of Civil Engineers: Reston, VA; 165–252.

Parker, G, Wilcock, PR, Paola, C *et al.* 2007. Physical basis for quasi-universal relations describing bankfull hydraulic geometry of single-thread gravel bed rivers. *J. Geophys. Res. Earth Surface* 112(F4). DOI: 10.1029/2006JF000549.

Pepin, E, Carretier, S, Guyot, JL, and Escobar, F 2010. Specific suspended sediment yields of the Andean rivers of Chile and their relationships to climate, slope, and vegetation. *Hydrol. Sci. J.* 55(7), 1190–1205. DOI:10.1080/02626667.2010.512868

Pondrelli, M, Rossi, AP, Marinangeli, L *et al.* 2008. Evolution and depositional environments of the Eberswalde fan delta, Mars. *Icarus* 197, 429–451. 10.1016/j.icarus.2008.05.018

Prancevic, JP, Lamb, MP, and Fuller, BM 2014. Incipient sediment motion across the river to debris-flow transition. *Geology* 42(3), 191–194.

Rice, MS, Bell III, J, Gupta, S *et al.* 2013. A detailed geologic characterization of Eberswalde crater, Mars. *Mars* 8, 15–59. http://dx.doi.org/10.1555/mars.2013.0002

Ryan, SE, Porth, LS, and Troendle, A 2005. Coarse sediment transport in mountain streams in Colorado and Wyoming, USA. *Earth Surface Processes and Landforms* 30(3), 269–288. DOI:10.1002/esp.1128.

Seizilles, G, Devauchelle, O, Lajeunesse, E, and Metivier, FM 2013. Width of laminar laboratory rivers. *Physical Review E* 87, 05224.

Stock, JD 2013. Waters divided: a history of alluvial fan research and a view of its future. In *Treatise on Geomorphology*, Vol. 9 (ed J. Shroder and E. Wohl). Elsevier: Amsterdam; 413–458.

Syvitski, JPM, Peckham, SD, Hilberman, R, and Mulder, T 2003. Predicting the terrestrial flux of sediment to the global ocean: a planetary perspective. *Sediment. Geol.* 162(1–2), 5–24.

Thomson, BJ, Bridges, NT, Milliken, R *et al.* 2011. Constraints on the origin and evolution of the layered mound in Gale Crater, Mars using Mars Reconnaissance Orbiter data. *Icarus* 214(2), 413–432. DOI: 10.1016/j.icarus.2011.05.002.

Trampush, SM, Huzurbazar, S, and McElroy, B 2014. Empirical assessment of theory for bankfull characteristics of alluvial channels. *Water Resour. Res.* 50, 9211–9220. DOI:10.1002/2014WR015597.

Venditti, JG and Church, M 2014. Morphology and controls on the position of a gravel-sand transition: Fraser River, *British Columbia. J. Geophys. Res. Earth Surf.* 119, 1959–1976. DOI:10.1002/2014JF003147.

Wohl, E and Merritt D 2005. Prediction of mountain stream morphology. *Water Resources Research* 41, W08419. DOI:10.1029/2004WR003779.

Williams, RME, Irwin III, RP, and Zimbelman, JR 2009. Evaluation of paleohydrologic models for terrestrial inverted channels: Implications for application to martian sinuous ridges. *Geomorphology* 107, 300–315.

Williams, RME, Grotzinger, JP, Dietrich, WE *et al.* 2013. Martian fluvial conglomerates at Gale crater. *Science* 340, 1068–1072. DOI:10.1126/science.1237317.

Wilson, L, Ghatan, G, Head, J, and Mitchell, K 2004. Mars outflow channels: a reappraisal of the estimation of water flow velocities from water depths, regional slopes, and channel floor properties. *J. Geophys. Res.* 109, E09003.
Wood, LJ 2006. Quantitative geomorphology of the Mars Eberswalde delta. *Geol. Soc. Am. Bull*, 118(5–6), 557–566, http://dx.doi.org/10.1130/B25822.1.
Woodsworth, R, Forget, F, Millour, E *et al.* 2013. Global modeling of the early Martian climate under a denser CO_2 atmosphere: Water cycle and ice evolution. *Icarus* 222(1), 1–19. DOI:10.1016/j.icarus.2012.09.036.
Yingst, RA, Kah, LC, Palucis, MC, Williams *et al.* 2013. Characteristics of pebble and cobble-sized clasts along the Curiosity rover traverse from Bradbury Landing to Rocknest. *J. Geophys. Res. Planets* 118(11), 2361–2380. DOI:10.1002/2013JE004435.

Discussion

Discussion by Colin Thorne and Mike Miles

The authors note that, "*A simple power law between channel width and discharge (derived from terrestrial rivers) may provide a more accurate estimate of discharge than mechanistic approaches.*" We agree. Hydraulic geometry equations predict width as a function of discharge and for application to Martian streams, the authors correctly reverse the dependence of the variables in deriving their regression equation (Williams 1983). This approach was pioneered in the UK during the 1980s (Wharton *et al.* 1989), and subsequently published as the "Channel-geometry Method" (Wharton 1995).

The coefficients in the best-fit regression equations for the two analyses are identical; the exponents differ. In most *hydraulic geometry* equations, the exponent is not significantly different from 0.5 (Soar and Thorne 2000), suggesting an exponent close to two in a power relationship for Q as a function of w. This is very close to Wharton's exponent (1.97). We would value the authors' interpretation of likely reasons for the lower exponent in their best-fit equation, which appears from Figure 28.10 to be due to inclusion of width data for large rivers with widths one or two orders of magnitude larger than those in Wharton's data set. We also suggest that adding confidence intervals to the graph in Figure 28.10 might add value by illustrating the degree of uncertainty in single and mean responses and would welcome the authors' view on this.

References

Soar, PJ and Thorne, CR 2001. *Channel Restoration Design for Meandering Rivers*. Final Report ERDC/CHL CR-01-1, US Army Research and Development Center: Vicksburg; 416 pp. plus appendices. http://libweb.wes.army.mil/uhtbin/cgisirsi/2fD4nuOSuv/ERDC_VBG/185360025/523/6327
Wharton, G 1995. The channel-geometry method: guidelines and applications. *Earth Surface Processes and Landforms* 20(7), 649–660.
Wharton, G, Arnell, NW, Gregory, KJ, and Gurnell AM. 1989. River discharge estimated from channel dimensions. *Journal of Hydrology* 106(3), 365–376.
Williams, GP 1983. Palaeohydrological methods and some examples from Swedish fluvial-environments. *I. Cobble and boulder deposits. Geografiska Annaler Series A* 65(3–4), 227–243.

Discussion by Jonathan Laronne

A variety of evidence has been available during the past decade that water was available on Mars, with even more convincing confirmation from the very recent months that at select locations it presently runs on the surface. Your paper is also based on several types of evidence, the most crucial is one of relatively close-up and very interesting photographs (Figures 28.2a–d) of a rock formation that has an obvious granular makeup and an appearance of a clastic deposit. In this photograph the particles are only slightly rounded, hence restating your words, if a meander is a meander, is this sufficiently convincing to ensure that it is fluvial gravel?

Discussion by Gordon Grant

Have you looked at the Froude numbers for your Martian rivers, not in a deterministic way, but as a check on your flow estimates? I would be suspicious, for example, if your results suggested supercritical flow where your channel dimensions are at least reasonably constrained.

Discussion by Mike Miles and Colin Thorne

Chinn and Mason (2015) summarize long-term hydrology data for the Onyx River, Antarctica – see our discussion of Session 16, Shaw and Healy (1980), and Head and Marchant (2014). Our analysis of these data indicates the 2-year return period daily and instantaneous flows (taken as representing bankfull discharge) are 2 and 3 m^3/s, respectively. Measurements on GoogleEarth indicate bankfull widths (in single-thread reaches) varying between 6 and 13 m. Figure 28.10 suggests that bankfull discharge in the Onyx River should be larger, ranging between 4 and 14 m^3/s. As Antarctic stream banks lack vegetation, which is known to result in wider channels when other bank properties are held constant, this is not unexpected.

This cursory analysis suggests that the hydraulic geometry relationships for unvegetated, high-latitude streams should be further investigated as possible analogues for ancient Martian streams and the effects of ice-rich permafrost on streambank stability (Church and Miles, 1982) and channel geometry should be better defined. Perhaps we should arrange a joint field trip to some polar streams (though probably on Earth rather than Mars!)?

References

Chinn, T and Mason, P 2015. The first 25 years of the hydrology of the Onyx River, Dry Valleys, Antarctica. *Polar Record* 52(1), 16–65. DOI:10.1017/S0032247415000212.

Church, MA and Miles, MJ 1982. Processes and mechanisms of riverbank erosion – discussion. In *Engineering Problems in the Management of Gravel-Bed Rivers* (ed. RE Hey). John Wiley & Sons: Chichester, UK; 259–268.

Head, JW and Marchant, DR 2014. The climate history of early Mars: insights from the Antarctic McMurdo Dry Valleys hydrologic system. *Antarctic Science* 26(6), 774–800.

Shaw, J and Healy, TR 1980. Morphology of the Onyx River system, McMurdo Sound region, Antarctica. *New Zealand Journal of Geology and Geophysics* 23(2), 223–238.

Discussion by John Buffington

Hydraulic geometry offers robust scaling of channels across many orders of magnitude in natural rivers, irrigation canals, and laboratory channels (Ferguson 1986). Although the authors' analysis focuses

(a)

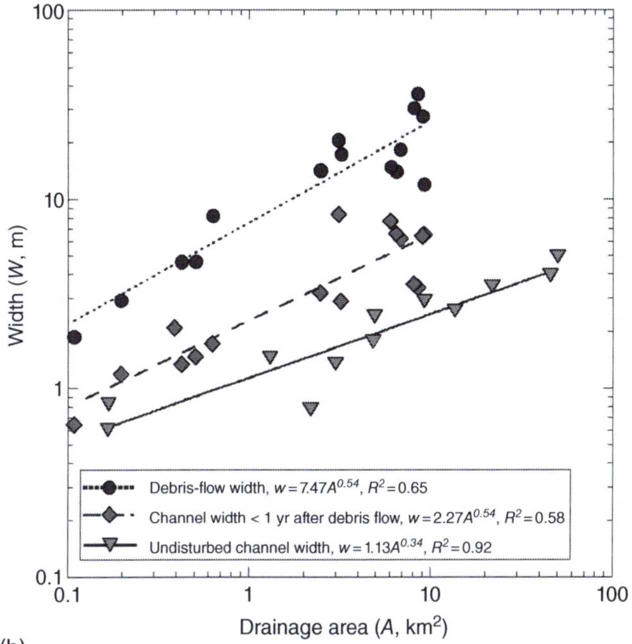

(b)

Figure 28.15 (a) Width as a function of drainage area for (i) post-fire debris flows (circles), (ii) channels carved into the debris-flow deposits via a mixture of hyperconcentrated and fluvial flows within the first year following disturbance (diamonds), and (iii) undisturbed fluvial channels in neighboring basins (inverted triangles). The study sites are mountain streams in confined, alluvial valleys (plane-bed, step-pool, and cascade morphologies, with gravel to boulder beds, and channel slopes > 1%) in the upper Boise River, central Idaho, USA. Debris flows were triggered by intense thunderstorms on bare, hydrophobic soils following fire. Subsequent thunderstorm and snowmelt runoff incised channels into the debris-flow deposits via hyperconcentrated and fluvial flows for bare, unvegetated banks. Undisturbed channels are produced by predominantly snowmelt runoff, with banks well-vegetated by shrubs and conifers, resulting in smaller channel widths for a given drainage area. (b) Photograph illustrating debris-flow width and width of channel subsequently carved into the debris-flow deposit.

on fluvial deposits, it is worth noting that hydraulic geometry relations can be developed for a variety of fluids, which may have relevance for some Martian channels given that differences in gravity and atmospheric conditions (pressure, temperature, and, thus, viscosity) may cause channel hydraulics and flow rheology (Froude and Reynolds numbers) to differ from those on Earth. Data from mountain rivers in Idaho provide an example of how channel width responds to differences in erosional processes, flow rheology, runoff characteristics, and degree of riparian vegetation within a given physiographic region (similar topography, geology, and climate) following fire disturbance (Figure 28.15). In all cases, the exponent of the hydraulic geometry equation is within the range typically observed for terrestrial alluvial and bedrock rivers (0.3–0.6; Montgomery and Gran 2001; Wohl 2004) but systematic differences in width are evident. The debris-flow widths are event-based responses and the post-debris-flow channel widths are transient responses following disturbance, both of which differ from the long-term bankfull conditions examined by the authors, but they nonetheless provide some sense of scaling for different formative processes across a range of rheological and environmental conditions.

References

Ferguson, RI 1986. Hydraulics and hydraulic geometry. *Progress in Physical Geography*, 10(1), 1–31.

Montgomery, DR and Gran KB 2001. Downstream variations in the width of bedrock channels. *Water Resources Research* 37(6), 1841–1846.

Wohl, E 2004. Limits of downstream hydraulic geometry. *Geology* 32(10), 897–900.

Reply by the Authors

The discussions touch on three important issues in the paper. (i) Are gravel deposits observed during the *Curiosity* mission convincingly of fluvial origin? (ii) Are the derived hydraulic expressions sensible, and, specifically, do they predict flows to be subcritical rather than supercritical in most circumstances? (iii) What is the significance of the specific power law relationship of bankfull discharge versus bankfull width (Equation (28.21)) and what processes might influence the relationship on Mars?

Jonathan Laronne comments that the gravels visible in our figures appear to be only slightly rounded and might this challenge the interpretation of their fluvial origin. For the outcrops shown in Figure 28.2 (and a nearby exposure), Williams *et al.* (2014) measured 535 gravels and concluded that the majority of the pebbles are subrounded to rounded (based on the qualitative Powers' roundness indices). Yingst *et al.* (2013) demonstrated that roundness measured at the outcrops shown in our Figure 28.2 was much greater than that found in previous missions on Mars. Williams *et al.* (2014) suggested the rounding required a minimum of several kilometers of transport, but was likely more given the reduced abrasion rates of less intense collisions of bedload transport on Mars. Szabo *et al.* (2015) employ other measures to document rounding of gravels observed by the *Curiosity* rover and conclude that rounding due to fluvial transport likely developed over distances of tens of kilometers by fluvial processes. Grotzinger *et al.* (2015) provide outcrop description, placing conglomerates in a fluvial deltaic system propagating from Gale Crater wall towards the center of the crater. Collectively these analyses argue strongly that these are fluvially transported gravels.

Gordon Grant suggests we review the Froude number implications of our derived hydraulic relationships, with the expectation that if the calculations predict falsely supercritical conditions it should be a concern. We calculated the Froude number by using Equations (28.12) and (28.8) (which are based on a threshold shear stress concept) and compared that to the Froude number *Fr* calculated from the observed bankfull velocity and depth values reported in Li *et al.* (2014). The regression yielded $Fr_{observed} = 0.97\ Fr_{\text{predicted}} + 0.02$, so a close comparison, but with considerable variance yielding only an R^2 of 0.38. None of the predictions exceeded a *Fr* number of 1.0 and only four of the observed values did. The *Fr* values calculated from Equations (28.17) and (28.15), expressions that use an empirical relationship for the bankfull shear velocity, showed essentially no correlation with observed *Fr* and predicted 24 of the rivers would have supercritical flows at bankfull. Hence, the performance is poorer but does not excessively overpredict supercritical conditions. The *Fr* derived from Equations (28.12) and (28.8) is independent of gravity (the gravity term in the velocity expression cancels the gravity term in the *Fr*) and thus the incidence of supercritical flow is not expected to be higher on Mars. The *Fr* derived from the empirical shear velocity relationship indicates that *Fr* is also only weakly dependent on gravity.

Colin Thorne and Mike Miles ask why the exponent in the plot of bankfull discharge versus bankfull width is not closer to the inverse of the common exponent of 0.5 when width is plotted against discharge. The least squares approach used in regressions typically leads to exponents that are not the inverse of each other when the variables are reversed. Also, many have reported discharge–width relationships in regional surveys, and Wharton (1995) summarized findings in his table 1. His value of 1.97 for British rivers is an exception. For the nine large data sets from western US rivers used in analyses by Osterkamp and Hedman (1977; cited in Wharton 1995), the exponents for the nine studies ranged from 1.4 to 1.97, but with six lying between 1.60 and 1.71, i.e. very similar to values reported in Equations 28.19 through 28.21. Many factors influence channel width, which Osterkamp and Hedman (1997), for example, explicitly explore empirically, and as John Buffington illustrates in his discussion of this paper. Our approach was to combine two recent data summaries with additional data from sources not included in these two summaries that represented smaller, steeper, and coarser bedded rivers. The data are all based on field measurements, including bankfull discharge. The streambed material ranged from sand to boulders and we purposely sought a wide range of conditions given the large uncertainty of Martian conditions. The data field is biased towards US rivers and to channels between 2 and 100 m wide. The largest rivers are lowland tropical sand-bedded rivers. The slope of the correlation reported in Figure 28.10 with a 95% confidence interval is 1.66 ± 0.04. Note that for the largest channel widths the data density is smaller and the range about the regression is about a factor of two. For the high data density the range about the regression is about 10. The width range for a given discharge is much less. The data from small Antarctic streams reported by Miles and Thorne fall within the data field shown in Figure 28.10. The range of variance of data, assuming measurements are accurate, reflect the varying factors that influence channel width for a given discharge, e.g., bank strength, sediment supply, bed grain size, flow dynamics, and process dominance. A conservative application of Equation 28.21 to Mars would be to assume that for a given calculated discharge the range could be a factor of ten larger or smaller. Despite this large uncertainty, such data provide key hydroclimatological information of the history of Mars.

Reference

Szabo, T, Domokos, G, Grotzinger, JP, and Jerolmack, DJ 2015. Reconstructing the transport history of pebbles on Mars. *Nature Communications* 6, 8366. DOI:10.1038/ncomms9366.

Index

Gravel-Bed Rivers: Processes and Disasters, First Edition. Edited by Daizo Tsutsumi and Jonathan B. Laronne.

e

f

g

Wiley The manufacturer's authorized representative according to the EU General Product Safety Regulation is Wiley-VCH GmbH, Boschstr. 12, 69469 Weinheim, Germany, e-mail: Product_Safety@wiley.com.

Printed and bound by CPI Group (UK) Ltd, Croydon, CR0 4YY
20/08/2026
02201740-0001